BACTERIAL STRESS RESPONSES

BACTERIAL STRESS RESPONSES

Edited by

Gisela Storz

Cell Biology and Metabolism Branch
National Institute of Child Health and Human Development
National Institutes of Health
Bethesda, Maryland

Regine Hengge-Aronis

Institut für Biologie—Mikrobiologie
Freie Universität Berlin
Berlin, Germany

ASM PRESS

Washington, D.C.

Library of Congress Cataloging-in-Publication Data

Bacterial stress responses / edited by Gisela Storz, Regine Hengge-Aronis.
 p. cm.
 Includes bibliographical references and index.
 ISBN 1-55581-192-2 (hc)
 1. Microorganisms—Physiology. 2. Stress (Physiology). 3. Adaptation (Physiology). I.
Storz, Gisela. II. Hengge-Aronis, Regine.
QR97.A1 B33 2000
571.2′93—dc21   99-086259

Cover photos: The two panels show similar areas of false-color, autoradiographic images of two polyacrylamide gels of [^{35}S]methionine-labeled proteins from extracts of *Escherichia coli* strain SC122. The cells were grown in defined rich medium and labeled during growth at 37°C (upper panel) or from 3 to 8 min after a shift from 37 to 42°C (bottom panel). Boxes surround heat shock proteins numbered identically in the two images (protein 2, GroEL; protein 3, DnaK; protein 6, HtpG; protein 7, HtpH; protein 8, ClpY, protein 9, LysU; protein 12, ClpB; protein 15, Lon). The arrow points to protein RpoH, the heat shock RNA polymerase factor σ32. Courtesy of Ruth A. VanBogelen and Frederick C. Neidhardt; gels produced in 1983.

To Ella, Toby, Felix, Lisa-Maria, and Manolis

CONTENTS

CONTRIBUTORS

Michael N. Alekshun
Center for Adaptation Genetics and Drug Resistance and Departments of Molecular Biology and Microbiology and of Medicine, Tufts University School of Medicine, Boston, Massachusetts 02111

L. Aravind
National Center for Biotechnology Information, National Library of Medicine, National Institutes of Health, Bethesda, Maryland 20894

Donna M. Bates
Department of Biomolecular Chemistry, University of Wisconsin Medical School, 1300 University Avenue, Madison, Wisconsin 53706

John R. Battista
Department of Biological Sciences, Louisiana State University and A & M College, Baton Rouge, Louisiana 70803

Volkmar Braun
Mikrobiologie/Membranphysiologie, Universität Tübingen, Auf der Morgenstelle 28, D-72076 Tübingen, Germany

Erhard Bremer
Laboratory of Microbiology, Department of Biology, University of Marburg, D-35032 Marburg, Germany

Jan A. M. de Bont
Division of Industrial Microbiology, Department of Food Technology and Nutritional Sciences, Wageningen Agricultural University, P.O. Box 8129, 6700 EV Wageningen, The Netherlands

David Dubnau
Public Health Research Institute, 455 First Avenue, New York, New York 10016

Ashlee M. Earl
Department of Biological Sciences, Louisiana State University and A & M College, Baton Rouge, Louisiana 70803

Steven E. Finkel
Department of Biological Sciences, University of Southern California, Los Angeles, California 90089

Patrick Forterre
Institut de Génétique et Microbiologie, Université Paris XI, 91405 Orsay Cedex, France

John W. Foster
Department of Microbiology and Immunology, University of South Alabama College of Medicine, Mobile, Alabama 36688

Patricia L. Foster
Department of Biology, Indiana University, Bloomington, Indiana 47405

Michael Y. Galperin
National Center for Biotechnology Information, National Library of Medicine, National Institutes of Health, Bethesda, Maryland 20894

Klaus Hantke
Mikrobiologie/Membranphysiologie, Universität Tübingen, Auf der Morgenstelle 28, D-72076 Tübingen, Germany

Regine Hengge-Aronis
Department of Biology—Microbiology, Freie Universität Berlin, 14195 Berlin, Germany

Masayori Inouye
Department of Biochemistry, Robert Wood Johnson Medical School, 675 Hoes Lane, Piscataway, New Jersey 08854

Ralph R. Isberg
Howard Hughes Medical Institute, Department of Molecular Biology and Microbiology, Tufts University School of Medicine, 136 Harrison Avenue, Boston, Massachusetts 02111

Masaaki Kanemori
HSP Research Institute, Kyoto Research Park, Shimogyo-ku, Kyoto 600-8813, Japan

Jasper Kieboom
Division of Industrial Microbiology, Department of Food Technology and Nutritional Sciences, Wageningen Agricultural University, P.O. Box 8129, 6700 EV Wageningen, The Netherlands

Patricia J. Kiley
Department of Biomolecular Chemistry, University of Wisconsin Medical School, 1300 University Avenue, Madison, Wisconsin 53706

Roberto Kolter
Department of Microbiology and Molecular Genetics, Harvard Medical School, Boston, Massachusetts 02115

Eugene V. Koonin
National Center for Biotechnology Information, National Library of Medicine, National Institutes of Health, Bethesda, Maryland 20894

Margareta Krabbe
Molecular, Cellular, and Developmental Biology, University of California, Santa Barbara, California 93106

Reinhard Krämer
Department of Chemistry, University of Cologne, D-50674 Cologne, Germany

Terry Ann Krulwich
Department of Biochemistry and Molecular Biology, Mount Sinai School of Medicine, New York, New York 10128

Robert A. LaRossa
Biochemical Science and Engineering, Central Research and Development, DuPont Company, Wilmington, Delaware 19880-0173

Stuart B. Levy
Center for Adaptation Genetics and Drug Resistance and Departments of Molecular Biology and Microbiology and of Medicine, Tufts University School of Medicine, Boston, Massachusetts 02111

Purificación López García
División de Microbiología, Facultad de Medicina, Campus de San Juan, Universidad Miguel Hernández, 03550 Alicante, Spain

David Low
Molecular, Cellular, and Developmental Biology, University of California, Santa Barbara, California 93106

Max Mergeay
Laboratory for Microbiology, Radioactive Waste and Clean-Up Division, Center of Studies for Nuclear Energy (SCK/CEN), and Environmental Technology, Flemish Institute for Technological Research (VITO), Mol, and Laboratoire de Génétique des Procaryotes, Institut de Biologie Moléculaire et Médicale, Université Libre de Bruxelles, Brussels, Belgium

Miyo T. Morita
HSP Research Institute, Kyoto Research Park, Shimogyo-ku, Kyoto 600-8813, Japan

Frederick C. Neidhardt
Department of Microbiology and Immunology, University of Michigan Medical School, Ann Arbor, Michigan 48109-0620

Thomas V. O'Halloran
Department of Chemistry and Department of Biochemistry, Molecular Biology and Cell Biology, Northwestern University, 2145 Sheridan Road, Evanston, Illinois 60208

Caryn E. Outten
Department of Chemistry, Northwestern University, 2145 Sheridan Road, Evanston, Illinois 60208

F. Wayne Outten
Department of Biochemistry, Molecular Biology and Cell Biology, Northwestern University, 2145 Sheridan Road, Evanston, Illinois 60208

Etana Padan
Division of Microbial and Molecular Ecology, Institute of Life Sciences, Hebrew University of Jerusalem, Jerusalem 91904, Israel

Thomas Patschkowski
Department of Biomolecular Chemistry, University of Wisconsin Medical School, 1300 University Avenue, Madison, Wisconsin 53706

Sangita Phadtare
Department of Biochemistry, Robert Wood Johnson Medical School, 675 Hoes Lane, Piscataway, New Jersey 08854

Chester W. Price
Department of Food Science and Technology, University of California, Davis, California 95616

Tracy L. Raivio
Department of Molecular Biology, Princeton University, Princeton, New Jersey 08544

William A. Rosche
Department of Biological Science, University of Tulsa, Tulsa, Oklahoma 74104

Peter Setlow
Department of Biochemistry, University of Connecticut Health Center, Farmington, Connecticut 06032

Thomas J. Silhavy
Department of Molecular Biology, Princeton University, Princeton, New Jersey 08544

Bradley T. Smith
Department of Biology, Massachusetts Institute of Technology, Cambridge, Massachusetts 02139

Abraham L. Sonenshein
Department of Molecular Biology and Microbiology, Tufts University School of Medicine, Boston, Massachusetts 02111

Gisela Storz
Cell Biology and Metabolism Branch, National Institute of Child Health and Human Development, National Institutes of Health, Bethesda, Maryland 20892

Mark D. Sutton
Department of Biology, Massachusetts Institute of Technology, Cambridge, Massachusetts 02139

Kürsad Turgay
Public Health Research Institute, 455 First Avenue, New York, New York 10016

Ruth A. VanBogelen
Molecular Biology Department, Parke-Davis Pharmaceutical Research, Division of Warner-Lambert Company, Ann Arbor, Michigan 48105

Tina K. Van Dyk
Biochemical Science and Engineering, Central Research and Development, DuPont Company, Wilmington, Delaware 19880-0173

Graham C. Walker
Department of Biology, Massachusetts Institute of Technology, Cambridge, Massachusetts 02139

Nathan Weyand
Molecular, Cellular, and Developmental Biology, University of California, Santa Barbara, California 93106

Owen White
The Institute for Genomic Research, 9712 Medical Center Drive, Rockville, Maryland 20850

Stephen C. Winans
Department of Microbiology, Cornell University, Ithaca, New York 14853

Kunitoshi Yamanaka
Department of Biochemistry, Robert Wood Johnson Medical School, 675 Hoes Lane, Piscataway, New Jersey 08854

Takashi Yura
HSP Research Institute, Kyoto Research Park, Shimogyo-ku, Kyoto 600-8813, Japan

Ming Zheng
Cell Biology and Metabolism Branch, National Institute of Child Health and Human Development, National Institutes of Health, Bethesda, Maryland 20892

Jun Zhu
Department of Microbiology, Cornell University, Ithaca, New York 14853

Erik R. Zinser
Department of Microbiology and Molecular Genetics, Harvard Medical School, Boston, Massachusetts 02115

PREFACE

Being exposed to stress has become an everyday experience in modern human life. But what is stress for bacteria? It is difficult to provide a universal definition for stress, since the perception of a stressful situation is highly dependent on the individual cell. Nevertheless, as a working definition, stress can be any deviation from optimal growth conditions that results in a reduced growth rate. Some adaptive or stress responses function so well, however, that growth is not impaired. Stress also can be defined as exposure to any environmental situation that results in damage of cellular components in the absence of a cellular response. The disadvantage of this definition is that something needs to be known about the physiology of the response. Finally, given the whole-genome expression data available for certain bacteria, stress may be defined as a situation that stimulates the expression of genes known to respond to a specific environmental condition. A limitation of this definition is that there might be novel stress conditions under which previously unidentified sets of genes are induced.

There are distinct levels of stress severity. Under minor stress, growth generally continues at the same rate and cells fully adapt to the new conditions. Under severe stress, the growth rate is reduced but cells can still adapt and tolerate the situation, albeit at a high price. Under extreme stress, growth ceases and cells devote their resources to survival. Under lethal stress, responses that may lead to the sacrifice of many individuals, but that may increase the survival of the population, are activated.

In the natural environment, bacterial cells usually have limited nutrients available to them and exist in environments that, at best, support slow growth. Bacteria are constantly faced with many different environmental assaults. Bacteria may plunge from one habitat to another drastically different habitat. Exposure to many different, constantly changing stresses is the normal lifestyle of bacteria. By investigating stress responses in the laboratory, microbiologists study responses that are physiologically more relevant than rapid growth in rich broth.

Research on stress responses has already provided many critical insights into bacterial cell function. First, the mechanisms by which bacteria regulate gene expression in response to environmental insults have become the paradigms for gene regulation. These mechanisms include two-component phosphorelays, sigma factors and anti-sigma factors, regulators that are allosterically regulated by small molecules or redox reactions, elaborate mechanisms that control translation or proteolysis of key regulators, and regulatory RNAs. The regulatory principles that operate in stress responses could hardly be more diverse. Second, studies of bacterial stress responses have given insights into cell physiology. An understanding of many different cellular processes (such as DNA repair and protein folding) has come from the studies of the functions of stress proteins.

In this book, we hope to summarize the current knowledge of bacterial stress responses, to synthesize information from different organisms and different systems, and to outline outstanding questions. The first section of the book focuses on specific stress responses, such as those to heat shock and oxidative stress. These environmental conditions provoke responses designed to cope with the specific stress situation, for example, induction of proteins that eliminate the stress agent or that repair cellular damage produced by the stress. These responses allow the cell to continue to grow, or at least survive, under the specific stress conditions but generally do not provide much protection against other stresses. The second section is devoted to general stress responses, such as spore formation and the σ^S- and σ^B-regulated responses. A general stress response can be induced by several different stress conditions but nevertheless can be considered a single response, usually governed by a single or a few crucial master regulators. The proteins induced have broadly stress-protective functions and often provide cross-protection against multiple stresses. The third section focuses on the connections between stress responses and pathogenesis. In their attempts to eliminate pathogenic bacteria, host organisms present stressful conditions for bacteria, including environments limited in essential nutrients such as iron, suboptimal physical conditions such as acidic pH, and attacks of the immune system such as oxidative bursts. The intriguing mechanisms by which pathogenic bacteria evade or cope with these stresses are just beginning to be revealed. The fourth section explores stress responses in bacteria that thrive in extreme environments. Throughout history, bacteria have been able to colonize some of the most extreme environments on earth. The characterization of these extremophiles is providing insights into survival under conditions such as very high temperature, radiation, and high concentrations of heavy metals. Given that their enzymes are active under extreme conditions, these bacteria are also of great value for biotechnology. The fifth section is devoted to general methods for the study of stress responses. Initially, our knowledge of stress responses was based on the analysis of single genes or proteins and their regulation. This approach is now being complemented by information gained from completed genomes and by multigene or multiprotein analyses. Multi-array approaches, together with the sensitivity of bacterial stress responses and the ease of bacterial genetic manipulation, also allow for the use of bacterial systems as monitors of stress.

Although much has been learned about bacterial stress responses, we are far from a complete picture. For most responses, it is not clear how the initial signal is perceived. For example, how are changes in osmolarity or pH sensed? What are the signals for starvation? It is also becoming increasingly apparent that the different stress responses are extensively interconnected. Future studies need to view the functioning of the cell as a single complex regulatory network. The physiological outcomes of many stress responses are not yet well understood. Many target genes are still unknown, and the molecular functions of some identified target genes are still ambiguous. We suspect that novel strategies to cope with stress, especially with extreme or potentially lethal stress, remain to be discovered. More needs to be learned about the living conditions of bacteria in their natural habitats. What are the environments encountered by pathogens? How are stress responses manifested in biofilms or in environments with many different competing bacterial species? Finally, the biotechnological uses of bacterial stress responses deserve greater attention. Can pollution be monitored by using bacteria with reporter fusions to stress genes as sensors? The principles of stress resistance discovered with certain bacteria may be used to make other bacterial strains or even plants more stress tolerant. Possibly, crucial components of stress responses can serve as targets for novel antibiotics. We hope that this book will help focus attention on these interesting problems, questions, and opportunities and stimulate further research on bacterial stress responses.

Gisela Storz
Regine Hengge-Aronis

ACKNOWLEDGMENTS

We thank all of the authors for their contributions and Greg Payne and Ken April of the ASM Press staff for their help in editing the book. We also appreciate the advice and tolerance of our laboratory members and families.

I. SPECIFIC STRESS RESPONSES

Bacterial Stress Responses
Edited by G. Storz and R. Hengge-Aronis
©2000 ASM Press, Washington, D.C.

Chapter 1

The Heat Shock Response: Regulation and Function

Takashi Yura, Masaaki Kanemori, and Miyo T. Morita

The heat shock response represents a ubiquitous protective and homeostatic cellular response to cope with heat-induced damage in proteins. Many heat shock proteins (HSPs) are molecular chaperones or ATP-dependent proteases and play major roles in protein folding, repair, and degradation under normal and stress conditions. In Escherichia coli, *induction of HSPs occurs primarily by transient increase in the master regulator σ^{32} (encoded by* rpoH), *which specifically directs RNA polymerase (RNAP) to transcribe from the heat shock promoters. The increase in σ^{32} level results primarily from two distinct events: (i) the translational induction due to temperature melting of the* rpoH *mRNA secondary structure, and (ii) transient stabilization of σ^{32}, which is normally unstable. Whereas this "classical" σ^{32} regulon provides major protection against cytoplasmic protein damage, the second regulon responding to high temperature is mediated by σ^E and provides functions to deal with protein misfolding in the extracytoplasmic compartment under extreme heat stress. The σ^{32} regulon is well conserved in gram-negative proteobacteria, although the details of the regulatory mechanisms vary considerably. In gram-positive bacteria, some major HSPs are under negative control by specific repressors (e.g., HrcA acting at CIRCE), whereas others are included in the σ^B-mediated general stress regulon.*

HISTORICAL PERSPECTIVE

Early work on the heat shock response in *Escherichia coli* contributed substantially to our basic understanding of not only the regulatory mechanism of stress responses but also the general issue of molecular chaperones and protein folding. Thus, the master regulator σ^{32} encoded by the *rpoH* (*htpR, hin*) gene (40, 58, 128) was the first of a series of minor σ factors to be discovered in *E. coli*, leading to the general concept that bacterial genes are under transcriptional control by multiple σ factors. In the course of studies on *rpoH*, one of the promoters for its transcription was found to be recognized by another σ factor, σ^E (or σ^{24}), which is activated under extreme heat stress (26, 119). Analysis of the latter system determined the basis for a compartment-specific (extracytoplasmic) stress response in bacteria (38, 74; see chapter 2).

The evolutionary conservation of major HSPs (6, 7) suggested their potentially important roles, while the diversity of agents that induce heat shock-like responses, as well as of phenotypes of some heat shock gene mutants, initially obscured functional significance of HSPs (32, 39, 89). The realization that many HSPs are molecular chaperones and help protein folding by limiting the nonproductive interactions awaited demonstration of HSP-assisted folding of specific proteins in vitro (25, 43). It also turned out that the GroE and DnaK proteins that are important in growth and DNA replication of bacteriophage λ (32) play major roles among various HSPs in bacterial growth as well (57), specifically by preventing protein aggregation (37). Through extensive work with GroE and DnaK in particular, roles of chaperones in protein folding including differential, cooperative, and overlapping functions within or among distinct HSP families are beginning to emerge (12, 28). No less important roles have been assigned to a set of HSPs that are ATP-dependent proteases or their subunits (35, 38), and additional HSPs are still being discovered, including interesting examples such as Hsp33 (46) and Hsp15 (55).

Given the basic and ubiquitous functions of HSPs both under normal and stress conditions, the regulatory mechanism of the heat shock response has been a subject of great interest. In general, heat induction of HSPs reflects increased cellular demands

Takashi Yura, Masaaki Kanemori, and Miyo T. Morita • HSP Research Institute, Kyoto Research Park, Shimogyo-ku, Kyoto 600-8813, Japan.

for HSPs at higher temperature. The amounts of HSPs are regulated primarily at the level of transcription, employing both positive control mediated by minor σ factors and negative control mediated by repressors. However, various mechanisms of posttranscriptional regulation, such as chaperone (and protease)-mediated feedback control, play important roles in modulating HSP synthesis. In the σ^{32} regulon of *E. coli* studied in detail, the amount of active σ^{32} is regulated at the levels of transcription, translation, stability, and activity, triggered by changes in the level of unfolded/misfolded proteins, mediated by the DnaK-DnaJ chaperones, and by ambient temperature (33, 39, 106, 111, 127). These elaborate regulatory mechanisms are interconnected and must be integrated through specific signaling pathways to adjust the level and activity of HSPs most effectively to meet the complex cellular requirements and to ensure maximal growth and survival under a variety of conditions.

We summarize here our current understanding of the heat shock regulation in *E. coli* which has been the most extensively studied, and discuss briefly some of the recent developments with other systems. We will highlight the regulatory mechanisms, particularly some initial events of the heat shock response. For discussion of the function of individual HSPs, more comprehensive or historical treatments of the subject, and discussion of the extracytoplasmic stress response and the heat shock response of hyperthermophiles, readers should consult chapters 2 and 12 as well as earlier reviews (e.g., 11, 33, 38, 39, 74, 89, 127).

E. COLI HEAT SHOCK REGULONS DEFINED BY THE MINOR σ FACTORS

Two major heat shock regulons whose transcription depends on a specific minor σ factor have been analyzed extensively in *E. coli*. These regulons are controlled separately from each other and from the rest of the genes by virtue of distinct promoters (see Table 1 for consensus sequences). Whereas the σ^{32} regulon plays a major role in coping with cytoplasmic protein damage, the σ^E regulon functions primarily to protect cells against extracytoplasmic or extreme heat stress. In addition, there are a number of heat-inducible genes that are controlled by other factors whose functions and regulations are less well understood. For example, induction of the *psp* operon, encoding several phage shock proteins at high temperature or following filamentous-phage infection, depends on σ^{54} and the PspF activator (75).

Table 1. Consensus sequences for selected promoters and operators involved in transcriptional control of heat shock genes

Promoter or operator	Sequence		References
Promoter			
(*E. coli*)	-35	-10	
σ^{70}	TTGACA (16–18 bp)	TATAAT	
σ^{32}	CTTGAA (13–17 bp)	CCCCAT-T	18, 38, 127
σ^E	GAACTT (16 bp)	TCTGA	20, 74
Operator			
CIRCE	TTAGCACTC (N9) GAGTGCTAA		85, 101, 131

The σ^{32} Regulon: Protection against Cytoplasmic Stress

This regulon consists of at least 30 HSPs (38, 74) (Table 2): they were mostly identified either by examining synthesis rates of individual proteins by SDS-polyacrylamide gel electrophoresis (59, 123) or by hybridizing cDNA (obtained with mRNA from heat-shocked cells) with membrane filters containing an ordered *E. coli* genomic library (15). Expression of this group of HSPs is transiently enhanced upon temperature upshift (e.g., 30 to 42°C) in the wild-type but not in the *rpoH* mutant defective in heat induction (88, 114, 125) due to deficiencies in σ^{32} that recognize the heat shock promoters (18, 27, 40).

The overall function of the σ^{32} regulon has been assessed by analysis of *rpoH* mutants. As expected from their temperature-sensitive phenotype, some of the HSPs are essential for growth at high temperature. In addition, pleiotropic phenotypes of *rpoH* mutants such as defects in proteolysis, cell division, phage growth, and plasmid DNA replication indicate the range of involvement of σ^{32} and HSPs in various cellular processes (33, 38, 127). The $\Delta rpoH$ mutants lacking σ^{32} can grow only at or below 20°C (57, 130). This extreme temperature sensitivity is markedly suppressed in the revertants able to grow up to 40°C by overproducing GroEL-GroES chaperones (also called chaperonins), and almost completely suppressed in the second-step revertants able to grow at 42°C by overproducing both GroEL-GroES and DnaK-DnaJ. This suggested the unique importance of GroE and DnaK among the HSPs that are primarily under σ^{32} control (57). Massive accumulation of protein aggregates was subsequently demonstrated in the $\Delta rpoH$ mutants but much less accumulation in the above revertants, suggesting that these chaperones play major and cooperative roles in preventing intracellular protein aggregation (37). Overproduction of GroE was also shown to suppress the temperature-sensitive phenotype of a variety of missense mutants, suggesting that GroE interacts with many proteins

Table 2. Heat-inducible proteins in *Escherichia coli*

Location (min)	Protein	Mol wt (kDa)	Function	Reference(s)
σ^{32} regulon				
0.3	HtpY	21	Unknown	71
0.3	DnaK	69	Chaperone	6
0.3	DnaJ	39	Chaperone	5
10.0	ClpP	24	Protease	67
10.0	ClpX	46	Protease	36
10.0	Lon	89	Protease	31
10.0	PpiD	70	Peptidyl-prolyl isomerase	21
10.0	HslA	65	Unknown	15
10.8	HtpG	70	Chaperone	7
19.2	HslC	80	Unknown	15
39.3	GapA	36	Dehydrogenase	14
39.8	HslK	49	Unknown	15
40.3	HtpX	32	Unknown	56
41.0	PrpA	24	Phosphatase	73
56.0	ClpB	84	Chaperone	53, 104
56.8	GrpE	26	Nucleotide exchange factor	63
69.0	σ^{70}	70	Sigma factor	13
69.2	FtsJ	26	Unknown	45
69.2	HflB (FtsH)	70	Metalloprotease	45, 115
75.0	HslO (Hsp33)	33	Chaperone (redox-controlled)	15, 46
75.0	HslP	30	Unknown	15
81.2	HtrM (RfaD)	34	Epimerase	93
83.2	IbpB (HtpE, HslS)	16	Chaperone	1, 16
83.2	IbpA (HtpN, HslT)	16	Chaperone	1, 16
89.0	ClpY (HtpI, HslU)	49	Protease	16
89.0	ClpQ (HtpO, HslV)	21	Protease	16
90.0	HtrC	21	Unknown	92
90.7	MetA (HTS)	36	Homoserine transsuccinylase	8
94.2	GroEL	60	Chaperone	87
94.2	GroES	16	Chaperone	113
94.2	HslW	22	Unknown	15
94.8	HslX	51	Unknown	15
94.8	HslY	45	Unknown	15
94.8	HslZ	37	Unknown	15
σ^{E} regulon				
3.9	DegP (HtrA)	50	Protease (periplasm)	64, 105
55.5	σ^{E}	22	Sigma factor	65, 94
74.9	FkpA	29	Peptidyl prolyl isomerase	20
77.5	σ^{32}	32	Sigma factor	58, 128
Others				
29.2	PspA	28	Unknown	75
29.7	HslE	60	Unknown	15
29.7	HslF	51	Unknown	15
29.7	HslG	41	Unknown	15
30.6	HslI (HtpH)	36	Unknown	15
30.6	HslJ	14	Unknown	15
69.2	HslM	31	Unknown	15
75.0	HslQ	24	Unknown	15
75.0	HslR (Hsp15)	15	Chaperone (RNA binding)	15, 55
93.5	LysU	60	Lysyl-tRNA synthetase	60

(117). Numerous studies on in vitro folding of specific proteins assisted by the GroEL or DnaK-DnaJ chaperones gave fully consistent results (25, 33, 43). The primary function of the chaperones is now understood in terms of their properties in binding to nonnative polypeptides that are generated during protein synthesis or by heat denaturation of existing proteins.

In addition to the chaperones important for protein folding, most if not all the ATP-dependent proteases and subunits are members of the σ^{32} regulon and are well conserved: Lon, two-component Clp proteases (the catalytic subunits ClpP and ClpQ [HslV], and the regulatory subunits ClpX, ClpY [HslU] and possibly ClpB that belong to the HSP100/ Clp family of chaperones [35, 38, 99]), and a membrane-bound metalloprotease, FtsH (HflB) (45, 115). The cytoplasmic proteases (and possibly FtsH) can degrade a variety of abnormal proteins as well as some specific substrates in vivo. Such roles for proteases are particularly vital during stresses that exacerbate the occurrence of damaged proteins (35). Interestingly, these same proteases can degrade σ^{32} in vivo and in vitro (45, 51, 52, 115, 120), suggesting that the bulk of heat shock proteases may also play active roles in modulating the heat shock response (see below).

The σ^{E} Regulon: Protection against Extracytoplasmic Stress

The discovery of σ^{E} (σ^{24}), the second heat shock σ factor in *E. coli* (26, 119), came from studies of *rpoH* transcription; one of the *rpoH* promoters (P3) was uniquely activated at very high temperature (45 to 50°C) whereas other promoters were virtually inactive. Under these conditions, the σ^{E}-promoted *rpoH* transcription effectively places the σ^{32} regulon under σ^{E} control, resulting in selective expression of cytoplasmic as well as periplasmic HSPs that are crucial for survival under extreme heat stress. σ^{E} is also autogenously regulated: transcription of *rpoE* (encoding σ^{E}) depends in part on RNAP-σ^{E} (94, 98), indicating that both σ^{32} and σ^{E} are members of this regulon (Table 2).

Other members of the regulon include a periplasmic protease DegP (HtrA) and a periplasmic peptidyl prolyl isomerase FkpA that can assist protein turnover or folding, respectively (20, 26, 74, 105). In addition, some of the proteins whose synthesis is enhanced upon σ^{E} overproduction are potential members of this regulon (94, 98). Contrary to the earlier reports, σ^{E} is essential for growth at all temperatures (23). The σ^{E} regulon thus appears to provide essential functions related to protein folding and turnover in the cell envelopes that are particularly important at extreme temperatures.

Not only heat or ethanol but misfolded outer membrane or periplasmic proteins can induce the σ^{E} regulon (69, 70). The σ^{E} activity is negatively controlled by a membrane-bound anti-σ factor, RseA, and another periplasmic protein, RseB, which binds to RseA (22, 72). Furthermore, part of the regulon (e.g., DegP) is under control of the Cpx two-

component signal transduction pathway that mediates the synthesis of periplasmic protein-folding factors such as DsbA (disulfide bond-forming enzyme), indicating that at least two partially overlapping regulatory systems are involved in the response to extracytoplasmic (periplasmic) stress (17, 20, 91). (See chapter 2 for further discussion.)

REGULATION OF THE σ^{32}-MEDIATED HEAT SHOCK RESPONSE

In a modest temperature upshift from 30 to 42°C, HSP synthesis increases almost immediately and reaches a maximum induction (10- to 15-fold) within 5 min (44, 59, 124), where HSPs represent over 20% of total proteins synthesized. The rapid induction is followed by a gradual decrease, during the adaptation phase, to attain a new steady-state level (two- to threefold of the preshift level) by 20 to 30 min. In a more extreme heat shock involving lethal temperature (e.g., 50°C), synthesis of most proteins ceases, presumably due to inactivation of σ^{70} (10), whereas that of HSPs under the control of σ^{32} or σ^{E} increases and continues as long as the cell can produce proteins. The rapid and dramatic change in transcription of the heat shock genes permits the cell to quickly attain new steady-state HSP levels. Although the difference in HSP levels between cells growing at different temperatures is generally small (two- to threefold between 30 and 42°C), the regulation of HSP levels is very tight.

Besides temperature changes, sublethal concentrations of ethanol and production of abnormal or heterologous proteins cause less rapid but prolonged induction of HSPs to appreciable extents (e.g., 34, 50, 121). Starvation for carbon source or amino acids; exposure to DNA-damaging agents, antibiotics, or heavy metals; oxidative stress; and phage infection can also induce at least some HSPs, although the induction is less striking and the underlying mechanisms are less well understood (89).

Temperature Regulation of the σ^{32} Level and Activity

The intracellular level of σ^{32} at low temperature (30°C) is very low, less than 50 (perhaps 10 to 30) molecules per cell (19, 107). The translational repression and protein instability effectively keep the σ^{32} level to a minimum when only low levels of HSPs are required. Upon shift to 42°C, the σ^{32} level increases 15- to 20-fold within 5 min and then declines to a new steady-state level, two- to threefold higher than the preshift level (107). Both the extent of σ^{32} increase and the time course are sufficient to account for the increased transcription of heat shock genes. Such a relatively modest heat shock (30 to 42°C) activates *rpoH* mRNA translation and transiently stabilizes σ^{32} (81, 107), whereas a severe heat shock (e.g., 50°C) can also activate *rpoH* transcription.

Upon downshift of temperature, transcription of the heat shock genes rapidly decreases as a result of decrease in σ^{32} activity (rather than σ^{32} level) caused by excessive HSPs, particularly the DnaK-DnaJ chaperones (108, 110). Within 5 min after a shift from 42 to 30°C, the σ^{32} activity drops 10-fold with only less than twofold decrease in the σ^{32} level. Transcription of the heat shock genes can therefore be regulated by controlling synthesis, degradation, or activity of σ^{32}, depending on the nature and extent of the temperature change. Analyses of each of these processes should be basic to our understanding of the early events of the heat shock response, including the nature of sensors and signaling pathways (see Fig. 1).

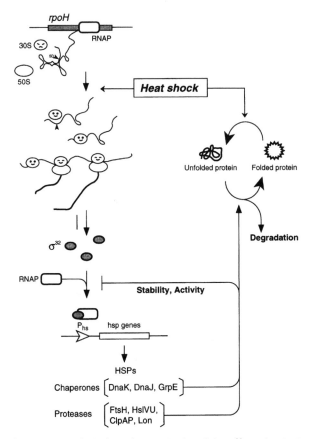

Figure 1. Hypothetical regulatory circuits of the σ^{32} regulon in *E. coli*. See text for explanation. RNAP, RNA polymerase core; P_{hs}, heat shock promoter.

Transcriptional Regulation of σ^{32} Synthesis

At least four promoters are involved in *rpoH* transcription; three of them (P1, P4, P5) are recognized by RNAP-σ^{70}, whereas one (P3) is transcribed by RNAP-σ^E (27, 80) (Fig. 2a). The P1 and P4 promoters are responsible for transcription under most conditions, whereas P5 requires cAMP and its receptor (CRP) for activity. The P3 promoter recognized by RNAP-σ^E is active only at very high temperatures (45 to 50°C) where σ^{70} is largely inactivated (26, 119). The P3- and P4-promoted transcription of *rpoH* also appears to be negatively modulated by DnaA, a key protein that controls chromosomal DNA replication (118). An additional negative control of P4–P5 transcription by the repressor that consists of a cAMP–CRP/CytR nucleoprotein complex was reported recently, implying a link between *rpoH* transcription and nucleoside metabolism (48).

Other inducers of the heat shock response, such as ethanol and DNA gyrase inhibitors, appear to induce HSP by inducing σ^{32} synthesis at least at the transcriptional level (27, 66, 80). Transcriptional regulation of *rpoH* is basically important in determining the basal level of HSP expression under a variety of steady-state conditions and specifically in maintaining the critical levels of *rpoH* mRNA when exposed to extreme heat stress, ethanol, or extracytoplasmic stress, perhaps as part of final efforts for cell survival. Despite the importance of *rpoH* transcriptional regulation, however, the potential regulatory roles of various promoters have yet to be fully explored.

Heat-Induced Synthesis of σ^{32} Mediated by the *rpoH* mRNA Secondary Structure

Several lines of evidence initially suggested that the heat induction of σ^{32} synthesis occurs at the translational level (41, 49, 81, 107): (i) synthesis of σ^{32} but not of *rpoH* mRNA increases markedly upon heat shock, (ii) expression of the *rpoH-lacZ* gene fusion (translational fusion) but not the operon fusion (transcriptional fusion) can be induced, (iii) increase in the synthesis of σ^{32} (or σ^{32}-β-galactosidase fusion protein) precedes that in the *rpoH* mRNA level, and (iv) heat induction of the fusion protein occurs even when RNA synthesis is inhibited by rifampin.

Subsequent deletion analysis of the *rpoH-lacZ* gene fusion identified two 5′-proximal regions of *rpoH* mRNA, called A and B, involved in thermoregulation (49, 81) (Fig. 2a). Region A (nt 6–20) is located downstream of the AUG initiation codon and enhances translation presumably through interaction with 16S rRNA (downstream box) (103). Region B

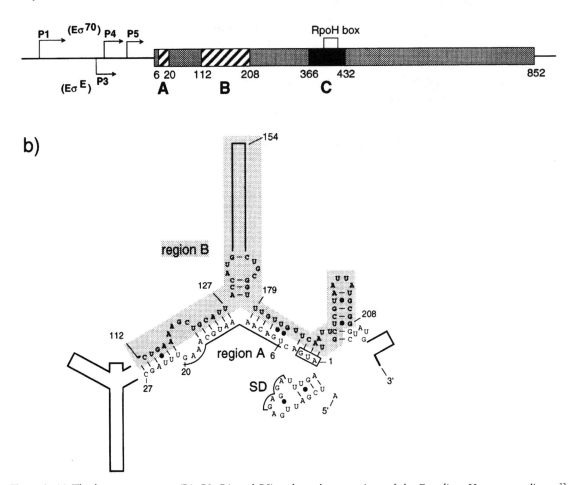

Figure 2. (a) The known promoters (P1, P3, P4, and P5) and regulatory regions of the *E. coli rpoH* gene encoding σ³². Regions A and B within the coding sequence are involved in the formation of the mRNA secondary structure shown in (b), whereas region C on the σ³² protein is implicated in the control of σ³² stability and/or in the binding with core RNAP (see text). The RpoH box is conserved among σ³² homologs but not in other σ factors (see Fig. 3). Numbers refer to nucleotides of the coding region.

(nt 112–208) is an internal segment that represses translation at low temperature (81, 129). Random point mutations causing high constitutive expression occurred mostly within these two regions (129) in agreement with the results of deletion analysis (81).

Computer prediction revealed that the 5′-portion of the *rpoH* mRNA might form a complex secondary structure by base pairing between region A (and the AUG codon) and part of region B (Fig. 2b) (81). Analysis of mutations predicted to disrupt a specific base pairing or compensatory mutations that should recover the base pairing revealed the importance of some of the base pairings for restricted translation at low temperature (30°C) (129). Although possible involvement of certain transacting factor(s) in modulating the thermoregulation was initially considered (129), extensive search for such factors was unsuccessful (Yuzawa and Morita, unpublished).

Meanwhile, the direct structural probing of the 5′-portion (ca. 240 nt) of *rpoH* mRNA confirmed the existence of the mRNA secondary structure and permitted the further pinning down of the minimal RNA segments required for regulation (78). Thus, appropriate stability of mRNA secondary structure as well as some of the base pairings seemed critical for repression at low temperature and induction upon heat shock (78). The temperature-melting profiles and toeprint analyses (using only purified wild type and several mutant mRNA, 30S ribosomes, and tRNA$_{Met}^f$) revealed strong correlation between the thermostability or the formation of mRNA-30S ribosome-tRNA$_{Met}^f$ complex in vitro and the expression levels in vivo at different temperatures (79). These results suggested that the *rpoH* mRNA alone, with no additional regulatory factors, acts as a thermosensor. The thermal melting of RNA structure at the initiation region should permit the ribosome entry followed by translation initiation (79).

The heat-induced σ^{32} synthesis had been thought to be transient and followed by a shutoff of *rpoH* translation during the adaptation phase of the heat shock response (107). Moreover, the DnaK chaperone team seemed to be required for the shutoff but not for the induction of σ^{32} synthesis (106). Subsequent work with the *rpoH-lacZ* gene fusion appeared to confirm these observations and implicated the involvement of a segment of σ^{32} (region C; Fig. 2a) both in the shutoff of synthesis and in the stability control (82). Contrary to such expectations, however, our recent studies suggest that the apparent "shutoff" observed previously resulted from the unexpectedly high instability of σ^{32} during the adaptation phase (52) (see below). It now seems evident that heat shock enhances the synthesis of σ^{32} by direct melting of the *rpoH* mRNA secondary structure and that the enhanced translation does not shut off and continues as long as the cells are maintained at high temperature (M. T. Morita et al., unpublished). Thus, the translational control of σ^{32} synthesis appears to be an event determined by the ambient temperature and is independent of the cellular level of chaperones (and proteases) and the state of protein folding.

Regulation of σ^{32} Degradation

Although σ^{32} is unstable during normal growth at 30°C (or at 42°C), marked stabilization occurs immediately upon shift to 42°C and continues for 4 to 5 min (107). This initial stabilization accounts for, at least in part, the almost instantaneous induction of HSP synthesis observed. Mutations in any member of the DnaK-DnaJ-GrpE chaperone team markedly stabilize σ^{32} under nonstress conditions (106, 112), indicating that this chaperone team is actively involved in in vivo turnover of σ^{32}. A membrane-associated metalloprotease FtsH was first implicated as a protease responsible for σ^{32} degradation (45, 115), which was confirmed by analyzing proteolysis with purified FtsH (115). However, a set of cytosolic proteases including HslVU, ClpAP, and Lon was subsequently shown to participate in turnover of σ^{32} in vivo (51) and in vitro (52, 120; Huang and Goldberg, personal communication). Although it is difficult to evaluate the relative contributions of different proteases in vivo, the cytosolic proteases appear to con-

tribute appreciably, perhaps comparable to FtsH (51).

Mutational analyses implicated the involvement of specific segments of σ^{32} for degradation. Thus, deletion or a frameshift mutation of an internal region (Region C, aa 122–144; Fig. 2a) markedly stabilizes the σ^{32}-β-galactosidase fusion protein (82). Importantly, a short segment of 9 residues (aa 132–140) within region C is uniquely conserved among the σ^{32} homologs (not found in other σ's) and thus denoted "RpoH box" (83) (see below). Moreover, synthetic oligopeptides (13-mers) containing these regions bind DnaK with high affinities (68). On the other hand, certain mutations in region C were shown to affect binding of σ^{32} to RNAP rather than binding to DnaK or stability of σ^{32} (2, 47). Although the C-terminus of σ^{32} was also suggested to be involved in initial attack by FtsH protease (9), other evidence does not support such a possibility (2, 82; M. T. Morita et al., unpublished). Whatever the mechanisms, DnaK and RNAP appear to compete with each other in binding to σ^{32} at some specific region(s): the chaperone-binding leads to σ^{32} degradation, whereas the RNAP binding stabilizes (and activates) σ^{32}. The protection of σ^{32} by RNAP from the protease's attack was observed with both FtsH (9, 116) and HslVU proteases (52). The finding that none of the members of the DnaK-chaperone team, in various combinations, affect the FtsH-mediated digestion of σ^{32} in vitro (9) may indicate that the role of DnaK-DnaJ chaperones in promoting σ^{32} degradation is simply to sequester σ^{32} away from RNAP. Obviously, further work is required to understand how σ^{32} is recognized by the proteolytic machinery and how the chaperones are involved in assisting σ^{32} degradation.

A new insight into the mechanism of heat shock regulation was recently obtained by studying σ^{32} degradation in vitro: both HslVU- and FtsH-mediated degradations of σ^{32} were found to exhibit unusual dependency on high temperature (52). The ratio of proteolytic activity of HslVU at 44°C to that of 35°C was about 15 with σ^{32}, in contrast to 2.5 or less with other substrates. Similar though less striking results were obtained with FtsH-mediated proteolysis. Such a high susceptibility of σ^{32} to the proteases at high temperature may involve a reversible conformational change of σ^{32} (52). Reexamination of in vivo turnover of σ^{32} in cells steadily growing at different temperatures suggested a clear inverse correlation between stability of σ^{32} and growth temperature: the higher the temperature, the lower the stability (52), consistent with the previous results of Tilly et al. (112). The high susceptibility of σ^{32} to proteases may explain the marked instability observed during the adaptation phase, which probably counterbalances the elevated synthesis of σ^{32}, since excessive accumulation of σ^{32} is deleterious to the cell.

Regulation of σ^{32} Activity

Whereas increased synthesis and stabilization of σ^{32} rather than activation of preformed σ^{32} largely account for HSP induction following temperature upshift, inactivation of σ^{32} partially contributes to the rapid shutoff of HSP synthesis early during the adaptation phase (107, 127). This inactivation of σ^{32} may be accompanied by the DnaK-DnaJ chaperone-mediated reactivation of partially heat-inactivated σ^{70}, as suggested by the in vitro experiments using purified RNAP, σ^{70}, and σ^{32} (10). Besides, reversible inactivation of σ^{32} is a predominant regulatory event following a temperature downshift or under the conditions in which excess HSPs are produced (108). Mutations in *dnaK*, *dnaJ*, or *grpE* result in a failure to inactivate σ^{32} (cited in reference 108). The involvement of DnaK-DnaJ chaperones in inactivation of σ^{32} was clearly shown by experiments with the $\Delta ftsH$ mutant (carrying the *sfhC* suppressor) in which the stability control was much reduced (109). These combined results indicate that the chaperones inhibit σ^{32} activity by inhibiting the function or formation of RNAP-σ^{32}.

Following the observation of physical interaction between DnaK (or DnaJ) and σ^{32} in crude lysates or in the purified system (30, 61), the mode of these interactions was studied extensively. Whereas DnaK binds weakly to σ^{32} and is dissociated by ATP, DnaJ binds to σ^{32} with much higher affinities and is insensitive to ATP. The binding of DnaK to σ^{32} is greatly facilitated by DnaJ and ATP, resulting in the formation of a (DnaK-ADP)-DnaJ-σ^{32} ternary complex. This complex is presumably involved in sequestering σ^{32} from the core RNAP or inactivating σ^{32} in vivo, since DnaK and DnaJ may compete with core RNAP for binding to σ^{32} and can synergistically inhibit σ^{32} activity in vitro (29, 62). GrpE then binds to the ternary complex and stimulates ADP release and complex dissociation by ATP rebinding. A binding and release cycle of the chaperone system therefore appears to play an important role in the modulation of σ^{32} activity (and perhaps stability) in vivo (29). Whether the chaperones associate only with free σ^{32} or can remove σ^{32} from RNAP-σ^{32} remains unsettled. As seems evident from our discussion in the previous section, the regions of σ^{32} implicated in degradation may well turn out to be important in the activity control as well, because of the intimate relationships between the two processes.

THE CHAPERONE (AND PROTEASE)-MEDIATED AND TEMPERATURE-DIRECTED SIGNALING PATHWAYS: BIPARTITE CONTROL

Heat shock enhances cellular levels of partially unfolded or nonnative proteins that preferentially bind to DnaK and/or DnaJ among other chaperones. This reduces the level of free DnaK-DnaJ chaperones, which in turn stabilize σ^{32} by shifting the equilibrium from the chaperone-bound form of σ^{32} to RNAP-σ^{32}, thus leading to HSP induction. By the time cells reach the adaptation phase, sufficient free chaperones accumulate and exert negative feedback control on the stability and activity (but not synthesis) of σ^{32} (see Fig. 1). It has been proposed that a pool of free DnaK (19) and/or DnaJ (11) acts as a "cellular thermometer," which monitors the state of unfolded proteins and adjusts the amount of active σ^{32} and expression of heat shock genes accordingly. This kind of model was recently supported by demonstrating that the DnaK-DnaJ chaperones are normally limiting in vivo and can serve as a highly sensitive stress sensor in response to protein misfolding: even small increases (1.5-fold) in the DnaK-DnaJ level reduced the level and activity of σ^{32} and caused faster shutoff of the heat shock response, whereas small decreases in the chaperone level caused inverse effects (116).

Production of various abnormal proteins in *E. coli* without temperature upshift can mimic the heat shock response and induce HSP synthesis (34). The level of unfolded proteins rather than protein degradation is important in such induction (90). The accumulation of secretory protein precursors, abnormal proteins containing amino acid analogs, or certain heterologous proteins can all stabilize σ^{32} (50, 121) but do not enhance the synthesis of σ^{32} (50). Like the DnaK-DnaJ chaperone mutants, those lacking FtsH (109) or the cytosolic proteases (51) exhibit enhanced HSP levels by stabilizing σ^{32}. Thus, a set of ATP-dependent proteases as well as the DnaK-DnaJ chaperones may collectively serve as a cellular thermometer in modulating the heat shock response (Fig. 1).

On the other hand, heat-induced synthesis of σ^{32} mediated by the *rpoH* mRNA secondary structure occurs by directly reflecting the ambient temperature and does not involve the DnaK-DnaJ-mediated negative control, because it occurs normally even in the mutants defective in these chaperones (81, 107) or the ATP-dependent protease (109). Contrary to the previous expectation, heat-induced synthesis of σ^{32} does not shut off during the adaptation phase (M. T. Morita et al., unpublished). In addition, the intracellular accumulation of abnormal proteins only stabilizes σ^{32} and cannot induce *rpoH* translation (50). Furthermore, a putative conformational change of σ^{32} reflecting ambient temperature appears to modulate its stability to avoid excessive σ^{32} at high temperature where σ^{32} synthesis continues at higher rates (52). The results of analysis of *rpoH* homologs from a number of proteobacteria are also consistent with the notion of "bipartite control" of σ^{32} levels (see below). Thus, two types of signaling pathways, chaperone (and protease)-mediated and temperature-directed, must have evolved to ensure the rapid and transient induction of HSP through enhancing the σ^{32} level in *E. coli*. These pathways and the effects on the σ^{32} level during the induction and adaptation phases are summarized in Table 3.

RpoH (σ^{32}) HOMOLOGS: CONSERVATION AND DIVERGENCY

A number of *rpoH* homologs have been isolated and characterized, not only from γ (including *E. coli*) but also from α and β (but not δ or ε) proteobacteria (Fig. 3). Most of the homologs can complement the temperature-sensitive growth of the *E. coli* $\Delta rpoH$ mutant at least partially, indicating that they are functional homologs of *E. coli* σ^{32} (see reference 83 for earlier references). A single *rpoH* gene was found from each species except that three homologs with distinct functional and regulatory features were obtained from *Bradyrhizobium japonicum* (85, 86). Sequence comparisons among various homologs revealed several conserved segments that may be of regulatory significance. First, they all contain the

highly and uniquely conserved segment called "RpoH box" (83) (see Figs. 2a and 3). Second, both the downstream box and mRNA secondary structures similar to that of *E. coli rpoH* were detected among the homologs from most γ but not α proteobacteria (83). In agreement with these results, immunoblotting with anti-*E. coli* σ^{32} permitted the identification of putative σ^{32} homologs in extracts of heat-shocked cells from most enteric bacteria tested (84, 95).

Heat shock experiments carried out with several γ and α proteobacteria revealed that both cellular levels and synthesis rates of RpoH homologs increase transiently in most cases (84, 95), although the σ^{32} level in *Klebsiella pneumoniae* hardly increased under the conditions used (95). The increased synthesis was observed with all members of the γ and α subgroups tested; it appears to occur primarily at the translation level in the γ subgroup as in *E. coli* (84), whereas it occurs at the transcription level by activating the σ^{32}-dependent promoter in the α subgroup including *Caulobacter* (96, 122), *Bradyrhizobium* (*rpoH*₃) (86), and *Agrobacterium* (K. Nakahigashi, unpublished). As to the control of RpoH protein stability, most homologs from the γ subgroup are normally unstable and stabilized upon heat shock as in *E. coli* (84). The RpoH of *Proteus mirabilis* is an exception in that it is rather stable at low temperature and little further stabilized upon heat shock. Furthermore, the *Agrobacterium* RpoH is quite stable at low temperature and destabilized during the adaptation phase of the heat shock response (K. Nakahigashi, unpublished data). Thus, the basic strategy of enhancing the RpoH level to cope with heat stress appears to be widely conserved, but detail mechanisms vary considerably.

HEAT SHOCK REGULATION IN GRAM-POSITIVE BACTERIA

Gram-positive bacteria use very different and diverse regulatory strategies depending on the specific heat shock genes and specific bacteria. The heat shock genes can be divided on the basis of *cis*-elements and the transcriptional activators or repressors used. In *B. subtilis* (low G+C gram-positive bacteria), they are divided into at least four classes. Class I genes (CIRCE/HrcA regulon) encode major chaperones DnaK-DnaJ-GrpE and GroEL-GroES, and their transcription depends on the vegetative σ factor σ^A (σ^{70}) and is negatively controlled by a repressor encoded by *hrcA*, the first gene of the *dnaK* operon. The HrcA repressor exerts its activity through binding to the operator, a very well conserved 9-bp inverted repeat with a 9-bp spacer, called CIRCE (100, 126, 131) (see Table 1). Class II is a large group of

Table 3. Bipartite signaling pathways for the σ^{32}-mediated heat shock response in *E. coli*

Sensor	Reaction affected	Effects on the σ^{32} level	
		Induction phase	Adaptation phase
Chaperone (and protease)-mediated			
DnaKJ chaperones[a] (and proteases)	Degradation	Increase	Decrease
Temperature-directed			
rpoH mRNA	Synthesis (translation)	Increase	Increase
σ^{32} protein	Degradation	(Decrease)[b]	Decrease

[a] The heat-induced decrease in the pool of free chaperones (and proteases) transiently stabilizes and increases the level and activity of σ^{32}.
[b] σ^{32} should become highly susceptible to proteases but would not actually decrease the σ^{32} level because of the expected shortage in free proteases.

γ

β

α

Eco 1 MTDKMQSL-ALAPVGN---LDSYIRAANAWPMLSADEERALAEKLHY 43
Cfr 1 MTKEMQNL-ALAPVGN---LESYIRAANAWPMLSADEERALAEKLHY 43
Ecl 1 MTKEMQTL-ALAPVGN---LESYIRAANTWPMLTAEEEKELAEKLHY 43
Sma 1 MTKEMQTL-ALVPQGS---LEAYIRAANAYPMLTAEEEERELAERLHY 43
Pmi 1 MTQEMQSL-ALVPQGS---IEAYIRAANTWPMLTAEEEKELAERLHY 43
Vch 1 MTNQAYPM-ALVSQDS---LDSYIRSVNGYPMLSADEERELAERLHY 43
Pae 1 MTTSLQPVHALVPGAN---LEAYHVSVNSIPLLSPEQERELAERLFY 44
Hin 1 MDKETQM-MLVPQGS---IEGYIRAANEYPMLTAEEEKELAERLYY 42
Bap 1 MINKVQIL-SVTPPGN---LDAYIRIANLWPMLSIEEEKKLTKRLRY 43
Xca 1 MNQITSTALVANNLPIPSALQS---LDAYIGAVHQIPVLSVDEEQNLARRFRD 50
Phy 1 MREATTMLPQPLASGVHSTAALPAAWQGSLPSPVSS---LSRYIQAVNRFPVLSEAEEHELARRFHE 64
Axy 1 MSLPLRLPETPAMKQPSTSLATSGNALALAIANPGALGT---IDAYISAVNRLPVLSAERETELGRRLRD 67
Atu 1 MARNSLPT-ITAGEAG---LNRYLDEIRKFPMLEPQEEYMLGKRYAE 43
Bj2 1 MARTAALP-VLNGESG---LSRYLAEIRKFPMLEPQQEYMLAKRWRE 43
Rca 1 MSSYANLP-APSPEQG---LNRYLQEIRKFPLLEPEEEYMLAKRWVD 43
Rsp 1 MSTYTSLP-APSPEQG---LNRYMQEIRKFPLLEPEEEYMLAKRWVD 43
Ccr 1 MAVNSLS-VMSPDGG---LSRYLTEIRKFPMLSKDEEFMLAQRWKE 43
Zmo 1 MATSSTLPAVVPALGGDQSLNHYLADIRKFPILKPEEEYMLAQRFQE 47
Rpr 1 MTNNINALAISSESG---FYSYLQKINKIPSLTQEEEFLLAKSYLE 42
Bj3 1 MQTSHEVARSASVAAAGAAVSAPFLSAYSAAIRRYELLEPGQEQQLARRWHE 52
Bj1 1 MFNNAALP-APSVDAG---LSKYLVEIRKFPLLTPEEELAYARRWRE 43

1.2

Eco 44 HGDLEAAKTLILSHLRFVVHIARNYAGYGLRQADLIQEGNIGLMKAVRRFNPEVGYVRLVSFAVHWIKAEI 113
Cfr 44 QGDLEAAKTLILSHLRFVVHVARNYAGYGLRQADLIQEGNIGLMKAVRRFNPEVGYVRLVSFAVHWIKAEI 113
Ecl 44 QGDLDAAKTLIILSHLRFVAHIARNYSGYGLPQADLIQEGNIGLMKAVRRFNPEVGYVRLVSFAVHWIKAEI 113
Sma 44 EGDLEAAKTLILSHLRFVIHVARSYSGYGLPQADLVQEGNIGLMKAVRHFNPEVGYVRLVSFAVHWIKAEI 113
Pmi 44 KGDIDAKGLILSHLRFVVHVARGYSGYGLPMADLVQEGNIGLMKAVKRFNPEMGVVRLVSFAVHWIKAEI 113
Vch 45 QQDLEAARQMVLAHLRFVIAKSYSGYGLAQADLIQEGNVGLMKAVRKFNPEVGMGVRLVSFAVHWLKAEI 114
Pae 44 HEDLDAAKKLIILSHLRFYIHVARSYSGYGLPQADLIQEQNIGLMKAVKRFNPEVGYVRLVSFAVHWIKSEI 112
Hin 43 NGDLDAAKTLIILSHLRFVIHISRNYSGYGLLQSDLVQEGNIGLMKAVRFNPEIGVVRLVSFAVHWIKSEI 113
Bap 44 ELDLDAARELVHSHLRFVVHSHLRFVVHARNYNGYGLPLGDLIQEGNIGLMKAVKRFPEMGVVRLVSFAVHWIRAEM 120
Xca 51 TNDLDAARKLVLANLRYVVMIARQYFGYGLPEADLIQEGNVGLLKAVRRFDPYKGVRFITFAAYWIKAEI 134
Phy 65 QEDLGAARELILSHLRLVVSVARQYLGYGLPHADLIQEGNVGLMQAVKKFDPERGVRLVSFAVHWIKASI 137
Axy 68 HGDRDAAHKLVTSHLRLVAKIAMGYRGYGLPIGEVVSEGNVGLMQAVKKFDPERGFRLATYAMWWIKASI 113
Atu 44 HDDRDAAHKLVTSHLRLVAKIAMGYRGYGLPISEVVSEGNVGLMQAVKRFDPEKGFRLATYAMWWIRAAI 113
Bj2 44 HQDPKAAHRLVTSHLRLAAKIAMGYRGYGLPQAEVISEANVGLMQAVVKFKFEPEKGFRLATYAMWWIRASI 113
Rca 44 HQDNRAAHRLVTSHLRLAAKIAMGYRGYGLPQAEVISEANVGLMQGVKKFPDERGFRLATYAIWWIRASI 113
Rsp 43 HQDPKAASRLVTSHLRLVAKIAMGYRQYGLPVSELISEGNIGLMQGVKKFPERGVRLATYAIWWIRASI 117
Ccr 48 ENDLQAANKLVTSHLKLVAAISSSYKTYGLPITELVSEGNIGLPITELVSEGNIGLPITELVSEGNIGL 113
Zmo 53 TRDRGAADALVTSHLRLAAKLARGYKGYGLPMVDLIAEANLGLVIAASRFEPGRGARFSTYAIWWIKAAI 122
Rpr 44 HDRDAAYHLVTSHLRLVAKIAMRYRGYGLPIAEVISEGNIGLMQAVRRFDPDRGVRLATYAMWWIRASI 113
Bj3
Bj1 44

2.1   2.2   2.3

Eco 114 HEYVLRNWRIVKVATTKAQRKLFFNLRKTKQRL----GWFNQDEVEMVARELGVTSKDVREMESRMA-A 177
Cfr 114 HEYVLRNWRIVKVATTKAQRKLFFNLRKTKQRL----GWFNQDEVEMVARELGVSSKDVREMESRMA-A 177
Ecl 114 HEYVLRNWRIVKVATTKAQRKLFFNLRKTKQRL----GWFNQDEVEMVARELGVTSKDVREMESRMA-A 177
Sma 114 HEYVLRNWRIVKVATTKSQRKLFFNLRKNKKRL----GWFNQDEVELVAKELGYSESDVREMESRMS-A 177
Pmi 114 HEYVLRNWRIVKIATTKAQRKLFFNLRKSKKRL----GWFNNGEVETVARELGYEPAEVREMESRLA-A 177
Vch 115 HEFILRNWRIVKIATTKAQRKLFFNLRSQKKRL----AWLNNEEVHRVAESLGVEPREVREMESRLT-G 178
Pae 113 HEFIKRNWRIVKVATTKAQRKLFFNLRKTKKRL----GWFNENEVDMVANELGVSKEDVIEMESRMS-A 177
Hin 114 HEYVLRNWRIVKVATTKSQRKLFFNLRKTKKRL----GWFNEEEIQIVARELGVSSRDVREMESRMS-A 177
Bap 121 HEFILRNWRIVKVATTKAQRKLFFNLRKSKTRL----GWLNASEVTAVAKDLNVSERRVEMRESRLS-G 185
Xca 135 NDYILRNWRLVKIATTKAQKKLFFNLRKLLGS----EPLTRAKADAIAETLAVKPEEVAEMHARFA-G 197
Phy 138 HEYIIRNWRIVKVATTKAQKKLFFNLRSMRPDG----QTLDPEQVEHIARELNVRREDVSEMEVRLS-G 201
Axy 114 QEYILRSWSLVKMGTTANQKRLFFNLRRLKGRIQAIDDGDLKPEHVKEIATKLQVSEEEVISMNRRLH- 181
Atu 114 QEYILRSWSLVKMGTTANQKRLFFNLRKAKSKINALDEGDLRPDQVATIAKRLGVTDQDVIDMNRRLG- 181
Bj2 114 QEYILRSWSLVKLGTTSAQKKLFFNLRRLKGRIQAIDDGDLKPEHVKEIATKLQVSEEEVISMNRRLH- 181
Rca 114 QEYILRSWSLVKLGTTSAQKKLFFNLRKAKLGAAEEGDLRPDAYAKIAHDLNVSEGDVIEMNRRLA-G 182
Rsp 114 QEYILRSWSLVKLGTTSAQKKLFFNLRKAKLGALEEGDLRPENVAIDLGVSETEVIDMNRRLS-G 182
Ccr 113 QEYILRSWSLVKMGTTAAQKKLFFSLNKVKHKITNLYSRAITTDDFAQIADELGVSVNEVSEMNTRIS-G 181
Zmo 118 QEYILRSWSLVKMGTTAAQKKLFFSLNKVKHKITNLYSRAITTDDFAQIADELGVSVNEVNMNRMAMG 187
Rpr 114 QEYILKSWSLVKMGTTAAQKKLFFSLNKVKHKITNLYSRAITTDDFAQIADELGVSVNEVSEMNTRIS-G 182
Bj3 123 HEYILRSWSLVKIGTTAASQKKLFFKLRRAKSAISALQDGDLRPEQVRLIAERLKVAERDYVAMDRARLR- 190
Bj1 114 QEYILRSWSLVKIAASASQKKLFFKLRRAKSAISALQDGDLRPEQVRLIAERLKVAERDYVAMDRARLR- 181

2.4   RpoH Box   3.1

Eco 178 QDMTFDLASDDDS-DSHAMAPVLYLQDK-SSNFADGIEEDNWEEQAANRLTDAMQGLDERSQDIIRARWL 245
Cfr 178 QDMTFDMSSDDES-DSQPMAPVLYLQDK-SSNFADGIEEDNWEDQAANKLTHAMEGLDERSQDIIRARWL 245
Ecl 178 QDMTFDMSADDASDSQPMAPVLYLQDK-SSNFADGIEEDNWEDQAANKLTFAMEGLDERSQDIIRARWL 246
Sma 178 QDMAFDMSADDSD-DPHPVAPVLFLEDK-SSDFAEGIEEDNWSNAADKLAYALEGLDERSQDIIRARWL 246
Pmi 178 QDMAFDMSADDSD-DPHPVAPVLFLEDK-SSDFADGIEEDNWDNHAADRLTLAIKTLDERSQDIIRARWL 246
Vch 178 QDAAFEMSAEDDENGMAYTAPVLYLEDK-HSDLADNLEAENWEAHTTQRLSMALASLDGSQHVIRARWL 246
Pae 179 QDAAFDPAADADD-ESAYQSPAHYLEDH-RYDPARQLEDADWSDSSSANLHEALEGLDERSRDILQQRWL 246
Hin 177 ADVGFDLPTDDAE--TETYSPALYLEDK-SSNFAAELENENFESQATEQLGAALQSLDARSQDIIRARWL 243
Bap 178 QDVAFNPSPEEHC-DSKTNSSIQYLEDK-SDNPSAGIEEDNWEEHAANKLSSALLRLDERSRHIIRARWL 245
Xca 185 RDIGFDASSDEDD-DHGPPSPVSYLVAN-EEDPSQAYERHDSEDNQLQLLREGMAGLDTRSRDIIVKRRWL 252
Phy 198 QDVALEAPIDSDE-EADWRAPLALYLPDP-SGTPEEAVAEAEAEERLSHEGLQQALTQDERSRAITRRWL 266
Axy 202 RDMSLENQDDDDD----SYAPIAYLSDDGRQEPTRVLERAARDQLQGSGLSDALEALDPRSRRIVEARWL 267
Atu 182 GDASLNAPIKASE-GESG-QWDWLVDDH-ESQEAVLIEQDELETRRRMLAKAMGVLNDRERRIFEARRL 246
Bj2 182 GDASLNAPIRD-D-GEAG-EWQDWDLVDNT-PNQEAMMAHEEYDHRRDALNGAMGVLNPRERRIFEAARL 247
Rca 183 SDASLNAQVGAGD-GESATQWQDWLEDED-ADQAEAYAEAELRPPNVLNDRERDILLMAARL 250
Rsp 183 SDASLNATIGS-D-GEGSTQWQDWLEDED-SDQAADYAERDELEIRRELLAQSMSVLNDREKDILVQRRL 249
Ccr 182 PDASLNAPLRA-D-GES--EWQDWLDEADEQVSGETRVALEDEEKSLRMSLLEEAMVELTDRERHILLTERRL 247
Zmo 188 GDSSLNITMRE-D-GEG--QMQDWLVDQE-PLQDQQIEEEESLVRHKLLIEAMDRLNDREHKIIVTERRL 252
Rpr 183 PDLSLNNSINSDD-VESG-ELIELLPELR-PTPEAMAINXVQDNWEQDNWEERNKLKNYTSKRLKLSALLNMQILNDRELRLILKDRKL 292
Bj3 191 GDMSLNARVGG---EESGTELEALLVDGA-VDAETMLADHEQTERRAKALRVALGGLAARERHVFEARL 256
Bj1 182 GDASLNVPIHD-E-DEGG-QTLDWLVDPA-PTCEITLAEEQEAKQRRLALANALANLANRIFTARWL 247

3.2   4.1

Eco 246 DEDNKSTLQELADRYGVSAERVRQLEKNAMKKLRAAIEA 284
Cfr 246 DEDNKSTLQELADRYGVSAERVRQLEKNAMKKLRAAIEA 284
Ecl 247 DEDNKSTLQELADRYGVSAERVRQLEKNAMKKLRAAIEA 285
Sma 247 DDDNKSTLQELADQYGVSAERVRQLEKNMKKMAIEA 285
Pmi 246 EDDNKSTLQELADKYGVSAERVRQLEKNAMKKLRLAIED 284
Vch 247 DDDNKTTLQDLAAKYGVSAERIRQLEKNAMSKLKEAVGEF 296
Pae 247 S-EEKATLHDLAEKYNVSAERIRQLETNALKKLKSRAYNF 284
Hin 244 D-IDNKATLHDLAAKYNVSAERIRQLEKNAMKKLLAVEA 281
Bap 246 DKNKKNTLQNIANNYGVSAERIQIEAANLLKKLAFVA 284
Xca 253 DSESKVTLQELANDEYGVSAERIRQIEANALKKNKALFVA 291
Phy 266 T-EKPATLHELAAEYGVSAERIRQIEAAALKKLRVWLSPQADAVL 309
Axy 268 QDDGGATLHELAQEFGVSAERIRQIEAAALKKRGNNLAA 306
Atu 249 A-EDPVTLEELSSEFDIRERVRQIEVRAFEKVQEAVQKEALEAA------RALRVVDA 300
Bj2 248 A-DEPMTLEDLAAEFGVSERVRQIEVRAFEKVGKTIARAE------QAALEAAH 299
Rca 251 R-DEPVTLEELSSQYDVSRERIRQIEVRAFEKLQARVALAKEKG------MSLPG 299
Rsp 250 T-DDPVTLEELSEGYGVSRERIRQIEVRAFEKLQAKMRELARSKG------MTIPA 298
Ccr 248 K-DDPTTLEELAAQYGVSRERIRQIEAAFEKTMREAIAK------NMVDA 295
Zmo 253 S-DNPKTLEELSQVYGVSRERVRQIEVRAFDKLQKAIMELAGDR------KLLPAMA 302
Rpr 250 T-DTPKTLDILSSEYNISRERIRQIENTAFEKIKKFILNNNREIA 292
Bj3 257 T-ECPVTLDQLRELSGVSRERVRQIEIRAFAKVKAAMLAAQDAP------RAAVCNV 307
Bj1 248 N-EESTTLEELAAEYGVSRERVRQIEERAFQKVKAAMLTSRHEANGPPSSRAKEMKQGVARA 308

4.2

genes (more than 50 members) that are positively controlled by a general stress σ factor, σ^B, and are activated by heat and other stresses, including starvation for glucose or oxygen (42). σ^B is a member of a subgroup of σ factors found in gram-positive bacteria, many of which (including σ^B) are regulated by a complex cascade of protein-protein interactions that also involve protein phosphorylation (see chapter 12). Class III genes (CtsR regulon) encode some of the highly conserved Clp proteins that are negatively regulated by the CtsR repressor and are important in the virulence and survival of several pathogens (24). Finally, Class IV includes the rest of the heat-inducible genes such as *htpG*, *ftsH*, and *lon* controlled by as yet undefined mechanisms. In the case of high G+C gram-positive bacteria such as *Streptomyces*, at least two regulatory mechanisms besides the CIRCE/HrcA system are known to be involved in the control of heat shock gene expression (85).

REGULATION OF THE CIRCE/HrcA REGULON

The CIRCE/HrcA regulon in *B. subtilis* is normally repressed by the HrcA repressor but can be heat-induced by inactivating the repressor. The question of how the HrcA repressor is transiently inactivated upon heat shock has been addressed in vivo and in vitro. Unlike the DnaK-DnaJ-mediated modulation of the σ^{32} regulon in *E. coli*, these operons seem to be modulated by GroE chaperones that can bind to, and facilitate folding of, HrcA and thereby modulate the repressor activity (76). Titration of GroE by stress-induced misfolded proteins results in lower HrcA repressor activity and thus higher expression of the regulon. Similarly, the GroE chaperones in *Bradyrhizobium* seem to be required for repression of the CIRCE-containing GroELS4 operon at low temperature (3). These results suggest that GroE plays an important role in monitoring protein folding in these bacteria. As might have been expected, both ethanol and puromycin that can cause formation of nonnative proteins induce the CIRCE regulon (77). On the other hand, the nonpolar $\Delta dnaK$ mutant of *Lacto-coccus lactis* exhibited elevated levels of HSP, raising the possibility that the DnaK-DnaJ chaperones could be involved in maturation of active HrcA repressor (54). Both the DnaK-DnaJ and GroEL-GroES chaperones are likely to play major roles in assisting protein folding in all these bacteria, but the mode of involvement of the two chaperone systems in the HrcA activity control may vary in different bacteria. Further work on these systems should shed new light on the mechanisms of chaperone-mediated feedback control of the HrcA repressor.

The role of HrcA repressor and CIRCE elements has been found (or implicated) not only in gram-positive and α proteobacteria but also in *Cyanobacteria* and more distantly related eubacteria such as *Chlamydia* and *Spirochaeta* (42, 101). However, their contribution to the heat shock response varies in different bacteria. Both the *dnaK* and *groE* operons are under HrcA/CIRCE control in *B. subtilis*, *L. lactis*, *Mycoplasma*, and possibly other members of the low G+C gram-positive bacteria as judged by the known distribution of CIRCE (42, 54, 101). In contrast, only the *groE* operons are under CIRCE/HrcA control in other groups such as α proteobacteria and the high G+C gram-positive bacteria (97, 101). Furthermore, the *groE* operons of certain proteobacteria including *Agrobacterium tumefaciens* (102; Nakahigashi et al., unpublished data), *Caulobacter crescentus* (4, 97), and *Bradyrhizobium japonicum* (3) are positively controlled by RpoH (σ^{32} homologs) and negatively controlled by HrcA, whereas *dnaK* and other heat shock genes appears to be primarily (or solely) under RpoH control. In addition to temperature regulation, the HrcA/CIRCE system was reported to be involved in the temporal (cell cycle) control of the *groE* expression in *C. crescentus* (4).

CONCLUSION AND FUTURE PERSPECTIVES

Extensive work on the heat shock response in *E. coli* and other bacteria brought us two major regulatory pathways that appear to be of general significance. One is the chaperone (and protease)-mediated feedback or autogenous control of the activity/sta-

---

Figure 3. Alignment of deduced amino acid sequences for σ^{32} homologs (provided by Kenji Nakahigashi). Multiple alignment was carried out with CLUSTAL W using 20 homologs from *Escherichia coli* (Eco, A94012; A00700), *Citrobacter freundii* (Cfr, S04697), *Enterobacter cloacae* (Ecl, D50829), *Serratia marcescens* (Sma, D50831), *Proteus mirabilis* (Pmi, D50830), *Vibrio cholerae* (Vch, U44432), *Pseudomonas aeruginosa* (Pae, U09560; S77322), *Haemophilus influenzae* (Hin, U32713), *Buchnera aphidicola* (Bap, U35400), *Xanthomonas campestis* (Xca, AFO42156), *Pseudomonas hydrogenothermophila* (Phy, IFO14978), *Alcaligenes xyloxydans* (Axy, ABO09990), *Agrobacterium tumefaciens* (Atu, D50828), *Bradyrhizobium japonicum* (Bj2, Y09502), *Rhodobacter capsulatus* (Rca, AFO017436), *Rhodobacter sphaeroides* (Rsp, U82397), *Caulobacter crescentus* (Ccr, U39791; U37792), *Zymomonas mobilis* (Zmo, D50832), *Rickettsia prowazekii* (Rpr, AJ235271), *Bradyrhizobium japonicum* (Bj3, Y09666; Bj1, U55047). Numbers below the sequences show generally conserved regions for all σ factors.

bility of the key regulatory factors such as σ^{32} in *E. coli* and HrcA in *B. subtilis*, which appear to represent ubiquitous and ancient strategies for tightly modulating the HSP synthesis to cope with heat and other stresses. Further mechanistic studies on each of these systems should provide opportunities for our deeper understanding of not only the central issue of heat shock regulation but also the physiological function of molecular chaperones and energy-dependent proteolytic apparatus that appear to be intimately interconnected. The other general pathway appears to be the direct sensing of high temperature by the regulatory elements or factors such as *rpoH* mRNA and σ^{32}. In view of the diverse heat shock regulatory factors and mechanisms that seem to be found in various bacteria as well as the known protein or RNA thermometers in various organisms, one or more thermometers are likely to be found that play key roles in the heat shock response for each species of bacteria.

The divergent regulatory mechanisms already apparent within proteobacteria or gram-positive bacteria would be bound to reflect their divergent ecological niches and should eventually be understood in those contexts. Finally, further understanding of the heat shock regulatory mechanisms should be important for our future endeavors trying to integrate the regulatory elements, factors, sensors, and signaling pathways with those involved in other stress responses to gain new insight into global intra- and intercellular regulatory mechanisms under physiological and pathological conditions.

Acknowledgments. We are grateful to a number of colleagues for helpful comments and to K. Nakahigashi for sequence alignment of *rpoH* homologs.

REFERENCES

1. **Allen, S. P., J. O. Polassi, J. K. Gierse, and A. M. Easton.** 1992. Two novel heat shock genes encoding proteins produced in response to heterologous protein expression in *Escherichia coli. J. Bacteriol.* **174:**6938–6947.

2. **Arsene, F., T. Tomoyasu, A. Mogk, C. Schirra, A. Schulze-Specking, and B. Bukau.** 1999. Role of region C in regulation of the heat shock gene-specific sigma factor of *Escherichia coli*, σ^{32}. *J. Bacteriol.* **181:**3552–3561.

3. **Babst, M., H. Hennecke, and H.-M. Fischer.** 1996. Two different mechanisms are involved in the heat-shock regulation of chaperonin gene expression in *Bradyrhizobium japonicum. Mol. Microbiol.* **19:**827–839.

4. **Baldini, R. L., M. Avedissian, and S. L. Gomes.** 1998. The CIRCE element and its putative repressor control cell cycle expression of the *Caulobacter crescentus groESL* operon. *J. Bacteriol.* **180:**1632–1641.

5. **Bardwell, J. C., K. Tilly, E. Craig, J. King, M. Zylicz, and C. Georgopoulos.** 1986. The nucleotide sequence of the *Escherichia coli* K12 *dnaJ+* gene. A gene that encodes a heat shock protein. *J. Biol. Chem.* **261:**1782–1785.

6. **Bardwell, J. C. A., and E. A. Craig.** 1984. Major heat shock gene of *Drosophila* and the *Escherichia coli* heat inducible *dnaK* gene are homologous. *Proc. Natl. Acad. Sci. USA* **81:**848–852.

7. **Bardwell, J. C. A., and E. A. Craig.** 1987. Eukaryotic M_r 83,000 heat shock protein has a homologue in *Escherichia coli. Proc. Natl. Acad. Sci. USA* **84:**5177–5181.

8. **Biran, D., N. Brot, H. Weissbach, and E. Z. Ron.** 1995. Heat shock-dependent transcriptional activation of the *metA* gene of *Escherichia coli. J. Bacteriol.* **177:**1374–1379.

9. **Blaszczak, A., C. Georgopoulos, and K. Liberek.** 1999. On the mechanism of FtsH-dependent degradation of the σ^{32} transcriptional regulator of *Escherichia coli* and the role of the DnaK chaperone machine. *Mol. Microbiol.* **31:**157–166.

10. **Blaszczak, A., M. Zylicz, C. Georgopoulos, and K. Liberek.** 1995. Both ambient temperature and the DnaK chaperone machine modulate the heat shock response in *Escherichia coli* by regulating the switch between σ^{70} and σ^{32} factors assembled with RNA polymerase. *EMBO J.* **14:**5085–5093.

11. **Bukau, B.** 1993. Regulation of the *Escherichia coli* heat-shock response. *Mol. Microbiol.* **9:**671–680.

12. **Bukau, B., and A. L. Horwich.** 1998. The Hsp70 and Hsp60 chaperone machines. *Cell* **92:**351–366.

13. **Burton, Z., R. R. Burgess, J. Lin, D. Moore, S. Holder, and C. A. Gross.** 1981. The nucleotide sequence of the cloned *rpoD* gene for the RNA polymerase sigma subunit from *E. coli* K12. *Nucleic Acids Res.* **9:**2889–2903.

14. **Charpentier, B., and C. Branlant.** 1994. The *Escherichia coli gapA* gene is transcribed by the vegetative RNA polymerase holoenzyme Eσ^{70} and by the heat shock RNA polymerase Eσ^{32}. *J. Bacteriol.* **176:**830–839.

15. **Chuang, S.-E., and F. R. Blattner.** 1993. Characterization of twenty-six new heat shock genes of *Escherichia coli. J. Bacteriol.* **175:**5242–5252.

16. **Chuang, S.-E., V. Burland, G. Plunkett III, D. L. Daniels, and F. R. Blattner.** 1993. Sequence analysis of four new heat-shock genes constituting the *hslTS/ibpAB* and *hslVU* operons in *Escherichia coli. Gene* **134:**1–6.

17. **Connolly, L., A. De Las Peñas, B. M. Alba, and C. A. Gross.** 1997. The response to extracytoplasmic stress in *Escherichia coli* is controlled by partially overlapping pathways. *Genes Dev.* **11:**2012–2021.

18. **Cowing, D. W., J. C. A. Bardwell, E. A. Craig, C. Woolford, R. W. Hendrix, and C. A. Gross.** 1985. Consensus sequence for *Escherichia coli* heat shock gene promoters. *Proc. Natl. Acad. Sci. USA* **82:**2679–2683.

19. **Craig, E. A., and C. A. Gross.** 1991. Is hsp70 the cellular thermometer? *Trends Biochem. Sci.* **16:**135–140.

20. **Danese, P., and T. Silhavy.** 1997. The σ^E and Cpx signal transduction systems control the synthesis of periplasmic protein-folding enzymes in *Escherichia coli. Genes Dev.* **11:**1183–1193.

21. **Dartigalongue, C., and S. Raina.** 1998. A new heat shock gene, *ppiD*, encodes a peptidyl-prolyl isomerase required for folding of outer membrane proteins in *Escherichia coli. EMBO J.* **14:**3968–3980.

22. **De Las Peñas, A., L. Connolly, and C. A. Gross.** 1997. The sigma E-mediated response to extracytoplasmic stress in *Escherichia coli* is transduced by RseA and RseB, two negative regulators of sigma E. *Mol. Microbiol.* **24:**373–385.

23. **De Las Peñas, A., L. Connolly, and C. A. Gross.** 1997. σ^E is an essential sigma factor in *Escherichia coli. J. Bacteriol.* **179:**6862–6864.

24. **Derre, I., G. Rapoport, and T. Msadek.** 1999. CtsR, a novel regulator of stress and heat shock response, controls *clp* and

molecular chaperone gene expression in gram-positive bacteria. *Mol. Microbiol.* **31:**117–131.

25. Ellis, R. J., and S. M. van der Vies. 1991. Molecular chaperones. *Annu. Rev. Biochem.* **60:**321–347.

26. Erickson, J. W., and C. A. Gross. 1989. Identification of the σ^E subunit of *Escherichia coli* RNA polymerase: a second alternate σ factor involved in high-temperature gene expression. *Genes Dev.* **3:**1462–1471.

27. Erickson, J. W., V. Vaughn, W. A. Walter, F. C. Neidhardt, and C. A. Gross. 1987. Regulation of the promoters and transcripts of *rpoH*, the *Escherichia coli* heat shock regulatory gene. *Genes Dev.* **1:**419–432.

28. Ewalt, K. L., J. P. Hendrick, W. A. Houry, and F. U. Hartl. 1997. In vivo observation of polypeptide flux through the bacterial chaperonin system. *Cell* **90:**491–500.

29. Gamer, J., G. Multhaup, T. Tomoyasu, J. S. McCarty, S. Rudiger, H.-J. Schonfeld, C. Schirra, H. Bujard, and B. Bukau. 1996. A cycle of binding and release of the DnaK, DnaJ and GrpE chaperones regulates activity of the *Escherichia coli* heat shock transcription factor σ^{32}. *EMBO J.* **15:**607–617.

30. Gamer, J., H. Bujard, and B. Bukau. 1992. Physical interaction between heat shock proteins DnaK, DnaJ, and GrpE and the bacterial heat shock transcription factor σ^{32}. *Cell* **69:**833–842.

31. Gayda, R. C., P. E. Stephens, R. Hewick, J. M. Schoemaker, W. J. Dreyer, and A. Markovitz. 1985. Regulatory region of the heat shock-inducible *capR* (*lon*) gene: DNA and protein sequences. *J. Bacteriol.* **162:**271–275.

32. Georgopoulos, C., D. Ang, A. Maddock, S. Raina, B. Lipinska, and M. Zylicz. 1990. Properties of the *Escherichia coli* heat shock proteins and their role in bacteriophage λ growth, p. 191–221. *In* R. I. Morimoto, A. Tissieres, and C. Georgopoulos (ed.), *Stress Proteins in Biology and Medicine.* Cold Spring Harbor Laboratory Press, Cold Spring Harbor, N.Y.

33. Georgopoulos, C., K. Liberek, M. Zylicz, and D. Ang. 1994. Properties of the heat shock proteins of *Escherichia coli* and the autoregulation of the heat shock response, p. 209–249. *In* R. I. Morimoto, A. Tissieres, and C. Georgopoulos (ed.), *The Biology of Heat Shock Proteins and Molecular Chaperones.* Cold Spring Harbor Laboratory Press, Cold Spring Harbor, N.Y.

34. Goff, S. A., and A. L. Goldberg. 1985. Production of abnormal proteins in *E. coli* stimulates transcription of *lon* and other heat shock genes. *Cell* **41:**587–595.

35. Gottesman, S. 1996. Proteases and their targets in *Escherichia coli*. *Annu. Rev. Genet.* **30:**465–506.

36. Gottesman, S., W. P. Clark, V. de Crecy-Lagard, and M. R. Maurizi. 1993. ClpX, and alternative subunit for the ATP-dependent Clp protease of *Escherichia coli*. *J. Biol. Chem.* **268:**22618–22626.

37. Gragerov, A., E. Nudler, N. Komissarova, G. A. Gaitanaris, M. E. Gottesman, and V. Nikiforov. 1992. Cooperation of GroEL/GroES and DnaK/DnaJ heat shock proteins in preventing protein misfolding in *E. coli*. *Proc. Natl. Acad. Sci. USA* **89:**10341–10344.

38. Gross, C. A. 1996. Function and regulation of the heat shock proteins, p. 1382–1399. *In* F. C. Neidhardt, R. Curtiss III, J. L. Ingraham, E. C. C. Lin, K. B. Low, B. Magasanik, W. S. Reznikoff, M. Riley, M. Schaechter, and H. E. Umbarger (ed.), *Escherichia coli and Salmonella: Cellular and Molecular Biology*, 2nd ed. ASM Press, Washington, D.C.

39. Gross, C. A., D. B. Straus, J. W. Erickson, and T. Yura. 1990. The function and regulation of heat shock proteins in *Escherichia coli*. *In* R. I. Morimoto, A. Tissieres, and C. Georgopoulos (ed.), *Stress Proteins in Biology and Medicine.* Cold Spring Harbor Laboratory Press, Cold Spring Harbor, N.Y.

40. Grossman, A. D., J. W. Erickson, and C. A. Gross. 1984. The *htpR* gene product of *E. coli* is a sigma factor for heat-shock promoters. *Cell* **38:**383–390.

41. Grossman, A. D., D. B. Straus, W. A. Walter, and C. A. Gross. 1987. σ^{32} synthesis can regulate the synthesis of heat shock proteins in *Escherichia coli*. *Genes Dev.* **1:**179–184.

42. Hecker, M., W. Schumann, and U. Volker. 1996. Heat-shock and general stress response in *Bacillus subtilis*. *Mol. Microbiol.* **19:**417–428.

43. Hendrick, J. P., and F.-U. Hartl. 1993. Molecular chaperone functions of heat-shock proteins. *Annu. Rev. Biochem.* **62:**349–384.

44. Herendeen, S. L., R. A. VanBogelen, and F. C. Neidhardt. 1979. Levels of major proteins of *Escherichia coli* during growth at different temperatures. *J. Bacteriol.* **139:**185–194.

45. Herman C., D. Thevenet, R. D'Ari, and P. Bouloc. 1995. Degradation of σ^{32}, the heat shock regulator in *Escherichia coli*, is governed by HflB. *Proc. Natl. Acad. Sci. USA* **92:**3516–3520.

46. Jacob, U., W. Muse, M. Eser, and J. C. A. Bardwell. 1999. Chaperone activity with a redox switch. *Cell* **96:**341–352.

47. Joo, D. M., A. Nolte, R. Calendar, Y.-N. Zhou, and D. J. Jin. 1998. Multiple regions on the *Escherichia coli* heat shock transcription factor σ^{32} determine core RNA polymerase binding specificity. *J. Bacteriol.* **180:**1095–1102.

48. Kallipolitis, B. H., and P. Valentin-Hansen. 1998. Transcription of *rpoH*, encoding the *Escherichia coli* heat-shock regulator σ^{32}, is negatively controlled by the cAMP-CRP/CytR nucleoprotein complex. *Mol. Microbiol.* **29:**1091–1099.

49. Kamath-Loeb, A. S., and C. A. Gross. 1991. Translational regulation of σ^{32} synthesis: requirement for an internal control element. *J. Bacteriol.* **173:**3904–3906.

50. Kanemori, M., H. Mori, and T. Yura. 1994. Induction of heat shock proteins by abnormal proteins results from stabilization and not increased synthesis of σ^{32} in *Escherichia coli*. *J. Bacteriol.* **176:**5648–5653.

51. Kanemori, M., K. Nishihara, H. Yanagi, and T. Yura. 1997. Synergistic roles of HslVU and other ATP-dependent proteases in controlling in vivo turnover of σ^{32} and abnormal proteins in *Escherichia coli*. *J. Bacteriol.* **179:**7219–7225.

52. Kanemori, M., H. Yanagi, and T. Yura. 1999. Marked instability of the σ^{32} heat shock transcription factor at high temperature: implications on the heat shock regulation. *J. Biol. Chem.* **274:**22002–22007.

53. Kitagawa, M., C. Wada, S. Yoshioka, and T. Yura. 1991. Expression of ClpB, an analog of the ATP-dependent protease regulatory subunit in *Escherichia coli*, is controlled by a heat shock σ factor (σ^{32}). *J. Bacteriol.* **173:**4247–4253.

54. Koch, B., M. Kilstrup, F. K. Vogensen, and K. Hammer. 1998. Induced levels of heat shock proteins in a *dnaK* mutant of *Lactococcus lactis*. *J. Bacteriol.* **180:**3873–3881.

55. Korber, P., T. Zander, D. Herschlag, and J. C. A. Bardwell. 1999. A new heat shock protein that binds nucleic acids. *J. Biol. Chem.* **274:**249–256.

56. Kornitzer, D., D. Teff, S. Altuvia, and A.B. Oppenheim. 1991. Isolation, characterization, and sequence of an *Escherichia coli* heat shock gene, *htpX*. *J. Bacteriol.* **173:**2944–2953.

57. Kusukawa, N., and T. Yura. 1988. Heat shock protein GroE of *Escherichia coli*: key protective roles against thermal stress. *Genes Dev.* **2:**874–882.

58. Landick, R., V. Vaughn, E. T. Lau, R. A. VanBogelen, J. W. Erickson, and F. C. Neidhardt. 1984. Nucleotide sequence of the heat shock regulatory gene of *E. coli* suggests its protein product may be a transcription factor. *Cell* **38:**175–182.

59. Lemaux, P. G., S. L. Herendeen, P. L. Bloch, and F. C. Neidhardt. 1978. Transient rates of synthesis of individual poly-

peptides in *E. coli* following temperature shifts. *Cell* 13:427–434.

60. Leveque, F., P. Plateau, P. Dessen, and S. Blanquet. 1990. Homology of *lysS* and *lysU*, the two *Escherichia coli* genes encoding distinct lysyl-tRNA synthetase species. *Nucleic Acids Res.* 18:305–312.

61. Liberek, K., T. P. Galitski, M. Zylicz, and C. Georgopoulos. 1992. The DnaK chaperone modulates the heat shock response of *Escherichia coli* by binding to the σ^{32} transcription factor. *Proc. Natl. Acad. Sci. USA* 89:3516–3520.

62. Liberek, K., and C. Georgopoulos. 1993. Autoregulation of the *Escherichia coli* heat shock response by binding to the σ^{32} transcription factor. *Proc. Natl. Acad. Sci. USA* 90:11019–11023.

63. Lipinska, B., J. King, D. Ang, and C. Georgopoulos. 1988. Sequence analysis and transcriptional regulation of the *Escherichia coli grpE* gene, encoding a heat shock protein. *Nucleic Acids Res.* 16:7545–7562.

64. Lipinska, B., S. Sharma, and C. Georgopoulos. 1988. Sequence analysis and regulation of the *htrA* gene of *Escherichia coli*: a sigma 32-independent mechanism of heat-inducible transcription. *Nucleic Acids Res.* 16:10053–10067.

65. Lonetto, M., K. L. Brown, K. E. Rudd, and M. J. Buttner. 1994. Analysis of the *Streptomyces coelicolor* sigmaE gene reveals the existence of a subfamily of eubacterial RNA polymerase sigma factors involved in the regulation of extracytoplasmic functions. *Proc. Natl. Acad. Sci. USA* 91:7573–7577.

66. Lopez-Sanchez, F., J. Ramírez-Santos, M. C., and Gomez-Eichelmann. 1997. In vivo effect of DNA relaxation on the transcription of gene *rpoH* in *Escherichia coli*. *Biochim. Biophys. Acta* 1353:79–83.

67. Maurizi, M. R., W. P. Clark, Y. Katayama, S. Rudikoff, J. Pumphrey, B. Bowers, and S. Gottesman. 1990. Sequence and structure of ClpP, the proteolytic component of the ATP-dependent Clp protease of *Escherichia coli*. *J. Biol. Chem.* 265:12536–12545.

68. McCarty, J. S., S. Rudiger, H. J. Schonfeld, J. Schneider-Mergener, K. Nakahigashi, T. Yura, and B. Bukau. 1996. Regulatory region C of the *E. coli* heat shock transcription factor, σ^{32}, constitutes a DnaK binding site and is conserved among eubacteria. *J. Mol. Biol.* 256:829–837.

69. Mecsas, J., P. E. Rouviere, J. W. Erickson, T. Donohue, and C. A. Gross. 1993. The activity of σ^E, an *Escherichia coli* heat-inducible σ-factor, is modulated by expression of outer membrane proteins. *Genes Dev.* 7:2619–2628.

70. Missiakas, D., J. M. Botton, and S. Raina. 1996. New components of protein folding in extracytoplasmic compartments of *Escherichia coli* SurA, FkpA and Skp/OmpH. *Mol. Microbiol.* 21:871–884.

71. Missiakas, D., C. Georgopoulos, and S. Raina. 1993. The *Escherichia coli* heat shock gene *htpY*: mutational analysis, cloning, sequencing, and transcriptional regulation. *J. Bacteriol.* 175:2613–2624.

72. Missiakas, D., M. P. Mayer, M. Lemaire, C. Georgopoulos, and S. Raina. 1997. Modulation of the *Escherichia coli* sigma E (RpoE) heat-shock transcription-factor activity by the RseA, RseB and RseC proteins. *Mol. Microbiol.* 24:355–371.

73. Missiakas, D., and S. Raina. 1997. Signal transduction pathways in response to protein misfolding in the extracytoplasmic compartments of *E. coli*: role of two new phosphoprotein phosphatases PrpA and PrpB. *EMBO J.* 16:1670–1685.

74. Missiakas, D., S. Raina, and C. Georgopoulos. 1996. Heat shock regulation, p. 481–501. *In* E. C. C. Lin and A. S. Lynch (ed.), *Regulation of Gene Expression in* Escherichia coli. R. G. Landes Company, Austin, Tex.

75. Model, P., G. Jovanovic, and J. Dworkin. 1997. The *Escherichia coli* phage-shock-protein (*psp*) operon. *Mol. Microbiol.* 24:255–261.

76. Mogk, A., G. Homuth, C. Scholz, L. Kim, F. X. Schmid, and W. Schumann. 1997. The GroE chaperonin machine is a major modulator of the CIRCE heat shock regulon of *Bacillus subtilis*. *EMBO J.* 16:4579–4590.

77. Mogk, A., A. Volker, S. Engelmann, M. Hecker, W. Schumann, and U. Volker. 1998. Nonnative proteins induce expression of the *Bacillus subtilis* CIRCE regulon. *J. Bacteriol.* 180:2895–2900.

78. Morita, M., M. Kanemori, H. Yanagi, and T. Yura. 1999. Heat-induced synthesis of σ^{32} in *Escherichia coli*: structural and functional dissection of *rpoH* mRNA secondary structure. *J. Bacteriol.* 181:401–410.

79. Morita, M. T., Y. Tanaka, T. Kodama, Y. Kyogoku, H. Yanagi, and T. Yura. 1999. Translational induction of heat shock transcription factor σ^{32}: evidence for a built-in RNA thermosensor. *Genes Dev.* 13:655–665.

80. Nagai, H., R. Yano, J. W. Erickson, and T. Yura. 1990. Transcriptional regulation of the heat shock regulatory gene *rpoH* in *Escherichia coli*: involvement of a novel catabolite-sensitive promoter. *J. Bacteriol.* 172:2710–2715.

81. Nagai, H., H. Yuzawa, and T. Yura. 1991. Interplay of two *cis*-acting mRNA regions in translational control of σ^{32} synthesis during the heat shock response of *Escherichia coli*. *Proc. Natl. Acad. Sci. USA* 88:10515–10519.

82. Nagai, H., H. Yuzawa, M. Kanemori, and T. Yura. 1994. A distinct segment of the σ^{32} polypeptide is involved in DnaK-mediated negative control of the heat shock response in *Escherichia coli*. *Proc. Natl. Acad. Sci. USA* 91:10280–10284.

83. Nakahigashi, K., H. Yanagi, and T. Yura. 1995. Isolation and sequence analysis of *rpoH* genes encoding σ^{32} homologs from gram-negative bacteria: conserved mRNA and protein segments for heat shock regulation. *Nucleic Acids Res.* 23:4383–4390.

84. Nakahigashi, K., H. Yanagi, and T. Yura. 1998. Regulatory conservation and divergence of σ^{32} homologs from gram-negative bacteria: *Serratia marcescens*, *Proteus mirabilis*, *Pseudomonas aeruginosa*, and *Agrobacterium tumefaciens*. *J. Bacteriol.* 180:2402–2408.

85. Narberhaus, F. 1999. Negative regulation of bacterial heat shock genes. *Mol. Microbiol.* 31:1–8.

86. Narberhaus, F., P. Krummenacher, H.-M. Fischer, and H. Hennecke. 1997. Three disparately regulated genes for σ^{32}-like transcription factors in *Bradyrhizobium japonicum*. *Mol. Microbiol.* 24:93–104.

87. Neidhardt, F. C., T. A. Phillips, R. A. VanBogelen, M. W. Smith, Y. Georgalis, and A. R. Subramanian. 1981. Identity of the B56.5 protein, the A-protein, and the *groE* gene product of *Escherichia coli*. *J. Bacteriol.* 145:513–520.

88. Neidhardt, F. C., and R. A. VanBogelen. 1981. Positive regulatory gene for temperature-controlled proteins in *Escherichia coli*. *Biochem. Biophys. Res. Commun.* 100:894–900.

89. Neidhardt, F. C., and R. A. VanBogelen. 1987. Heat shock response, p. 1334–1345. *In* F. C. Neidhardt, J. L. Ingraham, K. B. Low, B. Magasanik, M. Schaechter, and H. E. Umbarger (ed.), Escherichia coli *and* Salmonella typhimurium: *Cellular and Molecular Biology*. American Society for Microbiology, Washington, D.C.

90. Parsell, D. A., and R. T. Sauer. 1989. Induction of a heat shock-like response by unfolded protein in *Escherichia coli*: dependence on protein level not protein degradation. *Genes Dev.* 3:1226–1232.

91. Pogliano, J., A. S. Lynch, D. Belin, E. C. C. Lin, and J. Bechwith. 1997. Regulation of *Escherichia coli* envelope pro-

teins involved in protein folding and degradation by the Cpx two-component system. *Genes Dev.* 11:1169–1182.

92. Raina, S., and C. Georgopoulos. 1990. A new *Escherichia coli* heat shock gene, *htrC*, whose product is essential for viability only at high temperatures. *J. Bacteriol.* 172:3417–3426.

93. Raina, S., and C. Georgopoulos. 1991. The *htrM* gene, whose product is essential for *Escherichia coli* viability only at elevated temperatures, is identical to the *rfaD* gene. *Nucleic Acids Res.* 19:3811–3819.

94. Raina, S., D. Missiakas, and C. Georgopoulos. 1995. The *rpoE* gene encoding the σ^E (σ^{24}) heat shock sigma factor of *Escherichia coli*. *EMBO J.* 14:1043–1055.

95. Ramirez-Santos, J., and M. C. Gomez-Eichelmann. 1998. Identification of σ^{32}-like factors and *ftsX-rpoH* gene arrangements in enteric bacteria. *Can. J. Microbiol.* 44:565–568.

96. Reisenauer, A., C. D. Mohr, and L. Shapiro. 1996. Regulation of a heat shock σ^{32} homolog in *Caulobacter crescentus*. *J. Bacteriol.* 178:1919–1927.

97. Roberts, R. C., C. Toochinda, M. Avedissian, R. L. Baldini, S. L. Gomes, and L. Shapiro. 1996. Identification of a *Caulobacter crescentus* operon encoding *hrcA*, involved in negatively regulating heat-inducible transcription, and the chaperone gene *grpE*. *J. Bacteriol.* 178:1829–1841.

98. Rouviere, P. E., A. de las Peñas, J. Mecsas, C. Z. Lu, K. E. Rudd, and C. A. Gross. 1995. *rpoE*, the gene encoding the second heat-shock sigma factor, σ^E, in *Escherichia coli*. *EMBO J.* 14:1032–1042.

99. Schirmer, E. C., J. R. Glover, M. A. Singer, and S. Lindquist. 1996. HSP100/Clp proteins: a common mechanism explains diverse functions. *Trends Biochem. Sci.* 21:289–295.

100. Schulz, A., and W. Schumann. 1996. *hrcA*, the first gene of the *Bacillus subtilis dnaK* operon encodes a negative regulator of class I heat shock genes. *J. Bacteriol.* 178:1088–1093.

101. Segal, R., and E. Z. Ron. 1996. Regulation and organization of the *groE* and *dnaK* operons in Eubacteria. *FEMS Microbiol. Lett.* 138:1–10.

102. Segal, G., and E. Z. Ron. 1996. Heat shock activation of the *groESL* operon of *Agrobacterium tumefaciens* and the regulatory roles of the inverted repeat. *J. Bacteriol.* 178:3634–3640.

103. Sprengart, M. L., H. P. Fatscher, and E. Fuchs. 1990. The initiation of translation in *E. coli*: apparent base pairing between the 16S rRNA and downstream sequences of the mRNA. *Nucl. Acids Res.* 18:1719–1723.

104. Squires, C. L., S. Pedersen, B. M. Ross, and C. Squires. 1991. ClpB is the *Escherichia coli* heat shock protein F84.1. *J. Bacteriol.* 173:4254–4262.

105. Stauch, K. L., K. Johnson, and J. Bechwith. 1989. Characterization of *degP*, a gene required for proteolysis in the cell envelope and essential for growth of *Escherichia coli* at high temperature. *J. Bacteriol.* 171:2689–2696.

106. Straus, D., W. Walter, and C. A. Gross. 1990. DnaK, DnaJ, and GrpE heat shock proteins negatively regulate heat shock gene expression by controlling the synthesis and stability of σ^{32}. *Genes Dev.* 4:2202–2209.

107. Straus, D. B., W. A. Walter, and C. A. Gross. 1987. The heat shock response of *E. coli* is regulated by changes in the concentration of σ^{32}. *Nature* 329:348–351.

108. Straus, D. B., W. A. Walter, and C. A. Gross. 1989. The activity of σ^{32} is reduced under conditions of excess heat shock protein production in *Escherichia coli*. *Genes Dev.* 3:2003–2010.

109. Tatsuta T., T. Tomoyasu, B. Bukau, M. Kitagawa, H. Mori, K. Karata, and T. Ogura. 1998. Heat shock regulation in the

ftsH null mutant of *Escherichia coli*: dissection of stability and activity control mechanisms of σ^{32} in vivo. *Mol. Microbiol.* 30:583–593.

110. Taura, T., N. Kusukawa, T. Yura, and K. Ito. 1989. Transient shutoff of *Escherichia coli* heat shock protein synthesis upon temperature shift down. *Biochem. Biophys. Res. Commun.* 163:438–443.

111. Tilly, K., N. McKittrick, M. Zylicz, and C. Georgopoulos. 1983. The *dnaK* protein modulates the heat-shock response of *Escherichia coli*. *Cell* 34:641–646.

112. Tilly, K., J. Spence, and C. Georgopoulos. 1989. Modulation of stability of the *Escherichia coli* heat shock regulatory factor σ^{32}. *J. Bacteriol.* 171:1585–1589.

113. Tilly, K., R. A. VanBogelen, C. Georgopoulos, and F. C. Neidhardt. 1983. Identification of the heat-inducible protein C15.4 as the *groES* gene product in *Escherichia coli*. *J. Bacteriol.* 154:1505–1507.

114. Tobe, T., K. Ito, and T. Yura. 1984. Isolation and physical mapping of temperature-sensitive mutants defective in heat-shock induction of proteins in *Escherichia coli*. *Mol. Gen. Genet.* 195:10–16.

115. Tomoyasu, T., J. Gamer, B. Bukau, M. Kanemori, H. Mori, A. J. Rutman, A. B. Oppenheim, T. Yura, K. Yamanaka, H. Niki, S. Hiraga, and T. Ogura. 1995. *Escherichia coli* FtsH is a membrane-bound, ATP-dependent protease which degrades the heat-shock transcription factor σ^{32}. *EMBO J.* 14:2551–2560.

116. Tomoyasu, T., T. Ogura, T. Tatsuta, and B. Bukau. 1998. Levels of DnaK and DnaJ provide tight control of heat shock gene expression and protein repair in *Escherichia coli*. *Mol. Microbiol.* 30:567–581.

117. Van Dyk, T. K., A. A. Gatenby, and R. A. LaRossa. 1989. Demonstration by genetic suppression of interaction of GroE products with many proteins. *Nature* 342:451–453.

118. Wang, Q., and J. M. Kaguni. 1989. DnaA protein regulates transcription of the *rpoH* gene of *Escherichia coli*. *J. Biol. Chem.* 264:7338–7344.

119. Wang, Q., and J. M. Kaguni. 1989. A novel sigma factor is involved in expression of the *rpoH* gene of *Escherichia coli*. *J. Bacteriol.* 171:4248–4253.

120. Wawrzynow, A., D. Wojtkowiak, J. Marszalek, B. Banecki, M. Jonsen, B. Graves, C. Georgopoulos, and M. Zylicz. 1995. The ClpX heat-shock protein of *Escherichia coli*, the ATP-dependent substrate specificity component of the ClpP-ClpX protease, is a novel molecular chaperone. *EMBO J.* 14:1867–1877.

121. Wild, J., W. A. Walter, C. A. Gross, and E. Altman. 1993. Accumulation of secretory protein precursors in *E. coli* induces the heat shock response. *J. Bacteriol.* 175:3992–3997.

122. Wu, J., and A. Newton. 1996. Isolation, identification, and transcriptional specificity of the heat shock sigma factor σ^{32} from *Caulobacter crescentus*. *J. Bacteriol.* 178:2094–2101.

123. Yamamori, T., K. Ito, Y. Nakamura, and T. Yura. 1978. Transient regulation of protein synthesis in *Escherichia coli* upon shift-up of growth temperature. *J. Bacteriol.* 134:1133–1140.

124. Yamamori, T., and T. Yura. 1980. Temperature-induced synthesis of specific proteins in *Escherichia coli*: evidence for transcriptional control. *J. Bacteriol.* 142:843–851.

125. Yamamori, T., and T. Yura. 1982. Genetic control of heat-shock protein synthesis and its bearing on growth and thermal resistance in *Escherichia coli* K-12. *Proc. Natl. Acad. Sci. USA* 79:860–864.

126. **Yuan, G., and S.-L. Wong.** 1995. Isolation and characterization of *Bacillus subtilis groE* regulatory mutants: evidence for *orf39* in the *dnaK* operon as a repressor gene in regulating the expression of both *groE* and *dnaK*. *J. Bacteriol.* **177:**6442–6468.

127. **Yura, T., H. Nagai, and H. Mori.** 1993. Regulation of heat shock response in bacteria. *Annu. Rev. Microbiol.* **47:**321–350.

128. **Yura, T., T. Tobe, K. Ito, and T. Osawa.** 1984. Heat shock regulatory gene (*htpR*) of *Escherichia coli* is required for growth at high temeprature but is dispensable at low temperature. *Proc. Natl. Acad. Sci. USA* **81:**6803–6807.

129. **Yuzawa, H., H. Nagai, H. Mori, and T. Yura.** 1993. Heat induction of σ^{32} synthesis mediated by mRNA secondary structure: a primary step of the heat shock response in *Escherichia coli*. *Nucleic Acids Res.* **21:**5449–5455.

130. **Zhou, Y.-N., N. Kusukawa, J. W. Erickson, C. A. Gross, and T. Yura.** 1988. Isolation and characterization of *Escherichia coli* mutants that lack the heat shock sigma factor σ^{32}. *J. Bacteriol.* **170:**3640–3663.

131. **Zuber, U., and W. Schumann.** 1994. CIRCE, a novel heat shock element involved in regulation of heat shock operon *dnaK* of *Bacillus subtilis*. *J. Bacteriol.* **176:**1359–1363.

Bacterial Stress Responses
Edited by G. Storz and R. Hengge-Aronis
©2000 ASM Press, Washington, D.C.

Chapter 2

Sensing and Responding to Envelope Stress

TRACY L. RAIVIO AND THOMAS J. SILHAVY

The envelope of gram-negative bacteria is made up of the inner membrane (IM), periplasm (PP), and outer membrane (OM). Over the past 10 years, at least two responses have been identified in Escherichia coli *which are activated by a plethora of stresses specific to this compartment. One is regulated by an alternative sigma factor, σ^E, and the other one is regulated by the CpxAR two-component system. Recent study of these responses strongly suggests that the actual stress they perceive is misfolded proteins in the bacterial envelope. Further, although there are areas of overlap, each response appears to recognize a distinct type of misfolded protein and upregulate a unique set of protective factors, suggesting distinct physiological roles. This chapter summarizes the current state of knowledge concerning envelope stress responses and highlights future areas of interest in this field. We begin by introducing the envelope and its resident folding factors and then discuss the σ^E and Cpx envelope stress responses. Finally, we consider the possibility that yet more envelope stress responses remain to be characterized.*

The envelope consists of two lipid bilayers (inner or cytoplasmic membrane and outer membrane) of distinct composition (44, 74, 78). The viscous aqueous periplasmic space between these membranes is dense with proteins and peptidoglycan, and it is predicted that the diffusion rate in this compartment is close to 100-fold lower than that of the cytoplasm (15, 75). Further, conditions in the periplasm are oxidizing, favoring disulfide-bond formation, as opposed to the predominantly reducing environment of the cytoplasm (85). Finally, unlike the cytoplasm, the envelope is directly exposed to the environment.

Besides serving as a protective barrier, the envelope carries out numerous functions including solute transport, protein translocation, lipid biosynthesis, and oxidative phosphorylation (44, 74, 75, 76).

Extracellular appendages, such as pili and flagella, are also part of the envelope, and are required for attachment and motility, respectively (56, 57). The envelope is also centrally involved in the process of cell division (14).

These envelope-associated functions are carried out by distinct sets of proteins that are found in either the IM, PP, or OM. Accordingly, envelope proteins have a wide variety of physical properties. They may be either soluble, peripheral, or integral membrane proteins or covalently modified lipoproteins. Further, integral IM and OM proteins possess markedly different structural features (3, 37, 44, 74, 91).

Thus, the bacterial envelope is a physically and functionally complex compartment, containing a wide variety of proteins that are continually exposed to the changing conditions of the external milieu. It is not surprising, then, that this compartment also contains a distinct set of protein foldases and proteases, whose synthesis can be altered according to necessity.

ENVELOPE PROTEIN FOLDING AND DEGRADING FACTORS

Foldases

Although the field is still in its infancy, a number of foldases that reside in the bacterial envelope have recently been identified, and many of these have been shown to be essential for the correct folding of resident envelope proteins. These include disulfide-bond (dsb) oxidoreductases, petidyl-prolyl isomerases (ppiases), and a number of demonstrated and putative chaperones (Table 1).

Disulfide bond oxidoreductases

Despite the oxidizing environment, the efficient catalysis of correctly formed disulfide bonds in en-

Tracy L. Raivio and Thomas J. Silhavy • Department of Molecular Biology, Princeton University, Princeton, NJ 08544.

Table 1. Protein folding factors in the bacterial envelope

Folding factor	Function	Location	Link to envelope stress
Disulfide oxidoreductases			
DsbA/B	Disulfide bond formation	PP/IM	*dsbA* regulated by Cpx
DsbC/D	Disulfide bond isomerization	PP/IM	Null mutations induce σ^E
DsbE	Cytochrome biogenesis	IM	Null mutations induce σ^E
DsbG	?	PP	Null mutations induce σ^E
Peptidyl-prolyl isomerases			
SurA	OMP folding	PP	Null mutations confer envelope defects and induce σ^E
PpiD	OMP folding?	IM	Null mutations confer envelope defects
FkpA	OMP folding?	PP	*fkpA* regulated by σ^E
RotA/PpiA	?	PP	*rotA*/*ppiA* regulated by Cpx
Chaperones			
Pilus-specific chaperones	Folding and targeting of pilin subunits	PP	Misfolded pilus subunits activate Cpx
LolA/LolB	Incorporation of lipoprotein into the OM	PP/OM	?
Skp	OMP folding?	PP	?
DegP	Refolding of denatured envelope proteins at low temperatures	PP	Null mutants are temperature sensitive; regulated by both σ^E and Cpx

velope proteins requires the action of dsb proteins (85). These proteins function in pairs to catalyze the formation and isomerization of dsbs in envelope proteins. One partner uses its own reactive sulfhydryl groups to catalyze the formation or isomerization reaction in a substrate protein while the other functions to regenerate the initial redox state of the dsb system (6, 7, 25, 45, 67, 85). The main thiol-disulfide oxidase pair responsible for dsb formation in envelope proteins is DsbA/DsbB (6, 7, 45, 85). The DsbC/DsbD pair of thiol-disulfide oxidoreductases is needed to rearrange incorrectly formed dsbs in proteins containing more than one dsb (70, 85, 86, 93, 106).

In addition to DsbAB and DsbCD, two other thiol-disulfide oxidoreductases that reside in the bacterial envelope have been identified. DsbE is an IM protein that, in addition to DsbD, is required for cytochrome *c* biogenesis, and DsbG is a periplasmic protein whose function remains unknown (4, 24, 90, 101).

Peptidyl-prolyl isomerases

The second largest class of protein folding factors identified in the bacterial envelope to date is the ppiases (Table 1). These enzymes function to catalyze *cis-trans* isomerization around X-Pro peptide bonds. Four ppiases have thus far been described that reside in the bacterial envelope (Table 1). SurA and PpiD belong to the parvulin class of ppiases, so named for their similarity to the small *Escherichia coli* ppiase

parvulin (30, 51). FkpA belongs to the FK506-binding protein (FKBP) class of ppiases, which were initially identified as targets of the immunosuppressive drug FK506 (39). Finally, a fourth periplasmic ppiase, RotA/PpiA, is a cyclophilin ppiase, named for its binding of the drug cyclosporin (54).

Accumulating evidence suggests that these proteins facilitate envelope protein folding. Null mutations in many of these ppiase-encoding genes result in OM permeability defects and induction of the σ^E envelope stress response (30, 66, 88). Both events are indicative of perturbations in protein folding in the envelope. Further, multiple copies of some of these ppiases can alleviate certain envelope stresses (30, 66, 88). Finally, and most convincingly, *surA* nulls accumulate an unfolded species of OM porin, indicating that SurA may assist in the folding of OMPs at an early stage in assembly (88).

The apparent redundancy of the ppiases is striking. It will be interesting to determine why so many are required in a single compartment of the bacterium. Perhaps each facilitates the folding of a distinct subset of envelope proteins. Alternatively, it may be that they function under a different set of environmental parameters. Insight into the regulation of these genes has already suggested that this may be true in some cases as *fkpA* is specifically upregulated by stresses that induce the σ^E response, while *ppiA* expression is enhanced in the presence of Cpx-activating signals and *ppiD* is controlled both by the cytoplasmic heat shock and Cpx responses (27, 30, 77).

Chaperones

Although no general envelope chaperone has yet been described, several categories of chaperone molecules exist in this compartment (Table 1). The most abundant of these are the pilus-specific chaperones. They constitute a large class of periplasmic proteins, each of which is thought to facilitate folding and subsequent targeting of cognate pilus subunits to an OM assembly platform or usher where they are incorporated into the growing pilus on the cell surface (see references 40 and 99 for reviews).

A second type of envelope chaperone is responsible for the correct incorporation of many lipoproteins into the OM. LolA or p20 is a soluble periplasmic component that can extract modified OM lipoproteins out of the IM and shuttle them to their final destination (59). Final assembly in the OM requires that the lipoprotein be exchanged from LolA to LolB, another lipoprotein, which facilitates incorporation into the OM (60).

The periplasmic protein Skp (100) also appears to facilitate OMP biogenesis and has been proposed to be a general chaperone for this type of substrate. Several observations suggest that it may play a role in chaperoning OMPs in the bacterial envelope. Skp preferentially associates with denatured OmpF (19, 31), and multiple copies of *skp* are able to suppress the defective OM phenotypes associated with *surA* nulls (66). Finally, and perhaps most convincingly, it was recently found that coexpression of *skp* increased the efficiency of phage display and the functional yield of heterologous proteins that are secreted to the periplasm (13). The precise nature and stage of protein folding at which Skp function is required remain to be determined.

Finally, the periplasmic protease DegP appears to possess chaperone activity as well. While DegP functions predominantly to degrade misfolded envelope proteins at elevated temperatures, there is compelling evidence that at lower temperatures, when folding defects may be more easily repaired, it performs as a chaperone (94).

Proteases

In the cytoplasm stresses that cause protein denaturation upregulate the production of proteases that degrade the offending substrates (17, 23). Of the large number of proteases that have been identified as residents of the bacterial envelope, only DegP has been ascribed a central role in combating envelope stress to date. DegP is a periplasmic serine endoprotease that degrades abnormal or misassembled envelope proteins (22, 43, 50, 53, 92, 95). Moreover,

degP expression is under the control of two different envelope stress responses (29, 35, 52).

Of the remaining envelope proteases, only a few have been linked to a stress-combative role. DegQ and DegS are homologs of DegP (8, 102). Multiple copies of DegQ can suppress the temperature-sensitive phenotype of *degP* nulls, and *degS* null strains show a slow growth phenotype. OmpT and OmpP are OM serine endoproteases (46, 97). They may be involved in responding to envelope stress since OmpT has been shown to destroy abnormal envelope proteins, and the levels of both are elevated at increased temperature (5, 46, 89). Finally, Tsp is a periplasmic protease that degrades aberrantly translated, exported proteins that have been marked with a C-terminal peptide tag (47).

ENVELOPE STRESS RESPONSES

Two envelope stress responses have been characterized to date (Table 2). One is controlled through the alternative sigma factor σ^E, and the second is regulated by a classical two-component regulatory pair, made up of the sensor histidine kinase (HK) CpxA and the response regulator (RR) CpxR. A variety of stresses specific to the bacterial envelope activate one or both of these signal transduction pathways, resulting in upregulated expression of protein folding and degrading factors that reside in the envelope. Recent studies suggest that the actual stress that is perceived by each pathway corresponds to a distinct type of misfolded protein and that the regulon members function to refold or degrade the offending species.

The σ^E Response

History

σ^E was originally purified from fractions of RNA polymerase (RNAP) on the basis of its ability to initiate transcription from the *rpoH*P3 promoter in conjunction with core RNAP (35, 103). Since *rpoH* encodes the alternative sigma factor σ^H or σ^{32} that controls the cytoplasmic heat shock response, and *rpoH*P3 is the only promoter used to express this gene at extremely high temperature, Erickson and Gross proposed that σ^E holoenzyme might regulate a second heat shock regulon required for growth at very high temperatures (35). Concurrently, promoter mapping revealed that the promoter of *degP*, another gene which is essential at high temperatures, was similar to that of *rpoH*P3 (52) and was transcribed by RNAP σ^E holoenzyme in vitro (35).

A number of years later, the *rpoE* gene of *E. coli*, encoding σ^E, was identified (79, 87). These studies

Table 2. Similarities and differences between the σ^E and Cpx envelope stress responses

Characteristic	Envelope stress response	
	σ^E	Cpx
Inducing cues	Misfolded OM or PP proteins	Misfolded proteins; aggregates at IM?
Signal transduction mechanism	Relief of inhibitory interactions between the anti-sigma factor RseA and σ^E	Alterations in phosphotransfer reactions between the HK CpxA and the RR CpxR
Downstream targets	*degP fkpA rpoE rseA rseB rseC rpoH*	*degP yihE dsbA ppiA/rotA ppiD cpxRA cpxP*
Cellular role	Ensure correct OMP folding	Appropriately timed biogenesis of pili; others?
Autoregulation	Yes	Yes
Accessory regulatory molecules	RseB; exerts repressive effect from periplasm; shutoff molecule?	CpxP; exerts repressive effect from periplasm; shutoff molecule?
Overlap with other stress responses	Cpx, σ^H	σ^E, σ^H

demonstrated that *rpoE* nulls confer increased sensitivity to chemicals known to disrupt the OM and that expression of several proteins is induced upon overexpression of σ^E (79, 87). In fact, the *rpoE* gene is essential, and null mutations can only be obtained in conjunction with an unidentified suppressor mutation (32). Phylogenetic analysis demonstrated that σ^E belongs to the ECF class of alternative σ factors, all of which regulate extracytoplasmic functions (55). Cumulatively, all these observations led to the proposal of a second σ^E-controlled heat shock or stress response that appeared to be linked to the bacterial envelope.

σ^E activating signals

Studies of inducing signals have been extremely informative concerning the type of envelope stress recognized by the σ^E pathway and the physiological role of this response. Like the cytoplasmic heat shock response (17, 23), the σ^E pathway is activated by stresses such as heat and ethanol that denature proteins (36, 87). However, the σ^E pathway does not respond to aberrantly folded proteins in the cytoplasm as overproduced misfolded proteins in this compartment have no effect on the σ^E response (64).

In contrast, a bounty of observations point to the conclusion that the σ^E stress response monitors protein folding in the bacterial envelope. For example, a variety of mutations that impair protein folding in this compartment cause upregulated σ^E activity (66). Also, the overproduction of mutant envelope proteins known to be misfolded causes a similar upregulation of the σ^E response, and multiple copies of either *surA*, *fkpA*, or *skp* are able to counter this activation (66).

Not only is the σ^E pathway responsive to misfolded proteins in the envelope, it appears to be specifically activated by misfolded OMPs. Gross and coworkers showed strikingly that overexpression of OMPs leads to activation of the σ^E stress response,

while underexpression diminishes it (64). Similarly, mutations that cause OMP misfolding also activate the σ^E pathway. Further, they convincingly demonstrated that the activating signal is generated at a point after the fates of periplasmic and OM proteins diverge (64). Taken together, these observations argue that σ^E responds to a signal that is generated during OMP folding and assembly after secretion across the IM.

Despite this, production of mutant or abnormal periplasmic proteins can activate the σ^E pathway (66). Since the overexpression of wild-type periplasmic proteins does not activate the σ^E pathway (64), perhaps the mutant proteins possess some quality that wild-type periplasmic proteins never manifest. In support of this, it is known that one of these mutant periplasmic proteins, MalE31, becomes trapped in periplasmic inclusion bodies as an off-pathway aggregate (9). Alternatively, perhaps these mutant proteins activate the σ^E pathway by titrating folding factors required for OMP assembly or simply interfere with trafficking of OMPs to the OM.

The precise nature of the σ^E activating signal remains to be determined. Is there some specific feature of misfolded OMPs or abnormal periplasmic proteins that is recognized by the signal transduction machinery, or is protein misfolding sensed indirectly by monitoring levels of a critical folding factor? Future studies should be directed at answering these fundamental questions.

Transduction of envelope stress across the IM

In *E. coli*, the σ^E pathway accomplishes transduction of the envelope stress signal through a membrane-localized anti-sigma factor called RseA (33, 68). Gross and colleagues first noted the presence of *rseA* and *rseB* as two genes directly downstream of *rpoE* that shared homology with the MucAB negative regulators of the σ^E homolog AlgU in *Pseudomonas aeruginosa* (87). Null mutations in *rseA* lead to a 25-fold increase in activity of the σ^E pathway. Moreover, the σ^E regulon in these mutants can no longer be further induced by envelope stress, demonstrating that RseA is required for the sensation of activating cues and, further, that relief of the negative effect mediated by RseA is sufficient for full stimulation of the pathway. In agreement with these observations, overexpression of RseA was shown to repress the σ^E pathway (33, 68).

RseA contains a single transmembrane domain that orients an N-terminal cytoplasmic domain and a C-terminal periplasmic domain. Overexpression of the N terminus of RseA is sufficient to repress the σ^E regulon in vivo, and it was shown to physically interact with σ^E in vitro (33, 68). Thus, perhaps in the absence of inducing signals, the pathway is held in check by negative protein-protein interactions between RseA and σ^E that sequester σ^E at the IM in an inactive conformation (Fig. 1) (33, 68). Misfolded OMPs or periplasmic proteins could then act to relieve this negative interaction, leading to activation of the pathway (Fig. 1).

Recently, it was demonstrated that σ^E activating signals trigger DegS-mediated degradation of RseA (1). Thus, it appears that the inhibitory interaction between RseA and σ^E is relieved by proteolysis. Uncovering the mechanism of this regulated degradation promises to offer great insight into how protein misfolding is sensed in the bacterial envelope.

A second periplasmic signaling molecule may be involved. *rseB* is found in an operon with σ^E and *rseA* (33, 68). *rseB* null mutations cause a two- to threefold increase in σ^E activity, suggesting that RseB has a repressive role. RseB is localized in the periplasm, and overexpression downregulates the σ^E pathway. Moreover, RseB has been shown by two different approaches to interact directly with the periplasmic domain of RseA. However, *rseB* mutants can still be further induced by activating signals, demonstrating that removal of RseB is not sufficient for maximal activation of the pathway. Perhaps RseB functions to fine-tune the σ^E stress response. For example, RseB may be titrated away from RseA by inducing cues, priming RseA for activation (Fig. 1). Alternatively, RseB may function primarily as a shutoff molecule (82). Since the σ^E pathway is subject to autoactivation (87), rapid amplification of the response upon exposure to transitory envelope stresses could lead to an imbalance in envelope folding factors that is likely to be detrimental to the cell. Since RseB is also σ^E regulated, its increased expression could act as a safety valve, allowing for the rapid shutoff of the

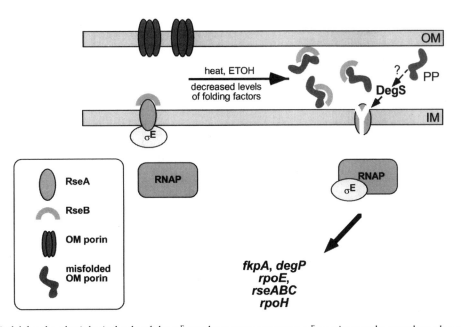

Figure 1. Model for the physiological role of the σ^E envelope stress response; σ^E monitors and responds to changes in OMP folding. Under nonstressed conditions, σ^E is maintained in an inactive conformation by interactions with the membrane-localized anti-sigma factor RseA. This negative interaction is further reinforced through periplasmic interactions between RseB and RseA. Envelope perturbations that affect OMP folding stimulate DegS-mediated degradation of RseA and alleviate the negative interaction between σ^E and RseA. These events cause σ^E to be released from RseA, allowing association with RNAP holoenzyme and activation of transcription from the promoters of downstream targets. These include the OMP folding and degrading factors FkpA and DegP, as well as *rpoH* and the *rpoE rseABC* operon. FkpA and DegP likely function to refold or degrade the misfolded OMPs. Autoregulation amplifies the response and allows for rapid shutoff once the misfolded OMPs have been cleared through overproduction of the negative regulators RseA and RseB.

pathway upon alleviation of the envelope stress. It will be interesting to determine whether RseA and RseB have additional roles in protein folding above and beyond their regulatory functions. Analysis of shutoff kinetics and other envelope proteins that interact with RseB would be useful in further clarifying the role of RseB.

The σ^E regulon

On the basis of analysis of protein profiles from strains overexpressing *rpoE*, it has been concluded that the σ^E regulon consists of at least 11 members (79, 87). Two of these include the envelope protein folding and degrading factors FkpA and DegP (27, 35, 52). Four other downstream targets of σ^E are the *rpoE* gene itself plus the downstream *rseA* and *rseB* and a fourth gene, *rseC*, in the operon of unknown function (33, 68, 79, 87). It may be that the signal amplification afforded by autoactivation is essential for escaping the onslaught of sudden and severe stresses to the bacterial envelope.

The σ^E response to envelope stress overlaps with other cellular stress responses. For example, σ^E activates a promoter upstream of *rpoH* (35, 103). The significance of this overlap is not currently understood. It may be that envelope stress, which would logically precede intracellular insults upon a change in environments, is used by the cell as an indicator of impending protein folding difficulties in the cytoplasm. Also, *degP* is a member of both the σ^E and Cpx envelope stress responses (29). This likely reflects the fact that DegP is required for the degradation of a wide variety of misfolded envelope proteins, including those that activate both the σ^E and Cpx signal transduction pathways (22, 29, 43, 95, 96). The challenge for the future will be to determine the remaining σ^E regulon members. Their identity will undoubtedly shed more light on the physiological role of the σ^E envelope stress response.

Physiological role of the σ^E stress response in *E. coli*

The nature of the activating signals and downstream targets suggests a distinct physiological role for the σ^E pathway in *E. coli* (Fig. 1). The σ^E pathway induces expression of *degP* and *fkpA* in response to misfolded OMP precursors or accumulated folding intermediates (27, 29, 35, 52, 64). Both these factors assist in alleviating protein folding difficulties in the bacterial envelope (22, 66, 92, 95). Moreover, the σ^E pathway is essential (32). Taken together, these observations suggest that the σ^E pathway plays a fun-

damental biological role in monitoring and responding to alterations in OMP biogenesis (Fig. 1).

The Cpx Envelope Stress Response

History

McEwen and Silverman first identified *cpxA* almost 20 years ago by virtue of mutations that affected conjugative pilus expression (61). Strains containing *cpxA* mutations were defective in conjugal transfer of the F plasmid because they no longer assembled a functional F pilus on the cell surface. Silverman's lab subsequently noted that *cpxA* mutations were associated with a variety of pleiotropic phenotypes, the majority of which reflected alterations in envelope-associated functions (62, 63, 80, 98). Cloning and characterization of the *cpx* locus showed that it encoded a typical two-component regulatory pair (38) consisting of the membrane-localized sensor HK CpxA (2, 104) and the cognate RR CpxR (34).

Studies in the Silhavy lab were the first to link the CpxRA two-component system with control of an envelope stress response. Gain-of-function mutations that mapped to the *cpx* locus, *cpx** or *cpxA** mutations, were able to suppress the toxic effects of mutant and abnormal envelope proteins (22), in large part through upregulated expression of the periplasmic protease *degP* (22, 29, 79). This upregulation is distinct from that conferred by the σ^E pathway, as signals that activate the σ^E pathway, such as OMP overexpression, have no effect on Cpx-mediated control of *degP* gene expression (29). Further, the Cpx pathway does not control expression from the *rpoH*P3 promoter, as σ^E does (29, 79). Thus, it was proposed that the Cpx two-component system controlled a distinct envelope stress response, which overlapped with that of σ^E at transcription of *degP* (22, 29, 79). Subsequent studies have borne out this hypothesis.

Cpx-activating signals

In general, Cpx-inducing cues fall into two broad categories: those of a general nature causing pleiotropic effects, and others that appear to be somewhat more specific in effect. What all of the activating signals have in common is that they are expected to perturb envelope protein biogenesis in some manner.

One rather broad environmental condition that influences Cpx signal transduction is pH. pH was first suggested to be involved in control of Cpx-mediated gene expression by studies examining the virulence of *Shigella* species. Nakayama and Watanabe demonstrated that, in *E. coli*, the *cpx* locus is required for the proper regulation of *virF*, a gene encoding a transcriptional activator of the *ipa* genes, in response to pH changes (73). Danese and Silhavy demonstrated that Cpx-mediated gene expression is dramatically elevated at increased pH and further that *cpx* mutants exhibit decreased survival under such conditions (28). Although it is unknown how alkaline pH induces the pathway, it is thought that Cpx activation at elevated pH may be the result of structural damage to the bacterial envelope caused by protein denaturation in this compartment (28).

A second rather general Cpx-inducing signal is generated in mutants that fail to synthesize phosphatidylethanolamine (PE) (65). Such mutants exhibit a wide variety of pleiotropic phenotypes that are partly due to activation of the Cpx signal transduction cascade. Interestingly, PE has been shown to facilitate IM protein folding, suggesting that it may act as a chaperone for some envelope proteins (10, 11). Accordingly, perhaps the Cpx pathway is activated by misfolded IM proteins in these mutants. Alternatively, PE$^-$ mutants may simply alter the conformation of the CpxA HK in the membrane such that it becomes constitutively active.

Last in this class of poorly understood Cpx-activating signals is accumulation of the enterobacterial common antigen (ECA) intermediate lipid II at the cytoplasmic membrane (26). ECA is a glycolipid found in the outer leaflet of the OM of members of the family *Enterobacteriaceae* (84). Interestingly, these mutations also cause a variety of envelope perturbations that are typical of bacteria that are defective in envelope protein folding and/or targeting, suggesting that the Cpx-activating effects could be due to a perturbation of protein folding and localization in the bacterial envelope (26). Analyses of inducing cues of a more specific nature support and extend this hypothesis.

One of the most salient clues to the function of the Cpx pathway derives from studies of P pilus assembly in uropathogenic *E. coli*. In the absence of their cognate chaperone, the interactive surfaces of P pilus subunits associate and aggregate at the periplasmic face of the IM, and this serves as a potent inducing signal for the Cpx pathway (43). Indeed, a second somewhat specific Cpx stimulus, overproduction of the novel OM lipoprotein NlpE, also appears to involve IM-associated, misfolded protein (92). It is tempting to speculate that mislocalization of the typically OM-associated, misfolded pilin subunits and NlpE to the IM is what activates the Cpx pathway.

Thus, like the σ^E pathway, the Cpx-activating signal consists of misfolded envelope protein and is generated after secretion across the IM. However, in contrast to the periplasmic and OM intermediates that induce the σ^E regulon, the Cpx response may be attuned to recognition of misfolded proteins having physically distinct locations and attributes.

Lastly, it should be noted that some Cpx-activating signals induce the σ^E pathway as well. For example, overexpression of the P pilus subunit PapG in the absence of the cognate PapD chaperone can activate both stress responses. σ^E activation, however, is approximately threefold while induction of the Cpx pathway by overproduction of misfolded pilin subunits can be greater than 10-fold (43). It may be that the σ^E pathway is responding to the relatively small pool of misfolded subunits that accumulate in the periplasm under these conditions, while the Cpx proteins sense the IM-associated aggregates. Also, the PapG subunit is distinct from others because it contains an adhesin domain in addition to the aggregative domain that is capped by the chaperone and required for pilus assembly (41). Accordingly, it may be that overexpression of the adhesin domain resembles the OMP-activating signal recognized by the σ^E pathway.

Similarly, lipid II accumulation also stimulates both envelope stress responses (26). It is possible that lipid II accumulation has effects on the folding and targeting of envelope proteins characteristic of those monitored by both the σ^E and Cpx signal transduction systems. At any rate, the precise nature of the Cpx-activating signal and how it is sensed remain areas ripe for future study.

Cpx-mediated signal transduction across the IM

Transduction of envelope stress by the Cpx pathway is accomplished by the CpxA sensor HK and the RR CpxR (34, 81, 104). Like many HKs, CpxA is localized to the IM by two transmembrane alpha helices that flank a single periplasmic domain (104). Its cytoplasmic domain contains the characteristic motifs that are required for the conserved autophosphorylation and phospho-transfer reactions possessed by this family of proteins (2, 38, 104). Likewise, CpxR contains an N-terminal phospho-accepting domain typical of the RR family of proteins (34, 38). The C-terminal domain of CpxR is homologous to the OmpR subclass of RRs and places it in the family of winged helix transcription factors (58).

Biochemical analysis indicates that the mechanism of signal transduction utilized by CpxA and CpxR is much the same as for other two-component systems. CpxA functions as an autokinase, a CpxR kinase, and a phosphatase of phosphorylated CpxR in vitro (Fig. 2) (81). Further, while a constitutively activated *cpxA** mutant can still catalyze phosphory-

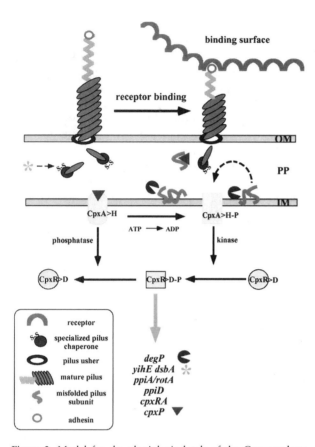

Figure 2. Model for the physiological role of the Cpx envelope stress response; the Cpx pathway mediates properly timed biogenesis of pili. In the absence of appropriate receptor molecules, the Cpx pathway is maintained in a predominantly off or kinase$^-$/phosphatase$^+$ state. Interaction with the periplasmic signaling molecule CpxP helps maintain this state. Attachment to an appropriate receptor impedes pilus growth by disrupting folding into the final structure on the cell surface, resulting in an accumulation of misfolded subunits at the periplasmic face of the IM. These aggregated, misfolded subunits titrate the repressive molecule CpxP and further activate CpxA via direct or indirect means, causing an elevation in the ratio of CpxA kinase:phosphatase activity and accumulation of phosphorylated CpxR in the cytoplasm. Phosphorylated CpxR activates expression of genes whose products are involved in pilus biogenesis, *dsbA* and *degP*, as well as those encoding the putative folding factors PpiA/RotA and PpiD. In addition, the response is subject to amplification by autoactivation of the *cpxRA* operon, ensuring a rapid commitment to pilus biogenesis when the conditions are appropriate. Further, the gene encoding the shutoff molecule CpxP is also upregulated, allowing for rapid downregulation of the pathway when adhesive organelles are no longer being made.

lation of both itself and CpxR, it is completely devoid of phosphatase activity. These observations suggest that it is the ratio of CpxA kinase to phosphatase activity that determines the level of activation of the pathway (Fig. 2) (81). Finally, like other RRs, phosphorylation of CpxR enhances its ability to bind to a consensus sequence that is found upstream of some Cpx-activated genes (77, 81).

Interestingly, as with the σ^E response, an accessory, negatively acting, periplasmic signaling molecule has been identified for the Cpx pathway (83). Like RseB of the σ^E response, overexpression of the small, periplasmic, Cpx-regulated molecule CpxP causes a three- to fivefold reduction in Cpx-mediated gene expression that is dependent on the CpxA sensory domain. Further, deletion of *cpxP* leads to activation of the Cpx response, consistent with a repressive role. However, *cpxP* null mutants can still be further induced by activating signals, demonstrating that removal of CpxP is not sufficient for full activation of the pathway. As mentioned earlier with respect to RseB, it has been suggested that these molecules perform a shutoff function, allowing for rapid turnoff of the pathway in question upon alleviation of envelope stress (82, 83).

Cpx signal transduction may also be affected by two phosphoprotein phosphatases that share homology with eukaryotic serine/threonine and tyrosine phosphatases (69). PrpA and PrpB were identified in a multicopy screen designed to detect factors that conferred enhanced transcription of a *degP-lacZ* reporter fusion. At present, it is unclear how these phosphatases exert their effect on the Cpx signal transduction cascade. It seems unlikely that they affect the phosphorylation status of CpxR directly, since it is the phosphorylated form of this RR that is responsible for up-regulated transcription of downstream target genes (29, 81). The situation is further complicated by the fact that PrpA and PrpB appear to affect multiple cellular signal transduction pathways (69).

Many questions concerning Cpx signal transduction remain unanswered. How are misfolded envelope proteins sensed by the CpxA sensor HK? What is the molecular mechanism behind transduction of envelope stress? How does CpxR activate transcription from promoters that are transcribed by at least three different forms of RNAP holoenzyme? The answers to these and other questions promise to add to our knowledge of both envelope stress as well as two-component signal transduction in general.

The Cpx regulon

To date, eight Cpx regulon members have been identified in *E. coli*. Of these, with the exception of the *cpxRA* genes themselves, all encode demonstrated or suspected envelope protein folding factors (Fig. 2). The first Cpx regulon member to be identified was the *degP* gene. The Silhavy lab noted that gain-of-function mutations that protected the cell from toxic envelope proteins and mapped to the *cpx* locus also conferred enhanced proteolysis of these proteins

(22). Accordingly, they analyzed expression of the gene encoding the periplasmic protease *degP* and found it to be under the control of the CpxRA proteins (29).

Expression of DsbA and an upstream gene in the same operon of unknown function, *yihE*, is also Cpx-regulated (27, 77). This finding was significant because it demonstrated that the Cpx response controls the expression of at least one envelope protein foldase, in addition to the DegP protease, thus supporting the notion that this pathway plays a general role in monitoring protein folding in the bacterial envelope. It will be of interest to determine the function of *yihE*. Does this gene, like DsbA, play a role in envelope protein folding?

A consensus CpxR binding site was identified by virtue of footprint analysis of the *degP* and *dsbA* promoter regions and used to identify the periplasmic ppiase encoding gene *rotA/ppiA* as another candidate for regulation by the Cpx pathway (77). Indeed, the *rotA/ppiA* promoter binds phosphorylated CpxR with high affinity and is transcriptionally regulated by the Cpx stress response, bringing the total number of Cpx regulon members to four.

In a search to identify random insertions of a promoterless *lacZ* element that were Cpx-regulated, Danese and Silhavy identified a fifth regulon member, *cpxP* (28). Found directly upstream of, and in the opposite orientation to, the *cpxRA* operon, *cpxP* was shown to encode a small, periplasmic protein that may play a role in resistance to stresses such as alkaline pH (28) in addition to feedback inhibition of the Cpx response (Fig. 2) (83). Further description of the role of CpxP in the periplasm will undoubtedly shed more light on the mechanism used by CpxA to monitor protein folding in the envelope.

The sixth reported Cpx regulon member is PpiD (30). Interestingly, regulation of *ppiD* expression represents another point of overlap between stress responses, as it is also under classical heat shock control by σ^H. Based on the presence of CpxR consensus binding sites in the *rpoH* promoter region, Pogliano et al. (77) have suggested that *rpoH* itself may be a Cpx regulon member; thus, the role of the Cpx response in *ppiD* regulation may not be straightforward.

Lastly, as with other stress responses, the *cpxRA* operon itself is a downstream target of the Cpx envelope stress response (83). Autoactivation may act as a trigger for switching on the Cpx response rapidly by allowing for massive signal amplification. This could be an essential feature for rescuing the cell upon abrupt environmental changes that would otherwise lead to death. Many such autoregulated responses appear to have a similar trigger function for

flipping on developmental and differentiative programs in other bacteria (38). Autoactivation comes into play once a threshold level of stimulus is present, ensuring that energetically expensive programs are not initiated inappropriately. Intriguingly, the Cpx response is tightly linked to one such program in *E. coli*, biogenesis of P pili (43, 82) (see below). It may be that this autoregulatory trigger feature of the Cpx response is required for the appropriately timed assembly of pili on the cell surface.

It is likely that the Cpx regulon extends beyond the eight genes described here. The presence of consensus CpxR binding sites has been noted upstream of other genes, including *rpoE*, *rpoH*, *groE*, and *skp* (77). Moreover, since known Cpx-regulated genes lack an obvious consensus CpxR binding motif, this suggests that an even greater number of potential downstream targets may exist. The identification of the remaining regulon members should be helpful in further defining the role of the Cpx envelope stress response and will hopefully broaden our views of the topics of envelope protein folding and targeting.

The physiological role of the Cpx envelope stress response

The Cpx pathway is not required for growth under nonstressed laboratory conditions (29). However, a number of observations point to a stress-combative role for this regulon. Firstly, activation of the Cpx pathway can rescue the cell from expression of otherwise toxic envelope proteins (22, 92). Secondly, *cpx* null mutants are impaired in their ability to withstand alkaline pH (28). Further, activation of the Cpx pathway can suppress the temperature-sensitive phenotype of *rpoE* null mutants (20). Thus, one physiological role of the Cpx pathway appears to be to protect the cell from the damage incurred by various envelope stresses.

The Cpx envelope stress response has also been co-opted to play a role in the virulence of pathogenic organisms. For example, in *Shigella* species, expression of the transcriptional activator VirF that is required for expression of the *ipa* genes that allow host cell invasion is controlled in response to pH by the Cpx two-component system (73). Thus, it seems that the Cpx proteins may sense pH changes in the host that signal a requirement for the Ipa cell invasion machinery.

Moreover, a wealth of recent observations have linked the Cpx envelope stress response with the process of pilus biogenesis in strains of uropathogenic *E. coli*. Misfolded, periplasmic pilus subunits serve as potent activators of the Cpx pathway, and this provides a mechanism for monitoring pilus assembly

(43). Indeed, it is thought that P pilus production in and of itself stimulates the Cpx response through the production of off-pathway, misfolded intermediates (43). It makes sense that the Cpx stress response is attuned to P pilus biogenesis since at least two Cpx-regulated factors are important for this process. DsbA is needed for the proper folding of the specialized pilus chaperone PapD, while DegP functions to degrade misfolded, off-pathway P pilin subunits (42, 43). Thus, a central role of the Cpx stress response seems to be to monitor and assist in assembly of P pili (Fig. 2).

Intriguingly, this role may extend beyond assembly to correctly timing the elaboration of pili as well (Fig. 2). It is expected that pilus-mediated attachment would confer an assembly block by impeding outward growth of the structure, causing a buildup of misfolded, aggregated subunits in the periplasm (72). Since misfolded pilin subunits activate the Cpx pathway, the signal transduction machinery could effectively sense the presence of an appropriate binding surface and concomitantly increase the number of pili on the bacterial surface by upregulating assembly factors. Interestingly, pili of the chaperone/usher type are widespread in gram-negative organisms and pathogens in particular (99). It will be interesting to determine the precise role of the Cpx response in the correctly timed biogenesis of P pili and to see whether it extends to other members of this family and perhaps other envelope-associated structures as well.

Other Envelope Stress Responses?

The Cpx and σ^E stress responses clearly play a role in protecting the cell from insults to the envelope that cause protein misfolding in this compartment. It seems possible that other responses exist which might also perceive envelope stresses via misfolded proteins. For example, although much has been learned about the regulation of envelope protein folding and degrading factors through the study of the σ^E and Cpx responses in recent years, our knowledge of this field is still in its infancy. Little is known concerning how, or if, central folding factors such as SurA, DsbB, DsbC, DsbD, or Skp are regulated. Further, it is unlikely that all of the envelope folding and targeting factors have been identified, much less studied at the level of regulation. Also, most convincingly, some studies argue strongly that other envelope stress responses exist. For instance, Silhavy and coworkers recently showed that expression of a small fragment of the integral IM protein TetA allowed for partial folding and correct localization of a toxic signal sequence mutant of the OM porin LamB (21). Remarkably,

this effect was independent of both the σ^E and Cpx envelope stress responses. Since it is highly unlikely that the mechanism of TetA'-mediated folding and localization of the mutant LamB is direct, these observations suggest that perhaps induction of an unidentified envelope stress response is involved.

Protein misfolding represents one symptom of envelope stress. However, insults to this compartment can lead to a plethora of other effects, including altered membrane structure and potential. It seems likely that distinct envelope stress responses must play roles in protecting the cell from these serious problems. Indeed, there is evidence suggesting that at least one such response does exist. The phage shock response induces the expression of the *psp* (phage shock proteins) operon upon exposure to a number of envelope perturbations (see reference 71 for a review). These include overexpression of a variety of OMPs, both mutant and wild-type (including the filamentous-phage OMP pIV, for which the response is named), extreme heat shock, ethanol, and hyperosmotic stress (12, 16, 18, 49). Because this response is also activated by media downshifts and uncoupling agents, it has been proposed that the inducing signal may be a perturbation in the level of ATP or its generation (105). In support of this, null mutations in the *psp* genes are less able to maintain their membrane potential, are slower to export proteins by the general secretory apparatus, and grow poorly in stationary phase at alkaline pH (48, 49, 105). Although the functions of the *psp*-encoded proteins are not fully understood, they are localized to the IM and PP, indicating some type of envelope function (see reference 71). Clearly there is some overlap with the σ^E and Cpx responses in that the *psp* response is also turned on by expression of mutant envelope proteins (12, 18, 49).

CONCLUSIONS

Many stresses to the bacterial envelope are perceived as protein folding problems by either or both the σ^E and Cpx stress responses. These pathways are parallel in many respects and share a number of common features (Table 2). For example, they both transduce the envelope stress signal across the IM by means of altered interactions between pairs of proteins found in the IM and cytoplasm. Further, both responses are capable of rapid amplification by means of auto-activation and subsequently to abrupt shutdown through accessory, repressive regulatory elements. Moreover, both pathways alleviate protein misfolding problems in the envelope by upregulating the production of folding and degrading agents localized to this compartment. In fact, one such factor, the DegP protease, is a downstream target of both stress responses.

However, each of these stress responses is also unique with respect to the type of misfolded protein that is sensed, and both pathways activate expression of a distinct, nonoverlapping set of downstream targets (Table 2). The σ^E pathway appears to have been specifically adapted to ensure proper folding of OMPs, while one function of the Cpx signal transduction cascade may be the correctly timed production and assembly of adhesive organelles. Future endeavors should be directed toward defining the mechanism of perceiving protein misfolding as well as identifying the remaining downstream targets of each response, with a view toward better understanding the physiological roles of the σ^E and Cpx envelope stress responses.

In addition, much remains to be learned concerning other mechanisms of envelope stress sensation and response. Perhaps a better understanding of how the many envelope protein folding catalysts are regulated will lead to the identification of other stress responses. What other physiological functions are disrupted, besides protein folding, upon stresses to this compartment, and how are alterations in these processes monitored and corrected? As our knowledge of envelope physiology and biogenesis expands, so too will our understanding of other responses that are mounted when this compartment is physically stressed.

Acknowledgments. We gratefully acknowledge the support of the American Heart Association (fellowship NJ-97-FW-04 to T.L.R.) and the National Institutes of Health (grant GM34821 to T.J.S.).

REFERENCES

1. **Ades, S. E., L. E. Connolly, B. M. Alba, and C. A. Gross.** 1999. The *Escherichia coli* σ^E-dependent extracytoplasmic stress response is controlled by the regulated proteolysis of an anti-σ factor. *Genes Dev.* **13**:2449–2461.

2. **Albin, R., R. Weber, and P. M. Silverman.** 1986. The Cpx proteins of *Escherichia coli* K12. Immunologic detection of the chromosomal *cpxA* gene product. *J. Biol. Chem.* **261**:4698–4705.

3. **Altenbach, C., T. Marti, H. G. Khorana, and W. L. Hubbell.** 1990. Transmembrane protein structure: spin labeling of bacteriorhodopsin mutants. *Science* **248**:1088–1092.

4. **Andersen, C. L., D. A. Matthey, D. Missiakas, and S. Raina.** 1997. A new *Escherichia coli* gene, *dsbG*, encodes a periplasmic protein involved in disulphide bond formation, required for recycling DsbA/DsbB and DsbC redox proteins. *Mol. Microbiol.* **26**:121–132.

5. **Baneyx, F., and G. Georgiou.** 1990. In vivo degradation of secreted fusion proteins by the *Escherichia coli* outer membrane protease OmpT. *J. Bacteriol.* **172**:491–494.

6. **Bardwell, J. C., J. O. Lee. G. Jander, N. Martin, D. Belin, and J. Beckwith.** 1993. A pathway for disulfide formation in vivo. *Proc. Natl. Acad. Sci. USA* **90**:1038–1042.

7. **Bardwell, J. C., K. McGovern, and J. Beckwith.** 1991. Identification of a protein required for disulfide bond formation in vivo. *Cell* **67**:581–589.

8. **Bass, S., G. Qimin, and A. Christen.** 1996. Multicopy suppressors of Prc mutant *Escherichia coli* include two HtrA (DegP) protease homologs (HhoAB), DksA, and a truncated RlpA. *J. Bacteriol.* **178**:1154–1161.

9. **Betton, J.-M., and M. Hofnung.** 1996. Folding of a mutant maltose-binding protein of *Escherichia coli* which forms inclusion bodies. *J. Biol. Chem.* **271**:8046–8052.

10. **Bogdanov, M., and W. Dowhan.** 1998. Phospholipid-assisted protein folding: phosphatidylethanolamine is required at a late step of the conformational maturation of the polytopic membrane protein lactose permease. *EMBO J.* **17**:5255–5264.

11. **Bogdanov, M., J. Sun, H. R. Kaback, and W. Dowhan.** 1996. A phospholipid acts as a chaperone in assembly of a membrane transport protein. *J. Biol. Chem.* **271**:11615–11618.

12. **Bosch, D., and J. Tommassen.** 1987. Effects of linker insertions on the biogenesis and functioning of the *Escherichia coli* outer membrane pore protein PhoE. *Mol. Gen. Genet.* **208**:485–489.

13. **Bothmann, H., and A. Pluckthun.** 1998. Selection for a periplasmic factor improving phage display and functional periplasmic expression. *Nature Biotech.* **16**:376–380.

14. **Bramhill, D.** 1997. Bacterial cell division. *Annu. Rev. Cell. Dev. Biol.* **13**:395–424.

15. **Brass, J. M., C. F. Higgins, M. Foley. P. A. Rugman, J. Birmingham, and P. B. Garland.** 1986. Lateral diffusion of proteins in the periplasm of *Escherichia coli*. *J. Bacteriol.* **165**:787–794.

16. **Brissette, J. L., M. Russel, L. Weiner, and P. Model.** 1990. Phage shock protein, a stress protein of *Escherichia coli*. *Proc. Natl. Acad. Sci. USA* **87**:862–866.

17. **Bukau, B.** 1993. Regulation of the *Escherichia coli* heat-shock response. *Mol. Microbiol.* **9**:671–680.

18. **Carlson, J. H., and T. J. Silhavy.** 1993. Signal sequence processing is required for the assembly of LamB trimers in the outer membrane of *Escherichia coli*. *J. Bacteriol.* **175**:3998–4007.

19. **Chen, R., and U. Henning.** 1996. A periplasmic protein (Skp) of *Escherichia coli* selectively binds a class of outer membrane proteins. *Mol. Microbiol.* **19**:1287–1294.

20. **Connolly, L., A. De Las Peñas, B. M. Alba, and C. A. Gross.** 1997. The response to extracytoplasmic stress in *Escherichia coli* is controlled by partially overlapping pathways. *Genes Dev.* **11**:2012–2021.

21. **Cosma, C. L., M. D. Crotwell, S. Y. Burrows, and T. J. Silhavy.** 1998. Folding-based suppression of extracytoplasmic toxicity conferred by processing-defective LamB. *J. Bacteriol.* **180**:3120–3130.

22. **Cosma, C. L., P. N. Danese, J. H. Carlson, T. J. Silhavy, and W. B. Snyder.** 1995. Activation of the Cpx two-component signal transduction pathway in *Escherichia coli* suppresses envelope associated stresses. *Mol. Microbiol.* **18**:491–505.

23. **Craig, E. A., B. D. Gambill, and R. J. Nelson.** 1993. Heat shock proteins: molecular chaperones of protein biogenesis. *Microbiol. Rev.* **57**:401–414.

24. **Crooke, H., and J. Cole.** 1995. The biogenesis of c-type cytochromes in *Escherichia coli* requires a membrane-bound protein, DipZ, with a protein disulphide isomerase domain. *Mol. Microbiol.* **15**:1139–1150.

25. **Dailey, F. E., and H. C. Berg.** 1993. Mutants in disulfide bond formation that disrupt flagellar assembly in *Escherichia coli*. *Proc. Natl. Acad. Sci. USA* **90**:1043–1047.

26. **Danese, P. N., G. R. Oliver, K. Barr, G. D. Bowman, P. D. Rick, and T. J. Silhavy.** 1998. Accumulation of the enterobac-

terial common antigen lipid II biosynthetic intermediate stimulates *degP* transcription in *Escherichia coli*. *J. Bacteriol.* **180**: 5875–5884.

27. **Danese, P. N., and T. J. Silhavy.** 1997. The σ^E and the Cpx signal transduction systems control the synthesis of periplasmic protein-folding enzymes in *Escherichia coli*. *Genes Dev.* **11**:1183–1193.

28. **Danese, P. N., and T. J. Silhavy.** 1998. CpxP, a stress-combative member of the Cpx regulon. *J. Bacteriol.* **180**:831–839.

29. **Danese, P. N., W. B. Snyder, C. L. Cosma, L. J. B. Davis, and T. J. Silhavy.** 1995. The Cpx two-component signal transduction pathway of *Escherichia coli* regulates transcription of the gene specifying the stress-inducible periplasmic protease, DegP. *Genes Dev.* **9**:387–398.

30. **Dartigalongue, C., and S. Raina.** 1998. A new heat-shock gene, *ppiD*, encodes a peptidyl-prolyl isomerase required for folding of outer membrane proteins in *Escherichia coli*. *EMBO J.* **17**:3968–3980.

31. **de Cock, H., U. Schafer, M. Potgeter, R. Demel, M. Muller, and J. Tommassen.** 1999. Affinity of the periplasmic chaperone Skp of *Escherichia coli* for phospholipids, lipopolysaccharides and non-native outer membrane proteins. *Eur. J. Biochem.* **259**:96–103.

32. **De Las Peñas, A., L. Connolly, and C. A. Gross.** 1997. σ^E is an essential sigma factor in *Escherichia coli*. *J. Bacteriol.* **179**: 6862–6864.

33. **De Las Peñas, A., L. Connolly, and C. A. Gross.** 1997. The σ^E-mediated response to extracytoplasmic stress in *Escherichia coli* is transduced by RseA and RseB, two negative regulators of σ^E. *Mol. Microbiol.* **24**:373–385.

34. **Dong, J. S., S. Iuchi, H. S. Kwan, Z. Lu, and E. C. C. Lin.** 1993. The deduced amino-acid sequence of the cloned *cpxR* gene suggests the protein is the cognate regulator for the membrane sensor, CpxA, in a two-component signal transduction system of *Escherichia coli*. *Gene* **136**:227–230.

35. **Erickson, J. W., and C. A. Gross.** 1989. Identification of the σ^E subunit of *Escherichia coli* RNA polymerase: a second alternate σ factor involved in high-temperature gene expression. *Genes Dev.* **9**:387–398.

36. **Erickson, J. W., V. Vaughn, W. A. Walter, F. C. Neidhardt, and C. A. Gross.** 1987. Regulation of the promoters and transcripts of *rpoH*, the *Escherichia coli* heat shock regulatory gene. *Genes Dev.* **1**:419–432.

37. **Hinkle, P. C., P. V. Hinkle, and H. R. Kaback.** 1990. Information content of amino acid residues in putative helix VIII of the *lac* permease from *Escherichia coli*. *Biochemistry* **29**: 10989–10994.

38. **Hoch, J. A., and T. J. Silhavy (ed.).** 1995. *Two-Component Signal Transduction*. American Society for Microbiology, Washington, D.C.

39. **Horne, S. M., and K. D. Young.** 1995. *Escherichia coli* and other species of the *Enterobacteriaceae* encode a protein similar to the family of Mip-like FK506-binding proteins. *Arch. Microbiol.* **163**:357–365.

40. **Hultgren, S. J., C. H. Jones, and S. Normark.** 1996. Bacterial adhesins and their assembly, p. 2730–2756. *In* F. C. Neidhardt, R. Curtiss III, J. L. Ingraham, E. C. C. Lin, K. B. Low, B. Magasanik, W. S. Reznikoff, M. Riley, M. Schaechter, and H. E. Umbarger (ed.), Escherichia coli *and* Salmonella: *Cellular and Molecular Biology*, 2nd ed. ASM Press, Washington, D.C.

41. **Hultgren, S. J., F. Lindberg, G. Magnusson, J. Kihlberg, J. M. Tennent, and S. Normark.** 1989. The PapG adhesin of uropathogenic *Escherichia coli* contains separate regions for re-

ceptor binding and for the incorporation into the pilus. *Proc. Natl. Acad. Sci. USA* **86**:4357–4361.

42. **Jacob-Dubuisson, F., J. Pinkner, Z. Xu, R. Striker, A. Padmanhaban, and S. J. Hultgren.** 1994. PapD chaperone function in pilus biogenesis depends on oxidant and chaperone-like activities of DsbA. *Proc. Natl. Acad. Sci. USA* **91**:11552–11556.

43. **Jones, C. H., P. N. Danese, J. S. Pinkner, T. J. Silhavy, and S. J. Hultgren.** 1997. The chaperone-assisted membrane release and folding pathway is sensed by two signal transduction systems. *EMBO J.* **21**:6394–6406.

44. **Kadner, R. J.** 1996. Cytoplasmic membrane, p. 58–87. *In* F. C. Neidhardt, R. Curtiss III, J. L. Ingraham, E. C. C. Lin, K. B. Low, B. Magasanik, W. S. Reznikoff, M. Riley, M. Schaecter, and H. E. Umbarger (ed.), Escherichia coli *and* Salmonella: *Cellular and Molecular Biology*, 2nd ed. ASM Press, Washington, D.C.

45. **Kamitani, S., Y. Akiyama, and K. Ito.** 1992. Identification and characterization of an *Escherichia coli* gene required for correctly folded alkaline phosphatase, a periplasmic enzyme. *EMBO J.* **11**:57–62.

46. **Kaufmann, A., Y. D. Stierhof, and U. Henning.** 1994. New outer membrane-associated protease of *Escherichia coli* K-12. *J. Bacteriol.* **176**:359–367.

47. **Keiler, K. C., and R. T. Sauer.** 1996. Sequence determinants of C-terminal substrate recognition by the Tsp protease. *J. Biol. Chem.* **271**:2589–2593.

48. **Kleerebezem, M., W. Crielaard, and J. Tommassen.** 1996. Involvement of stress protein PspA (phage shock protein A) of *Escherichia coli* in maintenance of the protonmotive force under stress conditions. *EMBO J.* **15**:162–171.

49. **Kleerebezem, M., and J. Tommassen.** 1993. Expression of the *pspA* gene stimulates efficient protein export in *Escherichia coli*. *Mol. Microbiol.* **7**:947–956.

50. **Kolmar, H., P. R. Waller, and R. T. Sauer.** 1996. The DegP and DegQ periplasmic endoproteases of *Escherichia coli*: specificity for cleavage sites and substrate conformation. *J. Bacteriol.* **178**:5925–5929.

51. **Lazar, S. W., and R. Kolter.** 1996. SurA assists the folding of *Escherichia coli* outer membrane proteins. *J. Bacteriol.* **178**: 1770–1773.

52. **Lipinska, B., S. Sharma, and C. Georgopoulos.** 1988. Sequence analysis and regulation of the *htrA* gene of *Escherichia coli*: a σ^{32}-independent mechanism of heat-inducible transcription. *Nucleic Acids Res.* **16**:10053–10067.

53. **Lipinska, B., M. Zylicz, and C. Georgopoulos.** 1990. The HtrA (DegP) protein, essential for *Escherichia coli* survival at high temperatures, is an endopeptidase. *J. Bacteriol.* **172**: 1791–1797.

54. **Liu, J., and C. T. Walsh.** 1990. Peptidyl-prolyl *cis-trans*-isomerase from *Escherichia coli*: a periplasmic homolog of cyclophilin that is not inhibited by cyclosporin. A. *Proc. Natl. Acad. Sci. USA* **87**:4028–4032.

55. **Lonetto, M., K. L. Brown, K. E. Rudd, and M. J. Buttner.** 1994. Analysis of the *Streptomyces coelicolor* sigmaE gene reveals the existence of a subfamily of eubacterial RNA polymerase sigma factors involved in the regulation of extracytoplasmic functions. *Proc. Natl. Acad. Sci. USA* **91**:7573–7577.

56. **Low, D., B. Braaten, and M. Van Der Woude.** 1996. Fimbriae, p. 146–157. *In* F. C. Neidhardt, R. Curtiss III, J. L. Ingraham, E. C. C. Lin, K. B. Low, B. Magasanik, W. S. Reznikoff, M. Riley, M. Schaechter, and H. E. Umbarger (ed.), Escherichia coli *and* Salmonella: *Cellular and Molecular Biology*, 2nd ed. ASM Press, Washington, D.C.

57. **Macnab, R. M.** 1996. Flagella and motility, p. 123–145. *In* F. C. Neidhardt, R. Curtiss III, J. L. Ingraham, E. C. C. Lin,

K. B. Low, B. Magasanik, W. S. Reznikoff, M. Riley, M. Schaechter, and H. E. Umbarger (ed.), Escherichia coli *and* Salmonella: *Cellular and Molecular Biology*, 2nd ed. ASM Press, Washington, D.C.

58. **Martinez-Hackert, E., and A. M. Stock.** 1997. Structural relationships in the OmpR family of winged-helix transcription factors. *J. Mol. Biol.* **269:**301–312.

59. **Matsuyama, S., T. Tajima, and H. Tokuda.** 1995. A novel periplasmic carrier protein involved in the sorting and transport of *Escherichia coli* lipoproteins destined for the outer membrane. *EMBO J.* **14:**3365–3372.

60. **Matsuyama, S.-I., N. Yokota, and H. Tokuda.** 1997. A novel outer membrane lipoprotein, LolB (HemM), involved in the LolA (p20)-dependent localization of lipoproteins to the outer membrane of *Escherichia coli*. *EMBO J.* **16:**6947–6955.

61. **McEwen, J., and P. M. Silverman.** 1980. Chromosomal mutations of *Escherichia coli* K-12 that alter the expression of conjugative plasmid function. *Proc. Natl. Acad. Sci. USA* **77:**513–517.

62. **McEwen, J., and P. M. Silverman.** 1980. Mutations in genes *cpxA* and *cpxB* of *Escherichia coli* K-12 cause a defect in isoleucine and valine synthesis. *J. Bacteriol.* **144:**68–73.

63. **McEwen, J., and P. M. Silverman.** 1982. Mutations in genes *cpxA* and *cpxB* alter the protein composition of *Escherichia coli* inner and outer membranes. *J. Bacteriol.* **151:**1553–1559.

64. **Mecsas, J., P. E. Rouviere, J. W. Erickson, T. J. Donohue, and C. A. Gross.** 1993. The activity of σ^E, an *Escherichia coli* heat-inducible σ-factor, is modulated by expression of outer membrane proteins. *Genes Dev.* **7:**2618–2628.

65. **Mileykovskaya, E., and W. Dowhan.** 1997. The Cpx two-component signal transduction pathway is activated in *Escherichia coli* mutant strains lacking phosphatidylethanolamine. *J. Bacteriol.* **179:**1029–1034.

66. **Missiakas, D., J.-M. Betton, and S. Raina.** 1996. New components of protein folding in extracytoplasmic compartments of *Escherichia coli* SurA, FkpA and Skp/OmpH. *Mol. Microbiol.* **21:**871–884.

67. **Missiakas, D., C. Georgopoulos, and S. Raina.** 1993. Identification and characterization of the *Escherichia coli* gene *dsbB*, whose product is involved in the formation of disulfide bonds in vivo. *Proc. Natl. Acad. Sci. USA* **90:**7084–7088.

68. **Missiakas, D., M. P. Mayer, M.Lemaire, C. Georgopoulos, and S. Raina.** 1997. Modulation of the *Escherichia coli* σ^E (RpoE) heat-shock transcription-factor activity by the RseA, RseB and RseC proteins. *Mol. Microbiol.* **24:**355–371.

69. **Missiakas, D., and S. Raina.** 1997. Signal transduction pathways in response to protein misfolding in the extracytoplasmic compartments of *E. coli*: role of two new phosphoprotein phosphatases PrpA and PrpB. *EMBO J.* **16:**1670–1685.

70. **Missiakas, D., F. Schwager, and S. Raina.** 1995. Identification and characterization of a new disulfide isomerase-like protein (DsbD) in *Escherichia coli*. *EMBO J.* **14:**3415–3424.

71. **Model, P., G. Jovanovic, and J. Dworkin.** 1997. The *Escherichia coli* phage-shock-protein (psp) operon. *Mol. Microbiol.* **24:**255–261.

72. **Mulvey, M. A., Y. S. Lopez-Boado, C. L. Wilson, R. Roth, W. C. Parks, J. Heuser, and S. J. Hultgren.** 1998. Induction and evasion of host defenses by type 1-piliated uropathogenic *Escherichia coli*. *Science* **282:**1494–1497.

73. **Nakayama, W.-I., and H. Watanabe.** 1995. Involvement of *cpxA*, a sensor of a two-component regulatory system, in the pH-dependent regulation of expression of *Shigella sonnei virF* gene. *J. Bacteriol.* **177:**5062–5069.

74. **Nikaido, H.** 1996. Outer membrane, p. 29–47. *In* F. C. Neidhardt, R. Curtiss III, J. L. Ingraham, E. C. C. Lin, K. B. Low, B. Magasanik, W. S. Reznikoff, M. Riley, M. Schaechter, and

H. E. Umbarger (ed.), Escherichia coli *and* Salmonella: *Cellular and Molecular Biology*, 2nd ed. ASM Press, Washington, D.C.

75. **Oliver, D. B.** 1996. Periplasm, p. 88–103. *In* F. C. Neidhardt, R. Curtiss III, J. L. Ingraham, E. C. C. Lin, K. B. Low, B. Magasanik, W. S. Reznikoff, M. Riley, M. Schaechter, and H. E. Umbarger (ed.), Escherichia coli *and* Salmonella: *Cellular and Molecular Biology*, 2nd ed. ASM Press, Washington, D.C.

76. **Park, J. T.** 1996. The murein sacculus, p. 48–57. *In* F. C. Neidhardt, R. Curtiss III, J. L. Ingraham, E. C. C. Lin, K. B. Low, B. Magasanik, W. S. Reznikoff, M. Riley, M. Schaechter, and H. E. Umbarger (ed.), Escherichia coli *and* Salmonella: *Cellular and Molecular Biology*, 2nd ed. ASM Press, Washington, D.C.

77. **Pogliano, J., A. S. Lynch, D. Belin, E. C. Lin, and J. Beckwith.** 1997. Regulation of *Escherichia coli* cell envelope proteins involved in protein folding and degradation by the Cpx two-component system. *Genes Dev.* **11:**1169–1182.

78. **Raetz, C. R. H., and W. Dowhan.** 1990. Biosynthesis and function of phospholipids in *Escherichia coli*. *J. Biol. Chem.* **265:**1235–1238.

79. **Raina, S., D. Missiakas, and C. Georgopoulos.** 1995. The *rpoE* gene encoding the sigma E (sigma 24) heat shock sigma factor of *Escherichia coli*. *EMBO J.* **14:**1043–1055.

80. **Rainwater, S., and P. M. Silverman.** 1990. The Cpx proteins of *Escherichia coli* K-12: evidence that *cpxA*, *ecfB*, *ssd*, and *eup* mutations all identify the same gene. *J. Bacteriol.* **172:**2456–2461.

81. **Raivio, T. L., and T. J. Silhavy.** 1997. Transduction of envelope stress in *Escherichia coli* by the Cpx two-component system. *J. Bacteriol.* **179:**7724–7733.

82. **Raivio, T. L., and T. J. Silhavy.** 1999. The σ^E and Cpx regulatory pathways: overlapping but distinct envelope stress responses. *Curr. Opin. Microbiol.* **2:**159–165.

83. **Raivio, T. L., D. L. Popkin, and T. J. Silhavy.** 1999. The Cpx envelope stress response is controlled by amplification and feedback inhibition. *J. Bacteriol.* **181:**5263–5272.

84. **Rick, P. D., and R. P. Silver.** 1996. Enterobacterial common antigen and capsular polysaccharides, p. 104–122. *In* F. C. Neidhardt, R. Curtiss III, J. L. Ingraham, E. C. C. Lin, K. B. Low, B. Magasanik, W. S. Reznikoff, M. Riley, M. Schaechter, and H. E. Umbarger (ed.), Escherichia coli *and* Salmonella: *Cellular and Molecular Biology*, 2nd ed. ASM Press, Washington, D.C.

85. **Rietsch, A., and J. Beckwith.** 1998. The genetics of disulfide bond metabolism. *Annu. Rev. Genet.* **32:**163–184.

86. **Rietsch, A., D. Belin, N. Martin, and J. Beckwith.** 1996. An in vivo pathway for disulfide bond isomerization in *Escherichia coli*. *Proc. Natl. Acad. Sci. USA* **93:**13048–13053.

87. **Rouviere, P. E., A. De Las Penas, J. Mecsas, C. Z. Lu, K. E. Rudd, and C. A. Gross.** 1995. *rpoE*, the gene encoding the second heat-shock sigma factor, σ^E, in *Escherichia coli*. *EMBO J.* **14:**1032–1042.

88. **Rouviere, P. E., and C. A. Gross.** 1996. SurA, a periplasmic protein with peptidyl-prolyl isomerase activity, participates in the assembly of outer membrane porins. *Genes Dev.* **10:**3170–3182.

89. **Rupprecht, K. R., G. Gordon, M. Lundrigan, R. C. Gayda, A. Markovitz, and C. Earhart.** 1983. *ompT: Escherichia coli* K-12 structural gene for protein a(3b). *J. Bacteriol.* **153:**1104–1106.

90. **Sambongi, Y., and S. J. Ferguson.** 1994. Specific thiol compounds complement deficiency in c-type cytochrome biogenesis in *Escherichia coli* carrying a mutation in a membrane-bound disulphide isomerase-like protein. *FEBS Lett.* **353:**235–238.

91. Schirmer, T. 1998. General and specific porins from bacterial outer membranes. *J. Struct. Biol.* **121**:101–109.

92. Snyder, W. B., L. J. B. Davis, P. N. Danese, C. L. Cosma, and T. J. Silhavy. 1995. Overproduction of NlpE, a new outer membrane lipoprotein, suppresses the toxicity of periplasmic LacZ by activation of the Cpx signal transduction pathway. *J. Bacteriol.* **177**:4216–4223.

93. Sone, M., Y. Akiyama, and K. Ito. 1997. Differential in vivo roles played by DsbA and DsbC in the formation of protein disulfide bonds. *J. Biol. Chem.* **272**:10349–10352.

94. Spiess, C., A. Beil, and M. Ehrmann. 1999. A temperature-dependent switch from chaperone to protease in a widely conserved heat shock protein. *Cell* **97**:339–347.

95. Strauch, K. L., and J. Beckwith. 1988. An *Escherichia coli* mutation preventing degradation of abnormal periplasmic proteins. *Proc. Natl. Acad. Sci. USA* **85**:1576–1580.

96. Strauch, K. L., K. Johnson, and J. Beckwith. 1989. Characterization of *degP*, a gene required for proteolysis in the cell envelope and essential for growth of *Escherichia coli* at high temperature. *J. Bacteriol.* **171**:2689–2696.

97. Sugimura, K., and N. Higashi. 1988. A novel outer-membrane-associated protease in *Escherichia coli*. *J. Bacteriol.* **170**:3650–3654.

98. Sutton, A., T. Newman, J. McEwen, P. M. Silverman, and M. Freundlich. 1982. Mutations in genes *cpxA* and *cpxB* of *Escherichia coli* K-12 cause a defect in acetohydroxyacid synthase I function *in vivo*. *J. Bacteriol.* **151**:976–982.

99. Thanassi, D. G., E. T. Saulino, and S. J. Hultgren. 1998. The chaperone/usher pathway: a major terminal branch of the general secretory pathway. *Curr. Opin. Microbiol.* **1**:223–231.

100. Thorne, B. M., and M. Muller. 1991. Skp is a periplasmic *Escherichia coli* protein requiring *secA* and SecY for export. *Mol. Microbiol.* **5**:2815–2817.

101. Throne-Holst, M., L. Thony-Meyer, and L. Hederstedt. 1997. *Escherichia coli ccm* in-frame deletion mutants can produce periplasmic cytochrome b but not cytochrome c. *FEBS Lett.* **410**:351–355.

102. Waller, P. R. H., and R. T. Sauer. 1996. Characterization of *degQ* and *degS*, *Escherichia coli* genes encoding homologs of the DegP protease. *J. Bacteriol.* **178**:1146–1153.

103. Wang, Q., and J. M. Kaguni. 1989. A novel sigma factor is involved in expression of the *rpoH* gene of *Escherichia coli*. *J. Bacteriol.* **171**:4248–4253.

104. Weber, R. F., and P. J. Silverman. 1988. The Cpx proteins of *Escherichia coli* K-12: Structure of the CpxA polypeptide as an inner membrane component. *J. Mol. Biol.* **203**:467–476.

105. Weiner, L., and P. Model. 1994. Role of an *Escherichia coli* stress-response operon in stationary-phase survival. *Proc. Natl. Acad. Sci. USA* **91**:2191–2195.

106. Zapun, A., D. Missiakas, S. Raina, and T. E. Creighton. 1995. Structural and functional characterization of DsbC, a protein involved in disulfide bond formation in *Escherichia coli*. *Biochemistry* **34**:5075–5089.

Bacterial Stress Responses
Edited by G. Storz and R. Hengge-Aronis
©2000 ASM Press, Washington, D.C.

Chapter 3

The Cold Shock Response

Sangita Phadtare, Kunitoshi Yamanaka, and Masayori Inouye

The cold shock response and adaptation have been extensively studied in Escherichia coli *and* Bacillus subtilis. *Cytoplasmic membranes, RNA/DNA, and ribosomes are suggested to be the cellular thermosensors. The major effects of cold shock are a decrease in membrane fluidity and the stabilization of secondary structures of RNA and DNA, which may affect the efficiency of translation, transcription, and DNA replication. The organisms overcome the membrane fluidity problem by decreasing the degree of saturation of fatty acids in the membrane phospholipids. In the case of* B. subtilis *and cyanobacteria, desaturases play an important role in alteration of the degree of saturation of fatty acids in the membrane phospholipids to restore the membrane-associated functions. Cold-inducible proteins overcome the deleterious effects of the cold shock on transcription and translation. CspA, the major cold shock protein of* E. coli, *is a β-barrel protein with two RNA-binding motifs. It is proposed to function as an RNA chaperone that facilitates translation at low temperature by preventing formation of secondary structures in RNA molecules. The expression of* cspA *at low temperature is regulated at the levels of transcription, mRNA stability, and translation. CspA-homologs are widely distributed in bacteria, and in some cases, these proteins are essential for survival even under optimal growth conditions.*

Temperature is one of the major stresses that all living organisms face. In contrast to the heat shock response (see chapter 1) which has been extensively studied from bacteria to humans, the cold shock response has caught the attention of researchers recently. The study of the cold shock response and adaptation is important for our understanding of how microorganisms are able to live under low temperature stress and for improvement of cold tolerance in plants and useful microbes. In addition, the cold shock response and adaptation have important health implications. As refrigeration is a commonly used method for extending the shelf life of food, it is necessary to understand the cold shock response of food-spoilage bacteria. One such example is *Listeria monocytogenes*, an opportunistic food-borne pathogen. This bacterium causes listeriosis, a disease primarily affecting pregnant woman, neonates, and immunocompromised patients, such as those with AIDS or undergoing corticosteroid therapy (5). The study of the cold shock response and adaptation is also important in bacteria such as *Lactobacillus*, which are widely used in the dairy industry (13).

The cold shock response and adaptation have been studied in detail using *Escherichia coli* and *Bacillus subtilis* as model systems (see reviews in references 35, 39, 49, 66, 68, 84, 94, 95). As a result of excretion from animals, enterobacteria such as *E. coli* encounter sudden drastic temperature downshift. Thus, the cold shock response and adaptation in this bacterium confer a selective advantage of quick adaptation to the new environment. In this chapter, we will discuss two important aspects of the cold shock response: how bacteria sense the change in temperature (i.e., what the cellular thermosensors are) and how they cope with the cold stress. We will also discuss cold shock proteins, especially CspA and its homologs, with respect to the regulation of their synthesis and their structure and function.

CELLULAR THERMOSENSORS FOR COLD SHOCK

The cytoplasmic membrane, nucleic acids, and ribosomes are implicated in sensing temperature changes. The cold shock response and adaptation in *E. coli* are summarized in Fig. 1.

Sangita Phadtare, Kunitoshi Yamanaka, and Masayori Inouye • Department of Biochemistry, Robert Wood Johnson Medical School, 675 Hoes Lane, Piscataway, NJ 08854.

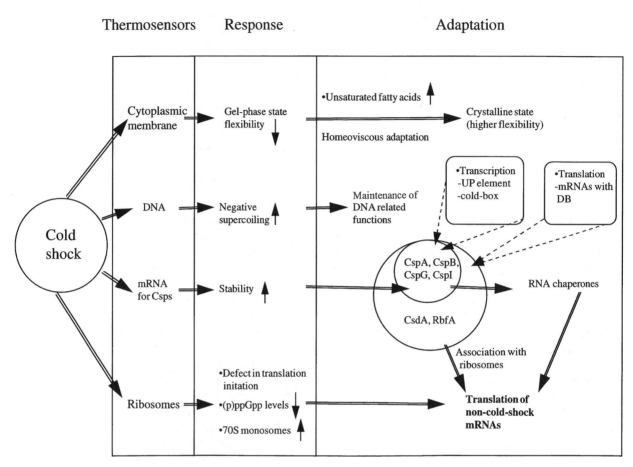

Figure 1. Model for cold shock response following the temperature downshift from 37 to 15°C in *E. coli*. Csps, cold shock proteins; DB, downstream box; (p)ppGpp, guanosine 5′-triphosphate-3′-diphosphate (pppGpp) and guanosine 5′-diphosphate-3′-diphosphate (ppGpp), collectively abbreviated as (p)ppGpp.

Cold Shock-Induced Membrane-Associated Changes

A highly conserved cold shock response observed among bacteria is the adjustment of membrane lipid composition. The usual liquid crystalline nature of the membrane is changed to a gel-phase state upon cold shock. The proportion of fluid (i.e., disordered lipid to ordered lipid) in the cell membrane plays an important role in membrane function. After the temperature downshift, the proportion of unsaturated fatty acids (UFAs) in the membrane lipids increases. Phospholipids with UFAs have lower melting points and a greater degree of flexibility than do phospholipids containing saturated fatty acids. This change compensates for the negative effect of low temperature on the physical state of the lipid bilayer. This is known as homeoviscous adaptation (80). In *E. coli*, the UFA produced after temperature downshift is *cis*-vaccenic acid (*cis*-11-octadecenic acid). The enzyme β-ketoacyl-acyl carrier protein synthase II converts palmitoleic acid to *cis*-vaccenic acid. Interestingly, the

synthesis of this enzyme is not induced upon cold shock, but the enzyme is activated at low temperature (28, 29).

In contrast, the desaturation system in *B. subtilis* is cold inducible. In this organism, temperature downshift results in increased synthesis and increased stability of the membrane-bound desaturase encoded by *des* (1). It has been shown that this enzyme inserts a double bond into the acyl chain of membrane phospholipids. This acyl-lipid desaturase is similar to those from cyanobacteria. In fact, *B. subtilis* is the only nonphotosynthetic bacterium in which the presence and cold induction of desaturase are reported. This also strengthens the theory of a close phylogenetic relationship between gram-positive bacteria and cyanobacteria (87). However, there are distinct differences between desaturases of these two bacteria. In *B. subtilis*, disruption of *des* caused a survival defect only during stationary phase at both 20 and 37°C. On the other hand, in the case of cyanobacteria such mutations caused cold sensitivity, as in these organisms desaturation of lipids is correlated with acclimatization of photosynthetic activity at low temperature. In cyanobacteria, the desaturases are bound to the thylakoid membrane. These are acyl-lipid desaturases that introduce double bonds into fatty acids that have been esterified to glycerolipids (61). They are the most efficient regulators of the extent of unsaturation of membrane lipids in response to temperature changes. In the case of *Synechococcus* sp., the two desaturase genes *desA* and *desB* are induced after temperature downshift from 38 to 22°C. Their expression is tightly controlled by a combination of mRNA synthesis and stabilization at low temperature (72). Introduction of the *desA* gene into the chilling-sensitive cyanobacterium *Anacystis nidulans* increased the cold resistance of this organism (89). These results suggest that unlike *B. subtilis*, desaturases are essential for cold shock adaptation of cyanobacteria.

By using *Synechococcus* PCC 6803 strain, it was demonstrated that the change in the fluidity of the membrane lipids is the first event that signals a change in temperature. In vivo, palladium-catalyzed hydrogenation of membrane lipids activated transcription of *desA*, without the temperature downshift. As this catalytic technique affects only the plasma membrane, this result implies that the membrane acts as a sensor and that transcription of *desA* is enhanced in response to the altered lipid saturation (88). It was also shown that the energy produced by photosynthesis is necessary for transcription of *desA*. Thus, the low temperature-induced desaturation of membrane lipids occurs only in the light but not in the dark. Another example of the membrane acting as a thermosensor is in the case of yeast, *Histoplasma capsulatum*, wherein the heat shock induction of mRNA transcripts is directly correlated with changes in the membrane fluidity (10). These results support the fact that cytoplasmic membrane helps in sensing the temperature changes.

Another interesting mechanism for maintaining a fluid, liquid-crystalline state of membrane lipids was observed in the case of *L. monocytogenes*, a gram-positive, food-borne pathogen (2). It is characterized by having a fatty acid profile dominated by branched-chain fatty acids, the major one being *anteiso* fatty acid. After shift to a lower temperature of 5°C, branching is changed from *iso* to *anteiso*. *Anteiso* fatty acids have a lower melting point than *iso* fatty acids. Deficiency of *anteiso* fatty acids results in cold sensitivity. Hence, it was suggested that *anteiso* fatty acids play a critical role in maintaining a fluid, liquid-crystalline state of the membrane lipids, while the role of fatty acid unsaturation in adaptation of this organism to cold temperatures is not clear (2).

Ribosomes as Thermosensors

Ribosomes have been suggested to be the sensors of temperature in bacteria by VanBogelen and Neidhardt (86). These authors have suggested that the state of ribosome signals the induction of the temperature responses. It was proposed (35) that after an increase in temperature, the speed of translation is higher than that of supply of charged tRNA, hence the A-site of the ribosome is empty. On other hand, after cold shock the translational efficiency reduces, so the A-site is blocked due to the high concentration of charged tRNA. This in turn lowers (p)ppGpp (guanosine 5′-triphosphate-3′-diphosphate [pppGpp] and guanosine 5′-diphosphate-3′-diphosphate [ppGpp] [collectively abbreviated as (p)ppGpp]) levels by inhibiting RelA-mediated synthesis of (p)ppGpp. Thus, the concentration of (p)ppGpp increases at high temperatures and decreases after a temperature downshift. Jones et al. (47, 49) have shown that artificially induced high levels of (p)ppGpp diminish the expression of cold shock proteins, while low concentration increases their production. Thus, (p)ppGpp affects the magnitude of the cold shock response. It has been suggested that upon cold shock, ribosomes become nonfunctional for cellular mRNAs, except for mRNAs for cold shock proteins (48). During the growth lag period after the temperature downshift, however, cold-unadapted ribosomes are converted to cold-adapted ribosomes by acquiring cold shock ribosomal factors such as RbfA and CsdA, which are produced during this period. These cold-adapted ribosomes are now able to translate non-cold shock mRNAs (48).

Nucleic Acids as Thermosensors

Supercoiling of DNA has been suggested as a transcriptional sensor for temperature changes by Wang and Syvanen (91). DNA is usually negatively supercoiled, and the extent of the supercoil is regulated mainly by DNA gyrase and DNA topoisomerase I. DNA gyrase introduces negative supercoiling into DNA (59). When *E. coli* cells are exposed to cold shock, negative supercoiling transiently increases. A similar effect was observed in the case of *B. subtilis* (52). Regulation of DNA supercoiling in the cells exposed to thermal stress is important in maintaining DNA-related functions, such as replication, transcription, and recombination. The changes in the linking number (i.e., the number of base pairs per helical turn) are often attributed to a change in superhelical density, which in turn influences transcription. Recognition of some σ^{70} promoters by RNA polymerase is dependent on relative orientation of the −35 and

−10 regions (91). This orientation can be controlled by changes in DNA twist brought about by environmental conditions. The cold shock-inducible *recA* promoter is one such twist-sensitive promoter. It has been proposed that DNA supercoiling may regulate UFA synthesis in *B. subtilis*, as the desaturation of fatty acids induced by temperature downshift is inhibited by novobiocin, an inhibitor of DNA gyrase (34).

H-NS is another cold shock-inducible protein of *E. coli*. It is a nucleoid-associated, DNA-binding protein and plays an important role in adaptation of *E. coli* to low temperature. It serves a role in structuring the chromosomal DNA. A mutation in H-NS resulted in cold sensitivity (16). These results suggest that DNA topology is important for survival of cells after temperature downshift.

RNA has been suggested to act as a thermosensor in the case of heat shock sigma factor σ^{32}, encoded by the *rpoH* gene, in *E. coli* (60) (refer to chapter 1). The secondary structure of its mRNA undergoes a partial melting at high temperature to enhance ribosome entry and initiation of translation without involvement of other cellular components. This means that intrinsic mRNA stability controls the synthesis of a transcriptional regulator. Other examples of RNA thermosensors have been reviewed by Storz (81) and include the *lcrF* mRNA of *Yersinia pestis* and mRNA encoding the *cIII* gene of bacteriophage λ. It has been shown that in *E. coli*, mRNA for the gene encoding its major cold shock protein CspA is dramatically stabilized after temperature downshift from 37 to 15°C in the absence of any protein synthesis (half-life of >20 min at 15°C as opposed to <12 s at 37°C) (30). This enhancement of mRNA stability is one of the important factors in the induction of *cspA* after temperature downshift. In this regard, it can be said that RNA acts as a thermosensor in the cold shock response of *E. coli*, though it is different from that in the heat shock response. The mechanism for mRNA stabilization is yet to be elucidated.

Proteins as Thermosensors

Recently the aspartate chemoreceptor (Tar) of *E. coli* has been described as a thermosensor. Its thermosensing properties are controlled by reversible methylation of the cytoplasmic signaling/adaptation domain of the protein. The fully methylated receptor senses decrease in temperature. The unmethylated and methylated (i.e., aspartate bound) receptors sense, as attractant stimuli, changes in temperature and act as warm or cold sensor (64). Methylation of a single residue converts a bacterial warm sensor to a cold sensor.

In addition, even though DNA has been suggested to act as thermosensor, it is possible that the proteins such as DNA gyrase, topoisomerase, and H-NS, which modulate DNA topology in response to temperature changes, are the thermosensors.

COLD SHOCK RESPONSE AND ADAPTATION

The cold shock response in bacteria is an immediate and transient response to the temperature downshift, which is followed by the low temperature adaptation that allows continued growth at low temperatures (65).

A major cold shock protein, CspA, was first identified in *E. coli*, and since then, its homologs have been reported from more than 60 species of gram-positive and gram-negative, psychrotrophic, psychrophilic, mesophilic, and thermophilic bacteria. CspA homologs have not been reported in archaea and cyanobacteria. In the case of mesophiles, after the temperature downshift, there is a growth lag period. During this period, cold shock proteins are induced dramatically, while the synthesis of all other proteins is arrested, resulting in a lag period of growth. Figure 2 shows the protein patterns of *E. coli* analyzed by two-dimensional gel electrophoresis at different time intervals before and after the cold shock (0.5 and 4 h). The production of cold shock proteins is transient and at the end of the growth lag period, their production is reduced to a new basal level (Fig. 2). After this period, synthesis of non-cold shock proteins and subsequently cell growth are resumed. Unlike mesophiles, CspA-like proteins are continuously synthesized during prolonged growth at low temperature in the case of psychrophilic and psychrotrophic bacteria. This probably enables these organisms to grow at temperatures near or below freezing. In these organisms, the synthesis of housekeeping proteins is never repressed following an abrupt temperature downshift, a large number of cold shock proteins are synthesized, the relative level of cold shock proteins is moderate even after a severe cold shock, and the synthesis of cold acclimation proteins is prolonged (6).

COLD SHOCK PROTEINS OF *E. COLI*

After a temperature downshift from 37 to 15°C, a number of proteins are induced in *E. coli* (Fig. 2B). These are of two types: the proteins (class I) that are expressed at an extremely low level at 37°C and are dramatically induced to very high levels after a shift

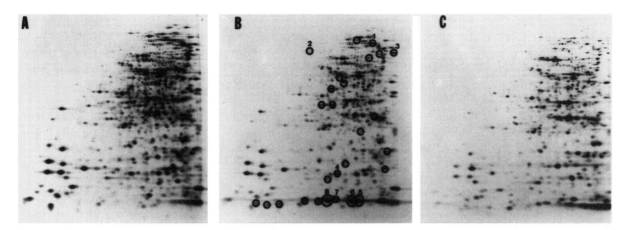

Figure 2. Two-dimensional analysis of total cellular proteins before and after cold shock. Cells were labeled with [^{35}S]methionine at 37°C (A) and at 0.5 h (B) and 4 h (C) after temperature downshift from 37 to 15°C. Cell lysates were analyzed by two-dimensional gel electrophoresis. The first dimension (isoelectric focusing) was carried out between the range of pH 3.5 (right side) to 10 (left side). Cold shock proteins are circled. Spot 1, polynucleotide phosphorylase; 2, CsdA; 3, NusA; 4, RbfA; 5, CspA; 6, CspG; 7, CspI; 8, CspB.

to lower temperatures and the proteins (class II) that are present at 37°C and are moderately induced after cold shock (<10-fold) (84). Class I proteins include CspA (32), CspB (54), CspG (62), CspI (92), CsdA (85), RbfA (15), NusA (27), and PNP (17). Class II proteins include RecA (90), IF-2 (40), H-NS (16), the α subunit of DNA gyrase (82), Hsc66 (55), HscB (55), and dihydrolipoamide transferase and pyruvate dehydrogenase (49) (Table 1).

At higher growth temperatures, misfolding of proteins and aggregation of misfolded peptides are major problems. Cells have heat shock-inducible systems to synthesize heat shock proteins. Some of them act as molecular chaperones by assisting in correct protein folding and proteolysis of abnormally folded polypeptides. GroEL is one of the major molecular chaperones that acts in association with other chaperones such as GroES. In contrast, protein misfolding

Table 1. Cold shock proteins of *E. coli* and *B. subtilis* and their functions

Organism and protein(s)	Function
E. coli	
CspA, CspB, CspG, CspI	Proposed RNA/DNA chaperones
CsdA	RNA unwinding activity
RbfA	Ribosomal binding factor
NusA	Transcriptional termination and antitermination
PNP	Ribonuclease
RecA	Recombination factor
IF-2	Initiation factor
H-NS	Structuring chromosomal DNA
α-subunit of DNA gyrase	DNA supercoiling
Hsc66, HscB	Cold shock molecular chaperones?
Trigger factor (TF)	Maintenance and repair of proteins?
Dihydrolipoamide transferase	Energy generation
Pyruvate dehydrogenase	Energy generation
B. subtilis	
CspB, CspC, and CspD	Proposed RNA chaperones
CheY	Chemotaxis
Hpr	Sugar uptake
Ribosomal proteins S6 and L7/L12	Translation
Peptidyl prolyl *cis/trans* isomerase	Protein folding
Cystein synthase	General metabolism
Ketol-acid reductoisomerase	General metabolism
Glyceraldehyde dehydrogenase	General metabolism
Triosephosphate isomerase	General metabolism

was not considered a major problem in the case of cold shock. But recent reports suggest that proper folding of proteins as well as refolding of cold-damaged proteins is important after cold shock (50). A peptidyl prolyl isomerase (trigger factor [TF]) that catalyzes the *cis/trans* isomerization of peptide bonds was identified in *E. coli*. This enzyme is induced upon cold shock at a modest level after a growth lag period of 2 to 3 h. Similar to other cold shock proteins, its synthesis was induced after temperature downshift from 37 to 10°C or exposure to choramphenicol. When stored at 4°C, cells with reduced levels of TF die faster; on the other hand, cells overexpressing TF show enhanced viability. At high temperature opposite effects were observed (i.e., cell viability was reduced by TF overproduction and increased by its deficiency). It was suggested that TF has a "maintenance and repair" function as it helps protein synthesis and folding to continue at low temperature (50). It may maintain preexisting proteins in functional form by promoting refolding of cold-damaged proteins. Its association with GroEL also supports its role as a molecular chaperone. It enhances affinity of GroEL for unfolded proteins and promotes degradation of certain polypeptides (50). Similarly, a trigger factor was identified in *B. subtilis* that along with another prolyl *cis/trans* isomerase (PPiB) is involved in protein folding at low temperature (37).

CspA Family of *E. coli*

The CspA family of *E. coli* consists of nine homologous proteins, CspA to CspI, but among them only CspA, CspB, CspG, and CspI are cold shock inducible (32, 54, 62, 92). Interestingly, *cspA* is dispensable at both normal and low growth temperatures (3). In a *cspA* deletion mutant, increased production of CspB and CspG was observed, suggesting that these proteins can compensate for CspA and probably have overlapping functions (3). It has been shown that CspA is differentially regulated from CspB, CspG, and CspI (21, 92). Induction of CspA is observed after a temperature shift from 37 to 30°C, and high levels of CspA production are seen between 24 and 10°C. CspB and CspG are produced only as temperature shifts below 20°C, the maximum induction being at 15°C. CspI is induced between 15 and 10°C. Recently, it has been shown that CspA, CspB, and CspG are induced at low temperature under conditions that completely block protein synthesis. These proteins can bypass the effect of protein synthesis inhibitors, such as kanamycin and chloramphenicol (18).

The elements involved in the transcription, mRNA stability, and translation in the cold shock-induction of CspA, CspB, CspG, and CspI of *E. coli* are illustrated in Fig. 3. One of the features of *cspA*, *cspB*, *cspG*, and *cspI* is the unusually long 5' untranslated region (5'-UTR). This region was shown to be important in expression of these genes.

The cold shock induction of *cspA* does not need any additional transcription factors unlike the induction of proteins by heat shock. It has been also suggested that *cspA* transcription is also enhanced by an AT-rich sequence (UP element) immediately upstream of the −35 region (31, 58) (Fig. 3). Deletion of the UP element resulted in diminished activity of the *cspA* promoter (58). Another important factor that contributes toward high promoter activity is the presence of a TGn motif immediately upstream of the −10 region (Fig. 3). It is reported that this motif together with the −10 region constitutes the extended −10 region and the −35 region is dispensable in the presence of this region (53). Interestingly, the 5'-UTR contains a highly conserved unique 11-base sequence called the cold box (44) (Fig. 3). The cold box sequence is a presumed transcriptional pausing site and is involved in the repression of *cspA* expression. As the cellular CspA concentration increases during the growth lag period after the temperature downshift, it starts to bind its own mRNA to destabilize the elongation complex of RNA polymerase, resulting in attenuation of transcription (3). Overproduction of the 5'-UTR at 15°C results in delayed induction of the cold shock response and in the prolonged synthesis of not only CspA but also CspB and CspG (23, 44). These effects are repressed by coproduction of the 5'-UTR together with CspA.

Recently, it has been shown that CspE negatively regulates the expression of CspA at the level of transcription (4). With the help of in vitro transcription assays, it was shown that CspE increases the efficiency of pausing by RNA polymerase. This is achieved by direct binding of CspE to the cold box region of *cspA*. This is consistent with the fact that *cspA* expression was derepressed at 37°C in a *cspE* deletion mutant (4).

Deletion analysis of the *cspA* 5'-UTR showed that this region is responsible for its extreme instability at 37°C and has a positive effect on mRNA stabilization at low temperature (58). *CspA* mRNA is dramatically stabilized (half-life more than 20 min) immediately following cold shock. This stabilization is transient and is lost once cells are adapted to low temperature. This in turn regulates the expression of *cspA*. A constitutive expression of *cspA* was observed when *cspA* mRNA was stabilized 150-fold due to a three-base substitution mutation within the 159-base 5'-UTR (24). This stabilization was found to be at least partially due to the resistance against RNase E

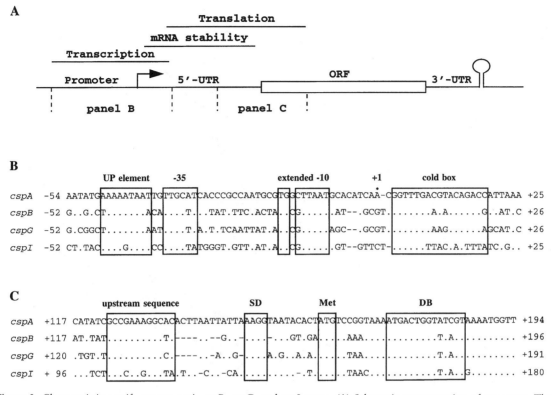

Figure 3. Characteristic motifs among *cspA*, *cspB*, *cspG*, and *cspI* genes. (A) Schematic representation of *csp* genes. They consist of their own promoter, 5'-UTR, ORF, 3'-UTR, and ρ-independent transcription terminator. The regions involved in the regulation at the levels of transcription, mRNA stability, and translation are indicated. (B) Sequence alignment of the region around the promoter. +1 indicates the transcription start site. Nucleotides identical to *cspA* are shown as dots. Gaps are indicated by dashes. (C) Sequence alignment of the region around the translation initiation. SD, Met, and DB indicate Shine-Dalgarno sequence, translation initiation codon, and downstream box, respectively. See details in the text.

degradation. This indicates that the *cspA* promoter is active at 37°C, but as the *cspA* mRNA is extremely unstable at this temperature (half-life less than 12 s), CspA is hardly detected. Recently, it has been shown that CspA is produced at 37°C during early exponential growth phase and its mRNA becomes unstable by mid- to late-exponential growth phase (8).

The expression of *cspA* is also regulated at the level of translation. The first in vivo evidence for the preference of a cold shock mRNA by cold-shocked ribosomes came from Brandi et al. (7). The preferential synthesis of cold shock proteins during the growth lag period suggests that their mRNAs, unlike most other cellular mRNAs, possess a mechanism to form the translation initiation complex at low temperature without the cold shock ribosome factors, such as RbfA. At low temperature, there is a defect in initiation of translation (25). As a result of this, 70S monosomes accumulate, with concomitant decrease in the number of polysomes (9). The Shine-Delgarno sequence and the initiation codon are not sufficient for the formation of translation initiation complex of RNAs encoding non-cold shock proteins.

mRNAs for cold shock proteins such as CspA, CspB, CspG, CspI, CsdA, and RbfA have an extra ribosome-binding site called the downstream box (DB) in the coding region (58) (Figs. 1 and 3). The DB sequence is complementary to a region in the penultimate stem of 16S rRNA and is located a few bases downstream of the initiation codon. The DB is considered to enhance translation initiation by facilitating the formation of translation preinitiation complex through binding to 16S rRNA. Recently, it has been shown that the DB in concert with the Shine-Dalgarno sequence plays a major role in the enhancement of translation initiation at low temperature (19, 20). In addition to the DB, the upstream sequence within the 5'-UTR of *cspA* mRNA has been recently shown to play a role in efficient translation (98).

It has been suggested that upon cold shock, ribosomes become nonfunctional for cellular mRNAs, except for mRNAs for these cold shock proteins (58). During the growth lag period, however, cold-unadapted ribosomes are converted to cold-adapted ribosomes by acquiring cold shock ribosomal factors such as RbfA and CsdA. These cold-adapted ribosomes are now able to translate non-cold shock mRNAs. One interesting phenomenon termed as "LACE" effect (low-temperature antibiotic effect of truncated cspA expression) was observed in *E. coli* (45). When a translatable truncated cspA gene is overexpressed at low temperature, cell growth is completely blocked. This is presumed to be caused by the entrapment of almost all the ribosomes by the truncated cspA mRNA. This truncated mRNA possesses the DB and thus is still able to form the preinitiation complex with nonadapted ribosomes at low temperature but is not able to synthesize the intact CspA. Furthermore, the production of truncated cspA mRNA is derepressed in the absence of autoregulation by the CspA production.

The CspA family of *E. coli* also contains non-cold shock-inducible proteins such as CspD, which is induced in the stationary phase and upon glucose starvation, and CspC and CspE, which are produced mainly at 37°C (96, 97). CspF and CspH have not yet been characterized. Judging from the chromosomal location of the csp genes, it has been proposed that the large CspA family probably resulted from a number of gene duplications and, after subsequent adaptation, resulted in specific groups of genes that respond to different environmental stresses (95).

COLD SHOCK PROTEINS FROM OTHER BACTERIA

In the case of *B. subtilis*, various proteins functioning at different levels of cellular physiology are induced after temperature downshift from 37 to 15°C (37). These proteins include CheY, Hpr, ribosomal proteins S6 and L7/L12, peptidyl prolyl *cis/trans* isomerase, cysteine synthase, ketol-acid reductoisomerase, glyceraldehyde dehydrogenase, and triosephosphate isomerase (Table 1). Similar to *E. coli*, *B. subtilis* also has a family of three cold shock-inducible, CspA-like proteins (38). The three cold shock proteins, CspB, CspC, and CspD, are essential for efficient growth at optimal temperature, for efficient adaptation to low temperatures, and for survival during stationary phase. A *cspB/cspC/cspD* triple deletion mutation is lethal, indicating that a minimum of one cold shock protein is essential for the viability of *B. subtilis* (38). There is a hierarchy of importance for these cold shock proteins, with CspB being the most essential. It is important at both low and high temperatures, while CspC functions mainly at low temperatures and CspD is needed mostly at optimal temperature (76). Recent evidence shows that CspB and CspD are stable during logarithmic growth at 37°C as well as after cold shock. CspC is less stable

at 37°C but has increased stability after cold shock. The in vitro proteolytic susceptibility of the cold shock proteins was strongly reduced in the presence of a nucleic acid ligand. This suggests that these proteins are stabilized in vivo by virtue of their binding to the substrate mRNAs at 37°C and especially after cold shock (76). Similar to *E. coli* cold shock genes, *B. subtilis* cold shock genes also have a long 5'-UTR region, which is supposed to be implicated in the autoregulation of these genes (38).

The cyanobacterium *Anabaena variabilis* has a family of RNA-binding proteins that are induced after temperature downshift (73). These belong to the RNA-binding domain family of proteins and are similar to RNA-binding proteins such as U1A snRNP from eukaryotes. This protein family is structurally different from the CspA family (36). Cyanobacteria do not have cold shock proteins, and *E. coli* does not have RNA-binding proteins. Interestingly, both of these proteins are RNA or single-stranded DNA binding proteins and are cold shock inducible. Hence, it has been suggested that this represents a very interesting case of convergent evolution in the sense that RNA-binding proteins could be the cyanobacterial counterparts of CspA-homologs from gram-positive and gram-negative bacteria (36, 39). Recently, a cold shock-inducible cyanobacterial RNA helicase, CrhC, was identified in *Anabaena* sp. (12). Even though the cold shock-specific expression of *crhC* is similar to that of the *E. coli* RNA helicase gene *csdA*, CrhC is not a CsdA homolog but is similar to *E. coli* RNA helicase RhlE. This suggests that the low temperature adaptation in bacteria requires an RNA helicase activity, but the helicases are not conserved between genera. The cold shock-induced expression of *crhC* is regulated at levels of transcription and translation. Similar to *cspA* of *E. coli*, *crhC* also contains an AT-rich UP element that functions as transcriptional activator and a cold box-like sequence that is considered to be involved in attenuation of transcription in *E. coli cspA*. Since this organism does not have cold shock proteins, it is not clear why it has a cold box-like sequence. Furthermore, similar to *E. coli* cold shock genes, *crhC* also contains a long, highly structured 5'-UTR and a downstream box that facilitates initiation of translation at low temperatures. It is suggested that cold shock-induced gene expression of CrhC may be regulated by these elements similar to that in the case of cold shock proteins. CrhC belongs to class I type of proteins, while another RNA helicase, CrhB, from this organism is also cold shock-inducible but belongs to class II type of proteins (12). In the case of two other cyanobacteria, *Synechococcus* sp. (70) and *Plectonema boryanum* (11), the heat shock protein ClpB (caseinolytic protease) was shown to be involved in the cold adaptation.

STRUCTURE OF COLD SHOCK PROTEINS

The three-dimensional structure of CspA from *E. coli* and CspB from *B. subtilis* has been resolved by X-ray crystallography and nuclear magnetic resonance analysis (26, 63, 74, 75, 78). The protein consists of five antiparallel β strands, β1 to β5, forming a β barrel structure with two β sheets. The two RNA-binding motifs, RNP1 and RNP2, are located on the β2 and β3 strands, respectively. The seven aromatic residues on the surface of CspA are involved in the hydrophilic interactions between protein and nucleic acid. Mutations of three phenylalanine residues from the aromatic cluster adversely affected the DNA binding in the case of CspA from *E. coli* (41). Similarly, in the case of CspB from *B. subtilis*, the nucleic acid binding as well as the protein stability was abolished by the mutations in the two RNP sites (77, 79). The unusually large nonpolar surface patch seems to have a functional role in single-stranded nucleic acid binding and in maintaining stability of the protein. Both proteins fold very rapidly in a two-state mechanism, without intermediates. They are moderately stable in solution. Two-state folding and an unusually native-like activated state of folding seem to be inherent properties of these proteins (43, 67, 71).

COLD SHOCK PROTEINS AS RNA CHAPERONES

CspA from *E. coli* binds RNA without apparent sequence specificity and with low binding affinity (46). It has been shown to increase translation of its own mRNA and render mRNA more susceptible to RNase degradation. Stable secondary structures in the RNA are resistant to degradation by RNases. CspA destabilizes these structures; hence it was described as an RNA chaperone (46). Such a function is probably crucial for efficient translation at low temperatures, as keeping RNA in a linear form is an essential prerequisite for efficient initiation of translation. CspB, CspG, and CspI are also speculated to function as RNA chaperones. In the case of *B. subtilis*, if the levels of cold shock proteins are experimentally reduced after temperature downshift, a decrease in overall protein synthesis is observed along with changes in the pattern of protein synthesis and significant reduction in growth of the organism (38). Hence, increased levels of cold shock proteins after cold shock appear to be important for compensating for higher stability of secondary structures in RNA at low temperatures. The structure of CspA is also ideal for its role as an RNA chaperone (38). The protein has overall negative surface charge while its nucleic acid-binding surface is positively charged and contains a number of aromatic residues. After binding to RNA by virtue of stacking of the aromatic side chains with RNA bases, the approach of other RNA for intramolecular or intermolecular base pairing will be prevented by charge repulsion. The nonspecific and weak binding of CspA to RNA/DNA is also important for the chaperone function, as the CspA binding would not hamper ribosome movement on mRNA. Hence, heat shock proteins function as protein chaperones at high temperatures, while cold shock proteins function as RNA chaperones at low temperatures (39).

Recently, it has been shown that CspB can interact with RNA/ssDNA sequence selectively. The preferred sequence is "UUUUU" residues. This suggests that CspB may have functions in addition to those of CspA. The other two CspA homologs, CspC and CspE, also show sequence-selective interaction with RNA/ssDNA. They prefer "AGGGAGGGA" and "AU-rich" regions, respectively. The in vivo significance of this finding remains to be elucidated (69).

CONCLUSIONS AND FUTURE PERSPECTIVES

Although a large body of information concerning the cold shock response and cold shock proteins has been accumulated, several questions are still unanswered. Unlike the heat shock response, a specific sigma factor has not been identified in the case of the cold shock response. Hence, it will be interesting to study what actual signal triggers the cold shock response. However, it has been shown that the stabilization of mRNAs for cold shock proteins, at least CspA homologs, plays a critical role in the cold shock response. It is also important to find the actual mechanism of the dramatic stabilization of the *cspA* mRNA after the temperature downshift. The secondary structure of the *cspA* mRNA should be determined, and the relationship between the structure and the translation efficiency should be elucidated.

In the case of the heat shock response, heat shock proteins, especially molecular chaperones such as DnaK, are known to be very well conserved from bacteria to humans (refer to chapter 1). In the case of the cold shock response, however, no such conserved cold shock proteins have been found so far. Therefore, whether there is a common cold shock protein(s) from bacteria to humans should be addressed. It also will be interesting to study why *E. coli* has so many CspA homologs. Further studies with multiple deletion strains of *cspA* homologs are necessary to answer this question. A number of functions, such as RNA chaperone and regulator of transcription and translation, have been indicated, but elaborate and extensive research is essential for thorough elucidation of functions of cold shock proteins. This lack of information has limited the design of strategies to identify the key regulatory factors responsible for survival of the organism after cold shock and for preventing cold adaptation of food-poisoning bacteria.

One of the important aspects of the study of the cold shock response and adaptation in bacteria is regarding health issues and economy of many industrial processes involving bacterial fermentations. As refrigeration is a commonly used method for extending the shelf life of food, it is necessary to understand the cold shock response and adaptation of food-spoilage bacteria, such as *Clostridium* (22), *Enterococcus* (83), *Listeria* (5), *Pseudomonas* (57), *Vibrio* (56), and *Yersinia* (33). Understanding the expression of cold shock proteins and designing means for decreasing their accumulation can potentially reduce the efficacy of the cold shock response and adaptation and prevent growth of bacteria at low temperatures. The knowledge acquired through the study of the cold shock response and adaptation in *E. coli* and *B. subtilis* as model systems can be applied to other organisms. It is reported that if *B. subtilis* cells are cold shocked before freezing, the viability during the freezing is enhanced. This is probably because the cold shock proteins somehow protect cells from the cold damage caused by freezing (93). On this basis, the food-spoilage bacteria can be sensitized to damage through direct freezing itself, and cold shock-induced cryotolerance may contribute to the development of processes that allow improved viability/activity of frozen or freeze-dried commercial lactic acid bacteria starter cultures (13, 51). In addition, *Rhizobium* isolated from arctic (psychrotrophic) legumes is of considerable interest in agricultural industry because of its potential to improve nitrogen fixation of legumes cultivated in temperate climate, where low temperature limits the efficiency of the symbiosis (14). Furthermore, a desaturase from cyanobacterium that plays an important role in maintaining the membrane fluidity after cold shock confers chilling resistance to tobacco plants (42). These instances imply that in the future the study of the cold shock response and adaptation is going to play a major role in biotechnology.

Acknowledgment. This work was supported by grant GM 19043 from the National Institutes of Health.

REFERENCES

1. **Aguilar, P. S., J. E. Cronan Jr., and D. de Mendoza.** 1998. A *Bacillus subtilis* gene induced by cold shock encodes a membrane phospholipid desaturase. *J. Bacteriol.* **180:**2194–2200.

2. **Annous, B. A., L. A. Becker, D. O. Bayles, D. P. Labeda, and B. J. Wilkinson.** 1997. Critical role of anteiso-$C_{15:0}$ fatty acid in the growth of *Listeria monocytogenes* at low temperatures. *Appl. Env. Microbiol.* **63:**3887–3894.

3. **Bae, W., P. G. Jones, and M. Inouye.** 1997. CspA, the major cold shock protein of *Escherichia coli*, negatively regulates its own gene expression. *J. Bacteriol.* **179:**7081–7088.

4. **Bae, W., S. Phadtare, K. Severinov, and M. Inouye.** 1999. Characterization of *Escherichia coli cspE*, whose product negatively regulates transcription of *cspA*, the gene for the major cold shock protein. *Mol. Microbiol.* **31:**1429–1441.

5. **Bayles, D. O., B. A. Annous, and B. J. Wilkinson.** 1996. Cold stress proteins induced in *Listeria monocytogenes* in response to temperature downshock and growth at low temperatures. *Appl. Env. Microbiol.* **62:**1116–1119.

6. **Berger, F., P. Normand, and P. Potier.** 1997. *capA*, a *cspA*-like gene that encodes a cold acclimation protein in the psychrotrophic bacterium *Arthrobacter globiformis* SI55. *J. Bacteriol.* **179:**5670–5676.

7. **Brandi, A., P. Pietroni, C. O. Gualerzi, and C. L. Pon.** 1996. Post-transcriptional regulation of CspA expression in *Escherichia coli*. *Mol. Microbiol.* **19:**231–240.

8. **Brandi, A., R. Spurio, C. O. Gualerzi, and C. L. Pon.** 1999. Massive presence of the *Escherichia coli* 'major cold-shock protein' CspA under non-stress conditions. *EMBO J.* **18:** 1653–1659.

9. **Broeze, R. J., C. J. Solomon, and D. H. Pope.** 1978. Effects of low temperature on in vitro and in vivo protein synthesis in *Escherichia coli* and *Pseudomonas fluorescens*. *J. Bacteriol.* **134:**861–874.

10. **Carratu, L., S. Franceschelli, C. L. Pardini, G. S. Kobayashi, I. Horvath, L. Vigh, and B. Maresca.** 1996. Membrane lipid perturbation modifies the set point of the temperature of heat shock response in yeast. *Proc. Natl. Acad. Sci. USA* **93:**3870–3875.

11. **Celerin, M., A. A. Gilpin, N. J. Schisler, A. G. Ivanov, E. Miskiewicz, M. Krol, and D. E. Laudenbach.** 1998. ClpB in a cyanobacterium: predicted structure, phylogenetic relationships, and regulation by light and temperature. *J. Bacteriol.* **180:**5173–5182.

12. **Chamot, D., W. C. Magee, E. Yu, and G. W. Owttrim.** 1999. A cold shock-induced cyanobacterial RNA helicase. *J. Bacteriol.* **181:**1728–1732.

13. **Chapot-Chartier, M. P., C. Schouler, A.-S. Lepeuple, J.-C. Gripon, and M.-C. Chopin.** 1997. Characterization of *cspB*, a cold-shock-inducible gene from *Lactobacillus lactis* and evidence for a family of genes homologous to the *Escherichia coli cspA* major cold shock gene. *J. Bacteriol.* **179:**5589–5593.

14. **Cloutier, J., D. Prevost, P. Nadeau, and H. Antoun.** 1992. Heat and cold shock protein synthesis in arctic and temperate strains of rhizobia. *Appl. Env. Microbiol.* **58:**2846–2853.

15. **Dammel, C. S., and H. F. Noller.** 1995. Suppression of a cold-sensitive mutation in 16S rRNA by overexpression of a novel ribosome-binding factor, RbfA. *Genes Dev.* **9:**626–637.

16. **Dersch, P., S. Kneip, and E. Bremer.** 1994. The nucleoid-associated DNA-binding protein H-NS is required for the efficient adaptation of *Escherichia coli* K-12 to a cold environment. *Mol. Gen. Genet.* **245:**255–259.

17. **Donovan, W. P., and S. R. Kushner.** 1986. Polynucleotide phosphorylase and ribonuclease II are required for cell viability and mRNA turnover in *Escherichia coli* K-12. *Proc. Natl. Acad. Sci. USA* **83:**120–124.

18. **Etchegaray, J.-P., and M. Inouye.** 1999. CspA, CspB, and CspG, the major cold shock proteins of *Escherichia coli*, are

induced at low temperature under conditions that completely block protein synthesis. *J. Bacteriol.* **181:**1827–1830.

19. **Etchegaray, J.-P., and M. Inouye.** 1999. A sequence downstream of the initiation codon is essential for cold shock induction of *cspB* of *Escherichia coli. J. Bacteriol.* **181:**5852–5854.

20. **Etchegaray, J.-P., and M. Inouye.** 1999. Translational enhancement by an element downstream of the initiation codon in *Escherichia coli. J. Biol. Chem.* **274:**10079–10085.

21. **Etchegaray, J.-P., P. G. Jones, and M. Inouye.** 1996. Differential thermoregulation of two highly homologous cold-shock genes, *cspA* and *cspB*, of *Escherichia coli. Genes Cells* **1:**171–178.

22. **Evans, R. I., N. J. Russell, G. W. Gould, and P. J. McClure.** 1997. The germinability of spores of a psychrotolerant, nonproteolytic strain of *Clostridium botulinum* is influenced by their formation and storage temperature. *J. Appl. Microbiol.* **83:**273–280.

23. **Fang, L., Y. Hou, and M. Inouye.** 1998. Role of the cold-box region in the 5′ untranslated region of the *cspA* mRNA in its transcript expression at low temperature in *Escherichia coli. J. Bacteriol.* **180:**90–95.

24. **Fang, L., W. Jiang, W. Bae, and M. Inouye.** 1997. Promoter-independent cold-shock induction of *cspA* and its derepression at 37°C by mRNA stabilization. *Mol. Microbiol.* **23:**355–364.

25. **Farewell, A., and F. C. Neidhardt.** 1998. Effect of temperature on in vivo protein synthetic capacity in *Escherichia coli. J. Bacteriol.* **180:**4704–4710.

26. **Feng, W., R. Tejero, D. E. Zimmerman, M. Inouye, and G. T. Montelione.** 1998. Solution NMR structure and backbone dynamics of the major cold-shock protein (CspA) from *Escherichia coli*: evidence for conformational dynamics in the single-stranded RNA-binding site. *Biochemistry* **37:**10881–10896.

27. **Friedman, D. I., E. R. Olson, C. Georgopoulos, K. Tilly, I. Herskowitz, and F. Banuett.** 1984. Interactions of bacteriophage and host macromolecules in the growth of bacteriophage lambda. *Microbiol Rev.* **48:**299–325.

28. **Garwin, J. L., and J. E. Cronan, Jr.** 1980. Thermal modulation of fatty acid synthesis in *Escherichia coli* does not involve de novo enzyme synthesis. *J. Bacteriol.* **141:**1457–1459.

29. **Garwin, J. L., A. L. Klages, and J. E. Cronan, Jr.** 1980. β-ketoacyl-acyl carrier protein synthase II of *Escherichia coli*. Evidence for function in the thermal regulation of fatty acid synthesis. *J. Biol. Chem.* **255:**3263–3265.

30. **Goldenberg, D., I. Azar, and A. B. Oppenheim.** 1996. Differential mRNA stability of the *cspA* gene in the cold-shock response of *Escherichia coli. Mol. Microbiol.* **19:**241–248.

31. **Goldenberg, D., I. Azar, A. B. Oppenheim, A. Brandi, C. L. Pon, and C. O. Gualerzi.** 1997. Role of *Escherichia coli cspA* promoter sequences and adaptation of transcriptional apparatus in the cold shock response. *Mol. Gen. Genet.* **256:**282–290.

32. **Goldstein, J., N. S. Pollitt, and M. Inouye.** 1990. Major cold shock protein of *Escherichia coli. Proc. Natl. Acad. Sci. USA* **87:**283–287.

33. **Goverde, R. L. J., J.-H. J. Huis in't Veld, J. G. Kusters, and F. R. Mooi.** 1998. The psychrotrophic bacterium *Yersinia enterocolitica* requires expression of *pnp*, the gene for polynucleotide phosphorylase, for growth at low temperature (5°C). *Mol. Microbiol.* **28:**555–569.

34. **Grau, R., D. Gardiol, G. C. Glikin, and D. de Mendoza.** 1994. DNA supercoiling and thermal regulation of unsaturated fatty acid synthesis in *Bacillus subtilis. Mol. Microbiol.* **11:**933–941.

35. **Graumann, P., and M. A. Marahiel.** 1996. Some like it cold: response of microorganisms to cold shock. *Arch. Microbiol.* **166:**293–300.

36. **Graumann, P., and M. A. Marahiel.** 1996. A case of convergent evolution of nucleic acid binding modules. *BioEssays* **18:**309–315.

37. **Graumann, P., K. Schroder, R. Schmid, and M. A. Marahiel.** 1996. Cold shock stress-induced proteins in *Bacillus subtilis. J. Bacteriol.* **178:**4611–4619.

38. **Graumann, P., T. M. Wendrich, M. H. W. Weber, K. Schroder, and M. A. Marahiel.** 1997. A family of cold shock proteins in *Bacillus subtilis* is essential for cellular growth and for efficient protein synthesis at optimal and low temperatures. *Mol. Microbiol.* **25:**741–756.

39. **Graumann, P. L., and M. A. Marahiel.** 1998. A superfamily of proteins that contain the cold-shock domain. *Trends Biochem. Sci.* **23:**286–290.

40. **Gualerzi, C. O., and C. L. Pon.** 1990. Initiation of mRNA translation in prokaryotes. *Biochemistry* **29:**5881–5889.

41. **Hillier, B. J., H. M. Rodriguez, and L. M. Gregoret.** 1998. Coupling protein stability and protein function in *Escherichia coli* CspA. *Folding and Design* **3:**87–93.

42. **Ishizaki-Nishizawa, O., T. Fujii, M. Azuma, K. Sekiguchi, N. Murata, T. Ohtani, and T. Toguri.** 1996. Low-temperature resistance of higher plants is significantly enhanced by nonspecific cyanobacterial desaturase. *Nature Biotech.* **14:**1003–1006.

43. **Jacob, M., T. Schindler, J. Balbach, and F. X. Schmid.** 1997. Diffusion control in an elementary protein folding reaction. *Proc. Natl. Acad. Sci. USA* **94:**5622–5627.

44. **Jiang, W., L. Fang, and M. Inouye.** 1996. The role of the 5′-end untranslated region of the mRNA for CspA, the major cold-shock protein of *Escherichia coli*, in cold-shock adaptation. *J. Bacteriol.* **178:**4919–4925.

45. **Jiang, W., L. Fang, and M. Inouye.** 1996. Complete growth inhibition of *Escherichia coli* by ribosome trapping with truncated *cspA* mRNA at low temperature. *Genes Cells* **1:**965–976.

46. **Jiang, W., Y. Hou, and M. Inouye.** 1997. CspA, the major cold-shock protein of *Escherichia coli*, is an RNA chaperone. *J. Biol. Chem.* **272:**196–202.

47. **Jones, P. G., M. Cashel, G. Glaser, and F. C. Neidhardt.** 1992. Function of a relaxed-like state following temperature downshifts in *Escherichia coli. J. Bacteriol.* **174:**3903–3914.

48. **Jones, P. G., and M. Inouye.** 1996. RbfA, a 30S ribosomal binding factor, is a cold-shock protein whose absence triggers the cold-shock response. *Mol Microbiol.* **21:**1207–1218.

49. **Jones, P. G., and M. Inouye.** 1994. The cold-shock response—a hot topic. *Mol. Microbiol.* **11:**811–818.

50. **Kandror, O., and A. L. Goldberg.** 1997. Trigger factor is induced upon cold shock and enhances viability of *Escherichia coli* at low temperatures. *Proc. Natl. Acad. Sci. USA* **94:**4978–4981.

51. **Kim, W. S., and N. W. Dunn.** 1997. Identification of a cold shock gene in lactic acid bacteria and the effect of cold shock on cryotolerance. *Curr. Microbiol.* **35:**59–63.

52. **Krispin, O., and R. Allmansberger.** 1995. Changes in DNA supertwist as a response of *Bacillus subtilis* towards different kinds of stress. *FEMS Microbiol. Lett.* **134:**129–135.

53. **Kumar, A., R. A. Malloch, N. Fujita, D. A. Smillie, A. Ishihama, and R. S. Hayward.** 1993. The minus 35-recognition region of *Escherichia coli* sigma 70 is inessential for initiation of transcription at an 'extended minus 10' promoter. *J. Mol. Biol.* **232:**406–418.

54. **Lee, S. J., A. Xie, W. Jiang, J.-P. Etchegaray, P. G. Jones, and M. Inouye.** 1994. Family of the major cold-shock protein, CspA (CS 7.4), of *Escherichia coli*, whose members show a high sequence similarity with the eukaryotic Y-box binding proteins. *Mol. Microbiol.* **11:**833–839.

55. Lelivelt, M. J., and T. H. Kawula. 1995. Hsc66, an Hsp70 homolog in *Escherichia coli*, is induced by cold shock but not by heat shock. *J. Bacteriol.* 177: 4900–4907.

56. McGovern, V. P., and J. D. Oliver. 1995. Induction of cold-responsive proteins in *Vibrio vulnificus. J. Bacteriol.* 177: 4131–4133.

57. Michel, V., I. Lehoux, G. Depret, P. Anglade, J. Labadie, and M. Hebraud. 1997. The cold shock response of the psychrotrophic bacterium *Pseudomonas fragi* involves four low-molecular-mass nucleic acid-binding proteins. *J. Bacteriol.* 179:7331–7342.

58. Mitta, M., L. Fang, and M. Inouye. 1997. Deletion analysis of *cspA* of *Escherichia coli*: requirement for the AT-rich UP element for *cspA* transcription, and the downstream box in the coding region for its cold shock induction. *Mol. Microbiol.* 26: 321–335.

59. Mizushima, T., K. Kataoka, Y. Ogata, R. Inoue, and K. Sekimizu. 1997. Increase in negative supercoiling of plasmid DNA in *Escherichia coli* exposed to cold shock. *Mol. Microbiol.* 23: 381–386.

60. Morita, M. T., Y. Tanaka, T. S. Kodama, Y. Kyogoku, H. Yanagi, and T. Yura. 1999. Translational induction of heat shock transcription factor σ^{32}: evidence for a built-in RNA thermosensor. *Genes Dev.* 13:655–665.

61. Murata, N., and H. Wada. 1995. Acyl-lipid desaturases and their importance in the tolerance and acclimatization to cold of cyanobacteria. *Biochem. J.* 308:1–8.

62. Nakashima, K., K. Kanamaru, T. Mizuno, and K. Horikoshi. 1996. A novel member of the *cspA* family of genes that is induced by cold-shock in *Escherichia coli. J. Bacteriol.* 178: 2994–2997.

63. Newkirk, K., W. Feng, W. Jiang, R. Tejero, S. D. Emerson, M. Inouye, and G. T. Montelione. 1994. Solution NMR structure of the major cold shock protein (CspA) from *Escherichia coli*: identification of a binding epitope for DNA. *Proc. Natl. Acad. Sci. USA* 91:5114–5118.

64. Nishiyama, S., T. Umemura, T. Nara, M. Homma, and I. Kawagishi. 1999. Conversion of a bacterial warm sensor to a cold sensor by methylation of a single residue in the presence of an attractant. *Mol. Microbiol.* 32:357–365.

65. Panoff, J.-M., D. Corroler, B. Thammavongs, and P. Boutibonnes. 1997. Differentiation between cold shock proteins and cold acclimation proteins in a mesophilic gram-positive bacterium, *Enterococcus faecalis* JH2-2. *J. Bacteriol.* 179: 4451–4454.

66. Panoff, J.-M., B. Thammavongs, M. Gueguen, and P. Boutibonnes. 1998. Cold stress responses in mesophilic bacteria. *Cryobiology* 36:75–83.

67. Perl, D., C. Welker, T. Schindler, K. Schroder, M. A. Marahiel, R. Jaenicke, and F. X. Schmid. 1998. Conservation of rapid two-state folding in mesophilic, thermophilic and hyperthermophilic cold shock proteins. *Nature Struct. Biol.* 5: 229–235.

68. Phadtare, S., J. Alsina, and M. Inouye. 1999. Cold-shock response and cold-shock proteins. *Curr. Opin. Microbiol.* 2:175–180.

69. Phadtare, S., and M. Inouye. 1999. Sequence selective interactions with RNA by CspB, CspC and CspE, members of the CspA family of *Escherichia coli. Mol. Microbiol.* 33:1004–1014.

70. Porankiewicz, J., and A. K. Clarke. 1997. Induction of the heat shock protein ClpB affects cold accclimation in the cyanobacterium *Synechococcus* sp. strain PCC 7942. *J. Bacteriol.* 179:5111–5117.

71. Reid, K. L., H. M. Rodriguez, B. J. Hillier, and L. M. Gregoret. 1998. Stability and folding properties of a model β-sheet protein, *Escherichia coli* CspA. *Protein Sci.* 7:470–479.

72. Sakamoto, T., and D. A. Bryant. 1997. Temperature-regulated mRNA accumulation and stabilization for fatty acid desaturase genes in the cyanobacterium *Synechococcus* sp. Strain PCC 7002. *Mol. Microbiol.* 23:1281–1292.

73. Sato, N. 1995. A family of cold-regulated RNA-binding protein genes in the cyanobacterium *Anabaena variabilis* M3. *Nucleic Acids Res.* 23:2161–2167.

74. Schindelin, H., W. Jiang, M. Inouye, and U. Heinemann. 1994. Crystal structure of CspA, the major cold shock protein of *Escherichia coli. Proc. Natl. Acad. Sci. USA* 91:5119–5123.

75. Schindelin, H., M. A. Marahiel, and U. Heinemann. 1993. Universal nucleic acid-binding domain revealed by crystal structure of the *Bacillus subtilis* major cold-shock protein. *Nature* 364:164–168.

76. Schindler, T., P. L. Graumann, D. Perl, S. Ma, F. X. Schmid, and M. A. Marahiel. 1999. The family of cold shock proteins of *Bacillus subtilis*. Stability and dynamics in vitro and in vivo. *J. Biol. Chem.* 274:3407–3413.

77. Schindler, T., D. Perl, P. Graumann, V. Sieber, M. A. Marahiel, and F. X. Schmid. 1998. Surface-exposed phenylalanines in the RNP 1/ RNP 2 motif stabilize the cold-shock protein CspB from *Bacillus subtilis. Proteins Struct. Funct. Genet.* 30: 401–406.

78. Schnuchel, A., R. Wiltscheck, M. Czisch, M. Herrler, G. Willimsky, P. Graumann, M. A. Marahiel, and T. A. Holak. 1993. Structure in solution of the major cold-shock protein from *Bacillus subtilis. Nature* 364:169–171.

79. Schroder, K., P. Graumann, A. Schnuchel, T. A. Holak, and M. A. Marahiel. 1995. Mutational analysis of the putative nucleic acid-binding surface of the cold-shock domain, CspB, revealed an essential role of aromatic and basic residues in binding of single-stranded DNA containing the Y-box motif. *Mol. Microbiol.* 16:699–708.

80. Sinensky, M. 1974. Homeoviscous adaptation: a homeostatic process that regulates the viscosity of membrane lipids in *Escherichia coli. Proc. Natl. Acad. Sci. USA* 71:522–525.

81. Storz, G. 1999. An RNA thermometer. *Genes Dev.* 13:633–636.

82. Sugino, A., C. Peebles, K. N. Kreuzer, and N. R. Cozzarelli. 1977. Mechanism of action of nalidixic acid: purification of *Escherichia coli* nalA gene product and its relationship to DNA gyrase and a novel nicking-closing enzyme. *Proc. Natl. Acad. Sci. USA* 74:4767–4771.

83. Thammavongs, B., D. Corroler, J.-M. Panoff, Y. Auffray, and P. Boutibonnes. 1996. Physiological response of *Enterococcus faecalis* JH2-2 to cold shock: growth at low temperatures and freezing/thawing challenge. *Lett. Appl. Microbiol.* 23:398–402.

84. Thieringer, H., P. G. Jones, and M. Inouye. 1998. Cold shock and adaptation. *BioEssays* 20:49–57.

85. Toone, W. M., K. E. Rudd, and J. D. Friensen. 1991. deaD, a new *Escherichia coli* gene encoding a presumed ATP-dependent RNA helicase can suppress a mutation in *rpsB*, the gene encoding ribosomal protein S2. *J. Bacteriol.* 173:3291–3302.

86. VanBogelen, R. A., and F. C. Neidhardt. 1990. Ribosomes as sensors of heat and cold shock in *Escherichia coli. Proc. Natl. Acad. Sci. USA* 87:5589–5593.

87. Viale, A. M., A. K. Arakaki, F. C. Soncini, and R. G. Ferreyra. 1994. Evolutionary relationships among eubacterial groups as inferred from GroEL (chaperonin) sequence comparisons. *Int. J. System. Bacteriol.* 44:527–533.

88. Vigh, L., D. A. Los, I. Horvath, and N. Murata. 1993. The primary signal in the biological perception of temperature: Pd-

catalyzed hydrogenation of membrane lipids stimulated the expression of the *desA* gene in *Synechocystis* PCC6803. *Proc. Natl. Acad. Sci. USA* **90:**9090–9094.

89. **Wada, H., Z. Gombos, and N. Murata.** 1990. Enhancement of chilling tolerance of a cyanobacterium by genetic manipulation of fatty acid desaturation. *Nature* **347:**200–203.

90. **Walker, G. C.** 1984. Mutagenesis and inducible responses to deoxyribonucleic acid damage in *Escherichia coli*. *Microbiol. Rev.* **48:**60–93.

91. **Wang, J.-Y., and M. Syvanen.** 1992. DNA twist as a transcriptional sensor for environmental changes. *Mol. Microbiol.* **6:**1861–1866.

92. **Wang, N., K. Yamanaka, and M. Inouye.** 1999. CspI, the ninth member of the CspA family of *Escherichia coli*, is induced upon cold shock. *J. Bacteriol.* **181:**1603–1609.

93. **Willimsky, G., H. Bang, G. Fischer, and M. A. Marahiel.** 1992. Characerization of *cspB*, a *Bacillus subtilis* inducible cold shock gene affecting cell viability at low temperatures. *J. Bacteriol.* **174:**6326–6335.

94. **Wolffe, A. P.** 1995. The cold-shock response in bacteria. *Science Progress* **78:**301–310.

95. **Yamanaka, K., L. Fang, and M. Inouye.** 1998. The CspA family in *Escherichia coli*: multiple gene duplication for stress adaptation. *Mol. Microbiol.* **27:**247–255.

96. **Yamanaka, K., and M. Inouye.** 1997. Growth-phase-dependent expression of *cspD*, encoding a member of the CspA family in *Escherichia coli*. *J. Bacteriol.* **179:**5126–5130.

97. **Yamanaka, K., T. Mitani, T. Ogura, H. Niki, and S. Hiraga.** 1994. Cloning, sequencing and characterization of multicopy suppressors of a *mukB* mutation in *Escherichia coli*. *Mol. Microbiol.* **13:**301–312.

98. **Yamanaka, K., M. Mitta, and M. Inouye.** 1999. Mutation analysis of the 5′ untranslated region of the cold shock *cspA* mRNA of *Escherichia coli*. *J. Bacteriol.* **181:**6284–6296.

Bacterial Stress Responses
Edited by G. Storz and R. Hengge-Aronis
©2000 ASM Press, Washington, D.C.

Chapter 4

Oxidative Stress

GISELA STORZ AND MING ZHENG

Oxidative stress caused by increased levels of superoxide anion ($O_2^{\bullet-}$), hydrogen peroxide (H_2O_2), or hydroxyl radical (HO•) can lead to the damage of all cellular components. Much has been learned about the bacterial defenses against oxidative stress from the studies of the Escherichia coli oxyR *and* soxRS *regulons. Here we review what is known about the antioxidant activities whose expression is controlled by OxyR and SoxRS and what has been discovered about the chemistry of OxyR and SoxR oxidation. We also discuss the responses to oxidative stress in other bacteria, how the oxidative stress responses overlap with other stress responses, and the potential roles of the* oxyR *and* soxRS *regulons in pathogenesis.*

Oxidative stress has been defined as a disturbance in the prooxidant-antioxidant balance in favor of prooxidants (93). Thus, conditions that lead to increased levels of reactive oxygen species or conditions that lead to the depletion of antioxidant molecules or enzymes constitute an oxidative stress. For aerobically growing bacterial cells, the autooxidation of components of the respiratory chain is the main source of endogenous $O_2^{\bullet-}$ and H_2O_2 (37, 53). Increased levels of the reactive oxygen species are also caused by exposure to radiation, metals, and redox-active drugs. In addition, plants, microorganisms, and animals all possess mechanisms to specifically generate oxidants as a defense against bacterial invasion. The reactive oxygen species are deleterious to cells since they can lead to protein, DNA, and membrane damage. Genes encoding antioxidant enzymes can be detected in the sequences of most completed genomes, showing that defenses against oxidative stress are critical to many organisms.

The existence of a regulated adaptive response to H_2O_2 was first discovered for *Escherichia coli* (21). Later a similar adaptive response was observed for *Salmonella enterica* serovar Typhimurium (16). If exponentially growing *E. coli* or serovar Typhimurium cells are treated with low doses of H_2O_2 (10 to 60 μM), they become resistant to what would otherwise be lethal doses of H_2O_2 (5 to 10 mM). *E. coli* cells treated with low doses of $O_2^{\bullet-}$-generating compounds such as plumbagin also become resistant to subsequent higher doses of these compounds (30). Two-dimensional gel analysis of *E. coli* and serovar Typhimurium cells treated with H_2O_2 showed that this oxidant induces the synthesis of ~40 proteins (80, 105). Similar two-dimensional gel analysis of cells exposed to plumbagin, paraquat, or menadione showed that these $O_2^{\bullet-}$-generating compounds also induce ~40 proteins, including some of the H_2O_2-inducible proteins (42, 107).

THE *E. COLI oxyR* AND *soxRS* REGULONS

Significant advances in understanding the adaptive responses to H_2O_2 and $O_2^{\bullet-}$ in *E. coli* came from the discoveries of the *soxRS* and *oxyR* regulators. The *oxyR* locus was defined by mutations that conferred increased resistance to H_2O_2 and led to elevated expression of a subset of H_2O_2-inducible proteins (16). The *soxRS* locus was defined by mutations that led to increased resistance to menadione, constitutive expression of a subset of $O_2^{\bullet-}$-inducible proteins, and elevated expression of reporter genes fused to the promoters of $O_2^{\bullet-}$-inducible genes (43, 103). Since the initial isolation of the *oxyR* and *soxR* mutants much has been learned about the members of the respective regulons (discussed below and summarized in Table 1) as well as the mechanism of regulation (discussed in a later section).

Gisela Storz and Ming Zheng • Cell Biology and Metabolism Branch, National Institute of Child Health and Human Development, National Institutes of Health, Bethesda, MD 20892.

Table 1. Antioxidant activities in *E. coli*[a]

Gene	Activity	Additional regulator(s)
oxyR regulon		
katG	Hydroperoxidase I	σ^s
ahpCF	Alkyl hydroperoxide reductase	
gorA	Glutathione reductase	σ^s
grxA	Glutaredoxin 1	
trxC	Thioredoxin 2	
fur	Ferric uptake repressor	SoxRS
dps	Nonspecific DNA binding protein	σ^s
oxyS	Regulatory RNA	
agn43	Outer membrane protein	Dam
fhuF	Ferric reductase	
soxRS regulon (also regulated by MarA and Rob)		
sodA	Manganese superoxide dismutase	ArcAB, FNR
nfo	Endonuclease IV	
zwf	Glucose-6-phosphate dehydrogenase	
fumC	Fumarase C	ArcAB, σ^s
acnA	Aconitase A	ArcAB, FNR, σ^s
tolC	Outer membrane protein	
fur	Ferric uptake repressor	OxyR
micF	RNA regulator of *ompF*	
acrAB	Multidrug efflux pump	
nfsA	Nitroreductase A	
fpr	Ferredoxin/flavodoxin reductase	
fldA	Flavodoxin	
fldB	Flavodoxin	
ribA	GTP cyclohydrolase	
Other defense activities		
katE	Hydroperoxidase II	σ^s
xthA	Exonuclease III	σ^s
polA	DNA polymerase I	
recA	RecA	
mutM (*fpg*)	Formamidopyrimidine glycosylase	FNR
nth	Endonuclease III	
mutY	Adenine glycosylase	
msrA	Methionine sulfoxide reductase	
hslO	Molecular chaperone	
sodB	Superoxide dismutase	
sodC	Copper-zinc superoxide dismutase	σ^s, FNR
iscS	NifS homolog	
iscU	NifU homolog	

[a] See text for references.

The *oxyR* Regulon

Many of the genes whose expression is activated by OxyR after treatment with H_2O_2 have clear antioxidant roles. The hydroperoxidase I (*katG*) and alkyl hydroperoxide reductase (*ahpCF*) protect against the toxic effects of peroxides by directly eliminating the oxidants (16). The induction of glutathione reductase (*gorA*), glutaredoxin 1 (*grxA*), and thioredoxin 2 (*trxC*) should help maintain the cellular thiol-disulfide balance (16, 90a, 98, 115). OxyR also induces the expression of the nonspecific DNA binding protein Dps, which protects against DNA damage and mutation (1, 73). Dps has been proposed to act by sequestering DNA since recent studies have shown

that Dps forms an extensive crystalline lattice in the presence of DNA in vitro and in vivo (108). Dps may also provide protection by sequestering iron since the crystal structure of Dps revealed that the protein is a ferritin homolog (41). OxyR induction of *fur*, which encodes a global repressor of ferric ion uptake, should prevent damage caused by HO• generated by H_2O_2 reacting with intracellular iron (the Fenton reaction) (116). OxyR activation also leads to high levels of a small RNA denoted OxyS (2). This unique RNA protects against mutagenesis, although the mechanism by which OxyS acts as an antimutator is unknown. The OxyS RNA also activates and represses the expression of numerous genes in *E. coli*. Since two of these target genes, *rpoS* and *fhlA*, encode transcriptional regulators, another role of OxyS is to integrate the adaptive response to H_2O_2 with other cellular stress responses. For both *rpoS* and *fhlA*, OxyS acts to repress translation (3, 113). The isolation of OxyR-regulated *lacZ* fusions indicates that OxyR also activates the expression of a coproporphyrinogen III oxidase (*hemF*), a regulator of capsular polysaccharide synthesis genes (*rcsC*), and a protein with homology to arylsulfatase enzymes (*f497*); however, the antioxidant roles of these proteins are not understood (82).

The OxyR protein acts as a repressor of its own gene as well as the Mu phage *mom* gene (encoding a DNA modification function) (7, 45, 97), the *E. coli agn43* gene (encoding antigen 43, a major phase-variable outer membrane protein) (48), and the *E. coli fhuF* gene (encoding a protein capable of reducing iron in cytoplasmic ferrioxamine B) (83; M. Zheng, B. Doan, K. Lewis, T. D. Schneider, and G. Storz, unpublished data). The role of OxyR repression of *fhuF* can be explained by the need to minimize ferric iron uptake to prevent the Fenton reaction. The roles of OxyR repression of *mom* and *agn43* are less clear. Recent studies have shown that the antigen 43 protein is required for biofilm formation by *E. coli* cells grown in glucose minimal medium, raising the possibility that biofilm formation is influenced by oxidative stress (P. N. Danese, L. A. Pratt, S. L. Dove, and R. Kolter, submitted for publication). Intriguingly, both the *mom* and *agn43* genes are subject to regulation by Dam methylation, and the OxyR DNA binding sites at both genes contain three GATC Dam recognition sequences (7, 45, 48, 97). Thus, OxyR competes with Dam methylation for binding to and regulating these promoters.

In addition to protecting against H_2O_2-induced damage, OxyR-regulated activities have been found to confer resistance to hypochlorous acid (HOCl) (28), organic solvents (31), and reactive nitrogen species (see below). For example, a screen for *E. coli* mutants with increased resistance to the organic solvent 1,2,3,4-tetrahydronaphthalene led to the isolation of an *ahpC* mutant (31).

The *soxRS* Regulon

Several of the activities whose expression is induced by SoxRS in response to treatment with $O_2^{\bullet-}$-generating compounds have understandable roles in a defense against oxidative stress. These include the manganese superoxide dismutase (*sodA*), the DNA repair enzyme endonuclease IV (*nfo*), and glucose-6-phosphate dehydrogenase (*zwf*), which increases the reducing power of the cell (43, 103). SoxRS activation also leads to increased levels of the $O_2^{\bullet-}$-resistant

isozymes of fumarase (*fumC*) and aconitase (*acnA*) (44, 63). Like OxyR activation of *fur*, SoxRS induction of *fur* should diminish the formation of HO• (116). Increased expression of the *tolC*-encoded outer membrane protein, the *acrAB*-encoded drug efflux pump, and the MicF regulatory RNA, which represses the expression of the OmpF outer membrane porin, should result in the exclusion of $O_2^{\bullet-}$-generating compounds (5, 15, 69). Nitroreductase A (*nfsA*), which catalyzes the divalent reduction of organic nitro compounds, may prevent the univalent reduction of quinones and dyes and thus diminish $O_2^{\bullet-}$ production due to redox cycling of these compounds (65). The roles of the SoxRS-induced flavodoxin A (*fldA*), flavodoxin B (*fldB*), and ferredoxin/flavodoxin reductase (*fpr*) in the $O_2^{\bullet-}$ defense response have not been characterized; however, the two proteins might function to maintain the reduced state of Fe-S clusters (64, 116; P. Gaudu and B. Weiss, submitted for publication). The protective roles of other members of the *soxRS* regulon, such as GTP cyclohydrolase II (*ribA*) and two proteins of unknown function (*inaA* and *pqi5*), are unclear (59, 60, 92).

In addition to protecting against $O_2^{\bullet-}$-induced damage, the *soxRS* regulon provides resistance against organic solvents (84) and reactive nitrogen species (see below). The regulon also protects against many different antibiotics and drugs (reviewed in chapter 22 and reference 78). In fact, activation of SoxRS might be considered a response to xenobiotics rather than just a defense against $O_2^{\bullet-}$.

ADDITIONAL ANTIOXIDANT DEFENSE ACTIVITIES IN *E. COLI*

The expression of some activities that clearly protect against H_2O_2 is not under the control of OxyR. These include hydroperoxidase II (*katE*) and the DNA repair enzymes exonuclease III (*xthA*), DNA polymerase I (*polA*), and RecA (54, 66). The genes encoding several other DNA repair enzymes, formamidopyrimidine glycosylase (*mutM* or *fpg*), endonuclease III (*nth*), and adenine glycosylase (*mutY*), which act on oxidative damage, also do not appear to be under the control of either OxyR or SoxRS (22). Peptide methionine sulfoxide reductase (*msrA*) and a molecular chaperone Hsp33 (*hslO*), whose activity is redox-regulated, probably protect against oxidative protein damage (55, 81). Strains carrying mutations in the *msrA* and *hslO* genes are sensitive to peroxides, but neither gene is regulated by OxyR. Two enzymes that eliminate $O_2^{\bullet-}$, but whose expression is not regulated by SoxRS, are the cytosolic iron superoxide dismutase (*sodB*) and the periplasmic copper-zinc superoxide dismutase (*sodC*) (33, 39). It is also conceivable that *E. coli* homologs of NifS (*iscS*) and NifU (*iscU*) proteins, which are required for Fe-S cluster assembly in *Azotobacter vinelandii*, are a defense against H_2O_2 and $O_2^{\bullet-}$ stress, but neither the expression nor the physiological role of *iscS* and *iscU* has been studied extensively in *E. coli* (114).

OVERLAP WITH OTHER REGULONS

A theme emerging from the studies of the *oxyR* and *soxRS* regulons is the existence of a significant overlap with other regulatory networks. As described in chapter 11, the *rpoS*-encoded σ^s subunit of RNA polymerase is important for the expression of a large group of genes that are induced when cells enter into stationary phase or encounter a number of different stresses including starvation, osmotic stress, and acid stress. Starved and stationary phase cells are intrinsically resistant to a variety of stress conditions including high levels of H_2O_2, and σ^s regulates the expression of several antioxidant genes including *katE*, *xthA*, and *sodC* (chapter 11 and references 39, 66). The SoxRS-regulated *pqi5* gene and the OxyR-regulated *katG*, *gorA*, and *dps* genes are also part of the σ^s regulon. To make the regulatory network even more complex, recent studies suggest that *oxyR* expression itself is under the control of σ^s (77). In addition, the OxyS RNA whose expression is activated by OxyR acts to repress *rpoS* translation (2, 113).

As reviewed in chapter 5, the FNR and ArcAB regulators play a central role in controlling cellular metabolism in response to oxygen levels. Thus it is not surprising that FNR and ArcAB modulate the expression of *mutM* and the SoxRS-regulated *sodA*, *acnA*, and *fumC* genes (18, 20, 61, 90), and FNR controls *sodC* expression (39).

In addition, the expression of almost all genes in the *soxRS* regulon is controlled by MarA and Rob. MarA was identified as part of an operon conferring multiple antibiotic resistance, and *marA* expression is induced by a variety of phenolic compounds including salicylate and 2,4-dinitrophenol and by certain antibiotics such as tetracycline (reviewed in chapter 22 and reference 78). Rob was detected by its ability to bind DNA near the bacterial origin of replication and is present at constitutively high levels in the cell (5,000 molecules) (95). The SoxS, MarA, and Rob proteins share the DNA binding motif characteristic of the AraC family of transcriptional activators. The transcription factors bind to and activate many, if not all, of the same promoters as monomers. The asymmetric "soxbox"/"marbox" has been defined by extensive promoter studies and contains an AYn-GCAYnnnnnnnYAA sequence at its core (71, 109). Interestingly, the boxes are oriented differently depending on whether they are located 7 or 15 bp upstream of the promoter −35 sequences (71, 109). Recent studies suggest that some promoter discrimination between SoxS and MarA is due to slight differences in binding to the individual sequences (72). However, each of the transcriptional activators is functional in the absence of the others, and the reasons for the extensive overlap between the *soxRS*, *mar*, and *rob* regulons are not understood. It is likely that many more overlaps and connections among the *E. coli* regulatory networks will be discovered. An important challenge will be to map and understand this connectivity.

CHEMISTRY OF OxyR AND SoxR OXIDATION

The cloning and purification of the OxyR and SoxR proteins allowed significant process to be made toward elucidating the mechanisms by which the activities of the OxyR and SoxR transcription factors are modulated by H_2O_2 and $O_2^{\bullet-}$, respectively (Fig. 1).

OxyR: Sensing H_2O_2 by a Thiol-Disulfide Switch

The 34-kDa OxyR protein shares homology with the LysR family of bacterial regulators and possesses an amino-terminal helix-turn-helix DNA binding domain common to this family (17). Transcriptional activation by OxyR is thought to occur via direct contacts with RNA polymerase, but the OxyR activation domain has not been defined yet (99, 100).

Early studies demonstrated that the tetrameric OxyR protein exists in two forms, reduced and oxidized, and only the oxidized form activates transcription (96). Toledano et al. then found that the two forms of OxyR make different DNA contacts: oxidized OxyR recognizes four ATAGnt elements in four adjacent major grooves while reduced OxyR recognizes ATAGnt elements present in two pairs of adjacent major grooves separated by one helical turn (102). Recent studies showed that oxidation leads to the formation of a disulfide bond between C199 and C208 (115). Mutation of either of these two conserved cysteine residues abolished the ability of the transcription factor to sense H_2O_2 in vivo and in vitro. Mass spectrometric analysis and thiol-disulfide titrations of purified OxyR showed that C199 and C208 are in a dithiol form in reduced OxyR and in a disulfide bonded form in oxidized OxyR. The redox potential of OxyR was determined to be −185

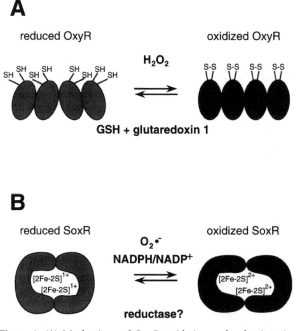

Figure 1. (A) Mechanisms of OxyR oxidation and reduction. Activation of OxyR by H_2O_2 is achieved by direct oxidation of the C199 and C208 thiols in OxyR to form an intramolecular disulfide bond (115). OxyR is deactivated upon reduction by glutaredoxin 1 with the consumption of glutathione (115). (B) Mechanisms of SoxR oxidation and reduction. Each SoxR dimer contains a pair of [2Fe-2S] clusters. Activation of the SoxR protein by $O_2{}^{\bullet-}$-generating compounds is achieved by direct oxidation of the [2Fe-2S] clusters possibly by $O_2{}^{\bullet-}$ or by changes in the NADPH/NADP$^+$ ratio (25, 27, 35, 36, 49, 52). The mechanism of SoxR reduction has not yet been elucidated but might involve a recently purified reductase (58).

mV, a value that ensures OxyR is in the dithiol form in the reducing environment of the cell (thiol-disulfide redox potential −280 mV). Thus direct oxidation of OxyR leading to the disulfide bond formation is the mechanism whereby cells sense H_2O_2 and induce the *oxyR* regulon.

Alkylation assays have allowed OxyR oxidation to be examined in more detail (6, 98a). In one alkylation assay, OxyR was treated with 4-acetamido-4′maleimidylstilbene-2,2′-disulfonic acid (AMS). Upon the addition of two 500-Da AMS adducts to reduced but not oxidized OxyR, the two forms can be separated on standard polyacrylamide gels (6). In a second alkylation assay, OxyR was treated with iodoacetamide, and oxidized and reduced OxyR proteins were separated on urea-containing polyacrylamide gels under nonreducing conditions (98a). These approaches showed that the OxyR protein is completely oxidized within 30 s in logarithmically growing wild-type cells and remains oxidized for approximately 5 min, paralleling the increase and decrease of mRNAs transcribed from OxyR-regulated

genes. The experiments also showed that the minimum concentration required to completely oxidize OxyR in vivo is 5 μM. This concentration agrees with the micromolar levels required for the OxyR-dependent activation of a *katG-lacZ* fusion in vivo (38). Given the millimolar levels of GSH inside the cell compared to the submicromolar levels of OxyR, the reaction between OxyR and H_2O_2 must be highly specific. An important direction for future studies will be to elucidate how the OxyR protein structure ensures such specificity.

The finding that OxyR was only activated/oxidized for a defined time raised the question as to how the protein was reduced (115). Two major disulfide-reduction systems are present in the *E. coli* cytosol: thioredoxin together with thioredoxin reductase and the tripeptide glutathione (GSH) together with glutathione reductase and the glutaredoxin proteins. The OxyR-regulated response was found to be prolonged in mutants lacking GSH or glutaredoxin 1 but not in mutants lacking thioredoxin or thioredoxin reductase. In vitro transcription assays showed that glutaredoxin 1 catalyzes the reduction of OxyR by GSH. These results indicate that oxidized OxyR is reduced by glutaredoxin 1 in the cell. Since the gene encoding glutaredoxin 1 (*grxA*) is itself regulated by OxyR, the OxyR response is autoregulated.

SoxR: Sensing $O_2{}^{\bullet-}$ by an Iron-Sulfur Cluster

Regulation of the *soxRS* regulon occurs by a two-stage process: the SoxR protein is first converted to an active form, which enhances *soxS* transcription, and increased levels of SoxS in turn activate expression of the regulon (85, 112). The constitutively expressed 17-kDa SoxR protein has homology to the mercury-dependent MerR regulator of *E. coli* and contains a helix-turn-helix DNA binding motif in the N-terminal domain (4, 111). SoxR is a homodimer in solution with two [2Fe-2S] centers per dimer (49, 50, 110). Site-directed mutagenesis studies showed that four conserved cysteine residues in the C-terminal domain of the SoxR polypeptide provide the ligands for the [2Fe-2S] clusters (8). The oxidation of the reduced [2Fe-2S]$^{1+}$ form of SoxR to a [2Fe-2S]$^{2+}$ form appears to be the mechanism of SoxR activation (25, 27, 35, 36, 49, 52). Evidence for this mechanism came from experiments in which the Fe-SoxR protein reduced with dithionite was found to regain transcriptional activity upon autooxidation (36). Electron paramagnetic resonance (EPR) spectroscopy of whole cells also showed that overproduced wild-type SoxR protein is oxidized within 2 min after cells are treated with $O_2{}^{\bullet-}$-generating compounds and that constitutively active mutant SoxR proteins are predominantly in the oxidized form even in the absence of stress (25, 35, 52). The redox potential measured for the purified wild-type protein in vitro is −285 mV, a value that ensures SoxR is reduced during normal growth (27, 36).

The nature of the oxidant that reacts with the [2Fe-2S] cluster in SoxR is still under debate. The *soxRS* regulon is induced by redox cycling reagents that generate $O_2 \cdot^-$. Thus the SoxR protein might be oxidized directly by $O_2 \cdot^-$. However, since the generation of $O_2 \cdot^-$ is accompanied by the consumption of cellular reductants, it is also possible that SoxR activity is modulated by alterations in NADPH, reduced flavodoxin, or reduced ferredoxin levels (64). On the one hand, Gort and Imlay observed very little induction of a *soxS-lacZ* fusion in a *sodA sodB* mutant that has elevated levels of $O_2 \cdot^-$, suggesting that SoxR senses some signal other than $O_2 \cdot^-$ (40). On the other hand, Liochev et al. detected significant SoxRS-dependent induction of fumarase C in a *sodA sodB* mutant strain (62). Assays of the *soxS* mRNA levels in different *sodA sodB* mutant strains should help resolve this discrepancy.

In vivo EPR studies showed that oxidized SoxR is rapidly reduced once the oxidative stress is removed (25); however, the mechanism of SoxR reduction/deactivation has not been elucidated. The high reducing power of ferridoxin (redox potential −380 mV) and flavodoxin (redox potential −450 mV) prompted Gaudu and Weiss to test whether these proteins can reduce SoxR (36). This is an intriguing possibility since both flavodoxin and ferrodoxin/flavodoxin reductase, which catalyze the reduction of ferridoxin and flavodoxin, are members of the *soxRS* regulon. However, neither protein effectively reduced SoxR in vitro (36). Recently, Kobayashi and Tagawa reported the purification of a SoxR reductase (58). The purified *E. coli* enzyme catalyzes SoxR-mediated NADPH reduction of cytochrome *c* and gives rise to a major band of molecular weight of 84 kD and minor bands of 50 kD and 150 kD in SDS-PAGE gels. It will be interesting to determine the sequence of the reductase and to test whether it reduces SoxR in vivo.

SoxR activity may also be modulated by the assembly/disassembly of the [2Fe-2S] clusters. The SoxR apoprotein binds to the *soxS* promoter but does not activate *soxS* expression. Hidalgo and Demple demonstrated that SoxR activity could be regenerated from the apoprotein by incubation with iron, inorganic sulfide, and a reducing agent or by incubation with the *A. vinelandii* NifS protein, which is involved in the formation of iron-sulfur clusters in nitrogenase (51). Exposure to monothiols such as glutathione led to the disruption of the SoxR [2Fe-2S] clusters under

aerobic conditions (24). In contrast, treatment with dithiols such as dithiothreitol and the dithiol enzyme thioredoxin promoted cluster assembly into apo-SoxR (26). More remains to be learned about the mechanism of SoxR [2Fe-2S] cluster assembly and disassembly, but these studies may reveal new levels of regulation.

OxyR AND SoxRS IN OTHER BACTERIA

OxyR homologs have been described in numerous organisms (Fig. 2A). The *Erwinia carotovora oxyR* gene was identified by DNA hybridization (12), the *Xanthomonas campestris* gene was cloned by complementation of H_2O_2 sensitivity of an *E. coli oxyR* mutant (67), and surprisingly, the *Haemophilus influenzae* gene was isolated in a screen for the expression of transferrin binding activity in *E. coli* (70). The OxyR homolog present in *Mycobacterium* species was first discovered by sequencing upstream of the mycobacterial *ahpC* gene (reviewed in reference 23). Although *oxyR* appears to be intact in *Mycobacterium leprae* and *Mycobacterium avium*, many *Mycobacterium* species carry mutations in the gene. The isolation of a H_2O_2-resistant mutant that shows constitutive expression of catalase, AhpC, and Dps allowed the identification of the *oxyR* gene in the obligate anaerobe *Bacteroides fragilis* (91; E. R. Rocha and C. J. Smith, Abstr. 99th Gen. Meet. Am. Soc. Microbiol. 1999, abstr. B/D-161, p. 60, 1999). The studies of *B. fragilis* indicate that aerobic growth is not a requirement for the presence of an *oxyR* regulon.

Interestingly, the arrangement of genes surrounding *oxyR* is different in several of the organisms listed above (Fig. 3): in *E. coli* and serovar Typhimurium, the *oxyS* gene is upstream of and on the opposite strand of *oxyR* (2); in *Mycobacterium*, *ahpC* is upstream of and on the opposite strand of *oxyR* (23); in *Xanthomonas*, *ahpC*, *ahpF*, and *oxyR* are adjacent and in the same orientation, with one transcript encoding *ahpC* and a second transcript encoding *ahpF* and *oxyR* (79); and in *B. fragilis*, the *dps* gene is upstream of and on the opposite strand of *oxyR* (Rocha and Smith, Abstr. 99th Gen. Meet. Am. Soc. Microbiol. 1999). Thus while the regulator and antioxidant activities have been conserved, the positions of the corresponding genes have been shuffled. Searches of the genomic databases indicate that many other bacteria contain OxyR homologs as defined by the presence of the redox-active C199 and C208 cysteines (Fig. 2A). It will be important to compare the activities, the target genes, and the gene configurations as additional homologs are characterized.

A

```
                                                                    C199        C208
Escherichia coli OxyR              EDHPWANRECVPMADLAGEKLLMLEDGHCLRDQAMGFCFEAGAD-EDT
Erwinia carotovora OxyR            QDHPWANRERVAMSDLAGEKLLMLEDGHCLRDQAMGFCFQAGAD-EDT
Xanthomonas campestris OxyR        EGHPLSRHDSMTLDDLSEQRLLLLEDGHCLRDQALDVCHLAGAL-EKS
Streptomyces viridosporus OxyR     LEHGLGGREGIPRKALRELNLLLLDEGHCLRDQALDICREAGSAGVAA
Brucella abortus OxyR              TNDHTVLASPMTQNHAALERLLLLEEGHCMRDQALAVCTLPSQR-QLV
Mycobacterium marinum OxyR         PGHPMADRHGVPVAALSELPLLLLDEGHCLRDQALDVCQNAGVRAELA
Acinetobacter calcoaceticus OxyR   K-HDKHSVNAHSLDDLDLSRLMLLEEGHCLRDHALSACPIGERK-NDN
Haemophilus influenzae OxyR        EHHPWAQESKLPMNQLNGQEMLMLDDGHCLRNQALDYCFTAGAK-ENS
Yersinia pestis OxyR               ADHPWANRERVEMHELAGEKLLMLEDGHCLRDQAMGFCFQAGAD-EDT
Actinobacillus actinomycetemcomitans OxyR  EQHPWANENSVSMSLLKDCEILMLDDGHCLRNQALGYCFTAGAR-ENA
Vibrio cholerae OxyR               CDHAWAARDEVDMLELKGKTVLALGDGHCLRDQALGFCFAAGAK-DDE
Pasteurella multocida OxyR         ENHPWANERTIAMNRLNGCEMLMLDDGHCLRDQTIGYCFSAGAK-ENA
Bordetella pertussis OxyR          HDHEWAQRKAIDAQDLKQQTMLLLGSGHCFRDQVLEVCPELSRFSASS
Pseudomonas aeruginosa OxyR        ADHPWTAKASIDSELLNDKSLLLLGEGHCFRDQVLEACPTVRKGDENK
Neisseria gonorrhoeae OxyR         KGHSFEELDAVSPRMLGEEQVLLLTEGNCMRDQVLSSCSELAAKQRIQ
Porphyromonas gingivalis OxyR      RCEPLFEQDVIRTTEVNPHRLWLLDEGHCFRDQLVRFCQMKGLHERQT
Conserved                                    *    *  *  *      *
```

B

```
                                               C119 C122 C124    C130
Escherichia coli SoxR          IHTLVALRDELDGCIGCGCLSRSDCPLRNPGD
Salmonella typhimurium SoxR    IHTLVALRDELDGCIGCGCLSRSDCPLRNPGD
Pseudomonas aeruginosa SoxR    IDKLLLLRDQLDGCIGCGCLSLQACPLRNPGD
Streptomyces violaceoruber SoxR IEQLLALRDGLEDCIGCGCLSVRDCPLTNPYD
Chromobacterium violaceum SoxR  IEALCRLRDQLDSCIGCGCLSLERCKLYNPDD
Bordetella pertussis SoxR       ILXLTQLRDQLDGCIGCGCLSRECPLRNPDD
Vibrio cholerae SoxR            IQQLNALKEDLSGCIGCGCLSLESCAIYNPKD
Conserved                       *   *   *    *  ****** *    *  ** *
```

Figure 2. (A) Sequence comparison of OxyR homologs from different organisms. Only the sequences surrounding the two critical cysteines (C199 and C208 in *E. coli* OxyR) are shown; however, the homologies extend across the protein sequences. (B) Sequence comparison of SoxR homologs from different organisms. Only the sequences near the four conserved cysteines, which provide ligands to the iron-sulfur cluster in *E. coli* SoxR, are shown; however, the homologies extend across the protein sequences.

The *soxR* and *soxS* genes are present in serovar Typhimurium and clearly function similarly to the *E. coli* homologs in conferring resistance to $O_2^{\bullet-}$-generating compounds and being required for induction of *sodA* (29). A search of the genome database indicates that SoxR homologs, which have the four-cysteine cluster required for activity in the *E. coli* protein, are present in *Pseudomonas aeruginosa*, *Streptomyces violaceoruber*, *Chromobacterium violaceum*, *Bordetella pertussis*, and *Vibrio cholerae* (Fig. 2B). It is interesting to note, however, that searches of the surrounding sequences did not reveal the presence of a *soxS* homolog near any of these genes. Further characterization of the responses to $O_2^{\bullet-}$ in

other bacteria may therefore reveal some important differences to the *E. coli* response.

ADDITIONAL OXIDATIVE STRESS REGULATORS

It is clear that oxidative stress regulators in addition to OxyR and SoxRS have yet to be identified. First, superoxide dismutase and catalase genes have been found in many organisms in which *oxyR* or *soxRS* genes are not detected. Second, even in *E. coli*, OxyR and SoxRS only control a subset of the ~80 H_2O_2- and $O_2^{\bullet-}$-inducible genes identified on two-

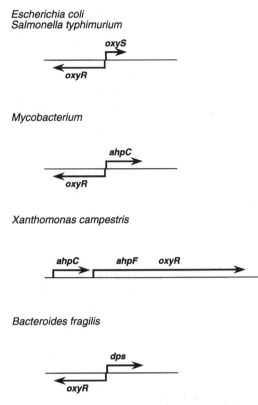

Figure 3. Gene organization around *oxyR* genes in *E. coli*, serovar Typhimurium, *Mycobacterium*, *Xanthomonas*, and *B. fragilis* (2, 23, 79; Rocha and Smith, Abstr. 99th Gen. Meet. Am. Soc. Microbiol. 1999).

dimensional gels. Two additional regulators that recently have been characterized are PerR and σ^R.

Bacillus subtilis PerR

Studies of the gram-positive bacterium *B. subtilis* showed that this organism possesses a regulatory response to peroxide stress (9, 13). H_2O_2 treatment as well as metal ion limitation led to increased expression of catalase (*katA*), alkyl hydroperoxide reductase (*ahpCF*), a Dps homolog (*mrgA*), and the heme biosynthesis operon (*hemAXCDBL*). Interestingly, the *katA*, *ahpC*, *mrgA*, and *hemA* promoters all contain a palindromic sequence with homology to a Fur box. Genetic studies suggested that these genes are coordinately regulated by a repressor denoted PerR. The identification of the *perR* gene came as a result of the completed *B. subtilis* genome, which revealed the presence of three distinct Fur homologs (*ygkL*, *yqfV*, and *ygaG*). The phenotype of mutants lacking *ygaG*, derepression of the peroxide regulon and resistance to H_2O_2, was identical to the phenotype of the previously identified *perR* mutants, demonstrating that *ygaG* is allelic to *perR* (9).

Like other Fur family members, the 21-kDa PerR protein has a helix-turn-helix DNA binding motif in its N-terminal region and contains two CXXC motifs in the C-terminal putative metal-binding domain. In vitro characterization of the PerR protein has not yet been reported; however, the mechanism of peroxide sensing by PerR is likely to be different from the mechanism of peroxide sensing by OxyR (9). On the basis of sequence homology and the effect of *perR* mutations, it has been speculated that PerR is a metalloprotein. The bound metal ion may play a role in peroxide sensing by contributing to the metal-catalyzed oxidation of PerR peptide. Alternatively, PerR activity may be modulated by the oxidation state of the bound metal ion.

The identification of *perR* gene in *B. subtilis* raises the possibility that other Fur-like proteins may play a role in oxidative stress responses (see also chapter 19). Intriguingly, *fur*-like genes are closely linked to antioxidant genes in *Mycobacterium tuberculosis*, *Streptomyces venezuelae*, and *Desulfovibrio vulgaris*. In addition, two *fur*-like genes have been identified in the genome of *Campylobacter jejuni*, and the disruption of one of these genes led to increased expression of the catalase (*katA*) and AhpC proteins as well as increased resistance to H_2O_2 (106).

Streptomyces coelicolor A3(2) σ^R

Another interesting regulatory system, composed of a sigma factor and its anti-sigma factor, was recently discovered in the gram-positive bacterium *Streptomyces coelicolor* A3(2) (57, 88). The first component of the regulatory system, an alternative sigma factor denoted σ^R (*sigR*), was initially purified as a component of stationary phase RNA polymerase (56). Deletion of the *sigR* gene caused sensitivity to the thiol-specific oxidant diamide and the $O_2^{\bullet-}$-generating compounds menadione and plumbagin, but not to H_2O_2 (88). This correlated with the inability to induce the operon encoding thioredoxin reductase and thioredoxin (*trxBA*) upon exposure to diamide. The second component of the regulatory system, RsrA, is a specific anti-sigma factor encoded immediately downstream of *sigR*. A *rsrA* null mutant substantially overexpresses the thioredoxin system, demonstrating that RsrA negatively regulates σ^R.

Biochemical studies showed that RsrA binds σ^R under reduced but not oxidized conditions, and binding prevents σ^R from activating *trxBA* transcription (57). The oxidized RsrA protein contains one or more intramolecular disulfide bonds, which can be reduced by the thioredoxin system. On the basis of these observations, Kang et al. proposed the following model (57). Under unstressed conditions, RsrA exists in its reduced form and binds σ^R. Exposure to oxidative stress leads to the formation of one or more intramolecular disulfide bonds in RsrA. This causes RsrA to lose affinity for σ^R, thereby allowing σ^R to bind to core RNA polymerase and induce *trxBA* transcription. Expression of the thioredoxin system in turn leads to reduction of RsrA. Reduced RsrA rebinds σ^R, thereby shutting off σ^R-dependent transcription and completing the regulatory loop. The posttranslational regulation of σ^R ensures a rapid and effective response to oxidative stress. In addition, since σ^R positively autoregulates its own expression via one of the two promoters, oxidative stress not only activates σ^R posttranslationally, but also induces σ^R synthesis.

The mechanism of RsrA action through reversible disulfide bond formation is reminiscent of OxyR. However, some important differences between the two regulatory systems can be noted. First, OxyR is a positive regulator, and the oxidized form of the protein is active. In contrast, RsrA is a negative regulator and the reduced form of the protein is active. Second, a single disulfide switch is sufficient to regulate OxyR activity, whereas more than one disulfide bond appears to be involved in RsrA action. The physiological role of the multiple disulfide bonds in RsrA is not clear, although one possibility is that they may be involved in sensing multiple levels of redox change in the cell. Finally, *E. coli* OxyR responds and confers resistance to disulfide and H_2O_2 stress while the *S. coelicolor* RsrA/σ^R regulon confers resistance to disulfide and $O_2^{\bullet-}$ stress. The presence of RsrA and σ^R homologs in other bacteria such as *M. tuberculosis* suggests that the same type of regulation exists in other organisms.

OVERLAP WITH THE *E. COLI* RESPONSE TO REACTIVE NITROGEN SPECIES

Given the reactions between reactive oxygen species and reactive nitrogen species, the damage caused by the two types of chemicals is intertwined. In addition, both $O_2^{\bullet-}$ and nitric oxide (NO•) are generated by macrophages as a defense against bacteria. Thus, it is worth considering the overlap between the responses to oxidative stress and nitrosative stress.

Hausladen et al. reported that the *E. coli oxyR* regulon is induced by and confers resistance to S-nitrosothiols (SNO) (47). It is not clear which of the OxyR-regulated activities is most important in conferring SNO resistance. However, Chen et al. recently demonstrated that in *Salmonella* serovar Ty-

phimurium, AhpC has antinitrosative activity that is independent of the AhpF subunit that is required for the antioxidative activity of AhpC (14). Some of the protective effect conferred by the *oxyR* regulon may also be due to its ability to restore thiol/disulfide homeostasis. The extent of OxyR activation by SNO as well as the mechanism of activation has not been resolved. Hausladen et al. claim that OxyR is fully activated by SNO modification of the crucial C199 residue (47). Zheng et al. observed significantly less activation by SNO than by H_2O_2 (115). Since S-nitrosothiols promote disulfide bond formation (94), OxyR activation by SNO may be due to a lowering of the thiol/disulfide ratio in the cell. Consistent with this proposal, SNO activation of OxyR is more readily observed in *gshA* mutant strains lacking GSH (47, 115). Additional experiments need to be carried out to examine OxyR reactivity with reactive nitrogen species.

The *soxRS* regulon can be induced by dissolved NO• gas (87), and an *E. coli* strain carrying a deletion of the *soxRS* locus is hypersusceptible to NO•-dependent killing by murine macrophages (86). Among the *soxRS*-regulated activities that might play a role in protecting against NO• are glucose-6-phosphate dehydrogenase, manganese superoxide dismutase, and endonuclease IV based on the macrophage sensitivity of *zwf*, *sodA*, and *nfo* mutant *E. coli* strains. Glucose-6-phosphate dehydrogenase has also recently been shown to be required for serovar Typhimurium resistance to H_2O_2 and GSNO (68). In vivo and in vitro EPR spectroscopy studies suggest that the mechanism of SoxR activation by NO• is the formation of dinitrosyl-iron-dithiol in which each sulfur atom in the [2Fe-2S] cluster is replaced by two NO• molecules (H. Ding and B. Demple, submitted for publication).

It is possible that the *oxyR* and *soxRS* regulons only play a secondary role in the defense against reactive nitrogen species. So far, the only bona fide NO• detoxifying enzyme to be identified is flavohemoglobin (*hmp*), which catalyzes the oxidation of NO• to nitrate (19, 34, 46, 74). The expression of *hmp* is induced by nitrosothiols and paraquat, but neither OxyR nor SoxR plays a role in this activation (76). Instead, recent studies showed that NO• induction of *hmp* occurs via the MetR transcription factor (75).

ROLES OF THE *oxyR* AND *soxRS* REGULONS IN PATHOGENESIS

The fact that a major portion of bacterial killing by macrophages is due to the respiratory burst raises the question whether the *oxyR* and *soxRS* regulons play a role in virulence. Nunoshiba et al. showed that *E. coli soxRS* mutants are hypersusceptible to killing by murine macrophages (86, 87). In addition, several of the genes that are part of the *oxyR* regulon in serovar Typhimurium have been found to be induced by the macrophage environment, including *ahpCF* and *dps* (32, 104).

However, *Salmonella* serovar Typhimurium *oxyR* and *soxS* mutants do not show increased sensitivity to macrophage killing (29, 89). The mutants also do not show attenuated virulence in BALB/c mice (29, 101). Thus, the *oxyR* and *soxRS* regulons of serovar Typhimurium do not appear to play a critical role in the virulence, and by extension, in protecting against the deleterious effects of the respiratory burst. Possibly, other regulons protect against the respiratory burst. In support of this conclusion, serovar Typhimurium strains carrying mutations in *slyA* (a transcriptional regulator), *rpoS*, and *recA* (a positive regulator of the SOS DNA repair pathway) show increased sensitivity to reactive oxygen species as well as attenuated virulence in mice (see chapter 11 and references 10, 11). It is also conceivable that the *oxyR* and *soxRS* regulons are important for virulence in organisms other than serovar Typhimurium.

PERSPECTIVES

Much has been learned about the responses to oxidative stress in *E. coli*, particularly with the identification and characterization of the SoxR, SoxS, and OxyR regulators. Further studies of the reactivity of SoxR and OxyR to $O_2^{\bullet-}$, H_2O_2, and reactive nitrogen species should give general insights into the chemistry of redox sensing. Another important direction for future studies will be to map the connectivity of the *oxyR* and *soxRS* regulons with other regulatory networks. It will also be interesting to compare the roles of SoxRS and OxyR in *E. coli* with the roles in other bacterial species. Finally, it is likely that several additional sensors of reactive oxygen species will be identified, and studies of these regulators will undoubtedly reveal unique mechanisms of redox regulation.

REFERENCES

1. **Altuvia, S., M. Almirón, G. Huisman, R. Kolter, and G. Storz.** 1994. The *dps* promoter is activated by OxyR during growth and by IHF and σ^s in stationary phase. *Mol. Microbiol.* 13: 265–272.

2. **Altuvia, S., D. Weinstein-Fischer, A. Zhang, L. Postow, and G. Storz.** 1997. A small, stable RNA induced by oxidative stress: role as a pleiotropic regulator and antimutator. *Cell* 90: 43–53.

3. **Altuvia, S., A. Zhang, L. Argaman, A. Tiwari, and G. Storz.** 1998. The *Escherichia coli* OxyS regulatory RNA represses *fhlA* translation by blocking ribosome binding. *EMBO J.* 17: 6069–6075.

4. **Amábile-Cuevas, C. F., and B. Demple.** 1991. Molecular characterization of the *soxRS* genes of *Escherichia coli*: two genes control a superoxide stress regulon. *Nucleic Acids Res.* 19: 4479–4484.

5. **Aono, R., N. Tsukagoshi, and M. Yamamoto.** 1998. Involvement of outer membrane protein TolC, a possible member of the *mar-sox* regulon, in maintenance and improvement of organic solvent tolerance of *Escherichia coli* K-12. *J. Bacteriol.* 180:938–944.

6. **Åslund, F., M. Zheng, J. Beckwith, and G. Storz.** 1999. Regulation of the OxyR transcription factor by hydrogen peroxide and the cellular thiol-disulfide status. *Proc. Natl. Acad. Sci. USA* 96:6161–6165.

7. **Bölker, M., and R. Kahmann.** 1989. The *Escherichia coli* regulatory protein OxyR discriminates between methylated and unmethylated states of the phage Mu *mom* promoter. *EMBO J.* 8:2403–2410.

8. **Bradley, T. M., E. Hidalgo, V. Leautaud, H. Ding, and B. Demple.** 1997. Cysteine-to-alanine replacements in the *Escherichia coli* SoxR protein and the role of the [2Fe-2S] centers in transcriptional activation. *Nucleic Acids Res.* 25:1469–1475.

9. **Bsat N., A. Herbig, L. Casillas-Martinez, P. Setlow, and J. D. Helmann.** 1998. *Bacillus subtilis* contains multiple Fur homologues: identification of the iron uptake (Fur) and peroxide regulon (PerR) repressors. *Mol. Microbiol.* 29:189–198.

10. **Buchmeier, N., S. Bossie, C.-Y. Chen, F. C. Fang, D. G. Guiney, and S. J. Libby.** 1997. SlyA, a transcriptional regulator of *Salmonella typhimurium*, is required for resistance to oxidative stress and is expressed in the intracellular environment of macrophages. *Infect. Immun.* 65:3725–3730.

11. **Buchmeier, N. A., C. J. Lipps, M. Y. H. So, and F. Heffron.** 1993. Recombination-deficient mutants of *Salmonella typhimurium* are avirulent and sensitive to the oxidative burst of macrophages. *Mol. Microbiol.* 7:933–936.

12. **Calcutt, M. J., M. S. Lewis, and A. Eisenstark.** 1998. The *oxyR* gene from *Erwinia carotovora*: cloning, sequence analysis and expression in *Escherichia coli*. *FEMS Microbiol. Lett.* 167:295–301.

13. **Chen, L., L. Keramati, and J. D. Helmann.** 1995. Coordinate regulation of *Bacillus subtilis* peroxide stress genes by hydrogen peroxide and metal ions. *Proc. Natl. Acad. Sci. USA* 92: 8190–8194.

14. **Chen, L., Q.-W. Xie, and C. Nathan.** 1998. Alkyl hydroperoxide reductase subunit C (AhpC) protects bacterial and human cells against reactive nitrogen intermediates. *Mol. Cell* 1: 795–805.

15. **Chou, J. H., J. T. Greenberg, and B. Demple.** 1993. Posttranscriptional repression of *Escherichia coli* OmpF protein in response to redox stress: positive control of the *micF* antisense RNA by the *soxRS* locus. *J. Bacteriol.* 175:1026–1031.

16. **Christman, M. F., R. W. Morgan, F. S. Jacobson, and B. N. Ames.** 1985. Positive control of a regulon for defenses against oxidative stress and some heat-shock proteins in *Salmonella typhimurium*. *Cell* 41:753–762.

17. **Christman, M. F., G. Storz, and B. N. Ames.** 1989. OxyR, a positive regulator of hydrogen peroxide-inducible genes in *Escherichia coli* and *Salmonella typhimurium*, is homologous to a family of bacterial regulatory proteins. *Proc. Natl. Acad. Sci. USA* 86:3484–3488.

18. Compan, I., and D. Touati. 1993. Interaction of six global transcription regulators in expression of manganese superoxide dismutase in *Escherichia coli* K-12. *J. Bacteriol.* **175**:1687–1696.

19. Crawford, M. J., and D. E. Goldberg. 1998. Role for the *Salmonella* flavohemoglobin in protection from nitric oxide. *J. Biol. Chem.* **273**:12543–12547.

20. Cunningham, L., M. J. Gruer, and J. R. Guest. 1997. Transcriptional regulation of the aconitase genes (*acnA* and *acnB*) of *Escherichia coli*. *Microbiology* **143**:3795–3805.

21. Demple, B., and J. Halbrook. 1983. Inducible repair of oxidative DNA damage in *Escherichia coli*. *Nature* **304**:466–468.

22. Demple, B., and L. Harrison. 1994. Repair of oxidative damage to DNA: enzymology and biology. *Annu. Rev. Biochem.* **63**:915–948.

23. Deretic, V., J. Song, and E. Pagán-Ramos. 1997. Loss of *oxyR* in *Mycobacterium tuberculosis*. *Trends Microbiol.* **5**:367–372.

24. Ding, H., and B. Demple. 1996. Glutathione-mediated destabilization *in vitro* of [2Fe-2S] centers in the SoxR regulatory protein. *Proc. Natl. Acad. Sci. USA* **93**:9449–9453.

25. Ding, H., and B. Demple. 1997. In vivo kinetics of a redox-regulated transcriptional switch. *Proc. Natl. Acad. Sci. USA* **94**:8445–8449.

26. Ding, H., and B. Demple. 1998. Thiol-mediated disassembly and reassembly of [2Fe-2S] clusters in the redox-regulated transcription factor SoxR. *Biochemistry* **37**:17280–17286.

27. Ding, H., E. Hidalgo, and B. Demple. 1996. The redox state of the [2Fe-2S] clusters in SoxR protein regulates its activity as a transcription factor. *J. Biol. Chem.* **271**:33173–33175.

28. Dukan, S., and D. Touati. 1996. Hypochlorous acid stress in *Escherichia coli*: resistance, DNA damage, and comparison with hydrogen peroxide stress. *J. Bacteriol.* **178**:6145–6150.

29. Fang, F. C., A. Vazquez-Torres, and Y. Xu. 1997. The transcriptional regulator SoxS is required for resistance of *Salmonella typhimurium* to paraquat but not for virulence in mice. *Infect. Immun.* **65**:5371–5375.

30. Farr, S. B., D. O. Natvig, and T. Kogoma. 1985. Toxicity and mutagenicity of plumbagin and the induction of a possible new DNA repair pathway in *Escherichia coli*. *J. Bacteriol.* **164**:1309–1316.

31. Ferrante, A. A., J. Augliera, K. Lewis, and A. M. Klibanov. 1995. Cloning of an organic solvent-resistance gene in *Escherichia coli*: the unexpected role of alkylhydroperoxide reductase. *Proc. Natl. Acad. Sci. USA* **92**:7617–7621.

32. Francis, K. P., P. D. Taylor, C. J. Inchley, and M. P. Gallagher. 1997. Identification of the *ahp* operon of *Salmonella typhimurium* as a macrophage-induced locus. *J. Bacteriol.* **179**:4046–4048.

33. Fridovich, I. 1995. Superoxide radical and superoxide dismutases. *Annu. Rev. Biochem.* **64**:97–112.

34. Gardner, P. R., A. M. Gardner, L. A. Martin, and A. L. Salzman. 1998. Nitric oxide dioxygenase: an enzymic function for flavohemoglobin. *Proc. Natl. Acad. Sci. USA* **95**:10378–10383.

35. Gaudu, P., N. Moon, and B. Weiss. 1997. Regulation of the *soxRS* oxidative stress regulon. Reversible oxidation of the Fe-S centers of SoxR *in vivo*. *J. Biol. Chem.* **272**:5082–5086.

36. Gaudu, P., and B. Weiss. 1995. SoxR, a [2Fe-2S] transcription factor, is active only in its oxidized form. *Proc. Natl. Acad. Sci. USA* **93**:10094–10098.

37. González-Flecha, B., and B. Demple. 1995. Metabolic sources of hydrogen peroxide in aerobically growing *Escherichia coli*. *J. Biol. Chem.* **270**:13681–13687.

38. González-Flecha, B., and B. Demple. 1997. Homeostatic regulation of intracellular hydrogen peroxide concentration in aerobically growing *Escherichia coli*. *J. Bacteriol.* **179**:382–388.

39. Gort, A. S., D. M. Ferber, and J. A. Imlay. 1999. The regulation and role of the periplasmic copper, zinc superoxide dismutase of *Escherichia coli*. *Mol. Microbiol.* **32**:179–191.

40. Gort, A. S., and J. A. Imlay. 1998. Balance between endogenous superoxide stress and antioxidant defenses. *J. Bacteriol.* **180**:1402–1410.

41. Grant, R. A., D. J. Filman, S. E. Finkel, R. Kolter, and J. M. Hogle. 1998. The crystal structure of Dps, a ferritin homolog that binds and protects DNA. *Nat. Struct. Biol.* **5**:294–303.

42. Greenberg, J. T., and B. Demple. 1989. A global response induced in *Escherichia* coli by redox-cycling agents overlaps with that induced by peroxide stress. *J. Bacteriol.* **171**:3933–3939.

43. Greenberg, J. T., P. Monach, J. H. Chou, P. D. Josephy, and B. Demple. 1990. Positive control of a global antioxidant defense regulon activated by superoxide-generating agents in *Escherichia coli*. *Proc. Natl. Acad. USA* **87**:6181–6185.

44. Gruer, M. J., and J. R. Guest. 1994. Two genetically-distinct and differentially-regulated aconitases (AcnA and AcnB) in *Escherichia coli*. *Microbiology* **140**:2531–2541.

45. Hattman, S., and W. Sun. 1997. *Escherichia coli* OxyR modulation of bacteriophage Mu *mom* expression in *dam*+ cells can be attributed to its ability to bind hemimethylated P$_{mom}$ promoter DNA. *Nucleic Acids Res.* **25**:4385–4388.

46. Hausladen, A., A. J. Gow, and J. S. Stamler. 1998. Nitrosative stress: metabolic pathway involving the flavohemoglobin. *Proc. Natl. Acad. Sci. USA* **95**:14100–14105.

47. Hausladen, A., C. T. Privalle, T. Keng, J. DeAngelo, and J. S. Stamler. 1996. Nitrosative stress: activation of the transcription factor OxyR. *Cell* **86**:719–729.

48. Henderson, I. R., and P. Owen. 1999. The major phase-variable outer membrane protein of *Escherichia coli* structurally resembles the immunoglobulin A1 protease class of exported protein and is regulated by a novel mechanism involving Dam and OxyR. *J. Bacteriol.* **181**:2132–2141.

49. Hidalgo, E., J. M. Bollinger, Jr., T. M. Bradley, C. T. Walsh, and B. Demple. 1995. Binuclear [2Fe-2S] clusters in the *Escherichia coli* SoxR protein and role of the metal centers in transcription. *J. Biol. Chem.* **270**:20908–20914.

50. Hidalgo, E., and B. Demple. 1994. An iron-sulfur center essential for transcriptional activation by the redox-sensing SoxR protein. *EMBO J.* **13**:138–146.

51. Hidalgo, E., and B. Demple. 1996. Activation of SoxR-dependent transcription *in vitro* by noncatalytic or NifS-mediated assembly of [2Fe-2S] clusters into Apo-SoxR. *J. Biol. Chem.* **271**:7269–7272.

52. Hidalgo, E., H. Ding, and B. Demple. 1997. Redox signal transduction: mutations shifting [2Fe-2S] centers of the SoxR sensor-regulator to the oxidized form. *Cell* **88**:121–129.

53. Imlay, J. A., and I. Fridovich. 1991. Assay of metabolic superoxide production in *Escherichia coli*. *J. Biol. Chem.* **266**:6957–6965.

54. Imlay, J. A., and S. Linn. 1988. DNA damage and oxygen radical toxicity. *Science* **240**:1302–1309.

55. Jakob, U., W. Muse, M. Eser, and J. C. A. Bardwell. 1999. Chaperone activity with a redox switch. *Cell* **96**:341–352.

56. Kang, J.-G. , M.-Y. Hahn, A. Ishihama, and J.-H. Roe. 1997. Identification of sigma factors for growth phase-related promoter selectivity of RNA polymerases from *Streptomyces coelicolor* A3(2). *Nucleic Acids Res.* **25**:2566–2573.

57. Kang, J.-G. , M. S. B. Paget, Y.-J. Seok, M.-Y. Hahn, J.-B. Bae, J.-S. Hahn, C. Kleanthous, M. J. Buttner, and J.-H. Roe. 1999. RsrA, an anti-sigma factor regulated by redox change. *EMBO J.* **18**:4292–4298.

58. Kobayashi, K., and S. Tagawa. 1999. Isolation of reductase for SoxR that governs an oxidative response regulon from *Escherichia coli*. *FEBS Lett.* **451**:227–230.

59. Koh, Y.-S., J. Choih, J.-H. Lee, and J.-H. Roe. 1996. Regulation of the *ribA* gene encoding GTP cyclohydrolase II by the *soxRS* locus in *Escherichia coli*. *Mol. Gen. Genet.* **251**:591–598.

60. Koh, Y.-S., and J.-H. Roe. 1995. Isolation of a novel paraquat-inducible (*pqi*) gene regulated by the *soxRS* locus in *Escherichia coli*. *J. Bacteriol.* **177**:2673–2678.

61. Lee, H.-S., Y.-S. Lee, H.-S. Kim, J.-Y. Choi, H. M. Hassan, and M.-H. Chung. 1998. Mechanism of regulation of 8-hydroxyguanine endonuclease by oxidative stress: roles of FNR, ArcA, and Fur. *Free Radic. Biol. Med.* **24**:1193–1201.

62. Liochev, S. I., L. Benov, D. Touati, and I. Fridovich. 1999. Induction of the *soxRS* regulon of *Escherichia coli* by superoxide. *J. Biol. Chem.* **274**:9479–9481.

63. Liochev, S. I., and I. Fridovich. 1992. Fumarase C, the stable fumarase of *Escherichia coli*, is controlled by the *soxRS* regulon. *Proc. Natl. Acad. Sci. USA* **89**:5892–5896.

64. Liochev, S. I., A. Hausladen, W. F. Beyer, Jr., and I. Fridovich. 1994. NADPH:ferredoxin oxidoreductase acts as a paraquat diaphorase and is a member of the *soxRS* regulon. *Proc. Natl. Acad. Sci. USA* **91**:1328–1331.

65. Liochev, S. I., A. Hausladen, and I. Fridovich. 1999. Nitroreductase A is regulated as a member of the *soxRS* regulon of *Escherichia coli*. *Proc. Natl. Acad. Sci. USA* **96**:3537–3539.

66. Loewen, P. C., B. Hu, J. Strutinsky, and R. Sparling. 1998. Regulation in the *rpoS* regulon of *Escherichia coli*. *Can. J. Microbiol.* **44**:707–717.

67. Loprasert, S., S. Atichartpongkun, W. Whangsuk, and S. Mongkolsuk. 1997. Isolation and analysis of the *Xanthomonas* alkyl hydroperoxide reductase gene and the peroxide sensor regulator genes *ahpC* and *ahpF-oxyR-orfX*. *J. Bacteriol.* **179**:3944–3949.

68. Lundberg, B. E., R. W. Wolf, Jr., M. C. Dinauer, Y. Xu, and F. C. Fang. 1999. Glucose 6-phosphate dehydrogenase is required for *Salmonella typhimurium* virulence and resistance to reactive oxygen and nitrogen intermediates. *Infect. Immun.* **67**:436–438.

69. Ma, D., M. Alberti, C. Lynch, H. Nikaido, and J. E. Hearst. 1996. The local repressor AcrR plays a modulating role in the regulation of *acrAB* genes of *Escherichia coli* by global stress signals. *Mol. Microbiol.* **19**:101–112.

70. Maciver, I., and E. J. Hansen. 1996. Lack of expression of the global regulator OxyR in *Haemophilus influenzae* has a profound effect on growth phenotype. *Infect. Immun.* **64**:4618–4629.

71. Martin, R. G., W. K. Gillette, S. Rhee, and J. L. Rosner. 1999. Structural requirements for marbox function in transcriptional activation of *mar/sox/rob* regulon promoters in *Escherichia coli*: sequence, orientation and spatial relationship to the core promoter. *Mol. Microbiol.* **34**:431–441.

72. Martin, R. G., W. K. Gillette, and J. L. Rosner. Promoter discrimination by the related transcriptional activators, MarA and SoxS: differential regulation by differential binding. *Mol. Microbiol.*, in press.

73. Martinez, A., and R. Kolter. 1997. Protection of DNA during oxidative stress by the nonspecific DNA-binding protein Dps. *J. Bacteriol.* **179**:5188–5194.

74. Membrillo-Hernández, J., M. D. Coopamah, M. F. Anjum, T. M. Stevanin, A. Kelly, M. N. Hughes, and R. K. Poole. 1999. The flavohemoglobin of *Escherichia coli* confers resistance to a nitrosalating agent, a "nitric oxide releaser," and paraquat and is essential for transcriptional responses to oxidative stress. *J. Biol. Chem.* **274**:748–754.

75. Membrillo-Hernández, J., M. D. Coopamah, A. Channa, M. N. Hughes, and R. K. Poole. 1998. A novel mechanism for upregulation of the *Escherichia coli* K-12 *hmp* (flavohae-moglobin) gene by the 'NO releaser,' S-nitrosoglutathione: nitrosation of homocysteine and modulation of MetR binding to the *glyA-hmp* intergenic region. *Mol. Microbiol.* **29**:1101–1112.

76. Membrillo-Hernández, J., S. O. Kim, G. M. Cook, and R. K. Poole. 1997. Paraquat regulation of *hmp* (flavohemoglobin) gene expression in *Escherichia coli* K-12 is SoxRS independent but modulated by σ^s. *J. Bacteriol.* **179**:3164–3170.

77. Michán, C., M. Manchado, G. Dorado, and C. Pueyo. 1999. In vivo transcription of the *Escherichia coli oxyR* regulon as a function of growth phase and in response to oxidative stress. *J. Bacteriol.* **181**:2759–2764.

78. Miller, P. F., and M. C. Sulavik. 1996. Overlaps and parallels in the regulation of intrinsic multiple-antibiotic resistance in *Escherichia coli*. *Mol. Microbiol.* **21**:441–448.

79. Mongkolsuk, S., S. Loprasert, W. Whangsuk, M. Fuangthong, and S. Atichartpongkun. 1997. Characterization of transcription organization and analysis of unique expression patterns of an alkyl hydroperoxide reductase C gene (*ahpC*) and the peroxide regulator operon *ahpF-oxyR-orfX* from *Xanthomonas campestris* pv. phaseoli. *J. Bacteriol.* **179**:3950–3955.

80. Morgan, R. W., M. F. Christman, F. S. Jacobson, G. Storz, and B. N. Ames. 1986. Hydrogen peroxide-inducible proteins in *Salmonella typhimurium* overlap with heat shock and other stress proteins. *Proc. Natl. Acad. Sci. USA* **83**:8059–8063.

81. Moskovitz, J., M. A. Rahman, J. Strassman, S. O. Yancey, S. R. Kushner, N. Brot, and H. Weissbach. 1995. *Escherichia coli* peptide methionine sulfoxide reductase gene: regulation of expression and role in protecting against oxidative damage. *J. Bacteriol.* **177**:502–507.

82. Mukhopadhyay, S., and H. E. Schellhorn. 1997. Identification and characterization of hydrogen peroxide-sensitive mutants of *Escherichia coli*: genes that require OxyR for expression. *J. Bacteriol.* **179**:330–338.

83. Müller, K., B. F. Matzanke, V. Schünemann, A. X. Trautwein, and K. Hantke. 1998. FhuF, an iron-regulated protein of *Escherichia coli* with a new type of [2Fe-2S] center. *Eur. J. Biochem.* **258**:1001–1008.

84. Nakajima, H., M. Kobayashi, T. Negishi, and R. Aono. 1995. *soxRS* gene increased the level of organic solvent tolerance in *Escherichia coli*. *Biosci. Biotech. Biochem.* **59**:1323–1325.

85. Nunoshiba, T., E. Hidalgo, C. F. Amábile-Cuevas, and B. Demple. 1992. Two-stage control of an oxidative stress regulon: the *Escherichia coli* SoxR protein triggers redox-inducible expression of the *soxS* regulatory gene. *J. Bacteriol.* **174**:6054–6060.

86. Nunoshiba, T., T. DeRojas-Walker, S. R. Tannenbaum, and B. Demple. 1995. Roles of nitric oxide in inducible resistance of *Escherichia coli* to activated murine macrophages. *Infect. Immun.* **63**:794–798.

87. Nunoshiba, T., T. DeRojas-Walker, J. S. Wishnok, S. R. Tannenbaum, and B. Demple. 1993. Activation by nitric oxide of an oxidative-stress response that defends *Escherichia coli* against activated macrophages. *Proc. Natl. Acad. Sci. USA.* **90**:9993–9997.

88. Paget, M. S. B., J.-G. Kang, J.-H. Roe, and M. J. Buttner. 1998. σ^R, an RNA polymerase sigma factor that modulates expression of the thioredoxin system in response to oxidative stress in *Streptomyces coelicolor* A3(2). *EMBO J.* **17**:5776–5782.

89. Papp-Szabò, E., M. Firtel, and P. D. Josephy. 1994. Comparison of the sensitivities of *Salmonella typhimurium oxyR* and *katG* mutants to killing by human neutrophils. *Infect. Immun.* **62**:2662–2668.

90. Park, S.-J., and R. P. Gunsalus. 1995. Oxygen, iron, carbon, and superoxide control of the fumarase *fumA* and *fumC* genes

of *Escherichia coli*: role of the *arcA, fnr,* and *soxR* gene products. *J. Bacteriol.* 177:6255–6262.

90a.Ritz, D., H. Patel, B. Doan, M. Zheng, F. Åslund, G. Storz, and J. Beckwith. Thioredoxin 2 is involved in the oxidative stress response in *Escherichia coli. J. Biol. Chem.,* in press.

91. Rocha, E. R., and C. J. Smith. 1998. Characterization of a peroxide-resistant mutant of the anaerobic bacterium *Bacteroides fragilis. J. Bacteriol.* 180:5906–5912.

92. Rosner, J. L., and J. L. Slonczewski. 1994. Dual regulation of *inaA* by the multiple antibiotic resistance (Mar) and superoxide (SoxRS) stress response systems of *Escherichia coli. J. Bacteriol.* 176:6262–6269.

93. Sies, H. (ed.). 1985. *Oxidative Stress.* Academic Press, London.

94. Singh, S. P., J. S. Wishnok, M. Keshive, W. M. Deen, and S. R. Tannenbaum. 1996. The chemistry of the *S*-nitrosoglutathione/glutathione system. *Proc. Natl. Acad. Sci. USA* 93:14428–14433.

95. Skarstad, K., B. Thöny, D. S. Hwang, and A. Kornberg. 1993. A novel binding protein of the origin of the *Escherichia coli* chromosome. *J. Biol. Chem.* 268:5365–5370.

96. Storz, G., L. A. Tartaglia, and B. N. Ames. 1990. Transcriptional regulator of oxidative stress-inducible genes: direct activation by oxidation. *Science* 248:189–194.

97. Sun, W., and S. Hattman. 1996. *Escherichia coli* OxyR protein represses the unmethylated bacteriophage Mu *mom* operon without blocking binding of the transcriptional activator C. *Nucleic Acids Res.* 24:4042–4049.

98. Tao, K. 1997. *oxyR*-dependent induction of *Escherichia coli* grx gene expression by peroxide stress. *J. Bacteriol.* 179:5967–5970.

98a. Tao, K. 1999. In vivo oxidation-reduction kinetics of OxyR, the transcriptional activator for an oxidative-stress regulon in *Escherichia coli. FEBS Lett.* 457:90–92.

99. Tao, K., N. Fujita, and A. Ishihama. 1993. Involvement of the RNA polymerase α subunit C-terminal region in cooperative interaction and transcriptional activation with OxyR protein. *Mol. Microbiol.* 7:859–864.

100. Tao, K., C. Zou, N. Fujita, and A. Ishihama. 1995. Mapping of the OxyR protein contact site in the C-terminal region of RNA polymerase α subunit. *J. Bacteriol.* 177:6740–6744.

101. Taylor, P. D., C. J. Inchley, and M. P. Gallagher. 1998. The *Salmonella typhimurium* AhpC polypeptide is not essential for virulence in BALB/c mice but is recognized as an antigen during infection. *Infect. Immun.* 66:3208–3217.

102. Toledano, M. B., I. Kullik, F. Trinh, P. T. Baird, T. D. Schneider, and G. Storz. 1994. Redox-dependent shift of OxyR-DNA contacts along an extended DNA-binding site: a mechanism for differential promoter selection. *Cell* 78:897–909.

103. Tsaneva, I. R., and B. Weiss. 1990. *soxR*, a locus governing a superoxide response regulon in *Escherichia coli* K-12. *J. Bacteriol.* 172:4197–4205.

104. Valdivia, R. H., and S. Falkow. 1996. Bacterial genetics by flow cytometry: rapid isolation of *Salmonella typhimurium* acid-inducible promoters by differential fluorescence induction. *Mol. Microbiol.* 22:367–378.

105. VanBogelen, R. A., P. M. Kelley, and F. C. Neidhardt. 1987. Differential induction of heat shock, SOS and oxidation stress regulons and accumulation of nucleotides in *Escherichia coli. J. Bacteriol.* 169:26–32.

106. van Vliet, A. H., M. L. Baillon, C. W. Penn, and J. M. Ketley. 1999. *Campylobacter jejuni* contains two Fur homologs: characterisation of iron-responsive regulation of peroxide stress defense genes by the PerR repressor. *J. Bacteriol.* 181:6371–6376.

107. Walkup, L. K. B., and T. Kogoma. 1989. *Escherichia coli* proteins inducible by oxidative stress mediated by the superoxide radical. *J. Bacteriol.* 171:1476–1484.

108. Wolf, S. G., D. Frenkiel, T. Arad, S. E. Finkel, R. Kolter, and A. Minsky. 1999. DNA protection by stress-induced biocrystallization. *Nature* 400:83–85.

109. Wood, T. I., K. L. Griffith, W. P. Fawcett, K.-W. Jair, T. D. Schneider, and R. E. Wolf, Jr. 1999. Interdependence of the position and orientation of SoxS binding sites in the transcriptional activation of the class I subset of *Escherichia coli* superoxide-inducible promoters. *Mol. Microbiol.* 34:414–430.

110. Wu, J., W. R. Dunham, and B. Weiss. 1995. Overproduction and physical characterization of SoxR, a [2Fe-2S] protein that governs an oxidative stress response regulon in *Escherichia coli. J. Biol. Chem.* 270:10323–10327.

111. Wu, J., and B. Weiss. 1991. Two divergently transcribed genes, *soxR* and *soxS*, control a superoxide response regulon of *Escherichia coli. J. Bacteriol.* 173:2864–2871.

112. Wu, J., and B. Weiss. 1992. Two-stage induction of the *soxRS* (superoxide response) regulon of *Escherichia coli. J. Bacteriol.* 174:3915–3920.

113. Zhang, A., S. Altuvia, A. Tiwari, L. Argaman, R. Hengge-Aronis, and G. Storz. 1998. The OxyS regulatory RNA represses *rpoS* translation and binds the Hfq (HF-I) protein. *EMBO J.* 17:6061–6068.

114. Zheng, L., V. L. Cash, D. H. Flint, and D. R. Dean. 1998. Assembly of iron-sulfur clusters. Identification of an *iscSUA-hscBA-fdx* gene cluster from *Azotobacter vinelandii. J. Biol. Chem.* 273:13264–13272.

115. Zheng, M., F. Åslund, and G. Storz. 1998. Activation of the OxyR transcription factor by reversible disulfide bond formation. *Science* 279:1718–1721.

116. Zheng, M., B. Doan, T. D. Schneider, and G. Storz. 1999. OxyR and SoxRS regulation of *fur. J. Bacteriol.* 181:4639–4643.

Bacterial Stress Responses
Edited by G. Storz and R. Hengge-Aronis
©2000 ASM Press, Washington, D.C.

Chapter 5

Mechanisms for Sensing and Responding to Oxygen Deprivation

THOMAS PATSCHKOWSKI, DONNA M. BATES, AND PATRICIA J. KILEY

Facultative bacteria have developed a variety of regulatory mechanisms to adapt to changes in the availability of oxygen. Several global regulatory proteins and their cofactors have been identified, and despite the metabolic diversity among different organisms, common features of oxygen-sensing proteins are emerging. Thus far, a heme group, a flavin moiety, and an Fe-S cluster have been assigned a function as cofactors in oxygen-sensing proteins. Oxygen concentrations seem to be sensed both directly and indirectly. Some regulators are distinguished by a conserved PAS domain whose specificity in signal sensing seems to vary according to the associated cofactor. This review summarizes the current knowledge about the oxygen-dependent expression of three well-studied metabolic pathways: anaerobic respiration in Escherichia coli, *anoxygenic photosynthesis in purple nonsulfur bacteria, and nitrogen fixation in both the free-living and symbiotic N$_2$-fixing bacteria. It will show that oxygen-dependent regulation of these pathways often consists of sophisticated overlapping regulatory circuits involving different types of oxygen sensory proteins resulting in stringent control of gene expression in response to oxygen availability.*

Given the prominent role of oxygen in the critical function of energy generation as well as in the generation of oxidative stress, it is not surprising that many organisms sense and adapt to changing oxygen levels in the environment. Such adaptive strategies are well illustrated in the lifestyles of many microbes where oxygen tension serves as an important environmental cue to trigger major changes in gene expression. For example, in response to environments poor in oxygen, some prokaryotes induce the synthesis of new energy-generating pathways such as anaerobic respiration (118), fermentation (9), or anoxygenic photosynthesis (128), providing alternative energy-generating mechanisms when O$_2$ is no longer available as a terminal electron acceptor. In addition, oxygen tension also serves to regulate other functions such as those in pathogenesis (77) and oxygen-sensitive assimilatory processes such as nitrogen fixation (30). In such organisms, oxygen availability plays a critical role in controlling the transcription of genes encoding these processes, ensuring that their expression is limited to anaerobic or microaerobic growth conditions. Much progress has been made in recent years in identifying several global regulatory proteins and their cofactors that function in oxygen-sensing pathways in bacteria. These global regulatory pathways will be highlighted in this review to illustrate the kinds of strategies uncovered thus far that allow cells to respond to changes in oxygen availability. The picture is emerging that cofactors known for their roles in oxygen metabolism also have a regulatory role in oxygen sensing (10, 25, 44, 102). In particular, three cofactors have been identified with well-known roles in O$_2$ binding (heme and flavins) or in electron transfer (heme, flavins, and Fe-S clusters). These findings have provided new insights for understanding the different ways that prokaryotes can sense changes in oxygen availability and also suggest that there are both direct and indirect mechanisms of oxygen sensing.

In this review, we have focused on the types of global regulatory pathways that control the expression of three major metabolic pathways in response to oxygen deprivation: anaerobic respiration in *Escherichia coli*, anoxygenic photosynthesis in the purple nonsulfur bacteria, and nitrogen fixation in both the free-living and symbiotic N$_2$-fixing bacteria. In considering how these pathways are regulated, a recurring theme to be presented is that oxygen regulation of these pathways often includes overlapping regulatory networks consisting of different types of oxy-

---

Thomas Patschkowski, Donna M. Bates, and Patricia J. Kiley • Department of Biomolecular Chemistry, University of Wisconsin Medical School, 1300 University Ave., Madison, WI 53706.

gen sensory proteins. While individual species of bacteria have evolved their own particular hierarchy of regulatory strategies, the evolving picture is that conservation of these regulatory strategies exists both in the classes of sensing cofactors and the type of proteins with which they are associated. With this review we do not intend to be all inclusive but rather to highlight the important features of some global regulators involved in oxygen-dependent gene regulation.

ANAEROBIC RESPIRATION IN *E. COLI*

As a facultative anaerobe, *E. coli* has a versatile metabolic lifestyle where it can respire or ferment several different carbon sources. For a comprehensive description of the energy-generating pathways of *E. coli*, the reader is referred to several recent reviews (13, 19, 33, 34, 118). In considering just the process of aerobic respiration, various carbon sources can be oxidized by *E. coli*, resulting in the production of reducing equivalents, such as NADH. The electrons derived from NADH are transferred to quinone by the action of NADH dehydrogenase (I and II), and then passed on to either cytochrome *o* or *d* oxidase, resulting in the four-electron reduction of O_2 to form two molecules of H_2O. Electrons may also be derived from the direct oxidation of compounds such as succinate, lactate, or glycerol through the action of their specific dehydrogenases. During oxidative phosphorylation, the majority of the energy released by the transport of electrons is coupled to the formation of an electrochemical gradient for H^+ across the cytoplasmic membrane that is the driving force for the synthesis of ATP via the F_1F_0-ATPase.

In the absence of O_2, *E. coli* can still respire, replacing O_2 with alternate electron acceptors such as nitrate, nitrite, fumarate, dimethylsulfoxide (DMSO), or trimethylamine N-oxide (TMAO) (reviewed in 118). The use of alternate electron acceptors requires the synthesis of their cognate terminal reductases. In addition, electrons are derived from the oxidation of compounds such as glycerol, formate, NADH, or H_2 under anaerobic conditions, and this requires the synthesis of specific dehydrogenases. Finally, in the absence of any terminal electron acceptor, ATP is generated via substrate-level phosphorylation utilizing fermentative pathways (9).

Upon the switch from aerobic to anaerobic conditions, many changes in gene expression occur in *E. coli* to adapt metabolic functions to changes in oxygen availability. These include the genes encoding the enzymes of the aerobic and anaerobic respiratory chains as well as enzymes of the citric acid cycle (reviewed in 50, 51, 74, 116, 117). *E. coli* possesses two global regulatory systems, ArcAB (arc for aerobic respiration control) and FNR (for fumarate and nitrate reduction), that function either independently or in conjunction with each other in the regulation of metabolic pathways in response to a lowering of O_2 tension (reviewed in 50, 51, 74, 116, 117). Whereas the ArcAB system mainly represses functions associated with aerobic respiration as well as several citric acid cycle enzymes, the transcriptional regulator FNR functions mainly as an activator of anaerobic pathways. Although the activity of ArcAB and FNR is limited to anaerobic conditions, their functions are controlled by distinct mechanisms. An additional hierarchy of regulation exists under anaerobic conditions in the presence of nitrate that is mediated by the two-component regulatory systems NarLX and NarPQ. Several excellent reviews have been published in this area and, therefore, nitrate regulation will not be considered further here (22, 110).

THE ArcAB MODULON

A major mechanism in prokaryotic signal transduction that facilitates an adaptive response to environmental stimuli is carried out by the so-called two-component regulatory systems (87, 88, 111). Typically such a system is composed of a membrane-anchored sensor-kinase and its cytoplasmically located cognate response-regulator. Upon stimulation by an environmental signal, the sensor-kinase undergoes autophosphorylation at a conserved histidine in its transmitter domain at the expense of ATP. The phosphoryl group is subsequently transferred to a conserved aspartate in the receiver domain of the response-regulator whose activity most often results in an altered pattern of gene expression in response to the specific stimulus.

ArcB and ArcA represent a two-component regulatory system that functions mainly as a repressor of aerobic respiratory pathways in *E. coli* in response to O_2 deprivation (reviewed in 56). ArcB is a sensor-kinase protein that resides within the cellular membrane, and ArcA is the DNA-binding response-regulator that functions to regulate transcription of target genes (56). Some of the metabolic genes that are regulated by the ArcAB system in *E. coli* are listed in Table 1. Several of the genes that are repressed by ArcA encode enzymes that function in the citric acid cycle that is operative as a cycle in *E. coli* only under aerobic conditions.

Upon O_2 deprivation, ArcB autophosphorylates at the conserved histidine in the transmitter domain and, via a complex phosphorelay process involving a

Table 1. Examples of metabolic enzymes that are regulated by FNR and/or ArcA[a]

Enzyme	Gene	FNR[b]	ArcA[b]
TCA cycle enzymes			
Aconitase (stationary phase)	acnA	−	−
Aconitase	acnB		−
Fumarase A	fumA	−	−
Fumarase B	fumB	+	+
Fumarase C	fumC		−
Citrate synthase	gltA		−
Isocitrate dehydrogenase	icd		−
Malate dehydrogenase	mdh		−
Succinate dehydrogenase	sdhCDAB	−	−
α-Ketoglutarate dehydrogenase	sucAB	−	−
Succinyl coenzyme A synthetase	sucCD	−	−
Respiratory chain dehydrogenases			
Formate dehydrogenase-N	fdnGHI	+	−
Glycerol-3-phosphate dehydrogenase (anaerobic enzyme)	glpACB	+	
Glycerol-3-phosphate dehydrogenase (aerobic enzyme)	glpD		−
Hydrogenase I	hyaA-F		+
L-Lactate dehydrogenase	ild		−
NADH dehydrogenase I	nuo	−	−
NADH dehydrogenase II	ndh	−	
Cytochrome d oxidase	cyd	−	+
Terminal reductases			
Cytochrome o oxidase	cyoABCDE	−	−
Cytochrome d oxidase	cydAB	−	+
Putative cytochrome oxidase	cyx		+
Dimethyl sulfoxide reductase	dmsABC	+	
Fumarate reductase	frdABCD	+	
Nitrate reductase	narGHJI	+	
Periplasmic nitrate reductase	napF	+	
NADH-dependent nitrite reductase	nirBDC	+	
Formate-linked nitrite reductase	nrfA-G	+	
Other enzymes			
Formate transport and pyruvate-formate lyase	focA-pfl	+	+
Pyruvate dehydrogenase complex and regulator	pdhR-aceEF-lpd	−	−

[a] Information in this table was obtained from reference 74 and updated by the addition of the following genes: acnA and acnB (20), fumA (86), fumB (114), fumC (86), sucAB and sucCD (21, 85), nuo (15), cyx (16), and cyd (18a).

[b] The − symbol refers to transcriptional repression by FNR or ArcA, the + symbol refers to the transcriptional activation by FNR or ArcA. The lack of a symbol indicates that FNR or ArcA has either no effect on transcription of the indicated gene or has not been tested. DNA sequences resembling FNR- and/or ArcA-binding sites have been identified in the promoter region for most of the listed genes. However, the expression of some of the genes might be regulated indirectly since they do not exhibit obvious binding sites.

central receiver domain and a secondary transmitter domain, transphosphorylates ArcA, thereby enabling it to function as a DNA-binding protein (35, 36). On the basis of the alignment of a variety of ArcA target promoters and DNase I protection assays, a consensus ArcA binding site, 5′-[A/T]GTTAATTA[A/T]-3′, has been proposed (75). The mechanism by which ArcB senses O_2 deprivation has yet to be determined, although it is interesting to note that ArcB has been identified as a member of the PAS domain superfamily (the abbreviation PAS stands for the initials of the first proteins to be identified in this family, the Dro-

sophila periodic clock protein, PER; vertebrate aryl hydrocarbon receptor nuclear translocator, ARNT; and Drosophila single-minded protein, SIM). PAS domains are signaling domains that are widely distributed in sensory proteins of prokaryotes and eukaryotes, where they monitor changes in light, redox potential, and oxygen (112). The finding that NADH and certain reduced products of fermentation, such as D-lactate and acetate, increase autophosphorylation of ArcB in vitro (55) suggests that the PAS domain of ArcB might respond to a reduced intermediate that accumulates when electron flow to O_2 is interrupted following a decrease in O_2 tension. Taken together these results suggest that ArcB may be monitoring the electron flux through the aerobic respiratory chain rather than O_2 itself.

THE FNR MODULON

The switch from aerobic to anaerobic growth conditions leads to increased expression of anaerobic respiratory pathways and simultaneously to reduced expression of certain aerobic respiratory functions (reviewed in 50, 74). This altered pattern in gene expression in response to changes in O_2 availability is partly achieved by the regulatory protein FNR. Mutations in the *fnr* gene have a pleiotropic effect (70), resulting in the inability to grow via anaerobic respiration utilizing the majority of the alternate electron acceptors. The genes that are activated by FNR in *E. coli* (Table 1) include those encoding the alternate reductases used during anaerobic respiration such as *narGHJI* (nitrate reductase), *frdABCD* (fumarate reductase), and *dmsABC* (DMSO/TMAO reductase) as well as some that are essential for fermentation such as *pfl* (pyruvate-formate lyase). In addition, a few genes are repressed by FNR, the most notable of which is *ndh* encoding the aerobic NADH dehydrogenase II. Active FNR protein exerts its function by binding to approximately 22 bp DNA sites where the consensus binding site is TTGAT-N_4-ATCAA (109). The position of this FNR binding site in the promoter region of target genes apparently determines whether FNR acts as an activator or repressor. Whereas FNR-activated promoters typically have binding sites centered at position −41.5 relative to the transcription start, the location of FNR sites in promoters repressed by FNR is more variable (50). In addition to the DNA binding function of FNR, regions of the protein have been identified that specifically promote transcription activation and possibly repression (49, 73, 124).

Oxygen Sensing by FNR In Vitro

While FNR is present in cells under both aerobic and anaerobic growth conditions, it functions as a transcription factor only when cells are grown anaerobically. Recent in vitro studies have shown that DNA binding by FNR is in fact O_2 sensitive, thus explaining why function of FNR is limited to anaerobic conditions. The anaerobically purified form of FNR is a dimer containing two $[4Fe-4S]^{2+}$ clusters that are necessary for dimerization and site-specific DNA binding under anaerobic conditions (48, 66, 67, 72). Upon exposure to air, the $[4Fe-4S]^{2+}$ clusters of FNR are converted to $[2Fe-2S]^{2+}$ clusters (Fig. 1), with a concomitant decrease in site-specific DNA binding (60, 66). The decrease in DNA binding following the $[4Fe-4S]^{2+}$ cluster conversion can be attributed to the dissociation of FNR dimers to monomers (72). Thus, it appears that the $[4Fe-4S]^{2+}$ cluster is acting as an O_2 sensor to regulate the DNA-binding activity of FNR through its oligomeric state (Fig. 1).

The $[4Fe-4S]^{2+}$ cluster of FNR has unusual redox properties, which may facilitate its function in O_2 sensing. Unlike the majority of $[4Fe-4S]^{2+}$ clusters that function in reversible electron transfer (e.g., ferredoxins) and are often easily reduced to the +1 oxidation state, reduction of the $[4Fe-4S]^{2+}$ cluster of FNR is difficult to achieve (67). Rather, the $[4Fe-4S]^{2+}$ cluster of FNR is more easily oxidized like the high-potential iron proteins (58, 59). However, unlike the high-potential iron proteins, the $[4Fe-4S]^{2+}$ cluster of FNR cannot be oxidized to a stable +3 state. Rather, it is rapidly converted to a $[2Fe-2S]^{2+}$ cluster. Such $[4Fe-4S]^{2+}$ cluster conversion may not be unique to FNR as recent data indicate that the activity of other Fe-S proteins may be regulated by O_2 in a similar manner (59).

On the basis of its similarity to the well-studied *E. coli* catabolic activator protein (CAP) (106), FNR is thought to consist of an N-terminal "allosteric" domain and a C-terminal DNA-binding domain. Such a conserved domain structure has been important in conceptualizing how the presence of the $[4Fe-4S]^{2+}$ cluster could be regulating dimerization and DNA-binding of FNR. The first 30 amino acids of FNR contain three of the four cysteine residues that are proposed to be involved in ligating the $[4Fe-4S]^{2+}$ cluster (see references 10 and 12 in reference 67). Although other models cannot be excluded, the current view (67) is that the ligation of one $[4Fe-4S]^{2+}$ cluster per subunit causes a long-range conformational change that increases dimerization at a region analogous to the CAP dimerization helices (62). The $[4Fe-4S]^{2+}$ cluster to $[2Fe-2S]^{2+}$ cluster conversion caused by O_2 has been proposed to reverse this conformational change, resulting in monomeric FNR protein (Fig. 1).

The Pathway for Air Inactivation of FNR from Anaerobic Cells

The finding that the $[4Fe-4S]^{2+}$ cluster of FNR is converted in vitro into a relatively stable $[2Fe-2S]^{2+}$ cluster was initially a surprise since apo-FNR has been typically isolated from aerobically grown cells (71). Nevertheless, the $[4Fe-4S]^{2+}$ to $[2Fe-2S]^{2+}$ cluster conversion also occurs in vivo since anaerobic

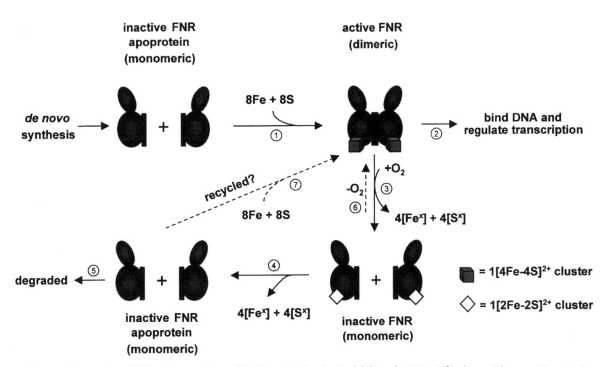

Figure 1. Proposed model for the regulation of FNR activity by the O_2 lability of a $[4Fe-4S]^{2+}$ cluster. The term "inactive" refers to the form of FNR that lacks site-specific DNA-binding activity. "Active" FNR is the form that contains the $[4Fe-4S]^{2+}$ cluster (4Fe-FNR) and is able to bind DNA site-specifically and activate transcription. The active anaerobically purified form of FNR contains two $[4Fe-4S]^{2+}$ clusters per dimer and presumably is formed in a concerted pathway during de novo synthesis (step 1). Ligation of the $[4Fe-4S]^{2+}$ cluster is required for site-specific DNA binding (step 2) of FNR because it increases dimerization. Exposure of anaerobically purified WT FNR to O_2 results in rapid conversion of $[4Fe-4S]^{2+}$ cluster to a $[2Fe-2S]^{2+}$ cluster and a corresponding loss in site-specific DNA binding (step 3). Although in vitro the $[2Fe-2S]^{2+}$ cluster form of FNR (2Fe-FNR) decays more slowly than 4Fe-FNR, little is known about the fate of 2Fe-FNR in vivo. It is likely that some 2Fe-FNR is converted to apo-FNR (step 4) and is possibly further degraded (step 5), but experimental data also suggest that 2Fe-FNR is reconverted to 4Fe-FNR under anaerobic conditions (step 6). Another open question is whether inactivated apo-FNR is recycled to form active 4Fe-FNR (step 7).

cells, exposed to air for 15 min, showed ~50% conversion of 4Fe-FNR to the 2Fe-form as judged by whole cell Mössbauer spectroscopy (97). These data support the notion that inactivation of FNR by exposure of anaerobic cells to air is mediated by $[4Fe-4S]^{2+}$ cluster conversion. Additionally, there is experimental evidence that 2Fe-FNR is reconverted to 4Fe-FNR under anaerobic conditions (97) (Fig. 1).

How FNR Is Kept in an Inactive State under Aerobic Conditions

In considering how FNR activity is regulated under steady-state aerobic conditions, the properties of an O_2 stable FNR mutant (FNR-L28H) have provided some key insights (4a). The L28H substitution has been shown both in vitro and in vivo to stabilize the $[4Fe-4S]^{2+}$ cluster, reducing its destruction in the presence of O_2. Mössbauer spectroscopy of aerobically grown cells overexpressing FNR-L28H were

shown to have as much 4Fe-FNR as those grown under anaerobic conditions (Bates et al., submitted). These data provide compelling evidence that assembly of the Fe-S cluster into FNR can occur under aerobic conditions. Therefore, the lack of WT-FNR activity under aerobic growth conditions cannot be simply explained by failure to assemble 4Fe-FNR. Rather, it seems more likely that, even under aerobic growth conditions, the cluster is assembled into WT-FNR but that this protein is rapidly inactivated due to the conversion of the $[4Fe-4S]^{2+}$ cluster to a $[2Fe-2S]^{2+}$ cluster when 4Fe-FNR is exposed to O_2 (Fig. 1). Since the steady-state amount of the $[4Fe-4S]^{2+}$ cluster should be influenced by the amount of O_2 present in a given culture, such a mechanism would be sufficient to explain FNR inactivation under aerobic growth conditions. The fate of 2Fe-FNR formed in aerobically grown cells remains to be determined, but it is likely that some 2Fe-FNR is converted to apo-FNR (Fig. 1).

Is the Effect of O_2 on FNR Activity Direct or Indirect In Vivo?

The concentration of FNR in either aerobically or anaerobically grown cells has been estimated to be 1 μM (119). Furthermore, FNR activity in cells was shown to depend on the dissolved O_2 concentration in the media over a range of 1 to 10 μM O_2 (8, 115). On the basis of cellular rates of O_2 diffusion and consumption, it has been estimated that the O_2 concentration within cells is equal to that of the surrounding medium until the external dissolved O_2 concentration drops below 0.2 μM (116). Thus, under growth conditions where the dissolved O_2 concentration is above 1 μM as in the experiments of Becker et al. (8) and Tseng et al. (115), the O_2 concentration in cells should have been sufficient to directly inactivate the ~1 μM levels of FNR. Although these data point to O_2 having a direct role in FNR inactivation in vivo, the observation that adding 50 mM ferricyanide to anaerobic cells also decreased FNR function suggested the possibility of an indirect O_2-sensing mechanism for FNR (120). Since the strong oxidant ferricyanide is not expected to equilibrate across the cell membrane, the fact that ferricyanide inactivated FNR was taken as evidence for a redox-sensitive intermediate in cells that communicates the presence of O_2 to FNR. However, recent data (60) show that ferricyanide can directly disassemble the [4Fe-4S]$^{2+}$ cluster of FNR in vitro, leading one to consider whether these in vivo results could alternatively be explained by even a small fraction of ferricyanide entering cells and directly oxidizing 4Fe-FNR. In addition, strains lacking the cytochrome o and d terminal oxidases show normal O_2 regulation of FNR-dependent genes, indicating that such a hypothetical intermediate that inactivates FNR cannot be generated by the aerobic electron transport chain (8).

A future challenge in this field is to provide more evidence that FNR responds directly to O_2 in vivo and to determine whether 2Fe-FNR is stable in whole cells or turned over to apo-FNR, the form previously purified from aerobic cells. While these properties of purified FNR from aerobic cells favor the idea that 2Fe-FNR is only a transient intermediate, recent data indicate that the [2Fe-2S]$^{2+}$ cluster of 2Fe-FNR is destabilized by some in vitro conditions, suggesting that most previous purification schemes would not have recovered 2Fe-FNR even if it was present in cells (K. Vogt and P. J. Kiley, unpublished results).

FNR Homologs

On the basis of their involvement in the regulation of oxygen-dependent functions and their amino acid sequence similarity to FNR (30), homologs have been identified in both gram-negative and gram-positive eubacteria where they have been implicated in the regulation of diverse anaerobic processes, such as nitrogen fixation, photosynthesis, and denitrification. In addition, FNR homologs may regulate virulence factors in pathogenic organisms such as *Bordetella pertussis* (3), *Actinobacillus pleuropneumoniae* (47), *Neisseria gonorrhoeae* (54), and *Listeria sp.* (123). Although all known FNR homologs have very similar DNA-binding regions, they can be divided into three groups on the basis of differences in the N-terminal region (30, 122). Many homologs fall into a group that has identical spacing of the N-terminal cysteine residues that ligate the [4Fe-4S]$^{2+}$ cluster in *E. coli* FNR. The second group of homologs (such as *Rhodobacter sphaeroides* FnrL and *Rhizobium leguminosarum* FnrN) has the essential N-terminal cysteine residues but with a spacing different from that of *E. coli* FNR. A third group lacks the N-terminal Cys cluster and thus may not directly sense O_2. Most of these FNR homologs are uncharacterized, but it will be interesting to see whether the regulation of these natural variants is similar to that of *E. coli* FNR or whether organisms have tailored the sensitivity of the proposed [4Fe-4S]$^{2+}$ clusters to the O_2 tensions representative of their particular ecological niches.

In summary, *E. coli* FNR apparently senses O_2 directly through the lability of a [4Fe-4S]$^{2+}$ cluster that causes its inactivation. In contrast, it appears that ArcB senses changes in oxygen concentrations indirectly, although more studies are needed to address this point. As global regulators of metabolic pathways in *E. coli*, the combined action of the ArcAB system and FNR ensures that the appropriate changes in gene expression take place to allow maximal energy generation and survival of this bacterium when O_2 becomes limiting.

BIOLOGICAL NITROGEN FIXATION

The element nitrogen is a major constituent of all organisms and exists in different oxidation states. Ammonia plays a key role in the nitrogen cycle and is produced during the decomposition of organic nitrogen compounds (ammonification). In well-aerated soils it can be oxidized to nitrate (nitrification), which, in turn, is readily reduced to molecular dinitrogen under anoxic conditions and released into the atmosphere (denitrification). This leads to a substantial loss of biologically available nitrogen sources in the soil. Atmospheric dinitrogen is recycled in a process called nitrogen fixation, which is carried out by certain prokaryotes (diazotrophs) either under free-living conditions and/or in symbiosis with certain plants. During biological nitrogen fixation, dinitrogen is reduced to ammonia by the enzyme complex nitrogenase consisting of the two components, dinitrogenase and dinitrogenase reductase (93). Since the enzyme nitrogenase is rapidly and irreversibly inactivated by O_2, the expression of nitrogen fixation (*nif*) genes is subject to stringent control, in response to oxygen availability in nitrogen-fixing bacteria.

One component common to this O_2-dependent control is the NifA protein, the key regulator of *nif* gene transcription (25, 30). NifA is a highly conserved protein in all nitrogen-fixing bacteria and consists of three domains. An N-terminal domain with a potential regulatory function is connected to a highly conserved central domain that interacts with RNA polymerase and the alternative sigma factor, σ^{54}, thereby catalyzing the formation of an open promoter complex (79). The transcription activation process is accompanied by the consumption of nucleoside triphosphates. The C-terminal domain contains a helix-turn-helix motif required for binding to upstream activator sequences (UAS) specific to the promoter regions of NifA-regulated genes. As described below, the expression and/or activity of NifA are regulated in response to O_2 and the concentration of fixed nitrogen, but the specific mechanism varies according to the organism (see Table 2) (25, 30).

FREE-LIVING NITROGEN FIXATION

NifA Activates the Transcription of *nif* Genes

Klebsiella pneumoniae and *Azotobacter vinelandii* are representatives of diazotrophs that fix nitrogen exclusively under free-living conditions. In both organisms *nifA* is cotranscribed with *nifL* and produced in equal amounts, which is achieved by coupling the translation of *nifA* to that of *nifL* (45, 46). Expression of the *nifLA* operon is independent of the O_2 concentration in these organisms. However, while expression of the *nifLA* operon is constitutive in *A. vinelandii* (11), its expression is induced upon nitrogen starvation in *K. pneumoniae* via the global nitrogen regulation system (Ntr) (26). To avoid wasteful synthesis of the O_2-sensitive nitrogenase, the activity of NifA is inhibited by the sensor protein NifL in response to fixed nitrogen and the external concentration of O_2 (25). NifA itself is not O_2-sensitive, since in a *nifL*⁻ mutant background, NifA activates *nif* gene expression even under aerobic conditions (11, 107). The inhibitory effect of NifL on NifA is most likely exerted by direct protein-protein interactions, and only recently, stoichiometric complex formation of purified NifL and NifA in vitro has been demonstrated and shown to be stimulated by adenosine nucleotides (78).

NifL Regulates the Activity of NifA

NifL consists of two domains tethered by a glutamine-rich linker. The N-terminal domain binds FAD as a prosthetic group and exhibits similarity to the PAS-domain family of proteins (53, 103). The redox state of the bound FAD has been shown in vitro to serve as a switch to regulate the inhibitor function of NifL from *A. vinelandii*. The oxidized form of NifL acts as an inhibitor of NifA activity, whereas the reduced form has no inhibitory effect (53). Although the physiological electron donors and acceptors of the NifL-FAD are still elusive, in vitro experiments indicate that the flavohaemoglobin (HMP) of *E. coli*, which is an NAD(P)H-dependent oxidoreductase (96), is able to function as an electron donor for oxidized NifL in the presence of NADH (76). On the basis of this experimental evidence, it has been proposed that NifL oxidation in vivo occurs directly and rapidly by O_2, whereas the return of NifL to the inactive reduced form occurs more slowly (76). Thus in the presence of O_2, NifL will largely be in the oxidized form and therefore active as an inhibitor of NifA.

The C-terminal domain of NifL, particularly that of *A. vinelandii*, exhibits homology to the transmitter domain of sensor-kinases of two-component regulatory systems. This domain is involved in detecting the nitrogen status of the cell and probably the presence of ADP. The mechanism of NifL functioning as a nitrogen sensor is still unclear, but there is some evidence for the participation of components or homologs of the global Ntr system (18, 52, 57). The detection of nitrogen/ADP by NifL is independent of its redox state. This was demonstrated with a truncated form of NifL of *A. vinelandii*, lacking part of the N-terminal FAD binding domain, which is still able to sense the nitrogen status (108). Therefore, the redox and nitrogen-sensing functions of NifL seem to be allocated to two different domains of the protein.

SYMBIOTIC NITROGEN FIXATION

The regulation of *nif* gene expression in symbiotically nitrogen-fixing organisms differs from that outlined for *K. pneumoniae* and *A. vinelandii*. A well-studied representative of this group is the bacterium *Sinorhizobium meliloti* (formerly *Rhizobium meliloti*), which induces the formation of specialized or-

Table 2. Major regulators involved in the oxygen-dependent expression of genes whose products are involved in nitrogen fixation or anoxygenic photosynthesis

Regulator	Cofactor[a]	Process regulated
Nitrogen fixation		
NifL	Flavin	Activity of NifA[b]
NifA[b]	None	*nif* gene expression
FixL	Heme	Activity of FixJ
FixJ	None	Expression of *nifA* and *fixK*
NifA[c]	FeS-cluster?	Expression of *nif* genes
FixK	Unknown	Expression of *fix* genes
FixT	Unknown	Expression of *fixK*
Anoxygenic photosynthesis[d]		
PpsR	Unknown	Expression of *puc*, *crt*, and *bch* genes
AppA	Flavin	Activity of PpsR
FnrL	FeS-cluster	Expression of *fnrL*, *hemA*, *hemZ*, *ccoNOQP*, *rdxBHIS*, *bch*, *dor*, *cta* genes
PrrB	Unknown	Activity of PrrA
PrrA	None	Expression of *puc*, *puf*, *puh*, *crt*, *bch* genes
TspO	Unknown	Expression of *crt* and *bch* genes

[a] For references, see text.
[b] NifA from *A. vinelandii* and *K. pneumoniae*.
[c] NifA from *S. meliloti*.
[d] Only regulators from *R. sphaeroides* are listed in this table.

gans on its host plant roots, called nodules, providing an ideal microaerobic environment for the nitrogen-fixing bacteria. Fixed nitrogen is made available to the host plant, which in return supplies the bacteria with fixed carbon compounds required for the high energy demand of the nitrogen fixation process (see also 5, 63, 104, 121).

NifA Is an Oxygen-Sensitive Regulator of *nif* Gene Expression

Thus far, 25 different genes required for symbiotic nitrogen fixation, organized in two clusters on an extrachromosomal megaplasmid (pSym), have been identified in *S. meliloti* (Fig. 2). Among the genes of cluster I are those encoding the polypeptides for nitrogenase (*nifHDK*) and the central regulator of *nif* gene transcription, NifA, which regulates the expression of all other genes of cluster I. In contrast to NifA of *K. pneumoniae* and *A. vinelandii*, rhizobial NifA proteins seem to be directly responsive to and inactivated by O_2. This property is dependent on an absolutely invariant $Cys-X_{11}-Cys-X_{19}-Cys-X_4-Cys$ motif residing in the C-terminal part of the central domain and an interdomain linker connecting it with the DNA-binding domain (31). The cysteine motif is missing in the NifA proteins of *K. pneumoniae* and *A. vinelandii*. Hitherto it has been proposed that the activity of rhizobial NifA proteins is regulated by the oxidation state of a bound metal ion. This assumption was based on the observation that NifA activity is inhibited by chelators in vivo and can be restored by the addition of Fe^{2+} ions (31). However, consid-

ering the identification of an Fe-S cluster in the O_2-responsive transcriptional regulator FNR of *E. coli*, it is tempting to propose a similar regulatory mechanism involving an O_2-sensitive Fe-S cluster for NifA of rhizobia.

FixK Regulates Expression of Symbiosis-Specific Genes

The FixK protein, encoded by a gene located in cluster II, activates the transcription of operons required for respiration in the microaerobic environment of the nodule (6). These include *fixN*, *fixO*, and *fixP*, encoding the three subunits of a heme-copper cytochrome cbb_3 oxidase with a high affinity for O_2 and the FixGHIS complex, which has been proposed to play a role in copper uptake/metabolism and assembly of the cbb_3 oxidase (98–100). Another target gene of FixK is *fixT*, whose gene product, FixT, negatively regulates *fixK* expression (32). Moreover, it has been reported that in a *fixK* mutant, *nifA* expression is elevated, suggesting that FixK negatively alters *nifA* expression. The FixK protein is homologous to FNR of *E. coli*, showing the same domain structure (see above). However, FixK lacks the cysteine residues located in the N-terminal domain of the FNR protein that are essential for the binding of the Fe-S cluster and therewith its O_2-responsive activity. In addition, cluster ligation of *E. coli* FNR is a prerequisite for the formation of the active dimeric protein (67). The missing cysteine motif makes it unlikely that FixK is able to respond directly to changes in the environmental O_2 concentration via an Fe-S cluster

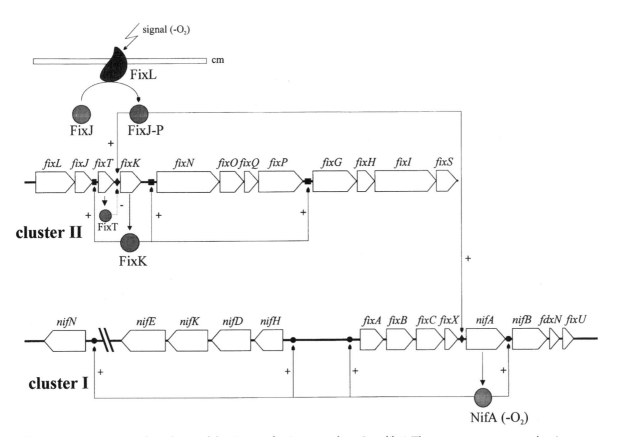

Figure 2. Organization and regulation of the nitrogen fixation genes from *S. meliloti.* The open arrows represent the nitrogen fixation genes located on the pSYM plasmid of *S. meliloti* and their direction of transcription. *nifHDKE, nifN,* and *nifB* are homologs to *nif* genes from *K. pneumoniae,* whereas *fix* refers to symbiosis specific genes also required for nitrogen fixation. Under microaerobic conditions, FixJ is converted into its active form by the membrane-bound FixL sensor-kinase and initiates transcription of *fixK* and *nifA.* NifA, in turn, activates transcription from promoters located upstream of *nifH, nifN, fixA,* and *nifB.* The transcription of the *fixNOQP* and *fixGHIS* operons and *fixT* is regulated by FixK. The way FixT downregulates expression of *fixK* is still elusive. cm, cytoplasmic membrane.

and raises the interesting question of the mechanism for FixK dimerization. The finding that FixK-regulated genes contain DNA sequence motifs similar to those bound by dimerized *E. coli* FNR implies that FixK acts as a dimer too. One possibility is that FixK activity is regulated by as yet unknown cofactor(s), whose abundance varies with the O_2 status of the cell and support dimerization. Alternatively, the activity of FixK could be O_2-independent and dimerization might be an intrinsic property of the FixK protein. Support for this possibility comes from the finding that the substitution of aspartate residue 154 in the putative dimerization helix of FNR to an alanine leads to an increase of apo-FNR dimerization (4, 71). An alanine residue is already present at the homologous position of FixK.

Another Level of Control is Achieved by the Two-Component System FixL and FixJ

The expression of *nifA* and *fixK* is strictly regulated in response to reduced O_2 tension by the reg-

ulatory proteins FixL and FixJ, which constitute a typical two-component regulatory system (23). The FixL protein of *S. meliloti* belongs to the family of PAS-domain–containing sensory proteins (44) and consists of an N-terminal membrane-anchoring domain, a central sensory domain, and a C-terminal transmitter domain. The sensory domain is a ferrous heme binding domain (37). The mechanism for regulating the O_2-responsive activity of FixL is dependent on the spin state of the heme iron in the sensory domain. Under microaerobic conditions, no oxygen is bound to the heme and the heme iron is in the high-spin state, thus allowing the reversible autophosphorylation of a histidine residue in the transmitter domain (39). Phosphorylated FixL transfers its phosphate residue to a conserved aspartate residue located in the N-terminal receiver domain of the cognate response-regulator FixJ. This phosphotransfer to FixJ is independent of the O_2 concentration (38). Phosphorylation of FixJ's receiver domain modulates the activity of its C-terminal output domain, thus enabling FixJ-P to activate transcription of *nifA* and *fixK*. If O_2 binds to the heme group of FixL, the heme iron is converted to the low-spin state (39). The transition between the high-spin and low-spin state results in a movement of the heme iron within the porphyrin plane, and it has been proposed that this motion results in a long-range conformational change in FixL, accounting for the switch in activity (39). A low-spin heme group inhibits the autophosphorylation rate in the FixL transmitter domain and leads to an increase of its intrinsic phosphatase activity upon FixJ-P. Dephosphorylated FixJ is no longer able to activate transcription of *nifA* and *fixK*. In addition to O_2, other high-field ligands like nitric oxide have been proposed to exert a regulatory effect on FixL activity by the same mechanism as O_2. In the light of the known deleterious effect of nitric oxide on the activity of nitrogenase this seems to be a compelling hypothesis.

Unity Despite Diversity

Homologs of the regulatory proteins FixL, FixJ, and FixK have also been identified in other rhizobia where they provide an excellent example for the "variation of a common theme," i.e., similar proteins can participate in different steps of regulatory circuits (30). The FixL proteins of *S. meliloti*, *Bradyrhizobium japonicum*, and *Azorhizobium caulinodans* are essential for symbiotic nitrogen fixation (1, 23, 64), whereas the loss of FixL leads only to a reduced fixation phenotype in *Rhizobium leguminosarum* bv. *viciae* VF39 and *Rhizobium etli* (24, 89). In *A. caulinodans* the O_2-responsive expression of *nifA* is regulated by FixK via FixL and FixJ (65). In contrast, the oxygen-responsive *nifA* expression in *B. japonicum* is independent of FixLJ and accomplished by another two-component regulatory system, RegSR (7).

Both *R. leguminosarum* bv. *viciae* VF39 and *B. japonicum* encode two FixK homologs. One resembles the FNR-type of *E. coli*, with the spacing of the N-terminal cysteine residues that ligate an $[4Fe-4S]^{2+}$ cluster being slightly different, and the second homolog resembles the FixK protein of *S. meliloti*, lacking this motif. Mutation of the FNR-type protein of *R. leguminosarum* bv. *viciae* VF39 (called FnrN) has a severe effect on its nitrogen fixation phenotype, whereas loss of the FixK-like regulator (called also FixK) has only minor effects (89). The opposite is true for the two homologs of *B. japonicum* (2, 82). Recent findings obtained with FnrN from *R. leguminosarum* bv. *viciae* VF39 suggest that the O_2-sensing mechanism of FnrN is probably identical to the mechanism proposed for *E. coli* FNR. Anaerobically purified FnrN is dimeric and exhibits UV/visible

spectral features similar to those obtained for *E. coli*, indicating that FnrN contains $[4Fe-4S]^{2+}$ clusters (T. Patschkowski and P. J. Kiley, unpublished results). Upon exposure to O_2, the $[4Fe-4S]^{2+}$ clusters decay to the $[2Fe-2S]^{2+}$ cluster form, and FnrN becomes monomeric.

These few examples demonstrate that although the principal regulators are conserved in different rhizobia, their use and position in the regulatory circuitry of the nitrogen fixation process can vary considerably according to the organism.

ANOXYGENIC PHOTOSYNTHESIS

Photosynthesis, the process during which light energy is converted into chemical energy in the form of ATP, is carried out by a variety of prokaryotic and eukaryotic organisms (phototrophs). Prokaryotic phototrophs are divided into oxygenic phototrophs and anoxygenic phototrophs on the basis of their ability to use H_2O as an electron donor for $NADP^+$ reduction, resulting in the production of molecular oxygen. The anoxygenic purple nonsulfur photosynthetic bacteria are extremely metabolically versatile, and most can grow aerobically, anaerobically in the light, fermentatively, or anaerobically in the dark in the presence of alternate electron acceptors. Reduction of the environmental O_2 tension leads to a differentiation of the cytoplasmic membrane into the intracytoplasmic membrane system (ICM) that houses the photosynthetic machinery. The photosynthetic apparatus consists of two types of light-harvesting (LH) complexes, designated B875 (LHI) and B800-850 (LHII), referring to their absorption maxima, and the reaction center (RC) as well as the supporting electron transport components that function in cyclic electron transfer (68). Each LH complex consists of two small, membrane-spanning polypeptides to which bacteriochlorophyll and carotenoids are noncovalently bound. The LH complexes absorb light energy and direct that energy to the RC, where the oxidation of a special pair of bacteriochlorophyll molecules is coupled with the reduction of a molecule of bacteriopheophytin within the RC. The subsequent electron flow via intermediate quinones, the bc_1 complex, and a periplasmically located cytochrome c_2 leads to rereduction of the RC and establishes a proton gradient that drives ATP synthesis (12). Purple non-sulfur bacteria photosynthesize only under anaerobic conditions, and the environmental O_2 concentration is the major signal in the regulation of photosynthetic gene expression. An additional layer of regulation is exerted by variations in the light intensity. The following section emphasizes the highly coordinated O_2-responsive regulation of photosynthesis gene expression in the anoxygenic phototrophs *R. sphaeroides* and *R. capsulatus*, which have been the main subjects of research in bacterial photosynthesis and its regulation. Photosynthesis gene expression is largely regulated by the combined action of three regulatory systems that act either as aerobic repressors or anaerobic activators (see Table 2) but involves additional levels of regulation that are still unfolding (for recent reviews see 91, 128).

Photosynthesis Gene Expression is Repressed under Aerobic Conditions

Most genes involved in photosynthesis are clustered in several operons on a 46-kbp region of the chromosomes of *R. sphaeroides* and *R. capsulatus*, called the photosynthesis gene cluster. Genes of this cluster encode enzymes involved in bacteriochlorophyll (*bch*) and carotenoid (*crt*) biosynthesis and polypeptides for the RC (*puf* and *puh*), LHI (*puf*), and LHII (*puc*). Moreover, the photosynthesis gene cluster codes for the O_2-responsive transcriptional regulators, PpsR (*R. sphaeroides*) and CrtJ (*R. capsulatus*), both members of PAS-domain-containing sensory proteins (112). Under aerobic conditions, the PpsR/CrtJ proteins function as a repressor of bacteriochlorophyll, carotenoid, and LHII gene expression (92, 95). A mutation in *ppsR/crtJ* leads to constitutive production of photopigments and the LHII complex. The C-terminal domain of PpsR/CrtJ contains a helix-turn-helix motif that has been shown by footprinting to bind to a conserved palindromic DNA sequence motif (TGT-N_{12}-ACA) either located in the promoter regions of PpsR/CrtJ-regulated genes or even further upstream (94). Binding of CrtJ as a dimer to such a sequence is redox-sensitive, with a higher binding affinity under oxidizing conditions (94). The mechanism of O_2-sensing by PpsR/CrtJ is still elusive, and no redox-sensitive cofactor has been identified yet. However, because in vivo PpsR is capable of responding to both changes in O_2 concentration and light intensity, it is tempting to speculate that different sensors may relay such changes to PpsR via a common redox mechanism.

The Flavoprotein AppA Decreases PpsR-Mediated Repression

The identification of the regulatory protein AppA in *R. sphaeroides* has shed some light on a potential sensing mechanism. AppA is a flavoprotein required for photosynthesis gene expression under anaerobic photosynthetic conditions (41). A null-mutation in *appA* results in low photosynthetic growth rates due to a decreased expression of photosynthesis genes (40). The inactivation of *ppsR* overrides this phenotype, and a *ppsR/appA* double mutant displays the same phenotype as a *ppsR* mutant (43). This suggested that AppA might interact with PpsR, thereby counteracting its repressive activity depending upon the redox state of AppA. The AppA protein has been proposed to directly sense the redox state of key molecules that would vary in response to altering O_2 tension or light intensity. The central domain of AppA, which appears to bind a heme, seems to be sufficient for the modulation of PpsR activity (41).

FNR Homologs as Global Oxygen-Responsive Regulators

In both *R. sphaeroides* and *R. capsulatus*, an *E. coli* FNR homolog, designated FnrL, has been identified (129, 130). Inactivation of the *fnrL* gene leads to an inability of both organisms to grow anaerobically in the dark in the presence of alternative electron acceptors, e.g., DMSO, indicating a role of FnrL in anaerobic respiration like FNR from *E. coli*. In contrast to *R. capsulatus*, a *fnrL* mutant strain of *R. sphaeroides* is also unable to grow under anaerobic photosynthetic conditions (130). This phenotype is not surprising since FNR consensus sequences have been identified in the promoter region of several *R. sphaeroides* genes, including genes involved in tetrapyrrole and bacteriochlorophyll biosynthesis or the structural genes for the LHII complex. A second FNR homolog of the FixK-type, designated NnrR, has been identified in different strains of *R. sphaeroides* where it plays a role in the expression of genes involved in anaerobic denitrification (69, 113). Consistent with this, NnrR seems to be dispensable for anaerobic photosynthetic growth.

Direct proof for *R. sphaeroides* FnrL-dependent expression under low O_2 concentrations has been shown for *hemA* (coding for δ-aminolevulinic acid synthase, catalyzing the first step in tetrapyrolle biosynthesis) and the *puc* operon (131). Therefore a *fnrL* mutant is unable to produce important components of the photosynthetic apparatus under photosynthetic conditions. In addition, FnrL regulates the expression of a cbb_3 type terminal cytochrome *c* oxidase (*ccoNOQP*) and probably the *rdxBHIS*-operon (81), which are both proposed to play an important role in regulation of gene expression by O_2 (see below). Homologs of the above described genes in *R. capsulatus* lack FNR consensus sequences in their promoter region, rendering them independent of FnrL. The mechanism of O_2-sensing by FnrL is probably identical to the mechanism proposed for *E. coli* FNR, since anaerobically purified FnrL exhibits UV/visible spectral features typical of a [4Fe-4S]$^{2+}$ cluster-containing protein. Exposure to O_2 results in a conversion of the [4Fe-4S]$^{2+}$ cluster to a [2Fe-2S]$^{2+}$ cluster (Patschkowski and Kiley, unpublished results). However, the expression of *hemA* in *R. sphaeroides* is elevated under aerobic conditions in an *fnrL* mutant background, implying that FnrL might be an active repressor under these conditions (83).

Role of the Electron Transport Chain in Controlling Photosynthesis Gene Expression

The *R. sphaeroides* regulators PrrB and PrrA (RegB/RegA in *R. capsulatus*) comprise a two-component regulatory system that is required for optimal anaerobic expression of the photosynthesis genes. PrrB and RegB are membrane-anchored sensor-kinases with a typical C-terminal transmitter domain (28, 80). PrrA and RegA are cytoplasmic response-regulators with a conserved receiver domain but an unusual C-terminal domain that lacks any similarity with other DNA-binding domains (29, 105). However, it has been recently shown that purified RegA binds directly to promoter sequences of specific target genes although so far no consensus DNA-binding site has been determined (15, 27). PrrB/PrrA and RegB/RegA regulate the anaerobic expression of genes coding for components of the photosynthetic apparatus as well as genes involved in bacteriochlorophyll and carotenoid biosynthesis. PrrA/RegA mutants are unable to grow photosynthetically, whereas mutations in *prrB/regB* result in poor cell growth under anaerobic photosynthetic conditions at high light intensities (28, 80). There is convincing experimental evidence that the minor effect of mutations in PrrB/RegB is due to cross-talk, i.e., other sensor-kinases are able to phosphorylate PrrA/RegA, thereby compensating for the loss of their cognate sensor-kinases (42). Additional members of the PrrB/PrrA and RegB/RegA regulatory systems are PrrC and SenC, which are encoded by genes located upstream of *prrA* and *regA*, respectively. Both proteins appear to be membrane-bound by the secondary structure predictions and exhibit homology to yeast proteins that are involved in the assembly of cytochrome *c* oxidase subunits (17, 28). Lack of PrrC/SenC leads to reduced induction of photosynthesis gene expression and, in the case of SenC, to a decrease in cytochrome *c* oxidase activity under aerobic conditions (17).

There are still open questions as to how the redox-sensing mechanism is accomplished by the Prr/Reg system, but recent findings in *R. sphaeroides* provide an interesting hypothesis upon which to base future work. Cells carrying mutations in the *ccoP* and *rdxB* genes exhibit an increased photosynthesis gene expression in the presence of O_2 (83). Another phenotype of these mutants is the unusual accumulation of the carotenoid, spheroidenone, under anaerobic photosynthetic growth conditions (83). This seems to be the result of an increased activity of the enzyme CrtA, which catalyzes the oxidation of spheroidene to spheroidenone. This suggests that the cbb_3 cytochrome *c* oxidase (encoded by the *ccoNOQP* operon) and the RdxBHIS complex function as O_2/redox sensors that transduce an inhibitory signal to the sensor-kinase PrrB under aerobic conditions to prevent photosynthesis gene expression. Electron flow through the cbb_3/RdxBHIS complex also seems to be necessary under anaerobic photosynthetic conditions to prevent the oxidation of spheroidene to spheroidenone, probably by keeping an as yet hypothetical cellular oxidant in its reduced form (84).

Signal transduction between the cbb_3/RdxBHIS complex and the PrrB sensor-kinase appears to be mediated by the CcoQ protein. Since CcoQ is dispensable for the activity of the cbb_3 oxidase, the lack of this protein seems to interrupt the signal exchange between cbb_3/RdxBHIS and PrrB (84). The FnrL-dependent expression of the *puc* operon is elevated several-fold under aerobic conditions in a *ccoP* mutant background, and it has been suggested that the activity of the FnrL protein might also be regulated by a signal emanating from the cbb_3 oxidase (83).

A Benzodiazepine Receptor Homolog Protein is Involved in Photosynthesis Gene Expression

An additional oxygen-responsive regulator, TspO, has been identified in *R. sphaeroides* (126). TspO is a tryptophan-rich protein with significant sequence homology to mammalian peripheral-type benzodiazepine receptors, and its activity can be complemented by a rat homolog in a *tspO* mutant strain (127). The TspO protein is located in the outer membrane and is most abundant in cells grown under photosynthetic conditions. A *tspO* mutant produces higher amounts of carotenoids and bacteriochlorophyll as compared to the wild type, especially if cells were grown semiaerobically or aerobically. This phenotype is due to an increased level of transcription of the corresponding biosynthetic genes. Under the same growth conditions the expression of the *puc* operon is also higher in a *tspO* mutant. So far, it is unknown how TspO responds to changes in O_2 concentrations and in which way such a signal leads to an altered pattern of gene transcription. However, recent in vivo data suggest a role of a heme biosynthetic pathway intermediate in the signal transduction process (125).

In summary, during the past few years important O_2-responsive regulators of photosynthesis gene expression have been identified. PpsR/CrtJ are aerobic repressors that seem to be involved exclusively in photosynthetic gene expression. The activity of PpsR is regulated by the flavoprotein AppA in response to O_2 and light, but it is still elusive what physiological signal is detected by AppA and how this signal is transferred to PpsR. The two-component regulatory systems PrrB/PrrA and RegB/RegA function maximally as transcriptional activators under anaerobic photosynthetic conditions. In addition, PrrB and PrrA have been shown to be involved in the expression of genes required for nitrogen fixation and carbon dioxide fixation in *R. sphaeroides* (61, 101). Therefore, PrrB/PrrA seems to represent a global regulatory system integrating the expression of three fundamental metabolic processes. Homologs of PrrB/PrrA were also found in the nitrogen-fixing bacterium *B. japonicum* (7), where they regulate the O_2-dependent expression of the transcriptional activator protein NifA (see above), and it is tempting to predict the presence of such regulators in other rhizobia. The activator FnrL exhibits a comparable versatility, at least in *R. sphaeroides*, where it is essential for the anaerobic expression of photosynthesis genes and components for anaerobic respiration.

CONCLUDING REMARKS

Major advances have been made in the identification of the protein components that are involved in O_2-sensing in a number of facultative bacteria. One unexpected and unifying theme for NifL, FixL, ArcB, PpsR, and CrtJ is that all five proteins are members of the PAS domain family of proteins, as inferred from their primary amino acid sequences (132). Members of the PAS domain superfamily are widespread throughout the biological world and in several cases are known to participate or have been implicated in the sensing of O_2, redox status, or light (112). The three-dimensional structure of a protein belonging to the PAS domain family, the photoactive yellow protein (PYP), has been solved (90). *Ectothiorhodospira halophila* PYP is a bacterial blue-light photoreceptor that contains a 4-hydroxylcinnamoyl chromophore (90). Since the PYP protein may be considered a PAS domain in its entirety, it has been proposed that the PYP structure represents a common fold for ligand binding in this family of sensory proteins (90). Consistent with this, the three-dimensional structure of the heme-binding domain of *B. japoni-*

cum FixL has recently been solved and exhibits a similar ligand fold as that of the PYP (44). Further work is needed to establish whether the other members of this group also bind a redox-sensitive ligand.

In conclusion, exciting progress has been made in understanding the different ways in which prokaryotic organisms sense and respond to changes in environmental oxygen concentrations. Such studies have thus far revealed that the regulatory or sensory proteins generally contain one of three cofactors for the sensing of O_2: a heme group, a flavin moiety, or an Fe-S cluster. While it is less clear how the flavin moiety is used to sense O_2, studies of FixL and FNR have shown that the use of heme involves direct binding of oxygen, whereas the use of an Fe-S cluster results in partial destruction of the cluster by O_2, respectively. Thus, it appears that the chemical properties of these molecules make them generally useful as direct biosensors of O_2. Nevertheless, there is also much exciting work ahead to determine how these elegant O_2 sensors interface with other cellular activities (such as cytochrome oxidases) to respond to O_2 tensions representative of their particular ecological niches.

Recent data also indicate that several O_2-sensing pathways such as ArcAB and PrrBA/RegBA may sense changes in O_2 indirectly. In these cases, the sensing pathways may be monitoring flux through the aerobic electron transport chain. Challenges for the future include identifying the cofactors that are involved in these pathways and determining how they may be interacting with the aerobic respiratory chain to sense changes in flux. Finally, it is worth noting that many organisms have multiple global regulators for O_2-sensing, indicating that control of the relevant target genes is critical in the adaptation and survival of facultative organisms. In addition, the fact that closely related organisms can apparently use different global regulators interchangeably suggests that at a gross level many of these systems can function in sensing similar changes in O_2 concentrations in the environment.

Acknowledgments. We thank Samuel Kaplan, Robert Gunsalus, Werner Klipp, Timothy Donohue, and Andreas Schlüter for helpful comments on this manuscript.

This work was supported by NIH grant GM-45844. P.J.K. was also a recipient of a Young Investigator Award from the National Science Foundation and a Shaw Scientist Award from the Milwaukee Foundation.

ADDENDUM IN PROOF

Georgellis et al. have shown that the fermentation products D-lactate, acetate, and pyruvate accelerate the autophosphorylation rate of the sensor kinase ArcB and the transphosphorylation rate of the response regulator ArcA without any change in the dephosphorylation of ArcA-P (D. Georgellis, O. Kwon, and E. C. C. Lin,

J. Biol. Chem. **274**:35950–35954, 1999). Garnerone et al. have shown that the negative effect of FixT on *fixK* transcription (and also on *nifA* transcription) is exerted indirectly by inhibiting the kinase activity of FixL, resulting in a decrease in the active form of FixJ (A. Garnerone, D. Cabanes, M. Foussard, P. Boistard, and J. Batut, *J. Biol. Chem.* **274**:32500–32506, 1999). A review addressing a similar subject has been published by Bauer et al. (C. Bauer, S. Elsen, and T. H. Bird, *Annu. Rev. Microbiol.* **53**:495–523, 1999).

REFERENCES

1. **Anthamatten, D., and H. Hennecke.** 1991. The regulatory status of the *fixL*- and *fixJ*-like genes in *Bradyrhizobium japonicum* may be different from that in *Rhizobium meliloti.* *Mol. Gen. Genet.* **225**:38–48.

2. **Anthamatten, D., B. Scherb, and H. Hennecke.** 1992. Characterization of a *fixLJ*-regulated *Bradyrhizobium japonicum* gene sharing similarity with the *Escherichia coli fnr* and *Rhizobium meliloti fixK* genes. *J. Bacteriol.* **174**:2111–2120.

3. **Bannan, J. D., M. J. Moran, J. I. MacInnes, G. A. Soltes, and R. L. Friedman.** 1993. Cloning and characterization of *btr*, a *Bordetella pertussis* gene encoding an FNR-like transcriptional regulator. *J. Bacteriol.* **175**:7228–7235.

4. **Bates, D. M., B. A. Lazazzera, and P. J. Kiley.** 1995. Characterization of FNR* mutant proteins indicates two distinct mechanisms for altering oxygen regulation of the *Escherichia coli* transcription factor FNR. *J. Bacteriol.* **177**:3972–3978.

4a. **Bates, D. M., C. V. Popescu, N. Khoroshilova, K. Vogt, H. Beinert, E. Munck, and P. J. Kiley.** 2000. Substitution of leucine 28 with histidine in the *Escherichia coli* transcription factor FNR results in increased stability of the $[4Fe-4S]^{2+}$ cluster to oxygen. *J. Biol. Chem.* **275**:6234–6240.

5. **Batut, J., and P. Boistard.** 1994. Oxygen control in *Rhizobium.* *Antonie Van Leeuwenhoek* **66**:129–150.

6. **Batut, J., M. L. Daveran-Mingot, M. David, J. Jacobs, A. M. Garnerone, and D. Kahn.** 1989. *fixK*, a gene homologous with *fnr* and *crp* from *Escherichia coli*, regulates nitrogen fixation genes both positively and negatively in *Rhizobium meliloti.* *EMBO J.* **8**:1279–1286.

7. **Bauer, E., T. Kaspar, H. M. Fischer, and H. Hennecke.** 1998. Expression of the *fixR-nifA* operon in *Bradyrhizobium japonicum* depends on a new response regulator, RegR. *J. Bacteriol.* **180**:3853–3863.

8. **Becker, S., G. Holighaus, T. Gabrielczyk, and G. Unden.** 1996. O_2 as the regulatory signal for FNR-dependent gene regulation in *Escherichia coli.* *J. Bacteriol.* **178**:4515–4521.

9. **Becker, S., D. Vlad, S. Schuster, P. Pfeiffer, and G. Unden.** 1997. Regulatory O_2 tensions for the synthesis of fermentation products in *Escherichia coli* and relation to aerobic respiration. *Arch. Microbiol.* **168**:290–296.

10. **Beinert, H., and P. J. Kiley.** 1999. Fe-S proteins in sensing and regulatory functions. *Curr. Opin. Chem. Biol.* **3**:152–157.

11. **Blanco, G., M. Drummond, P. Woodley, and C. Kennedy.** 1993. Sequence and molecular analysis of the *nifL* gene of *Azotobacter vinelandii.* *Mol. Microbiol.* **9**:869–879.

12. **Blankenship, R. E., M. T. Madigan, and C. E. Bauer (ed.).** 1995. *Anoxygenic Photosynthetic Bacteria.* Kluwer Academic Press, Dordrecht.

13. **Böck, A., and G. Sawers.** 1996. Fermentation, p. 262–282. *In* F. C. Neidhardt, R. Curtiss III, J. L. Ingraham, E. C. C. Lin, K. B. Low, B. Magasanik, W. S. Reznikoff, M. Riley, M. Schaecter, and H. E. Umbarger (ed.), Escherichia coli *and* Salmonella: *Cellular and Molecular Biology*, 2nd ed., vol. 1. ASM Press, Washington, D.C.

14. **Bongaerts, J., S. Zoske, U. Weidner, and G. Unden.** 1995. Transcriptional regulation of the proton translocating NADH dehydrogenase genes (*nuoA-N*) of *Escherichia coli* by electron acceptors, electron donors and gene regulators. *Mol. Microbiol.* **16**:521–534.

15. **Bowman, W. C., S. Du, C. E. Bauer, and R. G. Kranz.** 1999. In vitro activation and repression of photosynthesis gene transcription in *Rhodobacter capsulatus.* *Mol. Microbiol.* **33**:429–437.

16. **Brondsted, L., and T. Atlung.** 1996. Effect of growth conditions on expression of the acid phosphatase (*cyx-appA*) operon and the *appY* gene, which encodes a transcriptional activator of *Escherichia coli.* *J. Bacteriol.* **178**:1556–1564.

17. **Buggy, J., and C. E. Bauer.** 1995. Cloning and characterization of *senC*, a gene involved in both aerobic respiration and photosynthesis gene expression in *Rhodobacter capsulatus.* *J. Bacteriol.* **177**:6958–6965.

18. **Contreras, A., M. Drummond, A. Bali, G. Blanco, E. Garcia, G. Bush, C. Kennedy, and M. Merrick.** 1991. The product of the nitrogen fixation regulatory gene *nfrX* of *Azotobacter vinelandii* is functionally and structurally homologous to the uridylyltransferase encoded by *glnD* in enteric bacteria. *J. Bacteriol.* **173**:7741–7749.

18a. **Cotter, P. A., S. B. Melville, J. A. Albrecht, and R. P. Gunsalus.** 1997. Aerobic regulation of cytochrome *d* oxidase (*cydAB*) operon expression in *Escherichia coli*: roles of Fnr and ArcA in repression and activation. *Mol. Microbiol.* **25**:605–615.

19. **Cronan, J. E., and D. LaPorte.** 1996. Tricarboxylic acid cycle and glyoxylate bypass, p. 206–216. *In* F. C. Neidhardt, R. Curtiss III, J. L. Ingraham, E. C. C. Lin, K. B. Low, B. Magasanik, W. S. Reznikoff, M. Riley, M. Schaechter, and H. E. Umbarger (ed.), Escherichia coli *and* Salmonella: *Cellular and Molecular Biology*, 2nd ed., vol. 1. ASM Press, Washington, D.C.

20. **Cunningham, L., M. J. Gruer, and J. R. Guest.** 1997. Transcriptional regulation of the aconitase genes (*acnA* and *acnB*) of *Escherichia coli.* *Microbiology* **143**:3795–3805.

21. **Cunningham, L., and J. R. Guest.** 1998. Transcription and transcript processing in the *sdhCDAB-sucABCD* operon of *Escherichia coli.* *Microbiology* **144**:2113–2123.

22. **Darwin, A. J., and V. Stewart.** 1996. The NAR modulon systems: nitrate and nitrite regulation of anaerobic gene expression, p. 343–359. *In* E. C. C. Lin (ed.), *Regulation of Gene Expression in* Escherichia coli. R. G. Landes Company, Austin, Tex.

23. **David, M., M. L. Daveran, J. Batut, A. Dedieu, O. Domergue, J. Ghai, C. Hertig, P. Boistard, and D. Kahn.** 1988. Cascade regulation of *nif* gene expression in *Rhizobium meliloti.* *Cell* **54**:671–683.

24. **D'Hooghe, I., J. Michiels, K. Vlassak, C. Verreth, F. Waelkens, and J. Vanderleyden.** 1995. Structural and functional analysis of the *fixLJ* genes of *Rhizobium leguminosarum* biovar *phaseoli* CNPAF512. *Mol. Gen. Genet.* **249**:117–126.

25. **Dixon, R.** 1998. The oxygen-responsive NIFL-NIFA complex: a novel two-component regulatory system controlling nitrogenase synthesis in gamma-proteobacteria. *Arch. Microbiol.* **169**:371–380.

26. **Drummond, M., J. Clements, M. Merrick, and R. Dixon.** 1983. Positive control and autogenous regulation of the *nifLA* promoter in *Klebsiella pneumoniae.* *Nature* **301**:302–307.

27. **Du, S., T. H. Bird, and C. E. Bauer.** 1998. DNA binding characteristics of RegA. A constitutively active anaerobic activator of photosynthesis gene expression in *Rhodobacter capsulatus.* *J. Biol. Chem.* **273**:18509–18513.

28. **Eraso, J. M., and S. Kaplan.** 1995. Oxygen-insensitive synthesis of the photosynthetic membranes of *Rhodobacter sphaeroides*: a mutant histidine kinase. *J. Bacteriol.* **177:**2695–2706.

29. **Eraso, J. M., and S. Kaplan.** 1994. *prrA*, a putative response regulator involved in oxygen regulation of photosynthesis gene expression in *Rhodobacter sphaeroides*. *J. Bacteriol.* **176:**32–43.

30. **Fischer, H. M.** 1994. Genetic regulation of nitrogen fixation in rhizobia. *Microbiol. Rev.* **58:**352–386.

31. **Fischer, H. M., T. Bruderer, and H. Hennecke.** 1988. Essential and non-essential domains in the *Bradyrhizobium japonicum* NifA protein: identification of indispensable cysteine residues potentially involved in redox reactivity and/or metal binding. *Nucleic Acids Res.* **16:**2207–2224.

32. **Foussard, M., A.-M. Garnerone, F. Ni, E. Soupene, P. Boistard, and J. Batut.** 1997. Negative autoregulation of the *Rhizobium meliloti fixK* gene is indirect and requires a newly identified regulator, FixT. *Mol. Microbiol.* **25:**27–37.

33. **Fraenkel, D. G.** 1996. Glycolysis, p. 189–198. *In* F. C. Neidhardt, R. Curtiss III, J. L. Ingraham, E. C. C. Lin, K. B. Low, B. Magasanik, W. S. Reznikoff, M. Riley, M. Schaechter, and H. E. Umbarger (ed.), Escherichia coli *and* Salmonella*: Cellular and Molecular Biology*, 2nd ed., vol. 1. ASM Press, Washington, D.C.

34. **Gennis, R. B., and V. Stewart.** 1996. Respiration, p. 217–261. *In* F. C. Neidhardt, R. Curtiss III, J. L. Ingraham, E. C. C. Lin, K. B. Low, B. Magasanik, W. S. Reznikoff, M. Riley, M. Schaechter, and H. E. Umbarger (ed.), Escherichia coli *and* Salmonella*: Cellular and Molecular Biology*, 2nd ed., vol. 1. ASM Press, Washington, D.C.

35. **Georgellis, D., O. Kwon, P. De Wulf, and E. C. Lin.** 1998. Signal decay through a reverse phosphorelay in the Arc two-component signal transduction system. *J. Biol. Chem.* **273:**32864–32869.

36. **Georgellis, D., A. S. Lynch, and E. C. Lin.** 1997. In vitro phosphorylation study of the *arc* two-component signal transduction system of *Escherichia coli*. *J. Bacteriol.* **179:**5429–5435.

37. **Gilles-Gonzalez, M. A., G. S. Ditta, and D. R. Helinski.** 1991. A haemoprotein with kinase activity encoded by the oxygen sensor of *Rhizobium meliloti*. *Nature* **350:**170–172.

38. **Gilles-Gonzalez, M. A., and G. Gonzalez.** 1993. Regulation of the kinase activity of heme protein FixL from the two-component system FixL/FixJ of *Rhizobium meliloti*. *J. Biol. Chem.* **268:**16293–16297.

39. **Gilles-Gonzalez, M. A., G. Gonzalez, and M. F. Perutz.** 1995. Kinase activity of oxygen sensor FixL depends on the spin state of its heme iron. *Biochemistry* **34:**232–236.

40. **Gomelsky, M., and S. Kaplan.** 1995. *appA*, a novel gene encoding a *trans*-acting factor involved in the regulation of photosynthesis gene expression in *Rhodobacter sphaeroides* 2.4.1. *J. Bacteriol.* **177:**4609–4618.

41. **Gomelsky, M., and S. Kaplan.** 1998. AppA, a redox regulator of photosystem formation in *Rhodobacter sphaeroides* 2.4.1, is a flavoprotein. *J. Biol. Chem.* **273:**35319–35325.

42. **Gomelsky, M., and S. Kaplan.** 1995. Isolation of regulatory mutants in photosynthesis gene expression in *Rhodobacter sphaeroides* 2.4.1 and partial complementation of a PrrB mutant by the HupT histidine-kinase. *Microbiology* **141:**1805–1819.

43. **Gomelsky, M., and S. Kaplan.** 1997. Molecular genetic analysis suggesting interactions between AppA and PpsR in regulation of photosynthesis gene expression in *Rhodobacter sphaeroides* 2.4.1. *J. Bacteriol.* **179:**128–134.

44. **Gong, W. M., B. Hao, S. S. Mansy, G. Gonzalez, M. A. Gilles-Gonzalez, and M. K. Chan.** 1998. Structure of a biological oxygen sensor: a new mechanism for heme-driven signal transduction. *Proc. Natl. Acad. Sci. USA* **95:**15177–15182.

45. **Govantes, F., E. Andujar, and E. Santero.** 1998. Mechanism of translational coupling in the *nifLA* operon of *Klebsiella pneumoniae*. *EMBO J.* **17:**2368–2377.

46. **Govantes, F., J. A. Molina-Lopez, and E. Santero.** 1996. Mechanism of coordinated synthesis of the antagonistic regulatory proteins NifL and NifA of *Klebsiella pneumoniae*. *J. Bacteriol.* **178:**6817–6823.

47. **Green, J., and M. Baldwin.** 1997. HlyX, the FNR homologue of *Actinobacillus pleuropneumoniae*, is a [4Fe-4S]-containing oxygen-responsive transcription regulator that anaerobically activates FNR-dependent class I promoters via an enhanced AR1 contact. *Mol. Microbiol.* **24:**593–605.

48. **Green, J., B. Bennett, P. Jordan, E. T. Ralph, A. J. Thomson, and J. R. Guest.** 1996. Reconstitution of the [4Fe-4S] cluster in FNR and demonstration of the aerobic-anaerobic switch *in vitro*. *Biochem J.* **316:**887–892.

49. **Green, J., and F. A. Marshall.** 1999. Identification of a surface of FNR overlapping activating region 1 that is required for repression of gene expression. *J. Biol. Chem.* **274:**10244–10248.

50. **Guest, J. R., J. Green, A. S. Irvine, and S. Spiro.** 1996. The FNR modulon and FNR-regulated gene expression, p. 317–342. *In* E. C. C. Lin and A. S. Lynch (ed.), *Regulation of Gene Expression in* Escherichia coli. R. G. Landes Company, Austin, Tex.

51. **Gunsalus, R. P., and S. J. Park.** 1994. Aerobic-anaerobic gene regulation in *Escherichia coli*: control by the ArcAB and Fnr regulons. *Res. Microbiol.* **145:**437–450.

52. **He, L., E. Soupene, A. Ninfa, and S. Kustu.** 1998. Physiological role for the GlnK protein of enteric bacteria: relief of NifL inhibition under nitrogen-limiting conditions. *J. Bacteriol.* **180:**6661–6667.

53. **Hill, S., S. Austin, T. Eydmann, T. Jones, and R. Dixon.** 1996. *Azotobacter vinelandii* NIFL is a flavoprotein that modulates transcriptional activation of nitrogen-fixation genes via a redox-sensitive switch. *Proc. Natl. Acad. Sci. USA* **93:**2143–2148.

54. **Householder, T. C., W. A. Belli, S. Lissenden, J. A. Cole, and V. L. Clark.** 1999. *Cis*- and *trans*-acting elements involved in regulation of *aniA*, the gene encoding the major anaerobically induced outer membrane protein in *Neisseria gonorrhoeae*. *J. Bacteriol.* **181:**541–551.

55. **Iuchi, S.** 1993. Phosphorylation/dephosphorylation of the receiver module at the conserved aspartate residue controls transphosphorylation activity of histine kinase in sensor protein ArcB of *Escherichia coli*. *J. Biol. Chem.* **268:**23972–23980.

56. **Iuchi, S., and E. C. C. Lin.** 1995. Signal transduction in the Arc system for control of operons encoding aerobic respiratory enzymes, p. 223–231. *In* J. A. Hoch and T. J. Silhavy (ed.), *Two-Component Signal Transduction*. ASM Press, Washington, D.C.

57. **Jack, R., M. De Zamaroczy, and M. Merrick.** 1999. The signal transduction protein GlnK is required for NifL-dependent nitrogen control of nif gene expression in *Klebsiella pneumoniae*. *J. Bacteriol.* **181:**1156–1162.

58. **Johnson, M. K.** 1994. Iron-sulfur proteins, p. 1896–1913. *In* R. B. King (ed.), *Encyclopedia of Inorganic Chemistry*, vol. 4. John Wiley & Sons, Chichester.

59. **Johnson, M. K.** 1998. Iron-sulfur proteins: new roles for old clusters. *Curr. Opin. Chem. Biol.* **2:**173–181.

60. **Jordan, P. A., A. J. Thomson, E. T. Ralph, J. R. Guest, and J. Green.** 1997. FNR is a direct oxygen sensor having a biphasic response curve. *FEBS Lett.* **416:**349–352.

61. Joshi, H. M., and R. Tabita. 1996. A global two component signal transduction system that integrates the control of photosynthesis, carbon dioxide assimilation, and nitrogen fixation. *Proc. Natl. Acad. Sci. USA* **93**:14515–14520.

62. Joung, J. K., E. H. Chung, G. King, C. Yu, A. S. Hirsh, and A. Hochschild. 1995. Genetic strategy for analyzing specificity of dimer formation: *Escherichia coli* cyclic AMP receptor protein mutant altered in its dimerization specificity. *Genes Dev.* **9**:2986–2996.

63. Kaminski, P. A., J. Batut, and P. Boistard. 1998. A survey of symbiotic nitrogen fixation by rhizobia, p. 431–460. *In* H. P. Spaink, A. Kondorosi, and P. J. J. Hooykaas (ed.), *The Rhizobiaceae-Molecular Biology of Model Plant-Associated Bacteria*. Kluwer Academic Publishers, Dordrecht.

64. Kaminski, P. A., and C. Elmerich. 1991. Involvement of *fixLJ* in the regulation of nitrogen fixation in *Azorhizobium caulinodans*. *Mol. Microbiol.* **5**:665–673.

65. Kaminski, P. A., K. Mandon, F. Arigoni, N. Desnoues, and C. Elmerich. 1991. Regulation of nitrogen fixation in *Azorhizobium caulinodans*: identification of a *fixK*-like gene, a positive regulator of *nifA*. *Mol. Microbiol.* **5**:1983–1991.

66. Khoroshilova, N., C. Popescu, E. Münck, H. Beinert, and P. Kiley. 1997. Iron-sulfur cluster disassembly in the FNR protein of *Escherichia coli* by O_2: [4Fe-4S] to [2Fe-2S] conversion with loss of biological activity. *Proc. Natl. Acad. Sci. USA* **94**:6087–6092.

67. Kiley, P. J., and H. Beinert. 1999. Oxygen sensing by the global regulator, FNR: the role of the iron-sulfur cluster. *FEMS Microbiol. Lett.* **22**:341–352.

68. Kiley, P. J., and S. Kaplan. 1988. Molecular genetics of photosynthetic membrane biosynthesis in *Rhodobacter sphaeroides*. *Microbiol. Rev.* **52**:50–69.

69. Kwiatkowski, A. V., W. P. Laratta, A. Toffanin, and J. P. Shapleigh. 1997. Analysis of the role of the *nnrR* gene product in the response of *Rhodobacter sphaeroides* 2.4.1 to exogenous nitric oxide. *J. Bacteriol.* **179**:5618–5620.

70. Lambden, P. R., and J. R. Guest. 1976. Mutants of *Escherichia coli* unable to use fumarate as an anaerobic electron acceptor. *J. Gen. Microbiol.* **97**:145–160.

71. Lazazzera, B. A., D. M. Bates, and P. J. Kiley. 1993. The activity of the *Escherichia coli* transcription factor FNR is regulated by a change in oligomeric state. *Genes Dev.* **7**:1993–2005.

72. Lazazzera, B. A., H. Beinert, N. Khoroshilova, M. C. Kennedy, and P. J. Kiley. 1996. DNA binding and dimerization of the Fe-S containing FNR protein from *Escherichia coli* are regulated by oxygen. *J. Biol. Chem.* **271**:2762–2768.

73. Li, E., H. Wing, D. Lee, H. Wu, and S. Busby. 1998. Transcription activation by *Escherichia coli* FNR protein: similarities to, and differences from, the CRP paradigm. *Nucleic Acids Res.* **26**:2075–2081.

74. Lynch, A. S., and E. C. C. Lin. 1996. Responses to molecular oxygen, p. 1526–1538. *In* F. C. Neidhardt, R. Curtiss III, J. L. Ingraham, E. C. C. Lin, K. B. Low, B. Magasanik, W. S. Reznikoff, M. Riley, M. Schaechter, and H. E. Umbarger (ed.), Escherichia coli *and* Salmonella: *Cellular and Molecular Biology*, 2nd ed., vol. 1. ASM Press, Washington D.C.

75. Lynch, A. S., and E. C. C. Lin. 1996. Transcriptional control mediated by the ArcA two-component response regulator protein of *Escherichia coli*: characterization of DNA binding at target promoters. *J. Bacteriol.* **178**:6238–6249.

76. Macheroux, P., S. Hill, S. Austin, T. Eydmann, T. Jones, S. O. Kim, R. Poole, and R. Dixon. 1998. Electron donation to the flavoprotein NifL, a redox-sensing transcriptional regulator. *Biochem. J.* **332**:413–419.

77. Mahan, M. J., J. M. Slauch, and J. J. Mekalanos. 1996. Environmental regulation of virulence gene expression in *Escherichia*, *Salmonella*, and *Shigella* spp., p. 2803–2815. *In* F. C. Neidhardt, R. Curtiss III, J. L. Ingraham, E. C. C. Lin, K. B. Low, B. Magasanik, W. S. Reznikoff, M. Riley, M. Schaechter, and H. E. Umbarger (ed.), Escherichia coli *and* Salmonella: *Cellular and Molecular Biology*, 2nd ed., vol. 2. ASM Press, Washington D.C.

78. Money, T., T. Jones, R. Dixon, and S. Austin. 1999. Isolation and properties of the complex between the enhancer binding protein NIFA and the sensor NIFL. *J. Bacteriol.* **181**:4461–4468.

79. Morett, E., and M. Buck. 1989. In vivo studies on the interaction of RNA polymerase-sigma 54 with the *Klebsiella pneumoniae* and *Rhizobium meliloti nifH* promoters. The role of NifA in the formation of an open promoter complex. *J. Mol. Biol.* **210**:65–77.

80. Mosley, C. S., J. Y. Suzuki, and C. E. Bauer. 1994. Identification and molecular genetic characterization of a sensor kinase responsible for coordinately regulating light harvesting and reaction center gene expression in response to anaerobiosis. *J. Bacteriol.* **176**:7566–7573.

81. Mouncey, N. J., and S. Kaplan. 1998. Oxygen regulation of the *ccoN* gene encoding a component of the cbb_3 oxidase in *Rhodobacter sphaeroides* 2.4.1T: involvement of the FnrL protein. *J. Bacteriol.* **180**:2228–2231.

82. Nellen-Anthamatten, D., P. Rossi, O. Preisig, I. Kullik, M. Babst, H. M. Fischer, and H. Hennecke. 1998. *Bradyrhizobium japonicum* FixK2, a crucial distributor in the FixLJ-dependent regulatory cascade for control of genes inducible by low oxygen levels. *J. Bacteriol.* **180**:5251–5255.

83. O'Gara, J. P., and S. Kaplan. 1997. Evidence for the role of redox carriers in photosynthesis gene expression and carotenoid biosynthesis in *Rhodobacter sphaeroides* 2.4.1. *J. Bacteriol.* **179**:1951–1961.

84. Oh, J. I., and S. Kaplan. 1999. The cbb_3 terminal oxidase of *Rhodobacter sphaeroides* 2.4.1: structural and functional implications for the regulation of spectral complex formation. *Biochemistry* **38**:2688–2696.

85. Park, S. J., G. Chao, and R. P. Gunsalus. 1997. Aerobic regulation of the *sucABCD* genes of *Escherichia coli*, which encode alpha-ketoglutarate dehydrogenase and succinyl coenzyme A synthetase: roles of ArcA, Fnr, and the upstream *sdhCDAB* promoter. *J. Bacteriol.* **179**:4138–4142.

86. Park, S. J., and R. P. Gunsalus. 1995. Oxygen, iron, carbon, and superoxide control of the fumarase *fumA* and *fumC* genes of *Escherichia coli*: role of the *arcA*, *fnr*, and *soxR* gene products. *J. Bacteriol.* **177**:6255–6262.

87. Parkinson, J. S. 1993. Signal transduction schemes of bacteria. *Cell* **73**:857–871.

88. Parkinson, J. S., and E. C. Kofoid. 1992. Communication modules in bacterial signaling proteins. *Annu. Rev. Genet.* **26**:71–112.

89. Patschkowski, T., A. Schlüter, and U. B. Priefer. 1996. *Rhizobium leguminosarum* bv. viciae contains a second *fnr/fixK*-like gene and an unusual *fixL* homologue. *Mol. Microbiol.* **21**:267–280.

90. Pellequer, J. L., K. A. Wager-Smith, S. A. Kay, and E. D. Getzoff. 1998. Photoactive yellow protein: a structural prototype for the three-dimensional fold of the PAS domain superfamily. *Proc. Natl. Acad. Sci. USA* **95**:5884–5890.

91. Pemberton, J. M., I. M. Horne, and A. G. McEwan. 1998. Regulation of photosynthetic gene expression in purple bacteria. *Microbiology* **144**:267–278.

92. Penfold, R. J., and J. M. Pemberton. 1994. Sequencing, chromosomal inactivation, and functional expression in *Escherichia*

coli of *ppsR*, a gene which represses carotenoid and bacteriochlorophyll synthesis in *Rhodobacter sphaeroides*. *J. Bacteriol.* **176**:2869–2876.

93. Peters, J. W., K. Fisher, and D. R. Dean. 1995. Nitrogenase structure and function: a biochemical-genetic perspective. *Annu. Rev. Microbiol.* **49**:335–366.

94. Ponnampalam, S. N., and C. E. Bauer. 1997. DNA binding characteristics of CrtJ. *J. Biol. Chem.* **272**:18391–18396.

95. Ponnampalam, S. N., J. J. Buggy, and C. E. Bauer. 1995. Characterization of an aerobic repressor that coordinately regulates bacteriochlorophyll, carotenoid, and light harvesting-II expression in *Rhodobacter capsulatus*. *J. Bacteriol.* **177**: 2990–2997.

96. Poole, R. K., N. Ioannidis, and Y. Orii. 1994. Reactions of the *Escherichia coli* flavohaemoglobin (Hmp) with oxygen and reduced nicotinamide adenine dinucleotide: evidence for oxygen switching of flavin oxidoreduction and a mechanism for oxygen sensing. *Proc. R. Soc. Lond. B Biol. Sci.* **255**:251–258.

97. Popescu, C., D. Bates, E. Münck, and H. Beinert. 1998. Mössbauer spectroscopy as a tool for the study of activation/inactivation of the transcription regulator FNR in whole cells of *Escherichia coli*. *Proc. Natl. Acad. Sci. USA* **95**:13431–13435.

98. Preisig, O., D. Anthamatten, and H. Hennecke. 1993. Genes for a microaerobically induced oxidase complex in *Bradyrhizobium japonicum* are essential for a nitrogen-fixing endosymbiosis. *Proc. Natl. Acad. Sci. USA* **90**:3309–3313.

99. Preisig, O., R. Zufferey, and H. Hennecke. 1996. The *Bradyrhizobium japonicum fixGHIS* genes are required for the formation of the high-affinity *cbb₃*-type cytochrome oxidase. *Arch. Microbiol.* **165**:297–305.

100. Preisig, O., R. Zufferey, L. Thony-Meyer, C. A. Appleby, and H. Hennecke. 1996. A high-affinity *cbb₃*-type cytochrome oxidase terminates the symbiosis-specific respiratory chain of *Bradyrhizobium japonicum*. *J. Bacteriol.* **178**:1532–1538.

101. Qian, Y., and R. Tabita. 1996. A global signal transduction system regulates aerobic and anaerobic CO₂ fixation in *Rhodobacter sphaeroides*. *J. Bacteriol.* **178**:12–18.

102. Rodgers, K. R. 1999. Heme-based sensors in biological systems. *Curr. Opin. Chem. Biol.* **3**:158–167.

103. Schmitz, R. A. 1997. NifL of *Klebsiella pneumoniae* carries an N-terminally bound FAD cofactor, which is not directly required for the inhibitory function of NifL. *FEMS Microbiol. Lett.* **157**:313–318.

104. Schultze, M., and A. Kondorosi. 1998. Regulation of symbiotic root nodule development. *Annu. Rev. Genet.* **32**:33–57.

105. Sganga, M. W., and C. E. Bauer. 1992. Regulatory factors controlling photosynthetic reaction center and light-harvesting gene expression in *Rhodobacter capsulatus*. *Cell* **68**:945–954.

106. Shaw, D. J., D. W. Rice, and J. R. Guest. 1983. Homology between CAP and Fnr, a regulator of anaerobic respiration in *Escherichia coli*. *J. Mol. Biol.* **166**:241–247.

107. Sidoti, C., G. Harwood, R. Ackerman, J. Coppard, and M. Merrick. 1993. Characterisation of mutations in the *Klebsiella pneumoniae* nitrogen fixation regulatory gene *nifL* which impair oxygen regulation. *Arch. Microbiol.* **159**:276–281.

108. Söderback, E., F. Reyes-Ramirez, T. Eydmann, S. Austin, S. Hill, and R. Dixon. 1998. The redox- and fixed nitrogen-responsive regulatory protein NIFL from *Azotobacter vinelandii* comprises discrete flavin and nucleotide-binding domains. *Mol. Microbiol.* **28**:179–192.

109. Spiro, S. 1994. The FNR family of transcriptional regulators. *Antonie Van Leeuwenhoek* **66**:23–36.

110. Stewart, V., and R. S. Rabin. 1995. Dual sensors and dual response regulators interact to control nitrate- and nitrite-responsive gene expression in *Escherichia coli*, p. 233–252. *In* J. A. Hoch and T. J. Silhavy (ed.), *Two-Component Signal Transduction*. ASM Press, Washington, D.C.

111. Stock, J. B., A. M. Stock, and J. M. Mottonen. 1990. Signal transduction in bacteria. *Nature* **344**:395–400.

112. Taylor, B. L., and I. B. Zhulin. 1999. PAS domains: internal sensors of oxygen, redox potential, and light. *Microbiol. Mol. Biol. Rev.* **63**:479–506.

113. Tosques, I. E., J. Shi, and J. P. Shapleigh. 1996. Cloning and characterization of *nnrR*, whose product is required for the expression of proteins involved in nitric oxide metabolism in *Rhodobacter sphaeroides* 2.4.3. *J. Bacteriol.* **178**:4958–4964.

114. Tseng, C. P. 1997. Regulation of fumarase (*fumB*) gene expression in *Escherichia coli* in response to oxygen, iron and heme availability: role of the *arcA*, *fur*, and *hemA* gene products. *FEMS Microbiol. Lett.* **157**:67–72.

115. Tseng, C.-P., J. Albrecht, and R. P. Gunsalus. 1996. Effect of microaerophilic cell growth conditions on expression of the aerobic (*cyoABCDE* and *cydAB*) and anaerobic (*narGHJI*, *frdABCD*, and *dmsABC*) respiratory pathway genes in *Escherichia coli*. *J. Bacteriol.* **178**:1094–1098.

116. Unden, G., S. Becker, J. Bongaerts, G. Holighaus, J. Schirawski, and S. Six. 1995. O₂-sensing and O₂-dependent gene regulation in facultatively anaerobic bacteria. *Arch. Microbiol.* **164**:81–90.

117. Unden, G., S. Becker, J. Bongaerts, J. Schirawski, and S. Six. 1994. Oxygen regulated gene expression in facultatively anaerobic bacteria. *Antonie van Leeuwenhoek* **66**:3–22.

118. Unden, G., and J. Bongaerts. 1997. Alternative respiratory pathways of *Escherichia coli*: energetics and transcriptional regulation in response to electron acceptors. *Biochim. Biophys. Acta* **1320**:217–234.

119. Unden, G., and A. Duchene. 1987. On the role of cyclic-AMP and the FNR protein in *Escherichia coli* growing anaerobically. *Arch. Microbiol.* **147**:195–200.

120. Unden, G., M. Trageser, and A. Duchene. 1990. Effect of positive redox potentials (>+400mV) on the expression of anaerobic respiratory enzymes in *Escherichia coli*. *Mol. Microbiol.* **4**:315–319.

121. van Rhijn, P., and J. Vanderleyden. 1995. The *Rhizobium*-plant symbiosis. *Microbiol. Rev.* **59**:124–142.

122. Van Spanning, R. J., A. P. N. De Boer, W. N. M. Reijnders, H. V. Westerhoff, A. H. Stouthamer, and J. Van Der Oost. 1997. FnrP and NNR of *Paracoccus denitrificans* are both members of the FNR family of transcriptional activators but have distinct roles in respiratory adaptation in response to oxygen limitation. *Mol. Microbiol.* **23**:893–907.

123. Vega, Y., C. Dickneite, M. Ripio, R. Bockmann, B. Gonzalez-Zorn, S. Novella, G. Dominguez-Bernal, W. Goebel, and J. A. Vazquez-Boland. 1998. Functional similarities between the *Listeria monocytogenes* virulence regulator PrfA and cyclic AMP receptor protein: the PrfA* (Gly145Ser) mutation increases binding affinity for the target DNA. *J. Bacteriol.* **180**:6655–6660.

124. Williams, R., A. Bell, G. Sims, and S. Busby. 1991. The role of two surface exposed loops in transcription activation by the *Escherichia coli* CRP and FNR proteins. *Nucleic Acids Res.* **19**:6705–6712.

125. Yeliseev, A. A., and S. Kaplan. 1999. A novel mechanism for the regulation of photosynthesis gene expression by the TspO

outer membrane protein of *Rhodobacter sphaeroides* 2.4.1. *J. Biol. Chem.* **274**:21234–21243.

126. **Yeliseev, A. A., and S. Kaplan.** 1995. A sensory transducer homologous to the mammalian peripheral-type benzodiazepine receptor regulates photosynthetic membrane complex formation in *Rhodobacter sphaeroides* 2.4.1. *J. Biol. Chem.* **270**:21167–21175.

127. **Yeliseev, A. A., K. E. Krueger, and S. Kaplan.** 1997. A mammalian mitochondrial drug receptor functions as a bacterial "oxygen" sensor. *Proc. Natl. Acad. Sci. USA* **94**:5101–5106.

128. **Zeilstra-Ryalls, J., M. Gomelsky, J. M. Eraso, A. Yeliseev, J. O'Gara, and S. Kaplan.** 1998. Control of photosystem formation in *Rhodobacter sphaeroides*. *J. Bacteriol.* **180**:2801–2809.

129. **Zeilstra-Ryalls, J. H., K. Gabbert, N. J. Mouncey, S. Kaplan, and R. G. Kranz.** 1997. Analysis of the *fnrL* gene and its function in *Rhodobacter capsulatus*. *J. Bacteriol.* **179**:7264–7273.

130. **Zeilstra-Ryalls, J. H., and S. Kaplan.** 1995. Aerobic and anaerobic regulation in *Rhodobacter sphaeroides* 2.4.1: the role of the *fnrL* gene. *J. Bacteriol.* **177**:6422–6431.

131. **Zeilstra-Ryalls, J. H., and S. Kaplan.** 1998. Role of the *fnrL* gene in photosystem gene expression and photosynthetic growth of *Rhodobacter sphaeroides* 2.4.1. *J. Bacteriol.* **180**:1496–1503.

132. **Zhulin, I. B., B. L. Taylor, and R. Dixon.** 1997. PAS domain S-boxes in Archaea, Bacteria and sensors for oxygen and redox. *Trends Biochem. Sci.* **22**:331–333.

Bacterial Stress Responses
Edited by G. Storz and R. Hengge-Aronis
©2000 ASM Press, Washington, D.C.

Chapter 6

Coping with Osmotic Challenges: Osmoregulation through Accumulation and Release of Compatible Solutes in Bacteria

ERHARD BREMER AND REINHARD KRÄMER

Maintenance of turgor within physiologically acceptable boundaries is a key determinant for the growth of microorganisms. Bacteria that do not live permanently in hyperosmotic environments must adjust to fluctuations in the water content of their habitat. Osmotically instigated transmembrane water movements can cause either dehydration of the cytoplasm under hypertonic conditions or bursting of the cell in hypoosmotic environments. Microorganisms lack systems for active water transport and adjust cellular water content via osmosis by actively controlling the level of their intracellular solute pool. A limited group of organic osmolytes, the so-called compatible solutes, are used to balance environmental osmolality. These compounds can be accumulated to very high levels without disturbing the functioning of the cell and the most widely used are trehalose, proline, ectoine, and glycine betaine. Compatible solutes are amassed through synthesis or uptake from the environment, and their intracellular concentration is sensitively adjusted to the degree of osmotic stress. This process is controlled by an interplay between several finely tuned mechanisms that regulate either gene transcription or activity of the encoded enzymes and transporters used for compatible solute accumulation. The same compounds that protect the cell from the detrimental effects of hyperosmotic conditions become a threat to the cell's survival upon osmotic downshock. To reduce the driving force for water entry, the cell rapidly expels excess compatible solutes in a controlled manner through mechanosensitive channels. The amassing of organic osmolytes under hyperosmotic conditions and their expulsion when external osmolality drops allow the cell to control water content and turgor and thus sustain growth under unfavorable conditions.

Essentially all microorganisms other than intracellular parasites and symbionts face ever-changing environmental conditions. Their ability to thrive under often highly stressful circumstances depends on the sensing of environmental changes and responding to these challenges via highly integrated stress reactions. Two types of bacterial stress responses operate in microorganisms. General stress responses, which are frequently controlled through a single or a few master regulators, provide cross-protection against a wide variety of environmental insults regardless of the initial stimulus (54, 55) (see also chapters 11 and 12). While effective in ensuring the cell's survival, they may not be sufficient to allow growth when the cell is confronted with a particular stress condition for a prolonged period. In these circumstances, cells use specific stress responses that involve highly integrated networks of physiological and genetic adaptation mechanisms. There is also often a complex interplay between global regulators and cellular response systems directed against specific types of stress, thereby adding an additional level of sophistication to the cell's coordination of emergency and long-term stress reactions (57).

The availability of water is one of the most significant environmental parameters affecting the survival and growth of microorganisms (66, 121). Changes in the external osmolality immediately trigger fluxes of water along the osmotic gradient that could result either in swelling and bursting of the cell in hypotonic environments or in plasmolysis and dehydration under hypertonic conditions. The cell avoids these devastating alternatives by using active countermeasures to retain a suitable level of cytoplasmic water and turgor (30, 42, 76, 118, 149). Microorganisms control turgor by actively modulating

Erhard Bremer • Laboratory of Microbiology, Department of Biology, University of Marburg, D-35032 Marburg, Germany.   **Reinhard Krämer** • Department of Chemistry, University of Cologne, D-50674 Cologne, Germany.

the pool of osmotically active solutes in their cytoplasm, thereby allowing water content to be adjusted by osmosis. The physiological and molecular mechanisms through which this is accomplished are the focus of this chapter.

THE CELL AND THE SURROUNDING SOLVENT

When two compartments are separated by a semipermeable membrane, the solvent water flows from the compartment with high chemical potential of water to that with lower chemical potential. The osmotic pressure (Π) is defined as the hydrostatic pressure that, in an equilibrium state, balances the difference in chemical potential such that no movement of solvent occurs between the two compartments. The osmotic pressure (Π) of an aqueous solution is defined as $\Pi = (RT/V_w) \ln a_w$ and is thus directly related to the water activity (a_w, the mole fraction of water in a solution) and to V_w, the partial molal volume of water. Osmolality describes the osmotic pressure Π of a solution and is expressed in units of osmoles (moles of osmolyte) per kilogram of solvent. The term osmolarity, which is defined as the sum of the concentrations of osmotically active solutes (osmolytes) in solution, is a frequently used approximation for osmolality (149).

The total concentration of osmolytes within a cell is generally higher than in the environment, causing water to flow down its chemical potential into the cell. This influx increases the cellular volume, thereby pressing the cytoplasmic membrane toward the murein sacculus. The pressure difference across the inner membrane/cell wall complex is the turgor, $\Delta\Pi$ ($\Delta\Pi = \Pi_{in} - \Pi_{out}$). Turgor balances the difference in osmotic pressure between the cell interior and exterior and must be maintained through the growth cycle as the cell elongates (25). Turgor is considered essential for cell viability and critical for growth, and it is thought to provide the mechanical force for expansion of the cell wall (63, 80). Although turgor is quite difficult to quantify (149), values of 3 to 5 bars (0.3 to 0.5 MPa) for gram-negative bacteria and approximately 20 bars (2 MPa) for gram-positive organisms have been estimated (30, 66, 148). The much higher value in gram-positive bacteria is thought to reflect the large cytoplasmic solvent pool needed for expansion of the multilayered peptidoglycan.

WATER PERMEATION ACROSS THE CYTOPLASMIC MEMBRANE

Osmotically instigated water fluxes across the cytoplasmic membrane are accomplished by two distinct mechanisms. Simple diffusion of water across the lipid bilayer is usually sufficient to balance solute levels, but a much accelerated water transit is achieved by diffusion through water-selective channels. These so-called aquaporins are abundant in plant and animal cells that exhibit rapid transmembrane water movement (1, 102). They are also present in the lower eukaryote *Saccharomyces* (13) and have been detected in several bacterial species. The *Escherichia coli* aquaporin (AqpZ) (19) serves as a model for bacterial water channels and is a member of the ubiquitous major intrinsic protein family (MIP). Expression studies in *Xenopus* oocytes established the function of AqpZ as a water channel, and cryoelectron microscopy examinations have suggested that AqpZ mediates rapid and large water fluxes in both directions in response to sudden osmotic up- or downshifts (33). This indicates an important role for AqpZ in the cell's management of water flow. However, this channel is not essential (at least under laboratory conditions) since *aqpZ* deletion mutants exhibit only minor growth defects (20), and the physiological role of AqpZ in bacterial osmoregulation remains to be more fully explored.

MICROBIAL STRATEGIES FOR COPING WITH HYPEROSMOTIC ENVIRONMENTS

Microorganisms have developed two fundamentally different schemes for maintaining turgor under hyperosmotic conditions (42, 76, 109, 143). These are frequently referred to as the "salt-in" and "salt-out" strategies. The former has been adopted by microorganisms whose entire physiology has been adapted to a permanent life in high-osmolality surroundings; this group of bacteria accumulates large amounts of ions in their cytoplasm. Two phylogenetically unrelated groups of *Archaea* and *Bacteria*, the aerobic *Halobacteriales* and the *Haloanaerobiales*, colonize hypersaline environments, with salt concentrations ranging between 2 M and 5 M, as their preferred ecological niches (97, 109, 143). In their hypersaline habitats, K^+, Mg^{2+}, Na^+, and Ca^{2+} are generally the dominant cations, and Cl^-, SO_4^{2-}, and CO_3^{2-} are the major anions. These organisms selectively accumulate molar concentrations of K^+ and Cl^- in their cytoplasm and usually actively extrude the deleterious Na^+ ions (see chapter 8), resulting in a cytoplasmic ion composition that differs substantially from the surroundings. However, high concentrations of inorganic ions induce aggregation of macromolecules by enhancing hydrophobic interactions, and they restrict the availability of water for biochemical processes by salt ion hydration. These effects necessitated evolutionary changes in essentially all proteins to adjust the entire cell physiology to the high ion content of the cytoplasm (17, 35, 38, 117). As a result, halophilic microorganisms require high salt concentrations (frequently K^+) for most biochemical reactions, and charge repulsion of the many acid amino acids present in proteins of true halophiles leads to cell disintegration at lower salt concentrations. Ultimately, the salt-in strategy employed by the *Halobacteriales* and *Haloanaerobiales* severely limits their habitats and precludes the colonization of environments with moderate osmolalities.

A more versatile and flexible osmostress response is provided by the salt-out strategy. It is used by bacteria that live either in environments of moderate salinity or in ecological niches that are only periodically subjected to conditions of low water activity. This group of microorganisms avoids high ionic conditions in their cytoplasm and instead amasses large amounts of specific organic osmolytes that are highly congruous with cellular functions (30, 42, 69, 76, 109, 125, 143). These so-called compatible solutes (Fig. 1) can be accumulated up to molar concentrations in the cytoplasm and thus contribute significantly to the maintenance of turgor under hyperosmotic conditions. Because the salt-out adaptation mechanism does not require evolutionary adjustment of proteins and cellular processes to high salt concentrations, this response to osmotic stress is prevalent not only in *Bacteria* and *Archaea* (28, 32, 42, 76, 109, 118) but also in fungal, plant, animal, and human cells (18, 61, 100).

CHARACTERISTICS OF COMPATIBLE SOLUTES

By definition, compatible solutes are compounds that can be accumulated to very high levels (several moles per liter) without disturbing cellular physiology

(3, 16, 34, 135, 153). Because only a limited number of compounds meet these criteria, it is not surprising that certain compatible solutes have been widely adopted as osmostress protectants across the kingdoms (90, 154). The spectrum of compatible solutes identified comprises amino acids (e.g., glutamate, proline), amino acid derivatives (e.g., ectoine, proline betaine), small peptides (e.g., *N*-acetylglutaminylglutamine amide), methylamines (e.g., glycine betaines, carnitine) and their sulfonium analogs (e.g., dimethylsulfonium propionate), sulfate esters (e.g., choline-O-sulfate), polyols (e.g., glycerol, glycosylglycerol), and sugars (e.g., trehalose, sucrose). However, in the past few years, as more organisms have been examined, in particular those that thrive in extreme environments, a variety of novel organic osmolytes have been identified (32, 101). A given bacterium usually employs a selection of compatible solutes, and the composition of its compatible solute pool can vary in response to growth phase and growth conditions. Glycine betaine, ectoine, proline, and trehalose (Fig. 1) are probably the most widely used compatible solutes in the microbial world.

In general, compatible solutes are polar, highly soluble molecules, and they usually do not carry a net charge at physiological pH. Exceptions are negatively charged organic osmolytes (such as sulfotrehalose, diglycerolphosphate, and di-*myo*-1,1'-inositol phosphate) used by thermophilic *Archaea* (32, 101). K^+ is used as the counterion for these solutes; this ion is also accumulated transiently (often in combination with glutamate) by many nonhalophilic bacteria to initially offset the detrimental effects of an osmotic upshock (14, 30, 39, 152). The compatibility of organic osmolytes with cell functioning can be correlated with their effects on the behavior of water at interfaces and has been related to the so-called Hofmeister series (26, 153, 154). Compounds that break water structure and consequently destabilize proteins are called chaotropic, whereas those that increase wa-

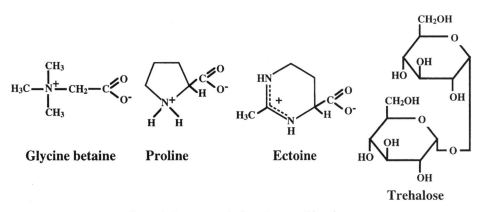

Figure 1. Structures of selected compatible solutes.

ter structure and stabilize proteins are termed kosmotropic. Compatible solutes are excluded from the immediate hydration shell of polypeptides, resulting in a preferential hydration of protein surfaces. The disruption of the ordered water structure around proteins by local or global unfolding becomes energetically unfavorable, and therefore the native conformations of proteins are stabilized (3, 123, 149, 154).

The accumulation of compatible solutes not only allows microbial cells to withstand a given osmolality, but it also extends their ability to colonize more saline ecological niches (42). Depending on the type, compatible solutes can also protect microorganisms against stresses other than dehydration. An example is the increased cold tolerance conferred on *Listeria monocytogenes* by the accumulation of glycine betaine from food sources (79). The beneficial effect of compatible solutes on membrane integrity and protein folding and stability has been demonstrated in in vitro experiments (129, 149, 153). These findings have heightened the biotechnological interest in these compounds as stabilizers in processes such as desiccation, freezing, and heating (23, 92). In addition, the common response of plant and microbial cells to high osmolality by producing compatible solutes (90, 100) has fostered interest in using bacterial systems for osmoprotectant synthesis as resources for the metabolic engineering of tolerance against high salinity and drought in commercially important crops (53, 62).

BIOSYNTHESIS OF COMPATIBLE SOLUTES: TREHALOSE, GLYCINE BETAINE, ECTOINE, AND PROLINE

Cell growth depends largely on the cell's ability to modulate its intracellular solvent pool in response to osmotic changes in the environment (125), and synthesis of compatible solutes plays an important role in this process. A considerable variety of compatible solutes can be produced by microorganisms (28, 32, 42, 101), but the biosynthetic pathways involved have been elucidated at the molecular level for only a few. Examples are the sugar trehalose, the trimethylammonium compound glycine betaine, the tetrahydropyrimidine ectoine, and the amino acid proline in *Bacteria* (Fig. 1).

The Disaccharide Trehalose

The sugar trehalose (Fig. 1) is an important stress compound and membrane-stabilizing agent in both prokaryotic and eukaryotic organisms (137). *E. coli* and *Salmonella enterica* serovar Typhimurium accumulate it via de novo synthesis as a compatible solute. Two enzymes encoded by the *otsAB* operon determine the osmoregulatory trehalose production (Fig. 2). OtsA, the trehalose-6-phosphate synthase, catalyzes the enzymatic condensation of the precursors glucose-6-phosphate and UDP-glucose; free trehalose is then generated from this intermediate by

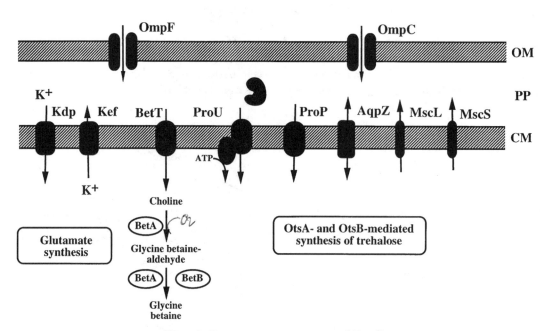

Figure 2. Osmostress response systems of *E. coli.*

rapid dephosphorylation via the *otsB*-encoded trehalose-6-phosphate (Tre-6-P) phosphatase. Transcription of the *otsAB* operon is induced in osmotically stressed cells and is also enhanced when unstressed cells enter stationary phase (54, 71). This pattern of *otsAB* expression reflects its control by RpoS, an important alternative transcription factor controlling (or contributing to) gene expression of a large regulon of stationary-phase and stress-induced genes (56, 58) (see chapter 11). However, stationary-phase levels of trehalose are at least severalfold lower than the levels reached in response to osmotic stress (43). This finding indicates that stationary-phase cells that are highly osmotolerant (88, 98) are not dependent on a massive amount of trehalose to withstand a high degree of osmotic stress (43). Increased levels of trehalose are also accumulated under carbon-starvation conditions, pointing to the role of this nonreducing disaccharide as a general stress protectant (129, 137).

While trehalose serves as a stress protectant under hyperosmotic conditions, it also functions as an energy source at moderate osmolalities (64). The role that it plays is determined by the osmotic strength of the growth medium. Under conditions of low osmolality, extracellular trehalose is taken up by the cells via a specific phosphotransferase system (PTS) involving the EIIBC^{Tre}(TreB)/EIIA^{Glc} complex, and it is released into the cytoplasm as Tre-6-P. A Tre-6-P hydrolase (TreC) cleaves this phosphorylated intermediate into glucose and glucose-6-phosphate for further metabolism. The genes (*treBC*) encoding the TreB transporter protein and the TreC enzyme are negatively controlled in response to the availability of trehalose in the growth medium by the TreR repressor protein, a member of the LacI family of prokaryotic DNA-binding proteins (52). Tre-6-P serves as the actual inducer. Since Tre-6-P is an intermediate in both the degradative and biosynthetic pathways, binding of the inducer to TreR must be sensitively controlled. The decision for one of the two pathways is governed by the intracellular ratio of Tre-6-P and trehalose, both of which can bind to TreR (52). While Tre-6-P prevents the binding of TreR to its operator, the TreR:trehalose complex retains DNA-binding capacity. As trehalose accumulates in osmotically stressed cells, this sugar competes (despite its lower affinity) with the inducer Tre-6-P for the available binding site on TreR and maintains this regulatory protein in a DNA-bound form that represses the expression of the catabolic *treBC* operon. A futile cycle of trehalose biosynthesis and degradation under hyperosmotic conditions is thus avoided.

Only de novo synthesized trehalose functions as a compatible solute in *E. coli* since this bacterium cannot take up free trehalose from the environment. Under high-osmolality growth conditions exogenously provided trehalose is hydrolyzed into two glucose molecules by a periplasmic trehalase (TreA) that are then transported into the cytoplasm via the glucose-specific PTS (50). Transcription of the *treA* structural gene is induced under hyperosmotic growth conditions and, like the *otsAB* genes, is regulated by RpoS (56, 137). The presence of TreA in the periplasm permits the cell to efficiently recapture exuded trehalose that was originally synthesized for osmoregulatory purposes and use it as a carbon source.

The Trimethylammonium Compound Glycine Betaine

The compound most widely used as an osmoprotectant by bacterial, plant, animal, and even human cells is glycine betaine (Fig. 1) (18, 30, 76, 100, 126). Two different routes for its production have been detected in bacteria. Some microorganisms have the ability to synthesize it de novo by a stepwise methylation of the amino acid glycine, with sarcosine and dimethylglycine as the intermediates and S-adenosyl methionine as the methyl donor (42). The second biosynthetic pathway, the enzymatic oxidation of choline to glycine betaine, is commonly used by both prokaryotic and eukaryotic cells, but the types of enzymes involved can vary.

The osmoregulatory choline to glycine betaine pathway has been characterized at the molecular level for the gram-negative and gram-positive model organisms, *E. coli* (85) and *B. subtilis* (12, 74). These bacteria cannot synthesize the precursor choline and acquire it from environmental sources such as decomposing plant and animal tissues. A single-component choline transporter (BetT) driven by the proton motive force (Table 1) is found in *E. coli* (Fig. 2), whereas two multicomponent, binding-protein-dependent ABC transporters (OpuB and OpuC) mediate high-affinity choline uptake in *B. subtilis* (Fig. 3). The first step in glycine betaine synthesis is performed by two different types of enzymes in *E. coli* and *B. subtilis*. *B. subtilis* uses a soluble, metal-containing, type III alcohol dehydrogenase (GbsB) (Fig. 3) to convert choline into glycine betaine aldehyde, whereas *E. coli* uses an FAD-containing, membrane-bound choline dehydrogenase (BetA) that can also oxidize glycine betaine aldehyde to glycine betaine at the same rate (Fig. 2). Both organisms possess evolutionarily well-conserved and highly salt-tolerant glycine betaine aldehyde dehydrogenases (BetB, GbsA) that convert the toxic intermediate into

Table 1. Uptake systems for compatible solutes

Organism	System	Type of mechanism	Substrate spectrum	Major substrate(s)[a]	Regulation at the level of:	
					Expression	Activity
E. coli	ProP	H[+] symport	Broad	GB, PB, Pro, Car, Ect, others	+	+
	ProU	ABC transporter[b]	Broad	GB, PB, Pro, others	+	+
	BetT	Secondary transport[c]	Narrow	Cho	+	+
B. subtilis	OpuA	ABC transporter	Medium	GB, PB, others	+	ND[d]
	OpuB	ABC transporter	Narrow	Cho	+	ND
	OpuC	ABC transporter	Broad	GB, PB, Cho, Car, others	+	ND
	OpuD	Na[+] symport	Narrow	GB, others	+	+
	OpuE	Na[+] symport	Narrow	Pro	+	−
C. glutamicum	BetP	Na[+] symport	Narrow	GB	+	+
	EctP	Na[+] symport	Broad	GB, Pro, Ect	−	+
	ProP	H[+] symport	Medium	Pro, Ect	+	+

[a] Car, carnitine; Cho, choline; Ect, ectoine; GB, glycine betaine; PB, proline betaine; Pro, proline.
[b] ABC, ATP-binding cassette.
[c] The cotransported ion (Na[+] or H[+]) is not known.
[d] ND, not determined.

the well-tolerated and metabolically inert glycine betaine (Fig. 2 and 3).

Glycine betaine production depends on the availability of choline. The presence of this precursor in the environment significantly affects the expression of genes encoding the systems for its uptake and enzymatic oxidation both in *E. coli* and *B. subtilis*. In *E. coli*, BetI, a member of the TetR family of bacterial regulators, serves as a specific choline-sensing repressor. It coordinates the transcription of the divergently oriented *betIBA* operon and the *betT* gene (127). Choline oxidation mediated by the choline dehydrogenase BetA (Fig. 2) can only occur in the presence

of oxygen. Under anaerobic conditions, the DNA-binding protein ArcA represses both the *betT* and *betIBA* promoters (87). The *bet* genes are therefore under the control of the two-component Arc (aerobic respiration control) system, which consists of the membrane-bound histidine kinase ArcB and the cytoplasmic response regulator ArcA. Increases in medium osmolality stimulate *betI* and *betIBA* expression, but the way in which osmotic changes are sensed and transmitted to the *bet* gene cluster is unknown.

B. subtilis possesses no BetI-related protein. Instead, a different type of choline-sensing regulatory

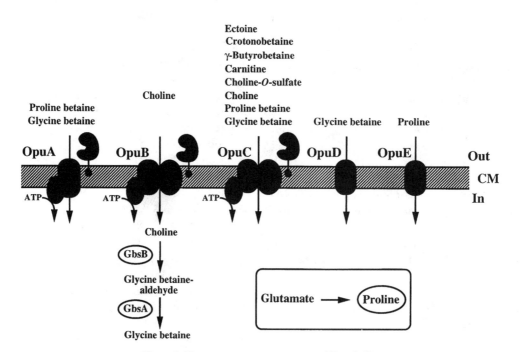

Figure 3. Osmostress response systems of *B. subtilis*.

protein (GbsR) is encoded by a gene that is transcribed divergently from the *gbsAB* glycine betaine biosynthetic gene cluster (G. Nau-Wagner, B. Kempf, J. Boch, and E. Bremer, unpublished data). GbsR negatively regulates both *gbsAB* and *opuB*, an operon encoding a high-affinity and highly substrate-specific choline transport system (Fig. 3). Expression of the *opuB* operon increases in response to elevations in medium osmolality (74), but like the choline-sensing BetI protein (87), the GbsR repressor is not involved in this regulatory process (Nau-Wagner et al., unpublished data). In contrast to the expression of the glycine betaine biosynthetic genes (*betBA*) in *E. coli*, *gbsAB* transcription is not greatly affected by osmotic stimuli.

The Tetrahydropyrimidine Ectoine

In addition to glycine betaine, ectoine (Fig. 1) is without doubt one of the most widely synthesized osmoprotectants in the microbial world (42). Many bacteria, including those that do not produce ectoine, such as *E. coli*, *B. subtilis*, and *Corynebacterium glutamicum*, have the ability to scavenge preformed ectoine from exogenous sources for osmoprotective purposes through transport processes (67, 68, 116). The pathway for ectoine biosynthesis was elucidated at the biochemical and molecular levels for two moderately halophilic eubacteria, the gram-positive *Marinococcus halophilus* (93) and the gram-negative *Halomonas elongata* (22). Disruption of the *ectABC* synthesis genes causes salt sensitivity in the resulting mutant strain (21). L-Aspartate-β-semialdehyde serves as the precursor for ectoine, and the consecutive action of three biosynthetic enzymes (EctBAC) is required for its production. Some microorganisms can modify ectoine by hydroxylation via a still uncharacterized pathway; in high-osmolality environments, these bacteria accumulate either a mixture of ectoine and hydroxyectoine or convert ectoine entirely into the hydroxylated derivative (23, 42).

The Amino Acid Proline

Proline plays a crucial role in osmotolerance in many bacteria. Often, it is taken up from environmental sources at times of osmotic stress (28, 29), and it is also synthesized in an osmotically controlled fashion by various bacteria, e.g., *Streptomyces* and *Bacillus* (147). Attesting to the effectiveness of this amino acid as an osmostress protectant, proline production is also widely used in plant cells as a defense against high salinity and drought (34). In bacteria, synthesis of proline for anabolic purposes usually proceeds from glutamate in three enzymatic steps involving γ-glutamyl kinase (ProB), glutamate-semialdehyde-dehydrogenase (ProA), and pyrroline-5-carboxylate-reductase (ProC). The activity of the first enzyme (ProB) of this pathway is frequently controlled by feedback inhibition through proline, thereby linking the production of this amino acid to the cell's biosynthetic requirements and preventing proline buildup. Substantially increased concentrations of proline are found in *proB* mutants of serovar Typhimurium and *E. coli*, where single amino acid substitutions cause a strong reduction in feedback inhibition of the γ-glutamyl kinase (27, 134). This confers enhanced osmotic stress tolerance in both species, although neither naturally synthesizes proline at elevated levels under hyperosmotic conditions.

While *E. coli* and serovar Typhimurium synthesize trehalose as their endogenous osmoprotectant (137), the soil bacterium *B. subtilis* increases proline production for this purpose. Upon a sudden osmotic upshock with 0.4 M NaCl, the intracellular proline concentration in *B. subtilis* is strongly increased via de novo synthesis from a basal level of 16 mM to approximately 500 mM, whereas the pool size of the other amino acids does not change (147). The intracellular proline content is proportionally linked to the osmolality of the growth medium over a considerable range of salt concentrations. A new proline biosynthetic pathway, which is different from that used for anabolic purposes, operates in high-osmolality stressed cells. It consists of two isoenzymes (ProH and ProJ) of the *proB*- and *proC*-encoded γ-glutamyl kinase and pyrroline-5-carboxylate-reductase, respectively, and also includes the *proA*-encoded glutamate-semialdehyde-dehydrogenase that functions in anabolic proline biosynthesis. Osmotic stress stimulates the expression of the operon encoding the ProHJ isoenzymes. Genetic disruption of the *proHJ* operon resulted in a strain that was proline prototrophic but no longer accumulated proline under high osmotic growth conditions; it exhibited a severe growth defect in hyperosmotic media (J. Brill, F. Spiegelhalter, and E. Bremer, unpublished data). This phenotype exemplifies the crucial role of compatible solute biosynthesis in the osmotic stress response of microorganisms.

TRANSPORT OF COMPATIBLE SOLUTES

In addition to accumulating compatible solutes by endogenous synthesis, a large variety of *Bacteria* and *Archaea* have developed the ability to acquire preformed osmoprotectants from exogenous sources. These compounds are released into ecosystems by primary microbial producers from osmotically downshocked cells; by decaying microbial, plant, and animal cells; by root exudates; and by mammals in their excretion fluids (e.g., urine).

Transporters for osmoprotectants (Table 1) have evolved to meet the special demands imposed by their physiological tasks. In natural ecosystems, the supply of osmoprotectants and their biosynthetic precursors is varying and generally very low, with concentrations usually in the nanomolar to micromolar range (78). Therefore, osmoprotectant transporters commonly exhibit very high affinity for their major substrates, and their capacity is geared to permit accumulation of compatible solutes to molar concentrations. In addition, they function effectively at high osmolality and at high ionic strength, conditions that frequently impair the performance of transporters for nutrients (128). To take advantage of the spectrum of osmoprotectants available in their habitat, microorganisms often possess several transport systems, some of which exhibit broad substrate specificity. Transporters for osmoprotectants have been most fully investigated at the molecular level in the gram-negative enteric bacteria *E. coli* and serovar Typhimurium and in the gram-positive soil bacteria *B. subtilis* and *C. glutamicum*.

Compatible Solute Uptake in the Gram-Negative Bacteria *E. coli* and Serovar Typhimurium

Two transport systems, ProP and ProU, are primarily responsible for the uptake of preformed compatible solutes in *E. coli* and serovar Typhimurium (Fig. 2). They were originally discovered as transporters for the uptake of proline under hyperosmotic growth conditions (hence their names), but subsequent studies revealed their broader substrate specificity (Table 1) and, in particular, their involvement in glycine betaine uptake (14, 30, 48, 77, 94).

ProP is a single-component system located in the cytoplasmic membrane (Table 1) and functions as an H^+ symporter for a wide variety of osmoprotectants structurally related to proline and glycine betaine (31, 96). ProP-mediated transport of compatible solutes is greatly enhanced in high-osmolality media as a result of increased *proP* expression and stimulated transport activity. Transcription of *proP* is directed from two promoters, P1 and P2, both of which are activated by osmotic upshifts (103). The activity of *proP*-P1 is normally repressed by cAMP-CRP complex, which binds to a site overlapping the −35 region. Activity of *proP*-P2 is dependent on the alternative sigma factor RpoS and the nucleoid-associated DNA-binding protein FIS (150, 151). Thus *proP* is a member of a growing class of RpoS-dependent genes that respond to both stationary-phase and high-osmolality signals

(56). Activity of the ProP protein is enhanced by hypertonicity both in vivo and in vitro (Table 1), suggesting that upon hyperosmotic upshock the ProP transporter functions as a sensor, transducer, and responder (124). ProP possesses a carboxyterminal extension with a high propensity for forming an alpha-helical coiled coil; this domain is thought to be critical for the osmoregulation of ProP transport activity (31).

The second osmoprotectant uptake system, ProU, is a multicomponent system and a member of the ABC superfamily of transporters (Table 1) in which substrate translocation across the cell membrane is dependent on ATP hydrolysis (30, 48, 94). It consists of three proteins. The periplasmic substrate-binding protein ProX recognizes glycine betaine and proline betaine with high affinity (47, 51) and delivers them to the integral inner membrane component ProW. The ProW-mediated substrate translocation across the cytoplasmic membrane depends on the hydrolysis of ATP by the inner membrane-associated ATPase ProV. ProU also serves as low-affinity transporter for a considerable range of other compatible solutes, but curiously none of them seems to bind to ProX (47, 51). A sudden osmotic upshock results in a rapid induction of *proU* (*proVWX*) transcription (up to 100-fold) to a level proportionally linked to the osmolality of the growth medium. This pattern of gene expression reflects the physiological function of ProU since it permits the number of ProU transporters to be sensitively adjusted to the degree of osmotic stress. Consequently, *proU* expression is kept at elevated levels as long as the osmotic stimulus persists (30, 48, 59, 94). This regulatory pattern distinguishes osmoregulation of *proU* from that of the *kdp* operon, which is only transiently induced (39, 84) and controlled through the two-component regulatory system KdpD and KdpE (70, 139, 144). The Kdp transport system mediates K^+ uptake as an immediate response to osmotic upshock, loss of turgor, and limiting K^+ concentrations, but it does not serve in the long-term adaptation to hyperosmotic environments. This contrast in regulatory patterns indicates that fundamentally different parameters are sensed to increase *proU* and *kdp* transcription in response to osmotic stimuli. No classical regulatory protein that specifically controls *proU* transcription has been identified. Genetic searches for such proteins have yielded mutants with alterations in DNA-binding proteins (e.g., H-NS, TopA, GyrAB, IHF, HU) that have pleiotropic effects on gene expression and DNA superstructure. The recovery of these types of mutants and the finding that DNA supercoiling is increased in high-osmolality-grown cells have suggested a model in which changes in DNA superstructure directly control the strength of *proU* expression (60, 112). However, the correlation between the level of *proU* expression and medium osmolality seems too finely tuned to be explained by DNA supercoiling alone, and it is unclear which features might make the *proU* control region so uniquely responsive to changes in DNA superstructure (30). The nucleoid-associated DNA-binding protein H-NS (5) plays an important role in governing *proU* expression and is one of the proteins that binds to an important transcriptional regulatory element (the silencer) within *proV*, the first gene of the *proU* operon (95, 111, 112). H-NS functions as a modulator rather than a regulator of *proU* because osmoregulation is not abolished in strains with lesions in *hns* (60). In vitro and in vivo studies have suggested a direct effect of K$^+$ glutamate on *proU* transcription, but the beneficial effects of these solutes might be indirect and reflect a nonspecific stimulation on the transcription apparatus of the cell. A model has been proposed in which the role of nucleoid-associated DNA-binding proteins (H-NS, HU, IHF), the intracellular accumulation of K$^+$ glutamate, and local changes in DNA supercoiling all play a part in the genetic control of *proU* (48). Despite intensive efforts in a number of laboratories, a full understanding of the molecular mechanisms controlling osmoregulated *proU* expression has not been achieved and remains a formidable challenge.

In gram-negative bacteria, the first step in uptake of osmoprotectants involves their movement across the outer membrane permeability barrier. The OmpC and OmpF porins form water-filled open channels and allow the passage of a wide variety of low-molecular-mass compounds (107). The levels of OmpC and OmpF vary in response to different demands and stresses, in particular to changes in medium osmolality (122). Hyperosmotic conditions favor the synthesis of OmpC and reduce the production of OmpF, whereas hypoosmotic conditions reverse this pattern. Glycine betaine can diffuse through both porins, but OmpC has a greater role in osmoprotectant permeation simply because it is synthesized at a much higher rate than OmpF when medium osmolality is high (40).

A two-component regulatory system comprising a membrane-bound sensor kinase (EnvZ) and a cytoplasmic response regulator (OmpR) serves to detect changes in environmental osmolality and to set the level of OmpF and OmpC synthesis (65, 122). However, these proteins do not serve as a general osmoregulatory module for other osmoadaptive cellular responses. The EnvZ protein monitors environmental osmolality and transmits information across the cytoplasmic membrane to the soluble transcription factor OmpR through phosphorylation and dephos-

phorylation reactions. The level of OmpR phosphorylation is critical for its DNA interactions with both the *ompC* and *ompF* regulatory regions. High levels of phosphorylated OmpR activate *ompC* expression but repress *ompF* transcription; low levels of phosphorylated OmpR can activate *ompF* expression but fail to increase *ompC* transcription. Despite extensive analysis, the nature of the signal to which the EnvZ sensor protein responds is not known. However, medium osmolality is not likely to be a direct signal since *ompF* expression is considerably increased when the high-osmolality growth medium is supplemented with glycine betaine without changing its osmotic strength (7).

Compatible Solute Uptake in the Gram-Positive Bacteria *B. subtilis* and *C. glutamicum*

The identification of osmoprotectant transporters in *B. subtilis* (Fig. 3) was facilitated by the availability of *E. coli* mutants defective in all osmoprotectant uptake systems (51) that permitted the cloning of transporter genes through a functional complementation approach (73, 75, 146). Five transport systems for osmoprotectants have been found: two are secondary transporters (OpuD and OpuE), and three are members of the ABC superfamily (OpuA, OpuB, and OpuC [Table 1]) (76). The latter are related to the *E. coli* ProU system, but the location of the substrate-binding proteins differs. While ProX is freely diffusable in the periplasm, the ligand-binding proteins from OpuA, OpuB, and OpuC are extracellular and tethered to the cytoplasmic membrane via a lipid modification at their amino terminus (Fig. 3) (77). The OpuA, OpuC, and OpuD transporters mediate high-affinity glycine betaine uptake (Fig. 1). OpuA and OpuD exhibit a restricted substrate specificity, whereas OpuC can transport a wide variety of compatible solutes with high affinity (72, 106). The OpuB transporter is closely related to OpuC, and both systems probably evolved through a gene duplication event (74). However, OpuB differs significantly from OpuC in that it essentially recognizes only one substrate, choline (Fig. 3). OpuB is part of the choline to glycine betaine biosynthetic pathway, and its structural genes are regulated along with the genes for the biosynthetic enzymes (GbsA and GbsB) by the GbsR repressor in response to the availability of choline in the growth medium. Proline uptake in *E. coli* and serovar Typhimurium under highly osmotic conditions is mediated through the ProP and ProU systems with moderate affinity (30), but *B. subtilis* possesses a dedicated, high-affinity transporter (OpuE) for this compound (Fig. 3) (146).

OpuE is unrelated to the proline transporter ProP, but surprisingly it does resemble PutP, a permease used by *E. coli* and serovar Typhimurium to acquire proline for catabolic purposes (146). Unlike *putP*, *opuE* is not induced by proline in the growth medium; instead, its transcription is upregulated in response to increase in medium osmolality (136, 146). Transcription of the *opuE* structural gene is sensitively adjusted to the osmolality of the environment and mediated by two overlapping and independently controlled promoters. The *opuE*-P1 promoter is controlled by the housekeeping sigma factor SigA, and *opuE*-P2 is recognized by the alternative transcription factor SigB (136, 146), the master regulator for a large stress regulon in *B. subtilis* that protects the cells against various environmental insults (54, 55) (see chapter 12).

The dual regulation of *opuE* exemplifies the complexity and sophistication of the cellular response to a single stress factor by illustrating the connection between the uptake of a specific osmoprotectant and the onset of a general stress defense reaction (F. Spiegelhalter and E. Bremer, unpublished data). Osmotic upshock triggers the rapid induction of *opuE* transcription, with the level of *opuE* expression proportionally linked to medium osmolality. Transcription initiating from *opuE*-P1 is sensitively adjusted to the degree of osmotic stress and is kept at elevated levels as long as the osmotic stimulus persists. In contrast, the activity of the SigB-controlled *opuE*-P2 promoter rises only transiently after an osmotic upshock and does not significantly contribute to the level of *opuE* expression in cells subjected to long-term osmotic stress. Thus at least two different signal transduction pathways must operate in *B. subtilis* to communicate osmotic changes in the environment to the transcription apparatus of the cell (136). Essentially all SigB-controlled genes can be induced by salt stress (54, 55), but the SigB regulon cannot adequately protect growing cells against the negative effects of high osmolality. In fact, the *sigB* gene can be deleted without affecting the growth of osmotically stressed cultures. However, SigB is crucial in protecting the cells against osmotic upshock under growth-restricting conditions (145). The linkage of this general stress response system to a particular proline transporter (OpuE) clearly shows the importance of compatible solute uptake in osmotically threatened and even nongrowing cells.

C. glutamicum is another gram-positive soil bacterium whose osmoregulatory systems have been studied in detail (116). It is a member of the high G/C group of gram-positive bacteria and is not phylogenetically closely related to *B. subtilis*, which is a representative of the low G/C group of gram-positive microorganisms. *C. glutamicum* synthesizes several compatible solutes such as trehalose, proline, and glutamine. Biosynthesis of trehalose and proline is regulated by osmotic stress conditions, but the type and the amount of the accumulated compounds depend highly on the available carbon and energy sources. High internal concentrations of glutamate (up to 1 M) are also found; however, in contrast to the situation observed in *E. coli*, this amino acid is not directly involved in osmoregulation in *C. glutamicum* (R. Krämer, unpublished data). *C. glutamicum* achieves osmoprotection primarily by the uptake of external compatible solutes, in particular, glycine betaine, ectoine, and proline (41, 116). It is equipped with three osmoregulated transporters: BetP, a high-affinity system specific for glycine betaine; ProP, a medium-affinity system for proline and ectoine; and EctP, a low-affinity system for all three compounds. All three systems are effectively regulated at the level of activity, and the former two are also controlled at the level of gene expression (113, 114). Recently, a carboxy-terminal domain of BetP was found to be directly involved in osmosensing (115). The BetP and EctP transporters from *C. glutamicum* (113) are related to the glycine betaine uptake systems OpuD from *B. subtilis* (73) and BetL from *L. monocytogenes* (133) and to the choline transporter BetT from *E. coli* (85); they are all members of the BCCT (betaine-choline-charnitine transporters) family of secondary transporters. In contrast to *B. subtilis*, *C. glutamicum* exclusively uses secondary transport systems for osmoprotectant acquisition (Table 1), although it does possess ABC-type transporters (83). An additional proline transport system (PutP) is present in *C. glutamicum*, but it is used for anabolic purposes and its physiological function is unrelated to osmoprotection (113); such a transporter is also present in *B. subtilis* (S. Moses and E. Bremer, unpublished data). *C. glutamicum* responds to hypoosmotic stress through the controlled release of osmoprotectants via mechanosensitive channels (130, 131).

OSMOTIC REGULATION OF OSMOPROTECTANT TRANSPORTER ACTIVITY

The activity of many transporters that take up and mechanosensitive channels that release (see below) compatible solutes is governed by environmental osmotic conditions. This inherent feature allows the cell to respond to osmotic fluctuations very rapidly by enhancing or diminishing the performance of existent proteins. Osmotically activated solute uptake systems have been detected in many bacterial species.

They comprise members of different families of transporters and include the *E. coli* K$^+$ uptake systems Kdp (a P-type ATPase) and Trk, the ABC transporter ProU from *E. coli*, as well as several secondary systems (e.g., ProP and BetT from *E. coli*, OpuD from *B. subtilis*, BetP from *C. glutamicum*, QacT of *Lactobacillus plantarum*, and transporters from *L. monocytogenes*) (6, 40, 46, 49, 73, 114, 115, 124, 138, 139, 144). The degree of hyperosmotic stress that leads to optimal activation of the transport systems closely parallels the degree of osmotolerance of the respective bacterial species, which is low for *E. coli* (approximately 0.2 to 0.3 M NaCl for ProP and ProU) and higher for more salt-tolerant species like *L. monocytogenes*, *Staphylococcus aureus*, *C. glutamicum* (0.7 to 0.8 M NaCl) (79, 81, 115, 119), and *L. plantarum* (1.2 M NaCl) (45, 46). Detailed molecular and biochemical studies on osmotic activation have been initiated for the ProP and BetP transporters from *E. coli* and *C. glutamicum*, respectively, and for the sensor kinase KdpD, a regulatory protein for the Kdp K$^+$ transport system (70, 115, 124, 139, 144). These studies focus on correlating the structure and function of the involved membrane proteins and have identified regions that seem to serve as sensory domains; however, the mechanism for recognizing the signal and transforming it into an activity change has not yet been deciphered.

OSMOREGULATION: STRATEGIES AND EVENTS

Within a given organism the stress responses discussed above are coordinated into a particular sequence of reactions resulting in cellular adaptation to unfavorable osmotic growth conditions. The best illustration of this coordination is provided by the intensively studied enteric bacteria *E. coli* and serovar Typhimurium (14, 30, 76, 149). The primary event after an osmotic upshift is the efflux of water by diffusion across the lipid bilayer and through aquaporins. This leads to many physical and structural changes in the cell that presumably trigger the osmosensing process. The first response in *E. coli* is a massive uptake of K$^+$, which is mediated by the Trk and Kdp systems (Fig. 2). This event occurs within seconds after the osmotic shift and is accompanied by putrescine efflux (6, 39, 141). Within a few minutes, glutamate begins to accumulate and serves to balance charges within the cell (30, 36, 99, 141, 152). The accumulation occurs via an increase in biosynthesis and a decrease of glutamate utilization. Changes in turgor contribute to the regulation in activity of both the Trk and the Kdp transporters, and the level of

turgor and the availability of K$^+$ appear to control expression of the *kdpFABC* operon via the KdpD-KdpE two-component regulatory module. An interesting regulatory model has been proposed that directly links osmotically mediated changes in turgor (or envelope tension) to the expression of the *kdp* operon (30, 39, 84). Signal perception induces autophosphorylation of the inner-membrane-embedded KdpD kinase; this phospor group is then transferred to the soluble response regulator KdpE (70), which then can activate *kdp* transcription. Transcription of the *kdp* operon is only transiently induced following an osmotic upshock, a regulatory pattern that is consistent with the observed restoration of turgor by an increase in the intracellular K$^+$ levels. K$^+$ availability also plays an important role in the genetic control of *kdp*: expression is transient when K$^+$ is freely available and permanent when K$^+$ is limiting. Furthermore, the threshold K$^+$ concentration for *kdp* expression is increased when the cells accumulate the osmoprotectants trehalose or glycine betaine (4). These findings challenge the validity of the turgor model, which predicts that the threshold concentration for K$^+$ will fall when accumulated compatible solutes contribute to turgor restoration and thereby lower the need for K$^+$ uptake via Kdp.

Because high cytoplasmic K$^+$ concentrations have negative effects on protein function and DNA-protein interactions, the massive accumulation of K$^+$ is an inadequate strategy for coping with prolonged osmotic stress. The initial increase in cellular K$^+$ content is followed by the accumulation of compatible solutes, allowing the cells to discharge large amounts of the initially acquired K$^+$ through specific and nonspecific efflux (Kef) systems (125, 141). The availability of compatible solutes in the environment determines whether synthesis or uptake will predominate in their cellular accumulation. A scarcity leads to the full expression of genes (*otsAB*) encoding enzymes for trehalose biosynthesis. Concomitantly, there is induction of genes coding for osmoprotectant transporters (ProP, ProU, and BetT) and enzymes for glycine betaine biosynthesis (BetBA), so that the cell also can take advantage of preformed osmoprotectants (or their biosynthetic precursors) that might be present in the environment. When exogenous osmoprotectants are abundant, uptake of these compounds is preferred over the endogenous synthesis of compatible solutes, and the expression of genes encoding the biosynthetic enzymes is repressed. Similarly, transcription of the osmoregulated genes encoding osmoprotectant transporters is reduced when the intracellular levels of compatible solutes are sufficient to counteract the osmotic stress (94). These observations imply a dedicated cellular and genetic control over the intracellular pool of compatible solutes through which synthesis and uptake of osmoprotectants are integrated into a finely tuned cellular stress response system.

The temporal sequence of events during the initial phase in osmoadaptation and the finding that many subsequent osmotic responses (e.g., induction of the *proU* operon) are dependent on the prior accumulation of K⁺ have led to the proposal that K⁺ and its counterion glutamate act as a second messenger (30, 39, 49). This concept is attractive, but it remains to be directly demonstrated whether the size of and temporary changes in the cellular K⁺/glutamate pool can serve as an internal signal for the onset of the second phase in osmoadaptation—the accumulation of compatible solutes.

Osmosensing

Cellular responses to osmotic challenges are triggered by the cell's perception of a stress situation. In contrast to the response systems, the mechanisms for sensing osmotic fluctuations and the parameters that are actually sensed are poorly understood. In principle, changes in osmotic conditions may be detected directly as changes in extracellular water activity or indirectly as changes in the cell wall/plasma membrane structure or in the composition of the cytoplasm. Consequently, numerous parameters may be recognized by an osmosensor (29, 149). Some of these stimuli are related to properties of the cytoplasm and the periplasm, such as hydrostatic pressure, osmolality, ionic strength, concentration of particular signal molecules, and macromolecular crowding. Other potential stimuli are related to membrane-based osmosensors, such as lateral pressure within the membrane, bilayer curvature, density and structure of the phospholipid head groups, and thickness of the membrane. There is already good evidence that integral membrane proteins can function as sensors, transducers, and also responders to osmotic shifts (115, 124).

Osmosensing can be fully understood only by considering the mechanical and elastic properties of the cell wall and plasma membrane. The peptidoglycan sacculus of the cell wall is more flexible than previously believed (37, 142). It is able to swell and shrink to a certain extent in response to the influx and efflux of water, and these changes could indirectly affect intrinsic parameters of the cytoplasmic membrane. The mechanical properties of the cell wall differ greatly in gram-negative and gram-positive bacteria and are reflected in substantially different values for turgor. In contrast, the mechanical properties of the plasma membrane can be assumed to be similar

in both groups of microorganisms, although the size of the cell must be taken into account. The strain at the surface of a spherical cell depends on its radius, i.e., larger cells experience a greater strain than smaller cells for a given stress (149). Further investigations are now needed to clarify the nature of the parameters that act as osmostress stimuli for the cell, the mechanisms by which these stimuli are recognized, and the pathway(s) by which the signal is transduced to further targets, such as regulatory proteins or the catalytic domains of transporters.

PROTECTION AGAINST EXTREME TURGOR: RELEASE OF COMPATIBLE SOLUTES

The stress response of bacterial cells to high-osmolality environments causes a massive intracellular buildup, often in molar concentrations, of compatible solutes. The same compounds that protect the cell from the detrimental effects of hyperosmotic conditions become a threat to the cell's survival in hypoosmotic habitats. In their natural ecosystems, bacteria are likely to experience osmotic downshocks caused by rain, flooding, and washout into freshwater sources (42, 108). Such conditions result in a rapid movement of water into the cell with a concomitant dramatic increase in turgor (15). To avoid bursting under this strain the cell must rapidly eliminate the accumulated organic osmolytes and ions (e.g., K⁺) to reduce the driving force for water entry. This cannot be achieved simply by breaking down compatible solutes since this process is rather slow and many of these compounds are metabolically inert in various bacterial species. Instead, the organic osmolytes and other cellular solutes are rapidly expelled in a controlled manner. The extent of solute loss is related to the severity of the hypoosmotic shock, allowing the cell to sensitively adjust turgor to the altered osmolality of its habitat (44, 118, 131, 132).

E. coli cells grown at high osmolality release potassium glutamate, trehalose, and glycine betaine upon osmotic downshock (36, 132). Expulsion of more than one molar glycine betaine in less than 200 ms (the time resolution limit of the experiments) was observed (2). This massive solute efflux apparently occurs without severely damaging the cells because they can reaccumulate glycine betaine to preshock concentrations when the osmolality of the growth medium is raised to the previous level. Electrophysiological studies with the patch-clamp technique have revealed that the cytoplasmic membrane of both *Bacteria* and *Archaea* contains gated channels with various levels of conductances (10, 89). Since their discovery in bacteria, mechanosensitive channels have been proposed to play a role in the sensing of and responding to osmotic changes and to serve as safety valves for the release of solutes when the environmental osmolality suddenly drops (9, 11, 110, 140). Recent genetic studies in *E. coli* have provided compelling support for this concept (15, 91).

Patch-clamp experiments have shown that *E. coli* contains at least three different species (MscL, MscM, and MscS) of efflux channels, with conductances ranging from 0.1 to 3 nS. These channels display little ion or solute specificity. Pioneering work by C. Kung and coworkers resulted in the isolation of the MscL protein, the identification of its structural gene (*mscL*), and the construction of mutants lacking MscL channel activity (140). Reconstitution of the MscL protein into artificial liposomes demonstrated that stretch forces in the lipid bilayer cause the channel to open, thereby establishing that mechanosensation is an inherent property of the MscL channel. Recently, Chang et al. succeeded in determining the crystal structure of the closed MscL channel from *Mycobacterium tuberculosis* at a resolution of 3.5 Å (24). The MscL protein has two transmembrane helices (TM1 and TM2) and a short periplasmic loop; both the N- and C-termini face the cytoplasm. The funnel-shaped channel is formed by a pentameric MscL configuration. All the transmembrane helices are slanted with respect to the plane of the membrane, and the five TM1 helices come together at their cytoplasmic ends, forming a gate that is held closed primarily by hydrophobic interactions between neighboring helices. It is thought that mechanical stretch pushes the cytoplasmic ends of the TM1 helices apart (9, 24) to form a large, nonspecific channel that can accommodate even the 12-kDa thioredoxin (2). The crystal structure of MscL (24) and mutant analysis of this integral membrane protein (11, 110) provide now a framework for understanding how a stretch in the lipid bilayer of the membrane can effect functional changes in a channel protein.

The search for the physiological function of MscL was originally inconclusive because a deletion mutant did not exhibit significant growth or survival phenotypes. The recent work of Levina et al. (91) not only characterized a new gene family (*yggB*) critical for MscS channel activity but also permitted new studies into the physiological role of MscL. Like the *mscL* deletion mutant, the *yggB* mutant did not have a severe growth phenotype. However, when both the *mscL* and *yggB* mutations were present in the same strain, these cells lost the ability to survive a severe osmotic downshock. Further physiological experiments with this double mutant revealed that the MscL and YggB channels open at a pressure change just below that which would cause cell disruption.

These studies establish that the rapid release of solutes is necessary after osmotic downshock to prevent the excessive buildup of turgor and that the MscL and YggB mechanosensitive channels are crucial for managing hypoosmotic stress. As expected for proteins with such important physiological functions, database searches and physiological studies have revealed homologs of MscL and YggB in a wide variety of microorganisms (91, 104). In addition to YggB, the KefA channel has MscS-like properties, but it does not play any major role in managing the transition from high to low osmolality (91). Likewise, the physiological function of the MscM channel is unknown, and its structural gene has not been identified.

Clearly, mechanosensitive channels play a primary role in osmolyte expulsion as a stress response to a sudden reduction in osmolality, but they are not the only means for removing compatible solutes from the cell. For instance, the efflux of glycine betaine and proline from osmotically downshocked *L. plantarum* cells is characterized by two kinetically distinct pathways. One, which acts very rapidly, fits the pattern of mechanosensitive channels, and the other, which exhibits substantially slower kinetics, is consistent with a carrier system (44). Compatible solute efflux carriers also appear to be present in enteric bacteria (81, 86). It should be kept in mind that a number of solute efflux systems exist in bacteria that are unrelated to osmotic stress, e.g., those for amino acids in *C. glutamicum* (82).

CONTRIBUTION OF MULTIPLE STRESS RESPONSES TO CELL SURVIVAL IN HYPEROSMOTIC ENVIRONMENTS

In addition to modulating the intracellular solute, bacteria alter the expression of a sizeable number of genes that are not directly involved in osmoprotectant uptake or synthesis. Osmotic induction of these genes has been observed during searches with reporter gene fusions randomly inserted into the chromosome and in two-dimensional gel electrophoresis studies. In *E. coli*, many of these genes are under the control of the alternative transcription factor RpoS, and their osmotic induction is either abolished or strongly reduced in *rpoS* mutants (56). The σ^s subunit of the RNA polymerase is itself subjected to posttranscriptional osmotic regulation (8, 105), and the increase in the cellular RpoS content provides a link between hyperosmotically induced and stationary-phase-associated processes (56, 57) (see chapter 11). Although the physiological function of most of these osmotically induced genes in stress resistance has not been determined, the genetic regulation of some has been studied in detail. These studies have revealed that RpoS-mediated gene expression is frequently governed by a complex interplay between global regulatory proteins such as H-NS, Lrp, CRP, IHF, and Fis, thus providing the cell with a finely tuned and interwoven stress response network (57). This network contributes significantly during stationary phase to the development of a broad stress resistance, including a high tolerance against osmotic challenges (88, 98). Increased osmotic sensitivity is observed in stationary-phase cells of *E. coli* lacking RpoS, but the cellular events involved in RpoS-mediated osmoadaptation are unknown. A similar phenomenon is observed in *B. subtilis* where a very large stress regulon that is controlled through the alternative sigma factor σ^B confers high stress resistance against a wide variety of environmental insults (54, 55) (see chapter 12). Mutants lacking σ^B are sensitive to a severe and growth-preventing osmotic upshock, and this osmosensitivity can be alleviated by glycine betaine (145). It is apparent that the general osmostress responses in both *E. coli* and *B. subtilis* add a level of osmoprotection that goes beyond compatible solute accumulation, and the underlying molecular and physiological mechanisms clearly deserve further exploration.

FUTURE PROSPECTS

The accumulation and release of compatible solutes have been clearly established as the principal physiological response of bacteria to hyper- and hypoosmotic challenges, and much has been learned about the systems for compatible solute synthesis and uptake. However, it is not yet possible to describe unambiguously in any bacterium the molecular events involved in perceiving osmotic changes and processing this information into a genetic signal that leads to altered gene expression and ultimately to physiological adaptation reactions. Considerable advances have been made in understanding these processes in yeast with respect to the synthesis of the compatible solute glycerol. Although two membrane-bound osmosensors (Sho1 and Sln1) and an elaborate signaling cascade involving MAP-kinases have been detected, the sensing process itself remains enigmatic (61, 120). This is also the case for the bacterial sensors EnvZ and KdpD, which control expression of the *ompF* and *ompC* porin genes and synthesis of the Kdp potassium transporter, respectively. There is genetic evidence that osmolality per se is not perceived by some osmoregulatory systems; the central problem that now must be urgently addressed is the precise nature

of the stimulus that is actually sensed in osmotically stressed cells.

It is well established that the activity of many transporters and channels that take up or release compatible solutes is governed by environmental osmotic conditions. We do not understand sufficiently how these membrane-bound proteins sense osmotic up- or downshifts, nor do we understand the nature of the signal and its transformation into an activity change. The successful functional reconstitution of the purified channel protein MscL and the osmoprotectant transporters ProP and BetP and that of the sensor kinase KdpD into artificial membrane provides important tools to further our understanding of the osmotic influence on protein activity.

It is now possible to approach the osmostress response of microorganisms on a genomic and cellular scale by analyzing the transcriptome and proteome. These techniques not only allow the detection of new osmoregulated genes but also permit the exploration of connections among individual osmostress response systems. In addition, interactions between osmotic and other types of stress responses can be analyzed, allowing an integrated view of the complex relationship between the cell and its everchanging environment.

Acknowledgments. We greatly appreciate the help of Vickie Koogle in editing the manuscript.

Work on osmoregulation in our laboratories was provided by the Deutsche Forschungsgemeinschaft and through the Fonds der Chemischen Industrie.

REFERENCES

1. **Agre, P., D. Brown, and S. Nielsen.** 1995. Aquaporin water channels: unanswered questions and unresolved controversies. *Curr. Opin. Cell Biol.* **7:**472–483.
2. **Ajouz, B., C. Berrier, A. Garrigues, M. Besnard, and A. Ghazi.** 1999. Release of thioredoxin via the mechanosensitive channel MscL during osmotic downshock of *Escherichia coli* cells. *J. Biol. Chem.* **273:**26670–26674.
3. **Arakawa, T., and S. N. Timasheff.** 1985. The stabilization of proteins by osmolytes. *Biophys. J.* **47:**411–414.
4. **Asha, H., and J. Growishankar.** 1993. Regulation of *kdp* operon expression in *Escherichia coli*: evidence against turgor as signal for transcriptional control. *J. Bacteriol.* **175:**4528–4537.
5. **Atlung, T., and H. Ingmer.** 1997. H-NS: a modulator of environmentally regulated gene expression. *Mol. Microbiol.* **24:** 7–17.
6. **Bakker, E. P.** 1993. Cell K$^+$ and K$^+$ transport systems in prokaryotes, p. 205–224. *In* E. P. Bakker (ed.), *Alkali Cation Transport Systems in Prokaryotes.* CRC Press Inc., Boca Raton, Fla.
7. **Barron, A., G. May, E. Bremer, and M. Villarejo.** 1986. Regulation of envelope protein composition during adaptation to osmotic stress in *Escherichia coli. J. Bacteriol.* **167:**433–438.
8. **Barth, M., C. Marschall, A. Muffler, D. Fischer, and R. Hengge-Aronis.** 1995. Role of the histone-like protein H-NS in growth phase-dependent and osmotic regulation of σ^s and

many σ^s dependent genes in *Escherichia coli*. *J. Bacteriol.* **177**: 3455–3464.

9. Batiza, A. F., I. Rayment, and C. Kung. 1999. Channel gate! Tension, leak and disclosure. *Structure* **7**:R99–R103.

10. Berrier, C. M., B. Besnard, B. Ajouz, A. Coloumbe, and A. Ghazi. 1996. Multiple mechanosensitive ion channels from *Escherichia coli*, activated at different thresholds of applied pressure. *J. Membr. Biol.* **151**:175–187.

11. Blount, P., M. J. Schroeder, and C. Kung. 1997. Mutations in a bacterial mechanosensitive channel change the cellular response to osmotic stress. *J. Biol. Chem.* **272**:32150–32157.

12. Boch, J., B. Kempf, R. Schmid, and E. Bremer. 1996. Synthesis of the osmoprotectant glycine betaine in *Bacillus subtilis*: characterization of the *gbsAB* genes. *J. Bacteriol.* **178**:5121–5129.

13. Bonhivers, M., J. M. Carbrey, S. J. Gould, and P. Agre. 1998. Aquaporins in *Saccharomyces*. Genetic and functional distinction between laboratory and wild-type strains. *J. Biol. Chem.* **273**:27565–27572.

14. Booth, I. R., and C. F. Higgins. 1990. Enteric bacteria and osmotic stress: intracellular potassium glutamate as secondary signal of osmotic stress? *FEMS Microbiol. Rev.* **75**:239–246.

15. Booth, I. R., and P. Louis. 1999. Managing hypoosmotic stress: aquaporins and mechanosensitive channels in *Escherichia coli*. *Curr. Opin. Microbiol.* **2**:166–169.

16. Braun, A. D. 1997. Microbial water stress. *Bacteriol. Rev.* **40**: 803–846.

17. Britton, K. L., T. J. Stillman, K. S. P. Yip, P. Forterre, P. C. Engel, and D. W. Rice. 1998. Insights into the molecular basis of salt tolerance from the study of glutamate dehydrogenase from *Halobacterium salinarum*. *J. Biol. Chem.* **273**:9023–9030.

18. Burg, M., E. Kwon, and D. Kültz. 1997. Regulation of gene expression by hypertonicity. *Annu. Rev. Physiol.* **59**:437–455.

19. Calamita, G., W. R. Bishai, G. M. Preston, W. B. Guggino, and P. Agre. 1995. Molecular cloning and characterization of AqpZ, a water channel from *Escherichia coli*. *J. Biol. Chem.* **270**:29063–29066.

20. Calamita, G., B. Kempf, M. Bonhivers, W. R. Bishai, E. Bremer, and P. Agre. 1998. Regulation of the *Escherichia coli* water channel gene *aqpZ*. *Proc. Natl. Acad. Sci. USA* **95**:3627–3631.

21. Cánovas, D., C. Vargas, F. Iglesias-Guerra, L. N. Csonla, D. Rhodes, A. Ventosa, and J. J. Nieto. 1997. Isolation and characterization of salt-sensitive mutants of the moderate halophile *Halomonas elongata* and cloning of the ectoine synthesis genes. *J. Biol. Chem.* **272**:25794–25801.

22. Cánovas, D., C. Vargas, M. I. Calderon, A. Ventosa, and J. J. Nieto. 1998. Characterization of the genes for the biosynthesis of the compatible solute ectoine in the moderately halophilic bacterium *Halomonas elongata* DSM 3043. *Syst. Appl. Microbiol.* **21**:487–497.

23. Cánovas, D., N. Borges, C. Vargas, A. Ventosa, J. J. Nieto, and H. Santos. 1999. Role of Nγ-acetyldiaminobutyrate as an enzyme stabilizer and an intermediate in the biosynthesis of hydroxyectoine. *Appl. Environ. Microbiol.* **65**:3774–3779.

24. Chang, G., R. H. Spencer, A. T. Lee, M. T. Barclay, and D. C. Rees. 1998. Structure of the MscL homolog from *Mycobacterium tuberculosis*: a gated mechanosensitive ion channel. *Science* **282**:2220–2226.

25. Chater, K. F., and H. Nikaido. 1999. Cell regulation: maintaining integrity and efficiency in microbial cells. *Curr. Opin. Microbiol.* **2**:121–125.

26. Collins, K. D., and M. W. Washabough. 1985. Hofmeister effect and the behaviour of water at interfaces. *Q. Rev. Biophys.* **18**:323–422.

27. Csonka, L. N. 1988. Regulation of cytoplasmic proline levels in *Salmonella typhimurium*: effect of osmotic stress on synthesis, degradation, and cellular retention of proline. *J. Bacteriol.* **170**:2374–2378.

28. Csonka, L. N. 1989. Physiological and genetic responses of bacteria to osmotic stress. *Microbiol. Rev.* **53**:121–147.

29. Csonka, L. N., and A. D. Hanson. 1991. Prokaryotic osmoregulation: genetics and physiology. *Annu. Rev. Microbiol.* **45**: 569–606.

30. Csonka, L. N., and W. Epstein. 1996. Osmoregulation, p. 1210–1223. *In* F. C. Neidhardt, R. Curtiss III, J. L. Ingraham, E. C. C. Lin, K. B. Low, B. Magasanik, W. S. Reznikoff, M. Riley, M. Schaechter, and H. E. Umbarger (ed.), Escherichia coli *and* Salmonella: *Cellular and Molecular Biology*, 2nd ed. ASM Press, Washington, D.C.

31. Culham, D. E., B. Lasby, A. G. Mrangoni, J. L. Milner, B. A. Steer, R. W. van Nues, and J. M. Wood. 1993. Isolation and sequencing of *Escherichia coli proP* reveals unusual structural features of the osmoregulatory proline/betaine transporter, ProP. *J. Mol. Biol.* **229**:268–276.

32. da Costa, M. S., H. Santos, and E. A. Galinski. 1998. An overview of the role and diversity of compatible solutes in *Bacteria* and *Archaea*. *Adv. Biochem. Eng. Biotechnol.* **61**:117–153.

33. Delamarche, C., D. Thomas, J.-P. Rolland, A. Froger, J. Gouranton, M. Svelto, P. Agre, and C. Calamita. 1999. Visualization of AqpZ-mediated water permeability in *Escherichia coli* by cryoelectron microscopy. *J. Bacteriol.* **181**:4193–4197.

34. Delauney A. J., and D. P. S. Verma. 1993. Proline biosynthesis and osmoregulation in plants. *Plant J.* **4**:215–223.

35. Dennis, P. P., and L. C. Shimmin. 1997. Evolutionary divergence and salinity-mediated selection in halophilic archaea. *Microbiol. Mol. Biol. Rev.* **61**:90–104.

36. Dinnbier, U., E. Limpinsel, R. Schmid, and E. P. Bakker. 1988. Transient accumulation of potassium glutamate and its replacement by trehalose during adaptation of growing cells of *Escherichia coli* K-12 to elevated sodium chloride concentrations. *Arch. Microbiol.* **150**:348–357.

37. Doyle, R. J., and R. E. Marquis. 1994. Elastic, flexible peptidoglycan and bacterial cell wall properties. *Trends Microbiol.* **2**:57–60.

38. Elcock, A. H., and J. A. McCammon. 1998. Electrostatic contributions to the stability of halophilic proteins. *J. Mol. Biol.* **280**:731–748.

39. Epstein, W. 1986. Osmoregulation by potassium transport in *Escherichia coli*. *FEMS Microbiol. Rev.* **39**:73–78.

40. Faatz, E., A. Middendorf, and E. Bremer. 1988. Cloned structural genes for the osmotically regulated binding-protein-dependent glycine betaine transport system (ProU) of *Escherichia coli*. *Mol. Microbiol.* **2**:265–279.

41. Farwick, M., R. M. Siewe, and R. Krämer. 1995. Glycine betaine uptake after hyperosmotic shift in *Corynebacterium glutamicum*. *J. Bacteriol.* **177**:4690–4695.

42. Galinski, E. A., and H. G. Trüper. 1994. Microbial behaviour in salt-stressed ecosystems. *FEMS Microbiol. Rev.* **15**:95–108.

43. Germer, J., A. Muffler, and R. Hengge-Aronis. 1998. Trehalose is not relevant for in vivo activity of σ^S-containing RNA polymerase in *Escherichia coli*. *J. Bacteriol.* **180**:1603–1606.

44. Glaasker, E., W. N. Konings, and B. Poolman. 1996. Glycinebetaine fluxes in *Lactobacillus plantarum* during osmostasis and hyper- and hypoosmotic shock. *J. Biol. Chem.* **271**: 10060–10065.

45. Glaasker, E., W. N. Konings, and B. Poolman. 1996. Osmotic regulation of intracellular solute pools in *Lactobacillus plantarum*. *J. Bacteriol.* **178**:575–582.

46. Glaasker, E., E. H. M. L. Heuberger, W. N. Konings, and B. Poolman. 1998. Mechanism of osmotic activation of the quaternary ammonium compound transporter (QacT) of *Lactobacillus plantarum. J. Bacteriol.* 180:5540–5546.

47. Gouesbet, G., M. Jebbar, R. Talibart, T. Bernard, and C. Blanco. 1994. Pipicolic acid is an osmoprotectant for *Escherichia coli* taken up by the general osmoporters ProU and ProP. *Microbiology* 140:2415–2422.

48. Gowrishankar, J., and D. Manna. 1996. How is osmotic regulation of transcription of the *Escherichia coli proU* operon achieved? *Genetica* 97:363–378.

49. Grothe, S., R. L. Krogsrud, D. J. McClellan, J. L. Milner, and J. M. Wood. 1986. Proline transport and osmotic stress response in *Escherichia coli* K-12. *J. Bacteriol.* 166:253–259.

50. Gutierrez, C., M. Ardourel, E. Bremer, A. Middendorf, W. Boos, and U. Ehmann. 1989. Analysis and DNA sequence of the osmoregulated *treA* gene encoding the periplasmic trehalase of *Escherichia coli* K12. *Mol. Gen. Genet.* 217:347–354.

51. Haardt, M., B. Kempf, E. Faatz, and E. Bremer. 1995. The osmoprotectant proline betaine is a major substrate for the binding-protein-dependent transport system ProU of *Escherichia coli* K-12. *Mol. Gen. Genet.* 246:783–786.

52. Hars, U., R. Horlacher, W. Boos, W. Welte, and K. Diederichs. 1998. Crystal structure of the effector-binding domain of the trehalose-repressor of *Escherichia coli*, a member of the LacI family, in its complex with inducer trehalose-6-phosphate and noninducer trehalose. *Prot. Sci.* 7:2511–2521.

53. Hayashi, H., A. L. Mustrady, P. Deshnium, M. Ida, and N. Murata. 1997. Transformation of *Arabidopsis thaliana* with the *codA* gene for choline oxidase; accumulation of glycine betaine and enhanced tolerance to salt and cold stress. *Plant J.* 12:133–142.

54. Hecker, M., W. Schumann, and U. Völker. 1996. Heat-shock and general stress response in *Bacillus subtilis. Mol. Microbiol.* 19:417–428.

55. Hecker, M., and U. Völker. 1998. Non-specific, general and multiple stress resistance of growth restricted *Bacillus subtilis* cells by the expression of the σ^B regulon. *Mol. Microbiol.* 29:1129–1136.

56. Hengge-Aronis, R. 1996. Back to log phase: σ^S as a global regulator in the osmotic control of gene expression in *Escherichia coli. Mol. Microbiol.* 21:887–893.

57. Hengge-Aronis, R. 1999. Interplay of global regulators and cell physiology in the general stress response of *Escherichia coli. Curr. Opin. Microbiol.* 2:148–152.

58. Hengge-Aronis, R., W. Klein, R. Lange, M. Rimmele, and W. Boos. 1991. Trehalose synthesis genes are controlled by the putative sigma-factor encoded by *rpoS* and are involved in stationary-phase thermotolerance in *Escherichia coli. J. Bacteriol.* 173:7918–7924.

59. Higgins, C. F., J. Cerney, D. A. Stirling, L. Sutherland, and I. R. Booth. 1987. Osmotic regulation of gene expression: ionic strength as an intracellular signal? *Trends Biochem. Sci.* 12:339–344.

60. Higgins, C. F., C. J. Dorman, D. A. Stirling, L. Wadell, I. R. Booth, G. May, and E. Bremer. 1988. A physiological role for DNA supercoiling in the osmotic regulation of gene expression in *S. typhimurium* and *E. coli. Cell* 52:569–584.

61. Hohmann, S. 1997. Shaping up: the response of yeast to osmotic stress, p. 101–145. *In* S. Hohmann and W. H. Mager (ed.), *Yeast Stress Responses.* Springer, Berlin, Germany.

62. Holmstrøm, K. O., B. Welin, A. Mandal, I. Kristiansdottir, T. H. Teeri, T. Lamark, A. R. Strøm, and E. T. Palva. 1994. Production of the *Escherichi coli* betaine-aldehyde dehydrogenase, an enzyme required for the synthesis of the osmoprotectant glycine betaine, in transgenic plants. *Plant J.* 6:749–758.

63. Höltje, V.-J. 1998. Growth of the stress-bearing and shape-maintaining murein sacculus of *Escherichia coli. Microbiol. Mol. Biol. Rev.* 62:181–203.

64. Horlacher, R., K. Uland, W. Klein, M. Ehrmann, and W. Boos. 1996. Characterization of a cytoplasmic trehalase of *Escherichia coli. J. Bacteriol.* 178:6250–6257.

65. Hsing, W., F. D. Russo, K. K. Bernd, and T. J. Silhavy. 1998. Mutations that alter the kinase and phosphatase activities of the two-component sensor EnvZ. *J. Bacteriol.* 180:4538–4546.

66. Ingraham, J. L., and A. G. Marr. 1996. Effect of temperature, pressure, pH, and osmotic stress on growth, p. 1570–1578. *In* F. C. Neidhardt, R. Curtiss III, J. L. Ingraham, E. C. C. Lin, K. B. Low, B. Magasanik, W. S. Reznikoff, M. Riley, M. Schaechter, and H. E. Umbarger (ed.), Escherichia coli *and* Salmonella: *Cellular and Molecular Biology*, 2nd ed. ASM Press, Washington, D.C.

67. Jebbar, M., R. Talibart, K. Gloux, T. Bernard, and C. Blanco. 1992. Osmoprotection of *Escherichia coli* by ectoine: uptake and accumulation characteristics. *J. Bacteriol.* 174:5027–5035.

68. Jebbar, M., C. von Blohn, and E. Bremer. 1997. Ectoine functions as an osmoprotectant in *Bacillus subtilis* and is accumulated via the ABC-transport system OpuC. *FEMS Microbiol. Lett.* 154:325–330.

69. Joset, F., R. Jeanjean, and M. Hagemann. 1996. Dynamics of the responses of cyanobacteria to salt stress: deciphering the molecular events. *Physiol. Plant* 96:738–744.

70. Jung, K., B. Tjaden, and K. Altendorf. 1997. Purification, reconstitution, and characterization of KdpD, the turgor sensor of *Escherichia coli. J. Biol. Chem.* 272:10847–10852.

71. Kaasen, I., P. Falkenberg, O. B. Styrvold, and A. R. Strom. 1992. Molecular cloning and physical mapping of the *otsBA* genes, which encode the osmoregulatory trehalose pathway of *Escherichia coli.* Evidence that transcription is activated by KatF (AppR). *J. Bacteriol.* 174:889–898.

72. Kappes, R. M., and E. Bremer. 1998. Responses of *Bacillus subtilis* to high osmolarity: uptake of carnitine, crotonobetaine and γ-butyrobetaine via the ABC transport system OpuC. *Microbiology* 144:83–90.

73. Kappes, R. M., B. Kempf, and E. Bremer. 1996. Three transport systems for the osmoprotectant glycine betaine operate in *Bacillus subtilis*: characterization of OpuD. *J. Bacteriol.* 178:5071–5079.

74. Kappes, R. M., B. Kempf, S. Kneip, J. Boch, J. Gade, J. Meier-Wagner, and E. Bremer. 1999. Two evolutionarily closely related ABC transporters mediate the uptake of choline for synthesis of the osmoprotectant glycine betaine in *Bacillus subtilis. Mol. Microbiol.* 32:203–216.

75. Kempf, B., and E. Bremer. 1995. OpuA, an osmotically regulated binding protein-dependent transport system for the osmoprotectant glycine betaine in *Bacillus subtilis. J. Biol. Chem.* 270:16701–16713.

76. Kempf, B., and E. Bremer. 1998. Uptake and synthesis of compatible solutes as microbial stress responses to high-osmolality environments. *Arch. Microbiol.* 170:319–330.

77. Kempf, B., J. Gade, and E. Bremer. 1997. Lipoprotein from the osmoregulated ABC transport system OpuA of *Bacillus subtilis*: purification of the glycine betaine binding protein and characterization of a functional lipidless mutant. *J. Bacteriol.* 179:6213–6220.

78. Kiene, R. P., L. P. Hoffmann Williams, and J. E. Walker. 1998. Seawater microorganisms have a high affinity glycine

betaine uptake system which also recognizes dimethylsulfoniopropionate. *Aquat. Microb. Ecol.* **15**:39–51.

79. Ko, R., L. T. Smith, and G. M. Smith. 1994. Glycine betaine confers enhanced osmotolerance and cryotolerance on *Listeria monocytogenes*. *J. Bacteriol.* **176**:426–431.

80. Koch, A. L. 1983. The surface stress theory of microbial morphogenesis. *Adv. Microbiol. Physiol.* **24**:301–336.

81. Koo, S.-P., C. F. Higgins, and I. R. Booth. 1991. Regulation of compatible solute accumulation in *Salmonella typhimurium*. *J. Gen. Microbiol.* **137**:2617–2625.

82. Krämer, R. 1994. Secretion of amino acids by bacteria: physiology and mechanism. *FEMS Microbiol. Rev.* **13**:75–94.

83. Kronemeyer, W., N. Peekhaus, R. Krämer, H. Sahm, and L. Eggeling. 1995. Structure of the *gluABC* cluster encoding the glutamate uptake system of *Corynebacterium glutamicum*. *J. Bacteriol.* **177**:1152–1158.

84. Laimins, L. A., D. B. Rhoads, and W. Epstein. 1981. Osmotic control of *kdp* operon expression in *Escherichia coli*. *Proc. Natl. Acad. Sci. USA* **78**:464–468.

85. Lamark, T., I. Kaasen, M. W. Eshoo, P. Falkenberg, J. McDougall, and A. R. Strøm. 1991. DNA sequence and analysis of the *bet* genes encoding the osmoregulatory choline-glycine betaine pathway of *Escherichia coli*. *Mol. Microbiol.* **5**:1049–1064.

86. Lamark, T., O. B. Styrvold, and A. R. Strøm. 1992. Efflux of choline and glycine betaine from osmoregulating cells of *Escherichia coli*. *FEMS Microbiol. Lett.* **96**:149–154.

87. Lamark, T., T. P. Røkenes, J. McDougall, and A. R. Strøm. 1996. The complex *bet* promoters of *Escherichia coli*: regulation by oxygen (ArcA), choline (BetI), and osmotic stress. *J. Bacteriol.* **178**:1655–1662.

88. Lange, R., and R. Hengge-Aronis. 1991. Identification of a central regulator of stationary-phase gene expression in *Escherichia coli*. *Mol. Microbiol.* **5**:49–59.

89. Le Dain, A. C., N. Saint, A. Kloda, A. Ghazi, and B. Martinac. 1998. Mechanosensitive ion channels of the archeon *Haloferax volcanii*. *J. Biol. Chem.* **273**:12116–12119.

90. Le Rudulier, D., A. R. Strøm, A. M. Dandekar, L. T. Smith, and R. C. Valentine. 1984. Molecular biology of osmoregulation. *Science* **224**:1064–1068.

91. Levina, N., S. Tötemeyer, N. R. Stokes, P. Louis, M. A. Jones, and I. R. Booth. 1999. Protection of *Escherichia coli* cells against extreme turgor by activation of MscS and MscL mechanosensitive channels: identification of genes required for MscS activity. *EMBO J.* **18**:1730–1737.

92. Lipper, K., and E. A. Galinski. 1992. Enzyme stabilization by ectoine-type compatible solutes: protection against heating, freezing and drying. *Appl. Microbiol. Biotechnol.* **37**:61–65.

93. Louis, P., and E. A. Galinski. 1997. Characterization of genes for the biosynthesis of the compatible solute ectoine from *Marinococcus halophilus* and osmoregulated expression in *Escherichia coli*. *Microbiology* **143**:1141–1149.

94. Lucht, J., and E. Bremer. 1994. Adaptation of *Escherichia coli* to high osmolarity environments: osmoregulation of the high-affinity glycine betaine transport system ProU. *FEMS Microbiol. Rev.* **14**:3–20.

95. Lucht, J. M., P. Dersch, B. Kempf, and E. Bremer. 1994. Interactions of the nucleoid associated DNA-binding protein H-NS with the regulatory region of the osmotically controlled *proU* operon of *Escherichia coli*. *J. Biol. Chem.* **269**:6578–6586.

96. MacMillan, S. V., D. A. Alexander, D. E. Culham, H. J. Kunte, E. V. Marshall, D. Rochon, and J. M. Wood. 1999. The ion coupling and organic substrate specificities of osmoregulatory transporter ProP in *Escherichia coli*. *Biochim. Biophys. Acta* **1420**:30–44.

97. Madigan, M. T., and A. Oren. 1999. Thermophilic and halophilic extremophiles. *Curr. Opin. Microbiol.* **2**:265–269.

98. McCann, P. M., J. P. Kidwell, and A. Matin. 1991. The putative σ factor KatF has a central role in development of starvation-mediated general resistance in *Escherichia coli*. *J. Bacteriol.* **173**:4188–4194.

99. McLaggan, D., J. Naprstek, E. T. Buurmann, and W. Epstein. 1994. Interdependence of K⁺ and glutamate accumulation during osmotic adaptation of *Escherichia coli*. *J. Biol. Chem.* **269**:1911–1917.

100. McNeil, S. D., M. L. Nuccio, and A. D. Hanson. 1999. Betaines and related osmoprotectants. Targets for metabolic engineering of stress resistance. *Plant Physiol.* **120**:945–949.

101. Martin, D. D., R. A. Ciulla, and M. F. Roberts. 1999. Osmoadaptation in archaea. *Appl. Environ. Microbiol.* **65**:1815–1825.

102. Maurel, C. 1997. Aquaporins and water permeability of plant membranes. *Annu. Rev. Plant Physiol. Plant Mol. Biol.* **48**:399–429.

103. Mellis, J., A. Wise, and M. Villarejo. 1995. Two different *Escherichia coli proP* promoters respond to osmotic and growth phase signals. *J. Bacteriol.* **177**:144–151.

104. Moe, P. C., P. Blount, and C. Kung. 1998. Functional and structural conservation in the mechanosensitive channel MscL implicates elements crucial for mechanosensation. *Mol. Microbiol.* **28**:583–592.

105. Muffler, A., D. D. Traulsen, R. Lange, and R. Hengge-Aronis. 1996. Posttranscriptional osmotic regulation of the σ^S subunit of RNA polymerase in *Escherichia coli*. *J. Bacteriol.* **178**:1607–1613.

106. Nau-Wagner, G., J. Boch, A. Le Good, and E. Bremer. 1999. High-affinity transport of choline-O-sulfate and its use as a compatible solute in *Bacillus subtilis*. *Appl. Environ. Microbiol.* **65**:560–568.

107. Nikaido, H. 1999. Microdermatology: cell surface in the interaction of microbes with the external world. *J. Bacteriol.* **181**:4–8.

108. Oren, A. 1990. Formation and breakdown of glycine betaine and trimethylamine in hypersaline environments. *Antonie Leeuwenhoek* **58**:291–298.

109. Oren, A. 1999. Bioenergetic aspects of halophilisms. *Microbiol. Mol. Biol. Rev.* **63**:334–348.

110. Ou, X., P. Blount, R. J. Hoffman, and C. Kung. 1998. One face of the transmembrane helix is crucial in mechanosensitivity channel gating. *Proc. Natl. Acad. Sci. USA* **95**:11471–11475.

111. Overdier, D. G., and L. N. Csonka. 1992. A transcriptional silencer downstream of the promoter in the osmotically controlled *proU* operon of *Salmonella typhimurium*. *Proc. Natl. Acad. Sci. USA* **89**:3140–3144.

112. Owen-Hughes, T. A., G. D. Pavitt, D. S. Santos, J. M. Sidebotham, C. J. Hulton, J. C. D. Hinton, and C. F. Higgins. 1992. The chromatin-associated protein H-NS interacts with curved DNA to influence DNA topology and gene expression. *Cell* **71**:255–265.

113. Peter, H., A. Bader, A. Burkovski, C. Lambert, and R. Krämer. 1997. Isolation of the *putP* gene from *Corynebacterium glutamicum* and characterization of a low-affinity uptake system for compatible solutes. *Arch. Microbiol.* **168**:143–151.

114. Peter, H., A. Burkovski, and R. Krämer. 1996. Isolation, characterization, and expression of the *Corynebacterium glutamicum betP* gene, encoding the transport system for the compatible solute glycine betaine. *J. Bacteriol.* **178**:5229–5234.

115. Peter, H., A. Burkovski, and R. Krämer. 1998. Osmosensing by N- and C-terminal extensions of the glycine betaine up-

take system BetP of *Corynebacterium glutamicum*. *J. Biol. Chem.* **273**:2567–2574.

116. Peter, H., B. Weil, A. Burkovski, R. Krämer, and S. Morbach. 1998. *Corynebacterium glutamicum* is equipped with four secondary carriers for compatible solutes: identification, sequencing, and characterization of the ectoine/proline uptake system, ProP, and the ectoine/proline/glycine betaine carrier, EctP. *J. Bacteriol.* **180**:6005–6012.

117. Pieper, U., G. Kapadia, M. Mevarech, and O. Herzberg. 1998. Structural features of halophilicity derived from the crystal structure of dihydropholate reductase from the Dead Sea halophilic archaeon, *Haloferax volcanii*. *Structure* **6**:75–88.

118. Poolman, B., and E. Glaasker. 1998. Regulation of compatible solute accumulation in bacteria. *Mol. Microbiol.* **29**:397–407.

119. Pourkomalian, B., and I. R. Booth. 1992. Glycine betaine transport by *Staphylococcus aureus*: evidence for two transport systems and their possible role in osmoregulation. *J. Gen. Microbiol.* **138**:2515–2518.

120. Posas, F., M. Takekawa, and H. Saito. 1998. Signal transduction by MAP kinase cascades in budding yeast. *Curr. Opin. Microbiol.* **1**:175–182.

121. Potts, M. 1994. Desiccation tolerance of prokaryotes. *Microbiol. Rev.* **58**:755–805.

122. Pratt, L. A., and T. J. Silhavy. 1996. From acids to *osmZ*: multiple factors influence synthesis of the OmpF and OmpC porins in *Escherichia coli*. *Mol. Microbiol.* **20**:911–917.

123. Qu, Y., C. L. Bolen, and D. W. Bolen. 1998. Osmolyte driven contraction of a random coil protein. *Proc. Natl. Acad. Sci. USA* **95**:9268–9273.

124. Racher, K. I., R. T. Voegele, E. V. Marshall, D. E. Culham, J. M. Wood, H. Jung, M. Bacon, M. T. Cairns, S. M. Ferguson, W.-J. Liang, and P. J. F. Henderson. 1999. Purification and reconstitution of an osmosensor: transporter ProP of *Escherichia coli* senses and responds to osmotic shifts. *Biochemistry* **38**:1676–1684.

125. Record, M. T., E. S. Courtenay, D. S. Cayley, and H. J. Guttman. 1998. Responses of *E. coli* to osmotic stress: large changes in amounts of cytoplasmic solutes and water. *Trends Biochem. Sci.* **23**:144–149.

126. Rhodes, D., and A. D. Hanson. 1993. Quaternary ammonium and tertiary sulfonium compounds in higher plants. *Annu. Rev. Plant Physiol. Plant Mol. Biol.* **44**:357–384.

127. Rokenes, T. P., T. Lamark, and A. R. Strom. 1996. DNA-binding properties of the BetI repressor protein of *Escherichia coli*: the inducer choline stimulates BetI-DNA complex formation. *J. Bacteriol.* **178**:1663–1670.

128. Roth, W. G., M. P. Leckie, and D. N. Dietzler. 1985. Osmotic stress drastically inhibits active transport of carbohydrates by *Escherichia coli*. *Biochem. Biophys. Res. Commun.* **126**:434–441.

129. Rudolph, A. S., J. H. Crowe, and L. M. Crowe. 1986. Effects of three stabilizing agents—proline, betaine and trehalose—on membrane phospholipids. *Arch. Biochem. Biophys.* **245**:134–143.

130. Ruffert, S., C. Berrier, R. Krämer, and A. Ghazi. 1999. Identification of mechanosensitive ion channels in the cytoplasmic membrane of *Corynebacterium glutamicum*. *J. Bacteriol.* **181**:1673–1676.

131. Ruffert, S., C. Lambert, H. Peter, V. F. Wendisch, and R. Krämer. 1997. Efflux of compatible solutes in *Corynebacterium glutamicum* mediated by osmoregulated channel activity. *Eur. J. Biochem.* **247**:572–580.

132. Schleyer, M., R. Schmid, and E. P. Bakker. 1993. Transient, specific and extremely rapid release of osmolytes from growing cells of *Escherichia coli* K-12 exposed to hypoosmotic shock. *Arch. Microbiol.* **160**:424–431.

133. Sleator, R. D., C. G. M. Gahan, T. Abee, and C. Hill. 1999. Identification and disruption of BetL, a secondary glycine betaine transport system linked to salt tolerance of *Listeria monocytogenes* LO28. *Appl. Environ. Microbiol.* **65**:2078–2083.

134. Smith, L. T. 1985. Characterization of γ-glutamyl kinase from *Escherichia coli* confers proline overproduction and osmotic tolerance. *J. Bacteriol.* **164**:1088–1093.

135. Somero, G. N. 1986. Protons, osmolytes, fitness of internal milieu for protein function. *Am. J. Physiol.* **251**:R197–R213.

136. Spiegelhalter, F., and E. Bremer. 1998. Osmoregulation of the *opuE* proline transport gene from *Bacillus subtilis*: contributions of the sigma A- and sigma B-dependent stress-responsive promoters. *Mol. Microbiol.* **29**:285–296.

137. Strøm, A. R., and I. Kaasen. 1993. Trehalose metabolism in *Escherichia coli*: stress protection and stress regulation of gene expression. *Mol. Microbiol.* **8**:205–210.

138. Styrvold, O. B., P. Falkenberg, B. Landfald, M. W. Eshoo, T. Bjornsen, and A. R. Strom. 1986. Selection, mapping, and characterization of osmoregulatory mutants of *Escherichia coli* blocked in the choline-glycine betaine pathway. *J. Bacteriol.* **165**:856–863.

139. Sugiura, A., K. Hirokawa, K. Nakashima, and T. Mizuno. 1994. Signal-sensing mechanism of the putative osmosensor KdpD in *Escherichia coli*. *Mol. Microbiol.* **14**:929–938.

140. Sukharev, S. I., P. Blount, B. Martinac, and C. Kung. 1997. Mechanosensitive channels of *Escherichia coli*: the MscL gene, protein, and activities. *Annu. Rev. Physiol.* **59**:633–657.

141. Stumpe, S., A. Schlösser, M. Schleyer, and E. P. Bakker. 1996. K$^+$ circulation across the prokaryotic cell membrane: K$^+$ uptake systems, p. 473–499. *In* W. N. Konings, H. R. Kaback, and J. S. Lolkema (ed.), *Transport Processes in Eukaryotic and Prokaryotic Organisms*. Elsevier, Amsterdam, The Netherlands.

142. Thwaites, J. J., and N. H. Mendelson. 1991. Mechanical behaviour of bacterial cell walls. *Adv. Microb. Physiol.* **32**:174–221.

143. Ventosa, A., J. J. Nieto, and A. Oren. 1998. Biology of moderately halophilic aerobic bacteria. *Microbiol. Mol. Biol. Rev.* **62**:504–544.

144. Voelkner, P., W. Puppe, and K. Altendorf. 1993. Characterization of KdpD protein, the sensor kinase of the K$^+$-translocating Kdp system of *Escherichia coli*. *Eur. J. Biochem.* **217**:1019–1026.

145. Völker, U., B. Maul, and M. Hecker. 1999. Expression of the σ^B-dependent general stress regulon confers multiple stress resistance in *Bacillus subtilis*. *J. Bacteriol.* **181**:3942–3948.

146. von Blohn, C., B. Kempf, R. M. Kappes, and E. Bremer. 1997. Osmostress response in *Bacillus subtilis*: characterization of a proline uptake system (OpuE) regulated by high osmolarity and the alternative transcription factor sigma B. *Mol. Microbiol.* **25**:175–187.

147. Whatmore, A. M., J. A. Chudek, and R. H. Reed. 1990. The effects of osmotic upshock on the intracellular solute pools of *Bacillus subtilis*. *J. Gen. Microbiol.* **136**:2527–2535.

148. Whatmore, A. M., and R. H. Reed. 1990. Determination of turgor pressure in *Bacillus subtilis*: possible role for K$^+$ in turgor regulation. *J. Gen. Microbiol.* **136**:2521–2526.

149. Wood, J. M. 1999. Osmosensing in bacteria: signals and membrane based sensors. *Microbiol. Mol. Biol. Rev.* **63**:230–262.

150. **Xu, J., and R. C. Johnson.** 1997. Cyclic AMP receptor protein functions as a repressor of the osmotically inducible promoter *proP* P1 in *Escherichia coli. J. Bacteriol.* **179:**2410–2417.

151. **Xu, J., and R. C. Johnson.** 1997. Activation of RpoS-dependent *proP* P2 transcription by the FIS protein in vitro. *J. Mol. Biol.* **270:**346–359.

152. **Yan, D., T. P. Ikeda, A. E. Shauger, and S. Kustu.** 1996. Glutamate is required to maintain the steady-state potassium pool in *Salmonella typhimurium. Proc. Natl. Acad. Sci. USA* **93:**6527–6531.

153. **Yancy, P. H.** 1994. Compatible and counteracting solutes, p. 81–109. *In* K. Strange (ed.), *Cellular and Molecular Physiology of Cell Volume Regulation.* CRC Press, Boca Raton, Fla.

154. **Yancy, P. H., M. E. Clark, S. C. Hand, R. D. Bowlus, and G. N. Somero.** 1982. Living with water stress: evolution of osmolyte systems. *Science* **217:**1214–1222.

Bacterial Stress Responses
Edited by G. Storz and R. Hengge-Aronis
©2000 ASM Press, Washington, D.C.

Chapter 7

Microbial Responses to Acid Stress

JOHN W. FOSTER

Microorganisms commonly live in widely fluctuating pH environments. As a result, bacteria have evolved adaptive strategies designed to minimize acid or alkaline-induced damage. Gram-negative and gram-positive neutralophiles utilize different as well as overlapping approaches for coping with acid stress. Some inducible systems attempt to alkalinize internal pH (e.g., F_oF_1 ATPases or specific amino acid decarboxylases/antiporters) while other systems involve complex global changes in the proteome that somehow protect crucial, acid-sensitive cellular components. pH-responsive regulatory mechanisms involved in these adaptations to acid environments include alternative sigma factors whose levels change in response to acid stress and specific signal transduction systems that sense an acidifying environment. These adaptive strategies are important to pathogens that encounter the acid environments of the stomach (e.g., Salmonella, Escherichia coli, and Helicobacter pylori) and dental caries (e.g., Streptococcus mutans). In addition, inducible acid tolerance mechanisms are vital to plant symbionts growing in acidic soils (e.g., Rhizobium). The studies described in this chapter indicate that the defensive responses of microbes to acid stress are complex interconnecting networks of regulators and metabolic processes designed to cope with the life-threatening consequences of an acidifying internal pH.

The ability to sense and respond to potentially lethal changes in the environment is a trait crucial to the survival of any microorganism. As a result, elegant regulatory networks designed to survive stress have evolved. A common, often self-imposed, environmental threat endured by bacteria is acid stress, defined as the combined biological effect of H^+ ion (i.e., pH) and weak acid concentrations. Although the permeability of the membrane toward protons is low, extreme low external pH (pH_O) will cause H^+ to leak across the membrane and acidify internal pH (pH_i).

Decreased pH_i, in turn, will have deleterious effects on biochemical reactions and macromolecular structures. However, even moderately acidic environmental conditions, normally tolerated by the cell, can be lethal if combined with weak acids (for example, organic or short-chain fatty acids). Weak acids include volatile fatty acids (VFA) such as butyrate, propionate, or acetate, which, in their protonated forms, can diffuse across the cell membrane and dissociate, lowering pH_i in the process. The more acidic the pH_O, the more undissociated weak acid will be available (based on pK_a values) to cross the membrane and affect pH_i. Thus, a pH_O 3 environment without any volatile fatty acids may present the same acid stress as a pH 4.5 environment containing 100 mM VFA. This chapter will deal specifically with systems having a demonstrable effect on the ability of microbes to survive acid stress. Studies examining the pH-regulated expression of genes not yet linked to acid tolerance will not be discussed (2, 4, 18, 41, 50, 51, 75, 77, 79, 93, 113, 124, 154, 157, 162).

Bacteria have evolved a variety of adaptive strategies designed to manage the broad range of acid stresses encountered in nature. This discussion will exclude the acidophiles and alkalophiles, two groups of organisms that have adapted unique bioenergetic approaches to solving the problems inherent with life at the extremes of pH (89–91, 105, 140). Focus, instead, will be on the tactics used mostly by neutralophiles to combat the biological effects of acid stress. Neutralophiles such as *Escherichia coli* and *Salmonella enterica* serovar Typhimurium have a remarkable ability to adapt to pH stress. They grow in conditions (minimal medium) between pH 5 and 8.5, a more than 3,000-fold range in H^+-ion concentration. However, if they are first allowed to adapt to moderate acid or alkaline conditions before their limits are tested, one finds cells can survive over a 1,000,000-fold range of H^+-ion concentration! This

John W. Foster • Department of Microbiology and Immunology, University of South Alabama College of Medicine, Mobile, AL 36688.

capability is important in non-host and pathogenic situations where pH fluctuates dramatically (33, 40, 129, 131). The principal defense against a 1 pH unit change in H^+ concentration above or below optimum growth pH involves housekeeping pH homeostasis systems. Greater deviations away from optimum growth pH elicit the inducible acid survival systems.

pH HOMEOSTASIS: THE GRAM-NEGATIVE APPROACH

Enteric organisms try to keep internal pH relatively constant at pH 7.6 to 7.8 even as external pH changes during growth (17, 132). As a result, the difference between pH_i and pH_o, known as ΔpH, also changes. A stable pH_i is maintained by pumps that either move protons into the cell at alkaline pH_o or extrude them at acidic pH_i. Two principal systems are recognized as potential generators of pH gradients. These systems are potassium-proton antiporters (87) and sodium-proton antiporters (23). As a simplified overview, shifts to acidic environments are handled by K^+/proton antiporters, resulting in alkalinization of the cytoplasm. Conversely, shifts to alkaline conditions result in acidification of the cytoplasm via an Na^+/proton antiporter (38, 171) (see chapter 8). A general perception, not always true, of these "housekeeping" pH-homeostasis systems is that it is the activity of a given porter, not its synthesis, that is controlled by pH (49, 172). In contrast, the adaptive acid stress response systems include inducible emergency pH homeostasis systems and a variety of other inducible pathways designed to deal with potential macromolecular damage caused by severe acid stress. A thought-provoking review of pH homeostasis mechanisms was published by Booth (20).

The main *E. coli* system responsible for adaptation to Na^+ and alkaline pH (in the presence of Na^+) is a 349-amino acid, membrane-bound, Na^+/H^+ antiporter encoded by *nhaA* (119). Mutants lacking *nhaA* will not grow under alkaline conditions in the presence of Na^+ (56). The exchange reaction is electrogenic, bringing in 2 H^+ for each Na^+ extruded, with the net result being acidification of the cytoplasm. NhaA antiporter activity is activated at the protein level by shifts to alkaline pH consistent with a role in pH homeostasis (156). Control of NhaA activity by pH has been localized to histidine residue 226. Site-directed mutagenesis of this residue changes the pH response profile of NhaA (132). Another residue, glycine 338, also plays a role in this pH response. For example, a G338S substitution creates a pH-relaxed antiporter constitutively active from pH 6 to pH 9. The *nhaA* gene is also subject to transcrip-

tional control by the NhaR regulator (126). NhaR senses intracellular Na^+ concentrations and induces *nhaA* under Na^+ or Li^+ stress. However, pH also modulates NhaR regulation of *nhaA*, with alkaline pH causing elevated expression of *nhaA* and a concomitant change in the NhaR footprint (27).

pH HOMEOSTASIS: THE GRAM-POSITIVE APPROACH

Gram-positive microorganisms such as the streptococci regulate cytoplasmic pH differently than do the enteric gram-negative organisms. While *E. coli* struggles to keep internal pH near 7.8 as the external pH acidifies, streptococci are more flexible, allowing pH_i to decrease along with the environment. Their only requirement is to maintain a ΔpH 0.5 to 1 unit higher than the external pH value (83). The F_oF_1 ATPase, rather than antiporters, appears to be the primary mechanism for extruding protons in the streptococci. Regulation of cytoplasmic pH by an F_oF_1 proton-translocating ATPase was first demonstrated in *Enterococcus faecalis* (84). A decrease in cytoplasmic pH causes an increase in the amount of this complex and enhanced acid tolerance through proton extrusion (85). However, the increase in ATPase is not due to transcriptional or significant translational control over the ATPase operon (5, 148). Regulation of ATPase levels apparently occurs at the assembly stage. When cytoplasmic pH is acceptable, unassembled F_1 and presumably F_o subunits are degraded and thus not assembled into an active membrane-associated complex (5). In contrast, assembled complexes are stable. Acidification of cytoplasmic pH in some way prevents degradation and stimulates subunit assembly into an active complex. Therefore, increasing the level of ATPase in response to external acidification allows the cell to compensate for concomitant increases in proton influx.

The medical significance of F_oF_1 ATPase as an acid tolerance mechanism is perhaps best illustrated by examining the organisms associated with tooth decay. Microbial tolerance to acid environments is of major importance in the ecology of dental plaque and in the pathogenesis of dental caries, which involves acid dissolution of tooth mineral (99). An important feature for organisms in this environment is to grow and metabolize carbohydrates at low pH values (22). Consequently, the relative acid tolerance of these organisms is often assessed by allowing dense cell preparations to acidify their medium via carbohydrate fermentation and observing how low the medium pH will drop (14). The more acid tolerant the organism, the lower the pH will drop before glycolysis ceases.

Organisms associated with dental caries, listed in order of decreasing acid resistance, include *Lactobacillus casei*, *Streptococcus mutans*, *Streptococcus sanguis*, and *Actinomyces viscosus*. Bender and Marquis (13) suggested that the increased acid resistance of *L. casei* relative to *A. viscosus* is due to the level of proton-translocating ATPase associated with the membranes (3.29 versus 0.06 U/mg of protein, respectively).

THE ACID TOLERANCE RESPONSE OF *SALMONELLA ENTERICA* SEROVAR TYPHIMURIUM

During its travels within a host, the enteric pathogen *Salmonella* serovar Typhimurium encounters extreme low pH in the stomach, confronts volatile fatty acids present in the intestine and feces, and must also endure low pH in the macrophage phagosome and phagolysosome environments (reviewed in reference 51). Since the infectious dose is fairly low for this organism (19), we predicted, and then demonstrated, the presence of inducible mechanisms of acid tolerance (47; J. W. Foster and H. K. Hall, Abstr. Annu. Meet. Am. Soc. Microbiol. 1989, poster H-24, p. 173). Serovar Typhimurium possesses both log-phase and stationary-phase acid tolerance response (ATR) systems that will protect cells at pH 3 for several hours (reviewed in reference 8, summarized in Fig. 1). The underlying concept is that growth in a moderately acid environment will trigger the synthesis of proteins that protect the cell from more extreme acid conditions. Exponential-phase cells grown at pH 7.7 (minimal media) and subsequently adapted to pH 5.8 produce an ill-defined pH homeostasis system that maintains internal pH as cells encounter potentially lethal pH 3 acid stress (49). Lowering adaptive pH further to pH 4.5 results in the induction of 50 acid shock proteins (ASPs) thought to contribute to acid survival (45, 46). Internal pH appears to be the inducing signal for some ASPs while others respond to external pH. Induction of acid tolerance also provides cross protection to a variety of other stresses such as high temperature, oxidative damage, and high osmolarity (96). It is interesting, too, that induction of acid tolerance can occur even at neutral pH if organic acid concentration is high, a condition that mimics the situation in some parts of the intestine (92). Although this is consistent with pH_i acidification being the inducing signal, it has been suggested that the system induced by organic acids may differ from that induced by lowering pH_i.

Several regulatory genes play a role in the log-phase ATR, including the alternate sigma factor σ^S,

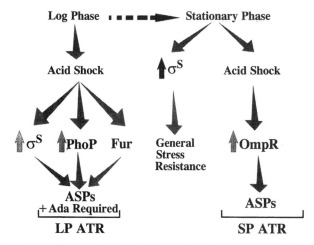

Figure 1. Comparison of log-phase and stationary-phase acid tolerance responses of *S. enterica* serovar Typhimurium. Acid shock (pH 4.5) of log-phase cells induces sigma S and PhoP, which in turn are required for the induction of subsets of acid shock proteins. Fur, not known to be induced by acid, is also required for induction of an ASP subset. Since *rpoS*, *phoP*, and *fur* mutants are acid sensitive, members of these ASPs are expected to be important for acid tolerance. In addition to the ASPs, the Ada protein is required but does not appear to be induced by acid or to regulate expression of other ASPs. Stationary phase will induce a general stress resistance dependent on sigma S, but acid-shocked stationary-phase cells also induce the regulator OmpR, which is required for enhanced acid tolerance known as the stationary-phase ATR.

encoded by *rpoS*; the iron regulator Fur; and the two-component signal transduction system PhoPQ. Each regulator controls the expression of a subset of ASPs: 10 ASPs for σ^S, five for Fur, and four for PhoP. Some members of each of these regulons must be important for the ATR since mutations in these regulators confer an acid-sensitive phenotype on cells. Learning how these regulators sense pH stress is an important part of understanding inducible acid stress survival. The transcription factor σ^S, for example, is barely detectable in log-phase cells (pH 7.7) due to rapid proteolytic turnover by the ClpXP protease (Webb et al., submitted; 141). Following acid shock, σ^S levels rise because a putative response regulator, MviA, signals decreased proteolytic turnover (10). Phosphorylation of MviA at residue 58 (aspartate) has been shown to occur in vitro (21), and site-directed mutations converting this residue to asparagine or glutamate reduce the ability of MviA to facilitate σ^S degradation (Moreno et al., submitted). On the basis of an "anti-sigma" activity demonstrated for MviA (Moreno et al., submitted; 170), it is believed that MviA directly interacts with σ^S, probably at σ^S residue 173 (11, 141), and aids in the recognition of σ^S by the ClpXP protease. An interaction between MviA(RssB) and σ^S has been demonstrated in vitro and in vivo using a bacterial two-hybrid system (Moreno and Foster, submitted). Phosphorylated MviA, in this model, would have a higher activity than dephosphorylated MviA. Although the amino terminus of MviA bears homology to the receiver modules of various response regulators, there is no obvious homology to known signal transduction output domains (16). Thus, it appears that MviA is not a transcriptional regulator but may represent a new class of response regulators whose role is to control the turnover of key regulatory molecules. A cognate sensor kinase has not been identified for MviA to date. However, acetyl phosphate appears to be an important in vivo phosphodonor. Acetyl phosphate was shown to phosphorylate RssB (the MviA homolog in *E. coli*) (21). In addition, mutants lacking the enzymes acetate kinase and phosphotransacetylase, and thus lacking acetyl phosphate, exhibit high log-phase levels of σ^S (Moreno et al., submitted; 21). How the system senses acid stress is not known.

The σ^S-dependent ATR system is required to survive acid stress imposed by volatile fatty acids and contributes to inorganic acid tolerance (7) although other systems also participate (9, 45, 94) (see PhoPQ below). Several σ^S-dependent genes encoding acid shock proteins have been identified through N-terminal sequencing, including genes *yciE* and *yajQ* of unknown function, *osmY* encoding a periplasmic protein of unknown function, and a periplasmic superoxide dismutase called *sodC_{II}* (*Salmonella* possesses two periplasmic SODs) (42). A mutation constructed in *sodC_{II}* did not affect acid tolerance, but its identification as an ASP illustrates that acid stress can trigger the synthesis of proteins that protect the cell against heterologous stresses (42).

The iron regulator Fur normally acts as a repressor of gene expression when bound to excess intracellular Fe^{2+} (6) but controls a subset of ASPs in an iron-independent manner (48, 68). Mutations have been identified in *fur* that produce acid-blind/iron-sensing and acid-sensing/iron-blind phenotypes (68) (Hall, Coy, and Foster, unpublished). This suggests that iron and acid are sensed separately by this protein. The iron-blind/acid-sensing mutation occurred in residue H90, part of the predicted iron-binding site of Fur. This mutant protein failed to repress iron-regulated genes but functioned normally in acid tolerance (68). In contrast, conversion of residue 144 from histidine to leucine did not affect the ability of Fur to repress iron-regulated genes but did interfere with its ability to function in acid tolerance (Hall, Coy, and Foster, unpublished). So it appears that the acid-sensing portion of Fur occurs at its carboxy terminus. Efforts are now under way to identify the acid-induced target genes regulated by Fur.

The third regulatory system affecting log-phase ATR involves the two-component PhoPQ system important for *Salmonella* virulence (43, 61, 62, 64, 109). In an elegant series of experiments, Groisman and colleagues have demonstrated that the membrane-bound PhoQ sensor-kinase senses periplasmic Mg^{2+} concentrations and phosphorylates the response regulator PhoP (53, 150). PhoP-P will then activate the expression of a series of target genes. The first inkling that this system was involved with acid tolerance came when PhoP was identified as an ASP. Subsequently, mutations in *phoP* and *phoQ* proved to make cells sensitive to inorganic acid in an *rpoS* mutant, but PhoPQ proved nonessential for log-phase acid tolerance in the presence of σ^S (9). This study also found that PhoQ, in addition to sensing magnesium, can probably sense pH even in the presence of high magnesium concentrations. Thus, in the macrophage phagolysosome environment where pH is acidic and Mg^{2+} levels are low, either one or both of these conditions may be involved in activating the PhoPQ system. PhoPQ controls the expression of another two-component system (PmrAB) involved in activating polymyxin resistance. Consistent with PhoPQ being an acid shock protein was the finding that an acidic environment can also stimulate PhoP-dependent polymyxin resistance (63).

Stationary-phase acid tolerance in serovar Typhimurium also involves σ^S-dependent and -independent systems. The σ^S-dependent stationary-phase system does not require acid induction, contrary to the σ^S-dependent log-phase system. Entry into stationary phase is enough to induce σ^S levels and acid tolerance. However, there is also an acid-induced σ^S-independent stationary-phase ATR that appears very different from the log-phase systems (95). The only regulator discovered to affect this system so far is the OmpR response regulator (Y. Park, J. W. Foster, and I. Bang, submitted). Interestingly, it does this independently from its known sensor-kinase, EnvZ.

Acidic pH_i is thought to negatively affect metabolic flow and damage macromolecules. Recent evidence that acid and alkaline stress induces DNA damage comes from finding that *ada* mutants of serovar Typhimurium are extremely acid and alkaline sensitive (9; White and Foster, unpublished). Ada is a DNA methyl transferase that repairs O^6-methyl guanine, O^4-methyl thymine, and methyl phosphotriesters. Although Ada is also a transcriptional regulator, its role in surviving pH stress does not appear to involve regulating ASP expression (9; B. White and J. Foster, unpublished).

The impact of low pH on the virulence of *Salmonella* is profound. A variety of genes appear to be induced in response to the low pH environment of the phagolysosome, including genes involved in virulence (3, 33, 162). In fact, macrophage vacuoles containing *Salmonella* acidify to between pH 4 and 5 within 60 min after formation (129), a process necessary for bacterial survival and replication within the macrophage. Studies also suggest that acid tolerance aids in the pathogenesis of the organism since several mutants defective in acid tolerance were attenuated (52, 166). Another possible link between low pH and virulence has been discovered in that serovar Typhimurium at low pH synthesizes an autoinducer similar to the system 2 autoinducer involved with induction of bioluminescence in *Vibrio*. The conditions causing autoinducer production are similar to what is encountered upon first interaction of an enteric pathogen with its host. However, no definitive role for autoinducer in pathogenesis has been established. Should such a link be established, the results will have a significant impact on our understanding of how pH influences virulence (8, 153).

ACID TOLERANCE, ACID RESISTANCE, AND ACID HABITUATION OF *SHIGELLA FLEXNERI* AND *E. COLI*

As with *Salmonella*, *E. coli* possesses log-phase and stationary-phase acid survival mechanisms. A problem that plagues the acid survival literature is the difficulty experienced when trying to compare various systems studied in different laboratories. Three systems have been named in *E. coli*: the acid tolerance response, acid habituation, and acid resistance. Problems comparing systems stem from the different strategies and media used to induce and test acid survival. For example, the ATR, which has been shown to protect log-phase cells for several hours at pH 3, appears to be very different from the process of acid habituation that is usually measured as protection for 7 min at pH 3 in complex media. However, as studies have progressed, it appears there are indeed numerous acid survival mechanisms, some of which have long-term dramatic effects while others have more subtle, yet significant, consequences. It is essential for the reader to be aware of the manner in which each system is tested. Key features of the *E. coli* acid resistance and acid habituation systems are summarized in Tables 1 and 2.

A direct comparison of the log-phase acid tolerance responses of serovar Typhimurium, *E. coli*, and *S. flexneri* indicates that, when tested in unsupplemented minimal media, serovar Typhimurium and *E. coli* exhibit similar levels of pH 3 acid tolerance (1 to 2 h challenge). Although detailed analyses of the ATR in *E. coli* have not been performed (97), it is thought that the mechanisms of the *E. coli* ATR will be similar to those described for serovar Typhimurium. *Shigella*, however, does not appear to possess a discernible ATR. One intriguing ATR study using *E. coli* has found that cyclopropane fatty acid formation in the membrane is extremely important for surviving acid stress (30). The gene responsible for the enzyme that carries out this postsynthetic modification of the lipid bilayer, *cfa* (CFA synthase), is induced by acid conditions and is RpoS dependent. Significantly, mutants defective in *cfa* are acid sensitive. Potential acid tolerance roles envisioned for CFAs include decreasing proton permeability of membranes or interacting with membrane proteins that influence proton traffic.

Acid-adapted *E. coli* and serovar Typhimurium, when placed in extreme acid environments, maintain pH_i closer to neutrality than do unadapted cells. This has been correlated in *E. coli* to changes in proton flux whereby adapted cells do not allow protons to flow into the cell as easily as unadapted cells do (Jordan, Oxford, and O'Byrne, submitted). Although an important finding, this study did not examine whether changes in proton movement required protein synthesis or was the result of activating preformed systems of pH homeostasis. The mechanism behind the acid-induced alterations in proton flux is not known. One possibility, protein synthesis-dependent increases in proton efflux via the F_oF_1 ATPase, has been demonstrated in *Streptococcus* (see above) but only suggested in the enteric gram-negative rods (reviewed in reference 69).

Much of the work examining the acid survival strategies of log-phase *E. coli* has been with a system referred to as acid habituation (57, 127). In contrast

Table 1. pH 2 acid resistance mechanisms of *E. coli*[a]

System	Mechanism	Regulation			Regulators	
		SP	SP/Acid	LP/Acid	SP	Acid
1	Unknown	++	++	−	RpoS, Crp	pH 8 inhib
2	Glt decarbox	+	++	++	RpoS, HNS	?
3	Arg decarbox	+	++	−	CysB, AdiY, HNS, IHF	?

[a] Acid resistance is tested using stationary-phase cells in minimal media, pH 2 to 2.5, and survival is measured after 2 to 4 h. SP, stationary phase; LP, log phase; decarbox, decarboxylation; inhib, inhibition.

Table 2. pH 3 acid habituation in *E. coli*[a]

Induction pH or heat shock	Inducing agent[b]	Extracellular components[c]	Mechanism	Regulator
Neutral pH	Glutamate	Yes	GAD-Dep?	
	Aspartate	Yes	?	
	Proline	Yes	?	
	Glucose	Yes	?	
	Fe^{2+}	Yes	?	
Acid pH	Low phosphate	Yes	PhoE Porin	CysB
	Cu^{2+}	?	?	
Heat shock	?	?	?	

[a] Acid habituation is usually tested in nutrient broth, pH 3 to 3.5, and survival is measured after 7 min.
[b] Inducing agents added to log-phase cells for 60 min.
[c] Filtered, dialyzed culture supernatants from adapted cells trigger acid habituation in sensitive cells.

to the ATR, acid habituation studies are generally performed in nutrient broth media, involving short exposures (7 min) to pH 3, and appear to involve mechanisms only peripherally related to those associated with the ATR (Table 2). The work reported by Rowbury and colleagues has revealed an amazing complexity to log-phase acid habituation (133). Many compounds have been shown to affect habituation, such as glucose, glutamate, aspartate, $FeCl_3$, KCl, and L-proline, which can induce habituation at neutral pH_O (Table 2). The control of these systems seems unrelated to Fur, CysB, and RelA although acid habituation induced at pH 5 is CysB dependent (135). Phosphate and cAMP also influence induction of acid habituation (134, 136). Another intriguing finding is that secreted proteins made during a pH 5 adaptation can induce acid habituation in log-phase cells grown in neutral pH, although the mechanism for this is unknown (137, 138). Similar extracellular proteins are thought to be important for signaling acid habituation at neutral pH (137, 138).

The most dramatic resistance to acid stress occurs in stationary-phase cells of *E. coli* and *S. flexneri* (146, 147). These systems, called acid resistance (AR) systems as opposed to the log-phase ATR or acid habituation, will protect cells at pH 2 for several hours (145). The seminal work in this area, published by Small and Slonczewski, revealed that there are σ^S-dependent and -independent systems of acid resistance present in stationary-phase cells. Their data also indicated that serovar Typhimurium was not capable of acid resistance at pH 2.5. A later study designed to dissect this phenomenon found three distinct and efficient stationary-phase systems of AR in *E. coli*, two of which are also present in *S. flexneri* (Table 1). AR system 1, present in both organisms, is RpoS dependent, CRP dependent, and glucose repressed. The system protects cells to pH 2.5 in minimal media and was originally thought to be induced by acid pH. More recent studies indicate that this may actually reflect the accumulation of an inhibitor produced

during growth at pH 8 that prevents the function, not the synthesis, of this system (29). AR system 1 also needs activation (not induction) by a brief exposure to glutamate prior to challenge at pH 2.5 (29). This is presumed to reflect the need for adequate intracellular glutamate levels for this system to function at extreme low pH. The mechanism by which this system affords protection to acid stress is not known.

E. coli acid resistance system 2, also present in *Shigella*, requires extracellular glutamate during pH 2 acid challenge (97, 98). Genetic studies indicate this system requires glutamate decarboxylase, an enzyme that converts intracellular glutamate to γ-amino butyric acid (GABA), and the glutamate/GABA antiporter *gadC* (29, 74, 146, 164). Inducible amino acid decarboxylase systems are thought to provide protection against acid stress through the consumption of intracellular protons (Fig. 2). In this model, protons leaking across the cell membrane during extreme acid stress can be consumed by amino acid decarboxylation reactions (e.g., glutamate decarboxylase to form GABA). However, to consume protons efficiently there must be a means to transport additional substrate rapidly into the cell and expel the decarboxylation product. This process is carried out by specific substrate/product antiporters such as the GadC antiporter in the case of the glutamate system. The net result of this process is intracellular alkalinization. However, as will be explained below, there is more to this mechanism than decarboxylation and antiport.

There are two glutamate decarboxylase (GAD) genes in *E. coli* designated *gadA* and *gadB* that map to different chromosomal locations but share 98% identity at the nucleotide level (149). The *gadA* gene is monocistronic, but the *gadB* gene forms an operon with *gadC*. Either one of the GAD enzymes will provide pH 2.5 resistance as long as the GadC antiporter is present, but both GAD isoforms are needed to protect the cell at pH 2 (29). Regulation of *gadA* and *gadBC* expression is complex. During growth in min-

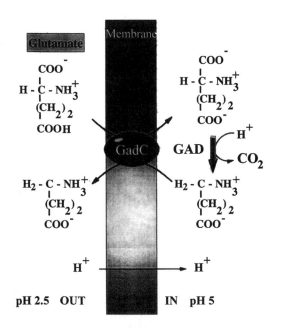

Figure 2. The glutamate decarboxylase/GadC antiporter system of acid resistance. GAD, glutamate decarboxylase; GABA, gamma amino butyric acid.

imal media, both genes are induced during entry into stationary phase or by acid pH in log phase (29, 168). Stationary phase induction requires the alternate sigma factor σ^S, but an unknown regulator, not σ^S, is involved in acid pH control (29). It is interesting to note that even though both *gad* transcriptional units respond to stationary phase and acid pH signals in minimal media, *gadBC* expression is primarily induced by stationary phase while *gadA* responds more to acid pH. This suggests that GadA is only needed when the cell anticipates, through sensing a progressively acidifying environment, an ultimate encounter with severe acid stress.

It is significant that *gadA* and *gadBC* are regulated differently during growth in Luria-Bertani broth (LB) as compared to minimal media. For some unknown reason *gad* genes, which are induced during log-phase growth in minimal media, are not induced by acid during log-phase growth in LB although acid will stimulate expression in stationary-phase LB cultures. In addition, the acid pH and stationary-phase controls observed during growth in LB both require σ^S in contrast to minimal media where acid control does not require this sigma factor (29, 36). Since only one promoter has been shown to function under all conditions, one wonders whether σ^{70} can substitute for σ^S under certain conditions and whether additional regulators are involved (36; Castanie-Cornet and Foster, unpublished).

As noted, *S. flexneri* also possesses the glutamate-dependent AR system 2. Although the *Shigella* system

was reported to be completely dependent on *rpoS*, the mutant cultures were not grown at an acid pH during adaptation or in minimal media (164). If the situation with *Shigella* is similar to that of *E. coli*, growth of a *Shigella rpoS* mutant at low pH (pH 5.5) in minimal media might also induce the *gad* system independently of σ^S and thereby restore glutamate-dependent acid resistance. It is noteworthy that mutations in the *hdeAB* operon also conveyed an acid-sensitive phenotype on *Shigella* (164). The products of this operon encode two small proteins of unknown function. Why these mutants lack glutamate-dependent acid resistance is not clear.

The third *E. coli* AR system, not present in *Shigella*, requires arginine during pH 2 acid challenge (97). It is an arginine decarboxylase (*adiA*)-dependent system that appears to function much like the GAD system (151). It converts intracellular arginine to agmatine, consuming a proton in the process. The antiporter for this system has not been identified. Optimum expression of *adiA* occurs at low pH under anaerobic conditions following growth in a complex medium. The positive regulator CysB is required for *adiA* expression and for arginine-dependent acid resistance, although it is probably not the component involved in sensing pH (98, 143). A gene downstream of *adiA*, called *adiY*, appears to be another positive regulator of *adiA* expression (152). AdiY bears significant homology to the AraC class of regulators, but its role in regulating *adiA* is unknown. There is some evidence that DNA supercoiling may play a role in pH control of this locus. The histone-like protein HNS thought to affect DNA supercoiling appears to repress *adiA* expression at pH 8 (144), and integration host factor (IHF), which is known to bend DNA, stimulates expression (151).

Curiously, not all amino acid decarboxylase/antiporter systems will protect *E. coli* from pH 2. The arginine and glutamate decarboxylase systems will provide this protection, but lysine and ornithine decarboxylase systems will not. Part of the reason for this lies in the pH optimum for each decarboxylase, with GAD having the most acidic pH optimum (pH 4.5)(44). Optima for the others are pH 5 (arginine decarboxylase), pH 5.7 (lysine decarboxylase), and pH 6.9 (ornithine decarboxylase) (155). At an external pH of 2.5, internal pH is estimated to be near 5 so that lysine and ornithine decarboxylases will either not operate or operate poorly. Several findings also suggest that the glutamate and arginine decarboxylase/antiporter systems are not, in and of themselves, capable of providing acid resistance. One striking result supporting this conclusion is that serovar Typhimurium possesses an outstanding arginine decarboxylase system, but this system will not protect the organism against pH 2. Furthermore, a major paradox that has not been resolved is that the consumption of protons via decarboxylation actually seems, on first glance, like a futile proton cycle. Close examination of Fig. 1 reveals that extracellular glutamic acid at pH 2.5 will be protonated at the γ-carboxyl group (resulting in a neutral charged molecule). As glutamate enters the cell, where internal pH will likely be above pH 5, that proton should be released. Then, during decarboxylation, only one proton will be consumed. This calculates, at best, to a net loss of 0 H^+! One possibility that might allow net consumption of protons would be if the pH of the GadC conduit is less acidic than the extracellular media so that the proton from the g-carboxyl group is released from glutamate into the periplasm *before* the amino acid gets into the cell. However, this may be unfavored energetically since a negatively charged glutamate would exchange for a neutral GABA. Clearly, accessory proteins are required for acid resistance dependent on glutamate (system 2) and arginine (system 3) in *E. coli*.

A particularly virulent form of *E. coli*, serotype O157:H7, has emerged as an important pathogen over the past decade. It is classified as an enterohemorrhagic *E. coli* that causes a variety of diseases including hemorrhagic colitis (130), hemolytic uremic syndrome (24, 34), and thrombotic thrombocytopenic purpura (101). *E. coli* O157:H7 is a major foodborne pathogen that threatens many aspects of the food industry, most notably the beef and poultry supply (120). Since disease caused by O157:H7 appears to require a very low infectious dose and has been associated with the organism's presence in acidic foods, it has been suggested that O157:H7 has an extraordinary acid defense system (1, 15, 60, 108). Several studies have shown this to be true. However, a fairly detailed comparison of true commensal strains of *E. coli* and O157:H7 clinical isolates has revealed little difference between the two groups, with the exception that arginine-dependent AR was more robust in the O157 strains (98). Two O157:H7 strains thought to be constitutively acid resistant have also been identified in the literature (25). However, both strains clearly exhibit the three inducible stationary-phase AR systems noted above, and no new systems were observed (Castanie-Cornet and Foster, unpublished observation). The overall levels of acid resistance for these strains, however, were elevated when compared with that of other O157:H7 strains.

It is assumed that the three stationary-phase AR systems help *E. coli*, including O157 strains, maintain a less acidic pH_i, yet this has never been unequivocally proven. A controversial and thought-provoking article by Diez-Gonzalez and Russell suggests that O157:H7 strains allow their internal pH to fall more than K-12 strains do to cope with an acidifying environment (39). Thus, it is proposed that O157 maintains a smaller ΔpH than K-12 does during acid stress. The lower pH_i of the pathogen would help protect the cell from acid stress because weak acids present in the medium will not deprotonate to the same degree and thus would not accumulate intracellularly. While this is an interesting concept, the model has not been confirmed. However, failure to maintain ΔpH cannot be the main acid protection mechanism since the three stationary-phase acid resistance systems noted above are tested in medium lacking any weak organic acids (29).

A recent study with *E. coli* O157 proposes another novel view of acid tolerance (K. Jordan, L. Oxford, and C. O'Byrne, submitted). These authors report that cells, adapted or not, will experience loss of viability, albeit at different rates, but that all cultures have a subpopulation of cells (physiological variants, not mutants) that will survive acid treatments (pH 3) for long periods. The acid-resistant nature of this subpopulation is not dependent on σ^S, but it is not known whether *gad* or *adi* systems might be important (J. Glover, L. Malcolm, F. Thompson-Carter, P. Carter, S. Jordan, S. Park, and I. Booth, unpublished).

HELICOBACTER PYLORI: UREASE-DEPENDENT AND -INDEPENDENT ACID TOLERANCE

H. pylori plays an etiological role in a variety of gastroduodenal diseases including gastritis, peptic ulcer disease, nonulcer dyspepsia, and gastric carcinoma (26). As a pathogen that has chosen to colonize the stomach, *H. pylori* has developed exceptional systems of acid tolerance. The major mechanism involves a powerful urease that the organism produces constitutively (104, 121). Urea is converted by urease to carbon dioxide and the alkaline product ammonia. Since urease has been found both intracellularly and bound to the cell surface, the manner by which this enzyme provides acid tolerance has met with some controversy. Scott et al. argue against a model whereby external urease somehow maintains a "cloud" of less acidic medium around the organism but clearly supports the idea that internal urease bolsters internal pH homeostasis (142). By measuring pH_i and by using an inhibitor of urease (fluorofamide) that poorly penetrates the cell, Scott et al. demonstrate that external urease is unimportant for the maintenance of internal pH (142). In addition, since external urease is inactive under pH conditions less than pH 4.5, the authors argue that external urease bound to the cell could not contribute to acid survival at pH 2. The data presented also suggest that there is an internal mechanism controlling urease activity at neutral pH. Internal urease does not become active until pH_O is below 6 and remains active through pH_O 2.5. This reflects the range of external pH at which pH_i acidification will begin to harm the organism. However, the pH optimum of urease is pH 7.5, equivalent to the pH_i when pH_O is neutral. Thus, one would expect cells grown at neutral pH to have high urease activity. Since internal urease is not active until cells encounter an acidic pH_O, there may be a novel mechanism that inhibits the intracellular activity of urease to protect the cell from inappropriately alkalinizing its internal pH when growing at neutral pH_O in the presence of urea.

Krishnamurthy et al. (86), however, present a counterargument in favor of external urease being most important for acid tolerance. Measuring pH 2 acid survival, they demonstrate that the same, poorly diffusible, urease inhibitor used in the above study caused a 6-log decrease in survival. Because it seems counterintuitive that external urease could contribute to acid resistance (e.g., *Yersinia enterocolitica* clearly uses internal urease) (37, 169) and because there are questions as to whether the urease inhibitor may still gain access to the cytoplasmic compartment, the resolution of this controversy may have to await the isolation of mutants that do not produce external urease.

Even though urease plays an important role in the survival of *Helicobacter* at extremely low pH, the organism also possesses a urease-independent ATR, although it is much less effective than urease-dependent acid resistance (17, 81). Three acid-induced genes have been associated with urease-independent acid tolerance in *H. pylori* (76, 81, 107). Expression of *cagA*, encoding a high molecular weight antigen, is maximal at pH 6 (82). However, *cagA* mutants are actually more acid resistant than wild type. The reasons for this are unknown, but it is hypothesized that the increased acid sensitivity of a *cagA*$^+$ strain provides a mechanism that helps drive the *cagA*$^+$ organism to a specific, less acidic part of the stomach and, in so doing, allows more intimate interaction with the host.

A second gene associated with urease-independent acid tolerance is *wbcJ*, involved in O-antigen synthesis (107). A *wbcJ* mutant was defective in the urease-independent ATR but possessed normal urease-dependent acid resistance. The manner in which WbcJ contributes to acid tolerance is not known. However, LPS may provide a barrier, reducing proton influx, or, alternatively, O-antigen composition may alter membrane permeability by changing surface charge. Loss of O-antigen expression may also affect conformation and function of various surface proteins needed for tolerance.

An unusual role for the chaperone Hsp70 (*dnaK*) in acid-induced stress tolerance of *H. pylori* has been proposed. Hueska et al. (76) demonstrate that brief exposure of *H. pylori* to acid conditions changes the glycolipid binding specificity of the organism to include sulfonated glycolipids that are present on eukaryotic cells. They further show that Hsp70 is induced by low pH, is expressed on the bacterial surface, and appears responsible for altered glycolipid binding. This altered binding is believed to facilitate association of the organism with stomach mucus and thus provide a survival advantage during periods of high acid stress.

Acid exposure in the stomach with subsequent acidification of the internal pH is thought to cause DNA damage. Evidence that repair of DNA damage may be important for acid survival of *H. pylori* includes the finding that *recA* mutants are 10-fold more sensitive to acid pH whether or not urea is present (158). Whether acid stress causes induction of a RecA-dependent SOS response is not known.

RHIZOBIUM: CALCIUM, ACID TOLERANCE, AND DEALING WITH THE ACID UNDERGROUND

Low soil pH restricts legume productivity in many parts of the world. Thus, the effects of low pH on the growth and survival of root nodule bacteria are of global importance. Different species of root nodule-forming *Rhizobium* display varying degrees of acid resistance as measured by their ability to grow (not just survive) at low pH (54, 55, 59). *Rhizobium leguminosarum* is moderately acid resistant, capable of growth down to pH 4.0 to 4.5. *Sinorhizobium meliloti*, however, is more acid sensitive, growing only down to pH 5.5. Studies designed to dissect the basis of acid resistance in these organisms have revealed that acid-resistant strains are better able to maintain internal pH when grown in acid conditions (116). Genes associated with acid resistance in *R. leguminosarum* are located on the second largest of four mega plasmids (31). Loss of this plasmid was shown to alter cell surface LPS structure, interfere with pH homeostasis at low pH, and confer an acid-sensitive growth phenotype. In a separate study, two loci (either chromosomal or plasmid) associated with acid resistance were identified using Tn5 insertional mutagenesis (32). The mutations interfered with membrane permeability but not proton extrusion, resulting in diminished pH homeostasis leading to the acid-sensitive phenotype.

Work with *S. meliloti* has shown this organism will respond to increasing concentrations of calcium in the growth medium by growing at progressively lower pH values (161). Transposon Tn5 insertion mutants with increased acid sensitivity have been isolated and grouped into two classes, those that are calcium repairable and those that are not (58, 161). How calcium participates in acid resistance and whether any of the systems are inducible are questions that have not been examined. One calcium-repairable, acid-sensitive mutation occurred in the *actA* gene whose product is a membrane-bound protein required to maintain pH_i above 7.0 when pH_O is below 6.5 (116). Mutants defective in *actA* fail to grow at pH 6 and are sensitive to Cu^{2+} and Zn^{2+}, although removal of these ions from media did not rescue acid tolerance (160). This is in contrast to an *actP* mutant that is acid sensitive only in the presence of Cu^{2+}. The ActP protein is a P-type ATPase that may be involved in Cu^{2+} transport. A unrelated two-component system, ActS (sensor)/ActR (regulator), predicted to sense pH has been identified (159). Mutants defective in either *actS* or *actR* do not exhibit inducible acid tolerance (55).

R. leguminosarum and *Bradyrhizobium japonicum* also exhibit acid tolerance responses induced by growth at pH 5.0 (115). Low pH-induced acid tolerance is best in stationary-phase cells, although significant induction also occurs in log-phase cells. Induction of acid tolerance requires protein synthesis and appears to be similar to the ATR systems described for serovar Typhimurium.

ATPASE-INDEPENDENT ACID TOLERANCE MECHANISMS IN GRAM-POSITIVE MICROORGANISMS

Listeria monocytogenes

An inducible ATR has been demonstrated for the food-borne pathogen *L. monocytogenes*, and the global effect of acid stress on protein synthesis has been reported (88, 113, 114, 122). The stress response sigma factor σ^B is involved in stationary-phase-induced acid tolerance, but its role in acid-induced log-phase acid tolerance has not been evaluated (165). A connection between the ATR and virulence has also been suggested. First, mutants with increased acid tolerance exhibit increased virulence in mice (114). Second, a mutant defective in low pH induction of tolerances against acid as well as other stresses (heat, ethanol) exhibited diminished virulence in mice (103). It should be noted that a *sigB* mutant lacking σ^B did not display a virulence defect, suggesting that σ^B-independent acid tolerance mechanisms may be employed during passage through the stomach (165) (see chapter 12).

Streptococcus

Adaptation to low pH by *S. mutans* was shown in that cells prepared from pH 5.0 cultures exhibited a lower pH drop resulting from glucose fermentation than did cells prepared from pH 7.0 cultures. This is due to a shift in the pH optima of glucose uptake and glycolysis itself as well as to an increase in the F_oF_1 H^+-translocating ATPase as described above (12, 70). Cells adapted to pH 5 are better able to maintain a pH gradient at lower pH values than are pH 7.5 unadapted cells (70). In addition, cells grown at pH 5.5 exhibit reduced proton permeability relative to cells grown at pH 7.5 (70). The combination of decreased membrane permeability and increased ability to expel protons improves internal pH homeostasis and enhances resistance to acid killing (12).

A related organism, *Streptococcus rattus*, also increases ATPase levels in response to acidification. In addition, it copes with acid stress by altering its fermentation end products from a mixture of formate, acetate, and ethanol at pH_O 7.0 to mostly lactate at pH_O 5.0 (110). At a pH_O of 5.0, the lower pK_a of lactate (pK_a 3.86) compared to acetate (pK_a 4.75) effectively lowers the amount of lipophilic undissociated weak acid available to permeate the membrane and cause intracellular toxic effects.

Some studies performed with *S. mutans* looked for genes, other than those encoding the proton-translocating ATPase, that are regulated by pH or that affect acid tolerance. The organisms induce several proteins in response to low pH, although their specific roles in acid tolerance have not been evaluated (71). Mutants of *S. mutans* with diminished acid tolerance have been isolated based on an inability to grow on trypticase soy agar containing 50 mM sodium acetate at pH 4.4 (167). One such insertion was found in the gene for diacylglycerol kinase, which generates phosphatidic acid (163). It is suspected that loss of this enzyme interferes with signal transduction of extracellular environmental signals due possibly to abnormal membrane function (167). It is not known whether diacylglycerol kinase constitutes part of the inducible or constitutive acid tolerance mechanisms for *S. mutans*. Another study using Tn917 mutagenesis found that a mutation in *fhs* encoding formyl-tetrahydrofolate synthetase conferred a pH 5 growth defect, although the physiological basis for this is unknown (65). The same study found that *ffh*, encoding a component of a signal recognition particle involved with protein translocation across the cytoplasmic membrane, also affected pH 5 growth and inducible acid tolerance (66). The *ffh* mutation appears to decrease the amount of H^+-translocating ATPase assembled in the membrane at pH 5. In a separate study, Jayaraman et al. (78) have shown that an operon including *hrc* (heat shock regulator), *grpE* (chaperone), and *dnaK* (ATP-dependent chaperone) is induced in response to acid shock. Presumably, the chaperoning properties of these gene products will protect cellular proteins from acid damage or will help refold damaged proteins, although these hypotheses were not tested directly.

Another finding that connects acid stress to DNA damage is the discovery of an acid-inducible DNA repair enzyme in *S. mutans* (67). Acid-adapted *S. mutans* exhibits increased survival after exposure to near UV irradiation (125). This occurs both in wild-type and *recA* mutants defective in the classic SOS DNA repair systems. An acid-inducible abasic site-specific endonuclease similar to *E. coli* exonuclease III has been purified 500-fold and is thought to contribute to acid-induced UV resistance (67).

Lactococcus lactis

L. lactis is a gram-positive microorganism that gains energy through carbohydrate fermentation, with the principal product being lactic acid. *L. lactis* is widely used in the dairy industry because the lactic acid produced acidifies food to a point that prevents spoilage. Because the organism spends a large part of its growth cycle at low pH, it has developed several acid survival strategies. Resting cells are better adapted to low pH than are dividing cells, indicating a stationary-phase-induced acid survival system (72). But, as with other microbes, acid resistance is significantly improved by preexposure to mild (pH 5 to 6) conditions (73, 128). A decrease in internal pH is a major signal in the induction of acid tolerance (117). Thirty-three acid-induced proteins, some of which are presumed to enable survival at low pH (such as the ATPase subunits), are produced in lactococcal cells when exposed to pH 5.5 (73, 118). Unfortunately, the identities and roles of these proteins are unknown. Consistent with the situation in streptococci, one acid tolerance system employed by *L. lactis* involves the F_oF_1 H^+-translocating ATPase. Mutants with reduced F_oF_1 H^+-translocating ATPase are more sensitive to acid (F. Tornita, S. Amachi, and A. Yokota, Fifth Symposium on Lactic Acid Bacteria, poster G38, 1996). However, elevated pH_i is not the sole reason for increased survival at acid pH_O. Cells with an induced ATR survived exposure to a lethal acidic pH_O better than did uninduced cells with a pH_i identical to the adapted cells (117). Clearly, mechanisms affecting consequenses of acid stress other than pH_i are important for survival. Insertional mutagenesis of *L. lactis* with selection for increased acid resistance in unadapted cells (minimal media) has revealed several genes associated with acid tolerance (R. Rallu, A. Gruss, S. D. Ehrlich, and E. Maguin, submitted). Mutations were identified in genes affecting glutamate transport (*glnP*), phosphate transport (*pstS* and *B*), purine metabolism (*deoB, guaA*), ppGpp synthesis (*relA*), as well as several genes with unknown functions.

In addition to minimal media ATR, some genera of lactococci, streptococci, and lactobacilli can utilize an arginine deiminase pathway to survive acid environments (28, 35, 102). This pathway is very different from the arginine decarboxylase system described above. The ADI system converts arginine to ornithine, ammonia, and CO_2 while concomitantly generating one mole of ATP per mole of arginine consumed (Fig. 3). The generation of ammonia alkalinizes the environment. The system is repressed by carbohydrates and induced by arginine, but it is not clear whether induction is influenced by environmental pH (35, 123). The genetic organization of this system (*arcABCTD*) has recently been examined and consists of *arcA* (encoding ADI), *arcB* (ornithine transcarbamylase), *arcC* (carbamate kinase), *arcT* (putative transaminase), and *arcD* (putative arginine/ornithine antiporter) (173).

Glutamate decarboxylase also plays an important role in lactococcal acid resistance. The gene arrangement in this organism, *gadCB*, is the reverse of that in *E. coli* but encodes the same proteins, a glutamate:GABA antiporter and glutamate decarboxylase, respectively (139). To protect cells from acid

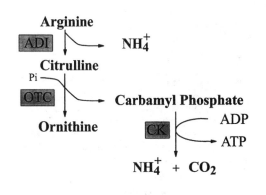

ADI Pathway

Figure 3. The arginine deiminase system. ADI, arginine deiminase; OTC, ornithine transcarbamylase; CK, carbamate kinase; and ArcD, putative arginine/ornithine antiporter.

challenge (pH 3.5), this system requires the presence of glutamate and Cl⁻ ion. A *gadB* mutant is very acid sensitive under these conditions, confirming a role for this system in acid survival. Maximal induction of *gadCB* occurs in stationary phase in the presense of Cl⁻ ion, low pH, and glutamate. The presence of choride ion is essential for induction with glutamate and low pH enhancing expression. Chloride-dependent activation of *gadCB* is controled by *gadR*, a regulatory gene located just upstream of *gadCB*, although *gadR* itself seems to be constitutively expressed.

CONCLUSIONS

In addition to the organisms discussed here, inducible acid tolerance has been demonstrated in several other species and, indeed, may be a universal phenomenon (80, 100, 106, 111, 112). Inducible acid tolerance is an extremely complex phenomenon of which little is known mechanistically, yet the importance of battling protons is widely recognized for its impact on pathogenesis, food microbiology, and various aspects of cell differentiation. Progress has been made in terms of identifying the genes and proteins brought to bear in response to pH stress, but how most of those proteins enable acid tolerance is unclear. Future work will surely focus on specific mechanisms of pH-regulated gene expression as well as processes by which acid or alkaline damage is prevented or repaired. Studies of the molecular responses to pH stress may expose novel antimicrobial targets and will provide insight into pathogenic processes leading to better vaccine development. The construction of hardier strains of root nodule bacteria that will better survive soil acidity and improve crop yields is another potential benefit. It is clear that the new insights provided by recent advances in molecular genetics have rekindled interest in this important aspect of communication between microorganisms and their environment.

Acknowledgments. I thank the many investigators who graciously shared their work prior to publication and the students who have made the journey both exciting and rewarding.

The work generated in my laboratory was supported by awards from the National Institutes of Health (GM48017), the National Science Foundation (DCB-89-04839), and the U.S. Department of Agriculture (97-35201-4751).

REFERENCES

1. **Abdul-Raouf, U. M., L. R. Beuchat, and M. S. Ammar.** 1993. Survival and growth of *Escherichia coli* 0157:H7 in ground, roasted beef as affected by pH, acidulants, and temperature. *Appl. Environ. Microbiol.* **59:**2364–2368.

2. **Abu Kwaik, Y., L. Y. Gao, O. S. Harb, and B. J. Stone.** 1997. Transcriptional regulation of the macrophage-induced gene (*gspA*) of *Legionella pneumophila* and phenotypic characterization of a null mutant. *Mol. Microbiol.* **24:**629–642.

3. **Alpuche-Aranda, C. M., J. A. Swanson, W. P. Loomis, and S. I. Miller.** 1992. *Salmonella typhimurium* activates virulence gene transcription within acidified macrophage phagosomes. *Proc. Natl. Acad. Sci. USA* **89:**10079–10083.

4. **Amaro, A. M., D. Chamorro, M. Seeger, R. Arredondo, I. Peirano, and C. A. Jerez.** 1991. Effect of External pH perturbations on in vivo protein synthesis by the acidophilic bacterium *Thiobacillus ferrooxidans*. *J. Bacteriol.* **173:**910–915.

5. **Arikado, E., H. Ishihara, T. Ehara, C. Shibata, H. Saito, T. Kakegawa, K. Igarashi, and H. Kobayashi.** 1999. Enzyme level of enterococcal F1Fo-ATPase is regulated by pH at the step of assembly. *Eur. J. Biochem.* **259:**262–268.

6. **Bagg, A., and J. B. Neilands.** 1987. Ferric uptake regulation protein acts as a repressor, employing iron (II) as a cofactor to bind the operator of an iron transport operon in *Escherichia coli. J. Biochem.* **26:**5471–5477.

7. **Baik, H. S., S. Bearson, S. Dunbar, and J. W. Foster.** 1996. The acid tolerance response of *Salmonella typhimurium* provides protection against organic acids. *Microbiology* **142:**3195–3200.

8. **Bassler, B. L., and M. R. Silverman.** 1995. Intercellular communication in marine Vibrio species: density-dependent regulation of the expression of bioluminescence, p. 431–445. *In* J. A. Hoch and T. J. Silhavy (ed.), *Two-Component Signal Transduction*. American Society of Microbiology, Washington, D.C.

9. **Bearson, B. L., L. Wilson, and J. W. Foster.** 1998. A low pH-inducible, PhoPQ-dependent acid tolerance response protects *Salmonella typhimurium* against inorganic acid stress. *J. Bacteriol.* **180:**2409–2417.

10. **Bearson, S. M. D., W. H. Benjamin, Jr., W. E. Swords, and J. W. Foster.** 1996. Acid shock induction of *rpoS* is mediated by the mouse virulence gene *mviA* of *Salmonella typhimurium*. *J. Bacteriol.* **178:**2572–2579.

11. **Becker, G., E. Klauck, and R. Hengge-Aronis.** 1999. Regulation of RpoS proteolysis in *Escherichia coli*: the response regulator RssB is a recognition factor that interacts with the turnover element in RpoS. *Proc. Natl. Acad. Sci. USA* **96:**6439–6444.

12. **Belli, W. A., and R. E. Marquis.** 1991. Adaptation of *Streptococcus mutants* and *Enterococcus hirae* to acid stress in continuous culture. *Appl. Environ. Microbiol.* **57:**1134–1138.

13. **Bender, G. R., and R. E. Marquis.** 1987. Membrane ATPase and acid tolerance of *Actinomyces viscous* and *Lactobacillus casei. Appl. Environ. Microbiol.* **53:**2124–2128.

14. **Bender, G. R., E. A. Thibodeau, and R. E. Marquis.** 1985. Reduction of acidurance of streptococcal growth and glycolysis by fluoride and gramicidin. *J. Dent. Res.* **64:**90–95.

15. **Benjamin, M. M., and A. R. Datta.** 1995. Acid tolerance of enterohemorrhagic *Escherichia coli. Appl. Environ. Microbiol.* **61:**1669–1672.

16. **Benjamin, W. H., Jr., X. Wu, and W. E. Swords.** 1996. The predicted amino acid sequence of the *Salmonella typhimurium* virulence gene *mviA*⁺ strongly indicates that MviA is a regulator protein of a previously unknown *S. typhimurium* response regulator family. *Infect. Immun.* **64:**2365–2367.

17. **Bijlsma, J. J., M. M. Gerrits, R. Imamdi, C. M. Vandenbroucke-Grauls, and J. G. Kusters.** 1998. Urease-positive, acid-sensitive mutants of *Helicobacter pylori*: urease-independent acid resistance involved in growth at low pH. *FEMS Microbiol. Lett.* **167:**309–313.

18. **Blankenhorn, D., J. Phillips, and J. L. Slonczewski.** 1999. Acid- and base-induced proteins during aerobic and anaerobic growth of *Escherichia coli* revealed by two-dimensional gel electrophoresis. *J. Bacteriol.* **181:**2209–2216.

19. **Blaser, M. J., and L. S. Newman.** 1982. A review of human salmonellosis. I. Infective dose. *Rev. Infect. Dis.* **4:**1096–1106.

20. **Booth, I. R.** 1999. The regulation of intracellular pH in bacteria, p. 19–27. *In* J. C. Derek and G. Gardner (ed.), *Bacterial Response to pH.* John Wiley & Sons, Ltd., Chichester, England.

21. **Bouche, S., E. Klauck, D. Fischer, M. Lucassen, K. Jung, and R. Hengge-Aronis.** 1998. Regulation of RssB-dependent proteolysis in *Escherichia coli*: a role for acetyl phosphate in a response-control process. *Mol. Microbiol.* **27:**787–795.

22. **Bowden, G. H., and I. R. Hamilton.** 1987. Environmental pH as a factor in the competition between strains of the oral streptococci *Streptococcus mutans, S. sanguis,* and "*S. mutans*" growing in continuous culture. *Can. J. Microbiol.* **33:**824–827.

23. **Brey, R. N., B. P. Rosen, and E. N. Sorensen.** 1979. Cation/proton antiport systems in *Escherichia coli. J. Biol. Chem.* **255:**39–44.

24. **Brynes, J. J., and J. L. Moake.** 1986. Thrombocytic thrombocytopenic purpura and the hemolytic-uremic syndrome: evolving concepts of pathogenesis and therapy. *Clin. Hematol.* **15:**413–442.

25. **Buchanan, R. L., and S. G. Edelson.** 1996. Culturing enterohemorrhagic *Escherichia coli* in the presence and absence of glucose as a simple means of evaluating the acid tolerance of stationary-phase cells. *Appl. Environ. Microbiol.* **62:**4009–4013.

26. **Buck, G. E.** 1990. *Campylobacter pylori* and gastroduodenal disease. *Clin. Microbiol. Rev.* **3:**1–12.

27. **Carmel, O., O. Rahav-Manor, N. Dover, B. Shaanan, and E. Padan.** 1997. The Na$^+$-specific interaction between the LysR-type regulator, NhaR, and the *nhaA* gene encoding the Na$^+$/H$^+$ antiporter of *Escherichia coli. EMBO J.* **16:**5922–5929.

28. **Casiano-Colón, A., and R. E. Marquis.** 1988. Role of the arginine deiminase system in protecting oral bacteria and an enzymatic basis for acid tolerance. *Appl. Environ. Microbiol.* **54:**1318–1324.

29. **Castanie-Cornet, M.-P., T. A. Penfound, D. Smith, J. F. Elliott, and J. W. Foster.** 1999. Control of acid resistance in *Escherichia coli. J. Bacteriol.* **181:**3525–3535.

30. **Chang, Y. Y., and J. E. Cronan.** 1999. Membrane cyclopropane fatty acid content is a major factor in acid resistance of *Escherichia coli. Mol. Microbiol.* **33:**249–259.

31. **Chen, H., E. Gartner, and B. Rolfe.** 1993. Involvement of genes on a megaplasmid in the acid-tolerant phenotype of *Rhizobium leguminosarum* biovar trifolii. *Appl. Environ. Microbiol.* **59:**1058–1064.

32. **Chen, H., A. Richardson, and B. Rolfe.** 1993. Studies of the physiological and genetic basis of acid tolerance in *Rhizobium leguminosarum* biovar trifolii. *Appl. Environ. Microbiol.* **59:**1798–1804.

33. **Cirillo, D. M., R. H. Valdivia, D. M. Monack, and S. Falkow.** 1998. Macrophage-dependent induction of the Salmonella pathogenicity island 2 type III secretion system and its role in intracellular survival. *Mol. Microbiol.* **30:**175–188.

34. **Cleary, T. G.** 1988. Cytotoxin producing *Escherichia coli* and the hemolytic uremic syndrome. *Pediatr. Clin. N. Am.* **35:**458–501.

35. **Curran, T. M., J. Lieou, and R. E. Marquis.** 1995. Arginine deiminase system and acid adaptation of oral streptococci. *Appl. Environ. Microbiol.* **61:**4494–4496.

36. **De Biase, D., A. Tramonti, F. Bossa, and P. Visca.** 1999. The response to stationary-phase stress conditions in *Escherichia coli*: role and regulation of the glutamic acid decarboxylase system. *Mol. Microbiol.* **32:**1198–1211.

37. **De Koning-Ward, T. F., and R. M. Robins-Browne.** 1995. Contribution of urease to acid tolerance in *Yersinia enterocolitica. Infect. Immun.* **63:**3790–3795.

38. **Dibrov, P. A.** 1991. The role of sodium ion transport in *Escherichia coli* genetics. *Biochim. Biophys. Acta* **1056:**209–224.

39. **Diez-Gonzalez, F., and J. B. Russell.** 1997. The ability of *Escherichia coli* 0157:H7 to decrease its intracellular pH and resist the toxicity of acetic acid. *Microbiology* **143:**1175–1180.

40. **Drasar, B. S., M. Shiner, and G. M. McLeod.** 1969. Studies on the intestinal flora. I. The bacterial flora of the gastrointestinal tract in healthy and achlorhydric persons. *Gastroenterology* **56:**71–79.

41. **Espeso, E. A., J. Tilburn, L. Sanchez-Pulido, C. V. Brown, A. Valencia, H. N. Arst, Jr., and M. A. Penalva.** 1997. Specific DNA recognition by the *Aspergillus nidulans* three zinc finger transcription factor PacC. *J. Mol. Biol.* **274:**466–480.

42. **Fang, F. C., M. A. DeGroote, J. W. Foster, A. J. Baumler, U. Ochsner, T. Testerman, S. Bearson, J. C. Giard, Y. Xu, G. Campbell, and T. Laessig.** 1999. Virulent *Salmonella typhimurium* has two periplasmic Cu, Zn-superoxide dismutases. *Proc. Natl. Acad. Sci. USA* **96:**7502–7507.

43. **Fields, P. I., E. A. Groisman, and F. Heffron.** 1989. A *Salmonella* locus that controls resistance to microbicidal proteins from phagocytic cells. *Science* **243:**1059–1062.

44. **Fonda, M. L.** 1985. L-Glutamate decarboxylase from bacteria. *Methods Enzymol.* **113:**11–16.

45. **Foster, J. W.** 1993. The acid tolerance response of *Salmonella typhimurium* involves transient synthesis of key acid shock proteins. *J. Bacteriol.* **175:**1981–1987.

46. **Foster, J. W.** 1991. *Salmonella* acid shock proteins are required for the adaptive acid tolerance response. *J. Bacteriol.* **173:**6896–6902.

47. **Foster, J. W., and H. K. Hall.** 1990. Adaptive acidification tolerance response of *Salmonella typhimurium. J. Bacteriol.* **172:**771–778.

48. **Foster, J. W., and H. K. Hall.** 1992. The effect of *Salmonella typhimurium* ferric-uptake regulator (*fur*) mutations on iron and pH-regulated protein synthesis. *J. Bacteriol.* **174:**4317–4323.

49. **Foster, J. W., and H. K. Hall.** 1991. Inducible pH homeostasis and the acid tolerance response of *Salmonella typhimurium. J. Bacteriol.* **173:**5129–5135.

50. **Foster, J. W., Y. K. Park, I. S. Bang, K. Karem, H. Betts, H. K. Hall, and E. Shaw.** 1994. Regulatory circuits involved with pH-regulated gene expression in *Salmonella typhimurium. Microbiology* **140:**341–352.

51. **Foster, J. W., and M. Spector.** 1995. How *Salmonella* survives against the odds. *Annu. Rev. Microbiol.* **49:**145–174.

52. **Garcia-del Portillo, F., J. W. Foster, and B. B. Finlay.** 1993. The role of acid tolerance response genes in *Salmonella typhimurium* virulence. *Infect. Immun.* **61:**4489–4492.

53. **Garcia Vescovi, E., F. C. Soncini, and E. A. Groisman.** 1996. Mg^{2+} as an extracellular signal: environmental regulation of salmonella virulence. *Cell* **84:**165–174.

54. **Glenn, A. R., and M. J. Dilworth.** 1994. The life of root nodule bacteria in the acidic underground. *FEMS Microbiol. Lett.* **123:**1–10.

55. **Glenn, A. R., W. G. Reeves, R. P. Tiwari, and M. J. Dilworth.** 1999. Acid tolerance in root nodule bacteria, p. 112–126. *In* J. C. Derek and G. Gardew (ed.), *Bacterial Responses to pH.* John Wiley & Sons, Ltd., Chichester, England.

56. **Goldberg, E. B., T. Arbel, J. Chen, R. Karpel, G. A. Mackie, S. Schuldiner, and E. Padan.** 1987. Characterization of a Na$^+$/

H⁺ antiporter gene of *Escherichia coli. Proc. Natl. Acad. Sci. USA* **84**:2615–2619.

57. Goodson, M., and R. J. Rowbury. 1989. Habituation to normal lethal acidity by prior growth of *Escherichia coli* at a sublethal acid pH value. *Lett. Appl. Microbiol.* **8**:77–79.

58. Goss, T. J., G. W. O'Hara, M. J. Dilworth, and A. R. Glenn. 1990. Cloning, characterization, and complementation of lesions causing acid sensitivity in Tn5-induced mutants of *Rhizobium meliloti* WSM419. *J. Bacteriol.* **172**:5173–5179.

59. Graham, P. H., K. J. Draeger, and M. L. Ferrey. 1994. Acid pH tolerance in strains of *Rhizobium* and *Bradyrhizobium*, and initial studies on the basis for acid tolerance of *Rhizobium tropici* UMR1899. *Can. J. Microbiol.* **40**:198–207.

60. Griffin, P. M., and R. V. Tauxe. 1991. The epidemiology of infections caused by *Escherichia coli* 0157:H7, other enterohemorrhagic *E. coli*, and the associated hemolytic uremic syndrome. *Epidemol. Rev.* **13**:60–98.

61. Groisman, E. A., E. Chiao, C. J. Lipps, and F. Heffron. 1989. *Salmonella typhimurium phoP* virulence gene is a transcriptional regulator. *Proc. Natl. Acad. Sci. USA* **86**:7077–7081.

62. Groisman, E. A., F. Heffron, and F. Solomon. 1992. Molecular genetic analysis of the *Escherichia coli phoP* locus. *J. Bacteriol.* **174**:486–491.

63. Groisman, E. A., J. Kayser, and F. C. Soncini. 1997. Regulation of polymyxin resistance and adaptation to low-Mg²⁺ environments. *J. Bacteriol.* **179**:7040–7045.

64. Gunn, J. S., E. L. Hohmann, and S. I. Miller. 1996. Transcriptional regulation of *Salmonella* virulence: a PhoQ periplasmic domain mutation results in increased net phospho transfer to PhoP. *J. Bacteriol.* **178**:6369–6373.

65. Gutierrez, J. A., P. J. Crowley, D. P. Brown, J. D. Hillman, P. Youngman, and A. S. Bleiweis. 1996. Insertional mutagenesis and recovery of interrupted genes of *Streptococcus mutans* by using transposon Tn917: preliminary characterization of mutants displaying acid sensitivity and nutritional requirements. *J. Bacteriol.* **178**:4166–4175.

66. Gutierrez, J. A., P. J. Crowley, D. G. Cvitkovitch, L. J. Brady, I. R. Hamilton, J. D. Hillman, and A. S. Bleiweis. 1999. *Streptococcus mutans ffh*, a gene encoding a homologue of the 54 kDa subunit of the signal recognition particle, is involved in resistance to acid stress. *Microbiology* **145**(Pt 2):357–366.

67. Hahn, K., R. C. Faustoferri, and R. G. Quivey, Jr. 1998. Induction of an AP endonuclease acitivity in *Streptococcus mutans* during growth at low pH. *Mol. Microbiol.* **31**:1489–1498.

68. Hall, H. K., and J. W. Foster. 1996. The role of Fur in the acid tolerance response of *Salmonella typhimurium* is physiologically and genetically separable from its role in iron aquisition. *J. Bacteriol.* **178**:5683–5691.

69. Hall, H. K., K. Karem, and J. W. Foster. 1995. Molecular responses of microbes to environmental pH stress. *Adv. Microb. Physiol.* **37**:229–272.

70. Hamilton, I. R., and N. D. Buckley. 1991. Adaptation by *Streptococcus mutants* to acid tolerance. *Oral Microbiol. Immunol.* **6**:65–71.

71. Hamilton, I. R., and G. Svensater. 1998. Acid-regulated proteins induced by *Streptococcus mutants* and other oral bacteria during acid shock. *Oral Microbiol. Immunol.* **13**:292–300.

72. Hartke, A., S. Bouche, X. Gansel, P. Boutibonnes, and Y. Auffray. 1994. Starvation-induced stress resistance in *Lactococcus lactis* subsp. *lactis* IL1403. *Appl. Environ. Microbiol.* **60**:3474–3478.

73. Hartke, A., S. Bouché, J. C. Giard, A. Benachour, P. Boutibonnes, and Y. Auffray. 1996. The lactic acid stress response of *Lactococcus lactis* subsp. *lactis. Curr. Microbiol.* **33**:194–199.

74. Hersh, B. M., F. T. Farooq, D. N. Barstad, D. L. Blankenshorn, and J. L. Slonczewski. 1996. A glutamate-dependent acid resistance gene in *Escherichia coli. J. Bacteriol.* **178**:3978–3981.

75. Heyde, M., and R. Portalier. 1990. Acid shock proteins of *Escherichia coli. FEMS Microbiol. Lett.* **69**:19–26.

76. Huesca, M., A. Goodwin, A. Bhagwansingh, P. Hoffman, and C. A. Lingwood. 1998. Characterization of an acidic-pH-inducible stress protein (hsp70), a putative sulfatide binding adhesin, from *Helicobacter pylori. Infect. Immun.* **66**:4061–4067.

77. Hyde, M., and R. Portalier. 1990. Acid shock proteins of *Escherichia coli. FEMS Microbiol. Lett.* **69**:19–26.

78. Jayaraman, G. C., J. E. Penders, and R. A. Burne. 1997. Transcriptional analysis of the *Streptococcus mutans hrcA, grpE* and *dnaK* genes and regulation of expression in response to heat shock and environmental acidification. *Mol. Microbiol.* **25**:329–341.

79. Karem, K., and J. W. Foster. 1993. The influence of DNA topology on the environmental regulation of a pH-regulated locus in *Salmonella typhimurium. Mol. Microbiol.* **10**:75–86.

80. Karem, K., J. W. Foster, and A. K. Bej. 1994. Adaptive acid tolerance response (ATR) in *Aeromonas hydrophila. Microbiology* **140**:341–352.

81. Karita, M., and M. J. Blaser. 1998. Acid-tolerance response in *Helicobacter pylori* and differences between CagA⁺ and CagA⁻ strains. *J. Infect. Dis.* **178**:213–219.

82. Karita, M., M. K. Tummuru, H. P. Wirth, and M. J. Blaser. 1996. Effect of growth phase and acid shock on *Helicobacter pylori cagA* expression. *Infect. Immun.* **64**:4501–4507.

83. Kashket, E. 1987. Bioenergetics of lactic acid bacteria: cytoplasmic pH and osmotolerance. *FEMS Microbiol. Rev.* **46**:233–244.

84. Kobayashi, H., N. Murakami, and T. Unemoto. 1982. Regulation of the cytoplasmic pH in *Streptococcus faecalis. J. Biol. Chem.* **257**:13246–13252.

85. Kobayashi, H., T. Suzuki, N. Kinoshita, and T. Unemoto. 1984. Amplification of *Streptococcus faecalis* proton-translocating ATPase by a decrease in cytoplasmic pH. *J. Bacteriol.* **158**:1157–1160.

86. Krishnamurthy, P., M. Parlow, J. B. Zitzer, N. B. Vakil, H. L. T. Mobley, M. Levy, S. H. Phadnis, and B. E. Dunn. 1998. *Helicobacter pylori* containing only cytoplasmic urease is susceptible to acid. *Infect. Immun.* **66**:5060–5066.

87. Kroll, R. G., and I. R. Booth. 1983. The relationship between intracellular pH, the pH gradient and potassium transport in *Escherichia coli. J. Biochem.* **216**:706–719.

88. Kroll, R. G., and R. A. Patchett. 1992. Induced acid tolerance in *Listeria monocytogenes. Lett. Appl. Microbiol.* **14**:224–227.

89. Krulwich, T. A. 1995. Alkaliphiles: 'basic' molecular problems of pH tolerance and bioenergetics. *Mol. Microbiol.* **15**:403–410.

90. Krulwich, T. A., A. A. Guffanti, and M. Ito. 1999. pH tolerance in *Bacillus*: alkaliphiles versus non-alkaliphiles, p. 167–179. *In* J. C. Derek and G. Gardew (ed.), *Bacterial Responses to pH.* John Wiley & Sons, Ltd., Chichester, England.

91. Krulwich, T. A., A. A. Guffanti, and D. Seto-Young. 1990. pH homeostasis and bioenergetic work in alkalophiles. *FEMS Microbiol. Rev.* **6**:271–278.

92. Kwon, Y. M., and S. C. Ricke. 1998. Induction of acid resistance of *Salmonella typhimurium* by exposure to short-chain fatty acids. *Appl. Environ. Microbiol.* **64**:3458–3463.

93. Lambert, L. A., K. Abshire, D. Blankenhorn, and J. L. Slonczewski. 1997. Proteins induced in *Escherichia coli* by benzoic acid. *J. Bacteriol.* **179**:7595–7599.

94. Lee, I. S., J. Lin, H. K. Hall, B. Bearson, and J. W. Foster. 1995. The stationary-phase sigma factor σˢ(RpoS) is required

for a sustained acid tolerance response in virulent *Salmonella typhimurium. Mol. Microbiol.* **17:**155–167.

95. **Lee, I. S., J. L. Slonczewski, and J. W. Foster.** 1994. A low-pH inducible stationary phase acid tolerance response in *Salmonella typhimurium. J. Bacteriol.* **176:**1422–1426.

96. **Leyer, G. J., and E. A. Johnson.** 1993. Acid adaptation induces cross-protection against environmental stresses in *Salmonella typhimurium. Appl. Environ. Microbiol.* **59:**1842–1847.

97. **Lin, J., I. S. Lee, J. Frey, J. L. Slonczewski, and J. W. Foster.** 1995. Comparative analysis of extreme acid survival in *Salmonella typhimurium, Shigella flexneri* and *Escherichia coli. J. Bacteriol.* **177:**4097–4104.

98. **Lin, J., M. P. Smith, K. C. Chapin, H. S. Baik, G. N. Bennett, and J. W. Foster.** 1996. Mechanisms of acid resistance in enterohemorrhagic *Escherichia coli. Appl. Environ. Microbiol.* **62:**3094–3100.

99. **Loesche, W. J.** 1986. Role of *Streptococcus mutans* in human dental decay. *Microbiol. Rev.* **50:**353–380.

100. **Lorca, G., R. Raya, M. Taranto, and G. de Valdez.** 1998. Adaptive acid tolerance response in *Lactobacillus acidophilus. Biotechnol. Lett.* **20:**239–241.

101. **Machin, S. J.** 1984. Clinical annotation: thrombotic thrombocytopenic purpura. *Br. J. Haematol.* **56:**191–197.

102. **Marquis, R. E., G. R. Bender, D. R. Murray, and A. Wong.** 1987. Arginine deiminase system and bacterial adaptation to acid environments. *Appl. Environ. Microbiol.* **53:**198–200.

103. **Marron, L., N. Emerson, C. G. Gahan, and C. Hill.** 1997. A mutant of *Listeria monocytogenes* LO28 unable to induce an acid tolerance response displays diminished virulence in a murine model. *Appl. Environ. Microbiol.* **63:**4945–4947.

104. **Marshall, B. J., L. J. Barrett, C. Prakash, R. W. McCallum, and R. Guerrant.** 1990. Urea protects *Helicobacter (Campylobacter) pylori* from the bactericidal effects of acid. *Gastroenterology* **99:**697–702.

105. **Matin, A.** 1999. pH homeostatis in acidophiles, p. 152–163. *In* J. C. Derek and G. Cardew (ed.), *Bacterial Responses to pH.* John Wiley & Sons, Ltd., Chichester, England.

106. **McDonald, L. C., H. P. Fleming, and H. M. Hassan.** 1990. Acid tolerance of *Leuconostoc mesenteroides* and *Lactobacillus plantarum. Appl. Environ. Microbiol.* **56:**2120–2124.

107. **McGowan, C. C., A. Necheva, S. A. Thompson, T. L. Cover, and M. J. Blaser.** 1998. Acid-induced expression of an LPS-associated gene in *Helicobacter pylori. Mol. Microbiol.* **30:**19–31.

108. **Miller, G. L., and C. W. Kaspar.** 1994. *Escherichia coli* 0157:H7 acid tolerance and survival in apple cider. *J. Food Prot.* **57:**460–464.

109. **Miller, S. I., A. M. Kukral, and J. J. Mekalanos.** 1989. A two-component regulatory system (*phoPphoQ*) controls *Salmonella typhimurium* virulence. *Proc. Natl. Acad. Sci. USA* **86:**5054–5058.

110. **Miyagi, A., H. Ohta, T. Kodama, K. Fukui, K. Kato, and T. Shimono.** 1994. Metabolic and energetic aspects of the growth response of *Streptococcus rattus* to environmental acidification in anaerobic continuous culture. *Microbiology* **140**(Pt. 8):1945–1952.

111. **Nojoumi, S. A., D. G. Smith, and R. J. Rowbury.** 1995. Tolerance to acid in pH 5.0-grown organisms of potentially pathogenic gram-negative bacteria. *Lett. Appl. Microbiol.* **21:**359–363.

112. **O'Brien, L. M., S. V. Gordon, I. S. Roberts, and P. W. Andrew.** 1996. Response of *Mycobacterium smegmatis* to acid stress. *FEMS Microbiol. Lett.* **139:**11–17.

113. **O'Driscoll, B., C. G. Gahan, and C. Hill.** 1997. Two-dimensional polyacrylamide gel electrophoresis analysis of

the acid tolerance response in *Listeria monocytogenes* LO28. *Appl. Environ. Microbiol.* **63:**2679–2685.

114. **O'Driscoll, B., C. G. M. Gahan, and C. Hill.** 1996. Adaptive acid tolerance response in *Listeria monocytogenes*: isolation of an acid-tolerant mutant which demonstrates increased virulence. *Appl. Environ. Microbiol.* **62:**1693–1698.

115. **O'Hara, G. W., and A. R. Glenn.** 1994. The adaptive acid tolerance response in root nodule bacteria and *Escherichia coli. Arch. Microbiol.* **161:**286–292.

116. **O'Hara, G. W., T. J. Goss, M. J. Dilworth, and A. R. Glenn.** 1989. Maintenance of intracellular pH and acid tolerance in *Rhizobium meliloti. Appl. Environ. Microbiol.* **55:**1870–1876.

117. **O'Sullivan, E., and S. Condon.** 1997. Intracellular pH is a major factor in the induction of tolerance to acid and other stresses in *Lactococcus lactis. Appl. Environ. Microbiol.* **63:**4210–4215.

118. **O'Sullivan, E., and S. Condon.** 1999. Relationship between acid tolerance, cytoplasmic pH, and ATP and H^+-ATPase levels in chemostat cultures of *Lactococcus lactis. Appl. Environ. Microbiol.* **65:**2287–2293.

119. **Padan, E.** 1999. The molecular mechanism of regulation of the NhaA Na^+/H^+ antiporter of *Escherichia coli*, a key transporter in the adaptation to Na^+ and H^+, p. 183–199. *In* D. Chadwick and G. Cardew (ed.), *Bacterial Responses to pH.* John Wiley & Sons, Ltd., Chichester, England.

120. **Padhye, N. V., and M. P. Doyle.** 1992. *Escherichia coli* 0157:H7: Epidemiology, pathogenesis, and methods of detection in food. *J. Food Prot.* **55:**555–565.

121. **Perez-Perez, G. I., A. Z. Olivares, T. L. Cover, and M. J. Blaser.** 1992. Characteristics of *Helicobacter pylori* variants selected for urease deficiency. *Infect. Immun.* **60:**3658–3663.

122. **Phan-Thanh, L., and T. Gormon.** 1997. Stress proteins in *Listeria monocytogenes. Electrophoresis* **18:**1464–1471.

123. **Poolman, B., A. J. Driessen, and W. N. Konings.** 1987. Regulation of arginine-ornithine exchange and the arginine deiminase pathway in *Streptococcus lactis* [published erratum appears in *J. Bacteriol.* 170:1415, 1988]. *J. Bacteriol.* **169:**5597–5604.

124. **Pratt, L. A., W. Hsing, K. E. Gibson, and T. J. Silhavy.** 1996. From acids to *osmZ*: multiple factors influence synthesis of the OmpF and OmpC porins in *Escherichia coli. Mol. Microbiol.* **20:**911–917.

125. **Quivey, R. G., Jr., R. C. Faustoferri, K. A. Clancy, and R. E. Marquis.** 1995. Acid adaptation in *Streptococcus mutans* UA159 alleviates sensitization to environmental stress due to RecA deficiency. *FEMS Microbiol. Lett.* **126:**257–262.

126. **Rahav-Manor, O., O. Carmel, R. Karpel, D. Taglicht, G. Glaser, S. Schuldiner, and E. Padan.** 1992. NhaR, a protein homologous to a family of bacterial regulatory proteins (LysR), regulates *nhaA*, the sodium proton antiporter gene in *Escherichia coli. J. Biol. Chem.* **267:**10433–10438.

127. **Raja, N., W. C. Goodson, C. Chui, D. G. Smith, and R. J. Rowbury.** 1991. Habitation to acid in *Escherichia coli*: conditions for habitation and its effects on plasmid transfer. *J. Appl. Bacteriol.* **70:**59–65.

128. **Rallu, F., A. Gruss, and E. Maguin.** 1996. *Lactococcus lactis* and stress. *Antonie Leeuwenhoek* **70:**243–251.

129. **Rathman, M., M. D. Sjaastad, and S. Falkow.** 1996. Acidification of phagosomes containing *Salmonella typhimurium* in murine macrophages. *Infect. Immun.* **64:**2765–2773.

130. **Remis, R. S., K. L. McDonald, L. W. Riley, N. D. Puhr, J. G. Wells, B. R. Davis, P. A. Blake, and M. L. Cohen.** 1984. Sporadic cases of hemorrhagic colitis associated with *Escherichia coli* 0157:H7. *Ann. Intern. Med.* **101:**728–742.

131. **Renberg, I., T. Korsman, and H. J. B. Birks.** 1993. Prehistoric increases in the pH of acid-sensitive Swedish lakes caused by land-use changes. *Nature* **362:**824–827.

132. **Rimon, A., Y. Gerchman, Y. Olami, S. Schuldiner, and E. Padan.** 1995. Replacements of histidine 226 of NhaA-Na$^+$/H$^+$ antiporter of *Escherichia coli*. Cysteine (H226C) or serine (H226S) retain both normal activity and pH sensitivity, aspartate (H226D) shifts the pH profile toward basic pH, and alanine (H226A) inactivates the carrier at all pH values. *J. Biol. Chem.* **270:**26813–26817.

133. **Rowbury, R. J.** 1999. Acid tolerance induced by metabolies and secreted proteins, and how tolerance can be counteracted, p. 93–106. *In* J. C. Derek and G. Gardew (ed.), *Bacterial Responses to pH*. John Wiley & Sons, Ltd., Chichester, England.

134. **Rowbury, R. J., and M. Goodson.** 1998. Glucose-induced acid tolerance appearing at neutral pH in log-phase *Escherichia coli* and its reversal by cyclic AMP. *J. Appl. Microbiol.* **85:**615–620.

135. **Rowbury, R. J., and M. Goodson.** 1997. Metabolites and other agents which abolish the CysB-regulated acid tolerance induced by low pH in log-phase *Escherichia coli*. *Recent Res. Dev. Microbiol.* **1:**1–12.

136. **Rowbury, R. J., and M. Goodson.** 1993. PhoE porin of *Escherichia coli* and phosphate reversal of acid damage and killing and of acid induction of the CadA gene product. *J. Appl. Bacteriol.* **74:**652–661.

137. **Rowbury, R. J., T. J. Humphrey, and M. Goodson.** 1999. Properties of an L-glutamate-induced acid tolerance response which involves the functioning of extracellular induction components. *J. Appl. Microbiol.* **86:**325–330.

138. **Rowbury, R. J., N. H. Hussain, and M. Goodson.** 1998. Extracellular proteins and other components as obligate intermediates in the induction of a range of acid tolerance and sensitisation responses in *Escherichia coli*. *FEMS Microbiol. Lett.* **166:**283–288.

139. **Sanders, J. W., K. Leenhouts, J. Burghoorn, J. R. Brands, G. Venema, and J. Kok.** 1998. A chloride-inducible acid resistance mechanism in *Lactococcus lactis* and its regulation. *Mol. Microbiol.* **27:**299–310.

140. **Schäfer, G.** 1999. How can archaea cope with extreme acidity?, p. 131–151. *In* D. Chadwick and G. Cardew (ed.), *Bacterial Response to pH*. John Wiley & Sons, Chichester, England.

141. **Schweder, T., K.-H. Lee, O. Lomovskaya, and A. Matin.** 1996. Regulation of *Escherichia coli* starvation sigma factor (σ^s) by ClpXP protease. *J. Bacteriol.* **178:**470–476.

142. **Scott, D. R., D. Weeks, C. Hong, S. Postius, K. Melchers, and G. Sachs.** 1998. The role of internal urease in acid resistance of *Helicobacter pylori*. *Gastroenterology* **114:**58–70.

143. **Shi, X., and G. N. Bennett.** 1994. Effects of *rpoA* and *cysB* mutations on acid induction of biodegradative agrinine decarboxylase in *Escherichia coli*. *J. Bacteriol.* **176:**7017–7023.

144. **Shi, X., B. C. Waasdorp, and G. N. Bennett.** 1993. Modulation of acid-induced amino acid decarboxylase gene expression by *hns* in *Escherichia coli*. *J. Bacteriol.* **175:**1182–1186.

145. **Small, P., D. Blankenhorn, D. Welty, E. Zinser, and J. L. Slonczewski.** 1994. Acid and base resistance in *Escherichia coli* and *Shigella flexneri*: Role of *rpoS* and growth pH. *J. Bacteriol.* **176:**1729–1737.

146. **Small, P. L., and S. R. Waterman.** 1998. Acid stress, anaerobiosis and *gadCB*: lessons from *Lactococcus lactis* and *Escherichia coli*. *Trends Microbiol.* **6:**214–216.

147. **Small, P. L. C.** 1998. *Shigella* and *Escherichia coli* strategies for survival at low pH. *Jpn. J. Med. Sci. Biol.* **51:**S81–S89.

148. **Smith, A. J., R. G. Quivey, Jr., and R. C. Faustoferri.** 1996. Cloning and nucleotide sequence analysis of the *Streptococcus mutans* membrane-bound, proton-translocating ATPase operon. *Gene* **183:**87–96.

149. **Smith, D. K., T. Kassam, B. Singh, and J. F. Elliott.** 1992. *Escherichia coli* has two homologous glutamate decarboxylase genes that map to distinct loci. *J. Bacteriol.* **174:**5820–5826.

150. **Soncini, F. C., E. G. Vescovi, F. Solomon, and E. A. Groisman.** 1996. Molecular basis of the magnesium deprivation response in *Salmonella typhimurium*: Identification of PhoP-regulated genes. *J. Bacteriol.* **178:**5092–5099.

151. **Stim, K. P., and G. N. Bennett.** 1993. Nucleotide sequence of the *adi* gene, which encodes the biodegradative acid-induced arginine decarboxylase of *Escherichia coli*. *J. Bacteriol.* **175:**1221–1234.

152. **Stim-Herndon, K. P., T. M. Flores, and G. N. Bennett.** 1996. Molecular characterization of *adiY*, a regulatory gene which affects expression of the biodegradative acid-induced arginine decarboxylase (*adiA*) of *Escherichia coli*. *Microbiology* **142:**1311–1320.

153. **Surette, M. G., M. B. Miller, and B. L. Bassler.** 1999. Quorum sensing in *Escherichia coli*, *Salmonella typhimurium*, and *Vibrio harveyi*: a new family of genes responsible for autoinducer production. *Proc. Natl. Acad. Sci. USA* **96:**1639–1644.

154. **Suziedlien, E., K. Suziedlis, V. Garbencit, and S. Normark.** 1999. The acid-inducible *asr* gene in *Escherichia coli*: transcriptional control by the *phoBR* operon. *J. Bacteriol.* **181:**2084–2093.

155. **Tabor, C. W., and H. Tabor.** 1985. Polyamines in microorganisms. *Microbiol. Rev.* **49:**81–99.

156. **Taglicht, D., E. Padan, and S. Schuldiner.** 1991. Overproduction and purification of a functional Na$^+$/H$^+$ antiporter coded by *nhaA* (*ant*) from *Escherichia coli*. *J. Biol. Chem.* **266:**11289–11294.

157. **Thomas, A. D., and I. R. Booth.** 1992. The regulation of expression of the porin gene *ompC* by acid pH. *Microbiology* **138:**1829–1835.

158. **Thompson, S. A., and M. J. Blaser.** 1995. Isolation of the *Helicobacter pylori recA* gene and involvement of the *recA* region in resistance to low pH. *Infect. Immun.* **63:**2185–2193.

159. **Tiwari, R. P., W. G. Reeve, M. J. Dilworth, and A. R. Glenn.** 1996. Acid tolerance in *Rhizobium meliloti* strain WSM419 involves a two-component sensor-regulator system. *Microbiology* **142**(Pt 7):1693–1704.

160. **Tiwari, R. P., W. G. Reeve, M. J. Dilworth, and A. R. Glenn.** 1996. An essential role for *actA* in acid tolerance of *Rhizobium meliloti*. *Microbiology* **142**(Pt 3):601–610.

161. **Tiwari, R. P., W. G. Reeve, and A. R. Glenn.** 1992. Mutations conferring acid sensitivity in the acid-tolerant strain *Rhizobium meliloti* WSM419 and *Rhizobium leguminosarum* biovar *viciae* WSM710. *FEMS Microbiol. Lett.* **100:**107–112.

162. **Valdivia, R. H., and S. Falkow.** 1996. Bacterial genetics by flow cytometry: rapid isolation of *Salmonella typhimurium* acid-inducible promoters by differential fluorescence induction. *Mol. Microbiol.* **22:**367–378.

163. **Walsh, J. P., L. Fahrner, and R. M. Bell.** 1990. sn-1,2-diacylglycerol kinase of *Escherichia coli*. Diacylglycerol analogues define specificity and mechanism. *J. Biol. Chem.* **265:**4374–4381.

164. **Waterman, S. R., and P. L. C. Small.** 1996. Identification of σ^S-dependent genes associated with the stationary-phase

acid-resistance phenotype of *Shigella flexneri. Mol. Microbiol.* 21:925–940.

165. Wiedmann, M., T. J. Arvik, R. J. Hurley, and K. J. Boor. 1998. General stress transcription factor sigmaB and its role in acid tolerance and virulence of *Listeria monocytogenes. J. Bacteriol.* 180:3650–3656.

166. Wilmes-Riesenberg, M. R., B. Bearson, J. W. Foster, and R. Curtiss III. 1996. Role of the acid tolerance response in virulence of *Salmonella typhimurium. Infect. Immun.* 64:1085–1092.

167. Yamashita, Y., T. Takehara, and H. K. Kuramitsu. 1993. Molecular characterization of a *Streptococcus mutans* mutant altered in environmental stress responses. *J. Bacteriol.* 175:6220–6228.

168. Yoshida, T., T. Yamashino, C. Ueguchi, and T. Mizuno. 1993. Expression of the *Escherichia coli* dimorphic glutamic acid decarboxylases is regulated by the nucleoid protein H-NS. *Biosci. Biotechnol. Biochem.* 57:1568–1569.

169. Young, G. M., D. Amid, and V. L. Miller. 1996. A bifunctional urease enhances survival of pathogenic *Yersinia enterocolitica* and *Morganella morganii* at low pH. *J. Bacteriol.* 178:6487–6495.

170. Zhou, Y., and S. Gottesman. 1998. Regulation of proteolysis of stationary-phase sigma factor RpoS. *J. Bacteriol.* 180:1154–1158.

171. Zilberstein, D., V. Agmon, S. Schuldiner, and E. Padan. 1984. *Escherichia coli* intracellular pH, membrane potential and cell growth. *J. Bacteriol.* 158:246–252.

172. Zilberstein, D., V. Agmon, S. Schuldiner, and E. Padan. 1982. The sodium/proton antiporter is part of the pH homeostasis mechanism in *Escherichia coli. J. Biol. Chem.* 257:3687–3691.

173. Zuniga, M., M. Champomier-Verges, M. Zagorec, and G. Perez-Martinez. 1998. Structural and functional analysis of the gene cluster encoding the enzymes of the arginine deiminase pathway of *Lactobacillus sake. J. Bacteriol.* 180:4154–4159.

Bacterial Stress Responses
Edited by G. Storz and R. Hengge-Aronis
©2000 ASM Press, Washington, D.C.

Chapter 8

Sodium Stress

ETANA PADAN AND TERRY ANN KRULWICH

All cells maintain an intracellular Na^+ concentration lower than the extracellular concentration. Na^+ stress is caused by an increase in the cytoplasmic concentration, which varies over a wide range; the harmful intracellular concentration in Escherichia coli *is above 20 mM and that of certain halophilic bacteria is suspected to be above 3 M. As yet, specific targets for Na^+ toxicity have been identified only in yeast. The difficulty in the identification may be the existing interplay between Na^+ and other physical-chemical stresses. Bacterial cells protect themselves from the adverse effect of Na^+ by primary and secondary Na^+ extrusion systems. This review focuses on the ion-specific bacterial Na^+/H^+ antiporters, which, similar to their eukaryotic counterparts, are very widely spread and play a primary role in homeostasis of Na^+ and H^+ in all cells. A new group of antiporters that, in addition to ions, have organic substrate (tetracycline) adds a new perspective to Na^+ metabolism and is also reviewed here.*

The best characterized prokaryotic ion-specific Na^+/H^+ antiporter is NhaA. NhaA is the main system responsible for adaptation to Na^+ and alkaline pH (in the presence of Na^+) in E. coli *and many other enteric bacteria. It is under a very intricate control. At the protein level it is regulated directly by pH, an environmental signal. At the gene level, its transcription is dependent on NhaR, a positive regulator of the LysR family, and Na^+, the other environmental signal. Na^+ affects directly the NhaR-nhaA interaction by changing the footprint of NhaR on nhaA in a pH-dependent fashion. It is suggested that in this novel Na^+-specific signal transduction and regulation, NhaR serves as both a specific Na^+ sensor and the transducer of the Na^+ signal. The expression of nhaA is also under*

global regulation of H-NS, which has been suggested to involve a change in the topology of the DNA in an unknown mechanism.

The TetA(L) and TetA(K) antiporters from the *Bacillus subtilis* chromosome and *Staphylococcus aureus* plasmids catalyze the efflux of a tetracycline-divalent metal complex in exchange for H^+. The TetA(L) protein was more recently shown to be a major, low-affinity $Na^+(K^+)/H^+$ antiporter in *B. subtilis* whose functions in Na^+ and alkali resistance can be replaced by TetA(K). An additional function of the monovalent cation/H^+ exchange mode of TetA(L) and TetA(K) may result in net K^+ uptake that may have an important physiological function of the Tet proteins with very significant selective advantage.

Comparison of the amino acid composition of halophilic proteins with those from nonhalophilic microorganisms revealed several properties common to halophilic proteins: (i) a large excess of the acidic amino acids, predominantly glutamate; (ii) a low content of the basic amino acids, lysine and arginine; and (iii) a low proportion of hydrophobic amino acids that is offset by a high frequency of amino acids such as Ser and Thr. Crystallization and atomic resolution of the crystal structure of several halophile enzymes have enabled a breakthrough toward understanding how the several properties mentioned above contribute to halotolerance of proteins. Thus, protein properties should be examined in the context of the three-dimensional structure of the protein. Furthermore, the more complete genome sequences become available, the more conserved structural features become amenable to assess by molecular genetic tools how proteins cope with high Na^+ concentration.

Etana Padan • Division of Microbial and Molecular Ecology, Institute of Life Sciences, The Hebrew University of Jerusalem, Jerusalem 91904, Israel.   **Terry Ann Krulwich** • Department of Biochemistry and Molecular Biology, Mount Sinai School of Medicine, New York, NY 10128.

WHAT IS Na+ STRESS?

Levels of Na+ Affecting Different Bacteria

Conventional bacteria generally grow optimally in media containing less than 1 M Na+ and are distinctly more tolerant to K+ than to Na+. The antipathy to high cytoplasmic levels of Na+, i.e., Na+ stress, persists in halotolerant bacteria that grow well in the low molar concentrations of Na+ and even in halophilic bacteria, many of which grow obligatorily at greater than 2 molar concentrations of Na+ (60). Even in halophilic bacteria whose adaptation to the high osmolarity involves a "salt in" approach, the high cytoplasmic salt is a K+ salt, whereas Na+ is substantially excluded (76). In spite of this evident exclusion of Na+ as a response to Na+ stress, the levels of cytoplasmic Na+ do not appear to be tightly regulated in the bacteria studied to date. Rather, there appears to be a complex interplay between the transmembrane gradients of Na+ and of H+ as well as K+. Thus, systematic definition of cytotoxic levels of Na+ will not only have to distinguish between bound and free Na+, but will have to take into account an intricate matrix of stress responses.

Using ^{23}Na+ nuclear magnetic resonance (NMR), Castle et al. (7) noted that *Escherichia coli* maintained substantially lower intracellular Na+ concentration ($[Na^+]_{in}$) over a range of extracellular Na+ concentration ($[Na^+]_{out}$) up to 285 mM. For example, $[Na^+]_{in}$ was 10 mM at $[Na^+]_{out}$ of 200 mM. The inwardly directed gradient of Na+ remained relatively constant over the range of $[Na^+]_{out}$ while the actual magnitude of $[Na^+]_{in}$ varied. Experiments were also conducted by Harel-Bronstein et al. (33) on wild-type *E. coli* that was grown to exponential phase on media at pH 7 containing 50 mM Na+. ^{22}Na+ was added and equilibrated, followed by 2 additional h of growth; at this point, $[Na^+]_{in}$ in the wild-type was 5 mM. Also with ^{22}Na+, $[Na^+]_{in}$ was assessed to be higher, about 30 mM, in wild-type *B. subtilis* cells growing in the presence of 100 mM added Na+ at pH 8.3 (11). In a halotolerant *Brevibacterium* sp. (71) and in halophilic *Salinivibrio costicola* (26), assayed by NMR protocols, the $[Na^+]_{in}$ was in the range of 5 to 25% of the $[Na^+]_{out}$; the data suggest that in addition to robust exclusion capacities, these organisms are probably better adapted than conventional bacteria to somewhat higher values of $[Na^+]_{in}$.

Basis for Na+ Stress: Specific Targets versus General Effects

Whereas specific enzymatic targets and their biochemical pathways have been implicated in Na+ toxicity in yeast (see references 27, 70, and 82), spe-

cific prokaryotic targets have not yet been identified. The greater toxicity of Na+ at elevated pH suggests that the lower electrochemical proton gradient at alkaline pH makes Na+ exclusion more problematic. Additionally, Na+ may more effectively compete with protons for some set of targets when the [H+] is low, leading to compromised function. The observation that Na+ toxicity is generally reduced under K+-replete conditions may reflect a cation preference with respect to optimal protein and nucleic acid folding. Na+ may support these functions less well or even interfere. There are significant adaptations in the amino acid sequences of proteins that function in halophiles that maintain high cytoplasmic salt concentrations (see below). It is not yet clear whether cytoplasmic proteins in cells with consistently but modestly elevated Na+ levels also have specific sequence features.

HOW CELLS PROTECT THEMSELVES FROM THE ADVERSE EFFECTS OF Na+

Na+ Extrusion Systems

Primary systems

Dimroth (15) has reviewed in detail the different types of primary Na+ extrusion systems except for the ABC-type Na+ systems that were described more recently (10, 109). Primary systems are those that catalyze active transport that is energized directly by energy conserved from light, respiration, ATP hydrolysis, or metabolic transformations carried out by the transport complex itself (as opposed to use of energizing transmembrane gradients set up by distinct proteins in the membrane). Among the primary Na+ efflux systems in bacteria are (i) ATP-driven systems that include V- and F-type ATPases and ABC-type transporters but not yet any P-type ATPases; (ii) metabolic energy-conserving enzyme-transport complexes, e.g., multisubunit, biotin-containing decarboxylases (Na+ pumps that are part of the pathway of CO_2 reduction to CH_4 by certain methanogenic bacteria); and (iii) respiration-coupled systems that use the energy made available from electron transport to energize active Na+ extrusion instead of the more common H+ coupling associated with more conventional respiratory chain complexes. These complexes are structurally distinct from the H+-coupled systems that catalyze the same overall redox reaction. Examples of these three categories of primary Na+ transporters, generally chosen from among those studied in most detail, are provided in Table 1. No Na+ efflux systems have yet been found among the light-driven primary transporters. Many of the pri-

Table 1. Examples of primary Na$^+$ efflux systems from bacteria

Energy source	Type or activity	Organism(s)	Reference(s)
ATP	F-ATPase	*Propionigenium modestum*	63
		Acetobacter woodii	87
		Ilyobacter tartaricus	72
	V-ATPase	*Enterococcus hirae*	51, 68
		Clostridium fervidus	43
	ABC-type transporter	*Bacillus subtilis*	10
		Bacillus firmus OF4	109
Metabolic conversion	Methylmalonyl-CoA decarboxylase	*Propionogenium modestum*	3
		Veillonella alcalescens	39
	Malonate decarboxylase	*Malonomonas rubra*	38
		Klebsiella pneumoniae	41
	Oxaloacetate decarboxylase	*Klebsiella pneumoniae*	15
		Salmonella enterica serovar Typhimurium	111
	N5-Methyltetrahydromethanopterin:CoM methyltransferase	*Methanobacterium thermoautotrophicum*	110
		Methanocarcina mazei Go1	64
Respiration	NADH:ubiquinone oxidoreductase	*Vibrio alginolyticus*	83, 89, 100
		Haemophilus influenzae	35
		Halomonas israelensis (Ba1)	56
	Cytochrome *bo*-type ubiquinol oxidase	*Vitreoscilla*	81

mary Na$^+$ efflux systems listed are probably not present to defend solely or primarily against Na$^+$ stress per se as opposed to other stresses or particular demands for growth. Thus, for example, the F-type ATPases that utilize Na$^+$ as coupling ion may function as ATP synthases, at least under some important conditions, rather than principally as Na$^+$ efflux systems (15).

Secondary systems

Secondary Na$^+$/H$^+$ antiporters evidently play a major role in Na$^+$ resistance as assessed by mutational disruption of established antiporter-encoding genes. In spite of the panoply of antiporters, a single mutation or deletion in an antiporter-encoding gene may produce a sodium-sensitive phenotype (11, 32, 45, 78, 108). The Na$^+$ and Li$^+$ sensitivity of the Δ*nhaA* strain of *E. coli*, NM81, was used as the basis for a screening approach that led to the cloning and characterization of many novel antiporter genes (80). The secondary antiporters are sometimes augmented, either routinely or under specific stress conditions, by primary Na$^+$ pumps of the types discussed above. The secondary antiporters catalyze efflux of Na$^+$ in exchange for external H$^+$. All three of the bacterial Na$^+$/H$^+$ antiporters that have been purified and studied in a reconstituted system, NhaA and NhaB from *E. coli* (85, 97) and TetA(L) from *B. subtilis* (12), have been shown to be electrogenic, with a H$^+$/Na$^+$ ratio greater than unity; the ratio has shown to be 2/1 for NhaA (97) and 1.5/1 for NhaB (85). The electrogenicity makes it possible for the antiport to

be energized by a transmembrane electrical potential and to acidify the cytoplasm relative to the external medium, a property that becomes important in connection with alkali resistance (59). Numerous homologs of NhaA and NhaB have been discovered, as reviewed elsewhere (79, 80). Additional systems that, on the basis of DNA sequence, belong to different structural groups and are still being characterized are NhaC (46, 47), NhaD (73), NapA (99, 108), MrpA (32, 45, 58), and TetA(L) (12, 30).

The Na$^+$/H$^+$ antiporters show individual properties with respect to how well they also use K$^+$ and/or Li$^+$ (79). In one instance, for ChaA from *E. coli*, Ca^{2+} is an important substrate (49, 74). The TetA(L) and TetA(K) antiporters from the *B. subtilis* chromosome and *Staphylococcus aureus* plasmids, respectively, are the first Na$^+$/H$^+$ antiporters that have well-established organic substrates. In fact, these proteins have long been known to be part of a large group of 14 structurally related membrane-spanning segment transporters that catalyze antibiotic efflux in exchange for H$^+$; these particular proteins catalyze the efflux of a tetracycline-divalent metal complex in exchange for the H$^+$ (65). The TetA(L) protein was more recently shown to be a major, low-affinity Na$^+$(K$^+$)/H$^+$ antiporter in *B. subtilis* whose functions in Na$^+$ and alkali resistance can be replaced by TetA(K) (11, 12, 30). An additional function of the monovalent cation/H$^+$ exchange mode of TetA(L) and TetA(K) is one in which the cytoplasmic cation, either Na$^+$ or K$^+$, is exchanged for a greater number of external K$^+$ instead of H$^+$. This mode of activity results in net K$^+$ uptake. It may be an important physiological function of the Tet proteins under some conditions (29). Moreover, the tetracycline-unrelated physiological roles of the Tet proteins could make antibiotic-resistant strains that have come to rely on the amplified Tet protein less likely to reduce the abundance of the protein simply upon elimination of the tetracycline from the environment. A possible rationale for the particular choice of Tet proteins to combine with monovalent cation/H^+ activity relates to the greater tetracycline toxicity under conditions of a high inwardly directed proton gradient that, in any event, a conventional Na^+/H^+ antiport would reduce (9). Thus, the activity of a Na^+/H^+ antiporter that would reduce the proton gradient across the membrane is anticipated to lower tetracycline toxicity even without tetracycline efflux capacity of its own, so the combination takes advantage of an existing common effect.

All the secondary Na^+ extruding antiporters noted above apparently function alone, without additional subunits, although they may function as homo-oligomers. Recently, an antiporter-encoding locus that was first discovered in alkaliphilic *Bacillus* C-125 (32) and implicated strongly in alkaliphile pH homeostasis has been found to be widespread among bacteria and is encoded as part of a 7-gene operon (40, 45, 57). The first gene, e.g., *mrpA* in *B. subtilis*, is apparently an antiporter-encoding gene although it may not be the only one in the operon (45). Expression of *mrpA* in an *mrp*-null mutant of *B. subtilis* does not restore antiport activity in the normal range, 5 mM $[Na^+]_{out}$. However, if the other *mrp* genes are expressed at very low levels, the expression of *mrpA* in *trans* restores significant activity. This makes it unlikely that a stoichiometric complex containing different protein products of the locus is required for MrpA-dependent antiport activity, but the possibility that the antiporter can function as a complex has not been ruled out. A complex was suggested for the *S. aureus* homolog that was studied in an antiporter-deficient *E. coli* mutant (40), but it is also possible that specific chaperones or assembly factors encoded by the operon are required for MrpA assembly or activity. Mutations in this locus render *B. subtilis* extremely sensitive to inhibition by Na^+ and also result in a defect in pH homeostasis under alkaline conditions (45, 57). This widespread locus may be a key element of the Na^+ stress response in many prokaryotes.

Passive mechanisms and import mechanisms

There may also be passive exclusion mechanisms for Na^+, e.g., involving properties of cell surface layers that bind or create a charged barrier or of the membrane lipids such that Na^+ penetration is reduced. There are also indications that some compatible solutes (31), artificial antifreeze proteins (42), and Na^+-binding proteins (48) can enhance Na^+ tolerance in bacteria, but the physiological significance of such phenomena is unclear. It is clear that for completion of the Na^+ cycle Na^+ import mechanisms are needed to get Na^+ into the cells. Na^+-coupled solute transporters serve this role in certain cases (59). The routes that operate in the absence of solutes are still unknown.

RESPONSES TO Na^+: EMERGING EXAMPLES OF SIGNAL TRANSDUCTION AND REGULATION

Na$^+$-Dependent Gene Expression in *E. coli*

nhaA of *E. coli* is induced specifically by an increase in either intracellular Na^+ or Li^+

Experiments with *nhaA'-'lacZ* protein fusions (54, 86) and Northern analysis (16, 77) showed that the environmental signals that turn on *nhaA* are Na^+ or Li^+ increases in the medium. Alkaline pH potentiates the effect of the ions, but neither an increase in osmolarity nor ionic strength induces *nhaA*. These results demonstrated for the first time that *E. coli* has a unique regulatory network responding specifically to Na^+ (and Li^+). Interestingly, a similar role has recently been assigned to Na^+ in the regulation of expression of the Na^+/ATPase of *Enterococcus hirae* (69) (see below).

To determine whether a change in $[Na^+]_{in}$ rather than $[Na^+]_{out}$ is the immediate signal, $[Na^+]_{in}$ was changed in the *nhaA'-'lacZ* strain while maintaining $[Na^+]_{out}$ constant. An increase in $[Na^+]_{in}$, achieved by deleting both antiporter genes, *nhaA* and *nhaB*, led to a high expression of *nhaA'-'lacZ* (16). A decrease in $[Na^+]_{in}$, obtained by introducing *nhaA* on a multicopy plasmid, led to a drastic reduction in *nhaA* expression (86). These results imply that $[Na^+]_{in}$ is the direct signal for induction of *nhaA*. The relative contributions of various possible Na^+ entry routes into the cells have not yet been determined.

The NhaR regulator

The regulator of *nhaA* is encoded by *nhaR*, which maps downstream of *nhaA* (53). A *nhaR* deletion strain is hypersensitive to Na^+ and Li^+ in spite of the presence of *nhaA* (86); NhaR is a positive regulator that works in *trans*, since *nhaR* supplied on a multicopy plasmid dramatically increases expression of the *nhaA'-'lacZ* protein fusion (86). The protein is homologous to the LysR family of positive regulators that are involved in the response of bacteria to various environmental stresses (13, 37, 86, 96).

To purify NhaR a His-tagged derivative of NhaR (His-tagged NhaR) was constructed. The His-tagged protein, which was found to be as active as the wild-type protein in vivo, was purified to homogeneity on a Ni-NTA affinity column. Separation by gel filtration on a HPLC column showed one homogenous peak at 72.5 kDa, suggesting that the 36.2-kDa His-tagged NhaR protein exists as a dimer in solution (6).

The cognate binding site and the Na+ specific footprint of NhaR on *nhaA*

A genetic approach undertaken previously (5) showed that NhaR may have an Na+ "sensor" activity; E134A, a point mutation in *nhaR* increased the affinity of the regulator to Na+ and conferred resistance to Li+. These results suggested that the interaction among Na+, NhaR, and the target regulatory sequences of *nhaA* is direct. Hence, a biochemical approach was undertaken to study this interaction with purified components.

DNA mobility shift and DNase I protection assays showed that both wild-type NhaR (86) and its purified His-tagged derivative (6) bind specifically to DNA sequences overlapping the promoter region of *nhaA* (6). The smallest fragment, overlapping all binding sequences of *nhaA*, was located between bp -120 and $+14$ (of the initiation codon) (Fig. 1). It contained three copies of the LysR consensus motif (T-N_{11}-A) (28, 93, 101). Na+ had no effect on the gel retardation patterns, suggesting that Na+ causes a change in the footprint rather than a change in the affinity. This behavior is characteristic to the LysR family of regulators (93, 101).

Na+ up to 100 mM had no effect on the DNase I footprint. This negative result, however, was found to be related to the limitations of the DNase I protection assay employed, i.e., it is not sensitive and is limited to the minor groove of the DNA (92). Thus, when dimethyl sulfate (DMS) methylation protection footprint analysis was used, both in vitro and in vivo experiments revealed a specific effect of Na+ (6). Whereas the NhaR-dependent protection of bases -24, -29, and -92 observed in vitro was not affected by either 100 mM K+ or 100 mM Na+, protection of base -60 was differentially affected by the two ions. Na+ but not K+ specifically exposed this base to methylation. The concentration of Na+ affecting the footprint was found to be within the range expected for the intracellular concentration (10 to 20 mM). Remarkably in line with the finding that the expression is potentiated in vivo by alkaline pH (54),

the Na+ interaction with base -60, but not with either bases -24 or -29, was dramatically affected by pH. At alkaline pH (7.5 to 8.5), but not at acidic pH (6.5), Na+ affected the NhaR/*nhaA* interaction at base -60 (6).

DMS methylation protection assays were also conducted in vivo. As expected from the in vitro results, bases -24 and -29 were protected but not affected by the type of ion; base -60 was protected in the presence of K+ but not in the presence of Na+ (Fig. 1B). Most interestingly, base -92, which in vitro was protected by the regulator but unaffected by the presence of Na+, in vivo was exposed to methylation by Na+ but not by K+. The Na+-specific effects on the footprint of NhaR on *nhaA* required NhaR since they were not observed in a $\Delta nhaR$ strain (6). Taken together this work suggests that NhaR is both the sensor and the transducer of the Na+ signal and undergoes a conformational change upon binding Na+. This change is expressed directly in a decrease in the binding of NhaR to G^{-60} in a pH-dependent fashion. This is also manifested in the change in NhaR binding to G^{-92} observed only in vivo. The G^{-92}/NhaR interaction suggests an involvement of another factor in vivo (6).

The Na+ specific promoter of *nhaA*

Two promoters, P_1 and P_2 located 31 bp and 172 bp upstream of the initiation codon, respectively, have been identified previously (54). Surprisingly, the binding region of NhaR on *nhaA* contains P_1 (Fig. 1). Therefore, it has become crucial to establish which is the Na+-dependent promoter of *nhaA*.

Primer extension assays (N. Dover and E. Padan, unpublished results) conducted on RNA isolated from cells grown in the absence of Na+ show a weak transcript band of a size expected from the P_1 promoter, which is accompanied by a transcript 8 bp shorter (Fig. 1A). The latter is most probably due to premature falloff of the reversed transcriptase. However, the possible involvement of another promoter cannot yet be ruled out. The level of both transcripts was drastically increased (more than 10-fold) when the assay was conducted on RNA isolated from Na+-induced cells. On the other hand, the transcript corresponding to P_2 was hardly detected, and it was not affected by the presence of Na+. These results suggest that P_1 is the Na+-specific promoter of *nhaA*.

Global regulation of *nhaA*, the involvement of H-NS

Recently, several global regulators have been shown to have an important impact on the expression

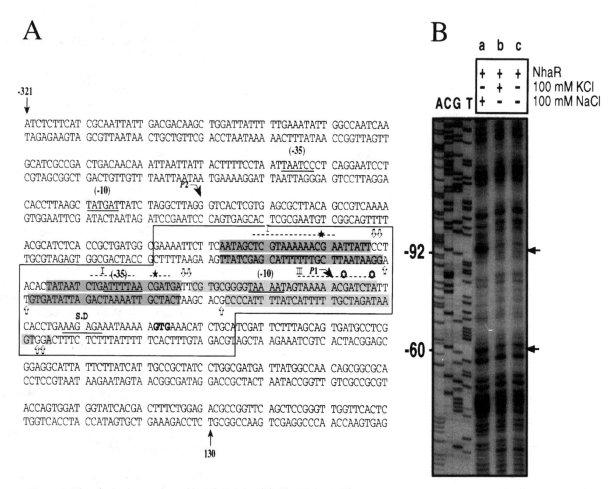

Figure 1. The *nhaA* sequence bound by NhaR is modified by Na⁺. (A) The upstream DNA sequences (DDBJ/EMBL/GenBank accession numbers are X17311, S67239, and J03897 for *nhaA* and L24072 for *nhaR*) containing the *cis*-regulatory sequences of *nhaA* are shown. The shortest fragment (bp −120 to 14) binding His-tagged NhaR in the gel retardation assay is delimited. The shaded sequences show the His-tagged NhaR domain protected from DNase I digestion. G⁻⁹² and G⁻⁶⁰ specifically affected by Na⁺ in the DMS methylation assay, in vivo or both in vivo and in vitro, respectively, are marked by dark stars. G⁻²⁴ and A⁻²⁹ protected by NhaR but not affected by Na⁺ in the DMS methylation protection assay are indicated by open stars. Open vertical arrows show the DNase I-hypersensitive sites. Three sequential consensus motifs of the lysR family (93), designated I, II, and III, are shown by interrupted lines above the *nhaA* sequence. Numbers in brackets relate to the indicated promoters P₁ and P₂ of *nhaA*. Other numbers relate to the first base (base 1) of the initiation codon GTG, in bold, while its upstream neighboring base is −1. (B) DMS methylation protection by NhaR in vivo. In all panels numbers at the left indicate position of bases in the promoter region relative to the first base of the initiation codon. The cells used were HB101 transformed with pGM42T, a plasmid harboring all upstream sequences of an inactive *nhaA* and wild-type *nhaR* (53). The cells were grown in the presence of the inducer (100 mM Na⁺) as indicated in the figure, exposed to DMS, plasmid DNA isolated and treated with piperidine, and the resulting fragments were analyzed by primer extension (7). Arrows, identified bases contacting NhaR and affected by Na⁺.

of many unlinked genes. Although the global regulator *rpoS* (see chapter 11) was not found to be involved in *nhaA* regulation, we have established a connection between the Na⁺-specific, NhaR-dependent regulation of *nhaA* and H-NS, a DNA-binding protein and a global regulator (16). Thus, the expression of *nhaA'-'lacZ* was derepressed in strains bearing *hns* mutation, and transformation with a low-copy-number plasmid carrying *hns*⁺ repressed expression and restored Na⁺ induction. The derepression in *hns*

strains was *nhaR* independent. Most interestingly, multicopy *nhaR*, which in an *hns*⁺ background acted only as a Na⁺-dependent positive regulator, acted as a repressor in an *hns* strain in the absence of Na⁺ but was activated in the presence of the ion. Hence, an interplay among *nhaR*, Na⁺, and *hns* in the regulation of *nhaA* was suggested (Fig. 2). Although the mechanism of regulation mediated by H-NS is not known, it has been suggested to involve a change in the topology of the DNA (103).

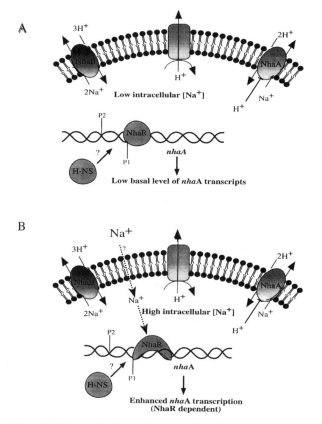

Figure 2. Na⁺ specifically regulates the NhaR-dependent transcription of *nha*A. A schematic model is shown. The NhaA (97) and NhaB (85) antiporters are shown with their respective Na⁺/H⁺ stoichiometries. P_1 and P_2 are the promoters of *nha*A (54). (A) A low level of basal transcription occurs when [Na]$_{in}$ is low. (B) Na⁺ stress leads to high transcription of *nha*A.

Na⁺-Dependent Gene Expression in Bacteria Other than *E. coli*

Na⁺ induces the Na⁺-ATPase in *E. hirae*

The gram-positive bacterium *E. hirae* has a V-type Na⁺-translocating ATPase that is encoded by the *ntp* operon (*ntpFIKECGABDHJ*) (98). Primer extension experiments identified the transcription start of this operon upstream of the *ntpF* gene. The Na⁺-pumping activity in whole cells, the Na⁺-stimulated ATPase activity in the membranes, and the amounts of the two major subunits (A and B) of the ATPase increased remarkably in cells grown on medium containing high concentrations of NaCl but not on medium containing KCl or sorbitol (55, 69). Chloramphenicol completely abolished the increases of the enzyme activity and the amounts of A and B subunits, suggesting that high concentrations of the external sodium ions increased the de novo synthesis of the enzyme. Finally, Western blot and Northern blot experiments revealed that the increase in the Na⁺-ATPase level with the external Na⁺ was further ac-

celerated by addition of an ionophore such as monensin, which rendered the cell membrane permeable to Na⁺. These results suggest that the transcription of the Na⁺-ATPase operon is regulated by the intracellular concentration of sodium ions (69).

Regulation of Na⁺ Extrusion Systems in *B. subtilis*

The response of *B. subtilis* to Na⁺ stress is part of a general stress response to heat shock, salt, ethanol, oxygen or nutrient starvation, and oxidative agents (36). The expression of most of the general stress response genes, encoding at least 42 distinct proteins, is dependent on σ^B, which is active in late-exponential and stationary phase (see chapter 12) (2, 107). It has been suggested that another "extracytoplasmic function" (ECF) sigma factor, σ^M, is also required for growth and survival at high [Na⁺]$_{out}$ and may be particularly important at early stages of growth (44). The connection between the general stress response and regulation of specific Na⁺ extrusion systems has yet to be made, and studies of how genes encoding Na⁺ extrusion systems are regulated have just recently begun. For unknown reasons mutants in the *mrp* operon that exhibit an extremely Na⁺-sensitive phenotype have large increases in the *mrp* mRNA abundance (45). By contrast, the strong upregulation of the multifunctional *tetA*(L) gene currently appears to occur primarily by a translational reinitiation mechanism that is mediated by tetracycline (95), although there may be a minor transcriptional component (1). Perhaps the levels of TetA(L) expression required to protect against the antibiotics are inhibitory to the cell, whereas the monovalent cation-related "housekeeping functions" of TetA(L) can be accomplished at levels much closer to the basal levels of *tetA*(L) expression.

Complex Relationship between Na⁺ and Other Stresses

Interplay with other physical-chemical stresses

The interplay of Na⁺ stress with elevated temperature is double-edged. On the one hand, the enhanced permeability of the coupling membrane to Na⁺ is anticipated to increase the cytotoxicity of the cation. On the other hand, since the proton permeability increases even more (104), Na⁺ is nonetheless the preferred or sole coupling ion above critical temperatures (the critical temperature for a given strain relates to the specific properties of the membrane lipids) (94, 102, 104). The relationship between Na⁺ stress and pH stress is also complex. Salt-induced enhancement of acid sensitivity has been noted in *E.*

coli (91), and a greater toxicity of Na$^+$ has been generally observed at alkaline versus neutral pH (11, 78). However, a key physiological adaptation to alkaline pH, i.e., acidification of the cytoplasm relative to the external medium using monovalent cation/proton antiporters (exchangers), works optimally with Na$^+$ in some bacteria and obligatorily with Na$^+$ in the "expert" alkaliphilic bacteria (58). For at least some Na$^+$/H$^+$ antiporters, activity is strongly related to [H$^+$]$_{in}$, with strong activation at high cytoplasmic pH (24, 30, 90, 93, 97).

Na$^+$ stress is also enhanced under conditions in which the membrane integrity or the specific proton-barrier function is compromised chemically. Under these conditions, e.g., treatment with ethanol or protonophoric uncouplers, the normal electrochemical gradient is reduced, Na$^+$ exclusion is challenged, and K$^+$ can be adversely lost from the cytoplasm. The induction of specific ATP-dependent Na$^+$ efflux systems can respond to the Na$^+$ accumulation, concomitantly increasing the failing membrane potential, thus energizing K$^+$ uptake. The interplay between K$^+$ sufficiency and Na$^+$ stress extends even beyond the conditions that imperil the cells by promoting both Na$^+$ accumulation and K$^+$ loss. Even in cells without any additional stresses, Na$^+$ toxicity is generally greater at suboptimal levels of K$^+$ (e.g., 11, 33) and, at least in *E. coli*, it has been suggested that both high K$^+$ and high osmolarity per se may lead to induction of specific Na$^+$ efflux pathways (106).

Relationship to virulence

An interplay between membrane transporters that mediate Na$^+$ flux or Na$^+$ levels themselves and virulence has been suggested by an increasing number of studies. It is worth noting a group of examples that may reflect a role of Na$^+$ in either virulence, its suppression, or host defenses. However, it should also be noted that it is not yet clear whether any such roles are direct. Some examples include the influence of Na$^+$ flux status on the expression of virulence genes in *Vibrio cholerae* (34). Aspects of the σ^B-mediated stress response in *Staphylococcus aureus* are also inhibited by elevated NaCl (8). An association of Na$^+$ stress with the intracellular life cycle of *Legionella pneumophila* is suggested by the finding that intracellular survival in macrophages and monocytes is reduced upon disruption of a gene encoding a homolog of oxaloacetate decarboxylase, a primary Na$^+$ efflux system (50). The presence of similar Na$^+$ pumps in additional prokaryotic pathogens is noted below (Table 1).

Adaptation to the stressor in more extreme conditions

Can proteins evolve to withstand high Na$^+$? Is the adaptation specific to Na$^+$? An answer to these questions should first be sought in organisms growing under extreme conditions of salt, i.e., in the halotolerant or halophilic bacteria (for recent reviews see 14, 19, 20, 75, 105). These bacteria possess three categories of proteins: (i) intracellular enzymes, which are not exposed to the salt concentration of the medium but rather to that of the intracellular environment; (ii) membrane-bound activities, including transport proteins, which are exposed to both the intracellular environment and the outer medium; and (iii) truly extracellular enzymes, exposed to the external hypersaline conditions.

Since in certain halophilic bacteria intracellular Na$^+$ concentration can reach about 3 M (75), all three categories of proteins in these bacteria are anticipated to have the capacity to endure high Na$^+$, with the extracellular enzymes representing the extreme. It is interesting to note that the integral membrane proteins, bacteriorhodopsin and halorhodopsin, do not require high salt for activity or stability (75). Are they unusual, or does the membrane shield its proteins, leaving only few residues exposed? Halotolerant bacteria maintain a low [Na$^+$]$_{in}$; therefore, only their membrane-bound and extracellular proteins function at high Na$^+$ (105). Although halotolerant organisms do not tolerate concentrations of Na$^+$ as high as the halophiles, they have the need and capacity to shift between low and high Na$^+$ within their life span. The mechanism underlying this capacity may thus be a key to the understanding of regulatory pathways specific to Na$^+$. In addition to being able to function at high salt, many of the enzymes from both halophiles and halotolerant bacteria require high salts for their stability. Are these two properties related in molecular terms, or do they represent different aspects of the structure/function relationship to Na$^+$?

Lipids of the cytoplasmic membranes that are exposed to a hypersaline environment or to a hypersaline cytoplasm exhibit unique properties that certainly affect the membrane permeability. This topic has recently been comprehensively reviewed (52, 104, 105) and therefore will not be discussed further here.

Extreme Halophiles

Structure and function of proteins

The devastating effects of high salt concentrations on proteins have been described (see above).

Hence the presence of molar concentrations of inorganic ions in many halophiles requires far-reaching adaptations of the intracellular enzymatic machinery. Comparison of the amino acid composition of halophilic proteins with those from nonhalophilic microorganisms revealed several properties common to halophilic proteins. This study was substantiated by cloning of the respective genes and comparison of their deduced products: (i) a large excess of the acidic amino acids, predominantly glutamate (4, 14, 19, 61, 88); (ii) a low content of the basic amino acids lysine and arginine; and (iii) a low proportion of hydrophobic amino acids that is offset by a high frequency of amino acids such as Ser and Thr (61).

Crystallization and atomic resolution of the crystal structure of several halophile enzymes have enabled a breakthrough toward understanding how the several properties mentioned above contribute to halotolerance of proteins. In turn, site-directed mutagenesis now makes it possible to test the ideas emerging from the structural analyses. The crystal structure of malate dehydrogenase from the archaebacterium *Haloarcula marismortui* (HmMDH) was solved at 3.0 Å (17). It revealed a large number of acidic residues distributed over the enzyme surface and an increased number of salt bridges relative to a nonhalophilic counterpart (17). The carboxylates sequester and coordinate a tight network of water and hydrated K^+ ions at the protein surface and form an unusually large number of internal salt bridges with strategically located basic amino acid residues. These bridges, protected by the protein from the solute ions, provide internal structural rigidity to the protein.

The second atomic structure of a halophilic protein was that of 2Fe-2S ferredoxin (HmFd) at 1.9-Å resolution (23). Comparison of the *H. marismortui* ferredoxin with the plant-type 2Fe-2S ferredoxin showed that the surface of the halophilic protein is coated with acidic residues, except for the vicinity of the iron-sulfur cluster, and that it contains two additional helices near the N-terminus that form a separate hyperacidic domain, postulated to provide extra surface caboxylates for solvation. Bound water molecules on the protein surface have on the average 40% more hydrogen bonds than in a typical nonhalophilic protein crystal structure. These water molecules are thus tightly bound within the hydration shell by protein-water and water-water hydrogen bonds and by hydration of interspersed K^+ ions (23). Surplus salt bridges have not been observed. This may be related to the fact that Fd is "less halophilic" than HmMDH (84).

The amino acid sequence of glutamate dehydrogenase (GluDH) from *Halobacterium salinarum* exhibits the above-described characteristics of halophilic proteins as well as large sequence similarity to mesophilic and hyperthermophilic GluDH, for which a three-dimensional structure was obtained. Although the 3D structure of the halophilic protein has not yet been solved, homology-based modeling has provided an insight into the molecular basis of salt tolerance (4). The model shows that the surface of the halophilic protein is coated with the acidic residues and contains an excess of serines and threonines. The surface also displays a significant reduction in hydrophobic character. It was suggested that this was not primarily due to loss of surface-exposed hydrophobic residues, as had previously been proposed, but to a reduction in surface-exposed lysine residues.

The structure of another halophile enzyme solved at 2.6-Å resolution is that of dihydrofolate reductase of *Haloferax volcanii* [Hy-DHFR] (84). This enzyme requires only moderate salt concentrations and retains its activity at monovalent cations as low as 0.5 M. Although there are no excess total negative charges in the protein, clusters of negative charges were identified at the surface of the protein. This study demonstrated the importance of examining charge distribution in the context of the 3D structure of the protein. An enrichment of arginine residues, compared with lysine residues, was also observed. Interestingly, arginine tends to bind more water molecules than lysine side chains (84).

In line with the high intracellular cation concentrations that characterize halobacteria, most proteins of these bacteria denature when suspended in solutions containing less than 1 to 2 M salt (84). The effect of salt is usually nonspecific to a particular ion, implying that the ionic strength is the effective factor. However, there are cases of Na^+ specificity (67, 75) or K^+ specificity (75). Notably, K^+ hydrates with less water than Na^+ does (14). The requirement for high salt may in part be needed to shield the negative charges on the protein surface. At low salt the repulsion caused by the negative charges results in protein instability. However, Lanyi and Stevenson (61, 62) argued that all the effects of salts cannot be due to charge-shielding action. Maximal electrostatic charge shielding would be reached in about 0.1 M salt, or 0.5 M at most, and in even much lower concentrations of divalent cations. The requirement for high salt concentrations for structural stability can also, to a large extent, be a consequence of the low content of hydrophobic residues; that is, high salt is needed to maintain the weak hydrophobic interactions in such circumstances (20).

Theoretical analysis further suggested that the acidic residues may also be important to prevent aggregation at high salt, which favors hydrophobic interactions, and do not contribute to intrinsic protein stability (21).

A high content of glutamate per se may be favorable as glutamate has the greatest water-binding ability of all amino acids, thus enabling the maximal maintenance of a hydration shell. The hydration, already noted, is thought to increase solubility and prevent the halophile proteins from aggregation (18).

Analysis of the phylogenetic record for halophilic protein-encoding genes suggests that the balance between fluctuating environmental salinity and optimal structure-function is delicate and precarious (14). Halophilic genes exhibit a high proportion of nonsynonymous nucleotide substitutions, and many of the resulting amino acid replacements involve the addition, removal, relocation, or rearrangement of acidic residues. The significance of these properties will become apparent when more completed genome sequences become available.

Protein-nucleic acid interaction

A major problem that so far has not been tackled in the study of extreme halophiles is the understanding of protein-nucleic acid interactions in the presence of 4 to 5 M K$^+$ and 2 M Na$^+$, their main internal ions (75). In this respect, the atomic resolution of halophilic ribosomal particles is of uttermost importance. The successful crystallization of ribosomal particles from *H. marismortui*, providing structural information at 2.9-Å resolution (22), may be expected to yield a wealth of new information on the structure-function relationships of halophilic proteins. This may include information on the protein-nucleic acid interactions that are essential for the performance of biological functions. The existence of meaningful protein-nucleic acid interactions in physiological concentrations of 4 to 5 M KCl constitutes as yet an unsolved enigma, worthy of intensive investigation.

Moderately halophilic and halotolerant aerobic bacteria

A comprehensive review of the taxonomy, ecology, physiology, and biochemistry of these bacteria has recently been published (75). Noted here are findings with respect to amino acid composition of proteins from this group in comparison with extreme halophiles and conventional bacteria.

Amino acid composition of proteins

The aerobic halophilic bacteria tolerate high salt concentrations, but organic osmotic solutes in the cytoplasm provide most of the osmotic balance. Therefore, one should expect that extracellular and

membrane-bound proteins of the halophilic bacteria may display halophilic characteristics with amino acid compositions similar to those found in the halophilic archaea, while there is little a priori reason to assume that the cytoplasmic proteins should show such adaptations. However, in view of reports of high apparent intracellular salt concentrations in certain species (75), some degree of halophilic characters may be expected in the intracellular enzymes as well, at least in particular organisms.

Comparative studies showed that ribosomal proteins from moderate halophiles often had a slightly higher content of acidic amino acids than did the comparable proteins from *E. coli* and other nonhalophiles. The average hydrophobicity of the ribosomal proteins of *S. costicola* and *Halomonas canadensis* did not differ greatly from that for *E. coli* and was much higher than that for *H. salinarum* (66).

Determinations of the abundance of different amino acids in the bulk protein of *Halomonas elongata* showed an excess of acidic amino acids and a low frequency of basic amino acids, with the values being intermediate between those for *E. coli* and the archaeon *Haloferax mediterranei*. Convergent evolution of amino acid usage was suggested to have led to this "halophilic" character of the *H. elongata* proteins (25). However, this convergent evolution did not lead to changes in the frequencies of the hydrophobic amino acids.

CONCLUSIONS AND FUTURE DIRECTIONS

Na$^+$ is one of the most common ions that plays a primary role in cell bioenergetics. However, it very often becomes a potent stressor to all cells. Hence, mechanisms counteracting Na$^+$ toxicity are essential to maintain normal cell physiology. There are many unknowns in both Na$^+$ metabolism and the mechanisms involved in its regulation. In most cells the target of Na$^+$ toxicity is still unknown. The emerging panoply of many secondary Na$^+$ extrusion systems is very intriguing but emphasizes the need for their biochemical and physiological characterization in more detail. Furthermore, to complete the Na$^+$ cycle, import, passive or active, of the ion must take place. This process is in most cases unknown. Most importantly, the multiplicity of systems involved stresses the importance of studying their regulation in order to understand how they operate in the Na$^+$ cycle.

The *nhaA*/NhaR Na$^+$-specific regulatory network is the only well-studied example that is specific to Na$^+$ and not to either osmolarity or ionic strength. This emerging Na$^+$-specific signal transduction, which is dependent on a direct interaction of Na$^+$

with NhaR, is most intriguing and raises the question as to how a small sodium ion selectively regulates gene expression. Identification of the Na$^+$-binding site is certainly one of the targets of this research, which may give general insights into selective ion sensing. The pattern of H-NS involvement in the *nhaA*/NhaR regulatory network is also an open question. In addition to the conventional approaches, it is foreseen that in the future the DNA chip technology will be used to look for additional Na$^+$-specific regulatory networks.

How proteins and nucleic acids evolved to withstand extreme Na$^+$ concentrations is certainly not only of an academic interest but also of a major biotechnological importance. From the very few crystals of halophile proteins, solved at atomic resolution, it is very clear that an interdisciplinary approach combining structural biology, molecular genetics, and analysis of conserved sequences in halophile genomes will be most rewarding.

Acknowledgments. The research in the laboratory of E.P. is supported by grants from the Israel Science Foundation administered by the Israel Academy of Sciences and Humanities, and the BMBF and the BMBF's International Bureau at the DLR (German-Israeli Project Cooperation on Future-Oriented Topics [DIP]). Thanks are also due to the Moshe Shilo Minerva Center for Marine Biogeochemistry and the Massimo and Adelina Della Pergolla Chair in Life Sciences. The research in the laboratory of T.A.K. is supported by research grants GM28454 and GM52837 from the National Institute of General Medical Sciences.

REFERENCES

1. **Bechhofer, D. H., and S. J. Stasinopoulos.** 1998. *tetA*(L) mutants of a tetracycline-sensitive strain of *Bacillus subtilis* with the polynucleotide phosphorylase gene deleted. *J. Bacteriol.* **180:**3470–3473.

2. **Bernhardt, J., U. Volker, A. Volker, H. Antelmann, R. Schmid, H. Mach, and M. Hecker.** 1997. Specific and general stress proteins in *Bacillus subtilis*—two-dimensional protein electrophoresis study. *Microbiology* **143:**999–1017.

3. **Bott, M., K. Pfister, P. Burda, O. Kalbermatter, G. Woehlke, and P. Dimroth.** 1997. Methylmalonyl-CoA decarboxylase from *Propionigenium modestum*—cloning and sequencing of the structural genes and purification of the enzyme complex. *Eur. J. Biochem.* **250:**590–599.

4. **Britton, K. L., T. J. Stillman, K. S. P. Yip, P. Forterre, P. C. Engel, and D. W. Rice.** 1998. Insights into the molecular basis of salt tolerance from the study of glutamate dehydrogenase from *Halobacterium salinarum*. *J. Biol. Chem.* **273:**9023–9030.

5. **Carmel, O., N. Dover, O. Rahav-Manor, P. Dibrov, D. Kirsch, S. Schuldiner, and E. Padan.** 1994. A single amino acid substitution (G134-Ala) in NhaR1 increases the inducibility by Na$^+$ of the product of *nhaA*, a Na$^+$/H$^+$ antiporter gene in *Escherichia coli*. *EMBO J.* **13:**1981–1989.

6. **Carmel, O., O. Rahav-Manor, N. Dover, B. Shaanan, and E. Padan.** 1997. The Na$^+$ specific interaction between the LysR-type regulator, NhaR and the *nha*A gene, encoding the Na$^+$/H$^+$ antiporter of *Escherichia coli*. *EMBO J.* **16:**5922–5929.

7. **Castle, A. M., R. M. Macnab, and R. G. Shulman.** 1986. Intracellular free Na$^+$ in *Escherichia coli* measured by NMR. *J. Biol. Chem.* **261:**3288–3294.

8. **Chan, P. F., S. J. Foster, E. Ingham, and M. O. Clements.** 1998. The *Staphylococcus aureus* alternative factor σ^B controls the environmental stress response but not starvation survival or pathogenicity in a mouse abscess model. *J. Bacteriol.* **180:**6082–6089.

9. **Cheng, J., K. Baldwin, A. A. Guffanti, and T. A. Krulwich.** 1996. Na$^+$/H$^+$ antiport activity conferred by *Bacillus subtilis tetA*(L), a 5′ truncation product of *tetA*(L), and related plasmid genes upon *Escherichia coli*. *Antimicrob. Agents Chemother.* **40:**852–857.

10. **Cheng, J., A. A. Guffanti, and T. A. Krulwich.** 1997. A two-gene ABC-type transport system that extrudes Na$^+$ in *Bacillus subtilis* is induced by ethanol or protonophore. *Mol. Microbiol.* **23:**1107–1120.

11. **Cheng, J., A. A. Guffanti, W. Wang, T. A. Krulwich, and D. H. Bechhofer.** 1996. Chromosomal *tetA*(L) gene of *Bacillus subtilis*: regulation of expression and physiology of a *tetA*(L) deletion strain. *J. Bacteriol.* **178:**2853–2860.

12. **Cheng, J., D. B. Hicks, and T. A. Krulwich.** 1996. The purified *Bacillus subtilis* tetracycline efflux protein TetA(L) reconstitutes both tetracycline-cobalt/H$^+$ and Na$^+$(K$^+$)/H$^+$ exchange. *Proc. Natl. Acad. Sci. USA* **93:**14446–14451.

13. **Christman, M., G. Storz, and B. N. Ames.** 1989. OxyR, a positive regulator of hydrogen peroxide-inducible genes in *Escherichia coli* and *Salmonella typhimurium* is homologous to a family of bacterial regulatory proteins. *Proc. Natl. Acad. Sci. USA* **86:**3484–3488.

14. **Dennis, P. P., and L. C. Shimmin.** 1997. Evolutionary divergence and salinity-mediated selection in halophilic archaea. *Microbiol. Mol. Biol. Rev.* **61:**90–104.

15. **Dimroth, P.** 1997. Primary sodium ion translocating enzymes. *Biochim. Biophys. Acta* **1318:**11–51.

16. **Dover, N., C. Higgins, O. Carmel, A. Rimon, E. Pinner, and E. Padan.** 1996. Na$^+$-induced transcription of *nhaA*, which encodes an Na$^+$/H$^+$ antiporter in *Escherichia coli*, is positively regulated by *nhaR* and affected by *hns*. *J. Bacteriol.* **178:**6508–6517.

17. **Dym, O., M. Mevarech, and J. L. Sussman.** 1995. Structural features that stabilize halophilic malate dehydrogenase from an archaebacterium. *Science* **267:**1344–1346.

18. **Ebel, C., B. Faou, B. Franzetti, B. Kernel, D. Madern, M. Pascu, C. Pfister, S. Richard, and G. Zaccai.** 1999. Molecular interactions in extreme halophiles—the solvation-stabilization hypothesis for halophilic proteins, p. 227–237. *In* A. Oren (ed.), *Microbiology and Biogeochemistry of Hypersaline Environments.* CRC Press, Boca Raton, Fla.

19. **Eisenberg, H.** 1995. Perspectives in biochemistry and biophysics. Life in unusual environments: progress in understanding the structure and function of enzymes from extreme halophilic bacteria. *Arch. Biochem. Biophys.* **318:**1–5.

20. **Eisenberg, H., M. Mevarech, and G. Zaccai.** 1992. Biochemical, structural and molecular genetic aspects of halophilism. *Adv. Prot. Chem.* **43:**1–62.

21. **Elcock, A. H., and J. A. McCammon.** 1998. Electrostatic contributions to the stability of halophilic proteins. *J. Mol. Biol.* **280:**731–748.

22. **Francheschi, F., I. Sagi, N. Böddeker, U. Evers, E. Amdt, C. Paulke, R. Hasenbank, M. Laschever, C. Glotz, J. Piefke, J. Müssig, S. Weinstein, and A. Yonath.** 1994. Crystallographic, biochemical and genetic studies on halophilic ribosomes. *Syst. Appl. Microbiol.* **16:**697–705.

23. **Frolow, F., M. Harel, J. L. Sussman, M. Mevarech, and M. Shoham.** 1996. Insights into protein adaptation to a saturated

salt environment from the crystal structure of a halophilic 2Fe-2S ferredoxin. *Nature Struct. Biol.* 3:452–458.

24. **Gerchman, Y., Y. Olami, A. Rimon, D. Taglicht, S. Schuldiner, and E. Padan.** 1993. Histidine-226 is part of the pH senosr of NhaA, a Na$^+$/H$^+$ antiporter in *Escherichia coli. Proc. Natl. Acad. Sci. USA* 90:1212–1216.

25. **Ghandbhir, M., I. Rasched, P. Marlière, and R. Mutzel.** 1995. Convergent evolution of amino acid usage in archaebacterial and eubacterial lineages adapted to high salt. *Res. Microbiol.* 146:113–120.

26. **Gilboa, H., M. Kogut, S. Chalamish, R. Regev, Y. Avi-Dor, and N. J. Russell.** 1991. Use of ^{23}Na nuclear magnetic resonance spectroscopy to determine the true intracellular concentration of free sodium in a halophilic eubacterium. *J. Bacteriol.* 173:7021–7023.

27. **Glaser, H.-U., D. Thomas, R. Gaxiola, F. Montrichard, Y. Surdin-Kerjan, and R. Serrano.** 1993. Salt tolerance and methionine biosynthesis in *Saccharomyces cerevisiae* involve a putative phosphatase gene. *EMBO J.* 12:3105–3110.

28. **Goethals, K., M. Van Montagu, and M. Holsters.** 1992. Conserved motifs in a divergent *nod* box of *Azorhizobium caulinodans* ORS571 reveal a common structure in promoters regulated by LysR-type proteins. *Proc. Natl. Acad. Sci. USA* 89:1646–1650.

29. **Guffanti, A. A., J. Cheng, and T. A. Krulwich.** 1998. Electrogenic antiport activities of the gram-positive Tet proteins include a Na$^+$(K$^+$)/K$^+$ mode that mediates net K$^+$ uptake. *J. Biol. Chem.* 273:26447–26454.

30. **Guffanti, A. A., and T. A. Krulwich.** 1995. Tetracycline/H$^+$ antiport and Na$^+$/H$^+$ antiport catalyzed by the *Bacillus subtilis* TetA(L) transporter expressed in *Escherichia coli. J. Bacteriol.* 177:4557–4561.

31. **Hagemann, M., and E. Zuther.** 1992. Selection and characterization of mutants of the cyanobacterium *Synechocystis* sp. PCC 6803 unable to tolerate high salt concentrations. *Arch. Microbiol.* 158:429–434.

32. **Hamamoto, T., M. Hashimoto, M. Hino, M. Kitada, Y. Seto, T. Kudo, and K. Horikoshi.** 1994. Characterization of a gene responsible for the Na$^+$/H$^+$ antiporter system of alkalophilic *Bacillus* species strain C-125. *Mol. Microbiol.* 14:939–946.

33. **Harel-Bronstein, M., P. Dibrov, Y. Olami, E. Pinner, S. Schuldiner, and E. Padan.** 1995. MH1, a second-site revertant of an *Escherichia coli* mutat lacking Na$^+$/H$^+$ antiporters (ΔnhaAΔnhaB), regains Na$^+$ resistance and a capacity to excrete Na$^+$ in a Δμ$_{H^+}$-independent fashion. *J. Biol. Chem.* 270:3816–3822.

34. **Hase, C. C., and J. J. Mekalanos.** 1999. Effects of changes in membrane sodium flux on viulence gene expression in *Vibrio cholerae. Proc. Natl. Acad. Sci. USA* 96:3183–3187.

35. **Hayashi, M., Y. Nakayama, and T. Unemoto.** 1996. Existence of Na$^+$-translocating NADH-quinone reductase in *Haemophilus influenzae. FEBS Lett.* 381:174–176.

36. **Hecker, M., W. Schumann, and U. Volker.** 1996. Heat-shock and general stress response in *Bacillus subtilis. Mol. Microbiol.* 19:417–428.

37. **Henikoff, S., G. Haughn, J. Calvo, and J. Wallace.** 1988. A large family of bacterial activator proteins. *Proc. Natl. Acad. Sci. USA* 85:6602–6606.

38. **Hilbi, H., I. Dehning, B. Schink, and P. Dimroth.** 1992. Malonate decarboxylase of *Malonomonas rubra*, a novel type of biotin-containing acetyl enzyme. *Eur. J. Biochem.* 207:117–123.

39. **Hilpert, W., and P. Dimroth.** 1983. Purification and characterization of a new sodium-transport decarboxylase. Methylmalonyl-CoA decarboxylase from *Veillonella alcalescens. Eur. J. Biochem.* 132:579–587.

40. **Hiramatsu, T., K. Kodama, T. Kuroda, T. Mizushima, and T. Tsuchiya.** 1998. A putative multisubunit Na$^+$/H$^+$ antiporter from *Staphylococcus aureus. J. Bacteriol.* 180:6642–6648.

41. **Hoenke, S., M. Schmid, and P. Dimroth.** 1997. Sequence of a gene cluster from *Klebsiella pneumoniae* encoding malonate decarboxylase and expression of the enzyme in *Escherichia coli. Eur. J. Biochem.* 246:530–538.

42. **Holmberg, N., G. Lilius, and L. Bulow.** 1994. Artificial antifeeze proteins can improve NaCl tolerance when expressed in *E. coli. FEBS Lett.* 349:354–358.

43. **Höner zu Bentrup, K., T. Ubbink-Kok, J. S. Lolkema, and W. N. Konings.** 1997. An Na$^+$-pumping V$_1$V$_0$-ATPase complex in the thermophilic bacterium *Clostridium fervidus. J. Bacteriol.* 179:1274–1279.

44. **Horsburgh, M. J., and A. Moir.** 1999. σ^M, an ECF RNA polymerase sigma factor of *Bacillus subtilis* 168, is essential for growth and survival in high concentrations of salt. *Mol. Microbiol.* 32:41–50.

45. **Ito, M., A. A. Guffanti, B. Oudega, and T. A. Krulwich.** 1999. *mrp*: a multigene, multifunctional locus in *Bacillus subtilis* with roles in resistance to cholate and to Na$^+$, and in pH homeostasis. *J. Bacteriol.* 181:2394–2402.

46. **Ito, M., A. A. Guffanti, J. Zemsky, D. M. Ivey, and T. A. Krulwich.** 1997. The role of the *nhaC*-encoded Na$^+$/H$^+$ antiporter of alkaliphilic *Bacillus firmus* OF4. *J. Bacteriol.* 179:3851–3857.

47. **Ivey, D. M., A. A. Guffanti, J. S. Bossewitch, E. Padan, and T. A. Krulwich.** 1991. Molecular cloning and sequencing of a gene from alkaliphilic *Bacillus firmus* OF4 that functionally complements an *Escherichia coli* strain carrying a deletion in the *nhaA* Na$^+$/H$^+$ antiporter gene. *J. Biol. Chem.* 266:23483–23489.

48. **Ivey, D. M., A. A. Guffanti, Z. Shen, N. Kudyan, and T. A. Krulwich.** 1992. The *cadC* gene product of alkaliphilic *Bacillus firmus* OF4 partially restores Na$^+$ resistance to an *Escherichia coli* strain lacking an Na$^+$/H$^+$ antiporter (NhaA). *J. Bacteriol.* 174:4878–4884.

49. **Ivey, D. M., A. A. Guffanti, J. Zemsky, E. Pinner, R. Karpel, S. Schuldiner, E. Padan, and T. A. Krulwich.** 1993. Cloning and characterization of a putative Ca^{2+}/H$^+$ antiporter gene from *Escherichia coli* upon functional complementation of Na$^+$/H$^+$ antiporter-deficient strains by the overexpressed gene. *J. Biol. Chem.* 268:11296–11303.

50. **Jain, B., B. C. Brand, P. C. Luck, M. Di Berardino, P. Dimroth, and J. Hacker.** 1996. An oxaloacetate decarboxylase homologue protein influences the intracellular survival of *Legionella pneumophila. FEMS Microbiol. Lett.* 145:273–279.

51. **Kakinuma, Y.** 1998. Inorganic cation transport and energy transduction in *Enterococcus hirae* and other streptococci. *Microbiol. Mol. Biol. Rev.* 62:1021–1045.

52. **Kamekura, M.** 1993. Lipids of extreme halophiles, p. 135–161. *In* R. H. Vreeland and L. I. Hockstein (eds.), *The Biology of Halophilic Bacteria.* CRC Press, Boca Raton, Fla.

53. **Karpel, R., Y. Olami, D. Taglicht, S. Schuldiner, and E. Padan.** 1988. Sequencing of the gene *ant* which affects the Na$^+$/H$^+$ antiporter activity of *Escherichia coli. J. Biol. Chem.* 263:10408–10414.

54. **Karpel, R., T. Alon, G. Glaser, S. Schuldiner, and E. Padan.** 1991. Expression of a sodium proton antiporter (NhaA) in *Escherichia coli* is inducted by Na$^+$ and Li$^+$ ions. *J. Biol. Chem.* 266:21753–21759.

55. **Kawano, M., K. Igarashi, and Y. Kakinuma.** 1998. The Na$^+$-responsive *ntp* operon is indispensable for homeostasis of K$^+$ and Na$^+$ in *Enterococcus hirae* at limited proton potential. *J. Bacteriol.* 180:4942–4945.

56. **Ken-Dror, S., J. K. Lanyi, B. Schobert, B. Silver, and Y. Avi-Dor.** 1986. An NADH:quinone oxidoreductase of the halotolerant bacterium Ba1 is specifically dependent on sodium ions. *Arch. Biochem. Biophys.* **244:**766–772.

57. **Kosono, S., S. Morotomi, M. Kitada, and T. Kudo.** 1999. Analyses of a *Bacillus subtilis* homologue of the Na^+/H^+ antiporter gene which is important for pH homeostasis of alkaliphilic *Bacillus* sp. C-125. *Biochim. Biophys. Acta* **1409:**171–175.

58. **Krulwich, T. A., A. A. Guffanti, and M. Ito.** 1999. pH tolerance in *Bacillus*: alkaliphile vs non-alkaliphile, p. 167–182. *In Mechanisms by Which Bacterial Cells Respond to pH* (Novartis Found. Symp. 221). Wiley, Chichester, England.

59. **Krulwich, T. A., M. Ito, R. Gilmour, and A. A. Guffanti.** 1997. Mechanisms of cytoplasmic pH regulation in alkaliphilic strains of *Bacillus. Extremophiles* **1:**163–169.

60. **Kushner, D. J.** 1988. What is the "true" internal environment of halophilic and other bacteria? *Can. J. Microbiol.* **34:**482–486.

61. **Lanyi, J. K.** 1974. Salt-dependent properties of proteins from extremely halophilic bacteria. *Bacteriol. Rev.* **38:**272–290.

62. **Lanyi, J. K., and J. Stevenson.** 1970. Studies of the electron transport chain of extremely halophilic bacteria. IV. Role of hydrophobic forces in the structure of menadione reductase. *J. Biol. Chem.* **245:**4074–4080.

63. **Laubinger, W., and P. Dimroth.** Characterization of the ATP synthase of *Propionogenium modestum* as a primary sodium pump. *Biochemistry* **27:**7531–7537.

64. **Lienard, T., B. Becher, M. Marschall, S. Bowien, and G. Gottschalk.** 1996. Sodium ion translocation by N5-methyltetrahydromethanopterin:coenzyme M methyltransferase from *Methanosarcina mazei* Gö1 reconstituted in either lipid liposomes. *Eur. J. Biochem.* **239:**857–864.

65. **Levy, S.** 1992. Active efflux mechanisms for antimicrobial resistance. *Antimicrob. Agents Chemother.* **36:**695–703.

66. **Matheson, A. T., M. Yaguchi, R. N. Nazar, L. P. Visentin, and G. E. Willick.** 1978. The structure of ribosomes from moderate and extreme halophilic bacteria, p. 481–500. *In S. R. Caplan and M. Ginzburg (ed.), Energetics and Structure of Halophilic Microorganisms.* Elsevier/North Holland Biomedical Press, Amsterdam, The Netherlands.

67. **Mohr, V., and H. Larsen.** 1963. On the structural transformations and lysis of *Halobacterium salinarium* in hypotonic and isotonic solutions. *J. Gen. Microbiol.* **31:**267–280.

68. **Murata, T., K. Takase, I. Uamato, K. Igarashi, and Y. Kakinuma.** 1997. Purification and reconstitution of Na^+-translocatng vacuolar ATPase from *Enterococcus hirae. J. Biol. Chem.* **272:**24885–24890.

69. **Murata, T., I. Yamato, K. Igarashi, and Y. Kakinuma.** 1996. Intracellular Na^+ regulates transcription of the *ntp* operon encoding a vacuolar-type Na^+ translocating ATPase in *Enterococcus hirae. J. Biol. Chem.* **271:**23661–23666.

70. **Murgufa, J. R., J. M. Belles, and R. Serrano.** 1996. The yeast *HAL2* nucleotidase is an in vivo target of salt toxicity. *J. Biol. Chem.* **271:**29029–29033.

71. **Nagata, S., K. Adachi, K. Shirai, and H. Sano.** 1995. ^{23}Na NMR spectroscopy of free Na^+ in the halotolerant bacterium *Brevibacterium* sp. and *Escherichia coli. Microbiology* **141:**729–736.

72. **Neumann, S., U. Matthey, G. Kaim, and P. Dimroth.** 1998. Purification and properties of the F_1F_0 ATPase of *Ilyobacter tartaricus*, a sodium ion pump. *J. Bacteriol.* **180:**3312–3316.

73. **Nozaki, K., T. Kuroda, T. Mizushima, and T. Tsuchiya.** 1998. A new Na^+/H^+ antiporter, NhaD, of *Vibrio parahaemolyticus. Biochim. Biophys. Acta* **1369:**213–220.

74. **Ohyama, T., K. Igarashi, and K. Kohayashi.** 1994. Physiological role of the *cha*A gene in sodium and calcium circulation at a high pH in *Escherichia coli. J. Bacteriol.* **176:**4311–4315.

75. **Oren, A.** Adaptation of halophilic archaea to life at high salt concentrations. *In A. Läuchli and U. Lüttge (ed.), Salinity: Environment-Plants-Molecules*, in press. Kluwer Academic Publishers, Dordrecht, The Netherlands.

76. **Oren, A.** 1999. Bioenergetic aspects of halophilism. *Microbiol. Mol. Biol. Rev.* **63:**334–348.

77. **Padan, E.** 1998. The molecular mechanism of regulation of the NhaA Na^+/H^+ antiporter of *Escherichia coli*, a paradigm for an adaptation to Na^+ and H^+, p. 163–175. *In A. Oren (ed.), Microbiology and Biogeochemistry of Hypersaline Environments.* CRC Press, Boca Raton, Fla.

78. **Padan, E., N. Maisler, D. Taglicht, R. Karpel, and S. Schuldiner.** 1989. Deletion of *ant* in *Escherichia coli* reveals its function in adaptation to high salinity and an alternative Na^+/H^+ antiporter system(s). *J. Biol. Chem.* **264:**20297–20302.

79. **Padan, E., and S. Schuldiner.** 1996. Bacterial Na^+/H^+ antiporters—molecular biology, biochemistry, and physiology, p. 501–531. *In W. N. Konings, H. R. Kaback, and J. Lolkema (ed.), The Handbook of Biological Physics*, vol. II. *Transport Processes in Membranes.* Elsevier Science, Amsterdam, The Netherlands.

80. **Padan, E., and S. Schuldiner.** 1994. Molecular physiology of Na^+/H^+ antiporters, key transporters in circulation of Na^+ and H^+ in cells. *Biochim. Biophys. Acta* **1185:**129–151.

81. **Park, C., J. Y. Moon, P. Cokie, and D. A. Webster.** 1996. Na(+)-translocating cytochrome *bo* terminal oxidase from *Vitreoscilla*: some parameters of its Na^+ pumping and orientation in synthetic vesicles. *Biochemistry* **35:**11895–11900.

82. **Peng, Z., and D. P. S. Verma.** 1995. A rice *HAL2*-like gene encodes a Ca^{2+}-sensitive 3′(2′),5′-diphosphonucleoside 3′(2′)-phosphohydrolase and complements yeast *met*22 and *Escherichia coli cys*Q mutations. *J. Biol. Chem.* **270:**29105–29110.

83. **Pfenninger-Li, X. D., S. P. J. Albracht, R. Van Belzen, and P. Dimroth.** 1996. NADH:ubiquinone oxidoreductase of *Vibrio alginolyticus*: purification, properties and reconstitution of the Na^+ pump. *Biochemistry* **35:**6233–6242.

84. **Pieper, U., G. Kapadia, M. Mevarech, and O. Herzberg.** 1998. Structural features of halophilicity derived from the crystal structure of dihydrofolate reductase from the Dead Sea halophilic archaeon, *Haloferax volcanii. Structure* **6:**75–88.

85. **Pinner, E., E. Padan, and S. Schuldiner.** 1994. Kinetic properties of NhaB, a Na^+/H^+ antiporter from *E. coli. J. Biol. Chem.* **269:**26274–26479.

86. **Rahav-Manor, O., O. Carmel, R. Karpel, D. Taglicht, G. Glaser, S. Schuldiner, and E. Padan.** 1992. NhaR, a protein homologous to a family of bacterial regulatory proteins (LysR), regulates *nha*A, the sodium proton antiporter gene in *Escherichia coli. J. Biol. Chem.* **267:**10433–10438.

87. **Reidlinger, J., and V. Müller.** 1994. Purification of ATP synthase from *Acetobacterium woodii* and identification as a Na(+)-translocating F_1F_O-type enzyme. *Eur. J. Biochem.* **226:**1079–1084.

88. **Reistad, R.** 1970. On the composition and nature of the bulk protein of extremely halophilic bacteria. *Arch. Mikrobiol.* **71:**353–360.

89. **Rich, P. R., B. Meinier, and F. B. Ward.** 1995. Predicted structure and possible ionmotive mechanisms of the sodium-linked NADH-ubiquinone oxidoreductase of *Vibrio alginolyticus. FEBS Lett.* **375:**5–10.

90. **Rimon, A., Y. Gerchman, Z. Kariv, and E. Padan.** 1998. A point mutation (G338S) and its suppresssor mutations affect both the pH response of the NhaA–Na^+/H^+ antiporter as well

as the growth phenotype of *Escherichia coli*. *J. Biol. Chem.* **273**:26470–26476.

91. **Rowbury, R. J.** 1997. Regulatory components, including integration host factor, CysB and H-NS, that influence pH responses in *Escherichia coli*. *Lett. Appl. Microbiol.* **24**:319–328.

92. **Sasse-Dwight, S., and J. Gralla.** 1991. Footprinting protein-DNA complexes in vivo. *Methods Enzymol.* **208**:146–168.

93. **Schell, M.** 1993. Molecular biology of the LysR family of transcriptional regulators. *Annu. Rev. Microbiol.* **47**:597–626.

94. **Speelmans, G., B. Poolman, T. Abee, and W. N. Konings.** 1993. Energy transduction in the thermophilic anaerobic bacterium *C. fervidus* is exclusively coupled to sodium ions. *Proc. Natl. Acad. Sci. USA* **90**:7975–7979.

95. **Stasinopoulos, S. J., G. A. Farr, and D. H. Bechhofer.** 1998. *Bacillus subtilis tetA*(L) gene expression: evidence for regulation by translational reinitiation. *Mol. Microbiol.* **30**:923–932.

96. **Storz, G., L. A. Tartaglia, and B. N. Ames.** 1990. Transcriptional regulator of oxidative stress-inducible genes: direct activation by oxidation. *Science* **248**:189–194.

97. **Taglicht, D., E. Padan, and S. Schuldiner.** 1993. Proton-sodium stoichiometry of NhaA, an electrogenic antiporter from *Escherichia coli*. *J. Biol. Chem.* **268**:5382–5387.

98. **Takase, K., S. Kakinuma, I. Yamato, K. Konishi, K. Igarashi, and Y. Kakinuma.** 1994. Sequencing and characterization of the *ntp* gene cluster for vacuolar-type Na$^+$-translocating ATPase of *Enterococcus hirae*. *J. Biol. Chem.* **269**:11037–11044.

99. **Tani, K., T. Watanabe, H. Matsuda, M. Nasu, and M. Kondo.** 1996. Cloning and sequencing of the spore germination gene of *Bacillus megaterium* ATCC12872: similarities to the NaH-antiporter coded by *nhaA*(*ant*) from *Enterococcus hirae*. *Microbiol. Immunol.* **40**:99–105.

100. **Tokuda, H., and T. Unemoto.** 1982. Characterization of the respiration-dependent Na$^+$ pump in the marine bacterium *Vibrio alginolyticus*. *J. Biol. Chem.* **257**:10007–10014.

101. **Toledano, M., I. Kullik, F. Trinh, P. Baird, T. Schneider, and G. Storz.** 1994. Redox dependent shift of OxyR-DNA contacts along an extended DNA-binding site: a mechanism for differential promoter selection. *Cell* **78**:897–909.

102. **Tolner, B., T. Ubbink-Kok, B. Poolman, and W. N. Konings.** 1995. Cation-selectivity of the L-glutamate transporters of *Escherichia coli*, *Bacillus stearothermophilus* and *Bacillus caldotenax*: dependence on the environment in which the proteins are expressed. *Mol. Microbiol.* **18**:123–133.

103. **Tupper, A., T. Owen-Hughes, D. Ussery, D. Santos, J. Ferguson, J. Sidebotham, J. Hinton, and C. Higgins.** 1994. The chromatin-associated protein H-NS alters DNA topology in vitro. *EMBO J.* **13**:258–268.

104. **van de Vossenberg, J. L. C. M., A. J. M. Driessen, and W. N. Konings.** 1998. The essence of being extremophilic: the role of the unique archaeal membrane lipids. *Extremophiles* **2**:163–170.

105. **Ventosa, A., J. J. Nieto, and A. Oren.** 1998. Biology of moderately halophilic aerobic bacteria. *Microbiol. Mol. Biol. Rev.* **62**:504–544.

106. **Verkhovskaya, M. L., B. Barquera, M. I. Verkhovsky, and M. Wikstrom.** 1998. The Na$^+$ and K$^+$ transport deficiency of an *E. coli* mutant lacking the NhaA and NhaB proteins is apparent and caused by impaired osmoregulation. *FEBS Lett.* **439**:271–274.

107. **Volker, U., S. Engelmann, B. Maul, S. Riethdorf, A. Volker, R. Schmid, H. Mach, and M. Hecker.** 1994. Analysis of the induction of general stress proteins of *Bacillus subtilis*. *Microbiology* **140**:741–752.

108. **Waser, M., D. Hess-Bienz, K. Davies, and M. Solioz.** 1992. Cloning and disruption of a putative NaH-antiporter gene of *Enterococcus hirae*. *J. Biol. Chem.* **267**:5396–5400.

109. **Wei, Y., A. A. Guffanti, and T. A. Krulwich.** 1999. Sequence analysis and functional studies of a chromosomal region of alkaliphilic *Bacillus firmus* OF4 encoding an ABC-type transporter with similarity of sequence and Na$^+$ exclusion capacity to the *Bacillus subtilis* NatAB transporter. *Extremophiles* **3**:113–120.

110. **Weiss, D. S., P. Gartner, and R. K. Thauer.** 1994. The energetic and sodium-ion dependence of N5-methyltetrahydromethanopterin:coenzyme M methyltransferase studied with cob(I)alamin as methyl acceptor and methylcob(III)alamin as methyl donor. *Eur. J. Biochem.* **226**:799–809.

111. **Woehlke, G., and P. Dimroth.** 1994. Anaerobic growth of *Salmonella typhimurium* on L(+)- and D(−)-tartrate involves an oxaloacetate decarboxylase Na$^+$ pump. *Arch. Microbiol.* **162**:233–237.

Bacterial Stress Responses
Edited by G. Storz and R. Hengge-Aronis
©2000 ASM Press, Washington, D.C.

Chapter 9

The SOS Response to DNA Damage

GRAHAM C. WALKER, BRADLEY T. SMITH, AND MARK D. SUTTON

The Escherichia coli SOS response is the best characterized regulatory network induced by DNA damage and has become the paradigm for how cells respond to damage to their genetic material. Three reviews in particular have attempted to trace the experimental and intellectual developments that gave rise to our current understanding of the SOS system (26, 112, 118) while other excellent reviews have covered important aspects of this field (35, 47, 60, 100, 113, 114, 121). The goal of this review is to provide a summary of our current understanding of the E. coli *SOS response to DNA damage while underscoring principles that may be useful in thinking about how other organisms may respond to this same type of stress. Because of space limitations, this review will not be comprehensive with respect to citations but rather will cite a limited number of older papers that played key roles in the development of the field as well as important more recent references.*

TRANSCRIPTIONAL CONTROL BY THE SOS SYSTEM: KEY DEVELOPMENTS AND CURRENT STATUS

Initial Recognition of the Existence of the SOS Regulatory System

Like all organisms, *Escherichia coli* is periodically exposed to any of a variety of environmental agents that damage its DNA. The idea that such damage might lead to the *recA*[+] *lexA*[+]-dependent induction of a variety of unlinked genes, whose products would enhance the survival of the cell, was first proposed by Radman (84). The term "SOS" was chosen to emphasize that this cellular response is to a distress signal, i.e., DNA damage (26). Radman's thinking was influenced by previous work, particularly Witkin's recognition of similarities between the UV induction of lambda prophage and filamentation in *E. coli* B, which led her to propose that a common regulatory system controlled these two diverse physiological functions (117).

Induction of SOS Gene Expression by DNA Damage-Induced Transcriptional Derepression

The most striking aspect of the SOS response of *E. coli* is the increase in the expression of more than 20 unlinked genes that occurs after DNA damage (26). As discussed below, recent work has made it clear that several posttranslational mechanisms are also involved in coordinating this bacterium's response to this form of stress. The LexA protein serves as a repressor (8, 61) of the various SOS loci by binding to a similar regulatory sequence, termed the SOS box, located near the promoter of each gene or operon. Most SOS-regulated loci, including the *recA* gene, are expressed at a basal level, even in an uninduced cell.

E. coli senses that it has experienced DNA damage by recognizing an internal signal. This signal consists of a RecA/ssDNA nucleoprotein filament that results from RecA protein polymerizing on single-stranded DNA that is generated when the cell initially attempts to replicate its damaged DNA (94) (Fig. 1). The subsequent signal transduction step occurs by an unusual mechanism in which interactions between LexA and this nucleoprotein filament (26) stimulate an otherwise latent capacity of LexA to autodigest (58, 59). This facilitated autodigestion cleaves LexA at its Ala[84]-Gly[85] bond and inactivates it as a repressor (33). Little and his colleagues proposed a mechanism for the cleavage in which an uncharged form of Lys[156] helps remove a proton from Ser[119], which then acts as a nucleophile to attack the Ala[84]-Gly[85] bond (99).

Graham C. Walker, Bradley T. Smith, and Mark D. Sutton • Department of Biology, Massachusetts Institute of Technology, Cambridge, MA 02139.

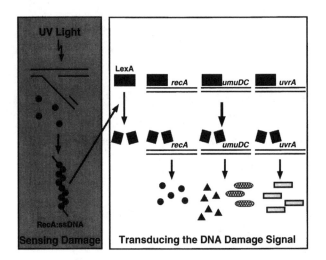

Figure 1. Sensing and transducing the DNA damage signal. The formation of the RecA/ssDNA nucleoprotein filament and the subsequent RecA/ssDNA-mediated autodigestion of the LexA repressor are the key events in the regulation of the SOS response. Autodigestion of LexA relieves its transcriptional repression of the approximately 20 unlinked genes that comprise the SOS response. Three genes that are components of this response are depicted here: *recA*, *uvrA*, and *umuDC*.

The RecA/ssDNA nucleoprotein filament similarly facilitates the cleavage of the repressors of lambda (91) and related phages and, as discussed below, the UmuD protein (26). The resulting decrease in the pool of LexA in the cell results in the increased expression of SOS-regulated genes (Fig. 1).

An SOS-signaling event has been been biochemically reconstituted using an in vitro system in which RecA protein, RecBCD enzyme, single-strand binding protein (SSB), and LexA repressor respond to double-strand breaks by derepressing transcription from an SOS-regulated promoter (1). In this model system, the single-stranded DNA necessary to form the RecA/ssDNA nucleoprotein filament is generated by the action of the RecBCD enzyme at the double-strand break.

Genes Controlled by the SOS Regulatory Circuitry

The *recA* gene was the first locus to be recognized as being under SOS control, and the *lexA* gene was subsequently recognized to be similarly controlled (26). Additional SOS-controlled genes, initially termed *din* for damage-inducible, were identified by screening randomly generated transcriptional fusions for their ability to be induced by DNA damage (41). Additional members of the SOS regulon

Table 1. LexA-regulated and -independent DNA damage-inducible genes[a]

Gene	Function and comments
LexA-regulated genes	
polB/*dinA*	DNA Pol II, involved in induced replisome reactivation/replication restart
dinB/*dinP*	DNA Pol IV, involved in untargeted mutagenesis
dinD/*pcsA*	*pcsA68* is a cold-sensitive mutant
dinF	Unknown, downstream of *lexA*
dinG	Putative ATP-dependent helicase
dinI	Inhibits RecA-coprotease activity, similar to TP110 ImpC
ftsK/*dinH*	Essential cell division gene; overexpression increases DNA damage resistance
lexA	Transcriptional repressor of the SOS genes
recA	Involved in homologous recombination, SOS regulation, and SOS mutagenesis
recN	Recombinational repair
ruvAB	Branch migration: RuvA binds Holliday junctions, RuvB is the helicase motor
sbmC	Resistance to Microcin B17
ssB	Single-stranded DNA-binding protein
sulA	Inhibition of cell division via inhibition of FtsZ polymerization
umuDC	DNA Pol V, required for SOS mutagenesis/translesion DNA synthesis
uvrA	Nucleotide excision repair: DNA damage recognition protein
uvrB	Nucleotide excision repair: required for incision step
uvrD	DNA helicase involved in excision and mismatch repair, and SOS induction
LexA-independent genes	
dinY	Mutant defective in Weigle reactivation of UV-irradiated bacteriophage λ
dnaA	Chromosomal replication initiator protein
dnaB	Replicative DNA helicase
dnaN	β-subunit of DNA Pol III, processivity factor
dnaQ	ε-subunit of DNA Pol III, proofreading activity
hga	2-Keto-4-hydroxyglutarate aldolase, required for respiration recovery post-UV
himA	Site-specific recombination
nrdAB	Ribonucleotide reductase
phr	DNA photolyase
recQ	Recombinational repair
sfiC	Inhibition of cell division

[a] Adapted from Friedberg et al. (26) and Koch and Woodgate (47). The reader is referred to those sources for the original references.

have been identified by a variety of techniques and are summarized in Table 1. Additional genes have been found to be induced in response to DNA damage in a recA$^+$-dependent but lexA$^+$-independent manner (Table 1), but, in general, the details of their regulation are not well understood and require further study.

Despite the relatively straightforward logic of SOS transcriptional control, it has become increasingly clear that the SOS circuitry is sophisticated and finely tuned to aid the cell in surviving insults to its genetic material. For example, as originally pointed out by Little (57), the repression of the lexA gene by the LexA protein has various effects on SOS induction (26). First, it extends the range, in terms of inducing signal, over which the cell can establish an intermediate state of induction and thus express only a subset of the total SOS responses. Second, since the affinity of the LexA protein for its operator is weak compared to its affinity for the operators of other SOS-regulated genes, the system is buffered against significant levels of induction by a very limited inducing signal. Third, by virtue of the fact that the expression of lexA is autoregulated, it speeds the return to the repressed state once the inducing signal starts to decrease.

Additional effects on SOS transcriptional control may be provided by double-strand DNA substrates for recombination competing with LexA for a binding site within the RecA/ssDNA nucleoprotein filament (125). This possibility is supported by recent experiments showing that LexA cleavage and DNA strand exchange functions of the RecA/ssDNA nucleoprotein filament are competitive reactions (32, 88). An additional intriguing twist to the circuitry for SOS transcriptional control has been the discovery that the SOS-controlled dinI gene (124) encodes an 81-amino acid protein that has the ability to inhibit the autodigestion of LexA and UmuD that is facilitated by the RecA/ssDNA nucleoprotein filament (123).

POSTTRANSCRIPTIONAL CONTROL IN THE SOS SYSTEM REVEALED BY ANALYSES OF umuDC

Over the past several years, studies of the SOS-regulated umuDC operon have revealed that, besides recA$^+$ lexA$^+$-mediated transcriptional control, a variety of posttranscriptional mechanisms play critically important roles in helping E. coli conduct an ordered and effective response to DNA damage. These studies have also offered fresh insights into the mechanisms cells use to tolerate DNA damage and into the unexpected universality of some of these mechanisms.

The umuDC Gene Products are Required for SOS Mutagenesis

The umuDC locus was originally identified by screening for E. coli mutants that were nonmutable after treatment with an agent such as UV light that is normally mutagenic (19, 40, 103). The observation that recessive loss-of-function mutations in umuD or umuC rendered E. coli nonmutable with agents such as UV indicated that some active cellular process involving the umuD and umuC gene products was necessary to process the damaged DNA in such a way that mutations would result. As discussed below, this active cellular process is now known to be translesion synthesis. The umuD and umuC genes were subsequently shown to be organized in an operon (19, 96) that is repressed by LexA and under SOS control (2).

Cleavage of UmuD Facilitated by RecA/ssDNA Nucleoprotein Filaments and Its Physiological Consequences

DNA sequencing of the umuDC operon (46, 82) led to the unexpected insight (82) that UmuD shared homology with carboxy-terminal domains of LexA and lambda repressor, both of which contain a latent autoprotease capacity. The UmuD protein was shown to resemble LexA and lambda repressor in undergoing a facilitated autodigestion upon interaction with the RecA/ssDNA nucleoprotein filament (13, 95). The cleavage removes the predicted N-terminal 24 amino acids of the protein to generate UmuD′ (Fig. 2). Recent evidence suggests that UmuD and LexA interact with the RecA/ssDNA nucleoprotein filament in a similar but not identical manner (49, 72). These differences may account, at least in part, for the fact that UmuD cleavage is much slower than LexA cleavage (13).

This cleavage of UmuD to UmuD′ was shown genetically to activate the umuD product for its role, together with UmuC, in SOS mutagenesis (74). An uncleaved UmuD or an uncleavable UmuD derivative (due to a missense mutation) is inactive in SOS mutagenesis. In contrast, the directly expressed UmuD′ protein (i.e., residues 25 to 139 of UmuD) is sufficient for the role of the umuD gene product in SOS mutagenesis (74). The cleavage of UmuD to UmuD′ was subsequently found to have also activated the umuD product for another potential physiological role, namely, acting together with UmuC to inhibit homologous recombination (7, 87, 101, 102).

Until recently, the intact UmuD protein was thought to simply be an inactive precursor that was subsequently converted to the active form, UmuD′, which acts together with UmuC in translesion syn-

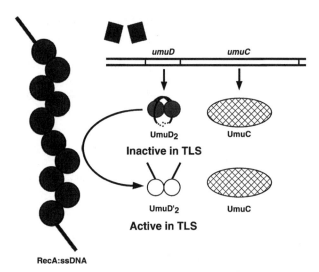

Figure 2. RecA-facilitated self-cleavage of UmuD to yield UmuD′ activates it for its role in translesion synthesis (TLS). Transcription of the *umuDC* operon is derepressed in response to DNA damage following cleavage of the LexA transcriptional repressor. Although UmuD and UmuC are unable to promote TLS, interaction of UmuD with RecA/ssDNA nucleoprotein filaments leads to self-cleavage of UmuD to yield UmuD′. The RecA-facilitated self-cleavage reaction that UmuD undergoes is similar to that which LexA undergoes and serves to activate the *umuD* gene product for its role in TLS.

thesis and possibly in a controlled inhibition of homologous recombination. However, evidence has now been presented that suggests that the uncleaved UmuD protein, acting together with UmuC, itself has an active physiological role. In particular, uncleaved UmuD, acting together with UmuC, has been proposed to function as a prokaryotic DNA damage checkpoint that serves to delay the resumption of DNA synthesis and cell growth after DNA damage, thereby allowing more time for accurate repair systems to remove or process the damage before replication is attempted (77).

Posttranslational Cleavage of UmuD Acts as a Molecular Switch That Regulates the Different Physiological Roles of the *umuDC* Gene Products

After a dose of 20 J/m², uncleaved UmuD accumulates and predominates over UmuD′ for 20 min, followed by a period during which UmuD′ predominates (77). The timing of the posttranslation modification of UmuD apparently permits the *umuDC* gene products to play two distinct and temporally separated roles in DNA damage tolerance. During the first phase, UmuD and UmuC play their DNA damage checkpoint role. During the second phase, UmuD′ and UmuC make it possible for the cell to replicate over any unrepaired or unrepairable damage

by enabling translesion synthesis (77). Thus, the SOS response not only induces physiologically useful functions in response to DNA damage, but ensures that they occur in a temporally ordered fashion.

Additional Mechanisms of Posttranscriptional Control of the *umuDC* Gene Products

The delay in the appearance of UmuD′ can be explained, at least in part, by the observation that, both in vivo (95) and in vitro (13), cleavage of UmuD mediated by the RecA/ssDNA nucleoprotein filament is slow compared to cleavage of LexA. The nature of the N-terminal 24 amino acids of UmuD plays an important role in determining the rate of this cleavage (67). However, other posttranscriptional controls are also important in modulating the various activities of the *umuDC* gene products.

One of these control systems is based on the differential protease sensitivities of the various forms of the *umuD* gene product. Both the UmuD and UmuD′ proteins form homodimers in solution (107, 122), and the UmuD•UmuD′ heterodimer is more stable than either of the homodimers (6). Both $UmuD_2$ and $UmuD'_2$ can form a complex with a molecule of UmuC (107, 122), and the heterodimer can do so as well (122). The different forms of the *umuD* gene product are differentially sensitive to ATP-dependent proteases found in *E. coli*. The $UmuD_2$ homodimer is sensitive to Lon, as is UmuC (23). The primary signal of the Lon-mediated proteolysis is located within the FPLF sequence that begins at residue 15 of UmuD, with a secondary site at the FPSP sequence that begins at residue 26 (28). In contrast, the UmuD′ in the UmuD•UmuD′ heterodimer is degraded by the ClpXP protease (23). This selective degradation of the UmuD′ in the UmuD•UmuD′ heterodimer may be another factor that enables cells to use the UmuD to UmuD′ conversion as a molecular timing device that enables the cell to switch from one mode of dealing with DNA damage (a checkpoint) to another (translesion synthesis) (77). The sequestration of UmuD′ in the UmuD•UmuD′ heterodimer had been proposed as a method by which the cell might prevent UmuD′ from enabling translesion synthesis at a time when the cell was beginning to shut off the SOS response (6). It now appears that the incorporation of UmuD′ into a UmuD•UmuD′ heterodimer is a way of targeting it for destruction by the ClpXP protease (29). Finally the SOS-induced DinI protein (123) is not only able to inhibit LexA cleavage, but UmuD cleavage as well. This may be yet another posttranscriptional mechanism that controls when the switch from UmuD predomination to UmuD′ predomination occurs following DNA damage.

STRUCTURES OF THE UmuD′₂ AND UmuD₂ HOMODIMERS: INSIGHTS INTO THE MECHANISM OF FACILITATED AUTODIGESTION AND THEIR DIFFERENT PHYSIOLOGICAL ROLES

Structure of UmuD′₂

The structure of a crystal of the UmuD′₂ protein was solved by Peat et al. (80, 81). The UmuD′ monomer, which was found to have a globular C-terminal domain, was arranged in a filament in which there were two different interfaces between UmuD′ monomers (Fig. 3). Peat et al. (80) assigned interface A as the UmuD′₂ dimer interface and proposed that interface B is involved in polymerizing UmuD′₂ into an extended filament that is found in solution. This latter proposal was supported by the observation that UmuD′ multimers could be detected in vitro following glutaraldehyde cross-linking (81). However, in an independent study employing nuclear magnetic resonance (NMR) Ferentz et al. (21) were able to show that the interface of the UmuD′₂ homodimer in solution closely resembled interface B seen in the crystals solved by Peat et al. (80, 81). Measurements of relaxation times revealed that the N-terminal 14 amino acids of each UmuD′ monomer (positions 25 to 39 using UmuD numbering) are free in solution, consistent with a number of them being disordered in the crystal structure. The conclusion that interface B is the correct interface in solution has been reinforced by a subsequent genetic study that showed that mutations affecting residues within interface B weakened the UmuD′₂ homodimer interface in solution (76). Furthermore, this NMR study showed that UmuD′₂ is a dimer rather than a filament in solution, even at the rather high (ca. 1 mM) concentrations needed for NMR analysis (21).

Implications of the UmuD′₂ Structure for UmuD Cleavage

The discovery of the correct interface for the UmuD′₂ homodimer had interesting implications for the mechanism of UmuD cleavage. On the basis of amino acid homology, Ser60 of UmuD corresponds to Ser119 of LexA (82), the putative nucleophile (99), while the Lys97 corresponds to Lys 156 of LexA. Thus, a similar mechanism of autodigestion for both LexA and UmuD was envisioned (13, 74). This model was strongly supported by the observation that a Ser60→Ala or Lys97→Ala mutation rendered the UmuD protein noncleavable (74). In the crystal structure solved by Peat et al. (80), these two residues were found at the end of a cleft in the UmuD′ C-terminal domain, with the Ser60 residue located only

2.8 Å from Lys97. Thus, it seemed likely that the cleft was the binding site for the amino terminus of UmuD and would serve to position the Cys24-Gly25 bond of UmuD close to the catalytic Ser60. The fact that mutations that impair UmuD cleavage are located along the side of this cleft (6, 80) lent support to this idea. However, in interface A, the one originally proposed to be the UmuD′₂ homodimer interface, the cleft is occluded in the homodimer, a fact that seemed to suggest that the UmuD′₂ homodimer would need to monomerize for cleavage to occur.

The discovery that the actual interface for the UmuD′₂ homodimer in solution was interface B (21) raised a new and interesting possibility, namely, that in the UmuD₂ homodimer, the amino terminal region of one UmuD′ monomer might lie along the exposed cleft of its partner so that its Cys24-Gly25 cleavage site would be near the Ser60 of its partner. Appropriate repositioning of the relative positions of the Ser60, Lys97, and the Cys24-Gly25 bond when UmuD₂ interacted with the RecA/ssDNA nucleoprotein filament could then result in cleavage. This hypothesis has been reinforced by the observation that UmuD autodigestion facilitated by the RecA/ssDNA nucleoprotein filament interactions can occur in *trans* between two different UmuD molecules (21, 65). These experiments were built on earlier work by Little and his colleagues with LexA and lambda *c*I repressor that demonstrated specific cleavage can take place in an intermolecular reaction (44). A UmuD derivative that is noncleavable because of a mutation at its Cys24-Gly25 cleavage site was mixed with a UmuD derivative that is noncleavable because of a mutation at its catalytic site. Only when both proteins were present did cleavage occur, demonstrating that the cleavage can occur in *trans* between the two proteins (65). It now seems plausible that, in vivo, this cleavage in *trans* occurs within the UmuD₂ homodimer.

The Mechanism of LexA and UmuD Cleavage Is Related to That of Signal Peptidases

Although it has been noted that the proposed mechanism for LexA and UmuD cleavage, in which a Lys activates a Ser to become a nucleophile, has some resemblance to the mechanism of action of beta-lactamases (59, 104, 105), the evolutionary origin of the autocleavage domain of LexA, various phage repressors, and UmuD had remained a mystery until recently. However, a highly intriguing possibility has been raised by the striking structural similarity between the structures of UmuD′ and those of signal peptidases. Signal peptidases cleave the signal peptides that are used to target secretion in both pro-

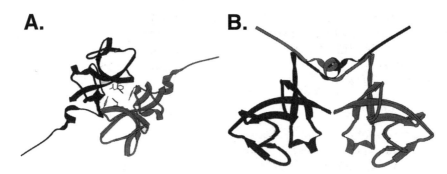

Figure 3. (A) The UmuD′$_2$ dimer interface proposed by Peat et al. (80) as seen in the crystal structure. (B) The correct UmuD′$_2$ dimer interface as observed by Ferentz et al. (21) using NMR techniques.

karyotes and eukaryotes. In the case of the *E. coli* signal peptidase, a conserved lysine residue (Lys145) has been proposed to act as a general base that abstracts the proton from the side chain hydroxyl group of the nucleophilic Ser90 (79) while analysis of the *Bacillus subtilis* Sip1 signal peptidase indicated that type 1 signal peptidases and LexA-like proteases were structurally and functionally related serine proteases (110). When the crystal structure of the *E. coli* signal peptidase in complex with beta-lactam inhibitor was solved, Paetzel et al. (78) made the remarkable observation that, although the overall main chain connectivity in UmuD′ and domain I of the signal peptidase differed in some regions, 68 common C$_\alpha$ atoms could be superimposed with a root mean square deviation of 1.6. It thus seems likely that the autodigestion module found in LexA and UmuD is evolutionarily related to the protease structural domain found in signal peptidases.

Relationship of the Structure of the UmuD$_2$ Homodimer to the UmuD′$_2$ Homodimer

The structure of the UmuD$_2$ homodimer is not yet known. However, some insights have been gained by analyzing the properties of a set of engineered functional derivatives of UmuD$_2$ and UmuD′$_2$, each of which contains a single cysteine at a defined position in the amino acid sequence. The relative ease of cross-linking of the monomers within each dimer was determined, either using an oxidizing agent to cross-link the monomers within the homodimer or by using a homobifunctional cysteine-specific cross-linking agent. In addition, the accessibility of each cysteine to an alkylating agent was assessed (30, 54; A. Guzzo, M. D. Sutton, I. Narumi, and G. C. Walker, unpublished results). Taken together, these analyses suggest that the UmuD$_2$ homodimer and UmuD′$_2$ homodimer share common elements in their interfaces but differ substantially in other elements of their structure. In particular, the amino acids between 25

and 39 are known from NMR studies to be free in solution in UmuD′$_2$ (21) yet appear to be constrained in UmuD$_2$. In UmuD$_2$, the single cysteines at positions 37 and 38 of one monomer seem to be very close to the corresponding cysteines in the other monomer and to be favorably situated for disulfide bond formation (30, 54). Other experiments suggest that the remaining part of the 25 to 39 region probably lies in the groove in the opposite monomer, with the cleavage site being located relatively close to Ser60, the catalytic serine (M. D. Sutton and G. C Walker, unpublished results). As discussed below, it seems likely that these differing structures of UmuD$_2$ and UmuD′$_2$ are responsible for their differing roles in DNA damage tolerance.

IN VIVO AND IN VITRO REQUIREMENTS FOR TRANSLESION SYNTHESIS

In Vivo Analyses of the Properties of Translesion Synthesis

Genetic studies had indicated that SOS mutagenesis, and hence the process of translesion synthesis that is responsible for such induced mutagenesis, requires not only the UmuD′ and UmuC proteins but also the RecA protein (4, 20, 74). Furthermore, limited genetic evidence indicated DNA polymerase III was required (9, 31) even when epsilon, the 3′ → 5′ proofreading subunit, was absent (98). In contrast, DNA Pol I and DNA Pol II were not required (5, 36). There is also an in vivo requirement for *groEL*$^+$ and *groES*$^+$ function (17), but this seems likely to reflect a requirement for chaperone function in UmuC folding or assembly (17, 83) and not a direct mechanistic role. The polymerization of RecA onto single-strand DNA generated when a Pol III holoenzyme complex stalls at a lesion has been proposed to create a RecA/ ssDNA nucleoprotein filament that targets the UmuD′C complex to the site of the lesion (24, 102).

Experiments in which plasmids carrying single defined lesions were transformed into SOS-induced cells revealed that *umuDC*-dependent translesion synthesis could occur over such lesions as abasic sites, UV-induced thymine-thymine cyclobutane dimers, and pyrimidine-pyrimidone (6-4) photoproducts (3, 24, 52, 53, 102). On the basis of indirect inferences from in vivo experiments, a model for translesion synthesis was proposed (10) that suggested misincorporation directly opposite the lesion occurs independently of UmuD′, UmuC, and RecA, but that continued replication required these three proteins (10, 11, 26). Experiments using *N*-2-acetylaminofluorene adducted plasmid DNA provided support for this model (48, 71), but these interpretations may need to be reevaluated in light of the recent biochemical studies involving UmuD′, UmuC, and RecA discussed below.

In Vitro Reconstitution of UmuD′/UmuC-Dependent Translesion Synthesis

Taking advantage of their knowledge of the genetic requirements of translesion synthesis, Echols and his colleagues (85) were able to reconstitute translesion synthesis in vitro using UmuD′, UmuC, RecA, DNA Pol III, single-stranded binding protein (SSB), and a linear ssDNA template with a synthetic abasic lesion at a defined location. In these experiments, the UmuC used had been purified in a denatured state and then renatured (115). Two groups subsequently extended these experiments using UmuC that had been purified in different manners (90, 107, 115). Tang et al. (107) purified the UmuD′$_2$C complex in its native state and showed that translesion synthesis could be observed in the presence of the UmuD′$_2$C complex, DNA polymerase III holoenzyme, RecA, and SSB. Reuven et al. (90) purified UmuC as a hybrid protein (MBP-UmuC) consisting of maltose-binding protein joined to the N-terminus of UmuC. They then showed that, when MBP-UmuC, UmuD′, DNA polymerase III holoenzyme, RecA, and SSB were present, translesion synthesis could be observed. In the course of their experiments, Tang et al. (107) made the unexpected observation that a significant amount of translesion synthesis could still be observed when they omitted the DNA polymerase III core, which contains the catalytic subunit. More detailed analyses discussed below have offered key insights into the mechanistic roles of UmuC and UmuD′ in this process.

DinB, an SOS-Induced *E. coli* Protein with Sequence Homology to UmuC

Before discussing the most recent insights into UmuC function, it is useful first to introduce the DinB protein of *E. coli*. The *dinB* gene was identified during the initial screen for SOS-regulated loci mentioned above (41) but only later was recognized as encoding a protein with sequence homology to UmuC (75). The function of the *dinB* gene had been found to be required for the untargeted mutagenesis that is observed when an undamaged lambda phage infects UV-irradiated cells (12, 45). It has also been observed that overexpression of DinB in *E. coli* cells causes a striking increase in the frequency of −1 frameshift mutations in homopolymeric runs (45), an observation consistent with earlier analyses of untargeted mutagenesis of lambda (119).

UmuC and DinB Are Members of a New Family of DNA Polymerases Which Are Widely Distributed in Nature

Important recent papers have now demonstrated that DinB is a DNA polymerase (DNA Pol IV) (111) and UmuC is a DNA polymerase (DNA Pol V) (see below) (89, 108). Furthermore, it has become clear that *umuC* and *dinB* are members of a superfamily of DNA polymerases, members of which are found in all three kingdoms of life (25, 120).

The first member of this superfamily determined to have a DNA polymerase activity was the product of the *REV1* gene of *Saccharomyces cerevisiae*. The *REV1* gene was identified in a manner similar to that in which the *umuD* and *umuC* genes of *E. coli* were identified, namely screening for yeast mutants that were nonmutable after exposure to a DNA-damaging agent (55). The Rev1 protein was found to share significant sequence homology with UmuC (51) and to possess an enzyme activity that catalyzes the unique incorporation of dCMP opposite abasic sites in a primer/template-dependent reaction (73). Although this activity was originally described as a deoxycytidyl transferase (73), it has now been recognized that Rev1 is more properly described as a DNA polymerase (50), albeit one with a limited polymerase activity (25, 120).

A *dinB* homolog, *RAD30*, was subsequently recognized in yeast (66, 92). *RAD30* was thought to play a role in a pathway of postreplication repair that is independent of *REV1*, *REV3*, and *REV7*. Its mechanism of action was revealed when Johnson et al. (38) discovered that the Rad30 protein was a DNA polymerase (DNA Pol η) that could catalyze the incorporation of nucleotides opposite UV-induced thymine-thymine cyclobutane dimers in an error-free manner, i.e., by incorporating deoxyadenosines opposite this lesion. The *E. coli* UmuC, *E. coli* DinB, *S. cerevisiae* Rev1, and *S. cerevisiae* Rad30 proteins have been found to define the four subfamilies of this superfamily of DNA polymerases (25, 120).

A human protein related to Rad30 has recently be shown to be a DNA polymerase that, similarly to *S. cerevisiae* DNA Pol η, is able to bypass a UV-induced pyrimidine dimer (63). The gene encoding this DNA polymerase was shown to be mutated in individuals with the inherited disorder, xeroderma pigmentosum variant (XP-V) (37, 64). Most xeroderma pigmentosum patients are defective in nucleotide excision repair whereas XP-V patients were known to be proficient. Thus, the demonstration that the XP-V DNA polymerase can carry out error-free translesion synthesis over thymine-thymine cyclobutane dimers provides an explanation for the nature of the defect in XP-V cells. Another human homolog of yeast *RAD30,* termed human *RAD30B,* has recently been identified (68), as has a human homolog of *E. coli dinB* (27). It seems highly likely that the products of these two human genes will also prove to be DNA polymerases.

E. coli DinB Is a DNA Polymerase

Recently, Wagner et al. (111) have shown that the SOS-inducible DinB protein of *E. coli* is a DNA polymerase (DNA Pol IV). DinB adds template-directed nucleotides in a highly distributive manner and lacks an intrinsic 3′ to 5′ proofreading exonuclease activity. It has the interesting ability to elongate misaligned (bulged) primer/template structures in such a manner that the replication product is one nucleotide shorter than would otherwise be expected. This property of DinB is consistent with the *dinB*⁺-dependent −1 frameshift mutations in homopolymeric runs observed in vivo (45, 119). Its behavior on other classes of substrates remains to be investigated.

E. coli UmuC Is a DNA Polymerase That Can Function in Translesion Synthesis

Two groups have recently presented evidence that UmuC is a DNA polymerase (DNA Pol V) (89, 108). Tang et al. (108) highly purified the native UmuD′₂C complex from a ΔpolB strain carrying a temperature-sensitive allele of *dnaE,* which encodes the catalytic subunit of DNA Pol III holoenzyme. The native UmuD′₂ complex was able to perform translesion synthesis at elevated temperature in the absence of added polymerase. In the presence of RecA, beta (the processivity clamp of DNA Pol III), the gamma complex (the clamp loader), and SSB, the polymerase activity observed with D′₂C was highly error-prone at both damaged and undamaged sites on DNA tem-

plates (108). Although the observed polymerase activity was shown to be a property of the UmuC protein and not due to contamination by DNA Pol II (the *polB* gene product) or Pol III, vastly stimulated translesion synthesis was observed in the presence of added DNA Pol III (108). Work by Reuven et al. (89) has shown that the MBP-UmuC fusion protein they purified had a polymerase activity on an undamaged template, but that UmuD′₂ and RecA are additionally necessary for translesion synthesis. However, in contrast to the findings of Tang et al. (108), the polymerase activity of UmuC in their experiments was not dependent on the DNA Pol III processivity clamp (beta) and the clamp loader (the gamma complex). It remains to be established whether the difference relates to the different experimental conditions used by the two groups. Thus, it appears that the UmuC has the polymerase catalytic activity, a result consistent with the properties of other members of the UmuC/DinB superfamily, but that the functions of the UmuD′ and RecA proteins are necessary for the polymerase to be able to carry out translesion synthesis.

Managing the Action of the UmuC Polymerase at the Replication Fork

The discovery that UmuC protein is a DNA polymerase raises the question of whether it acts independently of DNA Pol III or whether its action is somehow coordinated with that of DNA Pol III. Evidence that the action of the *umuDC* gene products is coordinated with the action of DNA Pol III is provided by the recent work of Sutton et al. (106), which showed that both the UmuD and UmuD′ proteins are capable of specific interactions with the alpha (catalytic), epsilon (proofreading), and beta (processivity) subunits of DNA polymerase III. These interactions, particularly those with the alpha subunit, could help account for the observation of Tang et al. (108) that UmuD′₂C apparently stabilized copurifying temperature-sensitive DNA Pol III in their experiments. Interestingly, the beta subunit interacted more strongly with UmuD than with UmuD′, while the alpha subunit interacted more strongly with UmuD′ than with UmuD. This suggested that these reciprocally different interactions are related to the differing physiological roles of the two forms of the *umuD* gene product, UmuD in a DNA damage checkpoint system and UmuD′ in translesion synthesis. A consideration of these issues led Sutton et al. (106) to propose that the *umuD* gene product is part of a system that manages the events occurring at a replication fork on a damaged DNA template and coordinates the action of the SOS-inducible UmuC polymerase with those of the replicative polymerase.

OTHER SOS-INDUCED COMPONENTS FORM PART OF A CELLULAR SYSTEM FOR MANAGING DAMAGED DNA

Previous reviews of the SOS system have focused on the mechanisms of induction and on the functions of individual genes induced as part of the SOS response. It now seems appropriate to view these SOS-induced functions as comprising a system for the management of damaged DNA. The proposed system for managing replication forks on damaged DNA can be seen as just one element of this global management system (Fig. 4). The coordinated action of the SOS gene products to manage cellular events occurring after DNA damage represents an emergent property of the system when one considers the entire SOS response in the context of the living cell. Other elements of the SOS response will be briefly discussed in this context.

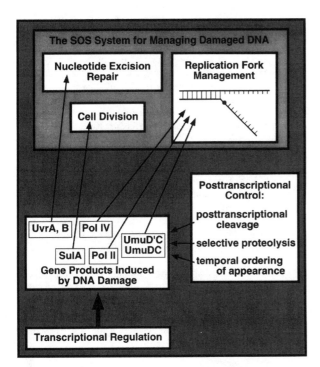

Figure 4. The SOS response to DNA damage. RecA binds to single-stranded DNA (ssDNA) that forms as a result of the cell's failed attempts to replicate damaged DNA. These RecA/ssDNA nucleo-protein filaments then facilitate the latent capacity of LexA to autodigest. LexA autodigestion inactivates it as a transcriptional repressor, thus resulting in derepression of SOS-regulated loci, among them being *sulA*, *umuDC*, *uvrA*, *uvrB*, and *polB*. SulA acts to inhibit cell division. UmuD together with UmuC function first to help regulate DNA synthesis. By inhibiting DNA Pol III, additional time is provided for error-free repair, such as nucleotide excision repair, prior to the resumption of DNA synthesis. After a defined amount of time that appears to be determined by the extent of DNA damage, UmuD is then activated for translesion synthesis by a RecA-facilitated autodigestion reaction that is mechanistically similar to that which LexA undergoes.

Management of Nucleotide Excision Repair

Nucleotide excision repair is a biochemically well-understood process that requires UvrA, UvrB, UvrC, UvrD, DNA polymerase I, and DNA ligase (26, 93). The *uvrA* (42), *uvrB* (22), and *uvrD* (97) genes were found many years ago to be induced as part of the SOS response, and it was originally assumed that their induction as part of the SOS response caused a quantitative increase in the cell's capacity to carry out nucleotide excision repair. It has subsequently been recognized that nucleotide excision repair can be separated into two subpathways, global repair and transcription-coupled repair. Global repair is the process by which most lesions are repaired irrespective of their location in the genome. In contrast, transcription-coupled repair is characterized by more repair of lesions in the transcribed strand than in the nontranscribed strand or in the rest of the genome (69). Crowley and Hanawalt (16) have now shown that SOS induction causes a qualitative change in the capacity of the cell to repair UV-induced lesions. Basal levels of the *uvr* proteins are sufficient for the removal of 6-4 photoproducts by both global and transcription-coupled repair and for the efficient removal of cyclobutane pyrimidine dimers by transcription-coupled repair. However, SOS induction is required for efficient global repair of cyclobutane pyrimidine dimers. Thus the cell is managing its nucleotide excision repair system during the SOS response not only by quantitatively increasing its capacity but by qualitatively altering its characteristics. It is interesting that eukaryotic cells similarly depend on damage-induced stress responses for efficient global repair of cyclobutane pyrimidine dimers (16).

Management of Cell Division after DNA Damage

While many of the proteins that are induced after DNA damage are involved directly in facets of DNA metabolism (excision repair, DNA replication, translesion synthesis), the product of the SOS-induced *sulA* gene acts to inhibit cell division (34). The SulA protein interacts with, and inhibits, the polymerization of the FtsZ protein (70, 109). Since the formation of the FtsZ ring at the site of septation is a crucial step in cell division, the induction of SulA by DNA damage results in a transient inhibition of division. Interestingly, the SOS-regulated locus originally identified as *dinH* (56) is the *ftsK*⁺ gene (116). FtsK is an essential cell division protein that is localized to the septum (116). Its N-terminus is required for cell division, and its C-terminus is involved in chromosomal localization within the cell (62). Increased expression of *ftsK* has been shown to confer increased resistance to DNA damage, suggesting that SOS induction of *ftsK*$^+$ is an additional mechanism that helps cells increase their resistance to DNA-damaging agents (116).

Additional Management of Replication Forks

In addition to the model of replication fork management discussed above, in which an *E. coli* cell is able to shift from DNA Pol III to the UmuD′$_2$C translesion DNA polymerase when the replication fork encounters a lesion in the template DNA, there now appears to be yet another SOS-induced DNA polymerase that is involved in the complex management of DNA replication after damage. Findings by Rangarajan et al. (86) indicate that DNA Pol II (encoded by the SOS-inducible *polB*/*dinA* gene) plays a key role in the reinitiation of replication after DNA-damaging treatments cause the replisome to pause. The rapid reinitiation of synthesis after damage, which has been termed induced replisome reactivation (43) or replication restart (18), requires DNA Pol II. After DNA Pol II restarts replication, it hands DNA synthesis duties back to DNA Pol III. However, if DNA Pol II is not present, UmuD′$_2$C can act to reinitiate replication (although less effectively than DNA Pol II). Interestingly, in the absence of both DNA pol II and UmuD′$_2$C, there is a third pathway of reinitiation that is even less efficient than UmuD′$_2$C and is not well understood.

Like DNA Pol II, the functions of proteins in the *recF* recombination pathway can also be viewed in a different light. In spite of their gene names, in an otherwise wild-type background the actual phenotypes of mutations in the *recF* or *recR* genes, while sensitizing the strains to UV light, do not impair recombination substantially. Work by Courcelle et al. (14) has shown that the primary function of the RecF and RecR proteins may be the reassembly and protection of the DNA in a replication fork that has been disrupted by DNA damage of an exogenous or endogenous nature. This protection could then allow the repair of the damage by excision repair prior to the reconstruction of the replisome and reinitiation of DNA replication (15). The level of complexity and redundancy with respect to the mechanisms employed by the cell to ensure its ability to continue replicating its chromosome in spite of DNA damage illustrates the sophisticated manner in which the cell is able to configure and then reconfigure the replication fork in response to DNA damage.

THE FUTURE

Despite the remarkable progress that has been made in understanding various mechanistic details of the SOS response of *E. coli*, many questions remain to be answered. Furthermore, we are just beginning to understand the ways that the various aspects of the SOS response alter cellular physiology after DNA damage and the ways that they act coordinately to increase the ability of a cell to survive damage to its genome. The fact that a remarkable series of advances was made even while this chapter was being written suggests that many more discoveries remain to be made, some perhaps having implications for eukaryotic biology, even for humans as well. Furthermore, the SOS response appears to be widely distributed among both gram-negative and gram-positive bacteria (26, 39, 47), so that it seems highly likely that investigations of the SOS response in other bacteria will offer insights into its importance in their individual lifestyles.

REFERENCES

1. **Anderson, D. G., and S. C. Kowalczykowski.** 1998. Reconstitution of an SOS response pathway derepression of transcription in response to DNA breaks. *Cell* **95:**975–979.
2. **Bagg, A., C. J. Kenyon, and G. C. Walker.** 1981. Inducibility of a gene product required for UV and chemical mutagenesis in *Escherichia coli. Proc. Natl. Acad. Sci. USA* **78:**5749–5753.
3. **Banerjee, S. K., R. B. Christensen, C. W. Lawrence, and J. E. LeClerc.** 1988. Frequency and spectrum of mutations produced by a single cis-syn thymine-thymine cyclobutane dimer in a single-stranded vector. *Proc. Natl. Acad. Sci. USA* **85:**8141–8145.
4. **Bates, H., and B. A. Bridges.** 1991. Mutagenic DNA repair in *Escherichia coli.* XIX. On the roles of RecA protein in ultraviolet light mutagenesis. *Biochimie* **73:**485–489.
5. **Bates, H., S. K. Randall, C. Rayssiguier, B. A. Bridges, M. F. Goodman, and M. Radman.** 1989. Spontaneous and UV-induced mutations in *Escherichia coli* K-12 strains with altered or absent DNA polymerase I. *J. Bacteriol.* **171:**2480–2484.
6. **Battista, J. R., T. Ohta, T. Nohmi, W. Sun, and G. C. Walker.** 1990. Dominant negative *umuD* mutations decreasing RecA-mediated cleavage suggest roles for intact UmuD in modulation of SOS mutagenesis. *Proc. Natl. Acad. Sci. USA* **87:**7190–7194.
7. **Boudsocq, F., M. Campbell, R. Devoret, and A. Bailone.** 1997. Quantitation of the inhibition of Hfr × F- recombination by the mutagenesis complex UmuD′C. *J. Mol. Biol.* **270:**201–211.
8. **Brent, R., and M. Ptashne.** 1981. Mechanism of action of the *lexA* gene product. *Proc. Natl. Acad. Sci. USA* **78:**4204–4208.
9. **Bridges, B. A., R. P. Mottershead, and S. G. Sedgwick.** 1976. Mutagenic DNA repair in *Escherichia coli. Mol. Gen. Genet.* **144:**53–58.
10. **Bridges, B. A., and R. Woodgate.** 1984. Mutagenic repair in *Escherichia coli*: The *umuC* gene product may be required for replication past pyrimidine dimers but not for the coding error in UV-mutagenesis. *Mol. Gen. Genet.* **196:**364–366.
11. **Bridges, B. A., and R. Woodgate.** 1985. The two-step model of bacterial UV mutagenesis. *Mutat. Res.* **150:**133–139.
12. **Brotcorne-Lannoye, A., and G. Maenhaut-Michel.** 1986. Role of RecA protein in untargeted UV mutagenesis of bacteriophage λ: evidence for the requirement for the *dinB* gene. *Proc. Natl. Acad. Sci. USA* **83:**3904–3908.

13. Burckhardt, S. E., R. Woodgate, R. H. Scheuermann, and H. Echols. 1988. UmuD mutagenesis protein of *Escherichia coli*: Overproduction, purification, and cleavage by RecA. *Proc. Natl. Acad. Sci. USA* **85**:1811–1815.

14. Courcelle, J., C. Carswell-Crumpton, and P. C. Hanawalt. 1997. *recF* and *recR* are required for the resumption of replication at DNA replication forks in *Escherichia coli*. *Proc. Natl. Acad. Sci. USA* **94**:3714–3719.

15. Courcelle, J., D. J. Crowley, and P. C. Hanawalt. 1999. Recovery of DNA replication in UV-irradiated *Escherichia coli* requires both excision repair and RecF protein function. *J. Bacteriol.* **181**:916–922.

16. Crowley, D. J., and P. C. Hanawalt. 1998. Induction of the SOS response increases the efficiency of global nucleotide excision repair of cyclobutane pyrimidine dimers, but not 6-4 photoproducts, in UV-irradiated *Escherichia coli*. *J. Bacteriol.* **180**:3345–3352.

17. Donnelly, C. E., and G. C. Walker. 1989. *groE* mutants of *Escherichia coli* are defective in *umuDC*-dependent UV mutagenesis. *J. Bacteriol.* **171**:6117–6125.

18. Echols, H., and M. F. Goodman. 1991. Fidelity mechanisms in DNA replication. *Annu. Rev. Biochem.* **60**:477–511.

19. Elledge, S. J., and G. C. Walker. 1983. Proteins required for ultraviolet light and chemical mutagenesis: identification of the products of the *umuC* locus of *Escherichia coli*. *J. Mol. Biol.* **164**:175–192.

20. Ennis, D. G., A. S. Levine, W. H. Koch, and R. Woodgate. 1995. Analysis of *recA* mutants with altered SOS functions. *Mutat. Res.* **336**:39–48.

21. Ferentz, A. E., T. Opperman, G. C. Walker, and G. Wagner. 1997. Dimerization of the UmuD′ protein in solution and its implications for regulation of SOS mutagenesis. *Nat. Struct. Biol.* **4**:979–983.

22. Fogliano, M., and P. F. Schendel. 1981. Evidence for the inducibility of the *uvrB* operon. *Nature* (London) **289**:196–198.

23. Frank, E. G., D. G. Ennis, M. Gonzalez, A. S. Levine, and R. Woodgate. 1996. Regulation of SOS mutagenesis by proteolysis. *Proc. Natl. Acad. Sci. USA* **93**:10291–10296.

24. Frank, E. G., J. Hauser, A. S. Levine, and R. Woodgate. 1993. Targeting of the UmuD, UmuD′ and MucA′ mutagenesis proteins to DNA by the RecA protein. *Proc. Natl. Acad. Sci. USA* **90**:8169–8173.

25. Friedberg, E. C., and V. L. Gerlach. 1999. Novel DNA polymerases offer clues to the molecular basis of mutagenesis. *Cell* **98**:413–416.

26. Friedberg, E. C., G. C. Walker, and W. Siede. 1995. *DNA Repair and Mutagenesis*. American Society for Microbiology, Washington, D.C.

27. Gerlach, V. L., L. Aravind, G. Gotway, R. A. Schultz, E. V. Koonin, and E. C. Friedberg. 1999. Human and mouse homologs of *E. coli* DinB (DNA polymerase IV), novel members of the UmuC/DinB superfamily. *Proc. Natl. Acad. Sci. USA* **96**:11922–11927.

28. Gonzalez, M., E. G. Frank, A. S. Levine, and R. Woodgate. 1998. Lon-mediated proteolysis of the *Escherichia coli* UmuD mutagenesis protein: in vitro degradation and identification of residues required for proteolysis. *Genes. Dev.* **12**:3889–3899.

29. Gonzalez, M., E. G. Frank, J. P. McDonald, A. S. Levine, and R. Woodgate. 1998. Structural insights into the regulation of SOS mutagenesis. *Acta Biochim. Pol.* **45**:163–172.

30. Guzzo, A., M. H. Lee, K. Oda, and G. C. Walker. 1996. Analysis of the region between amino acids 30 and 42 of intact UmuD by a monocysteine approach. *J. Bacteriol.* **178**:7295–7303.

31. Hagensee, M. E., T. L. Timme, S. Bryan, and R. E. Moses. 1987. DNA polymerase III of *Escherichia coli* is required for UV and ethyl methanesulfonate mutagenesis. *Proc. Natl. Acad. Sci. USA* **84**:4195–4199.

32. Harmon, F. G., W. M. Rehrauer, and S. C. Kowalczykowski. 1996. Interaction of *Escherichia coli* RecA protein with LexA repressor: II. Inhibition of DNA strand exchange by the uncleavable LexA S119A repressor argues that recombination and SOS induction are competitive processes. *J. Biol. Chem.* **271**:23874–23883.

33. Horii, T., T. Ogawa, T. Nakatani, T. Hase, H. Matsubara, and H. Ogawa. 1981. Regulation of SOS functions: purification of E. coli LexA protein and determination of its specific site cleaved by the RecA protein. *Cell* **27**:515–522.

34. Huisman, O., and R. D'Ari. 1981. An inducible DNA replication-cell division coupling mechanism in E. coli. *Nature* (London) **290**:797–799.

35. Humayun, M. Z. 1998. SOS and Mayday: multiple inducible mutagenic pathways in *Escherichia coli*. *Mol. Microbiol.* **30**: 905–910.

36. Iwasaki, H., A. Nakata, G. C. Walker, and H. Shinagawa. 1990. The *Escherichia coli polB* gene, which encodes DNA polymerase II, is regulated by the SOS system. *J. Bacteriol.* **172**:6268–6273.

37. Johnson, R. E., C. M. Kondratick, S. Prakash, and L. Prakash. 1999. hRAD30 mutations in the variant form of xeroderma pigmentosum. *Science* **285**:263–265.

38. Johnson, R. E., S. Prakash, and L. Prakash. 1999. Efficient bypass of a thymine-thymine dimer by yeast DNA polymerase, Polη. *Science* **283**:1001–1004.

39. Kaasch, M., J. Kaasch, and A. Quiñones. 1989. Expression of the *dnaN* and *dnaQ* genes of *Escherichia coli* is inducible by mitomycin C. *Mol. Gen. Genet.* **219**:187–192.

40. Kato, T., and Y. Shinoura. 1977. Isolation and characterization of mutants of *Escherichia coli* deficient in induction of mutations by ultraviolet light. *Mol. Gen. Genet.* **156**:121–131.

41. Kenyon, C. J., and G. C. Walker. 1980. DNA-damaging agents stimulate gene expression at specific loci in *Escherichia coli*. *Proc. Natl. Acad. Sci. USA* **77**:2819–2823.

42. Kenyon, C. J., and G. C. Walker. 1981. Expression of the *E. coli uvrA* gene is inducible. *Nature* (London) **289**:808–810.

43. Khidhir, M. A., S. Casaregola, and I. B. Holland. 1985. Mechanism of transient inhibition of DNA synthesis in ultraviolet-irradiated E. coli: inhibition is independent of *recA* whilst recovery requires RecA protein itself and an additional, inducible SOS function. *Mol. Gen. Genet.* **199**:133–140.

44. Kim, B., and J. W. Little. 1993. LexA and λ cI repressors as enzymes: specific cleavage in an intermolecular reaction. *J. Bacteriol.* **73**:1165–1173.

45. Kim, S.-R., G. Maenhaut-Michel, M. Yamada, Y. Yamamoto, K. Matsui, T. Sofuni, T. Nohmi, and H. Ohmori. 1997. Multiple pathways for SOS-mutagenesis in *Escherichia coli*: an overexpression of *dinB/dinP* results in strongly enhancing mutagenesis in the absence of any exogenous treatment to damage DNA. *Proc. Natl. Acad. Sci. USA* **94**:13792–13797.

46. Kitagawa, Y., E. Akaboshi, H. Shinagawa, T. Horii, H. Ogawa, and T. Kato. 1985. Structural analysis of the *umu* operon required for inducible mutagenesis in *Escherichia coli*. *Proc. Natl. Acad. Sci. USA* **82**:4336–4340.

47. Koch, H. K., and R. Woodgate. 1998. The SOS response, p. 107–134. *In* J. A. Nickoloff and M. F. Hoekstra (ed.), *DNA Damage and Repair*, vol. 1. *DNA Repair in Prokaryotes and Lower Eukaryotes*. Humana Press, Totowa, N.J.

48. Koffel-Schwartz, N., F. Coin, X. Veaute, and R. P. P. Fuchs. 1996. Cellular strategies for accommodating replication-hindering adducts in DNA: control by the SOS response in *Escherichia coli*. *Proc. Natl. Acad. Sci. USA* **93**:7805–7810.

49. **Konola, J. T., A. Guzzo, J.-B. Gow, G. C. Walker, and K. L. Knight.** 1998. Differential cleavage of LexA and UmuD mediated by *recA* Pro[67] mutants: implications for common LexA and UmuD binding sites on RecA. *J. Mol. Biol.* **276:**405–415.

50. **Kornberg, A., and T. A. Baker.** 1992. *DNA Replication.* W. H. Freeman and Company, New York, N.Y.

51. **Larimer, F., J. Perry, and A. Hardigree.** 1989. The *REV1* gene of *Saccharomyces cerevisiae*: isolation, sequence, and functional analysis. *J. Bacteriol.* **171:**230–237.

52. **Lawrence, C. W., A. Borden, S. K. Banerjee, and J. E. LeClerc.** 1990. Mutation frequency and spectrum resulting from a single abasic site in a single-stranded vector. *Nucleic Acids Res.* **18:**2153–2157.

53. **LeClerc, J. E., A. Borden, and C. W. Lawrence.** 1991. The thymine-thymine pyrimidine-pyrimidone (6-4) ultraviolet light photoproduct is highly mutagenic and specifically induces 3′ thymine-to-cytosine transitions in *Escherichia coli. Proc. Natl. Acad. Sci. USA* **88:**9685–9689.

54. **Lee, M. H., T. Ohta, and G. C. Walker.** 1994. A monocysteine approach for probing the structure and interactions of the UmuD protein. *J. Bacteriol.* **176:**4825–4837.

55. **Lemontt, J. F.** 1971. Mutants of yeast defective in mutation induced by ultraviolet light. *Genetics* **68:**21–33.

56. **Lewis, L. K., M. E. Jenkins, and D. W. Mount.** 1992. Isolation of DNA-damage inducible promoters in *E. coli*: regulation of *polB* (*dinA*), *dinG*, and *dinH* by LexA repressor. *J. Bacteriol.* **174:**3377–3385.

57. **Little, J. W.** 1983. The SOS regulatory system: control of its state by the level of *recA* protease. *J. Mol. Biol.* **167:**791–808.

58. **Little, J. W.** 1984. Autodigestion of *lexA* and phage λ repressors. *Proc. Natl. Acad. Sci. USA* **81:**1375–1379.

59. **Little, J. W.** 1993. LexA cleavage and other self-processing reactions. *J. Bacteriol.* **175:**4943–4950.

60. **Little, J. W., and D. W. Mount.** 1982. The SOS regulatory system of *Escherichia coli. Cell* **29:**11–22.

61. **Little, J. W., D. W. Mount, and C. R. Yanisch-Perron.** 1981. Purified *lexA* protein is a repressor of the *recA* and *lexA* genes. *Proc. Natl. Acad. Sci. USA* **78:**4199–4203.

62. **Liu, G., G. C. Draper, and W. D. Donachie.** 1998. FtsK is a bifunctional protein involved in cell division and chromosomal localization. *Mol. Microbiol.* **29:**893–903.

63. **Masutani, C., M. Araki, A. Yamada, R. Kusumoto, T. Nogimori, T. Maekawa, S. Iwai, and F. Hanaoka.** 1999. Xeroderma pigmentosum variant (XP-V) correcting protein from HeLa cells has a thymine dimer bypass DNA polymerase activity. *EMBO J.* **18:**3491–3501.

64. **Masutani, C., R. Kusumoto, A. Yamada, N. Dohmae, M. Yokoi, M. Yuasa, M. Araki, S. Iwai, K. Takio, and F. Hanaoka.** 1999. The XPV (xeroderma pigmentosum variant) gene encodes human DNA polymerase η. *Nature* (London) **399:**700–704.

65. **McDonald, J. P., E. G. Frank, A. S. Levine, and R. Woodgate.** 1998. Intermolecular cleavage by UmuD-like mutagenesis proteins. *Proc. Natl. Acad. Sci. USA* **95:**1478–1483.

66. **McDonald, J. P., A. S. Levine, and R. Woodgate.** 1997. The *Saccharomyces cerevisiae Rad30* gene, an homologue of *Escherichia coli dinB* and *umuC*, is DNA damage inducible and functions in a novel error-free postreplication repair mechanism. *Genetics* **147:**1557–1568.

67. **McDonald, J. P., E. E. Maury, A. S. Levine, and R. Woodgate.** 1998. Regulation of UmuD cleavage: role of the aminoterminal tail. *J. Mol. Biol.* **282:**721–730.

68. **McDonald, J. P., V. Rapic-Otrin, J. A. Epstein, B. C. Broughton, X. Wang, A. R. Lehmann, D. J. Wolgemuth, and R. Woodgate.** 1999. Novel human and mouse homologs of *Sac-*

charomyces cerevisiae DNA polymerase eta. *Genomics* **60:**20–30.

69. **Mellon, I., and P. C. Hanawalt.** 1989. Induction of the *Escherichia coli* lactose operon selectively increases repair of its transcribed DNA strand. *Nature* (London) **342:**95–98.

70. **Mukherjee, A., C. Cao, and J. Lutkenhaus.** 1998. Inhibition of FtsZ polymerization by SulA, an inhibitor of septation in *Escherichia coli. Proc. Natl. Acad. Sci. USA* **95:**2885–2890.

71. **Napolitano, R. L., I. B. Lambert, and R. P. P. Fuchs.** 1997. SOS factors involved in translesion synthesis. *Proc. Natl. Acad. Sci. USA* **94:**5733–5738.

72. **Nastri, H. G., A. Guzzo, C. Lange, G. C. Walker, and K. L. Knight.** 1997. Mutational analysis of the RecA protein L1 region identifies this area as a probable part of the co-protease substrate binding site. *Mol. Microbiol.* **25:**967–978.

73. **Nelson, J. R., C. W. Lawrence, and D. C. Hinkle.** 1996. Deoxycytidyl transferase activity of yeast *REV1* protein. *Nature* (London) **382:**729–731.

74. **Nohmi, T., J. R. Battista, L. A. Dodson, and G. C. Walker.** 1988. RecA-mediated cleavage activates UmuD for mutagenesis: mechanistic relationship between transcriptional derepression and posttranslational activation. *Proc. Natl. Acad. Sci. USA* **85:**1816–1820.

75. **Ohmori, H., E. Hatada, Y. Qiao, M. Tsuji, and R. Fukuda.** 1995. *dinP*, a new gene in *E. coli* whose product shows similarities to UmuC and its homologues. *Mutat. Res.* **347:**1–7.

76. **Ohta, T., M. D. Sutton, A. Guzzo, S. Cole, A. E. Ferentz, and G. C. Walker.** 1999. Mutations affecting the ability of the *Escherichia coli* UmuD′ protein to participate in SOS mutagenesis. *J. Bacteriol.* **181:**177–185.

77. **Opperman, T., S. Murli, B. T. Smith, and G. C. Walker.** 1999. A model for a *umuDC*-dependent prokaryotic DNA damage checkpoint. *Proc. Natl. Acad. Sci. USA* **96:**9218–9223.

78. **Paetzel, M., R. E. Dalbey, and N. C. J. Strynadka.** 1998. Crystal structure of a bacterial signal peptidase in complex with a β-lactam inhibitor. *Nature* (London) **396:**186–190.

79. **Paetzel, M., N. C. J. Strynadka, W. R. Tschantz, R. Casareno, P. R. Bullinger, and R. E. Dalbey.** 1997. Use of site-directed chemical modification to study an essential lysine in *Escherichia coli* leader peptidase. *J. Biol. Chem.* **272:**9994–10003.

80. **Peat, T. S., E. G. Frank, J. P. McDonald, A. S. Levine, R. Woodgate, and W. A. Hendrickson.** 1996. Structure of the UmuD′ protein and its regulation in response to DNA damage. *Nature* (London) **380:**727–730.

81. **Peat, T. S., E. G. Frank, J. P. McDonald, A. S. Levine, R. Woodgate, and W. A. Hendrickson.** 1996. The UmuD′ protein filament and its potential role in damage induced mutagenesis. *Structure* **4:**1401–1412.

82. **Perry, K. L., S. J. Elledge, B. Mitchell, L. Marsh, and G. C. Walker.** 1985. *umuDC* and *mucAB* operons whose products are required for UV light- and chemical-induced mutagenesis: UmuD, MucA, and LexA proteins share homology. *Proc. Natl. Acad. Sci. USA* **82:**4331–4335.

83. **Petit, M.-A., W. Bedale, J. Osipiuk, C. Lu, M. Rajagopalan, P. McInerney, M. F. Goodman, and H. Echols.** 1994. Sequential folding of UmuC by the Hsp70 and Hsp60 chaperone complexes of *Escherichia coli. J. Biol. Chem.* **269:**23824–23829.

84. **Radman, M.** 1974. Phenomenology of an inducible mutagenic DNA repair pathway in *Escherichia coli*: SOS repair hypothesis, p. 128–142. *In* L. Prakash, F. Sherman, M. Miller, C. Lawrence, and H. W. Tabor (ed.), *Molecular and Environmental Aspects of Mutagenesis.* Charles C Thomas, Springfield, Ill.

85. **Rajagopalan, M., C. Lu, R. Woodgate, M. O'Donnell, M. F. Goodman, and H. Echols.** 1992. Activity of the purified mu-

tagenesis proteins UmuC, UmuD', and RecA in replicative bypass of an abasic DNA lesion by DNA polymerase III. *Proc. Natl. Acad. Sci. USA* **89:**10777–10781.

86. **Rangarajan, S., R. Woodgate, and M. F. Goodman.** 1999. A phenotype for enigmatic DNA polymerase II: a pivotal role for Pol II in replication restart in UV-irradiated *Escherichia coli. Proc. Natl. Acad. Sci. USA* **96:**9224–9229.

87. **Rehrauer, W. M., I. Bruck, R. Woodgate, M. F. Goodman, and S. C. Kowalczykowski.** 1998. Modulation of RecA nucleoprotein function by the mutagenic UmuD'C protein complex. *J. Biol. Chem.* **273:**32384–32387.

88. **Rehrauer, W. M., P. E. Lavery, E. L. Palmer, R. N. Singh, and S. C. Kowalczykowski.** 1996. Interaction of *Escherichia coli* RecA protein with LexA repressor: I. LexA repressor cleavage is competitive with binding of a secondary DNA molecule. *J. Biol. Chem.* **271:**23865–23873.

89. **Reuven, N. B., G. Arad, A. Maor-Shoshani, and Z. Livneh.** 1999. The mutagenesis protein UmuC is a DNA polymerase activated by UmuD', RecA, and SSB and specialized for translesion synthesis. *J. Biol. Chem.* **274:**31763–31766.

90. **Reuven, N. B., G. Tomer, and Z. Livneh.** 1998. The mutagenesis proteins UmuD' and UmuC prevent lethal frameshifts while increasing base substitution mutations. *Mol. Cell* **2:**191–199.

91. **Roberts, J. W., and C. W. Roberts.** 1975. Proteolytic cleavage of bacteriophage lambda repressor in induction. *Proc. Natl. Acad. Sci. USA* **72:**147–151.

92. **Roush, A. A., M. Suarez, E. C. Friedberg, M. Radman, and W. Siede.** 1998. Deletion of the *Saccharomyces cerevisiae* gene *RAD30* encoding an *Escherichia coli* DinB homolog confers radiation sensitivity and altered mobility. *Mol. Gen. Genet.* **257:**686–692.

93. **Sancar, A.** 1996. DNA excision repair. *Annu. Rev. Biochem.* **65:**43–81.

94. **Sassanfar, M., and J. W. Roberts.** 1990. Nature of the SOS-inducing signal in *Escherichia coli*: the involvement of DNA replication. *J. Mol. Biol.* **212:**79–96.

95. **Shinagawa, H., H. Iwasaki, T. Kato, and A. Nakata.** 1988. RecA protein-dependent cleavage of UmuD protein and SOS mutagenesis. *Proc. Natl. Acad. Sci. USA* **85:**1806–1810.

96. **Shinagawa, H., T. Kato, T. Ise, K. Makino, and A. Nakata.** 1983. Cloning and characterization of the *umu* operon responsible for inducible mutagenesis in *Escherichia coli. Gene* **23:**167–174.

97. **Siegel, E. C.** 1983. The *Escherichia coli uvrD* gene is inducible by DNA damage. *Mol. Gen. Genet.* **191:**397–400.

98. **Slater, S. C., and R. Maurer.** 1991. Requirement for bypass of UV-induced lesions in single-stranded DNA of bacteriophage ϕX174 in *Salmonella typhimurium. Proc. Natl. Acad. Sci. USA* **88:**1251–1255.

99. **Slilaty, S. N., and J. W. Little.** 1987. Lysine-156 and serine-119 are required for LexA repressor cleavage: a possible mechanism. *Proc. Natl. Acad. Sci. USA* **84:**3987–3991.

100. **Smith, B. T., and G. C. Walker.** 1998. Mutagenesis and more: *umuDC* and the *Escherichia coli* SOS response. *Genetics* **148:**1599–1610.

101. **Sommer, S., A. Bailone, and R. Devoret.** 1993. The appearance of the UmuD'C protein complex in *Escherichia coli* switches repair from homologous recombination to SOS mutagenesis. *Mol. Microbiol.* **10:**963–971.

102. **Sommer, S., F. Boudsocq, R. Devoret, and A. Bailone.** 1998. Specific RecA amino acid changes affect RecA-UmuD'C interaction. *Mol. Microbiol.* **28:**281–291.

103. **Steinborn, G.** 1978. Uvm mutants of *Escherichia coli* K12 deficient in UV mutagenesis. I. Isolation of *uvm* mutants and

their phenotypical characterization in DNA repair and mutagenesis. *Mol. Gen. Genet.* **165:**87–93.

104. **Strynadka, N. C., S. E. Jensen, K. Johns, H. Blanchard, M. Page, A. Matagne, J.-M. Frère, and M. N. G. James.** 1994. Structural and kinetic characterization of a beta-lactamase inhibitor protein. *Nature* (London) **368:**657–660.

105. **Strynadka, N. C. J., H. Adachi, S. E. Jensen, K. Johns, A. Sielecki, C. Betzel, K. Sutoh, and M. N. G. James.** 1992. Molecular structure of the acyl-enzyme intermediate in β-lactam hydrolysis at 1.7 Angstrom resolution. *Nature* (London) **359:**700–705.

106. **Sutton, M. D., T. Opperman, and G. C. Walker.** 1999. The *Escherichia coli* SOS mutagenesis proteins UmuD and UmuD' interact physically with the replicative DNA polymerase. *Proc. Natl. Acad. Sci. USA* **96:**12373–12378.

107. **Tang, M., I. Bruck, R. Eritja, J. Turner, E. G. Frank, R. Woodgate, M. O'Donnell, and M. F. Goodman.** 1998. Biochemical basis of SOS-mutagenesis in *Escherichia coli*: reconstitution of in vitro lesion bypass dependent on the UmuD'$_2$C mutagenic complex and RecA protein. *Proc. Natl. Acad. Sci. USA* **95:**9755–9760.

108. **Tang, M., X. Shen, E. G. Frank, M. O'Donnell, R. Woodgate, and M. F. Goodman.** 1999. UmuD'$_2$C is an error-prone DNA polymerase, *Escherichia coli* Pol V. *Proc. Natl. Acad. Sci. USA* **96:**8919–8924.

109. **Trusca, D., S. Scott, C. Thompson, and D. Bramhill.** 1998. Bacterial SOS checkpoint protein SulA inhibits polymerization of purified FtsZ cell division protein. *J. Bacteriol.* **180:**3946–3953.

110. **van Dijl, J. M., A. de Jong, G. Venema, and S. Bron.** 1995. Identification of the potential active site of the signal peptidase SipS of *Bacillus subtilis. J. Bacteriol.* **270:**3611–3618.

111. **Wagner, J., P. Gruz, S.-R. Kim, M. Yamada, K. Matsui, R. P. P. Fuchs, and T. Nohmi.** 1999. The *dinB* gene encodes a novel *E. coli* DNA polymerase, DNA Pol IV, involved in mutagenesis. *Mol. Cell* **4:**281–286.

112. **Walker, G. C.** 1984. Mutagenesis and inducible responses to deoxyribonucleic acid damage in *Escherichia coli. Microbiol. Rev.* **48:**60–93.

113. **Walker, G. C.** 1985. Inducible DNA repair systems. *Annu. Rev. Biochem.* **54:**425–457.

114. **Walker, G. C.** 1996. The SOS response of *Escherichia coli*, p. 1400–1416. *In* F. C. Neidhardt, R. Curtiss III, J. L. Ingraham, E. C. C. Lin, K. B. Low, B. Magasanik, W. S. Reznikoff, M. Riley, M. Schaechter and H. E. Umbarger (ed.), *Escherichia coli and Salmonella: Cellular and Molecular Biology,* 2nd ed. ASM Press, Washington, D.C.

115. **Walker, G. C.** 1998. Skiing the black diamond slope: progress on the biochemistry of translesion synthesis. *Proc. Natl. Acad. Sci. USA* **95:**10348–10350.

116. **Wang, L., and J. Lutkenhaus.** 1998. FtsK is an essential cell division protein that is localized to the septum and induced as part of the the SOS response. *Mol. Microbiol.* **29:**731–740.

117. **Witkin, E. M.** 1967. The radiation sensitivity of *Escherichia coli* B: a hypothesis relating filament formation and prophage induction. *Proc. Natl. Acad. Sci. USA* **57:**1275–1279.

118. **Witkin, E. M.** 1976. Ultraviolet mutagenesis and inducible DNA repair in *Escherichia coli. Bacteriol. Rev.* **40:**869–907.

119. **Wood, R. D., and F. Hutchinson.** 1984. Non-targeted mutagenesis in unirradiated lambda phage in *Escherichia coli* host cells irradiated with ultraviolet light. *J. Mol. Biol.* **173:**293–305.

120. **Woodgate, R.** 1999. A plethora of lesion-replicating polymerases. *Genes Dev.* **13:**2191–2195.

121. **Woodgate, R., and A. S. Levine.** 1996. Damage inducible mutagenesis: recent insights into the activities of the Umu family of mutagenesis proteins. *Cancer Surveys* **28:**117–140.

122. **Woodgate, R., M. Rajagopalan, C. Lu, and H. Echols.** 1989. UmuC mutagenesis protein of *Escherichia coli:* purification and interaction with UmuD and UmuD'. *Proc. Natl. Acad. Sci. USA* **86:**7301–7305.

123. **Yasuda, T., K. Morimatsu, T. Horii, T. Nagata, and H. Ohmori.** 1998. Inhibition of *Escherichia coli* RecA coprotease activities by DinI. *EMBO J.* **17:**3207–3216.

124. **Yasuda, T., T. Nagata, and H. Ohmori.** 1996. Multicopy suppressors of the cold-sensitive phenotype of the *pcsA68* (*dinD68*) mutation in *Escherichia coli. J. Bacteriol.* **178:** 3854–3859.

125. **Yu, X., and E. H. Egelman.** 1993. The LexA repressor binds within the deep helical groove of the activated RecA filament. *J. Mol. Biol.* **231:**29–40.

Bacterial Stress Responses
Edited by G. Storz and R. Hengge-Aronis
©2000 ASM Press, Washington, D.C.

Chapter 10

Metalloregulatory Systems at the Interface between Bacterial Metal Homeostasis and Resistance

F. Wayne Outten, Caryn E. Outten, and Thomas V. O'Halloran

Many of the transition elements function as essential cofactors in metabolic pathways and are required for microbial growth. Yet the chemical properties of these metal ions can also lead to harmful side effects, including enzyme inhibition, biopolymer hydrolysis, and uncontrolled redox cycling within the cell. A wide range of stress-responsive genes are induced as metal ion concentrations increase from starvation to toxic levels. Metal-responsive regulatory proteins mediate the dilemma presented by transition metals by maintaining a balance between expression of nutrient uptake and sequestration systems on one hand and efflux systems on the other. Here we discuss several families of these metalloregulatory proteins and their associated metal stress response systems to illustrate a few transcriptional and homeostasis mechanisms common to metal stress.

Bacteria thrive in a multitude of environments and are exposed to a vast array of stresses. A common stress faced by bacteria is nutrient limitation. In complex communities of the natural environment, bacteria constantly struggle to acquire those nutrients that are essential to metabolism and reproduction. One subgroup of essential nutrients consists of transition metals such as iron (Fe), copper (Cu), zinc (Zn), nickel (Ni), and molybdenum (Mo). These elements play important roles in the proper metabolic functioning of the bacterial cell, and without certain minimal levels many necessary biochemical activities would cease. Complex and sometimes redundant metal uptake/sequestration systems are known in most organisms. The evolutionary pressure to develop and maintain such systems most likely derives from the fact that transition metal concentrations are often low in the natural environment and are thus growth-limiting nutrients in many situations (14).

The deleterious effects that occur under conditions of elevated metal concentration, however, present a significant dilemma. Redox-active metals such as iron and copper can create harmful byproducts through uncontrolled redox chemistry (54, 94). Metals can also stimulate hydrolytic reactions that degrade biopolymers. Excess levels of one metal can lead to competition with other metals during cellular transport or competition for specific sites in metal-utilizing enzymes (40, 72). Depending on the richness of the growth media, the window of optimal metal concentration in the medium can be quite narrow. Eukaryotic cells carefully maintain the number of transition metal ions per cell within an exceedingly narrow range to ensure key enzyme function without initiating cell damage (74, 96, 97). It is likely that bacteria adopt a similar strategy.

Defining the minimum and maximum levels of transition metals required by bacteria remains a key goal in metal homeostasis studies. Efforts to establish metal starvation and metal toxicity benchmarks have been complicated by changes in metal bioavailability in different media and by strain-dependent differences in response to metal stress. Some experiments have begun to examine this problem for specific metals, media, and strains (for example, see reference 35), but no systematic effort has addressed this issue.

This chapter focuses on a few of the known prokaryotic responses to transition metal stress, emphasizing themes that are common across diverse groups of organisms. Additional systems are known (for reviews see references 57 and 88) but not discussed here. Several stress response systems (listed in Table

F. Wayne Outten • Department of Biochemistry, Molecular Biology, and Cell Biology, Northwestern University, 2145 Sheridan Rd., Evanston, IL 60208. Caryn E. Outten • Department of Chemistry, Northwestern University, 2145 Sheridan Rd., Evanston, IL 60208. Thomas V. O'Halloran • Department of Chemistry and Department of Biochemistry, Molecular Biology, and Cell Biology, Northwestern University, 2145 Sheridan Rd., Evanston, IL 60208.

1) have been characterized at the genetic and biochemical levels. The expression of each of these systems is controlled by metal-sensing transcription factors (see Table 2) known as metalloregulatory proteins (60, 61). These systems were chosen to illustrate a few broad concepts and mechanisms that in some cases apply to a wide variety of stresses.

THE MerR FAMILY OF METAL STRESS RESPONSE REGULATORS

The mercury (Hg) resistance operon, mer, represents one paradigm for metal stress response in bacteria. The metal-responsive regulator of the mer operon, MerR, provides an example of a mechanism in which a single protein senses changes in metal ion concentration and directly alters gene expression accordingly. MerR represses transcription in the absence of Hg(II) and activates transcription when Hg(II) concentrations exceed a specific threshold (7, 12, 19, 27, 49, 60, 63, 76, 79, 98). Both events are accomplished while MerR is bound at a single site between the −10 and −35 promoter elements. Mechanisms for repression and metal-responsive activation that invoke protein-protein interactions between MerR and RNA polymerase (RNAP) and protein-induced distortions in DNA have been proposed.

In the DNA distortion mechanism, the key role of MerR is to alter the structure of the DNA, switching it from a very poor to a good substrate for RNAP. Promoters under the control of MerR have a suboptimal −10/−35 spacer region that is 2 bp longer than the consensus 17 bp (48). This places the −10 and −35 RNAP-binding sites in nonoptimal positions on the face of the DNA helix. In the absence of Hg, apo-MerR binds between the −10 and −35 elements and bends the DNA toward itself, producing two kinks in the DNA strand (5). This distortion forces the promoter region into an even less optimal conformation

that further represses transcription. Upon binding Hg(II), MerR relaxes these bends and also underwinds the DNA (5, 6). This underwinding event can bring the −10 and −35 elements closer to an optimal orientation for effective interaction with RNAP. This stabilization in turn leads to open complex formation and transcriptional activation (Fig. 1) of the Hg(II) detoxification system (5, 27). Allosteric control mutations in MerR that lead to transcriptional activation in the absence of Hg(II) (79) also exhibit the same types of DNA distortions in the absence of Hg(II), providing strong support for a role of distortion in altering transcription (65, 66).

Recent results indicate that RNAP open complex formation involves wrapping of the DNA strand around RNAP to form a left-handed superhelix (4, 21, 22, 78). The MerR DNA distortion mechanism conforms nicely with this model. MerR binds between the RNAP-induced bends upstream of −35 and the transcription bubble at the start site. Apo-MerR may bend the DNA away from RNAP, preventing DNA wrapping. The unbending and unwinding induced by Hg-MerR may then assist RNAP in the DNA wrapping process, as has been suggested for the homologous Zn-ZntR system described below (64).

An unusual feature of the MerR system is the apparently cooperative response of the switch mechanism to Hg(II) concentrations. Unlike most receptors following a Michaelis-Menten response, the activity of RNAP at the P_T promoter increases from 10% to 90% over a narrow change in Hg(II) concentration (62, 76, 81). This attribute of the system leads to rapid activation of the Hg(II) detoxification genes only after Hg(II) passes a critical threshold value. This response has many similarities to the square wave response of a switching device in electronic circuits, namely a transistor (62). While the mechanistic basis of this sensitivity phenomenon is not yet clear, it is likely a consequence of the ability of the repressor complex of MerR to sequester RNAP in the

Table 1. Prokaryotic metal-stress response systems

Target metal	Gene or operon	Organism	Stress response mechanism
Hg	mer	P. aeruginosa, etc.	Reduction of toxic Hg(II) to volatile Hg(0)
Cu	cop	P. syringae	Copper efflux
	pco	E. coli	Copper efflux
	cop	E. hirae	Copper efflux and uptake
Zn	znt	E. coli	Efflux of excess zinc
	znu	E. coli	Uptake of zinc under limiting conditions
Cd	czc	R. eutrophus	Efflux of excess cadmium
	cadA	S. aureus	Efflux of excess cadmium
Co	czc	R. eutrophus	Efflux of excess cobalt
Mo	mod	E. coli	Uptake of molybdate under limiting conditions
As	ars	E. coli	Reduction of metalloid followed by efflux

Table 2. Prokaryotic metalloregulatory proteins[a]

Family	Resistance to nonessential metals	Detoxification of essential metals	Metal acquisition/uptake
MerR	MerR (*P. aeruginosa*, etc.), Hg	ZntR (*E. coli*), Zn; PMTR (*P. mirabilis*), Zn; CoaR (*Synechocystis*), Co	
Two-component	SilRS (*Salmonella*), Ag; CzcRS (*Ralstonia*), Cd	CusRS (*E. coli*), Cu; PcoRS (*E. coli*), Cu; CopRS (*P. syringae*), Cu; CzcRS (*Ralstonia*), Zn, Co; LcoRS (*L. lactis*), Cu	
ArsR	ArsR (*E. coli*), As, Sb; CadC (*S. aureus*), Pb, Cd	ZntR (*S. aureus*), Co, Zn; CadC (*S. aureus*), Zn; CzrA (*S. aureus*), Zn	
Fur			Fur (*E. coli*, etc.), Fe; Zur (*E. coli*), Zn
Other regulators	CzcDI (*Ralstonia*), Cd	CopY (*E. hirae*), Cu; CzcDI (*Ralstonia*), Zn, Co	DtxR (*C. diphtheriae*), Fe; IdeR (*Mycobacterium*), Fe; SirR (*Staphylococcus*), Fe; CopY (*E. hirae*), Cu; ModE (*E. coli*, etc.), Mo

[a] Underlined systems are those in which a direct role in regulation is not yet clearly established.

closed complex. This leaves Hg(II) binding to MerR as the only diffusional step in the activation process (62, 76).

The proposed RNAP/apo-MerR complex is based on DNase I footprinting experiments that indicate that RNAP contacts regions upstream of the −10 element (5, 27). The stability of the closed RNAP/P_T/MerR complex could arise from MerR-induced distortions in DNA or from protein-protein interactions. It has been suggested that MerR forms stable contacts with RNAP in this poised state (19, 44), although it is difficult to rule out the possibility that distortions of the repressor-DNA complex lead RNAP to bind more tightly to this promoter. Little other evidence supporting a role for protein-protein interactions between RNAP and the repressor or activator forms of MerR is available.

The signal for activation is the binding of one Hg(II) per MerR dimer (63). The protein apparently discriminates against other cations by employing a trigonal, cysteine-thiolate coordination environment (101, 105). The metal binding site is most likely at the dimer interface, employing Cys residues from both subunits (37). Recent studies suggest that the core metal binding domain includes residues 80 to 128 located in the C-terminal portion of the protein (112).

In contrast to other resistance systems discussed below, the *mer* operon gene products chemically reduce their target metal. MerT, MerP, and MerC transport mercury into the cell. The cytosolic oxidoreductase, MerA, then reduces mercury from the highly toxic Hg(II) form to the volatile Hg(0) oxidation state. Hg(0) readily diffuses from the cell, thereby protecting cellular constituents from damage (89, 104).

MerR HOMOLOGS

Several MerR-like proteins that are specific for metals other than Hg have been identified (see Table 3). One such homolog, the *Escherichia coli* ZntR protein, mediates cellular responses to elevated Zn(II) levels. ZntR regulates transcription of the gene encoding a Zn(II)-exporting P-type ATPase, ZntA (10). DNase I and Cu-phenanthroline footprinting experiments show that ZntR causes the same DNA distortions that are seen with MerR (64). Recently, a cobalt-responsive homolog, CoaR, was identified in *Synechocystis* PCC 6803. CoaR was shown to regulate a P-type ATPase, CoaT, which is thought to efflux cobalt. Initial experiments seem to indicate that CoaR may also utilize DNA distortion mechanisms similar to those of MerR and ZntR (83).

Many stresses are mediated by other MerR homologs (Table 3). For example, SoxR, a superoxide sensor in *E. coli*, contains a pair of 2Fe-2S centers per dimer (see chapter 4). These centers are redox active, and the particular redox state of the 2Fe-2S centers regulates the ability of SoxR to activate transcription (25). SoxR is also activated by elevated concentrations of copper in the medium, but this is believed to occur indirectly through the generation of superoxide by copper (43). Another stress response regulator, BmrR, from *Bacillus subtilis*, contains an N-terminal domain homologous to the DNA-binding domain of MerR that has been characterized as a helix-turn-helix protein by X-ray crystallography (115). BmrR regulates multidrug efflux transporters and is induced by toxic compounds such as rhodamine (2). Similarly, TipAL, in *Streptomyces lividans*, is induced by the antibiotic thiostrepton and mediates antibiotic resistance (17). As these examples illustrate,

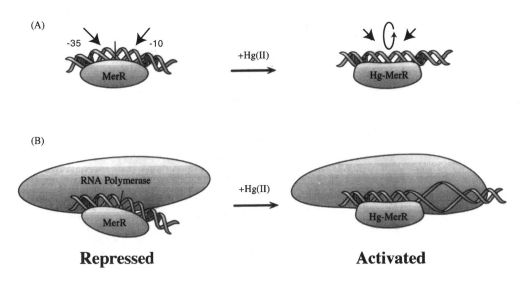

Figure 1. Mechanism of MerR-mediated repression/activation. (A) Apo-MerR bends DNA toward itself, producing two kinks (straight arrows) symmetrically spaced around the center of the operator (vertical line). Upon binding Hg(II), the bends are relaxed and the center of the operator is unwound (circular arrow). (B) Repression of transcription by apo-MerR followed by transcriptional activation by binding of Hg(II). See text for more detailed discussion.

the MerR stress-sensing mechanism has been adapted by numerous organisms to respond to stresses ranging from heavy metals to antibiotics.

TWO-COMPONENT FAMILY OF METAL-RESPONSIVE REGULATORS

Bacteria utilize two-component signal transduction systems for metal-responsive trancriptional control of a number of operons. Two-component signal transduction systems are common and facilitate responses to a variety of environmental stimuli such as changes in osmolarity and nitrogen limitation (39, 75, 82). While two-component systems can come in a variety of different forms, the typical two-component regulatory circuit consists of a sensor protein (S) and a receiver protein (R). The sensor is

usually membrane-bound and contains a highly conserved histidine kinase cytosolic domain at the C-terminus. Upon induction by the stimulus, the sensor protein autophosphorylates a conserved histidyl residue. The phosphoryl group is then transferred to an aspartyl residue on the cytosolic receiver protein (38). The activated receiver protein then goes on to effect transcription by binding to target promoters. In most cases the activated receiver protein activates transcription at the target promoter (58).

The *pco* (plasmid-encoded copper resistance) operon of *E. coli* (99, 106) and the *cop* copper resistance operon of *Pseudomonas syringae* (20) are regulated by two-component, metal-sensing systems. CopR and CopS form the regulatory circuit for the resistance operon *cop* (52) while PcoR and PcoS regulate the *pco* operon (11). A "copper box" binding sequence can be found just upstream of the consensus

Table 3. MerR homologs

Protein	Organism	% Identity (% of amino acids)[a]	Stress
MerR	*P. aeruginosa*	100 (144)	Hg(II)
ZntR	*E. coli*	34 (104)	Zn(II)
CoaR	*Synechocystis* PCC 6803	33 (72)	Co(II)
TipAL	*S. lividans*	32 (104)	Antibiotics
Mta	*B. subtilis*	32 (61)	Multidrug[b]
PMTR	*P. mirabilis*	30 (121)	Zn(II)
SoxR	*E. coli*	29 (125)	Superoxide
BmrR	*B. subtilis*	28 (75)	Multidrug

[a] Values were obtained by using the BLAST local alignment tool (NCBI) to compare homologs to the MerR sequence from Tn*501*.
[b] Underlining indicates that the specific inducer for the protein is still uncertain.

promoter elements of P_{copA} in the *cop* copper resistance plasmid. CopR, a close homolog of PcoR, has been shown to bind the *cop* copper box sequence (53). CopR activates transcription of the *copA*, *copB*, *copC*, and *copD* genes, which lie downstream of P_{copA} (50, 51). A similar sequence has been identified upstream of the weak consensus −35 element of the P_{pcoA} promoter (80). The *pcoRS* genes have been shown to be essential for copper-responsive transcription of the *pco* operon (11, 80). PcoS is thought to sense copper levels via a periplasmic loop between its two transmembrane domains. When Cu levels reach a critical level, PcoS presumably autophosphorylates, which leads to phosphoryl transfer to PcoR. PcoR may then activate transcription of the PcoABCD polycistron by binding to the "copper box" sequence.

Both *cop* and *pco* operons contain a core copper detoxification system consisting of four genes (*copABCD* in *Pseudomonas* and *pcoABCD* in *E. coli*). The four proteins of the *pco* operon (PcoABCD) share significant homology with their counterparts in the *cop* operon (76, 55, 60, and 38%, respectively) (11). Despite this homology, phenotypic differences have been observed between the two operons. For example, while null mutations in *copC* and *copD* still allow partial copper resistance (50), similar mutations in *pcoC* and *pcoD* completely abolish copper resistance (11). Another difference between the two systems is the presence of a seventh gene, *pcoE*, in the *pco* operon that is not found as part of the cloned *cop* operon. The mechanisms by which the *pco* and *cop* systems detoxify copper remain unclear.

On the basis of sequence homology between the *pco* and *cop* operons, PcoB and PcoD are proposed to be outer and inner membrane proteins, respectively (15). PcoA is a periplasmic protein with similarity to multicopper oxidases and has been shown to have oxidase activity (Huffman and O'Halloran, unpublished results). PcoC and PcoE are also periplasmic proteins. PcoC shares homology with CopC while PcoE is homologous to the silver resistance protein SilE (34). PcoC has recently been shown to bind one copper per monomer, in either the Cu(I) or Cu(II) oxidation state (Huffman et al., unpublished data).

Each gene in the *pcoABCD* polycistron has been shown to be essential for copper resistance (11). In contrast, no absolute requirement for *pcoE* in copper resistance was observed. When *pcoE* null mutants are challenged with copper stress, however, a marked lag phase is observed before growth resumes (Outten and O'Halloran, unpublished results). This suggests that PcoE may facilitate recovery from copper stress in stationary phase. Given that PcoE has 10 histidine residues spaced at regular intervals, it is likely to bind several copper atoms and may serve to decrease the concentration of free copper in the periplasm. PcoE may function to sequester copper from deleterious side reactions in the periplasm, thus providing an initial buffer against copper toxicity. It may also traffic copper between other resistance proteins. Expression of PcoE in the absence of PcoABCD leads to copper accumulation (55), which supports a role involving copper binding. This result also indicates that PcoE may require PcoABCD for proper function since the abnormal copper accumulation phenotype is not observed when PcoABCD are present (55). PcoE may additionally function as a copper chaperone, which facilitates copper transfer to an outer membrane protein for export; however, the exact role of PcoE remains to be established.

PcoE is under the regulatory control of a separate promoter, P_{pcoE}. Mutants that lack PcoRS still show normal, copper-responsive regulation of P_{pcoE} (Munson et al., unpublished data; 80). This led to the hypothesis that an unidentified chromosomal system was responsible for regulation of the plasmid-borne gene *pcoE* (80). A genetic screen has recently revealed a chromosomally encoded two-component signal transduction system, *cusRS* (for Cu-sensing) that regulates P_{pcoE} (Munson et al., unpublished data). Mutations in *cusRS* show loss of *pcoE* regulation but do not dramatically effect normal regulation of the other *pco* genes. It is unclear why PcoE is under the control of a separate, chromosomal regulatory circuit. This cross regulation may provide a means to link copper resistance with copper homeostasis.

The possibility of a wider role for CusRS in copper metabolism is supported by the identification of a chromosomal locus regulated by CusRS. Just upstream of CusR lies a divergent promoter and a series of three open reading frames designated CusCBA. The P_{cusC} promoter is copper-responsive in a CusRS-dependent manner (Munson et al., unpublished data). Several lines of evidence lead us to propose that CusCBA together form a tripartite chemiosmotic H^+/Cu antiporter complex.

CusC belongs to the outer membrane factor (OMF) family of transport proteins (68). The amino acid sequence of CusC contains a lipid attachment motif that is found in lipoproteins, such as NlpE, which are localized to the outer membrane (36, 91, 103). CusC may associate with a porin in the outer membrane or may function as a channel through the outer membrane, as has been postulated for other OMF family members (100).

CusB is a member of the membrane fusion protein (MFP) family (68) and has one predicted transmembrane domain at the N-terminal end of the amino acid sequence. Several members of this MFP family have been shown to be anchored by their N-terminus in the inner membrane with a periplasm-spanning domain that contacts the outer membrane or a target protein in the outer membrane (71, 113, 114). CusB-like proteins are thought to facilitate formation of a membrane fusion junction at the site of the transporter, thereby assisting in passage of the substrate from the cytoplasm to the extracellular space (90, 100). Formation of these tripartite junction complexes upon substrate transport has been demonstrated for the TolC/HlyD/HlyB secretion system in *E. coli*. HlyD, an MFP family member, and HlyB, its associated ATPase transport protein, are associated in the inner membrane while TolC, an OMF family member, is independently localized in the outer membrane (100). Upon binding to the substrate, the HlyA toxin, the HlyD/HlyB complex bridges the periplasm making contact with TolC through the MFP protein (HlyD). This forms a junction that spans the inner membrane, periplasm, and outer membrane. After secretion of the substrate, the inner and outer membrane components disengage, thereby resetting the transport complex (100).

In one model currently under evaluation, the workhorse of the putative *cus* transport complex would be CusA, a member of the resistance/nodulation/cell division (RND) family of antiporters (84). These large proteins, with 12 membrane-spanning regions, are found in the inner membrane. CusA is 34% identical to CzcA, a well-characterized RND protein from *Ralstonia*. Futhermore, hydropathy plots of CusA predict 12 transmembrane domains. RND transporters, such as CusA, utilize the proton gradient to provide energy for substrate transport, possibly copper efflux in this case.

Recent work with the *czc* metal/proton antiporter complex has given some insights into the functional structure of these RND transporters. One channel through the transporter CzcA harbors a conserved DDE motif that is essential for proton/cation antiport and could form a proton charge-relay system. Mutations of the DDE motif in CzcA do not abolish facilitated diffusion of metal cations through the transporter, indicating the possible existence of a second channel responsible for transport of the specific metal. Goldberg and colleagues have therefore suggested a two-channel model for proton/cation antiport (29). By analogy with *czc* and related systems, the CusCBA proteins may act together as a copper-transporting antiporter complex (see Fig. 2).

The nomenclature assigned to *cus* (CBA) was based on the previously characterized and homologous system, *czc* of *Ralstonia* (formerly *Alcaligenes eutrophus*, see chapter 26). The *czc* operon also con-

tains a two-component regulatory system, CzcRS. Recent work has shown that CzcRS are not absolutely required for transcription of the operon although they can modulate expression of target promoters (31). Two other genes on the *czc* operon, CzcD and CzcI, have been shown to influence expression of the operon. CzcD contains no significant homology with other histidine kinases of the two-component family, but it has been proposed to be a member of the cation diffusion facilitator (CDF) family (69). Van der Lelie and colleagues suggest that CzcD may act as a metal sensor in the membrane that transduces its signal to the receiver protein, CzcI, via a metal cation rather than a phosphate group. While CzcI contains putative metal-binding sites, it also contains a leader-peptidase signal sequence, which could localize it to the periplasm. If CzcI is located in the periplasm, it would not be suitable as a cytoplasmic signal transducer (102). The actual cellular location of czcI has not been determined experimentally, and while the CzcRS and CzcDI systems might interact, their role in signal transduction, if any, remains an open issue.

A plasmid-based silver resistance operon, *sil*, was recently identified in *Salmonella* (34). The *sil* operon contains the closest known homologs of CusRS, designated SilRS. The operon consists of two other transcripts. One transcript contains SilCBA, a proton/cation antiporter complex, and SilP, a P-type ATPase. The second transcript encodes SilE, a homolog of PcoE. Another copper-responsive two-component system, LcoRS, was found in *Lactococcus lactis* and shown to regulate a plasmid-based copper resistance determinant in a copper-responsive manner (42). The LcoR protein shows 34.3% sequence identity with PcoR and 34.4% identity with CopR.

THE ArsR FAMILY OF METAL STRESS RESPONSE REGULATORS

In the case of plasmid-borne arsenic resistance in *E. coli*, a repressor, ArsR, controls expression of an efflux system for this metalloid (16). ArsR is a repressor expressed at low basal levels in the cell. It binds as a dimer just upstream of the -35 region of the promoter and prevents transcription when arsenic concentrations in the media are below a toxic threshold (110). As levels of As(III) or Sb(III) in the medium rise, ArsR binds the inducing ion and dissociates from the promoter, allowing transcription initiation and upregulation of the other members of the operon, including another repressor, ArsD, and the efflux machinery, ArsA, ArsB, and ArsC. As ArsD protein levels increase, it binds to the promoter with about 2 orders of magnitude lower affinity than ArsR and represses

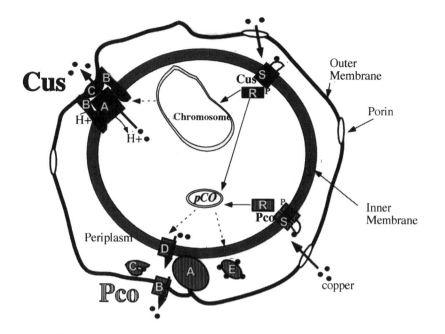

Figure 2. One hypothetical model of copper homeostasis and resistance circuits in *E. coli*. Both chromosomal elements (*cus*) and the plasmid system (*pco*) are shown. See text for detailed discussion of individual proteins and putative function assignments. Small solid circles indicate copper ions.

transcription (16). If arsenic concentrations continue to increase, ArsD will eventually dissociate from the promoter, allowing further increases in transcription of the *ars* operon. The need for a dual, high-affinity/low-affinity repressor system for the *ars* operon may stem from the toxicity of the ArsB integral membrane protein. As levels of ArsB protein rise, growth is inhibited (16). Therefore, the *ars* system must be carefully controlled lest the cure kill the patient.

Other metal-sensing homologs of ArsR have been identified. SmtB is an ArsR-like metalloregulatory protein in *Synechococcus* strain PCC7942 (53a). SmtB binds to and represses the *smt* metallothionein locus. In the presence of Zn(II), SmtB dissociates from the promoter and allows the transcription of the cyanobacterial metallothionein SmtA. A recent crystal structure of SmtB provides insights into the DNA-binding characteristics of the repressor and indicates several possible sites for Zn(II) binding (19a). A homolog of SmtB, ZiaR has been identified in *Synechococcus* strain PCC 6803. ZiaR is a repressor of the zinc export gene *ziaA* (100a).

The CadC regulatory protein of the *cad* metal resistance operon of *Staphylococcus aureus* plasmid pI258 represses expression of its target genes in the absence of metal (26). Apo-CadC binds in a region stretching between nucleotide positions −7 and +14 of the CadA promoter. Upon binding of metal, in this case either lead, cadmium, or zinc, repression by CadC is relieved and CadA transcription is activated.

The CadA gene encodes a P-type ATPase that transports the inducing metal (typically Cd[II]) from the cell.

Zn-responsive homologs of ArsR, namely CzrA and ZntR of *S. aureus* (the latter protein is not related to the *E. coli* ZntR), have also been identified (45, 109). Like the other family members, these two regulators are thought to be repressors, which, upon binding to their metal inducer, release from the operator DNA and allow transcription. ArsR homologs are also observed in eukaryotic systems, but their function is not understood. Clearly this metal-alleviated repression mechanism provides an alternative to activation and has been adopted in several metal stress responses.

THE Fur FAMILY OF METAL STRESS RESPONSE REGULATORS

The ferric uptake regulation (Fur) protein regulates transcription of a variety of iron uptake genes in bacteria. It has also been shown to repress transcription of other genes not directly involved in iron uptake, such as Shiga-like toxin, colicin I receptors, and Mn- and Fe-containing superoxide dismutases (see chapter 19). The mechanism of regulation for this protein remains controversial. Fur has been suggested to bind several divalent cations in vitro, such as Mn(II), Co(II), Cu(II), Cd(II), and Zn(II); however, it shows clear specificity for iron in vivo (8, 23). *E. coli* Fur binds one zinc tightly under all physiological conditions in an N_2S_2 coordination environment involving Cys 92 and Cys 95 (3, 30, 41). This form of the protein, Zn_1Fur, can only be converted to apo-Fur through denaturation of the protein (3). Upon addition of Fe(II) to Zn_1Fur, Fe(II) does not displace the Zn(II) but can bind in a second site, forming Fe(II),Zn(II)-Fur. Similarly, addition of excess Zn(II) to Zn_1Fur leads to formation of Zn_2Fur (3). Spectroscopic studies with Co(II) substituted into this second metal-binding site indicate that cobalt binding involves histidines and aspartates or glutamates, but no cysteines (1). DNA-binding studies on the two zinc forms and the apo form of Fur have shown that apo-Fur does not bind DNA, while Zn_1Fur, Fe(II), ZnFur, and Zn_2Fur bind the Fur operator with similar affinities (1; Althaus et al., unpublished results). Comparable results have been found with Zn_1Fur from *B. subtilis* (13). This study also found that addition of Fe(II) to *B. subtilis* Zn_1Fur did not significantly increase DNA-binding affinity or transcriptional repression. However, these studies were not conducted under strictly anaerobic conditions; therefore, the oxidation state of the iron ion and the cysteines is uncertain.

Since these results indicate that Fur binds specifically to the Fur operator without added Fe(II), a new model was proposed for the Fur repression mechanism. Fe(II) may bind Zn_1Fur and increase the binding affinity of the protein for the DNA, or cause a conformation change in the Zn_1Fur/DNA complex that enhances repression. Another possibility is that Fur does not sense intracellular iron by directly binding the metal ion. High-affinity DNA binding may instead require interaction with another component. It has also been proposed that Fur binds to a substrate that inhibits DNA binding, such as another protein or another metal ion, which is removed upon Fe(II) binding to the protein (13). These issues must be resolved to better understand the Fur mechanism of transcriptional regulation.

Studies of the *E. coli* Zur protein, a Zn-responsive Fur homolog, may provide further insight into the Fur mechanism. In vivo studies have shown that Zur represses transcription of the high-affinity zinc uptake system *znuABC* in response to elevated Zn levels (67) (see also chapter 19). The regulatory mechanism utilized by this protein is also yet to be determined.

OTHER METAL STRESS RESPONSE REGULATORS

The *cop* system of *Enterococcus hirae* represents one of the best-characterized bacterial copper ho-meostasis systems. The *cop* system also illustrates a common theme in copper homeostasis in both bacteria and higher eukaryotes, namely the copper transporting ATPase (77, 92). The *cop* operon consists of a copper-responsive repressor, CopY (95); a putative copper chaperone, CopZ; and two copper-transporting P-type ATPases, CopA and CopB. CopA transports copper into the cell while CopB effluxes copper from the cell (108). Apo-CopY binds to the *cop* promoter and represses expression until copper levels rise. Addition of Cu(I) (50 μM in vitro) abolishes CopY binding to the promoter and relieves the repression, allowing transcription initiation (95). While this mechanism explains CopB regulation, it is still unclear how the copper uptake transporter, CopA, is regulated.

CopZ shares homology with a family of proteins first shown to function as antioxidants in yeast (46). These proteins have recently been characterized as metallochaperones (47, 73). Metallochaperones are soluble, intracellular metal receptors that bind copper and guide it to target proteins, thereby protecting copper from intracellular chelators (74). The prototypical metallochaperone Atx1 binds Cu(I) and delivers it specifically to a cation transporter, Ccc2 (73). Another family member, CCS, also binds copper and delivers it to a specific target protein in the cytoplasm, namely superoxide dismutase (74, 86). The Cox17 protein is also thought to have a metallochaperone function in assembly of cytochrome c oxidase (9, 28, 93).

CopZ was initially classified as a regulatory protein that activates transcription (59). Recent work suggests that CopZ may deliver copper to the CopY repressor in vitro (18); however, no direct interaction between CopY and CopZ has been established. An alternative hypothesis is that the primary function of CopZ in vivo is to deliver Cu(I) to the copper efflux transporter, CopB, and not to CopY. Thus, the physiological roles or partners of CopZ have yet to be identified.

Molybdenum is required for several critical metabolic pathways in bacteria, including anaerobic respiration and nitrate reduction, yet, as a second row transition metal, molybdenum is only available in trace amounts in nature. To acquire this necessary trace metal, and avoid metal starvation stress, bacteria utilize molybdate uptake systems, some of which have K_m values for molybdate that reach as low as 50 nM (33). Molybdate uptake genes in *E. coli* and a variety of other microbes are regulated in part by the ModE repressor. ModE is a 28-kD DNA-binding protein with an N-terminal helix-turn-helix motif similar to that found in the LysR family of regulators (85), most of which are transcriptional activators.

Studies with molybdate uptake have shown that null mutations in the *modE* gene lead to molybdate-independent transcription of the molybdate uptake operon, *modABCD* (33). In vitro, ModE binds the operator region of the *modABCD* operon in three sites: −15 to −8, −4 to +4, and +8 to +14 (32). In the presence of molybdate, the affinity of ModE for these binding sites is enhanced. These experiments and other in vitro transcription/translation work indicate that ModE can act as a molybdate-sensing repressor that blocks expression of molybdate uptake genes at high molybdate concentrations.

Interestingly, ModE has also recently been shown to modulate transcription at other target promoters (87). The *hyc* and *nar* operons of *E. coli* encode for genes involved in production of the molybdoenzymes, formate hydrogenlyase and nitrate reductase, respectively. In the presence of molybdate, ModE can bind to the *hyc* promoter and the *narX-narK* intergenic region. Mutations in *modE* lead to suboptimal activation of *hyc, narX,* and *narK* (87). These results indicate that ModE functions as a secondary activator of the *hyc* and *nar* operons, somehow working in conjunction with the primary regulators, FhlA and NarL, to properly control transcription. ModE may therefore represent a global regulator of molybdate homeostasis that integrates elements of molybdate transport and metabolism by repressing some target promoters and contributing to the activation of others.

The DtxR family consists of iron-sensing repressors that regulate iron-uptake genes in *Corynebacterium diphtheriae, Mycobacterium tuberculosis,* and *Streptomyces.* The DtxR family is discussed in more detail in chapter 19. Here we note them as an example of a novel family of repressors that are involved in the bacterial stress response to metal starvation. There are also specific transport systems for nickel in bacteria (56). While nickel homeostasis was not discussed here, recent work has revealed a nickel-responsive metalloregulatory protein, NikR, which controls expression of the *nik* operon in *E. coli* (24).

FUTURE DIRECTIONS

The metalloregulatory systems presented here represent a few mechanisms utilized by bacteria to respond to metal starvation and metal toxicity. For a number of essential metals, such as copper, parts of the homeostasis circuit and its regulatory features have recently been elucidated. While a variety of metal-responsive copper efflux systems are now known, thus far only one copper uptake system (CopA of *Enterococcus*) has been identified in bac-

teria. Therefore our understanding of how bacteria balance efflux or detoxification with uptake to maintain optimal levels of transition metals is incomplete. Just how influx/efflux systems are integrated to accomplish this delicate task remains a key question. This integration may occur at the level of gene regulation as complex signal transduction networks function in concert to fine-tune the levels of transporters, chaperones, and other detoxification systems. In this sense, clues from the metal-responsive regulatory proteins in yeast (including Ace1, Mac1, and Amt1) may provide insights into the nature of the toxicity/homeostasis interface (70, 107, 111).

Another question arises upon close examination of these metal stress response systems: in cases where the metal substrate range for the sensing or detoxification proteins overlaps between both toxic and essential metals, can the bacterial cell ensure that homeostasis of essential metals (i.e., zinc) is not unduly disrupted by the presence of other, toxic metals (i.e., cadmium and lead)? Several of the systems shown (*pco, cop, znt, czc*) can be induced by both toxic and essential metals, and some can transport or detoxify both types of metals. Do bacteria have mechanisms for preventing disruption of essential metal homeostasis in the presence of chemically similar but toxic metals? Do transporters have different K_m values for each substrate that allow them to differentiate between toxic and essential metals? Are there adaptor proteins that modulate metal specificity for the transport and detoxification systems? Alternatively, is this disruption of essential metal homeostasis simply an added dimension of the toxicity of nonessential metals for which bacteria have little recourse? These questions pose a more difficult challenge in that they will force us to combine our understanding of metal sensing, metal transport, and overall intracellular metal physiology to formulate hypotheses.

Despite substrate overlap in some systems, many other metal stress response regulators are remarkably adept at discriminating among different metals. This presents the fascinating question of how the metal-sensing domains of response regulators distinguish between the various trace metals found in the environment. Characterization of the metal-binding domains of the regulatory, transporter, and chaperone proteins through spectroscopic means may provide us with insight into this conundrum. A deeper understanding of the ligand-metal interactions that take place in the metal-binding sites of these proteins should allow us to define criteria for proper binding sites for each metal. Furthermore, characterization of three-dimensional structures for these metal-sensing proteins should reveal how binding by different metals can influence the global properties of the protein.

Subtle changes in protein conformation may provide mechanisms by which the sensing protein can discern between different metal substrates.

The complexities of metal ion homeostasis and resistance in bacteria are beginning to be understood. As new species, microbial communities, and environments are discovered, other common themes will likely emerge from among the various bacterial strategies for responding to the dual nature of metal ion stress.

REFERENCES

1. Adrait, A., L. Jacquamet, L. Le Pape, A. Gonzalez de Peredo, D. Aberdam, J. L. Hazemann, J. M. Latour, and I. Michaud-Soret. 1999. Spectroscopic and saturation magnetization properties of the manganese- and cobalt-substituted Fur (ferric uptake regulation) protein from *Escherichia coli*. *Biochemistry* **38**:6248–6260.
2. Ahmed, M., C. M. Borsch, S. S. Taylor, N. Vazquez-Laslop, and A. A. Neyfakh. 1994. A protein that activates expression of a multidrug efflux transporter upon binding the transporter substrates. *J. Biol. Chem.* **269**:28506–28513.
3. Althaus, E. W., C. E. Outten, K. E. Olson, H. Cao, and T. V. O'Halloran. 1999. The ferric uptake regulation (Fur) repressor is a zinc metalloprotein. *Biochemistry* **38**:6559–6569.
4. Amouyal, M., and H. Buc. 1987. Topological unwinding of strong and weak promoters by RNA polymerase. A comparison between the lac wild-type and the UV5 sites of *Escherichia coli*. *J. Mol. Biol.* **195**:795–808.
5. Ansari, A. Z., J. E. Bradner, and T. V. O'Halloran. 1995. DNA-bend modulation in a repressor-to-activator switching mechanism. *Nature* **374**:371–375.
6. Ansari, A. Z., M. L. Chael, and T. V. O'Halloran. 1992. Allosteric underwinding of DNA is a critical step in positive control of transcription by Hg-MerR. *Nature* **355**:87–89.
7. Ansari, A. Z., and T. V. O'Halloran. 1994. An emerging role for allosteric modulation of DNA structure in transcription, p. 369–386. In R. C. Conaway and J. W. Conaway (ed.), *Transcription: Mechanisms and Regulation*. Raven Press, New York, N.Y.
8. Bagg, A., and J. B. Neilands. 1987. Ferric uptake regulation protein acts as a repressor, employing iron (II) as a cofactor to bind the operator of an iron transport operon in *Escherichia coli*. *Biochemistry* **26**:5471–5477.
9. Beers, J., D. M. Glerum, and A. Tzagoloff. 1997. Purification, characterization, and localization of yeast Cox17p, a mitochondrial copper shuttle. *J. Biol. Chem.* **272**:33191–33196.
10. Brocklehurst, K. R., J. L. Hobman, B. Lawley, L. Blank, S. J. Marshall, N. L. Brown, and A. P. Morby. 1999. ZntR is a Zn(II)-responsive MerR-like transcriptional regulator of zntA in *Escherichia coli*. *Mol. Microbiol.* **31**:893–902.
11. Brown, N. L., S. R. Barrett, J. Camakaris, B. T. Lee, and D. A. Rouch. 1995. Molecular genetics and transport analysis of the copper-resistance determinant (pco) from *Escherichia coli* plasmid pRJ1004. *Mol. Microbiol.* **17**:1153–1166.
12. Brown, N. L., J. Camakaris, B. T. Lee, T. Williams, A. P. Morby, J. Parkhill, and D. A. Rouch. 1991. Bacterial resistances to mercury and copper. *J. Cell Biochem.* **46**:106–114.
13. Bsat, N., and J. D. Helmann. 1999. Interaction of *Bacillus subtilis* Fur (ferric uptake repressor) with the dhb operator in vitro and in vivo. *J. Bacteriol.* **181**:4299–4307.
14. Butler, A. 1998. Acquisition and utilization of transition metal ions by marine organisms. *Science* **281**:207–210.
15. Cha, J. S., and D. A. Cooksey. 1991. Copper resistance in *Pseudomonas syringae* mediated by periplasmic and outer membrane proteins. *Proc. Natl. Acad. Sci. USA* **88**:8915–8919.
16. Chen, Y., and B. P. Rosen. 1997. Metalloregulatory properties of the ArsD repressor. *J. Biol. Chem.* **272**:14257–14262.
17. Chiu, M. L., M. Folcher, T. Katoh, A. M. Puglia, J. Vohradsky, B. S. Yun, H. Seto, and C. J. Thompson. 1999. Broad spectrum thiopeptide recognition specificity of the *Streptomyces lividans* TipAL protein and its role in regulating gene expression. *J. Biol. Chem.* **274**:20578–20586.
18. Cobine, P., W. A. Wickramasinghe, M. D. Harrison, T. Weber, M. Solioz, and C. T. Dameron. 1999. The *Enterococcus hirae* copper chaperone CopZ delivers copper(I) to the CopY repressor. *FEBS Lett.* **445**:27–30.
19. Comess, K. M., L. M. Shewchuk, K. Ivanetich, and C. T. Walsh. 1994. Construction of a synthetic gene for the metalloregulatory protein MerR and analysis of regionally mutated proteins for transcriptional regulation. *Biochemistry* **33**:4175–4186.
19a. Cook, W. J., S. R. Kar, K. B. Taylor, and L. M. Hall. 1998. Crystal structure of the cyanobacterial metallothionein repressor SmtB: a model for metalloregulatory proteins. *J. Mol. Biol.* **275**:337–346.
20. Cooksey, D. A. 1993. Copper uptake and resistance in bacteria. *Mol. Microbiol.* **7**:1–5.
21. Coulombe, B., and Z. F. Burton. 1999. DNA bending and wrapping around RNA polymerase: a "revolutionary" model describing transcriptional mechanisms. *Microbiol. Mol. Biol. Rev.* **63**:457–478.
22. Craig, M. L., W. C. Suh, and M. T. Record, Jr. 1995. HO and DNase I probing of E sigma 70 RNA polymerase—lambda PR promoter open complexes: Mg^{2+} binding and its structural consequences at the transcription start site. *Biochemistry* **34**:15624–15632.
23. de Lorenzo, V., S. Wee, M. Herrero, and J. B. Neilands. 1987. Operator sequences of the aerobactin operon of plasmid ColV-K30 binding the ferric uptake regulation (fur) repressor. *J. Bacteriol.* **169**:2624–2630.
24. De Pina, K., V. Desjardin, M. A. Mandrand-Berthelot, G. Giordano, and L. F. Wu. 1999. Isolation and characterization of the nikR gene encoding a nickel- responsive regulator in *Escherichia coli*. *J. Bacteriol.* **181**:670–674.
25. Ding, H., E. Hidalgo, and B. Demple. 1996. The redox state of the [2Fe-2S] clusters in SoxR protein regulates its activity as a transcription factor. *J. Biol. Chem.* **271**:33173–33175.
26. Endo, G., and S. Silver. 1995. CadC, the transcriptional regulatory protein of the cadmium resistance system of *Staphylococcus aureus* plasmid pI258. *J. Bacteriol.* **177**:4437–4441.
27. Frantz, B., and T. V. O'Halloran. 1990. DNA distortion accompanies transcriptional activation by the metal-responsive gene-regulatory protein MerR. *Biochemistry* **29**:4747–4751.
28. Glerum, D. M., A. Shtanko, and A. Tzagoloff. 1996. Characterization of COX17, a yeast gene involved in copper metabolism and assembly of cytochrome oxidase. *J. Biol. Chem.* **271**:14504–14509.
29. Goldberg, M., T. Pribyl, S. Juhnke, and D. H. Nies. 1999. Energetics and topology of CzcA, a cation/proton antiporter of the resistance-nodulation-cell division protein family. *J. Biol. Chem.* **274**:26065–26070.
30. Gonzalez de Peredo, A., C. Saint-Pierre, A. Adrait, L. Jacquamet, J. M. Latour, I. Michaud-Soret, and E. Forest. 1999. Identification of the two zinc-bound cysteines in the ferric uptake regulation protein from *Escherichia coli*: chemical modification and mass spectrometry analysis. *Biochemistry* **38**:8582–8589.

31. Grosse, C., G. Grass, A. Anton, S. Franke, A. N. Santos, B. Lawley, N. L. Brown, and D. H. Nies. 1999. Transcriptional organization of the czc heavy-metal homeostasis determinant from *Alcaligenes eutrophus. J. Bacteriol.* 181:2385–2393.

32. Grunden, A. M., W. T. Self, M. Villain, J. E. Blalock, and K. T. Shanmugam. 1999. An analysis of the binding of repressor protein ModE to modABCD (molybdate transport) operator/promoter DNA of *Escherichia coli. J. Biol. Chem.* 274:24308–24315.

33. Grunden, A. M., and K. T. Shanmugam. 1997. Molybdate transport and regulation in bacteria. *Arch. Microbiol.* 168: 345–354.

34. Gupta, A., K. Matsui, J. F. Lo, and S. Silver. 1999. Molecular basis for resistance to silver cations in Salmonella. *Nat. Med.* 5:183–188.

35. Gupta, A., M. Maynes, and S. Silver. 1998. Effects of halides on plasmid-mediated silver resistance in *Escherichia coli. Appl. Environ. Microbiol.* 64:5042–5045.

36. Gupta, S. D., B. T. Lee, J. Camakaris, and H. C. Wu. 1995. Identification of cutC and cutF (nlpE) genes involved in copper tolerance in *Escherichia coli. J. Bacteriol.* 177:4207–4215.

37. Helmann, J. D., B. T. Ballard, and C. T. Walsh. 1990. The MerR metalloregulatory protein binds mercuric ion as a tricoordinate, metal-bridged dimer. *Science* 247:946–948.

38. Hess, J. F., R. B. Bourret, and M. I. Simon. 1988. Histidine phosphorylation and phosphoryl group transfer in bacterial chemotaxis. *Nature* 336:139–143.

39. Hoch, J. A., and T. J. Silhavy (ed.). 1995. *Two-Component Signal Transduction.* ASM Press, Washington, D.C.

40. Hutchens, T. W., M. H. Allen, C. M. Li, and T. T. Yip. 1992. Occupancy of a C2-C2 type 'zinc-finger' protein domain by copper. Direct observation by electrospray ionization mass spectrometry. *FEBS Lett.* 309:170–174.

41. Jacquamet, L., D. Aberdam, A. Adrait, J. L. Hazemann, J. M. Latour, and I. Michaud-Soret. 1998. X-ray absorption spectroscopy of a new zinc site in the fur protein from *Escherichia coli. Biochemistry* 37:2564–2571.

42. Khunajakr, N., C. Q. Liu, P. Charoenchai, and N. W. Dunn. 1999. A plasmid-encoded two-component regulatory system involved in copper-inducible transcription in *Lactococcus lactis. Gene* 229:229–235.

43. Kimura, T., and H. Nishioka. 1997. Intracellular generation of superoxide by copper sulphate in *Escherichia coli. Mutat. Res.* 389:237–242.

44. Kulkarni, R. D., and A. O. Summers. 1999. MerR cross-links to the alpha, beta, and sigma 70 subunits of RNA polymerase in the preinitiation complex at the merTPCAD promoter. *Biochemistry* 38:3362–3368.

45. Kuroda, M., H. Hayashi, and T. Ohta. 1999. Chromosome-determined zinc-responsible operon czr in *Staphylococcus aureus* strain 912. *Microbiol. Immunol.* 43:115–125.

46. Lin, S. J., and V. C. Culotta. 1995. The ATX1 gene of *Saccharomyces cerevisiae* encodes a small metal homeostasis factor that protects cells against reactive oxygen toxicity. *Proc. Natl. Acad. Sci. USA* 92:3784–3788.

47. Lin, S. J., R. A. Pufahl, A. Dancis, T. V. O'Halloran, and V. C. Culotta. 1997. A role for the *Saccharomyces cerevisiae* ATX1 gene in copper trafficking and iron transport. *J. Biol. Chem.* 272:9215–9220.

48. Lund, P. A., and N. L. Brown. 1989. Regulation of transcription in *Escherichia coli* from the mer and merR promoters in the transposon Tn501. *J. Mol. Biol.* 205:343–353.

49. Lund, P. A., S. J. Ford, and N. L. Brown. 1986. Transcriptional regulation of the mercury-resistance genes of transposon Tn501. *J. Gen. Microbiol.* 132:465–480.

50. Mellano, M. A., and D. A. Cooksey. 1988. Induction of the copper resistance operon from *Pseudomonas syringae. J. Bacteriol.* 170:4399–4401.

51. Mellano, M. A., and D. A. Cooksey. 1988. Nucleotide sequence and organization of copper resistance genes from *Pseudomonas syringae* pv. tomato. *J. Bacteriol.* 170:2879–2883.

52. Mills, S. D., C. A. Jasalavich, and D. A. Cooksey. 1993. A two-component regulatory system required for copper-inducible expression of the copper resistance operon of *Pseudomonas syringae. J. Bacteriol.* 175:1656–1664.

53. Mills, S. D., C. K. Lim, and D. A. Cooksey. 1994. Purification and characterization of CopR, a transcriptional activator protein that binds to a conserved domain (cop box) in copper-inducible promoters of *Pseudomonas syringae. Mol. Gen. Genet.* 244:341–351.

53a. Morby, A. P., J. S. Turner, J. W. Huckle, and N. J. Robinson. 1993. SmtB is a metal-dependent repressor of the cyanobacterial metallothionein gene smtA: identification of a Zn inhibited DNA-protein complex. *Nucleic Acids Res.* 21:921–925.

54. Mukhopadhyay, C. K., B. Mazumder, P. F. Lindley, and P. L. Fox. 1997. Identification of the prooxidant site of human ceruloplasmin: a model for oxidative damage by copper bound to protein surfaces. *Proc. Natl. Acad. Sci. USA* 94:11546–11551.

55. Munson, G. P. 1997. Copper resistance of enteric bacteria: molecular and genetic analysis of resistance, homeostasis, and transcriptional regulation. Ph.D. thesis. Northwestern University, Evanston, Ill.

56. Navarro, C., L. F. Wu, and M. A. Mandrand-Berthelot. 1993. The nik operon of *Escherichia coli* encodes a periplasmic-binding protein-dependent transport system for nickel. *Mol. Microbiol.* 9:1181–1191.

57. Nies, D. H. 1999. Microbial heavy-metal resistance. *Appl. Microbiol. Biotechnol.* 51:730–750.

58. Ninfa, A. J. 1996. Regulation of gene transcription by extracellular stimuli, p. 1246–1262. *In* F. C. Neidhardt, R. Curtiss III, J. L. Ingraham, E. C. C. Lin, K. B. Low, B. Magasanik, W. S. Reznikoff, M. Riley, M. Schaechter, and H. E. Umbarger (ed.), *Escherichia coli and Salmonella: Cellular and Molecular Biology*, 2nd ed. ASM Press, Washington D.C.

59. Odermatt, A., and M. Solioz. 1995. Two trans-acting metal-loregulatory proteins controlling expression of the copper-ATPases of *Enterococcus hirae. J. Biol. Chem.* 270:4349–4354.

60. O'Halloran, T., and C. Walsh. 1987. Metalloregulatory DNA-binding protein encoded by the merR gene: isolation and characterization. *Science* 235:211–214.

61. O'Halloran, T. V. 1989. Metalloregulatory proteins: metal-responsive molecular switches governing gene expression, p. 105–146. *In* H. Sigel and A. Sigel (ed.), *Metal Ions in Biological Systems.* Marcel Dekker, Inc., New York, N.Y.

62. O'Halloran, T. V. 1993. Transition metals in control of gene expression. *Science* 261:715–725.

63. O'Halloran, T. V., B. Frantz, M. K. Shin, D. M. Ralston, and J. G. Wright. 1989. The MerR heavy metal receptor mediates positive activation in a topologically novel transcription complex. *Cell* 56:119–129.

64. Outten, C. E., F. W. Outten, and T. V. O'Halloran. 1999. DNA distortion mechanism for transcriptional activation by ZntR, a Zn(II)-responsive MerR homologue in *E. coli. J. Biol. Chem.* 274:37517–37524.

65. Parkhill, J., A. Z. Ansari, J. G. Wright, N. L. Brown, and T. V. O'Halloran. 1993. Construction and characterization of a mercury-independent MerR activator (MerRAC): transcriptional activation in the absence of Hg(II) is accompanied by DNA distortion. *EMBO J.* 12:413–421.

66. Parkhill, J., B. Lawley, J. L. Hobman, and N. L. Brown. 1998. Selection and characterization of mercury-independent activation mutants of the Tn501 transcriptional regulator, MerR. *Microbiology* **144:**2855–2864.

67. Patzer, S. I., and K. Hantke. 1998. The ZnuABC high-affinity zinc uptake system and its regulator Zur in *Escherichia coli. Mol. Microbiol.* **28:**1199–1210.

68. Paulsen, I. T., J. H. Park, P. S. Choi, and M. H. Saier, Jr. 1997. A family of gram-negative bacterial outer membrane factors that function in the export of proteins, carbohydrates, drugs and heavy metals from gram-negative bacteria. *FEMS Microbiol. Lett.* **156:**1–8.

69. Paulsen, I. T., and M. H. Saier, Jr. 1997. A novel family of ubiquitous heavy metal ion transport proteins. *J. Membr. Biol.* **156:**99–103.

70. Pena, M. M., J. Lee, and D. J. Thiele. 1999. A delicate balance: homeostatic control of copper uptake and distribution. *J. Nutr.* **129:**1251–1260.

71. Pimenta, A. L., J. Young, I. B. Holland, and M. A. Blight. 1999. Antibody analysis of the localisation, expression and stability of HlyD, the MFP component of the *E. coli* haemolysin translocator. *Mol. Gen. Genet.* **261:**122–132.

72. Predki, P. F., and B. Sarkar. 1992. Effect of replacement of "zinc finger" zinc on estrogen receptor DNA interactions. *J. Biol. Chem.* **267:**5842–5846.

73. Pufahl, R. A., C. P. Singer, K. L. Peariso, S. J. Lin, P. J. Schmidt, C. J. Fahrni, V. C. Culotta, J. E. Penner-Hahn, and T. V. O'Halloran. 1997. Metal ion chaperone function of the soluble Cu(I) receptor Atx1. *Science* **278:**853–856.

74. Rae, T. D., P. J. Schmidt, R. A. Pufahl, V. C. Culotta, and T. V. O'Halloran. 1999. Undetectable intracellular free copper: the requirement of a copper chaperone for superoxide dismutase. *Science* **284:**805–808.

75. Raivio, T. L., and T. J. Silhavy. 1997. Transduction of envelope stress in *Escherichia coli* by the Cpx two-component system. *J. Bacteriol.* **179:**7724–7733.

76. Ralston, D. M., and T. V. O'Halloran. 1990. Ultrasensitivity and heavy-metal selectivity of the allosterically modulated MerR transcription complex. *Proc. Natl. Acad. Sci. USA* **87:**3846–3850.

77. Rensing, C., M. Ghosh, and B. P. Rosen. 1999. Families of soft-metal-ion-transporting ATPases. *J. Bacteriol.* **181:**5891–5897.

78. Rivetti, C., M. Guthold, and C. Bustamante. 1999. Wrapping of DNA around the *E. coli* RNA polymerase open promoter complex. *EMBO J.* **18:**4464–4475.

79. Ross, W., S. J. Park, and A. O. Summers. 1989. Genetic analysis of transcriptional activation and repression in the Tn21 mer operon. *J. Bacteriol.* **171:**4009–4018.

80. Rouch, D. A., and N. L. Brown. 1997. Copper-inducible transcriptional regulation at two promoters in the *Escherichia coli* copper resistance determinant pco. *Microbiology* **143:**1191–1202.

81. Rouch, D. A., J. Parkhill, and N. L. Brown. 1995. Induction of bacterial mercury- and copper-responsive promoter: functional differences between inducible systems and implications for their use in gene-fusions for in vivo metal biosensors. *J. Ind. Microbiol.* **14:**249–253.

82. Russo, F. D., and T. J. Silhavy. 1993. The essential tension: opposed reactions in bacterial two-component regulatory systems. *Trends Microbiol.* **1:**306–310.

83. Rutherford, J. C., J. S. Cavet, and N. J. Robinson. 1999. Cobalt-dependent transcriptional switching by a dual-effector MerR-like protein regulates a cobalt-exporting variant CPx-type ATPase. *J. Biol. Chem.* **274:**25827–25832.

84. Saier, M. H., Jr., R. Tam, A. Reizer, and J. Reizer. 1994. Two novel families of bacterial membrane proteins concerned with nodulation, cell division and transport. *Mol. Microbiol.* **11:**841–847.

85. Schell, M. A. 1993. Molecular biology of the LysR family of transcriptional regulators. *Annu. Rev. Microbiol.* **47:**597–626.

86. Schmidt, P. J., T. D. Rae, R. A. Pufahl, T. Hamma, J. Strain, T. V. O'Halloran, and V. C. Culotta. 1999. Multiple protein domains contribute to the action of the copper chaperone for superoxide dismutase. *J. Biol. Chem.* **274:**23719–23725.

87. Self, W. T., A. M. Grunden, A. Hasona, and K. T. Shanmugam. 1999. Transcriptional regulation of molybdoenzyme synthesis in *Escherichia coli* in response to molybdenum: ModE-molybdate, a repressor of the modABCD (molybdate transport) operon is a secondary transcriptional activator for the hyc and nar operons. *Microbiology* **145:**41–55.

88. Silver, S. 1998. Genes for all metals—a bacterial view of the periodic table. The 1996 Thom Award Lecture. *J. Ind. Microbiol. Biotechnol.* **20:**1–12.

89. Silver, S., and L. T. Phung. 1996. Bacterial heavy metal resistance: new surprises. *Annu. Rev. Microbiol.* **50:**753–789.

90. Skvirsky, R. C., S. Reginald, and X. Shen. 1995. Topology analysis of the colicin V export protein CvaA in Escherichia coli. *J. Bacteriol.* **177:**6153–6159.

91. Snyder, W. B., L. J. Davis, P. N. Danese, C. L. Cosma, and T. J. Silhavy. 1995. Overproduction of NlpE, a new outer membrane lipoprotein, suppresses the toxicity of periplasmic LacZ by activation of the Cpx signal transduction pathway. *J. Bacteriol.* **177:**4216–4223.

92. Solioz, M., A. Odermatt, and R. Krapf. 1994. Copper pumping ATPases: common concepts in bacteria and man. *FEBS Lett.* **346:**44–47.

93. Srinivasan, C., M. C. Posewitz, G. N. George, and D. R. Winge. 1998. Characterization of the copper chaperone Cox17 of *Saccharomyces cerevisiae*. *Biochemistry* **37:**7572–7577.

94. Storz, G., and J. A. Imlay. 1999. Oxidative stress. *Curr. Opin. Microbiol.* **2:**188–194.

95. Strausak, D., and M. Solioz. 1997. CopY is a copper-inducible repressor of the *Enterococcus hirae* copper ATPases. *J. Biol. Chem.* **272:**8932–8936.

96. Suhy, D. A., and T. V. O'Halloran. 1996. Metal-responsive gene regulation and the zinc metalloregulatory model. *Met. Ions Biol. Syst.* **32:**557–578.

97. Suhy, D. A., K. D. Simon, D. I. Linzer, and T. V. O'Halloran. 1999. Metallothionein is part of a zinc-scavenging mechanism for cell survival under conditions of extreme zinc deprivation. *J. Biol. Chem.* **274:**9183–9192.

98. Summers, A. O. 1992. Untwist and shout: a heavy metal-responsive transcriptional regulator. *J. Bacteriol.* **174:**3097–3101.

99. Tetaz, T. J., and R. K. Luke. 1983. Plasmid-controlled resistance to copper in *Escherichia coli*. *J. Bacteriol.* **154:**1263–1268.

100. Thanabalu, T., E. Koronakis, C. Hughes, and V. Koronakis. 1998. Substrate-induced assembly of a contiguous channel for protein export from *E. coli*: reversible bridging of an inner-membrane translocase to an outer membrane exit pore. *EMBO J.* **17:**6487–6496.

100a. Thelwell, C., N. J. Robinson, and J. S. Turner-Cavet. 1998. An SmtB-like repressor from Synechocystis PCC 6803 regulates a zinc exporter. *Proc. Natl. Acad. Sci. USA* **95:**10728–10733.

101. Utschig, L. M., J. W. Bryson, and T. V. O'Halloran. 1995. Mercury-199 NMR of the metal receptor site in MerR and its protein-DNA complex. *Science* **268:**380–385.

102. van der Lelie, D., T. Schwuchow, U. Schwidetzky, S. Wuertz, W. Baeyens, M. Mergeay, and D. H. Nies. 1997. Two-component regulatory system involved in transcriptional control of heavy-metal homoeostasis in *Alcaligenes eutrophus*. *Mol. Microbiol.* 23:493–503.

103. von Heijne, G. 1989. The structure of signal peptides from bacterial lipoproteins. *Protein Eng.* 2:531–534.

104. Wakatsuki, T. 1995. Metal oxidoreduction by microbial cells. *J. Ind. Microbiol.* 14:169–177.

105. Watton, S. P., J. G. Wright, F. M. MacDonnell, J. W. Bryson, M. Sabat, and T. V. O'Halloran. 1990. Trigonal mercuric complex of an aliphatic thiolate—a spectroscopic and structural model for the receptor-site in the Hg(II) biosensor MerR. *J. Am. Chem. Soc.* 112:2824–2826.

106. Williams, J. R., A. G. Morgan, D. A. Rouch, N. L. Brown, and B. T. Lee. 1993. Copper-resistant enteric bacteria from United Kingdom and Australian piggeries. *Appl. Environ. Microbiol.* 59:2531–2537.

107. Winge, D. R., L. T. Jensen, and C. Srinivasan. 1998. Metal-ion regulation of gene expression in yeast. *Curr. Opin. Chem. Biol.* 2:216–221.

108. Wunderli-Ye, H., and M. Solioz. 1999. Copper homeostasis in *Enterococcus hirae*. *Adv. Exp. Med. Biol.* 448:255–264.

109. Xiong, A., and R. K. Jayaswal. 1998. Molecular characterization of a chromosomal determinant conferring resistance to zinc and cobalt ions in *Staphylococcus aureus*. *J. Bacteriol.* 180:4024–4029.

110. Xu, C., and B. P. Rosen. 1997. Dimerization is essential for DNA binding and repression by the ArsR metalloregulatory protein of *Escherichia coli*. *J. Biol. Chem.* 272:15734–15738.

111. Yamaguchi-Iwai, Y., M. Serpe, D. Haile, W. Yang, D. J. Kosman, R. D. Klausner, and A. Dancis. 1997. Homeostatic regulation of copper uptake in yeast via direct binding of MAC1 protein to upstream regulatory sequences of FRE1 and CTR1. *J. Biol. Chem.* 272:17711–17718.

112. Zeng, Q., C. Stalhandske, M. C. Anderson, R. A. Scott, and A. O. Summers. 1998. The core metal-recognition domain of MerR. *Biochemistry* 37:15885–15895

113. Zgurskaya, H. I., and H. Nikaido. 1999. AcrA is a highly asymmetric protein capable of spanning the periplasm. *J. Mol. Biol.* 285:409–420.

114. Zgurskaya, H. I., and H. Nikaido. 1999. Bypassing the periplasm: reconstitution of the AcrAB multidrug efflux pump of *Escherichia coli*. *Proc. Natl. Acad. Sci. USA* 96:7190–7195.

115. Zheleznova, E. E., P. N. Markham, A. A. Neyfakh, and R. G. Brennan. 1997. Preliminary structural studies on the multi-ligand-binding domain of the transcription activator, BmrR, from *Bacillus subtilis*. *Protein Sci.* 6:2465–2468.

II. GENERAL STRESS RESPONSES

Chapter 11

The General Stress Response in *Escherichia coli*

REGINE HENGGE-ARONIS

The general stress response of Escherichia coli *is characterized by numerous alterations in cellular physiology and even morphology that enhance survival by increasing cellular stress resistance, i.e., by preventing cellular damage rather than by repairing it. This response is triggered by many different stresses including starvation (which results in stationary phase), high osmolarity, high or low temperature, and acidic pH. These conditions result in the accumulation of σ^S (RpoS), a sigma subunit of RNA polymerase, which acts as the master regulator of this response. Many of the more than 50 σ^S-controlled genes confer stress tolerance, whereas others mediate structural and morphological rearrangements or redirect metabolism. In addition, a number of virulence genes have recently been found to be part of the σ^S regulon. Surprisingly, σ^S and the "housekeeping" σ^{70} recognize similar promoter sequences. σ^S-controlled promoter regions often feature multiple binding sites for additional regulators such as cAMP-CRP or the histone-like proteins H-NS, Lrp, IHF, and Fis. Recent data suggest that these regulators are crucial for determining sigma factor specificity. The regulation of σ^S has also been studied intensively. Induction of σ^S is mainly due to posttranscriptional control with different stresses acting independently at different steps. rpoS translation is probably stimulated by rearrangements of rpoS mRNA secondary structure that render the ribosomal binding site accessible. trans-acting factors involved are the RNA-binding protein Hfq and two small regulatory RNAs, OxyS, and DsrA. In addition, σ^S levels are regulated by proteolysis, for which the ClpXP protease and the response regulatory RssB are essential. RssB is a recognition factor that interacts with σ^S in a phosphorylation-dependent manner, which suggests that environmental stress modulates σ^S proteolysis by controlling RssB affinity for σ^S. Taken together, the σ^S-dependent general stress response provides an intriguing model system to study transcriptional and translational control of gene expression, regulated proteolysis, and signal transduction and integration in a complex genetic network that underlies the drastic changes in cell physiology that occur in response to many different stresses.*

The concept of the general stress response in *Escherichia coli* has only recently emerged from the analysis of the molecular processes in stationary-phase cells. When *E. coli* became the model organism during the times of early molecular biology, it became rapidly accepted practice to use almost exclusively exponentially growing cells. This provided researchers with a relatively simple experimental system that allowed the basic concepts of molecular biology such as the paradigm of local gene regulation exemplified in the *lac* operon to be established (85). When it became apparent that bacterial cells responded to alterations in their environment by activating small or large groups of genes under the control of a common regulatory protein, the concept of global gene regulation was developed. Such responses allow cells to cope with specific stress situations. Currently, the interconnections between global regulatory circuits are being studied, with regulatory networks becoming apparent that ultimately coordinate metabolism, growth, and division of the entire cell in response to environmental signals.

While it had been known for a long time that stationary-phase cells are morphologically and physiologically distinct from rapidly growing cells, only the concepts of global regulation and regulatory networks provided the theoretical framework for the molecular analysis of stationary-phase processes. This allowed the identification of a master regulator of the stationary-phase response, σ^S (108), and many stationary phase-inducible genes (140, 152, 160, 176). More recently, it was observed that σ^S and σ^S-dependent genes are also induced in response to

Regine Hengge-Aronis • Institut für Biologie—Mikrobiologie, Freie Universität Berlin, 14195 Berlin, Germany.

many other stresses such as high osmolarity, nonoptimal temperature, or acid pH (73, 78, 112, 126, 130). Cells subject to these stresses exhibit phenotypes (such as multiple stress resistance or smaller and ovoid cell shape) that have been thought to be typical for stationary-phase cells (78, 112, 151). The current view thus has σ^S as the master regulator of the general stress response, which is elicited by any of a number of different stresses, is usually accompanied by a reduction or cessation of growth, and provides the cells with the ability to survive the currently experienced stress as well as additional stresses not yet encountered (cross-protection).

THE MASTER REGULATOR σ^S

Identification of the *rpoS*-Encoded Sigma Factor σ^S as a Central Regulator of Stationary Phase-Inducible Genes

The *rpoS* gene was independently discovered by several groups in different contexts (summarized in reference 99). It was identified as a gene involved in near-UV resistance (169); as a regulator for the *katE*-encoded catalase (116, 150), exonuclease III (*xthA*) (150), and an acidic phosphatase (167); and finally as a starvation-inducible gene encoding a central regulator for stationary phase-inducible genes (108). Only then it became clear that all these studies had been dealing with the same gene (108, 168), which in fact encoded a sigma factor (132, 135, 166), for which the name σ^S or RpoS was proposed (108). Since then, the number of σ^S-controlled genes identified has steadily increased to currently more than 50 (see references 74 and 115 for compilations of σ^S-regulated genes).

Basic Properties of σ^S

σ^S is the product of the *rpoS* gene (located in counterclockwise orientation at 61.76 min between coordinates 2864.6 and 2865.6 of the physical map of *E. coli* [149]). The gene is part of the *nlpD-rpoS* operon, which exhibits basic expression from a polycistronic mRNA starting from two closely spaced promoter sites upstream of *nlpD* (which encodes a lipoprotein of unclear function) (110). In addition, *rpoS* is expressed from a promoter located within *nlpD*, and this shorter transcript further accumulates under certain stationary-phase conditions (106).

σ^S acts as a sigma subunit of RNA polymerase and is a close relative of the vegetative ("housekeeping") sigma factor σ^{70} (RpoD) to the extent that both recognize the similar promoter sequences (see below). σ^S is thus considered a second primary sigma

factor rather than an alternative sigma factor (118). Whereas σ^S is dispensable during exponential growth (and, in fact, is hardly present in growing wild-type cells), it is essential in stationary phase or when cells are otherwise stressed (108, 124), i.e., under conditions that prevail in the natural environments of *E. coli*. *rpoS* mutants are devoid of the typical properties associated with the general stress response: they remain rod-shaped and sensitive to multiple stresses, they are unable to accumulate glycogen and trehalose, and they die off rapidly in stationary phase (77, 107, 108, 124).

The form of σ^S that is considered to be wild type (and is present in the widely used strains MC4100 and W3110) has a molecular mass of 37.8 kDa (therefore, σ^S is sometimes also designated σ^{38}). However, σ^S variants of different sizes have been found in long-term stationary-phase cultures in the laboratory as well as in natural isolates (84, 92, 187). It seems that for not yet understood reasons, the *rpoS* locus exhibits high variability under conditions similar to the natural living conditions of *E. coli*.

σ^S in Other Bacterial Species

σ^S has been identified in a number of gram-negative bacteria such as other enteric bacteria or various *Pseudomonas* species. In these bacteria, σ^S also seems to act as a regulator of stationary phase- and stress-inducible genes (58, 103, 140). In *Legionella pneumophila*, it is required for survival within host amoeba (71). In certain plant pathogens such as *Erwinia carotovora* and *Ralstonia solanacearum*, σ^S seems to be involved in the control of virulence genes and the production of exoenzymes (7, 55, 131). So far, σ^S has not been found in species that belong to the α-branch of the proteobacteria. Gram-positive bacteria do not have a σ^S-like sigma factor, although they exhibit a general stress response that physiologically is similar to that of *E. coli*. This response, however, is under the control of a genuine alternative sigma factor, σ^B (see chapter 12). In fact, σ^B controls the expression of genes that are homologous to the σ^S-dependent genes *katE*, *dps*, *proP*, and *osmC* in *E. coli* (8, 50, 171, 172).

THE σ^S REGULON: PHYSIOLOGICAL FUNCTIONS OF σ^S-CONTROLLED GENES

Most of our knowledge of the physiology of the general stress response is based on the analysis of starving (i.e., stationary-phase) cells. Although σ^S itself is also induced in response to many other stresses (see below), this is not necessarily so for all σ^S-

dependent genes since their expression can be modulated by various coregulators that may or may not be present or active under different stress conditions. Therefore, the physiological consequences of the general stress response, i.e., an induction of σ^S, need not necessarily always be the same under different stress conditions. In addition, certain stress conditions often not only trigger the general stress response but at the same time also induce σ^S-independent responses. Thus, stationary-phase cells are also subject to oxygen control via the Arc system, and similar multiple overlapping responses can be observed in hyperosmotically shocked or low pH-exposed cells. Given this complexity, it is clear that we are still far from a deep understanding of the physiology of stressed cells.

The general stress response can serve both as a rapid emergency response and as a long-term program of adaptation to starvation and other stresses that involve dramatic changes in cellular physiology and morphology (for a summary, see reference 74). The emergency character is apparent in the extremely rapid and transient induction of σ^S and σ^S-dependent genes at the beginning of the classic diauxic lag phase of glucose/lactose-grown cells, even before β-galactosidase is induced (54). On the other hand, entry into stationary phase triggered by complete exhaustion of an essential nutrient is characterized by an adaptation phase that lasts approximately 4 h. During this period, starvation proteins are induced according to a temporal program, and multiple stress resistance gradually develops (66, 88). σ^S is crucial for this genetic program in early stationary phase and thus also sets the stage for the survival of long-term starvation (108, 124). In contrast to the developmental programs of sporulating bacteria, this genetic program does not involve a definite commitment but rather allows reversal at any point in case environmental conditions should improve. With these properties, the σ^S-mediated general stress response is the key to the remarkable adaptive flexibility of enteric bacteria.

In contrast to specific stress responses, which allow bacterial cells to cope with a single acute stress situation by eliminating the stress agent and/or repairing damage that has already occurred, the physiological function of the general stress response is predominantly preventive. It renders the cells broadly stress-resistant in a way that damage is avoided rather than has to be repaired. Even though a large number of σ^S-controlled genes probably still await identification, and functions have not yet been assigned to all known members of the σ^S regulon, the overall function of the general stress response becomes apparent from the relatively large number of genes whose expression enhances stress tolerance. By contrast, genes encoding proteins with direct repair functions seem to be scarce in the σ^S regulon. Beyond its role in stress resistance and survival, σ^S has been shown to control also some virulence genes and therefore may play a role in pathogenicity (see below). Moreover, σ^S may be involved in biofilm formation (1).

Genes That Confer Stress Resistance

In the laboratory, *E. coli rpoS* mutants are rapidly killed by oxidative stress (e.g., by hydrogen peroxide), by shift to high temperature (above 50°C), and upon exposure to high osmolarity, to acid pH, or to ethanol (52, 108, 112, 124, 159). Similar findings have been reported under more "natural" conditions, such as in seawater (134) or when inoculated in food sources (37).

σ^S-controlled genes involved in oxidative stress resistance have recently been reviewed (49). Crucial genes are *xthA*, which encodes exonuclease III, a component involved in DNA repair (45, 150), and *dps*, the structural gene for a histone-like protein that forms large clusters with DNA (4). The protective function of Dps is based on its ferritin-like structure that may act as a "trap" for iron and thus prevent the formation of hydroxyl radicals via the Fenton reaction in close vicinity of the DNA (64). Moreover, both catalases present in *E. coli* (HPI and HPII, encoded by *katG* and *katE*) are under σ^S control (83, 114), and the same has been shown for glutathione reductase (*gor*) (21). Most recently, a σ^S-controlled periplasmic superoxide dismutase (CuZnSOD, encoded by *sodC*) has been found to significantly contribute to H_2O_2 resistance, specifically during entry into stationary phase. In pathogens, this enzyme may play a role in protection against the oxidative burst of host macrophages (161).

The disaccharide trehalose is a general stress protectant in bacteria, plants, and lower animals (41, 170). In *E. coli*, it serves as an osmoprotectant (162); it is crucial for desiccation resistance (177) and also seems to play a minor role in thermotolerance (77). The *otsBA* operon, which encodes the enzymes involved in the production of trehalose, is under the control of σ^S (77, 95). Under conditions of high osmolarity, trehalose is made de novo (provided alternative compatible solutes such as glycine betaine or proline are not present in the medium). If present, external trehalose is degraded to glucose by a periplasmic trehalase, which is equally σ^S-dependent for expression (27, 77, 162) (also see chapter 6). In addition to trehalose, the partially σ^S-dependent *ecp-htrE* operon (encoding homologs of PapD and PapC, i.e., of proteins involved in the postsecretional assembly of pili) has been implicated in thermotolerance and osmotic resistance (145). Taken together, however, the genes identified so far cannot fully account for the strong σ^S-dependent stationary-phase thermotolerance, since null mutations in *otsBA* or *ecphtrE* confer much weaker phenotypes than that observed for *rpoS* mutants.

The σ^S-dependent acid resistance is incompletely understood, in part because it is tightly interwoven with non–σ^S-dependent mechanisms of acid tolerance (see also chapter 7). In *Shigella*, a mutation in *gadC* confers acid sensitivity. *gadC* may encode a glutamate/γ-aminobutyrate (GABA) antiporter, which together with glutamate decarboxylase (*gadB*) appears to play a role in pH homeostasis (175). In *E. coli*, the *gad* system is also involved in extreme acid resistance (pH 2.5). Either one of two glutamate decarboxylases (encoded by the σ^S-controlled *gadA* and *gadB* genes) can function in concert with the glutamate/GABA antiporter (encoded by *gadC*) (35) (also see chapter 7). In addition, the *hdeAB* operon, which encodes two small periplasmic proteins of unknown function, is important for acid resistance (175).

Recently, σ^S-dependent resistance against ethanol has been described that is mediated by the product of the *uspB* gene. UspB is probably a transmembrane protein that may somehow counteract the membrane-deleterious effects of ethanol (53).

In summary, our knowledge of stress-protective genes under σ^S control is incomplete. Moreover, even for most of the known genes, the molecular modes of action of their gene products are unknown or a matter of speculation. More studies will be necessary to really understand the physiology of σ^S-dependent stress tolerance.

Genes Affecting Programmed Cell Death in Stationary Phase

Whereas stress resistance as described above is effective at the single cell level, programmed cell death of part of a bacterial population may increase long-term survival, i.e., starvation tolerance of a population. Because of lysis, dead cells may provide nutrients that allow cryptic growth of a stationary-phase cell population (186), which thereby could be less affected by the detrimental consequences of stasis (48). The σ^S/OmpR-controlled entericidin locus (*ecnAB*) encodes an antidote/toxin system consisting of two similar lipoproteins that may induce programmed cell death in starving cells (23).

Another gene, *ssnA*, seems to promote actively the rapid decrease in viable cell numbers of a stationary-phase culture grown in rich medium. The

mechanisms of action of its gene product remains unknown, but in contrast to the *ecn* genes, *ssnA* is negatively regulated by σ^S, i.e., increased σ^S levels would protect against SsnA-mediated cell death (182).

Genes Affecting the Cell Envelope and Overall Morphology

σ^S is also involved in generating the smaller and more spherical or ovoid morphology typical for stationary-phase *E. coli* cells (107). This is due to σ^S control of the morphogen *bolA*, which appears to encode a regulator for the expression of penicillin-binding protein 6 (*dacC*) (2, 3, 24, 107). In addition, one of the *ftsQ* promoters is σ^S-controlled, and growth phase- and growth rate-dependent expression of the *ftsQAZ* operon may play a role in reductive cell division that results in smaller and ovoid cell shape (15).

The cell envelope is also affected by stationary phase. The σ^S-regulated genes *osmB*, *osmC*, and *osmE* have initially been identified using a *phoA* fusion approach selective for extracytoplasmic proteins (29, 39, 69, 93). Whereas *osmC* and *osmE* functions remain unknown, *osmB* encodes an outer membrane lipoprotein that has been implicated in cell aggregation (94). Also the expression of thin aggregative fimbriae (encoded by the *csg* and *agf* genes in *E. coli* and *Salmonella enteritis* serovar Typhimurium, respectively) is under σ^S control and affects cell-cell contacts, thereby producing a distinct colony morphology (10, 148). The role of *osmY*, which encodes a strongly inducible periplasmic protein, is still mysterious, although *osmY* has been used for many detailed regulatory studies (46, 105, 176, 179, 184, 185). Alterations in the membrane include the formation of cyclopropane fatty acids that depends on the σ^S-controlled *cfa* gene (173).

Genes That Redirect Metabolism

In stationary phase, major metabolic readjustments are required. These include decreased Krebs cycle activity combined with increased glycolytic activity, which reduces the danger of damage by oxygen radicals. However, these changes are not σ^S-dependent but require the Arc system as a global regulator (139). Nevertheless, σ^S is involved in the control of some adjustments. The σ^S-controlled acetyl-CoA synthase (*acs*) is important for the reutilization of acetate that is produced and excreted during rapid growth (156). Pyruvate metabolism may be altered due to σ^S-dependent induction of pyruvate oxidase (*poxB*), which produces acetate and CO_2 from pyruvate (36). σ^S also strongly influences trehalose metabolism. In growing nonstressed cells, growth on trehalose occurs via a phosphotransferase system (PTS) (27). Upon osmotic upshift, this PTS (specifically EII^{Tre}) is inhibited and external trehalose is degraded to glucose by a σ^S-controlled periplasmic trehalase (*treA*), and glucose is then taken up via the PTS^{Glu} (26, 68). In parallel, trehalose is synthesized de novo by trehalose-6-phosphate (Tre6P) synthase and Tre6P phosphatase, which are encoded by the σ^S-dependent *otsBA* operon, and now serves as a compatible solute and osmoprotectant (77, 95, 162). Cytoplasmic trehalase (*treF*) is similarly controlled and may serve in the degradation of the accumulated trehalose when external conditions improve again (80).

Virulence Genes

For pathogenic bacteria, the host organism provides a stressful environment. Upon infection, bacteria may have to deal with severe acid, high osmolarity, starvation, and oxidative stress. It is therefore not surprising that virulence genes, especially those involved in the initial phase of colonization or those that allow survival in a specific intracellular compartment, exhibit a strong regulatory overlap with stress-regulated genes (see chapter 20).

Salmonella rpoS mutants as well as σ^S-overproducing strains exhibit reduced virulence (19, 40, 51, 178). Although part of this phenotype may be due to general stress sensitivity or inappropriate regulation of stress tolerance and therefore reduced viability in the host, there is now increasing evidence that σ^S also controls specific virulence genes. A well-documented case is that of the *spv* genes in *Salmonella* that are encoded by a virulence plasmid and are necessary for *Salmonella* to grow in deep lymphoid organs such as spleen and liver (67). The *spv* operon requires both a local activator, SpvR, and the global regulator σ^S for expression (72, 100, 137, 146).

Pathogenic *E. coli* strains also make use of σ^S as a regulator for virulence genes. The *esp* genes, which encode a type III secretion system crucial for cellular adhesion of EHEC and EPEC strains, have been reported to be under σ^S control (22). Curli fimbriae (termed thin aggregative fimbriae in *Salmonella*), which have been implicated in adhesion to eukaryotic tissue and in bacterial aggregation, require σ^S for expression (141, 148). On the other hand, *E. coli* type 1 fimbriae are negatively controlled by σ^S (47).

In *Legionella pneumophila*, σ^S is induced in stationary phase, but it is not required for stationary phase-associated acid, hydrogen peroxide, or salt resistance of cells grown in vitro. *rpoS* mutants also can grow in and kill macrophages. By contrast, σ^S is es-

sential for survival in an environmental protozoan host, *Acanthamoeba castellanii* (71). Intracellular growth of *Legionella* (either in amoeba or in macrophages) results in distinct features such as smaller morphology and increased resistance against antibiotics and host defense mechanisms (16), but a potential role of *rpoS* in these phenotypes has not been studied.

Yersinia enterocolitica rpoS mutants do not seem to be affected in virulence (13, 82). However, a mutation in *nlpD*, which may be polar on *rpoS*, has recently been found to affect survival of *Y. enterocolitica* in mice (44). A *rpoS* mutant of *Pseudomonas aeruginosa* produces less exotoxin A and alginate but shows increased synthesis of pyocyanin and pyoverdine. The latter may explain enhanced virulence of the mutant in a rat chronic lung infection model (163). Information on *rpoS* in some plant pathogens, together with evidence for an effect on virulence, has now also become available (7, 55). Taken together, these findings may be just the tip of an iceberg, and further studies on the involvement of σ^S in virulence gene regulation are certainly warranted.

Secondary Regulatory Genes

Several σ^S-controlled genes have regulatory functions and thereby establish regulatory cascades within the σ^S regulon. Those for which specific target genes are known include *appY*, *rob*, and *bolA*. AppY is an AraC-like regulator that controls expression of the *hya* (encoding hydrogenase I) and *cbdAB-appA* operons (encoding a third cytochrome oxidase and an acid phosphatase) (12). Besides being regulated by σ^S, AppY is also under oxygen control mediated by the ArcB/ArcA two-component system (12, 31). The secondary regulator thus serves as the point of integration of additional environmental signals that thereby affect only a subset of σ^S-dependent genes.

BolA is involved in controlling cell shape (see above). Rob is a close homolog of SoxS and MarA, which respond to oxidative stress and certain antibiotics, respectively (9, 97). Rob, SoxS, and MarA can bind to a common motif present in the promoter regions of several target genes (see below), which thereby are activated also in response to stresses to which σ^S-dependent genes normally do not respond.

ROLE OF σ^S IN TRANSCRIPTIONAL INITIATION AT STRESS-RESPONSIVE PROMOTERS

σ^S is not a "classical" alternative sigma factor (as, for example, the heat shock factor σ^{32}) that transiently accumulates and more or less outcompetes σ^{70}, thereby reprogramming the transcription machinery to recognize alternative promoter sequences. Even under conditions of maximal expression, the cellular σ^S level always remains lower than that of σ^{70} (91), and moreover, these two very similar sigmas have the same basic in vitro promoter recognition specificity. Nevertheless, they clearly control different genes in vivo, since *rpoS* mutations strongly reduce or even eliminate the expression of many σ^S-regulated target genes, i.e., σ^{70} cannot compensate for the absence of σ^S (for review, see reference 74). The explanation for this promoter specificity paradox is probably provided by the complex interplay of σ^S and σ^{70} holoenzyme forms of RNA polymerase (Eσ^S and Eσ^{70}, respectively) with a number of additional global regulators, many of which are histone-like, i.e., chromatin-associated proteins (summarized in reference 76).

Formation of Active σ^S-Containing RNA Polymerase under Stress Conditions

The cellular σ^S levels observed in stationary phase or otherwise stressed cells reach approximately 30% of the σ^{70} level (91). Surprisingly, the affinity for core polymerase (E) turned out to be significantly lower for σ^S than for σ^{70} (101), indicating that some factor or factors crucial for the formation of active σ^S-containing RNA polymerase (Eσ^S) in stressed cells have so far escaped our attention. The same problem has recently also been recognized for the replacement of vegetative RNA polymerase (Eσ^A) by the forespore-specific Eσ^F during sporulation of *Bacillus subtilis* (119).

It could be that the formation of σ^{70}-containing holoenzyme (Eσ^{70}) is somehow actively inhibited. A recently discovered stationary phase-inducible σ^{70}-binding protein, Rsd, may play such a role, even though its relatively low level of abundance seems to argue against a direct stoichiometric mechanism of action (90). On the other hand, the formation of active Eσ^S may be actively stimulated. Genetic data demonstrate that the *crl* gene promotes the expression of σ^S-dependent genes without increasing σ^S levels but probably does not code for a transcriptional activator (144). Thus, the Crl protein may be involved in activation of σ^S, possibly by promoting holoenzyme formation. In addition, covalent modification of RNA polymerase core in stationary phase (142) may also favor σ^S holoenzyme formation or activity.

Interaction of Eσ^S and Eσ^{70} with Promoter DNA

In vitro experiments have clearly indicated that Eσ^S and Eσ^{70} can recognize the same promoter se-

quences, and under standard in vitro transcription conditions, even yield more or less similar amounts of transcript with many templates (46, 135, 165, 166). A SELEX in vitro evolution approach yielded completely identical consensus sequences (R. Gourse, personal communication). These findings are consistent with the strong similarity of the promoter recognizing regions 2.4 and 4.2 in the two sigma factors.

Nevertheless, the actual details of promoter binding are slightly different. In general, Eσ^S seems to tolerate deviations from the consensus better than Eσ^{70} (R. Gourse, personal communication). Genetic studies have suggested that recognition in the -35 part of the promoter is less strictly required for Eσ^S than for Eσ^{70} (79, 165). These results correlate well with in vitro data that demonstrate that footprint patterns and open complex formation observed with the two holoenzymes at specific promoters are somewhat different (122, 165). In addition, Eσ^S contacts (as probed with the chemical nuclease FeBABE tethered to a number of amino acid positions in σ^S and σ^{70}) seem to be relatively compact and centered between positions -25 and $+2$ on the promoter DNA (38). By contrast, Eσ^{70} contacts are scattered along a more extended promoter sequence including strong interaction between the -35 region of a promoter and region 4.2 in σ^{70} (30). A region 2.5 involved in a contact with an extended -10 promoter region has been defined for σ^{70} (17). Genetic and biochemical evidence indicates that a region 2.5 is also important for σ^S (20, 38). Interestingly, a key amino acid in region 2.5 is different in the two sigmas (K173 in σ^S, which corresponds to E458 in σ^{70}).

Even though these differences may seem minor, they appear to be crucial for differential promoter recognition in vivo as described below.

Complex Molecular Architecture of Stress-Activated Promoters

A fair number of promoter regions that are controlled by σ^S in vivo have now been analyzed in some detail. Several genetic studies have demonstrated that in addition to their dependency on σ^S, these genes are often differentially controlled by other global regulators, which include cAMP-CRP, leucine-responsive regulatory protein (Lrp), integration host factor (IHF), FIS, and H-NS. Depending on the location of specific binding sites, these abundant and often histone-like factors can act positively or negatively. Target gene-specific combinations of these regulators can result in distinct regulatory patterns that are superimposed upon the regulation of σ^S (75). If the additional regulator is itself under environmental control, this can efficiently restrict the expression of the target gene to very specific conditions (104, 122, 180) (see below). Some of the additional regulators, most notably Lrp, can also bind to multiple sites in a single promoter region, which may result in a wrapping of the DNA around multiple Lrp molecules (122). In general, the regulatory factors involved also bend DNA. All this indicates that in vivo complex nucleoprotein structures are formed at these stress-activated promoters.

The σ^S/σ^{70} Promoter Specificity Paradox: Role of Histone-Like and Other Global Regulators

The finding that $E\sigma^S$ and $E\sigma^{70}$ recognize many promoter sequences equally well in vitro but differentially control genes in vivo has provided strong evidence that the current concept of recognition specificity of sigma factors that is based on specific interactions of certain regions of sigma with certain nucleotides of the promoter is incomplete. At certain promoters, in vitro preference for $E\sigma^S$ can be somewhat improved by variations in salts or superhelical density of the template or the addition of trehalose (46, 101, 102). This indicates that the two holoenzymes have slightly different preferences with respect to certain physical parameters. For instance, promoter sequences, which in vivo are σ^S-dependent, may have evolved to function optimally in the presence of high concentrations of potassium glutamate (which accumulates in osmotically stressed cells). Local superhelical density may be relevant for $E\sigma^S$ preference rather than average supercoiling density of the whole chromosome or large domains thereof. The stimulatory effect of trehalose does not seem to be relevant in vivo (60).

There is increasing evidence that the additional regulators mentioned above play a crucial role in establishing sigma factor specificity at stress-inducible promoters (summarized in reference 76). At many stress-inducible promoters, H-NS protein strongly inhibits transcription by $E\sigma^{70}$, whereas $E\sigma^S$ can overcome this inhibition. This becomes apparent when expression of a gene, which is strongly reduced in strains containing a mutation in *rpoS* alone, is partially or even completely restored when a secondary mutation in *hns* is introduced (10, 18, 183). A similar differential effect on expression mediated by $E\sigma^S$ and $E\sigma^{70}$ was also observed for other regulatory proteins. These include cAMP-CRP, Lrp, and IHF, which act as negative modulators at the *osmY* promoter (105; F. Colland, M. Barth, R. Hengge-Aronis, and A. Kolb, submitted). At the *osmC*p1 promoter, Lrp interferes specifically with activation by $E\sigma^S$ (63). In in vitro experiments with the *csiD* promoter, cAMP-CRP and Lrp activate more strongly in conjunction with $E\sigma^S$ than with $E\sigma^{70}$, whereas H-NS displays a differential negative effect with the two holoenzymes (122).

Taken together, these examples provide initial evidence that certain histone-like proteins and other global regulators can crucially contribute to generating sigma factor specificity. In vivo, RNA polymerase holoenzyme does not find naked DNA but has to deal with complex three-dimensional nucleoprotein structures, in which promoter DNA could be bent, distorted, or only partially accessible. Within this context, minor differences in the mode of binding and open complex formation observed for $E\sigma^S$ and $E\sigma^{70}$ could translate into ultimately large differences in their abilities to initiate transcription at specific promoters. In addition, sigma factor specificity may not only depend on protein-DNA contacts but could be modulated by multiple interactions that also include protein-protein contacts.

REGULATION OF THE CELLULAR σ^S LEVEL

In rapidly growing cells not subject to any particular stress, σ^S is hardly detectable. However, σ^S rapidly accumulates in response to many different stress conditions. These include starvation for sources of carbon, nitrogen, or phosphorus or for amino acids; shift to high osmolarity or continuous growth under high osmolarity conditions; shift to high or low temperature (42°C or below 30°C, respectively); shift to acid pH; and probably high cell density (19, 59, 89, 91, 109, 126, 130, 158). How can so many different conditions be integrated to affect a single parameter, i.e., the cellular σ^S content? A first step toward an answer to this question was the observation that *rpoS* is regulated at multiple levels, which include transcription, translation, and σ^S proteolysis, with different stresses affecting different levels of control (19, 109, 117, 123, 126, 130, 158). This also suggests that there is no single common cellular signal reflecting all the different stress conditions. Posttranscriptional control seems particularly important, with most stresses mentioned above influencing the efficiency of *rpoS* mRNA translation or the rate of σ^S degradation (Fig. 1). In general, potentially lethal stress conditions, which may require the most rapid response, affect σ^S proteolysis, whereas a more gradual deterioration of environmental conditions tends to stimulate *rpoS* transcription or translation. The complexity of σ^S regulation is illustrated by an ever increasing number of *cis*-acting elements and *trans*-acting factors involved in this control (summarized in Table 1 and described below).

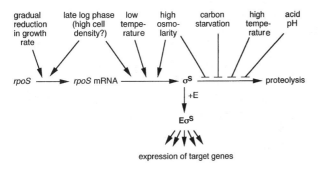

Figure 1. Different stresses activate the general stress response by affecting either transcription, translation, or proteolysis of the master regulator σ^S. (This figure is a modified version of an already published figure [74].)

Regulation of *rpoS* Transcription

Although σ^S levels are very low during exponential growth, polycistronic *nlpD-rpoS* mRNA as well as a monocistronic *rpoS* transcript originating from two closely spaced promoters upstream of the *nlpD* gene ($nlpD_{P1+P2}$) and from a promoter within the *nlpD* gene (*rpoSp*), respectively, are synthesized (10, 106, 110, 164). Studies with transcriptional *rpoS* ::*lacZ* fusions indicate that the major *rpoS* transcript originates at *rpoSp*. This transcript further accumulates during late log phase and entry into stationary phase when cells are grown in rich medium. Growth in minimal medium, however, produces little or no increase (108, 109, 123, 133, 153). *rpoS* transcription is not autoregulated (R. Lange and R. Hengge-Aronis, unpublished results).

cAMP-CRP is a negative regulator of *rpoS* transcription. *crp* and *cya* mutants exhibit increased levels of σ^S and increased activity of transcriptional *rpoS* ::*lacZ* fusions, and both are reversed by the addition of cAMP to a *cya* mutant (109). On the other hand, certain *rpoS* ::*lacZ* fusions showed reduced expression in a *cya* background for reasons that have not

been clarified (123). A putative cAMP-CRP binding site is present upstream of *rpoSp*, but its actual role has not been investigated. Besides CRP, no other regulatory protein has been found to be involved in *rpoS* transcription. However, small molecules can influence *rpoS* expression. σ^S levels correlate with guanosine-3′,5′-bispyrophosphate (ppGpp), and it was shown that ppGpp acts at the level of *rpoS* transcription, probably by positively affecting transcriptional elongation (59, 106). In addition, homoserine lactone seems to play an uncharacterized positive role in *rpoS* transcription (81). Finally, a positive role has been suggested for polyphosphate, but this role may be indirect since polyphosphate did not stimulate in vitro transcription of *rpoS* (155).

Taken together, the regulation of *rpoS* transcription and transcript stability has not been studied in detail. This is probably due to the fact that most environmental signals that induce σ^S do not affect *rpoS* transcript levels. Therefore, translational and posttranslational control of *rpoS* has been analyzed in much more detail.

Regulation of *rpoS* Translation

First evidence for a posttranscriptional control of *rpoS* was provided by strikingly different patterns of expression observed for transcriptional and translational *rpoS* ::*lacZ* fusions (109, 117, 123). Subsequently, it was demonstrated that translation of *rpoS* mRNA is stimulated by shift to high osmolarity or low temperature as well as during late exponential phase (provided cultures reach a certain minimal optical density) (109, 130, 158). After the onset of starvation, *rpoS* translation is reduced again (as is protein synthesis in general), and the continuing increase in σ^S levels is then due to enhanced stability of σ^S in cells that have ceased to grow and divide (109).

The translational initiation region (TIR) of the *rpoS* mRNA, which includes the ribosomal binding

Table 1. *cis*-acting elements and *trans*-acting factors in the control of $\sigma^{S\,a}$

Element or factor	*rpoS* transcription	*rpoS* translation	σ^S proteolysis
cis-acting elements	$nlpD_{p1+p2}$, $rpoS_p$	Translational initiation region (TIR) in a specific secondary structure	Turnover element (crucial amino acids: K173, E174, V177)
trans-acting factors			
Macromolecular regulators	CRP-cAMP (−)	Hfq (+)	ClpXP (+)
		DsrA RNA (+)	RssB (+)
		OxyS RNA (−)	RssA (−)
		H-NS (−)	H-NS (+)
		LeuO (−)	LrhA (+)
		DnaK (+)	DnaK (−)
Small molecules	ppGpp (+), homoserine lactone (+)	UDP-glucose (−)	Acetyl phosphate (+)

^{a} Plus and minus signs indicate whether a factor stimulates or inhibits the corresponding step in σ^S regulation (for further details and references, see text).

site (or SD sequence), the initiation codon, and a putative downstream box, is predicted to form extensive secondary structures, in which much of the TIR would be base-paired and nonaccessible for ribosomes (33, 34, 42, 128, 189). An obvious hypothesis is that the stress conditions mentioned above trigger an alteration of this mRNA secondary structure, which should improve accessibility of the TIR, but this has yet to be demonstrated directly. There is recent genetic and biochemical evidence that the secondary structure involving the TIR has to be in a finely balanced semistable state for translation to be possible (increasing the stability of single base pairs by site-directed mutagenesis without affecting overall structure can virtually eliminate translation [S. Bouché, D. Traulsen, and R. Hengge-Aronis, unpublished results]).

The RNA-binding protein Hfq is required for *rpoS* translation (33, 128, 129). This protein, which is also known as host factor I (HF-I), was first identified as an essential host-provided component of phage Qβ RNA replicase (56, 57). Hfq binds to *rpoS* mRNA in vitro (D. Traulsen and R. Hengge-Aronis, unpublished results). Hfq does not resemble RNA helicases, but by just binding it may contribute to maintaining the secondary structure around the *rpoS* TIR in the semistable state required for translation to be inducible. It is not yet clear whether and how Hfq activity is controlled by environmental signals.

H-NS protein, even though known as a DNA-binding protein, plays an uncharacterized inhibitory role in *rpoS* translation. It could be that H-NS binds to *rpoS* mRNA (StpA, a close homolog of H-NS, even acts as a RNA chaperone) (43). It may also bind to Hfq (the two proteins in fact copurify) (96) and perhaps influence Hfq activity or availability. Further work will have to clarify these issues. Data with translational *rpoS*::*lacZ* fusions indicate that the DnaK chaperone also may be involved in controlling *rpoS* translation, but the actual mechanism has not been studied (126).

Two small regulatory RNAs play a role in *rpoS* translational control. One is DsrA RNA, which is induced upon temperature downshift and is required for low temperature induction of σ^S (157, 158). It is partially complementary to the region of *rpoS* mRNA that base pairs with the TIR, i.e., DsrA could stimulate *rpoS* translation by an anti-antisense mechanism (111, 120). The LysR-like regulator LeuO inhibits expression of DsrA, and thereby can negatively influence *rpoS* translation, especially at low temperature. However, in wild-type cells, *leuO* is strongly repressed or even "silenced" by H-NS under laboratory conditions, and therefore its physiological role is unclear (98). The second small regulatory RNA, OxyS,

is induced in response to oxidative stress (hydrogen peroxide). It interferes with *rpoS* translation possibly by forming a translationally incompetent complex with Hfq and *rpoS* mRNA (189) (see chapter 4). Thus, OxyS may interfere with unnecessary induction of σ^S and the general stress response under conditions that can be dealt with by inducing the stress-specific OxyR-dependent response alone.

Finally, a metabolic product, UDP-glucose, negatively influences σ^S levels (25) and has been found to interfere with *rpoS* translation (A. Muffler and R. Hengge-Aronis, unpublished results). Its actual mechanism of action is under investigation.

With the exception of the low temperature-induced DsrA RNA and the hydrogen peroxide-induced OxyS RNA, it has not yet been possible to connect any of these *trans*-acting factors to specific environmental signals that affect *rpoS* translation. Thus, the mechanisms that stimulate *rpoS* translation in response to high osmolarity or high cell density have yet to be clarified.

Although the wealth of data described above support *rpoS* translational regulation, its mere existence has been questioned (188). This was probably due to a misunderstanding based on the still relatively common narrow view of σ^S as a stationary phase-specific sigma factor. In fact, *stationary-phase* stimulation of *rpoS* translation has never been described or claimed to exist. On the contrary, like for most other mRNAs, the translation of *rpoS* mRNA is reduced after the onset of starvation (109, 188), probably simply because of a shortage of amino acyl tRNAs. If cells are not exposed to low temperature or high osmolarity, translational induction of *rpoS* can only be observed during late log phase, provided the growth medium contains sufficient carbon source, so that the culture reaches a certain minimal density. This is a late exponential phase phenomenon that may be related to cell density but is not connected to starvation or stationary phase (109).

Regulation of σ^S Proteolysis

In exponentially growing cells, there is readily measurable synthesis of σ^S, but actual σ^S levels are very low because of continuous proteolysis. Under these conditions, σ^S half-lives are in the 1 to 4 min range (109, 126, 130, 154, 164). However, σ^S is rapidly stabilized when cells experience carbon starvation or are shifted to high osmolarity, high temperature, or acidic pH (19, 109, 126, 130, 164). Again, the question arises of how so many different signals can be integrated to rapidly affect a single process, i.e., σ^S proteolysis.

The protease involved is ClpXP (154), a complex ATP-dependent processive protease of proteasome-like shape (65, 174). At least during the onset of starvation, the cellular level of this protease does not change (154), indicating that some other more specific process controls the rate of σ^S degradation. A second protein absolutely essential for σ^S proteolysis is the response regulator RssB (also known as SprE or MviA, the latter in *Salmonella*) (19, 127, 143). With a relatively large output domain not homologous to any protein of known function, RssB seems to be a very special response regulator. Recently, RssB was found to bind directly to σ^S. Thus, RssB is a direct recognition factor that acts in the initiation of σ^S degradation, probably by presenting σ^S to ClpXP protease (20) (Fig. 2). RssB may be specific for σ^S, because another ClpXP substrate, λO protein, does not require it for degradation (190). Moreover, the finding that phosphorylation of the RssB receiver domain (at aspartate 58) strongly stim-

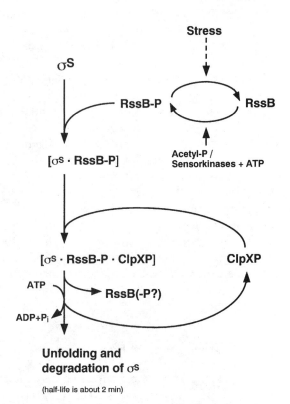

Figure 2. Model of σ^S recognition by the ClpXP proteolytic machinery. Under nonstress conditions, σ^S is recognized by phosphorylated RssB and is transferred to the ClpXP complex where it is degraded rapidly and completely. RssB is not codegraded (E. Klauck and R. Hengge-Aronis, unpublished results). Stress conditions probably result in dephosphorylation of RssB, which results in reduced affinity for σ^S and consequently its stabilization. It is not yet known whether all stresses that affect σ^S proteolysis act by dephosphorylating RssB. Also, the details of the signal transduction pathway(s) that affect RssB phosphorylation remain to be elucidated.

ulates σ^S binding suggests that environmental signals control σ^S proteolysis by altering RssB affinity for σ^S via changes in the phosphorylation of the RssB receiver domain (20). The components of the upper part of this complex signal transduction pathway(s) still await identification. In particular, a cognate sensor kinase has not been found, and the *E. coli* genome sequence does not provide any obvious candidate. It could be that phosphorylation of RssB is due to acetyl phosphate and cross-talk from other sensor kinases (28), with dephosphorylation being the stress-regulated process in the control of RssB activity. So far it is not clear whether all signals that affect σ^S proteolysis are transmitted via RssB.

σ^S proteolysis is also dependent on a *cis*-acting element (the "turnover element"), whose existence was originally derived from the finding that hybrid proteins with relatively large N-terminal σ^S moieties are degraded just as σ^S itself, whereas shorter hybrids are stable (130, 154). Recently, the core of the turnover element has been localized to K173, E174, and V177 of σ^S. Moreover, this element constitutes the site of recognition by RssB, with K173 being absolutely crucial for σ^S degradation in vivo and RssB binding in vitro (20). Strikingly, K173 is not only located just downstream of the −10 promoter recognizing region 2.4 in σ^S, but as part of a region 2.5, seems to be itself involved in transcription initiation (20). It may recognize an extended −10 region in analogy to the role of the corresponding region in σ^{70} (17, 30). Consistent with this overlap of σ^S regions crucial for transcription initiation and for proteolysis, RssB can also inhibit σ^S activity. This becomes apparent when proteolysis is slowed down or eliminated (e.g., in stationary phase or in a *clpP* mutant background) and the cellular RssB:σ^S ratio is increased to quasistoichiometry (20a, 190). In other words, RssB can act like an anti-sigma factor, and it is conceivable, that in the course of evolution, RssB originally was an anti-sigma factor, which has then been recruited to serve as a specific substrate recognition factor for the ClpXP proteolytic machinery.

While the role of RssB is now becoming clear, a few more factors have been implicated in σ^S proteolysis, but so far their mechanisms of action have remained obscure. RssA, which is encoded by a gene located just upstream of *rssB*, seems to somehow reduce σ^S proteolysis (127; G. Kampmann and R. Hengge-Aronis, unpublished results). In stationary phase, and perhaps also in heat-shocked cells, the DnaK chaperone may protect σ^S against proteolysis by an unknown mechanism (126, 147). On the other hand, the histone-like protein H-NS appears to promote σ^S turnover, but this effect could well be indirect, with H-NS affecting the expression of some other component involved in the control of σ^S proteolysis (18, 183). LrhA, a LysR-like transcriptional regulator, which genetic studies have placed upstream of RssB in the signal transduction cascade, also seems to affect σ^S proteolysis in such an indirect way (61).

Even though σ^S turnover has become one of the best-characterized examples of regulated proteolysis in prokaryotes, the signal transduction and integration mechanisms involved seem so complex that they will keep researchers busy perhaps for years to come.

THE σ^S-MEDIATED GENERAL STRESS RESPONSE AS PART OF THE OVERALL CELLULAR REGULATORY NETWORK

At many points in the general stress response regulatory network, connections to other regulatory circuits are now becoming apparent. These can be seen as first glimpses of the overall cellular regulatory network that ultimately governs and integrates metabolism, survival, growth, and proliferation of bacterial cells during good and bad times. Due to hierarchical, often cascade-like regulatory dependencies and complex composite promoter structures, there are numerous possibilities to tie together signal transduction and regulatory pathways of the global regulatory circuits involved.

There are many examples of genes that can be activated either by σ^S or by some other global regulator (the latter acting together with σ^{70}-containing RNA polymerase). These genes are thus recruited to different regulons, depending on the stress imposed. The targets of this regulation can be the promoters of structural genes, as in the case of *dps*, *katG*, and *gor*, which are under the control of σ^S (and therefore activated by stresses that induce σ^S) or of OxyR/σ^{70} (and thereby respond to hydrogen peroxide stress) (5, 21, 83). Alternatively, the common target of σ^S and another global regulator can be a secondary regulator, with the result that a "subregulon" is under common control. This is exemplified by the AppY regulator, which is under oxygen control via the Arc two-component system besides being under general stress control via σ^S (11, 12, 31, 32; L. Brøndsted and T. Atlung, submitted). Even more complex is the case of a group of stress-responsive genes that includes *zwf*, *nfo*, *sodA*, *micF*, *fpr*, and *fumC*, which can be activated by either one of three closely homologous regulators, SoxS, MarA, and Rob, which all bind to the same region in the promoters of the target genes (9, 86, 87, 113). These direct regulators are in turn activated by SoxR, MarR, and σ^S, which respond to superoxide anion, antibiotics, and multiple stresses, respectively (70, 97, 138). Here, σ^S is a member of one of several regulatory cascades that operate in parallel, and ultimately all activate the same set of structural genes.

In contrast to the cases just described, in which either σ^S or some other global regulator(s) activate certain target genes, regulatory arrangements have also been found in which both σ^S and another global regulator are required for expression. Such double dependencies can efficiently restrict target gene expression to very specific situations. Several genes have been found to require cAMP-CRP besides σ^S for expression, with *csiD* being the best-studied (36, 121, 122). Due to having high cAMP levels as a prerequisite, *csiD* expression occurs only in carbon-starved cells (122). A similar example is the *proP*p2 promoter, which requires both Fis and σ^S for activity. Fis is strongly upregulated in freshly diluted cells, but its cellular level then declines throughout the growth cycle (14, 136). Consequently, *proP*p2 is active only (i) during a narrow time window in late exponential phase where Fis and σ^S are both present or (ii) in exponentially growing cells that are exposed to some stress that induces σ^S (e.g., high osmolarity) (180, 181).

Finally, there are numerous possibilities for σ^S and other global regulators to influence each other. Besides controlling many other genes, cAMP-CRP and H-NS contribute to keeping σ^S levels low in nonstressed growing cells. While cAMP-CRP reduces *rpoS* transcription (108, 109), H-NS negatively affects *rpoS* translation and σ^S stability (18, 183). In both cases, it is not clear whether these effects are direct or indirect. Also, it has yet to be shown whether signals that control σ^S levels are transmitted through CRP or H-NS. Alternatively, CRP or H-NS may just be required for the expression of some other factors involved in σ^S regulation, whose activity would then be controlled by stress signals independently of CRP or H-NS. Putative (stress?) signals that affect H-NS levels or activity remain unknown. However, H-NS virtually silences some *E. coli* genes, which nevertheless are stably maintained. This indicates that there must be unknown conditions that downregulate H-NS levels or activity, which in turn would probably result in massive accumulation of σ^S and therefore strong activation of the general stress response. In addition, H-NS as well as CRP interferes with activation of some target promoters by Eσ^{70}, and thereby contribute to produce σ^S specificity at these promoters (10, 18, 122, 183).

Another complex interplay between σ^S and other global regulators involves OxyR protein and the small OxyS RNA. OxyS is itself part of the oxidative stress-induced OxyR regulon (which overlaps with the σ^S regulon also at the target gene level, i.e., in the control of *dps*, *katG*, and *gor*; see above) (6). OxyS can directly inhibit *rpoS* translation (by forming a complex with *rpoS* mRNA and Hfq). This may avoid the induction of the complete σ^S-dependent general stress response under conditions where the oxidative stress response is sufficient (189). On the other hand, σ^S can negatively or positively affect OxyR expression in stationary phase or after osmotic upshift, respectively (62, 125). The physiological implications of this complex interplay among σ^S, OxyR, and OxyS are not yet apparent.

PERSPECTIVES

Even though the general stress response in *E. coli* has been under intense investigation, a number of crucial features remain to be understood. Many σ^S-dependent genes still await identification and/or functional characterization before we can fully appreciate the physiological implications of the general stress response. Concerning regulation of σ^S levels, the upstream parts of the signal transduction pathways involved are virtually unknown, and therefore the question of integration of multiple stress signals has not yet been answered. How can high osmolarity or a putative cell-density signal in late log phase trigger rearrangements in *rpoS* mRNA secondary structure and thereby stimulate translation initiation? σ^S degradation is a prototypical example of regulated proteolysis in prokaryotes. The identification of the response regulator RssB as an essential direct recognition factor for σ^S proteolysis was crucial, but what stresses affect RssB activity, i.e., phosphorylation of its receiver domain, and how do they do so? How does RssB bind σ^S, and does it also interact with ClpX and/or ClpP? The analysis of σ^S as a transcription factor is also providing valuable new insights. Abundant DNA-binding proteins, e.g., H-NS, play a crucial role in establishing σ^S specificity at stress-regulated promoters; this indicates that sigma specificity can be more than recognition of the nucleotides that constitute the $-35/-10$ promoter regions. RNA polymerase does not "see" naked linear DNA, but rather complex nucleoprotein structures in which promoter DNA can be bent or otherwise distorted or only partially accessible. In such a context, protein-protein interactions may also be important for generating sigma factor specificity (and thereby coupling of gene expression to environmental signal transduction), a situation that is strongly reminiscent of regulation at eukaryotic promoters. A complete description of all these interactions and their consequences at certain model promoters is a challenge for the future. Finally, the mechanism and selective pressure

underlying the strong genetic variability of *rpoS* in stationary phase (186) (see also chapter 15), as well as a putative role of σ^S in stationary-phase "adaptive" mutagenesis (see chapter 16), promise to be rewarding subjects for future studies.

Acknowledgments. Research has been supported by the Deutsche Forschungsgemeinschaft (Priority Program "Regulatory Networks in Bacteria"; SFB 156-A8; Gottfried-Wilhelm-Leibniz Program), the State of Baden-Württemberg (Landesforschungspreis), and the Fonds der Chemischen Industrie.

REFERENCES

1. **Adams, J. L., and R. J. C. MacLean.** 1999. Impact of *rpoS* deletion on *Escherichia coli* biofilms. *Appl. Environ. Microbiol.* **65:**4285–4287.
2. **Aldea, M., T. Garrido, C. Hernández-Chico, M. Vicente, and S. R. Kushner.** 1989. Induction of a growth-phase-dependent promoter triggers transcription of *bolA*, an *Escherichia coli* morphogene. *EMBO J.* **8:**3923–3931.
3. **Aldea, M., C. Hernandez-Chico, A. G. de la Campa, S. R. Kushner, and M. Vicente.** 1988. Identification, cloning and expression of *bolA*, an *ftsZ*-dependent morphogene of *Escherichia coli*. *J. Bacteriol.* **170:**5169–5176.
4. **Almirón, M., A. Link, D. Furlong, and R. Kolter.** 1992. A novel DNA binding protein with regulatory and protective roles in starved *Escherichia coli*. *Genes Dev.* **6:**2646–2654.
5. **Altuvia, S., M. Almirón, G. Huisman, R. Kolter, and G. Storz.** 1994. The *dps* promoter is activated by OxyR during growth and by IHF and σ^S in stationary phase. *Mol. Microbiol.* **13:**265–272.
6. **Altuvia, S., D. Weinstein-Fischer, A. Zhang, L. Postow, and G. Storz.** 1997. A small, stable RNA induced by oxidative stress: roles as a pleiotropic regulator and antimutator. *Cell* **90:**43–53.
7. **Andersson, R. A., E. T. Palva, and M. Pirhonen.** 1999. The response regulator ExpM is essential for the virulence of *Erwinia carotovora* subsp. *carotovora* and acts negatively on the sigma factor RpoS (σ^S). *Mol. Plant-Microbe Interact.* **12:**575–584.
8. **Antelmann, H., S. Engelmann, R. Schmid, A. Sorokin, A. Lapidus, and M. Hecker.** 1997. Expression of a stress- and starvation-induced *dps/pexB*-homologous gene is controlled by the alternative sigma factor σ^B in *Bacillus subtilis*. *J. Bacteriol.* **179:**7251–7256.
9. **Ariza, R. R. Z., Z. Li, N. Ringstad, and B. Demple.** 1995. Activation of multiple antibiotic resistance and binding of stress-inducible promoters by *Escherichia coli* Rob protein. *J. Bacteriol.* **177:**1655–1661.
10. **Arnqvist, A., A. Olsén, and S. Normark.** 1994. σ^S-dependent growth-phase induction of the *csgBA* promoter in *Escherichia coli* can be achieved in vivo by σ^{70} in the absence of the nucleoid-associated protein H-NS. *Mol. Microbiol.* **13:**1021–1032.
11. **Atlung, T., and L. Brøndsted.** 1994. Role of the transcriptional activator AppY in regulation of the *cyx appA* operon of *Escherichia coli* by anaerobiosis, phosphate starvation and growth phase. *J. Bacteriol.* **176:**5414–5422.
12. **Atlung, T., K. Knudsen, L. Heerfordt, and L. Brøndsted.** 1997. Effect of σ^S and the transcriptional activator AppY on induction of the *Escherichia coli hya* and *cbdAB-appA* operons in response to carbon and phosphate starvation. *J. Bacteriol.* **179:**2141–2146.

13. Badger, J. L., and V. L. Miller. 1995. Role of RpoS in survival of *Yersinia enterocolitica* to a variety of environmental stresses. *J. Bacteriol.* **177:**5370–5373.

14. Ball, C. A., R. Osuna, K. C. Ferguson, and R. C. Johnson. 1992. Dramatic changes in Fis levels upon nutrient upshift in *Escherichia coli. J. Bacteriol.* **174:**8043–8056.

15. Ballesteros, M., S. Kusano, A. Ishihama, and M. Vicente. 1998. The *ftsQ*1p gearbox promoter of *Escherichia coli* is a major sigma S-dependent promoter in the *ddlB-ftsA* region. *Mol. Microbiol.* **30:**419–430.

16. Barker, J., H. Scaife, and M. Brown. 1995. Intraphagocytic growth induces an antibiotic-resistant phenotype of *Legionella pneumophila. Antimicrob. Agents Chemother.* **39:**2684–2688.

17. Barne, K. A., J. A. Bown, S. J. W. Busby, and S. D. Minchin. 1997. Region 2.5. of the *Escherichia coli* RNA polymerase σ^{70} subunit is responsible for the recognition of the "extended −10" motif at promoters. *EMBO J.* **16:**4034–4040.

18. Barth, M., C. Marschall, A. Muffler, D. Fischer, and R. Hengge-Aronis. 1995. A role for the histone-like protein H-NS in growth phase-dependent and osmotic regulation of σ^S and many σ^S-dependent genes in *Escherichia coli. J. Bacteriol.* **177:**3455–3464.

19. Bearson, S. M. D., W. H. Benjamin, Jr., W. E. Swords, and J. W. Foster. 1996. Acid shock induction of RpoS is mediated by the mouse virulence gene *mviA* of *Salmonella typhimurium. J. Bacteriol.* **178:**2572–2579.

20. Becker, G., E. Klauck, and R. Hengge-Aronis. 1999. Regulation of RpoS proteolysis in *Escherichia coli*: the response regulator RssB is a recognition factor that interacts with the turnover element in RpoS. *Proc. Natl. Acad. Sci. USA* **96:**6439–6444.

20a. Becker, G., E. Klauck, and R. Hengge-Aronis. *Mol. Microbiol.*, in press.

21. Becker-Hapak, M., and A. Eisenstark. 1995. Role of *rpoS* in the regulation of glutathione oxidoreductase (*gor*) in *Escherichia coli. FEMS Microbiol. Lett.* **134:**39–44.

22. Beltrametti, F., A. U. Kresse, and C. A. Guzmán. 1999. Transcriptional regulation of the *esp* genes of enterohemorrhagic *Escherichia coli. J. Bacteriol.* **181:**3409–3418.

23. Bishop, R. E., B. K. Leskiw, R. S. Hodges, C. M. Kay, and J. H. Weiner. 1998. The entericidin locus of *Escherichia coli* and its implications for programmed bacterial cell death. *J. Mol. Biol.* **280:**583–596.

24. Bohannon, D. E., N. Connell, L. K., A. Tormo, M. Espinosa-Urgel, M. M. Zambrano, and R. Kolter. 1991. Stationary-phase-inducible "gearbox" promoters: differential effects of *katF* mutations and role of σ^{70}. *J. Bacteriol.* **173:**4482–4492.

25. Böhringer, J., D. Fischer, G. Mosler, and R. Hengge-Aronis. 1995. UDP-glucose is a potential intracellular signal molecule in the control of expression of σ^S and σ^S-dependent genes in *Escherichia coli. J. Bacteriol.* **177:**413–422.

26. Boos, W., U. Ehmann, E. Bremer, A. Middendorf, and P. Postma. 1987. Trehalase of *Escherichia coli. J. Biol. Chem.* **262:**13212–13218.

27. Boos, W., U. Ehmann, H. Forkl, W. Klein, M. Rimmele, and P. Postma. 1990. Trehalose transport and metabolism in *Escherichia coli. J. Bacteriol.* **172:**3450–3461.

28. Bouché, S., E. Klauck, D. Fischer, M. Lucassen, K. Jung, and R. Hengge-Aronis. 1998. Regulation of RssB-dependent proteolysis in *Escherichia coli*: a role for acetyl phosphate in a response regulator-controlled process. *Mol. Microbiol.* **27:**787–795.

29. Bouvier, J., S. Gordia, G. Kampmann, R. Lange, R. Hengge-Aronis, and C. Gutierrez. 1998. Interplay between global regulators of *Escherichia coli*: effect of RpoS, H-NS and Lrp on

30. Bown, J. A., J. T. Owens, C. F. Meares, N. Fujita, A. Ishihama, S. J. W. Busby, and S. D. Minchin. 1999. Organisation of open complexes at *Escherichia coli* promoters: location of promoter DNA sites close to region 2.5 of the σ^{70} subunit of RNA polymerase. *J. Biol. Chem.* **274:**2263–2270.

31. Brøndsted, L., and T. Atlung. 1994. Anaerobic regulation of the hydrogenase 1 (*hya*) operon of *Escherichia coli. J. Bacteriol.* **176:**5423–5428.

32. Brøndsted, L., and T. Atlung. 1996. Effect of growth conditions on expression of the acid phosphatase (*cyx-appA*) operon and the *appY* gene, which encodes a transcriptional activator of *Escherichia coli. J. Bacteriol.* **178:**1556–1564.

33. Brown, L., and T. Elliott. 1996. Efficient translation of the RpoS sigma factor in *Salmonella typhimurium* requires host factor I, an RNA-binding protein encoded by the hfq gene. *J. Bacteriol.* **178:**3763–3770.

34. Brown, L., and T. Elliott. 1997. Mutations that increase expression of the *rpoS* gene and decrease its dependence of *hfq* function in *Salmonella typhimurium. J. Bacteriol.* **179:**656–662.

35. Castanie-Cornet, M.-P., T. A. Penfound, D. Smith, J. F. Elliott, and J. W. Foster. 1999. Control of acid resistance in *Escherichia coli. J. Bacteriol.* **181:**3525–3535.

36. Chang, Y.-Y., A.-Y. Wang, and J. E. Cronan, Jr. 1994. Expression of *Escherichia coli* pyruvate oxidase (PoxB) depends on the sigma factor encoded by the *rpoS* (*katF*) gene. *Mol. Microbiol.* **11:**1019–1028.

37. Cheville, A. M., K. W. Arnold, C. Buchrieser, C. M. Cheng, and C. W. Kaspar. 1996. *rpoS* regulation of acid, heat, and salt tolerance in *Escherichia coli* O157:H7. *Appl. Environ. Microbiol.* **62:**1822–1824.

38. Colland, F., N. Fujita, D. Kotlarz, A. Ishihama, and A. Kolb. 1999. Positioning of σ^S, the stationary phase σ factor, in *Escherichia coli* RNA polymerase-promoter open complexes. *EMBO J.* **18:**4049–4059.

39. Conter, A., C. Menchon, and C. Gutierrez. 1997. Role of DNA supercoiling and RpoS sigma factor in the osmotic and growth phase-dependent induction of the gene *osmE* of *Escherichia coli* K12. *J. Mol. Biol.* **273:**75–83.

40. Coynault, C., V. Robbe-Saule, and F. Norel. 1996. Virulence and vaccine potential of *Salmonella typhimurium* mutants deficient in the expression of the RpoS (σ^S) regulon. *Mol. Microbiol.* **22:**149–160.

41. Crowe, J. H., F. A. Hoekstra, and L. M. Crowe. 1992. Anhydrobiosis. *Annu. Rev. Physiol.* **54:**579–599.

42. Cunning, C., L. Brown, and T. Elliott. 1998. Promoter substitution and deletion analysis of upstream region required for *rpoS* translational regulation. *J. Bacteriol.* **180:**4564–4570.

43. Cusick, M. E., and M. Belfort. 1998. Domain structure and RNA annealing activity of the *Escherichia coli* regulatory protein StpA. *Mol. Microbiol.* **28:**847–857.

44. Darwin, A. J., and V. L. Miller. 1999. Identification of *Yersinia enterocolitica* genes affecting survival in an animal host using signature-tagged transposon mutagenesis. *Mol. Microbiol.* **32:**51–62.

45. Demple, B., J. Halbrook, and S. Linn. 1983. *Escherichia coli xth* mutants are hypersensitive to hydrogen peroxide. *J. Bacteriol.* **153:**1079–1082.

46. Ding, Q., S. Kusano, M. Villarejo, and A. Ishihama. 1995. Promoter selectivity control of *Escherichia coli* RNA polymerase by ionic strength: differential recognition of osmoregulated promoters by $E\sigma^D$ and $E\sigma^S$ holoenzymes. *Mol. Microbiol.* **16:**649–656.

47. **Dove, S. L., and C. J. Dorman.** 1994. The site-specific recombination system regulating expression of the Type 1 fimbrial subunit gene of *Escherichia coli* is sensitive to changes in DNA supercoiling. *Mol. Microbiol.* **14:**957–988.

48. **Dukan, S., and T. Nyström.** 1998. Bacterial senescence: stasis results in increased and differential oxidation of cytoplasmic proteins leading to developmental induction of the heat shock regulon. *Genes Dev.* **12:**3431–3441.

49. **Eisenstark, A., M. J. Calcutt, M. Becker-Hapak, and A. Ivanova.** 1999. Role of *Escherichia coli rpoS* and associated genes in defense against oxidative damage. *Free Rad. Biol. Med.* **21:** 975–993.

50. **Engelmann, S., C. Lindner, and M. Hecker.** 1995. Cloning, nucleotide sequence, and regulation of *katE* encoding a σ^B-dependent catalase in *Bacillus subtilis. J. Bacteriol.* **177:**5598–5605.

51. **Fang, R. C., S. J. Libby, N. A. Buchmeier, P. C. Loewen, J. Switala, J. Harwood, and D. G. Guiney.** 1992. The alternative σ factor KatF (RpoS) regulates *Salmonella* virulence. *Proc. Natl. Acad. Sci. USA* **89:**11978–11982.

52. **Farewell, A., K. Kvint, and T. Nyström.** 1998. *uspB,* a new σ^S-regulated gene in *Escherichia coli* which is required for stationary-phase resistance to ethanol. *J. Bacteriol.* **180:**6140–6147.

53. **Farewell, A., K. Kvint, and T. Nyström.** 1998. uspB, a new σ^S-regulated gene in *Escherichia coli* which is required for stationary-phase resistance to ethanol. *J. Bacteriol.* **180:**6140–6147.

54. **Fischer, D., A. Teich, P. Neubauer, and R. Hengge-Aronis.** 1998. The general stress sigma factor σ^S of *Escherichia coli* is induced during diauxic shift from glucose to lactose. *J. Bacteriol.* **180:**6203–6206.

55. **Flavier, A. B., M. A. Schell, and T. P. Denny.** 1998. An RpoS (σ^S) homologue regulates acylhomoserine lacton-dependent autoinduction in *Ralstonia solanacearum. Mol. Microbiol.* **28:** 475–486.

56. **Franze de Fernandez, M. T., L. Eoyang, and J. T. August.** 1968. Factor fraction required for the synthesis of bacteriophage Qβ RNA. *Nature* (London) **219:**588–590.

57. **Franze de Fernandez, M. T., W. S. Hayward, and J. T. August.** 1972. Bacterial proteins required for replication of phage Qβ ribonucleic acid. Purification and properties of host factor I, a ribonucleic acid-binding protein. *J. Biol. Chem.* **247:**824–831.

58. **Fujita, M., K. Tanaka, H. Takahashi, and A. Amemura.** 1994. Transcription of the principal sigma-factor genes, *rpoD* and *rpoS,* in *Pseudomonas aeruginosa* is controlled according to the growth phase. *Mol. Microbiol.* **13:**1071–1077.

59. **Gentry, D. R., V. J. Hernandez, L. H. Nguyen, D. B. Jensen, and M. Cashel.** 1993. Synthesis of the stationary-phase sigma factor σ^S is positively regulated by ppGpp. *J. Bacteriol.* **175:** 7982–7989.

60. **Germer, J., A. Muffler, and R. Hengge-Aronis.** 1998. Trehalose is not relevant for in vivo activity of σ^S-containing RNA polymerase in *Escherichia coli. J. Bacteriol.* **180:**1603–1606.

61. **Gibson, K. E., and T. J. Silhavy.** 1999. The LysR homolog LrhA promotes RpoS degradation by modulating activity of the response regulator SprE. *J. Bacteriol.* **181:**563–571.

62. **González-Flecha, B., and B. Demple.** 1997. Transcriptional regulation of the *Escherichia coli* oxyR gene as a function of cell growth. *J. Bacteriol.* **179:**6181–6186.

63. **Gordia, S., and C. Gutierrez.** 1996. Growth-phase-dependent expression of the osmotically inducible gene *osmC* of *Escherichia coli* K-12. *Mol. Microbiol.* **19:**729–736.

64. **Grant, R. A., D. H. Gilman, S. E. Finkel, R. Kolter, and J. M. Hogle.** 1998. The crystal structure of Dps, a ferritin

65. **Grimaud, R., M. Kessel, B. Beuron, A. C. Steven, and M. R. Maurizi.** 1998. Enzymatic and structural similarities between the *Escherichia coli* ATP-dependent proteases, ClpXP and ClpAP. *J. Biol. Chem.* **273:**12476–12481.

66. **Groat, R. G., J. E. Schultz, E. Zychlinski, A. T. Bockman, and A. Martin.** 1986. Starvation proteins in *Escherichia coli:* kinetics of synthesis and role in starvation survival. *J. Bacteriol.* **168:**486–493.

67. **Gulig, P. A., H. Danbar, D. G. Guiney, A. J. Lax, F. Norel, and M. Rhen.** 1993. Molecular analysis of *spv* virulence genes of the *Salmonella* virulence plasmids. *Mol. Microbiol.* **7:**825–830.

68. **Gutierrez, C., M. Ardourel, E. Bremer, A. Middendorf, W. Boos, and U. Ehmann.** 1989. Analysis and DNA sequence of the osmoregulated *treA* gene encoding the periplasmic trehalase of *Escherichia coli* K12. *Mol. Gen. Genet.* **217:**347–354.

69. **Gutierrez, C., J. Barondess, C. Manoil, and J. Beckwith.** 1987. The use of transposon Tn*phoA* to detect genes for cell envelope proteins subject to a common regulatory stimulus. *J. Mol. Biol.* **195:**289–297.

70. **Hächler, H., S. P. Cohen, and S. B. Levy.** 1991. *marA,* a regulated locus which controls expression of chromosomal multiple antibiotic resistance in *Escherichia coli. J. Bacteriol.* **173:** 5532–5538.

71. **Hales, L. M., and H. A. Shuman.** 1999. *Legionella pneumophila rpoS* is required for growth within *Acanthamoeba castellanii. J. Bacteriol.* **181:**4879–4889.

72. **Heiskanen, P., S. Taira, and M. Rhen.** 1994. Role of *rpoS* in the regulation of *Salmonella* plasmid virulence (*spv*) genes. *FEMS Microbiol. Lett.* **123:**125–130.

73. **Hengge-Aronis, R.** 1996. Back to log phase: σ^S as a global regulator in the osmotic control of gene expression in *Escherichia coli. Mol. Microbiol.* **21:**887–893.

74. **Hengge-Aronis, R.** 1996. Regulation of gene expression during entry into stationary phase, p. 1497–1512. *In* F. C. Neidhardt, R. Curtiss III, J. L. Ingraham, E. C. C. Lin, K. B. Low, B. Magasanik, W. S. Reznikoff, M. Riley, M. Schaechter, and H. E. Umbarger (ed.), Escherichia coli *and* Salmonella*: Cellular and Molecular Biology,* 2nd ed. ASM Press, Washington, D.C.

75. **Hengge-Aronis, R.** 1999. Integration of control devices: a global regulatory network in *Escherichia coli,* p. 169–193. *In* S. Baumberg (ed.), *Prokaryotic Gene Expression.* Oxford University Press, Oxford, England.

76. **Hengge-Aronis, R.** 1999. Interplay of global regulators in the general stress response of *Escherichia coli. Curr. Opin. Microbiol.* **2:**148–152.

77. **Hengge-Aronis, R., W. Klein, R. Lange, M. Rimmele, and W. Boos.** 1991. Trehalose synthesis genes are controlled by the putative sigma factor encoded by *rpoS* and are involved in stationary phase thermotolerance in *Escherichia coli. J. Bacteriol.* **173:**7918–7924.

78. **Hengge-Aronis, R., R. Lange, N., Henneberg, and D. Fischer.** 1993. Osmotic regulation of *rpoS*-dependent genes in *Escherichia coli. J. Bacteriol.* **175:**259–265.

79. **Hiratsu, K., H. Shinagawa, and K. Makino.** 1995. Mode of promoter recognition by the *Escherichia coli* RNA polymerase holoenzyme containing the σ^S subunit: identification of the recognition sequence of the *fic* promoter. *Mol. Microbiol.* **18:** 841–850.

80. **Horlacher, R., K. Uhland, W. Klein, M. Ehrmann, and W. Boos.** 1996. Characterization of a cytoplasmic trehalase of *Escherichia coli. J. Bacteriol.* **178:**6250–6257.

homolog that binds and protects DNA. *Nature Struct. Biol.* **5:** 294–303.

81. **Huisman, G. W., and R. Kolter.** 1994. Sensing starvation: a homoserine lactone-dependent signaling pathway in *Escherichia coli. Science* **265:**537–539.

82. **Iriarte, M., I. Stanier, and G. R. Cornelis.** 1995. The *rpoS* gene from *Yersinia enterocolitica* and its influence on expression of virulence factors. *Infect. Immun.* **63:**1840–1847.

83. **Ivanova, A., C. Miller, G. Glinsky, and A. Eisenstark.** 1994. Role of *rpoS* (*katF*) in *oxyR*-independent regulation of hydroperoxidase I in *Escherichia coli. Mol. Microbiol.* **12:**571–578.

84. **Ivanova, A., M. Renshaw, R. V. Guntaka, and A. Eisenstark.** 1992. DNA base sequence variability in *katF* (putative sigma factor) gene of *Escherichia coli. Nucleic Acids Res.* **20:**5479–5480.

85. **Jacob, R., and J. Monod.** 1961. Genetic regulatory mechanisms in the synthesis of proteins. *J. Mol. Biol.* **3:**318–356.

86. **Jair, K.-W., X. Yu, K. Skarstad, B. Thöny, N. Fujita, A. Ishihama, and R. E. Wolf, Jr.** 1996. Transcriptional activation of promoters of the superoxide and multiple antibiotic resistance regulons by Rob, a binding protein of the *Escherichia coli* origin of chromosomal replication. *J. Bacteriol.* **178:** 2507–2513.

87. **Jair, K. W., R. G. Martin, J. L. Rosner, N. Fujita, A. Ishihama, and R. E. Wolf.** 1995. Purification and regulatory properties of MarA protein, a transcriptional activator of *Escherichia coli* multiple antibiotic and superoxide resistance promoters. *J. Bacteriol.* **177:**7100–7104.

88. **Jenkins, D. E., J. E. Schultz, and A. Matin.** 1988. Starvation-induced cross-protection against heat or H_2O_2 challenge in *Escherichia coli. J. Bacteriol.* **170:**3910–3914.

89. **Jishage, M., and A. Ishihama.** 1995. Regulation of RNA polymerase sigma subunit synthesis in *Escherichia coli*: intracellular levels of σ^{70} and σ^{38}. *J. Bacteriol.* **177:**6832–6835.

90. **Jishage, M., and A. Ishihama.** 1998. A stationary phase protein in *Escherichia coli* with binding activity to the major σ subunit of RNA polymerase. *Proc. Natl. Acad. Sci. USA* **95:** 4953–4958.

91. **Jishage, M., A. Iwata, S. Ueda, and A. Ishihama.** 1996. Regulation of RNA polymerase sigma subunit synthesis in *Escherichia coli*: intracellular levels of four species of sigma subunit under various growth conditions. *J. Bacteriol.* **178:**5447–5451.

92. **Jishage, M. I., and A. Ishihama.** 1997. Variation in RNA polymerase sigma subunit composition within different stocks of *Escherichia coli* W3110. *J. Bacteriol.* **179:**959–963.

93. **Jung, J. U., C. Gutierrez, F. Martin, M. Ardourel, and M. Villarejo.** 1990. Transcription of *osmB*, a gene encoding an *Escherichia coli* lipoprotein, is regulated by dual signals. *J. Biol. Chem.* **265:**10574–10581.

94. **Jung, J. U., C. Gutierrez, and M. R. Villarejo.** 1989. Sequence of an osmotically inducible lipoprotein gene. *J. Bacteriol.* **171:** 511–520.

95. **Kaasen, I., P. Falkenberg, O. B. Styrvold, and A. R. Strøm.** 1992. Molecular cloning and physical mapping of the *otsBA* genes, which encode the osmoregulatory trehalose pathway of *Escherichia coli*: evidence that transcription is activated by KatF(AppR). *J. Bacteriol.* **174:**889–898.

96. **Kajitani, M., and A. Ishihama.** 1991. Identification and sequence determination of the host factor gene for bacteriophage Qβ. *Nucleic Acids Res.* **19:**1063–1066.

97. **Kakeda, M., C. Ueguchi, H. Yamada, and T. Mizuno.** 1995. An *Escherichia coli* curved DNA-binding protein whose expression is affected by the stationary phase specific sigma factor σ^S. *Mol. Gen. Genet.* **248:**629–634.

98. **Klauck, E., J. Böhringer, and R. Hengge-Aronis.** 1997. The LysR-like regulator LeuO in *Escherichia coli* is involved in the translational regulation of *rpoS* by affecting the expression of the small regulatory DsrA-RNA. *Mol. Microbiol.* **25:**559–569.

99. **Kolter, R.** 1999. Growth in studying the cessation of growth. *J. Bacteriol.* **181:**697–699.

100. **Kowarz, L., C. Coynault, V. Robbe-Saule, and F. Norel.** 1994. The *Salmonella typhimurium katF* (*rpoS*) gene: cloning, nucleotide sequence, and regulation of *spvR* and *spvABCD* virulence plasmid genes. *J. Bacteriol.* **176:**6852–6860.

101. **Kusano, S., Q. Q. Ding, N. Fujita, and A. Ishihama.** 1996. Promoter selectivity of *Escherichia coli* RNA polymerase Eσ^{70} and Eσ^{38} holoenzymes: effect of DNA supercoiling. *J. Biol. Chem.* **271:**1998–2004.

102. **Kusano, S., and A. Ishihama.** 1997. Stimulatory effect of trehalose on formation and activity of *Escherichia coli* RNA polymerase Eσ^{38} holoenzyme. *J. Bacteriol.* **179:**3649–3654.

103. **Lafiti, A., M. Foglino, K. Tanaka, P. Williams, and A. Lazdunski.** 1996. A hierarchical quorum-sensing cascade in *Pseudomonas aeruginosa* links the transcriptional activators LasR and RhlR (VsmR) to expression of the stationary-phase sigma factor RpoS. *Mol. Microbiol.* **21:**1137–1146.

104. **Landini, P., L. I. Hajec, L. H. Nguyen, R. R. Burgess, and M. R. Volkert.** 1996. The leucine-responsive regulatory protein (Lrp) acts as a specific repressor for σ^S-dependent transcription of the *Escherichia coli aidB* gene. *Mol. Microbiol.* **20:**947–955.

105. **Lange, R., M. Barth, and R. Hengge-Aronis.** 1993. Complex transcriptional control of the σ^S-dependent stationary phase-induced and osmotically regulated *osmY* (*csi-5*) gene suggests novel roles for Lrp, cyclic AMP (cAMP) receptor protein-cAMP complex and integration host factor in the stationary phase response of *Escherichia coli. J. Bacteriol.* **175:**7910–7917.

106. **Lange, R., D. Fischer, and R. Hengge-Aronis.** 1995. Identification of transcriptional start sites and the role of ppGpp in the expression of *rpoS*, the structural gene for the σ^S subunit of RNA-polymerase in *Escherichia coli. J. Bacteriol.* **177:** 4676–4680.

107. **Lange, R., and R. Hengge-Aronis.** 1991. Growth phase-regulated expression of *bolA* and morphology of stationary phase *Escherichia coli* cells is controlled by the novel sigma factor σ^S (*rpoS*). *J. Bacteriol.* **173:**4474–4481.

108. **Lange, R., and R. Hengge-Aronis.** 1991. Identification of a central regulator of stationary-phase gene expression in *Escherichia coli. Mol. Microbiol.* **5:**49–59.

109. **Lange, R., and R. Hengge-Aronis.** 1994. The cellular concentration of the σ^S subunit of RNA-polymerase in *Escherichia coli* is controlled at the levels of transcription, translation and protein stability. *Genes Dev.* **8:**1600–1612.

110. **Lange, R., and R. Hengge-Aronis.** 1994. The *nlpD* gene is located in an operon with *rpoS* on the *Escherichia coli* chromosome and encodes a novel lipoprotein with a potential function in cell wall formation. *Mol. Microbiol.* **13:**733–743.

111. **Lease, R. A., M. E. Cusick, and M. Belfort.** Riboregulation in *Escherichia coli*: DsrA RNA acts by RNA:RNA interaction at multiple loci. *Proc. Natl. Acad. Sci. USA*, in press.

112. **Lee, I. S., J. Lin, H. K. Hall, B. Bearson, and J. W. Foster.** 1995. The stationary-phase sigma factor σ^S (RpoS) is required for a sustained acid tolerance response in virulent *Salmonella typhimurium. Mol. Microbiol.* **17:**155–167.

113. **Li, Z. Y., and B. Demple.** 1996. Sequence specificity for DNA binding by *Escherichia coli* SoxS and Rob proteins. *Mol. Microbiol.* **20:**937–945.

114. **Loewen, P. C.** 1992. Regulation of bacterial catalase synthesis, p. 97–115. *In* J. Scandalios (ed.), *Molecular Biology of*

Free Radical Scavenging Systems. Cold Spring Harbor Laboratory, Cold Spring Harbor, N.Y.

115. Loewen, P. C., B. Hu, J. Strutinsky, and R. Sparling. 1998. Regulation in the *rpoS* regulon of *Escherichia coli*. *Can. J. Microbiol.* 44:707–717.

116. Loewen, P. C., and B. L. Triggs. 1984. Genetic mapping of *katF*, a locus that with *katE* affects the synthesis of a second catalase species in *Escherichia coli*. *J. Bacteriol.* 160:668–675.

117. Loewen, P. C., I. von Ossowski, J. Switala, and M. R. Mulvey. 1993. KatF (σ^S) synthesis in *Escherichia coli* is subject to posttranscriptional regulation. *J. Bacteriol.* 175:2150–2153.

118. Lonetto, M., M. Gribskov, and C. A. Gross. 1992. The σ^{70} family: sequence conservation and evolutionary relationships. *J. Bacteriol.* 174:3843–3849.

119. Lord, M., D. Barillà, and M. D. Yudkin. 1999. Replacement of vegetative σ^A by sporulation-specific σ^F as a component of the RNA polymerase holoenzyme in sporulating *Bacillus subtilis*. *J. Bacteriol.* 181:2346–2350.

120. Majdalani, N., C. Cunning, D. Sledjeski, T. Elliott, and S. Gottesman. 1998. DsrA RNA regulates translation of RpoS message by an anti-antisense mechanisms, independent of its action as an antisilencer of transcription. *Proc. Natl. Acad. Sci. USA* 95:12462–12467.

121. Marschall, C., and R. Hengge-Aronis. 1995. Regulatory characteristics and promoter analysis of *csiE*, a stationary phase-inducible σ^S-dependent gene under positive control of cAMP-CRP in *Escherichia coli*. *Mol. Microbiol.* 18:175–184.

122. Marschall, C., V. Labrousse, M. Kreimer, D. Weichart, A. Kolb, and R. Hengge-Aronis. 1998. Molecular analysis of the regulation of *csiD*, a carbon starvation inducible gene in *Escherichia coli* that is exclusively dependent on σ^S and requires activation by cAMP-CRP. *J. Mol. Biol.* 276:339–353.

123. McCann, M. P., C. D. Fraley, and A. Matin. 1993. The putative σ factor KatF is regulated posttranscriptionally during carbon starvation. *J. Bacteriol.* 175:2143–2149.

124. McCann, M. P., J. P. Kidwell, and A. Matin. 1991. The putative σ factor KatF has a central role in development of starvation-mediated general resistance in *Escherichia coli*. *J. Bacteriol.* 173:4188–4194.

125. Michán, C., M. Manchado, G. Dorado, and C. Pueyo. 1999. In vivo transcription of the *Escherichia coli* *oxyR* regulon as a function of growth phase and in response to oxidative stress. *J. Bacteriol.* 181:2759–2764.

126. Muffler, A., M. Barth, C. Marschall, and R. Hengge-Aronis. 1997. Heat shock regulation of σ^S turnover: a role for DnaK and relationship between stress responses mediated by σ^S and σ^{32} in *Escherichia coli*. *J. Bacteriol.* 179:445–452.

127. Muffler, A., D. Fischer, S. Altuvia, G. Storz, and R. Hengge-Aronis. 1996. The response regulator RssB controls stability of the σ^S subunit of RNA polymerase in *Escherichia coli*. *EMBO J.* 15:1333–1339.

128. Muffler, A., D. Fischer, and R. Hengge-Aronis. 1996. The RNA-binding protein HF-I, known as a host factor for phage Qβ RNA replication, is essential for the translational regulation of *rpoS* in *Escherichia coli*. *Genes Dev.* 10:1143–1151.

129. Muffler, A., D. D. Traulsen, D. Fischer, R. Lange, and R. Hengge-Aronis. 1997. The RNA-binding protein HF-I plays a global regulatory role which is largely, but not exclusively, due to its role in expression of the σ^S subunit of RNA polymerase in *Escherichia coli*. *J. Bacteriol.* 179:197–300.

130. Muffler, A., D. D. Traulsen, R. Lange, and R. Hengge-Aronis. 1996. Posttranscriptional osmotic regulation of the σ^S subunit of RNA polymerase in *Escherichia coli*. *J. Bacteriol.* 178:1607–1613.

131. Mukherjee, A., Y. Cui, W. Ma, Y. Liu, A. Ishihama, A. Eisenstark, and A. K. Chatterjee. 1998. RpoS (Sigma-S) controls expression of *rsmA*, a global regulator of secondary metabolites, hairpin and extracellular proteins in *Erwinia carotovora*. *J. Bacteriol.* 180:3629–3634.

132. Mulvey, M. R., and P. C. Loewen. 1989. Nucleotide sequence of *katF* of *Escherichia coli* suggest KatF protein is a novel σ transcription factor. *Nucleic Acids Res.* 17:9979–9991.

133. Mulvey, M. R., J. Switala, A. Borys, and P. C. Loewen. 1990. Regulation of transcription of *katE* and *katF* in *Escherichia coli*. *J. Bacteriol.* 172:6713–6720.

134. Munro, P. M., G. N. Flatau, R. L. Clement, and M. J. Gauthier. 1995. Influence of the RpoS (KatF) sigma factor on maintenance of viability and culturability of *Escherichia coli* and *Salmonella typhimurium* in seawater. *Appl. Environ. Microbiol.* 61:1853–1858.

135. Nguyen, L. H., D. B. Jensen, N. E. Thompson, D. R. Gentry, and R. R. Burgess. 1993. In vitro functional characterization of overproduced *Escherichia coli* *katF/rpoS* gene product. *Biochemistry* 32:11112–11117.

136. Ninneman, O., C. Koch, and R. Kahmann. 1992. The *E. coli* *fis* promoter is subject to stringent control and autoregulation. *EMBO J.* 11:1075–1083.

137. Norel, F., V. Robbe-Saule, M. Y. Popoff, and C. Coynault. 1992. The putative sigma factor KatF (RpoS) is required for the transcription of the *Salmonella typhimurium* virulence gene *spvB* in *Escherichia coli*. *FEMS Microbiol. Lett.* 99:271–276.

138. Nunoshiba, T., E. Hidalgo, C. F. Amábile-Cuevas, and B. Demple. 1992. Two-stage control of an oxidative stress regulon: the *Escherichia coli* SoxR protein triggers redox-inducible expression of the *soxS* regulatory gene. *J. Bacteriol.* 174:6054–6060.

139. Nyström, T., C. Larsson, and L. Gustafsson. 1996. Bacterial defense against aging: role of the *Escherichia coli* ArcA regulator in gene expression, readjusted energy flux and survival during stasis. *EMBO J.* 15:3219–3228.

140. O'Neal, C. R., W. M. Gabriel, A. Turk, S. J. Libby, F. Fang, and M. P. Spector. 1994. RpoS is necessary for both the positive and negative regulation of starvation survival genes during phosphate, carbon, and nitrogen starvation in *Salmonella typhimurium*. *J. Bacteriol.* 176:4610–4616.

141. Olsén, A., A. Arnqvist, M. Hammar, S. Sukupolvi, and S. Normark. 1993. The RpoS sigma factor relieves H-NS-mediated transcriptional repression of *csgA*, the subunit gene of fibronectin binding curli in *Escherichia coli*. *Mol. Microbiol.* 7:523–536.

142. Ozaki, M., A. Wada, N. Fujita, and A. Ishihama. 1991. Growth phase-dependent modification of RNA polymerase in *Escherichia coli*. *Mol. Gen. Genet.* 230:17–23.

143. Pratt, L. A., and T. J. Silhavy. 1996. The response regulator, SprE, controls the stability of RpoS. *Proc. Natl. Acad. Sci. USA* 93:2488–2492.

144. Pratt, L. A., and T. J. Silhavy. 1998. Crl stimulates RpoS activity during stationary phase. *Mol. Microbiol.* 29:1225–1236.

145. Raina, S., D. Missiakas, L. Baird, S. Kumar, and C. Georgopoulos. 1993. Identification and transcriptional analysis of the *Escherichia coli* *htrE* operon which is homologous to *pap* and related pilin operons. *J. Bacteriol.* 175:5009–5021.

146. Robbe-Saule, V., F. Schaeffer, L. Kowarz, and F. Norel. 1997. Relationships between H-NS, σ^S, SpvR and growth phase in the control of spvR, the regulatory gene of the *Salmonella* plasmid virulence operon. *Mol. Gen. Genet.* **256:** 333–347.

147. Rockabrand, D., K. Livers, T. Austin, R. Kaiser, D. Jensen, R. Burgess, and P. Blum. 1998. Roles of DnaK and RpoS in starvation-induced thermotolerance of *Escherichia coli. J. Bacteriol.* **180:**846–854.

148. Römling, U., W. D. Sierralta, K. Eriksson, and S. Normark. 1998. Multicellular and aggregative behaviour of *Salmonella typhimurum* strains is controlled by mutations in the *agfD* promoter. *Mol. Microbiol.* **28:**249–264.

149. Rudd, K. E. 1998. Linkage map of *Escherichia coli* K-12, edition 10: the physical map. *Microbiol. Mol. Biol. Rev.* **62:** 985–1019.

150. Sak, B. D., A. Eisenstark, and D. Touati. 1989. Exonuclease III and the catalase hydroperoxidase II in *Escherichia coli* are both regulated by the *katF* product. *Proc. Natl. Acad. Sci. USA* **86:**3271–3275.

151. Santos, J. M., P. Freire, M. Vicente, and C. M. Arraiano. 1999. The stationary-phase morphogene *bolA* from *Escherichia coli* is induced by stress during early stages of growth. *Mol. Microbiol.* **32:**789–798.

152. Schellhorn, H. E., J. P. Audia, L. I. C. Wei, and L. Chang. 1998. Identification of conserved, RpoS-dependent stationary phase genes of *Escherichia coli. J. Bacteriol.* **180:**6283–6291.

153. Schellhorn, H. E., and V. L. Stones. 1992. Regulation of *katF* and *katE* in *Escherichia coli* K-12 by weak acids. *J. Bacteriol.* **174:**4769–4776.

154. Schweder, T., K.-H. Lee, O. Lomovskaya, and A. Matin. 1996. Regulation of *Escherichia coli* starvation sigma factor (σ^S) by ClpXP protease. *J. Bacteriol.* **178:**470–476.

155. Shiba, T., K. Tsutsumi, H. Yano, Y. Ihara, A. Kameda, K. Tanaka, K. H. Takahashi, M. Munekata, N. N. Rao, and A. Kornberg. 1997. Inorganic polyphosphate and the induction of *rpoS* expression. *Proc. Natl. Acad. Sci. USA* **94:** 11210–11215.

156. Shin, S., S. G. Song, D. S. Lee, J. G. Pan, and C. Park. 1997. Involvement of *iclR* and *rpoS* in the induction of *acs*, the gene for acetyl coenzyme A synthetase of *Escherichia coli* K-12. *FEMS Microbiol. Lett.* **146:**103–108.

157. Sledjeski, D., and S. Gottesman. 1995. A small RNA acts as an antisilencer of the H-NS-silenced *rcsA* gene of *Escherichia coli. Proc. Natl. Acad. Sci. USA* **92:**2003–2007.

158. Sledjeski, D. D., A. Gupta, and S. Gottesman. 1996. The small RNA, DsrA, is essential for the low temperature expression of RpoS during exponential growth in *E. coli. EMBO J.* **15:**3993–4000.

159. Small, P., D. Blankenhorn, D. Welty, E. Zinser, and J. L. Slonczewski. 1994. Acid and base resistance in *Escherichia coli* and *Shigella flexneri*: role of *rpoS* and growth pH. *J. Bacteriol.* **176:**1729–1737.

160. Spector, M. P., and C. L. Cubitt. 1992. Starvation-inducible loci of *Salmonella typhimurium*: regulation and roles in starvation survival. *Mol. Microbiol.* **6:**1467–1476.

161. Strohmeier-Gort, A., D. M. Ferber, and J. A. Imlay. 1999. The regulation and role of the periplasmic copper, zinc superoxide dismutase of *Escherichia coli. Mol. Microbiol.* **32:** 179–191.

162. Strøm, A. R., and I. Kaasen. 1993. Trehalose metabolism in *Escherichia coli*: stress protection and stress regulation of gene expression. *Mol. Microbiol.* **8:**205–210.

163. Suh, S.-J., L. Silo-Suh, D. E. Woods, D. J. Hassett, S. E. H. West, and D. E. Ohman. 1999. Effect of *rpoS* mutation on the stress response and expression of virulence factors in *Pseudomonas aeruginosa. J. Bacteriol.* **181:**3890–3897.

164. Takayanagi, Y., K. Tanaka, and H. Takahashi. 1994. Structure of the 5′upstream region and the regulation of the *rpoS* gene of *Escherichia coli. Mol. Gen. Genet.* **243:**525–531.

165. Tanaka, K., S. Kusano, N. Fujita, A. Ishihama, and H. Takahashi. 1995. Promoter determinants for *Escherichia coli* RNA polymerase holoenzyme containing σ^{38} (the *rpoS* gene product). *Nucleic Acids Res.* **23:**827–834.

166. Tanaka, K., Y. Takayanagi, N. Fujita, A. Ishihama, and H. Takahashi. 1993. Heterogeneity of the principal sigma factor in *Escherichia coli*: the *rpoS* gene product, σ^{38}, is a second principal sigma factor of RNA polymerase in stationary phase *Escherichia coli. Proc. Natl. Acad. Sci. USA* **90:**3511–3515.

167. Touati, E., E. Dassa, and P. L. Boquet. 1986. Pleiotropic mutations in *appR* reduce pH 2.5 acid phosphatase expression and restore succinate untilization in CRP-deficient strains of *Escherichia coli. Mol. Gen. Genet.* **202:**257–264.

168. Touati, E., E. Dassa, J. Dassa, P. L. Boquet, and D. Touati. 1991. Are *appR* and *katF* the same *Escherichia coli* gene encoding a new sigma transcription initiation factor? *Res. Microbiol.* **142:**29–36.

169. Tuveson, R. W., and R. B. Jonas. 1979. Genetic control of near-UV (300–400 nm) sensitivity independent of the *recA* gene in strains of *Escherichia coli* K12. *Photochem. Photobiol.* **30:**667–676.

170. van Laere, A. 1989. Trehalose, reserve and/or stress metabolite? *FEMS Microbiol. Rev.* **63:**201–210.

171. Völker, U., K. K. Andersen, H. Antelmann, K. M. Devine, and M. Hecker. 1998. One of the two OsmC homologs in *Bacillus subtilis* is part of the σ^B-dependent general stress regulon. *J. Bacteriol.* **180:**4212–4218.

172. von Blohn, C., B. Kempf, R. M. Karres, and E. Bremer. 1997. Osmostress response in *Bacillus subtilis*: characterization of a proline uptake system (*opuE*) regulated by high osmolarity and the alternative transcription factor sigma B. *Mol. Microbiol.* **25:**175–187.

173. Wang, A.-Y., and J. E. Cronan, Jr. 1994. The growth phase-dependent synthesis of cyclopropane fatty acids in *Escherichia coli* is the result of an RpoS (KatF)-dependent promoter plus enzyme instability. *Mol. Microbiol.* **11:**1009–1017.

174. Wang, J. M., J. A. Hartling, and J. M. Flanagan. 1997. The structure of ClpP at 2.3 Å resolution suggests a model for ATP-dependent proteolysis. *Cell* **91:**447–456.

175. Waterman, S. R., and P. L. C. Small. 1996. Identification of σ^S-dependent genes associated with the stationary-phase acid resistance phenotype of *Shigella flexneri. Mol. Microbiol.* **21:** 925–940.

176. Weichart, D., R. Lange, N. Henneberg, and R. Hengge-Aronis. 1993. Identification and characterization of stationary phase-inducible genes in *Escherichia coli. Mol. Microbiol.* **10:**407–420.

177. Welsh, D. T., and R. A. Herbert. 1999. Osmotically induced intracellular trehalose, but not glycine betaine accumulation promotes desiccation tolerance in *Escherichia coli. FEMS Microbiol. Lett.* **174:**57–63.

178. Wilmes-Riesenberg, M. R., J. W. Foster, and R. Curtiss III. 1997. An altered *rpoS* allele contributes to the avirulence of *Salmonella typhimurium* LT2. *Infect. Immun.* **65:**203–210.

179. Wise, A., R. Brems, V. Ramakrishnan, and M. Villarejo. 1996. Sequences in the −35 region of *Escherichia coli rpoS*-dependent genes promote transcription by Eσ^S, *J. Bacteriol.* **178:**2785–2793.

180. Xu, J., and R. C. Johnson. 1995. Fis activates the RpoS-dependent stationary phase expression of *proP* in *Escherichia coli. J. Bacteriol.* **177:**5222–5231.

181. **Xu, J. M., and C. Johnson.** 1997. Activation of RpoS-dependent *proP* P2 transcription by the Fis protein in vitro. *J. Mol. Biol.* **270:**346–359.

182. **Yamada, M., A. A. Talukder, and T. Nitta.** 1999. Characterization of the *ssnA* gne, which is involved in the decline of cell viability at the beginning of stationary phase in *Escherichia coli. J. Bacteriol.* **181:**1838–1846.

183. **Yamashino, T., C. Ueguchi, and T. Mizuno.** 1995. Quantitative control of the stationary phase-specific sigma factor, σ^S, in *Escherichia coli*: involvement of the nucleoid protein H-NS. *EMBO J.* **14:**594–602.

184. **Yim, H. H., R. L. Brems, and M. Villarejo.** 1994. Molecular characterization of the promoter of *osmY*, an *rpoS* dependent gene. *J. Bacteriol.* **176:**100–107.

185. **Yim, H. H., and M. Villarejo.** 1992. *osmY*, a new hyperosmotically inducible gene, encodes a periplasmic protein in *Escherichia coli. J. Bacteriol.* **174:**3637–3644.

186. **Zambrano, M. M., and R. Kolter.** 1996. GASPing for life in stationary phase. *Cell* **86:**181–184.

187. **Zambrano, M. M., D. A. Siegele, M. Almirón, A. Tormo, and R. Kolter.** 1993. Microbial competition: *Escherichia coli* mutants that take over stationary phase cultures. *Science* **259:**1757–1760.

188. **Zgurskaya, H. I., M. Keyhan, and A. Matin.** 1997. The σ^S level in starving *Escherichia coli* cells increases solely as a result of its increased stability, despite decreased synthesis. *Mol. Microbiol.* **24:**643–651.

189. **Zhang, A., S. Altuvia, A. Tiwari, L. Argaman, R. Hengge-Aronis, and G. Storz.** 1998. The OxyS regulatory RNA represses *rpoS* translation and binds the Hfq (HF-I) protein. *EMBO J.* **17:**6061–6068.

190. **Zhou, A. N., and S. Gottesman.** 1998. Regulation of proteolysis of the stationary-phase sigma factor RpoS. *J. Bacteriol.* **180:**1154–1158.

Bacterial Stress Responses
Edited by G. Storz and R. Hengge-Aronis
©2000 ASM Press, Washington, D.C.

Chapter 12

Protective Function and Regulation of the General Stress Response in *Bacillus subtilis* and Related Gram-Positive Bacteria

CHESTER W. PRICE

The general stress response of gram-positive bacteria contributes to their survival in the natural environment, their persistence in foods, and perhaps to their pathogenicity in humans. In Bacillus subtilis *this response is controlled by the σ^B transcription factor, which is activated by diverse growth-limiting stresses to direct the synthesis of more than 100 general stress proteins. Loss of σ^B function leads to increased sensitivity to multiple stresses, including acid, heat, osmotic, and oxidative stress. The known functions of genes controlled by σ^B are consistent with a role in protecting cellular DNA, protein, and lipid against the deleterious effects of these stresses, and specifically against the reactive oxygen species generated by unbalanced metabolism. Examples include the ClpP protease and the ClpC ATPase, which are thought to sort, repair, or degrade damaged proteins; a Dps homolog, which is thought to bind and protect DNA; and OpuE, which transports proline as a compatible osmoprotectant. Many general stress genes are under the dual control of σ^B and another σ factor, which provides a second regulatory input. σ^B itself is activated in response to two broad classes of stresses: (i) energy stress, including starvation for glucose, phosphate, or oxygen, and (ii) environmental stress, including acid, ethanol, heat, or salt stress. According to the current model, these two classes of stress are conveyed to σ^B by separate signal transduction pathways, each terminating in a differentially regulated PP2C phosphatase and converging on the RsbV regulator of σ^B. In unstressed cells the phosphorylated RsbV anti-anti-σ is unable to complex the RsbW anti-σ, which is then free to bind and inactivate σ^B. Either PP2C phosphatase, when triggered by its particular class of stress, can dephosphorylate RsbV and thereby activate σ^B. With the discovery of σ^B and its associated regulators in the human pathogens* Listeria monocytogenes *and* Staphylococcus aureus, *the role of the general stress response in their survival and pathogenesis can be fully explored.*

It is now well established that the σ^B transcription factor of *Bacillus subtilis* is activated by growth-limiting environmental or metabolic stress to direct the synthesis of at least 100 general stress proteins. The synthesis of these proteins significantly increases resistance to a variety of stresses, including acid, alkali, ethanol, heat, osmotic, or oxidative stress (11, 42, 45, 138), and their function may become particularly important when sporulation is prevented by inimical conditions. Indeed, the number of genes known to be under σ^B control is now rapidly approaching the estimated number of sporulation genes (124), and their expression can absorb between 25 and 35% of new protein synthesis under growth-limiting conditions (20). These observations, coupled with the complexity of the signal transduction pathway that controls σ^B activity (46, 132, 147), imply that the general stress regulon and the sporulation process are both highly important for survival in the natural environment. However, sporulation and the general stress response produce dramatically different physiological states and are elicited in response different external and internal signals. The discovery of σ^B and its distinctive regulatory proteins in gram-positive human pathogens (15, 81, 82, 92, 143, 146) underscores the significance of *B. subtilis* as an accessible model to explore the physiological role and regulation of the general stress response.

R. Losick and his colleagues laid the foundation for the field with their pioneering discovery of σ^B in *B. subtilis*, which was the first alternative σ factor

Chester W. Price • Department of Food Science and Technology, University of California, Davis, CA 95616.

found in bacteria. This group also provided the best-characterized σ^B-dependent promoter to date, that for the *ctc* gene, which is widely employed for studies of σ^B regulation. Haldenwang and Losick (55, 56) took a biochemical approach to identify σ^B, using the so-called "0.4 kb" gene and the *ctc* gene as templates for in vitro transcription reactions. The 0.4 kb gene was of unknown function but was chosen because it was more highly expressed in sporulating cells (117). Subsequently, a mutation within the 0.4 kb gene was found to influence the sporulation process, and the gene was renamed *spoVG* (111). *ctc* is also a gene of unknown function that lies adjacent to *spoVG*. Notably, σ^B could direct RNA polymerase core enzyme to transcribe *spoVG* and *ctc* in vitro whereas σ^A, the major σ factor of *B. subtilis*, could not (55, 56, 70, 95, 101). On the basis of these and other results, it was widely assumed that σ^B was involved in sporulation (83). Two research groups tested this notion by isolating the σ^B structural gene, *sigB*, and constructing null mutants (22, 41, 68). Significantly, these null mutants proved to have no obvious growth or sporulation phenotype under standard laboratory conditions. Furthermore, although *ctc* expression was found to be σ^B-dependent in vivo, expression of *spoVG* proved to be σ^B-independent (68). The physiological role of σ^B therefore remained mysterious.

The mystery was solved by the contributions of three different research groups, working from two different directions. In their studies of stress responses in *B. subtilis*, Hecker and his colleagues identified a common set of proteins whose synthesis was induced by a variety of environmental and metabolic stresses (57, 58, 139). Because the synthesis of this common set of proteins was overlaid upon that of proteins specifically evoked by the stress employed, and because these common proteins were elicited by diverse stresses, they were termed general stress proteins (57, 61). At the same time, the Haldenwang lab and my laboratory were studying the signals that activated σ^B. Igo and Losick (68, 69) had already established that σ^B activity increased sharply at the *ctc* promoter upon entry into the stationary growth phase, and that the sequences required for this increase were congruent with those required for σ^B activity. Subsequent work in the Haldenwang lab and my lab found that this stationary-phase increase also occurred at other σ^B-dependent promoters (23, 26, 72) and identified components of the signaling pathway that conveyed these stationary signals to σ^B (17, 18, 25). Benson and Haldenwang (19) then showed that σ^B activity was also induced by heat shock during logarithmic growth, and that the synthesis of a small set of proteins distinct from the established heat shock proteins was in fact dependent on σ^B. In a sep-

arate study Boylan et al. (23) found that salt or ethanol stresses were also very effective inducers of σ^B activity during logarithmic growth and that these environmental stress signals were conveyed to σ^B via the same signaling pathway that communicated stationary-phase stress signals.

On the basis of the sum of these results, Boylan et al. (23) proposed that σ^B controls a general stress response that is induced when cells encounter a variety of growth-limiting conditions. These investigators further suggested that this general stress response allows the cells to counter stresses to which they have not yet been exposed, and that this alternative might become of critical importance under conditions that do not support the sporulation process. Hecker and his colleagues independently reached the same conclusions based on their discovery that many of the general stress proteins they previously identified indeed required σ^B for full expression (137). The proposed physiological role for σ^B in general stress response was then firmly established by the subsequent discovery of clear phenotypes for *sigB* null mutants and by the functional characterization of key general stress proteins, as discussed in the following two sections.

LOSS OF σ^B FUNCTION LEADS TO MULTIPLE STRESS SENSITIVITY

Even with the understanding that σ^B controlled the synthesis of a number of general stress proteins, it was not straightforward to establish the phenotype of a *sigB* null mutant. This reflects two features of the general stress response in *B. subtilis*. On one hand, many genes under σ^B control are also under the control of another σ factor, presenting substantial regulatory redundancy (see reference 60). And on the other hand, the σ^B regulon affords greater protection if it is first induced by a mild stress before exposure to a more severe challenge (138, 139). Loss of σ^B function is now known to cause sensitivity to multiple stresses, including acid, alkali, heat, osmotic, and oxidative stress (see Table 1).

Protection of starved cells against oxidative stress was the first role attributed to the σ^B regulon, and sensitivity to this stress remains one of the strongest phenotypes of a *sigB* null mutant. The oxidative stress response of *B. subtilis* was known to have at least two components, one specifically induced by oxidative stress and the other generally induced by starvation (36). With the discovery that the stationary-phase catalase KatE was under σ^B control (43), Hecker and his colleagues tested the contribution of KatE and of σ^B to oxidative stress resistance

Table 1. *sigB* null mutants are multiply deficient in stress survival

| Stress class | Without preadaptation[a] | | With preadaptation[b] | | Reference(s) |
	Stress	Relative viability loss of ΔsigB (fold)	Prestress	Relative viability loss of ΔsigB (fold)	
Heat	37 → 54°C	1,000	37 → 48°C	5,000	138
Salt	10% NaCl	50–200	4% NaCl	500–2,000	138
Ethanol	9% Ethanol	100	ND[c]	ND	138
Acid (log)	pH 4.3	50–100	pH 5.2	50–100	138
Acid (stat)	pH 5.0	10–20	ND	ND	45
Alkali (stat)	pH 9.0	10–20	ND	ND	45
Oxidative	10 mM H_2O_2	10	Glucose exhaustion	1,000	11, 42
Freeze/thaw	−20°C	ND	Glucose exhaustion	500	138
Desiccation	Lyophilization	10	ND	ND	Unpublished; cited in reference 138

[a] Fold viability decrease when cells were subjected to the stress shown.
[b] Fold viability decrease when cells were first subjected to the prestress shown, then to the stress displayed in the previous column.
[c] ND = not determined.

(42). Although loss of *katE* function was not deleterious to either the specific or the starvation-induced resistance, loss of *sigB* function diminished the latter by 10- to 100-fold. With the finding that loss of *sigB* function also decreased resistance to organic hydroperoxides (11), the role of the general stress response in oxidative stress resistance was clearly established.

Following the observation that many genes under σ^B control encoded products that appeared to be associated with the cell envelope, Gaidenko and Price (45) examined the contribution of σ^B to survival under conditions known to tax envelope function. These investigators found that the absence of σ^B caused a 10- to 20-fold decrease in survival of stationary-phase cells held under acidic or alkaline conditions for 5 days, and a significant decrease in yield when cells were grown in medium containing between 7 and 9% ethanol. These effects of a *sigB* null mutation were in sharp contrast to the absence of any effect on stationary-phase survival at neutral pH in rich medium.

In a comprehensive study, Völker and Hecker (138) extended these findings to logarithmically growing cells, showing that mutants lacking σ^B manifested a 50- to 100-fold decrease in survival within 2 h of either acid or ethanol shock. Notably, these authors further demonstrated that a *sigB* null mutant was 1,000-fold more sensitive than wild type to heat stress and 50- to 200-fold more sensitive to salt stress, and they cited unpublished data that implicated σ^B function in resistance to desiccation stress. Lastly, this study showed that the σ^B regulon could afford significantly increased cross-protection against some of these same stresses when first induced with an unrelated, nonlethal stress.

The pleiotropic phenotype of a *sigB* null mutant (Table 1) underscores the importance of the σ^B regulon for the survival of *B. subtilis* in the soil and water environment, where it frequently encounters pH, osmotic, temperature, and desiccation stresses, and where it must protect its cellular contents against oxidative damage while persisting in a nongrowing state. This pleiotropic phenotype is consistent with the known physiological roles of well-characterized σ^B-dependent genes, described in the following section.

σ^B-DEPENDENT GENES AFFECT A RANGE OF CELLULAR FUNCTIONS

Well before any phenotype for a *sigB* null mutant had been established, various investigators endeavored to isolate genes whose expression was dependent on σ^B, with the notion that a genetic and biochemical characterization of such genes would provide important clues to the physiological role of σ^B. This has indeed proven to be the case, and has provided an outline of how expression of the σ^B regulon promotes general stress resistance.

Thus far four different methods have been employed to identify σ^B-dependent genes. The first of these was in vitro transcription studies, which were highly successful in the case of *ctc*, whose transcription remains the best characterized of any σ^B-dependent gene (68, 69, 94, 106, 107, 125). However, with the advent of authentic *sigB* null mutants, it was found that two other genes transcribed by σ^B-containing RNA polymerase in vitro were unlikely to be under σ^B control in vivo. The previously mentioned *spoVG* gene (55, 56, 95, 101) was found not to require σ^B for its expression in vivo (68). Likewise, expression of the *aprE* gene encoding subtilisin (145) was found to be σ^B-independent in vivo (44). A third gene originally identified by in vitro transcription studies, *cdd* (120, 141), has yet to be tested for its σ^B dependence in vivo. In contrast, the use of *sigB* null mutants clearly demonstrated that expression of both *ctc* (68) and the *sigB* operon itself (72) required σ^B function. The basis of the discrepancy between the in vitro and in vivo results for *spoVG* and *aprE* has not been further investigated, but these examples encouraged subsequent investigations to seek σ^B-dependent genes only in the context of the cellular environment.

Consequently, the second approach employed was a genetic screen that identified Tn*917-lacZ* fusions requiring σ^B for all or part of their expression (26). This genetic approach ultimately identified eight new genes controlled by σ^B, called *csb* genes (3, 4, 26, 130, 131), and led to the important conclusion that σ^B controls a large stationary-phase regulon (24). The third approach used reverse genetics, determining the amino terminal sequences of general stress proteins separated on two-dimensional gels and matching these to the predicted products of sequenced genes. This approach has led to the identification of the largest number of σ^B-dependent genes to date and to the pivotal discovery that key stress genes are indeed controlled by σ^B (12, 47, 78–80, 98, 102, 112). The fourth approach relied on searching the *B. subtilis* genome for promoters that resembled the σ^B consensus, then screening this limited set of candidate genes by hybridization with RNA isolated from wild-type and *sigB* mutant cells (103). The sum of all of these approaches has identified nearly 100 σ^B-dependent genes thus far, and the actual number may be considerably greater (60). As a result of the variety of approaches and aims of the different studies, these genes have been characterized to different degrees.

The 42 genes listed in Table 2 are the best characterized with regard to their σ^B dependency. They fall into 28 transcription units for which (i) the 5′ end of a σ^B-dependent message has been shown to lie the expected distance downstream from a consensus σ^B recognition sequence, or (ii) critical bases within the putative recognition sequence (94, 106, 107, 125) have been shown by site-directed mutagenesis to be required for σ^B-dependent fusion expression. These genes are therefore likely to be directly transcribed by σ^B-containing holoenzyme in vivo. Among the gene products listed in Table 2, the biochemical functions of some have been clearly established, the importance of others in stress resistance has been demonstrated genetically, and the role of still others is suggested by similarity to proteins of known function. However, the roles of many of the newly identified σ^B-dependent genes remain completely unknown, suggesting that additional effects on cellular physiology are still to be discovered. The stress protective roles of the more evident gene products are discussed in following sections.

The ClpP Protease and ClpC Chaperone Belong to the σ^B Regulon

The σ^B-dependent genes with the most clear relation to stress resistance are those encoding the ClpP ATP-dependent protease (47, 97) and its ClpC ATPase regulatory subunit (78, 80, 98). ClpC belongs to the Hsp104 family of heat shock proteins, members of which have a dual function: they manifest chaperone activity on their own and are also involved in targeting dispensable or damaged proteins for degradation by a protease moiety such as ClpP (52, 122, 142). ClpC could therefore serve to sort damaged proteins, restoring the less impaired and presenting the terminally damaged for destruction by the ClpP protease (52). Consistent with this role, growth of *clpC* and *clpP* null mutants is impaired at elevated temperatures (80, 97, 98).

In *Escherichia coli*, *clpP* and *clpB* (a *clpC* homolog) both belong to the heat shock regulon controlled by σ^{32} (see chapter 1). In contrast, heat shock genes in *B. subtilis* fall into at least three different classes (59). Class I genes are those specifically induced by heat shock and include *dnaK*, *dnaJ*, *grpE*, *groES*, and *groEL*. Expression of these genes is controlled by σ^A-containing RNA polymerase and the CIRCE/HrcA system, where CIRCE is an inverted repeat found in Class I heat shock promoters (149, 150) and HrcA is a repressor that binds to CIRCE (114, 148). The heat shock induction of Class II genes is wholly under σ^B control (59). Class II genes therefore encode the general stress proteins whose synthesis is induced by either environmental stresses, such as heat, or by metabolic stresses, such as energy starvation. These genes are frequently under the additional control of a second σ factor. It is common for this second promoter to have σ^A-like recognition sequences, but dependence of the second promoter on σ^F (104), σ^H (37, 131), and σ^X (66) has also been reported. Class III genes include the *clpP* and *clpC* genes under discussion here and are distinguished by the independence of their heat shock response from

Table 2. σ^B-dependent genes for which σ^B consensus sequences have been experimentally associated with expression

Gene[a]	Function from biochemical or genetic analysis	Nearest homolog (accession no.)	E value[b]	Nearest homolog of known function	E value[b]
bmrU (102)	Not known	*B. subtilis* YerQ (CAB12492)	5e-32	None	
bmr	Multidrug efflux (100)	*B. subtilis* Blt (P39843)	1e-89	*B. subtilis* Bmr (P33449)	e-175
bmrR	Transcriptional regulator of *bmr* (1)	*B. subtilis* YdfL (BAA19380)	1e-14	*B. subtilis* BmrR (P39075)	e-160
ctc (56, 68, 69, 94)	Not known	*Borrelia burgdorferi ctc* (AE001177.1)	6e-14	*E. coli* ribosomal protein L25 (P02426)	3e-04
clpP (47)	Clp energy-dependent protease (127)	*E. coli* ClpP protease (P19245)	2e-73	*B. subtilis* ClpP protease (P80244)	e-106
csbA (26)	Not known	None	None	None	None
csbB (4)	Not known	*Streptomyces coelicolor* putative sugar transferase (AL031182.1)	2e-69	Bacteriophage SfII bactoprenol glucosyl transferase (AAC39272)	1e-48
csbC (*yxcC*) (3)	Not known	*B. subtilis* YwtG hypothetical transporter (Z92954.1)	e-116	*Lactobacillus brevis* XylT xylose symporter (O52733)	4e-99
csbD (*ywmG*) (3)	Not known	*E. coli* YjbJ (P32691)	4e-03	None	None
csbX (50)	Not known	*B. subtilis* YoaB hypothetical transporter (AF027868.1)	e-128	*Klebsiella pneumoniae* DalT D-arabinitol transporter (AF045245.1)	5e-83
bofC	Sporulation control; forespore checkpoint for σ^K (49)	None	None	None	None
ctsR (78)	Repressor of Class III heat shock genes *clpC*, *clpE*, and *clpP* (34, 77)	*L. monocytogenes* first frame in ClpC operon (AAC44443)	4e-39	*B. subtilis* CtsR (P37568)	5e-71
yacH	Not known	*L. monocytogenes* second frame in ClpC operon (AAC44444)	2e-19	None	None
yacI	Not known	*L. monocytogenes* third frame in ClpC operon (AAC44445)	2e-97	*Limulus polyphemus* arginine kinase (P51541)	7e-20
clpC	Clp ATPase (127, 128)	*L. monocytogenes* ClpC homolog (AAC44446)	>e-180	*B. subtilis* ClpC (P37571)	>e-180
sms	UV and methyl methanesulfonate resistance (79)	*L. monocytogenes* RadA/Sms homolog (AAC33293)	e-162	*E. coli* RadA/Sms (P24554)	e-104
comY (*yacK*)	UV resistance and genetic competence (79)	*S. coelicolor* putative DNA binding protein (CAB40852)	2e-94	None	None
dps (*pexB*) (12)	Oxidative stress resistance (12)	*B. subtilis* MgrA metalloregulation protein (P37960)	6e-35	*B. subtilis* MgrA metalloregulation protein (P37960)	6e-35
gsiB (88)	Not known	*Arabidopsis thaliana* late embryogenesis protein (Q07187)	3e-18	None	None
gspA (10)	Not known	*Haemophilus influenzae* hypothetical protein (P43974)	1e-22	*Neisseria gonorrhoeae* LgtC glycosyl transferase (AAA68011)	5e-19
gtaB (119, 130)	UDP-glucose pyrophosphorylase (105)	*Bacillus anthracis* pXO-94 frame on pXO1 virulence plasmid (AAD32398)	5e-93	*B. subtilis* UDP-glucose pyrophosphorylase (Q05852)	e-157

Continued on following page

Table 2. *Continued*

Gene[a]	Function from biochemical or genetic analysis	Nearest homolog (accession no.)	E value[b]	Nearest homolog of known function	E value[b]
katB (*katE*) (43)	Stationary-phase catalase 2 (43)	*B. firmus* KatA (AAF11546)	>e-180	*B. subtilis* KatB (P42234)	>e-180
katX (104)	Spore germination and out-growth catalase (14)	*Deinococcus radiodurans* KatA (BAA09937)	>e-180	*B. subtilis* KatX (P94377)	>e-180
nadE (*outB*) (13)	NAD synthetase (99)	*E. coli* NadE NAD synthetase (P18843)	4e-77	*B. subtilis* NadE (P08164)	e-155
opuE (140)	Osmoregulated proline transport (140)	*B. subtilis* YcgO (BAA08956)	e-148	*B. subtilis* OpuE (O06493)	>e-180
rsbV (72)	Anti-anti-σ for σ^B (6, 38)	*Bacillus licheniformis* RsbV (AAC29508)	2e-39	*B. subtilis* RsbV (P17903)	1e-55
rsbW	Anti-σ for σ^B (6, 16)	*B. licheniformis* RsbW (AAC29509)	2e-72	*B. subtilis* RsbW (P17904)	3e-89
sigB	σ^B structural gene (22, 41, 56)	*B. licheniformis* σ^B (AAC29510)	e-127	*B. subtilis* σ^B (AAA22713)	e-145
rsbX	PP2C phosphatase for RsbS-P (147)	*B. licheniformis* RsbX (AAC29511)	4e-67	*B. subtilis* RsbX (P17906)	e-111
trxA (112)	Essential gene (112)	*L. monocytogenes* thioredoxin (probable) (CAB40815)	9e-37	*E. coli* thioredoxin (AAA24534)	3e-24
ycnH (103)	Not known	*E. coli* GabD NADP-dependent succinate semialdehyde dehydrogenase (P25526)	e-126	*E. coli* GabD NADP-dependent succinate semialdehyde dehydrogenase (P25526)	e-126
ydaD (102)	Not known	*B. subtilis* YhdF (O07575)	e-113	*Bacillus megaterium* Dhg2 glucose 1-dehydrogenase 2 (P39483)	1e-38
ydaE	Not known	None	None	None	None
ydaG (102)	Not known	*D. radiodurans* putative general stress protein (AAF10718)	5e-04	None	None
ydaP (102)	Not known	*Lactobacillus plantarum* pyruvate oxidase (P37063)	1e-96	*L. plantarum* pyruvate oxidase (P37063)	1e-96
yflA (103)	Not known	*Treponema pallidum* putative D-alanine-glycine permease (AAC65402)	7e-65	*Pseudoalteromonas haloplanktis* Na-linked D-alanine-glycine permease (P30144)	3e-54
yjbC (103)	Not known	None	None	None	None
ykzA (136)	Not known	*B. subtilis* YklA (CAA05593)	2e-26	*Xanthamonas campestris* Ohr (organic hydro-peroxide resistance) (AAC38562)	7e-18
ytxG (131)	Not known	*B. subtilis* YoxC (P28670)	4.8e-10	None	None
ytxH	Not known	*B. subtilis* YhaH (CAA74413)	1.4e-10	None	None
ytxJ	Not known	None	None	None	None
yvyD (37)	Not known	*Staphylococcus carnosus* hypothetical protein (P47995)	8e-38	Spinach chloroplast r-protein S30 (P19954)	7e-18

[a] Genes are listed alphabetically by the first gene in the transcription unit; within transcription units genes are listed according to their order. Transcription units are separated by horizontal lines.

[b] E value from BLAST 2.0 (8) using the BLOSUM 62 matrix (62). E values of <1e-3 are listed as "none."

both CIRCE and σ^B control (59). Analysis of the promoter regions for *clpP* (47) and for the *clpC* operon (78) has demonstrated dual control from an upstream σ^B-dependent promoter and a downstream σ^A-like promoter. The σ^A-like promoter responds to multiple environmental stresses and can largely compensate for the loss of σ^B function, explaining the apparent σ^B independence of the heat shock response. For both *clpP* and *clpC*, transcription from this downstream promoter is controlled by the CtsR repressor, encoded by the first gene in the *clpC* operon (34, 77).

Although ClpP and ClpC are considered to be heat shock proteins that repair or recycle proteins damaged by excessive temperature, they might manifest similar recycling or repair activities toward proteins damaged by salt, ethanol, or oxidative stress. Consistent with this notion, *clpP* null mutants are more sensitive to salt or ethanol stress (47), and *clpC* null mutants are more sensitive to salt stress (80), particularly when growing in minimal medium.

Contribution of the σ^B Regulon to a General Oxidative Stress Resistance

Protection of cellular protein, DNA, and lipid against reactive oxygen species following growth-limiting energy or environmental stress is thought to be a major role of the σ^B regulon (60). In addition to ClpP and ClpC, the known or suspected functions of a number of σ^B-dependent genes are consistent with this role.

First, the KatB (43) and KatX (104) catalases are specifically expressed under growth-limiting conditions. Although neither is essential for survival of oxidative stress in starved cells (42, 104), each likely contributes to removing deleterious molecules that form when cellular metabolism becomes unbalanced.

Second, protein oxidation may be reversed in part by the action of a putative thioredoxin encoded by *trxA* (112). In contrast to the case in *E. coli*, in *B. subtilis trxA* appears to be an essential gene, perhaps because there is no glutathione system in *B. subtilis* (112). Like *clpP* and *clpC*, *trxA* is a Class III heat shock gene with a σ^B-dependent and σ^A-like promoter, but it is not yet clear whether *trxA* expression is also under CtsR control (34, 77).

Third, DNA protection and repair may involve the products of the fifth and sixth genes in the *clpC* operon, *sms* and *comY* (79), and the product of the *dps* gene (12). The *sms* product is important for methyl methane-sulfonate resistance, and both the *sms* and *comY* products are important for UV resistance and for the development of genetic competence (79). The predicted *dps* product belongs to the *dps*

family of DNA-binding proteins, and a *dps* null mutant fails to develop starvation-induced resistance to oxidative stress, leading Antelmann et al. (12) to suggest that *B. subtilis* Dps has a DNA protective function similar to that demonstrated for *E. coli* Dps/PexB (5, 87). However, the similarity between the predicted sequences of *B. subtilis* and *E. coli* Dps is not strong (E value = 5e-7), and there is presently no experimental evidence that *B. subtilis* Dps binds and protects DNA or that it is synthesized in sufficient amounts to perform such a function. The closest homolog to Dps is *B. subtilis* MgrA, another member of the *dps* family that is specifically induced in response to H_2O_2 addition and has been shown to confer oxidative stress resistance by forming a multimeric complex with DNA (29). Both MgrA and Dps may therefore fulfill equivalent DNA protective roles, with MgrA representing the peroxide stress-inducible regulon controlled by PerR (27) and Dps representing the general stress and starvation inducible regulon controlled by σ^B. This contrasts with the situation in *E. coli*, where expression of a single *dps* gene is under the dual control of the specific peroxide regulator OxyR and the general stress regulator σ^S (9). It is possible that *B. subtilis dps* may also afford some DNA protection during acid stress, because the σ^B regulon is known to contribute to acid stress resistance (45, 138) and expression of a *Salmonella enterica dps* homolog is rapidly induced by acid shock (129).

Fourth, although a general stress protection of lipids against oxidative damage has not been demonstrated experimentally, the observation that a large proportion of σ^B-dependent genes identified by transposon mutagenesis do encode products known or suspected to affect envelope function is consistent with this notion (45). This observation is also consistent with modifications to the *B. subtilis* cell membrane that are known to occur in response to osmotic shock (85).

Contribution of the σ^B Regulon to Osmotic Stress Resistance

One way that *B. subtilis* responds to osmotic stress is by synthesizing or importing compatible solutes such as proline and glycine betaine (75). The *opuE* locus, encoding a proline transporter, has been shown to be under the dual control of an osmotically regulated σ^A-like promoter and a separate σ^B-dependent promoter (140). Following a detailed analysis of *opuE* expression, Spiegelhalter and Bremer (121) concluded that the σ^B-dependent promoter was transiently induced immediately following an osmotic shock while the σ^A-like promoter was responsible for increased *opuE* transcription over an extended period. These results suggest that σ^B is not responsible for resistance to steady-state osmotic stress but may contribute to resistance immediately following osmotic shock. In the same study, Spiegelhalter and Bremer also found that the σ^B-dependent promoter was solely responsible for *opuE* expression in response to heat or ethanol shock. The physiological role of proline uptake during heat or ethanol stress remains to be established.

The predicted sequences of two other σ^B-dependent genes also suggest a function in osmotic stress response. One is GsiB (88), which resembles plant desiccation proteins (123) and is among the most abundant products of the general stress response (71). The other is the YtxH, which weakly resembles other plant desiccation proteins and is encoded by the second gene of the *ytxG* operon (131). However, the importance of *gsiB* and *ytxH* in resistance to osmotic stress and desiccation has not been experimentally tested.

Two other gene products, which by their sequence similarity to *E. coli* OsmC might be expected to have a role in osmotic stress response, may instead be involved in resistance to organic hydroperoxides. *E. coli osmC* is a gene of unknown function that is induced by osmotic stress (54). Its transcription is controlled by a σ^{70}-like promoter and by a second promoter dependent on the *E. coli* stress and starvation factor σ^S (51). By contrast, *B. subtilis* has two *osmC* homologs: *yklA*, which is transcribed from a σ^A-like promoter that is apparently not subject to osmotic control, and *ykzA*, which responds to osmotic and other stresses via a σ^B-dependent promoter (136). Because the predicted YklA and YkzA products are more closely related to *Xanthomonas campestris* Ohr (93) than they are to *E. coli* OsmC, Völker et al. (136) suggested that YklA and YkzA may help detoxify organic hydroperoxides, but this remains to be tested. As was the case for *dps* and *mgrA*, *yklA* and *ykz* appear to be duplicated genes that together exhibit the same pattern of expression as their *E. coli osmC* counterpart, which is a single cistron under dual control.

Do Members of the σ^B Regulon Mediate the Efflux of Toxic Compounds?

The *bmr* locus consists of three genes in the order *bmrU-bmr-bmrR*, where *bmrU* encodes a protein of unknown function, *bmr* encodes a efflux transporter for structurally diverse drugs, and *bmrR* encodes a transcriptional activator that binds in a drug-dependent manner to a promoter in the *bmrU-bmr* interval (1, 100). Notably, Petersohn et al. (102) have

detected a σ^B-dependent promoter immediately preceding *bmrU*, and a Northern analysis is consistent with extension of this σ^B-dependent transcript into *bmr* and *bmrR*. The possibility that *bmr* and *bmrR* are under σ^B control is particularly intriguing in light of the finding that expression of a σ^B-like factor in *Mycobacterium tuberculosis* is induced by cycloserine, rifampin, and streptomycin (89). However, Neyfakh and his colleagues (76) have suggested that the natural solute transported by *B. subtilis bmr* could be a toxic compound that accumulates as a result of normal metabolism. If this proves to be the case, the drug efflux capability of *bmr* would then be an extension of this basic function.

The Presumed Roles of Less Well Characterized Genes in the σ^B Regulon Suggest Two Added Cellular Functions

In addition to the genes listed in Table 2, the expression of at least 30 other genes is also thought to be under σ^B control. Work in the Hecker lab has shown that the product or the message of each of these genes is synthesized under stress conditions in a σ^B-dependent manner, and a presumptive σ^B recognition sequence lies within 200 bp of the coding region (102, 103). Although the biological relevance of the proposed σ^B recognition sequences has not yet been demonstrated, most or all of these additional genes will likely prove to be members of the σ^B regulon. On the basis of sequence analysis, many appear to lie in operons, increasing the number of genes whose expression is potentially under σ^B control (102, 103).

Two new themes emerge when these less well characterized genes are also considered. First, on the basis of similarity of their predicted products to known proteins, Hecker and his colleagues suggested that four of them encode NAD(P)-dependent dehydrogenases which help maintain redox balance during stress (102, 103). Second, the products of many σ^B-dependent loci identified by genetic means are either known or suspected to influence cell envelope function (45). In support of this observation, a significant portion of the genes identified by a more recent genomic approach also appears to encode products that influence the cell wall or membrane (103). In this study, candidate genes were first identified by searching the genome for σ^B consensus recognition sequences, then the stress-induced expression of each candidate was tested for σ^B dependence by hybridization with RNA probes isolated from wild-type and mutant strains.

σᴮ IS CONTROLLED BY A SIGNAL TRANSDUCTION NETWORK THAT RELIES ON SERINE AND THREONINE PHOSPHORYLATION

σᴮ is a stable protein (109) whose activity is negatively controlled by the association of a specific anti-σ factor, RsbW (6, 16), where *rsb* stands for regulator of sigma-B. Upon release from RsbW, σᴮ associates with RNA polymerase core enzyme to direct the transcription of the general stress genes (6, 16). This release occurs in response to two broad classes of stress: (i) internal or energy stresses, such as those caused by starvation for glucose, phosphate, or oxygen or by addition of oxidative uncouplers to the growth medium; and (ii) external or environmental stresses, such as acid, ethanol, heat, or salt stress (7, 19, 23, 135, 137).

According to the model shown in Fig. 1, these two classes of stress signals are channeled to RsbW—and ultimately to σᴮ—by two branches of a signal transduction network that converge on RsbV, an anti-anti-σ factor that regulates the action of RsbW (6, 38). This network functions by a "partner switching" mechanism, in which key protein-protein interactions are controlled by serine phosphorylation and dephosphorylation (6, 16, 38, 147). It is therefore distinct from the two-component mechanism typically used in bacterial signal transduction. Only one other σ factor in *B. subtilis* is known to be regulated by the partner switching mechanism, and that is the σᶠ factor, which governs forespore-specific gene expression during the sporulation process (see chapter 13). Indeed, the term "partner switching" was coined by Alper et al. (7) to describe the distinctive means by which σᶠ activity is controlled. Although the σᴮ and σᶠ regulatory networks respond to different sets of external and internal signals, analysis of the σᴮ network has benefited from its considerable parallel with the genetic (35, 113) and biochemical (7, 39, 40, 90) analysis of the σᶠ system.

The key elements of the σᴮ network are largely encoded within the eight-gene *sigB* operon. As shown in Fig. 2, a σᴬ-like promoter controls the entire eight-gene operon while an internal, σᴮ-dependent promoter is responsible for the autocatalytic induction of the downstream four genes (72, 144). Three of these downstream genes encode products that function in the common part of the signal transduction network—RsbV, RsbW, and σᴮ. The fourth downstream gene encodes a product, RsbX, which appears to function in a negative feedback loop that damps continued environmental signaling. The next three sections will describe (i) the interactions among these common elements, (ii) the components of the energy stress branch, and (iii) the components of the environmental stress branch. These latter two sections include a discussion of how energy and environmental signals might enter their respective branches.

RsbV and RsbW Are the Primary Regulators of σᴮ Activity

Genetic analysis established that RsbW is a negative regulator of σᴮ activity, that RsbV is a positive regulator, and that RsbV requires RsbW in order to exert its positive function (17, 18, 25). This analysis also indicated that RsbV and RsbW control σᴮ activity at the posttranslational level, and that RsbW is responsible for rendering σᴮ inactive in unstressed cells. In vitro and in vivo biochemical analyses then established the mechanism by which this regulation is achieved (6, 16, 38).

The crux of the mechanism is the control of the phosphorylation state of the RsbV anti-anti-σ factor, which governs the ability of RsbW to bind its alternate partners—either RsbV or σᴮ. In unstressed cells RsbV is phosphorylated on a conserved serine residue and cannot bind the RsbW anti-σ factor, which is then free to complex σᴮ and prevent its association with RNA polymerase (6, 16, 38, 147). However, when RsbV is dephosphorylated, it complexes RsbW and forces it to switch partners, thereby releasing σᴮ (6, 38). The central role of RsbV in this signal transduction pathway is underscored by the phenotype of an *rsbV* null mutant, which is incapable of activating σᴮ in response to either energy or environmental stresses (17, 23, 25, 135).

The primary elements controlling the phosphorylation state of RsbV have now been identified. RsbW, in addition to its role as an anti-σ factor, also possesses a serine kinase activity that specifically phosphorylates and inactivates its RsbV antagonist (6, 38, 147). RsbV-P is dephosphorylated by one of two homologous PP2C phosphatases: RsbP, which is required for the transmission of energy stress signals (132), or RsbU, which is required for the transmission of environmental stress signals (135, 147). Thus the phosphorylation state of RsbV is controlled by the opposing activities of the RsbW kinase and two complementary PP2C phosphatases. How these opposing activities respond to energy and environmental stress therefore controls the induction of the general stress regulon.

The Energy Signaling Branch

A null *rsbP* mutant abolishes activation of σᴮ in response to energy stress but has little effect on response to environmental stress (132). Moreover, pu-

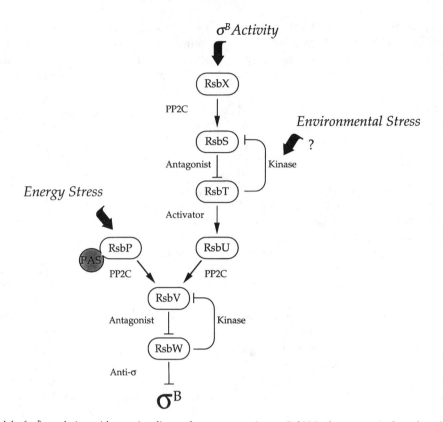

Figure 1. Model of σ^B regulation with two signaling pathways converging on RsbV-P, the antagonist form found in unstressed cells. The energy-stress signaling pathway terminates with the RsbP phosphatase, which contains a PAS domain in its amino terminal half. This PAS domain is thought to sense the overall energy level of the cell and communicate this information to the PP2C phosphatase domain. In contrast, the environmental-stress signaling pathway terminates with the RsbU phosphatase, which is activated by direct interaction with the RsbT switch protein. Whether RsbT binds and activates RsbU is regulated by upstream signaling components. When activated by its particular class of stress, either the RsbP or the RsbU phosphatase removes the serine phosphate from RsbV-P. Dephosphorylated RsbV then binds the RsbW anti-σ factor, which releases σ^B to activate its target general stress genes. The RsbX phosphatase provides a feedback loop that responds to increasing σ^B levels and prevents continued environmental signaling.

rified RsbP has a phosphatase activity specific for RsbV-P in vitro. RsbP therefore possesses the genetic and biochemical properties expected for an essential, energy-responsive phosphatase. There appear to be two separate routes by which energy stress signals might enter this branch of the σ^B network. These two routes are not mutually exclusive, and the relative contribution of each remains to be tested.

The first route emerged from the in vitro studies of Alper et al. (6), who proposed that the activity of the RsbW kinase reflects the energy level of the cell. When ATP levels fall during energy stress (6, 7, 135), RsbW activity is also thought to fall, reducing its capacity to phosphorylate RsbV. In this view, RsbV and RsbW would themselves be sufficient to sense energy stress. If RsbV and RsbW were solely responsible for

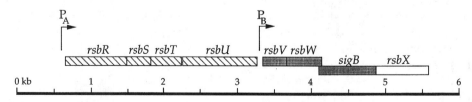

Figure 2. The structural gene for σ^B (*sigB*) lies in an eight-gene operon. The other genes (termed *rsb* for regulator of sigma-B) encode products that posttranslationally regulate σ^B activity. The products of the four *crosshatched* genes comprise the environmental signaling branch of the pathway that communicates with the common regulators encoded by the three *shaded* genes. Expression of the *sigB* and *rsbX* genes may be transcriptionally coupled. The activity of the RsbX phosphatase may therefore provide an indirect measure of σ^B levels in the cell. Transcription of the *sigB* operon is controlled by two promoters, the σ^A-like promoter P_A and an internal, σ^B-dependent promoter P_B.

sensing the decrease in energy levels, the essential RsbP phosphatase could be unregulated and mainly provide a fixed reference against which the activity of the RsbW kinase rises and falls. However, analysis of the predicted RsbP sequence suggests that its activity is directly regulated by energy stress, and this regulation may therefore provide a second route by which energy signals enter the system (132).

Inspection of the predicted RsbP sequence suggests two domains: a C-terminal catalytic domain homologous to other PP2C phosphatases, and an N-terminal PAS domain similar to that found in a wide variety of proteins that monitor changes in redox potential, oxygen tension, and the overall energy level of the cell (126). In bacteria these changes are interrelated with changes in electron transport and proton motive force, and it is thought that monitoring any single parameter can provide a warning of energy stress that precedes a measurable decrease in the ATP pool. Specificity for the parameter sensed relies in part on the association of a particular cofactor with the PAS domain (126). Well-characterized examples include the redox-sensing aerotaxis transducer Aer of *E. coli* (21, 108) and the nitrogen fixation regulatory protein NifL of *Azotobacter vinelandii* (63), both of which are flavoproteins. Modulation of the redox state of a similar chromophore bound to the amino terminal PAS domain of RsbP could conceivably regulate activity of the carboxyl terminal PP2C phosphatase domain. Alternatively, the PAS domain could serve to control the interactions of RsbP with other proteins, as has been shown to be the case in a number of eukaryotic regulators (67, 84).

Clearly, the importance of each of these proposed routes of signal entry must be tested experimentally. Moreover, the existence of these routes does not eliminate the possible involvement of additional elements that remain to be discovered on the energy stress signaling branch.

The Environmental Signaling Branch

Genetic analysis indicates that RsbU has a signaling function that is the reciprocal of RsbP. That is, a null *rsbU* mutant abolishes activation of σ^B in response to environmental stress but has no effect on activation in response to energy stress (134, 135). Moreover, RsbU specifically dephosphorylates RsbV-P in vitro and is activated by upstream members of an environmental signaling cascade (73, 74, 147).

Two of these upstream members, RsbS and RsbT, are homologs of the downstream regulators RsbV and RsbW, and the available evidence indicates that they likewise function by a partner switching mechanism (73, 147). According to the model shown

in Fig. 1, the ability of RsbT to bind its alternate partners—either RsbS or the RsbU phosphatase—depends on the phosphorylation state of the RsbS antagonist. When RsbS is unphosphorylated in unstressed cells, it binds and inactivates RsbT. However, when RsbS is phosphorylated, it cannot bind RsbT, which is then free to bind and activate the RsbU phosphatase. In contrast to the downstream regulators, in which RsbV is phosphorylated in the default or inactive signaling state, here RsbS is unphosphorylated in the default state.

This difference in the default phosphorylation state of the antagonist protein reflects the fact that the output of the downstream regulators is negative, leading to the inactivation of σ^B by the association of RsbW. In contrast, the output of the upstream regulators is positive, leading to the activation of RsbU by the association of RsbT, which can be considered a regulatory subunit of the phosphatase (74, 147). This model is derived from a combined genetic and biochemical analysis of the RsbS, RsbU, and RsbT regulators that included assays of the activities and interactions of wild-type and mutant proteins in vitro and in the yeast two-hybrid system (73, 74, 147). However, the strong inference that the in vivo phosphorylation state of RsbS governs the partner switch of RsbT between its alternate partners RsbS and RsbU remains to be confirmed with biochemical experiments in *B. subtilis*.

Control of the phosphorylation state of RsbS is one means by which external stresses might enter the environmental signaling branch. According to the model shown in Fig. 1, the phosphorylation state of RsbS depends on the tension between RsbT, which possesses a serine kinase activity directed toward its RsbS antagonist, and RsbX, a PP2C phosphatase that specifically dephosphorylates RsbS-P in vitro (147). Genetic experiments have established that RsbX phosphatase activity is not required for the stress response but instead provides a negative feedback loop that serves to damp σ^B activity and prevent continued environmental signaling (118, 133). In contrast, the kinase activity of RsbT is essential for the environmental stress response. Kang et al. (74) constructed a mutant RsbT protein that lacked kinase activity but retained the capacity to activate the RsbU phosphatase both in vitro and in vivo. Therefore, the inability of the mutant RsbT to respond to environmental stress reflected its loss of kinase activity and not its inability to activate RsbU. Modulation of the kinase activity of RsbT therefore becomes a good candidate for the mechanism by which environmental stresses enter the system.

If modulation of the kinase activity of RsbT is indeed how multiple external signals enter the environmental signaling pathway, the question becomes how this modulation is achieved. Specifically, is there a single modulator that monitors a common cellular target—such as the cell membrane—for perturbation by diverse stresses, or are there multiple modulators, each responding to a different stress? Although these questions are just beginning to be addressed, analysis of the RsbR regulator suggests multiple routes of signal entry.

RsbR has the genetic and biochemical properties expected for a protein that modulates the kinase activity of RsbT. Previous work had shown that null alleles in *rsbS*, *rsbT*, *rsbU*, *rsbV*, *rsbW*, and *rsbX* all have strong regulatory phenotypes (17, 25, 72, 73, 135). In marked contrast, an *rsbR* null has only a partial phenotype (2). This phenotype is partial in two ways. First, response to salt and heat stress is affected by an *rsbR* null whereas response to ethanol stress is not. Second, even those responses that are affected by an *rsbR* null are only diminished two- to threefold. These findings indicate that response to ethanol is at least partly separable from the response to salt and heat stress. One explanation for these results is that RsbR function is redundant with other regulators, and this may in fact be the case. This possibility is examined in the section that follows this discussion of RsbR function.

A biochemical analysis identified an intriguing activity for purified RsbR protein and found that this activity is controlled by phosphorylation of RsbR. The carboxy terminal half of RsbR shares 30% sequence identity with the entire RsbS antagonist protein, the substrate of the RsbT kinase (2, 147). In vitro analysis with purified proteins has shown that (i) RsbR itself has no kinase activity but instead greatly stimulates the phosphorylation of RsbS by RsbT; (ii) in addition to phosphorylating RsbS, RsbT also specifically phosphorylates RsbR on two threonine residues, one of which corresponds to the position of the phosphorylated serine in RsbS; and (iii) these phosphorylation events abrogate the ability of RsbR to stimulate phosphorylation of RsbS by RsbT (46). These results, summarized in Fig. 3, are consistent with a model in which RsbR modulates the kinase activity of RsbT directed toward its RsbS antagonist in vivo, either specifically in response to environmental signals or as part of a feedback mechanism to prevent continued signaling. These alternatives can be distinguished by determining the in vivo phosphorylation state of RsbR before and after stress.

Other Proteins May Play a Role in the Environmental Stress Response

By moving the eight-gene *sigB* operon and a reporter fusion into *E. coli*, Scott et al. (116) were able

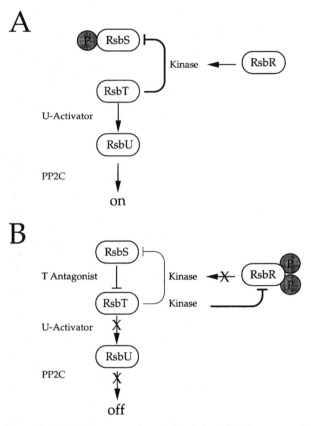

Figure 3. (A) RsbR increases the activity of the RsbT kinase toward its RsbS antagonist. RsbS-P releases RsbT, initiating a cascade of events that ultimately activates σ^B. (B) RsbR is also phosphorylated by the RsbT kinase, abrogating the ability of RsbR to stimulate the phosphorylation of RsbS by RsbT. According to this model, RsbR could function in two different ways. On one hand, if RsbR were unphosphorylated in unstressed cells, its eventual phosphorylation following stress would serve as a feedback mechanism to damp environmental signaling. This feedback mechanism would be superimposed on the mechanism relying on RsbX phosphatase activity. On the other hand, if RsbR were phosphorylated in unstressed cells, its dephosphorylation could trigger the environmental stress response. The phosphatase for RsbR-P has not yet been identified.

to reconstitute a qualitatively functional environmental stress pathway in a heterologous host. However, heat and ethanol stress failed to activate σ^B in this system, leading these authors to conclude that the environmental stress pathway required specific *B. subtilis* regulators not encoded in the *sigB* operon. These additional regulators might include Obg, a small GTP-binding protein. Because Obg is an essential protein, Scott and Haldenwang (115) used depletion studies to show that Obg is directly or indirectly required for the environmental stress response. Another class of potential regulatory proteins is represented by the four newly discovered RsbR homologs YkoB, YojH, YqhA, and YtvA (S. Akbar, C. M. Kang, T. A. Gaidenko, M. O'Reilly, K. M. Devine, and C. W. Price, unpublished data). The carboxy terminal halves of these homologs and RsbR are 50% identical, and it is this common portion that resembles the entire RsbS antagonist protein. Moreover, either singly or in combination, null mutations in these new genes modestly affect activation of σ^B by environmental stress. In contrast, loss of a fifth potential RsbR homolog (encoded by the adjacent yetI and yezB loci) was found to elicit no obvious phenotype. Together these data suggest roles for Obg and the RsbR homologs in regulating σ^B activity, but their mechanism of action has yet to be determined.

RELATIONSHIP OF THE GENERAL STRESS RESPONSE TO OTHER REGULATORY NETWORKS

A sophisticated coordination of adaptive responses in growth-arrested B. subtilis cells is becoming increasingly evident (for a recent review see reference 96). Such responses include chemotaxis and motility, the synthesis of degradative enzymes and antibiotics, the development of natural competence for DNA uptake, and the sporulation process. Because the general stress regulon appears to influence such diverse aspects of cellular physiology, and because expression of the regulon can occupy such a significant proportion of cellular resources under growth-limiting conditions, it is likely that σ^B regulation will prove to be integrated in the expanding web of stationary-phase interactions. Even with our current limited understanding, there are two indications of how this integration might be accomplished.

First, the products of some genes under σ^B control are themselves known to be regulatory proteins. These include ClpP and ClpC, which play essential roles in adaptive responses such as motility, competence development, and sporulation (97, 98, 127), and CtsR, a repressor for a subset of Class III heat shock genes (34, 77). The product of another gene under σ^B control, YvyD, has some resemblance to a negative regulator of σ^{54} activity in gram-negative bacteria, and loss of YvyD function causes a modest increase in activity of a B. subtilis σ^{54} homolog (37). The activation of σ^B in response to energy or environmental stress could therefore send ancillary messages to either up- or downregulate other adaptive systems. The converse circumstance—of another adaptive system sending a signal to σ^B—may be provided by the observation that σ^B activity is elevated at least three- to five-fold in a σ^H null mutant (107). σ^H is known to be essential for sporulation (124), but it also directs the transcription of a number of genes with other roles, including stress response genes (see reference 131 and references therein). The means by

which the loss of σ^H function is communicated to σ^B is not known.

Second, it is apparent that many genes in the σ^B regulon are also under the control of another σ factor. This allows genes to be shared among different stimulons and expressed in response to different combinations of environmental signals. Examples include (i) clpP, clpC, and trxA, which also have a σ^A-like promoter subject to heat shock control (47, 78, 112); (ii) opuE, with its additional σ^A-like promoter subject to osmotic control (121, 140); and (iii) the ytxG operon and the yvyD gene, each of which has a second, σ^H-dependent promoter (37, 131). Two genes in the σ^B regulon are also known or suspected to be under control of ECF σ factors. One is csbB (4), which is partly controlled by σ^X (66), and the other is ydjB (103), which may be partly controlled by σ^W (65). In other bacteria ECF σ factors control a wide variety of mostly extracytoplasmic functions, such as the export or import of ions and macromolecules (53, 91) and the folding and degradation of extracytoplasmic proteins (see chapter 2).

Even in the absence of an obvious regulatory connection to coordinate the action of key adaptive responses, complementary mechanisms appear to have evolved for systems that respond to the same signal. One example is the presumably different roles of σ^B and σ^M in survival of salt stress. σ^M is an ECF σ whose function is required for balanced growth and survival under conditions of high, steady-state osmostress, such as in nutrient broth containing 1.4 M NaCl (64). In contrast, σ^B function is required for full viability following addition of 1.7 M NaCl to cells growing exponentially in a synthetic medium (138) (see Table 1). Considering that σ^B activity is only transiently induced when cells are shocked by salt addition (23, 88, 121), and that σ^B activity is not induced by steady-state osmostress (unpublished data cited in reference 64), it appears that the action of the σ^B regulon is chiefly involved in negotiating the growth-limiting transition of an osmotic shock.

PHYSIOLOGICAL ROLES OF THE σ^B REGULON IN OTHER BACTERIA

With the discovery of σ^B and its associated regulatory proteins in Listeria monocytogenes (15, 143) and Staphylococcus aureus (81, 146), the contribution of the general stress response to the pathogenesis of gram-positive bacteria can now be explored. This contribution has two distinct facets. On one hand, if the general stress response governs persistence in the natural environment, in foods, and in medical facilities, then it indirectly influences pathogenesis by increasing the bacterial load to which potential hosts are exposed. On the other hand, the general stress response could directly influence the interaction of the pathogen with host tissues, as it does for some gram-negative bacteria (see chapter 11).

Listeria monocytogenes

The available evidence suggests that the σ^B regulon is not directly involved in the interaction of *L. monocytogenes* with its host but rather that σ^B function is important for environmental persistence. *L. monocytogenes* is the agent of a food-borne infection with a high fatality rate among susceptible populations (110). The organism is widespread in nature and is a particular danger in refrigerated, salty foods because it grows well at low temperatures and it manifests high osmotolerance. The σ^B structural gene and part of the *sigB* operon were independently isolated by two research groups (15, 143), and the entire *sigB* operon with a gene order identical to that of *B. subtilis* has since been characterized (K. J. Boor, personal communication). Notably, loss of σ^B function had no obvious effect on the spread of *L. monocytogenes* to mouse liver or spleen following intragastric or intraperitoneal infection (143). However, it is not clear how well this animal model mimics food-borne infection in humans. For example, a *sigB* null mutant manifested a 1,000- to 5,000-fold decrease in survival when stationary-phase *L. monocytogenes* was exposed to pH 2.5 (143), suggesting a role in surviving a long stomach passage. It is also likely that *sigB* contributes to environmental survival in other ways. For example, a *sigB* null mutant was found to be impaired in its ability to use betaine and carnitine as osmoprotectants (15), suggesting a role in maintaining the population on salty foods. Considering the diverse phenotypes manifest by *sigB* mutants in *B. subtilis* (Table 1), it is likely that additional roles remain to be discovered for the σ^B regulon in *Listeria* species. Given the induction of σ^B activity by cold stress in *L. monocytogenes* (15), it will be particularly interesting to test the contribution of σ^B to psychrotrophic growth, a property integral to the emergence of this organism as a serious food-borne pathogen.

Staphylococcus aureus

The *sigB* regulon of *S. aureus* appears to be important for stress response, and its possible direct role in pathogenicity has been suggested but not fully tested. *S. aureus* causes a variety of human illnesses, including food-borne intoxications, soft-tissue infections, septicemia, and toxic shock syndrome. Many of these outcomes require products formed during stationary phase, when the *S. aureus* σ^B protein accumulates (28). The organization and expression of the *S. aureus sigB* region appears to differ in three significant ways from that of *B. subtilis*. First, the *S. aureus* region seems to contain only the genes corresponding to *rsbU*, *rsbV*, *rsbW*, and *sigB* of *B. subtilis* (81, 146). It is not presently known whether the *S. aureus* signal transduction network lacks the RsbX feedback phosphatase as well as the environmental stress regulators RsbR, RsbS, and RsbT, or whether their functions have been assumed by other proteins. A second difference comes from a Northern analysis that revealed a more complex transcriptional organization of the *S. aureus sigB* region (81). And a third difference is that expression of the *S. aureus sigB* region (and of at least one additional σ^B-dependent gene) is constitutive in a synthetic medium but inducible in complex medium (48). However, the *S. aureus* homolog of RsbW has been shown to have anti-σ factor activity in vitro (92), implying that the basic partner switching mechanism of regulation is retained in this organism.

The involvement of *S. aureus* σ^B in pathogenesis was suggested by the discovery that the P3 promoter of *sarA* is dependent on σ^B in vivo and in vitro (33, 86). *sarA* encodes a DNA-binding protein that modulates expression of the global virulence regulator *agr* (31). However, in a rabbit endocarditis model, Cheung et al. (30) found that a P3-GFP promoter fusion was silent when borne in *S. aureus* strain RN6390, suggesting that the P3 promoter contributes little to *sarA* expression in vivo. Chan et al. (28) then used a mouse subcutaneous abscess model in an apparent direct test of the role of σ^B during infection and found no difference in pathogenicity between *S. aureus* strain 8325-4 and a congenic *sigB* null mutant. In contrast, these same authors found that the *sigB* mutant had significantly reduced tolerance to such environmental stresses as acid shock in exponential growth and oxidative stress in stationary phase, and also had a mild defect in recovery from heat shock, but was indistinguishable from the parent strain with regard to salt or ethanol stress. These authors therefore concluded that σ^B controls an environmental stress response in *S. aureus* but does not contribute to its pathogenicity in the mouse model. However, these results are difficult to interpret because the parent strain 8325-4 employed in this study carried an 11-bp deletion in the gene for the *rsbU* homolog (48, 82). In *B. subtilis* the RsbU phosphatase is responsible for transmitting environmental signals to the RsbV and RsbW regulators (135, 147). Because some stress signals were evidently still communicated to *S. aureus* σ^B in the study of Chan et al. (28) despite the absence of RsbU function, the possibility arises that environmental stress transmission is accomplished differently in this organism. Until the signaling role of *S. aureus* RsbU is further defined, it may be premature to conclude that σ^B has no direct role in pathogenicity.

Mycobacterium tuberculosis

An intriguing *sigB*-like gene called *sigF* has also been discovered in *M. tuberculosis* (32). The *sigF* gene is more highly expressed during anaerobic growth, nitrogen depletion, oxidative stress, cold shock, and entry into stationary phase (32, 89). Moreover, *sigF* is preceded by a gene whose predicted product is similar to the RsbW anti-σ factor of *B. subtilis* σ^B and to the related SpoIIAB anti-σ of *B. subtilis* σ^F, which controls forespore-specific gene expression during sporulation (7, 90). Because expression of *M. tuberculosis sigF* is increased when cells are exposed to antibiotics, Michele et al. (89) speculated that *sigF* may control genes that are important for mycobacterial resistance to chemotherapy.

SUMMARY AND PROSPECTS

The scope and function of the *B. subtilis* general stress regulon are beginning to emerge, and the framework of the signal transduction network that controls σ^B activity has been established. Although there is some overlap in the kinds of general stress genes controlled by σ^B in *B. subtilis* and σ^S in *E. coli*, there are already indications of meaningful distinctions. Chief among these are the additional levels of local control superimposed on σ^S promoters by different DNA-binding proteins, dividing the σ^S regulon into subclasses of coregulated genes (see chapter 11). In contrast, most genes in the σ^B regulon appear to respond primarily to the global control of σ^B activity via the signal transduction network shown in Fig. 1. This undoubtedly reflects the different cellular architecture, ecological niches, and evolutionary histories of the two organisms. Moreover, the global mechanisms regulating σ^B and σ^S activities are distinctly different. If σ^B and its regulatory network are to provide potential targets for control of gram-positive pathogens, the channels by which this network senses and transduces external and internal signals require further investigation. Of particular interest are (i) biochemical and structural studies of the partner switching mechanism that forms the backbone of the σ^B signal transduction network, (ii) identification of upstream elements through which energy and environmental stresses might enter their respective branches of the backbone, and (iii) how the σ^B reg-

ulatory network exchanges signals with other adaptive response networks. Expanding this investigation to other gram-positive bacteria will reveal whether σ^B and its partner switching regulatory mechanism are found largely in *B. subtilis* and its close relatives, such as *L. monocytogenes* and *S. aureus*, or whether it is widespread among these medically and commercially important organisms.

Acknowledgments. I thank W. G. Haldenwang and M. Hecker for communicating results prior to publication and R. Losick and M. Hecker for their helpful comments on the manuscript.

Work from my laboratory was supported by Public Health Service grant GM42077 from the National Institute of General Medical Sciences.

REFERENCES

1. **Ahmed, M., C. M. Borsch, S. S. Taylor, N. Vazquez-Laslop, and A. A. Neyfakh.** 1994. A protein that activates expression of a multidrug efflux transporter upon binding the transporter substrates. *J. Biol. Chem.* **269:**28506–28513.
2. **Akbar, S., C. M. Kang, T. A. Gaidenko, and C. W. Price.** 1997. Modulator protein RsbR regulates environmental signalling in the general stress pathway of *Bacillus subtilis. Mol. Microbiol.* **24:**567–578.
3. **Akbar, S., S. Y. Lee, S. A. Boylan, and C. W. Price.** 1999. Two genes from *Bacillus subtilis* under the sole control of general stress transcription factor σ^B. *Microbiology* **145:**1069–1078.
4. **Akbar, S., and C. W. Price.** 1996. Isolation and characterization of *csbB*, a gene controlled by *Bacillus subtilis* general stress transcription factor σ^B. *Gene* **177:**123–128.
5. **Almiron, M., A. J. Link, D. Furlong, and R. Kolter.** 1992. A novel DNA-binding protein with regulatory and protective roles in starved *Escherichia coli. Genes Dev.* **6:**2646–2654.
6. **Alper, S., A. Dufour, D. A. Garsin, L. Duncan, and R. Losick.** 1996. Role of adenosine nucleotides in the regulation of a stress-response transcription factor in *Bacillus subtilis. J. Mol. Biol.* **260:**165–177.
7. **Alper, S., L. Duncan, and R. Losick.** 1994. An adenosine nucleotide switch controlling the activity of a cell type-specific transcription factor in *B. subtilis. Cell* **77:**195–205.
8. **Altschul, S. F., T. L. Madden, A. A. Schaffer, J. Zhang, Z. Zhang, W. Miller, and D. J. Lipman.** 1997. Gapped BLAST and PSI-BLAST: a new generation of protein database search programs. *Nucleic Acids Res.* **25:**3389–3402.
9. **Altuvia, S., M. Almiron, G. Huisman, R. Kolter, and G. Storz.** 1994. The *dps* promoter is activated by OxyR during growth and by IHF and σ^S in stationary phase. *Mol. Microbiol.* **13:**265–272.
10. **Antelmann, H., J. Bernhardt, R. Schmid, and M. Hecker.** 1995. A gene at 333 degrees on the *Bacillus subtilis* chromosome encodes the newly identified σ^B-dependent general stress protein GspA. *J. Bacteriol.* **177:**3540–3545.
11. **Antelmann, H., S. Engelmann, R. Schmid, and M. Hecker.** 1996. General and oxidative stress responses in *Bacillus subtilis*: cloning, expression, and mutation of the alkyl hydroperoxide reductase operon. *J. Bacteriol.* **178:**6571–6578.
12. **Antelmann, H., S. Engelmann, R. Schmid, A. Sorokin, A. Lapidus, and M. Hecker.** 1997. Expression of a stress- and starvation-induced *dps/pexB*-homologous gene is controlled by the alternative sigma factor σ^B in *Bacillus subtilis. J. Bacteriol.* **179:**7251–7256.

13. Antelmann, H., R. Schmid, and M. Hecker. 1997. The NAD synthetase NadE (OutB) of *Bacillus subtilis* is a σ^B-dependent general stress protein. *FEMS Microbiol. Lett.* **153:**405–409.

14. Bagyan, I., L. Casillas-Martinez, and P. Setlow. 1998. The *katX* gene, which codes for the catalase in spores of *Bacillus subtilis*, is a forespore-specific gene controlled by σ^F, and KatX is essential for hydrogen peroxide resistance of the germinating spore. *J. Bacteriol.* **180:**2057–2062.

15. Becker, L. A., M. S. Cetin, R. W. Hutkins, and A. K. Benson. 1998. Identification of the gene encoding the alternative sigma factor σ^B from *Listeria monocytogenes* and its role in osmotolerance. *J. Bacteriol.* **180:**4547–4554.

16. Benson, A. K., and W. G. Haldenwang. 1993. *Bacillus subtilis* σ^B is regulated by a binding protein (RsbW) that blocks its association with core RNA polymerase. *Proc. Natl. Acad. Sci. USA* **90:**2330–2334.

17. Benson, A. K., and W. G. Haldenwang. 1992. Characterization of a regulatory network that controls σ^B expression in *Bacillus subtilis*. *J. Bacteriol.* **174:**749–757.

18. Benson, A. K., and W. G. Haldenwang. 1993. Regulation of σ^B levels and activity in *Bacillus subtilis*. *J. Bacteriol.* **175:**2347–2356.

19. Benson, A. K., and W. G. Haldenwang. 1993. The σ^B-dependent promoter of the *Bacillus subtilis sigB* operon is induced by heat shock. *J. Bacteriol.* **175:**1929–1935.

20. Bernhardt, J., U. Völker, A. Völker, H. Antelmann, R. Schmid, H. Mach, and M. Hecker. 1997. Specific and general stress proteins in *Bacillus subtilis*—a two-dimensional protein electrophoresis study. *Microbiology* **143:**999–1017.

21. Bibikov, S. I., R. Biran, K. E. Rudd, and J. S. Parkinson. 1997. A signal transducer for aerotaxis in *Escherichia coli*. *J. Bacteriol.* **179:**4075–4079.

22. Binnie, C., M. Lampe, and R. Losick. 1986. Gene encoding the σ^{37} species of RNA polymerase sigma factor from *Bacillus subtilis*. *Proc. Natl. Acad. Sci. USA* **83:**5943–5947.

23. Boylan, S. A., A. R. Redfield, M. S. Brody, and C. W. Price. 1993. Stress-induced activation of the σ^B transcription factor of *Bacillus subtilis*. *J. Bacteriol.* **175:**7931–7937.

24. Boylan, S. A., A. R. Redfield, and C. W. Price. 1993. Transcription factor σ^B of *Bacillus subtilis* controls a large stationary-phase regulon. *J. Bacteriol.* **175:**3957–3963.

25. Boylan, S. A., A. Rutherford, S. M. Thomas, and C. W. Price. 1992. Activation of *Bacillus subtilis* transcription factor σ^B by a regulatory pathway responsive to stationary-phase signals. *J. Bacteriol.* **174:**3695–3706.

26. Boylan, S. A., M. D. Thomas, and C. W. Price. 1991. Genetic method to identify regulons controlled by nonessential elements: isolation of a gene dependent on alternate transcription factor σ^B of *Bacillus subtilis*. *J. Bacteriol.* **173:**7856–7866.

27. Bsat, N., A. Herbig, L. Casillas-Martinez, P. Setlow, and J. D. Helmann. 1998. *Bacillus subtilis* contains multiple Fur homologues: identification of the iron uptake (Fur) and peroxide regulon (PerR) repressors. *Mol. Microbiol.* **29:**189–198.

28. Chan, P. F., S. J. Foster, E. Ingham, and M. O. Clements. 1998. The *Staphylococcus aureus* alternative sigma factor σ^B controls the environmental stress response but not starvation survival or pathogenicity in a mouse abscess model. *J. Bacteriol.* **180:**6082–6089.

29. Chen, L., and J. D. Helmann. 1995. *Bacillus subtilis* MgrA is a Dps(PexB) homologue: evidence for a metalloregulation of an oxidative stress gene. *Mol. Microbiol.* **18:**295–300.

30. Cheung, A. L., C. C. Nast, and A. S. Bayer. 1998. Selective activation of *sar* promoters with the use of green fluorescent protein transcriptional fusions as the detection system in the rabbit endocarditis model. *Infect. Immun.* **66:**5988–5993.

31. Chien, Y., and A. L. Cheung. 1998. Molecular interactions between two global regulators, *sar* and *agr*, in *Staphylococcus aureus*. *J. Biol. Chem.* **273:**2645–2652.

32. DeMaio, J., Y. Zhang, C. Ko, D. B. Young, and W. R. Bishai. 1996. A stationary-phase stress-response sigma factor from *Mycobacterium tuberculosis*. *Proc. Natl. Acad. Sci. USA* **93:**2790–2794.

33. Deora, R., T. Tseng, and T. K. Misra. 1997. Alternative transcription factor σ^{SB} of *Staphylococcus aureus*: characterization and role in transcription of the global regulatory locus *sar*. *J. Bacteriol.* **179:**6355–6359.

34. Derre, I., G. Rapoport, and T. Msadek. 1999. CtsR, a novel regulator of stress and heat shock response, controls *clp* and molecular chaperone gene expression in gram-positive bacteria. *Mol. Microbiol.* **31:**117–131.

35. Diederich, B., J. F. Wilkinson, T. Magnin, M. Najafi, J. Errington, and M. D. Yudkin. 1994. Role of interactions between SpoIIAA and SpoIIAB in regulating cell-specific transcription factor σ^F of *Bacillus subtilis*. *Genes Dev.* **8:**2653–2663.

36. Dowds, B. C., P. Murphy, D. J. McConnell, and K. M. Devine. 1987. Relationship among oxidative stress, growth cycle, and sporulation in *Bacillus subtilis*. *J. Bacteriol.* **169:**5771–5775.

37. Drzewiecki, K., C. Eymann, G. Mittenhuber, and M. Hecker. 1998. The *yvyD* gene of *Bacillus subtilis* is under dual control of σ^B and σ^H. *J. Bacteriol.* **180:**6674–6680.

38. Dufour, A., and W. G. Haldenwang. 1994. Interactions between a *Bacillus subtilis* anti-sigma factor (RsbW) and its antagonist (RsbV). *J. Bacteriol.* **176:**1813–1820.

39. Duncan, L., S. Alper, F. Arigoni, R. Losick, and P. Stragier. 1995. Activation of cell-specific transcription by a serine phosphatase at the site of asymmetric division. *Science* **270:**641–644.

40. Duncan, L., and R. Losick. 1993. SpoIIAB is an anti-σ factor that binds to and inhibits transcription by regulatory protein σ^F from *Bacillus subtilis*. *Proc. Natl. Acad. Sci. USA* **90:**2325–2329.

41. Duncan, M. L., S. S. Kalman, S. M. Thomas, and C. W. Price. 1987. Gene encoding the 37,000-dalton minor sigma factor of *Bacillus subtilis* RNA polymerase: isolation, nucleotide sequence, chromosomal locus, and cryptic function. *J. Bacteriol.* **169:**771–778.

42. Engelmann, S., and M. Hecker. 1996. Impaired oxidative stress resistance of *Bacillus subtilis sigB* mutants and the role of *katA* and *katE*. *FEMS Microbiol. Lett.* **145:**63–69.

43. Engelmann, S., C. Lindner, and M. Hecker. 1995. Cloning, nucleotide sequence, and regulation of *katE* encoding a σ^B-dependent catalase in *Bacillus subtilis*. *J. Bacteriol.* **177:**5598–5605.

44. Ferrari, E., D. J. Henner, M. Perego, and J. A. Hoch. 1988. Transcription of *Bacillus subtilis* subtilisin and expression of subtilisin in sporulation mutants. *J. Bacteriol.* **170:**289–295.

45. Gaidenko, T. A., and C. W. Price. 1998. General stress transcription factor σ^B and sporulation transcription factor σ^H each contribute to survival of *Bacillus subtilis* under extreme growth conditions. *J. Bacteriol.* **180:**3730–3733.

46. Gaidenko, T. A., X. Yang, Y. M. Lee, and C. W. Price. 1999. Threonine phosphorylation of modulator protein RsbR governs its ability to regulate a serine kinase in the environmental stress signaling pathway of *Bacillus subtilis*. *J. Mol. Biol.* **288:**29–39.

47. Gerth, U., E. Krüger, I. Derre, T. Msadek, and M. Hecker. 1998. Stress induction of the *Bacillus subtilis clpP* gene encoding a homologue of the proteolytic component of the Clp

protease and the involvement of ClpP and ClpX in stress tolerance. *Mol. Microbiol.* **28:**787–802.

48. Gertz, S., S. Engelmann, R. Schmid, K. Ohlsen, J. Hacker, and M. Hecker. 1999. Regulation of σ^B-dependent transcription of *sigB* and *asp23* in two different *Staphylococcus aureus* strains. *Mol. Gen. Genet.* **261:**558–566.

49. Gomez, M., and S. M. Cutting. 1997. BofC encodes a putative forespore regulator of the *Bacillus subtilis* σ^K checkpoint. *Microbiology* **143:**157–170.

50. Gomez, M., and S. M. Cutting. 1997. Identification of a new σ^B-controlled gene, *csbX*, in *Bacillus subtilis*. *Gene* **188:**29–33.

51. Gordia, S., and C. Gutierrez. 1996. Growth-phase-dependent expression of the osmotically inducible gene *osmC* of *Escherichia coli* K-12. *Mol. Microbiol.* **19:**729–736.

52. Gottesman, S., M. R. Maurizi, and S. Wickner. 1997. Regulatory subunits of energy-dependent proteases. *Cell* **91:**435–438.

53. Gross, C. A. 1996. Function and regulation of the heat shock proteins, p. 1382–1399. *In* F. C. Neidhardt, R. Curtiss III, J. L. Ingraham, E. C. C. Lin, K. B. Low, B. Magasanik, W. S. Reznikoff, M. Riley, M. Schaechter, and H. E. Umbarger (ed.), *Escherichia coli and Salmonella: Cellular and Molelcular Biology*, 2nd ed. ASM Press, Washington, D.C.

54. Gutierrez, C., and J. C. Devedjian. 1991. Osmotic induction of gene *osmC* expression in *Escherichia coli* K12. *J. Mol. Biol.* **220:**959–973.

55. Haldenwang, W. G., and R. Losick. 1979. A modified RNA polymerase transcribes a cloned gene under sporulation control in *Bacillus subtilis*. *Nature* **282:**256–260.

56. Haldenwang, W. G., and R. Losick. 1980. Novel RNA polymerase sigma factor from *Bacillus subtilis*. *Proc. Natl. Acad. Sci. USA* **77:**7000–7004.

57. Hecker, M., C. Heim, U. Völker, and L. Wölfel. 1988. Induction of stress proteins by sodium chloride treatment in *Bacillus subtilis*. *Arch. Microbiol.* **150:**564–566.

58. Hecker, M., A. Richter, A. Schroeter, L. Wölfel, and F. Mach. 1987. [Synthesis of heat shock proteins following amino acid or oxygen limitation in *Bacillus subtilis* RelA+ and RelA strains]. *Z. Naturforsch.* [C] **42:**941–947.

59. Hecker, M., W. Schumann, and U. Völker. 1996. Heat-shock and general stress response in *Bacillus subtilis*. *Mol. Microbiol.* **19:**417–428.

60. Hecker, M., and U. Völker. 1998. Non-specific, general and multiple stress resistance of growth-restricted *Bacillus subtilis* cells by the expression of the σ^B regulon. *Mol. Microbiol.* **29:**1129–1136.

61. Hecker, M., and U. Völker. 1990. General stress proteins in *Bacillus subtilis*. *FEMS Microbiol. Ecol.* **74:**197–214.

62. Henikoff, S., and J. G. Henikoff. 1992. Amino acid substitution matrices from protein blocks. *Proc. Natl. Acad. Sci. USA* **89:**10915–10919.

63. Hill, S., S. Austin, T. Eydmann, T. Jones, and R. Dixon. 1996. *Azotobacter vinelandii* NifL is a flavoprotein that modulates transcriptional activation of nitrogen-fixation genes via a redox-sensitive switch. *Proc. Natl. Acad. Sci. USA* **93:**2143–2148.

64. Horsburgh, M. J., and A. Moir. 1999. σ^M, an ECF RNA polymerase σ factor of *Bacillus subtilis* 168, is essential for growth and survival in high concentrations of salt. *Mol. Microbiol.* **32:**41–50.

65. Huang, X., A. Gaballa, M. Cao, and J. D. Helmann. 1999. Identification of target promoters for the *Bacillus subtilis* extracytoplasmic function σ factor, σ^W. *Mol. Microbiol.* **31:**361–371.

66. Huang, X., and J. D. Helmann. 1998. Identification of target promoters for the *Bacillus subtilis* σ^X factor using a consensus-directed search. *J. Mol. Biol.* **279:**165–173.

67. Huang, Z. J., I. Edery, and M. Rosbash. 1993. PAS is a dimerization domain common to *Drosophila period* and several transcription factors. *Nature* **364:**259–262.

68. Igo, M., M. Lampe, C. Ray, W. Schafer, C. P. Moran, Jr., and R. Losick. 1987. Genetic studies of a secondary RNA polymerase sigma factor in *Bacillus subtilis*. *J. Bacteriol.* **169:**3464–3469.

69. Igo, M. M., and R. Losick. 1986. Regulation of a promoter that is utilized by minor forms of RNA polymerase holoenzyme in *Bacillus subtilis*. *J. Mol. Biol.* **191:**615–624.

70. Johnson, W. C., C. P. Moran, Jr., and R. Losick. 1983. Two RNA polymerase sigma factors from *Bacillus subtilis* discriminate between overlapping promoters for a developmentally regulated gene. *Nature* **302:**800–804.

71. Jurgen, B., T. Schweder, and M. Hecker. 1998. The stability of mRNA from the *gsiB* gene of *Bacillus subtilis* is dependent on the presence of a strong ribosome binding site. *Mol. Gen. Genet.* **258:**538–545.

72. Kalman, S., M. L. Duncan, S. M. Thomas, and C. W. Price. 1990. Similar organization of the *sigB* and *spoIIA* operons encoding alternate sigma factors of *Bacillus subtilis* RNA polymerase. *J. Bacteriol.* **172:**5575–5585.

73. Kang, C. M., M. S. Brody, S. Akbar, X. Yang, and C. W. Price. 1996. Homologous pairs of regulatory proteins control activity of *Bacillus subtilis* transcription factor σ^B in response to environmental stress. *J. Bacteriol.* **178:**3846–3853.

74. Kang, C. M., K. Vijay, and C. W. Price. 1998. Serine kinase activity of a *Bacillus subtilis* switch protein is required to transduce environmental stress signals but not to activate its target PP2C phosphatase. *Mol. Microbiol.* **30:**189–196.

75. Kempf, B., and E. Bremer. 1998. Uptake and synthesis of compatible solutes as microbial stress responses to high-osmolality environments. *Arch. Microbiol.* **170:**319–330.

76. Klyachko, K. A., and A. A. Neyfakh. 1998. Paradoxical enhancement of the activity of a bacterial multidrug transporter caused by substitutions of a conserved residue. *J. Bacteriol.* **180:**2817–2821.

77. Krüger, E., and M. Hecker. 1998. The first gene of the *Bacillus subtilis clpC* operon, *ctsR*, encodes a negative regulator of its own operon and other class III heat shock genes. *J. Bacteriol.* **180:**6681–6688.

78. Krüger, E., T. Msadek, and M. Hecker. 1996. Alternate promoters direct stress-induced transcription of the *Bacillus subtilis clpC* operon. *Mol. Microbiol.* **20:**713–723.

79. Krüger, E., T. Msadek, S. Ohlmeier, and M. Hecker. 1997. The *Bacillus subtilis clpC* operon encodes DNA repair and competence proteins. *Microbiology* **143:**1309–1316.

80. Krüger, E., U. Völker, and M. Hecker. 1994. Stress induction of *clpC* in *Bacillus subtilis* and its involvement in stress tolerance. *J. Bacteriol.* **176:**3360–3367.

81. Kullik, I., and P. Giachino. 1997. The alternative sigma factor σ^B in *Staphylococcus aureus*: regulation of the *sigB* operon in response to growth phase and heat shock. *Arch. Microbiol.* **167:**151–159.

82. Kullik, I., P. Giachino, and T. Fuchs. 1998. Deletion of the alternative sigma factor σ^B in *Staphylococcus aureus* reveals its function as a global regulator of virulence genes. *J. Bacteriol.* **180:**4814–4820.

83. Lewin, B. 1987. *Genes*, 3rd ed. Wiley, New York, N.Y.

84. Lindebro, M. C., L. Poellinger, and M. L. Whitelaw. 1995. Protein-protein interaction via PAS domains: role of the PAS domain in positive and negative regulation of the bHLH/PAS

dioxin receptor-Arnt transcription factor complex. *EMBO J.* **14:**3528–3539.

85. **López, C. S., H. Heras, S. M. Ruzal, C. Sánchez-Rivas, and E. A. Rivas.** 1998. Variations of the envelope composition of *Bacillus subtilis* during growth in hyperosmotic medium. *Curr. Microbiol.* **36:**55–61.

86. **Manna, A. C., M. G. Bayer, and A. L. Cheung.** 1998. Transcriptional analysis of different promoters in the *sar* locus in *Staphylococcus aureus. J. Bacteriol.* **180:**3828–3836.

87. **Martinez, A., and R. Kolter.** 1997. Protection of DNA during oxidative stress by the nonspecific DNA-binding protein Dps. *J. Bacteriol.* **179:**5188–5194.

88. **Maul, B., U. Völker, S. Riethdorf, S. Engelmann, and M. Hecker.** 1995. σ^B-dependent regulation of *gsiB* in response to multiple stimuli in *Bacillus subtilis. Mol. Gen. Genet.* **248:**114–120.

89. **Michele, T. M., C. Ko, and W. R. Bishai.** 1999. Exposure to antibiotics induces expression of the *Mycobacterium tuberculosis sigF* gene: implications for chemotherapy against mycobacterial persistors. *Antimicrob. Agents Chemother.* **43:**218–225.

90. **Min, K. T., C. M. Hilditch, B. Diederich, J. Errington, and M. D. Yudkin.** 1993. σ^F, the first compartment-specific transcription factor of *B. subtilis*, is regulated by an anti-σ factor that is also a protein kinase. *Cell* **74:**735–742.

91. **Missiakas, D., and S. Raina.** 1998. The extracytoplasmic function sigma factors: role and regulation. *Mol. Microbiol.* **28:**1059–1066.

92. **Miyazaki, E., J. M. Chen, C. Ko, and W. R. Bishai.** 1999. The *Staphylococcus aureus rsbW* (orf159) gene encodes an anti-sigma factor of SigB. *J. Bacteriol.* **181:**2846–2851.

93. **Mongkolsuk, S., W. Praituan, S. Loprasert, M. Fuangthong, and S. Chamnongpol.** 1998. Identification and characterization of a new organic hydroperoxide resistance (*ohr*) gene with a novel pattern of oxidative stress regulation from *Xanthomonas campestris* pv. *phaseoli. J. Bacteriol.* **180:**2636–2643.

94. **Moran, C. P., Jr., W. C. Johnson, and R. Losick.** 1982. Close contacts between sigma 37-RNA polymerase and a *Bacillus subtilis* chromosomal promoter. *J. Mol. Biol.* **162:**709–713.

95. **Moran, C. P., Jr., N. Lang, C. D. Banner, W. G. Haldenwang, and R. Losick.** 1981. Promoter for a developmentally regulated gene in *Bacillus subtilis. Cell* **25:**783–791.

96. **Msadek, T.** 1999. When the going gets tough: survival strategies and environmental signaling networks in *Bacillus subtilis. Trends Microbiol.* **7:**201–207.

97. **Msadek, T., V. Dartois, F. Kunst, M. L. Herbaud, F. Denizot, and G. Rapoport.** 1998. ClpP of *Bacillus subtilis* is required for competence development, motility, degradative enzyme synthesis, growth at high temperature and sporulation. *Mol. Microbiol.* **27:**899–914.

98. **Msadek, T., F. Kunst, and G. Rapoport.** 1994. MecB of *Bacillus subtilis*, a member of the ClpC ATPase family, is a pleiotropic regulator controlling competence gene expression and growth at high temperature. *Proc. Natl. Acad. Sci. USA* **91:**5788–5792.

99. **Nessi, C., A. M. Albertini, M. L. Speranza, and A. Galizzi.** 1995. The *outB* gene of *Bacillus subtilis* codes for NAD synthetase. *J. Biol. Chem.* **270:**6181–6185.

100. **Neyfakh, A. A., V. E. Bidnenko, and L. B. Chen.** 1991. Efflux-mediated multidrug resistance in *Bacillus subtilis*: similarities and dissimilarities with the mammalian system. *Proc. Natl. Acad. Sci. USA* **88:**4781–4785.

101. **Ollington, J. F., W. G. Haldenwang, T. V. Huynh, and R. Losick.** 1981. Developmentally regulated transcription in a cloned segment of the *Bacillus subtilis* chromosome. *J. Bacteriol.* **147:**432–442.

102. **Petersohn, A., H. Antelmann, U. Gerth, and M. Hecker.** 1999. Identification and transcriptional analysis of new members of the σ^B regulon in *Bacillus subtilis. Microbiology* **145:**869–880.

103. **Petersohn, A., J. Bernhardt, U. Gerth, D. Höper, T. Koburger, U. Völker, and M. Hecker.** 1999. Identification of σ^B-dependent genes in *Bacillus subtilis* using a promoter consensus directed search and oligonucleotide hybridization. *J. Bacteriol.* **181:**5718–5724.

104. **Petersohn, A., S. Engelmann, P. Setlow, and M. Hecker.** 1999. The *katX* gene of *Bacillus subtilis* is under the dual control of σ^B and σ^F. *Mol. Gen. Genet.* **262:**173–179.

105. **Pooley, H. M., D. Paschoud, and D. Karamata.** 1987. The *gtaB* marker in *Bacillus subtilis* 168 is associated with a deficiency in UDP-glucose pyrophosphorylase. *J. Gen. Microbiol.* **133:**3481–3493.

106. **Ray, C., R. E. Hay, H. L. Carter, and C. P. Moran, Jr.** 1985. Mutations that affect utilization of a promoter in stationary-phase *Bacillus subtilis. J. Bacteriol.* **163:**610–614.

107. **Ray, C., M. Igo, W. Shafer, R. Losick, and C. P. Moran, Jr.** 1988. Suppression of *ctc* promoter mutations in *Bacillus subtilis. J. Bacteriol.* **170:**900–907.

108. **Rebbapragada, A., M. S. Johnson, G. P. Harding, A. J. Zuccarelli, H. M. Fletcher, I. B. Zhulin, and B. L. Taylor.** 1997. The Aer protein and the serine chemoreceptor Tsr independently sense intracellular energy levels and transduce oxygen, redox, and energy signals for *Escherichia coli* behavior. *Proc. Natl. Acad. Sci. USA* **94:**10541–10546.

109. **Redfield, A. R., and C. W. Price.** 1996. General stress transcription factor σ^B of *Bacillus subtilis* is a stable protein. *J. Bacteriol.* **178:**3668–3670.

110. **Rocourt, J., and P. Cossart.** 1997. *Listeria monocytogenes*, p. 337–352. *In* M. P. Doyle, L. R. Beuchat, and T. J. Montville (ed.), *Food Microbiology—Fundamentals and Frontiers.* ASM Press, Washington, D.C.

111. **Rosenbluh, A., C. D. Banner, R. Losick, and P. C. Fitz-James.** 1981. Identification of a new developmental locus in *Bacillus subtilis* by construction of a deletion mutation in a cloned gene under sporulation control. *J. Bacteriol.* **148:**341–351.

112. **Scharf, C., S. Riethdorf, H. Ernst, S. Engelmann, U. Völker, and M. Hecker.** 1998. Thioredoxin is an essential protein induced by multiple stresses in *Bacillus subtilis. J. Bacteriol.* **180:**1869–1877.

113. **Schmidt, R., P. Margolis, L. Duncan, R. Coppolecchia, C. P. Moran, Jr., and R. Losick.** 1990. Control of developmental transcription factor σ^F by sporulation regulatory proteins SpoIIAA and SpoIIAB in *Bacillus subtilis. Proc. Natl. Acad. Sci. USA* **87:**9221–9225.

114. **Schulz, A., and W. Schumann.** 1996. *hrcA*, the first gene of the *Bacillus subtilis dnaK* operon encodes a negative regulator of class I heat shock genes. *J. Bacteriol.* **178:**1088–1093.

115. **Scott, J. M., and W. G. Haldenwang.** 1999. Obg, an essential GTP binding protein of *Bacillus subtilis*, is necessary for stress activation of transcription factor σ^B. *J. Bacteriol.* **181:**4653–4660.

116. **Scott, J. M., N. Smirnova, and W. G. Haldenwang.** 1999. A Bacillus-specific factor is needed to trigger the stress-activated phosphatase/kinase cascade of σ^B induction. *Biochem. Biophys. Res. Commun.* **257:**106–110.

117. **Segall, J., and R. Losick.** 1977. Cloned *Bacillus subtilis* DNA containing a gene that is activated early during sporulation. *Cell* **11:**751–761.

118. Smirnova, N., J. Scott, U. Voelker, and W. G. Haldenwang. 1998. Isolation and characterization of *Bacillus subtilis sigB* operon mutations that suppress the loss of the negative regulator RsbX. *J. Bacteriol.* **180:**3671–3680.

119. Soldo, B., V. Lazarevic, P. Margot, and D. Karamata. 1993. Sequencing and analysis of the divergon comprising *gtaB*, the structural gene of UDP-glucose pyrophosphorylase of *Bacillus subtilis* 168. *J. Gen. Microbiol.* **139:**3185–3195.

120. Song, B. H., and J. Neuhard. 1989. Chromosomal location, cloning and nucleotide sequence of the *Bacillus subtilis cdd* gene encoding cytidine/deoxycytidine deaminase. *Mol. Gen. Genet.* **216:**462–468.

121. Spiegelhalter, F., and E. Bremer. 1998. Osmoregulation of the *opuE* proline transport gene from *Bacillus subtilis*: contributions of the σ^A- and σ^B-dependent stress-responsive promoters. *Mol. Microbiol.* **29:**285–296.

122. Squires, C., and C. L. Squires. 1992. The Clp proteins: proteolysis regulators or molecular chaperones? *J. Bacteriol.* **174:**1081–1085.

123. Stacy, R. A., and R. B. Aalen. 1998. Identification of sequence homology between the internal hydrophilic repeated motifs of group 1 late-embryogenesis-abundant proteins in plants and hydrophilic repeats of the general stress protein GsiB of *Bacillus subtilis*. *Planta* **206:**476–478.

124. Stragier, P., and R. Losick. 1996. Molecular genetics of sporulation in *Bacillus subtilis*. *Annu. Rev. Genet.* **30:**297–241.

125. Tatti, K. M., and C. P. Moran, Jr. 1984. Promoter recognition by σ^{37} RNA polymerase from *Bacillus subtilis*. *J. Mol. Biol.* **175:**285–297.

126. Taylor, B. L., and I. B. Zhulin. 1999. PAS domains: internal sensors of oxygen, redox potential, and light. *Microbiol. Mol. Biol. Rev.* **63:**479–506.

127. Turgay, K., J. Hahn, J. Burghoorn, and D. Dubnau. 1998. Competence in *Bacillus subtilis* is controlled by regulated proteolysis of a transcription factor. *EMBO J.* **17:**6730–6738.

128. Turgay, K., L. W. Hamoen, G. Venema, and D. Dubnau. 1997. Biochemical characterization of a molecular switch involving the heat shock protein ClpC, which controls the activity of ComK, the competence transcription factor of *Bacillus subtilis*. *Genes Dev.* **11:**119–128.

129. Valdivia, R. H., and S. Falkow. 1996. Bacterial genetics by flow cytometry: rapid isolation of *Salmonella typhimurium* acid-inducible promoters by differential fluorescence induction. *Mol. Microbiol.* **22:**367–378.

130. Varón, D., S. A. Boylan, K. Okamoto, and C. W. Price. 1993. *Bacillus subtilis gtaB* encodes UDP-glucose pyrophosphorylase and is controlled by stationary-phase transcription factor σ^B. *J. Bacteriol.* **175:**3964–3971.

131. Varón, D., M. S. Brody, and C. W. Price. 1996. *Bacillus subtilis* operon under the dual control of the general stress transcription factor σ^B and the sporulation transcription factor σ^H. *Mol. Microbiol.* **20:**339–350.

132. Vijay, K., M. S. Brody, E. Fredlund, and C. W. Price. 2000. A PP2C phosphatase containing a PAS domain is required to convey signals of energy stress to the σ^B transcription factor of *Bacillus subtilis*. *Mol. Microbiol.* **35:**180–188.

133. Voelker, U., T. Luo, N. Smirnova, and W. Haldenwang. 1997. Stress activation of *Bacillus subtilis* σ^B can occur in the absence of the σ^B negative regulator RsbX. *J. Bacteriol.* **179:**1980–1984.

134. Voelker, U., A. Voelker, and W. G. Haldenwang. 1996. Reactivation of the *Bacillus subtilis* anti-σ^B antagonist, RsbV, by stress- or starvation-induced phosphatase activities. *J. Bacteriol.* **178:**5456–5463.

135. Voelker, U., A. Voelker, B. Maul, M. Hecker, A. Dufour, and W. G. Haldenwang. 1995. Separate mechanisms activate σ^B of *Bacillus subtilis* in response to environmental and metabolic stresses. *J. Bacteriol.* **177:**3771–3780.

136. Völker, U., K. K. Andersen, H. Antelmann, K. M. Devine, and M. Hecker. 1998. One of two *osmC* homologs in *Bacillus subtilis* is part of the σ^B-dependent general stress regulon. *J. Bacteriol.* **180:**4212–4218.

137. Völker, U., S. Engelmann, B. Maul, S. Riethdorf, A. Völker, R. Schmid, H. Mach, and M. Hecker. 1994. Analysis of the induction of general stress proteins of *Bacillus subtilis*. *Microbiology* **140:**741–752.

138. Völker, U., and M. Hecker. 1999. Expression of the σ^B-dependent general stress regulon confers multiple stress resistance in *Bacillus subtilis*. *J. Bacteriol.* **181:**3942–3948.

139. Völker, U., H. Mach, R. Schmid, and M. Hecker. 1992. Stress proteins and cross-protection by heat shock and salt stress in *Bacillus subtilis*. *J. Gen. Microbiol.* **138:**2125–2135.

140. von Blohn, C., B. Kempf, R. M. Kappes, and E. Bremer. 1997. Osmostress response in *Bacillus subtilis*: characterization of a proline uptake system (OpuE) regulated by high osmolarity and the alternative transcription factor σ^B. *Mol. Microbiol.* **25:**175–187.

141. Wang, P. Z., and R. H. Doi. 1984. Overlapping promoters transcribed by *Bacillus subtilis* sigma 55 and sigma 37 RNA polymerase holoenzymes during growth and stationary phases. *J. Biol. Chem.* **259:**8619–8625.

142. Wawrzynow, A., B. Banecki, and M. Zylicz. 1996. The Clp ATPases define a novel class of molecular chaperones. *Mol. Microbiol.* **21:**895–899.

143. Wiedmann, M., T. J. Arvik, R. J. Hurley, and K. J. Boor. 1998. General stress transcription factor σ^B and its role in acid tolerance and virulence of *Listeria monocytogenes*. *J. Bacteriol.* **180:**3650–3656.

144. Wise, A. A., and C. W. Price. 1995. Four additional genes in the *sigB* operon of *Bacillus subtilis* that control activity of the general stress factor σ^B in response to environmental signals. *J. Bacteriol.* **177:**123–133.

145. Wong, S. L., C. W. Price, D. S. Goldfarb, and R. H. Doi. 1984. The subtilisin E gene of *Bacillus subtilis* is transcribed from a sigma 37 promoter in vivo. *Proc. Natl. Acad. Sci. USA* **81:**1184–1188.

146. Wu, S., H. de Lencastre, and A. Tomasz. 1996. Sigma-B, a putative operon encoding alternate sigma factor of *Staphylococcus aureus* RNA polymerase: molecular cloning and DNA sequencing. *J. Bacteriol.* **178:**6036–6042.

147. Yang, X., C. M. Kang, M. S. Brody, and C. W. Price. 1996. Opposing pairs of serine protein kinases and phosphatases transmit signals of environmental stress to activate a bacterial transcription factor. *Genes Dev.* **10:**2265–2275.

148. Yuan, G., and S. L. Wong. 1995. Isolation and characterization of *Bacillus subtilis groE* regulatory mutants: evidence for orf39 in the *dnaK* operon as a repressor gene in regulating the expression of both *groE* and *dnaK*. *J. Bacteriol.* **177:**6462–6468.

149. Yura, T., and K. Nakahigashi. 1999. Regulation of the heat-shock response. *Curr. Opin. Microbiol.* **2:**153–158.

150. Zuber, U., and W. Schumann. 1994. CIRCE, a novel heat shock element involved in regulation of heat shock operon *dnaK* of *Bacillus subtilis*. *J. Bacteriol.* **176:**1359–1363.

Bacterial Stress Responses
Edited by G. Storz and R. Hengge-Aronis
©2000 ASM Press, Washington, D.C.

Chapter 13

Bacterial Sporulation: a Response to Environmental Signals

ABRAHAM L. SONENSHEIN

Endospore formation in Bacillus subtilis, *myxospore formation in* Myxococcus xanthus, *and sporulation of aerial hyphae of* Streptomyces coelicolor *have proved to be useful paradigms for understanding bacterial differentiation. Although the three systems differ considerably in the detailed biochemistry and morphology of differentiation, they share important features. In each case, sporulation is induced by nutritional limitation and cell crowding and leads to formation of a metabolically limited (or even dormant) cell that has increased resistance to certain environmental stresses (e.g., heat, desiccation, organic solvents). Thus, these systems fall into the category of general stress responses but differ from other such responses in that the precipitating stresses (nutritional limitation, cell crowding) are not overcome by the response. Furthermore, once committed to the differentiation pathway, the cells cannot revert to active growth until after having completed spore formation.*

Bacterial sporulation has the unusual property of endowing cells with resistance to environmental changes they have not yet encountered and may never encounter. Generally, adaptive responses to sublethal changes in temperature, doses of oxidizing compounds, or changes in pH induce expression of genes whose products provide protection against more potent doses of the same type of insult. Sporulation, by contrast, is a response to one type of stress (nutritional limitation) but leads to resistance to several potentially damaging agents, while failing to counteract the problem that caused the induction. Even though exposure to the deleterious agents is not known to induce sporulation, the levels of resistance achieved by spores are much greater than those achieved by agent-specific induction in most other bacteria. It must be that the likelihood that a dormant spore will encounter dangerous conditions in its nat-

ural niche is high enough that the ability to achieve a long-term, protected state provides a selective advantage. From the long-range perspective, sporulation is more like general than specific stress responses in that the formation of a spore allows the cell to survive a large variety of potential environmental challenges. Unlike the case of the general stress response controlled by σ^B in *Bacillus subtilis* (see chapter 12), however, induction of spore formation has an absolute requirement for nutritional limitation. Moreover, sporulation becomes, at some point, an irreversible pathway; general stress responses are rapidly reversible by elimination of the stress. Sporulation and other stress responses may have additive effects, since exposure of sporulating cells to mild heat stress leads to a small increase in spore heat resistance (120). But σ^B is not needed for sporulation in *B. subtilis* (10, 36, 57), and no sporulation-essential gene is needed for the general stress response (58).

Four modes of bacterial spore formation have been recognized. Three of them have been the subject of intensive study, especially in a small number of paradigm species. (i) Endospore formers (including *Bacillus, Clostridium, Sporosarcina, Thermoactinomyces*, and *Sporolactobacillus*) are almost exclusively gram-positive bacteria (132); they differentiate by intracellular division and engulfment of the developing spore protoplast within a mother cell (106) (Fig. 1A, B). The model endosporulator is *B. subtilis*, the most highly studied bacterial species other than *Escherichia coli* (133). (ii) Actinomycetes are gram-positive examples of spore-formers that do so without the benefit of a mother cell. In this case, a long, multinucleate hypha divides into numerous mononucleate cells, many of which then differentiate into arthrospores or conidia (22) (Fig. 1D). The best studied actinomycetes with respect to genetics and differentiation are *Streptomyces coelicolor* and *Streptomyces griseus*.

Abraham L. Sonenshein • Department of Molecular Biology and Microbiology, Tufts University School of Medicine, Boston, MA 02111.

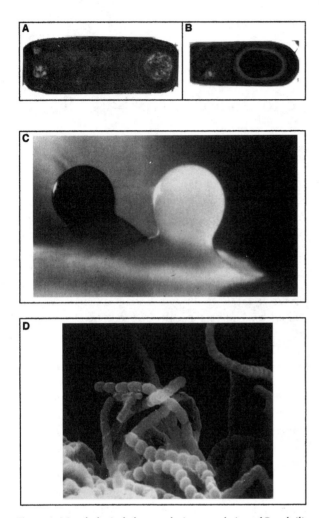

Figure 1. Morphological changes during sporulation of *B. subtilis*. (A) A cell at the transition from stage II to stage III. A division septum has formed near one pole of the cell and has begun to migrate around the forespore compartment. When engulfment is completed, the forespore will be a separate cell, surrounded by a double membrane and fully contained within the mother cell compartment. (B) A late sporulating cell. After completion of engulfment, peptidoglycan-like cortex material has been deposited between the two forespore membranes, and coat proteins have begun to assemble in two layers around the forespore. The electron micrographs of parts A and B were prepared by A. Brown Cormier, Tufts University Electron Microscopy Facility. (C) Fruiting bodies of *Myxococcus fulvus*, photographed by H. Reichenbach (Braunschweig, Germany). The diameter of the fruiting body at its widest point is about 100 μm. This picture appeared previously in reference 38 and is reprinted with permission of Wiley-Liss, Inc., a subsidiary of John Wiley & Sons, Inc. (D) Aerial mycelia of *S. coelicolor* strain J1501 grown for 7 days on a rich sporulation agar medium. In this scanning electron micrograph, provided by J. Willey (Hofstra University), the hyphae have divided into individual cells, each of which will form a spore.

(iii) The gram-negative myxobacteria form fruiting bodies (sporangioles [Fig. 1C]) within which some of the cells round up, becoming microcysts or myxospores (37). The paradigmatic organism is *Myxococcus xanthus*. In all of the above cases, the model

species were chosen because of their ease of growth in the laboratory, their relatively efficient differentiation, and their susceptibility to genetic manipulation. (iv) The fourth mode is exosporulation, a budding-like phenomenon about which relatively little is known. It can be found in certain species of cyanobacteria and purple bacteria.

Bacillus spores are metabolically dormant and have very high resistance to heat, desiccation, acids, bases, organic solvents, oxidizing agents, radiation, and proteolytic and saccharolytic enzymes (122, 143; see also chapter 14). Myxospores resist drying and have moderate heat and UV resistance (37). *Streptomyces* spores have greatly reduced metabolic activity and have substantial resistance to desiccation but only moderately increased resistance to heat compared to vegetative cells (22).

While it is clearly advantageous to any organism to become resistant to potentially lethal conditions, the formation of spores comes at a price. In *M. xanthus*, sporulation occurs within a fruiting body in which many fewer than half of the cells reach the stage of heat-resistant spores (37, 64). Most of the remaining cells undergo lysis. In *S. coelicolor*, only those cells that are included within an aerial mycelium can sporulate; the majority of the cell mass remains in a nonsporulating surface mycelium and may be cannibalized to provide nutrients to the aerial hyphae (20). In *B. subtilis*, half of the cellular DNA and more than half of the cell mass are discarded during the conversion of a predivisional cell to a spore. Moreover, once committed to sporulate, a cell cannot revert to growth until it has completed differentiation (about 8 h at 37°C in *Bacillus*) and has undergone spore germination (about 1 h for *Bacillus*), a second differentiation process. If the nutritional environment changes for the better during these rather lengthy processes, nonsporulating competitor bacteria in the same niche will have at least a temporary growth advantage and may crowd out the spore-formers.

COMMON THEMES

All spore-forming bacteria that have been studied in detail induce spore formation under conditions of nutrient limitation and cell crowding. Although the details differ considerably, all sporulation-competent cells sense the disappearance of nutrients that support rapid growth and simultaneously monitor the local concentration of the same or related species. For population density- or quorum-sensing, the cells depend on species-specific signal compounds produced during growth. The accumulation of these compounds in the extracellular environment serves as a measure of the cell concentration. A common metabolic signal for all types of spore formation discussed here may be a drop in the intracellular concentration of GTP or an increase in ppGpp or both. In all cases, the initial nutritional and cell population density signals lead to cascading induction of differentiation-specific genes through a dependent temporal sequence of gene expression. At least in *Bacillus*, this gene expression is also ordered in space and is responsive to structural changes in the differentiating cell.

Sporulation-Associated Morphological Changes in *B. subtilis*

The defining morphological event for endospore formation is division of the cytoplasm into forespore and mother cell compartments (83). In rod-shaped endospore-formers, this division event is asymmetrically oriented with respect to the long axis of the cell. As a result, the forespore is much smaller than the mother cell. (By convention, cells that have completed this septation event are said to have reached stage II of endosporulation. Stage I corresponds to predivisional alignment of the chromosomes (two copies) along the long axis of the cell.) In *Sporosarcina ureae*, a spore-forming coccus, the division event is centrally located and creates subcellular compartments of equal volume (155).

To divide asymmetrically, the cell must redirect its control system for septum placement. The primary event in septation is formation of a ring of FtsZ protein at the site of eventual septation (9). Other proteins then associate with FtsZ, creating the septation machine (septasome). In growing bacterial cells, the septasome forms only at mid-cell because the inhibitory MinCD protein complex sits at potential polar septation sites (30). At the onset of sporulation, inhibition of polar septation is relieved at the same time that medial assembly of septasomes becomes inhibited. Although septasomes form at both poles of an early sporulating cell, only one of the machines actually forms a septum; the other subsequently dissociates (82). "Disporic" mutants are unable to restrict septation to a single pole (106).

After sporulation-associated septation, whether polar (as in *B. subtilis*) or medial (as in *S. ureae*), the cytoplasmic membrane of the mother cell compartment engulfs the nascent forespore (Fig. 1A). When the encirclement is complete, the tips of the mother cell membrane fuse, creating a forespore compartment that is surrounded by a double membrane and is wholly contained within the cytoplasm of the mother cell. Completion of engulfment defines stage III of endosporulation. At stage IV, spore cortex, a

peptidoglycan closely related to the normal cell wall, assembles between the two forespore membranes. This structure gives rigidity to the spore, and its formation contributes to the dehydration of the spore cytoplasm (see chapter 14).

The final steps in endospore formation (stages V to VII) are the assembly of multiple layers of spore coat proteins (Fig. 1B), maturation of the spore, and release by lysis of the mother cell. Spore coats are synthesized and assembled in the mother cell cytoplasm. They attach to the outer forespore membrane through several scaffolding proteins (32, 114). Recent reviews detail how these protective layers are made and how they contribute to the resistance properties of endospores (33 and chapter 14).

Sporulation-Associated Signal Transduction in *B. subtilis*

Initiation of sporulation in *B. subtilis* can be seen as a two-stage process during which the cell first attempts to maintain growth despite high population density and nutrient limitation and subsequently determines that it is obliged to sporulate because it is unable to find any alternative nutrient sources (131, 132). During the initial phase (referred to here, for simplicity, as the stationary-phase response), the cell responds to limitation of favored carbon or nitrogen or phosphorus sources (29, 119, 134) by inducing a variety of adaptive responses whose goal is to seek out and assimilate alternative sources of food. These responses include chemotaxis and motility, secretion of degradative enzymes, activation of transport systems, induction of catabolic pathways, synthesis of antibiotics and toxins, and induction of genetic competence. If this search is successful, the cells will continue to grow, albeit slowly, and will not sporulate en masse. If no alternative food sources can be found or when those secondary sources eventually become limiting, the cell turns on genes whose purpose is to commit it to the sporulation pathway. Complete starvation does not lead to efficient sporulation because formation of the spore is an energy-intensive, biosynthetic process.

There is considerable but incomplete evidence that nutritional limitation is sensed intracellularly, at least in part, through a drop in the GTP pool or the accumulation of ppGpp or both. First, all known protocols for inducing sporulation lead to at least a transient drop in intracellular GTP (86). Second, stationary-phase gene expression and sporulation can be induced in growing cells by treating them with an inhibitor of guanine nucleotide synthesis (93, 95). Third, the Obg protein, a GTP-binding homolog of eukaryotic Ras proteins, appears to be a key factor in transmitting a GTP signal in early stationary phase (74, 142). Fourth, amino acid deprivation, a condition known to induce the stringent response, also induces sporulation. On the other hand, a *relA* mutant, which is incapable of converting GTP to ppGpp, sporulates well, albeit with a delayed time frame (86, 148; A. L. Sonenshein, unpublished results). There is as yet no clear mechanistic connection between changes in guanine nucleotide levels and regulation of sporulation.

Curiously, both the initial, adaptive response and the sporulation-specific response to nutritional limitation depend on the same transcription factor, phosphorylated Spo0A protein (Spo0A~P) (54), but seem to be sensitive to different concentrations of this transcription factor. Spo0A~P accumulates as a result of increases in the rate of transcription of the *spo0A* gene and modification of the gene product by phosphorylation. The pathway of phosphorylation, known as the Spo0A phosphorelay, involves several protein kinases; a phosphorylated intermediate, Spo0F; and a phosphotransferase enzyme, Spo0B (14) (Fig. 2). During the first phase of the response, an initial level of Spo0A~P is achieved, thanks to low-level transcription of *spo0A* and *spo0F* from σ^A-type promoters and the relatively low activity of two kinases, KinB and KinC (73, 80, 139). This level of phosphorelay activity appears to be activated by one or more peptides that accumulate in proportion to the cell population density (48, 79, 130, 144) and by a metabolic signal sensed in the cytoplasmic

membrane by KinB (27, 139). The putative GTP-derived signal may also influence the phosphorelay (142). The amount of Spo0A~P that accumulates initially is sufficient to repress the synthesis of AbrB, a general repressor of early stationary-phase genes (136). Because AbrB is an unstable protein, arresting its synthesis leads to rapid derepression of AbrB-repressed genes (43). In fact, only a few genes are actually induced by removal of AbrB alone; most other genes are under additional control by one or more other regulators that sense nutrient availability. One such regulator is CodY, a protein that represses genes involved in transport of dipeptides (127) and γ-aminobutyrate (40), motility (E. Olmsted and L. Márquez-Magaña, Abstr. 96th Gen. Meet. Am. Soc. Microbiol. 1996), histidine degradation (42), genetic competence (121), and acetate utilization (S. H. Fisher, personal communication). Thus, the cell activates expression of many stationary phase genes only when two conditions are met simultaneously—the population has reached a critical density and critical nutrients are becoming limiting for growth.

Most of the genes repressed by AbrB (and, thus, indirectly activated by Spo0A~P) are not required for sporulation (136). Among the required genes is *spo0H*, the gene that codes for σ^H (147). As the rate of synthesis of σ^H increases in early stationary-phase cells, σ^H protein also becomes more stable (6, 52). Accumulation of σ^H activates several genes critical for sporulation (and many that are dispensable for sporulation). First, σ^H stimulates transcription of

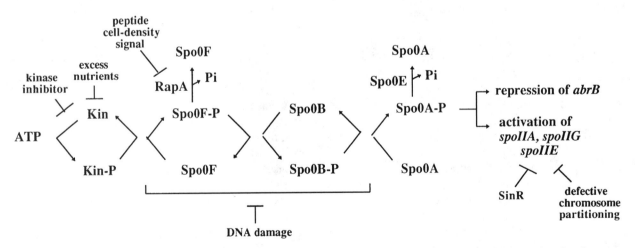

Figure 2. The *B. subtilis* Spo0A phosphorelay. Three different histidine protein kinases (KinA, KinB, KinC) autophosphorylate and then transfer their phosphate groups to Spo0F. Through the intermediary of a phosphotransferase, Spo0B, the phosphate is finally transferred to an aspartate residue on Spo0A. Spo0A-phosphate is active as a DNA-binding transcription factor, having both negative and positive effects on gene expression. See references 14, 54, and 80 for additional details. The activity of the phosphorelay is mitigated by excess nutrients; by an inhibitor of KinA; by RapA, a phosphatase for Spo0F~P; by Spo0E, a phosphatase for Spo0A~P; and by a signal indicating DNA damage. RapA is inactivated by a peptide that accumulates in the medium as a culture reaches high density. The positive regulatory activity of Spo0A~P is in competition with repression by SinR and Soj, a protein also involved in regulating chromosome partitioning. See text for details. This figure is a variation of a figure that appeared in reference 132 and is reprinted with permission of the publisher.

spo0A and *spo0F* from sporulation-specific promoters (113). Second, it activates transcription of *kinA* (6, 113), which encodes a kinase able to stimulate the Spo0A phosphorelay to high level (14). Third, σ^H recognizes the promoter of the *spoIIA* operon, one of the earliest expressed sporulation-specific operons (11). The following sequence of events would account for the onset of sporulation once the concentration of active σ^H reaches a threshold level. By stimulating expression of *kinA*, *spo0F*, and *spo0A*, σ^H makes possible a substantial increase in the concentration of Spo0A~P. By rising to these higher levels, Spo0A~P is able to bind to the promoters of the *spoIIA*, *spoIIE*, and *spoIIG* operons (8, 11, 12, 153), acting as a direct positive regulator of transcription directed by σ^A-containing (*spoIIE, spoIIG*) or σ^H-containing (*spoIIA*) forms of RNA polymerase. The *spoIIA* and *spoIIG* operons encode σ^F and σ^E, the σ factors responsible for early forespore-specific and mother cell-specific transcription, respectively (see below), while SpoIIE is a phosphatase needed for activation of σ^F. In addition, Spo0A~P and σ^H are needed for polar septation, presumably because each of these transcription factors stimulates expression of one or more genes whose products are required for assembly of the polar septasome (Spo0A~P) and for septation itself (σ^H) (82).

At least two proteins delay the onset of sporulation even when σ^H and Spo0A are being activated. RapA is a phosphatase for Spo0F~P that is induced simultaneously with KinA (96, 105). Since KinA and RapA produce and degrade, respectively, the same compound (Spo0F~P), they are direct competitors. Only when RapA itself is inactivated by a small peptide product of the *phrA* gene can high-level accumulation of Spo0A~P occur (104). But another protein also interferes with accumulation of Spo0A~P. Spo0E, a phosphatase specific for Spo0A~P, is expressed from an AbrB-repressed promoter and, thus, appears at the same time as KinA, RapA, and increased Spo0A (101). How Spo0E-dependent dephosphorylation of Spo0A~P is overcome by a cell that proceeds to sporulate is not known.

The net effect of this series of interactions is that signals reflecting population density and nutritional limitation turn on a variety of stationary-phase responses. One of those responses, sporulation, is delayed relative to the others, because inhibitory proteins interfere with the accumulation of Spo0A~P to a level sufficient to activate transcription of the *spoIIA*, *spoIIE*, and *spoIIG* operons, as well as the as yet unidentified gene whose product directs septasome formation to polar sites. These sporulation-specific events only occur after two phosphatases, RapA and Spo0E, have been inhibited, leading to suf-

ficient accumulation of Spo0A~P. Surprisingly, additional proteins also interfere with the commitment to sporulation. SinR acts as a direct repressor of the *spo0A*, *spoIIA*, *spoIIE*, and *spoIIG* promoters (16, 87). Relief of this repression is due to the activity of SinI, a small protein that binds directly to SinR (7), but the timing of inactivation of SinR is uncertain. Additional negative regulation of an overlapping set of genes is mediated by Soj protein (A. D. Grossman, personal communication). Interestingly, Soj and its inhibitor, Spo0J, are also involved in partitioning chromosomes in conjunction with both medial and polar septation (61). Furthermore, the activity of KinA is controlled by an inhibitor protein, KipI, whose synthesis is regulated by nitrogen availability (146).

Whether the eventual elevation of Spo0A~P to levels sufficient to activate sporulation genes is the result of specific signals or simply a passive process that reflects the passage of time is not yet known. It seems more likely that a specific nutritional signal is needed, namely, one that indicates that the attempt to locate and metabolize secondary nutritional sources was unsuccessful. Interestingly, mutations in most genes that encode enzymes of the Krebs citric acid cycle interfere with sporulation (53). To some extent this outcome is due to a reduction in the generation of energy and reducing power, but in several cases other bases for sporulation inefficiency have been found. The absence of isocitrate dehydrogenase, for instance, is associated with a block at stage I; Spo0A~P-dependent genes are expressed normally in isocitrate dehydrogenase mutant cells, but the polar septum does not form (62). This blockage has been attributed to accumulation of citrate and isocitrate and, as a consequence, a reduction in extracellular pH and depletion of certain divalent cations (90), implying that polar septation has special requirements for pH and cations. An aconitase-defective mutant is blocked at stage 0, being unable to express at normal levels the genes that depend directly on Spo0A~P for their transcription (25). This effect appears to be due both to the absence of the enzymatic activity of aconitase and to the absence of the protein itself, which has a second activity as an RNA-binding protein that responds to availability of iron (1).

Other types of signals are also integrated by the Spo0A phosphorelay. A cell that is unable to complete a round of DNA replication or to repair DNA damage cannot activate the expression of Spo0A~P-dependent sporulation genes (47, 60). The mechanisms by which such signals are sensed are unknown.

A curious finding is that when sporulation-competent cells are induced to sporulate under the most favorable conditions known, only 50 to 75% of the cells are converted to spores. The remainder are mostly blocked at the initiation step. There is considerable evidence that entry into sporulation is a threshold-dependent phenomenon (24, 119), an explanation that fits well with the autocatalytic nature of key early sporulation events. For example, a cell that accumulates just enough Spo0A~P to reduce AbrB, even transiently, to some critical level will increase the concentration and activity of σ^H, one of whose functions is to stimulate synthesis of phosphorelay components, leading to rapid amplification of Spo0A~P. In addition, σ^H stimulates synthesis of peptide factors that protect phosphorelay components against dephosphorylation (A. D. Grossman, personal communication). Thus, two cells that have only slightly different concentrations of AbrB may have very different fates.

Coupling of Morphological Changes to Transcriptional Regulation

Once the cell becomes able to accumulate a high enough concentration of Spo0A~P to permit polar septation and transcription of the *spoIIA*, *spoIIE*, and *spoIIG* operons, a cascade of regulatory events is set in motion (135). This cascade is, at one level, two series of sequential substitutions of σ factors initiated by σ^A and σ^H in the predivisional cell (Fig. 3); at

another level, it is a remarkable example of the interplay of structural components of the sporulating cell and the transcriptional machinery. Just before polar septation, the predivisional cell contains RNA polymerase molecules that are associated with σ^A and σ^H. In addition, the cell contains σ^F and pro-σ^E, neither of which is in an active form. σ^F is held in an inactive state by SpoIIAB, an anti-σ factor (35); pro-σ^E is the inactive precursor of σ^E (77). All four active or inactive σ factors are found in both compartments immediately after septation. Yet, the forespore is the unique locus of σ^F activation, and the mother cell is the unique site of σ^E-dependent transcription.

σ^F is activated by a pathway that starts with association of SpoIIE with the septal membrane (5, 84). Upon septation, SpoIIE acts as a phosphatase for SpoIIAA~P (3, 34, 41); SpoIIAA then binds to SpoIIAB, causing the latter to release σ^F in an active form (2). The pathway is complicated by the fact that SpoIIAB is the kinase for SpoIIAA (31, 92); were there no other intervention, an equilibrium would be reached between SpoIIA~P and SpoIIAA-SpoIIAB that would disfavor σ^F activation. Two factors bias the outcome toward σ^F activation. First, the association of SpoIIAB with σ^F, on the one hand, and with SpoIIAA, on the other hand, is influenced at least in vitro by the relative concentrations of adenine nucleotides (2). When ATP is relatively high, SpoIIAB is

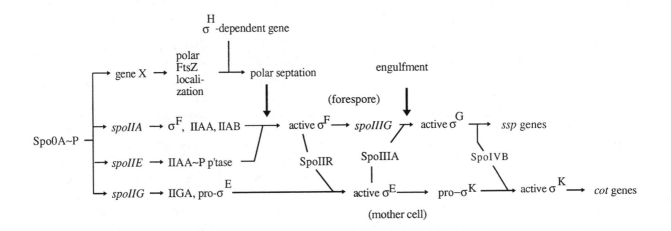

Figure 3. The sporulation sigma factor cascade. At the onset of stationary phase, activation of Spo0A by phosphorylation (see Fig. 2) leads to expression of the *spoIIA*, *spoIIG*, and *spoIIE* operons. An additional Spo0A~P-dependent gene of unknown identity is required for asymmetric septation. σ^F, a product of the *spoIIA* operon, interacts with core RNA polymerase to direct transcription of the early class of forespore-specific genes. Activation of σ^F requires its release from an inhibitory complex with SpoIIAB, a process that depends on SpoIIA, after the latter is dephosphorylated by SpoIIE. One of the early forespore-specific genes is *spoIIIG*, which codes for σ^G. When activated, a step that requires the mother cell-expressed *spoIIIA* operon, σ^G directs transcription of late forespore-specific genes, including those that encode internal (Ssp) proteins of the spore. σ^E is encoded in the *spoIIG* operon as an inactive precursor; cleavage/activation depends on an early forespore protein, SpoIIR. Upon activation, σ^E-containing RNA polymerase transcribes genes for early mother cell-specific proteins. Among these proteins is the precursor of σ^K. Cleavage/activation of pro-σ^K depends on a late forespore protein (SpoIVB), as well as on other mother cell proteins. When activated, σ^K recognizes promoters for late mother cell genes, including spore coat protein genes. This figure previously appeared in reference 132 and is reprinted with permission of the publisher.

preferentially bound to σ^F; when ADP is relatively high, SpoIIAB is bound instead to SpoIIAA. Second, SpoIIAB that is released from a complex with σ^F is subject to proteolysis, a process that prevents reversal of σ^F activation (44).

A major unanswered question is why this pathway for σ^F activation occurs only in the forespore compartment. One theory is that a critical protein, such as SpoIIE, is present in equal numbers of molecules on both faces of the nascent septum. After septation is completed, the concentration of that protein (given the difference in volume of the two compartments) would be much higher in the forespore. Recent experiments, however, introduce several complicating factors that bring into the question the prevailing view of σ^F activation. First, the fraction of total SpoIIAA that is dephosphorylated suggests that dephosphorylated SpoIIAA is present in both compartments (72). Second, dephosphorylation of SpoIIAA may not be sufficient for σ^F activation; polar septation is still required to mediate some other aspect of σ^F-SpoIIAB-SpoIIAA-SpoIIE interaction (72). Third, expression of a mutant form of SpoIIE that retains phosphatase activity but is not membrane-associated supports compartment-specific activation of σ^F and sporulation in a surprisingly large fraction of the cell population, implying that factors other than association of SpoIIE with the septum help to establish compartment specificity (4).

After septation, the inhibitory N-terminal domain of pro-σ^E is removed by an interaction at the septal membrane, leading to cleavage by a membrane-bound protease encoded by *spoIIGA* (55). The cleavage depends, however, on SpoIIR, the membrane-associated product of a σ^F-dependent gene (55, 65, 85). Thus, activation of σ^E only occurs after σ^F has become active in the forespore, a mechanism that controls the timing of σ^E activation. But why doesn't σ^E also become active in the forespore? The answer is that it is not there, apparently because it is selectively degraded in the forespore compartment (63, 111).

σ^E-dependent gene expression in the mother cell is needed for the next morphological change, engulfment of the forespore. That event, when completed, is followed by activation of σ^G in the forespore. The mechanistic relationship, if any, between completion of engulfment and σ^G activation is not clear; their temporal relationship may only be coincidental. The gene for σ^G, *spoIIIG*, is initially transcribed by σ^F-containing RNA polymerase, but the gene product is not immediately active. Substantial evidence implicates SpoIIAB as an inactivating protein for σ^G (68, 115), but this idea seems in some ways paradoxical. If SpoIIAB is still present and active in the forespore

at the time of *spoIIIG* transcription, why does it not inhibit σ^F from transcribing the *spoIIIG* gene? But it is certain that protein products of the *spoIIIA* operon, expressed in the mother cell under control of σ^E (59), associate with the forespore membrane and are necessary for activation of σ^G (68). A current theory is that the SpoIIIA membrane complex provides a signal to SpoIIAB (or to another anti-σ^G factor), inducing dissociation of the σ^G–anti-σ^G complex. Once activated, σ^G stimulates further transcription of *spoIIIG* and activates transcription of genes, such as the *ssp* family members, that encode internal proteins of the spore.

The final step in the sigma cascade is the proteolytic activation of pro-σ^K in the mother cell. The *sigK* gene is initially transcribed by σ^E-containing RNA polymerase, leading to synthesis of pro-σ^K exclusively in the mother cell. (Among endospore-formers, *B. subtilis* has the unique property of carrying an inserted prophage within its *sigK* gene. As a result, in this species only, a site-specific recombination event that removes the inserted DNA must occur in the mother cell genome before an intact *sigK* gene can be transcribed [76].) Processing of pro-σ^K requires several mother cell proteins and, in addition, depends on SpoIVB, which is synthesized in the forespore (26). The *spoIVB* gene is transcribed by σ^G-containing RNA polymerase. Again the membrane is the site of regulation and processing. Pro-σ^K and its protease bind to the outer surface of the forespore double membrane (154); SpoIVB is thought to bind to the inner surface of the forespore membranes and, through a transmembrane protein-protein interaction, stimulate cleavage of pro-σ^K. Once activated, σ^K recognizes promoters for late mother cell genes, including spore coat protein genes.

In summary, sporulation in *B. subtilis* is initiated by nutrient limitation in a population at high density. The cell's first response to these signals is to search for alternative nutrient sources while preparing itself to sporulate. If restoration of growth is unsuccessful, the cell expresses genes for sporulation-specific σ factors, whose subsequent activation leads inexorably to an ordered and tightly coupled sequence of morphological and regulatory events, culminating in the production of one metabolically dormant spore per cell. The spore is protected from its environment by a tough outer shell of lamellar coats, by a rigid cortex, by dehydration of the cytoplasm, and by the coating of its genome by a family of small, nonspecific DNA-binding proteins (122). The spore can remain dormant and viable for decades (69) or, as some claim, for millennia (15). Spores germinate only when they are exposed to specific mixtures of germinant chemicals (94). The coat lamellae become loosened, the cytoplasm becomes rehydrated, enzymatic activity is resuscitated, and the cell begins to expand its mass, breaking out of the spore envelope.

Aggregation and Morphological Changes in *M. xanthus*

The myxobacteria are characterized by their gliding motility and the elegant fruiting bodies they construct. Ironically, *M. xanthus*, the best-studied member of the family, and its close relatives form the least elaborate fruiting bodies, consisting essentially of a mound of cells (Fig. 1C) within which individual cells differentiate into environmentally resistant spores by acquiring new characteristics, including distinct surface proteins. Molecular genetic studies of producers of more elaborate fruiting bodies, such as *Stigmatella* and *Chondromyces* (see reference 37), are still at an embryonic state.

Aggregation leading to fruiting body development is totally dependent on the motility of myxobacteria. Unlike other motile bacteria, however, myxobacteria do not have flagella. Their motility is mediated in a not-fully-explained manner by a mechanism that allows the bacteria to glide across a surface. In fact, *M. xanthus* shows two kinds of motility—adventurous (movement of individual cells) and social (movement of clusters of cells, including aggregation to fruiting bodies) (51). Each is determined by large, independent sets of genes, but both types depend on a small GTP-binding protein, MglA (50). MglA is, as one would anticipate, required for fruiting body formation and sporulation (50). Social motility also depends on type IV pili (150), on fibrillar protein-carbohydrate complexes (38, 128), and on lipopolysaccharide O-antigen (13). The directionality of gliding is determined by the *frz* gene products, whose sequences are similar to those of the chemotaxis proteins of other bacteria (123).

While formation of the fruiting body is not an essential step in differentiation, it helps to ensure that sporulation will usually take place when a critical mass of like cells is present in a very localized environment, a useful attribute when sporulation is inefficient (64). At a defined time after formation of the fruiting body, individual cells are converted from rods to spherical spores. Few details of the morphological changes associated with this transformation are available. In one study, it was shown that only an inner ring of nonmotile cells within the fruiting body is able to express developmentally regulated genes (118). Formation of myxospores is accompanied by intracellular accumulation of trehalose, thickening of the cell wall, and deposition of several coat proteins on the myxospore surface (37, 64). By raising the myxospores above the surface of the local habitat, the fruiting body helps to ensure that the spores will be dispersed by the wind or passing animals to new and perhaps more nutritionally supportive environments (64).

Developmental Signal Transduction and Regulation of Gene Expression in *M. xanthus*

As in *B. subtilis*, the initiation of development is signaled by intercellular communication and requires nutrient limitation, approaching complete starvation (64). The initial signal can be depletion of the carbon, nitrogen, or phosphorus source or of amino acids and seems to involve synthesis of ppGpp (88). Mutants defective in the ppGpp synthetase, RelA, are unable to initiate fruiting body formation or differentiation (49) and cells forced to express *relA* inappropriately express developmental genes even when the environment permits growth (126).

As in other bacteria, the response of *M. xanthus* to nutrient limitation includes synthesis of extracellular degradative enzymes and antibiotics that could potentially allow the bacteria to maintain the growing state (37). How the cell differentiates between this response and commitment to sporulation is less clear than in *Bacillus*.

At least five intercellular signals, defined by extracellular complementation studies, have been implicated in myxococcal development (Table 1). The earliest of these appears to be the B signal, in the absence of which cells fail to aggregate or differentiate or express genes normally turned on during the first hour after nutrient limitation. The signal itself has not been identified; the only genetic locus (*bsgA*) in which B signal mutations arise encodes a putative protease of the Lon family (46). A model for B signaling is that BsgA normally degrades a negative regulator of a gene required for aggregation and differentiation. A compound dependent on the function of that gene must be secreted into the medium. This view of B signaling is complicated by the fact that not all B signal mutants participate well in intercellular complementation, at least in a particular experimental regimen (75).

The A signal can be any one of several amino acids at a specific low, extracellular concentration (10 μM). This signal can also be generated by extracellular hydrolysis of peptides or by extracellular proteases that catalyze such hydrolysis using cellular surface proteins or components of the medium as substrates (109). In the absence of A signal, cells form a flat film but fail to create a focal point of aggregation and are blocked in expression of genes normally turned on between hours 1 and 2 of development.

Table 1. Intercellular signaling in *M. xanthus*[a]

Signal type	Biochemical nature of signal	Time of action during development	Genes implicated in signal production	Protein products of signaling genes
B	Unknown	0–1 h	*bsgA*	Lon protease
A	Certain amino acids or peptides containing those amino acids	1–2 h	*asgA* *asgB*	Two-component system Transcription factor
D	Unknown	1–2 h	*dsgA*	Translation factor IF-3
E	Branched-chain fatty acids	3–5 h	*esgA*	Branched-chain keto acid dehydrogenase
C	Unknown	5–6 h	*csgA*	Short-chain alcohol dehydrogenase

[a] For details, see text.

Generation of A signal by nutrient-limited cells seems to rely on AsgA, a two-component–type signal transduction system that controls expression of one or more as yet unidentified genes (110); on AsgB, a putative transcription factor (107); and on AsgC, the major σ factor of *M. xanthus* RNA polymerase (28). Amino acids or mixtures of amino acids that stimulate development at low concentration (10 μM) also inhibit development at high concentration (1 mM). A model that links these various elements would postulate that nutrient limitation (causing available amino acids to drop below 1 μM) activates the AsgA/AsgB signal transduction pathway and the stringent response, leading to accumulation of ppGpp, which interacts with the σ subunit of RNA polymerase to stimulate, in conjunction with AsgA/AsgB, the expression of genes coding for intracellular and (or) extracellular proteases. Extracellular A signal (i.e., amino acids) is thought to be recognized by another two-component system, consisting of the membrane-bound sensor kinase, SasS (C. Yang and H. B. Kaplan, personal communication), and the transcription factor, SasR (108). A second level of control appears to be imposed by SasN, which may act as an inhibitor of either SasS or SasR (151). Although A signal is required for aggregation and differentiation, the only genes shown to date to respond to A signal are not essential for fruiting body formation.

The coupling of quorum-sensing to availability of nutrients (by virtue of the fact that the two signals are identical) allows each cell to determine individually whether to produce A signal. The sensing of A signal, however, is a collective response since it reflects the average concentration of A signal in the local environment. In other words, A signal produced by a single cell would diffuse away and never reach a high enough local concentration to induce that cell to differentiate.

An A signal-independent pathway is also required at an early stage of development. Mutations in the *sdeK* gene block development after formation of loose aggregates, but expression of the *sdeK* gene, although dependent on ppGpp, is independent of A signal (45). The putative product of *sdeK* is a sensor histidine kinase with homology to *B. subtilis* KinA, suggesting that it initiates a signal transduction pathway that acts, in conjunction with the A signal pathway, to move cells forward in development (45).

The D signal is also needed between the first and second hours of differentiation. The signal itself has not been identified, and the only locus known to be responsible for its production is, surprisingly, the gene for translation initiation factor IF-3 (23). Current thinking is that a reduction in IF-3 activity slows the expression of one or more key proteins. It was recently suggested, however, that the phenotype of a D signal mutant may be due to reduced production of A signal (145). Curiously, the absence of D signal, but not of A signal, can be overcome by addition of a mixture of fatty acids (116).

E signal mutants are defective in branched chain keto acid dehydrogenase complex and can be rescued by addition of short, branched chain fatty acids (138). Branched chain keto acids are precursors of branched chain fatty acids. Extracellular complementation of E signal mutants requires cell-cell contact, suggesting that a surface-bound enzyme, such as the keto acid dehydrogenase, is responsible for the activity. The E signal is required between hours 3 and 5 of development.

The C signal, which is required after the fifth hour, appears to be an enzyme of the short chain alcohol dehydrogenase family (81), but its substrate is unknown. Extracellular complementation requires cell-cell contact and that both partners be motile. The inference is that cell surface structures implicated in social motility are responsible for holding aggregated cells together. As noted above, several surface macromolecules appear to be needed for cell cohesion and fruiting body formation. Recent work suggests that CsgA plays an interesting and critical role in the cell's decision to commit itself to sporulation (125).

Accumulation of ppGpp due to amino acid depriva-tion both induces CsgA synthesis and shuts off syn-thesis of SocE, an inhibitor of sporulation. As a result, growth is arrested and sporulation ensues. Complex, interdependent regulation of the *csgA*, *socE*, and *relA* genes is thought to be a major determinant of the choice between growth and sporulation, tipping the balance one way or the other (125).

The pathway from the signaling events described above to the mature myxospore is not very clear. One protein implicated in later development is FruA, a putative transcription factor needed for expression of several late genes, including the gene for a major spore surface protein (100). The roles of various RNA polymerase σ factors have also been surmised, but their number and specificities remain to be clar-ified. σ^D accumulates at the transition from exponen-tial phase to stationary phase and is required for max-imal viability of stationary-phase cells and efficiency of sporulation (141). σ^{54}, a member of the RpoN family of sigma factors, is essential for growth of *M. xanthus* (71) and is probably responsible for tran-scription of the *sdeK* (45), *sasN* (152), *pilA* (150), and *mbhA* (70) genes.

Morphological Changes in *S. coelicolor*

The first step in differentiation in *S. coelicolor* is the lifting up of hyphae from the surface mycelium to initiate growth perpendicular to the substrate sur-face. Elongation of the gravity-defying aerial hyphae gives a hairy appearance to colonies. The aerial hy-phae, each of which is a single, multinucleate cell, subsequently lay down evenly spaced division septa and segregate their chromosomes in such a way that each hypha generates many mononucleate cells. These cells remain attached in a long, coiled, chain and then undergo sporulation (Fig. 1D). The chains of spores have a grayish brown cast that distinguishes them from the white, presporulation hyphae. The mature spores are eventually released and dispersed. Variants that fail to form aerial hyphae are known as "bald" (*bld*) mutants, and those that form aerial hy-phae but fail to sporulate are known as "white" (*whi*) mutants (Table 2).

Unlike the case for bacteria that grow as individ-ual cells, *S. coelicolor* does not absolutely need FtsZ for growth but does require FtsZ during sporulation (91). Proper chromosome partitioning is dependent on ParB. In the absence of ParB, about 10% of the cells in the aerial hyphae are anucleate and lyse (71a). Other cells seem to have an extra copy of the chro-mosome, as would be expected when there is imper-fect segregation.

Table 2. Properties of *Streptomyces* mutants defective in differentiation[a]

Genetic locus	Function of gene product
Defective in aerial mycelium formation	
amfR	Positive regulator of aerial mycelium genes
bldA	Leucyl tRNA
bldB	DNA-binding protein
bldC	Unknown
bldD	DNA-binding protein
bldG	Unknown
bldK	Oligopeptide permease
bld261	Unknown
brgA	ADP-ribosylation
citA[b]	Citrate synthase
sapB	Surfactant
Sporulation-defective mutants	
sigF	Sigma factor
whiA	Unknown
whiB	Unknown
whiG	Sigma factor
whiH	DNA-binding protein
whiI	DNA-binding protein

[a] For fuller explanation, see text.
[b] Aerial mycelium formation restored in buffered medium.

Developmental Signal Transduction and Sequential Gene Expression in *S. coelicolor*

The relationship between stationary-phase re-sponses and sporulation in *Streptomyces* seems to be similar to that in *Bacillus*. An initial signal of nutrient limitation leads to formation of aerial hyphae, accu-mulation of storage compounds, and synthesis of ex-tracellular degradative enzymes and antibiotics (19, 22). It is only at a later stage in stationary phase that the aerial hyphae undergo sporulation. *S. coelicolor*, the paradigm species, is difficult to grow as a dis-persed culture in liquid medium and does not spor-ulate well under such conditions.

The mechanism by which *Streptomyces* cells sense nutritional limitation is not really understood. There is some evidence that GTP and ppGpp are im-portant nutritional signaling molecules. First, *S. coe-licolor* and *Streptomyces griseus* both contain homo-logs of the *B. subtilis* Obg protein (102). As in *B. subtilis*, this protein is essential for growth in *S. coe-licolor*. In addition, mutants defective in ppGpp syn-thesis are unable to sporulate or make antibiotics, at least under some growth conditions (17, 89).

Bacillus and *Myxococcus* cultures grow as indi-vidual cells but differentiate when the population density reaches a critical level. Myxobacteria respond to this critical density by forming an even more fo-cused mass of cells before sporulating. Actinomy-cetes, such as *S. coelicolor*, grow as multinucleate mycelia with only occasional crosswalls interrupting the cytoplasm. Thus, the mycelium is, in effect, a multicellular mass even when it is growing. Nonetheless, actinomycetes produce and sense extracellular compounds that indicate the local concentration of like bacteria. Whether the purpose of the signaling is to establish the existence of a quorum or to measure the passage of time is unclear. These signals are mostly uncharacterized, but a few of them have been studied in detail.

Production of antibiotics early during stationary phase is interesting with respect to both human health and actinomycete biology. Each antibiotic biosynthetic pathway has its own regulatory protein (usually a transcription activator) whose accumulation is the initiating event for antibiotic production (18). Synthesis of the positive regulators is controlled by global control mechanisms that link stationary phase, aerial mycelium development, and cell population density. An extracellular factor (A-factor) induces the antibiotic production pathway and sporulation in *S. griseus* (56). A-factor is a γ-butyrolactone, a family of chemicals structurally related to the homoserine lactones that serve as quorum indicators in many gram-negative bacteria. A-factor is sensed intracellularly by ArpA (103), a direct or indirect repressor of the antibiotic pathway activator, and by AdpB, a repressor of the aerial mycelium activator, AmfR (140). Thus, uptake of A-factor leads to expression of both antibiotic synthesis and aerial mycelium-inducing genes. Analogous mechanisms are thought to regulate antibiotic synthesis and sporulation in other *Streptomyces* species (18).

The *bld* mutants are defective in both sporulation and antibiotic synthesis. A surprising aspect of the *bld* mutations is that most of them can be bypassed in two unexpected ways. First, many *bld* mutants are conditionally defective in aerial mycelium formation and sporulation. That is, they are defective in glucose-containing medium but not in media containing less rapidly utilized carbon sources (18). The basis for this conditionality is unknown, but the conditionality has been suggested to indicate that wild-type cells have two different pathways for inducing aerial mycelium formation, one of which is dependent on *bld* genes and the other is repressed by glucose. Second, the *bld* mutants can cross-complement each other extracellularly. Inoculation of a petri dish with two different *bld* mutants (or a mutant and a wild-type strain) can allow aerial mycelium formation at the edges of the streaks where the two strains most closely approach each other (149). On the basis of their complementation pattern, the *bld* mutants can be assigned to a strict hierarchy, implying that a series of ordered signals is being exchanged (99). Only a

few of the *bld* gene products have been identified, however.

All *bld* mutants lack SapB, a protein needed for aerial mycelium development (149), as if a primary purpose of the *bld* gene pathway is to stimulate synthesis of SapB at the appropriate time and place. Addition of purified SapB can restore aerial hypha production to all *bld* mutants (149). SapB is thought to coat the hyphae and to have surfactant properties, breaking the tension that holds the substrate mycelium to the surface and allowing the hyphae to extend away from the plane of the surface.

The *bldA* gene is turned on in stationary phase and encodes a tRNA for a rarely used leucine codon (78). Since regulators for antibiotic synthesis and aerial mycelium development contain codons recognized by this tRNA, it is presumed that differential translation of mRNA explains in part the timing of stationary-phase events (22). The extracellular signal that bypasses the need for BldA is unknown. BldB and BldD appear to be DNA-binding proteins (39, 112), but their targets are unknown. Mutations in *bldB* also cause aberrations in regulation of carbon source utilization (112).

The *bldK* locus encodes an oligopeptide permease (98), suggesting that a peptide serves as one of the developmental signals. A candidate peptide has been identified (97). Some mutations in the *brgA* gene also cause a bald phenotype; *brgA* mutants are defective in ADP-ribosylation of certain target proteins, but the roles of the target proteins have not been elucidated (124).

A newly identified *bld* mutation inactivates citrate synthase, as a result of which the extracellular pH drops because of failure to metabolize organic acids produced during fermentation (P. H. Viollier and C. J. Thompson, unpublished data). Buffering the medium allowed the mutant to differentiate (Viollier and Thompson, unpublished data). Other mutants blocked early in the *bld* pathway (e.g., *bldA*, *bldB*, *bldC*, *bldG*, *bld261*, *brgA*) are also unable to maintain neutral extracellular pH, but, with the exception of *bld261*, they cannot be suppressed by addition of buffer (137). Like other *bld* mutants, the citrate synthase mutant can be complemented extracellularly by later blocked mutants (Viollier and Thompson, unpublished data). In this case, the extracellular complementation probably reflects the ability of later-blocked mutants to metabolize compounds that inhibit the differentiation of the citrate synthase mutant.

The ability to utilize acidic fermentation products also depends on cyclic AMP. Wild-type cells make the switch from acid-producing to acid-utilizing metabolism as the cells enter stationary phase, but mutants lacking adenyl cyclase remain stuck in the acidogenic phase and cannot form aerial hyphae (137). The implication of these results is that aerial hypha formation depends on induction of the Krebs citric acid cycle, a process that depends in turn on a cAMP-related signal.

Conversion of aerial hyphae to coiled chains of spores depends on the *whi* genes, as well as on other proteins, such as FtsZ (91). The best characterized of the *whi* gene products is WhiG, an RNA polymerase σ factor of the motility regulon family (21). A *whiG* mutant produces aerial hyphae that are uncoiled and that neither septate nor sporulate. WhiG appears to be at the head of a cascade of gene expression, since it is required for expression of *whiH* and *whiI* (117). The *whiI* gene may encode a DNA-binding protein similar to response regulators of the two-component family (20), but there is no direct evidence for its function. WhiH may also be a regulatory protein since it has similarity to repressors of the GntR family (117).

There may, in fact, be at least two gene expression pathways operating simultaneously, since *whiA* and *whiB* mutants have early blocks in sporulation but are not dependent on WhiG (20, 129). In addition, the detailed phenotypes of the various mutants differ. The *whiG* mutant makes straight aerial hyphae, while the *whiA* and *whiB* mutants make coiled but unusually long hyphae (21). By contrast, *whiH* and *whiI* mutants seem to be blocked later, inasmuch as they produce aerial hyphae of normal length with a few septa (117). The sequences of WhiA and WhiB have not yet revealed their functions.

Postseptation sporulation events are at least in part dependent on σ^F, a homolog of *Bacillus* σ^F family members (67). One of the promoters dependent on σ^F drives transcription of a gene whose product is responsible for the gray-brown color of the spore (66).

COMMON THEMES REVISITED

The three modes of spore formation discussed here are characteristic of evolutionarily distant bacteria. These bacteria differ so much in their lifestyles, and their modes of sporulation are so different in the details of the morphological changes that occur, that one assumes that ability to sporulate evolved independently in each case. Yet, certain common characteristics are striking and raise the possibility that all modes of sporulation are derived from a common ancestral mode. Alternatively, one might conclude that all bacteria use a common set of mechanisms to respond to similar environmental stresses. In all three

cases, the precipitating event is nutritional limitation, but response to this signal is tempered by the requirement that the bacterial mass be sufficiently high to allow a reasonable number of spores to be produced in processes that are considerably less than 100% efficient. GTP or ppGpp or both may be the critical intracellular signaling compound in all the cases, and peptides appear to play critical roles as extracellular signals. All the sporulators respond to nutritional limitation initially by inducing enzymes, antibiotics, and transport systems that would allow them to make use of any available secondary nutritional sources before committing themselves to spore formation. In all the cases, sporulation genes are turned on in a temporally regulated, cascading manner that is at least partly determined by successive replacements of RNA polymerase σ factors. From the biologist's point of view, however, the beauty of each system lies in the distinguishing details, the gliding and aggregation of myxobacteria and their elegantly designed fruiting bodies, the mycelial growth of actinomycetes and their clever way of raising up their spore-forming hyphae to maximize dispersal, and the truly remarkable interrelationship between morphological structures and transcriptional regulation in endospore-formers. We have begun to see how spore-forming bacteria deal with a type of stress that is likely to be common in their natural habitats, but many of the most interesting details remain to be uncovered.

Acknowledgments. I thank R. Losick and H. Kaplan for helpful comments on the manuscript; W. Champness and D. Kaiser for providing preprints of their reviews; S. Fisher, A. Grossman, L. Márquez-Magaña, L. Shimkets, and C. J. Thompson for allowing me to cite unpublished results; and J. Willey, M. Dworkin, and H. Reichenbach for providing photographs of differentiating cells. The preparation of this chapter was greatly aided by several excellent review articles by W. Champness (18), K. F. Chater and R. Losick (20, 22), M. Dworkin (37), D. Kaiser (64), L. Plamann and H. Kaplan (108), and P. Hartzell and P. Youderian (51).

Unpublished work from my laboratory was funded by grant GM42219 from the U.S. Public Health Service.

REFERENCES

1. **Alén, C., and A. L. Sonenshein.** 1999. *Bacillus subtilis* aconitase is an RNA-binding protein. *Proc. Natl. Acad. Sci. USA* **96:** 10412–10417.
2. **Alper, S., L. Duncan, and R. Losick.** 1994. An adenosine nucleotide switch controlling the activity of a cell type-specific transcription factor in *B. subtilis*. *Cell* **77:**195–205.
3. **Arigoni, F., L. Duncan, S. Alper, R. Losick, and P. Stragier.** 1996. SpoIIE governs the phosphorylation state of a protein regulating transcription factor sigma F during sporulation in *Bacillus subtilis*. *Proc. Natl. Acad. Sci. USA* **93:**3238–3242.
4. **Arigoni, F., A.-M. Guérout-Fleury, I. Barak, and P. Stragier.** 1999. The SpoIIE phosphatase, the sporulation septum and the establishment of forespore-specific transcription in *Bacillus subtilis*: a reassessment. *Mol. Microbiol.* **31:**1407–1415.
5. **Arigoni, F., K. Pogliano, C. D. Webb, P. Stragier, and R. Losick.** 1995. Localization of protein implicated in establishment

of cell type to sites of asymmetric division. *Science* 270:637–640.

6. **Asai, K., F. Kawamura, H. Yoshikawa, and H. Takahashi.** 1995. Expression of *kinA* and accumulation of σ^H at the onset of sporulation in *Bacillus subtilis*. *J. Bacteriol.* 177:6679–6683.

7. **Bai, U., I. Mandec-Mulec, and I. Smith.** 1993. SinI modulates the activity of SinR, a developmental switch protein of *Bacillus subtilis*, by protein-protein interaction. *Genes Dev.* 7:139–148.

8. **Baldus, J. M., B. D. Green, P. Youngman, and C. P. Moran, Jr.** 1994. Phosphorylation of *Bacillus subtilis* transcription factor SpoOA stimulates transcription from the *spoIIG* promoter by enhancing binding to weak 0A boxes. *J. Bacteriol.* 176:296–306.

9. **Beall, B., and J. Lutkenhaus.** 1991. FtsZ in *Bacillus subtilis* is required for vegetative septation and for asymmetric septation during sporulation. *Genes Dev.* 5:447–455.

10. **Binnie, C., M. Lampe, and R. Losick.** 1986. Gene encoding the σ^{37} species of RNA polymerase sigma factor from *Bacillus subtilis*. *Proc. Natl. Acad. Sci. USA* 83:5943–5947.

11. **Bird, T., D. Burbulys, J. J. Wu, M. A. Strauch, J. A. Hoch, and G. B. Spiegelman.** 1992. The effect of supercoiling on the in vitro transcription of the *spoIIA* operon from *Bacillus subtilis*. *Biochimie* 74:627–634.

12. **Bird, T. H., J. K. Grimsley, J. A. Hoch, and G. B. Spiegelman.** 1993. Phosphorylation of Spo0A activates its stimulation of in vitro transcription from the *Bacillus subtilis spoIIG* operon. *Mol. Microbiol.* 9:741–749.

13. **Bowden, M. G., and H. B. Kaplan.** 1998. The *Myxococcus xanthus* lipopolysaccharide O-antigen is required for social motility and multicellular development. *Mol. Microbiol.* 30:275–284.

14. **Burbulys, D., K. A. Trach, and J. A. Hoch.** 1991. Initiation of sporulation in *B. subtilis* is controlled by a multicomponent phosphorelay. *Cell* 64:545–552.

15. **Cano, R. J., and M. K. Borucki.** 1995. Revival and identification of bacterial spores in 25-40 million year old Dominican amber. *Science* 268:1060–1064.

16. **Cervin, M. A., R. J. Lewis, J. A. Brannigan, and G. B. Spiegelman.** 1998. The *Bacillus subtilis* regulator SinR inhibits *spoIIG* promoter transcription in vitro without displacing RNA polymerase. *Nucleic Acids Res.* 26:3806–3812.

17. **Chakraburtty, R., and M. Bibb.** 1997. The ppGpp synthetase gene (*relA*) of *Streptomyces coelicolor* A3(2) plays a conditional role in antibiotic production and morphological differentiation. *J. Bacteriol.* 179:5854–5861.

18. **Champness, W.** 2000. Actinomycete development, antibiotic production, and phylogeny: questions and challenges, p. 11–31. *In* Y. V. Brun and L. J. Shimkets (ed.), *Prokaryotic Development*. American Society for Microbiology, Washington, D.C.

19. **Chater, K. F.** 1993. Genetics of differentiation in *Streptomyces*. *Annu. Rev. Microbiol.* 47:685–713.

20. **Chater, K. F.** 2000. Developmental decisions during sporulation in the aerial mycelium in *Streptomyces*, p. 33–48. *In* Y. V. Brun and L. J. Shimkets (ed.), *Prokaryotic Development*. American Society for Microbiology, Washington, D.C.

21. **Chater, K. F., C. J. Bruton, K. A. Plaskitt, M. J. Buttner, C. Méndez, and J. Helmann.** 1989. The developmental fate of *S. coelicolor* hyphae depends crucially on a gene product homologous with the motility sigma factor of *B. subtilis*. *Cell* 59:133–143.

22. **Chater, K. F., and R. Losick.** 1997. Mycelial life cycle of *Streptomyces coelicolor* A3(2) and its relatives, p. 149–182. *In* J. A.

Shapiro and M. Dworkin (ed.), *Bacteria as Multicellular Organisms*. Oxford University Press, New York, N.Y.

23. **Cheng, Y. L., L. V. Kalman, and D. Kaiser.** 1994. The *dsg* gene of *Myxococcus xanthus* encodes a protein similar to translation initiation factor IF3. *J. Bacteriol.* 176:1427–1433.

24. **Chung, J. D., G. Stephanopoulos, K. Ireton, and A. D. Grossman.** 1994. Gene expression in single cells of *Bacillus subtilis*: evidence that a threshold mechanism controls the initiation of sporulation. *J. Bacteriol.* 176:1977–1984.

25. **Craig, J. E., M. J. Ford, D. C. Blaydon, and A. L. Sonenshein.** 1997. A null mutation in the *Bacillus subtilis* aconitase gene causes a block in Spo0A-phosphate dependent gene expression. *J. Bacteriol.* 179:7351–7359.

26. **Cutting, S., A. Driks, R. Schmidt, B. Kunkel, and R. Losick.** 1991. Forespore-specific transcription of a gene in the signal transduction pathway that governs pro-sigma K processing in *Bacillus subtilis*. *Genes Dev.* 5:456–466.

27. **Dartois, V., T. Djavakhishvili, and J. A. Hoch.** 1997. KapB is a lipoprotein required for KinB signal transduction and activation of the phosphorelay to sporulation in *Bacillus subtilis*. *Mol. Microbiol.* 26:1097–1108.

28. **Davis, J. M., J. Mayor, and L. Plamann.** 1995. A missense mutation in *rpoD* results in an A-signalling defect in *Myxococcus xanthus*. *Mol. Microbiol.* 18:943–952.

29. **Dawes, I. W., and J. Mandelstam.** 1970. Sporulation of *Bacillus subtilis* in continuous culture. *J. Bacteriol.* 103:529–535.

30. **de Boer, P. A., R. E. Crossley, and L. I. Rothfield.** 1989. A division inhibitor and a topological specificity factor coded for by the minicell locus determine proper placement of the division septum in *E. coli*. *Cell* 56:641–649.

31. **Diederich, B., J. F. Wilkinson, T. Magnin, M. Najafi, J. Erringston, and M. D. Yudkin.** 1994. Role of interactions between SpoIIAA and SpoIIAB in regulating cell-specific transcription factor sigma F of *Bacillus subtilis*. *Genes Dev.* 8:2653–2663.

32. **Driks, A., S. Roels, B. Beall, C. Moran, and R. Losick.** 1994. Subcellular localization of proteins involved in the assembly of the spore coat of *Bacillus subtilis*. *Genes Dev.* 8:234–244.

33. **Driks, A., and P. Setlow.** 2000. Morphogenesis and properties of the bacterial spore, p. 191–218. *In* Y. V. Brun and L. J. Shimkets (ed), *Prokaryotic Development*. American Society for Microbiology, Washington, D.C.

34. **Duncan, L., S. Alper, F. Arigoni, R. Losick, and P. Stragier.** 1995. Activation of cell-specific transcription by a serine phosphatase at the site of asymmetric division. *Science* 270:641–644.

35. **Duncan, L., and R. Losick.** 1993. SpoIIAB is an anti-σ factor that binds to and inhibits transcription by regulatory protein σ^F from *Bacillus subtilis*. *Proc. Natl. Acad. Sci. USA* 90:2325–2329.

36. **Duncan, M. L., S. S. Kalman, S. M. Thomas, and C. W. Price.** 1987. Gene encoding the 37,000-dalton minor sigma factor of *Bacillus subtilis* RNA polymerase: isolation, nucleotide sequence, chromosomal locus, and cryptic function. *J. Bacteriol.* 169:771–778.

37. **Dworkin, M.** 1996. Recent advances in the social and developmental biology of the myxobacteria. *Microbiol. Rev.* 60:70–112.

38. **Dworkin, M.** 1999. Fibrils as extracellular appendages of bacteria: their role in contact-mediated cell-cell interactions in *Myxococcus xanthus*. *BioEssays* 21:590–595.

39. **Elliot, M., F. Damji, R. Passantino, K. Chater, and B. Leskiw.** 1998. The *bldD* gene of *Streptomyces coelicolor* A3(2); a reg-

ulatory gene involved in morphogenesis and antibiotic production. *J. Bacteriol.* **180**:1549–1555.

40. Ferson, A. E., L. V. Wray, Jr., and S. H. Fisher. 1996. Expression of the *Bacillus subtilis gabP* gene is regulated independently in response to nitrogen and amino acid availability. *Mol. Microbiol.* **22**:693–701.

41. Feucht, A., T. Magnin, M. D. Yudkin, and J. Errington. 1996. Bifunctional protein required for asymmetric cell division and cell-specific transcription in *Bacillus subtilis. Genes Dev.* **10**: 794–803.

42. Fisher, S. H., K. Rohrer, and A. E. Ferson. 1996. Role of CodY in regulation of the *Bacillus subtilis hut* operon. *J. Bacteriol.* **178**:3779–3784.

43. Furbass, R., M. Gocht, P. Zuber, and M. A. Marahiel. 1991. Interaction of AbrB, a transcriptional regulator from *Bacillus subtilis* with the promoters of the transition state-activated genes *tycA* and *spoVG. Mol. Gen. Genet.* **225**:347–354.

44. Garsin, D. 1999. Ph.D. thesis. Harvard University, Cambridge, Massachusetts.

45. Garza, A. G., J. S. Pollack, B. Z. Harris, A. Lee, I. M. Keseler, E. F. Licking, and M. Singer. 1998. SdeK is required for early fruiting body development in *Myxococcus xanthus. J. Bacteriol.* **180**:4628–4637.

46. Gill, R. E., M. Karlok, and D. Benton. 1993. *Myxococcus xanthus* encodes an ATP-dependent protease which is required for developmental gene transcription and intercellular signalling. *J. Bacteriol.* **175**:4538–4544.

47. Grossman, A. D. 1995. Genetic networks controlling the initiation of sporulation and the development of genetic competence in *Bacillus subtilis. Annu. Rev. Genet.* **29**:477–508.

48. Grossman, A. D., and R. Losick. 1988. Extracellular control of spore formation in *Bacillus subtilis. Proc. Natl. Acad. Sci. USA* **85**:4369–4373.

49. Harris, B. Z., D. Kaiser, and M. Singer. 1998. The guanosine nucleotide (p)ppGpp initiates development and A-factor production in *Myxococcus xanthus. Genes Dev.* **12**:1022–1035.

50. Hartzell, P., and D. Kaiser. 1991. Function of MglA, a 22-kilodalton protein essential for gliding in *Myxococcus xanthus. J. Bacteriol.* **173**:7615–7624.

51. Hartzell, P. L., and P. Youderian. 1995. Genetics of gliding motility and development in *Myxococcus xanthus. Arch. Microbiol.* **164**:309–323.

52. Healy, J., J. Weir, I. Smith, and R. Losick. 1991. Post-transcriptional control of a sporulation regulatory gene encoding transcription factor sigma H in *Bacillus subtilis. Mol. Microbiol.* **5**:477–487.

53. Hederstedt, L. 1993. The Krebs citric acid cycle, p. 181–197. *In* A. L. Sonenshein, J. A. Hoch, and R. Losick (eds.), *Bacillus subtilis and Other Gram-Positive Bacteria: Biochemistry, Physiology, and Molecular Genetics.* American Society for Microbiology, Washington, D.C.

54. Hoch, J. A. 1993. Regulation of the phosphorelay and the initiation of sporulation in *Bacillus subtilis. Annu. Rev. Microbiol.* **47**:441–465.

55. Hofmeister, A., A. Londoño-Vallejo, E. Harry, P. Stragier, and R. Losick. 1995. Extracellular signal protein triggering the proteolytic activation of a developmental transcription factor in *B. subtilis. Cell* **83**:219–241.

56. Horinouchi, S., and T. Beppu. 1994. A-factor as a microbial hormone that controls cellular differentiation and secondary metabolism in *Streptomyces griseus. Mol. Microbiol.* **12**:859–864.

57. Igo, M. , M. Lampe, C. Ray, W. Schafer, C. P. Moran, Jr., and R. Losick. 1987. Genetic studies of a secondary RNA po-

lymerase sigma factor in *Bacillus subtilis. J. Bacteriol.* **169**: 3464–3469.

58. Igo, M. M., and R. Losick. 1986. Regulation of a promoter that is utilized by minor forms of RNA polymerase holoenzyme in *Bacillus subtilis. J. Mol. Biol.* **191**:615–624.

59. Illing, N., and J. Errington. 1991. The *spoIIIA* operon of *Bacillus subtilis* defines a new temporal class of mother-cell-specific sporulation genes under the control of the σ^E form of RNA polymerase. *Mol. Microbiol.* **5**:1927–1940.

60. Ireton, K., and A. D. Grossman. 1994. A developmental checkpoint couples the initiation of sporulation to DNA replication in *Bacillus subtilis. EMBO J.* **13**:1566–1573.

61. Ireton, K., N. W. Gunther IV, and A. D. Grossman. 1994. *spo0J* is required for normal chromosome segregation as well as the initiation of sporulation in *Bacillus subtilis. J. Bacteriol.* **176**:5320–5329.

62. Jin, S., P. A. Levin, K. Matsuno, A. D. Grossman, and A. L. Sonenshein. 1997. Deletion of the *Bacillus subtilis* isocitrate dehydrogenase gene causes a block at stage I of sporulation. *J. Bacteriol.* **179**:4725–4732.

63. Ju, J., T. Luo, and W. G. Haldenwang. 1995. Forespore expression and processing of the SigE transcription factor in wild-type and mutant *Bacillus subtilis. J. Bacteriol.* **180**:1673–1681.

64. Kaiser, D. 2000. Cell-interactive sensing of the environment, p. 263–275. *In* Y. V. Brun and L. J. Shimkets (ed.), *Prokaryotic Development.* American Society for Microbiology, Washington, D.C.

65. Karow, M. L., P. Glaser, and P. J. Piggot. 1995. Identification of a gene, *spoIIR*, that links the activation of sigma E to the transcriptional activity of sigma F during sporulation in *Bacillus subtilis. Proc. Natl. Acad. Sci. USA* **92**:2012–2016.

66. Kelemen, G. H., P. Brian, K. Flärdh, L. Chamberlin, K. F. Chater, and M. J. Buttner. 1998. Developmental regulation of transcription of *whiE*, a locus specifying the polyketide spore pigment in *Streptomyces coelicolor* A3(2). *J. Bacteriol.* **180**:2515–2521.

67. Kelemen, G. H., G. L. Brown, J. Kormanec, L. Potúcková, K. F. Chater, and M. J. Buttner. 1996. The positions of the sigma factor genes, *whiG* and *sigF*, in the heirarchy controlling the development of spore chains in the aerial hyphae of *Streptomyces coelicolor* A3(2). *Mol. Microbiol.* **21**:593–603.

68. Kellner, E. M., A. Decatur, and C. P. Moran, Jr. 1996. Two-stage regulation of an anti-sigma factor determines developmental fate during bacterial endospore formation. *Mol. Microbiol.* **21**:913–924.

69. Kennedy, M. J., S. L. Reader, and L. M. Swierczynski. 1994. Preservation records of microorganisms: evidence for the tenacity of life. *Microbiology* **140**:2513–2529.

70. Keseler, I. M., and D. Kaiser. 1995. An early A-signal-dependent gene in *Myxococcus xanthus* has a σ^{54}-like promoter. *J. Bacteriol.* **177**:4638–4644.

71. Keseler, I. M., and D. Kaiser. 1997. Sigma 54, a vital protein for *Myxococcus xanthus. Proc. Natl. Acad. Sci. USA* **94**:1979–1984.

71a. Kim, H.-J., M. J. Calcutt, F. J. Schmidt, and K. F. Chater. Chromosome partitioning during sporulation of *Streptomyces coelicolor* A3(2) involves an *oriC*-linked *parAB* locus. *J. Bacteriol.*, in press.

72. King, N., O. Dreesen, P. Stragier, K. Pogliano, and R. Losick. 1999. Septation, dephosphorylation and the activation of σ^F during sporulation in *Bacillus subtilis. Genes Dev.* **13**:1156–1167.

73. Kobayashi, K., K. Shoji, T. Shimizu, K. Nakano, T. Sato, and Y. Kobayashi. 1995. Analysis of a suppressor mutation *ssb* (*kinC*) of *sur0B20* (*spoOA*) mutation in *Bacillus subtilis* reveals

that *kinC* encodes a histidine protein kinase. *J. Bacteriol.* **177:** 176–182.

74. Kok, J., K. A. Trach, and J. A. Hoch. 1994. Effects on *Bacillus subtilis* of a conditional lethal mutation in the essential GTP-binding protein Obg. *J. Bacteriol.* **176:**7155–7160.

75. Kroos, L., and D. Kaiser. 1987. Expression of many developmentally regulated genes in *Myxococcus* depends on a sequence of cell interactions. *Genes Dev.* **1:**840–854.

76. Kunkel, B., R. Losick, and P. Stragier. 1990. The *Bacillus subtilis* gene for the developmental transcription factor σ^K is generated by excision of a dispensable DNA element containing a sporulation recombinase gene. *Genes Dev.* **4:**525–535.

77. LaBell, T. L., J. E. Trempy, and W. G. Haldenwang. 1987. Sporulation-specific sigma factor sigma 29 of *Bacillus subtilis* is synthesized from a precursor protein, P31. *Proc. Natl. Acad. Sci. USA* **84:**1784–1788.

78. Lawlor, E. J., H. A. Baylis, and K. F. Chater. 1987. Pleiotropic morphological and antibiotic deficiencies result from mutations in a gene encoding a tRNA-like product in *Streptomyces coelicolor* A3(2). *Genes Dev.* **1:**1305–1310.

79. Lazazzera, B. A., and A. D. Grossman. 1998. The ins and outs of peptide signalling. *Trends Microbiol.* **6:**288–294.

80. LeDeaux, J. R., N. Yu, and A. D. Grossman. 1995. Different roles for KinA, KinB, and KinC in the initiation of sporulation in *Bacillus subtilis*. *J. Bacteriol.* **177:**861–863.

81. Lee, K., and L. J. Shimkets. 1996. Suppression of a signaling defect during *Myxococcus xanthus* development. *J. Bacteriol.* **178:**977–984.

82. Levin, P. A., and R. Losick. 1996. Transcription factor Spo0A switches the localization of the cell division protein FtsZ from a medial to a bipolar pattern in *Bacillus subtilis*. *Genes Devel.* **10:**478–488.

83. Levin, P. A., and R. Losick. 2000. Asymmetric division and cell fate during sporulation in *Bacillus subtilis*, p. 167–189. *In* Y. V. Brun and L. J. Shimkets (ed.), *Prokaryotic Development*. American Society for Microbiology, Washington, D.C.

84. Levin, P. A., R. Losick, P. Stragier, and F. Arigoni. 1997. Localization of the sporulation protein SpoIIE in *Bacillus subtilis* is dependent upon the cell division protein FtsZ. *Mol. Microbiol.* **25:**839–846.

85. Londoño-Vallejo, J. A., and P. Stragier. 1995. Cell-cell signaling pathway activating a developmental transcription factor in *Bacillus subtilis*. *Genes Dev.* **9:**503–508.

86. Lopez, J. M., A. Dromerick, and E. Freese. 1981. Response of guanosine 5′-triphosphate concentration to nutritional changes and its significance for *Bacillus subtilis* sporulation. *J. Bacteriol.* **146:**605–613.

87. Mandic-Mulec, I., L. Doukhan, and I. Smith. 1995. The *Bacillus subtilis* SinR protein is a repressor of the key sporulation gene *spo0A*. *J. Bacteriol.* **177:**4619–4627.

88. Manoil, C., and D. Kaiser. 1980. Guanosine pentaphosphate and guanosine tetraphosphate accumulation and induction of *Myxococcus xanthus* fruiting body development. *J. Bacteriol.* **141:**305–315.

89. Martinez-Costa, H., P. Arias, N. M. Romero, V. Parro, R. P. Mellado, and F. Malpartida. 1996. A *relA/spoT* homologous gene from *Streptomyces coelicolor* A3(2) controls antibiotic biosynthesis genes. *J. Biol. Chem.* **271:**10627–10634.

90. Matsuno, K., T. Blais, A. W. Serio, T. Conway, T. M. Henkin, and A. L. Sonenshein. 1999. Metabolic imbalance and sporulation in an isocitrate dehydrogenase mutant of *Bacillus subtilis*. *J. Bacteriol.* **181:**3382–3391.

91. McCormick, J., E. P. Su, A. Driks, and R. Losick. 1994. Growth and viability of *Streptomyces coelicolor* mutant for the cell division gene *ftsZ*. *Mol. Microbiol.* **14:**243–254.

92. Min, K.-T., C. M. Hilditch, B. Diederich, J. Errington, and M. D. Yudkin. 1993. σ^F, the first compartment-specific transcription factor of *B. subtilis*, is regulated by an anti-σ factor that is also a protein kinase. *Cell* **74:**735–742.

93. Mitani, T., J. E. Heinze, and E. Freese. 1977. Induction of sporulation in *Bacillus subtilis* by decoyinine or hadacidin. *Biochem. Biophys. Res. Commun.* **77:**1118–1125.

94. Moir, A., and D. A. Smith. 1990. The genetics of bacterial spore germination. *Annu. Rev. Microbiol.* **44:**531–553.

95. Mueller, J. P., G. Bukusoglu, and A. L. Sonenshein. 1992. Transcriptional regulation of *Bacillus subtilis* glucose starvation-inducible genes: control of *gsiA* by the ComP-ComA signal transduction system. *J. Bacteriol.* **174:**4361–4373.

96. Mueller, J. P., and A. L. Sonenshein. 1992. Role of the *Bacillus subtilis gsiA* gene in regulation of early sporulation gene expression. *J. Bacteriol.* **174:**4374–4383.

97. Nodwell, J. R., and R. Losick. 1998. Purification of an extracellular signaling molecule involved in production of aerial mycelium by *Streptomyces coelicolor*. *J. Bacteriol.* **180:**1334–1337.

98. Nodwell, J. R., K. McGovern, and R. Losick. 1996. An oligopeptide permease responsible for the import of an extracellular signal governing aerial mycelium formation in *Streptomyces coelicolor*. *Mol. Microbiol.* **22:**881–893.

99. Nodwell, J. R., M. Yang, D. Kuo, and R. Losick. 1999. Extracellular complementation and the identification of additional genes involved in aerial mycelium formation in *Streptomyces coelicolor*. *Genetics* **151:**569–584.

100. Ogawa, M., S. Fujitani, X. Mao, S. Inouye, and T. Komano. 1996. FruA, a putative transcription factor essential for the development of *Myxococcus xanthus*. *Mol. Microbiol.* **22:** 757–767.

101. Ohlsen, K. L., J. K. Grimsley, and J. A. Hoch. 1994. Deactivation of the sporulation transcription factor Spo0A by the Spo0E protein phosphatase. *Proc. Natl. Acad. Sci.* **91:**1756–1760.

102. Okamoto, S., and K. Ochi. 1998. An essential GTP-binding protein functions as a regulator for differentiation in *Streptomyces coelicolor*. *Mol. Microbiol.* **30:**107–119.

103. Onaka, H., and S. Horinouchi. 1997. DNA-binding activity of the A-factor receptor protein and its recognition DNA sequences. *Mol. Microbiol.* **24:**991–1000.

104. Perego, M. 1998. Kinase-phosphatase competition regulates *Bacillus subtilis* development. *Trends Microbiol.* **6:**366–370.

105. Perego, M., and J. A. Hoch. 1996. Cell-cell communication regulates the effects of protein aspartate phosphatases on the phosphorelay controlling development in *Bacillus subtilis*. *Proc. Natl. Acad. Sci. USA* **93:**1549–1553.

106. Piggot, P. J., and J. G. Coote. 1976. Genetic aspects of bacterial endospore formation. *Bacteriol. Rev.* **40:**908–962.

107. Plamann, L., J. M. Davis, B. Cantwell, and J. Mayor. 1994. Evidence that *asgB* encodes a DNA-binding protein essential for growth and development of *Myxococcus xanthus*. *J. Bacteriol.* **176:**2013–2020.

108. Plamann, L., and H. B. Kaplan. 1999. Cell-density sensing during early development in *Myxococcus xanthus*, p. 67–82. *In* G. M. Dunny and S. C. Winans (ed.), *Cell-Cell Signaling in Bacteria*. American Society for Microbiology, Washington, D.C.

109. Plamann, L., A. Kuspa, and D. Kaiser. 1992. Proteins that rescue A-signal-defective mutants of *Myxococcus xanthus*. *J. Bacteriol.* **174:**3311–3318.

110. Plamann, L., Y. Li, B. Cantwell, and J. Mayor. 1995. The *Myxococcus xanthus asgA* gene encodes a novel signal trans-

duction protein required for multicellular development. *J. Bacteriol.* **177**:2014–2020.

111. **Pogliano, K., A. E. Hofmeister, and R. Losick.** 1997. Disappearance of the sigma E transcription factor from the forespore and the SpoIIE phosphatase from the mother cell contributes to establishment of cell-specific gene expression during sporulation in *Bacillus subtilis*. *J. Bacteriol.* **179**:3331–3341.

112. **Pope, M. K., B. Green, and J. Westpheling.** 1998. The *bldB* gene encodes a small protein required for morphogenesis, antibiotic production and catabolite control in *Streptomyces coelicolor*. *J. Bacteriol.* **180**:1556–1562.

113. **Predich, M., G. Nair, and I. Smith.** 1992. *Bacillus subtilis* early sporulation genes *kinA*, *spo0F*, and *spo0A* are transcribed by the RNA polymerase containing sigma H. *J. Bacteriol.* **174**:2771–2778.

114. **Price, K. D., and R. Losick.** 1999. A four-dimensional view of assembly of a morphogenetic protein during sporulation in *Bacillus subtilis*. *J. Bacteriol.* **181**:781–790.

115. **Rather, P. N., R. Coppolecchia, H. DeGrazia, and C. P. Moran, Jr.** 1990. Negative regulator of sigma G-controlled gene expression in stationary-phase *Bacillus subtilis*. *J. Bacteriol.* **172**:709–715.

116. **Rosenbluh, A., and E. Rosenberg.** 1989. Autocide AMI rescues development in *dsg* mutants of *Myxococcus xanthus*. *J. Bacteriol.* **171**:1513–1518.

117. **Ryding, N. J., G. H. Kelemen, C. A. Whatling, K. Flärdh, M. J. Buttner, and K. F. Chater.** 1998. A developmentally regulated gene encoding a repressor-like protein is essential for sporulation in *Streptomyces coelicolor* A3(2). *Microbiology* **29**:343–357.

118. **Sager, B., and D. Kaiser.** 1993. Spatial restriction of cellular differentiation. *Genes Dev.* **7**:1645–1653.

119. **Schaeffer, P. J. Millet, and J.-P. Aubert.** 1965. Catabolite repression of bacterial sporulation. *Proc. Natl. Acad. Sci. USA* **54**:704–711.

120. **Sedlak, M., V. Vinter, J. Adamec, J. Vohradsky, Z. Voburka, and J. Chaloupka.** 1993. Heat shock applied early in sporulation affects heat resistance of *Bacillus megaterium* spores. *J. Bacteriol.* **175**:8049–8052.

121. **Serror, P., and A. L. Sonenshein.** 1996. CodY is required for nutritional repression of *Bacillus subtilis* genetic competence. *J. Bacteriol.* **178**:5910–5915.

122. **Setlow, P.** 1995. Mechanisms for the prevention of damage to DNA in spores of *Bacillus* species. *Annu. Rev. Microbiol.* **49**:29–54.

123. **Shi, W., and D. R. Zusman.** 1995. The *frz* signal transduction system controls multicellular behavior in *Myxococcus xanthus*, p. 419–430. *In* J. A. Hoch and T. J. Silhavy (eds.), *Two-Component Signal Transduction*. American Society for Microbiology, Washington, D.C.

124. **Shima, J., A. Penyige, and K Ochi.** 1996. Changes in patterns of ADP-ribosylated proteins during differentiation of *Streptomyces coelicolor* A3(2) and its developmental mutants. *J. Bacteriol.* **178**:3785–3790.

125. **Shimkets, L. J.** 2000. Growth, sporulation, and other tough decisions, p. 277–284. *In* Y. V. Brun and L. J. Shimkets (ed.), *Prokaryotic Development*. American Society for Microbiology, Washington, D.C.

126. **Singer, M., and D. Kaiser.** 1995. Ectopic production of guanosine penta- and tetraphosphate can initiate early developmental gene expression in *Myxococcus xanthus*. *Genes Dev.* **9**:1633–1644.

127. **Slack, F. J., P. Serror, E. Joyce, and A. L. Sonenshein.** 1995. A gene required for nutritional repression of the *Bacillus subtilis* dipeptide permease operon. *Mol. Microbiol.* **15**:689–702.

128. **Smith, D. R., and M. Dworkin.** 1997. A mutation that affects fibril protein, development, cohesion and gene expression in *Myxococcus xanthus*. *Microbiology* **143**:3683–3692.

129. **Soliveri, J., K. L. Brown, M. J. Buttner, and K. F. Chater.** 1992. Two promoters for the *whiB* sporulation gene of *Streptomyces coelicolor* A3(2), and their activities in relation to development. *J. Bacteriol.* **174**:6215–6220.

130. **Solomon, J. M., B. A. Lazazzera, and A. D. Grossman.** 1996. Purification and characterization of an extracellular peptide factor that affects two different developmental pathways in *Bacillus subtilis*. *Genes Dev.* **10**:2014–2024.

131. **Sonenshein, A. L.** 1989. Metabolic regulation of sporulation and other stationary-phase phenomena, p. 109–130. *In* I. Smith, R. A. Slepecky, and P. Setlow (ed.), *Regulation of Prokaryotic Development*. American Society for Microbiology, Washington, D.C.

132. **Sonenshein, A. L.** 2000. Endospore-forming bacteria: an overview, p. 133-150. *In* Y. V. Brun and L. J. Shimkets (ed.), *Prokaryotic Development*. American Society for Microbiology, Washington, D.C.

133. **Sonenshein, A. L., J. A. Hoch, and R. Losick.** 1993. Bacillus subtilis *and Other Gram-Positive Bacteria: Biochemistry, Physiology, and Molecular Genetics*. American Society for Microbiology, Washington, D.C.

134. **Sterlini, J. M., and J. Mandelstam.** 1969. Commitment to sporulation in *Bacillus subtilis* and its relationship to development of antibiotic resistance. *Biochem. J.* **113**:29–37.

135. **Stragier, P., and R. Losick.** 1996. Molecular genetics of sporulation in *Bacillus subtilis*. *Annu. Rev. Genet.* **30**:297–341.

136. **Strauch, M. A., and J. A. Hoch.** 1993. Transition-state regulators: sentinels of *Bacillus subtilis* post-exponential gene expression. *Mol. Microbiol.* **7**:337–342.

137. **Susstrunk, U., J. Pidoux, S. Taubert, A. Ullmann, and C. J. Thompson.** 1998. Pleiotropic effects of cAMP on germination, antibiotic synthesis and morphological development in *Streptomyces coelicolor*. *Mol. Microbiol.* **30**:33–46.

138. **Toal, D. R., S. W. Clifton, B. A. Roe, and J. Downard.** 1995. The *esg* locus of *Myxococcus xanthus* encodes the E1 alpha and E1 beta subunits of a branched chain keto-acid dehydrogenase. *Mol. Microbiol.* **16**:177–189.

139. **Trach, K. A., and J. A. Hoch.** 1993. Multisensory activation of the phosphorelay initiating sporulation in *Bacillus subtilis*: identification and sequence of the protein kinase of the alternate pathway. *Mol. Microbiol.* **8**:69–79.

140. **Ueda, K., C.-W. Hsheh, T. Tosaki, H. Shinkawa, T. Beppu, and S. Horinouchi.** 1998. Characterization of an A-factor-responsive repressor for *amfR* essential for onset of aerial mycelium formation in *Streptomyces griseus*. *J. Bacteriol.* **180**:5085–5093.

141. **Ueki, T., and S. Inouye.** 1998. The new sigma factor, SigD, essential for stationary phase is also required for multicellular differentiation in *Myxococcus xanthus*. *Genes Cells* **3**:371–385.

142. **Vidwans, S. J., K. Ireton, and A. D. Grossman.** 1995. Possible role for the essential GTP-binding protein Obg in regulating the initiation of sporulation in *Bacillus subtilis*. *J. Bacteriol.* **177**:3308–3311.

143. **Waites, W. M., D. Kay, I. W. Dawes, D. A. Wood, S. C. Warren, and J. Mandelstam.** 1970. Sporulation in *Bacillus subtilis*. Correlation of biochemical events with morphological changes in asporogenous mutants. *Biochem. J.* **118**:667–676.

144. **Waldburger, C., D. Gonzalez, and G. H. Chambliss.** 1993. Characterization of a new sporulation factor in *Bacillus subtilis. J. Bacteriol.* **175:**6321–6327.

145. **Wall, D., P. E. Kolenbrander, and D. Kaiser.** 1999. The *Myxococcus xanthus pilQ* (*sglA*) gene encodes a secretin homolog required for type IV pilus biogenesis, social motility, and development. *J. Bacteriol.* **181:**24–33.

146. **Wang, L., R. Grau, M. Perego, and J. A. Hoch.** 1997. A novel histidine kinase inhibitor regulating development in *Bacillus subtilis. Genes Dev.* **11:**2569–2579.

147. **Weir, J., M. Predich, E. Dubnau, G. Nair, and I. Smith.** 1991. Regulation of *spo0H*, a gene coding for the *Bacillus subtilis* sigma H factor. *J. Bacteriol.* **173:**521–529.

148. **Wendrich, T. M., and M. A. Marahiel.** 1997. Cloning and characterization of a *relA/spoT* homologue from *Bacillus subtilis. Mol. Microbiol.* **26:**65–79.

149. **Willey, J., J. Schwedock, and R. Losick.** 1993. Multiple extracellular signals govern the production of a morphogenetic protein involved in aerial mycelium formation by *Streptomyces coelicolor. Genes Dev.* **7:**895–903.

150. **Wu, S. S., and D. Kaiser.** 1997. Regulation of expression of the *pilA* gene in *Myxococcus xanthus. J. Bacteriol.* **179:**7748–7758.

151. **Xu, D., C. Yang, and H. B. Kaplan.** 1998. *Myxococcus xanthus sasN* encodes a regulator that prevents developmental gene expression during growth. *J. Bacteriol.* **180:**6215–6223.

152. **Yang, C., and H. B. Kaplan.** 1997. *Myxococcus xanthus sasS* encodes a sensor histidine kinase required for early developmental gene expression. *J. Bacteriol.* **179:**7759–7767.

153. **York, K., T. J. Kenney, S. Satola, C. P. Moran, Jr., H. Poth, and P. Youngman.** 1992. Spo0A controls the σ^A-dependent activation of *Bacillus subtilis* sporulation-specific transcription unit *spoIIE. J. Bacteriol.* **174:**2648–2658.

154. **Zhang, B., A. Hofmeister, and L. Kroos.** 1998. The prosequence of pro-sigmaK promotes membrane association and inhibits RNA polymerase core binding. *J. Bacteriol.* **180:**2434–2441.

155. **Zhang, L., M. L. Higgins, and P. J. Piggot.** 1997. The division during bacterial sporulation is symmetrically located in *Sporosarcina ureae. Mol. Microbiol.* **25:**1091–1098.

Bacterial Stress Responses
Edited by G. Storz and R. Hengge-Aronis
©2000 ASM Press, Washington, D.C.

Chapter 14

Resistance of Bacterial Spores

PETER SETLOW

Dormant endospores of Bacillus *and* Clostridium *species are much more resistant than their growing cell counterparts to wet and dry heat, UV and γ-radiation, oxidizing agents and other chemicals, enzymes, mechanical disruption, desiccation, and high pressures. Although all the reasons for these different resistance properties are not known, factors important in one or more spore resistance properties include: (i) the relative dehydration of the spore core, (ii) the mineralization of the spore core, (iii) the protection of spore DNA by its saturation with a unique group of spore-specific proteins, (iv) DNA repair, (v) the decreased permeability of the spore core to chemicals, and (vi) the encasement of the spore in the rigid spore coats. Spores of myxobacterial and* Streptomyces *species are generally much less resistant than are spores of* Bacillus *and* Clostridium *species but are more resistant than their corresponding growing cells—in particular to wet heat, desiccation, UV radiation, mechanical disruption, and at least some chemicals. Although relatively little is known about the factors that contribute to the resistance of spores of myxobacterial and* Streptomyces *species, factors that have been identified include these spores' high level of the disaccharide trehalose (not present in spores of* Bacillus *and* Clostridium *species), which contributes to spore heat and desiccation resistance, and the spore coats, which may contribute to resistance to mechanical disruption and enzymes.*

At some point all bacteria find themselves in environments in which nutrients are insufficient to support an increase in their biomass and possibly even to allow continued high-level energy production. Faced with these starvation conditions, bacterial species initiate a variety of different responses as out-

lined in other chapters in this book (see chapters 11, 12, 13, 15, and 17). One common response is to initiate a developmental pathway leading to conversion of the growing cell into a spore that is metabolically dormant or quiescent and can thus survive in environments of low or no nutrients. Since formation of these spores is designed to generate a cell that can survive for long periods with little or no endogenous metabolism and with no foreknowledge of when nutrients will return, it is not surprising that these bacterial spores are invariably more resistant than their growing cell counterparts to a variety of stress conditions that might be encountered in the environment. Although the degree of this elevated spore resistance varies among different classes of spore-formers including myxobacterial, *Streptomyces*, and *Bacillus*/*Clostridium* species, this elevated spore resistance is another salient feature of spores of these organisms. However, the physiological stimulus for sporulation appears to be starvation, as there is no evidence that other stresses (e.g., heat or desiccation) induce sporulation.

The purpose of this review is to examine not only specific resistance properties of spores, but more importantly the detailed mechanisms bringing about this resistance. Some aspects of resistance of spores of myxobacterial and *Streptomyces* species will be discussed briefly. However, the review will concentrate on the resistance of spores of *Bacillus* and *Clostridium* species, since there is a fairly large body of information on this topic. In the discussion of this latter topic statements are meant to refer to the situation with spores of both *Bacillus* and *Clostridium* species, unless otherwise noted. However, in general, there is much more detailed knowledge about spores of *Bacillus* species than of *Clostridium* species.

Peter Setlow • Department of Biochemistry, University of Connecticut Health Center, Farmington, CT 06032.

SPORES OF *BACILLUS AND CLOSTRIDIUM* SPECIES

Endospores are formed by the members of the gram-positive *Bacillus* and *Clostridium* genera and their close relatives in response to nutrient limitation (see chapter 13). While other environmental stresses (heat, desiccation, etc.) are not known to trigger spore formation, there are data indicating that in *Bacillus subtilis* a number of genes in sporulation regulons are also members of stress regulons (25, 39, 122; A. Petersohn, S. Engelmann, P. Setlow, and M. Hecker, submitted for publication; see also chapter 12). However, the significance of the overlap between these regulons is not clear. The products of the sporulation process, the dormant spores, have a large pool of free adenine nucleotides, but less than 1% is ATP and the energy charge ([ATP] + 0.5[ADP]/[ATP] + [ADP] + [AMP]) is only ~ 0.1 (55, 112, 116). As is not surprising given their minimal ATP levels, these spores exhibit no detectable metabolism of exogenous or even endogenous metabolites, despite the presence of significant levels of endogenous metabolites as well as the enzymes for their utilization (116). A major reason for the dormancy of these spores is the very low level of hydration of the spore core, the site of most spore enzymes (see below). In addition to their dormancy, these spores are also extremely resistant to a wide variety of stress conditions, much more resistant than their corresponding vegetative cells and significantly more resistant than spores of myxobacterial or *Streptomyces* species (Ta-

ble 1) (29–31, 35, 40, 52, 59, 92, 93, 95, 107, 115, 116, 120).

Spore Structure

There are many factors involved in spore resistance to different stress conditions, and these will be discussed in detail below. However, since spore structure plays a major role in a number of the spore's resistance properties, we will first briefly outline the major structural features and components of spores (Fig. 1) and their role in spore resistance. Beginning at the outer spore surface there is a loosely fitting structure, termed the exosporium, whose size varies greatly between spores of different species (24). This structure appears to have no counterpart in growing cells, but the individual components of the exosporium have not been well defined. However, in so far as is known, the exosporium plays no role in spore resistance properties, although it may well play some important role in the spores' natural environments. Underlying the exosporium is the spore coat. This structure is composed primarily of protein although the number of proteins in the coat and the morphological complexity of the coat vary greatly between species (23, 24). The coat proteins are sporulation-specific gene products that are synthesized at defined times in the mother cell compartment of the sporulating cell under the control of two mother cell-specific sigma factors for RNA polymerase, σ^E and σ^K, as well as the regulatory GerE protein; several comprehensive reviews of coat protein gene expres-

Table 1. Comparison of the heat and desiccation resistance of growing cells and spores of *B. subtilis*, *M. xanthus*, and *Streptomyces* species[a]

Organism and stage	Heat resistance		Desiccation resistance	
	Temp (°C)	Time to obtain 90% killing	Time in dry storage at low humidity (days)	% Killing
B. subtilis[b]				
Growing cells	65	<15 s	1	2
Spores	65	105 h	30	100
M. xanthus[c]				
Growing cells	60	15 s	6	0
Spores	60	82 min	6	50
S. albidoflavus[d]				
Growing cells	55	25 min	15	<1
Spores	55	75 min	15	82
S. griseus[d]				
Growing cells	60	<5 min	28	23
Spores	60	~150 min		

[a] Values have been calculated from data in references 20, 30, 52, 64, 107, and 120.
[b] Data are for growing cells in the log phase of growth.
[c] Data are for spores isolated from fruiting bodies.
[d] Data are for vegetative mycelia (growing cells) and spores formed in aerial hyphae.

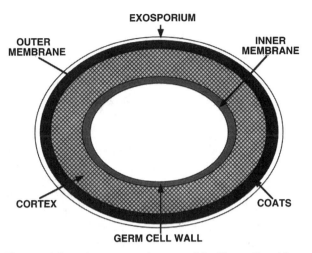

Figure 1. Schematic structure of a spore of *Bacillus or Clostridium* species. The sizes of the various layers are not drawn exactly to scale, and the size of the exosporium layer varies widely in shape and size across species.

sion and coat assembly in *B. subtilis* have recently appeared (23, 24). In contrast to the exosporium, the coat does play a role in spore resistance, as the coat provides a barrier against peptidoglycan lytic enzymes such as lysozyme (23, 24, 81), and may be important in preventing digestion of spores by foraging organisms (58) (i.e., the coats provide "eat" resistance). The coats also appear to play a role in the resistance of spores to some chemicals (4, 7–9, 124), although the precise coat proteins involved and exactly how this resistance mechanism operates are not clear.

Beneath the coat structure is the outer spore membrane. During spore formation this membrane is almost certainly a complete membrane and thus a potential permeability barrier that could play a role in spore resistance to chemicals. However, it is not clear if the outer spore membrane is in fact a complete membrane in the dormant spore. Indeed, spores missing some particular coat proteins are lysozyme sensitive (23), suggesting that the outer membrane is not a significant permeability barrier.

Underlying the outer spore membrane is a large layer of peptidoglycan (PG), the majority of which is termed the cortex; underlying this structure is the germ cell wall. The germ cell wall likely has a PG structure similar to that in growing cells, but the cortex has a number of unique features, including the presence of muramic acid lactam and muramic acid-alanine in the PG backbone and a much lower degree of cross-linking than in growing cell PG (1, 11, 87, 88,127, 128). The spore cortex likely plays an important if indirect role in spore resistance, as this structure appears essential in the dehydration of the spore core (1, 10, 45, 87, 89). However, the precise mechanism whereby the cortex causes core dehydration during sporulation is not known. The germ cell wall is not thought to play any role in spore resistance but becomes the cell wall of the germinated spore, since the spore cortex is degraded during spore germination.

Underlying the spore cortex is the inner spore membrane. This membrane is a major permeability barrier in the dormant spore, restricting or slowing the entry of hydrophilic molecules into the spore core, as well as blocking egress of the huge depot of hydrophilic molecules in the core (36, 98). However, at present the reason for the extremely low permeability of the inner spore membrane is not clear.

The central region of the spore is the core, the site of spore DNA, ribosomes, and most enzymes. Five major physiological differences between the spore core and the protoplast of a growing bacterium are (i) the absence from the spore core of common "high-energy compounds" (e.g., ATP, NADH) found at high levels in growing cells (112); (ii) the high levels of divalent ions (predominantly Ca^{2+}) in the core (74); (iii) the high levels (up to 10% of dry weight) of pyridine-2,6-dicarboxylic acid (dipicolinic acid [DPA]) (Fig. 2A) in the core (74); much, if not all of this DPA is likely chelated with divalent cations and DPA is absent from growing cells; (iv) a pH in the spore core 1 to 1.5 units lower than in a growing cell (56, 57); and (v) the low level of hydration of the

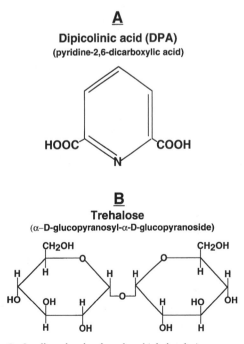

Figure 2. Small molecules found at high levels in spores of (A) *Bacillus* or *Clostridium* and (B) myxobacterial and *Streptomyces* species. Note that trehalose is not a reducing sugar and thus lacks a reactive aldehyde group.

core compared to the hydration level in growing cells (5, 35). In a growing bacterium 75 to 80% of the wet weight is water, and this is also largely the case in the spore layers external to the core (35). However, in the spore core itself only 25 to 35% of the wet weight is water (35). Although neither how the low core hydration is achieved, nor the level of free water in the core, nor the precise state of divalent cations and DPA in the core is known, it is clear that these physiological properties of the spore core play a major role in several aspects of spore resistance as discussed below. One group of small molecules that are notable by their absence from the spore core are free sugars such as trehalose (Fig. 2B) (112, 116). This latter disaccharide is accumulated at high levels by spores of myxobacterial and *Streptomyces* species and most likely plays a role in the resistance of these spores (19, 64, 66). However, spores of *Bacillus* and *Clostridium* species have no significant levels of any small mono- or oligosaccharides (112).

Small Acid-Soluble Proteins

There is one other group of molecules unique to the spore core that plays a major role in spore resistance. These are the small, acid-soluble proteins (SASP) of the α/β-type, so named for the major proteins (SASP-α and $-\beta$) of this type in *B. subtilis* (24, 61, 113, 116, 117). The α/β-type SASP are synthesized during sporulation only within the developing forespore, and comprise 3 to 8% of total spore protein in spores of *Bacillus* and *Clostridium* species. These proteins are the products of the multigene *ssp* family with from 4 to 7 members. In *Bacillus* species two of these genes (*sspA* and *B* in *B. subtilis*) are expressed at a high level, with the remainder expressed at a much lower level (113). The sequences of α/β-type SASP are very highly conserved both within and between various *Bacillus* species and their close relatives, and this sequence conservation extends to the α/β-type SASP of *Clostridium* species. The genes encoding α/β-type SASP are under the control of RNA polymerase with the forespore-specific sigma factor σ^G (80, 121; see also chapter 13).

The α/β-type SASP are DNA-binding proteins, binding only to double-stranded DNA (110). There is sufficient α/β-type SASP in spores to saturate the spore chromosome, and there is strong evidence that the spore DNA is indeed saturated with α/β-type SASP (34, 85, 97, 116; M. A. Ross and P. Setlow, unpublished data). Formation of the α/β-type SASP/ DNA complex changes both DNA structure and topology appreciably and plays an important role in spore DNA resistance to a variety of stresses (70, 77, 79, 114, 115–117). The α/β-type SASP structure

also changes dramatically in the complex, going from a largely random coil in the free protein to a highly α-helical conformation when complexed with DNA (C. S. Hayes and P. Setlow, unpublished data; S. C. Mohr and P. Setlow, unpublished data); formation of the complex also protects the protein against a variety of types of damage (43, 44, 106). A number of facts are known about the α/β-type SASP/DNA complex, including (i) α/β-type SASP bind on the outside of the DNA helix (42); (ii) the DNA in the complex is stiffened and straightened but has the same length as free DNA (42); (iii) ~4 bp are bound per α/β-type SASP (110); and (iv) some of the spectroscopic properties of the DNA in the complex are characteristic of DNA in an A-like helix (70). However, the precise structures of the proteins and DNA in this complex are not known. As a consequence, the precise molecular mechanism(s) whereby α/β-type SASP binding brings about resistance of spore DNA to a variety of stresses is not known.

The major DNA binding protein found in growing cells of *Bacillus* species (termed HBsu) is highly homologous to the HU protein of *Escherichia coli* but is essential in *B. subtilis* (67, 68). HBsu is present in spores at levels similar to those in growing cells (M. A. Ross, N. M. Magill, and P. Setlow, unpublished data), and studies in vitro have shown that HU can modulate the effects of α/β-type SASP on DNA properties (M. A. Ross and P. Setlow, unpublished data). Spores also have low levels of a large number of minor SASP unrelated to α/β-type SASP but with rather high isoelectric points (3); as yet no specific functions have been ascribed to these minor proteins.

Temporal Development of Resistance

As will be discussed in detail below, the dormant spore is extremely resistant to a variety of stresses. These resistance properties are acquired by the developing spore at defined times during spore formation. UV resistance is acquired first in parallel with synthesis of α/β-type SASP, and full wet heat resistance is acquired several hours later when the final level of spore core dehydration is achieved (69, 112). Chemical resistance is acquired at various times between these two time points as is γ-radiation resistance (49, 69). As yet the precise times of acquisition of dry heat and desiccation resistance are not known.

While the dormant spore has no metabolism and is extremely resistant to many treatments, these properties are lost in the first minutes of spore germination when the spore "returns to life." In these first minutes the strong permeability barrier limiting access to and egress from the spore core is breached; the core takes up water to achieve the hydration level of a growing cell; hydrogen ions, metal ions, and DPA are excreted; and the α/β-type SASP are degraded (46, 71, 112). Of the spore resistance properties, wet heat resistance is lost first, and UV resistance is lost last. Subsequently, the germinated spore is converted back into a vegetative cell through spore outgrowth.

TYPES OF *BACILLUS* AND *CLOSTRIDIUM* SPORE RESISTANCE

As noted above, spores of *Bacillus* and *Clostridium* species are much more resistant than their growing cell counterparts, and there is a huge amount of data on resistance of spores from many different species to many different stresses. There are also a number of studies attempting to correlate physiological or biochemical properties of spores of different species with the level of resistance of these spores. However, these latter studies are limited and always subject to the concern that the property being measured may only be a reflection of some other unknown spore property that correlates directly with spore resistance. Unfortunately, for most *Bacillus* and *Clostridium* species genetic manipulation of spore properties that might be correlated with spore resistance remains relatively to extremely difficult. The exception is *B. subtilis* in which facile genetic and molecular genetic analyses as well as the complete genomic sequence (51) have greatly facilitated direct assessment of the roles of specific spore components or properties in spore resistance. Consequently, the discussion below will focus on knowledge gained primarily from studies with *B. subtilis*, although relevant data from other species will also be mentioned. However, all available evidence indicates that the basic mechanisms of spore resistance are similar in spores of all *Bacillus* and *Clostridium* species.

As will be seen from the sections below, spores of *Bacillus* and *Clostridium* species are extremely resistant to acute stresses, e.g., high temperatures, very high levels of toxic chemicals, etc. It seems likely that this resistance of spores to acute stresses is a reflection of the phenomenal resistance of spores to "normal" levels of stress, and this in turn is reflected in the spore's tremendous ability to survive for long periods under relatively mild conditions. There are good data indicating that spores can survive in the dry state for hundreds if not thousands of years (37, 48, 119), and other data that have been interpreted as indicating that spores can survive for millions of years (12, 86). However long spores can survive (and it clearly is a long time), it appears likely that this long-term survival is a reflection of the spore's extreme dormancy as well as the spore's resistance to the everyday stresses of its environment.

Wet Heat Resistance

Undoubtedly the property most associated with spores of *Bacillus* and *Clostridium* species is their wet heat resistance (Table 1). When heat resistance is defined as a D_t value, the time at temperature t for killing of 90% of the cells or spores in a population, then spores require ~40°C higher temperatures than growing cells to obtain similar D values (35). A number of factors have been identified that contribute to spore wet heat resistance. First, the higher the temperature optimum for growth of an organism, the higher the wet heat resistance of spores of the organism; consequently, spores of thermophiles (e.g., *Bacillus stearothermophilus*) have higher heat resistance than do spores of mesophiles (e.g., *B. subtilis*). Presumably, the reason for this factor is simply the higher intrinsic heat resistance of proteins in thermophiles than in mesophiles. Second, for one organism, the higher the temperature at which spores are prepared, the higher is the spore's heat resistance (5, 35). The precise mechanism for this effect is not clear, although there are some data suggesting that spore core hydration levels fall as the sporulation temperature for an organism is increased (5, 35). Although this latter phenomenon is not understood, it may explain at least in part the increased spore wet heat resistance as sporulation temperature is increased, because decreasing spore core hydration is associated with increased spore heat resistance (see below). However, the possible role of other factors, including the heat shock response, in the increased spore heat resistance as sporulation temperatures increase has not been thoroughly studied. Third is the spore's degree of mineralization; both the amount of minerals and their identity are important, with increasing levels of mineral ions, predominantly Ca^{2+}, being associated with increased spore wet heat resistance (34, 59, 60). The precise mechanism of this effect, however, is not clear, although models to explain it have been presented (35, 59). Fourth, and most important, is the degree of hydration of the spore core (35, 60, 126). When the heat resistance and core hydration of spores of a variety of species are compared, a good correlation between decreasing spore core hydration and increasing spore wet heat resistance is observed. However, the correlation is by no means perfect, undoubtedly reflecting the role of factors besides core hydration. In vitro studies have shown that relatively dehydrated proteins are stabilized tremendously against heat, and presumably this is the reason for much of the wet heat resistance of spores (reviewed in 126). Note, however, that the spore core is not dry, having 25 to 35% of its wet weight as water, but the amount of free water in the spore core is not known. We also do not yet know the details of the process that causes the reduction in spore core hydration, although this almost certainly requires the action of the spore cortex as it is formed during sporulation. Interestingly, although DPA makes up an enormous percentage of the core's dry weight, DPA appears to play no direct role in spore heat resistance (35, 60).

One current limitation in our understanding of spore wet heat resistance is the identity of the target for spore killing by this treatment, as well as the spore injury caused by wet heat that can sometimes be "repaired" during spore germination and outgrowth. Likely candidates for these targets are one or more spore proteins, and while spore enzymes are extremely well protected against the denaturation as well as covalent modification (e.g., deamidation) that can be caused by elevated temperatures, inactivation of some spore enzymes accompanies and in a few cases appears to precede spore killing (44, 125). Possible candidates for proteins whose inactivation by wet heat might lead to spore killing or injury are the components of the spore germination apparatus, in particular, the molecules that sense the presence of germinants. These are almost certainly proteins and are likely to be outside the spore core and thus removed from the protective effects of core dehydration (71). Indeed, a spore whose germination machinery was inactivated would appear dead, although it might be recovered by artificial means (e.g., germination with lysozyme). However, spores appear to have multiple receptors for germinants (71), all of which would need to be inactivated to give a phenotypically dead spore. Unfortunately, while there are data consistent with one or more spore proteins as the target for spore killing by wet heat (6), neither the identity of this protein or proteins, nor proof that the target is a protein, nor the exact mechanism whereby wet heat inactivates some key proteins is yet known. However, spore killing by wet heat is not through oxidative damage promoted by elevated temperatures (109). In addition, the killing target is not spore DNA, as wet heat killing of spores generates no mutations in the survivors, no DNA damage (in particular, the base loss by depurination that might be expected at the elevated temperatures used for spore killing), and mutations in DNA repair genes do not decrease spore wet heat resistance (30, 105, 108). The protection of spore DNA specifically from heat damage is due to the fifth factor involved in spore heat resistance, which is the spore DNA's saturation with α/β-type SASP that block DNA depurination both in vitro and in vivo (30). Indeed, spores lacking the majority of their α/β-type SASP (termed $\alpha^-\beta^-$ spores) exhibit decreased spore wet heat resistance and a high percentage of mutations in the survivors of wet heat treatment (30, 63, 105). In addition, mutation of key DNA repair genes decreases the wet heat resistance of $\alpha^-\beta^-$ spores even further (108).

As noted above, the available evidence suggests that a protein(s) is the killing target of wet heat treatment of spores. If so, one might expect that one or more products of the heat shock response regulon might be able to reduce spore killing by wet heat by renaturing or degrading damaged proteins during subsequent spore germination. Although there are a few observations consistent with this idea (5, 99), it has by no means been proven and is one area that requires further study.

Dry Heat and Desiccation Resistance

Spores are again much more resistant to dry heat than are growing cells, and as mentioned above, it is likely that the resistance of dry spores to elevated temperatures is a reflection of the tremendously long survival times of dry spores at moderate temperatures. The target for spore killing by dry heat is at least in part DNA, as there is a large percentage of mutations in survivors of dry heat treatment of spores as well as much DNA damage, and loss of DNA repair capacity decreases dry heat resistance of spores (17, 82, 107, 108, 133). The major protective factor increasing spore resistance to dry heat appears to be the binding of α/β-type SASP to DNA, as $\alpha^-\beta^-$ spores exhibit dry heat resistance very similar to that of growing cells, and as with wet heat, α/β-type SASP binding protects DNA against depurination caused by dry heat (107).

As is not surprising given their dry heat resistance, spores are also extremely resistant to desiccation and can survive multiple cycles of hydration and desiccation with little if any killing (31, 92). However, spores incubated in very high vacuum appear to accumulate slowly some type of DNA damage that is normally repaired very efficiently during spore germination (73). All the factors responsible for spore desiccation resistance are not known, but α/β-type SASP as well as DNA repair during spore germination are again important in protecting spore DNA from damage during freeze drying (31, 108). Although spore components besides α/β-type SASP are also likely important in spore desiccation resistance, these factors do not include the high levels of sugars such as trehalose that are almost certainly involved in the desiccation resistance of spores of myxobacterial and

Streptomyces species (19, 64, 66), since such sugars are absent from spores of *Bacillus* species (112).

Radiation Resistance

Spores are significantly more resistant to γ-radiation than are growing cells. Although the factors responsible for the elevated spore γ-radiation resistance have not been explored in detail, one likely candidate is the reduced spore core hydration, which should reduce production of toxic radicals (e.g., hydroxyl radicals) by radiolysis of water in the spore core (40, 92, 116; Ross et al., unpublished data). However, this has by no means been proven, and spore γ-radiation resistance is not affected by the presence or absence of α/β-type SASP, which protect spore DNA from hydroxyl radical attack on the DNA backbone (116).

Although the increase in spore UV resistance over that of growing cells varies somewhat depending on the wavelength of the UV light (53, 76, 132), at 254 nm (the most effective wavelength) spores are 10 to 50 times more resistant than growing cells (115–117). The reason for the difference in spore and growing cell resistance to 254 nm UV radiation is a difference in the UV photochemistry of DNA in spores and growing cells. While the major DNA photoproduct generated by 254 nm UV irradiation of growing cells is a *cis*, syn-cyclobutane-type thymine dimer (TT, Fig. 3A) formed between adjacent thymines on the same DNA strand, TT is not formed by UV irradiation of spores (21). Rather, the major photoproduct is a thyminyl-thymine adduct termed spore photoproduct (SP, Fig. 3B), which is also formed between adjacent thymines in the same DNA strand. TT and SP are formed with similar efficiency as a function of UV fluence in growing cells and spores. However, TT formation in vitro is more efficient, since it appears that the high level of DPA in spores acts as a photosensitizer for SP formation (115, 117). While SP is the predominant lesion formed in spores by 254 nm radiation, other photoproducts appear to be formed at longer wavelengths; however, the identity of these other photoproducts is not clear (76, 132).

Both TT and SP are potentially lethal lesions, and spores and growing cells have repair systems for dealing with these lesions (117). Growing cells repair TT almost exclusively by nucleotide excision repair (NER), and NER also operates to repair SP lesions early in spore germination (76, 115, 117). However, spores have an additional mechanism for SP repair, which is the enzyme SP lyase (Spl) that monomerizes SP to two thymines without cleavage of the DNA backbone (72, 76). This enzyme is synthesized only during sporulation in the developing spore and op-

A

Cyclobutane-type thymine-thymine dimer (TT)

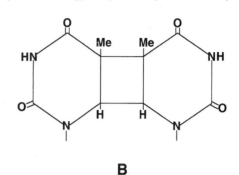

B

Thyminyl-thymine adduct or spore photoproduct (SP)

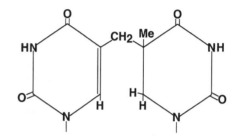

Figure 3. Structures of (A) cyclobutane-type thymine-tymine dimer and (B) spore photoproduct. Me denotes a methyl group, and the bonds to the N atoms at the bottoms of the pyrimidine rings are from the sugar residues in DNA.

erates during spore germination (33, 76). Action of Spl does not require energy or light, although the protein does exhibit limited amino acid sequence homology to DNA photolyases, which use light to monomerize TT in DNA to two thymines (72, 76). Purified Spl is an iron sulfur protein (91), but the mechanism of action of this enzyme is not yet clear. Mutations inactivating either the NER or Spl repair system reduce spore UV resistance two- to fourfold, while spores of double mutants exhibit 20 to 50 fold lower UV resistance (76, 115).

The reason for the unique UV photochemistry of DNA in spores is the saturation of spore DNA with α/β-type SASP (100, 115, 117). α⁻β⁻ spores exhibit UV resistance slightly below that of growing cells, and UV radiation generates much TT in α⁻β⁻ spores (62, 103). In vitro studies have shown that binding of α/β-type SASP to DNA in dilute buffer results in SP formation upon UV irradiation and suppression of formation of both TT and other cyclobutane dimers, as well as the various 6,4-photoproducts (32, 78). α/β-type SASP also bind to DNA in *E. coli* and result in formation of SP upon UV irradiation (101). However, α/β-type SASP with changes in residues conserved throughout evolution are no longer capable of altering spore DNA photochemistry either in vitro or in vivo in either *B. subtilis* or *E. coli* (32, 78, 101, 122). In contrast to the clear role for α/β-type SASP in spore UV resistance, spore core hydration levels appear to play no significant role in spore UV resistance (89, 100); indeed, developing spores become UV resistant well before full spore core dehydration (116).

Chemical Resistance

Spores are much more resistant than growing cells to a wide variety of chemicals, including acids, alcohols, alkylating agents, aldehydes, bases, enzymes, halogen gases, phenols, and peroxides (4, 7–9, 24, 49, 93, 124). For many of these chemicals the specific factors in spore resistance have not been identified. However, the spore coat is thought to play a role in resistance to enzymes such as lysozyme and chemicals such as halogens and oxidizing agents (4, 8, 9, 24, 81, 124). The relative impermeability of the spore core also undoubtedly plays a major role in spore chemical resistance (8, 36). For a number of these chemicals the target for their killing of spores is also not known. However, for peroxides, formaldehyde, and alkylating agents some of the details about the mechanisms of spore resistance are known.

Peroxides

The resistance of growing cells to both hydrogen peroxide and alkyl hydroperoxides is due in large part to the detoxification of these chemicals by catalases and alkyl hydroperoxide reductases (13, 22). These enzymes are present in dormant spores but play no role in dormant spore resistance to peroxides, undoubtedly because of the lack of enzyme activity in the dormant spore due to the spore core's low level of hydration (2, 13, 116).

The target of spore killing by peroxides it is not known, but it is not spore DNA as peroxide killing of spores causes neither mutagenesis nor DNA damage, and mutation of DNA repair genes does not decrease spore peroxide resistance (102, 104, 106). Peroxide killing of spores is associated with and accompanied by inactivation of spore core enzymes (84). However, it is not yet known if this is the cause of spore killing or only an effect of spore killing by damage to some other spore molecule, possibly the spore's inner membrane.

It appears likely that two factors in spore peroxide resistance are the presence of spore coats and the impermeability of spores, in particular the inner spore membrane (4, 8, 36). A third factor is the relative dehydration of the spore core, as spore hydrogen peroxide resistance decreases as spore core hydration increases (89); presumably low core hydration slows oxidative or other chemical reactions in the core. A fourth factor is spore core mineralization, as increased core levels of Ca^{2+} are associated with increased peroxide resistance (59); however, the mechanism of this effect is not known. The fifth factor is the protection of spore DNA from oxidative damage by the binding of α/β-type SASP; $\alpha^-\beta^-$ spores exhibit increased peroxide sensitivity, a high level of mutants in survivors of killing by peroxides, and DNA damage in spores treated with peroxides (102, 104, 108). However, the precise nature of this DNA damage is not known. The binding of α/β-type SASP also protects DNA from peroxide-induced damage in vitro (102, 104, 110). The methionine residues in α/β-type SASP are extremely sensitive to oxidation in vitro, as these proteins are essentially unstructured in solution, and oxidation of one highly conserved methionine abolishes binding of α/β-type SASP to DNA (43; Hayes and Setlow, unpublished data; Mohr and Setlow, unpublished data). However, the binding of α/β-type SASP to DNA in vitro results in protection of crucial methionine residues against oxidation, and these residues are also not oxidized in spores (43). Consequently, it appears that the protection against oxidative damage in an α/β-type SASP/DNA complex is mutual.

Spores are also resistant to hypochlorite (7, 94). As with peroxide resistance, spore hypochlorite resistance appears due in part to the spore coats, possibly in part to the impermeability of spores to this hydrophilic chemical, and certainly to the protection of spore DNA from damage by the binding of α/β-type SASP (7, 8, 94).

Formaldehyde

Spores are much more resistant to formaldehyde and other aldehydes than are growing cells (49, 54, 95), and it is likely that spore impermeability again plays a role in formaldehyde resistance. As with peroxides, the target for spore killing by formaldehyde is not DNA, and again the protection of spore DNA against reaction with formaldehyde is due to the saturation of the DNA with α/β-type SASP (54).

Alkylating Agents

Spores are much more resistant than growing cells to alkylating agents, including ethylene oxide, which is used for some sterilization applications (93, 111). Spore core impermeability to these hydrophilic chemicals appears to play a significant role in spore resistance to alkylating agents (93). However, in contrast to formaldehyde and peroxides, the target for spore killing by alkylating agents is spore DNA, as survivors of ethylmethane sulfonate (EMS) treatment of spores have a high level of mutations and DNA damage consistent with DNA base alkylation, and a DNA repair mutation decreases spore EMS resistance (111). Also in contrast to results with peroxides and formaldehyde, $\alpha^- \beta^-$ spores exhibit identical EMS resistance to wild-type spores, indicating that α/β-type SASP binding does not protect spore DNA against alkylation (111). Indeed, α/β-type SASP binding does not protect DNA from base alkylation by alkylating agents in vitro (110).

Other Resistance

In addition to the resistance properties discussed above, there are also a number of other spore resistance properties whose mechanisms are poorly understood, if at all. These include the following: (i) resistance to disruption, which is likely due to the mechanical rigidity of the spore coat structures (23, 24); (ii) resistance to osmotic stress, likely due to the impermeability of the spore's inner membrane and the maintenance of spore core volume by the spore cortex; (iii) resistance to extremely high pressures (40, 41, 75, 131); (iv) resistance to multiple cycles of freezing and thawing (31, 92); and (v) resistance to predation, as spores have been reported to survive passage through food vacuoles of protozoa (58); presumably this is largely because the spore is resistant to enzymes due to the presence of the spore coat (23, 24).

SPORES OF MYXOBACTERIA AND *STREPTOMYCES* SPECIES

Both gram-positive *Streptomyces* species and many gram-negative myxobacteria (most notably *Myxococcus xanthus*) can also form spores (14–16, 26–29; see also chapter 13). While gram-positive, *Streptomyces* are not particularly closely related to *Bacillus*/*Clostridium* species, and neither myxobacteria nor *Streptomyces* species form endospores. Rather, cells of these latter genera differentiate into round spores in either fruiting body formation (myxobacteria) or in an aerial mycelium (*Streptomyces*). However, conditions have been developed for obtaining spores of both types of organisms in submerged culture (20, 38, 47, 50, 52). The spores of myxobacteria and *Streptomyces* species differ from those of *Bacillus*/*Clostridium* species in structure, but even more so in physiological properties. In contrast to spores of *Bacillus*/*Clostridium* species, which have no detectable metabolism or ATP, spores of *M. xan-*

thus and *Streptomyces* species are only metabolically quiescent, with significant levels of ATP, overall respiration rates 10 to 15% those of vegetative cells, and significant metabolism of endogenous compounds (14, 26, 28, 29, 65, 118). DPA is also not present in spores of *M. xanthus* or *Steptomyces* species.

While there are often significant differences in resistance of spores prepared in submerged culture versus those prepared in fruiting bodies or aerial mycelia, spores of *M. xanthus* and various *Streptomyces* species are significantly more resistant than growing cells of these organisms to mild (55 to 65°C) heat, UV radiation, desiccation, lysozyme, and mechanical disruption (Table 1) (26, 29, 38, 47, 52, 120). There are also reports that spores are more resistant than growing cells to some chemicals (e.g., sodium dodecylsulfate) but not others (28, 52, 130). For the most part (but see below), the reasons for the increased spore resistance to these various treatments are not known. However, it seems likely that these spores' proteinaceous coat structures, which at least in some cases are significantly mineralized, are responsible for resistance to mechanical disruption and lysozyme, as they are in spores of *Bacillus* and *Clostridium* species (83, 96, 129). In addition, it has been suggested, but not proven, that pigments accumulated by spores of *M. xanthus* contribute to their UV resistance (120). It does not appear that spores of *M. xanthus* and *Streptomyces* species are particularly dehydrated, since they continue significant metabolism, and this is supported by at least one measurement of *Streptomyces* spore hydration levels (90). Thus spore dehydration seems unlikely to be a major cause of these spores' resistance properties.

One factor that does seem likely to be important in the resistance of spores of *M. xanthus* and *Streptomyces* species to both mild heat and desiccation is the accumulation of large amounts of the disaccharide trehalose, routinely 25 to 50% of the dry weight of spores prepared by fruiting body formation or on aerial mycelia (19, 64–66). A variety of in vitro studies have shown that trehalose is extremely effective in protecting proteins from denaturation and membranes from disruption, in particular upon desiccation (18). Since the trehalose content of spores of *Streptomyces* species can be altered significantly by modification of culture conditions or media, this has allowed assessment of the effects of spore trehalose levels on spore resistance (19, 64, 65). Several studies have thus shown that an increase in spore trehalose content correlates with an increase in spore desiccation resistance, which is certainly consistent with the effects of trehalose on protein and membrane resistance to desiccation in vitro. However, it is not clear that by manipulating culture conditions to modulate spore trehalose levels that some other spore parameter that modulates spore resistance has not been altered. There is also one study that indicates that spore trehalose levels correlate with increasing resistance to a high temperature but not to lower temperatures (19). Given the advances in the genetics and genomics of *Streptomyces*, it would seem possible to block spore trehalose formation by mutation. This should allow comparison of the resistance properties of spores with high and no trehalose that have been prepared under the same conditions.

FUTURE DIRECTIONS

Although much has been learned about the resistance of spores of various bacterial species, there remain a number of significant unanswered questions. For *Bacillus*/*Clostridium* spores these include the following: (i) How is the core dehydration that is essential for spore resistance to a number of treatments established and maintained? While the spore cortex is almost certainly required for core dehydration, how exactly do changes in cortical structure result in core dehydration? (ii) How much free water is present in the spore core? While the amount of core water per gram wet weight is known for spores of a number of species, the amount of this water that is free, if any, is not known. The knowledge of this parameter may contribute to our detailed understanding of both spore wet heat resistance as well as spore dormancy. (iii) What are the precise changes in the structure of the spore's inner membrane that severely restrict the permeability of this membrane? (iv) What are the factors involved in spore resistance to γ-radiation? Given the increasing use of γ-radiation in food processing/sterilization applications, it is somewhat surprising that there has been so little work analyzing the role of specific spore parameters such as water content, mineralization, and so forth, in spore γ-radiation resistance. (v) What is the state of DPA and divalent cations in the spore core? Despite the identification of DPA in spores more than 40 years ago and the knowledge of the spore's high level of divalent cations, the precise state of these two components is not known. Since DPA is a metal ion chelator, it is supposed that much of the spore's depot of divalent cations is complexed with DPA. However, this has not been proven. In view of the role of core mineralization in spore resistance to several stresses, knowledge of the state of both DPA and divalent cations would undoubtedly increase our knowledge of the roles of these compounds in spore resistance. (vi) Exactly how does α/β-type SASP binding alter spore DNA structure, thus resulting in spore DNA resistance? Whereas the general features of the α/β-type SASP/DNA complex have been elucidated, the specific details of the structure of this complex remain unknown. Also unknown are the roles that other factors (e.g., DPA, HBsu, other proteins, etc.) might play in modulating the structure and thus the properties of the α/β-type SASP/DNA complex in spores.

For spores of myxobacterial and *Streptomyces* species it is obvious that much less is known of the details of resistance of these spores compared to the state of knowledge about resistance of spores of *Bacillus*/*Clostridium* species. This is undoubtedly because of the role for spores of a number of *Bacillus*/*Clostridium* species in food spoilage and food-borne diseases, and the lack of such a role for spores of myxobacterial and *Streptomyces* species. Consequently, a large number of fundamental questions about the latter types of spores remain unanswered. Some of these include the following: (i) What is the state of hydration of these spores? (ii) What is the reason for the metabolic quiescence of these spores? Why is trehalose used so slowly in these dormant spores and much more rapidly upon spore germination? (iii) Does trehalose accumulation fully explain the desiccation resistance of these dormant spores? (iv) What are the factors involved in spore heat resistance besides the spore's trehalose depot? It would seem that detailed studies using some of the techniques proven to be valuable in the study of spores of *Bacillus* species could provide a wealth of new information about spores of *M. xanthus* and *Streptomyces* species.

Acknowledgments. I am grateful for advice from Jan Westpheling, Kieth Chater, and Larry Shimkets.

Work in my laboratory has been supported by grants from the Army Research Office and the National Institutes of Health (GM-19698).

REFERENCES

1. **Atrih, A., P. Zollner, G. Allmaier, and S. F. Foster.** 1996. Structural analysis of *Bacillus subtilis* 168 endospore peptidoglycan and its role during differentiation. *J. Bacteriol.* **178:**6173–6183.
2. **Bagyan, I., L. Casillas-Martinez, and P. Setlow.** 1998. The *katX* gene which codes for the catalase in spores of *Bacillus subtilis* is a forespore specific gene controlled by σ^F, and KatX is essential for hydrogen peroxide resistance of the germinating spore. *J. Bacteriol.* **180:**2057–2062.
3. **Bagyan, I., B. Setlow, and P. Setlow.** 1998. New small, acid soluble proteins unique to spores of *Bacillus subtilis:* identification of the coding genes and studies of the regulation and function of two of these genes. *J. Bacteriol.* **180:**6704–6712.
4. **Bayliss, C. E., and W. M. Waites.** 1976. The effect of hydrogen peroxide on spores of *Clostridium bifermentans. J. Gen. Microbiol.* **96:**401–407.
5. **Beaman, T. C., and P. Gerhardt.** 1986. Heat resistance of bacterial spores correlated with protoplast dehydration, min-

eralization, and thermal adaptation. *Appl. Environ. Microbiol.* **52:**1242–1246.

6. **Belliveau, B. H., T. C. Beaman, S. Pankratz, and P. Gerhardt.** 1992. Heat killing of bacterial spores analyzed by differential scanning calorimetry. *J. Bacteriol.* **174:**4463–4474.

7. **Bloomfield, S. F., and M. Arthur.** 1992. Interaction of *Bacillus subtilis* spores with sodium hypochlorite, sodium dichloroisocyanurate and chloramine-T. *J. Appl. Bacteriol.* **72:**166–172.

8. **Bloomfield, S. F., and M. Arthur.** 1994. Mechanisms of inactivation and resistance of spores to chemical biocides. *J. Appl. Bacteriol.* **76:**91S–104S.

9. **Bloomfield, S. F., and R. Megid.** 1994. Interaction of iodine with *Bacillus subtilis* spores and spore formers. *J. Appl. Bacteriol.* **76:**492–499.

10. **Buchanan, C. E., and A. Gustafson.** 1992. Mutagenesis and mapping of the gene for a sporulation-specific penicillin-binding protein in *Bacillus subtilis*. *J. Bacteriol.* **174:**5430–5435.

11. **Buchanan, C. E., A. O. Henriques, and P. J. Piggot.** 1994. Cell wall changes during bacterial endospore formation, p. 167–186. *In* J.-M. Ghuysen and R. Hackenbeck (ed.), *Bacterial Cell Wall*. Elsevier, Amsterdam.

12. **Cano, R. J., and M. K. Borucki.** 1995. Revival and identification of bacterial spores in 25- to 40-million-year-old Dominican amber. *Science* **268:**1060–1064.

13. **Casillas-Martinez, L., and P. Setlow.** 1997. Alkyl hydroperoxide reductase, catalase, MrgA, and superoxide dismutase are not involved in resistance of *Bacillus subtilis* spores to heat or oxidizing agents. *J. Bacteriol.* **179:**7420–7425.

14. **Champness, W.** 2000. Actinomycete development, antibiotic production, and phylogeny: questions and challenges, p. 11–31. *In* Y. V. Brun and L. J. Shimkets (ed.), *Prokaryotic Development*. American Society for Microbiology, Washington, D.C.

15. **Chater, K. F.** 1984. Morphological and physiological differentiation in *Streptomyces*, p. 89–115. *In* R. Losick and L. Shapiro (ed.), *Microbial Development*. Cold Spring Harbor Laboratory, Cold Spring Harbor, N.Y.

16. **Chater, K. F.** 1991. Saps, hydrophobins and aerial growth. *Curr. Biol.* **1:**318–320.

17. **Chiasson, L. P., and S. Zamenhof.** 1966. Studies on induction of mutations by heat in spores of *Bacillus subtilis*. *Can. J. Microbiol.* **12:**43–46.

18. **Crowe, J. H., J. F. Carpenter, and L. M. Crowe.** 1998. The role of vitrification in anhydrobiosis. *Annu. Rev. Physiol.* **60:**73–103.

19. **Cruz-Martin, M., A. Diaz, M. B. Manzanal, and C. Hardisson.** 1986. Role of trehalose in the spores of *Streptomyces*. *FEMS Microbiol. Lett.* **35:**49–54.

20. **Daza, A., J. F. Martin, A. Dominguez, and J. A. Gil.** 1989. Sporulation of several species of *Streptomyces* in submerged cultures after nutritional downshift. *J. Gen. Microbiol.* **135:**2483–2491.

21. **Donnellan, J. E., Jr., and R. B. Setlow.** 1965. Thymine photoproducts but not thymine dimers are found in ultraviolet irradiated bacterial spores. *Science* **149:**308–310.

22. **Dowds, B. C.** 1994. The oxidative stress response in *Bacillus subtilis*. *FEMS Microbiol. Lett.* **124:**255–264.

23. **Driks, A.** 1999. The *Bacillus subtilis* spore coat. *Microbiol. Mol. Biol. Rev.* **63:**1–20.

24. **Driks, A., and P. Setlow.** 2000. Morphogenesis and properties of the bacterial spore, p. 191–218. *In* Y. V. Brun and L. J. Shimkets (ed.), *Prokaryotic Development*. American Society for Microbiology, Washington, D.C.

25. **Drzewiecki, K., C. Eymann, G. Mittenhuber, and M. Hecker.** 1998. The *yvyD* gene of *Bacillus subtilis* is under dual control of σ^B and σ^H. *J. Bacteriol.* **180:**6674–6680.

26. **Dworkin, M.** 1996. Recent advances in the social and developmental biology of the myxobacteria. *Microbiol. Rev.* **60:**70–102.

27. **Dworkin, M.** 2000. Introduction to the myxobacteria p. 221–242. *In* Y. V. Brun and L. J. Shimkets (ed.), *Prokaryotic Development*. American Society for Microbiology, Washington, D.C.

28. **Elias, M., and F. J. Murillo.** 1991. Induction of germination in *Myxococcus xanthus* fruiting body spores. *J. Gen. Microbiol.* **137:**381–388.

29. **Ensign, J. C.** 1978. Formation, properties and germination of Actinomycetes spores. *Annu. Rev. Microbiol.* **32:**185–219.

30. **Fairhead, H., B. Setlow, and P. Setlow.** 1993. Prevention of DNA damage in spores and in vitro by small, acid-soluble proteins from *Bacillus* species. *J. Bacteriol.* **175:**1367–1374.

31. **Fairhead, H., B. Setlow, W. M. Waites, and P. Setlow.** 1994. Small, acid-soluble proteins bound to DNA protect *Bacillus subtilis* spores from killing by freeze-drying. *Appl. Environ. Microbiol.* **60:**2647–2649.

32. **Fairhead, H., and P. Setlow.** 1992. Binding of DNA to α/β-type small, acid-soluble proteins from spores of *Bacillus* or *Clostridium* species prevents formation of cytosine dimers, cytosine-thymine dimers and dipyrimidine photoadducts upon ultraviolet irradiation. *J. Bacteriol.* **174:**2874–2880.

33. **Fajardo-Cavazos, P., C. Salazar, and W. L. Nicholson.** 1993. Molecular cloning and characterization of the *Bacillus subtilis* spore photoproduct lyase (*spl*) gene, which is involved in the repair of UV radiation-induced DNA damage during spore germination. *J. Bacteriol.* **175:**1735–1744.

34. **Francesconi, S. C., T. J. MacAlister, B. Setlow, and P. Setlow.** 1988. Immunoelectron microscopic localization of small, acid-soluble spore proteins in sporulating cells of *Bacillus subtilis*. *J. Bacteriol.* **170:**5963–5967.

35. **Gerhardt, P., and R. E. Marquis.** 1989. Spore thermoresistance mechanisms, p. 43–63. *In* I. Smith, R. A. Slepecky, and P. Setlow (ed.), *Regulation of Prokaryotic Development*. American Society for Microbiology, Washington, D.C.

36. **Gerhardt, P., R. Scherrer, and S. H. Black.** 1972. Molecular sieving by dormant spore structures, p. 68–74. *In* H. O. Halvorson, R. Hanson, and L. L. Campbell (ed.). *Spores V*. American Society for Microbiology, Washington, D.C.

37. **Gest, H., and J. Mandelstam.** 1987. Longevity of microorganisms in natural environments. *Microbiol. Sci.* **4:**69–71.

38. **Glazebrook, M. A., J. L. Doull, C. Stuttard, and L. C. Vining.** 1990. Sporulation of *Streptomyces venezuelae* in submerged cultures. *J. Gen. Microbiol.* **136:**581–588.

39. **Gomez, A., and S. M. Cutting.** 1997. Identification of a new σ^B-controlled gene, *asbX*, in *Bacillus subtilis*. *Gene* **188:**29–33.

40. **Gould, G. W.** 1983. Mechanisms of resistance and dormancy, p. 397–444. *In* G. W. Gould and A. Hurst (ed.), *The Bacterial Spore*, vol. 2. Academic Press, London.

41. **Gould, G. W., and A. J. H. Sole.** 1972. Role of pressure in the stabilization and destabilization of bacterial spores. *Symp. Soc. Exp. Biol.* **26:**147–147.

42. **Griffith, J., A. Makhov, L. Santiago-Lara, and P. Setlow.** 1994. Electron microscopic studies of the interaction between a *Bacillus* α/β-type small, acid-soluble spore protein with DNA: protein binding is cooperative, stiffens the DNA and induces negative supercoiling. *Proc. Natl. Acad. Sci. USA.* **91:**8224–8228.

43. **Hayes, C. S., B. Illades-Aguiar, L. Casillas-Martinez, and P. Setlow.** 1998. In vitro and in vivo oxidation of methionine residues in small, acid-soluble spore proteins from *Bacillus* species. *J. Bacteriol.* **180:**2694–2700.

44. Hayes, C. S., and P. Setlow. 1997. Analysis of deamidation of small acid-soluble spore proteins from *Bacillus subtilis* in vitro and in vivo. *J. Bacteriol.* **179**:6020–6027.

45. Imae, Y., and J. L. Strominger. 1976. Relationship between cortex content and properties of *Bacillus sphaericus* spores. *J. Bacteriol.* **126**:907–913.

46. Johnstone, K. 1994. The trigger mechanism of spore germination. *J. Appl. Bacteriol.* **76**:17S–24S.

47. Kendrick, K. E., and J. C. Ensign. 1983. Sporulation of *Streptomyces griseus* in submerged culture. *J. Bacteriol.* **155**:357–336.

48. Kennedy, M. J., S. L. Reader, and L. M. Swierczynski. 1994. Preservation records of micro-organisms: evidence of the tenacity of life. *Microbiol.* **140**:2513–2529.

49. Knott, A. G., A. D. Russell, and B. N. Dancer. 1995. Development of resistance to biocides during sporulation of *Bacillus subtilis*. *J. Appl. Bacteriol.* **79**:492–498.

50. Koepsel, R., and J. C. Ensign. 1984. Microcycle sporulation of *Streptomyces viridochromogenes*. *Arch. Microbiol.* **140**:9–14.

51. Kunst, F., N. Ogasawara, I. Moszer, A. M. Albertini, G. Alloni, V. Azevedo, M. G. Bertero, P. Bessieres, A. Bolotin, S. Borchert, R. Borriss, L. Boursier, A. Brans, M. Braun, S. C. Brignell, S. Bron, S. Brouillet, C. V. Bruschi, B. Caldwell, V. Capuano, N. M. Carter, S. K. Choi, J. J. Codani, I. F. Connerton, and A. Danchin, et al. 1997. The complete genome sequence of the gram-positive bacterium *Bacillus subtilis*. *Nature* **390**:249–256.

52. Lee, K. J., and Y. T. Rho. 1993. Characteristics of spores formed by surface and submerged cultures of *Streptomyces albidoflavus* SMF301. *J. Gen. Microbiol.* **139**:3131–3137.

53. Lindberg, C., and G. Horneck. 1991. Action spectra for survival and spore photoproduct formation of *Bacillus subtilis* irradiated with short-wavelength (200–300 nm) UV at atmosphere pressure and in vacuo. *J. Photochem. Photobiol. B. Biol.* **11**:69–80.

54. Loshon, C. A., P. C. Genest, B. Setlow, and P. Setlow. 1999. Formaldehyde kills spores of *Bacillus subtilis* by DNA damage, and small, acid-soluble spore proteins of the α/β-type protect spores against this DNA damage. *J. Appl. Microbiol.* **87**:8–14.

55. Loshon, C. A., and P. Setlow. 1993. Levels of small molecules in dormant spores of *Sporosarcina* species and comparison with levels in spores of *Bacillus* and *Clostridium* species. *Can. J. Microbiol.* **39**:259–262.

56. Magill, N. G., A. E. Cowan, D. E. Koppel, and P. Setlow. 1994. The internal pH of the forespore compartment of *Bacillus megaterium* decreases by about 1 pH unit during sporulation. *J. Bacteriol.* **176**:2252–2258.

57. Magill, N. G., A. E. Cowan, M. A. Leyva-Vazquez, M. Brown, D. E. Koppel, and P. Setlow. 1996. Analysis of the relationship between the decrease in pH and accumulation of 3-phosphoglyceric acid in developing forespores of *Bacillus* species. *J. Bacteriol.* **178**:2204–2210.

58. Manasherob, R., E. Ben-Dov, A. Zaritsky, and Z. Barak. 1998. Germination, growth and sporulation of *Bacillus thuringiensis* subsp. *israelensis* in excreted food vacuoles of the protozoan *Tetrahymena pyriformis*. *Appl. Environ. Microbiol.* **64**:1750–1758.

59. Marquis, R. E., and S. Y. Shin. 1994. Mineralization and responses of bacterial spores to heat and oxidative agents. *FEMS Microbiol. Rev.* **14**:375–380.

60. Marquis, R. E., J. Sim, and S. Y. Shin. 1994. Molecular mechanisms of resistance to heat and oxidative damage. *J. Appl. Bacteriol.* **76**:40S–48S.

61. Mason, J. M., R. H. Hackett, and P. Setlow. 1988. Studies on the regulation of expression of genes coding for small, acid-soluble proteins of *Bacillus subtilis* spores using *lacZ* gene fusions. *J. Bacteriol.* **170**:239–244.

62. Mason, J. M., and P. Setlow. 1986. Evidence for an essential role for small, acid-soluble, spore proteins in the resistance of *Bacillus subtilis* spores to ultraviolet light. *J. Bacteriol.* **167**:174–178.

63. Mason, J. M., and P. Setlow. 1987. Different small, acid-soluble proteins of the α/β-type have interchangeable roles in the heat and ultraviolet radiation resistance of *Bacillus subtilis* spores. *J. Bacteriol.* **169**:3633–3637.

64. McBride, M. J., and J. C. Ensign. 1987. Effects of intracellular trehalose content on *Streptomyces griseus* spores. *J. Bacteriol.* **169**:4995–5001.

65. McBride, M. J., and J. C. Ensign. 1987. Metabolism of endogenous trehalose by *Streptomyces griseus* spores and by spores or cells of other Actinomycetes. *J. Bacteriol.* **169**:5002–5007.

66. McBride, M. J., and D. R. Zusman. 1989. Trehalose accumulation in vegetative cells and spores of *Myxococcus xanthus*. *J. Bacteriol.* **171**:6383–6386.

67. Micka, B., N. Groch, U. Heinemann, and M. A. Marahiel. 1991. Molecular cloning, nucleotide sequence and characterization of the *Bacillus subtilis* gene encoding the DNA binding protein HBsu. *J. Bacteriol.* **173**:3191–3198.

68. Micka, B., and M. A. Marahiel. 1992. The DNA-binding protein HBsu is essential for normal growth and development in *Bacillus subtilis*. *Biochimie* **74**:641–650.

69. Milhaud, P., and G. Balassa. 1973. Biochemical genetics of bacterial sporulation: IV Sequential development of resistance to chemical and physical agents during sporulation of *Bacillus subtilis*. *Mol. Gen. Genet.* **125**:241–250.

70. Mohr, S. C., N. V. H. A. Sokolov, C. He, and P. Setlow. 1991. Binding of small acid-soluble spore proteins from *Bacillus subtilis* changes the conformation of DNA from B to A. *Proc. Natl. Acad. Sci. USA* **88**:77–81.

71. Moir, A., and D. A. Smith. 1990. The genetics of bacterial spore germination. *Annu. Rev. Microbiol.* **44**:531–553.

72. Munakata, N., and C. S. Rupert. 1972. Genetically controlled removal of "spore photoproduct" from deoxyribonucleic acid of ultraviolet-irradiated *Bacillus subtilis* spores. *Mol. Gen. Genet.* **104**:258–263.

73. Munakata, N., M. Saiton, N. Takahashi, K. Hieda, and F. Morihoshi. 1997. Induction of unique tandem-base change mutations in bacterial spores exposed to extreme dryness. *Mutat. Res.* **390**:189–195.

74. Murrell, W. G. 1967. The biochemistry of the bacterial endospore. *Adv. Microbial Physiol.* **1**:133–251.

75. Nakayama, A., Y. Yano, S. Kobayashi, M. Ishikawa, and K. Sakai. 1996. Comparison of pressure resistance of spores of six *Bacillus* strains with their heat resistance. *Appl. Environ. Microbiol.* **62**:3897–3900.

76. Nicholson, W. L., and P. Fajardo-Cavazos. 1997. DNA repair and the ultraviolet radiation resistance of bacterial spores: from the laboratory to the environment. *Recent Res. Devel. Microbiol.* **1**:125–140.

77. Nicholson, W. L., B. Setlow, and P. Setlow. 1990. Binding of DNA in vitro by a small, acid-soluble spore protein and its effect on DNA topology. *J. Bacteriol.* **172**:6900–6096.

78. Nicholson, W. L., B. Setlow, and P. Setlow. 1991. Ultraviolet irradiation of DNA complexed with α/β-type small, acid-soluble proteins from spores of *Bacillus* or *Clostridium* species makes spore photoproduct but not thymine dimers. *Proc. Natl. Acad. Sci. USA* **88**:8288–8292.

79. Nicholson, W. L., and P. Setlow. 1990. Dramatic increase in the negative superhelicity of plasmid DNA in the forespore

compartment of sporulating cells of *Bacillus subtilis*. *J. Bacteriol.* **172:**7–14.

80. **Nicholson, W. L., D. Sun, B. Setlow, and P. Setlow.** 1989. Promoter specificity of sigma-G-containing RNA polymerase from sporulating cells of *Bacillus subtilis*: identification of a group of forespore-specific promoters. *J. Bacteriol.* **171:**2708–2718.

81. **Nishihara, T., Y. Takubo, E. Kawamata, T. Koshikawa, J. Ogaki, and M. Kondo.** 1989. Role of outer coat in resistance of *Bacillus megaterium* spore. *J. Biochem.* **106:**270–273.

82. **Northrop, J., and R. A. Slepecky.** 1967. Sporulation mutations induced by heat in *Bacillus subtilis*. *Science* **155:**838–839.

83. **Otani, M., S. Kozuka, C. Xu, C. Umezawaa, K. Sano, and S. Inouye.** 1998. Protein W, a spore-specific protein in *Myxococcus xanthus*, formation of a large electron dense particle in a spore. *Mol. Microbiol.* **30:**57–66.

84. **Palop, A., G. C. Rutherford, and R. E. Marquis.** 1996. Hydroperoxide inactivation of enzymes within spores of *Bacillus megaterium* ATCC19213. *FEMS Microbiol. Lett.* **142:**283–287.

85. **Pogliano, K., E. Harry, and R. Losick.** 1995. Visualization of the subcellular location of sporulation proteins in *Bacillus subtilis* using immunofluorescence microscopy. *Mol. Microbiol.* **18:**459–470.

86. **Poinar, H. N., M. Hoss, J. L. Bada, and S. Paabo.** 1996. Amino acid racemization and the preservation record of ancient DNA. *Science* **272:**864–866.

87. **Popham, D. L., J. Helin, C. E. Costello, and P. Setlow.** 1996. Analysis of the peptidoglycan structure of *Bacillus subtilis* endospores. *J. Bacteriol.* **178:**6451–6458.

88. **Popham, D. L., B. Illades-Aguiar, and P. Setlow.** 1995. The *Bacillus subtilis dacB* gene, encoding penicillin-binding protein 5*, is part of a three-gene operon required for proper spore cortex synthesis and spore core dehydration. *J. Bacteriol.* **177:**4721–4729.

89. **Popham, D. L., S. Sengupta, and P. Setlow.** 1995. Heat, hydrogen peroxide, and UV resistance of *Bacillus subtilis* spores with increased core water content and with or without major DNA binding proteins. *Appl. Environ. Microbiol.* **61:**3633–3638.

90. **Quiros, L. M., and J. A. Salas.** 1996. Intracellular water volume and intracellular pH of *Streptomyces antibioticus* spores. *FEMS Microbiol. Lett.* **141:**245–249.

91. **Rebeil, R., Y. Sun, L. Chooback, M. Pedraza-Reyes, C. Kinsland, T. P. Begley, and W. L. Nicholson.** 1998. Spore photoproduct lyase from *Bacillus subtilis* spores is a novel iron-sulfur DNA repair enzyme which shares features with proteins such as class III anaerobic ribonucleotide reductases and pyruvate-formate lyases. *J. Bacteriol.* **180:**4879–4885.

92. **Roberts, T. A., and A. D. Hitchins.** 1969. Resistance of spores, p. 611–670. *In* G. W. Gould and A. Hurst (ed.), *The Bacterial Spore*. Academic Press, New York, N.Y.

93. **Russell, A. D.** 1982. *The Destruction of Bacterial Spores*. Academic Press, London, United Kingdom.

94. **Sabli, M. Z. H., P. Setlow, and W. M. Waites.** 1996. The effect of hypochlorite on spores of *Bacillus subtilis* without small acid soluble proteins. *Lett. Appl. Microbiol.* **22:**405–407.

95. **Sagripanti, J.-L., and A. Bonifacio.** 1996. Comparative sporicidal effects of liquid chemical agents. *Appl. Environ. Microbiol.* **62:**545–551.

96. **Salas, J. A., A. Guijarro, and C. Hardisson.** 1983. High calcium content in *Streptomyces* spores and its release as an early event in spore germination. *J. Bacteriol.* **155:**1316–1323.

97. **Sanchez-Salas, J.-L., M. L. Santiago-Lara, B. Setlow, M. D. Sussman, and P. Setlow.** 1992. Properties of mutants of *Bacillus megaterium* and *Bacillus subtilis* which lack the protease that degrades small, acid-soluble proteins during spore germination. *J. Bacteriol.* **174:**807–814.

98. **Scherrer, R., T. C. Beaman, and P. Gerhardt.** 1971. Macromolecular sieving by the dormant spore of *Bacillus cereus*. J. Bacteriol. **108:**868–873.

99. **Sedlak, M., V. Vinter, J. Adames, J. Vohradsky, Z. Voburka, and J. Chaloupka.** 1993. Heat shock applied early in sporulation affects heat resistance of *Bacillus megaterium* spores. *J. Bacteriol.* **175:**8049–8052.

100. **Setlow, B., R. H. Hackett, and P. Setlow.** 1982. Noninvolvement of the spore cortex in acquisition of low-molecular-weight proteins and UV light resistance during *Bacillus sphaericus* sporulation. *J. Bacteriol.* **149:**494–498.

101. **Setlow, B., A. R. Hand, and P. Setlow.** 1991. Synthesis of a *Bacillus subtilis* small, acid-soluble spore protein in *Escherichia coli* causes cell DNA to assume some characteristics of spore DNA. *J. Bacteriol.* **173:**1642–1653.

102. **Setlow, B., C. A. Setlow, and P. Setlow.** 1997. Killing bacterial spores by organic hydroperoxides. *J. Indust. Microbiol.* **18:**384–388.

103. **Setlow, B., and P. Setlow.** 1987. Thymine containing dimers as well as spore photoproduct are found in ultraviolet-irradiated *Bacillus subtilis* spores that lack small acid-soluble proteins. *Proc. Natl. Acad. Sci. USA* **84:**421–423.

104. **Setlow, B., and P. Setlow.** 1993. Binding of small, acid-soluble spore proteins to DNA plays a significant role in the resistance of *Bacillus subtilis* spores to hydrogen peroxide. *Appl. Environ. Microbiol.* **59:**3418–3423.

105. **Setlow, B., and P. Setlow.** 1994. Heat inactivation of *Bacillus subtilis* spores lacking small, acid-soluble spore proteins is accompanied by generation of abasic sites in spore DNA. *J. Bacteriol.* **176:**2111–2113.

106. **Setlow, B., and P. Setlow.** 1995. Binding to DNA protects α/β-type small, acid-soluble spore proteins of *Bacillus* and *Clostridium* species against digestion by their specific protease as well as other proteases. *J. Bacteriol.* **177:**4149–4151.

107. **Setlow, B., and P. Setlow.** 1995. Small, acid-soluble proteins bound to DNA protect *Bacillus subtilis* spores from killing by dry heat. *Appl. Environ. Microbiol.* **61:**2787–2790.

108. **Setlow, B., and P. Setlow.** 1995. Role of DNA repair in *Bacillus subtilis* spore resistance. *J. Bacteriol.* **178:**3486–3495.

109. **Setlow, B., and P. Setlow.** 1998. Heat killing of *Bacillus subtilis* spores in water is not due to oxidative damage. *Appl. Environ. Microbiol.* **64:**4109–4112.

110. **Setlow, B., D. Sun, and P. Setlow.** 1992. Studies of the interaction between DNA and α/β-type small, acid-soluble spore proteins: a new class of DNA binding protein. *J. Bacteriol.* **174:**2312–2322.

111. **Setlow, B., K. J. Tautvydas, and P. Setlow.** 1998. Small, acid-soluble spore proteins of the α/β-type do not protect the DNA in *Bacillus subtilis* spores against base alkylation. *Appl. Environ. Microbiol.* **64:**1958–1962.

112. **Setlow, P.** 1983. Germination and outgrowth, p. 211–254. *In* A. Hurst and G. W. Gould (ed.), *The Bacterial Spore*, vol. 2. Academic Press, London.

113. **Setlow, P.** 1988. Small acid-soluble, spore proteins of *Bacillus* species: structure, synthesis, genetics, function and degradation. *Annu. Rev. Microbiol.* **42:**319–338.

114. **Setlow, P.** 1992. DNA in dormant spores of *Bacillus* species is in an A-like conformation. *Molec. Microbiol.* **6:**563–567.

115. **Setlow, P.** 1992. I will survive: protecting and repairing spore DNA. *J. Bacteriol.* **174:**2737–2741.

116. **Setlow, P.** 1994. Mechanisms which contribute to the long-term survival of spores of *Bacillus* species. *J. Appl. Bacteriol.* **76:**49S–60S.

117. Setlow, P. 1995. Mechanisms for the prevention of damage to the DNA in spores of *Bacillus* species. *Annu. Rev. Microbiol.* **49:**29–54.

118. Smith, B. A., and M. Dworkin. 1980. Adenylate energy charge during fruiting body formation by *Myxococcus xanthus. J. Bacteriol.* **142:**1007–1009.

119. Sneath, P. H. A. 1962. Longevity of micro-organisms. *Nature* **195:**643–646.

120. Sudo, S. Z., and M. Dworkin. 1969. Resistance of vegetative cells and microcysts of *Myxococcus xanthus. J. Bacteriol.* **98:** 883–887.

121. Sun, D., P. Stragier, and P. Setlow. 1989. Identification of a new σ-factor involved in compartmentalized gene expression during sporulation of *Bacillus subtilis. Genes Dev.* **3:**141–149.

122. Tovar-Rojo, F., and P. Setlow. 1991. Analysis of the effects of mutant small, acid-soluble spore proteins from *Bacillus subtilis* on DNA in vivo and in vitro. *J. Bacteriol.* **173:**4827–4835.

123. Varon, D., M. S. Brody, and C. W. Price. 1996. *Bacillus subtilis* operon under the dual control of the general stress transcription factor σ^B and the sporulation transcription factor σ^F. *Mol. Microbiol.* **20:**339–350.

124. Waites, W. M. 1985. Inactivation of spores with chemical agents, p. 383–396. *In* G. J. Dring, D. J. Ellar, and G. W. Gould (ed.), *Fundamental and Applied Aspects of Bacterial Spores.* Academic Press, London.

125. Warth, A. D. 1980. Heat stability of *Bacillus cereus* enzymes within spores and in extracts. *J. Bacteriol.* **143:**27–34.

126. Warth, A. D. 1985. Mechanisms of heat resistance, p. 209–225. *In* G. J. Dring, D. J. Ellar, and G. W. Gould (ed.), *Fundamental and Applied Aspects of Bacterial Spores.* Academic Press, London.

127. Warth, A. D., and J. L. Strominger. 1969. Structure of the peptidoglycan of bacterial spores: occurrence of the lactam of muramic acid. *Proc. Natl. Acad. Sci. USA* **64:**528–535.

128. Warth, A. D., and J. L. Strominger. 1972. Structure of the peptidoglycan from spores of *Bacillus subtilis. Biochemistry* **11:**1389–1396.

129. Wenk, M., and E. M. Mayr. 1998. *Myxococcus xanthus* spore coat protein S, a stress-induced member of the β-, γ-crystallin superfamily, gains stability from binding of calcium ions. *Eur. J. Biochem.* **255:**604–610.

130. Whitmore, T. N., and S. Denny. 1992. The effect of disinfectants on a geosmin-producing strain of *Streptomyces griseus. J. Appl. Bacteriol.* **72:**160–165.

131. Wuytack, E. Y., S. Boven, and C. W. Michiels. Comparative study of pressure-induced germination of *Bacillus subtilis* spore at low and high pressures. *Appl. Environ. Microbiol.* **64:**3220–3224.

132. Xue, Y., and W. L. Nicholson. 1996. The two major spore DNA repair pathways, nucleotide excision repair and spore photoproduct lyase, are sufficient for the resistance of *Bacillus subtilis* spores to artificial UV-C and UV-B but not to solar radiation. *Appl. Environ. Microbiol.* **62:**2221–2227.

133. Zamenhof, S. 1960. Effects of heating dry bacteria and spores on their phenotype and genotype. *Proc. Natl. Acad. Sci. USA* **46:**101–105.

Bacterial Stress Responses
Edited by G. Storz and R. Hengge-Aronis
©2000 ASM Press, Washington, D.C.

Chapter 15

Long-Term Survival and Evolution in the Stationary Phase

STEVEN E. FINKEL, ERIK R. ZINSER, AND ROBERTO KOLTER

Microbes have evolved diverse strategies to exploit nutrient resources and to cope with environmental stresses, thus ensuring their survival. What are the mechanisms that underlie the evolution of these survival strategies? During the past decade work from several laboratories has begun to directly address this question under a variety of experimental conditions. Our laboratory has focused its attention on the mechanisms of survival during long-term stationary-phase incubation. A major concept that has emerged from these studies is the phenomenon we have termed the growth advantage in stationary phase (GASP) phenotype. When microbial populations die as a consequence of starvation, almost invariably there are some survivors that are mutants better able to exploit the scarce resources made available from the dying cells. It has become clear from studies of stationary-phase cultures that the surviving populations are highly dynamic and that this flux provides a constant source of genetic diversity. Moreover, we are able to observe newly evolving activities that confer increased fitness on cells during times of intense competition. This is a slightly modified view of the concept of cryptic growth initially proposed by Postgate and Hunter (27). Instead of a homogeneous population of survivors feeding off the dying cells, the survivors are fitter mutants, and thus this survival strategy is at the same time an evolutionary strategy.

The bacterial life cycle is often depicted as consisting of three phases: the lag phase, the exponential or logarithmic growth phase, and the stationary phase. The lag phase is a time when bacteria, after a period of starvation, reencounter nutrients and begin to synthesize the metabolic machinery that will allow growth and entry into exponential phase. Exponen-

tial phase is a time when cells find themselves under the rare conditions of balanced growth afforded by an environment where no particular nutrient is limiting. As nutrients are depleted, growth slows and cultures enter stationary phase (23). As traditionally defined, stationary phase begins when the optical density or colony-forming units (CFU) in a bacterial culture cease to increase. Work by many investigators has examined the physiological changes occurring during the transition from exponential phase to the first few hours or days of stationary phase. Our investigations have focused on what happens to bacterial cultures during much longer periods of incubation in stationary phase, lasting days, weeks, or even years (7, 8, 38).

In most natural environments bacteria spend the majority of their existence under conditions of starvation (24). Therefore, knowledge gained from studying starved bacterial populations in the laboratory is likely to provide key insights into survival mechanisms in the "real world." Studies of the entry into stationary phase by nonsporulating bacteria, e.g., *Escherichia coli*, have revealed that such cells display significant morphological and physiological differentiation in response to starvation (16, 17; see also chapter 11). Additional studies of stationary-phase cultures have gone beyond characterization of cellular changes and the molecular mechanisms that underlie the regulation of these changes. Of particular interest are the studies on the population dynamics of cultures under long-term incubation conditions. These experiments have been performed with one of three culture regimens: constant or batch culture, continuous incubation in chemostats, or serial transfer protocols; the latter two are reviewed elsewhere (3, 21). In this chapter we will concentrate only on work that was performed with batch culture.

Steven E. Finkel • Department of Biological Sciences, University of Southern California, Los Angeles, CA 90089. **Erik R. Zinser and Roberto Kolter** • Department of Microbiology and Molecular Genetics, Harvard Medical School, Boston, MA 02115.

A SHORT HISTORY OF LONG-TERM STATIONARY-PHASE INCUBATION EXPERIMENTS

Studies of the long-term survival of bacteria grown in batch culture date back to the beginning of the century (31). Many of these studies have been little noticed since the advent of molecular genetics. It is interesting to note that at least as early as 1935 the standard view of the bacterial life cycle was divided not into the three familiar phases, but actually into five: lag phase, exponential phase, stationary phase, death phase, and the "period of prolonged decrease." This last phase was seen as a period of relatively slow decline in the number of surviving cells over long periods (11).

A study of long-term survival performed during the late 1930s (33) evaluated the survival patterns of more than a dozen different bacterial species, including *E. coli* and *Salmonella* (see Table 1). In this study, the survival kinetics of *Sarcina lutea* and *Serratia marcescens* were determined for cells incubated in nutrient broth cultures for 2 years without the addition of nutrients. Quite strikingly, these cultures retained significant numbers of viable cells even after such prolonged incubation. Using more detailed enumeration of CFUs during the 13th month of incubation, the authors noted fluctuations in viable counts that they attributed to fluctuations in the growth of subpopulations within the culture, alluding to the potential dynamics of the aging populations. Similar survival kinetics have been observed over the decades by investigators who maintained bacterial strains in batch culture, as well as in agar slants or stabs (5, 25). The heterogeneity of colony morphologies recovered from old stabs (see below) also suggests potential dynamics among cultures kept on solid media.

THE GASP PHENOTYPE

While studying long-term survival of *E. coli* in batch culture, it was observed that cells from aged LB cultures, incubated for at least 10 days, would outcompete cells of the same initial strain that had been incubated for only 1 day (see Fig. 1). These cells from aged cultures exhibited what came to be called the growth advantage in stationary phase, or GASP, phenotype (37, 39). The ability of these cells to outcompete their parent is conferred by novel mutations. Aged and parental populations are distinguished using neutral genetic markers such as resistance to the antibiotics nalidixic acid or streptomycin, valine-resistant growth on glucose (20), or ability to grow on β-glucosides (28). It has been shown that GASP-conferring alleles can be transduced into fresh strain backgrounds, and the transductants are able to take over cultures during stationary-phase incubation (see section below, "Genetics of GASP").

Virtually all *E. coli* K-12 derived cell lineages will exhibit GASP takeovers after extended incubation in stationary phase under a variety of growth conditions. In addition, the GASP phenomenon is not restricted to laboratory strains of *E. coli*. Several laboratories have observed the GASP phenotype in a range of microorganisms, including clinical isolates of *E. coli*; other gram-negative bacteria such as *Shigella dysenteriae*, *Enterobacter cloacae*, *Pseudomonas putida*, and *Brucella abortus*; gram-positive bacteria such as *Staphylococcus aureus*, *Enterococcus faecalis*, and *Bacillus globigii*; and the acid-fast bacterium *Mycobacterium smegmatis* (4, 6, 12, 18, 32, 35; E. R. Zinser and R. Kolter, unpublished data). The ability of *E. coli* to express the GASP phenotype was first detected in aerated liquid cultures grown in rich medium. Since then, the GASP phenomenon has been

Table 1. Representative counts of the numbers of bacteria per millilter in aging cultures[a]

| Bacterium | Avg. no. of bacteria/ml | | | pH of culture at end of 1 yr |
	Immediately after inoculation	After 48 h of incubation	After 1 yr of incubation	
Sarcina lutea	6×10^3	1×10^9	5.2×10^6	8.60
Serratia marcescens	8×10^3	2×10^9	4.8×10^6	8.78
Bacillus subtilis	4×10^3	1×10^7	3×10^5	8.92
Staphylococcus aureus	7×10^3	1.8×10^9	1.3×10^6	8.62
Aerobacter aerogenes	5×10^3	1.1×10^9	1.6×10^6	8.75
Escherichia coli	9×10^3	1.7×10^9	3×10^5	8.91
Klebsiella pneumoniae	6×10^3	2.5×10^9	7.8×10^5	8.81
Pseudomonas aeruginosa	4×10^3	1.4×10^9	1.1×10^6	8.87
Rhodococcus rhodochrous	1×10^3	7×10^5	4.3×10^5	7.86
Micrococcus tetragena	2×10^3	3×10^8	2.1×10^6	8.83

[a] Based on data from reference 33.

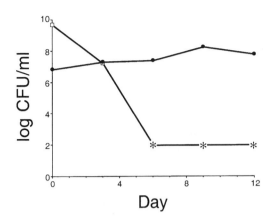

Figure 1. The GASP phenotype. Cells from 10-day-old cultures (filled circles) are mixed as a minority into a culture of 1-day-old cells (open squares). Asterisk indicates that cell numbers are below the level of detection (<100 CFU/ml).

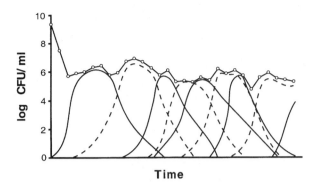

Figure 2. "Waves of takeover" of GASP mutants within a single culture over time. Open circles indicate the total CFU/ml of the culture. Solid and dashed curves model the appearance and subsequent disappearance of different GASP mutant subpopulations.

observed under many other culture conditions, including minimal media supplemented with glucose or amino acid mixtures, anoxic LB, and semisolid media, such as LB agar plates or stabs (36, 40; S. E. Finkel and R. Kolter, unpublished data; Zinser and Kolter, unpublished data).

Our current model of the GASP phenomenon can be summarized as follows: (i) Initially isogenic cells growing in batch culture have equal competitive fitnesses within a given environment. (ii) As these cells grow and multiply within a culture, new mutant alleles spontaneously arise within individual cells. (iii) These new alleles can affect a cell in one of three ways: they can have no effect on relative fitness, be deleterious, or can confer a competitive advantage on the cell. (iv) Cells containing advantageous alleles that permit growth during stationary phase (GASP mutants) will increase their numbers relative to the rest of the population as the culture ages. (v) Cells with a particular GASP phenotype will dominate until they are themselves outcompeted by cells containing additional GASP alleles. This results in "waves of takeover" of more fit GASP competitors over time in a given culture (see Fig. 2). Implicit in this view of the GASP phenomenon is the fact that population takeovers will occur during stationary phase under any incubation condition so long as population sizes are large enough and mutation rates are high enough to permit the appearance of GASP-conferring mutations. However, the incubation condition defines the alleles that can confer GASP, and there is no reason to expect that an allele that is advantageous under one condition will be able to confer GASP under a different stationary-phase incubation condition.

The initial observation that cells from 10-day-old cultures outcompete cells from fresh overnight cultures has been extended, and a more careful ex-

amination of stationary-phase cultures incubated for extended periods indicates that new "waves" of GASP mutants are constantly arising and taking over batch culture populations (7, 8). Cells from 20-day-old cultures will outcompete cells from 10-day-old cultures, and cells from 30-day-old cultures outcompete those from 20-day-old cultures. This growth advantage phenomenon has been observed in cultures aged for as long as 120 days when competed against cells aged up to 90 days (7). However, after about 120 to 150 days of incubation in LB, cells no longer are able to take over fresh overnight LB cultures of the parental strain. We believe that this is due to the fact that after this long period of incubation, the environment under which these cells have been selected, i.e., the environment where they can express their growth advantage, is so different from the environment present in an overnight culture that they no longer have a fitness advantage in the overnight culture conditions. However, this does not mean that cell populations in these older cultures are not dynamic or that new fitter competitors are not being selected under those conditions, taking over the cultures. Quite the contrary, not only do some cells in these cultures remain viable for extraordinarily long periods without the addition of nutrients, but the populations continue to be dynamic (7, 8; Finkel and Kolter, unpublished data).

When cultures incubated for many months are periodically sampled for CFUs on LB agar, differences in colony morphology, referred to as colony morphotypes, are observed (7). While the overall numbers of viable cells (as determined by colony-forming ability) remain roughly constant, from one to seven different colony morphotypes have been observed from a single culture at any given time. For example, a culture that maintained a homogeneous colony morphology for the first 4 months of incubation had three different morphotypes at day 150, four types at day 180, and five types by day 240 (Fig. 3). Assays for genetic markers and the use of DNA fingerprinting techniques (RFLP analysis) confirmed that all cells in the culture were descended from the same parent. We have recorded the type and number of colony morphotypes from four initially isogenic cultures for more than 900 days, and changes continue to be observed (see Fig. 3). The fact that cells of the same colony morphology do not necessarily share the same genotype, thus underrepresenting the extent of genetic diversity, and that the type and number of colony morphotypes are different at each sampling, suggests to us that there is both a considerable amount of turnover and generation of genetic diversity in these long-term stationary-phase cultures.

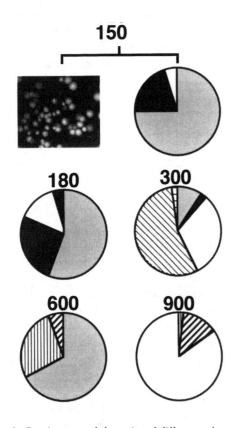

Figure 3. Coexistence and dynamics of different colony morphotypes during long-term incubation observed in a single culture. The photograph shows different colony morphotypes of cells harvested from a 150-day-old culture (7). Note the "opaque" and "mini" colonies as compared to the normal colony phenotype. Pie charts reflect the population composition (percent) of each morphotype over the course of incubation (shown in days.) Each color or pattern represents a different colony morphotype.

Not only is there diversity of GASP mutations within a culture but also between different cultures seeded from a single inoculum. This was determined directly by demonstrating that initially isogenic cells from different 1-month-old cultures behaved differently when competed against each other (Fig. 4). Sometimes, cells from one culture outcompeted cells from one particular parallel culture but lost in competition with another (7). This suggests that different mutations appeared in each culture over time and were selected or outcompeted depending on their relative fitness. That is, a GASP mutant that had a competitive advantage in one culture had no fitness gain when it was competed in another same-aged culture (see Fig. 4; compare Fig. 4B with Fig. 4C). That these mixes had different outcomes despite the fact that conditions were initially identical in all cultures indicates that each culture evolved along an independent path.

There are two factors that contribute to the genetic heterogeneity between cultures. First, different mutations may arise in different populations (e.g., different culture tubes) even when those populations are initially isogenic and the culture environments are initially identical. Second, as different GASP alleles appear in the populations, the resulting physiological changes may lead to alterations in the culture environments. Therefore, different culture tubes will contain not only a different spectrum of genetic changes, but also a different chemical environment which leads to different selective pressures.

The ability of some cells to survive for long periods in a nutrient-limiting environment has also been demonstrated by studying cells that have survived in nutrient agar "stabs" or "slants." In these cases, viable cells have been recovered from stabs that have been sealed for more than four decades (5, 25). In all cases where phenotypic and genotypic characterizations have been performed on the survivors of long-term incubations, it has been discovered that mutants with new phenotypes and increased fitness have arisen in these environments. This confirms that long-term cell viability, population dynamism, and the generation of genetic diversity in the laboratory are not reserved only for bacteria growing in liquid culture.

The fact that the GASP phenotype is expressed under a broad range of conditions suggests that the process of microbial evolution during extended starvation may be a generalizable phenomenon. That is, perhaps the stresses experienced by cells incubated for extended periods in stationary-phase culture approximate conditions that the cells may encounter in natural environments. While this still remains to be proven, hints from studies of natural *E. coli* isolates lend credence to this hypothesis (see below).

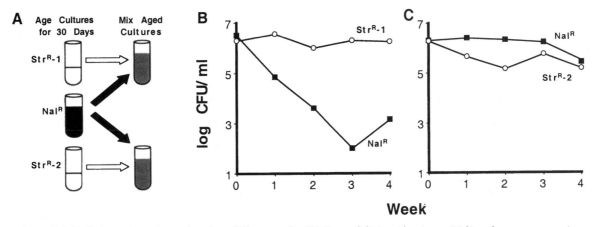

Figure 4. Initially isogenic strains evolve along different paths. (A) One nalidixic acid-resistant (Nalʳ) and two streptomycin-resistant (Strʳ) cultures were incubated at 37°C for 30 days. The Nalʳ culture was then mixed (1:1) with cells from either of the Strʳ cultures, which were derived from the same colony, hence initially isogenic. (B & C) A representative pair of mixed cultures. Nalʳ cells (filled squares), Strʳ cells (open circles.)

GENETICS OF GASP

As stated earlier, the GASP phenotype is not a physiological adaptation to starvation stress but is due to the acquisition of beneficial mutations. Two lines of evidence support the genetic basis for GASP. First, the GASP phenotype of aged cultures is stably inherited and is not reversed by allowing the cells to exit stationary phase and grow prior to competition. Second, a strain can be made to express the GASP phenotype by crossing in chromosomal DNA from aged cells into unaged cells by transduction. Hence, single mutations are sufficient for expression of the GASP phenotype.

The first GASP-conferring mutation to be described was an allele of *rpoS* (*rpoS819*) (39). The product of *rpoS*, σˢ, regulates the transcription of numerous genes upon the onset of starvation (16; see also chapter 11). Transduction of the *rpoS819* allele of the GASP mutant into an otherwise wild-type strain was sufficient to confer the GASP phenotype. This allele has a small duplication near the 3′ end of the gene, resulting in a protein with the last four amino acids replaced by 39 new amino acids (39). Additional GASP alleles of *rpoS* have been identified from aged *E. coli* cultures that include missense and frameshift mutations (14). Although some mutations have been shown to affect protein stability, the effect of the other mutant alleles remains to be determined. However, because of the location of these mutations, we can infer that some affect the activity of the protein and others may affect expression of *rpoS*. However, all alleles identified so far share a partial loss-of-function phenotype in their ability to activate the transcription of downstream positively regulated genes. The allelic state of *rpoS* in clinical and envi-

ronmental isolates of *E. coli* is highly variable (14, 18, 35; Zinser and Kolter, unpublished data). Interestingly, many natural isolates contain alleles that display similar partial loss of function as those GASP-conferring alleles isolated from the survivors of prolonged incubation in the laboratory. This suggests that the selection conditions imposed during prolonged incubation of cultures in stationary phase may have some resemblance to the selection conditions in several natural settings.

Since a null allele of *rpoS* does not confer a GASP phenotype over a wild-type parent (39), it is inferred that there is a selection for a partial but not complete loss of σˢ activity. Three points are worthy of special attention with respect to the ability of partial loss-of-function alleles of *rpoS* to confer the GASP phenotype. First, strains harboring these alleles can express the GASP phenotype only in the conditions under which they are selected, i.e., aerated rich medium. Second, the physiological basis of the growth advantage afforded during stationary phase by the presence of the partial loss-of-function *rpoS* allele remains unknown. Third, not only are *rpoS* mutations not essential for cells to express the GASP phenotype but mutations in other loci can also confer the GASP phenotype (see below). In fact, *rpoS* is completely dispensable for GASP. Cells that are cultured anaerobically will express the GASP phenotype without acquiring mutations in *rpoS* and, most strikingly, a 10-day-old *rpoS*-null strain cultured aerobically will express the GASP phenotype when competed with a fresh overnight culture of its initially isogenic *rpoS*-null parent (Finkel and Kolter, unpublished data; Zinser and Kolter, unpublished data).

Mutations in genes other than *rpoS* can also confer the GASP phenotype. This is most clearly demonstrated by the fact that multiple rounds of population takeover occur in cultures starved for several weeks. More specifically, aged cultures of an initially *rpoS819* strain express the GASP phenotype over their *rpoS819* parents. One particular isolate from an aged *rpoS819* culture (strain ZK1141) has new GASP-conferring mutations in addition to *rpoS819*. These loci, designated *sgaA*, *sgaB*, and *sgaC* (for stationary-phase growth advantage), mapping at 14′, 20′, and 74′, respectively, are each capable of conferring the GASP phenotype to the *rpoS819* parent when individually transduced (40). Each of these GASP-conferring mutations can provide successively higher fitnesses for the bacteria in the starved cultures (Fig. 5).

Beneficial mutations exert their effects by altering the physiological state of the cell such that a competitive advantage relative to the parent is manifested. Our laboratory has begun an analysis of the physiological changes associated with GASP-conferring mutations, in an attempt to identify the changes responsible for the selective advantages of the mutants. We speculated that the reason why GASP mutants are able to grow and take over the population during stationary phase is because they grow faster than the parental cells on nutrients that become available from dying cells in the culture. Since, in our system, cultures enter stationary phase

as a result of carbon depletion (36), the only source of carbon that the cells compete for during stationary phase comes from the dying cells: as they die, they release utilizable carbon into the medium.

The majority of the nutrients released by the dying cells are expected to be amino acids, as they account for over half the dry weight of an *E. coli* cell (26). Significantly, all four GASP-conferring mutations isolated thus far, *rpoS*, *sgaA*, *sgaB*, and *sgaC*, confer faster growth on a complex mixture of amino acids (Casamino Acids) (40). Each of these mutants was also able to grow faster on several amino acids as the sole source of carbon. The *rpoS819* allele conferred an advantage during growth on alanine, serine, threonine, glutamate, glutamine, and asparagine as the sole carbon source. The *sgaA* GASP allele conferred growth advantages on aspartate, glutamate, glutamine, asparagine, and proline. The *sgaB* GASP allele conferred growth advantages on alanine, serine, and threonine, and the *sgaC* GASP allele conferred growth advantages on alanine, serine, threonine, and proline. Hence, all four GASP alleles have a pleiotropic phenotype of enhanced amino acid catabolism, consistent with the model that the cells are selected for the ability to scavenge amino acids released by the dying cells for the purposes of growth and cell maintenance during starvation conditions.

It has been shown that each of these GASP mutations confers an increased ability of the mutant cell to catabolize some amino acids obtained from the medium. However, it is likely that other GASP mutations will confer different catabolic activities. With further study it might be possible to find GASP mutations that allow the enhanced catabolism of different cellular components, including, for example, the complex carbohydrates and lipids derived from cell wall and membrane components.

Is There a Mechanism Underlying the Acquisition of Mutations Conferring the GASP Phenotype?

The fact that all strains of bacteria tested thus far are able to express the GASP phenotype begs the question of whether specific mechanisms underlie the ability of cells to acquire and express GASP mutations. Although no such mechanism has yet been identified, several lines of evidence suggest that the mutation frequency during stationary phase is modulated, possibly resulting in a transient increase in the appearance of mutations as a stress response.

Mutations seem to appear at a higher than expected frequency in aged cultures. For example, three new GASP-conferring mutations were identified after 2 weeks of stationary-phase incubation of *E. coli* strain ZK819 (40). The doubling time of GASP mu-

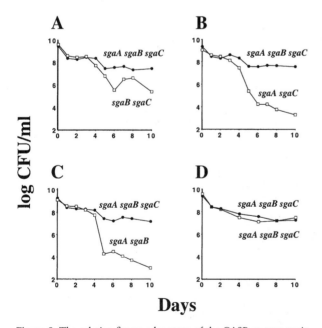

Figure 5. The relative fitness advantage of the GASP mutant strain ZK1141 is conferred by the additive affect of three novel GASP alleles: *sgaA*, *sgaB*, and *sgaC*. (A–C) Reconstructed strains that carry only two of the three new mutations cannot outcompete the ZK1141 parent. (D) A reconstructed triple mutant strain (*sgaA sgaB sgaC*) has equal fitness compared to ZK1141. ZK1141 (filled circles), reconstructed strains (open squares.)

tant cells within these cultures has been estimated at about 16 to 24 h after day 3 of incubation (Finkel and Kolter, unpublished data). Given population densities of $\sim 10^7$ CFU/ml and reported replication error rates of $\sim 10^{-10}$ under exponential growth conditions, we would expect to see very few mutations fixed in the population—in fact, too few to account for the appearance and dominance of three beneficial mutations over this time.

Explanations for this phenomenon include a relaxation of replication fidelity mechanisms, possibly resulting from a decrease in the rate of mismatch repair or an increase in the kinds of adaptive mutation events characterized during stationary phase (9, 10, 15, 22, 29, 30; see also chapter 16). Other possible sources of mutation are "nonstandard" replication events, such as the *oriC*-independent form of replication proposed by Kogoma (19), or the induction of error-prone mechanisms similar to those induced during the SOS response (13, 34). The status of the SOS response mechanism during long-term stationary-phase incubation has not been characterized.

Other mechanisms that may increase genetic diversity are the movement of transposable elements within the chromosome and chromosomal rearrangements such as deletions, inversions, duplications, and amplifications. Evidence for movement of transposable elements and other genomic rearrangements has been obtained from analyses of cells rescued from old stabs (5, 25) as well as from aged batch cultures (2, 7). Of particular interest is a report that chromosome copy number may be greater than one during stationary phase; under several culture conditions chromosome ploidy during stationary-phase incubation was estimated to be as high as eight per cell (1). A plethora of possibilities arise when thinking about the presence of multiple chromosome copies and the effects of increased mutation frequency or chromosomal rearrangements during stationary phase.

Initial Survival Mechanisms Required for the Expression of GASP

It is possible that, in addition to mutations that are selected to become GASP alleles, there exists a group of activities that the cell must possess to express mutations conferring the growth advantage phenotype. These activities, which might be essential to allow cells to survive long enough to acquire mutations that can confer a GASP phenotype, can be broadly placed into three categories: (i) activities that allow the appearance of new mutations, possibly modulating mutation rates (as discussed above); (ii) those activities that are required for the cell to maintain viability, at some level, until a GASP mutation

conferring a competitive advantage appears; and (iii) activities that are necessary for a GASP mutation to manifest its growth advantage phenotype. Genetic screens and selections for mutants that either do not express the GASP phenotype or express GASP in an accelerated fashion should identify these mechanisms. Such a screen identified the locus encoding the primary respiratory NADH dehydrogenase (*nuo* genes) as required for the ZK1141 strain to express the GASP phenotype, but the physiological basis for the requirement of this activity is not clear (37).

PERSPECTIVES

The phenomenon we have been studying in the laboratory, the appearance of fitter mutant competitors, is essentially a demonstration of "evolution in a test tube." We are able to observe, under a given set of environmental conditions and genetic backgrounds, evolution in real time. The appearance of cells containing GASP mutations leads to populations or communities that are highly dynamic, constantly altering their environment and, inevitably, themselves. Given that in natural environments the organisms we study rarely encounter nutritional conditions as replete as in the laboratory under exponential growth conditions, we feel that our long-term culture system better approximates the competitive and selective conditions encountered by many bacteria in their natural setting. Further study of the mechanisms in place that allow this generation of genetic diversity will continue to shed light on mechanisms of bacterial stress response while, at the same time, possibly allowing us to begin to understand how bacterial diversity originates.

REFERENCES

1. **Akerlund, T., K. Nordstrom, and R. Bernander.** 1995. Analysis of cell size and DNA content in exponentially growing and stationary-phase batch cultures of *Escherichia coli. J. Bacteriol.* **177:**6791–6797.
2. **Andersson, D. I., E. S. Slechta, and J. R. Roth.** 1998. Evidence that gene amplification underlies adaptive mutability of the bacterial *lac* operon. *Science* **282:**1133–1135.
3. **Dykhuizen, D. E.** 1990. Experimental studies of natural selection in bacteria. *Annu. Rev. Ecol. Syst.* **21:**373–398.
4. **Eberl, L., M. Givskov, C. Sternberg, S. Moller, G. Christiansen, and S. Molin.** 1996. Physiological responses of *Pseudomonas putida* KT2442 to phosphate starvation. *Microbiology* **142:**155–163.
5. **Eisenstark, A., C. Miller, J. Jones, and S. Leven.** 1992. *Escherichia coli* genes involved in cell survival during dormancy: role of oxidative stress. *Biochem. Biophys. Res. Commun.* **188:**1054–1059.
6. **Espinosa-Urgel, M.** 1995. Bacterial stationary phase: new physiological and molecular aspects related to the *rpoS* gene.

Ph.D. thesis. Universidad Complutense de Madrid, Madrid, Spain.

7. **Finkel, S. E., and R. Kolter.** 1999. Evolution of microbial diversity during prolonged starvation. *Proc. Natl. Acad. Sci. USA* **96:**4023–4027.

8. **Finkel, S. E., E. Zinser, S. Gupta, and R. Kolter.** 1997. Life and death in stationary phase, p. 3–16. *In* S. J. W. Busby, C. M. Thomas, and N. L. Brown (ed.), *Molecular Microbiology.* Springer-Verlag, Berlin, Germany.

9. **Foster, P.** 1993. Adaptive mutation: the uses of adversity. *Annu. Rev. Microbiol.* **47:**467–504.

10. **Foster, P. L., and W. A. Rosche.** 1999. Mechanisms of mutation in nondividing cells. Insights from the study of adaptive mutation in *Escherichia coli. Ann. N. Y. Acad. Sci.* **870:**133–145.

11. **Gay, F. P.** 1935. Bacterial growth and reproduction. *In* F. P. Gay, G. W. Bachman, R. H. Benham, and L. Buchbinder (ed.), *Agents of Disease and Host Resistance, Including the Principles of Immunology, Bacteriology, Mycology, Protozoology, Parasitology and Virus Disease.* Charles C Thomas, Springfield, Ill.

12. **Goodlow, R. J., W. Braun, and L. A. Mika.** 1951. Metabolite-resistance and virulence of smooth *Brucella* variants isolated after prolonged cultivation. *Proc. Soc. Exp. Biol. Med.* **76:**786–788.

13. **Goodman, M. F.** 1998. Purposeful mutations. *Nature* **395:**221–223.

14. **Gupta, S.** 1997. Mutations that confer a competitive advantage during starvation. M.A. thesis. Harvard University, Cambridge, Mass.

15. **Harris, R. S., G. Feng, K. J. Ross, R. Sidhu, C. Thulin, S. Longerich, S. K. Szigety, P. J. Hastings, M. E. Winkler, and S. M. Rosenberg.** 1999. Mismatch repair is diminished during stationary-phase mutation. *Mutat. Res.* **437:**51–60.

16. **Hengge-Aronis, R.** 1996. Regulation of gene expression during entry into stationary phase, p. 1497–1512. *In* F. C. Neidhardt, R. Curtiss III, J. L. Ingraham, E. C. C. Lin, K. B. Low, B. Magasanik, W. S. Reznikoff, M. Riley, M. Schaechter, and H. E. Umbarger (ed.), *Escherichia coli and Salmonella: Cellular and Molecular Biology,* 2nd ed. ASM Press, Washington, D.C.

17. **Huisman, G. W., D. A. Siegele, M. M. Zambrano, and R. Kolter.** 1996. Morphological and physiological changes during stationary phase, p. 1672–1682. *In* F. C. Neidhardt, R. Curtiss III, J. L. Ingraham, E. C. C. Lin, K. B. Low, B. Magasanik, W. S. Reznikoff, M. Riley, M. Schaechter, and H. E. Umbarger (ed.), *Escherichia coli and Salmonella: Cellular and Molecular Biology,* 2nd ed. ASM Press, Washington, D.C.

18. **Jishage, M., and A. Ishihama.** 1997. Variation in RNA polymerase sigma subunit composition within different stocks of *Escherichia coli* W3110. *J. Bacteriol.* **179:**959–963.

19. **Kogoma, T.** 1997. Stable DNA replication: interplay between DNA replication, homologous recombination, and transcription. *Microbiol. Mol. Biol. Rev.* **61:**212–238.

20. **Lawther, R. P., D. H. Calhoun, C. W. Adams, C. A. Hauser, J. Gray, and G. W. Hatfield.** 1981. Molecular basis of valine resistance in *Escherichia coli* K-12. *Proc. Natl. Acad. Sci. USA* **78:**922–925.

21. **Lenski, R. E., J. A. Mongold, P. D. Sniegowski, M. Travisano, F. Vasi, P. J. Gerrish, and T. M. Schmidt.** 1998. Evolution of competitive fitness in experimental populations of *E. coli:* what makes one genotype a better competitor than another? *Antonie Leeuwenhoek* **73:**35–47.

22. **Lombardo, M. J., J. Torkelson, H. J. Bull, G. J. McKenzie, and S. M. Rosenberg.** 1999. Mechanisms of genome-wide hypermutation in stationary phase. *Ann. N.Y. Acad. Sci.* **870:**275–289.

23. **Madigan, M. T., J. M. Martinko, and J. Parker.** 1999. *Brock Biology of Microorganisms.* Prentice-Hall, New York, N.Y.

24. **Morita, R. Y.** 1993. Bioavailability of energy and the starvation state, p. 1–23. *In* S. Kjelleberg (ed.), *Starvation in Bacteria.* Plenum Press, New York, N.Y.

25. **Naas, T., M. Blot, W. M. Fitch, and W. Arber.** 1994. Insertion sequence-related genetic variation in resting *Escherichia coli* K–12. *Genetics* **136:**721–730.

26. **Neidhardt, F. C., and H. E. Umbarger.** 1996. Chemical composition of *Escherichia coli,* p. 13–28. *In* F. C. Neidhardt, R. Curtiss III, J. L. Ingraham, E. C. C. Lin, K. B. Low, B. Magasanik, W. S. Reznikoff, M. Riley, M. Schaechter, and H. E. Umbarger (ed.), *Escherichia coli and Salmonella: Cellular and Molecular Biology,* 2nd ed. ASM Press, Washington, D.C.

27. **Postgate, J. R., and J. R. Hunter.** 1962. The survival of starved bacteria. *J. Gen. Microbiol.* **29:**233–263.

28. **Reynolds, A. E., J. Felton, and A. Wright.** 1981. Insertion of DNA activates the cryptic *bgl* operon of *E. coli. Nature* **293:**625–629.

29. **Rosche, W. A., and P. L. Foster.** 1999. The role of transient hypermutators in adaptive mutation in *Escherichia coli. Proc. Natl. Acad. Sci. USA* **96:**6862–6867.

30. **Rosenberg, S. M., C. Thulin, and R. S. Harris.** 1998. Transient and heritable mutators in adaptive evolution in the lab and in nature. *Genetics* **148:**1559–1566.

31. **Shearer, C.** 1917. On the toxic action of dilute pure sodium chloride solutions on the meningococcus. *Proc. R. Soc.* **89:**440.

32. **Smeulders, M. J., J. Keer, R. A. Speight, and H. D. Williams.** 1999. Adaptation of *Mycobacterium smegmatis* to stationary phase. *J. Bacteriol.* **181:**270–283.

33. **Steinhaus, E. A., and J. M. Birkeland.** 1939. Studies on the life and death of bacteria. I. The senescent phase in aging cultures and the probable mechanisms involved. *J. Bacteriol.* **38:**249–261.

34. **Walker, G. C.** 1996. The SOS response of *Escherichia coli,* p. 1400–1416. *In* F. C. Neidhardt, R. Curtiss III, J. L. Ingraham, E. C. C. Lin, K. B. Low, B. Magasanik, W. S. Reznikoff, M. Riley, M. Schaechter, and H. E. Umbarger (ed.), *Escherichia coli and Salmonella: Cellular and Molecular Biology,* 2nd ed. ASM Press, Washington, D.C.

35. **Waterman, S. R., and P. L. C. Small.** 1996. Characterization of the acid resistance phenotype and *rpoS* alleles of shiga-like toxin-producing *Escherichia coli. Infect. Immun.* **64:**2808–2811.

36. **Zambrano, M. M.** 1993. *Echerichia coli* mutants with a growth advantage in stationary phase. Ph.D. thesis. Harvard University, Cambridge, Massachusetts.

37. **Zambrano, M. M., and R. Kolter.** 1993. *Escherichia coli* mutants lacking NADH dehydrogenase-I have a competitive disadvantage in stationary phase. *J. Bacteriol.* **175:**5642–5647.

38. **Zambrano, M. M., and R. Kolter.** 1996. GASPing for life in stationary phase. *Cell* **86:**181–184.

39. **Zambrano, M. M., D. A. Siegele, M. Almirón, A. Tormo, and R. Kolter.** 1993. Microbial competition: *Escherichia coli* mutants that take over stationary phase cultures. *Science* **259:**1757–1760.

40. **Zinser, E. R., and R. Kolter.** 1999. Mutations enhancing amino acid catabolism confer a growth advantage in stationary phase. *J. Bacteriol.* **181:**5800–5807.

Bacterial Stress Responses
Edited by G. Storz and R. Hengge-Aronis
©2000 ASM Press, Washington, D.C.

Chapter 16

Mutation under Stress: Adaptive Mutation in *Escherichia coli*

WILLIAM A. ROSCHE AND PATRICIA L. FOSTER

When bacteria encounter a foreign environment, they may not be able to establish themselves in this new niche unless the population contains a pre-adapted variant. The process of random mutation is believed to be the source of such variants. But when a bacterium finds itself in a new environment, can it mutate its genome to provide a new adapted variant? Can a bacterium sense its environment and direct mutations to relieve the stress it finds itself in? A paper published by Cairns et al. (9) made us question the assumption that the mutational process is random. Cairns et al. challenged us to ask whether mutations could be directed, allowing adaptation to the environment. In this chapter we will review the evidence for the spontaneous nature of mutations and describe the experimental systems used to study bacterial populations in stressful environments. We will discuss the mechanisms that cells use to achieve apparent adaptive mutations and give a detailed example of one mechanism.

A variety of names have been given to the process by which mutations arise in static populations of cells. For clarity, we use the term "adaptive mutation" to mean the process that, during selection, produces mutations that relieve the selective pressure whether or not nonselected mutations are also occurring. "Adaptive mutations," meaning the sequence changes themselves, is used in the same sense as it is used by evolutionists to distinguish beneficial from neutral or deleterious mutations.

MUTATIONS ARE RANDOM

In 1943 Luria and Delbrück (51) published a paper describing an experiment that has come to be called a fluctuation test. This test was used to show that mutations occurred randomly during the growth of a bacterial culture. A fluctuation test consists of a large number of identical cultures, each started with an inoculum so small that the chance of it containing a preexisting mutant is low. These cultures are grown under nonselective conditions and then plated onto a selective medium. Mutants that arise during the nonselective growth of the culture can replicate and produce a clone of identical descendants, each of which will produce a colony after plating. Occasionally a rare mutation will arise early in the growth of the culture, producing a large number of mutant progeny (called a "jackpot"). Luria and Delbrück realized that the high variation among the cultures due to the jackpots proved that the mutations were occurring randomly during the growth of the culture and not after the selection had been applied.

ARE MUTATIONS INDUCED BY SELECTION?

The first questioning of the random mutation doctrine began with the work of Ryan in the 1950s (65). While trying to confirm the Luria-Delbrück experiment, Ryan observed that histidine utilization (His$^+$) revertants continued to appear for 10 days after His$^-$ cells were plated onto medium without histidine. Ryan was able to show that the delayed appearance of mutants was not due to slowly growing mutants (65). Nor could the rate of cell turnover account for the number of mutants observed if the mechanism of mutation was the same during growth as during selection (66). Thus, Ryan concluded that even nondividing cells must perform some limited DNA synthesis and suggested that this synthesis has an increased error rate (67).

Twenty years later, Shapiro (68) published a study reexamining the process by which mutations are produced during prolonged selection. The strain he used has the regulatory region of the arabinose

William A. Rosche • Department of Biological Science, University of Tulsa, Tulsa, OK 74104. **Patricia L. Foster** • Department of Biology, Indiana University, Bloomington, IN 47405.

(*ara*) operon separated from the *lacZ* gene by a *Mu* bacteriophage. The strain is Ara⁻ and Lac⁻, but excision of the prophage can fuse *araB* to *lacZ*, resulting in a cell that can use lactose if arabinose is present as an inducer. On lactose-arabinose minimal medium mutants began appearing after 5 to 6 days of incubation. The rate of accumulation reached its maximum at 20 days postplating and then declined. Loss of the prophage was not detected in actively growing cultures, but Ara-Lac⁺ fusions were observed at a frequency of 1 per 10^8 cells during selection on lactose-arabinose plates. Merely starving the cells in buffer did not stimulate fusion formation, indicating that the mutation process was responding to some environmental trigger besides starvation. However, more recent experiments have shown that aerobic starvation can induce fusion formation (28, 53, 56, 70).

As mentioned above, in 1988 Cairns et al. (9) published a paper asking the scientific community to rethink the origin of mutants. In that paper they showed that Lac⁺ mutations accumulated with time when Lac⁻ cells were under selection for lactose utilization. Two days after parallel cultures of Lac⁻ cells were plated on medium that contained lactose as the sole carbon and energy source, Lac⁺ colonies appeared with the expected Luria-Delbrück distribution. These mutants had arisen spontaneously during the generation of the culture (i.e., they were due to growth-dependent mutations). However, as in the Ryan experiments, Lac⁺ colonies continued to appear with time, and after day 2, the distribution of the number of mutants per plate became narrower, indicating that the mutations were occurring after the cells were plated on lactose.

To test whether the stress of lactose selection was causing an increase in mutation rate, Cairns et al. looked for the appearance of mutants with a different, rapidly expressed phenotype, valine resistance (Val^r). At various times during lactose selection the plates were overlaid with agar containing glucose (to provide a carbon source) and valine and the number of Val^r mutants determined. Surprisingly, the number of new Val^r mutants went down with time. Cairns et al. concluded that a population of cells during selection for lactose utilization was not at the same time accumulating other mutations. Lactose selection appeared to be "directing" the mutations to the genes needed to allow growth, i.e., selective conditions did not cause a general increase in mutation rates (9).

Cairns et al. (9) also used a "delayed overlay" experiment to show that the mutations that arose during selection were not induced by mere starvation but rather required the presence of the selecting agent. In this experiment, Lac⁻ *Escherichia coli* cells were plated onto medium without a carbon source

(i.e., the cells were starving). At various times after plating (usually 4-day intervals), the plates were overlaid with agar containing lactose. Lac⁺ mutants did not accumulate before the overlay nor was there a "burst" of mutants immediately following the overlay. Instead, there was a steady increase in the appearance of Lac⁺ mutants after the lactose was added. Thus, the continuous presence of lactose was needed for the mutations to occur.

GENERAL MECHANISMS TO EXPLAIN ADAPTIVE MUTATION

Many hypotheses have been proposed to explain the apparent adaptiveness of the mutations that arise after selection (see Table 1). Most of these can be thought of as "trial-and-error" mechanisms, an idea originally proposed by Cairns et al. (9). The basic idea is that variants arise at random in a population during selection, but these are transitory and only those that allow the cell to divide are immortalized as mutations. Several of these hypotheses have been tested in strain FC40 and more or less eliminated. This does not mean that they might not be relevant in other cases of adaptive mutation.

Cairns et al. (9) proposed that during selection cells might produce highly variable mRNA molecules. If a mutant transcript was translated into a functional protein that allowed growth under the selective condition, the transcript would be reverse transcribed and inserted into the chromosome. This mechanism is attractive because the genome would be preserved from deleterious mutations while a large number of variants were "tried" by the cell. However, this mechanism requires the enzyme reverse transcriptase, and Bridges (6) has shown that curing *E. coli* B of its retron (a retroviral-like element that encodes reverse transcriptase) has no effect on the production of mutations during selection. Thus, there is no evidence to support the reverse transcription mechanism in *E. coli*.

Trial-and-error mechanisms could also involve variant DNA molecules. Cells could produce extra copies of the genes needed to relieve the selection, and these extra copies could accumulate mutations while the cell retained a "master copy" of the gene (9, 22). If a variant were produced that allowed growth under the selective condition, that copy would be retained and incorporated into the genome. In the best-studied case of adaptive mutations, *E. coli* FC40, the mutations that arise after selection are recombination dependent (see below), and it has been proposed that this requirement could mean that gene amplification is required (1, 27). However, gene am-

Table 1. Summary of the various models for adaptive mutation

Model	Limitations	Supporting data	Refuting data	Conclusion	Reference(s)
Variant RNAs	Requires reverse transcriptase	None	Absence of reverse transcriptase has no effect	Interesting but unlikely to be important in *E. coli*	6, 9
Slow repair	MMR or other repair pathway must be limiting	Levels of MMR proteins decline in stationary phase	MMR repair capacity remains sufficient for the amount of DNA synthesis	May be important in some cases	3, 18, 25, 41, 71
Hypermutable state	Severe consequences for the cell	Multiple mutations found in a subpopulation	Most Lac⁺ adaptive mutations do not come from hypermutators	Accounts for only minority of single mutations but most multiple mutations	3, 24, 39, 61, 75
Gene amplification	Requires that the cells proliferate or that the amplification is unstable	Sectoring colonies of *S. enterica* serovar Typhimurium	Unsupported by physical evidence in *E. coli*	Important in *S. enterica* serovar Typhimurium but less so in *E. coli*	1, 23, 27, 30, Rosche and Foster, unpublished data
Error-prone DNA synthesis		None	None	Still unknown	
Mutagenic transcription		Correlation between transcription and mutation rates in some genes	Lactose is still required for adaptive Lac⁺ mutations when the *lac* allele is constitutive	Could be of general importance in mutagenesis	2, 8, 13, 14, 20, 27, 64, 76, 77
Recombination		Lac⁺ reversion in FC40	Accounts for only some cases of adaptive mutation	Well characterized in FC40 but generality still unknown	26

plification probably only accounts for a minority of the mutations in FC40 (23) (see below).

Another trial-and-error mechanism is Stahl's slow repair model (71). Cells in stationary phase or in a nutrient-poor environment may not have enough energy to synthesize all the DNA repair enzymes they need. As a result, the cells repair their genome "here and there" (71). If a mutation is not repaired but allows growth, it would be immortalized when the cell divides. It has been shown that cells in stationary phase have reduced levels of the enzymes that perform postreplicative methyl-directed mismatch repair (MMR) (18, 41). However, MMR⁻ cells have an increased rate of mutation during selection, just as they do during nonselected growth (25, 27, 45, 60). Therefore, although reduced, the levels of MMR proteins in stationary phase cells must remain at a level that is adequate for the amount of DNA synthesis that occurs; if this were not so, eliminating MMR would have little effect (25, 60).

Hall (39) proposed that during selection most cells in the population do not mutate, but a minority transiently experience a mutation rate so high that they die unless a useful mutation occurs. A specific prediction of this "hypermutable state model" is that

nonselected mutations should occur at a higher frequency among cells that bear adaptive mutations than among cells that do not (39). This prediction has been confirmed in five cases (3, 24, 39, 61, 75). However, whether transient hypermutators are important contributors to a population of mutants depends on the proportion of cells that are hypermutators and the degree to which their mutation rate is elevated (7, 58). Using a broad mutational screen, loss of motility, we determined the frequency of nonselected mutations in starved Lac⁻ cells, in selected Lac⁺ revertants, and in selected Lac⁺ revertants carrying another nonselected mutation in *E. coli* FC40. These frequencies were used to solve an algebraic model of hypermutation developed by Cairns (7, 61). The calculations revealed that during lactose selection about 0.1% of the cells were hypermutators and their mutation rate was increased about 200-fold, estimates remarkably similar to ones predicted by Ninio on theoretical grounds (58). Although these hypermutators accounted for nearly all multiple mutations, they accounted for only about 10% of the adaptive Lac⁺ mutations. Thus 90% of the adaptive Lac⁺ mutations were not arising from hypermutators. Surprisingly, the proportion of Lac⁺ clones bearing second mutations increased linearly with time during lactose selection, suggesting that most of the hypermutators did not die during the 5 days of the experiment (61). These results indicate that transient hypermutators do not make a substantial contribution to the population of cells bearing a single mutation but are responsible for nearly all cells bearing more than one mutation. Therefore, when the solution to a particular challenge requires several mutations, transient mutators may have a significant evolutionary advantage.

It has often been hypothesized that "directed mutation" could be achieved if transcription were mutagenic because, then, mutation rates would be enhanced only in genes induced by selective conditions (14, 20). Although gratuitous transcription of *lac* does not produce Lac$^+$ mutants in the absence of lactose (8, 27), transcription may nonetheless be mutagenic. Transcription produces single-stranded DNA, which may be more easily damaged than double-stranded DNA (14). For example, cytosine deaminates to uracil at a 100-fold higher rate in single-stranded than in double-stranded DNA, although the half-life is still 200 years (36). In addition, because transcribed genes are subjected to more repair than nontranscribed genes, they have a greater chance of accumulating polymerase errors (40).

Correlations between transcription and spontaneous mutation rates have been reported for *Saccharomyces cerevisiae* (13), *E. coli* (2, 77) and *Bacillus subtilis* (64). In the case of one gene, *leuB*, the spontaneous mutation rate was roughly proportional to the level of its transcript (77). But starvation for amino acids does not appear to be globally mutagenic (39, 64, 76). Because starvation for one amino acid appears to increase the mutation rates of genes required for the biosynthesis of other amino acids (76), the mutational process is "directed" to classes of genes, not the particular gene under selection.

ADAPTIVE MUTATION IN *E. COLI* FC40

Although not the strain originally used by Cairns et al., FC40 has become a paradigm of adaptive mutation. It is an attractive model system because the large number of adaptive mutations that appear during lactose selection allows obvious artifacts to be easily eliminated. However, it was clear early on that not all aspects of adaptive mutation in FC40 were generalizable to other systems (21). In particular, the recombination dependence of adaptive mutation in FC40 has not been found to be common to other cases of adaptive mutation. Nonetheless, studies of FC40 have revealed at least one mechanism by which

mutations can arise in cells during selection. Because FC40 is the best characterized system for the study of adaptive mutation, the recombination-dependent mechanism is given here as a case study.

E. coli FC40 cannot utilize lactose (Lac$^-$) but reverts to lactose utilization (Lac$^+$) when lactose is its sole carbon and energy source. Two days after plating FC40 cells on minimal lactose plates, Lac$^+$ colonies start appearing at a constant rate of about 1 per 10^7 cells per day. The number of mutants that appear 2 days after plating has a Luria-Delbrück distribution, indicating that these mutations occurred randomly during the growth of the culture. Mutants continue to accumulate and, after a week, 95% of all the Lac$^+$ colonies that appear are due to mutations that occurred after plating. To produce this number of mutations at the normal growth-dependent mutation rate every cell would have to be replicating about once an hour during lactose selection, a rate of replication comparable to that achieved by cells during nonselective growth in minimal medium (8, 23).

Although mutation to Lac$^+$ in FC40 appeared to be adaptive as originally defined, it has been shown that another gene (*tetA*, encoding tetracycline resistance) mutates at approximately the same rate and with the same genetic requirements when it is located near the *lac* operon on the episome (24). Thus, Lac$^+$ mutations in FC40 are not adaptive as originally defined. However, we continue to call the Lac$^+$ mutations that occur during lactose selection adaptive mutations.

Properties of the Lac$^-$ Allele in FC40

The Lac$^-$ allele in FC40 is derived from a fusion of the *lacI* gene to the *lacZ* gene. Although the fusion eliminates the coding sequence for the last four residues of *lacI*, all of *lacP* and *lacO*, and the first 23 residues of *lacZ*, it is Lac$^+$. Constitutive transcription is initiated from the *lacI*^{q} promoter (57). For ease of genetic manipulation the Lac fusion is located on F'128, a conjugal plasmid. Mutant versions of this fusion constructed by J. H. Miller and coworkers have been used for a variety of mutational studies (e.g., see reference 55).

The Lac$^-$ allele carried by FC40, Φ(*lacI33-lacZ*), has an ICR-191–induced +1 frameshift at the 320th codon of *lacI*, changing CCC to CCCC (10). The allele is slightly leaky, producing about 2 Miller units of β-galactosidase, an amount insufficient to allow FC40 to grow on lactose as a sole carbon source (23). The frameshift is polar on *lacY* so FC40 is also lactose permease-defective (P. L. Foster, unpublished data). However, stationary-phase reversion of FC40 to Lac$^+$ requires lactose (8), suggesting that the mutational process requires the energy provided by the residual amount of β-galactosidase and permease activity.

Characteristics of Adaptive Mutation in FC40

The most important characteristics of adaptive mutation in FC40 are summarized in Table 2. A defining property is the spectrum of sequence changes that give a Lac⁺ phenotype. Mutations that occur during the nonselective growth of the culture (called growth-dependent mutations) are predominantly deletions, duplications, and some simple frameshifts (32, 62). In contrast, the majority of the mutations that occur during lactose selection are simple −1 frameshifts in runs of iterated bases. About 75% are at one site (the same site that was originally mutated by ICR-191) (32, 62).

In FC40, the mutations that arise during selection are recombination dependent. Loss of *E. coli*'s major recombinase, RecA, completely abolishes adaptive mutation (8, 22). RecA protein binds to single-stranded DNA and promotes homologous recombination (46). Adaptive mutations are also abolished if the cells lack RecBCD, the enzyme that initiates double-strand end repair (30, 42). Surprisingly, mutations in the *ruvAB* and *recG*, genes that encode partially redundant enzyme systems required during the late stages of recombination, have opposite effects on adaptive mutations in FC40 (35, 43).

Growth-dependent and adaptive mutations share two important characteristics. (i) Both growth-dependent and adaptive mutations probably are the result of polymerase errors made by *E. coli*'s major replicative polymerase, DNA polymerase III (Pol III)

(29). Although DNA polymerase II is active in stationary-phase cells, it makes very few errors (17, 29). (ii) Both growth-dependent and adaptive mutations can be corrected by MMR (25, 27, 60).

A Model for the Mechanism of Adaptive Mutation in FC40

The following model can account for all the characteristics of adaptive mutation in *E. coli* FC40. Nicking at the episome's conjugal origin, *oriT*, is known to initiate recombination (11), thus the initiating event for adaptive mutation to Lac⁺ in FC40 is likely to be a nick at *oriT*. This nicking occurs even in the absence of a conjugal signal and persists in stationary-phase cells (19, 37). Kuzminov (49) proposed that a double-strand end is created when a replication fork initiated at one of the episome's vegetative origins collapses at the nick at *oriT* (Fig. 1A and B). The exonuclease and helicase activities of RecBCD create an invasive 3′ end that initiates RecA-catalyzed recombination (Fig. 1C and D), accompanied by the restoration of a replication fork (Fig. 1E). Uncorrected replication errors produced by the reassembled fork at this point produce the adaptive mutations. Indeed DNA synthesis at a restored replication fork may be particularly error-prone (15, 54, 72). The new fork differs from a normal fork in that it is accompanied by a four-stranded recombination intermediate (a Holliday junction). Translocation of the Holliday junction toward the fork (e.g., by RuvAB) would create a tract of doubly unmethylated DNA in which polymerase errors would be randomly repaired by MMR (Fig. 1F, left). This tract would thus contain a higher than normal number of muta-

Table 2. Characteristics of mutation in *E. coli* FC40

Characteristic	Mutation		Reference(s)
	Growth dependent	Adaptive	
Differences			
Sequence changes	Deletions, duplications, frameshifts	Predominately −1-bp frameshifts	32, 62
Recombination functions	Not required	RecA, RecBCD	8, 22, 30, 42
Holliday junction processing	Not required	RuvAB required, RecG inhibits, RuvC required	35, 43
SOS functions	Not required	*umuDC* not required, one unknown SOS gene required for maximum response[a]	8, 35; Foster, unpublished data
Conjugal functions	Not required	Required	33, 34, 38
Commonalities			
DNA polymerase	Probably Pol III	Probably Pol III	29
Methyl-directed mismatch repair	Mutator	Mutator	25, 27, 60

[a] Could be *dinB*.

Fork Collapse

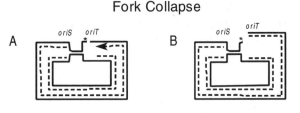

Recombination-Dependent Reestablishment of the Replication Fork

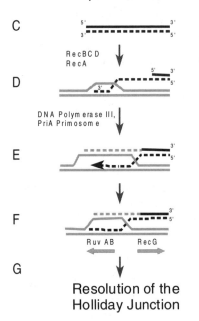

Resolution of the Holliday Junction

Figure 1. A model for recombination-dependent mutation. Panels A and B, collapse of the replication fork at *oriT*; panels C and D, RecABCD catalyzed recombination; panel E, restoration of the replication fork; panel F, translocation of the Holliday junction in opposite directions by RuvAB and RecG; panel G, resolution of the Holliday junction by RuvC (not shown). See text for details. Dashed lines are newly synthesized DNA; * at *oriT* indicates TraI. Reprinted with permission from reference 26, © 1999 Annual Reviews.

tions. Translocation of the junction away from the fork, or resolution of the Holliday junction before DNA synthesis begins (e.g., by RecG), would preserve the hemimethylated state of the DNA, allowing polymerase errors to be correctly repaired (Fig. 1F, right). This model not only accounts for the opposite effects of RuvAB and RecG on adaptive mutation, but also for the fact that the mutational spectrum bears the mark of MMR deficiency (35).

Alternative models to explain adaptive mutations in FC40 have been proposed (reviewed in referencess 26, 31). Some evidence indicates that conjugation can be mutagenic (12, 48), so it is possible that the adaptive Lac⁺ mutations are produced by conjugal replication initiated by nicking at *oriT* (33, 38, 59). But this hypothesis provides no clear role for

recombination. Because gene amplification is recombination dependent, amplification is an attractive hypothesis to explain adaptive mutation in FC40. True revertants might arise during replication of amplified copies of the *lac* region, which would then be resolved in favor of the Lac⁺ copy when the cells began to grow (27). This hypothesis has been extended to include a growing subpopulation of cells carrying amplified Lac⁻ alleles (63). Amplification of the *lac* region and subsequent colony sectoring were detected in *Salmonella enterica* serovar Typhimurium cells carrying the episome from FC40 (1). However, we found no evidence for a growing subpopulation of FC40 cells during lactose selection, and only a few Lac⁺ colonies prove to have amplified Lac⁻ alleles (23). Amplification of the *lac* region during 3 days of lactose selection was also not detectable by Southern hybridization (30; Rosche and Foster, unpublished data). Thus, amplification of the Lac⁻ allele appears to play only a minor role in adaptive reversion of FC40, although it may be more important in *S. enterica* serovar Typhimurium.

These various models are not necessarily mutually exclusive. Thus, multiple mechanisms may contribute to adaptive mutation, and each may be more or less important in different strains and under different conditions.

STRESS REGULATION OF CONJUGAL FUNCTIONS

As mentioned above, the high rate of adaptive mutation of the *lac* allele, the *tetA* gene, and perhaps all genes carried on the F' episome is dependent on the expression of the conjugal functions encoded by the *tra* operon. According to the model we have presented, the important Tra function is nicking at *oriT*. The Tra proteins form a large membrane-spanning complex, so changes in the expression of any *tra* gene might affect the nicking function. The rate of mutagenesis of genes carried on the episome could respond to, and be regulated by, environmental conditions via modulation of the expression of the *tra* operon or the levels of its gene products.

One important regulator of *tra* expression is the FinO protein. FinO binds to and stabilizes an antisense RNA that regulates *tra* expression by inhibiting the expression of *traJ*, a positive regulator of the operon (19). However, the F factor carried in laboratory strains of *E. coli* K-12 is defective in *finO*. Nevertheless, several additional factors regulate expression of the *tra* operon even in *finO* mutants and may be even more powerful regulators of *tra* functions on *finO⁺* episomes in nature.

Transcription from P_{traY}, the major promoter of the *tra* operon, decreases in stationary-phase cells, although the levels of some of the Tra proteins remain high. *oriT* nicking is at its maximum in early stationary-phase cells and then decreases as the levels of TraI (the nickase) decline. But a low level of nicking was detectable even in late stationary-phase cells (37). Mutations in *rpoS*, which encodes the stationary-phase sigma factor, increase the stability of some Tra proteins (37), but RpoS does not appear to play a role in adaptive mutation in FC40 (23).

Expression of the *tra* operon is responsive to the cell's energy level. *tra* expression requires cyclic AMP (cAMP) and is probably mediated through the cAMP-receptor protein (CRP or CAP) (19, 47). Adaptive mutations in FC40 are completely inhibited if *cyaA* (encoding adenylate cyclase) is defective (P. L. Foster, unpublished data). ArcA, a chromosomal-encoded regulatory protein, is a potent activator of P_{traY}. ArcA and B constitute a two-component signal transduction system. ArcB apparently senses the redox level of the cell and, as oxygen levels fall, phosphorylates itself and then ArcA. ArcA-P directly or indirectly activates or represses about 15 other genes, allowing a smooth transition from aerobic to anaerobic conditions (which probably overlaps with the transition from exponential growth into stationary phase) (52). ArcA's function as activator of P_{traY} is known as Sfr (sex factor regulation) and is genetically and physiologically separable from its Arc (anaerobic respiration control) function. Sfr function does not require ArcB, although it can be modulated by ArcB, and deletion of the *arcA* locus eliminates the expression of *traY* under aerobic conditions (44, 69). A less studied host factor, CpxA, may modify Sfr activity or may act independently to modify *tra* expression (19). The effects of ArcA and CpxA on adaptive mutation in FC40 are not known.

Adaptive mutation of FC40 is reduced under anaerobic conditions (1; P. Srinivasan and R. G. Fowler, unpublished data). It was concluded that this reduction was due to a reduced energy yield from lactose metabolism (1). However, it is more likely that adaptive mutations were inhibited because the *tra* operon is downregulated during anaerobiosis. It has long been known that conjugation declines some 2 to 3 orders of magnitude under anaerobic conditions in minimal medium (71a). These same conditions reduce *tra* expression, as measured by the number of sex pili, about 50% in wild-type cells (44). Curiously, some *arcA* and *arcB* mutations cause increased piliation during anaerobiosis, although conjugation itself was not measured in these mutants (44). Clearly regulation of *tra* expression is not well understood and needs further investigation.

RESTING ORGANISMS IN A STRUCTURED ENVIRONMENT (ROSE)

Taddei et al. described a phenomenon they called ROSE, an acronym for resting organisms in a structured environment (74). They define an unstructured environment as one in which each individual cell influences the environment equally. Liquid medium with constant agitation (a "normal" laboratory setting for *E. coli*) would be an example. In a structured environment the cells are clustered so that they influence neighboring cells. Examples include colonies on agar surfaces, biofilms, and cellular aggregates (74). ROSE mutagenesis is distinct from adaptive mutation in FC40 because it has different genetic requirements (74).

If cells modulate their mutation rates when stressed, allowing for adaptation to the stressful environment, one way they might do this is by inducing the SOS system (see chapter 9). The SOS response, which is normally induced by DNA damage, affects both cell division and the accuracy of DNA replication. Using an epigenetic switch to monitor SOS induction, Taddei et al. showed that most cells in aging colonies were induced (74). This induction was abolished in a strain mutant in the *cyaA* gene (74), demonstrating that SOS induction in aging colonies was influenced by cAMP levels. ROSE is also dependent on two genes involved in repair of damaged bases: *uvrB*, which encodes an enzyme required to repair ultraviolet light damage, and *polA*, which encodes *E. coli*'s DNA repair polymerase (73).

The SOS response is controlled by LexA. In the absence of DNA damage, LexA binds to LexA-binding sites and represses the expression of about 20 SOS genes. In response to DNA damage, RecA becomes a coprotease and facilitates the self-cleavage of LexA, allowing expression of the SOS genes. Repression by LexA can also be modulated by the pH of the medium (16). LexA cleavage is inhibited at low pH, and the affinity of the repressor for its binding site is decreased at high pH. Therefore, in the absence of DNA damage the induction of SOS genes may be modulated by the cell's metabolic state.

CONCLUSIONS AND PERSPECTIVES

The research on adaptive mutation has revealed mutagenic mechanisms that may be of evolutionary significance. The recombinational mechanism utilized by FC40 could be an important source of spontaneous mutation in cells that are not undergoing genomic replication. Recombination is often, perhaps always, associated with DNA synthesis (50), and this synthesis appears to have a high error rate (15, 72). Thus, recombination has the potential to increase variation both by recombining existing alleles and by creating new ones.

Conjugal plasmids are common in natural isolates of enteric bacteria (5). On an evolutionary time scale, F and related plasmids frequently recombine and are passed among the major groups of *E. coli* and *Salmonella* (4, 5). Because F can recombine with its host's chromosome, it can acquire chromosomal genes that would then be exposed to its high mutation rate and be free to diverge from the chromosomal copies. These diverged alleles could then spread though the population and even to other species. Indeed, some organisms (e.g., *Pseudomonas* spp.) carry whole metabolic pathways on conjugal plasmids.

Mutagenesis of the episomal genes and the conjugal transfer of F′ episomes may be controlled by environmental conditions. *tra* expression, necessary for adaptive mutation in *E. coli* FC40, is responsive to the energy level of the cell. Anaerobic growth reduces adaptive mutation in FC40, probably due to downregulation of *tra* functions. The structural nature of the environment, whether an agar plate in the laboratory or a biofilm in nature, may also influence the mutagenic process. Both cAMP levels and the pH of the environment are important to the regulation of the SOS regulon. It is possible that there are yet to be discovered environmental cues that the cell uses to modulate mutation rates.

The case of the Ara-Lac⁺ fusion shows that the movement of insertion elements can, in principle, be responsive to environmental conditions. Insertion elements not only activate and inactivate genes, they provide sequence homology that allows large-scale genomic rearrangements. These rearrangements may provide gene duplications that can then evolve new functions for a cell.

Finally, if a subpopulation of nutritionally deprived cells enters into a state of transient hypermutation, they could be a source of multiple variant alleles. Whether an evolutionary strategy, or simply a manifestation of pathology, transient hypermutation could provide a mechanism for rapid adaptive evolution under adverse conditions.

REFERENCES

1. **Andersson, D. I., E. S. Slechta, and J. R. Roth.** 1998. Evidence that gene amplification underlies adaptive mutability of the bacterial *lac* operon. *Science* **282**:1133–1135.
2. **Beletskii, A., and A. S. Bhagwat.** 1996. Transcription-induced mutations: increases in C to T mutations in the nontranscribed strand during transcription in *Escherichia coli*. *Proc. Natl. Acad. Sci. USA* **93**:13919–13924.
3. **Boe, L.** 1990. Mechanism for induction of adaptive mutations in *Escherichia coli*. *Mol. Microbiol.* **4**:597–601.
4. **Boyd, E. F., and D. L. Hartl.** 1997. Recent horizontal transmission of plasmids between natural populations of *Escherichia coli* and *Salmonella enterica*. *J. Bacteriol.* **179**:1622–1627.
5. **Boyd, E. F., C. S. Hill, S. M. Rich, and D. L. Hartl.** 1996. Mosaic structure of plasmids from natural populations of *Escherichia coli*. *Genetics* **143**:1091–1100.
6. **Bridges, B. A.** 1995. Starvation-associated mutation in *E. coli* strains with and without reverse transcriptase. *Mutat. Res.* **347**:13–15.
7. **Cairns, J.** 1998. Mutation and cancer: the antecedents to our studies of adaptive mutation. *Genetics* **148**:1433–1440.
8. **Cairns, J., and P. L. Foster.** 1991. Adaptive reversion of a frameshift mutation in *Escherichia coli*. *Genetics* **128**:695–701.
9. **Cairns, J., J. Overbaugh, and S. Miller.** 1988. The origin of mutants. *Nature* (London) **335**:142–145.
10. **Calos, M. P., and J. H. Miller.** 1981. Genetic and sequence analysis of frameshift mutations induced by ICR-191. *J. Mol. Biol.* **153**:39–66.
11. **Carter, J. R., D. R. Patel, and R. D. Porter.** 1992. The role of *oriT* in *tra*-dependent enhanced recombination between mini-F-*lac-oriT* and lambda p*lac*5. *Genet. Res.* **59**:157–165.
12. **Christensen, R. B., J. R. Christensen, and C. W. Lawrence.** 1985. Conjugation-dependent enhancement of induced and spontaneous mutation in the *lacI* gene of *E. coli*. *Mol. Gen. Genet.* **201**:35–37.
13. **Datta, A., and S. Jinks-Robertson.** 1995. Association of increased spontaneous mutation rates with high levels of transcription in yeast. *Science* **268**:1616–1619.
14. **Davis, B. D.** 1989. Transcriptional bias: a non-Lamarckian mechanism for substrate-induced mutations. *Proc. Natl. Acad. Sci. USA* **86**:5005–5009.
15. **Demerec, M.** 1963. Selfer mutants of *Salmonella typhimurium*. *Genetics* **48**:1519–1531.
16. **Dri, A.-M., and P. L. Moreau.** 1994. Control of the LexA regulon by pH: evidence for a reversible inactivation of the LexA repressor during the growth cycle of *Escherichia coli*. *Mol. Microbiol.* **12**:621–629.
17. **Escarceller, M., J. Hicks, G. Gudmundsson, G. Trump, D. Touati, S. Lovett, P. L. Foster, K. McEntee, and M. F. Goodman.** 1994. Involvement of *Escherichia coli* DNA polymerase II in response to oxidative damage and adaptive mutation. *J. Bacteriol.* **176**:6221–6228.
18. **Feng, G., H.-C.T. Tsui, and M. E. Winkler.** 1996. Depletion of the cellular amounts of the MutS and MutH methyl-directed mismatch repair proteins in stationary-phase *Escherichia coli* K-12 cells. *J. Bacteriol.* **178**:2388–2396.
19. **Firth, N., K. Ippen-Ihler, and R. A. Skurray.** 1996. Structure and function of the F factor and mechanism of conjugation, p. 2377–2412. *In* F. C. Neidhardt, R. Curtiss III, J. L. Ingraham, E. C. C. Lin, K. B. Low, B. Magasanik, W. S. Reznikoff, M. Riley, M. Schaechter, and H. E. Umbarger (ed.), *Escherichia coli and Salmonella: Cellular and Molecular Biology*, 2nd ed. ASM Press, Washington, D.C.
20. **Fitch, W. M.** 1982. The challenges to Darwinism since the last centennial and the impact of molecular studies. *Evolution* **36**:1133–1143.
21. **Foster, P. L.** 1992. Directed mutation: between unicorns and goats. *J. Bacteriol.* **174**:1711–1716.
22. **Foster, P. L.** 1993. Adaptive mutation: the uses of adversity. *Annu. Rev. Microbiol.* **47**:467–504.
23. **Foster, P. L.** 1994. Population dynamics of a Lac⁻ strain of *Escherichia coli* during selection for lactose utilization. *Genetics* **138**:253–261.

24. Foster, P. L. 1997. Nonadaptive mutations occur on the F' episome during adaptive mutation conditions in *Escherichia coli*. *J Bacteriol*. **179:**1550–1554.

25. Foster, P. L. 1999. Are adaptive mutations due to a decline in mismatch repair? The evidence is lacking. *Mutat. Res*. **436:** 179–184.

26. Foster, P. L. 1999. Mechanisms of stationary phase mutation: a decade of adaptive mutation. *Annu. Rev. Genet*. **33:**57–88.

27. Foster, P. L., and J. Cairns. 1992. Mechanisms of directed mutation. *Genetics* **131:**783–789.

28. Foster, P. L., and J. Cairns. 1994. The occurrence of heritable *Mu* excisions in starving cells of *Escherichia coli*. *EMBO J*. **13:** 5240–5244.

29. Foster, P. L., G. Gudmundsson, J. M. Trimarchi, H. Cai, and M. F. Goodman. 1995. Proofreading-defective DNA polymerase II increases adaptive mutation in *Escherichia coli*. *Proc. Natl. Acad. Sci. USA* **92:**7951–7955.

30. Foster, P. L., and W. A. Rosche. 1999. Increased episomal replication accounts for the high rate of adaptive mutation in *recD* mutants of *Escherichia coli*. *Genetics* **152:**15–30.

31. Foster, P. L., and W. A. Rosche. 1999. Mechanism of mutation in nondividing cells; insights from the study of adaptive mutation in *Escherichia coli*. *Ann. N.Y. Acad. Sci*. **870:**133–145.

32. Foster, P. L., and J. M. Trimarchi. 1994. Adaptive reversion of a frameshift mutation in *Escherichia coli* by simple base deletions in homopolymeric runs. *Science* **265:**407–409.

33. Foster, P. L., and J. M. Trimarchi. 1995. Adaptive reversion of an episomal frameshift mutation in *Escherichia coli* requires conjugal functions but not actual conjugation. *Proc. Natl. Acad. Sci. USA* **92:**5487–5490.

34. Foster, P. L., and J. M. Trimarchi. 1995. Conjugation is not required for adaptive reversion of an episomal frameshift mutation in *Escherichia coli*. *J. Bacteriol*. **177:**6670–6671.

35. Foster, P. L., J. M. Trimarchi, and R. A. Maurer. 1996. Two enzymes, both of which process recombination intermediates, have opposite effects on adaptive mutation in *Escherichia coli*. *Genetics* **142:**25–37.

36. Frederico, L. A., T. A. Kunkel, and B. R. Shaw. 1990. A sensitive genetic assay for the detection of cytosine deamination: determination of rate constants and the activation energy. *Biochemistry* **29:**2532–2537.

37. Frost, L. S., and J. Manchak. 1998. F-phenocopies: characterization of expression of the F transfer region in stationary phase. *Microbiology* **144:**2579–2587.

38. Galitski, T., and J. R. Roth. 1995. Evidence that F plasmid transfer replication underlies apparent adaptive mutation. *Science* **268:**421–423.

39. Hall, B. G. 1990. Spontaneous point mutations that occur more often when they are advantageous than when they are neutral. *Genetics* **126:**5–16.

40. Hanawalt, P. C. 1994. Transcription-coupled repair and human disease. *Science* **266:**1957–1958.

41. Harris, R. S., G. Feng, K. J. Ross, R. Sidhu, C. Thulin, S. Longerich, S. K. Szigety, M. E. Winkler, and S. M. Rosenberg. 1997. Mismatch repair protein MutL becomes limiting during stationary-phase mutation. *Genes Dev*. **11:**2426–2437.

42. Harris, R. S., S. Longerich, and S. M. Rosenberg. 1994. Recombination in adaptive mutation. *Science* **264:**258–260.

43. Harris, R. S., K. J. Ross, and S. M. Rosenberg. 1996. Opposing roles of the Holliday junction processing systems of *Escherichia coli* in recombination-dependent adaptive mutation. *Genetics* **142:**681–691.

44. Iuchi, S., D. Furlong, and E. C. C. Lin. 1989. Differentiation of arcA, arcB, and cpxA mutant phenotypes of *Escherichia coli*

by sex pilus formation and enzyme regulation. *J. Bacteriol*. **171:**2889–2893.

45. Jayaraman, R. 1992. Cairnsian mutagenesis in *Escherichia coli*: genetic evidence for two pathways regulated by *mutS* and *mutL* genes. *J. Genet*. **71:**23–41.

46. Kowalczykowski, S.D., D.A. Dixon, A.K. Eggleston, S.C. Lauder, and W.M. Rehrauer. 1994. Biochemistry of homologous recombination in *Escherichia coli*. *Microbiol. Rev*. **58:**401–465.

47. Kumar, S., and S. Srivastava. 1983. Cyclic AMP and its receptor protein are required for expression of transfer genes of conjugative plasmid F in *Escherichia coli*. *Mol. Gen. Genet*. **190:**27–34.

48. Kunz, B. A., and B. W. Glickman. 1983. The infidelity of conjugal DNA transfer in *Escherichia coli*. *Genetics* **105:**489–500.

49. Kuzminov, A. 1995. Collapse and repair of replication forks in *Escherichia coli*. *Mol. Microbiol*. **16:**373–384.

50. Kuzminov, A., and F. W. Stahl. 1999. Double-strand end repair via the RecBC pathway in *Escherichia coli* primes DNA replication. *Genes Dev*. **13:**345–356.

51. Luria, S. E., and M. Delbrück. 1943. Mutations of bacteria from virus sensitivity to virus resistance. *Genetics* **28:**491–511.

52. Lynch, A. S., and E. C. C. Lin. 1996. Responses to molecular oxygen, p. 1526–1538. *In* F. C. Neidhardt, R. Curtiss III, J. L. Ingraham, E. C. C. Lin, K. B. Low, B. Magasanik, W. S. Reznikoff, M. Riley, M. Schaechter, and H. E. Umbarger (ed.), Escherichia coli *and* Salmonella: *Cellular and Molecular Biology*, 2nd ed. ASM Press, Washington, DC.

53. Maenhaut-Michel, G., and J. A. Shapiro. 1994. The roles of starvation and selection in the emergence of *araB-lacZ* fusion clones. *EMBO J*. **13:**5229–5244.

54. Marians, K. J., H. Hiasa, D. R. Kim, and C. S. McHenry. 1998. Role of the core DNA polymerase III subunits at the replication fork; α is the only subunit required for processive replication. *J. Biol. Chem*. **273:**2452–2457.

55. Miller, J. H. 1985. Mutagenic specificity of ultraviolet light. *J. Mol. Biol*. **182:**45–65.

56. Mittler, J. E., and R. E. Lenski. 1990. New data on excisions of Mu from *E. coli* MCS2 cast doubt on directed mutation hypothesis. *Nature* (London) **344:**173–175.

57. Müller-Hill, B., and J. Kania. 1974. *Lac* repressor can be fused to β-galactosidase. *Nature* (London) **249:**561–562.

58. Ninio, J. 1991. Transient mutators: a semiquantitative analysis of the influence of translation and transcription errors on mutation rates. *Genetics* **129:**957–962.

59. Radicella, J. P., P. U. Park, and M. S. Fox. 1995. Adaptive mutation in *Escherichia coli*: a role for conjugation. *Science* **268:**418–420.

60. Reddy, M., and J. Gowrishankar. 1997. A genetic strategy to demonstrate the occurrence of spontaneous mutations in nondividing cells within colonies of *Escherichia coli*. *Genetics* **147:** 991–1001.

61. Rosche, W. A., and P. L. Foster. 1999. The role of transient hypermutators in adaptive mutation in *Escherichia coli*. *Proc. Natl. Acad. Sci. USA* **96:**6862–6867.

62. Rosenberg, S. M., S. Longerich, P. Gee, and R. S. Harris. 1994. Adaptive mutation by deletions in small mononucleotide repeats. *Science* **265:**405–407.

63. Roth, J. R., N. Benson, T. Galitski, K. Haack, J. G. Lawrence, and L. Miesel. 1996. Rearrangements of the bacterial chromosome: formation and applications, p. 2256–2276. *In* F. C. Neidhardt, R. Curtiss III, J. L. Ingraham, E. C. C. Lin, K. B. Low, B. Magasanik, W. S. Reznikoff, M. Riley, M. Schaechter, and H. E. Umbarger (ed.), Escherichia coli *and* Salmonella:

Cellular and Molecular Biology, 2nd ed. ASM Press, Washington, D.C.

64. **Rudner, R., A. Murray, and N. Huda.** 1999. Is there a link between mutation rates and the stringent reponse in *Bacillus subtilis. Ann. N. Y. Acad. Sci.* **870:**418–422.

65. **Ryan, F. J.** 1955. Spontaneous mutation in non-dividing bacteria. *Genetics* **40:**726–738.

66. **Ryan, F. J.** 1959. Bacterial mutation in stationary phase and the question of cell turnover. *J. Gen. Microbiol.* **21:**530–549.

67. **Ryan, F. J., D. Nakada, and M. J. Schneider.** 1961. Is DNA replication a necessary condition for spontaneous mutation? *Z. Vererbungsl.* **92:**38–41.

68. **Shapiro, J. A.** 1984. Observations on the formation of clones containing *araB-lacZ* cistron fusions. *Mol. Gen. Genet.* **194:**79–90.

69. **Silverman, P. M., S. Rother, and H. Gaudin.** 1991. Arc and Sfr functions of the *Escherichia coli* K-12 arcA gene product are genetically and physiologically separable. *J. Bacteriol.* **173:**5648–5652.

70. **Sniegowski, P. D.** 1995. A test of the directed mutation hypothesis in *Escherichia coli* MCS2 using replica plating. *J. Bacteriol.* **177:**1119–1120.

71. **Stahl, F. W.** 1988. A unicorn in the garden. *Nature* (London) **335:**112–113.

71a.**Stallions, D. R., and R. Curtiss III.** 1972. Bacterial conjugation under anaerobic conditions. *J. Bacteriol.* **111:**294–295.

72. **Strathern, J. N., B. K. Shafer, and C. B. McGill.** 1995. DNA synthesis errors associated with double-strand-break repair. *Genetics* **140:**965–972.

73. **Taddei F., J. A. Halliday, I. Matic, and M. Radman.** 1997. Genetic analysis of mutagenesis in aging *Escherichia coli* colonies. *Mol. Gen. Genet.* **256:**277–281.

74. **Taddei, F., I. Matic, and M. Radman.** 1995. cAMP-dependent SOS induction and mutagenesis in resting bacterial populations. *Proc. Natl. Acad. Sci. USA* **92:**11736–11740.

75. **Torkelson, J., R. S. Harris, M.-J. Lombardo, J. Nagendran, C. Thulin, and S. M. Rosenberg.** 1997. Genome-wide hypermutation in a subpopulation of stationary-phase cells underlies recombination-dependent adaptive mutation. *EMBO J.* **16:**3303–3311.

76. **Wright, B. E.** 1997. Does selective gene activation direct evolution? *FEBS Lett.* **402:**4–8.

77. **Wright, B. E., A. Longacre, and J. M. Reimers.** 1999. Hypermutation in derepressed operons of *E. coli* K12. *Proc. Natl. Acad. Sci. USA* **96:**5089–5094.

Chapter 17

Regulation of Competence in *Bacillus subtilis* and Its Relation to Stress Response

DAVID DUBNAU AND KÜRSAD TURGAY

Competence, the ability to be genetically transformed, can be viewed as a response to stress. The regulation of competence in Bacillus subtilis *is complex since it involves multiple inputs and control at the levels of both transcription and protein stability. A quorum-sensing module that responds to two extracellular signaling molecules ensures that competence development is initiated in response to increasing population density, as a culture approaches stationary phase. The product of this quorum-sensing mechanism is the small protein ComS. ComK is the critical factor required for the transcription of genes needed for DNA uptake. ComK binds to MecA and is then degraded by a proteolytic machine composed of ClpC and ClpP. ComS interacts with MecA, causing the release of ComK, which is thereby protected from degradation. In addition, multiple controls are exerted at the promoter of* comK. *ComK is itself needed for* comK *transcription, as are the response regulator DegU and the transition state regulators AbrB and SinR. High concentrations of AbrB act negatively at the* comK *promoter, as does CodY. These proteins respond to a variety of extra and intracellular signals, and interactions at the* comK *promoter act to integrate these multiple signals. In B.* subtilis, *competence is only one of several stress responses that may be initiated at stationary phase. The signal transduction network that controls competence also regulates the other responses, some of which (like sporulation) are temporally incompatible with competence. Competence is widespread among bacteria, although its regulation varies markedly. The biological role of competence is a matter of controversy, and some current hypotheses are discussed.*

Natural competence, the subject of this review, refers to the ability of a bacterial cell to be genetically

transformed. When *Bacillus subtilis* is grown in minimal salts medium supplemented with amino acids and glucose, competence develops in about 10% of the cell population and reaches a maximum 2 h after the onset of stationary phase. Competence is the endpoint of a highly regulated developmental program, resulting in cells that are physiologically distinct from sporulating cells, from noncompetent cells in stationary phase, and from growing cells. In addition to their ability to bind and internalize macromolecular DNA, competent cells are in a nondividing state, are arrested for DNA replication, and have greatly reduced rates of stable RNA and protein synthesis. The regulation of competence has been reviewed (15, 74).

It is worth considering why a discussion of competence belongs in a book on bacterial stress response. Signaling pathways and adaptive strategies for dealing with stress have classically been elucidated by imposing adverse conditions on growing cultures in the laboratory. These conditions include heat shock, cold shock, DNA damage, and exposure to high salt, ethanol, or oxidative damage. However, the transition from logarithmic growth to stationary phase, when the population density is elevated and environmental signals or limitations of essential nutrients lead to a decrease in the rate of growth, can also be regarded as a period of stress in the growth cycle of a bacteria culture. This is biologically relevant, because *B. subtilis*, a soil organism, is certainly exposed in the wild to extreme environmental fluctuations and may hover for prolonged periods between growing and nongrowing states. Depending on conditions, this type of transition state stress induces a variety of responses in *B. subtilis*, including competence, the decision to sporulate, the synthesis of degradative enzymes and antibiotics, and increased motility (see chapters 12 and 13). In agreement with

David Dubnau and Kürsad Turgay • Public Health Research Institute, 455 First Ave., New York, NY 10016.

the theory that these are bona fide stress responses, several proteins that regulate the classical stress responses are also intimately involved in the control of competence and sporulation.

In recent years it has become more apparent that all the relevant regulatory pathways form an interconnected network that integrates the sensing, signaling, and subsequent behavior of bacterial populations. In *B. subtilis* different knots in this fabric of interwoven signal pathways are represented by the general stress proteins ClpC, ClpP, and ClpX, as well as by the transcriptional regulators ComA, ComK, SinR, AbrB, CodY, and DegU and the phosphoprotein phosphatases. These proteins are pleiotropic in their associated phenotypes, affecting sporulation, degradative enzyme synthesis, motility, competence, and the heat shock and SOS responses.

WHY TRANSFORMATION?

Several of the *B. subtilis* stationary-phase responses named above have plausible roles in dealing with stress. For instance, under conditions of nutrient depletion, extracellular degradative enzymes can increase the food supply, motility can permit escape to a more hospitable environment, and sporulation (dormancy) is a strategy of last resort. What might be the role of transformation as a stress response? Three notions have been advanced, and they are not mutually exclusive. The "transformation for food" idea holds that DNA is taken up by competent cells as a nutritional source (65, 66). This may be the case in some organisms and under certain conditions. However, in *B. subtilis* an extracellular nuclease and efficient uptake pathways for nucleolytic products are available, and it is not obvious why the elaborate and exquisitely regulated competence pathway would evolve to serve this redundant function. In *Neisseria* and *Haemophilus*, DNA uptake requires the presence of a specific recognition sequence found in closely related strains (4, 13). What would be the selective advantage of such a finicky, cannibalistic feeding mechanism? Additionally, some transformation systems, as in *B. subtilis*, take up a single strand and expel the complement into the medium as degradation products. Why throw away half the steak? It is more reasonable to consider that competence is a mechanism for the acquisition of genetic information, not mere meat.

The "transformation for repair" hypothesis suggests that exogenous DNA is taken up as a template to restore damaged DNA (64). This is certainly plausible and supported by the fact that the competence and SOS repair regulons are coinduced in *B. subtilis*

(22, 47). The third notion, "transformation for genetic variation," is also attractive. Transformation (and recombination in general) can contribute to genetic diversity in several ways. Although all genetic variation ultimately derives from mutation, transformation can generate new fitness-enhancing allelic combinations, and the horizontal transfer of novel genes between species might facilitate dramatic evolutionary change.

These arguments imply that a competent cell has a selective advantage when stressed, since it would have a greater probability of survival due to the uptake of fitness-enhancing alleles or novel genes. Transformation can therefore be regarded as a means for stressed cells to explore "genetic information space." The repair hypothesis is certainly also plausible and is not mutually exclusive with the notion that transformation promotes genetic diversity. But all of this remains speculative, and the debate concerning the role of transformation will continue. In fact, the signals and pathways that induce competence are species-specific. This may be telling us that competence serves a fixed set of needs that are advantageous for different organisms under specific conditions. Alternatively, it may serve a variety of needs, with each species adapting the ability to transport DNA to meet its particular requirements. Whichever of the three hypotheses discussed above is valid, competence may properly be regarded as a stress response.

HOW IS COMPETENCE REGULATED IN *B. SUBTILIS*? A TALE OF TWO MODULES

In *B. subtilis*, as noted above, the regulatory pathway of competence intersects with those of other stress responses, including sporulation, motility, the SOS, and heat shock responses and the secretion of macromolecule-degrading enzymes. It is important to bear in mind that because these pathways intersect at multiple points, few if any regulatory genes are specific to one stress response. In this presentation, for convenience, we will often describe competence as if it were an independent pathway, but interactions with other parts of the network will be a repeated theme.

It is useful to consider the regulation of competence as consisting of two modules (Fig. 1). Mod-

Figure 1. Simplified schematic of competence regulation.

ule 1 is a quorum-sensing machine. Its major input is population-density information mediated by the extracellular accumulation of a pair of pheromones. Its output is a small protein, ComS. Module 2 accepts ComS as its major input while its output is the synthesis of ComK. ComK is a transcription factor that is necessary and sufficient for the expression of the late competence genes (19, 79) and the SOS pathway (22). These late genes encode DNA binding and transport proteins, located in the cell surface layers. The transcription of *comK* is itself ComK-dependent. Because of this positive feedback loop, the increase in ComK during competence development is explosive. We will describe the operation of these modules and how they are influenced by secondary input signals.

Module 1: Quorum Sensing

The operation of module 1

This module is summarized in Fig. 2, and quorum-sensing systems in general are reviewed in chapter 18. ComS, the output of module 1, is a 46-residue protein required for the synthesis of ComK and hence for the development of competence. *comS* is embedded within the *srfA* operon (8, 23) and is transcribed from the *srfA* promoter. The phosphorylated form of a transcription factor, ComA, is required for optimal transcription of *srfA(comS)* and ComA~P has been shown to bind upstream from the *srfA* promoter (54, 55, 67). ComA is a response regulator protein, and its cognate histidine kinase, ComP, is encoded immediately upstream from *comA* on the *B. subtilis* chromosome. Upstream from *comP* is a small open reading frame, *comX*, and a larger

one, *comQ*, that overlaps *comX*. The Grossman lab has established that *comX* is translated as a protein of 55 amino acids that is processed and exported as a derivatized decapeptide, derived from the C-terminus of pre-ComX (49). The synthesis of active ComX requires ComQ. Mature ComX accumulates in the medium as a culture approaches stationary phase, acting as a primary input signal for module 1. The N-terminal 380 amino acids of the histidine kinase ComP comprise a largely hydrophobic anchor with 6 to 8 membrane-spanning segments and 2 large extracellular loops (62, 81). It is likely that ComX interacts directly with ComP, initiating a phosphorylation cascade that culminates in the accumulation of ComA~P. A second exported factor, CSF (competence and sporulation factor), also acts as a signaling molecule for module 1 (75). CSF is a pentapeptide (ERGMT) that is processed by cleavage from the *phrC* gene product and exported by an unknown mechanism. Although ComX probably acts extracellularly, CSF is internalized by an oligopeptide permease (42) and appears to prevent the dephosphorylation of ComA~P by inhibiting a phosphoprotein phosphatase (RapC) in the cytosolic compartment (59). It is not clear whether this inhibition of RapC is direct (60). The two pheromone signaling pathways therefore converge, elevating the intracellular concentration of ComA~P (73), thereby increasing the rate of transcription of *srfA(comS)*.

Additional module 1 inputs

The transition state regulatory protein SinR positively regulates ComS synthesis, possibly on a posttranscriptional level (45). Polynucleotide phosphorylase (PnpA) is required for ComS synthesis and also appears to act posttranscriptionally (48). AbrB and DegU~P act negatively on *srfA(comS)* transcription (18, 31), possibly as direct repressors. Finally, CodY acts as a repressor at the promoter of *srfA* (71). The activities and amounts of these proteins are regulated by specific mechanisms, discussed below.

Module 2: Action, Sequestration, and Degradation of ComK

ComK as a transcription factor

As noted above, the input and output of module 2 are, respectively, ComS and ComK. Before discussing the internal workings of module 2, it is pertinent to describe the DNA-binding activity of ComK. ComK acts not only at its own promoter but at those of all the known late competence genes (79, 80). It also activates the promoters of the recombination/

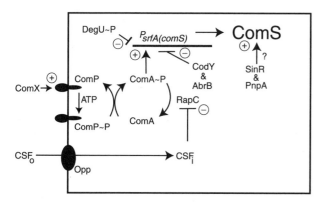

Figure 2. Module 1, the quorum-sensing mechanism. The output of this module is ComS. The principal (quorum-sensing) inputs are the pheromones ComX and CSF. Plus and minus signs represent positive and negative effects. The box represents the cell membrane. The positive effects of SinR and PnpA on the yield of ComS are not understood.

repair genes *recA*, *addAB*, and *dinA*, establishing a link between the regulation of competence and the SOS regulon (21, 22). A consensus sequence for ComK binding (AAAAN$_5$TTTT) has been defined (24). This palindromic sequence is always present in two copies, separated by an integral number of helical turns, placing the recognition sequences on the same face of the helix. However, the number of turns in the spacer varies dramatically in different promoters. For instance, in the recombination/repair gene binding sites, the AT boxes are repeated after two helical turns; in the *comC*, *comE*, *comG*, and *comF* late competence gene sites, repetition occurs after three turns; and in the *comK* site, after four turns. This variation may reflect the action of additional proteins at these sites. For instance, in the case of *comG*, ComK is sufficient in vitro to drive transcription, and genetic evidence suggests that this is true of all the so-called late competence genes that have the three-turn spacing (19). *comK* transcription, on the other hand, is modulated by at least three other proteins that bind to its promoter, and the unique spacing found in this promoter may accommodate the binding of these proteins. In agreement with the paired palindromic nature of all known ComK-binding sites, it has been shown that ComK binds to the *comK* promoter as a tetramer, probably consisting of a pair of dimers (24).

Three proteins (SinR, AbrB, DegU), in addition to ComK itself and ComS, are known to be required for the synthesis of ComK. All but ComS act at the level of transcription, because their need is bypassed when *comK* is driven by *P-xyl* (19). In this respect they are distinct from ComS, which is still required in the *P-xyl comK* strain because it acts posttranslationally. SinR and DegU also act as modulators of module 1, as noted above.

MecA and ClpC

In addition to the transcriptional regulation of *comK*, ComK is controlled posttranslationally. Two key players in this control are MecA and ClpC (originally called MecB). The genes encoding these proteins were identified by mutations that permitted the expression of late competence proteins during exponential growth and in complex media that ordinarily do not support competence (10, 68). These were shown to be loss-of-function mutations, and *mecB* was subsequently characterized as the *B. subtilis clpC* gene (52).

ClpC belongs to the extensively studied and widespread family of HSP100 proteins (70). On the basis of sequence analysis, these proteins can be classified into several subfamilies. *B. subtilis*, for instance, possesses single members of the ClpY, ClpX, and ClpE subfamilies in addition to ClpC. ClpC contains two ATP-binding sites, whereas members of the ClpX family contain a single site. Several of the HSP100 proteins have been shown to act as chaperones, and all of them appear to be ATPases. As their name implies, HSP100 family members are often heat shock proteins. ClpC is no exception (37, 52), and the inactivation of *clpC* in *B. subtilis* results in decreased thermotolerance (37, 52). Inactivation of *clpX* has a profound negative effect on competence development (J. Hahn and D. Dubnau, unpublished data). The mechanism of this effect is not understood. In contrast, inactivation of *clpE* (5) or *clpY* (J. Hahn, unpublished data) confers no apparent competence phenotypes.

Several of the *Escherichia coli* HSP100 proteins (ClpA, ClpX) form multisubunit complexes with the serine protease ClpP (14, 33), also a heat shock protein in *B. subtilis* (12, 51). These complexes, capable of protein degradation, are barrel-like structures resembling the more complex eukaryotic proteasome (33). A hexamer ring of ClpA subunits abuts one or both surfaces of two 7-membered rings of ClpP. The ClpP rings enclose a space, on the walls of which 14 ClpP proteolytic sites are exposed. The chaperone activity of ClpA unfolds a substrate protein, which is translocated into the enclosed space of the ClpP rings where it is degraded. Since ClpC and ClpP are both heat shock proteins in *B. subtilis*, it is reasonable to propose that they work like ClpAP in *E. coli* to degrade heat-damaged proteins. This background information suggests that the chaperone activity of ClpC, its role in proteolysis, or both activities are required for the regulation of competence.

The sequestration model

The inactivation of either *mecA* or *clpC* results in greatly enhanced expression of the late competence genes and of *comK* (10, 34, 35, 68). On the other hand, overexpression of the *mecA* or *clpC* gene products has distinct effects. When MecA is overexpressed, the transcription of *comK* is completely suppressed, and this phenotype is the same in a *clpC* null mutant background (34). In contrast, overexpression of ClpC has no detectable competence phenotype. These genetic data led to the prediction that MecA is an effector protein and that ClpC acts to potentiate the action of MecA. When the intracellular MecA concentration is high, as in the overexpressing strain, the need for ClpC is bypassed. A complication in this system derives from the dependence of *comK* transcription on ComK. Thus the observed negative effect of MecA on *comK* transcription might indicate that MecA represses the *comK* promoter or that it inactivates ComK posttranscriptionally. In vitro experiments with purified MecA revealed that MecA could bind to ComK, whereas no evidence for binding to the *comK* promoter was obtained (34). It was shown that ClpC, in the presence of ATP, formed a ternary complex with MecA and ComK and increased the affinity of MecA for ComK (78). When sequestered in this way by binding to MecA/ClpC, ComK was no longer available to bind to the *comK* promoter. As expected, the in vitro addition of ComS could reverse this binding, causing the release of ComK. Following release, ComK was once again able to bind to its *comK* target.

These genetic and biochemical data led to the sequestration model for ComK regulation (78). It was proposed that during exponential growth, any ComK present would be bound in a ternary complex with MecA and ClpC, and thereby inactivated as a transcription factor. Because *comK* transcription requires ComK binding to its own promoter, the synthesis of ComK would be limited by this sequestration mechanism during exponential growth. However, when ComS, the output of module 1, reached a sufficient intracellular concentration, ComK would be released to drive its own transcription, leading to the rapid activation of the late competence promoters.

Regulation of ComK degradation

The sequestration model seemed capable of explaining all the known physiological and genetic observations concerning ComK regulation. However, further experiments suggested that this was not the whole story. Measurements of the content of ClpC, MecA, and ComK in fully competent cells revealed that ComK was present in vast excess. In fact, there were more than 20,000 ComK tetramers per competent cell and only about 1,000 MecA dimers and ClpC hexamers (77). This observation posed a problem for the sequestration model in explaining the escape from the competent state. It had been known for many years that competent cells exhibit a long delay before resuming growth after dilution into fresh medium (56). *mecA* null mutants lose viability when they reach stationary phase, whereas *mecA comK* double mutants do not (17). From this it was inferred that the expression of ComK in the *mecA* background prevented the resumption of growth because MecA was needed to inactivate ComK. This was confirmed by the finding that the introduction of a cloned copy of MecA by transformation into a competent *mecA* culture was able to restore the viability of the transformants. The sequestration model requires that ComK be rebound by the MecA/ClpC

complex, perhaps as the ComS signal is inactivated. However, the discovery that ComK is present in large excess in competent cells precluded this escape mechanism and suggested that ComK might be degraded during escape. This was tested in vivo, and it was shown that ComK is indeed degraded by a mechanism that requires both MecA and ClpC (77). Overexpression of ComS, which would lead to dissociation of ComK from the MecA/ComK/ClpC ternary complex, delayed the degradation of ComK.

These observations suggested that MecA might be targeting ComK for degradation by ClpCP. Inactivation of the *B. subtilis clpP* gene revealed a complication, because this resulted in the accumulation of excess MecA, with a consequent marked drop in ComK synthesis. This difficulty was bypassed by expressing *comK* from a conditional promoter (*P-xyl*) that responds to xylose in the medium. This construct freed ComK synthesis from its dependence on ComK, thus insulating *comK* transcription from the effect of excess MecA. When the *clpP* mutation was introduced into the *P-xyl comK* background, ComK accumulated dramatically, supporting the postulated role of ClpP in the degradation of ComK (77). In vitro experiments using purified proteins confirmed that in the presence of ATP, ComK was targeted by MecA for degradation by ClpCP. Under these conditions MecA was also slowly degraded, consistent with the in vivo data. When ComS was added, ComK was protected from degradation, whereas the rate of MecA degradation increased and ComS was also proteolyzed (77).

These observations occasioned a modification of the original sequestration model (Fig. 3). It appears that during exponential growth, the targeting of ComK to the ClpCP proteasome by the binding of both to MecA results in the degradation of ComK. As ComS accumulates in the cell, ComK is released, and the rate of MecA degradation increases, possibly accelerating the net release of ComK. ComK is thus protected and late competence genes are transcribed. ComS degradation by ClpCP ensues, perhaps acting as a timing mechanism; as the ComS concentration decreases, ComK is retargeted for degradation, and the cells are able to escape from the competent state. It is worth noting that the inhibition of ComK synthesis that occurs when MecA is overexpressed in a *clpC* mutant demonstrates that sequestration in the absence of degradation can apparently also inhibit competence, at least under the artificial conditions of MecA hyperaccumulation.

Protein-protein interactions and the domain structure of MecA

ClpC was shown to be an ATPase (78). When ComK or ComS was added, a stimulation of this

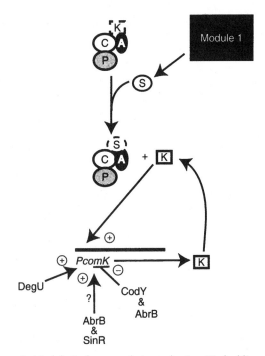

Figure 3. Module 2, the proteolytic mechanism. Dashed lines represent proteolysis. Plus and minus signs represent positive and negative effects. The positive effects of AbrB and SinR on the transcription of *comK* are not understood.

ATPase activity was noted, but only when MecA was also present. From this it was inferred that ComS probably acted by binding to MecA. It had already been established that ComK and ClpC also were able to bind to MecA. These interactions have been confirmed by cross-linking and surface plasmon resonance binding experiments with both purified proteins and with crude extracts (57, 61).

Controlled proteolysis of MecA in solution, followed by N-terminal sequence analysis of proteolytic products, suggested that MecA consists of N- and C-terminal domains, separated by an exposed protease-sensitive linker (61). These domains were cloned individually and expressed as His-tagged products. Surface plasmon resonance binding studies revealed that the NTD of MecA interacted with ComS and ComK, whereas the CTD interacted only with ClpC (61). Cross-linking and gel filtration analysis demonstrated that MecA, its NTD, and its CTD all behaved as homodimers in solution, suggesting that dimerization surfaces were present on both domains.

The role of MecA as an adapter protein

MecA acts in this system as an adapter protein. By binding to both ComK and to ClpC it adapts the general stress-related proteolytic machinery (ClpCP) to the degradation of a transcription factor. It also

couples the output of module 1 (ComS) to the turnoff of this proteolytic mechanism, resulting in the activation and protection of ComK. We do not know whether this is a general mechanism or whether the role of MecA in competence is unique. However, several considerations suggest that it may be general. For instance, database searches have revealed MecA orthologs in *Streptococcus pneumoniae*, *Streptococcus mutans*, *Streptococcus pyogenes*, and *Staphylococcus aureus*. A MecA ortholog was also identified in *Listeria monocytogenes* (E. Borezée and P. Berche, personal communication). Finally, a MecA ortholog from *Bacillus firmus* has been reported (16). Of these organisms only *S. pneumoniae* and *S. mutans* have been shown to be competent. It is possible that MecA targets proteins other than ComK for degradation in these systems and that additional signaling molecules equivalent to ComS are at work. In *B. subtilis* we have identified a close paralog of MecA (YpbH) in the database, of unknown function. Finally, our preliminary data (M. Persuh and D. Dubnau) suggest that MecA may play a role in the regulation of sporulation and of autolysins in *B. subtilis*, independent of its role in the regulation of ComK.

How does the regulation of degradation work?

Although the structure of the ClpCP complex has not yet been addressed, we assume that it forms a multisubunit barrel-like particle similar to that of ClpAP and that the chaperone activity of ClpC unfolds ComK and translocates the unwound protein strand into the lumen of the complex for degradation. Because the NTD domain of MecA binds to ComK whereas the CTD binds to ClpC, and ClpC somehow increases the affinity of MecA for ComK (78), these domains must communicate with one another. It is possible that the binding of ClpC displaces the CTD from a position in which it sterically interferes with the binding of the NTD to ComK. Alternatively the binding of ClpC to the CTD may cause a conformational change that is propagated to the NTD, increasing the affinity of the latter for ComK.

The binding of the NTD of MecA to ComK and ComS must present the latter two proteins to ClpC for unwinding. In this connection it is interesting that the DnaJ cochaperone has been reported to stabilize the binding of a substrate protein to DnaK by closing the DnaK binding pocket (40). This process is accompanied by stimulation of the DnaK ATPase activity. MecA works analogously; it apparently transfers ComK and ComS to ClpCP for degradation and stimulates ATP hydrolysis by ClpC. Perhaps similar mechanisms are used by both systems. ComS releases ComK from binding to MecA. Because ComS and

ComK both bind to the NTD of MecA, this may occur by competition for overlapping sites or by a ComS-induced conformational change in MecA.

Additional module 2 inputs

DegU is a response regulator protein required for the synthesis of degradative enzymes (38). For this, DegU receives a phosphoryl group from the histidine kinase DegS. However, for competence expression, *degU* is needed but *degS* is not. It has been suggested that the unphosphorylated form of DegU is required for competence (38, 53). In fact, certain mutations in *degU* or *degS* that result in hyperphosphorylation of DegU strongly inhibit the synthesis of ComK, probably acting to repress *srfA(comS)* transcription (18). Because DegU is needed for *comK* transcription, and because it is no longer needed when ComK is overproduced in a *mecA* mutant (68), it was postulated that DegU acts as a "priming protein," required only at the onset of competence induction when the concentration of ComK is low (19). Recently it has been established that DegU binds at the *comK* promoter, to a site between the two AT boxes, and that it acts by increasing the affinity of ComK for its own promoter (L. W. Hamoen, A. F. Van Werkhoven, G. Venema, and D. Dubnau, submitted). These results are in accord with the priming hypothesis.

AbrB and SinR act positively on *comK* transcription in unknown ways. CodY and AbrB act negatively at the *comK* promoter, perhaps by sterically blocking ComK binding (71). CodY represses *comK* and *srfA* transcription in response to rapid growth rate, and this repression no longer operates after the transition to stationary phase (11, 71).

Coordination with Other Pathways

CodY, SinR, and AbrB each acts to modulate the synthesis of ComK in response to a distinct signaling pathway (Fig. 4). The requirement for unphosphorylated DegU or for low concentrations of DegU~P may mean that competence is vetoed when conditions exist that phosphorylate this response regulator. The signals that determine the level of phosphorylation of DegU are not understood (but see references 39 and 69). This implies that competence and degradative enzyme synthesis may be mutually exclusive. AbrB, a repressor of sporulation, is downregulated when the response regulator SpoOA is phosphorylated by its own phosphorelay cascade (76) (Fig. 4). As noted above, AbrB plays both positive and negative roles in competence and must be present within

a narrow concentration range for ComK to be synthesized (20). As SpoOA is phosphorylated, the concentration of AbrB in the cell decreases and the repression at *comK* is therefore relieved. When the amount of AbrB decreases further, *comK* transcription will decrease and the sporulation pathway may be activated. A related mechanism appears to operate in the case of SinR (Fig. 4). This protein is required for competence but inhibits sporulation (72). SinR is antagonized by binding to SinI, which is itself synthesized in response to SpoOA~P (1). Thus, conditions that activate sporulation will inhibit competence via this pathway. It has been reported that excess SinR also represses the synthesis of ComS (45). Thus, like AbrB, SinR may also act within a certain concentration range to permit competence development, and both proteins may serve to ensure the proper relative timing of competence and sporulation. From the above discussion it is clear that the strengths of the signals that activate the SpoOA phosphorelay will influence the probability that a cell will enter the competence or sporulation pathways. These signals include starvation, DNA replication, bacterial population density, and the tricarboxylic acid cycle (27–30). All of the inputs described above serve as checks on the regulation of *comK* transcription via the degradative mechanism described above. They make competence responsive to nutritional and cell cycle signals that modulate the signals flowing through modules 1 and 2, and they serve to temporally coordinate competence, sporulation, and probably other stationary-phase stress responses (Fig. 4).

As noted previously, at low concentrations CSF contributes to competence induction by inhibiting RapC, the ComA-specific phosphatase. However, at concentrations above about 20 nM, CSF stimulates sporulation while inhibiting competence by an unknown mechanism. The stimulation of sporulation may be due to the inhibition of RapB, a phosphorelay-specific phosphatase (59). Phosphorylated ComA not only activates the transcription of *srfA* but also the transcription of the *rapA* operon, which encodes another phosphorelay-specific phosphatase. This may serve to delay the phosphorylation of Spo0A and consequently the initiation of sporulation (58). The *rapC* operon is also positively activated by phosphorylated ComA (41). This establishes a negative feedback since ComA~P induces the synthesis of a ComA~P-specific phosphatase. Since *phrC* is cotranscribed with *rapC*, ComA~P also serves to further increase the concentration of CSF, which negatively affects competence development at high concentration while stimulating sporulation. The CSF concentration is further increased since *phrC* is driven by a second, sigma H-dependent promoter (3), and sigma

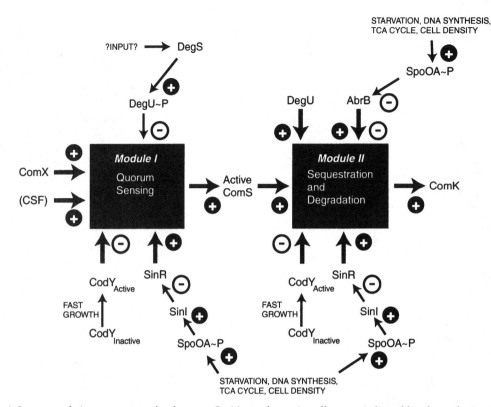

Figure 4. Inputs regulating competence development. Positive and negative effects are indicated by plus and minus signs.

H activity increases as the cells reach stationary phase. Like the SinR and AbrB pathways described above, these mechanisms probably serve a timing function, ensuring that competence precedes sporulation.

ComS, the output of module 1, is a small open reading frame embedded in the *srfA* operon. *srfA* encodes the surfactin synthetase. Synthesis of the secondary metabolite surfactin therefore also responds to ComA~P, as does transcription of *rapA*, *rapC*, and *degQ*. It is likely that additional genes are activated by ComA~P (41). These stationary phase responses are in effect branching outputs from module 1.

Mutations or deletions of the three module 2 competence switch proteins ClpC, ClpP, and MecA have pleiotropic effects on the physiology of *B. subtilis*, affecting motility, heat tolerance, and sporulation (10, 35, 37, 45, 52, 63). The effect of *mecA* and *clpC* mutation on motility is partly understood. The ComK-driven expression of *comF* leads to readthrough into *flgM* (44). FlgM is an anti-sigma factor, inhibiting sigma D-dependent gene expression, and flagellar synthesis and chemotaxis require sigma factor D activity. This explains the negative effect on motility and chemotaxis of *mecA* and *clpC* mutations, since these cause overproduction of ComK and hence enhanced read-through. This is only part of the story,

since inactivation of *comK* only partially suppresses the effect of *mecA* mutation on motility (44).

Mutations in *clpC* reduce sporulation in *B. subtilis* (10), and mutations in *clpP* also block sporulation at an early stage (51). In contrast, mutations in *mecA* lead to a glucose-resistant sporulation phenotype (10, 51) while overproduction of MecA inhibits sporulation (35). Although the mechanisms of these effects on sporulation are not understood, they serve to further illustrate the interdependence of these various stress response systems.

clpC and *clpP* are class 3 stress response genes (26). As such, they are driven by both sigma factor A and sigma factor B forms of RNA polymerase. This establishes a link between the processes and pathways controlled by *clpC* and *clpP*, and the general stress response *sigB* regulon (26) (see chapter 12). In addition, the class 3 genes are negatively regulated by CtsR, encoded as the first open reading frame of the *clpC* operon (6, 36). CtsR is itself degraded by ClpCP under stress conditions (E. Krüger, E. Witt, S. Ohlmeier, R. Hanschke, and M. Hecker, submitted). Consequently this protease plays a key role in the regulation of its own synthesis and of other class 3 genes. This demonstrates once again the intimate relationship of competence regulation to the entire stress response regulatory network.

Competence as a Cellular Differentiation System

Two cell types in competence

In *B. subtilis* competence is expressed under optimal conditions in about 10% of the cells. Competent and noncompetent cells are characterized by dramatically different transcriptional programs. Competent cells transcribe *comK* and therefore all the late competence genes, as well as *recA*, *addAB*, *dinA*, *nucA*, and probably other genes not yet identified. In addition, DNA replication and bulk RNA and protein synthesis are markedly decreased in the competent cells (7, 50, 56). Upon dilution into fresh growth medium, the noncompetent cells resume growth after a brief lag. They elongate, nuclear bodies duplicate and segregate, Z rings form, and the cells divide. We have shown microscopically that these processes are arrested in the competent cells for about 2 h (J. Hahn, B. J. Haijema, and D. Dubnau, unpublished). For these studies the competent cells were distinguished using a fusion of *comK* to green fluorescent protein or by immunofluorescence with anti-ComK antisera. It is clear that there is a profound physiological distinction between the two cell types coexisting in the same culture. What is the role of this heterogeneity and how does it arise?

The role of cell type heterogeneity in competence

As noted above, the regulation of the various stationary-phase stress responses relies on a shared set of gene products. Yet in at least some cases these responses are mutually exclusive in a given cell. For instance, sporulation is not compatible with the other stress responses. The same is apparently true of competence and motility (44). Perhaps heterogeneity permits a population of cells to explore several survival strategies simultaneously. By restricting these choices to different cells in a given population, the probability that at least some cells will survive is enhanced. It will be interesting to determine whether similar heterogeneity is present with respect to each of the stress responses. For instance, do all cells synthesize degradative enzymes? Are those that do distinct from those that express competence or motility?

How does the population heterogeneity with respect to competence arise?

ComK is clearly a determinative molecule, in the sense that cells in which ComK is synthesized become competent. How is ComK synthesis restricted to just a few cells? When *mecA* is inactivated, nearly 100% of the cells in a culture become competent (17). This suggests that in the wild-type strain, ComK may be protected by ComS from degradation by the MecA/ClpC degradative machinery in only those cells fated for competence. In fact, when ComS is overexpressed, nearly 100% of the cells in a culture also become competent, as determined microscopically using a ComK-GFP fusion (M. Albano and D. Dubnau, unpublished data). So it is attractive to consider that the concentration of ComS determines whether a given cell will become competent. However, this possibility must be evaluated cautiously because it was suggested by data obtained using nonphysiological conditions, and ComS overproduction may simply overwhelm the normal control mechanism. Whatever the initial determinative molecule, it is an open question whether the heterogeneity is regulated, perhaps by a feedback mechanism, or whether it is determined stochastically.

COMPETENCE IN OTHER BACTERIA

Transformability has been reported in a wide variety of eubacterial species, both gram positive and gram negative (46). However, regulatory studies have been largely restricted to only a few of these. Competence in *Neisseria gonorrhoeae* appears to be expressed constitutively (2). In *S. pneumoniae*, competence is expressed during growth, in response to a quorum-sensing system that responds to a peptide pheromone (25). As in *B. subtilis*, the extracellular accumulation of this peptide initiates a phosphorylation cascade by interacting with a membrane-localized histidine kinase. However, the downstream segment of the streptococcal regulatory pathway differs from that in *B. subtilis*, since the phosphorylated response regulator (ComE~P) activates the transcription of a competence sigma factor gene. This sigma factor apparently initiates transcription of the DNA uptake genes (43). Competence in the gram-negative bacterium *H. influenzae* is regulated differently, apparently under the positive control of cAMP and of the regulatory gene *sxy* (32, 82). Although the DNA uptake proteins are quite similar in all systems that have been studied (9), the regulatory systems that govern the synthesis of these proteins obviously differ. Probably different species need to develop competence at specific times in their life cycles and in response to particular signals.

PERSPECTIVES

Although much is known about competence regulation in *B. subtilis*, several areas deserve future attention, and several will now be mentioned. One question concerns the mechanism that determines the cellular heterogeneity in competence development. Another subject deserving attention is the molecular mechanism of transmembrane signal transduction during quorum sensing. How does ComP respond to the presence of ComX, and how does this response result in the transfer of phosphate to ComA? The molecular details of MecA ClpC, ClpP, and ComK interactions and the roles of ATP in these interactions, as well as the secrets of ComS interaction with MecA, are all fascinating and amenable subjects for experimentation. The complex protein-protein and protein-DNA interactions at the *comK* promoter are unique and deserving of attention. Finally, our understanding of the spatial and temporal interrelationships of the various stationary-phase stress responses is incomplete. Can competent cells efficiently go on to sporulate? Is the competent subpopulation motile or nonmotile? Do the competent cells secrete degradative enzymes and antibiotics? And what are the mechanisms that govern these mutually exclusive or coordinate responses?

Acknowledgments. We thank all the members of our lab for constant and useful input during and prior to the preparation of this review.

The work carried out by our group was supported by National Institutes of Health grants GM57720 and GM43756.

REFERENCES

1. **Bai, U., I. Mandic-Mulec, and I. Smith.** 1993. SinI modulates the activity of SinR, a developmental switch protein of *Bacillus subtilis*, by protein-protein interaction. *Genes Dev.* **7:**139–148.

2. **Biswas, G. D., T. Sox, and P. F. Sparling.** 1977. Factors affecting genetic transformation of *Neisseria gonorrhoeae. J. Bacteriol.* **129:**983–992.

3. **Carter, H. L. D., K. M. Tatti, and C. P. Moran, Jr.** 1990. Cloning of a promoter used by sigma H RNA polymerase in *Bacillus subtilis. Gene* **96:**101–105.

4. **Danner, D. B., R. A. Deich, K. L. Sisco, and H. O. Smith.** 1980. An eleven-base-pair sequence determines the specificity of DNA uptake in *Haemophilus* transformation. *Gene* **11:**311–318.

5. **Derre, I., G. Rapoport, K. Devine, M. Rose, and T. Msadek.** 1999. ClpE, a novel type of HSP100 ATPase, is part of the CtsR heat shock regulon of *Bacillus subtilis. Mol. Microbiol.* **32:**581–593.

6. **Derré, I., G. Rapoport, and T. Msadek.** 1999. CtsR, a novel regulator of stress and heat shock response, controls *clp* and molecular chaperone gene expression in gram-positive bacteria. *Mol. Microbiol.* **31:**117–131.

7. **Dooley, D. C., C. T. Hadden, and E. W. Nester.** 1971. Macromolecular synthesis in *Bacillus subtilis* during development of the competent state. *J. Bacteriol.* **108:**668–679.

8. **D'Souza, C., M. M. Nakano, and P. Zuber.** 1994. Identification of *comS*, a gene of the *srfA* operon that regulates the establishment of genetic competence in *Bacillus subtilis. Proc. Natl. Acad. Sci. USA* **91:**9397–9401.

9. **Dubnau, D.** 1999. DNA uptake in bacteria. *Annu. Rev. Microbiol.* **53:**217–244.

10. **Dubnau, D., and M. Roggiani.** 1990. Growth medium-independent genetic competence mutants of *Bacillus subtilis. J. Bacteriol.* **172:**4048–4055.

11. **Fisher, S. H., K. Rohrer, and A. E. Ferson.** 1996. Role of CodY in regulation of the *Bacillus subtilis* hut operon. *J. Bacteriol.* **178:**3779–3784.

12. **Gerth, U., E. Krüger, I. Derré, T. Msadek, and M. Hecker.** 1998. Stress induction of the *Bacillus subtilis clpP* gene encoding a homologue of the proteolytic component of the Clp protease and the involvement of ClpP and ClpX in stress tolerance. *Mol. Microbiol.* **28:**787–802.

13. **Goodman, S. D., and J. J. Scocca.** 1988. Identification and arrangement of the DNA sequence recognized in specific transformation of *Neisseria gonorrhoeae. Proc. Natl. Acad. Sci. USA* **85:**6982–6986.

14. **Grimaud, R., M. Kessel, F. Beuron, A. C. Steven, and M. R. Maurizi.** 1998. Enzymatic and structural similarities between the *Escherichia coli* ATP-dependent proteases, ClpXP and ClpAP. *J. Biol. Chem.* **273:**12476–12481.

15. **Grossman, A. D.** 1995. Genetic networks controlling the initiation of sporulation and the development of genetic competence in *Bacillus subtilis. Ann. Rev. Genet.* **29:**477–508.

16. **Guo, D., and B. E. Tropp.** 1998. Cloning of the *Bacillus firmus* OF4 cls gene and characterization of its gene product. *Biochim. Biophys. Acta* **1389:**34–42.

17. **Hahn, J., J. Bylund, M. Haines, M. Higgins, and D. Dubnau.** 1995. Inactivation of *mecA* prevents recovery from the competent state and the partitioning of nucleoids in *Bacillus subtilis. Mol. Microbiol.* **18:**755–767.

18. **Hahn, J., and D. Dubnau.** 1991. Growth stage signal transduction and the requirements for *srfA* induction in the development of competence. *J. Bacteriol.* **173:**7275–7282.

19. **Hahn, J., A. Luttinger, and D. Dubnau.** 1996. Regulatory inputs for the synthesis of ComK, the competence transcription factor of *Bacillus subtilis. Mol. Microbiol.* **21:**763–775.

20. **Hahn, J., M. Roggiani, and D. Dubnau.** 1995. The major role of SpoOA in genetic competence is to downregulate *abrB*, an essential competence gene. *J. Bacteriol.* **177:**3601–3605.

21. **Haijema, B. J., L. W. Hamoen, J. Kooistra, G. Venema, and D. van Sinderen.** 1995. Expression of the ATP-dependent deoxyribonuclease of *Bacillus subtilis* is under competence-mediated control. *Mol. Microbiol.* **15:**203–211.

22. **Haijema, B. J., D. van Sinderen, K. Winterling, J. Kooistra, G. Venema, and L. W. Hamoen.** 1996. Regulated expression of the *dinR* and *recA* genes during competence development and SOS induction in *Bacillus subtilis. Mol. Microbiol.* **22:**75–85.

23. **Hamoen, L. W., H. Eshuis, J. Jongbloed, G. Venema, and D. van Sinderen.** 1995. A small gene, designated *comS*, located within the coding region of the fourth amino acid-activation domain of *srfA*, is required for competence development in *Bacillus subtilis. Mol. Microbiol.* **15:**55–63.

24. **Hamoen, L. W., A. F. Van Werkhoven, J. J. E. Bijlsma, D. Dubnau, and G. Venema.** 1998. The competence transcription factor of *Bacillus subtilis* recognizes short A/T-rich sequences arranged in a unique, flexible pattern along the DNA helix. *Genes Dev.* **12:**1539–1550.

25. **Håverstein, L. S., and D. A. Morrison.** 1999. Quorum sensing and peptide pheromones in streptococcal competence for genetic transformation, p. 9–26. *In* G. M. Dunny and S. C. Winans (ed.), *Cell-Cell Signaling in Bacteria.* American Society for Microbiology, Washington, D.C.

26. **Hecker, M., and U. Volker.** 1998. Non-specific, general and multiple stress resistance of growth-restricted *Bacillus subtilis* cells by the expression of the sigmaB regulon. *Mol. Microbiol.* **29:**1129–1136.

27. Ireton, K., and A. D. Grossman. 1992. Coupling between gene expression and DNA synthesis early during development in *Bacillus subtilis*. *Proc. Natl. Acad. Sci.* **89:**8808–8812.

28. Ireton, K., and A. D. Grossman. 1994. A developmental checkpoint couples the initiation of sporulation to DNA replication in *Bacillus subtilis*. *EMBO J.* **13:**1566–1573.

29. Ireton, K., S. Jin, A. D. Grossman, and A. L. Sonenshein. 1995. Krebs cycle function is required for activation of the SpoOA transcription factor in *Bacillus subtilis*. *Proc. Natl. Acad. Sci. USA* **92:**2845–2849.

30. Ireton, K., D. Z. Rudner, K. Jaacks-Siranosian, and A. D. Grossman. 1993. Integration of multiple developmental signals in *Bacillus subtilis* through the SpoOA transcription factor. *Genes Dev.* **7:**283–294.

31. Jaacks, K. J., J. Healy, R. Losick, and A. D. Grossman. 1989. Identification and characterization of genes controlled by the sporulation-regulatory gene *spoOH* in *Bacillus subtilis*. *J. Bacteriol.* **171:**4121–4129.

32. Karudapuram, S., and G. J. Barcak. 1997. The *Haemophilus influenzae* dprABC genes constitute a competence-inducible operon that requires the product of the *tfoX* (*sxy*) gene for transcriptional activation. *J. Bacteriol.* **179:**4815–4820.

33. Kessel, M., M. R. Maurizi, B. Kim, E. Kocsis, B. L. Trus, S. K. Singh, and A. C. Steven. 1995. Homology in structural organization between *E. coli* ClpAP protease and the eukaryotic 26 S proteasome. *J. Mol. Biol.* **250:**587–594.

34. Kong, L., and D. Dubnau. 1994. Regulation of competence-specific gene expression by Mec-mediated protein-protein interaction in *Bacillus subtilis*. *Proc. Natl. Acad. Sci. USA* **91:**5793–5797.

35. Kong, L., K. J. Siranosian, A. D. Grossman, and D. Dubnau. 1993. Sequence and properties of *mecA*, a negative regulator of genetic competence in *Bacillus subtilis*. *Mol. Microbiol.* **9:**365–373.

36. Kruger, E., and M. Hecker. 1998. The first gene of the *Bacillus subtilis clpC* operon, *ctsR*, encodes a negative regulator of its own operon and other class III heat shock genes. *J. Bacteriol.* **180:**6681–6688.

37. Krüger, E., U. Völker, and M. Hecker. 1994. Stress induction of *clpC* in *Bacillus subtilis* and its involvement in stress tolerance. *J. Bacteriol.* **176:**3360–3367.

38. Kunst, F., T. Msadek, J. Bignon, and G. Rapoport. 1994. The DegS/DegU and ComP/ComA two-component systems are part of a network controlling degradative enzyme synthesis and competence in *Bacillus subtilis*. *Res. Microbiol.* **145:**393–402.

39. Kunst, F., and G. Rapoport. 1995. Salt stress is an environmental signal affecting degradative enzyme synthesis in *Bacillus subtilis*. *J Bacteriol.* **177:**2403–2407.

40. Laufen, T., M. P. Mayer, C. Beisel, D. Klostermeier, A. Mogk, J. Reinstein, and B. Bukau. 1999. Mechanism of regulation of hsp70 chaperones by DnaJ cochaperones. *Proc. Natl. Acad. Sci. USA* **96:**5452–5457.

41. Lazazzera, B. A., T. Palmer, J. Quisel, and A. D. Grossman. 1999. Cell density control of gene expression and development in *Bacillus subtilis*, p. 27–46. *In* G. M. Dunny and S. C. Winans (ed.), *Cell-Cell Signaling in Bacteria*. ASM Press, Washington, D.C.

42. Lazazzera, B. A., J. M. Solomon, and A. D. Grossman. 1997. An exported peptide functions intracellularly to contribute to cell density signaling in *B. subtilis*. *Cell* **89:**917–925.

43. Lee, M. S., and D. A. Morrison. 1999. Identification of a new regulator in *Streptococcus pneumoniae* linking quorum sensing to competence for genetic transformation. *J. Bacteriol.* **181:**5004–5016.

44. Liu, J., and P. Zuber. 1998. A molecular switch controlling competence and motility: competence regulatory factors ComS, MecA, and ComK control sigmaD-dependent gene expression in *Bacillus subtilis*. *J. Bacteriol.* **180:**4243–4251.

45. Liu, L., M. Nakano, O. H. Lee, and P. Zuber. 1996. Plasmid-amplified *comS* enhances genetic competence and suppresses *sinR* in *Bacillus subtilis*. *J. Bacteriol.* **178:**5144–5152.

46. Lorenz, M. G., and W. Wackernagel. 1994. Bacterial gene transfer by natural genetic transformation in the environment. *Microbiol Rev.* **58:**563–602.

47. Love, P. E., M. J. Lyle, and R. E. Yasbin. 1985. DNA-damage-inducible (*din*) loci are transcriptionally activated in competent *Bacillus subtilis*. *Proc. Natl. Acad. Sci. USA* **82:**6201–6205.

48. Luttinger, A., J. Hahn, and D. Dubnau. 1996. Polynucleotide phosphorylase is necessary for competence development in *Bacillus subtilis*. *Mol. Microbiol.* **19:**343–356.

49. Magnuson, R., J. Solomon, and A. D. Grossman. 1994. Biochemical and genetic characterization of a competence pheromone. *Cell* **77:**207–216.

50. McCarthy, C., and E. W. Nester. 1967. Macromolecular synthesis in newly transformed cells of *Bacillus subtilis*. *J. Bacteriol.* **94:**131–40.

51. Msadek, T., V. Dartois, F. Kunst, M. L. Herbaud, F. Denizot, and G. Rapoport. 1998. ClpP is required for competence development, motility, degradative enzyme synthesis, growth at high temperature and sporulation. *Mol. Microbiol.* **27:**899–914.

52. Msadek, T., F. Kunst, and G. Rapoport. 1994. MecB of *Bacillus subtilis* is a pleiotropic regulator of the ClpC ATPase family, controlling competence gene expression and survival at high temperature. *Proc. Natl. Acad. Sci. USA* **91:**5788–5792.

53. Msadek, T., F. Kunst, and G. Rapoport. 1995. A signal transduction network in *Bacillus subtilis* includes the DegS/DegU and ComP/ComA two-component systems. *In* J. A. Hoch and T. J. Silhavy (ed.), *Two-Component Signal Transduction*. ASM Press, Washington, D.C.

54. Nakano, M. M., L. Xia, and P. Zuber. 1991. Transcription initiation region of the *srfA* operon which is controlled by the *comP-comA* signal transduction system in *Bacillus subtilis*. *J. Bacteriol.* **173:**5487–5493.

55. Nakano, M. M., and P. Zuber. 1993. Mutational analysis of the regulatory region of the *srfA* operon in *Bacillus subtilis*. *J. Bacteriol.* **175:**3188–3191.

56. Nester, E. W., and B. A. D. Stocker. 1963. Biosynthetic latency in early stages of deoxyribonucleic acid transformation in *Bacillus subtilis*. *J. Bacteriol.* **86:**785–796.

57. Ogura, M., L. Liu, M. Lacelle, M. M. Nakano, and P. Zuber. 1999. Mutational analysis of ComS: evidence for the interaction of ComS and MecA in the regulation of competence development in *Bacillus subtilis*. *Mol. Microbiol.* **32:**799–812.

58. Perego, M. 1998. Kinase-phosphatase competition regulates *Bacillus subtilis* development. *Trends Microbiol.* **6:**366–370.

59. Perego, M. 1997. A peptide export-import control circuit modulating bacterial development regulates protein phosphatases of the phosphorelay. *Proc. Natl. Acad. Sci. USA* **94:**8612–8617.

60. Perego, M. 1999. Self-signaling by Phr peptides modulates *Bacillus subtilis* development, p. 243–258. *In* G. M. Dunny and S. C. Winans (ed.), *Cell-Cell Signaling in Bacteria*. ASM Press, Washington, D.C.

61. Persuh, M., K. Turgay, I. Mandic-Mulec, and D. Dubnau. 1999. The N and C-terminal domains of MecA recognize different partners in the competence molecular switch. *Mol. Microbiol.* **33:**886–894.

62. Piazza, F., P. Tortosa, and D. Dubnau. 1999. Mutational analysis and membrane topology of ComP, a quorum-sensing sensor histidine kinase of *Bacillus subtilis* controlling competence development. *J. Bacteriol.* **181:**4540–4548.

63. Rashid, H. R., A. Tamakoshi, and J. Sekiguchi. 1996. Effects of *mecA* and *mecB* (*clpC*) mutations on expression of *sigD*, which encodes an alternative sigma factor, and autolysin operons and on flagellin synthesis in *Bacillus subtilis*. *J. Bacteriol.* **178:**4861–4869.

64. Redfield, R. J. 1988. Evolution of bacterial transformation: is sex with dead cells ever better than no sex at all? *Genetics* **119:**213–221.

65. Redfield, R. J. 1993. Genes for breakfast: the have-your-cake-and-eat-it-too of bacterial transformation. *J. Hered.* **84:**400–404.

66. Redfield, R. J., M. R. Schrag, and A. M. Dean. 1997. The evolution of bacterial transformation: sex with poor relations. *Genetics* **146:**27–38.

67. Roggiani, M., and D. Dubnau. 1993. ComA, a phosphorylated response regulator protein of *Bacillus subtilis*, binds to the promoter region of *srfA*. *J. Bacteriol.* **175:**3182–3187.

68. Roggiani, M., J. Hahn, and D. Dubnau. 1990. Suppression of early competence mutations in *Bacillus subtilis* by *mec* mutations. *J. Bacteriol.* **172:**4056–4063.

69. Ruzal, S. M., and C. Sanchez-Rivas. 1998. In *Bacillus subtilis* DegU-P is a positive regulator of the osmotic response. *Curr. Microbiol.* **37:**368–372.

70. Schirmer, E. C., J. R. Glover, M. A. Singer, and S. Lindquist. 1996. HSP100/Clp proteins: a common mechanism explains diverse functions. *Trends Biochem. Sci.* **21:**289–296.

71. Serror, P., and A. L. Sonenshein. 1996. CodY is required for nutritional repression of *Bacillus subtilis* genetic competence. *J. Bacteriol.* **178:**5910–5915.

72. Smith, I. 1993. Regulatory proteins that control late-growth development, p. 785–800. *In* A. L. Sonenshein, J. A. Hoch, and R. Losick (ed.), *Bacillus subtilis and Other Gram-Positive Bacteria: Biochemistry, Physiology, and Molecular Genetics.* American Society for Microbiology, Washington, D.C.

73. Solomon, J., R. Magnuson, A. Srivastava, and A. D. Grossman. 1995. Convergent sensing pathways mediate response to two extracellular competence factors in *Bacillus subtilis*. *Genes Dev.* **9:**547–558.

74. Solomon, J. M., and A. D. Grossman. 1996. Who's competent and when: regulation of natural competence in bacteria. *Trends Genet.* **12:**150–155.

75. Solomon, J. M., B. A. Lazazzera, and A. D. Grossman. 1996. Purification and characterization of an extracellular peptide factor that affects two developmental pathways in *Bacillus subtilis*. *Genes and Dev.* **10:**2014–2024.

76. Strauch, M., V. Webb, G. Spiegelman, and J. A. Hoch. 1990. The SpoOA protein of *Bacillus subtilis* is a repressor of the *abrB* gene. *Proc. Natl. Acad. Sci. USA* **87:**1801–1805.

77. Turgay, K., J. Hahn, J. Burghoorn, and D. Dubnau. 1998. Competence in *Bacillus subtilis* is controlled by regulated proteolysis of a transcription factor. *EMBO J.* **17:**6730–6738.

78. Turgay, K., L. W. Hamoen, G. Venema, and D. Dubnau. 1997. Biochemical characterization of a molecular switch involving the heat shock protein ClpC, that controls the activity of ComK, the competence transcription factor of *Bacillus subtilis*. *Genes Dev.* **11:**119–128.

79. van Sinderen, D., A. Luttinger, L. Kong, D. Dubnau, G. Venema, and L. Hamoen. 1995. *comK* encodes the competence transcription factor, the key regulatory protein for competence development in *Bacillus subtilis*. *Mol. Microbiol.* **15:**455–462.

80. van Sinderen, D., A. ten Berge, B. J. Hayema, L. Hamoen, and G. Venema. 1994. Molecular cloning and sequence of comK, a gene required for genetic competence in *Bacillus subtilis*. *Mol. Microbiol.* **11:**695–703.

81. Weinrauch, Y., R. Penchev, E. Dubnau, I. Smith, and D. Dubnau. 1990. A *Bacillus subtilis* regulatory gene product for genetic competence and sporulation resembles sensor protein members of the bacterial two-component signal-transduction systems. *Genes Dev.* **4:**860–872.

82. Williams, P. M., L. A. Bannister, and R. J. Redfield. 1994. The *Haemophilus influenzae sxy-1* mutation is in a newly identified gene essential for competence. *J. Bacteriol.* **176:**6789–6794.

Bacterial Stress Responses
Edited by G. Storz and R. Hengge-Aronis
©2000 ASM Press, Washington, D.C.

Chapter 18

Roles of Cell-Cell Communication in Confronting the Limitations and Opportunities of High Population Densities

STEPHEN C. WINANS AND JUN ZHU

As recently as 10 years ago, few bacteria were known to monitor their population densities and to modify their patterns of behavior in response to cell-density signals. Since that time, so many examples of this behavior have been described that one is tempted to conclude that cell-cell signaling is the rule rather than the exception among prokaryotes. Two broad categories of pheromones seem to account for most of the known examples. First, many gram-positive organisms use oligopeptides as pheromones, which in some cases are posttranslationally modified prior to their release. Some of these peptides are reimported into the cells and bind to cytoplasmic receptors, while others remain outside the bacteria and bind to membrane-spanning two-component kinases. Proteobacteria seldom if ever use peptide signals and instead communicate using acylhomoserine lactone pheromones. These systems require at least two proteins, one a pheromone synthase and the other a pheromone receptor, which also acts as a transcriptional regulator. Many members of this family of regulatory systems have been described, including several that play critical roles in pathogenic or symbiotic associations with plant, animal, or human hosts. New insights into these regulatory systems will inevitably lead to novel approaches to direct benefical bacterial processes and to control harmful ones.

The bicentennial of the publication of *An Essay on the Principle of Population* was celebrated in 1998 (40). In this essay, Thomas Malthus argued that the majority of humanity must forever live under conditions of extreme poverty, since any increase in the production of food or other necessities will inevitably result in a compensatory increase in human population size. More recently, this message was echoed in a report called *The Limits to Growth*, which argued

that the earth will imminently be unable to provide sufficient resources to sustain ever increasing human populations (42). When applied to bacterial populations, these ideas certainly seem self-evident. All microbiologists accept the idea that bacteria generally replicate until their nutrients are depleted. While in principle it ought to be possible for a solitary bacterium to exhaust its nutrients, nutrient limitation is more often a consequence of growing populations of bacteria that compete for finite resources. It has recently become clear that under such conditions, many groups of bacteria can exchange chemical signals to help them monitor their population densities, a phenomenon sometimes referred to as quorum sensing (24). This information is used to trigger novel and varied patterns of behavior, including the formation of an endospore or other resting structures, the release of toxins to limit interspecific competition, the exchange of genetic information by transformation or conjugation, and the induction of pathogenesis genes. This review describes a selected subset of the known mechanisms that bacteria use for cell-cell signaling and the consequent changes in bacterial behavior. This area has also been reviewed in two books (15, 17).

SIGNAL-DEPENDENT COMPETENCE AND SPORULATION BY *BACILLUS SUBTILIS*

Most types of bacteria are able to survive for extended intervals without steady supplies of nutrients. Under such conditions some groups of bacteria form metabolically inert spores that are able to endure inimical environments for indefinite periods of time. Probably the most thoroughly studied example of this phenomenon is formation of structures called

Stephen C. Winans and Jun Zhu • Department of Microbiology, Cornell University, Ithaca, NY 14853.

endospores by the soil bacterium *B. subtilis* (see chapter 13). During the transition from logarithmic growth to stationary phase, these bacteria become hypermotile and chemotactic and release a variety of hydrolytic enzymes, all in an effort to obtain new sources of nutrients. Some strains also release diverse antibiotics, presumably to limit interspecific competition. Also during the transition to stationary phase, the bacterium becomes competent for the uptake and homologous integration of naked DNA released during lysis of other *Bacillus* cells (see chapter 17). Under some conditions, *B. subtilis* can also form endospores. Endospore formation is initiated by an asymmetric cell division, subsequent engulfment of the smaller cell by the larger cell, synthesis of a spore coat and cortex around the engulfed cell (the forespore), and eventual sacrificial lysis of the mother cell.

In the past decade, various studies indicated that many of these processes occur preferentially at high population densities and are stimulated by the exchange of chemical signals between bacteria (27, 66). The intricate process of endospore formation is initiated by a phosphorelay composed of two histidine protein kinases (KinA and KinB), two intermediate phosphoproteins (Spo0F and Spo0B), and a response regulator (Spo0A) that is a global transcriptional regulator of genes directing many stationary-phase responses, including sporulation and the production of hydrolytic enzymes and antibiotics (Fig. 1). The phosphorelay is regulated at numerous points. Most important for this discussion is that Spo0F~P can be dephosphorylated by two specific phosphatases (RapA and RapB), and these phosphatases are regulated by diffusible peptides (54, 56). PhrA, the RapA-specific pheromone, is synthesized as a 44-amino-acid precursor, which is exported from the bacterial cytoplasm by the general secretory pathway. During export, a 25-amino-acid signal sequence is removed, leaving a 19-amino-acid proinhibitor, which is further processed by an uncharacterized enzyme to a 5-amino-acid mature pheromone. The RapB-specific peptide CSF (or PhrC) has similar properties (see below). These pheromones may plausibly diffuse from cells that produce them and accumulate only at high cell densities, although there is some disagreement about this critical point (see below). The peptides can be reimported into the cytoplasm of these bacteria via an ABC-type oligopeptide permease, where they bind to RapA and RapB proteins and inhibit phosphatase activity (56). The net result of this complex regulatory circuit is that transcription of several genes required for sporulation is stimulated by these pheromones, which accumulate only at high cell densities.

Doubts have been expressed about whether the purpose of this system is really to estimate cell den-

sities. Perego and colleagues in particular favor a model in which these peptides act in a cell-autonomous fashion, in that each bacterium reimports the pheromones that it produces rather than releasing them to the environment (54, 55). Two observations strongly suggest that these peptides mediate cell-cell communication. First, exogenously added peptides are efficiently recognized by the bacteria (54). Second, and most important, a strain that is unable to synthesize PhrA and therefore is unable to sporulate can be rescued by cocultivation with a strain that produces PhrA (56). This proves that the processed form of PhrA can be released from one population of cells at concentrations sufficient to be taken up by a separate population. Furthermore, sporulation seems to be an appropriate response to the Malthusian scenario of competition-induced scarcity. At low cell densities, starving bacteria might best enter stationary phase and await new nutrient sources, whereas at high cell densities, any new nutrients would likely be depleted by competing bacteria, and the more drastic solution of suspended animation via sporulation would therefore be a better survival strategy.

B. subtilis is able to import naked DNA from its environment and integrate this DNA into its genome. This process occurs during the transition from logarithmic phase to stationary phase and requires expression of approximately 40 genes (39). The regulation of these genes is controlled by two different peptide pheromones, CSF (which also controls the RapB phosphatase of the sporulation phosphorelay) and ComX. Both pheromones influence the phosphorylation state of the response regulator ComA, which regulates several operons, including one that encodes a key regulator (ComS) of competence genes (Fig. 2). CSF acts by a mechanism that is highly reminiscent of PhrA (see above), in that it is synthesized as a 40-amino-acid precursor, exported by the general secretory pathway, processed to a 5-amino-acid mature pheromone, and is imported by an oligopeptide permease. The CSF peptide inhibits the activity of a phosphatase (RapC) that dephosphorylates ComA~P (Fig. 2). By this mechanism, high cell densities stabilize the phosphorylated form of ComA, which in turn leads to the induction of genetic competence. The chromosome of *B. subtilis* contains no fewer than 9 additional genes that resemble *rapA* and *rapC* (39, 55), so additional two-component regulatory systems may be regulated in this fashion by dedicated protein phosphatases.

Expression of competence genes is further stimulated by a second peptide called ComX. Surprisingly, ComX acts by a different mechanism (39). It is synthesized as a 55-amino-acid precursor and is

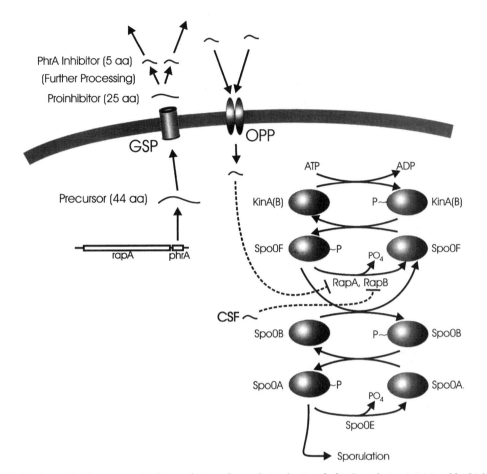

Figure 1. Role of peptide pheromones in the regulation of sporulation by *B. subtilis*. Sporulation is initiated by high concentrations of Spo0A~P, which is the final protein in a phosphorelay consisting of KinA, Spo0F, and Spo0B. Spo0F can be dephosphorylated by the RapA and RapB phosphatases, thereby inhibiting sporulation. RapA and RapB are themselves inhibited by the diffusible peptides PhrA and CSF, which are exported by the general secretory pathway (GSP) and imported by an oligopeptide permease (OPP), encoded by the *spo0K* operon.

processed by proteolysis and covalent modification to a 10-amino-acid form. It is exported from the cytoplasm by mechanism that is independent of the general secretory pathway. Processing and export may require the ComQ protein, which is encoded just upstream of *comX* (Fig. 2). Unlike CSF, ComX is not reimported into the cells but rather is thought to bind to the sensory domain of a transmembrane histidine protein kinase (ComP). Binding of the pheromone stimulates phosphorylation of the response regulator ComA (Fig. 2). Thus, ComX acts by stimulating phosphorylation of ComA, while CSF acts by inhibiting the dephosphorylation of the same protein.

Like sporulation, induction of genetic competence at high cell density makes sense teleologically, since these conditions would favor the accumulation of homologous naked DNA. Furthermore, only genetically similar or identical cells are likely to produce CSF, so competence requires high populations of cells whose DNA is similar to the recipient's DNA

(essential for homologous recombination), while cells do not become competent in the presence of unrelated bacteria.

OTHER EXAMPLES OF CELL-CELL SIGNALING VIA PEPTIDE PHEROMONES

Peptide pheromones are widely used by gram-positive bacteria to gauge their population densities. Many of these systems resemble the ComX system described above, in that they are composed of a pheromone, a specific pheromone export and processing system, a transmembrane two-component kinase that serves as a pheromone receptor, and a cognate response regulator (Fig. 3).

One such system regulates the competence genes of *Streptococcus pneumoniae*, which, like the competence proteins of *B. subtilis*, directs the uptake of homologous DNA from the environment and its ho-

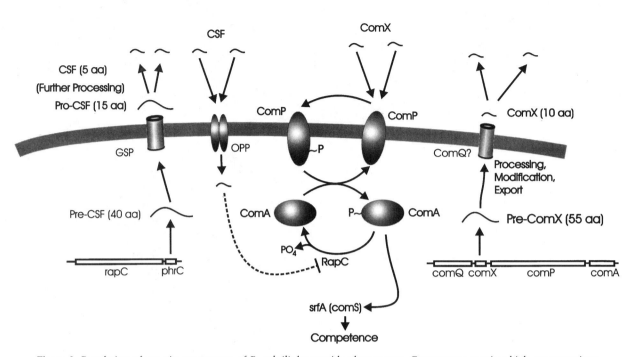

Figure 2. Regulation of genetic competence of *B. subtilis* by peptide pheromones. Competence requires high concentrations of ComA~P, a protein that is phosphorylated by ComP and dephosphorylated by RapC. The ComX peptide stimulates ComP-mediated phosphorylation, while the CSF peptide inhibts RapC-mediated dephosphorylation. CSF is exported by the general secretory pathway (GSP) and imported by an oligopeptide permease (OPP), while ComX is exported by an uncharacterized permease, possibly composed of ComQ, and is thought to act from the outside of the cell by binding to the extracellular domain of ComP.

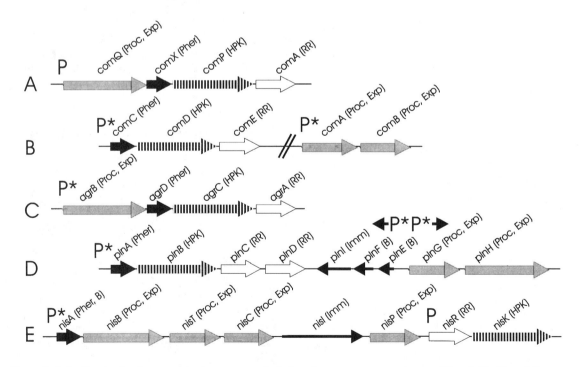

Figure 3. A family of cell-cell communication cassettes, including the competence regulators of *B. subtilis* (A) and *S. pneumoniae* (B), the pathogenesis regulators of *S. aureus* (C), and the bacteriocin synthesis regulons of *L. plantarum* (D) and *L. lactis* (E). Each system contains a peptide pheromone (Pher; solid arrows), a transmembrane histidine protein kinase (HPK; striped arrows), a response regulator (RR; white arrows), and one or more proteins required for pheromone processing and export (Proc, Exp; gray arrows). The bacteriocin regulons also contain bacteriocin structure genes (B) and immunity genes (Imm). Positively autoregulated promoters are designated "P*" while other promoters are designated "P."

mologous integration (30). Also like *B. subtilis*, *S. pneumoniae* becomes genetically competent during late exponential phase of growth and is stimulated by high population densities. Signaling is mediated by the products of the *comCDE* and *comAB* operons. The pheromone is encoded by the *comC* gene as a 40-amino-acid precursor and is exported and processed by the products of the *comAB* operon to a mature form, designated CSP, which is 17 amino acids in length (30). It is detected by the ComB histidine protein kinase, which phosphorylates the response regulator ComE. The *comCDE* operons of several strains have been sequenced, and a variety of different pheromones have been identified. In general, a particular streptococcal species responds to its cognate pheromone but not to heterologous pheromones (29). This has an obvious adaptive significance, since it would cause genetic competence to be induced by genetically identical or very similar bacteria, but not by more distantly related strains of the same genus, whose DNA might be too divergent for homologous recombination.

Another example of a peptide pheromone is found in the pathogen *Staphylococcus aureus*, which causes a variety of skin and lung infections (48). This bacterium induces the expression of a battery of exotoxins during late log phase growth and utilizes a cell-density detection system to control this induction. In this system, the AgrD protein is synthesized as a 46-amino-acid precursor, which is thought to be exported and processed by the AgrB protein (Fig. 4). The mature peptide is 8 amino acids in length and contains a covalent bond between the amino terminus and an internal cysteine residue (31, 41). It is detected by the AgrC two-component kinase, which phosphorylates AgrA. The phosphorylated form of AgrA positively regulates the *agrBDCA* operon, resulting in positive autoregulation of this system, and also positively regulates a divergent gene that encodes a regulatory RNA (RNA III), which positively effects expression of other pathogenesis genes (44).

Additional examples of peptide pheromones are found in the production of proteinaceous antimicrobial compounds called bacteriocins. Bacteriocins themselves are generally small peptides (30 to 60 amino acid residues) that are synthesized as precursor proteins, processed and exported from the bacteria, and act as broad-specificity antimicrobials (2, 35, 46, 47). These compounds are synthesized by a broad variety of gram-positive bacteria and are economically important in retarding the spoilage of fermented foods. Two classes of bacteriocins have been recognized. Class I bacteriocins, often called lantibiotics, are heavily modified after translation to include lanthionine, didehydroalanine, and didehydrobutyrine

residues, and frequent thioether bridges. Class II bacteriocins are similar in size and function but lack posttranslational modifications. Both classes of bacteriocins are synthesized as precursor molecules containing amino terminal signal sequences that are recognized by a dedicated export/processing system and cleaved at a Gly-Gly dipeptide motif. Bacteria that produce either class of bacteriocins also have immunity proteins, some of which probably act by causing active efflux of the peptide.

These peptides are generally produced during late exponential phase of growth, and their production in early exponential phase can be stimulated by addition of a cell-free supernatant from stationary-phase cultures, suggesting the existence of a diffusible pheromone (14). In the case of class I bacteriocins, the pheromone is the bacteriocin itself (37). For example, the *nis* operon of *Lactococcus lactis* (Fig. 4) encodes a precursor of the lantibiotic nisin (*nisA*). This peptide is processed and exported by the NisB, NisC, NisP, and NisT proteins. It is toxic toward a wide variety of other bacteria, and its toxicity toward *L. lactis* is prevented by the NisI immunity protein. In addition to being a bacteriocin, nisin acts as a pheromone and stimulates activity of the NisK histidine proteins kinase, which phosphorylates the NisR response regulator (36).

Class II bacteriocins are regulated in a similar manner, except that the bacteriocin is not used as a pheromone. For example, *Lactobacillus plantarum* synthesizes several Type II bacteriocins, two of which are encoded by the *plnE* and *plnF* genes (Fig. 3). These peptides are exported and processed by the PlnG and PlnH proteins. The peptide pheromone for this system is encoded by a separate gene (*plnA*) and is processed and exported by PlnG and PlnH and detected by the histidine protein kinase PlnB, which phosphorylates two response regulators, PlnC and PlnD (14). Although the pheromone does not have bacteriocin activity, its similarity to bacteriocins in size and processing suggests a common evolutionary origin.

CELL-DENSITY REGULATION USING ACYL-HOMOSERINE LACTONE PHEROMONES

The peptide-type pheromones described above are widespread among gram-positive bacteria but have not been documented among other groups. Many bacterial genera within the proteobacteria communicate via diffusible nonpeptide pheromones called autoinducers, which consist of homoserine lactone rings conjugated to acyl side chains of variable length, oxidation state, and saturation (20). Many of

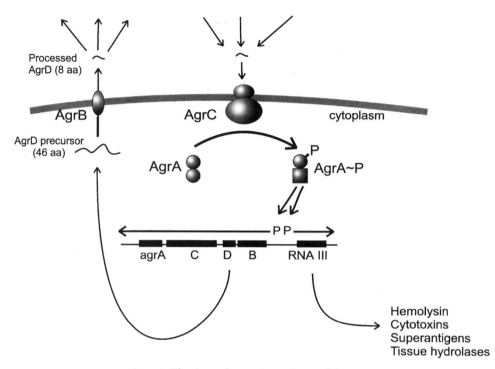

Figure 4. The Agr pathogenesis regulators of *S. aureus*.

these systems are composed of just two proteins, one of which serves as an autoinducer synthase and the other serves as an autoinducer receptor and an autoinducer-dependent transcriptional regulator (22).

The first known proteins of this type are the LuxR and LuxI proteins of *Vibrio fischeri* that regulate the bioluminescence (*lux*) operon of this marine bacterium (16). LuxR is a positive regulator of the *luxICDABEG* operon, which encodes the autoinducer synthase (*luxI*), the luciferase proteins (*luxA* and *luxB*), and other proteins that synthesize luciferase substrates (*luxC, D, E,* and *G* [see reference 16]). At low cell densities, this operon is weakly expressed, and the bacteria are nonluminescent, while at high cell densities, intracellular concentrations of the autoinducer 3-oxohexanoylhomoserine lactone (3-oxo-C6-HSL) rise sufficiently to bind LuxR, and the *lux* operon is strongly expressed. Under these conditions, the bacteria also synthesize greatly increased levels of the autoinducer, resulting in a positive feedback loop.

For many years, cell-cell communication via autoinducers was thought to be limited to *V. fischeri* and its relative *Vibrio harveyi* (see below). However, in the early 1990s it was demonstrated that the pathogen *Pseudomonas aeruginosa* uses a similar pair of proteins, designated LasR and LasI, to regulate the expression of elastolytic proteins that play a role in pathogenesis (25, 52). Since that time, many members of the proteobacteria have been found to use similar regulatory systems to regulate diverse groups

of genes (3). In one study, 106 isolates from eight genera of plant-associated bacteria were examined for release of acyl-HSLs, and each genus included at least some isolates that did so (9).

GENERAL FEATURES OF LuxR-LuxI-TYPE REGULATORY SYSTEMS

Considerable progress has been made by many laboratories in understanding the properties of this family of proteins. For example, the LuxI, RhlI, and TraI proteins (of *V. fischeri, P. aeruginosa,* and *Agrobacterium tumefaciens,* respectively), which synthesize autoinducer-type pheromones, have all been studied biochemically (33, 43, 51, 61). These proteins appear to carry out two reactions: (i) formation of a homoserine lactone ring at the expense of S-adenosyl-methionine (AdoMet), and (ii) acylation of the amine of homoserine lactone at the expense of acylated acyl carrier protein (acyl-ACP). Although AdoMet is generally thought of as a source of methyl groups, it is also a source of homoserine groups in various reactions (8). In autoinducer synthesis, the homoserine lactone group is created by intramolecular nucleophilic attack by the carboxyl group of AdoMet upon its γ-methylene group. This ring is somewhat unstable to hydrolysis, especially at alkaline pH, and this may provide an ecologically significant "fuse" that slowly inactivates the pheromone

(70). Although the acyl groups of autoinducers vary in chain length, oxidation state, and saturation, virtually all of these acyl groups would be available to synthase enzymes as ACP conjugates that are diverted from fatty acid biosynthesis. Autoinducers are generally thought to diffuse readily across the bacterial envelope, although diffusion rates have been measured only for the *V. fischeri* autoinducer (34). This diffusion is thought to cause intracellular concentrations to be similar to extracellular concentrations and to rise only in the presence of large numbers of bacteria.

LuxR-type proteins are thought to contain two modules: an amino terminal module (approximately 180 amino acid residues) that binds autoinducers and also mediates multimerization, and a carboxyl terminal module (approximately 70 amino acid residues) that binds particular DNA sites near target promoters and also makes stimulatory contacts with RNA polymerase (RNAP) (reference 63). High-level expression of the amino terminal domain of LuxR in *Escherichia coli* caused that strain to accumulate radiolabeled autoinducer (28), indicating that this fragment is sufficient for autoinducer binding. These same fragments exert a dominant negative effect on wild-type LuxR (11), suggesting that these fragments bind to full-length LuxR, forming inactive heterodimers. Expression of the carboxyl terminal fragment of LuxR caused constitutive expression of a target promoter (10), although these experiments were done using strains that grossly overexpressed LuxR. This LuxR fragment and RNAP cooperatively bound a target promoter in vitro (64). The fact that this fragment acts constitutively in vivo was taken to suggest that the amino terminal fragment plays an inhibitory role that is neutralized by autoinducer (10). However, the extremely high expression level of this fragment could help explain its constitutive activity. The finding that the amino terminal module mediates oligomerization suggests a positive role rather than an inhibitory role.

A small number of biochemical studies of LuxR-type proteins have been reported. The first, described above, examined properties of a carboxyl terminal fragment of LuxR (64). Another report analyzed a fusion protein composed of LasR and glutathione S-transferase. This fusion protein was described as binding to a DNA fragment containing the cognate binding site in an autoinducer-dependent fashion (69). However, UV cross-linking was required to detect binding and the binding site was not identified by footprinting. The ExpR of *Erwinia chrysanthemi* bound to a variety of promoters, including several that appeared not to be regulated by this protein in vivo (45, 60). Binding in gel shift assays was inhibited

by the cognate autoinducer (3-oxo-C6-HSL), although, surprisingly, binding appeared to be stimulated by this ligand in DNase I footprinting assays (45).

The TraR protein of *A. tumefaciens* was also purified as a native protein and characterized biochemically (72). Overexpression of this protein in *E. coli* in the absence of the cognate autoinducer (3-oxo-C8-HSL) resulted in the accumulation of insoluble inclusion bodies, while addition of autoinducer during growth caused much of the TraR protein to accumulate in a soluble and active form. Purified TraR-autoinducer complexes bound specifically to the predicted TraR binding site and activated transcription of two divergent promoters that flank this site (72). Binding and transcription were not stimulated by adding autoinducer, since the TraR preparation already contained saturating amounts of this ligand. Autoinducer bound TraR in a 1:1 mole ratio but was removable by extensive dialysis in the presence of mild detergent. The resulting preparation of apo-TraR bound to *tra* box DNA more weakly than TraR-autoinducer complexes. Apo-TraR eluted from a gel filtration column as a monomer, while TraR-autoinducer complexes eluted as a dimer (Zhu and Winans, unpublished data). Autoinducer also increased the intracellular abundance of TraR by decreasing TraR turnover rates about 20-fold (72).

EXAMPLES OF REGULATORS THAT USE ACYL-HSL PHEROMONES

While the LuxR-LuxI family of proteins probably share many properties, at least some members have unusual and surprising properties. For example, as described above, *P. aeruginosa* contains the LasR and LasI proteins and the autoinducer 3-oxo-C12-HSL, which regulate genes that encode various elastolytic proteases (*lasA* and *lasB*) as well as exotoxin A (*toxA*) and an exotoxin A-specific export system (*xcp*) (Fig. 5) (see reference 57). LasR-autoinducer complexes also activate *lasI* (reminiscent of LuxR activating *luxI*), resulting in a positive feedback loop (62). A third protein, RsaL, inhibits the transcription of *lasI* (13). Expression of *lasR* is positively regulated by Vfr (an ortholog of CRP) and by the LemA/GacA two-component proteins (1, 59). Surprisingly, this organism expresses a second pair of LuxR-LuxI type proteins consisting of the RhlR and RhlI proteins and the pheromone C4-HSL (57). RhlR-autoinducer complexes regulate the *rhlAB* operon, which directs production of a rhamnolipid surfactant (50), as well as the *rpoS* stationary-phase sigma factor (38) and the *rhlI* gene (68), causing a second example of positive

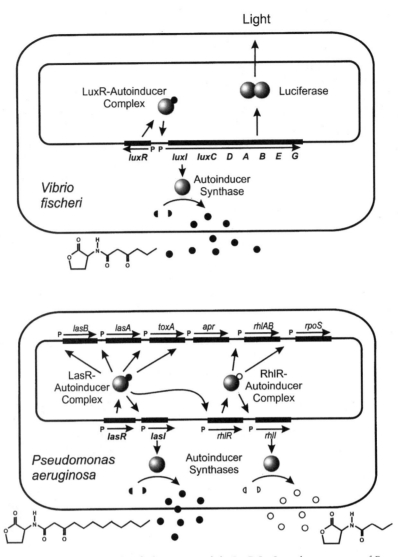

Figure 5. The LuxR-LuxI regulatory system of *V. fischeri* (top) and the LasR-LasI regulatory system of *P. aeruginosa* (bottom), both of which use acylhomoserine lactone pheromones. The LuxI and LasI proteins synthesize 3-oxo-C6-HSL and 3-oxo-C12-HSL, respectively, at the expense of AdoMet and acyl-ACP. The *V. fischeri* pheromone passively diffuses across the cell envelope, while that of *P. aeruginosa* is actively exported by an efflux pump. Both pheromones bind to their respective receptors (LuxR and LasR), and the resulting complexes activate transcription of target promoters. LasR-pheromone complexes activate the *rhlR* gene, which encodes a second LuxR-type protein that binds C4-HSL, which is synthesized by RhlI, thus creating a regulatory cascade involving two pheromones. RhlR-autoinducer complexes activate the *rhlAB*, which directs synthesis of a rhamnolipid surfactant, *rpoS*, which encodes a stationary-phase sigma factor, and the *rhlI* gene, among others.

feedback. These two regulatory systems appear to be interdigitated in two opposing ways: LasR-autoinducer complexes activate transcription of *rhlR*, resulting in a regulatory cascade (38, 58), while the pheromone 3-oxo-C12-HSL appears to act as an antagonist for the second autoinducer C4-HSL (58). It was recently shown that *P. aeruginosa* actively exports one of its autoinducers (3-oxo-C12-HSL) but not the other (C4-HSL) via an efflux pump (53).

Although *P. aeruginosa* is flagellated and motile, it is also known to adhere to solid surfaces, forming biofilms approximately 100 microns thick. These biofilms allow colonization of many solid surfaces, in-

cluding lung epithelium and indwelling catheters. These biofilms contain clumps of pillar- or mushroom-shaped structures consisting of an extracellular polysaccharide matrix in which the bacteria are embedded. The maturation of biofilms by *P. aeruginosa* was recently reported to require the LasR-LasI regulatory system, in that a LasI mutant failed to form a normal biofilm (12). Instead, this mutant produced much thinner, homogeneous biofilms that lacked water channels. Biofilms of wild-type bacteria are resistant to washing with a strong detergent, while biofilms of the of a *lasI* mutant were quickly dislodged. This defect was fully suppressed by exogenous 3-oxo-

C12-HSL. LasR must therefore regulate one or more genes that are required for biofilm formation, although these target genes have not been identified.

The TraR/TraI system of *A. tumefaciens* also has several unexpected properties. TraR, when complexed with 3-oxo-C8-HSL, activates several promoters that direct transcription of *tra* genes, which direct interbacterial transfer of the Ti plasmid (67). The *traR* gene of the octopine-type Ti plasmid is itself induced by the opine octopine and the activator OccR, thus explaining older data indicating that the Ti plasmid conjugation is stimulated by this opine (21). Induction of *traR* in nopaline-type Ti plasmids requires the opines agrocinopine A and B, which inactivate the AccR repressor (7). OccR and AccR are unrelated, indicating that induction of the two *traR* genes by opines evolved independently. TraR-autoinducer complexes also activate the *traR* and *traI* genes, thereby creating two positive autoregulatory loops (23). TraR activity is antagonized by two proteins, TraM and TrlR. Interestingly, TraR activates transcription of *traM*, providing a negative autoregulatory loop (19). TrlR closely resembles TraR but is truncated at its carboxyl terminus and lacks a DNA-binding module (49, 71). TrlR may act by binding to TraR, forming inactive heterodimers. Expression of the *trlR* gene is induced by the opine mannopine, and this opine therefore antagonizes octopine for *tra* gene induction. TraR function can also be antagonized by many autoinducer analogs, including several found in the rhizosphere (70). This suggests that Agrobacteria exposed to these analogs may not induce *tra* genes even in conditions that would otherwise favor *tra* gene expression.

Perhaps no cell-cell communication system has more unexpected properties than that of *V. harveyi*. Like *V. fischeri*, *V. harveyi* is a marine bioluminescent microbe that uses an acyl homoserine lactone (3-hydroxy-C4-HSL) as a signaling molecule to regulate its *lux* operon (4). However, the molecular mechanisms used by these organisms could not be more different and represent a remarkable example of convergent evolution. First, the pheromone described above is only one of two chemical signals, the other of which has not been characterized chemically. The synthases for both signals have been identified (6, 65). Remarkably, the 3-hydroxy-C4-HSL synthase (LuxLM) does not resemble members of the LuxI family (though it does resemble the synthase for a second autoinducer made by *V. fischeri* [see reference 26]). The synthase for the second signal, LuxS, does not resemble any other characterized protein, though homologous genes have been identified in several other organisms. Each of the two signals is detected by a transmembrane histidine protein kinase (LuxN

and LuxQ, respectively) reminiscent of those described above that detect peptide pheromones (6). These kinases are part of a phosphorelay system in that each contains a carboxyl terminal receiver module that passes phosphoryl groups to an intermediary protein (LuxU), which in turn phosphorylates a response regulator (LuxO) that directly regulates the *lux* operon (18). 3-hydroxy-C4-HSL appears to be a species-specific pheromone, while in contrast, many genera of eubacteria, including *E. coli* and *Salmonella*, release the second pheromone (5). It is possible that all these bacteria can detect this signal and use it for cross-species communication.

CONCLUSIONS AND PERSPECTIVES

This review began with Thomas Malthus' ideas about high population densities as a cause of world poverty. Clearly, bacteria are accustomed to prolonged periods of scarcity and are better equipped than humans to endure such conditions. Perhaps the most extreme adaptation to life in the idiophase is to form a resting stage such as an endospore, and we have seen how *B. subtilis* uses intercellular signals at the outset of sporulation. However, sporulation is only one solution to the stresses of high population densities. High cell densities are more likely to elicit cooperative behavior, whether genetic exchange, pathogenesis, bioluminescence, or biological warfare. These responses echo the philosopher John Locke, who argued that the default state of human existence is independent and solitary but that individuals may voluntarily form communities for mutual protection and economic prosperity. Both humans and bacteria form coalitions to accomplish goals that would be unattainable by individuals. Cooperative behavior among humans or bacteria requires accurate estimation of population size. Whereas humans require an act of Congress to take a regular census (for example, Article 1 of the U.S. Constitution), bacteria have evolved a number of mechanisms to count their numbers by simpler, essentially olfactory mechanisms.

While bacteria have been signaling each other ever since the Precambrian era, our appreciation of these processes began within the past 20 years and is still extremely rudimentary. The number of examples of this form of communication will continue to grow explosively, and each new discovery will help us to delineate familiar patterns as well as uncover unimaginable surprises. It will be important to study a small number of these systems in more detail, as each of the known systems is embedded within a highly intricate regulatory circuitry. It will be important to see how different organisms have adapted these proteins to their unique needs. Future work will also have to address fundamental questions about these regulators, for example, to understand exactly how these pheromones alter the properties of their receptors.

Future experiments may conceivably focus on reconstituting these regulatory systems in heterologous cells, including eukaryotic cells, to facilitate the controlled gene expression. As we learn more about the fundamentals of these systems, it will also be possible to interfere with these processes to prevent undesirable bacterial processes and to enhance those processes that we find beneficial. For example, it may be possible to increase the quality and shelf life of fermented foods by encouraging the production to bacteriocins (35, 46). Conversely, a few laboratories have begun the task of identifying pheromone antagonists that could help in preventing human or plant diseases (32, 70). Clearly this is an important area for clinicians, and we anticipate new classes of antagonists in the next few years.

Acknowledgments. We are grateful for the cooperative spirit of our many colleagues that enlivens the field of cell-cell communication.

Research in our laboratory is supported by National Institutes of Health grant GM42893 and by National Science Foundation grant MB-9904917.

REFERENCES

1. **Albus, A. M., E. C. Pesci, L. J. Runyen-Janecky, S. E. West, and B. H. Iglewski.** 1997. Vfr controls quorum sensing in *Pseudomonas aeruginosa. J. Bacteriol.* **179:**3928–3935.
2. **Baba, T., and O. Schneewind.** 1998. Instruments of microbial warfare: bacteriocin synthesis, toxicity and immunity. *Trends Microbiol.* 6:66–71.
3. **Bainton, N. J., B. W. Bycroft, S. R. Chhabra, P. Stead, L. Gledhill, P. J. Hill, C. E. Rees, M. K. Winson, G. P. C. Salmond, G. S. A. B. Stewart, and P. Williams.** 1992. A general role for the *lux* autoinducer in bacterial cell signalling: control of antibiotic biosynthesis in *Erwinia. Gene* **116:**87–91.
4. **Bassler, B. L.** 1999. A multichannel two-component signaling relay controls quorum sensing in *Vibrio harveyi*, p. 259–273. *In* G. M. Dunny and S. C. Winans (ed.), *Cell-Cell Signaling in Bacteria.* American Society for Microbiology, Washington, D.C.
5. **Bassler, B. L., E. P. Greenberg, and A. M. Stevens.** 1997. Cross-species induction of luminescence in the quorum-sensing bacterium *Vibrio harveyi. J. Bacteriol.* **179:**4043–4045.
6. **Bassler, B. L., M. Wright, and M. R. Silverman.** 1994. Multiple signalling systems controlling expression of luminescence in *Vibrio harveyi*: sequence and function of genes encoding a second sensory pathway. *Mol. Microbiol.* **13:**273–286.
7. **Beck von Bodman, S., G. T. Hayman, and S. K. Farrand.** 1992. Opine catabolism and conjugal transfer of the nopaline Ti plasmid pTiC58 are coordinately regulated by a single repressor. *Proc. Natl. Acad. Sci. USA* **89:**643–647.
8. **Bodley, J. W., and S. A. Veldman.** 1990. Biosynthesis of diphthamide: ADP-ribose acceptor for diphtheria toxin, p. 21–30. *In* J. Moss and M. Vaughan (ed.), *ADP-Ribosylating Toxins and G Proteins: Insights into Signal Transduction.* American Society for Microbiology, Washington, D.C.
9. **Cha, C., P. Gao, Y. C. Chen, P. D. Shaw, and S. K. Farrand.** 1998. Production of acyl-homoserine lactone quorum-sensing signals by gram-negative plant-associated bacteria. *Mol. Plant-Microbe Interact.* **11:**1119–1129.
10. **Choi, S. H., and E. P. Greenberg.** 1991. The C-terminal region of the *Vibrio fischeri* LuxR protein contains an autoinducer-independent *lux* gene activating domain. *Proc. Natl. Acad. Sci. USA* **88:**11115–11119.
11. **Choi, S. H., and E. P. Greenberg.** 1992. Genetic evidence for multimerization of LuxR, the transcriptional activator of *Vibrio fischeri* luminescence. *Mol. Mar. Biol. Biotech.* **1:**408–413.
12. **Davies, D. G., M. R. Parsek, J. P. Pearson, B. H. Iglewski, J. W. Costerton, and E. P. Greenberg.** 1998. The involvement of cell-to-cell signals in the development of a bacterial biofilm. *Science* **280:**295–298.
13. **de Kievit, T., P. C. Seed, J. Nezezon, L. Passador, and B. H. Iglewski.** 1999. RsaL, a novel repressor of virulence gene expression in *Pseudomonas aeruginosa. J. Bacteriol.* **181:**2175–2184.
14. **Diep, D. B., L. S. Havarstein, and I. F. Nes.** 1995. A bacteriocin-like peptide induces bacteriocin synthesis in *Lactobacillus plantarum* C11. *Mol. Microbiol.* **8:**631–639.
15. **Dunny, G. M., and S. C. Winans** (ed.). 1999. *Cell-Cell Signaling in Bacteria.* American Society for Microbiology, Washington, D.C.
16. **Engebrecht, J., K. Nealson, and M. Silverman.** 1983. Bacterial bioluminescence: isolation and genetic analysis of functions from *Vibrio fischeri. Cell* **32:**773–781.
17. **England, R., G. Hobbs, N. Bainton, and D. M. Roberts** (ed.). 1999. *Microbial Signalling and Communication.* Cambridge University Press, Cambride, United Kingdom.
18. **Freeman, J. A., and B. L. Bassler.** 1999. Sequence and function of LuxU: a two-component phosphorelay protein that regulates quorum sensing in *Vibrio harveyi. J. Bacteriol.* **181:**899–906.
19. **Fuqua, C., M. Burbea, and S. C. Winans.** 1995. Activity of the *Agrobacterium* Ti plasmid conjugal transfer regulator TraR is inhibited by the product of the *traM* gene. *J. Bacteriol.* **177:**1367–1373.
20. **Fuqua, C., and A. Eberhard.** 1999. Signal generation in autoinduction systems: synthesis of acylated homoserine lactones by LuxI-type proteins, p. 211–230. *In* G. M. Dunny and S. C. Winans (ed.), *Cell-Cell Signaling in Bacteria.* American Society for Microbiology, Washington, D.C.
21. **Fuqua, C., and S. C. Winans.** 1996. Localization of OccR-activated and TraR-activated promoters that express two ABC-type permeases and the *traR* gene of Ti plasmid pTiR10. *Mol. Microbiol.* **20:**1199–1210.
22. **Fuqua, C., S. C. Winans, and E. P. Greenberg.** 1996. Census and consensus in bacterial ecosystems: the LuxR-LuxI family of quorum-sensing transcriptional regulators. *Annu. Rev. Microbiol.* **50:**727–751.
23. **Fuqua, W. C., and S. C. Winans.** 1994. A LuxR-LuxI type regulatory system activates *Agrobacterium* Ti plasmid conjugal transfer in the presence of a plant tumor metabolite. *J. Bacteriol.* **176:**2796–2806.
24. **Fuqua, W. C., S. C. Winans, and E. P. Greenberg.** 1994. Quorum sensing in bacteria: the LuxR/LuxI family of cell density responsive transcriptional regulators. *J. Bacteriol.* **176:**269–275.
25. **Gambello, M. J., and B. H. Iglewski.** 1991. Cloning and characterization of the *Pseudomonas aeruginosa lasR* gene: a transcriptional activator of elastase expression. *J. Bacteriol.* **173:**3000–3009.

26. Gilson, L., A. Kuo, and P. V. Dunlap. 1995. AinS and a new family of autoinducer synthesis proteins. *J. Bacteriol.* **177:** 6946–6951.

27. Grossman, A. D., and R. Losick. 1988. Extracellular control of spore formation in *Bacillus subtilis. Proc. Natl. Acad. Sci. USA* **85:**4369–4373.

28. Hanzelka, B. L., and E. P. Greenberg. 1995. Evidence that the N-terminal region of the *Vibrio fischeri* LuxR protein constitutes an autoinducer-binding domain. *J. Bacteriol.* **177:**815–817.

29. Havarstein, L. S., R. Hakenbeck, and P. Gaustad. 1997. Natural competence in the genus *Streptococcus:* evidence that streptococci can change pherotype by interspecies recombinational exchanges. *J. Bacteriol.* **179:**6589–6594.

30. Havarstein, L. S., and D. A. Morrison. 1999. Quorum sensing and peptide pheromones in streptococcal competence for genetic transformation, p. 9–26. *In* G. M. Dunny and S. C. Winans (ed.), *Cell-Cell Signaling in Bacteria.* American Society for Microbiology, Washington, D.C.

31. Ji, G., R. C. Beavis, and R. P. Novick. 1995. Cell density control of staphylococcal virulence mediated by an octapeptide pheromone. *Proc. Natl. Acad. Sci. USA* **92:**12055–12059.

32. Ji, G., R. Beavis, and R. P. Novick. 1997. Bacterial interference caused by autoinducing peptide variants. *Science* **276:** 2027–2030.

33. Jiang, Y., M. Camara, S. R. Chhabra, K. R. Hardie, B. W. Bycroft, A. Lazdunski, G. P. Salmond, G. S. Stewart, and P. Williams. 1998. In vitro biosynthesis of the *Pseudomonas aeruginosa* quorum-sensing signal molecule N-butanoyl-L-homoserine lactone. *Mol. Microbiol.* **28:**193–203.

34. Kaplan, H. B., and E. P. Greenberg. 1985. Diffusion of autoinducer is involved in regulation of the *Vibrio fischeri* luminescence system. *J. Bacteriol.* **163:**1210–1214.

35. Kleerebezem, M., W. M. de Vos, and O. P. Kuipers. 1999. The lantibiotics nisin and subtilin act as extracellular regulators of their own biosynthesis, p. 159–174. *In* G. M. Dunny and S. C. Winans (ed.), *Cell-Cell Signaling in Bacteria.* American Society for Microbiology, Washington, D.C.

36. Kuipers, O. P., M. M. Beerthuyzen, P. G. de Ruyter, E, J. Luesink, and W. M. de Vos. 1995. Autoregulation of nisin biosynthesis in *Lactococcus lactis* by signal transduction. *J. Biol. Chem.* **270:**27299–27304.

37. Kuipers, O. P., M. M. Beerthuyzen, R. J. Siezen, and W. M. De Vos. 1993. Characterization of the nisin gene cluster *nisABTCIPR* of *Lactococcus lactis.* Requirement of expression of the *nisA* and *nisI* genes for development of immunity. *Eur. J. Biochem.* **216:**281–291.

38. Latifi, A., M. Foglino, K. Tanaka, P. Williams, and A. Lazdunski. 1996. A hierarchical quorum-sensing cascade in *Pseudomonas aeruginosa* links the transcriptional activators LasR and RhlR to expression of the stationary phase sigma factor RpoS. *Mol. Microbiol.* **21:**1137–1146.

39. Lazazzera, B. A., T. Palmer, J. Quisel, and A. D. Grossman. 1999. Cell density control of gene expression in *Bacillus subtilis,* p. 27–46. *In* G. M. Dunny and S. C. Winans (ed.), *Cell-Cell Signaling in Bacteria.* American Society for Microbiology, Washington, D.C.

40. Malthus, T. R. 1976. *An Essay on the Principle of Population.* Prometheus Books, Amherst, N.Y.

41. Mayville, P., G. Ji, R. Beavis, H. Yang, M. Goger, R. P. Novick, and T. W. Muir. 1999. Structure-activity analysis of synthetic autoinducing thiolactone peptides from *Staphylococcus aureus* responsible for virulence. *Proc. Natl. Acad. Sci. USA* **96:** 1218–1223.

42. Meadows, D. H., D. L. Meadows, J. Randers, and W. W. Behrens III. 1972. *The Limits to Growth.* Universe Books, New York, N.Y.

43. Moré, M. I., L. D. Finger, J. L. Stryker, C. Fuqua, A. Eberhard, and S. C. Winans. 1996. Enzymatic synthesis of a quorum-sensing autoinducer using defined substrates. *Science* **272:** 1655–1658.

44. Morfeldt, E., L. Janzon, S. Arvidson, and S. Lofdahl. 1988. Cloning of a chromosomal locus (*exp*) which regulates the expression of several exoprotein genes in *Staphylococcus aureus. Mol. Gen. Genet.* **211:**435–440.

45. Nasser, W., M. L. Bouillant, G. Salmond, and S. Reverchon. 1998. Characterization of the *Erwinia chrysanthemi expI-expR* locus directing the synthesis of two N-acyl-homoserine lactone signal molecules. *Mol. Microbiol.* **29:**1391–1405.

46. Nes, I. F., and V. G. H. Eijsink. 1999. Regulation of group II peptide bacteriocin synthesis by quorum-sensing mechanisms, p. 175–192. *In* G. M. Dunny and S. C. Winans (ed.), *Cell-Cell Signaling in Bacteria.* American Society for Microbiology, Washington, D.C.

47. Nissen-Meyer, J., and I. F. Nes. 1997. Ribosomally synthesized antimicrobial peptides: their function, structure, biogenesis, and mechanism of action. *Arch. Microbiol.* **167:**67–77.

48. Novick, R. P. 1999. Regulation of pathogenicity in *Staphylococcus aureus* by a peptide-based density-sensing system, p. 129–146. *In* G. M. Dunny and S. C. Winans (ed.), *Cell-Cell Signaling in Bacteria.* American Society for Microbiology, Washington, D.C.

49. Oger, P., K. S. Kim, R. L. Sackett, K. R. Piper, and S. K. Farrand. 1998. Octopine-type Ti plasmids code for a mannopine-inducible dominant-negative allele of *traR,* the quorum-sensing activator that regulates Ti plasmid conjugal transfer. *Mol. Microbiol.* **27:**277–288.

50. Oshsner, U. A., A. K. Koch, A. Fiechter, and J. Reiser. 1994. Isolation and characterization of a regulatory gene affecting rhamnolipid biosurfactant synthesis in *Pseudomonas aeruginosa. J. Bacteriol.* **176:**2044–2054.

51. Parsek, M. R., D. L. Val, B. L. Hanzelka, J. E. Cronan, Jr., and E. P. Greenberg. 1999. Acyl homoserine-lactone quorum-sensing signal generation. *Proc. Natl. Acad. Sci. USA* **96:**4360–4365.

52. Passador, L., J. M. Cook, M. J. Gambello, L. Rust, and B. H. Iglewski. 1993. Expression of *Pseudomonas aeruginosa* virulence genes requires cell-to-cell communication. *Science* **260:** 1127–1130.

53. Pearson, J. P., C. Van Delden, and B. H. Iglewski. 1999. Active efflux and diffusion are involved in transport of *Pseudomonas aeruginosa* cell-to-cell signals. *J. Bacteriol.* **181:**1203–1210.

54. Perego, M. 1997. A peptide export-import control circuit modulating bacterial development egulates protein phosphatases of the phosphorelay. *Proc. Natl. Acad. Sci. USA* **94:**8612–8617.

55. Perego, M. 1999. Self-signalling by Phr peptides molulates *Bacillus subtilis* development, p. 243–258. *In* G. M. Dunny and S. C. Winans (ed.), *Cell-Cell Signaling in Bacteria.* American Society for Microbiology, Washington, D.C.

56. Perego, M., and J. A. Hoch. 1996. Cell-cell communication regulates the effects of protein aspartate phosphatases on the phosphorelay controlling development in *Bacillus subtilis. Proc. Natl. Acad. Sci. USA* **93:**1549–1553.

57. Pesci, E. C., and B. H. Iglewski. 1999. Quorum sensing in *Pseudomonas aeruginosa,* p. 147–155. *In* G. M. Dunny and S. C. Winans (ed.), *Cell-Cell Signaling in Bacteria.* American Society for Microbiology, Washington, D.C.

58. Pesci, E. C., J. P. Pearson, P. C. Seed, and B. H. Iglewski. 1997. Regulation of *las* and *rhl* quorum sensing in *Pseudomonas aeruginosa. J. Bacteriol.* **179:**3127–3132.

59. Reimmann, C., M. Beyeler, A. Latifi, H. Winteler, M. Foglino, A. Lazdunski, D. Haas. 1997. The global activator GacA of *Pseudomonas aeruginosa* PAO positively controls the production of the autoinducer N-butyryl-homoserine lactone and the formation of the virulence factors pyocyanin, cyanide, and lipase. *Mol. Microbiol.* **24:**309–319.

60. Reverchon, S., M. L. Bouillant, G. Salmond, and W. Nasser. 1998. Integration of the quorum-sensing system in the regulatory networks controlling virulence factor synthesis in *Erwinia chrysanthemi. Mol. Microbiol.* **29:**1407–1418.

61. Schaefer, A. L., D. L. Val, B. L. Hanzelka, J. E. Cronan, Jr., and E. P. Greenberg. 1996. Generation of cell-to-cell signals in quorum sensing: acyl homoserine lactone synthase activity of a purified *Vibrio fischeri* LuxI protein. *Proc. Natl. Acad. Sci. USA* **93:**9505–9509.

62. Seed, P. C., L. Passador, and B. H. Iglewski. 1995. Activation of the *Pseudomonas aeruginosa* lasI gene by LasR and the *Pseudomonas* autoinducer PAI: an autoinduction regulatory hierarchy. *J. Bacteriol.* **177:**654–659.

63. Stevens, A. M., and E. P. Greenberg. 1999. Transcription activation by LuxR, p. 231–242. *In* G. M. Dunny and S. C. Winans (ed.), *Cell-Cell Signaling in Bacteria.* American Society for Microbiology, Washington, D.C.

64. Stevens, A. M., K. M. Dolan, and E. P. Greenberg. 1994. Synergistic binding of the *Vibrio fischeri* LuxR transcriptional activator domain and RNA polymerase to the lux promoter region. *Proc. Natl. Acad. Sci. USA* **91:**12619–12623.

65. Surette, M. G., M. B. Miller, and B. L. Bassler. 1999. Quorum sensing in *Escherichia coli, Salmonella typhimurium,* and *Vibrio harveyi*: a new family of genes responsible for autoinducer production. *Proc. Natl. Acad. Sci. USA* **96:**1639–1644.

66. Waldburger, C., D. Gonzalez, and G. H. Chambliss. 1993. Characterization of a new sporulation factor in *Bacillus subtilis. J. Bacteriol.* **175:**6321–6327.

67. Winans, S. C., J. Zhu, and M. I. Moré. 1999. Cell density-dependent gene expression by *Agrobacterium tumefaciens* during colonization of crown gall tumors, p. 117–128. *In* G. M. Dunny and S. C. Winans (ed.), *Cell-Cell Signaling in Bacteria.* American Society for Microbiology, Washington, D.C.

68. Winson, M. K., M. Camara, A. Latifi, M. Foglino, S. R. Chhabra, M. Daykin, M. Bally, V. Chapon, G. P. Salmond, B. W. Bycroft, A. Lazdunski, G. S. A. B. Stewart, and P. Williams. 1995. Multiple N-acyl-L-homoserine lactone signal molecules regulate production of virulence determinants and secondary metabolites in *Pseudomonas aeruginosa. Proc. Natl. Acad. Sci. USA* **92:**9427–9431.

69. You, Z., J. Fukushima, T. Ishiwata, B. Chang, M. Kurata, S. Kawamoto, P. Williams, and K. Okuda. 1996. Purification and characterization of LasR as a DNA-binding protein. *FEMS Microbiol. Lett.* **142:**301–307.

70. Zhu, J., J. W. Beaber, M. I. Moré, C. Fuqua, A. Eberhard, and S. C. Winans. 1998. Analogs of the autoinducer 3-oxo-octanoyl-homoserine lactone strongly inhibit activity of the TraR protein of *Agrobacterium tumefaciens. J. Bacteriol.* **180:**5398–5405.

71. Zhu, J., and S. C. Winans. 1998. Activity of the quorum-sensing regulator TraR of *Agrobacterium tumefaciens* is inhibited by a truncated, dominant defective TraR-like protein. *Mol. Microbiol.* **27:**289–297.

72. Zhu, J., and S. C. Winans. 1999. Autoinducer binding by the quorum-sensing regulator TraR increases affinity for target promoters in vitro and decreases TraR turnover rates in whole cells. *Proc. Natl. Acad. Sci USA.* **96:** 4832–4837.

III. PATHOGENIC RESPONSES

Bacterial Stress Responses
Edited by G. Storz and R. Hengge-Aronis
©2000 ASM Press, Washington, D.C.

Chapter 19

The Art of Keeping Low and High Iron Concentrations in Balance

KLAUS HANTKE AND VOLKMAR BRAUN

Aerobic bacteria transport iron(III) mostly with the help of siderophores, low molecular weight iron chelators, while commensals and pathogens often use directly the iron sources heme, transferrin, and lactoferrin of their host. Gram-negative bacteria have in their outer membrane specific receptors for these iron-carrying molecules. Transport of substrate through the receptors requires energy of the cytoplasmic membrane which is mediated by the Ton system to the outer membrane receptors. In the periplasm, an ABC transporter with a binding protein takes over to bring the substrate into the cytoplasm. The crystal structures of the FhuA and the FepA siderophore receptors are very similar: a β-barrel with 22 β-strands is closed by a globular N-terminal cork-like structure. Removal of the cork, including the known interaction site of TonB, revealed that FhuA still transports dependent on TonB. Peptide mapping disclosed for the first time an interaction of the binding protein in close proximity to the ATPase subunit in the membrane component of the ABC transporter. In gram-negative bacteria regulation of iron transport is achieved by the corepressor Fe(II) and the repressor protein Fur, which seems to recognize three GATAAT hexanucleotides. Since iron is a main cause of oxidative stress, Fur regulation is intimately connected to SoxS/R and OxyR regulation. Regulators of the Fur protein family have been shown also to control zinc uptake. In gram-positive bacteria the DtxR protein family serves as iron regulator, and Fur-like proteins regulate oxidative stress defense genes. The only exception is Bacillus subtilis *where one of three Fur-like proteins regulates iron uptake.*

Certain siderophore transport systems are only expressed in the presence of the respective siderophore in the medium. This has been studied in detail in the iron citrate transport system of Escherichia coli. *TonB-dependent binding of iron-citrate to the FecA outer membrane receptor triggers via FecR the activation of the sigma factor FecI. The genes encoding the transport system are only transcribed with the activated FecI.*

Aerobic bacteria living in a neutral or alkaline environment encounter the problem of how to acquire sufficient amounts of iron and transition metals since these metals are highly insoluble under these conditions. The same problem faces bacteria that invade eukaryotic organisms. These hosts solubilize iron mainly by binding it to the high-affinity binding proteins transferrin in the serum as well as lactoferrin in secretory fluids and lactoferrin stored in macrophages and deposited at sites of infection. These host binding proteins reduce even further the free iron available for bacteria—the free Fe^{3+} concentration in equilibrium with transferrin and lactoferrin is in the order of 1 ion per liter. In nature, the abundant Fe^{3+} forms insoluble hydroxyaquo complexes with a free Fe^{3+} concentration of 10^{-18} M. Bacteria require a Fe^{3+} concentration of 10^{-7} M for growth.

To bridge the huge gap between supply and demand, bacteria synthesize siderophores with a very high affinity and specificity for Fe^{3+}. The free Fe^{3+} concentration in equilibrium with siderophores is in the order of 10^{-24} M. Symbiotic or parasitic bacteria utilize the iron sources presented by their hosts: iron-loaded transferrin and lactoferrin, heme, and heme-containing proteins. Bacteria also overcome the extreme iron shortage by synthesizing elaborate iron-transport systems. Several of these systems—at least seven in the *Escherichia coli* K-12 laboratory strain—are synthesized under conditions of iron shortage and occasionally also in response to the type of iron source present in the culture medium. Hence, iron is not taken up as a metal ion, but as a Fe^{3+}–siderophore complex. In the cytoplasm,

Klaus Hantke and Volkmar Braun • Mikrobiologie/Membranphysiologie, Universität Tübingen, Auf der Morgenstelle 28, D-72076 Tübingen, Germany.

Fe^{3+} is released from the siderophores by reduction to Fe^{2+}.

Bacteria and fungi produce a large variety of siderophores, which can be grouped into three structural classes: the hydroxamates, the catecholates, and the hydroxycarboxylates. In Fig. 1, ferrichrome produced by fungi is shown as a representative of the Fe^{3+} siderophores of the hydroxamate type. *E. coli* produces the catechol enterobactin (also called enterochelin), which is a cyclic trimer of 2,3-dihydroxy-N-benzoyl-L-serine. Citrate is a simple representative of the hydroxycarboxylate class of siderophores.

Although cells require iron, free iron in the presence of oxygen is also a dangerous metal for the cell. Toxic radicals, generated via the Fenton reaction, destroy DNA and membranes. Therefore, iron uptake is strictly controlled by the cell, and the intracellular iron must be bound and stored in a way that makes it inaccessible to oxygen.

Iron uptake and regulation, especially in pathogenic bacteria, have attracted great interest since iron-transport systems in certain cases determine virulence and since toxins unrelated in their activity to iron metabolism are regulated by iron. Under low-iron stress, many pathogenic bacteria induce the synthesis of toxins and express additional virulence factors such as adhesins (63). Since most of these topics have been reviewed in the past few years (14, 44), we will concentrate on recent developments on how bacteria respond to low-iron stress. An increasing number of variations of well-known transport systems with very similar regulatory principles are being found in different pathogenic bacteria, and our focus will be on the progress made in the detection of new systems. In addition, we will discuss major advances that have been made in the characterization of high-affinity iron-transport systems and in the regulation of iron uptake and metabolism.

TRANSPORT OF IRON

Transport of Fe^{3+} across the Outer Membrane of Gram-Negative Bacteria

In contrast to the diffusion of substrates through the porins in the outer membrane of gram-negative bacteria, iron is actively transported. The various Fe^{3+} siderophores have molecular masses of around 700 Da, which is too large to diffuse through the porin channels. In addition, the concentration of the Fe^{3+} siderophores would not suffice to support growth (at least 10^5 Fe^{3+} ions per cell are needed per generation) if they were to pass the outer membrane by diffusion. Therefore, the Fe^{3+} siderophores are extracted from the medium and concentrated at the bacterial cell surface by binding to receptor proteins. Input of energy is required to release the Fe^{3+} siderophores from the active center of the receptor proteins and to transport them across the outer membrane into the periplasm, where they bind to binding proteins that deliver them to ABC transporters in the cytoplasmic membrane (Fig. 2).

In humans, Fe^{3+} is bound to extracellular transferrin and lactoferrin and deposited in intracellular ferritin. Large amounts of Fe^{3+} are also contained in heme, hemoglobin, and hemopexin. Bacteria that live in humans can use these iron sources and bind them to specific receptor proteins at the cell surface (Fig. 2). Iron is released by an unknown mechanism from transferrin and lactoferrin, and heme is taken up as such (88, 89). Mobilization of heme from hemoglo-

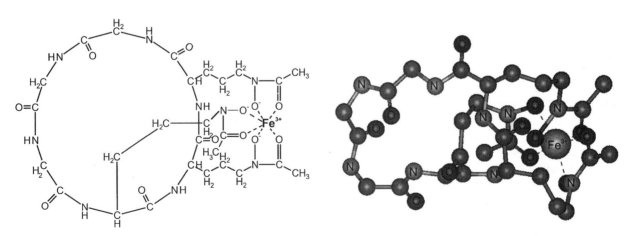

Figure 1. Chemical structure of ferrichrome (left), which shows the octahedral binding and the location of Fe^{3+} as revealed by the crystal structure of ferrichrome (right) (93). Ferrichrome is a cyclic hexapeptide composed of three Nδ-acetyl-Nδ-hydroxy-L-ornithine and three glycine residues. In the crystal structure H is not shown to allow a better view on the backbone structure.

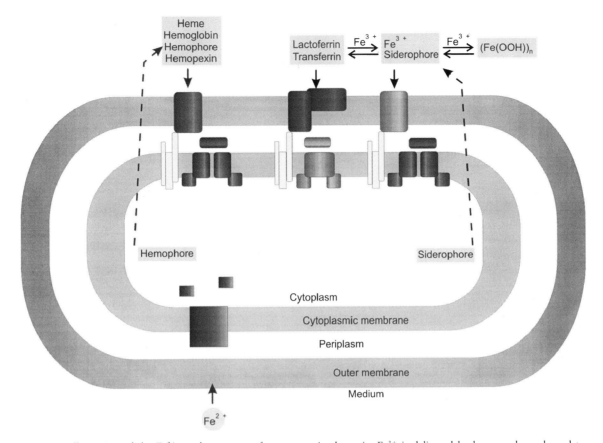

Figure 2. Illustration of the Fe^{3+}-uptake systems of gram-negative bacteria. Fe^{3+} is delivered by heme or heme bound to hemoglobin or hemopexin, transferrin or lactoferrin of the host organisms, or siderophores synthesized by the transporting strains or provided by other bacteria and fungi. The siderophores incorporate free Fe^{3+} in equilibrium with the iron hydroxyaquo complex or with transferrin and lactoferrin. The outer membrane receptor proteins are specific for the respective iron donor. Active transport across the outer membrane is energized by the proton gradient across the cytoplasmic membrane. Presumably, conformational energy is transferred from the cytoplasmic membrane to the outer membrane by a protein complex consisting of the TonB, which interacts with the receptor; ExbB, proximal to TonB; and ExbD, proximal to ExbB (shown in white). In the periplasm, Fe^{3+}, heme, or Fe^{3+}-siderophores bind to binding proteins that are specific for Fe^{3+}, heme, or Fe^{3+}-siderophores of the hydroxamate, catecholate, or hydroxycarboxylate type. The binding proteins in the periplasm deliver the iron compounds to transport proteins in the cytoplasmic membrane, which translocate the iron compounds into the cytoplasm at the expense of ATP. The respective ATPases are associated with the cytoplasmic membrane on the inside. Fe^{2+} presumably diffuses across the outer membrane and is actively transported across the cytoplasmic membrane by proteins shown in the lower part.

bin and from hemopexin may occur at the receptor proteins or by secreted proteins (Fig. 2). The latter, termed hemophores (36), deliver the heme to receptor proteins at the cell surface, as has been shown for HasA of *Serratia marcescens* (36), HasAp of *Pseudomonas aeruginosa* (62), and HuxA of *Haemophilus influenzae* (21).

As far as is known, the mechanism of iron transport across the outer membrane is the same for Fe^{3+}-siderophores, heme, and iron released from transferrin and lactoferrin. The most advanced studies have been performed on the FhuA and the FepA receptor proteins of *E. coli* K-12 and include the determination of their crystal structures (19, 30, 64). Insights gained on the structure and function of these two

proteins may apply to all the iron-transport systems and will therefore be discussed in some detail below.

The FhuA outer membrane receptor protein specifically transports ferrichrome, the structurally closely related antibiotic albomycin, and the antibiotic rifamycin CPG 4832, and it serves as receptor for the infection of the phages T1, T5, ϕ80, and UC-1 and for the uptake of colicin M and microcin 25. The multiple functions of FhuA make it a particularly attractive and suitable transport protein to study. With the exception of infection by phage T5, all FhuA activities depend on the TonB, ExbB, and ExbD proteins, which together form the Ton system thought to be involved in the transfer of energy from the cytoplasmic membrane to the outer membrane transporters (11, 12, 50, 69, 79). Binding of phages T1 and φ80 to FhuA requires energy and the activity of the Ton system, which suggests a Ton-induced FhuA conformation that differs from the Ton-independent conformation (42).

The crystal structure of the FhuA protein consists of a β-barrel composed of 22 antiparallel β-strands (Fig. 3). The β-barrel is closed on the periplasmic side by a domain designated the "cork" (30) or "plug" (64) that comprises residues 1 to 159, which form a four-stranded β-sheet and four short α-helices. Ferrichrome binds in a pocket slightly above the external outer membrane surface to which the cork and the barrel contribute. Two of the eight aromatic residues that line the ferrichrome binding pocket, Tyr244 and Trp246, are lacking in the deletion mutant FhuAΔ236-248, which no longer binds and transports ferrichrome but still functions as a phage and colicin receptor (51). Binding of ferrichrome causes a strong structural transition across the entire membrane-embedded portion of FhuA. A 1.7-Å translation close to the ferrichrome binding site of the cork domain is enhanced by a 17-Å movement of Glu19, located in the periplasmic pocket, away from its former α-carbon position. Despite this large conformational change, ferrichrome binding does not

open a channel in FhuA. The channel opening is thought to occur by interaction of FhuA with TonB, a concept derived from suppression of missense mutations in the "TonB box" of FhuA by amino acid replacements at residues 158 and 160 of TonB (81), from stabilization of overexpressed TonB by over-expressed FhuA (40), from chemical cross-linking of FhuA to TonB (70), and from an alteration in the proteinase degradation pattern of TonB in response to the proton gradient across the cytoplasmic membrane and the presence of a transport substrate, presumably bound to the transport protein (58). The mutations in FhuA that are suppressed by mutations in TonB are located close to the N-terminal end (residues 7 to 11, termed the "TonB box"); although not seen in the crystal structure, these residues are most likely in the periplasm, where most of TonB also resides. All receptors that require the Ton system for their transport activity and all group B colicins that need TonB for entry into sensitive cells contain a TonB box near the N terminal end (12). Alteration of the conformation of the N-terminus of FhuA may allow interaction with TonB. This in fact has been shown in that binding of TonB to FhuA is enhanced in the presence of ferricrocin, a ferrichrome derivative, and colicin M (70).

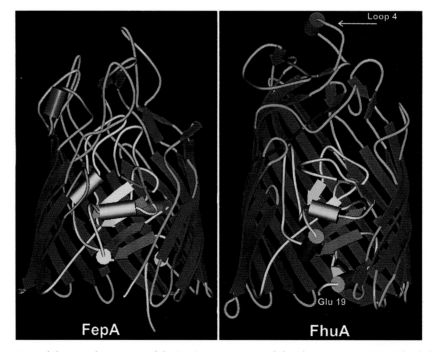

Figure 3. Comparison of the crystal structures of the FepA protein (19) and the FhuA protein (64), not loaded with substrate, as arranged in the outer membrane: top, cell surface; bottom, periplasm. The β-barrel composed of 22 antiparallel β-strands in each protein is shown in blue (FhuA residues 160 to 612, FepA residues 156 to 614); in the β-barrel of the deletion mutant FhuAΔ236-248 residues 236 to 248 (in red) are missing; the N-proximal cork or plug that closes the channel formed by the β-barrels in each protein is shown in yellow (FhuA residues 18 to 160, FepA residues 11 to 155). In FepA the carboxy-terminal residues 615 to 724 are not shown, and in FhuA residues 613 to 714 are not shown to allow a better view of the cork structures.

The crystal structure of FhuA indicates that the cork domain completely closes the β-barrel of FhuA and suggests that a conformational change in the cork must occur to open a channel through which ferrichrome, albomycin, and rifamycin CGP 4832 can be translocated into the periplasm. Interaction of TonB with the TonB box of wild-type FhuA might impose the structural change in the cork that is required to open a channel in the cork, which otherwise tightly closes the channel lumen of FhuA. This structural change may or may not completely induce a structural transition in the ferrichrome binding pocket for the release of ferrichrome. To release ferrichrome, the barrel changes its structure in response to TonB, and this change may only partially be caused by structural changes in the cork. It is not known whether passage of ferrichrome through the FhuA channel is unidirectional from the cell surface into the periplasm or whether ferrichrome released from its binding site can also escape into the surrounding medium. Removal of the cork should result in a permanently open channel through which ferrichrome and the antibiotics diffuse into the periplasm.

Studies with the FhuAΔ5-160 deletion mutant, which lacks residues 5 to 160 of mature FhuA, surprisingly revealed additional site(s) of interaction between FhuA and TonB (12). Since the TonB box and the cork are lacking in FhuAΔ5-160, the mutant protein should not display TonB-dependent activities. In contrast to this expectation, FhuAΔ5-160 functions similarly to wild-type FhuA in all FhuA activities and still requires TonB activity. Since FhuAΔ5-160 does not contain the cork, TonB does not open a channel and therefore must have an additional function. Interaction of TonB with FhuAΔ5-160 might release ferrichrome from its binding sites at the barrel, which would also have to occur with wild-type FhuA. Due to the lack of the cork-binding sites, 10% of the ferrichrome that binds to wild-type FhuA binds to FhuAΔ5-160. Release from the binding site seems to be sufficient since FhuAΔ5-160 mediates uptake of ferrichrome with about half the rate of FhuA wild-type. Interaction of TonB with periplasmic sites of FhuA induces structural changes in FhuA that are transmitted across the outer membrane to the cell surface where ferrichrome resides. This long-range conformational change implies that the FhuA β-barrel is not a rigid structure but must be sufficiently flexible to transmit a signal. As expected, when the cork is deleted, a low level of TonB-independent unspecific permeability of FhuAΔ5-160 for ferrichrome, certain antibiotics, and maltodextrins is observed.

TonB can activate FhuAΔ5-160 exclusively through the β-barrel, and this reveals sites of interaction between TonB and FhuA in addition to the TonB box. It is conceivable that these sites are located in the periplasm. Since the binding sites of the TonB-dependent phages T1 and φ80 have been localized to three subregions of amino acid residues 316 to 355 (52), which comprise the most prominent loop (loop 4) at the cell surface, interaction with TonB may well be propagated through the FhuA barrel up to loop 4, well above the outer bilayer boundary. In FhuAΔ5-160, the conformational change of loop 4 caused by TonB interaction is transmitted through the β-barrel, and this may also be the case in wild-type FhuA.

Phage T5 infects cells without the help of the Ton system and is inactivated by an isolated outer membrane fragment (96) that contains FhuA (15). Binding of phage T5 to the outer membrane fragment triggers release of the DNA from the phage head (32). The release of DNA from phage T5 by synthetic hexapeptides of loop 4 (52) supports the theory that phage T5 binding to loop 4 is the minimal requirement for phage T5 infection. FhuA incorporated into a planar lipid bilayer opens a channel that displays such a high conductance that the estimated channel diameter would accommodate phage DNA with a 2-nm cross section (8, 76). Ferrichrome entrapped in liposomes containing FhuA in the membrane flows out of liposomes upon addition of phage T5 (61), and cryoelectron microscopy has revealed phage DNA inside the liposomes (57). The straight tail fiber of the phage traverses into the liposomes and is shortened to 50% of its normal length. These data demonstrate the opening of a channel in FhuA that might be initiated by binding of phage T5 to loop 4, and they suggest penetration of the straight tail fiber into the channel. In FhuAΔ5-160, interaction of phage T5 with loop 4 should be sufficient to trigger DNA release, and no cork would have to be removed to allow penetration of the phage DNA. However, the multiple interactions of amino acid side chains of the cork and the barrel render it energetically unlikely that the entire cork is pushed out of the barrel of wild-type FhuA upon phage binding. Therefore, it is unclear whether phage DNA goes through a FhuA channel or along the side of FhuA across the outer membrane. Colicin M is also too large (molecular weight 27,000) to go through the FhuA channel without removal of the entire cork.

The crystal structure of FepA (Fig. 3) is similar to the crystal structure of FhuA (19). Since no comparison has been made between FepA occupied with ferric enterobactin, a siderophore of the catecholate type, and FepA free of enterobactin, it is not known whether FepA undergoes a structural transition upon substrate binding similar to that observed in FhuA. However, electron-spin resonance spectroscopy has revealed changes in FepA conformation upon binding of ferric enterobactin and in response to energization via TonB (49). The TonB box of FepA that comprises residues 12 to 18 is seen in the crystal structure that starts with residue 11. The TonB box forms an extended conformation without obvious secondary structure elements. It is likely that the structure of the FepA TonB box also applies to the FhuA TonB box and the other TonB-dependent outer membrane transporters and the TonB-dependent colicins. Interaction of TonB with the TonB box of transporters, hitherto mainly based on genetic evidence, was directly shown by cysteine scanning mutagenesis of the TonB box of the vitamin B_{12} transporter BtuB and residue 160 (20).

The extracellular human iron carrier proteins transferrin and lactoferrin serve as iron sources of *Neisseria gonorrhoeae*, *Neisseria meningitidis*, and *H. influenzae* without involvement of siderophores (Fig. 2). Utilization of iron bound to transferrin requires the two outer membrane proteins TbpA and TbpB. TbpA contains a TonB box close to the N-terminal end. For *N. gonorrhoeae* it has been shown that cleavage of TbpB by added trypsin depends on the energization of TbpA by the Ton system. In *tonB* mutants, in unenergized cells, and in cells devoid of TbpA, trypsin readily degrades TbpB (22). These experiments demonstrate that energization promotes interaction of TbpA with TbpB, most likely through a conformational change in TbpA caused by the Ton system.

Transport of Fe^{3+} across the Cytoplasmic Membrane

All Fe^{3+}-transport systems hitherto studied belong to the family of ABC transporters, which consist of a binding protein contained in the periplasm of gram-negative bacteria or linked by a lipid anchor to the cytoplasmic membrane of gram-positive bacteria (85), one or two integral proteins in the cytoplasmic membrane, and two copies of an ATPase associated at the inner side of the cytoplasmic membrane with the integral membrane transport proteins (Fig. 2). The respective proteins that transport ferrichrome are designated FhuD, FhuB, and FhuC. Ferrichrome binds to FhuD with an apparent K_d of 1 μM, renders FhuD resistant to proteases, and changes its tryptophan fluorescence (80). FhuD traps ferrichrome in the periplasm and delivers it to FhuB in the cytoplasmic membrane.

Interaction of FhuD with FhuB has been determined with synthetic 10- and 20-residue overlapping peptides that cover the entire sequence of FhuB (659 residues). Binding of the peptides to isolated FhuD was measured with the ELISA technique (65). The peptides bound to nine segments of FhuB of which, according to a transmembrane topology model of FhuB (37), four segments represent transmembrane regions and five segments represent loops. Only two loops are located in the periplasm. In addition, the 10-residue peptides were transferred into the periplasm through the FhuAΔ322-355 deletion protein, which forms a permanently open channel in the outer membrane that is at least threefold larger than the porin channels (51). Those FhuB peptides that in the ELISA assay reacted with FhuD inhibited ferrichrome transport, most likely by interference with the interaction of FhuD with FhuB. The same experimental approach identified two regions of FhuD that interact with FhuB (C. Dangelmayr, H. Killmann, and V. Braun, unpublished data). These results demonstrate for the first time for any ABC transporter the interacting regions between a periplasmic binding protein and the transport protein in the cytoplasmic membrane. They show that the sites of interaction are not confined to periplasmic loops of the integral membrane protein but rather involve transmembrane regions and cytoplasmic loops. They can be reconciled by the proposal that FhuB forms a channel that is lined with transmembrane regions. The cytoplasmic loops and perhaps the periplasmic loops, as defined by the model, fold into the FhuB channel. FhuD loaded with ferrichrome inserts into the FhuB channel and thereby delivers ferrichrome into the channel.

One site of interaction is very close to the proposed binding site of the FhuC ATPase (55, 86). As a consequence, FhuD may interact directly with FhuC or through a short interconnecting peptide between the two binding sites on FhuB. It is no longer necessary to assume that binding of the substrate-loaded binding protein to the transmembrane transport protein(s) initiates a signal across the cytoplasmic membrane that triggers ATP hydrolysis catalyzed by the traffic ATPase (6, 24). Rather, the substrate-loaded binding proteins and the ATPases come in close contact to trigger ATP hydrolysis to provide the energy required for substrate transport across the cytoplasmic membrane. We consider these results obtained with the FhuABCD transport system as representative of other bacterial iron ABC transporters (14) and ABC transporters in general (9).

The ferrichrome-transport system is typical for other Fe^{3+}-siderophore transport systems and most probably also applies to heme transport and iron transport across the cytoplasmic membrane (14). The first transport system for presumably free Fe^{3+} across the cytoplasmic membrane was characterized in *S. marcescens* (5). The number of the *sfuABC* genes, the nucleotide sequences, and the subcellular location of the encoded proteins are typical for ABC transporters. It has been shown that iron released from transferrin and lactoferrin at the cell surface uses this kind of transport system across the cytoplasmic membrane (68). The crystal structure of the periplasmic HitA iron-binding protein of *H. influenzae* (17) shows a remarkable similarity to the conformation of one lobe of transferrin, and the iron-binding amino acid side chains of both proteins are largely identical. From these results, it can be inferred that the binding proteins for Fe^{3+}, Fe^{3+}-siderophores, and heme display a similar and characteristic three-dimensional structure.

REGULATION

Regulation by the Fur Protein Family

Iron uptake is regulated in all gram-negative bacteria studied to date by the repressor protein Fur (ferric iron uptake regulation), with ferrous iron as corepressor (14, 44). In *E. coli* K-12 it has been shown that more than 50 genes are regulated by Fur. The Fur protein of *E. coli* consists of two domains: an N-terminal DNA-binding domain, which contains an unusual helix-turn-helix motif, and a C-terminal domain, the beginning of which has a cluster of histidine residues that may bind the regulatory ferrous iron (44). A careful analysis has revealed that Fur unexpectedly contains two metal-binding sites, which have been characterized by several spectroscopic techniques including electron paramagnetic resonance, nuclear magnetic resonance, Mössbauer, and X-ray absorption spectroscopy. Per monomer, 0.5 to 0.8 zinc atoms were determined; these zinc atoms are bound tetrahedrally to two cysteines (most likely positions 92 and 95) and two nitrogen/oxygen atoms (2, 48). The second binding site for Fe^{2+} important for regulation may be occupied in vitro by Mn^{2+}, Co^{2+}, or some other divalent cations. The metal is bound in this site by at least two histidines and one carboxylate in an axially distorted octahedral environment. However, no cysteines contribute to metal binding (2) in this site, confirming earlier work cited in reference 14.

By footprinting experiments, binding sites of Fur have been defined in the promoter region of different iron-regulated genes; a Fur-binding site consensus sequence, in *E. coli* GATAATGATAATCATTATC, called the "Fur box," seems to be similar to sites in other enteric bacteria (25, 26). Further in vitro studies with synthetic binding sites have indicated that Fur recognizes the short hexamer GATAAT and not the 19-bp palindromic sequence (29), as was generally assumed. However, at least three hexamers are necessary for productive binding of Fur to DNA. The orientation of the hexamers seems not to be important. This unusual binding behavior may also explain why Fur at high concentrations polymerizes and wraps around the DNA starting at the Fur binding sites (59).

In *E. coli*, a proportion of manganese-resistant mutants have been found to be mutated in *fur* (43). The selection procedure was also used to select for *fur* mutants in other bacteria; however, this method was successful only in some species. One reason for failure may have been that in *Pseudomonas* (46), *Synechococcus* (35), *N. gonorrhoeae* (91), and *Vibrio anguillarum* (98), knockout mutants of *fur* are not viable, possibly because Fur has, besides its iron regulatory function, an influence on genes that defend the cells against oxidative damage. In *E. coli*, a *recA fur* double mutant is not viable (92), most likely because some genes of oxidative response regulons are activated by Fur. SodB, the manganese-dependent superoxide dismutase in *E. coli*, is an example of this positive regulation by Fur. The regulation may be indirect since no Fur-binding sites have been found in the promoter region of *sodB*. In *E. coli*, Fur has a similar positive regulatory influence on *acnA*, which encodes one of the two aconitases, and on *fumA*, which encodes one of the three fumarase genes (38). For *acnA*, the regulation is more complex since the gene is regulated by Crp (cAMP receptor protein), ArcA, and SoxRS (superoxide regulatory system). In all cases where a positive regulation by Fur is observed, it is not clear whether its influence is direct or indirect. Some oxidative stress genes are negatively regulated (i.e., repressed) by Fur and iron, as has been shown for *sodA*, which encodes the manganese superoxide dismutase of *E. coli* (72). *sodA* is also negatively regulated by the global regulator ArcA (aerobic respiration control), which couples the expression of *sodA* to aerobic metabolism and the intracellular iron pool. A similar multifactorial regulation has been observed for the 8-hydroxyguanine endonuclease by Fur, Fnr, and ArcA (60).

Recently another important link of Fur to oxidative stress response has been revealed. The oxidative stress regulator OxyR activates expression of Fur and binds to the *fur* promoter (99). In addition, stimulation of the SoxRS system with paraquat induces synthesis of a transcript that encodes FldA and Fur (99). The gene *fldA* upstream of *fur* encodes flavodoxin A, which can replace ferredoxin in some reduction systems. In some organisms it is known that flavodoxin and flavins are induced under low-iron stress. However, in *E. coli* flavodoxin was assumed to be synthesized constitutively (67), but this now has to be revised since oxidative stress seems to induce flavodoxin synthesis (99).

Under slightly acidic growth conditions, *E. coli* and many other enterobacteria induce protective systems that allow the cells to survive in an otherwise deadly acidic surrounding. The acid response regulation is a major factor determining the survival and infectivity of enterobacteria passing the acidic gut (27). Also, Fur has an influence on acid response regulation, which has been studied mainly in *Salmonella enterica* serovar Typhimurium and other pathogenic enterobacteria. Eight acid shock proteins are regulated by Fur in an unusual iron-independent manner, but nothing is known about the molecular mechanism (31).

Fur-Related Regulators with Other Functions

Proteins that regulate oxidative stress genes

Fur-like proteins in different gram-positive bacteria regulate peroxide stress genes. Recently a link of Fur regulation was also observed in gram-negative bacteria. The PerR protein in *Bacillus subtilis* acts as a repressor of the catalase *katA* and alkyl hydroperoxide reductase *ahpC* genes (18). The proteins are induced by H_2O_2, and manganese is a corepressor of PerR.

Another group of *fur*-like genes, namely *furA* in *Mycobacterium marinum* and *furS* in *Streptomyces reticuli*, are cotranscribed with the catalase-peroxidase gene *katG* (74, 100). In *S. reticuli*, iron represses the synthesis of the catalase-peroxidase (100), while in *M. marinum*, H_2O_2 has no inducing effect on *furA katG* expression (74). The regulatory influence of manganese was not tested in either case.

Proteins of the Fur family recognize different divalent cations and have fulfilled different functions during evolution: regulation of iron transport and metabolism, regulation of Zn^{2+} transport and metabolism (see below), and regulation of the oxidative stress response in connection with (possibly only divalent) cations such as iron or manganese. These different activities of Fur-like proteins make it difficult to predict from sequence similarities alone the in vivo activities of the proteins. A sequence comparison of most of the known Fur proteins is shown as an unrooted phylogenetic tree in Fig. 4. Fur proteins acting as iron-transport regulators in pseudomonads and enteric bacteria form a defined subgroup, while the other Fur-like proteins seem to be very diverse. It is obvious that the oxidative stress responsive proteins FurA from *M. marinum*, FurS from *S. reticulum*, and PerR from *B. subtilis* are not closely related. Also, the two identified Zur proteins for zinc regulation from *B. subtilis* and *E. coli* are not closely related. The observed regulation of heme biosynthesis by Irr in

Bradyrhizobium japonicum (see below) may also be one aspect of an oxidative response regulation since catalases contain heme. However, further characterization of these regulons is necessary to learn which metals act as corepressor and which genes are regulated by these proteins.

Fur-related proteins regulate zinc-uptake systems

From the sequence of the *E. coli* genome, the gene product of *yjbK* was found to have 27% sequence identity with Fur (7). The characterization of the high-affinity zinc-uptake system *znuABC* in *E. coli* led to the discovery that the *yjbK*-encoded protein is the regulator of this transport system. The repressor was named Zur (for zinc uptake regulator) (75). This was one of the first examples that showed that the Fur protein family is not specific for Fe^{2+} as corepressor. Another example of a Fur-like protein with zinc as corepressor has been found in *B. subtilis* (34), where Zur regulates two operons, one that may encode a low-affinity zinc-transport system and one a high-affinity zinc-transport system. Zinc uptake and excretion seem to be thoroughly controlled processes since a MerR-type regulator, ZntR, has been recently identified in *E. coli* as a regulator of the *zntA* gene, which encodes a zinc/cadmium exporter (16). Genes with gene products significantly similar to Zur have been identified in *Vibrio cholerae* as well as in *Pseudomonas aeruginosa* in a screening for genes relevant for virulence. This may have its reason in the high affinity of serum albumin for zinc, which makes zinc a scarce element for bacteria in serum. In horse serum, free Zn^{2+} is present at a concentration of 2×10^{-4} μM (66). The concentration of zinc in serum is lowered during infection, and this may inhibit bacterial growth in serum. The same phenomenon is observed for iron.

The Fur-like gene product Irr regulates heme biosynthesis

In *B. japonicum*, a *fur* gene and a *fur*-like gene, *irr*, have been identified. While the *fur* gene seems to regulate iron transport, the *irr* gene product seems to inhibit porphyrin production under low-iron growth conditions, connecting porphyrin biosynthesis to iron metabolism (41). A new interesting aspect of the Irr-mediated regulation is the rapid degradation of Irr in the cell when the medium changes from low-iron to iron-replete (41). This opens up the possibility that a divalent cation activates a specific protease.

Since all known Fur-like proteins interact with divalent cations, it remains to be elucidated how Irr represses heme biosynthesis under low-iron growth

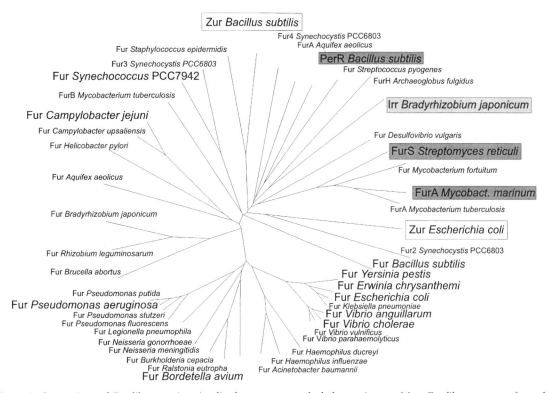

Figure 4. Comparison of Fur-like proteins visualized as an unrooted phylogenetic tree. Most Fur-like sequences from the National Center for Biotechnology Information database (May 1999) are included. Fur-like proteins with a better-defined function are printed in larger type, and a reference is given: Fur *Bacillus subtilis* (18), Fur *Bordetella avium* (70), Fur *Campylobacter jejuni* (94), Fur *Erwinia chrysanthemi* (33), Fur *Escherichia coli* (14), Fur *Pseudomonas aeruginosa* (46), Fur *Synechococcus* sp. (35), Fur *Vibrio anguillarum* (98), Fur *Vibrio cholerae* (56), and Fur *Yersinia pestis* (87). Fur-like proteins that act as zinc regulators (boxed) are Zur from *E. coli* (75) and from *B. subtilis* (34), and those acting mainly as oxidative stress regulators are (dark gray box) PerR from *B. subtilis* (18), FurS from *Streptomyces reticuli* (100), and FurA from *Mycobacterium marinum* (74); Irr from *B. japonicum* (41) regulates heme biosynthesis (light gray box).

conditions. One corepressor candidate is zinc. For the zinc-Zur regulon in *E. coli,* Zur-dependent genes are derepressed by iron (75).

Regulation by DtxR-Like Proteins

DtxR, the diphtheria toxin regulator from *Corynebacterium diphtheriae,* was the first iron-dependent regulator identified in gram-positive bacteria (10, 83). DtxR regulates not only the phage-encoded diphtheria toxin, but also the iron-uptake genes of this organism (82–84). Proteins similar to DtxR regulate iron uptake and the biosynthesis of ferrioxamine B in *Streptomyces* (39).

The nucleotide sequence of the genome of *Mycobacterium tuberculosis* revealed a DtxR-like protein, designated IdeR. The cloned protein was studied in *E. coli,* where it binds in an iron-dependent manner to the promoter region of transformed DtxR-dependent genes (28). An *ideR* mutant in *Mycobacterium smegmatis* has been characterized, and regulation of siderophore biosynthesis by IdeR has been demonstrated (28).

The staphylococcal protein SirR, which has 28% amino acid identity to DtxR, has recently been identified as a regulator of a putative iron-transport system (47). Proteins of the DtxR-IdeR-SirR family have to date only been characterized in gram-positive bacteria. They have been shown to fulfill the same function as Fur in gram-negative bacteria by regulating iron uptake and metabolism and the virulence factor diphtheria toxin. However, homologs of the Fur protein have also been identified in gram-positive bacteria, but they seem to regulate oxidative stress response genes or the uptake and metabolism of other divalent cations (see above). The only known exception to this is in *B. subtilis,* where a Fur homolog regulates siderophore biosynthesis and siderophore-iron uptake (18). *B. subtilis* and *E. coli* contain open reading frames (YQHN BACSU and b817, respectively) that belong to the DtxR family; however, no functions have been assigned to these open reading frames.

The crystal structures of DtxR and IdeR and mutants thereof give a detailed picture of this repressor protein family. Unfortunately, no crystal structure is available for a Fur protein, and one has to rely on more indirect studies of the topology of Fur. The DtxR protein family has no obvious sequence similarity to Fur, but the structural studies disclosed similarities in their domain organization: (i) the N-terminal domain (residues 1 to 73) carries the DNA-binding site, which has been defined by a crystal structure of the repressor–DNA complex (77), and (ii) the dimerization domain (residues 74 to 140) carries two metal-binding sites, which may in vivo bind Fe^{2+} or certain other divalent cations. Met10 in the N-terminal domain of IdeR is a ligand of the second metal-binding site (78). How metal binding induces the structural change in the repressor to allow binding to DNA is not understood. In the IdeR crystal structure, the C-terminal domain (residues141 to 230) is too disordered to be included and its function is unknown.

Transmembrane Transcriptional Control of the Ferric-Citrate-Transport Genes

Under iron-limiting growth conditions and in the presence of ferric citrate in the medium, synthesis of a ferric-citrate-transport system is induced in *E. coli* K-12. The system consists of the outer membrane protein FecA, the Ton energy-transfer device (TonB, ExbB, and ExbD), the periplasmic binding protein FecB, the cytoplasmic membrane proteins FecC and FecD, and the FecE ATPase. The *fec* locus consists of seven genes, *fecIRABCDE*, arranged and transcribed in this order. The *fecIR* genes are two regulatory genes (13). FecA plays a dual role: it transports ferric citrate across the outer membrane into the periplasm, and it is required for initiation of *fecABCDE* gene transcription. Mutants of any of the *fecA, tonB, exbB,* and *exbD* genes involved in transport across the outer membrane are no longer inducible. This does not reflect the need to transport ferric citrate into the periplasm since missense mutations in *fecA* have been isolated that lead to constitutive induction of *fec* transport gene transcription but do not transport ferric citrate (45). In addition, FecA contains an extended N-terminus of 40 residues that is not present in other Fe^{3+}-siderophore receptors of *E. coli* K-12. Deletion of this portion of FecA blocks induction but does not affect transport (53). These results indicate that the transport activity of FecA is not required for induction and imply that ferric citrate only needs to bind to FecA to induce transcription from the cell surface. This conclusion is corroborated by the full induction of mutants that cannot transport iron across the cytoplasmic membrane. A structural change in the N-terminus of FecA upon binding of ferric citrate, similar to the structural change in FhuA

caused by ferrichrome, may be the physical basis of the transcription initiation signal.

After the signal has reached the periplasm through FecA, it is transmitted across the cytoplasmic membrane by the FecR protein. FecR spans the cytoplasmic membrane once and thus can transmit information across it (97). FecR receives the signal by direct interaction of its C-proximal end in the periplasm with the N-terminal end of FecA (S. Enz, U. Stroeher, and V. Braun, unpublished data). Cells that synthesize only the cytoplasmic N-terminal portion of FecR constitutively transcribe the *fec* transport genes (73). This N-terminal fragment might assume the conformation that it adopts in the complete FecR protein upon ferric citrate induction. FecR activates FecI, a sigma factor, by an unknown mechanism; when activated, FecI binds to a promoter upstream of the *fecA* gene and initiates transcription by RNA polymerase (4).

Transcription of the *fecIR* regulatory genes is not induced by ferric citrate. Rather, transcription is derepressed under iron-limiting growth conditions. If the cells contain sufficient iron, the Fe^{2+}-loaded Fur repressor represses transcription of *fecIR* and *fecABCDE* (3). The hierarchical order of transcription regulation, in that under iron deprivation, the FecIR regulatory proteins are synthesized first and then the cells respond only if the suitable iron source is contained in the culture medium via FecIR, represents an optimal and economical device.

Evidence for a regulatory system similar to the ferric-citrate regulatory mechanism has been found in *Pseudomonas putida* (54). Induction by the iron-loaded siderophores pseudobactin BN7 and BN8 is clearly related to the outer membrane receptor protein and occurs independent of its transport activity. The genome sequence of *Pseudomonas aeruginosa* reveals at least two sets of genes homologous to *fecI* and *fecR*, and the heme-uptake system of *S. marcescens* seems to be regulated by FecIR homologs (C. Wandersman, personal communication).

FUTURE DIRECTIONS

There are additional control mechanisms of Fe^{3+}-siderophore biosynthesis and transport induced by Fe^{3+}-siderophores, not described in this chapter, that have not been completely characterized (95). The most complex regulation has been found in *Vibrio anguillarum* and involves, in addition to Fe^{2+}-Fur, a leucine zipper protein and an antisense RNA (23).

In eukaryotic cells low iron or oxidative stress leads to an enzymatically inactive, [4Fe-4S]-free cytosolic aconitase IRP1, which binds to iron-responsive elements (IRE) on the mRNA. An IRE contains roughly a 10-bp-long stem and a loop of 6 bases. Depending on the encoded protein, the binding to the aconitase may enhance mRNA stability by binding to the 3'-untranslated mRNA as it is found with the transferrin receptor mRNA, or it may hinder translation of ferritin mRNA by binding to the IRE in the 5' untranslated region. Aconitases are well conserved in evolution, the aconitase A from *E. coli* is 53% identical to the human aconitase IRP1. It is of interest that the apo-aconitases of *E. coli* bind and stabilize aconitase mRNA (90). This leads to an increase in aconitase synthesis in vitro. Most recently IRE-like elements have been demonstrated in *B. subtilis* on the mRNA of the main cytochrome oxidase and on an iron uptake system (1). An enzymatically inactive aconitase was able to bind mRNA, and the mutant sporulated more efficiently than an aconitase null mutant (1). Future experiments have to show to what extent the translational control contributes to the overall regulation of the iron and oxidative stress response.

Acknowledgments. Our work was supported by the Deutsche Forschungsgemeinschaft (Schwerpunkt, Regulationsnetzwerke, SFB 323, Graduiertenkolleg Mikrobiologie) and the Fonds der Chemischen Industrie.

We thank Karen A. Brune for critically reading the manuscript and M. Braun and H. Killmann for computer drawings of Fig. 1, 2, and 3.

REFERENCES

1. **Alen, C., and A. L. Sonenshein.** 1999. *Bacillus subtilis* aconitase is an RNA-binding protein. *Proc. Natl. Acad. Sci. USA* **96:** 10412–10417.

2. **Althaus, E. W., C. E. Outten, K. E. Olson, H. Cao, and T. V. O'Halloran.** 1999. The ferric uptake regulation (Fur) repressor is a zinc metalloprotein. *Biochemistry* **38:**6559–6569.

3. **Angerer, A., and V. Braun.** 1998. Iron regulates transcription of the *Escherichia coli* ferric citrate transport genes directly and through the transcription initiation proteins. *Arch. Microbiol.* **169:**483–490.

4. **Angerer, A., S. Enz, M. Ochs, and V. Braun.** 1995. Transcriptional regulation of ferric citrate transport in *Escherichia coli* K-12. FecI belongs to a new subfamily of sigma-type factors that respond to extracytoplasmic stimuli. *Mol. Microbiol.* **19:** 163–174.

5. **Angerer, A., S. Gaisser, and V. Braun.** 1990. Nucleotide sequences of the *sfuA, sfuB,* and *sfuC* genes of *Serratia marcescens* suggest a periplasmic-binding-protein-dependent iron transport mechanism. *J. Bacteriol.* **172:**572–578.

6. **Bishop, L., H. R. Agbayani, S. V. Ambudkar, P. C. Maloney, and G. F. L. Ames.** 1989. Reconstitution of a bacterial periplasmic permease in proteoliposomes and demonstration of ATP hydrolysis concomitant with transport. *Proc. Natl. Acad. Sci. USA* **86:**6953–6957.

7. **Blattner, F. R., V. Burland, G. Plunkett III, H. J. Sofia, and D. L. Daniels.** 1993. Analysis of the *Escherichia coli* genome. IV. DNA sequence of the region from 89.2 to 92.8 minutes. *Nucleic. Acids. Res.* **21:**5408–5417.

8. **Bonhivers, M., A. Ghazi, P. Boulanger, and L. Letellier.** 1996. FhuA, a transporter of the *Escherichia coli* outer membrane, is converted into a channel upon binding of bacteriophage T5. *EMBO J.* **15:**1850–1856.

9. **Boos, W., and J. M. Lucht.** 1996. Periplasmic binding protein-dependent ABC transporters, p. 1175–1209. *In* F. C. Neidhardt, R. Curtiss III, J. L. Ingraham, E. C. C. Lin, K. B. Low, B. Magasanik, W. S. Reznikoff, M. Riley, M. Schaechter, and H. E. Umbarger (ed.), *Escherichia coli and Salmonella: Cellular and Molecular Biology,* 2nd ed. ASM Press, Washington, D.C.

10. **Boyd, J., M. N. Oza, and J. R. Murphy.** 1990. Molecular cloning and DNA sequence analysis of a diphtheria tox iron-dependent regulatory element (*dtxR*) from *Corynebacterium diphtheriae. Proc. Natl. Acad. Sci. USA* **87:**5968–5972.

11. **Bradbeer, C.** 1993. The proton motive force drives the outer membrane transport of cobalamin in *Escherichia coli. J. Bacteriol.* **175:**3146–3150.

12. **Braun, M., H. Killmann, and V. Braun.** 1999. The β-barrel domain of FhuAΔ5-160 is sufficient for TonB-dependent FhuA activities of *Escherichia coli. Mol. Microbiol.* **33:**1037–1049.

13. **Braun, V.** 1997. Surface signaling: novel transcription initiation mechanism starting from the cell surface. *Arch. Microbiol.* **237:**325–331.

14. **Braun, V., K. Hantke, and W. Köster.** 1998. Bacterial iron transport: mechanisms, genetics, and regulation, p. 67–145. *In* A. Sigel and H. Sigel (ed.), *Metal Ions in Biological Systems,* vol. 35. *Iron Transport and Storage in Microorganisms, Plants, and Animals.* Marcel Dekker, New York, N.Y.

15. **Braun, V., K. Schaller, and H. Wolff.** 1973. A common receptor protein for phage T5 and colicin M in the outer membrane of *Escherichia coli* B. *Biochim. Biophys. Acta* **323:** 87–97.

16. **Brocklehurst, K. R., J. L. Hobman, B. Lawley, L. Blank, S. J. Marshall, N. L. Brown, and A. P. Morby.** 1999. ZntR is a Zn(II)-responsive MerR-like transcriptional regulator of zntA in *Escherichia coli. Mol. Microbiol.* **31:**893–902.

17. **Bruns, C. M., A. J. Nowalk, A. S. Arvai, M. A. McTigue, K. G. Vaughan, T. A. Mietzner, and D. E. McRee.** 1997. Structure of *Haemophilus influenzae* Fe(+3)-binding protein reveals convergent evolution within a superfamily. *Nat. Struct. Biol.* **4:**919–924.

18. **Bsat, N., A. Herbig, L. Casillas-Martinez, P. Setlow, and J. D. Helmann.** 1998. *Bacillus subtilis* contains multiple Fur homologues: identification of the iron uptake (Fur) and peroxide regulon (PerR) repressors. *Mol. Microbiol.* **29:**189–198.

19. **Buchanan, S. K., B. S. Smith, L. Venkatramani, D. Xia, L. Esser, M. Palnitkar, R. Chakraborty, D. van der Helm, and J. Deisenhofer.** 1999. Crystal structure of the outer membrane active transporter FepA from *Escherichia coli. Nat. Struct. Biol.* **6:**56–63.

20. **Cadieux, N., and R. J. Kadner.** 1999. Site-directed disulfide bonding reveals an interaction site between energy-coupling protein TonB and BtuB, the outer membrane cobalamin transporter. *Proc. Natl. Acad. Sci. USA* **96:**10673–10678.

21. **Cope, L. D., R. Yogev, U. Muller Eberhard, and E. J. Hansen.** 1995. A gene cluster involved in the utilization of both free heme and heme:hemopexin by *Haemophilus influenzae* type b. *J. Bacteriol.* **177:**2644–2653.

22. **Cornelissen, C. N., J. E. Anderson, and P. F. Sparling.** 1997. Energy-dependent changes in the gonococcal transferrin receptor. *Mol. Microbiol.* **26:**25–35.

23. **Crosa, J. H.** 1997. Signal transduction and transcriptional and posttranscriptional control of iron-regulated genes in bacteria. *Microbiol. Mol. Biol. Rev.* **61:**319–336.

24. Davidson, A. L., H. A. Shuman, and H. Nikaido. 1992. Mechanism of maltose transport in *Escherichia coli*: transmembrane signaling by periplasmic binding proteins. *Proc. Natl. Acad. Sci. USA* **89**:2360–2364.

25. De Lorenzo, V., M. Herrero, F. Giovannini, and J. B. Neilands. 1988. Fur (ferric uptake regulation) protein and CAP (catabolite-activator protein) modulate transcription of *fur* gene in *Escherichia coli*. *Eur. J. Biochem.* **173**:537–546.

26. De Lorenzo, V., S. Wee, M. Herrero, and J. B. Neilands. 1987. Operator sequences of the aerobactin operon of plasmid ColV-K30 binding the ferric uptake regulation (*fur*) repressor. *J. Bacteriol.* **169**:2624–2630.

27. Diez-Gonzalez, F., T. R. Callaway, M. G. Kizoulis, and J. B. Russell. 1998. Grain feeding and the dissemination of acid-resistant *Escherichia coli* from cattle. *Science* **281**:1666–1668.

28. Dussurget, O., M. Rodriguez, and I. Smith. 1996. An IdeR mutant of *Mycobacterium smegmatis* has derepressed siderophore production and altered oxidative stress response. *Mol. Microbiol.* **22**:535–544.

29. Escolar, L., J. Perez-Martin, and V. De Lorenzo. 1998. Binding of the fur (ferric uptake regulator) repressor of *Escherichia coli* to arrays of the GATAAT sequence. *J. Mol. Biol.* **283**:537–547.

30. Ferguson, A. D., E. Hofmann, J. W. Coulton, K. Diederichs, and W. Welte. 1998. Siderophore-mediated iron transport: crystal structure of FhuA with bound lipopolysaccharide. *Science* **282**:2215–2220.

31. Foster, J. W., and M. Moreno. 1999. Inducible acid tolerance mechanisms in enteric bacteria. *Novartis Found. Symp.* **221**:55–69; discussion **70–4**:55–69.

32. Frank, H., M.-L. Zarnitz, and W. Weidel. 1963. Über die Rezeptorsubstanz für den Phagen T5. Elektronenoptische Darstellung und Längenbestimmung der aus T5/R5-Komplexen freigesetzten DNA. *Z. Naturforsch.* **18b**:281–284.

33. Franza, T., C. Sauvage, and D. Expert. 1999. Iron regulation and pathogenicity in *Erwinia chrysanthemi* 3937: role of the Fur repressor protein. *Mol. Plant-Microbe Interact.* **12**:119–128.

34. Gaballa, A., and J. D. Helmann. 1998. Identification of a zinc-specific metalloregulatory protein, Zur, controlling zinc transport operons in *Bacillus subtilis*. *J. Bacteriol.* **180**:5815–5821.

35. Ghassemian, M., and N. A. Straus. 1996. Fur regulates the expression of iron stress genes in the cyanobacterium *Synechococcus* sp. strain PCC 7942. *Microbiol.* **142**:1469–1476.

36. Ghigo, J.-M., S. Ltoffe, and C. Wandersman. 1997. A new type of hemophore-dependent heme acquisition system of *Serratia marcescens* reconstituted in *Escherichia coli*. *J. Bacteriol.* **179**:3572–3579.

37. Groeger, W., and W. Koster. 1998. Transmembrane topology of the two FhuB domains representing the hydrophobic components of bacterial ABC transporters involved in the uptake of siderophores, haem and vitamin B12. *Microbiology* **144**:2759–2769.

38. Gruer, M. J., and J. R. Guest. 1994. Two genetically-distinct and differentially-regulated aconitases (AcnA and AcnB) in *Escherichia coli*. *Microbiology* **140**:2531–2541.

39. Günter-Seeboth, K. and T. Schupp. 1995. Cloning and sequence analysis of the *Corynebacterium diphtheriae* dtxR homologue from *Streptomyces lividans* and *S. pilosus* encoding a putative iron repressor protein. *Gene* **166**:117–119.

40. Günter, K., and V. Braun. 1990. In vivo evidence for FhuA outer membrane receptor interaction with the TonB inner membrane protein of *Escherichia coli*. *FEBS Lett.* **274**:85–88.

41. Hamza, I., S. Chauhan, R. Hassett, and M. R. O'Brian. 1998. The bacterial irr protein is required for coordination of heme biosynthesis with iron availability. *J. Biol. Chem.* **273**:21669–21674.

42. Hancock, R. E., and V. Braun. 1976. Nature of the energy requirement for the irreversible adsorption of bacteriophages T1 and phi 80 *Escherichia coli*. *J. Bacteriol.* **125**:409–415.

43. Hantke, K. 1987. Selection procedure for deregulated iron transport mutants (fur) in *Escherichia coli* K-12: fur not only affects iron metabolism. *Mol. Gen. Genet.* **210**:135–139.

44. Hantke, K., and V. Braun. 1997. Control of bacterial iron transport by regulatory proteins, p. 11–44. *In* S. Silver and W. Walden (eds.), *Metal Ions in Gene Regulation.* Chapman and Hall, New York, N.Y.

45. Härle, C., I. Kim, A. Angerer, and V. Braun. 1995. Signal transfer through three compartments: transcription initiation of the *Escherichia coli* ferric citrate transport system from the cell surface. *EMBO J.* **14**:1430–1438.

46. Hasset, D. J., P. A. Sokol, M. L. Howell, J. F. Ma, H. T. Schweizer, U. Ochsner, and M. L. Vasil. 1996. Ferric uptake regulator (Fur) mutants of *Pseudomonas aeruginosa* demonstrate defective siderophore mediated iron uptake, altered aerobic growth, and decreased superoxide dismutase and catalase activities. *J. Bacteriol.* **178**:3996–4003.

47. Hill, P. J., A. Cockayne, P. Landers, J. A. Morrissey, C. M. Sims, and P. Williams. 1998. SirR, a novel iron-dependent repressor in *Staphylococcus epidermidis*. *Infect. Immun.* **66**:4123–4129.

48. Jacquamet, L., D. Aberdam, A. Adrait, J. L. Hazemann, J. M. Latour, and I. Michaud-Soret. 1998. X-ray absorption spectroscopy of a new zinc site in the fur protein from *Escherichia coli*. *Biochemistry* **37**:2564–2571.

49. Jiang, X., M. A. Payne, Z. Cao, S. B. Foster, J. B. Feix, S. M. Newton, and P. E. Klebba. 1997. Ligand-specific opening of a gated-porin channel in the outer membrane of living bacteria. *Science* **276**:1261–1264.

50. Kadner, R. J. 1990. Vitamin B$_{12}$ transport in *Escherichia coli*: energy coupling between membranes. *Mol. Microbiol.* **4**:2027–2033.

51. Killmann, H., R. Benz, and V. Braun. 1993. Conversion of the FhuA transport protein into a diffusion channel through the outer membrane of *Escherichia coli*. *EMBO J.* **12**:3007–3016.

52. Killmann, H., G. Videnov, G. Jung, H. Schwarz, and V. Braun. 1995. Identification of receptor binding sites by competitive peptide mapping: phages T1, T5, and Φ80 and colicin M bind to the gating loop of FhuA. *J. Bacteriol.* **177**:694–698.

53. Kim, I., A. Stiefel, S. Plantör, A. Angerer, and V. Braun. 1997. Transcription induction of the ferric citrate transport genes via the N-terminus of the FecA outer membrane protein, the Ton system and the electrochemical potential of the cytoplasmic membrane. *Mol. Microbiol.* **23**:333–344.

54. Koster, M., W. van Klompenburg, W. Bitter, J. Leong, and P. J. Weisbeek. 1994. Role for the outer membrane ferric-siderophore receptor PupB in signal transduction across the bacterial cell envelope. *EMBO J.* **13**:2805–2813.

55. Köster, W., and B. Böhm. 1992. Point mutations in two conserved glycine residues within the integral membrane protein FhuB affect iron(III)hydroxamate transport. *Mol. Gen. Genet.* **232**:399–407.

56. Lam, M. S., C. M. Litwin, P. A. Carroll, and S. B. Calderwood. 1994. *Vibrio cholerae fur* mutations associated with loss of repressor activity: implications for the structural-functional relationships of fur. *J. Bacteriol.* **176**:5108–5115.

57. Lambert, O., L. Plancon, J. L. Rigaud, and L. Letellier. 1998. Protein-mediated DNA transfer into liposomes. *Mol. Microbiol.* 30:761–765.

58. Larsen, R. A., M. G. Thomas, and K. Postle. 1999. Protonmotive force, ExbB and ligand-bound FepA drive conformational changes in TonB. *Mol. Microbiol.* 31:1809–1824.

59. Le Cam, E., D. Frechon, M. Barray, A. Fourcade, and E. Delain. 1994. Observation of binding and polymerization of Fur repressor onto operator-containing DNA with electron and atomic force microscopes. *Proc. Natl. Acad. Sci. USA* 91:11816–11820.

60. Lee, H. S., Y. S. Lee, H. S. Kim, J. Y. Choi, H. M. Hassan, and M. H. Chung. 1998. Mechanism of regulation of 8-hydroxyguanine endonuclease by oxidative stress: roles of FNR, ArcA, and Fur. *Free Radic. Biol. Med.* 24:1193–1201.

61. Letellier, L., K. P. Locher, L. Plancon, and J. P. Rosenbusch. 1997. Modeling ligand-gated receptor activity. FhuA-mediated ferrichrome efflux from lipid vesicles triggered by phage T5. *J. Biol. Chem.* 272:1448–1451. (Erratum, 272:8836.)

62. Letoffe, S., V. Redeker, and C. Wandersman. 1998. Isolation and characterization of an extracellular haem-binding protein from *Pseudomonas aeruginosa* that shares function and sequence similarities with the *Serratia marcescens* HasA haemophore. *Mol. Microbiol.* 28:1223–1234.

63. Litwin, C. M., and S. B. Calderwood. 1993. Role of iron in regulation of virulence genes. *Clin. Microbiol. Rev.* 6:137–149.

64. Locher, K. P., B. Rees, R. Koebnik, A. Mitschler, L. Moulinier, J. P. Rosenbusch, and D. Moras. 1998. Transmembrane signaling across the ligand-gated FhuA receptor: crystal structures of free and ferrichrome-bound states reveal allosteric changes. *Cell* 95:771–778.

65. Mademidis, A., H. Killmann, W. Kraas, I. Flechsner, G. Jung, and V. Braun. 1997. ATP-dependent ferric hydroxamate transport system in *Escherichia coli*: periplasmic FhuD interacts with a periplasmic and a transmembrane/cytoplasmic region of the integral membrane protein FhuB, as revealed by peptide mapping. *Mol. Microbiol.* 26:1109–1123.

66. Magneson, G. R., J. M. Puvathingal, and W. J. Ray, Jr. 1987. The concentrations of free $Mg2^+$ and free $Zn2^+$ in equine blood plasma. *J. Biol. Chem.* 262:11140–11148.

67. Mayhew, S. G., and M. L. Ludwig. 1975 Flavodoxins and electron-transferring flavoproteins, p. 57–109. *In* P. B. Boyer (ed.), *The Enzymes*, vol. 12. Academic Press, New York, N.Y.

68. Mietzner, T. A., S. B. Tencza, P. Adhikari, K. G. Vaughan, and A. J. Nowalk. 1998. Fe(III) periplasm-to-cytosol transporters of gram-negative pathogens. *Curr. Top. Microbiol. Immunol.* 225:113–135.

69. Moeck, G. S., and J. W. Coulton. 1998. TonB-dependent iron acquisition: mechanisms of siderophore-mediated active transport. *Mol. Microbiol.* 28:675–681.

70. Moeck, G. S., J. W. Coulton, and K. Postle. 1997. Cell envelope signaling in *Escherichia coli*. Ligand binding to the ferrichrome-iron receptor Fhua promotes interaction with the energy-transducing protein TonB. *J. Biol. Chem.* 272:28391–28397.

71. Murphy, E. R., A. Dickenson, K. T. Militello, and T. D. Connell. 1999. Genetic characterization of wild-type and mutant *fur* genes of *Bordetella avium*. *Infect. Immun.* 67:3160–3165.

72. Niederhoffer, E. C., C. M. Naranjo, K. L. Bradley, and J. A. Fee. 1990. Control of *Escherichia coli* superoxide dismutase (*sodA* and *sodB*) genes by the ferric uptake regulation (*fur*) locus. *Mol. Microbiol.* 9:53–63.

73. Ochs, M., S. Veitinger, I. Kim, D. Welz, A. Angerer, and V. Braun. 1995. Regulation of citrate-dependent iron transport of *Escherichia coli*: FecR is required for transcription activation by FecI. *Mol. Microbiol.* 15:119–132.

74. Pagan-Ramos, E., J. Song, M. McFalone, M. H. Mudd, and V. Deretic. 1998. Oxidative stress response and characterization of the *oxyR-ahpC* and *furA-katG* loci in *Mycobacterium marinum*. *J. Bacteriol.* 180:4856–4864.

75. Patzer, S. I., and K. Hantke. 1998. The ZnuABC high-affinity zinc uptake system and its regulator Zur in *Escherichia coli*. *Mol. Microbiol.* 28:1199–1210.

76. Plancon, L., M. Chami, and L. Letellier. 1997. Reconstitution of FhuA, an *Escherichia coli* outer membrane protein, into liposomes. Binding of phage T5 to FhuA triggers the transfer of DNA into the proteoliposomes. *J. Biol. Chem.* 272:16868–16872.

77. Pohl, E., R. K. Holmes, and W. G. Hol. 1998. Motion of the DNA-binding domain with respect to the core of the diphtheria toxin repressor (DtxR) revealed in the crystal structures of apo- and holo-DtxR. *J. Biol. Chem.* 273:22420–22427.

78. Pohl, E., R. K. Holmes, and W. G. J. Hol. 1999. Crystal structure of the iron-dependent regulator (IdeR) from *Mycobacterium tuberculosis* shows both metal binding sites fully occupied. *J. Mol. Biol.* 285:1145–1156.

79. Postle, K. 1993. TonB protein and energy transduction between membranes. *J. Bioenerg. Biomembr.* 25:591–601.

80. Rohrbach, M. R., V. Braun, and W. Köster. 1995. Ferrichrome transport in *Escherichia coli* K-12: altered substrate specificity of mutated periplasmic FhuD and interaction of FhuD with the integral membrane protein FhuB. *J. Bacteriol.* 177:7186–7193.

81. Schöffler, H., and V. Braun. 1989. Transport across the outer membrane of *Escherichia coli* K12 via the FhuA receptor is regulated by the TonB protein of the cytoplasmic membrane. *Mol. Gen. Genet.* 217:378–383.

82. Schmitt, M. P. 1997. Transcription of the *Corynebacterium diphtheriae hmuO* gene is regulated by iron and heme. *Infect. Immun.* 65:4634–4641.

83. Schmitt, M. P., and R. K. Holmes. 1991. Iron-dependent regulation of diphtheria toxin and siderophore expression by the cloned *Corynebacterium diphtheriae* repressor gene *dtxR* in C. *diphtheriae* C7 strains. *Infect. Immun.* 59:1899–1904.

84. Schmitt, M. P., B. G. Talley, and R. K. Holmes. 1997. Characterization of lipoprotein IRP1 from *Corynebacterium diphtheriae*, which is regulated by the diphtheria toxin repressor (DtxR) and iron. *Infect. Immun.* 65:5364–5367.

85. Schneider, R., and K. Hantke. 1993. Iron-hydroxamate uptake systems in *Bacillus subtilis*: identification of a lipoprotein as a part of a binding proteindpendent transport system. *Mol. Microbiol.* 8:111–121.

86. Schultz-Hauser, G., W. Köster, H. Schwartz, and V. Braun. 1992. Iron(III)hydroxamate transport in *Escherichia coli* K-12. FhuB-mediated membrane association of the FhuC protein and negative complementation of *fhuC* mutants. *J. Bacteriol.* 174:2305–2311.

87. Staggs, T. M., J. D. Fetherston, and R. D. Perry. 1994. Pleiotropic effects of a *Yersinia pestis fur* mutation. *J. Bacteriol.* 176:7614–7624.

88. Stojiljkovic, I., and K. Hantke. 1992. Hemin uptake system of *Yersinia enterocolitica*: similarities with other TonB-dependent systems in gram-negative bacteria. *EMBO J.* 11:4359–4367.

89. Stojiljkovic, I., and K. Hantke. 1994. Transport of haemin across the cytoplasmic membrane through a haemin-specific periplasmic binding-protein-dependent transport system in *Yersinia enterocolitica*. *Mol. Microbiol.* 13:719–732.

90. Tang, Y., and J. R. Guest. 1999. Direct evidence for mRNA binding and post-transcriptional regulation by *Escherichia coli* aconitases. *Microbiology* 145:3069–3079.

91. Thomas, C. E., and P. F. Sparling. 1996. Isolation and analysis of a Fur mutant of *Neisseria gonorrhoeae*. *J. Bacteriol.* **178:** 4224–4232.

92. Touati, D., M. Jacques, B. Tardat, L. Bouchard, and S. Despied. 1995. Lethal oxidative damage and mutagenesis are generated by iron in *fur* mutants of *Escherichia coli*: protective role of superoxide dismutase. *J. Bacteriol.* **177:**2305–2314.

93. van der Helm, D., J. R. Baker, D. L. Eng-Wilmot, M. B. Hossain, R. A. Loghry. 1980. Crystal structure of ferrichrome and a comparison with the structure of ferrichrome A. *J. Am. Chem. Soc.* **102:**4224–4231.

94. van Vliet, A. H. M., K. G. Wooldridge, and J. M. Ketley. 1998. Iron-responsive gene regulation in a *Campylobacter jejuni fur* mutant. *J. Bacteriol.* **180:**5291–5298.

95. Venturi, V., P. J. Weisbeek, and M. Koster. 1995. Gene regulation of siderophore-mediated iron acquisition in *Pseudomonas*: not only the Fur repressor. *Mol. Microbiol.* **17:**603–610.

96. Weidel, W., and E. Kellenberger. 1955. The *E. coli* B-receptor for the phage T5, II. Electron microscopic studies. *Biochim. Biophys. Acta* **17:**1–9.

97. Welz, D., and V. Braun. 1998. Ferric citrate transport of *Escherichia coli*: functional regions of the FecR transmembrane regulatory protein. *J. Bacteriol.* **180:**2387–2394.

98. Wertheimer, A. M., M. E. Tolmasky, L. A. Actis, and J. H. Crosa. 1994. Structural and functional analyses of mutant Fur proteins with impaired regulatory function. *J. Bacteriol.* **176:**5116–5122.

99. Zheng, M., B. Doan, T. D. Schneider, and G. J. Storz. 1999. OxyR and SoxRS regulation of fur. *J. Bacteriol.* **181:**4639–4643.

100. Zou, P., I. Borovok, Lucana, D. O., D. Müller, and H. Schrempf. 1999. The mycelium-associated *Streptomyces reticuli* catalase-peroxidase, its gene and regulation by FurS. *Microbiology* **145:**549–559.

Bacterial Stress Responses
Edited by G. Storz and R. Hengge-Aronis
©2000 ASM Press, Washington, D.C.

Chapter 20

Identification and Analysis of Proteins Expressed by Bacterial Pathogens in Response to Host Tissues

RALPH R. ISBERG

Bacterial pathogens regulate gene expression in response to the environment encountered within host tissues. Studies indicate that bacterial colonization within hosts induces expression of genes that encode stress response and iron sequestration proteins. In addition, specific regulatory circuits tailored for a single group of pathogens are induced, resulting in virulence properties that are associated with distinctive pathways of disease. A variety of strategies have been used to identify genes that are specifically expressed in response to host tissues. These strategies have included isolating promoters that are regulated in response to host cells or to growth within animals. Each strategy has had biases that affected the nature of the genes identified. The relative strengths of these strategies are discussed, as is the nature of the genes identified with these approaches.

Bacterial pathogens encounter a number of different environments during growth within their hosts. Bacteria face a number of challenges that either are due to direct attack by the immune system or result from bacterial entry into tissue sites that discourage microbial growth. In either case, the bacterium must respond nimbly to changes that take place during the encounter with the host in various tissue sites. Research on the physiology of bacterial growth in culture has shown that microorganisms are extremely economical in the choice of the genes that are expressed in response to environmental conditions, fine-tuning the array of expressed proteins to the chemical composition and physical state of the growth medium (34). Regulatory circuits must exist that allow efficient colonization when the organism leaves an environmental site and enters the host or when the microorganism moves to different organ

sites within the host (43, 48). This chapter will detail key strategies that have been used to identify genes that are specifically expressed in response to encounters with host cells or tissue sites.

The rationale behind identifying host- or host cell-expressed genes is that it should allow identification of proteins necessary for pathogenesis that cannot be identified by other means. Although there is no strong evidence that this is the most efficient strategy for identifying such proteins, a few of the newly identified genes have been shown to be of central importance in causing disease in animal models (50). The majority of previously uncharacterized genes that have been isolated in this fashion, however, have not been demonstrated to be required for pathogenesis (32). More direct strategies, such as signature-tagged mutagenesis (35), allow identification of such factors. Therefore, this approach should be used with the intention of (i) gaining insight into the global regulatory systems encoded by the bacterium that are necessary to promote disease, (ii) obtaining evidence for particular microenvironments during various stages of disease, and (iii) identifying factors required for pathogenesis that may be functionally redundant. This last point is poorly appreciated and generally underinvestigated, as many factors required for growth within the host may be of such great importance that the microorganism has a variety of back-up strategies. Furthermore, disease in some animal models may result from multiple parallel pathways that allow spread of the microorganism simultaneously via different host tissues.

Many of these studies have also attempted to identify simple culture conditions that specifically stimulate expression of host-induced genes, in hopes of gaining insight into the chemical composition of

Ralph R. Isberg • Howard Hughes Medical Institute, Department of Molecular Biology and Microbiology, Tufts University School of Medicine, 136 Harrison Ave., Boston, MA 02111.

the site of colonization of the microorganism (33). The concept that in vitro culture conditions can be used to mimic host tissue cues is traceable to observations made over 50 years ago during attempts to produce *Corynebacterium diphtheriae* toxin in culture (55). Maximal diphtheria toxin was found to require growth of the bacteria in an iron-depleted medium, while addition of iron to the growth medium greatly reduced yields of toxin in culture supernatants. It is now known that the transcriptional iron regulator DtxR (70) controls this response. As *C. diphtheriae* produces sufficient toxin during colonization of the human throat to result in a potentially lethal disease, the obvious conclusion was that the microorganism grows in host tissues containing vanishingly small amounts of iron. Subsequent work on a number of virulent organisms, such as *Vibrio cholerae*, *Yersinia* species, and tissue-invasive *Escherichia coli* species, has demonstrated the importance in pathogenesis of genes induced by iron limitation and molecules involved in iron sequestration (9, 24). On the basis of this early success, a major goal of studies involving the identification of genes induced by host tissues has been to catalog all the regulatory circuits that respond to host tissues and to determine the sites within the host that cause the induction of such genes. This chapter will review strategies used to directly identify genes on the basis of their ability to be expressed either after contact with host cells or during growth in an animal infection model. The strengths and weaknesses of these strategies will be analyzed, as will the major themes that have emerged from such studies.

USE OF TISSUE CULTURE MODELS TO ANALYZE DIFFERENTIALLY EXPRESSED GENES

The simplest model systems for analyzing microbial gene expression in response to host tissues involve studying the interaction of bacteria with cultured mammalian cells. The advantages of using cell line models are that gene expression by a relatively large population of bacteria can be assayed, separation of bacteria from cultured cells is often straightforward, and bacterial responses to different cell types are easily compared (21). The disadvantages are that cultured cells cannot fully mimic the conditions found within an animal, and there may be large numbers of bacterial proteins that are expressed within an animal in response to humoral, rather than cellular, factors. There is no easy way to reproduce these latter conditions, although genes whose expression levels are altered by the presence of mucus in the culture medium have been identified (77).

Protein Labeling

Bacterial proteins that are specifically expressed after encounter with cultured cell monolayers have been successfully identified with differential labeling strategies (2, 11). Such an approach relies on biosynthetically incorporating radioactive label into bacterial proteins in the absence of labeling mammalian host cell proteins. Three major strategies have been used to eliminate labeling of mammalian cells. The first involves biosynthetically labeling bacteria that have been allowed to bind mammalian cells killed by a fixative (25). The second strategy eliminates protein synthesis in the mammalian cell by adding a specific inhibitor of the eukaryotic translation machinery (2). The third utilizes a labeled biosynthetic precursor that can be incorporated into proteins by bacteria, but not by mammalian cells (11).

Each of the above-noted strategies has distinct advantages and shortcomings. Labeling bacteria bound to fixed mammalian cell monolayers is the simplest strategy, but it has several detractions. First, proteins that are expressed only after host cells internalize the bacteria cannot be detected. Second, proteins that require interplay between the bacterium and a live host cell for their synthesis will not be labeled. Finally, fixation may destroy an important macromolecule on the mammalian host cell surface that is recognized by the bacterium as an inducing signal. For that reason, this has not been a popular strategy. The most popular labeling strategy has been to introduce bacteria to host cells that have been treated with cycloheximide to eliminate host protein synthesis (2, 8, 10) and has been used with success in *Salmonella enterica* serovar Typhimurium and *Legionella pneumophila*. This technique overcomes most of the concerns of using fixed cells but has serious drawbacks as well (1). The absence of protein synthesis within host cells is likely to alter greatly the display of eukaryotic proteins encountered by the bacterium that could result in either the lack of induction of important proteins or ectopic expression of bacterial proteins not normally synthesized after contact with host cells.

The use of a labeled amino acid precursor that is specifically utilized by the bacterium appears to be the most promising strategy for looking at bulk protein synthesis by the bacterium after contact with mammalian cells. The best attempt at this approach was labeling serovar Typhimurium-infected macrophages with ^{3}H-DAP, which can be used as a lysine precursor (11). The specific activity and incorporation efficiency of this particular label were sufficiently low, however, that the total number of proteins visible on two-dimensional electrophoresis gels was far less than could be detected after labeling with ^{35}S-methionine in the presence of cycloheximide (10). Even so, 34 proteins were found to be preferentially expressed by serovar Typhimurium after incubation with cultured macrophages, as compared to medium-incubated bacteria (11).

LacZ Reporter Constructions

Analysis of the total array of proteins expressed by bacteria in response to interaction with cultured cells has had significant technical problems, and once a protein band has been identified on a gel, it is potentially problematic to obtain sufficient amounts for protein microsequencing. For these reasons, the most common approach has been to use a variety of reporter constructs. In this strategy, transcriptional units are fused to genes encoding either selectable markers or assayable proteins, and the relative level of transcriptional activation of the resulting fusion is measured by growth in various culture conditions. The sensitivity of this approach was first clearly demonstrated with fusions between the *E. coli lacZ* gene and the *Yersinia pestis yopK* gene, which is believed to encode a protein necessary for modulating the translocation of secreted proteins into the host cell cytoplasm (60). After the bacteria were incubated with mouse peritoneal macrophages, *Y. pestis* harboring the fusion was shown to have approximately 10 times the level of β-galactosidase activity relative to bacteria exposed to media alone. Although the investigators may have incorrectly attributed the increased expression to phagocytosis rather than to extracellular adhesion of the bacteria to the macrophage surface (37, 60), this result firmly established that mammalian cells are able to modulate the expression of select bacterial genes.

Fusions to *lacZ* were also used to attempt to identify genes encoded by *Listeria monocytogenes* that are expressed during growth of the bacterium within phagocytic cells but have little expression in culture medium (45). Random fusions were constructed in the genome by transposon insertion of Tn917lac containing a promoterless *lacZ* gene, and the resulting individual strains were then assayed for β-galactosidase activity after a cell line was infected. Insertions in five different genes passed the screening procedure, three of which affected nucleotide metabolism (45). The main detraction of this work was that if an insertion mutation affected the survival or intracellular growth of the bacterium, and if either growth or survival was required for induction of the

fusion, then such insertions could not be recovered from this mutant pool. For this reason, workers searching for genes that have increased expression after contact with host cells have favored making fusions that result in a meridiploid strain, in which the fusion to the reporter gene is introduced into a strain that has an intact copy of the wild-type gene.

Auxotrophic Complementation

The most intensively used strategy that allows identification of promoters preferentially expressed after bacteria interact with host cells has been to set up a reporter fusion system that allows direct selection for the desired strains. The two selections that have been pursued involve using reporters that allow complementation of an auxotrophic mutation (62) or confer antibiotic resistance on the bacterium (52).

In the complementation strategy, which can be used with pathogens that grow intracellularly, a promoterless cassette containing a bicistronic operon consisting of a gene encoding a biosynthetic enzyme and *lacZ* is fused to random DNA fragments from the pathogenic organism (Fig. 1). The fusion library

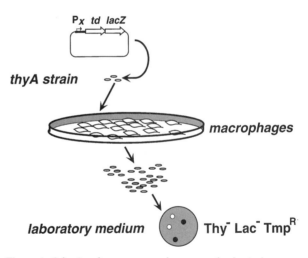

Figure 1. Selection for promoters that respond selectively to encounter with phagocytic cells. A gene bank is constructed by inserting a random pool of DNA fragments (noted as P_x) upstream of the promoterless bicistronic operon consisting of the phage T4 *td* gene (encoding thymidylate synthetase) and *lacZ*. The gene bank is transformed into a *thyA* (thymine-requiring) bacterial strain, and the resulting pool of bacteria is internalized by macrophages. Bacteria harboring promoters active intracellularly are selected by their ability to grow in the absence of thymine. To identify strains harboring plasmids containing promoters activated by host cells, the bacteria that are competent for intracellular replication are plated on bacteriological media containing thymidine and either trimethoprim (to select for organisms with low thymidylate synthetase activity) or X-Gal (to screen for β-galactosidase activity). Bacteria resistant to trimethoprim (Tmp^R) or having low β-galactosidase activity (Lac$^-$) are retained as containing isolated promoters that are selectively activated within phagocytic cells.

is then introduced into a mutant strain that is defective for production of the biosynthetic gene, and the resulting pool of strains is introduced onto a monolayer of cultured mammalian cells. After internalization by the host cells, the mutant bacterial strain is unable to grow, while strains harboring fusions with active promoters grow intracellularly. To identify promoters preferentially expressed when the bacteria are in an intracellular compartment, the enriched pool is plated on bacteriological medium containing an indicator for β-galactosidase activity. Strains that survive the intracellular enrichment and have the lowest levels of β-galactosidase activity on bacteriological media are retained as having cloned promoters that are more active when the bacteria are in an intracellular compartment compared to growth on agar plates. If expression of the biosynthetic gene can be used in a counterselection procedure, then it is possible to select for lack of expression of the promoter in an extracellular environment by plating on media containing the appropriate components. This strategy was used to identify more than 30 promoters from *L. pneumophila* that were preferentially expressed in phorbol ester-differentiated U937 cells relative to growth on bacteriological plates (62). In this procedure, random *L. pneumophila* DNA fragments were fused to a cassette encoding the thymidylate synthetase gene from phage T4 and *lacZ* and introduced into an *L. pneumophila thyA* strain, which is unable to grow intracellularly or extracellularly in the absence of thymine. After enriching for growth in U937 cells in the absence of thymidine, the organisms were plated on bacteriological media containing thymine and either bromochloroindolyl-β-D-galactoside (X-Gal), an indicator for β-galactosidase activity, or trimethroprim, which selects against thymidylate synthetase activity. Strains that either had the lowest β-galactosidase activity or grew well in the presence of trimethoprim on bacteriological plates were retained as having fusions to promoters with preferential intracellular activity.

A similar approach was taken with a reporter cassette having a promoterless gene for the antibiotic resistance element encoding chloramphenicol transacetylase (*cat*) upstream of *lacZY* (52). A pool of serovar Typhimurium strains harboring random insertions of DNA upstream of the cassette was introduced onto a lawn of RAW 264.7 macrophages and then subjected to media containing chloramphenicol, which penetrates mammalian cells and selects against strains harboring fusions to promoters inactive in an intracellular environment. Surviving bacteria were then plated on MacConkey lactose, and strains with the lowest apparent β-galactosidase activity were saved. It is not clear how many such fusions were

analyzed, but eight such fusions have been characterized (32, 33) and were shown to be appropriately regulated.

The primary disadvantage of the above enrichments is that these strategies systematically eliminate bacterial strains harboring fusions that have any appreciable expression on bacteriological media. As a result, these procedures strongly select against promoters that are active extracellularly but are significantly upregulated on encountering a host cell. In fact, the most important virulence factors that are regulated in response to conditions within the host may have significant expression levels when bacteria are grown outside the host. This is a singular problem when identifying promoters expressed within the animal and is known to have prevented the isolation of promoters responsive to the major regulators of *V. cholerae* virulence (13, 15; see below). The use of the *cat* reporter system could overcome this problem by a simple change in the selection procedure. Bacterial strains that grow intracellularly in the presence of high concentrations of chloramphenicol and can grow extracellularly in lower concentrations of the antibiotic may contain *cat* under the control of a promoter that is partially active extracellularly.

GFP Reporter Constructions

The use of flow cytometry combined with green fluorescence protein (GFP) reporters may allow the identification of promoters that have significant extracellular expression and enhanced activity after bacterial interaction with host cells. In this procedure, a population of bacterial strains harboring promoter fusions to the *gfp* gene are pooled and subjected to fluorescence-activated cell sorting (75, 76). The amount of fluorescence exhibited by each cell in the population is a function of promoter strength, and bacterial cells can be collected on the basis of this property. Theoretically, a population of bacteria exhibiting any fluorescence intensity, and thus harboring promoters of any strength, may be collected for further analysis. To identify promoters expressed in serovar Typhimurium specifically after contact with host cells, bacteria harboring random transcriptional fusions to the *gfp* gene were pooled and subjected to differential fluorescence induction (DFI). In this procedure, bacteria grown in culture media were sorted, and those cells having the lowest fluorescence intensity were collected. This pool was then introduced into the RAW macrophage line and then sorted once again, isolating bacteria that had the highest fluorescence intensity and greatest promoter strength after the bacteria were internalized. Although the nature of this enrichment is similar to that used with complementation or antibiotic resistance reporters, DFI allows pools of bacteria that have significant expression extracellularly but enhanced expression during interaction with host cells to be reliably selected. This is because the flow cytometer provides quantitation of the degree of fluorescence of each bacterium in a population, so that strains having some amount of extracellular expression can be saved (76).

Analysis of cDNA

Two strategies have been used that are derived from techniques developed to analyze transcription in eukaryotic cells (49). These approaches involve the isolation and amplification of cDNA generated from bacterial mRNA and have not been popular for two reasons. First, the turnover rate of bacterial mRNA is higher than that in eukaryotes, making it difficult to reproducibly isolate uniform populations of transcripts, particularly in pathogens for which few molecular techniques have been established. Second, the isolation of cDNA in eukaryotes is greatly aided by priming with an oligo(dT) track that is complementary to the polyA found at the 3′ of eukaryotic mRNA. No such easy choice of primers is available that allows efficient cDNA synthesis off a bacterial mRNA template. Neither of these problems is insurmountable, and isolation of cDNAs corresponding to transcripts that are differentially expressed after bacterial association with host cells is a potentially powerful strategy to analyze genes turned on at very specific times during this interaction.

Subtractive hybridization was used to identify a *Mycobacterium avium* transcript that was specifically induced by growth within human peripheral blood monocytes (58, 59). In this strategy, cDNA clone banks were isolated from bacteria grown in broth or within monocytes for 5 days. The bank from the broth-grown bacteria was used in excess to remove cDNA clones from the bank derived from the monocyte-grown bacteria, and the remaining clones that resisted subtraction were used to probe a cosmid gene bank (58, 59). Differential display-PCR (49) was used to isolate a gene from *L. pneumophila* that was expressed preferentially a few hours after internalization of the bacterium by phorbol ester–transformed U937 cells (4). In this strategy, a series of arbitrarily chosen primer pairs were used to PCR amplify cDNAs derived from bacteria grown on agar plates or within U937 cells. By displaying the electrophoretic profiles of each population that was subjected to amplification by the identical pairs, PCR products that were unique to intracellularly grown bacterium were identified. As was true for the subtractive hybridization method described above, only a single

transcript was identified. The reason for the small number of transcripts identified by these strategies may be a consequence of the relative difficulty in producing bacterial cDNA. It seems just as likely that certain technical details of the approaches pursued in these studies may have significantly reduced the potential number of candidate genes that could have been isolated. Perhaps by changing some of the parameters used by these investigators, much larger numbers of genes could be identified that are preferentially expressed after association with host cells.

USE OF ANIMAL INFECTION MODELS TO ANALYZE DIFFERENTIALLY EXPRESSED GENES

The identification of genes preferentially expressed in an animal raises issues that are more complex than those encountered by isolating genes that are induced by cell association. Genes that are candidates for being host regulated certainly do not all respond to the same tissue site, yet determining which site within the host causes induction of each gene is difficult. Furthermore, quantitation of induction levels of each gene within the host is problematic.

IVET

The primary strategy for isolating host-regulated promoters has been to use reporters that either allow direct selection within the animal or allow screening by colony phenotype after fusion pools are introduced into animals. The vectors used for identifying host-induced genes are primarily derived from the in vivo expression technology (IVET) plasmids (51) or from the recombinase-based screening strain, also called RIVET (14). In the original IVET strategy, random fusions between serovar Typhimurium DNA and a bicistronic *purAlacZ* reporter were constructed and integrated by homologous recombination into a serovar Typhimurium *purA* strain. Although wild-type serovar Typhimurium strains replicate within multiple lymphoid glands, as well as the liver and the spleen, *purA* mutants are unable to do so (Fig. 2). Therefore, replication in these sites requires the *purA* gene to be fused to a promoter that is active within host tissues. Promoters that are preferentially activated within the animal (called *ivi* for in vivo induced) will show a *lac⁻* colony phenotype, indicating lack of expression on bacteriological media (51). In an alternate strategy, the *cat* fusions described above were used, this time with simultaneous injection and feeding of chloramphenicol into the animal as the se-

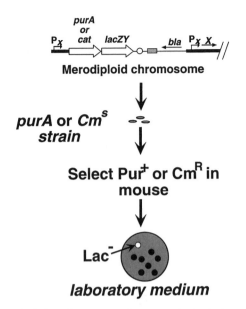

Figure 2. Isolation of promoters that respond selectively to infection of a mouse. A gene bank is constructed by inserting a random pool of DNA fragments (noted as P$_x$) upstream of the promoterless bicistronic operon consisting of either the *purA* gene and *lacZ* or *cat* and *lacZ*. The resulting pools are transformed either into a *purA* strain or a chloramphenicol-sensitive strain and inoculated into mice. Bacteria harboring promoters active within the mouse are selected by their ability to complement the growth defect that is a result of the *purA* mutation or by resistance to chloramphenicol. Replication-competent organisms are recovered from mouse tissues and plated on lactose MacConkey agar to identify strains having lowest β-galactosidase activity (Lac). Such strains have promoters selectively expressed within the host.

lective agent. Enrichments with the two vectors have been used after either oral or intraperitoneal inoculations of mice with serovar Typhimurium and after isolation of bacteria from regional intestinal lymph nodes as well as the spleen (32).

Recombinase-Based Reporters

The related recombinase-based strategy (14, 15) has been used with *V. cholerae* and *Staphylococcus aureus* (15, 50). In this particular strategy, random fragments of *V. cholerae* DNA are placed upstream of a promoterless gene encoding the TnpR site-specific recombinase. Activation of the recombinase by a promoter results in excision of a drug resistance cassette, which can be detected on selective media (see Fig. 3). Strains are then chosen on the basis of the ability of the fusion to promote excision of the cassette selectively within the animal and not in bacteriological medium. Originally, this particular strategy failed to identify virulence regulatory circuits in *V. cholerae* that are known to be activated in the intestine of the infant mouse because the levels of ex-

pression of these circuits are sufficiently high to drive excision of the cassette in bacteriological medium. For instance, promoters sensitive to the ToxR regulatory circuit, which controls the production of cholera toxin and the TCP pili, were not recovered from this enrichment, even though they are clearly activated within the animal. The solution to this problem was to highly attenuate the reporter *tnpR* gene by introducing a defective ribosome binding site. This then allowed the ToxR protein to control the appropriate fusions to *tnpR* in a fashion that is induced specifically within the host animal, without causing activation of *tnpR* in broth culture (S. H. Lee, D. L. Hava, M. K. Waldor, and A. Camilli, submitted). Similar strategies may allow a less biased identification of tissue-induced genes than has been used to date.

GENES INDUCED BY CONTACT WITH HOST CELLS

The best-characterized regulatory circuit that responds directly to extracellular adhesion to host cells is encoded by pathogenic *Yersinia* species. Each of the three major pathogenic species (*Y. enterocolitica, Y. pseudotuberculosis,* and *Y. pestis*) harbors a large plasmid necessary for pathogenesis that encodes a variety of proteins called Yops that are either translocated into target mammalian cells or secreted extracellularly by a specialized Type III secretion system (22; for review of Type III secretion, see reference 40). Exit of Yops from the bacterial cell is highly regulated by contact with mammalian cells, as is transcription of several Yops (57, 68). With transcriptional fusions to either *lacZ*, *luxA*, or *gfp*, large increases in expression after cell contact have been observed for reporters fused to the genes of the exported proteins YopH (tyrosine phosphatase), YopE (cytoskeletal antagonist), YopK (see above), and LcrQ (global regulator). Binding to the extracellular surface of target cells is clearly necessary and sufficient to induce expression of at least some of these genes. Induction of YopE and YopH does not require bacterial internalization by host cells, and bacteria bound randomly to the extracellular matrix surrounding cultured cells show no enhanced expression of Yops (56). The demonstration that induction of gene expression is specific for bacteria directly in contact with host cells was greatly facilitated by the use of the *lux* and *gfp* reporters. The expression of these proteins can be directly visualized by microscopy, and with such fusions, bacteria glow brightly only if they are in contact with mammalian cells (43, 57).

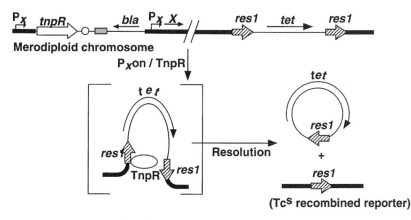

Figure 3. Recombinase-based strategy to identify promoters activated in response to host tissues. A gene bank is constructed by inserting a random pool of DNA fragments (noted as P_x) upstream of the promoterless *tnpR* gene encoding a site-specific recombinase located on a plasmid unable to replicate within the pathogen. The plasmids constructed in this fashion are allowed to integrate into a tester strain, using the homology provided by the inserted DNA. The tester strain has an expressed drug resistance gene (noted as *tet*) flanked by a direct repetition of *res* sites, which are targets of the TnpR site-specific recombinase. If P_x contains an active promoter, then the *tnpR* gene is expressed, allowing excision of the drug resistance element and loss of the ability to confer resistance to tetracycline. Promoters that are active exclusively within hosts are unable to express *tnpR* in bacteriological media and result in strains that are tetracycline resistant when grown in culture but are tetracycline sensitive during infection of the animal.

The bacteria must have a method for transmitting a signal from the bacterial cell surface to the regulatory circuit that controls expression of the Yops. A variety of mutations disrupt this signaling and cause either constitutive expression of the Yops or prevent induction of expression of these proteins. Many of these mutations also result in unregulated secretion of the Yops, and it has been proposed that these lesions identify factors that form a gate, associated with the Type III translocation apparatus, that responds to contact of the bacterial cell surface with the mammalian cell membrane (22). By the most popular model, binding of the bacterium to the host cell leads to opening of the gate, and the open position leads to induction of transcription of several Yops. Based on the mutational analysis, a fully closed position requires the function of several proteins, most notably YopN, TyeA, LcrG, LcrH, and probably YopD, that presumably act to ensure that a negative regulator maintains its cytoplasmic localization within the bacterium and is not secreted. The best candidate for such a regulator is the LcrQ protein, which is secreted in response to inducing conditions in vitro but behaves genetically like a repressor (57). Unfortunately, there is no information on how this protein exerts its repressive effects, and it may do so in an unconventional manner. These results are similar to what was seen previously with regulation of flagellum biosynthesis in serovar Typhimurium, which also involves a Type III secretion system (18). In the case of flagellum biosynthesis, assembly intermediates that lead to the formation of an uncapped channel in the membrane result in the secretion of FlgM protein, an anti-sigma factor (41).

There have been a number of documented accounts of bacterial gene expression induced after contact with host cell surfaces (chapter 2) (81), but no detailed models exist for how such induction could occur in any organism other than *Yersinia* species. The direct channel formed between *Yersinia* and the mammalian cell makes it easy to imagine how a host cell could communicate with the bacterium, and many microorganisms having similar Type III secretion systems may have expression of translocated proteins regulated in an identical fashion. Not clear is how bacterial gene expression could be regulated in the absence of direct channel formation between the bacterium and host cell. A solution to this problem has been forwarded by Silhavy, Hultgren, and coworkers (44). They have proposed that in gram-negative organisms, the environment of the periplasm may be able to transmit signals from the bacterial cell surface into the cytoplasm through a two-component regulatory system. By this formulation, disruption of polymerization of a bacterial organelle, such as a pilus, occurs after host cell contact. Depolymerization then results in accumulation of pilin subunits in the periplasm. The absence of binding of pilin to a dedicated chaperone in the periplasm results in recognition of these subunits by either the *cpx* two-component global regulator or a cytoplasmic membrane sensor that normally shuts off sigma E-dependent promoters. This allows large banks of genes to be turned on or off in response to cellular adhesion.

GENES INDUCED BY AN INTRACELLULAR ENVIRONMENT

It is relatively straightforward to understand how entry into an intracellular niche causes gene expression patterns of the bacterium to change relative to extracellular growth. Uptake of a microorganism into a vacuole or entry into the cytosol places it in an environment that is drastically different from culture medium. Potentially, the organism encounters lowered pH, a variety of stress response-inducing conditions, limiting nutrients, and altered osmolarity relative to bacteriological media. Responses to all of these changes are controlled by a number of well-characterized regulatory circuits. It should be noted that each microorganism represents a different challenge for uncovering these regulatory circuits, as intravacuolar pathogens differ significantly from one another in the nature of the compartments they occupy, and intracytoplasmic microorganisms clearly encounter different environmental cues than do those remaining within a vacuole.

L. monocytogenes

The only microorganism for which there is information regarding signals within the host cell cytosol is *L. monocytogenes*. This gram-positive rod breaks out of its phagosome shortly after it is internalized by macrophages and grows within the host cell cytoplasm (23). A number of bacterial genes have been identified that are expressed in response to an intracellular environment (45). As noted above, none of the genes was shown to be required for intracellular growth, although the nature of the screen used to identify such genes probably selected against such mutants. Three clearly had auxotrophic mutations in genes involved in nucleotide metabolism (*purH*, *purD*, and *pyrE*). The fact that these mutants, as well as most auxotrophs screened in *L. monocytogenes* (73), are replication proficient indicates that the concentration of nucleotides within the host cell cytoplasm is probably limiting but sufficient to support intracellular growth. To identify other genes induced after the bacteria are internalized, directed fusions to *lacZ* and *cat* reporters were constructed in merodiploid strains. With this approach, it was demonstrated that genes important for intracellular growth were activated after bacterial internalization into the J774 macrophage-like cell line (26, 53). The listeriolysin O structural gene *hlyA* and the *actA* gene were expressed at higher levels in cells than in L broth, with the induction levels exhibited by *actA* being well over 200-fold (53). Listeriolysin is required for the organism to break out of the phagosome and enter into the cytosol (61), whereas ActA is involved in nucleating actin and promoting actin-based motility, resulting in cell-to-cell spread of the bacterium (23).

Salmonella Serovar Typhimurium

In contrast to intracytoplasmic pathogens, considerable effort has been invested in identifying intracellular regulatory circuits of intravacuolar pathogens. The most intensively investigated intravacuolar bacterium has been *Salmonella* serovar Typhimurium. In mouse models, this organism causes diseases that are both localized within the intestine as well as systemic due to entry of bacteria into the liver and spleen. The bacteria-laden vacuoles within both macrophages and epithelial-like cultured cells have been characterized in some detail (27). Although these compartments are acidified and contain some proteins found in lysosomes, they are clearly nonlysosomal (27), and there is some evidence that proteins that prevent maturation of the compartment into a lysosome are required for bacterial survival within macrophages (74).

Initial studies using two-dimensional gel electrophoresis indicated that multiple regulatory circuits that control the expression of stress response proteins were induced after serovar Typhimurium uptake into macrophages (8). In particular, enhanced synthesis of heat shock proteins, such as DnaK and GroEL, was observed for intracellularly localized bacteria relative to those grown in culture. A second serovar Typhimurium global regulatory circuit activated within macrophages was shown to be controlled by the PhoP/PhoQ two-component system (5). Through fusions to the PhoP-regulated *pagA* (*ugd*), *pagB* (*pmrC*), and *pagC* genes, it was shown that acidification of the phagosome was required to activate this system, since addition of a lysosomotropic base to infected macrophages prevented increased expression of *pagC*. Maximal activation of *pagC* occurred several hours after uptake, when the phagosome was maximally acidified (5). A similar connection between acidification and the PhoP/PhoQ system was made by analyzing serovar Typhimurium localized within epithelial-like MDCK cells (28). Transcriptional fusions of *lacZ* to preselected operons indicated that promoters responsive to acid stress as well as magnesium and iron starvation are induced in response to internalization of serovar Typhimurium by the epithelial-like MDCK cell line (28). It was later demonstrated that the particular operon chosen to sense magnesium starvation, *mgtCB*, is under the control of the PhoP/PhoQ regulon (65). Several PhoP-activated genes appear to be involved in promoting bacterial growth in Mg^{2+}-limiting conditions (65), and Mg^2 starvation causes induction of the PhoP regulon (29, 66). Although these results do not directly demonstrate that intracellular starvation for divalent cations controls bacterial gene expression, there appears to be a close correlation between the bank of genes activated intracellularly and those activated by Mg^{2+}-limiting conditions. In fact, it has been proposed that sensing limiting amounts of Mg^{2+} ions within the vacuole is the single key signal for inducing *Salmonella* gene expression after uptake by host cells (30).

Two major studies have identified serovar Typhimurium promoters activated within cultured phagocytes. The most significant conclusion obtained from these studies is that the majority of promoters identified in this fashion are PhoP-activated (33, 76). Mahan and coworkers used the *cat-lacZ* fusion cassette, described above, to identify promoters that conferred chloramphenicol resistance during growth in the RAW 264.7 macrophage-like cell line but that were Lac⁻ when grown on bacteriological medium (32). PhoP-activated genes, such as *pmrB* (polymyxin resistance) and *mgtA/mgtB*, were identified, as were genes regulated by iron concentration (*fhuA, cirA,* and *entF*), consistent with previous work (5, 28). In addition, two genes of unknown function and regulation were identified (32). There has been no detailed study, as yet, that evaluates the relative contribution of these genes to the survival of serovar Typhimurium within cultured macrophages.

Valdivia and Falkow used random fusions to *gfp* followed by cell sorting to identify promoters active within RAW 264.7 macrophages (76). Eight of the 14 fusions identified in this fashion were to promoters that were PhoP-activated. Although no information is available on the contribution of these eight genes to survival in macrophages, two (a lipoprotein and a gene of no known sequence homology) clearly are important for full virulence in a mouse model. In addition, a second regulatory circuit of critical importance to virulence, controlled by the SsrA/SsrB two-component regulator, was shown to respond to the intravacuolar environment. This was demonstrated by showing that one of the fusions that responded to macrophage uptake was within the *ssaH* gene, a component of a Type III secretion system found in the serovar Typhimurium pathogenicity island SPI-2 (35) that is regulated by SsrA/SsrB. Several of the genes regulated by this two-component regulator have been shown to be required for intramacrophage survival. Mutations in components of the SPI-2 secretion system, as well as in putative effector molecules translocated into target cells, result in poor bacterial survival and lack of proliferation in cultured macrophages (19, 36). It has been demonstrated that one of these effector molecules alters endocytic trafficking within the infected macrophage (74). SsrA/SsrB may sense nutrient deprivation, as a potential target of the SPI-2 system that is regulated by this two-component regulator accumulates within the bacterium during entry into stationary phase (6).

L. pneumophila

L. pneumophila is found in a very different vacuole than is serovar Typhimurium, so it may be presumed that the signals to which this organism responds are different. *L. pneumophila* resides in a considerably less acidic compartment than does serovar Typhimurium (39) that is devoid of lysosomal markers (38). Nevertheless, two-dimensional gel analysis of *L. pneumophila* introduced into cycloheximide-treated cultured macrophages indicated that the bacteria express increased amounts of at least 13 proteins that respond to a variety of stress conditions, such as heat shock or oxidizing conditions (2). One of these proteins, GspA, was analyzed in detail and found not to be important for intracellular growth (3). Unfortunately, it is not clear from these studies whether the majority of bacteria were grown in culture conditions that allowed them to enter into a vacuole that was competent for intracellular replication (12). If most of the bacteria were internalized into a lysosomal compartment rather than a replication vacuole, then this may explain why induction of stress-induced proteins was so prominent. A second approach, using differential display PCR, identified an open reading frame called *eml* that is induced after uptake into cultured U937 cells. Mutations in and around this gene result in delayed kinetics of intracellular replication (4). It is not known what conditions other than entry into cultured cells cause induction of this locus.

The Td-*lacZ* fusion strategy, described above, allowed the isolation of *L. pneumophila* promoters that respond to an intracellular environment (62). As the enrichment procedure required that an active fusion result in intracellular growth, promoters selected in this fashion must have been active in bacteria localized within replication vacuoles. Among the promoters identified was one that drives a bicistronic operon encoding a glutaredoxin-like protein and AhpC, a subunit of alkyl hydrogen peroxide reductase (72). AhpC has been implicated in protecting bacteria from NO killing (16), and the gene encoding this protein was originally identified in serovar Typhimurium as being regulated by oxidative stress via the OxyR system (16). A similar regulatory response for this cluster has not been demonstrated for *L. pneumophila*. In addition, a two-component regulator, as well as at least one gene homologous to antibiotic efflux transporters, was identified, indicating that there may be a detoxification response by the bacterium after entry into the phagocytic vacuole. In no case were any of the loci identified in this procedure shown to be required for intracellular growth, implying that many of the bacterial genes induced within the vacuole perform many functions. Alternatively, conditions within the vacuoles of cultured cells may not provide growth conditions that are as restrictive as vacuoles found in tissue-localized macrophages.

GENES INDUCED IN TISSUE SITES

Serovar Typhimurium

The studies to date that have made the most significant contribution to the characterization of bacterial genes induced within host tissues have been performed with serovar Typhimurium (32, 51). By using the IVET vectors containing random fragments of serovar Typhimurium DNA, strains were isolated that had a *purA* gene selectively induced or that were chloramphenicol resistant within the mouse after either intragastric or intraperitoneal infection. These strains were then pooled and analyzed along with the strains described above that expressed chloramphenicol resistance specifically after introduction into RAW 264.7 macrophages. More than 100 different genes were recovered and sequenced from these selections (32), although only about 25 of them have been described in tables (20, 32, 33). The criterion for choosing which fusions are significant is not totally clear. The fusions described to date have either been to genes that were previously demonstrated to be important in virulence or in macrophage survival, had DNA sequences that had G + C contents that differed significantly from the majority of *Salmonella* genes (20), or showed some interesting homology to adhesins (32). As seen with macrophage-induced genes, several of the loci identified were either PhoP-regulated, such as *pmrB, mgtB,* and *spvB,* or were regulated by iron concentration. Interestingly, a fusion to PhoP itself was recovered, consistent with previous data showing that *phoPQ* is autoregulated (67). Among fusions to genes that had not been previously identified as contributing to serovar Typhimurium virulence, none was shown to be essential for pathogenicity, as insertion mutations isolated downstream from the promoters identified in this enrichment showed no effect on LD$_{50}$ levels of serovar Typhimurium after intragastric or intraperitoneal inoculations of the mouse (32).

Two interesting observations came from these studies. First, 10 of the promoters identified in the fashion described above controlled genes that were in regions of the serovar Typhimurium chromosome either absent in other enteric pathogens or missing from other *Salmonella* serovars (20). It was postulated that these regions are sequences involved in adaptation of the bacterium to its particular animal host. Consistent with this theory, a large deletion in at least one of these regions caused severe attenuation of virulence (20). These regions may be pathogenicity islands (31), containing genes that are specifically carried by the organism to allow colonization of a host. Whether the isolation of genes that are induced within the animal is an efficient strategy to identify such regions is arguable, however, since only 10 out of more than 100 insertions allowed identification of unique serovar Typhimurium sequences. In contrast, it is estimated that 30% of the serovar Typhimurium genome differs greatly from that of *E. coli* (47). A second observation is that many of the genes identified that were induced within an animal and were also required for full virulence were shown to be induced within cultured cells (28, 33). This suggests that in the case of serovar Typhimurium, a large number of the genes important for virulence that are induced within the animal can be identified by using enrichments in cultured cells rather than in animals.

There were clearly several genes induced within the animal that were missed in the exhaustive IVET search. For instance, the large group of genes in SPI-2 that are induced within macrophages (described above; 19, 76) were not recovered. The explanation for this inability to recover some fusions may be that some of the important genes that are induced during growth within the animal are expressed at sufficiently high levels in laboratory media to prevent their isolation. As mentioned above, this has clearly been demonstrated with the recombinase-based RIVET strategy in *V. cholerae* (15).

Y. enterocolitica

Enteropathogenic *Yersinia* causes a disease in mice that superficially resembles that of serovar Typhimurium. Within 3 days after oral inoculation, bacteria can be found growing in Peyer's patches and regional lymph nodes (22). Eventually, bacteria enter the liver and spleen, followed by inevitable death of the animal. The organ localization of bacteria is similar to that of serovar Typhimurium, but there is little evidence that pathogenic *Yersinia* grows within macrophages in the host. For this reason, it may be expected that the panel of genes identified in this organism that are specifically induced within the animal would be different from those identified in serovar Typhimurium.

Using a *cat* reporter system similar to that described for serovar Typhimurium, a fusion gene bank to *Y. enterocolitica* chromosomal DNA was constructed to identify genes activated within the animal after orogastric inoculation. Strains were selected that were able to multiply in the Peyer's patch in the presence of chloramphenicol but were unable to survive in the presence of the antibiotic on laboratory medium. As was shown for serovar Typhimurium, stress response proteins such as glutaredoxin, HflX, and a variety of DNA repair proteins showed increased expression in the animal (79). In addition, several genes that respond to iron starvation were induced. Interestingly, no PhoP-activated genes were isolated by this procedure, and as yet there is no evidence for the PhoP regulatory system playing a role in *Yersinia* pathogenesis. One explanation for this difference is that the PhoP system is specifically involved in promoting bacterial survival within macrophages by controlling synthesis of proteins that protect the bacterium against an environment that is limiting in divalent cations. As factors translocated into host cells by the *Yersinia* Type III secretion system prevent phagocytosis by macrophages, the organism never encounters an environment that causes induction of PhoP-activated genes.

V. cholerae

As mentioned above, *V. cholerae* promoters that are activated during colonization of the mouse intestine were identified with a recombinase-based strategy (Fig. 3) (15). *V. cholerae* colonizes the small intestine of the mouse with a pilus called TCP that is required for pathogenesis (63). Organisms that successfully colonize this site elicit cholera toxin, which induces unregulated fluid secretion from cells found within the intestinal crypts (71, 78). Both TCP and cholera toxin are expressed preferentially during infection and are regulated by a complex circuit involving the ToxR/ToxS regulator and the ToxT transcriptional activator (7, 80). The original recombinase-based strategy to identify host-induced promoters resulted in the identification of promoters that control genes encoding proteins involved in motility and biosynthetic pathways, as well as five unknown genes. Only one of these genes when knocked out had a subtle effect on colonization of the mouse intestine, and none was regulated by ToxR/ToxS.

Detailed analysis of one of the fusions to a previously uncharacterized gene revealed one of the strengths of using the recombinase-based system. When the recombinase is activated within the animal, the loss of drug resistance leaves a permanent genotypic change in the strain. This can be followed spatially and temporally, so that the exact time and site within the animal that provides conditions that induce expression can be determined. This was demonstrated with a fusion to one of the previously identified *V. cholerae* promoters that was preferentially activated within the mouse intestine (48). When the promoter for *vieB*, which encodes a putative polypeptide having sequence similarity to two-component response regulators, was fused to *tnpR*, it was found that the recombinase was activated shortly after the organism entered into the small bowel. The sites where recombination took place appeared to be in both the proximal and mid-bowel.

In the most remarkable use of this system, the attenuated *tnpR* gene having a defective ribosome binding site was fused to the genes for cholera toxin and TCP (S. H. Lee, D. L. Hava, M. K. Waldor, and A. Camilli, submitted). It was demonstrated that induction of expression of the TCP pilus occurred in a biphasic fashion in the small intestine, with 20% of the population being induced immediately after entry into the intestine, and the rest of the population being induced 4 h later. Cholera toxin showed no such immediate induction and instead was induced 4 to 5 h after inoculation. To emphasize the power of this strategy, the amounts of toxin produced within the mouse intestine were below the limits of detection, so the only way the time of induction could be determined was by using the recombinase assay. Interestingly, induction of cholera toxin expression within the intestine required the expression of functional TCP pilus, suggesting that the bacteria must bind the intestinal mucosa to induce expression of the toxin gene. This is the first indication that adhesion by *V. cholerae* causes signaling back to the bacterium to induce expression of a gene critical for virulence. It is possible that the ToxR/ToxS regulator, which is a transmembrane complex necessary for toxin expression within the intestine (Lee et al., submitted), may sense increased concentrations of a component in the bacterial periplasm that accumulates in response to adhesion.

S. aureus

Perhaps the most successful isolation of host-induced genes for identifying factors required for virulence within the animal was with *S. aureus* (50). In this study, the recombinase system as developed in *V. cholerae* was used to identify strains growing within the mouse kidney that had undergone recombination as a result of a cloned promoter selectively activating the *tnpR* gene. Forty-five such fusions were identified that promoted recombination within the mouse and not in laboratory media. Of these fusions, several affected known genes required for virulence, such as capsule biosynthesis and the gene encoding the quorum-sensing regulator AgrA. The latter is known to be required for the activation of a bank of secreted virulence factors in *S. aureus* (17, 54). Eleven of the genes that either had no homology to genes in database searches or had not been demonstrated to be involved in virulence were analyzed further. Deletion mutations isolated in each of these genes showed that at least six of these genes were required for efficient colonization of kidney, with significant attenuation resulting from elimination of any one of these genes. No other hunt for genes induced within the animal has been so successful in identifying previously uncharacterized genes required for virulence.

PERSPECTIVES AND FUTURE APPROACHES

For most of the successful attempts at identifying large numbers of genes induced within the bacterium in response to host cells or tissues, the use of promoter trap vectors that result in fusion to transcriptional reporters has been critical. However, the development of nucleic acid microarrays, which have been primarily used to analyze eukaryotic gene expression, may change the way researchers approach this issue (42, 64, 69). In the most relevant strategy for this purpose, a small chip matrix with thousands of different immobilized nucleic acid fragments at individual known addresses is used as a target for a pair of differentially marked fluorescent probes that are prepared under different biological conditions. Sites on the matrix that preferentially hybridize to probes prepared under one condition can be detected by enhanced fluorescence emission wavelength of one of the probes at this site. Typically, this is displayed as a three-color array, with preferential hybridization to either probe revealed as the color of emission of that particular probe. Equivalent hybridization to the two probes is displayed as a third color. This approach is most feasible with organisms for which the complete nucleotide sequence is known, in which a variety of growth conditions can be compared, and for which it is relatively easy to obtain large amounts of mRNA probes. For this reason, the most successful applications of this approach have been with *Saccharomyces cerevisiae*, which satisfies all of these criteria (46).

Although the complete nucleotide sequences of several bacterial pathogens have now been identified, some technical problems involved in isolating sufficient amounts of mRNA or preparing an appropriate cDNA probe pool have delayed application of array technology. Progress has been made in applying this technology to *Bacillus subtilis* and *Pseudomonas*

aeruginosa, and it should be possible to amplify selectively message-specific cDNA by using PCR primers of appropriate complexity, so the outlook is bright for using arrays to analyze the interaction of pathogens with host cells. Technically, by preparing mRNA-derived probes from bacteria at various times after introduction onto mammalian cell monolayers, and by comparing each to probes made from bacteria grown in culture, it should be possible to catalog the complete expression program of a pathogen after contact with host cells. Applying this technology to bacteria found within tissues of an infected animal will be more technically difficult because of the small amount of sample available. Presumably a variation on using a PCR primer pool that selectively amplifies the complete array of mRNA from the pathogen of choice will be able to solve this problem.

No matter what the technology that is chosen to identify host-encoded genes, the most vexing technical problem will be to determine how the detection system should be tuned. Considerable evidence exists that most of the schemes used to identify host-induced genes have eliminated genes that have small, but detectable, amounts of expression in laboratory media. As a result, the schemes have enriched for genes that have the lowest levels of expression in culture (62; Lee et al., submitted). These may not be the most interesting genes to analyze, and in fact, very few previously unidentified genes isolated by these enrichments have been demonstrated to play a profound role in virulence. For instance, the original vector system failed to identify *V. cholerae* genes that respond to the most important virulence regulators in the intestinal environment. This appears to be a clear warning signal, that the detection scheme must be calibrated in such a way to optimize the isolation of desired genes, allowing the isolation of genes expressed in laboratory media but which respond to host inducing conditions. On the positive side, the relative failure to show clearly a virulence defect when newly identified genes are mutated may be a result of this strategy uncovering functionally redundant genes, although this has yet to be formally demonstrated. As pointed out above, this is a potential strength of identifying host-induced genes, in that isolation of genes that are potentially important for virulence does not rely on observing a virulence defect in mutant strains.

A common theme that emerges from these studies is that there are a few regulatory circuits that seem to be activated within most pathogens, and in addition, there seems to be at least one central regulatory circuit that is specific for the pathogen, facilitating the unique lifestyle of that microorganism. For instance, in the case of serovar Typhimurium, genes involved in stress response, protection from oxidizing conditions, and response to iron limitation are all found to be regulated by interaction with host tissues. Such genes are also known to be activated in *Y. enterocolitica*, *L. pneumophila*, and *C. diphtheriae*. In addition, a large group of genes is activated by the PhoP/PhoQ system in serovar Typhimurium, both during mouse infection and within cultured macrophages. Although many other bacterial pathogens have the identical regulatory circuit, there is little evidence to support such a critical role for PhoP in virulence. This regulatory circuit may allow the organism to survive within an intracellular environment that is not identical to that experienced by other pathogens. In the future, much of the excitement will lie in the ability to identify such unique and critical regulatory circuits, as well as the ability to identify redundant pathways necessary for virulence in animal models. Presumably, the development of array technology, tunable vector systems, and the use of GFP technology should facilitate these discoveries.

Acknowledgments. I thank Andrew Camilli for reviewing the manuscript, sharing submitted manuscripts, and contributing Fig. 3.

Work in my laboratory is supported by grant RO1-AI23538 from the National Institute of Allergy and Infectious Diseases, and a Howard Hughes Medical Institute Investigator Award.

REFERENCES

1. **Abu Kwaik, Y.** 1998. Induced expression of the *Legionella pneumophila* gene encoding a 20-kilodalton protein during intracellular infection. *Infect. Immun.* 66:203–212.

2. **Abu Kwaik, Y., B. I. Eisenstein, and N. C. Engleberg.** 1993. Phenotypic modulation by *Legionella pneumophila* upon infection of macrophages. *Infect. Immun.* 61:1320–1329.

3. **Abu Kwaik, Y., L. Y. Gao, O. S. Harb, and B. J. Stone.** 1997. Transcriptional regulation of the macrophage-induced gene (*gspA*) of *Legionella pneumophila* and phenotypic characterization of a null mutant. *Mol. Microbiol.* 24:629–642.

4. **Abu Kwaik, Y., and L. L. Pederson.** 1996. The use of differential display-PCR to isolate and characterize a *Legionella pneumophila* locus induced during the intracellular infection of macrophages. *Mol. Microbiol.* 21:543–556.

5. **Alpuche Aranda, C. M., J. A. Swanson, W. P. Loomis, and S. I. Miller.** 1992. *Salmonella typhimurium* activates virulence gene transcription within acidified macrophage phagosomes. *Proc. Natl. Acad. Sci. USA* 89:10079–10083.

6. **Beuzön, C. R., G. Banks, J. Deiwick, M. Hensel, and D. W. Holden.** 1999. pH-dependent secretion of SseB, a product of the SPI-2 type III secretion system of *Salmonella typhimurium*. *Mol. Microbiol.* 33:806–816.

7. **Brown, R. C., and R. K. Taylor.** 1995. Organization of *tcp*, *acf*, and *toxT* genes within a ToxT-dependent operon. *Mol. Microbiol.* 16:425–439.

8. **Buchmeier, N. A., and F. Heffron.** 1990. Induction of Salmonella stress proteins upon infection of macrophages. *Science* 248:730–732.

9. **Buchrieser, C., R. Brosch, S. Bach, A. Guiyoule, and E. Carniel.** 1998. The high-pathogenicity island of *Yersinia pseudotuberculosis* can be inserted into any of the three chromosomal asn tRNA genes. *Mol. Microbiol.* 30:965–978.

10. **Burns-Keliher, L., C. A. Nickerson, B. J. Morrow, and R. Curtiss III.** 1998. Cell-specific proteins synthesized by *Salmonella typhimurium*. *Infect. Immun.* 66:856–861.

11. **Burns-Keliher, L. L., A. Portteus, and R. Curtiss III.** 1997. Specific detection of *Salmonella typhimurium* proteins synthesized intracellularly. *J. Bacteriol.* 179:3604–3612.

12. **Byrne, B., and M. S. Swanson.** 1998. Expression of *Legionella pneumophila* virulence traits in response to growth conditions. *Infect. Immun.* 66:3029–3034.

13. **Camilli, A.** 1996. Noninvasive techniques for studying pathogenic bacteria in the whole animal. *Trends Microbiol.* 4:295–296.

14. **Camilli, A., D. T. Beattie, and J. J. Mekalanos.** 1994. Use of genetic recombination as a reporter of gene expression. *Proc. Natl. Acad. Sci. USA* 91:2634–268.

15. **Camilli, A., and J. J. Mekalanos.** 1995. Use of recombinase gene fusions to identify *Vibrio cholerae* genes induced during infection. *Mol. Microbiol.* 18:671–683.

16. **Chen, L., Q. W. Xie, and C. Nathan.** 1998. Alkyl hydroperoxide reductase subunit C (AhpC) protects bacterial and human cells against reactive nitrogen intermediates. *Mol. Cell.* 1:795–805.

17. **Chien, Y., and A. L. Cheung.** 1998. Molecular interactions between two global regulators, *sar* and *agr*, in *Staphylococcus aureus*. *J. Biol. Chem.* 273:2645–2652.

18. **Chilcott, G. S., and K. T. Hughes.** 1998. The type III secretion determinants of the flagellar anti-transcription factor, FlgM, extend from the amino-terminus into the anti-sigma28 domain. *Mol. Microbiol.* 30:1029–1040.

19. **Cirillo, D. M., R. H. Valdivia, D. M. Monack, and S. Falkow.** 1998. Macrophage-dependent induction of the *Salmonella* pathogenicity island 2 type III secretion system and its role in intracellular survival. *Mol. Microbiol.* 30:175–188.

20. **Conner, C. P., D. M. Heithoff, S. M. Julio, R. L. Sinsheimer, and M. J. Mahan.** 1998. Differential patterns of acquired virulence genes distinguish *Salmonella* strains. *Proc. Natl. Acad. Sci. USA* 95:4641–4645.

21. **Conner, C. P., D. M. Heithoff, and M. J. Mahan.** 1998. In vivo gene expression: contributions to infection, virulence, and pathogenesis. *Curr. Top. Microbiol. Immunol.* 225:1–12.

22. **Cornelis, G. R., A. Boland, A. P. Boyd, C. Geuijen, M. Iriarte, C. Neyt, M. P. Sory, and I. Stainier.** 1998. The virulence plasmid of *Yersinia*, an antihost genome. *Microbiol. Mol. Biol. Rev.* 62:1315–1352.

23. **Cossart, P., and M. Lecuit.** 1998. Interactions of *Listeria monocytogenes* with mammalian cells during entry and actin-based movement: bacterial factors, cellular ligands and signaling. *EMBO J.* 17:3797–3806.

24. **Crosa, J. H.** 1997. Signal transduction and transcriptional and posttranscriptional control of iron-regulated genes in bacteria. *Microbiol. Mol. Biol. Rev.* 61:319–336.

25. **Finlay, B. B., F. Heffron, and S. Falkow.** 1989. Epithelial cell surfaces induce *Salmonella* proteins required for bacterial adherence and invasion. *Science* 243:940–943.

26. **Freitag, N. E., and K. E. Jacobs.** 1999. Examination of *Listeria monocytogenes* intracellular gene expression by using the green fluorescent protein of *Aequorea victoria*. *Infect. Immun.* 67:1844–1852.

27. **Garcia-del Portillo, F., and B. B. Finlay.** 1995. Targeting of *Salmonella typhimurium* to vesicles containing lysosomal membrane glycoproteins bypasses compartments with mannose 6-phosphate receptors. *J. Cell Biol.* 129:81–97.

28. **Garcia-del Portillo, F., J. W. Foster, M. E. Maguire, and B. B. Finlay.** 1992. Characterization of the micro-environment of *Salmonella typhimurium*-containing vacuoles within MDCK epithelial cells. *Mol. Microbiol.* 6:3289–3297.

29. Garcia Vescovi, E., F. C. Soncini, and E. A. Groisman. 1996. Mg^{2+} as an extracellular signal: environmental regulation of *Salmonella* virulence. *Cell* 84:165–174.

30. Groisman, E. A. 1998. The ins and outs of virulence gene expression: Mg^{2+} as a regulatory signal. *Bioessays* 20:96–101.

31. Groisman, E. A., and H. Ochman. 1997. How *Salmonella* became a pathogen. *Trends Microbiol.* 5:343–349.

32. Heithoff, D. M., C. P. Conner, P. C. Hanna, S. M. Julio, U. Hentschel, and M. J. Mahan. 1997. Bacterial infection as assessed by in vivo gene expression. *Proc. Natl. Acad. Sci. USA* 94:934–939.

33. Heithoff, D. M., C. P. Conner, U. Hentschel, F. Govantes, P. C. Hanna, and M. J. Mahan. 1999. Coordinate intracellular expression of *Salmonella* genes induced during infection. *J. Bacteriol.* 181:799–807.

34. Hengge-Aronis, R. 1999. Interplay of global regulators and cell physiology in the general stress response of *Escherichia coli. Curr. Opin. Microbiol.* 2:148–152.

35. Hensel, M., J. E. Shea, C. Gleeson, M. D. Jones, E. Dalton, and D. W. Holden. 1995. Simultaneous identification of bacterial virulence genes by negative selection. *Science* 269:400–403.

36. Hensel, M., J. E. Shea, S. R. Waterman, R. Mundy, T. Nikolaus, G. Banks, A. Vazquez-Torres, C. Gleeson, F. C. Fang, and D. W. Holden. 1998. Genes encoding putative effector proteins of the type III secretion system of *Salmonella* pathogenicity island 2 are required for bacterial virulence and proliferation in macrophages. *Mol. Microbiol.* 30:163–174.

37. Holmstrom, A., J. Petterson, R. Rosqvist, S. Hakansson, F. Tafazoli, M. Fallman, K. E. Magnusson, H. Wolf-Watz, and A. Forsberg. 1997. YopK of *Yersinia pseudotuberculosis* controls translocation of Yop effectors across the eukaryotic cell membrane. *Mol. Microbiol.* 24:73–91.

38. Horwitz, M. A. 1992. Interactions between macrophages and *Legionella pneumophila. Curr. Top. Microbiol. Immunol.* 181:265–282.

39. Horwitz, M. A., and F. R. Maxfield. 1984. *Legionella pneumophila* inhibits acidification of its phagosome in human monocytes. *J. Cell Biol.* 99:1936–1943.

40. Hueck, C. J. 1998. Type III protein secretion systems in bacterial pathogens of animals and plants. *Microbiol. Mol. Biol. Rev.* 62:379–433.

41. Hughes, K. T., and K. Mathee. 1998. The anti-sigma factors. *Ann. Rev. Microbiol.* 52:231–286.

42. Iyer, V. R., M. B. Eisen, D. T. Ross, G. Schuler, T. Moore, J. C. F. Lee, J. M. Trent, L. M. Staudt, J. Hudson, Jr., M. S. Boguski, D. Lashkari, D. Shalon, D. Botstein, and P. O. Brown. 1999. The transcriptional program in the response of human fibroblasts to serum. *Science* 283:83–87.

43. Jacobi, C. A., A. Roggenkamp, A. Rakin, R. Zumbihl, L. Leitritz, and J. Heesemann. 1998. In vitro and in vivo expression studies of *yopE* from *Yersinia enterocolitica* using the *gfp* reporter gene. *Mol. Microbiol.* 30:865–882.

44. Jones, C. H., P. N. Danese, J. S. Pinkner, T. J. Silhavy, and S. J. Hultgren. 1997. The chaperone-assisted membrane release and folding pathway is sensed by two signal transduction systems. *EMBO J.* 16:6394–6406.

45. Klarsfeld, A. D., P. L. Goossens, and P. Cossart. 1994. Five *Listeria monocytogenes* genes preferentially expressed in infected mammalian cells: *plcA, purH, purD, pyrE* and an arginine ABC transporter gene, *arpJ. Mol. Microbiol.* 13:585–597.

46. Lashkari, D. A., J. L. DeRisi, J. H. McCusker, A. F. Namath, C. Gentile, S. Y. Hwang, P. O. Brown, and R. W. Davis. 1997. Yeast microarrays for genome wide parallel genetic and gene expression analysis. *Proc. Natl. Acad. Sci. USA* 94:13057–13062.

47. Lawrence, J., and J. Roth. 1996. Selfish operons: horizontal transfer may drive the evolution of gene clusters. *Genetics* 143:1843–1860.

48. Lee, S. H., M. J. Angelichio, J. J. Mekalanos, and A. Camilli. 1998. Nucleotide sequence and spatiotemporal expression of the *Vibrio cholerae vieSAB* genes during infection. *J. Bacteriol.* 180:2298–2305.

49. Liang, P., D. Bauer, L. Averboukh, P. Warthoe, M. Rohrwild, H. Muller, M. Strauss, and A. B. Pardee. 1995. Analysis of altered gene expression by differential display. *Methods Enzymol.* 254:304–321.

50. Lowe, A. M., D. T. Beattie, and R. L. Deresiewicz. 1998. Identification of novel staphylococcal virulence genes by in vivo expression technology. *Mol. Microbiol.* 27:967–976.

51. Mahan, M. J., J. M. Slauch, and J. J. Mekalanos. 1993. Selection of bacterial virulence genes that are specifically induced in host tissues. *Science* 259:686–688.

52. Mahan, M. J., J. W. Tobias, J. M. Slauch, P. C. Hanna, R. J. Collier, and J. J. Mekalanos. 1995. Antibiotic-based selection for bacterial genes that are specifically induced during infection of a host. *Proc. Natl. Acad. Sci. USA* 92:669–673.

53. Moors, M. A., B. Levitt, P. Youngman, and D. A. Portnoy. 1999. Expression of listeriolysin O and ActA by intracellular and extracellular *Listeria monocytogenes. Infect. Immun.* 67:131–139.

54. Novick, R. P., and T. W. Muir. 1999. Virulence gene regulation by peptides in staphylococci and other gram-positive bacteria. *Curr. Opin. Microbiol.* 2:40–45.

55. Pappenheimer, A. M., and D. M. Gill. 1973. Diphtheria. *Science* 182:353–358.

56. Persson, C., R. Nordfelth, A. Holmstrom, S. Hakansson, R. Rosqvist, and H. Wolf-Watz. 1995. Cell-surface-bound *Yersinia* translocate the protein tyrosine phosphatase YopH by a polarized mechanism into the target cell. *Mol. Microbiol.* 18:135–150.

57. Pettersson, J., R. Nordfelth, E. Dubinina, T. Bergman, M. Gustafsson, K. E. Magnusson, and H. Wolf-Watz. 1996. Modulation of virulence factor expression by pathogen target cell contact. *Science* 273:1231–1233.

58. Plum, G., M. Brenden, J. E. Clark-Curtiss, and G. Pulverer. 1997. Cloning, sequencing, and expression of the *mig* gene of *Mycobacterium avium*, which codes for a secreted macrophage-induced protein. *Infect. Immun.* 65:4548–4557.

59. Plum, G., and J. E. Clark-Curtiss. 1994. Induction of *Mycobacterium avium* gene expression following phagocytosis by human macrophages. *Infect. Immun.* 62:476–483.

60. Pollack, C., S. C. Straley, and M. S. Klempner. 1986. Probing the phagolysosomal environment of human macrophages with a Ca^{2+}-responsive operon fusion in *Yersinia pestis. Nature* 322:834–836.

61. Portnoy, D. A., and S. Jones. 1994. The cell biology of *Listeria monocytogenes* infection (escape from a vacuole). *Ann. N.Y. Acad. Sci.* 730:15–25.

62. Rankin, S., and R. R. Isberg. 1993. Identification of *Legionella pneumophila* promoters regulated by the macrophage intracellular environment. *Infect. Agents Dis.* 2:269–271.

63. Rhine, J. A., and R. K. Taylor. 1994. TcpA pilin sequences and colonization requirements for O1 and O139 *Vibrio cholerae. Mol. Microbiol.* 13:1013–1020.

64. Schena, M., D. Shalon, R. W. Davis, and P. O. Brown. 1995. Quantitative monitoring of gene expression patterns with a complementary DNA microarray. *Science* 270:467–470.

65. Soncini, F. C., E. Garcia Vescovi, F. Solomon, and E. A. Groisman. 1996. Molecular basis of the magnesium deprivation response in *Salmonella typhimurium*: identification of PhoP-regulated genes. *J. Bacteriol.* 178:5092–5099.

66. Soncini, F. C., and E. A. Groisman. 1996. Two-component regulatory systems can interact to process multiple environmental signals. *J. Bacteriol.* **178**:6796–6801.

67. Soncini, F. C., E. G. Vescovi, and E. A. Groisman. 1995. Transcriptional autoregulation of the *Salmonella typhimurium* phoPQ operon. *J. Bacteriol.* **177**:4364–4371.

68. Sory, M. P., and G. R. Cornelis. 1994. Translocation of a hybrid YopE-adenylate cyclase from *Yersinia enterocolitica* into HeLa cells. *Mol. Microbiol.* **14**:583–594.

69. Spellman, P. T., G. Sherlock, M. Q. Zhang, V. R. Iyer, K. Anders, M. B. Eisen, P. O. Brown, D. Botstein, and B. Futcher. 1998. Comprehensive identification of cell cycle-regulated genes of the yeast *Saccharomyces cerevisiae* by microarray hybridization. *Mol. Biol. Cell.* **9**:3273–3297.

70. Sun, L., J. vanderSpek, and J. R. Murphy. 1998. Isolation and characterization of iron-independent positive dominant mutants of the diphtheria toxin repressor DtxR. *Proc. Natl. Acad. Sci. USA* **95**:14985–14990.

71. Sundaram, U., R. G. Knickelbein, and J. W. Dobbins. 1991. Mechanism of intestinal secretion: effect of cyclic AMP on rabbit ileal crypt and villus cells. *Proc. Natl. Acad. Sci. USA* **88**:6249–6253.

72. Tartaglia, L. A., G. Storz, and B. N. Ames. 1989. Identification and molecular analysis of oxyR-regulated promoters important for the bacterial adaptation to oxidative stress. *J. Mol. Biol.* **210**:709–719.

73. Thompson, R. J., H. G. Bouwer, D. A. Portnoy, and F. R. Frankel. 1998. Pathogenicity and immunogenicity of a *Listeria monocytogenes* strain that requires D-alanine for growth. *Infect. Immun.* **66**:3552–3561.

74. Uchiya, K., M. A. Barbieri, K. Funato, A. H. Shah, P. D. Stahl, and E. A. Groisman. 1999. A *Salmonella* virulence protein that inhibits cellular trafficking. *EMBO J.* **18**:3924–3933.

75. Valdivia, R. H., and S. Falkow. 1996. Bacterial genetics by flow cytometry: rapid isolation of *Salmonella typhimurium* acid-inducible promoters by differential fluorescence induction. *Mol. Microbiol.* **22**:367–378.

76. Valdivia, R. H., and S. Falkow. 1997. Fluorescence-based isolation of bacterial genes expressed within host cells. *Science* **277**:2007–2011.

77. Wang, J., S. Lory, R. Ramphal, and S. Jin. 1996. Isolation and characterization of *Pseudomonas aeruginosa* genes inducible by respiratory mucus derived from cystic fibrosis patients. *Mol. Microbiol.* **22**:1005–1012.

78. Weiser, M. M., and H. Quill. 1975. Intestinal villus and crypt cell responses to cholera toxin. *Gastroenterology* **69**:479–482.

79. Young, G. M., and V. L. Miller. 1997. Identification of novel chromosomal loci affecting *Yersinia enterocolitica* pathogenesis. *Mol. Microbiol.* **25**:319–328.

80. Yu, R. R., and V. J. DiRita. 1999. Analysis of an autoregulatory loop controlling ToxT, cholera toxin, and toxin-coregulated pilus production in *Vibrio cholerae*. *J. Bacteriol.* **181**:2584–2592.

81. Zhang, J. P., and S. Normark. 1996. Induction of gene expression in *Escherichia coli* after pilus-mediated adherence. *Science* **273**:1234–1236.

Bacterial Stress Responses
Edited by G. Storz and R. Hengge-Aronis
©2000 ASM Press, Washington, D.C.

Chapter 21

Environmental Control of Pilus Gene Expression

Margareta Krabbe, Nathan Weyand, and David Low

This review focuses on the mechanisms by which environmental and cellular factors regulate the expression of adhesive appendages on bacteria known as fimbriae or pili. The investigation of pilus regulatory mechanisms over the past decade has shown that there is a complex network of protein-protein and protein-DNA interactions that integrate pilus expression with environmental response pathways. A number of pili are subject to phase variation control mechanisms by which pilus expression is skewed between active and inactive expression states. Type 1 and MR/P pili are regulated by a DNA inversion mechanism that is orchestrated by site-specific recombination between inverted repeats flanking the DNA switch region containing the fimbrial promoter. The frequency of off to on switching of type 1 pili is controlled by leucine via the global regulator Lrp, with additional regulatory control exerted by histone-like nucleoid structuring protein (H-NS) and integration host factor. The pap family of pilus operons is regulated by DNA methylation-dependent phase variation. Methylation of two GATC sites within the upstream regulatory region controls the binding of Lrp. Switch frequency is controlled by carbon source via catabolite gene activator protein, with additional thermoregulatory control exerted by H-NS. Numerous other pilus operons are regulated by a wide variety of environmental signals including pH, ammonium, and osmolarity.

Many gram-negative bacteria express multiple types of pili that enable them to adhere to various cell surfaces within the environments that they colonize. The regulation of pilus expression is highly complex, with input from a number of local and global regulatory factors. The focus of this review is on the mechanisms by which pilius expression in gram-negative bacteria is regulated by environmental stimuli.

INVERSION-CONTROLLED FIMBRIAL OPERONS

Type 1 Pili (*fim*)

Most *Escherichia coli* cells express type 1 pili, also known as common pili, that bind to mannose residues on epithelial cells (45). Type 1 pili have been shown to play roles in host-to-host transmission of bacteria (8, 135) and in colonization of the bladder by uropathogenic *E. coli* (UPEC) (27, 119). Type 1 pili are regulated by the inversion of a 314-bp DNA fragment known as the *fim* switch, containing the *fimA* promoter (Fig. 1). In cells expressing type 1 pili ("on" phase), the *fimA* promoter is in the same orientation as the *fimA* gene coding for the type 1 pilin structural gene. In the "off" phase the *fim* switch is inverted to an orientation opposite that of phase on cells, which places the *fimA* promoter in orientation opposite that of *fimA*.

Inversion of the *fim* switch occurs by a conservative site-specific recombination mechanism. Two *fim*-encoded proteins, FimB and FimE, share significant protein sequence identity with phage-encoded integrases, which facilitate insertion of lysogenic viruses into the bacterial chromosome. In addition, three DNA bending proteins—leucine-responsive regulatory protein (Lrp), histone-like nucleoid structuring protein (H-NS), and integration host factor (IHF)—play roles in site-specific recombination of the *fim* switch (regulatory information is summarized in Table 1).

Margareta Krabbe, Nathan Weyand, and David Low • Molecular, Cellular, and Developmental Biology, University of California, Santa Barbara, CA 93106.

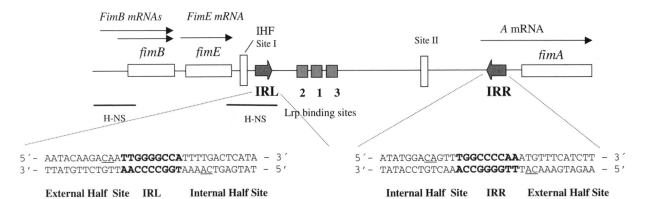

Figure 1. Regulatory sequences in the *E. coli fim* operon. The 314-bp *fim* switch is shown flanked by inverted repeat left (IRL) and inverted repeat right (IRR) sequences. The DNA sequences encompassing IRL and IRR are shown, including the conserved "CA" base pairs of the half sites, which are underlined. The binding sites for H-NS, Lrp, and IHF are shown by boxes and lines. The *fimB*, *fimE*, and *fimA* transcripts are shown by arrows. The "on" *fim* switch orientation is shown (6–8, 27, 32, 39, 44, 45, 47, 104, 157).

FimB and FimE

Although both FimB and FimE are recombinases, FimB promotes recombination to both the off and on *fim* orientations with equal frequency (about 1/1,000 per cell per generation), whereas FimE promotes recombination only from on to off at a very high frequency (0.7 per cell per generation) (113). Further analysis showed that FimE and FimB could each promote inversion of the *fim* switch in vitro, maintaining the switch specificities of these two recombinases observed in vivo (59). In vitro DNA binding analysis of FimB and FimE to the inverted repeat left (IRL) and inverted repeat right (IRR) regions of *fim* suggested that both proteins recognized a consensus sequence "ANNAGACANTTNGG" (59) (see Fig. 1). FimB has been shown to bind to IRL and IRR in vivo (40). Mutation of the totally conserved "CA" bases (underlined) reduced the affinities of FimB and FimE for the *fim* switch and significantly reduced *fim* switching in vivo (59). These results raise the possibility that the switch specificities of FimB and FimE are due to differences in DNA binding affinities. For example, if the affinity of FimE for one of the half sites at IRL or IRR was low when the switch was off, this could greatly reduce its ability to facilitate off-to-on switching. However, it was found that FimE has the highest affinity for IRR in the off orientation compared to the other three sites, yet FimE does not act on the off-phase *fim* switch. Thus, there is not a simple relationship between DNA binding affinities and the specificity by which *fim* switch inversion occurs.

Further studies indicate that FimE specificity occurs by at least two different mechanisms: FimE preferentially acts to promote recombination when the *fim* switch is in the on orientation and *fimE* activity is regulated by the orientation of the switch (99). DNA sequences within IRR and IRL as well as sequences outside the *fim* switch are required for specificity of FimE for the on orientation of the *fim* switch. Reciprocal exchange of either the internal or external IRR and IRL half sites reversed FimE specificity. However, in the latter case this was observed only in the absence of additional external flanking sequences, which contain IHF site 1 (Fig. 1). These results are consistent with the hypothesis that IHF might inhibit potential off to on switching promoted by FimE, thus maintaining FimE specificity. In any case, the specific interactions of FimE with the on-phase *fim* switch are likely to involve *trans*-acting factors such as IHF (see below).

The hypothesis that *fimE* activity might be controlled by the orientation of the *fim* switch was initially put forth by Pallesen et al. (137). This hypothesis was recently tested by mutating IRR to lock the switch in the on and off orientations and measuring FimE activity. The results showed that *fimE*-mediated recombination was detected only when the switch was in the on orientation. Merodiploid analysis showed that control of *fimE* occurred via a *cis*-acting mechanism, ruling out the possible involvement of an antisense *fimE* mRNA, which could be generated by *fim* transcription in the off switch (Fig 1). On the basis of these results it was speculated that regulation of *fimE* activity might vary as a result of mRNA instability, since the *fim* switch in the off and on orientations could alter the 3′ end of *fimE* mRNA (99). In summary, the activity of FimE is regulated by at least two mechanisms operating at the levels of expression and/or enzyme activity of the recombinase, and preference for the on-phase orientation of the *fim* switch as a substrate for recombination.

Table 1. Regulatory and environmental factors that control pilus expression

Regulatory grouping	Fimbrial operon(s) included[a]	Environmental stimuli[b]	Regulators or alleles[c]
Inversion	*fim* (EC[d], ST, SE), *mrp* (PM)	Alanine, isoleucine, leucine, valine (58), temperature (58, 95, 134), carbon source (46), rich medium (58), solid medium (25, 83, 161), standing culture (46), anaerobic incubation (25), aeration (4), stationary phase (42)	Lrp (11, 61, 74, 152), H-NS (95, 134, 157), IHF (11, 39, 47), RpoS (40, 42), *leuX* (149, 150), *topA, gyrA, gyrB* (41, 80)
Methylation	*pap* (EC), *sfa* (EC), *daa* (EC), *afa* (EC), *fae* (EC), *f165* (EC), *foo* (EC), *clp* (EC), *prf* (EC), F17? (EC), *pef* (ST)	Alanine, leucine (28, 110), temperature (28, 67, 158, 189), carbon source (7, 9, 28, 158), aeration (158), growth rate (158), solid vs. liquid growth medium (28, 158), osmolarity/NaCl (158), pH (28, 125)	Lrp (7, 81, 128, 180), Dam (13, 17, 81, 181), CAP (66), H-NS (67, 117, 189), CpxR? (145), IHF (7), RpoD (188), RpoS (125), RimJ (191), RNase E (117)
K99	*fan* (EC)	Alanine (86), leucine (18), temperature (178), carbon source (85), growth rate (176), pH (182), monoamines (139)	Lrp (10, 18, 108), CAP (84, 108)
Curli	*csg* (EC), *sef17* (SE), *agf* (ST)	Temperature (33, 133, 186), solid vs. liquid growth medium (154), growth phase (154, 186), osmolarity/NaCl (131, 154), iron-limiting conditions (154), pH (186)	RpoS (1, 132), Crl (1, 133, 143), OmpR (153, 183), H-NS (1)
Colonization factor antigens and 987P fimbriae	*cfa* (EC), *coo* (EC), *agg* (EC), *fas* (*fap*) (EC), *afr* (EC)	Temperature (21, 44, 142), iron (93), growth medium (179), growth rate (179), O₂ (179), ammonium (44), pH (44), carbon source (44), osmolarity/NaCl (43)	H-NS (89, 121), CAP (44), AraC homologs (Rns [121, 142], CfaR/CfaD [89, 156], FasH/FapR [44, 96], AggR [123], AfaR [20], CsvR [31, 192])
Type IV pili including TCP	*pil* (PA, MX, NG, NM), *tcp* (VC), *bfp* (EC), *lng* (EC)	Isoleucine (100), calcium (144), oxygen (32), carbon dioxide (184), high nutrients (193), temperature (100, 138, 144, 163), osmolarity/NaCl (71, 111, 116), ammonium (100, 111, 144), anaerobic growth (100, 115), log phase (144), stationary phase (115), growth medium (113), pH (36, 100, 116), cAMP (163), presence of amino acids (116), bile (69, 159), bacterial cell contact (169), iron-limiting conditions (100), urea (100), CAB agar (64), CDC anaerobe blood agar (62)	ToxT/TcpN (35), ToxR (35), BfpT/PerA (111), SigF (6), CAP (163), RegF (32), RpoN (87, 167), AphA (164)

[a] Pilus operon loci are listed if references are cited in the table that deal with regulation of the specific pilus operon by a factor(s) or environmental stimulus.
[b] The environmental stimuli listed do not affect all regulatory family members in all cases. Please refer to the specific references to identify the family member(s) affected by a given environmental signal.
[c] This list does not include operon-encoded regulatory proteins.
[d] EC, *Escherichia coli*; SE, *Salmonella enteritidis*; ST, *Salmonella enterica* serovar Typhimurium; PM, *Proteus mirabilis*; PA, *Pseudomonas aeruginosa*; MX, *Myxococcus xanthus*; NG, *Neisseria gonorrhoeae*; NM, *Neisseria meningitidis*; VC, *Vibrio cholerae*.

Lrp and Response to Amino Acids

In addition to its essential role in methylation-dependent pilus phase variation systems such as *pap* (18), Lrp also plays a role in type 1 pilus phase variation since both FimB- and FimE-catalyzed switching is greatly reduced in its absence (10, 60). In contrast to *pap*, which is not regulated by aliphatic amino acids such as leucine (18), phase switching of type 1 pili is stimulated 15-fold by leucine (58). Lrp binds cooperatively to three binding sites near IRL of the *fim* switch (Fig. 1). On the basis of mutation analysis,

binding of Lrp to sites 1 and 2 is essential for Lrp-promoted phase switching (60), whereas binding of Lrp to site 3 inhibits recombination (151). Addition of leucine specifically decreases the affinity of Lrp for site 3, consistent with the enhancement of *fim* switching by leucine. The mechanism by which binding of Lrp at sites 1 and 2 enhances recombination is not clear, though it seems likely that Lrp plays a structural role. This could occur by DNA bending, since a dimer of Lrp has been reported to bend DNA 130° (19).

H-NS

In a search for genes that affect type 1 fimbrial switching, Kawula and Orndorff found that mutations in the gene coding for H-NS, a 16-kDa abundant DNA-binding protein (2), increased type 1 fimbrial inversion up to 100-fold (95). H-NS specifically binds within the regulatory regions of the *fimB* gene (38) as well as the 3′ end of *fimE* encompassing IRL (157) (Fig. 1). H-NS has a fivefold repressive effect on *fimB* transcription in vivo and specifically represses *fimB* transcription in vitro (38). However, the increased level of FimB in an *hns* null isolate was not enough to account for the high switch frequencies observed in the absence of H-NS. Furthermore, bacteria containing the *hns-1* mutation (95) (a single base-pair mutation in the promoter that reduces the H-NS levels in the cell to approximately one-half that of the wild-type levels) result in an aberrantly high switch rate, yet *fimB* expression was unaffected.

These data suggest that H-NS may play a direct role in *fim* switching. This could occur via interactions of H-NS with *fim* DNA or with regulatory proteins such as IHF, Lrp, or an unidentified factor. Recently, mutational analysis of *hns* showed that the C-terminal DNA-binding domain of H-NS, although involved in regulation of *fimB* expression, was not required for the effects of H-NS on *fim* switching (37). Thus, it seems likely that H-NS binds to at least one regulatory protein to regulate *fim* recombination. This hypothesis was tested by measuring the effects of H-NS on *fim* switching in vitro (134) with a system developed previously (59). Although it was concluded that H-NS does not directly affect *fim* switch frequency, it was not determined whether Lrp and IHF still played roles in the recombination process observed in vitro. In fact, under conditions of FimB overexpression in vivo, *fim* recombination occurs in the absence of Lrp and IHF (40). Thus, the high levels of FimB present in the in vitro assay might alter the recombination process to become independent of Lrp and IHF as well. Further work needs to be done to test this hypothesis under in vitro conditions in which Lrp and IHF influence *fim* switching.

Temperature Response

H-NS represses *fimE* transcription more than *fimB* transcription as temperature is increased from 30 to 37°C. This could account for the observed 10-fold increase in the fraction of phase on cells at 37°C compared to 28°C, since on to off switching would be preferentially downregulated at 37°C (58). Since FimE-mediated switching occurs at much higher rates than that of FimB, relatively small changes in FimE activity can have large effects on phase switching.

Integration Host Factor

IHF is a small heterodimeric protein that binds to the DNA minor groove (148) with a consensus of 5′-AATCAANNNANTTA-3′ (65), bends DNA by more than 140° (98), and plays important roles in regulating biological processes such as site-specific recombination, DNA replication, and gene expression (54). Initial studies showed that IHF regulates type 1 pilus switching (39, 47). Since the *E. coli* isolate CSH50 used in this work was later shown to contain an insertion sequence mutation within *fimE* (12), further work was carried out to assess the role of IHF in FimE-dependent *fim* switching. Introduction of mutations in the *ihfA* and *ihfB* genes (187) coding for IHF reduced *fim* switch rates by 100-fold for FimB- and 15,000-fold for FimE-controlled switching (11). IHF was shown to bind to two DNA sites, IHF site 1 just outside of IRL and IHF site 2 internal to the *fim* switch (Fig. 1). Mutation of the IHF-binding sites showed that FimB-promoted switching was reduced about 100- to 200-fold, similar to the reduction in switch frequency observed in the absence of IHF. In contrast, FimE-promoted recombination was reduced 15,000-fold in the absence of IHF but only 300-fold in cells containing mutations in IHF-binding sites 1 and 2. These results indicate either that an additional unidentified IHF site is important for FimE-promoted switching or that a majority of the effect of IHF is indirect, such as the regulation of other proteins playing roles in *fim* recombination.

Other Regulatory Factors

A number of other regulatory factors have been reported to affect type 1 pilus expression. Analysis of UPEC isolates showed that the majority of isolates shut off type 1 pilus expression when grown on a solid agar surface, whereas they underwent phase variation in liquid medium (83). It was also reported that cell extracts from cells grown in broth retarded the migration of a DNA fragment containing the *fimB* promoter, whereas extracts from cells grown on agar did not (160). It was hypothesized that some factor(s), for example, an inducer of *fimB* transcription, is present in broth- but not agar-grown cells.

The stationary-phase sigma factor RpoS appears to repress transcription at the *fimA* and *fimB* promoters (42). These results suggest that type 1 pilus expression is maximal during logarithmic growth and is reduced upon entry of cells into stationary phase. However, the effect of such regulation may be counteracted by the fact that the growth of a bacterial culture incubated without agitation (standing culture) results in the selective outgrowth of type 1 piliated cells (46). This occurs by formation of a pellicle at the liquid-air interface because of the hydrophobic nature of the pilus, giving bacteria that express type 1 pili a growth advantage. Thus, the physiologic relevance of stationary-phase repression of type 1 piliation is not clear.

Regulation of type 1 pilus expression also occurs at the translational level via the *leuX* gene coding for tRNA^{Leu5} (150). LeuX has been shown to be required for expression of type 1 pili in addition to other cellular properties such as motility and serum resistance (149). Results showed that LeuX operated at the level of FimB translation, since replacement of the five TTG codons recognized by LeuX with CTG codons recognized by LeuV resulted in synthesis of type 1 pili in the absence of LeuX. It was reasoned that *fimE* was not significantly affected by LeuX because of the presence of only two TTG codons. It is possible that this type of regulation based on codon usage might be involved in the environmental control of type 1 pilus expression, since regulation of LeuX levels could alter type 1 piliation.

Glucose was reported to repress type 1 pilus expression when *E. coli* isolate CSH50 (*fimE$^-$*) was grown in static culture but not in shaking broth culture. Eisenstein and Dodd (46) showed that this effect of glucose was not mediated by the catabolite gene activator protein (CAP), but was due to the prevention of outgrowth of piliated cells in static culture that normally occurs as a result of pellicle formation as discussed above.

MR/P Pili

Proteus mirabilis expresses pili denoted as mannose-resistant/Proteus-like (MR/P) by a mechanism that shares common features with type 1 pilus expression in *E. coli* (196). A putative site-specific recombinase, MrpI, sharing sequence identity with FimB and FimE appears to regulate the inversion of a 252-bp DNA element containing the *mrp* promoter. In the on position the *mrp* promoter drives expression of genes, including *mrpA*, the structural gene for the pilus. Expression of MR/P pili is inhibited by aeration and by growth on agar (4). Recent evidence suggests that there is an inverse relationship between MR/P pilus expression and flagellation (102), which might involve the Lrp global regulator (74). Such a mechanism is appealing since it would prevent useless expression of flagella when cells are bound to a surface and are thus prevented from moving.

METHYLATION-CONTROLLED FIMBRIAL OPERONS

The *pap* Paradigm

The *pap* operon present in many UPEC strains (130, 141) encodes proteins for the regulation and synthesis of Pap pili on the bacterial surface. These pili mediate the attachment of *E. coli* to uroepithelial cells in the human host. The expression of Pap pili is subject to phase variation in which transcription of the *pap* operon and expression of Pap pili from individual bacteria occur (on phase) or do not occur (off phase). The transcription from the *pBA* and *pI* promoters (Fig. 2) is sensitive to environmental signals such as temperature and carbon source (see below). The synthesis of Pap pili was also recently suggested to be coupled to the CpxAR two-component pathway of stress response in *E. coli* (145). The *pap* operon encodes nine proteins required for expression and assembly of Pap pili, together with two transcription regulatory proteins, PapI and PapB (82, 126, 127).

Methylation by Deoxyadenosine Methylase (Dam) Regulates *pap* Transcription

The *pap* operon is regulated by the 32-kDa *E. coli* deoxyadenosine methylase (Dam) (5), which is required for the formation of specific DNA methylation patterns at two target GATC sequences located in the regulatory region between the *papI* and *papBA* promoter sequences. In phase on cells the GATC site proximal to the *papBA* promoter (GATC^{prox}) is methylated whereas GATC^{dist}, located about 100-bp upstream, is not methylated (Fig. 2). In phase off cells the methylation pattern is reversed: GATC^{dist} is methylated and GATC^{prox} is nonmethylated (13). *pap* DNA methylation patterns directly regulate *pap* transcription (17) by affecting protein-DNA binding (see below). Dam levels are thus critical for Pap expression, as evidenced by the fact that overexpression and underexpression of Dam block Pap pilus production (13).

Factors That Regulate *pap* DNA Methylation Patterns

The formation of *pap* DNA methylation patterns by Dam is dependent on a complex interplay between global regulatory proteins and *pap*-encoded proteins. One central player, the 19-kDa leucine-responsive regulatory protein (Lrp) (19, 29, 124), binds cooperatively to two sets of *pap* DNA sites designated sites 1, 2, and 3 overlapping GATC^{prox} and sites 4, 5, and 6 overlapping GATC^{dist}. All six binding sites in the

A.

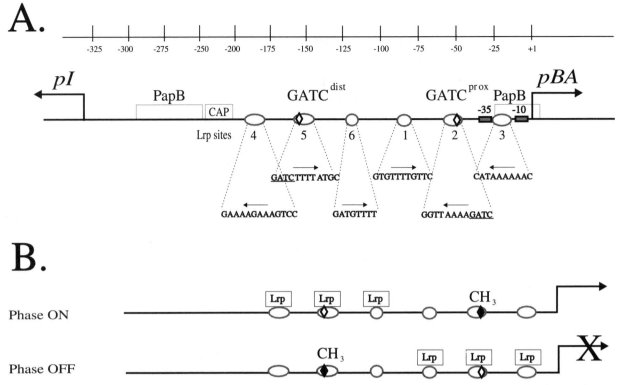

Figure 2. Regulatory sequences of the *E. coli pap* operon. (A) The *pap* regulatory region is shown with the divergently transcribed promoters *pBA* and *pI*. The two GATC sites subject to methylation by *E. coli* Dam are GATC^{prox} and GATC^{dist} (the adenines in GATC are located at −154 and −52 base pairs relative to the *pBA* transcription start site). The Lrp-binding sites are shown as ovals, and the DNA-binding sequences of each Lrp-binding site are depicted (128). The directionality of the Lrp sites (using a consensus sequence 5′-Gnn(n)TTTt-3′) is indicated with arrows above the sequence. PapB-binding sites at −285 to −234 (site 1) and −27 to +6 (site 2) are shown as boxes. The third site in the PapB-coding region at +92 to +125 (site 3) is not shown (53). The CAP-binding site at −215 is also shown. GATC sites are depicted as diamonds. (B) The DNA methylation patterns of phase on and phase off cells are shown with the inferred binding of Lrp to the *pap* regulatory DNA region. Open diamonds represent nonmethylated GATC sites whereas filled diamonds represent methylated GATC sites.

regulatory region of *pap* share the consensus sequence GN$_{2-3}$TTT (128) (see Fig. 2).

Lrp is essential for activation as well as repression of *pap* (180). Activation of *pap* transcription requires binding of Lrp at *pap* sites 4, 5, and 6 (128) and methylation of the GATC^{prox} site. Conversely, repression of transcription requires binding of Lrp at sites 1, 2, and 3 and methylation of the GATC^{dist} site (17). Lrp is required for *pap* transcription (17), but unlike many other Lrp-regulated operons, leucine does not alter *pap* expression (18). Two of the six Lrp-binding sites directly overlap the regulatory GATC sequences. Binding of Lrp at sites 1, 2, and 3 blocks methylation of GATC^{prox} whereas binding of Lrp at sites 4, 5, and 6 blocks methylation of GATC^{dist} (Fig. 2) (129). Lrp binds to these sites in vitro as well as in vivo, blocking Dam methylation of the GATC sites (175).

In conjunction with Lrp, the PapI coregulatory protein (8.8 kDa) is required to form the on-phase *pap* DNA methylation pattern and to activate *pap* transcription. The PapI protein acts by increasing the affinity of Lrp for *pap* sites 4, 5, and 6, thus favoring

the phase on DNA methylation pattern. PapI expression, in turn, is regulated by the global regulator CAP. The expression of PapI (and therefore Pap pili) is sensitive to the cAMP levels in the cell, and growth in media containing poor carbon sources leads to high intracellular cAMP levels, expression of PapI, and increased switching to the phase on state. In contrast, growth of cells in glucose reduces the rate of switching to the on phase by over 60-fold, presumably as a result of lowered cAMP levels, which reduces the PapI level (14). cAMP-CAP is required for transcription of the *papBA* promoter as well as the *papI* promoter (66).

The 11-kDa PapB protein (3) indirectly activates *pap* transcription by inducing PapI transcription via the *papI* promoter. PapB binds to *pap* DNA as an oligomer in the minor groove (194, 195). There are three binding sites for PapB in the intergenic *papI-B* region (see Fig. 2), a high-affinity site near the *papI* promoter and lower-affinity sites overlapping the *papBA* promoter and within the coding region for PapB (53). It was hypothesized that this arrangement of *papB*-binding sites results in autoregulation since high levels of PapB repress transcription because of the occupancy of the low-affinity site overlapping the *papBA* promoter (53).

The global regulator protein H-NS (15.5 kDa) was first implicated in the regulation of *pap* gene expression by Uhlin and coworkers (67). A mutation in a gene designated as *drdX*, shown later to be identical to *hns*, blocked the normal silencing of Pap pilus expression in response to a temperature downshift. H-NS binds specifically within the *pap* regulatory region and blocks methylation of the *pap* GATC sites at low temperature ($<26°C$) (189). Although the PapI and PapB regulatory proteins are not required for thermoregulation, DNA sequences within the *papB* gene are required (189).

Biochemical Mechanisms That Control Pap Phase Variation

In phase off cells, the GATC^{prox} located within *pap* sites 1, 2, and 3 is protected from methylation, which requires Lrp (13, 16). Lrp binds in vitro to promoter proximal *pap* sites 1, 2, and 3 with higher affinity than to sites 4, 5, and 6, which are distal to the *papBA* transcription start site. Binding of Lrp to sites 1, 2, and 3 blocks transcription from *papBA* in vitro (189), consistent with the hypothesis that in phase off cells Lrp binds to *pap* sites 1, 2, and 3, blocking methylation of GATC^{prox}.

PapI is essential for the movement of Lrp from *pap* sites 1, 2, and 3 to sites 4, 5, and 6 and for protection of the GATC^{dist} sequence from methylation (90, 129). PapI binds to Lrp-*pap* complexes (90), increasing the affinity of Lrp for sites 4, 5, and 6 by reducing dissociation of Lrp from sites 4, 5, 6 (M. Krabbe, unpublished data).

The frequency of switching from off to on is low even under optimal laboratory conditions (about 1/10,000 per cell per generation). PapI levels regulate the formation of the phase on methylation state, and thus any factor that controls PapI expression should also control *pap* switch rates. A case in point is cAMP, which binds to CAP and activates *papI* transcription (66). A mutation that was found to upregulate cAMP levels in cells resulted in a higher frequency of off to on switching (L. Kaltenbach and D. Low, unpublished data).

The on-to-off switch rate of about 3/100 per cell per generation is about 100-fold higher than the off to on rate (14). Thus, Pap pilus expression is biased toward the off phase, which has also been reported for type 1 pili (12, 58). Analysis of different *pap* operons has shown that there is variability in Pap pilus switch frequencies. For example, the *pap$_{21}$* operon of UPEC strain C1212 (97) has an on-to-off switch fre-

quency of about 3/1,000 per cell per generation, significantly lower than that for *pap$_{17}$* reported above. The mechanism by which different Pap-switch frequencies are achieved is unknown but likely involves alterations in the binding sites for regulatory proteins such as Lrp.

Thermoregulation of *pap* Expression

The transcription of the Pap pilus operon is repressed at temperatures below 26°C (14, 68, 190). This thermoregulation is dependent on H-NS and independent of PapI and PapB (67, 189). In cells grown at 23°C both *pap* regulatory GATC sequences were protected from methylation in an H-NS-dependent manner. H-NS specifically bound to the regulatory region of *pap* and protected both *pap* GATC sites from methylation by Dam (189). However, it has not yet been determined whether the DNA-binding domain of *hns* is required for thermoregulation.

Total repression of *pap* transcription occurs within one cell generation following a shift from 37 to 23°C, yet at this point 20% of cells still have a phase on DNA methylation pattern, indicating that Lrp is bound at GATC^{dist}. Thus, H-NS appears to block *papBA* transcription in phase on cells and does not simply inhibit binding of Lrp at GATC^{dist} and formation of the phase on DNA methylation pattern (189).

The fact that H-NS represses *pap* transcription at 23°C but has a positive effect on the off-to-on switch frequency at 37°C suggests that either the level of H-NS or the DNA- or protein-binding properties of H-NS are altered in response to a temperature downshift. H-NS expression is induced at low temperature (10°C) (88, 101), but the level of H-NS protein at 23 and 37°C is similar (189). There are indications that H-NS is present in different isoforms in the cell (37, 174). It is not known whether the relative levels of H-NS isoforms are different at 23 and 37°C. It is also possible that H-NS has a higher affinity for *pap* DNA and/or a coregulatory protein at 23°C compared to 37°C, as was recently hypothesized for regulation of *Shigella* virulence genes by H-NS (50). Such an alteration in the affinity of H-NS for DNA could result from temperature-dependent alterations in regions of bent DNA that were reported to occur in the *Yersinia* pYV plasmid by S. Minnich and coworkers (152).

Other Pilus Operons within the Pap Family

Comparison of the *pap* regulatory DNA region with other genes in the database shows that 16 pilus operons share GATC sequences that are spaced 102 to 103 base pairs apart, and most contain *papI* and *papB* homologs (Fig. 3A). Additional DNA sequences around the GATC^{prox} and GATC^{dist} sites are identical and comprise "GATC boxes." The GATC box sequence is defined here as "CGATCTTTT" (177). The two GATC boxes are in opposite orientations (Fig. 3A). Notably, many of these operons appear to share conserved regions within the CAP-binding site as well as putative Lrp-binding sites in similar positions to those of *pap* (Fig. 3B). Both *pef* and *fae* are thermoregulated, and neither operon is induced by leucine or alanine.

The *pef* and *fae* operons, though sharing important features with *pap*, have distinct regulatory features (Fig. 3A). In contrast to *pap*, the PapI homologs FaeA and PefI negatively regulate *fae* and *pef* transcription, respectively. PapI also represses *pef* and *fae* transcription, indicating that the regulatory difference between these operons is likely due to *cis*-acting sites within the regulatory DNAs of these operons. For example, both *fae* and *pef* contain an additional GATC near GATC^{prox}. In contrast to Fae and Pap pili, which are expressed well in laboratory media, Pef pili are not (125). Induction of Pef pilus expression occurs only at low pH (optimal pH of 5.1), with only a fraction of cells expressing Pef pili (125). The mechanism by which the pH signal is integrated into the Dam- and Lrp-dependent genetic switch is not known.

The *daa* and *sfa* operons encoding F1845 and S pili, respectively, were shown to be under Dam and Lrp control similar to *pap* (177). In addition to sharing GATC box motifs as well as PapB and PapI homologs, they share sequence similarity with *pap* at the CAP-binding site (Fig. 3A), suggesting that both *daa* and *sfa* are under catabolite repression control. Recent data indicate that Daa expression is also translationally regulated (107). Like *pap*, neither *daa* nor *sfa* is regulated by leucine or alanine.

The PrfI and PrfB genes (*papI* and *papB* homologs) of P-related fimbriae (*prf*, see Fig. 3A) were shown to complement mutations in the *sfaB* and *sfaC* regulatory genes both in vivo and in vitro (118). These results indicate that *prf* pili are under methylation-dependent control similar to *sfa*. Pili related to S and F1C, known as Sfr pili ([S and F1C-related [140]]), have been described and are likely within the family of methylation-dependent pilus operons (not shown in Fig. 3A).

Clp (CS31A), F165$_1$, and F165$_2$ pilus operons share common features, including conserved GATC box motifs and a PapB homolog (Fig. 3A). All three operons are repressed by the aliphatic amino acids leucine and alanine (28, 110). Thus, it is likely that Clp, F165$_1$, and F165$_2$ pilus operons share some

regulatory feature that enhances their sensitivity to aliphatic amino acids, in contrast with other methylation-controlled operons such as *pap*. R. Matthews and coworkers hypothesized that the sensitivity to leucine may occur when the effective intracellular concentration of Lrp is low compared to the affinity of Lrp for regulatory DNA-binding sites (49). In addition, F165$_1$ pili are also repressed by glucose and low temperature (28), which are likely to be mediated by cAMP-CAP (via a conserved CAP-binding site within F165$_1$) (Fig. 3A) and H-NS (which has been shown to be required for *pap* thermoregulation) (67, 189), respectively.

Similar to Clp, F165$_1$, and F165$_2$, the K99 pilus (*fan* operon) is also repressed by leucine and alanine (30, 86). K99 pili are regulated by Lrp, contain a *papB* homolog but lack a *papI* homolog, contain conserved GATC box motifs, do not undergo phase variation, and are not under Dam control (18). Thus, although *fan* is regulated by Lrp, it is not a member of the Dam-controlled pilus operon family.

F17 pili are expressed by enterotoxigenic *E. coli* (ETEC) and invasive *E. coli* (48, 105) that cause diarrhea and septicemia in domestic animals. F17 pili share common features with Pap and type 1 pili (104). Analysis of the regulatory regions of F17a and F17d pili shows that there is some similarity to the conserved GATC box motifs of Dam-controlled operons, but the GATC^{dist} sequence is present at the position normally occupied by GATC^{prox} (Fig. 3A). These results suggest the possibility that F17 pili may be under Dam control. It is not known whether the F17 pilus operons contain PapI or PapB homologs.

Other Methylation-Controlled Pilus Operons

Elegant work carried out in the laboratory of J. Casadesus has shown that the Dam methylase inhibits the expression of both the F conjugation pilus in *E. coli* and pili encoded by the pSLT virulence plasmid in *Salmonella enterica* serovar Typhimurium (172, 173). In the absence of Dam, F plasmid pilus expression increases significantly. Thus, Dam controls bacterial mating. Dam appears to enhance expression of the *finP* antisense RNA coded for by both pSLT and the F plasmid, which, in turn, represses pilus expression. Torreblanca and coworkers hypothesized that methylation of a GATC sequence within the −10 promoter hexamer site of the *finP* promoter enhances binding of RNA polymerase or inhibits binding of a repressor (173). These results suggest that environmental factors could control F and pSLT pilus expression via regulation of repressor expression. Alternatively, Dam levels might also be under regulatory control, which could alter expression of F

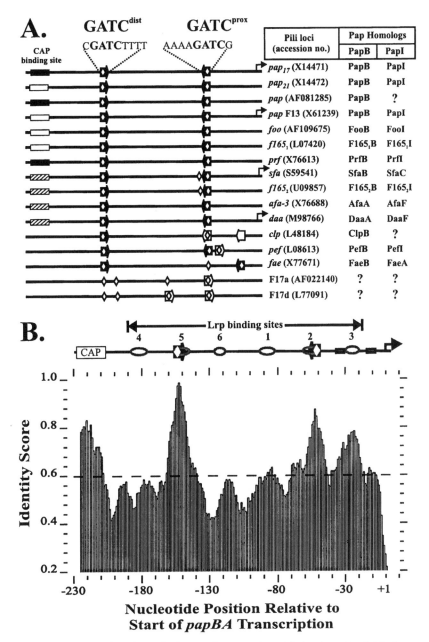

Figure 3. DNA sequence alignment and analysis of pilus regulatory regions. (A) A schematic of 16 pilus operons that have similarly spaced GATC sequences (targets for Dam methylation) is presented. All GATC sequences within the pilus operons are depicted by diamonds. Arrows pointing right depict the GATC box containing GATC^{dist} (sequence 5′-CGATCTTTT-3′) while left-pointing arrows represent the inverted GATC box containing GATC^{prox} (sequence 5′-AAAAGATCG-3′). Black arrows represent exact matches to the pap_{17} GATC boxes, while open arrows represent sequences that contain a single mismatch. The pap_{17} CAP-binding site (5′-TTTATTTGATGTGTATCACA-3′) is depicted by a black rectangle as are identical CAP-binding sites in other pilus operons. CAP sites containing a single mismatch to the pap_{17} CAP site appear as open rectangles, while sites with fewer than 5 mismatches out of 20 appear as hatched rectangles. The start sites of transcription for pap_{17}, pap F13, sfa, and daa are depicted with small right arrows. The pilus loci (followed by accession numbers) and operon-encoded homologs of PapB and PapI are listed at the right of the figure. A "?" indicates that the presence of a particular homolog is not known. (B) The pilus operon regulatory regions (~ 230 bp) depicted in panel A (with the exception of F17a and F17d) were aligned by using PileUp from Genetics Computer Group (GCG), Inc., Madison, Wis. PileUp uses a simplification of the progressive alignment method of Feng and Doolittle (51). The multiple sequence alignment created by the PileUp program was used as input to PlotSimilarity (GCG). A bar graph of the average identity of windows of 10 nucleotides was plotted at the middle position of each window. The overall average identity of all 14 sequences was 60%, as shown by the dotted horizontal line. The strongest regions of identity in the pilus operons occur around the GATC^{dist} and GATC^{prox} sites. The next highest regions of identity include the CAP-binding site (rectangle), Lrp-binding site 3 (Lrp-binding sites are numbered gray ovals), and the −35 and −10 regions of the pBA promoter (gray filled boxes). The peaks are aligned with corresponding sequence features of the pap_{17} operon, shown at the top.

and pSLT pili as well as the *pap* pilus family (Fig. 3A). Work by Marinus and others has shown that *dam* has at least five promoters and is regulated by growth rate (106, 146, 147). Recent work showing that Dam is required for the pathogenesis of *Salmonella* supports the hypothesis that Dam expression may be controlled by environmental factors (61, 75).

Finally, the outer membrane protein Ag43, which facilitates flocculation caused by autoaggregation of bacteria, is under phase-variation control by a mechanism involving DNA methylation. Ag43 is negatively regulated by OxyR, an important stress response protein, and positively by Dam (73, 76–78, 109). The mechanism by which Ag43 is regulated is distinct from that of *pap* since there is no conservation of GATC box motifs with 102- to 103-bp spacing, PapI homolog, or control by Lrp. Instead, methylation of GATC sites in the *agg* regulatory region appears to control binding of OxyR.

The Curli Family

Curli and thin aggregative fimbriae are homologous fibers expressed by *E. coli* and *Salmonella* spp., respectively. The divergently transcribed operons are denoted *agfDEFG* and *agfBA(C)* in serovar Typhimurium and *csgDEFG* and *csgBA(C)* in *E. coli*. In addition, *Salmonella enteritidis* expresses thin aggregative fimbriae (Sef-17) (26). The regulator CsgD (transcriptional regulator belonging to the LuxR family) represses transcription of the *csgBA* operon in *E. coli* (70). Serovar Typhimurium curli fibers are optimally expressed under conditions of low temperature, stationary-phase growth, and low osmolarity (154). Both RpoS and the transcriptional regulator OmpR are required for fimbrial expression. However, mutations in *hns* relieve the requirement for RpoS (131, 132). Although pilus expression is normally shut off at 37°C on laboratory media, pili are expressed under iron starvation conditions. The mechanisms regulating expression of the Curli family have not been reported.

COLONIZATION FACTOR ANTIGENS

ETEC expresses more than 20 distinct pili that were originally designated colonization factor antigens (CFAs), including CS1, CS2, and CFA/I (155). The amino acid sequences of these pili are highly conserved, but they have different antigenic properties and likely bind to different receptor structures on host cells (155).

CS1 and CS2 pili are regulated by the Rns transcription regulator, a member of the AraC family that positively regulates pilin expression (103, 155). Rns is 95% identical to CfaR (also known as CfaD), which regulates the related CFA/I pili expression. Rns complements mutations in *cfaR* (22) as well as *virF*, which regulates *Shigella* invasion gene expression (142). Notably, *virF* does not complement *rns*, indicating that there is some difference in mechanisms by which Rns and VirF activate transcription. CfaR, however, complements *rns* null mutants in CS1 pili as well as a mutation in *aggR* coding for a regulator of aggregative adhesive fimbriae (AAF/I) (22, 123). Rns shares the highest sequence identity with AraC in the carboxy-terminal DNA binding domain. The amino terminus of AraC, required for dimerization, appears to be an effector site for small molecules such as arabinose, which binds to AraC (57, 165). The UreR regulator in the *rns* family responds to urea, but otherwise no effectors are known for any *rns*-like regulator (57). Two binding sites for Rns were identified in the *coo* promoter for CS1 pili, at −112 and −44 relative to the transcriptional start site (120). Mutation analysis indicated that Rns activates transcription directly and that both sites identified in vitro were important in vivo.

Thermoregulation

CFA/I fimbriae of ETEC strains are thermoregulated, with pili expressed at 37°C but not at 20°C. H-NS negatively regulates both CFA/I and CS1 pilus expression and is required for the thermoregulatory phenotype (89). The CfaD positive regulator was not required for CFA/I expression in an H-NS⁻ strain, indicating that CfaD antagonizes the inhibitory effect of H-NS. The negative effect of H-NS on both *coo* and CFA/I transcription requires promoter regulatory DNA sequences as well as DNA within downstream DNA-coding regions (89, 121) similar to H-NS-mediated repression of *proU* (52, 136).

Certain ETEC strains express 987P pili (122). The AraC-like regulator FasH (FapR) is an activator of 987P pilus expression, is controlled by nitrogen availability and pH, and is positively regulated by cAMP-CAP and ammonium (44). 987P pilus expression is also thermoregulated and osmoregulated by H-NS. The activation of 987P pilus expression by ammonium correlates with the colonization by porcine ETEC strains of the distal small intestine, which contains high ammonium levels (44).

TYPE IV PILI

Type IV pili are a diverse group of adhesins that share common features such as conserved leader peptides and N-terminal sequences (112, 166). They are divided into two classes. Class A pili contain an N-methylated phenylalanine (N-CH$_3$Phe) at the amino terminus and are expressed by *Neisseria* spp., *Pseudomonas aeruginosa*, and *Myxococcus xanthus*, among others. Class B type IV pili expressed by enteric pathogens, including *Vibrio cholerae* and diarrheagenic *E. coli* strains, lack N-CH$_3$Phe. Ishimoto and Lory showed that *P. aeruginosa* type IV pili are transcribed by RNA polymerase core containing sigma factor RpoN (σ^{54}) (87). Transcription is regulated by a classic two-component regulatory system: PilS appears to be the sensor, phosphorylating and activating PilR, a σ^{54}-dependent transcription activator that binds upstream of the pilin promoter (15, 166). Type IV pilus expression in *Neisseria gonorrhoeae* and *Neisseria meningitidis* is also regulated by a two-component regulatory system (PilA-PilB) (168, 170). However, it is not clear whether RpoN plays a role in transcription despite the fact that a conserved σ^{54} promoter is present (23, 56). The expression of these pili has been shown to vary in response to a number of environmental signals (100) and appears to involve the PilA-PilB proteins. Another protein, RegF, a homolog of the *E. coli* stringent starvation protein (SspA), was shown to repress transcription of the *pilE* pilin gene. PilC, which is involved in assembly of the gonococcal pilus, was reported to be induced by oxygen (32). Type IV pili expressed by *M. xanthus*, controlled by the two-component PilR-PilS system, are induced by high-nutrient conditions and are developmentally regulated (193).

Among type IVB pili are toxin coregulated pili (TCP) expressed by *V. cholerae* (55, 72, 138), bundle-forming pili (BFP) expressed by enteropathogenic *E. coli* (EPEC) strains (63, 91), and long pilus (Longus) of ETEC (64). TCP are regulated indirectly by the global regulatory protein ToxR, which is responsive to temperature, osmolarity, pH, and cAMP. ToxR, in turn, positively regulates ToxT, an AraC-like protein that directly activates TCP (24, 35, 72, 94, 138, 163, 164). ToxT is temperature controlled and negatively regulated by bile (159). The reader is referred to excellent reviews for additional information on TCP regulation (34, 114, 162).

Bundle-forming pili of EPEC are controlled by ammonium level, growth medium, calcium, and temperature (144). The BfpA pilin is positively regulated by BfpT, an AraC-like regulatory protein. The mechanism by which ammonium represses BFP expression is not known but requires a factor present in EPEC strains but absent in *E. coli* K-12. The ammonium response appears to be physiologically relevant since EPEC colonizes the proximal small intestine, which has a lower ammonium concentration than the distal small intestine (144).

TCP mediates interbacterial adherence and function as an essential colonization factor (79, 171). TCP also acts as a receptor for the CTXΦ phage (185), which encodes cholera toxin. A *V. cholerae* pathogenicity island (VPI) was shown to contain a second phage, VPIΦ (92), encoding TCP. The *tcpA* gene appears to encode a coat protein for this VPIΦ phage. These exciting results suggest a possible origin of type IV pili as coat proteins for bacteriophage.

PERSPECTIVES

In this review we have focused on pilus regulatory systems for which we have some knowledge of environmental regulatory factors. This is not an all-encompassing review; there are many additional pilus regulatory systems that we have not discussed, including those outside of the gram-negative eubacteria. Indeed, since many pilus operons are only induced under specific environmental conditions, it is highly likely that many pilus systems remain undiscovered.

It is clear that many if not all pilus regulatory systems are connected to important environmental response pathways and that pilus expression is tightly coregulated with many other cellular functions. For example, many pilus operons are regulated by H-NS, which functions as a thermostat to shut off pilus expression in response to temperature and regulates many other nonpilus genes as well. It seems that we have only scratched the surface with regard to understanding the biochemical mechanisms by which pilus expression is regulated. Important unanswered questions related to environmental response include the following. (i) Do pilus operons controlled by AraC-like regulators respond to small molecules analogous to arabinose? (ii) What signals control many of the two-component regulatory systems that have been identified? (iii) What are the true signals to which bacteria respond in their environments, and how are multiple signal inputs integrated into pilus expression output? (iv) Why are many pilus operons regulated by phase variation? (v) What signals alter on-off switch rates, and what role(s) does phase variation play in the colonization and spread of bacteria?

Acknowledgments. We thank the National Institutes of Health (grant AI23348 to D.L.) and the National Science Foundation (grant MCB-9305166) for support of work on pilus gene expression and methylation-dependent gene regulation in our laboratory. Margareta Krabbe was supported by a postdoctoral grant from the Swedish Foundation for International Cooperation in Research and Higher Education (STINT).

We extend special thanks to M. Mahan (University of California, Santa Barbara) for helpful discussion and comments on the manuscript.

REFERENCES

1. **Arnqvist, A., A. Olsen, and S. Normark.** 1994. Sigma S-dependent growth-phase induction of the *csgBA* promoter in *Escherichia coli* can be achieved in vivo by sigma 70 in the absence of the nucleoid-associated protein H-NS. *Mol. Microbiol.* **13:**1021–1032.

2. **Atlung, T., and H. Ingmer.** 1997. H-NS: a modulator of environmentally regulated gene expression. *Mol. Microbiol.* **24:**7–17.

3. **Baga, M., M. Goransson, S. Normark, and B. E. Uhlin.** 1985. Transcriptional activation of a *pap* pilus virulence operon from uropathogenic *Escherichia coli*. *EMBO J.* **4:**3887–3893.

4. **Bahrani, F. K., G. Massad, C. V. Lockatell, D. E. Johnson, R. G. Russell, J. W. Warren, and H. L. Mobley.** 1994. Construction of an MR/P fimbrial mutant of *Proteus mirabilis*: role in virulence in a mouse model of ascending urinary tract infection. *Infect. Immun.* **62:**3363–3371.

5. **Barras, F., and M. G. Marinus.** 1989. The great GATC: DNA methylation in *E. coli*. *Trends Genet.* **5:**139–143.

6. **Bhaya, D., N. Watanabe, T. Ogawa, and A. R. Grossman.** 1999. The role of an alternative sigma factor in motility and pilus formation in the cyanobacterium *Synechocystis* sp. strain PCC6803. *Proc. Natl. Acad. Sci. USA* **96:**3188–3193.

7. **Bilge, S. S., J. M. Apostol, Jr., K. J. Fullner, and S. L. Moseley.** 1993. Transcriptional organization of the F1845 fimbrial adhesin determinant of *Escherichia coli*. *Mol. Microbiol.* **7:**993–1006.

8. **Bloch, C. A., and P. E. Orndorff.** 1990. Impaired colonization by and full invasiveness of *Escherichia coli* K1 bearing a site-directed mutation in the type 1 pilin gene. *Infect. Immun.* **58:**275–278.

9. **Blomberg, L., and P. L. Conway.** 1991. Influence of raffinose on the relative synthesis rate of K88 fimbriae and the adhesive capacity of *Escherichia coli* K88. *Microb. Pathog.* **11:**143–147.

10. **Blomfield, I. C., P. J. Calie, K. J. Eberhardt, M. S. McClain, and B. I. Eisenstein.** 1993. Lrp stimulates phase variation of type 1 fimbriation in *Escherichia coli* K-12. *J. Bacteriol.* **175:**27–36.

11. **Blomfield, I. C., D. H. Kulasekara, and B. I. Eisenstein.** 1997. Integration host factor stimulates both FimB- and FimE-mediated site-specific DNA inversion that controls phase variation of type 1 fimbriae expression in *Escherichia coli*. *Mol. Microbiol.* **23:**705–717.

12. **Blomfield, I. C., M. S. McClain, J. A. Princ, P. J. Calie, and B. I. Eisenstein.** 1991. Type 1 fimbriation and *fimE* mutants of *Escherichia coli* K-12. *J. Bacteriol.* **173:**5298–5307.

13. **Blyn, L. B., B. A. Braaten, and D. A. Low.** 1990. Regulation of *pap* pilin phase variation by a mechanism involving differential dam methylation states. *EMBO J.* **9:**4045–4054.

14. **Blyn, L. B., B. A. Braaten, C. A. White-Ziegler, D. H. Rolfson, and D. A. Low.** 1989. Phase-variation of pyelonephritis-associated pili in *Escherichia coli*: evidence for transcriptional regulation. *EMBO J.* **8:**613–620.

15. **Boyd, J. M., and S. Lory.** 1996. Dual function of PilS during transcriptional activation of the *Pseudomonas aeruginosa* pilin subunit gene. *J. Bacteriol.* **178:**831–839.

16. **Braaten, B. A., L. B. Blyn, B. S. Skinner, and D. A. Low.** 1991. Evidence for a methylation-blocking factor (*mbf*) locus involved in *pap* pilus expression and phase variation in *Escherichia coli*. *J. Bacteriol.* **173:**1789–1800.

17. **Braaten, B. A., X. Nou, L. S. Kaltenbach, and D. A. Low.** 1994. Methylation patterns in *pap* regulatory DNA control pyelonephritis-associated pili phase variation in *E. coli*. *Cell* **76:**577–588.

18. **Braaten, B. A., J. V. Platko, M. W. van der Woude, B. H. Simons, F. K. de Graaf, J. M. Calvo, and D. A. Low.** 1992. Leucine-responsive regulatory protein controls the expression of both the *pap* and *fan* pili operons in *Escherichia coli*. *Proc. Natl. Acad. Sci. USA* **89:**4250–4254.

19. **Calvo, J. M., and R. G. Matthews.** 1994. The leucine-responsive regulatory protein, a global regulator of metabolism in *Escherichia coli*. *Microbiol. Rev.* **58:**466–490. (Erratum, **59:**323, 1995.)

20. **Cantey, J. R., R. K. Blake, J. R. Williford, and S. L. Moseley.** 1999. Characterization of the *Escherichia coli* AF/R1 pilus operon: novel genes necessary for transcriptional regulation and for pilus-mediated adherence. *Infect. Immun.* **67:**2292–2298.

21. **Caron, J., L. M. Coffield, and J. R. Scott.** 1989. A plasmid-encoded regulatory gene, *rns*, required for expression of the CS1 and CS2 adhesins of enterotoxigenic *Escherichia coli*. *Proc. Natl. Acad. Sci. USA* **86:**963–967.

22. **Caron, J., and J. R. Scott.** 1990. A *rns*-like regulatory gene for colonization factor antigen I (CFA/I) that controls expression of CFA/I pilin. *Infect. Immun.* **58:**874–878.

23. **Carrick, C. S., J. A. Fyfe, and J. K. Davies.** 1997. The normally silent sigma54 promoters upstream of the *pilE* genes of both *Neisseria gonorrhoeae* and *Neisseria meningitidis* are functional when transferred to *Pseudomonas aeruginosa*. *Gene* **198:**89–97.

24. **Chiang, S. L., R. K. Taylor, M. Koomey, and J. J. Mekalanos.** 1995. Single amino acid substitutions in the N-terminus of *Vibrio cholerae* TcpA affect colonization, autoagglutination, and serum resistance. *Mol. Microbiol.* **17:**1133–1142.

25. **Clegg, S., L. S. Hancox, and K. S. Yeh.** 1996. *Salmonella typhimurium* fimbrial phase variation and FimA expression. *J. Bacteriol.* **178:**542–545.

26. **Collinson, S. K., L. Emody, T. J. Trust, and W. W. Kay.** 1992. Thin aggregative fimbriae from diarrheagenic *Escherichia coli*. *J. Bacteriol.* **174:**4490–4495.

27. **Connell, I., W. Agace, P. Klemm, M. Schembri, S. Marild, and C. Svanborg.** 1996. Type 1 fimbrial expression enhances *Escherichia coli* virulence for the urinary tract. *Proc. Natl. Acad. Sci. USA* **93:**9827–9832.

28. **Daigle, F., C. M. Dozois, M. Jacques, and J. Harel.** 1997. Mutations in the f165(1)A and f165(1)E fimbrial genes and regulation of their expression in an *Escherichia coli* strain causing septicemia in pigs. *Microb. Pathog.* **22:**247–252.

29. **D'Ari, R., R. T. Lin, and E. B. Newman.** 1993. The leucine-responsive regulatory protein: more than a regulator? *Trends Biochem. Sci.* **18:**260–263.

30. **de Graaf, F. K., P. Klaasen-Boor, and J. E. van Hees.** 1980. Biosynthesis of the K99 surface antigen is repressed by alanine. *Infect. Immun.* **30:**125–128.

31. **de Haan, L. A., G. A. Willshaw, B. A. van der Zeijst, and W. Gaastra.** 1991. The nucleotide sequence of a regulatory gene present on a plasmid in an enterotoxigenic *Escherichia coli* strain of serotype O167:H5. *FEMS Microbiol. Lett.* **67:**341–346.

32. **De Reuse, H., and M. K. Taha.** 1997. RegF, an SspA homologue, regulates the expression of the *Neisseria gonorrhoeae* *pilE* gene. *Res. Microbiol.* **148:**289–303.

33. **Dibb-Fuller, M. P., E. Allen-Vercoe, C. J. Thorns, and M. J. Woodward.** 1999. Fimbriae- and flagella-mediated association with and invasion of cultured epithelial cells by *Salmonella enteritidis*. *Microbiology* **145:**1023–1031.

34. **DiRita, V. J.** 1992. Co-ordinate expression of virulence genes by ToxR in *Vibrio cholerae*. *Mol. Microbiol.* **6:**451–458.

35. **DiRita, V. J., M. Neely, R. K. Taylor, and P. M. Bruss.** 1996. Differential expression of the ToxR regulon in classical and E1 Tor biotypes of *Vibrio cholerae* is due to biotype-specific

control over *toxT* expression. *Proc. Natl. Acad. Sci. USA* **93**: 7991–7995.

36. **DiRita, V. J., C. Parsot, G. Jander, and J. J. Mekalanos.** 1991. Regulatory cascade controls virulence in *Vibrio cholerae. Proc. Natl. Acad. Sci. USA* **88**:5403–5407.

37. **Donato, G. M., and T. H. Kawula.** 1999. Phenotypic analysis of random *hns* mutations differentiate [sic] DNA-binding activity from properties of *fimA* promoter inversion modulation and bacterial motility. *J. Bacteriol.* **181**:941–948.

38. **Donato, G. M., M. J. Lelivelt, and T. H. Kawula.** 1997. Promoter-specific repression of *fimB* expression by the *Escherichia coli* nucleoid-associated protein H-NS. *J. Bacteriol.* **179**:6618–6625.

39. **Dorman, C. J., and C. F. Higgins.** 1987. Fimbrial phase variation in *Escherichia coli*: dependence on integration host factor and homologies with other site-specific recombinases. *J. Bacteriol.* **169**:3840–3843.

40. **Dove, S. L., and C. J. Dorman.** 1996. Multicopy *fimB* gene expression in *Escherichia coli*: binding to inverted repeats in vivo, effect on *fimA* gene transcription and DNA inversion. *Mol. Microbiol.* **21**:1161–1173.

41. **Dove, S. L., and C. J. Dorman.** 1994. The site-specific recombination system regulating expression of the type 1 fimbrial subunit gene of *Escherichia coli* is sensitive to changes in DNA supercoiling. *Mol. Microbiol.* **14**:975–988.

42. **Dove, S. L., S. G. Smith, and C. J. Dorman.** 1997. Control of *Escherichia coli* type 1 fimbrial gene expression in stationary phase: a negative role for RpoS. *Mol. Gen. Genet.* **254**:13–20.

43. **Edwards, R. A., L. H. Keller, and D. M. Schifferli.** 1998. Improved allelic exchange vectors and their use to analyze 987P fimbria gene expression. *Gene* **207**:149–157.

44. **Edwards, R. A., and D. M. Schifferli.** 1997. Differential regulation of *fasA* and *fasH* expression of *Escherichia coli* 987P fimbriae by environmental cues. *Mol. Microbiol.* **25**:797–809.

45. **Eisenstein, B. I.** 1981. Phase variation of type 1 fimbriae in *Escherichia coli* is under transcriptional control. *Science* **214**: 337–339.

46. **Eisenstein, B. I., and D. C. Dodd.** 1982. Pseudocatabolite repression of type 1 fimbriae of *Escherichia coli. J. Bacteriol.* **151**:1560–1567.

47. **Eisenstein, B. I., D. S. Sweet, V. Vaughn, and D. I. Friedman.** 1987. Integration host factor is required for the DNA inversion that controls phase variation in *Escherichia coli. Proc. Natl. Acad. Sci. USA* **84**:6506–6510.

48. **el Mazouari, K., E. Oswald, J. P. Hernalsteens, P. Lintermans, and H. De Greve.** 1994. F17-like fimbriae from an invasive *Escherichia coli* strain producing cytotoxic necrotizing factor type 2 toxin. *Infect. Immun.* **62**:2633–2638.

49. **Ernsting, B. R., J. W. Denninger, R. M. Blumenthal, and R. G. Matthews.** 1993. Regulation of the *gltBDF* operon of *Escherichia coli*: how is a leucine-insensitive operon regulated by the leucine-responsive regulatory protein? *J. Bacteriol.* **175**: 7160–7169.

50. **Falconi, M., B. Colonna, G. Prosseda, G. Micheli, and C. O. Gualerzi.** 1998. Thermoregulation of *Shigella* and *Escherichia coli* EIEC pathogenicity. A temperature-dependent structural transition of DNA modulates accessibility of *virF* promoter to transcriptional repressor H-NS. *EMBO J.* **17**:7033–7043.

51. **Feng, D. F., and R. F. Doolittle.** 1987. Progressive sequence alignment as a prerequisite to correct phylogenetic trees. *J. Mol. Evol.* **25**:351–360.

52. **Fletcher, S. A., and L. N. Csonka.** 1995. Fine-structure deletion analysis of the transcriptional silencer of the *proU* operon of *Salmonella typhimurium. J. Bacteriol.* **177**:4508–4513.

53. **Forsman, K., M. Goransson, and B. E. Uhlin.** 1989. Autoregulation and multiple DNA interactions by a transcriptional

regulatory protein in *E. coli* pili biogenesis. *EMBO J.* **8**:1271–1277.

54. **Friedman, D. I.** 1988. Integration host factor: a protein for all reasons. *Cell* **55**:545–554.

55. **Fullner, K. J., and J. J. Mekalanos.** 1999. Genetic characterization of a new type IV-A pilus gene cluster found in both classical and El Tor biotypes of *Vibrio cholerae. Infect. Immun.* **67**:1393–1404.

56. **Fyfe, J. A., C. S. Carrick, and J. K. Davies.** 1995. The *pilE* gene of *Neisseria gonorrhoeae* MS11 is transcribed from a sigma 70 promoter during growth in vitro. *J. Bacteriol.* **177**: 3781–3787.

57. **Gallegos, M.-T., R. Schleif, A. Bairoch, K. Hofmann, and J. L. Ramos.** 1997. AraC/XylS family of transcriptional regulators. *Microbiol. Mol. Biol. Rev.* **61**:393–410.

58. **Gally, D. L., J. A. Bogan, B. I. Eisenstein, and I. C. Blomfield.** 1993. Environmental regulation of the *fim* switch controlling type 1 fimbrial phase variation in *Escherichia coli* K-12: effects of temperature and media. *J. Bacteriol.* **175**:6186–6193.

59. **Gally, D. L., J. Leathart, and I. C. Blomfield.** 1996. Interaction of FimB and FimE with the *fim* switch that controls the phase variation of type 1 fimbriae in *Escherichia coli* K-12. *Mol. Microbiol.* **21**:725–738.

60. **Gally, D. L., T. J. Rucker, and I. C. Blomfield.** 1994. The leucine-responsive regulatory protein binds to the *fim* switch to control phase variation of type 1 fimbrial expression in *Escherichia coli* K-12. *J. Bacteriol.* **176**:5665–5672.

61. **Garcia-Del Portillo, F., M. G. Pucciarelli, and J. Casadesus.** 1999. DNA adenine methylase mutants of *Salmonella typhimurium* show defects in protein secretion, cell invasion, and M cell cytotoxicity. *Proc. Natl. Acad. Sci. USA* **96**:11578–11583.

62. **Giron, J. A., O. G. Gomez-Duarte, K. G. Jarvis, and J. B. Kaper.** 1997. Longus pilus of enterotoxigenic *Escherichia coli* and its relatedness to other type-4 pili—a minireview. *Gene* **192**:39–43.

63. **Giron, J. A., A. S. Ho, and G. K. Schoolnik.** 1991. An inducible bundle-forming pilus of enteropathogenic *Escherichia coli. Science* **254**:710–713.

64. **Giron, J. A., M. M. Levine, and J. B. Kaper.** 1994. Longus: a long pilus ultrastructure produced by human enterotoxigenic *Escherichia coli. Mol. Microbiol.* **12**:71–82.

65. **Goodrich, J. A., M. L. Schwartz, and W. R. McClure.** 1990. Searching for and predicting the activity of sites for DNA binding proteins: compilation and analysis of the binding sites for *Escherichia coli* integration host factor (IHF). *Nucleic Acids Res.* **18**:4993–5000.

66. **Goransson, M., P. Forsman, P. Nilsson, and B. E. Uhlin.** 1989. Upstream activating sequences that are shared by two divergently transcribed operons mediate cAMP-CRP regulation of pilus-adhesin in *Escherichia coli. Mol. Microbiol.* **3**: 1557–1565.

67. **Goransson, M., B. Sonden, P. Nilsson, B. Dagberg, K. Forsman, K. Emanuelsson, and B. E. Uhlin.** 1990. Transcriptional silencing and thermoregulation of gene expression in *Escherichia coli. Nature* **344**:682–685.

68. **Goransson, M., and B. E. Uhlin.** 1984. Environmental temperature regulates transcription of a virulence pili operon in *E. coli. EMBO J.* **3**:2885–2888.

69. **Gupta, S., and R. Chowdhury.** 1997. Bile affects production of virulence factors and motility of *Vibrio cholerae. Infect. Immun.* **65**:1131–1134.

70. **Hammar, M., Z. Bian, and S. Normark.** 1996. Nucleator-dependent intercellular assembly of adhesive curli organelles in *Escherichia coli. Proc. Natl. Acad. Sci. USA* **93**:6562–6566.

71. Hase, C. C., and J. J. Mekalanos. 1999. Effects of changes in membrane sodium flux on virulence gene expression in *Vibrio cholerae*. *Proc. Natl. Acad. Sci. USA* **96:**3183–3187.

72. Hase, C. C., and J. J. Mekalanos. 1998. TcpP protein is a positive regulator of virulence gene expression in *Vibrio cholerae*. *Proc. Natl. Acad. Sci. USA* **95:**730–734.

73. Hasman, H., T. Chakraborty, and P. Klemm. 1999. Antigen-43-mediated autoaggregation of *Escherichia coli* is blocked by fimbriation. *J. Bacteriol.* **181:**4834–4841.

74. Hay, N. A., D. J. Tipper, D. Gygi, and C. Hughes. 1997. A nonswarming mutant of *Proteus mirabilis* lacks the Lrp global transcriptional regulator. *J. Bacteriol.* **179:**4741–4746.

75. Heithoff, D. M., R. L. Sinsheimer, D. A. Low, and M. J. Mahan. 1999. An essential role for DNA adenine methylation in bacterial virulence [see comments]. *Science* **284:**967–970.

76. Henderson, I. R., M. Meehan, and P. Owen. 1997. Antigen 43, a phase-variable bipartite outer membrane protein, determines colony morphology and autoaggregation in *Escherichia coli* K-12. *FEMS Microbiol. Lett.* **149:**115–120.

77. Henderson, I. R., M. Meehan, and P. Owen. 1997. A novel regulatory mechanism for a novel phase-variable outer membrane protein of *Escherichia coli*. *Adv. Exp. Med. Biol.* **412:**349–355.

78. Henderson, I. R., and P. Owen. 1999. The major phase-variable outer membrane protein of *Escherichia coli* structurally resembles the immunoglobulin A1 protease class of exported protein and is regulated by a novel mechanism involving Dam and *oxyR*. *J. Bacteriol.* **181:**2132–2141.

79. Herrington, D. A., R. H. Hall, G. Losonsky, J. J. Mekalanos, R. K. Taylor, and M. M. Levine. 1988. Toxin, toxin-coregulated pili, and the *toxR* regulon are essential for *Vibrio cholerae* pathogenesis in humans. *J. Exp. Med.* **168:**1487–1492.

80. Higgins, C. F., C. J. Dorman, D. A. Stirling, L. Waddell, I. R. Booth, G. May, and E. Bremer. 1988. A physiological role for DNA supercoiling in the osmotic regulation of gene expression in *S. typhimurium* and *E. coli*. *Cell* **52:**569–584.

81. Huisman, T. T., and F. K. de Graaf. 1995. Negative control of *fae* (K88) expression by the 'global' regulator Lrp is modulated by the 'local' regulator FaeA and affected by DNA methylation. *Mol. Microbiol.* **16:**943–953.

82. Hultgren, S. J., S. Normark, and S. N. Abraham. 1991. Chaperone-assisted assembly and molecular architecture of adhesive pili. *Annu. Rev. Microbiol.* **45:**383–415.

83. Hultgren, S. J., W. R. Schwan, A. J. Schaeffer, and J. L. Duncan. 1986. Regulation of production of type 1 pili among urinary tract isolates of *Escherichia coli*. *Infect. Immun.* **54:**613–620.

84. Inoue, O. J., J. H. Lee, and R. E. Isaacson. 1993. Transcriptional organization of the *Escherichia coli* pilus adhesin K99. *Mol. Microbiol.* **10:**607–613.

85. Isaacson, R. E. 1980. Factors affecting expression of the *Escherichia coli* pilus K99. *Infect. Immun.* **28:**190–194.

86. Isaacson, R. E. 1983. Regulation of expression of *Escherichia coli* pilus K99. *Infect. Immun.* **40:**633–639.

87. Ishimoto, K. S., and S. Lory. 1989. Formation of pilin in *Pseudomonas aeruginosa* requires the alternative sigma factor (RpoN) of RNA polymerase. *Proc. Natl. Acad. Sci. USA* **86:**1954–1957.

88. Jones, P. G., and M. Inouye. 1994. The cold-shock response—a hot topic. *Mol. Microbiol.* **11:**811–818.

89. Jordi, B. J., B. Dagberg, L. A. de Haan, A. M. Hamers, B. A. van der Zeijst, W. Gaastra, and B. E. Uhlin. 1992. The positive regulator CfaD overcomes the repression mediated by histone-like protein H-NS (H1) in the CFA/I fimbrial operon of *Escherichia coli*. *EMBO J.* **11:**2627–2632.

90. Kaltenbach, L. S., B. A. Braaten, and D. A. Low. 1995. Specific binding of PapI to Lrp-*pap* DNA complexes. *J. Bacteriol.* **177:**6449–6455.

91. Kaper, J. B., T. K. McDaniel, K. G. Jarvis, and O. Gomez-Duarte. 1997. Genetics of virulence of enteropathogenic *E. coli*. *Adv. Exp. Med. Biol.* **412:**279–287.

92. Karaolis, D. K., S. Somara, D. R. Maneval, Jr., J. A. Johnson, and J. B. Kaper. 1999. A bacteriophage encoding a pathogenicity island, a type-IV pilus and a phage receptor in cholera bacteria [see comments]. *Nature* **399:**375–379.

93. Karjalainen, T. K., D. G. Evans, D. J. Evans, Jr., D. Y. Graham, and C. H. Lee. 1991. Iron represses the expression of CFA/I fimbriae of enterotoxigenic *E. coli*. *Microb. Pathog.* **11:**317–323.

94. Kaufman, M. R., C. E. Shaw, I. D. Jones, and R. K. Taylor. 1993. Biogenesis and regulation of the *Vibrio cholerae* toxin-coregulated pilus: analogies to other virulence factor secretory systems. *Gene* **126:**43–49.

95. Kawula, T. H., and P. E. Orndorff. 1991. Rapid site-specific DNA inversion in *Escherichia coli* mutants lacking the histonelike protein H-NS. *J. Bacteriol.* **173:**4116–4123.

96. Klaasen, P., and F. K. de Graaf. 1990. Characterization of FapR, a positive regulator of expression of the 987P operon in enterotoxigenic *Escherichia coli*. *Mol. Microbiol.* **4:**1779–1783.

97. Klemm, P., I. Orskov, and F. Orskov. 1982. F7 and type 1-like fimbriae from three *Escherichia coli* strains isolated from urinary tract infections: protein chemical and immunological aspects. *Infect. Immun.* **36:**462–468.

98. Kosturko, L. D., E. Daub, and H. Murialdo. 1989. The interaction of *E. coli* integration host factor and lambda *cos* DNA: multiple complex formation and protein-induced bending. *Nucleic Acids Res.* **17:**317–334.

99. Kulasekara, H. D., and I. C. Blomfield. 1999. The molecular basis for the specificity of *fimE* in the phase variation of type 1 fimbriae of *Escherichia coli* K-12. *Mol. Microbiol.* **31:**1171–1181.

100. Larribe, M., M. K. Taha, A. Topilko, and C. Marchal. 1997. Control of *Neisseria gonorrhoeae* pilin gene expression by environmental factors: involvement of the *pilA/pilB* regulatory genes. *Microbiology* **143:**1757–1764.

101. La Teana, A., A. Brandi, M. Falconi, R. Spurio, C. L. Pon, and C. O. Gualerzi. 1991. Identification of a cold shock transcriptional enhancer of the *Escherichia coli* gene encoding nucleoid protein H-NS. *Proc. Natl. Acad. Sci. USA* **88:**10907–10911.

102. Latta, R. K., A. Grondin, H. C. Jarrell, G. R. Nicholls, and L. R. Berube. 1999. Differential expression of nonagglutinating fimbriae and MR/P pili in swarming colonies of *Proteus mirabilis*. *J. Bacteriol.* **181:**3220–3225.

103. Levine, M. M., P. Ristaino, R. B. Sack, J. B. Kaper, F. Orskov, and I. Orskov. 1983. Colonization factor antigens I and II and type 1 somatic pili in enterotoxigenic *Escherichia coli*: relation to enterotoxin type. *Infect. Immun.* **39:**889–897.

104. Lintermans, P., P. Pohl, F. Deboeck, A. Bertels, C. Schlicker, J. Vandekerckhove, J. Van Damme, M. Van Montagu, and H. De Greve. 1988. Isolation and nucleotide sequence of the F17-A gene encoding the structural protein of the F17 fimbriae in bovine enterotoxigenic *Escherichia coli*. *Infect. Immun.* **56:**1475–1484.

105. Lintermans, P. F., P. Pohl, A. Bertels, G. Charlier, J. Vandekerckhove, J. Van Damme, J. Schoup, C. Schlicker, T. Korhonen, H. De Greve, et al. 1988. Characterization and purification of the F17 adhesin on the surface of bovine en-

teropathogenic and septicemic *Escherichia coli*. *Am. J. Vet. Res.* **49**:1794–1799.

106. **Lobner-Olesen, A., E. Boye, and M. G. Marinus.** 1992. Expression of the *Escherichia coli dam* gene. *Mol. Microbiol.* **6:** 1841–1851.

107. **Loomis, W. P., and S. L. Moseley.** 1998. Translational control of mRNA processing in the F1845 fimbrial operon of *Escherichia coli*. *Mol. Microbiol.* **30**:843–853.

108. **Lo-Tseng, T., J. Lee, and R. E. Isaacson.** 1997. Regulators of *Escherichia coli* K99 region 1 genes. *Adv. Exp. Med. Biol.* **412**:303–310.

109. **Marinus, M. G., and N. R. Morris.** 1975. Pleiotropic effects of a DNA adenine methylation mutation (*dam*-3) in *Escherichia coli* K12. *Mutat. Res.* **28**:15–26.

110. **Martin, C.** 1996. The *clp* (CS31A) operon is negatively controlled by Lrp, ClpB, and L-alanine at the transcriptional level. *Mol. Microbiol.* **21**:281–292.

111. **Martinez-Laguna, Y., E. Calva, and J. L. Puente.** 1999. Autoactivation and environmental regulation of *bfpT* expression, the gene coding for the transcriptional activator of *bfpA* in enteropathogenic *Escherichia coli*. *Mol. Microbiol.* **33:** 153–166.

112. **Mattick, J. S., C. B. Whitchurch, and R. A. Alm.** 1996. The molecular genetics of type-4 fimbriae in *Pseudomonas aeruginosa*—a review. *Gene* **179**:147–155.

113. **McClain, M. S., I. C. Blomfield, and B. I. Eisenstein.** 1991. Roles of *fimB* and *fimE* in site-specific DNA inversion associated with phase variation of type 1 fimbriae in *Escherichia coli*. *J. Bacteriol.* **173**:5308–5314.

114. **Mekalanos, J. J.** 1992. Environmental signals controlling expression of virulence determinants in bacteria. *J. Bacteriol.* **174**:1–7.

115. **Mellies, J., T. Rudel, and T. F. Meyer.** 1997. Transcriptional regulation of pilC2 in *Neisseria gonorrhoeae*: response to oxygen availability and evidence for growth-phase regulation in *Escherichia coli*. *Mol. Gen. Genet.* **255**:285–293.

116. **Miller, V. L., and J. J. Mekalanos.** 1988. A novel suicide vector and its use in construction of insertion mutations: osmoregulation of outer membrane proteins and virulence determinants in *Vibrio cholerae* requires *toxR*. *J. Bacteriol.* **170:** 2575–2583.

117. **Morschhauser, J., B. E. Uhlin, and J. Hacker.** 1993. Transcriptional analysis and regulation of the *sfa* determinant coding for S fimbriae of pathogenic *Escherichia coli* strains. *Mol. Gen. Genet.* **238**:97–105.

118. **Morschhauser, J., V. Vetter, L. Emody, and J. Hacker.** 1994. Adhesin regulatory genes within large, unstable DNA regions of pathogenic *Escherichia coli*: cross-talk between different adhesin gene clusters. *Mol. Microbiol.* **11**:555–566.

119. **Mulvey, M. A., Y. S. Lopez-Boado, C. L. Wilson, R. Roth, W. C. Parks, J. Heuser, and S. J. Hultgren.** 1998. Induction and evasion of host defenses by type 1-piliated uropathogenic *Escherichia coli*. *Science* **282**:1494–1497.

120. **Munson, G. P., and J. R. Scott.** 1999. Binding site recognition by Rns, a virulence regulator in the AraC family. *J. Bacteriol.* **181**:2110–2117.

121. **Murphree, D., B. Froehlich, and J. R. Scott.** 1997. Transcriptional control of genes encoding CS1 pili: negative regulation by a silencer and positive regulation by Rns. *J. Bacteriol.* **179**: 5736–5743.

122. **Nagy, B., H. W. Moon, and R. E. Isaacson.** 1977. Colonization of porcine intestine by enterotoxigenic *Escherichia coli*: selection of piliated forms in vivo, adhesion of piliated forms to epithelial cells in vitro, and incidence of a pilus antigen among porcine enteropathogenic *E. coli*. *Infect. Immun.* **16**:344–352.

123. **Nataro, J. P., D. Yikang, D. Yingkang, and K. Walker.** 1994. AggR, a transcriptional activator of aggregative adherence fimbria I expression in enteroaggregative *Escherichia coli*. *J. Bacteriol.* **176**:4691–4699.

124. **Newman, E. B., and R. Lin.** 1995. Leucine-responsive regulatory protein: a global regulator of gene expression in *E. coli*. *Annu. Rev. Microbiol.* **49**:747–775.

125. **Nicholson, B., and D. Low.** 2000. DNA methylation-dependent regulation of Pef expression in *Salmonella typhimurium*. *Mol. Microbiol.* **35**:728–742.

126. **Norgren, M., S. Normark, D. Lark, P. O'Hanley, G. Schoolnik, S. Falkow, C. Svanborg-Eden, M. Baga, and B. E. Uhlin.** 1984. Mutations in *E. coli* cistrons affecting adhesion to human cells do not abolish Pap pili fiber formation. *EMBO J.* **3**:1159–1165.

127. **Normark, S., D. Lark, R. Hull, M. Norgren, M. Baga, P. O'Hanley, G. Schoolnik, and S. Falkow.** 1983. Genetics of digalactoside-binding adhesin from a uropathogenic *Escherichia coli* strain. *Infect. Immun.* **41**:942–949.

128. **Nou, X., B. Braaten, L. Kaltenbach, and D. A. Low.** 1995. Differential binding of Lrp to two sets of *pap* DNA binding sites mediated by Pap I regulates Pap phase variation in *Escherichia coli*. *EMBO J.* **14**:5785–5797.

129. **Nou, X., B. Skinner, B. Braaten, L. Blyn, D. Hirsch, and D. Low.** 1993. Regulation of pyelonephritis-associated pili phase-variation in *Escherichia coli*: binding of the PapI and the Lrp regulatory proteins is controlled by DNA methylation. *Mol. Microbiol.* **7**:545–553.

130. **O'Hanley, P., D. Low, I. Romero, D. Lark, K. Vosti, S. Falkow, and G. Schoolnik.** 1985. Gal-Gal binding and hemolysin phenotypes and genotypes associated with uropathogenic *Escherichia coli*. *N. Engl. J. Med.* **313**:414–420.

131. **Olsen, A., A. Arnqvist, M. Hammar, and S. Normark.** 1993. Environmental regulation of curli production in *Escherichia coli*. *Infect. Agents Dis.* **2**:272–274.

132. **Olsen, A., A. Arnqvist, M. Hammar, S. Sukupolvi, and S. Normark.** 1993. The RpoS sigma factor relieves H-NS-mediated transcriptional repression of *csgA*, the subunit gene of fibronectin-binding curli in *Escherichia coli*. *Mol. Microbiol.* **7**:523–536.

133. **Olsen, A., A. Jonsson, and S. Normark.** 1989. Fibronectin binding mediated by a novel class of surface organelles on *Escherichia coli*. *Nature* **338**:652–655.

134. **Olsen, P. B., M. A. Schembri, D. L. Gally, and P. Klemm.** 1998. Differential temperature modulation by H-NS of the *fimB* and *fimE* recombinase genes which control the orientation of the type 1 fimbrial phase switch. *FEMS Microbiol. Lett.* **162**:17–23.

135. **Orndorff, P. E., and C. A. Bloch.** 1990. The role of type 1 pili in the pathogenesis of *Escherichia coli* infections: a short review and some new ideas. *Microb. Pathog.* **9**:75–79.

136. **Overdier, D. G., and L. N. Csonka.** 1992. A transcriptional silencer downstream of the promoter in the osmotically controlled *proU* operon of *Salmonella typhimurium*. *Proc. Natl. Acad. Sci. USA* **89**:3140–3144.

137. **Pallesen, L., O. Madsen, and P. Klemm.** 1989. Regulation of the phase switch controlling expression of type 1 fimbriae in *Escherichia coli*. *Mol. Microbiol.* **3**:925–931.

138. **Parsot, C., and J. J. Mekalanos.** 1990. Expression of ToxR, the transcriptional activator of the virulence factors in *Vibrio cholerae*, is modulated by the heat shock response. *Proc. Natl. Acad. Sci. USA* **87**:9898–9902.

139. **Paul, P. S., D. H. Francis, and D. A. Benfield.** 1997. *Mechanisms in the Pathogenesis of Enteric Diseases.* Plenum Press, New York, N.Y.

140. Pawelzik, M., J. Heesemann, J. Hacker, and W. Opferkuch. 1988. Cloning and characterization of a new type of fimbria (S/F1C-related fimbria) expressed by an *Escherichia coli* O75:K1:H7 blood culture isolate. *Infect. Immun.* **56**:2918–2924.

141. Plos, K., S. I. Hull, R. A. Hull, B. R. Levin, I. Orskov, F. Orskov, and C. Svanborg-Eden. 1989. Distribution of the P-associated-pilus (*pap*) region among *Escherichia coli* from natural sources: evidence for horizontal gene transfer. *Infect. Immun.* **57**:1604–1611.

142. Porter, M. E., S. G. Smith, and C. J. Dorman. 1998. Two highly related regulatory proteins, *Shigella flexneri* VirF and enterotoxigenic *Escherichia coli* Rns, have common and distinct regulatory properties. *FEMS Microbiol. Lett.* **162**:303–309.

143. Pratt, L. A., and T. J. Silhavy. 1998. Crl stimulates RpoS activity during stationary phase. *Mol. Microbiol.* **29**:1225–1236.

144. Puente, J. L., D. Bieber, S. W. Ramer, W. Murray, and G. K. Schoolnik. 1996. The bundle-forming pili of enteropathogenic *Escherichia coli*: transcriptional regulation by environmental signals. *Mol. Microbiol.* **20**:87–100.

145. Raivio, T. L., and T. J. Silhavy. 1999. The sigmaE and Cpx regulatory pathways: overlapping but distinct envelope stress responses. *Curr. Opin. Microbiol.* **2**:159–165.

146. Rasmussen, L. J., A. Lobner-Olesen, and M. G. Marinus. 1995. Growth-rate-dependent transcription initiation from the *dam* P2 promoter. *Gene* **157**:213–215.

147. Rasmussen, L. J., M. G. Marinus, and A. Lobner-Olesen. 1994. Novel growth rate control of *dam* gene expression in *Escherichia coli*. *Mol. Microbiol.* **12**:631–638.

148. Rice, P. A., S. Yang, K. Mizuuchi, and H. A. Nash. 1996. Crystal structure of an IHF-DNA complex: a protein-induced DNA U-turn. *Cell* **87**:1295–1306.

149. Ritter, A., G. Blum, L. Emody, M. Kerenyi, A. Bock, B. Neuhierl, W. Rabsch, F. Scheutz, and J. Hacker. 1995. tRNA genes and pathogenicity islands: influence on virulence and metabolic properties of uropathogenic *Escherichia coli*. *Mol. Microbiol.* **17**:109–121.

150. Ritter, A., D. L. Gally, P. B. Olsen, U. Dobrindt, A. Friedrich, P. Klemm, and J. Hacker. 1997. The Pai-associated leuX specific tRNA5(Leu) affects type 1 fimbriation in pathogenic *Escherichia coli* by control of FimB recombinase expression. *Mol. Microbiol.* **25**:871–882.

151. Roesch, P. L., and I. C. Blomfield. 1998. Leucine alters the interaction of the leucine-responsive regulatory protein (Lrp) with the *fim* switch to stimulate site-specific recombination in *Escherichia coli*. *Mol. Microbiol.* **27**:751–761.

152. Rohde, J. R., X. S. Luan, H. Rohde, J. M. Fox, and S. A. Minnich. 1999. The *Yersinia enterocolitica* pYV virulence plasmid contains multiple intrinsic DNA bends which melt at 37 degrees C. *J. Bacteriol.* **181**:4198–4204.

153. Romling, U., Z. Bian, M. Hammar, W. D. Sierralta, and S. Normark. 1998. Curli fibers are highly conserved between *Salmonella typhimurium* and *Escherichia coli* with respect to operon structure and regulation. *J. Bacteriol.* **180**:722–731.

154. Romling, U., W. D. Sierralta, K. Eriksson, and S. Normark. 1998. Multicellular and aggregative behaviour of *Salmonella typhimurium* strains is controlled by mutations in the *agfD* promoter. *Mol. Microbiol.* **28**:249–264.

155. Sakellaris, H., and J. R. Scott. 1998. New tools in an old trade: CS1 pilus morphogenesis. *Mol. Microbiol.* **30**:681–687.

156. Savelkoul, P. H., G. A. Willshaw, M. M. McConnell, H. R. Smith, A. M. Hamers, B. A. van der Zeijst, and W. Gaastra. 1990. Expression of CFA/I fimbriae is positively regulated. *Microb. Pathog.* **8**:91–99.

157. Schembri, M. A., P. B. Olsen, and P. Klemm. 1998. Orientation-dependent enhancement by H-NS of the activity of the type 1 fimbrial phase switch promoter in *Escherichia coli*. *Mol. Gen. Genet.* **259**:336–344.

158. Schmoll, T., M. Ott, B. Oudega, and J. Hacker. 1990. Use of a wild-type gene fusion to determine the influence of environmental conditions on expression of the S fimbrial adhesin in an *Escherichia coli* pathogen. *J. Bacteriol.* **172**:5103–5111.

159. Schuhmacher, D. A., and K. E. Klose. 1999. Environmental signals modulate ToxT-dependent virulence factor expression in *Vibrio cholerae*. *J. Bacteriol.* **181**:1508–1514.

160. Schwan, W. R., H. S. Seifert, and J. L. Duncan. 1994. Analysis of the *fimB* promoter region involved in type 1 pilus phase variation in *Escherichia coli*. *Mol. Gen. Genet.* **242**:623–630.

161. Schwan, W. R., H. S. Seifert, and J. L. Duncan. 1992. Growth conditions mediate differential transcription of *fim* genes involved in phase variation of type 1 pili. *J. Bacteriol.* **174**:2367–2375.

162. Skorupski, K., and R. K. Taylor. 1997. Control of the ToxR virulence regulon in *Vibrio cholerae* by environmental stimuli. *Mol. Microbiol.* **25**:1003–1009.

163. Skorupski, K., and R. K. Taylor. 1997. Cyclic AMP and its receptor protein negatively regulate the coordinate expression of cholera toxin and toxin-coregulated pilus in *Vibrio cholerae*. *Proc. Natl. Acad. Sci. USA* **94**:265–270.

164. Skorupski, K., and R. K. Taylor. 1999. A new level in the *Vibrio cholerae* ToxR virulence cascade: AphA is required for transcriptional activation of the *tcpPH* operon. *Mol. Microbiol.* **31**:763–771.

165. Soisson, S. M., B. MacDougall-Shackleton, R. Schleif, and C. Wolberger. 1997. Structural basis for ligand-regulated oligomerization of AraC. *Science* **276**:421–425.

166. Strom, M. S., and S. Lory. 1993. Structure-function and biogenesis of the type IV pili. *Annu. Rev. Microbiol.* **47**:565–596.

167. Taha, M. K., and D. Giorgini. 1995. Phosphorylation and functional analysis of PilA, a protein involved in the transcriptional regulation of the pilin gene in *Neisseria gonorrhoeae*. *Mol. Microbiol.* **15**:667–677.

168. Taha, M. K., D. Giorgini, and X. Nassif. 1996. The *pilA* regulatory gene modulates the pilus-mediated adhesion of *Neisseria meningitidis* by controlling the transcription of *pilC1*. *Mol. Microbiol.* **19**:1073–1084.

169. Taha, M. K., P. C. Morand, Y. Pereira, E. Eugene, D. Giorgini, M. Larribe, and X. Nassif. 1998. Pilus-mediated adhesion of *Neisseria meningitidis*: the essential role of cell contact-dependent transcriptional upregulation of the PilC1 protein. *Mol. Microbiol.* **28**:1153–1163.

170. Taha, M. K., M. So, H. S. Seifert, E. Billyard, and C. Marchal. 1988. Pilin expression in *Neisseria gonorrhoeae* is under both positive and negative transcriptional control. *EMBO J.* **7**:4367–4378.

171. Taylor, R. K., V. L. Miller, D. B. Furlong, and J. J. Mekalanos. 1987. Use of *phoA* gene fusions to identify a pilus colonization factor coordinately regulated with cholera toxin. *Proc. Natl. Acad. Sci. USA* **84**:2833–2837.

172. Torreblanca, J., and J. Casadesus. 1996. DNA adenine methylase mutants of *Salmonella typhimurium* and a novel *dam*-regulated locus. *Genetics* **144**:15–26.

173. Torreblanca, J., S. Marques, and J. Casadesus. 1999. Synthesis of FinP RNA by plasmids F and pSLT is regulated by DNA adenine methylation. *Genetics* **152**:31–45.

174. Ussery, D. W., J. C. Hinton, B. J. Jordi, P. E. Granum, A. Seirafi, R. J. Stephen, A. E. Tupper, G. Berridge, J. M. Sidebotham, and C. F. Higgins. 1994. The chromatin-associated protein H-NS. *Biochimie* 76:968–980.

175. van der Woude, M., W. B. Hale, and D. A. Low. 1998. Formation of DNA methylation patterns: nonmethylated GATC sequences in *gut* and *pap* operons. *J. Bacteriol.* 180:5913–5920.

176. van der Woude, M. W., P. A. Arts, D. Bakker, H. W. van Verseveld, and F. K. de Graaf. 1990. Growth-rate-dependent synthesis of K99 fimbrial subunits is regulated at the level of transcription. *J. Gen. Microbiol.* 136:897–903.

177. van der Woude, M. W., B. A. Braaten, and D. A. Low. 1992. Evidence for global regulatory control of pilus expression in *Escherichia coli* by Lrp and DNA methylation: model building based on analysis of *pap. Mol. Microbiol.* 6:2429–2435.

178. van der Woude, M. W., M. Braster, H. W. van Verseveld, and F. K. de Graaf. 1990. Control of temperature-dependent synthesis of K99 fimbriae. *FEMS Microbiol. Lett.* 56:183–188.

179. van der Woude, M. W., F. K. de Graaf, and H. W. van Verseveld. 1989. Production of the fimbrial adhesin 987P by enterotoxigenic *Escherichia coli* during growth under controlled conditions in a chemostat. *J. Gen. Microbiol.* 135:3421–3429.

180. van der Woude, M. W., L. S. Kaltenbach, and D. A. Low. 1995. Leucine-responsive regulatory protein plays dual roles as both an activator and a repressor of the *Escherichia coli pap* fimbrial operon. *Mol. Microbiol.* 17:303–312.

181. van der Woude, M. W., and D. A. Low. 1994. Leucine-responsive regulatory protein and deoxyadenosine methylase control the phase variation and expression of the *sfa* and *daa* pili operons in *Escherichia coli. Mol. Microbiol.* 11:605–618.

182. van Verseveld, H. W., P. Bakker, T. van der Woude, C. Terleth, and F. K. de Graaf. 1985. Production of fimbrial adhesins K99 and F41 by enterotoxigenic *Escherichia coli* as a function of growth-rate domain. *Infect. Immun.* 49:159–163.

183. Vidal, O., R. Longin, C. Prigent-Combaret, C. Dorel, M. Hooreman, and P. Lejeune. 1998. Isolation of an *Escherichia coli* K-12 mutant strain able to form biofilms on inert surfaces: involvement of a new *ompR* allele that increases curli expression. *J. Bacteriol.* 180:2442–2449.

184. Voss, E., and S. R. Attridge. 1993. In vitro production of toxin-coregulated pili by *Vibrio cholerae* El Tor. *Microb. Pathog.* 15:255–268.

185. Waldor, M. K., and J. J. Mekalanos. 1996. Lysogenic conversion by a filamentous phage encoding cholera toxin [see comments]. *Science* 272:1910–1914.

186. Walker, S. L., M. Sojka, M. Dibb-Fuller, and M. J. Woodward. 1999. Effect of pH, temperature and surface contact on the elaboration of fimbriae and flagella by *Salmonella* serotype Enteritidis. *J. Med. Microbiol.* 48:253–261.

187. Weisberg, R. A., M. Freundlich, D. Friedman, J. Gardner, N. Goosen, H. Nash, A. Oppenheim, and J. Rouviere-Yaniv. 1996. Nomenclature of the genes encoding IHF [letter]. *Mol. Microbiol.* 19:642.

188. Weyand, N. J., and D. A. Low. 2000. Regulation of Pap phase variation. Lrp is sufficient for the establishment of the phase off *pap* DNA methylation pattern and repression of *pap* transcription *in vitro. J. Biol. Chem.* 275:3192–3200.

189. White-Ziegler, C. A., M. L. Angus Hill, B. A. Braaten, M. W. van der Woude, and D. A. Low. 1998. Thermoregulation of *Escherichia coli pap* transcription: H-NS is a temperature-dependent DNA methylation blocking factor. *Mol. Microbiol.* 28:1121–1137.

190. White-Ziegler, C. A., L. B. Blyn, B. A. Braaten, and D. A. Low. 1990. Identification of an *Escherichia coli* genetic locus involved in thermoregulation of the *pap* operon. *J. Bacteriol.* 172:1775–1782.

191. White-Ziegler, C. A., and D. A. Low. 1992. Thermoregulation of the *pap* operon: evidence for the involvement of RimJ, the N-terminal acetylase of ribosomal protein S5. *J. Bacteriol.* 174:7003–7012.

192. Willshaw, G. A., H. R. Smith, M. M. McConnell, and B. Rowe. 1991. Cloning of regulator genes controlling fimbrial production by enterotoxigenic *Escherichia coli. FEMS Microbiol. Lett.* 66:125–129.

193. Wu, S. S., and D. Kaiser. 1997. Regulation of expression of the *pilA* gene in *Myxococcus xanthus. J. Bacteriol.* 179:7748–7758.

194. Xia, Y., K. Forsman, J. Jass, and B. E. Uhlin. 1998. Oligomeric interaction of the PapB transcriptional regulator with the upstream activating region of pili adhesin gene promoters in *Escherichia coli. Mol. Microbiol.* 30:513–523.

195. Xia, Y., and B. E. Uhlin. 1999. Mutational analysis of the PapB transcriptional regulator in *Escherichia coli*. Regions important for DNA binding and oligomerization. *J. Biol. Chem.* 274:19723–19730.

196. Zhao, H., X. Li, D. E. Johnson, I. Blomfield, and H. L. Mobley. 1997. In vivo phase variation of MR/P fimbrial gene expression in *Proteus mirabilis* infecting the urinary tract. *Mol. Microbiol.* 23:1009–1019.

Bacterial Stress Responses
Edited by G. Storz and R. Hengge-Aronis
©2000 ASM Press, Washington, D.C.

Chapter 22

Bacterial Drug Resistance: Response to Survival Threats

MICHAEL N. ALEKSHUN AND STUART B. LEVY

Antibiotic resistance emerges in bacteria in response to survival threats. Among large populations of bacteria, certain few survive: namely, those able to resist the antibiotic's inhibitory activity. Resistance arises in many instances via the acquisition of a gene(s) or a genetic element(s) such as a transposon, plasmid, or a fragment of chromosomal DNA bearing a resistance trait from another organism. These genes can specify enzymes that inactivate an antibiotic, modify the target to which these drugs bind, or catalyze the removal of the drugs from the cell. Spontaneous chromosomal mutations also play a role in antibiotic resistance and result in an altered target(s) or an altered accumulation of the drug. Genes already present in the microbial flora are reorganized, altered, combined, or shuffled to effect a response to life-threatening events. Many of these genes specify mechanisms that are specific for a single antibiotic or class of drug while others, namely, reduced expression of nonspecific hydrophilic channels and increased expression of multidrug efflux systems, can produce resistance to multiple structurally unrelated drugs. In certain circumstances, a single global transcriptional regulator can coordinate the regulation of both decreased influx and increased efflux as well as activate the expression of cytoprotective enzymes to ensure survival of the bacterial cell.

Over the past 50 years, clinicians have learned to rely on natural and chemical substances to control infectious diseases. Beginning with Alexander Fleming's discovery of penicillin in 1928, the introduction into clinical practice of sulfonamides in the 1930s, and the champion efforts of Chain and Florey in the late 1930s and early 1940s in finding ways to produce enough penicillin for use in war victims, and eventually for the broader public, antibiotics have become a regular part of the physician's armamentarium (186).

In the mid-1980s, the surgeon general of the United States stated that, with all the antibiotic power then, the war against infectious diseases had been won. Many pharmaceutical companies left the field. Only a few years later, at the end of the 1980s, the first vancomycin-resistant *Enterococcus* appeared. If any organism was to be the beacon of the potential for the future demise of antibiotics, this one probably was it. Steadily since then, penicillin-resistant pneumococci and multidrug-resistant *Mycobacterium tuberculosis*—all types of bacteria, gram-positive and gram-negative—have emerged with resistance to antibiotics. As we reach the end of the 1990s, we note that the decade is clearly one of multidrug resistance (187, 188).

While the different structures and activities of antibiotics have revealed the variety of ways we can inhibit microorganisms, resistance to these agents has shown us the flexibility of the bacterial targets. The appearance of antibiotic resistance determinants has unveiled a plethora of genetic elements in bacteria that they can exchange (Fig. 1) and that allow them to survive antibiotic treatments. The existence of such genetic variety has shed light on the facility and ability of microbes to survive—a feature that has evolved and been maintained over the millions of years that bacteria have inhabited the earth.

The relatively short history (~50 years) of antibiotic use has uncovered a large assortment of resistance genes, which clearly suggests that genes for resistance have long existed, probably not so much to deal with antibiotics as to provide other necessary functions to the host bacterium. During the present half century, the genes have served bacteria well by helping them to survive the onslaught of antibiotics

Michael N. Alekshun and Stuart B. Levy • Center for Adaptation Genetics and Drug Resistance and Departments of Molecular Biology and Microbiology and of Medicine, Tufts University School of Medicine, Boston, MA 02111.

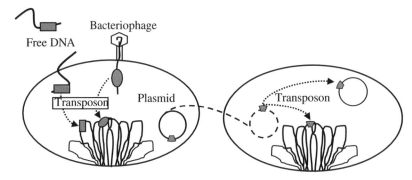

Figure 1. Mechanisms of antibiotic resistance gene acquisition and transfer. Susceptible organisms can acquire resistance genes via free DNA through transformation or following infection by a bacteriophage. During conjugation, plasmids are copied within a donor and transferred to a recipient. These plasmids can exist as extrachromosomal elements or integrate into the host's chromosome or resident plasmids. Once inside the cell, the resistance genes on transposons can move to the chromosome or resident plasmids through transposition.

and other antibacterial substances. The numbers of mechanisms of resistance (Table 1) surpass the numbers of families of different antibiotics we use: they may affect the drug itself, the drug target, or how the drug accumulates within the cell.

As modern medical technology improves, more people are being put into situations of increased susceptibility to infections, including those caused by microorganisms that intact host defense systems generally keep in check. These "opportunistic" organisms have, as added unwanted features, intrinsic resistance to the antibiotics being used to treat the more classic pathogens.

This chapter will review the genetics and mechanisms of antibiotic resistance, both as structural genes and as regulatory genes. It will attempt to describe, in a variety of ways, how bacteria are able to evade and survive the threats of antibiotics. Clearly, if any one human development has had a broad ecologic effect, antibiotics rank among the top, since they have affected vast numbers of species and genera making up the microbial world.

GENETICS

Plasmids

Plasmids bearing antibiotic resistance genes, originally termed resistance (R) factors (e.g., R222), are extrachromosomal DNA structures that replicate independent of the chromosome and vary in size from 1 to greater than 100 kb. In most cases, plasmids carry genes that are accessory to those borne on the host chromosome. Self-transmissible plasmids are readily disseminated among different bacteria through a process called conjugation (Fig. 1). In this

process, a donor cell attaches to a recipient via a sex pilus, which is specified by genes on the self-transmissible plasmid. The plasmid is then replicated within the donor, and a copy of it is passed on to the recipient. Thus, both the donor and the recipient now contain copies of the resistance plasmid and are able to pass it on to new hosts. A unique feature of some self-transmissible plasmids and conjugative transposons (see below) is that they are able to promote the transfer of non-self-conjugatable coresident plasmids (so-called mobilizable plasmids), which make use of the conjugative machinery for movement.

Some strains can contain 30 or more different plasmids that may exist in low (1 to 3 per cell) copy number, while others are found at intermediate (greater than 10 but less than 100) or high (>100) copy numbers. The coexistence of different plasmids is determined by compatibility traits; plasmids with the same origin of replication are incompatible and thus cannot coexist within the same organism. Plasmids may bear genes for many different traits that render them profoundly resistant to antibiotics (for additional information see reference 367).

Transposable Elements

A transposable element represents any genetic unit that can catalyze its own movement from one DNA element to another (28, 55, 167) (Fig. 1). Transposons may be found on plasmids, bacteriophages, and chromosomal DNA (Fig. 1) and can exist as many different types: insertion sequences (IS), composite and complex transposons, the bacteriophages Mu and D108, and integrons (167) (Fig. 2). Another type, the conjugative transposon, has both plasmid- and phage-like properties (64, 136, 334, 341).

Table 1. Known antibiotic resistance mechanisms

Antibiotic class	Resistance mechanism	Genetic location[a]	Reference(s)[b]
Tetracyclines	Increased efflux	P, Tn, Chr	185, 228
	Decreased influx	Chr	108, 268, 320
	Ribosome protection	P, Tn, Chr	64, 335
	Drug inactivation	Tn, Chr (?)	178, 232, 335
	Altered ribosome	Chr	327
β-Lactams	Drug degradation	P, Chr, Tn	202
	Altered PBP(s)	Chr, P	92, 153
	Increased efflux	Chr	268
	Decreased influx	Chr	268
	Altered CihA, CpoA, LytA	Chr	135
	Tolerance (Altered LytA)	Chr	380
Aminoglycosides	Drug modification	P, Tn	81, 314, 347
	Altered ribosome	Chr	397
	Increased efflux	Chr	7, 236
Glycopeptides	Altered target(s)	Chr, Tn	23, 292, 350, 402
Macrolides,	Increased efflux	P, Chr	180, 185, 397
lincosamides,	Altered targets	P, Chr	91, 324, 397
and	Drug inactivation	P	180, 250, 397
streptogramins		P	
Fluoroquinolones	Increased efflux	Chr, P	67, 147, 265
	Altered targets	Chr	147
Chloramphenicol	Increased efflux	Tn, Chr, P	88, 227, 252
	Decreased influx	Chr, P	318, 319
	Drug inactivation	P, Tn, Chr	250, 285, 348
	Drug degradation	Chr	248
	Altered ribosome	Chr	281
Florfenicol	Increased efflux	Chr, Tn, P	20, 40, 166
Rifampin	Altered RpoB	Chr	344, 395, 407
	Drug modification	Tn, P	77, 245, 407
	Increased efflux	P, Chr	59, 280
Trimethoprim and	Altered DHFR	Chr, P, Tn	148
sulfonamides	Substituted drug targets	Tn, P	136, 148
	Altered DHPS	Chr	148
	Overexpressed target	Chr	148
Nitroimidazoles	Altered NADPH	Chr	123, 125
	Nitroreductase	Chr	96
	Decreased accumulation	?	254
	Drug inactivation (*Bacteroides* spp., *nim* genes)	Chr, P	384
Fusidic acid	Drug degradation	Chr	392
	Altered EF-G	Chr	62
	Sequestration by CAT$_I$	P	250
	Decreased permeability	P	210
	Unknown (*S. aureus*)	P	305
Fosfomycin	Drug modification	P, Tn	19, 118
	Decreased accumulation	Chr	163
	Altered pyruvyl transferase	Chr	391
	Unknown (*S. epidermidis*)	?	98
Mupirocin	Altered *ileS*	Chr, P	157, 313

[a] P, plasmid; Chr, chromosome; Tn, transposable element (integron [gene cassette], transposon).
[b] Where relevant, reviews containing pertinent references are provided.

Acquisition of a transposable element has important consequences. It not only delivers the genes that it carries, but it may have polar effects on the cell by positively or negatively affecting the expression of genes that are downstream of its insertion (167).

Insertion sequences

Simple insertion sequences are relatively small genetic elements (28, 55, 167) ranging in size from 768 bp (IS*1*) to 2.1 kb (IS*21*) and specifying only the products required for their movement: terminal in-

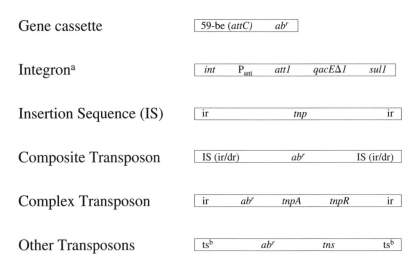

Figure 2. Prokaryotic transposable sequences (not drawn to scale). The bacteriophages are not shown. Abbreviations: 59-be (base element) is also known as *attC* (329); *bla*, a β-lactamase; *int*, an integrase; *att1*, attachment site; P$_{ant}$, anterior promoter; *qacEΔ1*, a deletion derivative of *qacE* that specifies resistance to quaternary ammonium compounds (289); *sul1*, specifies resistance to the sulfonamides (136); IS, insertion sequence; ir/dr, inverted repeat/direct repeat; *tnp*, a transposase; *ab^r*, antibiotic resistance gene(s); ts, terminal sequences. a, For simplicity, a class 1 integron is shown (136). b, The terminal sequences of the mobile elements classified as other transposons (see text) are longer than those found in complex transposons.

verted repeats required for recombination and a transposase that catalyzes transposition (28, 55, 167) (Fig. 2). IS*1* is an integral part of the *Escherichia coli* and *Shigella* chromosomes and many plasmids (28).

Composite (compound) transposable elements

Composite transposons are genes flanked by insertion sequences (Fig. 2). Examples include the chloramphenicol resistance Tn*9* (2.5 kb and flanked by direct repeats of IS*1*), the tetracycline resistance Tn*10* (9.3 kb and flanked by inverted repeats of IS*10*), and the kanamycin resistance Tn*903* (3.1 kb and flanked by inverted repeats of IS*903*) (28, 55, 167). Tn*5* (flanked by inverted repeats of IS*50* and specifying kanamycin, bleomycin, and streptomycin resistances) and Tn*2671* (flanked by copies of IS*1* and specifying resistance to chloramphenicol, fusidic acid, streptomycin, sulfonamide, and mercury) are composite transposons bearing multiple resistance traits (28, 55, 167).

Complex transposons: Tn*1*, Tn*3*, and their relatives

Complex transposons do not terminate in insertion sequences but are instead flanked by shorter terminal inverted repeats (28) (Fig. 2). They specify, at least, one antibiotic resistance mechanism and two additional proteins, TnpA (an essential transposase) and TnpR (a nonessential site-specific recombinase, but its absence decreases the frequency of transposition), that control transposition (28, 55, 167) (Fig.

2). These transposons are large (>5 kb), and their mechanism of transposition is more complex than that of elements bearing flanking insertion sequences (for reviews, see references 55 and 167).

Tn*1* and Tn*3*, the paradigms of this family, originated in gram-negative bacteria. They specify TEM-2 and TEM-1 β-lactamases (Table 2), respectively (28, 167). Tn*21* (~20 kb) is also of a gram-negative origin and specifies resistance to streptomycin, spectinomycin, and mercury (citations in reference 28). Complex transposons that originated in gram-positive bacteria include Tn*551* (specifying erythromycin resistance) and Tn*4451* (specifying chloramphenicol resistance) (28).

Complex transposons with Tn*3*-like direct repeat termini and plasmids formed from these elements are unstable, a property attributable to TnpR (citations in reference 167). Tn*1721*, however, is a complex transposon that specifies tetracycline resistance and is a stable genetic element; at one end, a TnpR resolution site has been deleted (citations in reference 167).

Some transposable elements are distantly related to the Tn*3* family. One relative, Tn*917* of *Enterococcus faecalis*, confers resistance to erythromycin (167). Other complex transposons, such as Tn*7* (specifying trimethoprim, streptomycin, and spectinomycin resistances) and Tn*554* (specifying erythromycin and spectinomycin resistances), display target specificity; in *E. coli* the Tn*7* insertion site is *att*Tn*7* (28). While the ends of Tn*7* are imperfect inverted repeats, Tn*554* does not possess terminal repeats (28).

Table 2. Classification of β-lactamases[a]

Group[b]	Class[c]	Representative member[d]	Location[e]	Resistance profile and relevant properties[f]
1	C	AmpC		
		Basal low level	P/Chr	Ceph
		Constitutive high level	P/Chr	Ext Ceph, PEN
2				
2a	A	Gram-positive bacteria	P	SUL, CLA, PEN
2b and 2be	A	Gram-negative bacteria	Chr	SUL, CLA, ATM, CLX, CAZ
2b and 2be	A	TEM-1/2 and SHV-1	P	CLA[g], TZB[g], PEN, Ceph, Ure[h]
2be	A	TEM-3 (CTX-1), *Klebsiella* spp.	P	CLA, TZB, Ext Ceph, Mono, Oxi, ESBL
2br	A	TEM-30, IRT-3, TRC-1, OXY-2	P	PEN, Ceph[i], IR
2c	A	PSE-1	P/Chr	CLA, SUL, PEN, CAR
2d	D	PSE-2 (OXA-10)	P/Chr	CLA[j], SUL, PEN, CLX, OXA, Ceph
2e	A	*P. vulgaris* Ceph	Chr	CLA, SUL, CLX, Ceph[k]
2f	A	IMI-1, NMC-A, Sme-1	Chr	CLA, PEN, Ceph, Carb, NM
3	B	IMP-1[l], CcrA (CfiA)	Chr/Te	CLA, TZB, EDTA (Met), Carb
		CcrA (CfiA)	P	EDTA (Met), Carb
4	ND	SAR-2 of *B. cepacia*	Chr	PEN

[a] The table was adapted from references 54, 202, and 203. An updated listing of extended spectrum β-lactamases can be obtained elsewhere (http://www.lahey.org/studies/web.htm [22 December 1999, last date accessed]).

[b] The classification of β-lactams according to groups has been described by Bush et al. (54).

[c] The classification of β-lactams according to molecular class has been described previously (14). ND, not determined.

[d] Alternative designations are given in parentheses.

[e] P, plasmid; Chr, chromosomal; Te, transmissible element.

[f] ATM, hydrolyzes aztreonam; CAR, hydrolyzes carbenicillin; Carb, hydrolyzes carbapenems; CAZ, hydrolyzes ceftazidime; CLA, inhibited by clavulanic acid; CLX, hydrolyzes cloxacillin; Ceph, cephalosporinase; Ext Ceph, extended-spectrum cephalosporinase; ESBL, extended-spectrum β-lactamase; EDTA, inhibited by EDTA; IR, inhibitor resistant; Met, metallo-β-lactamase, requires a divalent cation for function; Mono, hydrolyzes monobactams; NM, non-metallo-β-lactamase; OXA, hydrolyzes oxacillin; Oxi, hydrolyzes oximino-β-lactams; PEN, penicillinase; SUL, inhibited by penicillanic acid sulfones; TZB, inhibited by tazobactam; Ure, hydrolyzes ureidopenicillins.

[g] Overexpression affects CLA/TZB sensitivity (54, 203).

[h] Resistance is apparent only when organisms are analyzed at a high inoculum (203).

[i] The cephalosporinase activity of the inhibitor-resistant β-lactamases is lower than that of the parental enzymes (203).

[j] Intermediate level of clavulanate sensitivity.

[k] An inducible cephalosporinase (54).

[l] IMP-1 has been reported as chromosomal (54) and as plasmid-specified (203). In a clinical isolate of *P. aeruginosa*, IMP-1 is plasmid-specified and is located within an integron as a gene cassette (177).

Transposing bacteriophages

The Mu and D108 bacteriophages employ replicative transposition for the synthesis of new phage in both lytic and lysogenic growth phases and are thus classified as transposons (reference 167 and citations therein). The products of the Mu *A* (transposase) and *B* genes comprise the transposition machinery, and mutations in both genes affect transposition (citations in reference 167). Insertion of Mu into a genome is random (28), and since Mu and D108 are lytic phages, infection of a strain with multiple resistances may eventually result in lysis providing a pool of resistance traits.

Other transposable elements

Some other transposons that do not terminate in insertion sequences are not composite transposons. Since their terminal sequences are much longer than those of complex transposons (Fig. 2), they are grouped as a separate unclassified collection of transposons (167). Tn*4* specifies ampicillin, streptomycin, sulfonamide, and mercury resistances, and Tn*1691*

confers resistance to gentamicin, streptomycin, sulfonamide, chloramphenicol, and mercury (167).

Integrons and gene cassettes

Integrons are specialized genetic elements that aid in the acquisition and dissemination of a gene cassette(s) (136). They may be found on plasmids or the chromosome and, by themselves, cannot promote self-transposition (329). Functionally, an integron is composed of three parts: an *int* gene specifying a site-specific recombinase (IntI, the integrase); an *attI* site, the point where the gene cassette(s) is inserted by the integrase, and a promoter (P_{ant}) used to drive expression of the gene cassette (Fig. 2) (136). Most class 1 integrons contain an integrase gene at their 5' end and a *sul1* gene, specifying sulfonamide resistance, at their 3' end. These two elements are separated by antibiotic resistance gene cassettes and *qacEΔ1*, a deletion derivative of *qacE* (289). Both QacE and QacEΔ1 specify resistance to quaternary ammonium compounds (Qacs) in *Klebsiella aerogenes* (289) (see below). Most integrons are found on transposons or have remnants of transposable elements in their genetic organization (136). While the vast majority of integrons have been found in gram-negative bacteria, these elements have also been identified in *Mycobacterium fortuitum* and *Corynebacterium glutamicum* (citations in reference 329).

A gene cassette, commonly a gene (or group of genes) specifying an antibiotic resistance mechanism(s), encompasses a promoterless resistance trait(s) (except for the chloramphenicol resistance genes *cmlA* from *Pseudomonas aeruginosa* [31] and *cmlA2* from *Enterobacter aerogenes* [298] [see section below], which contain their own promoters) and a recombination site, known as a 59-base element (59-be [316] or *attC* [329]) proximal to the resistance gene, that is recognized by the integrase (316) (Fig. 2). The expression of a gene within a cassette is determined by its proximity to P_{ant} (70): the closer it is to P_{ant}, the higher the level of resistance. A gene cassette can exist by itself in a circularized form (136), but a recombinational event between the *attI* site of the integron and the 59-base element of the gene cassette is required for movement of the resistance genes (136, 316). While integrons are required for the movement of gene cassettes, by definition, a "cassette-free" integron is still an integron (citations in reference 136).

Conjugative transposons

Conjugative transposons are unique in that they behave like plasmids and phages (64, 334, 341). They can exist as covalently closed circles and can be transferred by conjugation like plasmids (Fig. 1), but integrate and excise like bacteriophages (64, 334, 341). They do not, however, replicate in the closed circular form (62, 330, 337). Additionally, conjugative transposons are able to promote the transfer of coresident plasmids (62, 330, 337). In one study, transfer of Tn*1545* (specifying resistance to tetracycline, kanamycin, and erythromycin) from *E. faecalis* to *Listeria monocytogenes* occurred in the digestive tracts of gnotobiotic (germ-free) mice (90).

Some conjugative transposons can be found in both gram-positive and gram-negative bacteria, e.g., Tn*916*, Tn*1545*, and their derivatives (64, 334), which carry TetM-mediated resistance to tetracycline (see below). Tn*5253*, another gram-positive conjugative transposon found in *Streptococcus pneumoniae*, specifies resistance to tetracycline, kanamycin, and erythromycin (citations in reference 334).

Most tetracycline-resistant clinical *Bacteroides* isolates have a conjugative transposon that specifies the resistance determinant (334 and references therein), but the conjugative transposons in *Bacteroides* spp. are unrelated to Tn*916* yet they carry TetQ

(64, 334), a TetM homolog (citations in reference 192 and see below). Additionally, *Bacteroides* cells harboring a conjugative transposon promote self-transfer of it following exposure to low levels of a tetracycline (reference 334 and citations therein).

SPECIFIC ANTIBIOTIC RESISTANCES

Acquired Mechanisms of Resistance

Drug modification

β-Lactams. The penicillin-binding proteins (PBPs) play an important role in the assembly of the bacterial cell well, as they catalyze the various stages of peptidoglycan biosynthesis, and are found in all *Eubacteria* (393). An individual organism may contain three to eight different PBPs (393). The *β*-lactam antibiotics bind to PBPs and prevent the cross-linking of the bacterial cell wall.

β-Lactamases hydrolyze the *β*-lactam ring of antibiotics containing this structure, such as the penams (penicillins), cephems (cephalosporins), penems (carbapenems), and monobactams. These enzymes constitute the most common mechanism of resistance to these drugs (202) and are classified into different groups according to their biochemical properties (54) and molecular class (14) (Table 2). Many of the commonly encountered *β*-lactamases have a serine residue within their active sites and are sensitive to *β*-lactamase inhibitors such as clavulanic acid and the penicillanic acid sulfones (sulbactam and tazobactam).

Most gram-negative bacteria (particularly the enterobacteria), with the notable exceptions of *Salmonella* spp. and *Helicobacter pylori*, have a chromosomal *β*-lactamase (*ampC*; group 1, class C enzymes), which is more active against cephalosporins than penicillins (Table 2). Under normal circumstances, these chromosomal enzymes are poorly expressed and thus do not play a major role in high-level resistance phenotypes (154).

Mutations that result in constitutive high-level AmpC expression confer resistance to both penicillins (except for temocillin, a penicillin that is stable against most transferable and chromosomal gram-negative *β*-lactamases) and broad-spectrum cephalosporins (Table 2) (203). Mutations within an *Enterobacter cloacae ampC* gene expressed in *E. coli* confer resistance to cefepime and cefpirome (fourth-generation cephalosporins) (247). In *E. cloacae* and *Proteus rettgeri*, overexpression of AmpC coupled with decreased porin expression engenders carbapenem resistance (312). Resistance to both carbapenems and oxyimino-*β*-lactams is attributed to AmpC overexpression and porin deficiency in *Klebsiella pneumoniae* (220) (also see section on altered transport below). (In *P. aeruginosa*, loss of the OprD porin alone is sufficient to confer carbapenem resistance [see below]). Even more troublesome is the movement of *ampC*, with this extended spectrum of activity, to a transferable plasmid, an event that is responsible for its clinical dissemination (153, 203).

β-Lactamases (groups 2b, 2be, and 2e; all class A enzymes [Table 2]) are found in the chromosomes of *Citrobacter diversus*, *Proteus vulgaris*, *Klebsiella* spp., and *Bacteroides* spp. (203). Plasmid-specified, and in some instances gene cassette-encoded (316), β-lactamases are common mediators of clinically significant levels of β-lactam resistance (154). In gram-negative bacteria, the TEM-1, TEM-2, TEM-13, and SHV-1 (e.g., *K. pneumoniae*) (group 2b and 2be, class A enzymes) β-lactamases are commonly found (54, 154, 203) (Table 2).

In gram-positive organisms, the group 2a, class A β-lactamases are responsible for resistance (54, 203). They are penicillinases that are sensitive to both clavulanate and penicillanic acid sulfones (54, 203) (Table 2).

The plasmid-specified narrow-spectrum β-lactamases in both gram-negative and gram-positive bacteria are ineffective against broad-spectrum cephalosporins, monobactams, oxyimino-β-lactams, and cephamycins. Mutations within these genes, particularly those derived from TEM and SHV, give rise to the extended-spectrum β-lactamases (ESBL) (group 2be, class A enzymes) (152, 294), which are especially prevalent in nosocomial pathogens such as *K. pneumoniae* (124, 154) (Table 2). These ESBL enzymes confer broad β-lactam resistances but are sensitive to temocillin, carbapenems, and cephamycins and to inhibition by clavulanate and tazobactam (54, 203) (Table 2).

A new class of β-lactamase that is inhibitor resistant is emerging (group 2br, class A enzymes) (58, 169). These enzymes, which are related to TEM-1, TEM-2, and SHV, have arisen through mutations that decrease sensitivity to clavulanic acid and the penicillanic acid sulfones but do not alter the spectrum of β-lactamase activity (203) (Table 2). (For many of these enzymes, the mutations that confer inhibitor resistance lower the spectrum of β-lactamase activity [203].) The inhibitor-resistant enzymes are termed inhibitor-resistant TEM (IRT) β-lactamases (58). (A previous designation for these enzymes was TEM resistant to β-lactamase inhibitors (TRI) [390].) While the expression of these enzymes has been observed predominantly in the *Enterobacteriaceae*, in organisms such as *E. coli*, *K. pneumoniae*, *Citrobacter freundii*, and *Proteus mirabilis* (58), there is an ex-

ample of an inhibitor-resistant β-lactamase in *Nocardia brasiliensis* (54). More recently, an inhibitor-resistant β-lactamase (OXY-2) derived from a chromosomal enzyme has been identified in a clinical isolate of *Klebsiella oxytoca* (351).

Another class of β-lactamase, the metallo-enzymes (group 3, class B enzymes) (Table 2), has a requirement, as its name implies, for the divalent cation Zn^{2+} (75). They are either chromosomal (406) or are found on transmissible elements (151). As a group, the metallo-β-lactamases mediate resistance to all β-lactams except for the monobactams (54, 151, 203, 406). More importantly, however, the metallo-β-lactamases confer resistance to the carbapenems such as imipenem, panipenem, and meropenem (54, 151, 203, 406) (Table 2). (Of note, particular metallo-β-lactamases from *Aeromonas* spp. confer resistance only to the carbapenems [357].) All serine metallo-β-lactamases are sensitive to inhibition by clavulanate and tazobactam (Table 2).

IMI-1 and NMC-A of *E. cloacae* and Sme-1 of *Serratia marcescens* (group 2f, class A enzymes) are non–metallo-enzymes that confer carbapenem resistance (54) (Table 2).

Chloramphenicol. Chloramphenicol is an uncharged bacteriostatic agent that binds the 50S ribosomal subunit and inhibits the transpeptidation step of protein synthesis. Its bone marrow toxicity has limited its use in clinical practice. The most prevalent mechanism of chloramphenicol resistance is mediated by an acquired, i.e., plasmid-specified, chloramphenicol acetyltransferase (CAT) that is widespread in gram-positive and gram-negative bacteria and specified by a large number of related genes (348). In gram-negative bacteria, CAT expression is constitutive, but in gram-positive and some gram-negative organisms, CAT synthesis is inducible following exposure to the drug (reviewed in reference 348) via translational attenuation (reviewed in 208). In some hosts, e.g., *Staphylococcus aureus*, resistance is exclusively mediated by plasmids bearing the resistance genes (210).

There are four classes or type variants of CAT enzymes (CAT$_I$ [specified by Tn9], CAT$_{II}$, CAT$_{III}$, and XAT [250]; the XATs have also been termed CATB enzymes [46]) (reviewed in references 250 and 348). The enzymes belonging to the first three classes are specified by the *catA* genes and, as of 1997, 23 nonredundant sequences were documented (250). The CAT$_I$ enzymes bind fusidic acid with high affinity and thereby confer resistance to this drug as well (see below), whereas the CAT$_{III}$ proteins have only a low affinity for fusidate (citations in reference 250). The

CAT$_{II}$ proteins are inhibited by thiol-reactive chemicals (citations in reference 250).

The newest family of chloramphenicol acetyltransferases are the xenobiotic acetyltransferases (XATs) that are apparently structurally unrelated to the more common CAT enzymes, i.e., those specified by *catA* (250 and references therein). The CATB designation has also been given to this family of enzymes, to differentiate them from the *catA*-specified proteins (46), but this terminology is not ideal since other XATs, i.e., SatA and Vat/VatB (see below), acetylate antibiotics other than chloramphenicol (citations in reference 250). With respect to SatA from *Enterococcus faecium*, the purified protein acetylates virginiamycin M1 (a streptogramin) but does not acetylate chloramphenicol (unpublished data cited in reference 250).

The first *xat* gene, identified in *Agrobacterium tumefaciens*, is chromosomal and was termed *catB* (373). Homologs have also been identified in *E. coli* (specified on Tn2424 [285]), *P. aeruginosa* PAO1 (a partial chromosomal *orf* of unknown function) and in a clinical isolate of *P. aeruginosa* (plasmid-encoded and integron/gene cassette-specified [177]), *S. aureus* (the plasmid-specified *vgb*, *vat*, and *vatB* genes), *Bacillus sphaericus* (a partial *orf* of unknown function), *E. aerogenes* (the integron/plasmid-specified *catB3*), *E. faecium* (the plasmid-specified *satA* gene), *Morganella morganii* (specified on Tn840), and *S. marcescens* (citations in reference 250).

An alignment of the XAT proteins has demonstrated that the SatA, Vat, VatB, and the *B. sphaericus* (the partial sequence reported) enzymes are more related to themselves than they are to the other proteins (250). The identification of genes specifying XAT proteins in both stable (chromosomes) and mobile (plasmids/transposons/integrons) genetic elements suggests that all are related but, at present, the origin of the progenitor is unknown (46).

A second degradative mechanism, mediated by the chromosomal *cml*^{r} locus of *Streptomyces venezuelae* (249), a chloramphenicol-producing organism, specifies a chloramphenicol 3′-O-phosphotransferase that is now termed Cpt (248). Cpt converts chloramphenicol to a 3′-O-phosphoryl ester derivative (248), which lacks antibacterial activity, and confers a modest level (12.5 μg/ml) of chloramphenicol resistance in *S. venezuelae* but does not engender chloramphenicol resistance in *E. coli* (249).

Fusidic acid. Fusidic acid, produced by the fungus *Fusidium coccineum* (citations in reference 392), affects translation by interfering with the function of elongation factor G (EF-G) and inhibiting translocation of the ribosome (370). Resistance to fusidic acid

in *Streptomyces lividans* 66 and other *Streptomyces* spp. (citations in reference 392) is attributable to an esterase specified by the *fusH* locus (392). This enzyme is secreted by the bacterium and converts fusidic acid to its inactive lactone derivative (392). FusH is specific for fusidic acid and helvolic acid, a structural analog of fusidic acid produced by *Aspergillus fumigatus* (citations in reference 392), but is ineffective against other fusidic acid derivatives (392).

Aminoglycosides. The aminoglycosides are hydrophilic bactericidal antibiotics that bind the 30S ribosomal subunit and promote premature translational termination and mRNA misreading. In gram-negative bacteria, the abberant proteins generated following exposure to the aminoglycosides are thought to compromise the integrity of the bacterial membrane, an event that ultimately leads to cell death. The aminoglycosides freely transverse the outer membrane, but to cross the cytoplasmic membrane, an energy-dependent component (an efflux process driven by active electron transport) is required (368) (see below). Thus, these drugs are ineffective in the treatment of infections caused by obligate anaerobes such as *Bacteroides fragilis* and *Clostridium perfringens*, since decreased uptake of the drug is observed under aerobic conditions (368).

The mechanisms of aminoglycoside resistance are remarkably diverse, but most resistance genes specify enzymes that catalyze antibiotic modification at many different sites on a particular aminoglycoside (81, 237, 314, 347, 404). There are three general types of aminoglycoside-inactivating proteins (81, 237, 347, 404): acetyltransferases (AAC) use acetyl CoA while phosphotransferases (APH) and adenyltransferases (ANT) use ATP to catalyze acetylation, phosphorylation, and adenylation reactions (81, 347). While most of these enzymes are specific for a single type of modification, there are examples of proteins in the streptococci, staphylococci, and enterococci that possess both acetyl- and phosphotransferase activities (citations in references 4 and 286; for reviews, see references 81 and 347).

An aminoglycoside adenyltransferase (*aadB*) (56) and an aminoglycoside acetyltransferase (*aacA*) (115) have been identified on integrons (316). AacA engenders amikacin, netilmicin, and tobramycin resistance (citations in reference 46), and AadB confers resistance to kanamycin, gentamicin, and tobramycin (citations in reference 56).

Macrolides-lincosamides-streptogramins. The macrolides, lincosamides, and type A and B streptogramins (MLS antibiotics) are structurally dissimilar agents that bind to the 50S ribosomal subunit and inhibit translation. Specifically, the macrolides and type B streptogramins inhibit translocation of the translational apparatus, and the type A streptogramins interfere with the delivery of charged tRNA molecules to the donor and acceptor sites in the ribosome. The lincosamides are thought to mimic the 3′-end(s) of tRNA and to interfere with peptide bond formation by binding to the ribosome in the region that contains the peptidyl transferase activity (106 and references therein). This interaction is thought to promote the premature dissociation of tRNAs from the ribosome (106 and references therein). By themselves, the MLS antibiotics function as bacteriostatic agents; however, the type A and B streptogramins can function synergistically (e.g., Synercid [a registered trademark of Rhône-Poulenc Rorer Inc.] is a mixture of dalfopristin-quinupristin) in a presumed bactericidal manner whereby binding of the A component to the large ribosomal proteins L10 and L11 increases the affinity of L24 for the B component (97). The MLS antibiotics are grouped together since resistance to one agent usually confers resistance to the others.

Enzymatic inactivation is a prominent mechanism of resistance for the MLS antibiotics. The erythromycin resistance esterase (ere) and streptogramin/virginamycin B hydrolase genes (sbh, vgb, and vgbB) specify hydrolytic enzymes. The mgt (macrolide glycosylation), mph (macrolide phosphotransferase), lin (lincosamide nucleotidyltransferase), saa/satA (streptogramin A acetyltransferase), and vat/vatB (virginamycin factor A acetylation) gene products all catalyze independent mechanisms of antibiotic modification (180, 397). (Because the lin designation has been previously used for a completely unrelated resistance mechanism, these genes are now termed lun for lincomycin nucleotidyltransferase [324]. The sat[A] designation has been changed to vat[D] [324].) SatA/Vat(D) and Vat/VatB are members of the XAT family of enzymes, which include a group of chloramphenicol acetyltransferases (see above), that use many structurally unrelated hydroxyl-containing compounds as acyl acceptors (250).

Rifampin. Rifampin (referred to as rifampicin in countries outside the United States) is a semisynthetic derivative of a natural product, rifamycin, produced by *Nocardia* (*Amycolatopsis*) *mediterranei*. Rifampin interacts with the β-subunit of RNA polymerase, specified by the *rpoB* gene (395), and inhibits transcription. The drug either prevents exit of the nascent transcript (224) or destabilizes the formation of RNA-DNA oligonucleotide hybrids (340).

Rifampin is a first-line antituberculosis drug, and it is also used to treat *Mycobacterium leprae*, the agent of leprosy. It has a place along with other drugs in the treatment of coagulase-positive and -negative staphylococci and of opportunistic infections caused by organisms belonging to the *Mycobacterium avium* complex that commonly occur in patients with AIDS.

While chromosomal mutations in *rpoB* that confer high-level resistance are common (see below), an inducible mechanism of rifampin inactivation was first described in *Rhodococcus erythropolis* and then demonstrated using cell-free extracts in *Mycobacterium smegmatis* (77). Subsequently, studies demonstrated that pathogenic *Nocardia* spp. glycosylate and phosphorylate rifampin (245, 407) and several fast-growing *Mycobacteria* spp. ribosylate the drug (78). In *P. aeruginosa*, *arr-2* is specified by an integron, is homologous to *arr* from *M. smegmatis* (329), and presumably mediates rifampin resistance through an ADP-ribosylase (383).

Tetracycline. The tetracycline family of antibiotics bind the 30S ribosomal subunit and prevent access of aminoacyl tRNA to the acceptor site on the ribosome. The *tet*(X) resistance determinant, originally identified in two related *B. fragilis* transposons, specifies an enzyme capable of inactivating tetracycline (citations in references 335 and 353). The resistance mechanism was identified on the basis of its ability to confer tetracycline resistance in the heterologous host *E. coli* (citations in references 335 and 353). Since the mechanism of inactivation requires both oxygen and NADPH, it produced no resistance in its anaerobic *B. fragilis* host. This finding implies that it originated elsewhere (citations in references 335 and 353), but curiously, so far no other sources of the gene have been identified.

The degradation of tetracycline and many of its derivatives has been demonstrated in *Xylaria digitata*, a fungus (232). This activity was found to be dependent on both temperature (no degradation was observed at 37°C) and oxygen and was inhibited by sodium azide and potassium arsenate, but not by potassium cyanide (232). Subsequently, the modification of oxytetracycline in *E. coli* was reported (178). While the precise nature of drug modification was not determined, a number of the altered compounds retained antibacterial activity (178). None of the degradative mechanisms of tetracycline resistance have emerged as a clinical problem.

Nitroimidazoles. The antibiotics discussed so far are intact active antibacterial compounds that readily exert their bacteriostatic/bactericidal properties. Alternatively, the nitroimidazoles, i.e., metronidazole (Flagyl [Rhône-Poulenc Rorer Inc.]), are anaerobicidal prodrugs that require metabolic activation by a susceptible organism (reduction of their nitro group, an oxygen-sensitive reaction, to a nitro anion radical and then the corresponding hydroxylamine, both of which are toxic to bacteria). They are active against both bacteria and parasites (citations in reference 336). Use of the nitroimidazoles is limited to infections caused by strict microaerophiles, e.g., *H. pylori*, or obligate/strict anaerobes and, in some instances, is preferred over other therapeutic agents that may allow the anaerobic flora to flourish, i.e., to prevent *Clostridium difficile*-associated pseudomembranous colitis.

Metronidazole is inactivated by strains of *E. faecalis*, *E. coli*, and *K. aerogenes* (254 and references therein). With respect to the former, metronidazole is inactivated both by whole cells and cell extracts prepared from cultures grown either anaerobically or aerobically (254). In all instances, the biochemical basis of resistance has not yet been determined.

Fosfomycin. Fosfomycin (phosphonomycin) interferes with formation of the bacterial cell wall by irreversibly inhibiting pyruvyl transferase and, subsequently, synthesis of peptidoglycan by blocking the formation of *N*-acetylmuramic acid (163). In clinical isolates of *S. marcescens*, fosfomycin resistance was found on a plasmid encoding transposon Tn*2921* (118). Subsequent studies with crude cell extracts determined that this resistance was conferred by the *fosA*-specified glutathione *S*-transferase, a gene found only in gram-negative bacteria (19). The purified enzyme catalyzes the formation of a fosfomycin-glutathione adduct that inactivates the antibiotic (18). The *fosB* gene, responsible for fosfomycin resistance in *Staphylococcus epidermidis* and *S. aureus*, has also been localized to a plasmid, but its mechanism of action is unknown (98).

Altered transport: decreased uptake, increased efflux

Bacteria can achieve drug resistance by limiting the amount of drug that diffuses into the cell or by actively effluxing antibiotics that enter the cell. Decreased influx is achieved by downregulating the expression of porins. In *K. pneumoniae*, *opmK36* specifies a porin that is homologous to the *E. coli* OmpC protein (citations in reference 143). Loss of OmpK35, an OmpF homolog, and OmpK36 results in resistance to cefoxitin and extended-spectrum cephalosporins (citations in reference 143). Inactivation of *opmK36* by the transposition of an insertion sequence into the gene has occurred both in vitro and in patients during the course of antimicrobial chemotherapy (143).

Increased efflux results from the synthesis of proteins that exhibit relatively nonspecific substrate profiles (multidrug efflux, see below) or that are specific for a particular class of antibiotic. Specific antibiotic efflux systems can be differentiated from the multidrug resistance proteins in their ability to be saturated by their respective drugs and not by others.

Tetracycline. Decreased permeability was initially considered the major factor in tetracycline resistance. Preliminary studies with plasmid-specified tetracycline resistance determinants, however, were the first to associate antibiotic resistance with active drug efflux (191, 226). Subsequent experiments demonstrated an energy-dependent saturable tetracycline efflux process (226) linked to a 43-kDa inner membrane protein (Tet) (190, 191), which consisted of two equally sized and evolutionarily related domains, α and β (157, 330). Efflux is specified by Tet proteins that show varying degrees of similarity, although those in gram-negative bacteria (e.g., classes A–E) appear to have 12 putative transmembrane spanning segments (TMS), and those in gram-positive organisms (e.g., TetK and TetL) have 14 TMS (185).

There are six different groups of tetracycline efflux proteins (228). The 11 proteins of group 1, TetA–E, G, H, I, J, Z1, and Tet30, and one in group 4 [TetA(P)] consist of 12 putative α-helical TMS. Group 1 proteins have a large cytoplasmic loop that connects helices 6 and 7, but this segment is not present in TetA(P) (citations in reference 228). Members of group 2 are found mainly in gram-negative bacteria, and those of group 3 are specified by chromosomal genes (228 and references therein). The group 5 and 6 proteins are unusual; the former, found in *M. smegmatis*, has 10 or 11 putative TMS and the latter, an unclassified protein from *Corynebacterium striatum*, is composed of two separate peptides (TetA and TetB) that use ATP to confer resistance to both tetracycline and oxacillin, a finding that indicates that it may be a MDR protein (reference 228 and citations therein).

Except for the group 6 protein, the mechanism of tetracycline efflux is drug-divalent cation/H^+ antiport; i.e., for every tetracycline-divalent cation complex exported from the cell, a proton is brought into the cell, and this process is energized by proton motive force (PMF) (405). Thus, tetracycline efflux is a net electroneutral reaction.

Fluoroquinolones. Following the initial studies describing tetracycline efflux, norfloxacin (a completely synthetic, hydrophilic fluoroquinolone) efflux was reported in susceptible and fluoroquinolone-resistant *E. coli* mutants (67). This efflux in *E. coli* was saturatable with other hydrophilic fluoroquinolones (ciprofloxacin, enoxacin, and ofloxacin) but not with their hydrophobic members (nalidixic acid and oxolinic acid), tetracycline, or chloramphenicol (67). A similar type of efflux was found in *S. aureus* and was attributed to NorA (160), a single protein that also mediates resistance to other drugs (259) (see below). Despite the early findings with norfloxacin, the majority of fluoroquinolone efflux in *E. coli* appears to be associated with the AcrAB/TolC multidrug efflux system (265, 280) (see below).

A salicylate-inducible mechanism of fluoroquinolone resistance has been found in ciprofloxacin-susceptible and -resistant *S. aureus* isolates (132). This weak acid also increased the frequency at which ciprofloxacin resistant mutants were isolated (132). Although the mechanism of resistance has not yet been identified, it is thought that induction may proceed through a system similar to that found with *mar* in *E. coli* (see below) and may involve NorA (132).

Macrolides-lincosamides-streptogramins. Efflux catalyzed by specific or MDR transport proteins also accounts for many instances of macrolide resistance (see references in references 180, 324, 362, and 397). ErpA (erythromycin resistance permeability) is constitutively expressed in an isolate of *S. epidermidis* and confers a low level of resistance to 14 (e.g., erythromycin and clarithromycin)- and 15 (e.g., azithromycin)-membered macrolides (180). The *Streptococcus pyogenes* MefA and *S. pneumoniae* MefE [macrolide (M)-type efflux] proteins, members of the major facilitator superfamily of drug efflux proteins (see below), are presumed to be PMF dependent and engender resistance to 14- and 15-membered macrolides (397). Since their sequences are 98% similar, they have been placed in a single gene class designated *mef*(A) (324). The *mef* genes are widely distributed among gram-positive bacteria (citations in reference 324).

MsrA (macrolide streptogramin resistance), which was initially described in *S. epidermidis* (180, 397), mediates macrolide streptogramin (MS)-type efflux of 14- and 15-membered macrolides and type B streptogramins via an ATP-dependent mechanism. The staphylococcal plasmid-specified Vga and VgaB proteins, which confer resistance to the streptogramins (97), are homologs of MsrA and are presumed to function in a similar manner (97). (The Vga and VgaB proteins are members of the ABC family of drug efflux proteins [324] [see below].) In *Burkholderia pseudomallei*, AmrAB confers clinical levels of resistance to clarithromycin and erythromycin but not to clindamycin (a lincosamide) (242) (see below).

Erythromycin is also a substrate for many MDR efflux systems (see below and Table 4).

Chloramphenicol and florfenicol. Many nonenzymatic chloramphenicol resistance mechanisms (excluding the MDR systems) have been identified in both gram-positive and gram-negative bacteria (Table 3). Most appear to be specific for chloramphenicol alone, while others (e.g., from *Salmonella enterica* serovar Typhimurium DT104 [a human and animal pathogen] and *Pasteurella piscicida* [a pathogen of fish]) confer resistance to both chloramphenicol and florfenicol (Table 3).

A single-step mutation in the *E. coli* chromosomal *cmlA* locus confers chloramphenicol resistance in a nonenzymatic manner (319, 320). This locus was independently rediscovered (35, 271) and was termed *cmr* and later *mdfA* (94). *cmlA/cmr/mdfA* is homologous to *yjiO*, which specifies an Orf in *E. coli* (30, 33) that confers resistance to tetraphenylphosphonium (TPP) and low-level resistance to ethidium bromide (EtBr) (94). (In addition to conferring chloramphenicol resistance, CmlA/Cmr/MdfA is also thought to function as an MDR protein [94, 235] [see below].)

Efflux of chloramphenicol (94, 235, 271) by a drug/H^+ antiport mechanism has been demonstrated for the *E. coli* CmlA/Cmr/MdfA protein (235). While the AcrAB multidrug efflux system is responsible for most of the endogenous chloramphenicol efflux in both wild-type *E. coli* (10, 12) and Mar mutants (280), the *E. coli* CmlA/Cmr/MdfA system presumably plays a role in chloramphenicol resistance under certain conditions. Earlier reports of an energy-dependent efflux of chloramphenicol described a unsaturatable mechanism in both susceptible and multidrug-resistant *E. coli* mutants (227).

The chromosomal *S. lividans* 1326 *cml*^{R} locus specifies an approximately 38.8-kDa protein with 12 putative transmembrane spanning regions that confers a high-level (200 µg/ml) chloramphenicol resistance (87). It is 24% identical to the *E. coli* TetA efflux protein and 52% identical to the plasmid-specified *Rhodococcus fascians* Cmr protein (87) (see below). Cml^R is not homologous to the plasmid (R26)-specified (see below) nonenzymatic chloramphenicol resistance gene (87) (see below).

The chloramphenicol 3'-O-phosphotransferase from *S. venezuelae* (see above) was identified on a cloned fragment of genomic DNA that also specified an Orf (ORF1) with homology to Cml^R (39.8% identity) from *S. lividans* 1326, Cmr (42.4% identity) from *R. fascians* (see below), as well as other known efflux proteins: TetL from *Bacillus stearothermophilus*, NorA from *S. aureus*, and Mmr from *Strepto-*

Table 3. Nonenzymatic chloramphenicol and chloramphenicol/florfenicol resistance determinants

Organism	Current designation[a]	Location[b]	Substrate(s)[c]	Reference(s)
E. coli	CmlA/Cmr/MdfA	Chr	MDR/IPTG	94, 271, 318
	YjiO	Chr	TPP, EtBr, CAM (?)	94
	Cml	P	CAM	88, 252
S. lividans 1326	Cml^R	Chr	CAM	97
S. venezuelae M252	ORF1	G	Phos CAM (?)	248
P. aeruginosa	CmlA	Tn	CAM	31
E. aerogenes	CmlA2	Int	CAM	298
R. fascians	Cmr	P	CAM	85
S. enterica serovar Typhimurium DT104	FloR (bovine)	Chr	CAM/FLR	20
	Flo$_{St}$ (cattle)	Chr	CAM/FLR	40
	CmlA-like (human)	Chr	Cm/FLR (?)	43
P. piscicida	Flo$_{Pp}$	P	CAM/FLR	166

[a] See text.

[b] Chr, chromosomal; Int, integron-specified; Tn, transposon-mediated; P, plasmid-specified; G, genomic DNA (It was not indicated whether the gene was plasmid-, transposon-, integron-, or chromosomally specified [248]).

[c] CAM, chloramphenicol; FLR, florfenicol; Phos CAM, 3'-O-phospho-chloramphenicol; MDR, multidrug resistant; IPTG, isopropyl-β-thiogalactoside.

myces coelicolor (248). ORF1 by itself, however, does not play a role in chloramphenicol resistance in either *S. lividans* M252 (a chloramphenicol-hypersusceptible host) (248) or in *E. coli* (249). It has been suggested that ORF1 may act as a specific 3'-O-phospho-chloramphenicol efflux system that functions in the presence of the phosphotransferase (249).

Many instances of nonenzymatic chloramphenicol resistance also involve plasmid-specified (e.g., R factors R26 and R55-1) proteins, and initial experiments with R factors from clinical *E. coli* isolates were the basis for future work (252). Further experiments identified these R factors in other gram-negative organisms (113, 252). *cml* of the *E. coli* plasmid R26 specifies a 33.8-kDa (31 kDa in minicells) hydrophobic protein (89) located in the bacterial cytoplasmic membrane (113). It is presumed that these determinants promote drug efflux in a manner analogous to the tetracycline efflux proteins (113). Fluorinated chloramphenicol analogs (see below) like florfenicol can induce these nonenzymatic resistance mechanisms but do not produce high-level resistance to themselves (88) or the fluoroquinolones (31). Thus, these plasmid-mediated putative efflux systems, like the CAT enzymes, appear to be specific only for chloramphenicol.

A transposon-specified (Tn*1696*) gene in *P. aeruginosa* (31, 50) (confusingly also called *cmlA*) (31 and references therein) and an integron-specified (In*40*) gene in *E. aerogenes* BM2688 (termed *cmlA2*) (298) also specify nonenzymatic chloramphenicol resistance mechanisms. Each specifies an approximately 45-kDa protein (31, 298). CmlA shows homology to both gram-positive and gram-negative tetracycline efflux proteins, as well as QacA (a known MDR protein from *S. aureus* [see below]) and Mmr, and is closely

related to the chromosomal *cmlA*-like gene of serovar Typhimurium DT104 (43). CmlA is thought to comprise 12 transmembrane-spanning segments (31 and references therein).

Another chloramphenicol-inducible plasmid-mediated resistance mechanism (again a duplicated term, *cmr*) was identified in *R. fascians* (85). The *R. fascians* Cmr protein is approximately 40.3 kDa and contains 12 transmembrane-spanning segments. Like Cml^R from *S. lividans*, it confers a high level (200 μg/ml) of chloramphenicol resistance (85). Cmr is 52% identical to the chromosomal *S. lividans* 1326 Cml^R protein (85) but is not similar to the *E. coli* Cmr protein.

Susceptible *Haemophilus influenzae* exhibits an energy-dependent and saturable chloramphenicol uptake system (52). This activity is negatively affected by puromycin and can be competitively inhibited by fluorinated chloramphenicol analogs (52). These data and the similarity of some of the nonenzymatic chloramphenicol resistance proteins to other known efflux systems, particularly TetC, TetA, and QacA, suggest that they mediate resistance in a common manner.

Resistance to both chloramphenicol and florfenicol (a fluorinated derivative of chloramphenicol approved for use in the treatment of bovine respiratory pathogens) was initially attributed to the plasmid-specified *flo*$_{Pp}$ (*flo*$_{Pasteurella piscicida}$) gene from *P. piscicida* (166). Similar determinants have now been identified on integrons in the chromosomes of serovar Typhimurium DT104 human (*cmlA*-like) (43), bovine (*floR*) (20), and cattle *flo*$_{St}$ (*flo*$_{serovar Typhimurium}$) (40) isolates. The *flo*$_{St}$ locus is widely distributed among *Salmonella* isolates as well as in florfenicol-resistant *E. coli*, suggesting that resistance to this drug is an emerging problem (40).

At some level, the nonenzymatic chloramphenicol and chloramphenicol/florfenicol resistance mechanisms are homologous to each other. The exception appears to be the *E. coli* plasmid-specified Cml protein, an observation that was previously noted (85). This finding suggests that most of these proteins are derived from a common ancestor that diverged into two lineages: one with the ability to confer chloramphenicol resistance (CmlA/Cmr/MdfA [*E. coli*], Cml [*E. coli*], CmlA [*P. aeruginosa*] and CmlA2 [*E. aerogenes*]) and possibly multidrug resistance (CmlA/Cmr/MdfA/YjiO [95, 235] [see below]), and the other with the ability to engender resistance to both chloramphenicol and florfenicol (Flo$_{St}$/FloR/Flo$_{Pp}$) (Fig. 3).

The original communication that described the serovar Typhimurium DT104 CmlA-like protein in a human isolate (43) did not address whether this protein could also confer resistance to florfenicol. Its similarity to the florfenicol-chloramphenicol resistance proteins of animal origin suggests that it may mediate cross-resistance to florfenicol. If the CmlA-like protein were able to confer resistance to both chloramphenicol and florfenicol, then one might expect it to have originated in cattle, since florfenicol is not used in human therapy.

The *R. fascians* Cmr and the *S. lividans* Cml^R proteins, the *P. aeruginosa* CmlA and the *E. aerogenes* CmlA2 proteins, and the serovar Typhimurium DT104 CmlA-like/FloR/Flo$_{St}$ and *P. piscicida* Flo$_{Pp}$ proteins form distinct nodes on a phylogenetic tree, demonstrating their close relationship to each other (Fig. 3). In one representation, the *R. fascians* Cmr and the *S. lividans* Cml^R proteins branch together with the *E. coli* CmlA/Cmr/MdfA protein, and this node further diverges into two separate lineages (Fig. 3). The *E. coli* plasmid-specified Cml protein did not fit on this or other phylogenetic representations, which implies that it may have evolved from a different ancestor to perform a common function.

A rational system for the nomenclature of proteins that specify chloramphenicol or chloramphenicol/florfenicol resistance(s) does not exist. As a start, it would be convenient to designate those that confer resistance to chloramphenicol alone, such as CmlA, or both drugs, FloA, together with the genus and species of the organism from which they were isolated in a manner similar to that currently used for some of the chloramphenicol/florfenicol efflux proteins (see above). The use of subscripts, however, can be cumbersome and so a nomenclature system, like that for tetracycline (192, 193) or macrolide and MLS (324) resistance determinants would be welcomed.

Rifampin. Mutations within *rpoB* are, by far, the most common mechanism of rifampin resistance. However, a putative rifampin efflux protein termed Ptr was first described in *Streptomyces pristinaespiralis* (32). This system is chromosomally specified; is homologous to QacA and EmrB, proteins that are known to confer resistance to multiple antibiotics (see below); and engenders resistance to both rifampin and streptogramins (32). A plasmid found in *Pseudomonas fluorescens* was also found to mediate rifampin resistance via efflux (59). Rifampin-susceptible *E. coli* and *Pseudomonas putida* bearing this plasmid accumulated less rifampin (59). Rifampin resistance in these hosts was affected by the presence of potassium cyanide, an uncoupler, and was accompanied by changes in outer membrane proteins (59).

Aminoglycosides. Many of the antibiotics that have been discussed to this point have both hydrophobic and hydrophilic properties. This characteristic is a major factor that governs the ability of both specific and nonspecific drug efflux systems to recognize and remove these antibiotics from the cell (see below). Since the aminoglycosides are purely hydrophilic drugs, it was not anticipated that they could serve as substrates of efflux proteins. Recent findings have demonstrated otherwise. AmrAB from *B. pseudomallei* (242) and MexXY (also termed MexGH and AmrAB) from *P. aeruginosa* (7, 236, 399) confer aminoglycoside resistance via efflux. When these findings are considered, it is probable that an energy-dependent active efflux process is a major determinant in the net penetration of the aminoglycosides through the cytoplasmic membrane (H. Nikaido, personal communication).

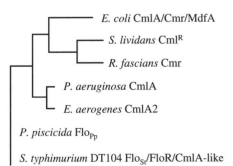

E. coli CmlA/Cmr/MdfA

S. lividans Cml^R

R. fascians Cmr

P. aeruginosa CmlA

E. aerogenes CmlA2

P. piscicida Flo$_{Pp}$

S. typhimurium DT104 Flo$_{St}$/FloR/CmlA-like

Figure 3. The evolutionary relationship between the nonenzymatic proteins that confer resistance to chloramphenicol and chloramphenicol and florfenicol. This figure was produced using Blocks (142) and PHYLIP (101) at the Blocks website (http://www.blocks.fhcrc.org/blocks/ [21 December 1999, last date accessed]).

Altered drug targets

β-Lactams. Methicillin resistance in *S. aureus* (MRSA) results from the acquisition of a new PBP (PBP-2′ specified by *mecA*) with a very low affinity for all β-lactam antibiotics (153). This low-affinity PBP can take the place functionally of all other PBPs (153) and thereby provide β-lactam resistance.

The most prevalent mechanism of β-lactam resistance in *S. pneumoniae* (pneumococci) involves the production of altered PBPs. The pneumococci can achieve a 1,000-fold resistance to benzylpenicillin through the production of altered high molecular weight PBPs (135). While there are many examples of mutational events that give rise to antibiotic resistance through altered targets (see above), clinical β-lactam resistance in the pneumococci involves the development of mosaic PBPs, which are formed from the acquisition of pieces of PBP genes from intrinsically resistant commensal streptococci (92 and reviewed in references 65, 354, and 379).

Imipenem was specifically designed to target extended-spectrum β-lactamases (see below), and most organisms containing an ESBL remain sensitive to the drug (155). Altered PBP-4 expression in *P. aeruginosa* (27) and PBP-3 expression in *L. monocytogenes* (297), however, are associated with imipenem resistance.

Glycopeptides. The antibacterial activity of the glycopeptides, e.g., vancomycin and teicoplanin, is linked to their binding of the C-terminal dipeptide (D-alanine–D-alanine [D-Ala–D-Ala]) of peptidoglycan precursors. This association prevents the transglycosylation and transpeptidation of the bacterial cell wall. Since the glycopeptides are very large antibiotics, they penetrate poorly into gram-negative organisms (402).

The enterococci are notoriously resistant to many commonly used antimicrobial agents. In circumstances of high-level multidrug-resistant enterococcal infections, vancomycin is the drug of last resort, but the emergence of vancomycin-resistant enterococci (VRE) has seriously impaired that option (181). There are five glycopeptide resistance genotypes among the enterococci and all except for one (*vanC*) are acquired resistance traits (215). The *vanA* (of chromosomal, transferable/nontransferable plasmid, and transposon origin) and *vanB* (of chromosomal, composite transposon, and plasmid origin) determinants are widely distributed in the enterococci and are transferable on plasmids by conjugation. While VanA is inducible by both vancomycin and teicoplanin and confers high-level resistances to both drugs, VanB is inducible only by vancomycin, but

when induced, it provides resistance to both drugs (23, 153, 215). The *vanC* determinant is chromosomal; is specific to *Enterococcus gallinarum*, *Enterococcus casseliflavus*, and *Enterococcus flavescens* (citations in references 105 and 215); and is constitutively expressed but engenders only a low level of resistance to vancomycin alone (23, 153, 215). The chromosomal VanD of *E. faecium* is constitutively expressed (292) and, while it cannot be transferred by conjugation, it is an acquired trait (215); it confers a substantial level of vancomycin resistance (not as high as that of VanA or VanB [215]) but a lower level of resistance to teicoplanin (292). The newly identified VanE determinant acquired by an *E. faecalis* isolate is similar to VanC-mediated glycopeptide resistance in being inducible by vancomycin and in conferring resistance to low levels of vancomycin but not to teicoplanin (105).

While more than one gene product is responsible for the expression of vancomycin resistance (23), the end result is the same—the composition of the bacterial cell wall is altered whereby D-Ala–D-Ala is replaced with either D-Ala–D-Lac (VanA, B, and D) or D-Ala–D-Ser (VanC). This new structure has a greatly reduced affinity for the glycopeptide(s). With respect to the VanA, B, C, and D phenotypes, a D,D-dipeptidase (a hydrolase termed VanX$_{A, B, C, or D}$) cleaves the normally produced D-Ala–D-Ala and a D,D-carboxypeptidase (termed VanY$_{A, B, C, or D}$) removes D-Ala from existing precursor (acyl-D-Ala–D-Ala) molecules (23). In one instance of the VanC phenotype in *E. gallinarum*, and possibly *E. faecalis*, both activities are specified within a single locus (*vanXY$_C$*) (105 and references therein). For the VanA, B, and D phenotypes, a cytoplasmic dehydrogenase (termed VanH$_{A, B, or D}$) synthesizes lactate from pyruvate (23) whereas a membrane bound serine racemase (termed VanT) produces D-Ser in the expression of VanC resistance (21). *vanA, B, C,* and *D* specify ligases that join D-Ala with a new terminal residue, either lactate or serine. In addition to the production of D-Ala–D-Ser, strains with the VanC phenotype have the ability to synthesize D-Ala–D-Ala (references in 215). This property may explain why these isolates exhibit only a low level of vancomycin resistance (215).

Curiously, some clinical isolates are dependent on the presence of vancomycin for growth (109). These vancomycin-dependent enterococci are unable to complete the biosynthesis of normal peptidoglycan precursors and therefore require the drug for the production of alternative synthetic pathways (citations in references 153 and 215).

Macrolides-lincosamides-streptogramins. As with enzymatic modification of the MLS antibiotics, an alteration in the target(s) of the MLS group can confer cross-resistance. (See reference 324 for the nomenclature of MLS target alteration determinants.) The Erm (erythromycin resistance methylase or erythromycin ribosome methylation) family of proteins (324) catalyze the most common mode of MLS resistance by methylating a specific adenine nucleotide within the 23S rRNA (see citations in reference 397). The enzyme responsible for the mechanism is specified by *ermB* (also *ermAM*). At present, there are more than 21 different classes of *erm* genes (324).

Chloramphenicol. Ribosomes purified from different chloramphenicol-resistant *Bacillus subtilis* mutants were of two classes: those that exhibited a decreased affinity for chloramphenicol and those with decreased affinity for both chloramphenicol and erythromycin (281). While the precise mechanism of resistance is not known, alterations in large ribosomal subunit proteins L3, L5, and L6 have been implicated (281).

Tetracycline. While tetracycline efflux is a well-known resistance mechanism (see above), resistance conferred by ribosomal protection is probably more widespread (323, 335). A number of determinants, including class M, O, P, Q, S, T, W, and Z tetracycline resistance determinants and the *tet* and *otrA* loci (192, 323), specify an ~68-kDa cytoplasmic protein that protects the ribosome from the inhibitory effects of tetracycline in whole cells and in cell-free systems (citations in reference 353). These ribosome protection proteins are homologs of elongation factor G (citations in reference 47). The binding of Tet M to the ribosome requires GTP, and it competes with elongation factor G for ribosome binding (79). TetM cannot, however, substitute for either elongation factor G or Tu in a reconstituted system (47). The ribosome protection proteins presumably confer tetracycline resistance by promoting the release of the drug from the ribosome in a GTP-dependent manner (47).

Substituted drug targets

Sulfonamides and trimethoprim. Sulfonamides (SUL) and trimethoprim (TMP) are competitive inhibitors of folate metabolism; TMP targets dihydrofolate reductase (DHFR) and the SUL act on dihydropteroate synthase (DHPS). DHFR is specified by *folA*, which is found in many different organisms (O. Sköld and G. Swedberg, personal communication). In gram-negative bacteria, *dfr* (*dhfr*) is an alternative designation (148). Two TMP-resistant DHFR variants have been identified in gram-positive bacteria (citations in reference 60); the S1 variant is transposon-

encoded (Tn4003) and specified by *dfrA* in *S. aureus*, *Staphylococcus haemolyticus*, *S. epidermidis*, and *Staphylococcus hominis* and the S2 type variant is plasmid-encoded and is specified by *dfrB* in *S. haemolyticus* MUR313 and *dfrD* in *L. monocytogenes* (citations in reference 60). The human homolog of DHFR is naturally TMP insensitive. DHPS is specified by *folP* (this designation should not be confused with that for the proteins that confer resistance to florfenicol and chloramphenicol [see above]), and it also is very widespread in bacteria (O. Sköld and G. Swedberg, personal communication). *folP* has also been termed *dhps* in *E. coli* and *Neisseria meningitidis* and *sulA* in *S. pneumoniae*; it is not found in human cells (148).

Chromosomal TMP resistance is attributed to the presence of Tn7, which encodes a gene cassette-specified (*dhfrI*) TMP-insensitive DHFR (28). While Tn7 has a preferred site of attachment on bacterial chromosomes, it is frequently a plasmid-specified trait. Additionally, there are many different plasmid-specified and nonallelic TMP resistance mechanisms in gram-negative bacteria and relatively few descriptions of resistance in gram-positive organisms (148). Some plasmid-specified TMP resistance genes, i.e., *dhfrI, Ib, V, VI, VII,* and *XV*, confer high-level resistances ($\gg 1$ mg/ml) (3, 148); others, i.e., *dhfrIII, IIIb, IIIc, IV,* and *IX*, confer lower levels of TMP resistance (< 1 mg/ml) (148). The *dhfrIIa, IIb,* and *IIc* genes are reported to be TMP resistance determinants (3, 148). Since they are unaffected by TMP, even at extremely high concentrations (3, 148), they probably have other functions within the cell (O. Sköld and G. Swedberg, personal communication).

Conversely, in contrast to the large number of trimethoprim resistance determinants, only two plasmid-encoded genes, *sulI* and *sulII*, specify SUL-insensitive DHPSs (148). *sulI* is present on most class 1 integrons of clinical origin and is located immediately adjacent to an integron's gene cassette (136). *sulI* is also found on transposons related to Tn21, and *sulIII*, a *sulI* deletion derivative, in *M. fortuitum* has been described (148). *sulII* is found on both conjugative and nonconjugative plasmids (148).

Resistance mediated by chromosomal mutation

Quinolones. DNA gyrase and topoisomerase IV are essential to cell viability. DNA gyrase introduces negative supercoils into DNA to relieve the topological stress that is generated during transcription and DNA replication. Topoisomerase IV catalyzes the decatenation of newly replicated daughter chromosomes. The quinolones and their broad spectrum (34) fluorinated derivatives, the fluoroquinolones, are synthetic bactericidal agents that target both DNA gyrase and topo IV (147).

In gram-negative bacteria, DNA gyrase (specified by *gyrAB*) is the primary target and topo IV (specified by *parCE*) is the secondary target of the quinolones; in gram-positive bacteria the target preference is reversed (147). In *S. pneumoniae*, trovafloxacin, pefloxacin, levofloxacin, norfloxacin, and ciprofloxacin preferentially target ParC (111, 389), while gatifloxacin and sparfloxacin (a hydrophobic fluoroquinolone) target GyrA (111). Other fluoroquinolones may initially act on either ParC or GyrA (389). Thus, high-level resistance to the quinolones can occur through mutations within either of these two targets; however, depending on which target is affected, different phenotypes are manifested. For instance, the inhibition of DNA gyrase is a bactericidal event and resistance mediated by mutations within this enzyme is a dominant trait, while *parC*$^+$/*parC*^{R} alleles in a heterozygote are codominant and topo IV inhibition is bacteriostatic in *E. coli* (165).

Rifampin. Since rifampin presumably interacts directly with RpoB, resistance is conferred by mutations that prevent this interaction. While most point mutations (both missense and nonsense) that confer high-level rifampin resistance in *E. coli*, *M. tuberculosis*, and other organisms occur within the center of RpoB in a portion of the protein termed the Rif-region (395), mutations outside this area have also been described (199, 344).

Aminoglycosides. Streptomycin is an aminocyclitol glycoside antibiotic that targets both the 16S rRNA (*rrs*) and small ribosomal protein (30S associated) S12 (*rpsL*) in *E. coli*. Since *E. coli* has multiple copies of the 16S rRNA, single streptomycin-resistant *rrs* mutations are recessive. However, in organisms of low rRNA copy number, e.g., *M. tuberculosis* and *M. leprae* (a single copy) and *M. smegmatis* (two copies), one *rrs* mutation is sufficient to confer a high level of resistance. Clinical *Mycobacterium* spp. bearing an A→G transition mutation at nucleotide 1408 (numbering based on the sequence of the *E. coli* 16S rRNA) in the 16S rRNA are resistant to kanamycin and apramycin (9, 304). While most streptomycin-resistant *M. tuberculosis* isolates have mutations in either *rrs* or *rpsL*, others do not, suggesting the presence of alternative resistance mechanisms in these strains (251).

E. coli cells bearing a single A→G transition mutation at position 1408 of a plasmid-specified 16S rRNA produce functional ribosomes and exhibit a >64- to 512-fold increase in resistance to aminoglycosides such as kanamycin A (512-fold), amikacin (256-fold), neomycin (128-fold), and neamine (>64-fold) (317). Footprinting of mutant 30S subunits in the presence of neomycin demonstrated a lack of antibiotic binding (317).

β-Lactams. In some in vitro derived β-lactam-esistant pneumococci, the resistance mechanism involves a mutation in one of two chromosomal loci, *ciaH* (specifying a histidine protein kinase) and *cpoA* (specifying a putative glycosylase) (135). These data suggest that, in addition to the PBPs, the β-lactams have other cellular targets (citations in reference 135). When these mutations were introduced into a sensitive host, there was only a modest, 1.5- to 2-fold, increase in the level of β-lactam resistance, but strains bearing these mutations exhibited a transformation-defective phenotype (citations in reference 135). Both loci may affect the cell's ability to sense damage to its cell wall and, by conferring a low level of resistance, promote antibiotic tolerance (references in reference 135) (see below).

Macrolides-lincosamides-streptogramins. Point mutations within the 23S rRNA can decrease the affinity of the MLS antibiotics for their target (91, 324, 397). However, this mode of resistance, like that of streptomycin, is applicable only to organisms with low rRNA multiplicity (397).

Glycopeptides. In the United States, vancomycin is the drug of choice for the treatment of methicillin-resistant *S. aureus* (MRSA) (261). However, an initial report describing the emergence of glycopeptide-intermediate-resistant *S. aureus* (GISA) in Japan (145) was soon followed by reports of GISA in other parts of the world, including the United States and Europe. (Strains were originally termed vancomycin-intermediate-resistant *S. aureus* [VISA].) The composition of the cell wall in laboratory generated high-level glycopeptide-resistant *S. aureus* mutants is unique (see citations in reference 350), and this property has been attributed to a deficiency in muropeptide cross-linking that is presumably due to the lack of PBP-4 (350 and references within). It has been postulated, for at least part of the resistance mechanism, that there is an increase in the number of muropeptide monomers within the altered cell wall of a resistant organism (350). This excess of monomeric muropeptide binds the glycopeptides before they reach their normal target (see reference 350 for a model of this proposal).

Tetracycline. The interaction of tetracycline with the 30S ribosomal subunit involves the 16S rRNA and many ribosomal proteins (citations in reference 327). There are seven *rrn* operons in *E. coli*, and this multiplicity has confounded the ability to isolate mutations in the 16S rRNA that might confer tetracycline resistance in this organism. Clinical tetracycline resistance in *Propionibacterium* spp. (*Propionibacterium acnes* contains three *rrn* operons [citations in reference 327]) is associated with a mutation in the 16S rRNA (327). A single base change (corresponding to nucleotide 1058 of the *E. coli* 16S rRNA) confers a 16- to 64-fold increase in resistance to drugs in the tetracycline family (327). An *E. coli* strain with all seven chromosomal *rrn* operons inactivated, but with one plasmid-specified *rrn* operon (24), will be a useful tool in determining whether similar mutational events in this host result in tetracycline resistance.

Fusidic acid. Fusidic acid resistance in *S. aureus* can result from mutations in the chromosomal *fusA* locus (62), which specifies EF-G. The mutant EF-G has a decreased affinity for fusidic acid relative to the wild-type protein (62). Decreased permeability associated with the plasmid (pUB101)-specified *fusB* locus may also contribute to fusidic acid resistance (210). An alternative but yet unknown mechanism of fusidic acid resistance in *S. aureus*, presumed to be *fusA* and pUB101 independent, is inducible by salicylate, acetyl salicylate, benzoate, and ibuprofen (305).

Another mechanism of resistance includes the ability of the type I CAT to bind and thereby sequester the drug from its normal target (citations in reference 250 and also see the discussion of CAT above and of MDR in *Yersinia pestis* below).

Nitroimidazoles. The most common mechanisms of nitroimidazole resistance are attributed to chromosomal mutations. In particular, mutations within the *H. pylori rdxA* locus, specifying oxygen-insensitive NADPH nitroreductase (125), the enzyme responsible for activation of the prodrug, confer nitroimidazole resistance. In some resistant *H. pylori* isolates, however, mutations in *rdxA* were not found, suggesting the presence of another resistance mechanism(s) (158). Decreased accumulation of the nitroimidazoles in *Bacteroides* spp. and *Clostridium* spp. is another resistance mechanism described whose basis (influx, efflux) is not known (citations in reference 96).

In contrast to high-level metronidazole resistance, clinical isolates of *Bacteroides* spp. with the Nim (5-nitroimidazole resistance) phenotype display moderate levels of metronidazole resistance and high levels of resistance to other 5-nitroimidazoles (citations in reference 384). The *nimB* gene is chromosomal while the *nimA*, *C*, and *D* alleles are plasmid encoded, suggesting that they may be readily disseminated (citations in reference 384). The mechanism by which the *nim* genes confer resistance is currently unknown (citations in reference 384).

Fosfomycin. In addition to the plasmid-specified resistance mechanisms described above, there are two chromosomally specified modes of fosfomycin resistance. The hexose phosphate uptake (*uhp*; functional only when induced by glucose-6-phosphate) and L-α-glycerophosphate transport (*glpT*) systems are specific fosfomycin uptake pathways (163). A mutation that abolishes the function of either system will confer resistance (163). In addition to decreased uptake, mutations in the chromosomal pyruvyl transferase, the target of fosfomycin, that decrease the affinity for the drug will also engender resistance (391).

Sulfonamides and trimethoprim. In both laboratory and clinical isolates, chromosomal mutations confer SUL resistance (121, 148, 282), e.g., point mutations in *E. coli* (citations in reference 148), *Campylobacter jejuni* (121), and *S. aureus* (137). The resistant DHPS in *C. jejuni* exhibited a 1,000-fold reduced affinity for sulfathiazole (121). In *N. meningitidis* (102, 174, 310) and *S. pneumoniae* (207, 221, 282) amino acid duplications within the *folP* coding sequence confer the resistance phenotype. An altered *floP* in *S. pyogenes* engenders SUL resistance (365). In *N. meningitidis*, the acquisition of SUL resistance has adverse biological effects; however, this growth disadvantage is overcome by the accumulation of compensatory mutations (103). (A resistant organism will often accumulate additional mutations to overcome particular defects that result from a mutation(s) that confers antibiotic resistance [15].) With respect to *S. pneumoniae*, an unidentified factor(s) in addition to an altered DHPS also contributes SUL resistance (282).

Chromosomal TMP resistance may be attributed to mutations that decrease the affinity of DHFR for the drug, that result in DHFR overproduction, or that create thymine auxotrophs (DHFR plays an intermediate role in the generation of 5,10-methylenetetrahydrofolate, a metabolite that participates in thymidine-5′-phosphate biosynthesis) (citations in reference 148). DHFR overproduction and synthesis of an altered enzyme both contribute to clinical TMP resistance in *E. coli* and *H. influenzae* (citations in reference 148). In a clinical *Pseudomonas cepacia* isolate, resistance is linked to a DHFR with reduced affinity for TMP (48).

Mupirocin. Mupirocin (pseudomonic acid A), produced by *P. fluorescens*, is an isoleucine analog that binds to bacterial isoleucyl-tRNA synthetase, specified by *ileS* (also termed *ileS-2* or *mupA* [citations in reference 283]), to inhibit protein synthesis (citations in reference 364). Mupirocin is highly active against all staphylococci and streptococci (except group D streptococci) and moderately effective against some gram-negative bacteria, including *H. influenzae* and *Neisseria gonorrhoeae* (364 and references therein). Since mupirocin is unstable in human tissue, its chief use has been limited to topical ointments (citations in reference 283).

Resistance to mupirocin follows chromosomal mutations in *ileS* that confer low-level resistance (313) and high-level resistance that is attributed to plasmid-specified determinants (93, 122). In *Methanobacterium thermoautotrophicum* (157), and possibly *S. aureus* (93, 122), a single amino acid point difference in the isoleucyl-tRNA synthetase results in the production of an enzyme with a reduced affinity for mupirocin (157).

MULTIPLE ANTIBIOTIC RESISTANCE IN BACTERIA

Acquired Mechanisms

Bacteria are ingenious in the manner in which they accumulate different antibiotic resistance genes. For some, these resistance traits may reside on the chromosome, on plasmids, and on integrons; their location can facilitate their transmission to other susceptible organisms. Several clinical examples in both gram-negative and gram-positive bacteria illustrate this situation.

Multidrug-resistant plague

In the past, *Y. pestis*, the etiologic agent of plague, has been susceptible to many of the antibiotics that are used in the treatment of other gram-negative infections; however, instances of tetracycline resistance have been described (citations in reference 114). A report describing plasmid-mediated MDR in *Y. pestis* (114) caused great concern. The plasmid was transferable by conjugation to both *E. coli* and avirulent *Y. pestis* (114). It was a true cornucopia of resistance traits: a TEM-1 β-lactamase; a type I chloramphenicol acetyltransferase that conferred resistance to chloramphenicol in *Y. pestis* and resistance to fusidic acid in *E. coli*; two aminoglycoside resistance mechanisms, an APH for kanamycin resistance and an ANT for streptomycin-spectinomycin resistances; a tetracycline efflux pro-

tein specified by *tet*(D); and a sulfonamide resistance gene (114).

Multiple antibiotic resistance in food organisms

Many studies have raised concerns that the rise in antibiotic resistance, particularly in certain zoonotic organisms, can be attributed to the widespread use of antibiotics in agricultural settings and animal husbandry (187, 241, 296, 371, 374, 394).

A multiple-antibiotic-resistant *Lactococcus lactis* subsp. *lactis* strain K-214 was isolated from cheese (lactococci are commonly used as starter cultures for subsequent fermentative processes) (293). This strain bears an ~29.9-kb plasmid that specifies the following resistance traits: a gene specifying a streptomycin adenyltransferase (ANT) and a CAT_I enzyme (both were presumably acquired from *S. aureus*) and a gene conferring tetracycline resistance that is almost identical to *tet*(S) (ribosome protection) from *L. monocytogenes* (293). All three resistance phenotypes were transferred to *E. faecalis* by electroporation (293). In addition, this plasmid specifies a putative macrolide efflux protein (similar to MefA from *S. pyogenes*) that confers resistance to erythromycin and other macrolides in *E. coli* (293). These findings are a further demonstration of the exchange of antibiotic resistance genes among pathogenic organisms and nonpathogenic hosts.

S. enterica serovar Typhimurium DT104

The number of human infections caused by *S. enterica* serovar Typhimurium definitive phage type 104 (DT104) is on the rise (123). Serovar Typhimurium DT104 has also been found in cattle (most commonly), farm workers, domestic pets, sheep, pigs, goats, poultry (chickens and turkeys), as well as many food products and processed foods (378 and references therein). Its presence in food is thought to play a major role in the infectious process in humans (references 322 and 378 and citations therein).

Although serovar Typhimurium DT104 is no more invasive than other nontyphoidal salmonellas, it can be resistant to seven or more different antibiotics. All the resistance determinants, except for that which engenders trimethoprim resistance, are chromosomal (377). Ciprofloxacin resistance is attributed to chromosomal mutations in the gene that specifies DNA gyrase (241, 322), and ampicillin and streptomycin resistances are due to a gene cassette-specified β-lactamase and an integron-specified adenyltransferase, respectively (citations in reference 40). Sulfonamide resistance is an attribute of *sul1* (citations in reference 40), which is present on most clinical class

1 integrons of clinical origin (136). Chloramphenicol (and florfenicol) resistance is due to a CmlA-like/FloR/Flo$_{St}$ protein (see above). A nonconjugative but mobilizable plasmid confers resistance to both trimethoprim and sulfonamides in this host (citations in reference 378).

It has been hypothesized that resistance to trimethoprim arose following the use of the drug in cattle to combat multidrug-resistant serovar Typhimurium DT104 (378). Likewise, resistance to gentamicin (in serovar Typhimurium DT104 204c) and ciprofloxacin may be linked to the use of apramycin (an aminoglycoside) and enrofloxacin (a fluoroquinolone) in veterinary medicine (241, 378).

Intrinsic and Overexpressed Chromosomally Mediated Mechanisms

As compared to gram-positive bacteria, gram-negative organisms generally display a higher level of intrinsic resistance to a range of toxic compounds. This property is in part attributed to the dual membranous structure of the gram-negative cell envelope (139, 266, 268) and to endogenous efflux systems (185). The gram-negative outer membrane is composed mainly of phospholipids, a single outer layer of lipopolysaccharide (LPS), and water-filled channels termed porins. Negatively charged LPS and the porins, the predominant nonspecific entry point for most small hydrophilic drugs, function as a selective barrier to chemicals with anionic or hydrophobic properties (139, 266, 267). This is exemplified by the low permeability of *P. aeruginosa* to many antibacterials (16, 412).

Although most gram-negative porins do not discriminate among the antibiotics that pass through them (e.g., OmpF is a nonspecific channel for many hydrophilic antibiotics), there are examples of specific antibiotic uptake pathways. The Cir and Fiu porins of *E. coli*, normally involved in siderophore transport, are responsible for the influx of catechol-containing cephalosporins (76) and β-lactams (270) into the cell (139). The OprD porin of *P. aeruginosa* (there are 17 OprD homologs specified within the *P. aeruginosa* genome [data cited in reference 275]) normally functions in basic amino acid uptake (382); its expression is increased when cells are grown on either arginine, histidine, glutamate, or alanine, as a sole carbon source (274); and it is a specific channel for imipenem (381) (a carbapenem β-lactam [326]). Thus, a decrease in porin expression will afford the cell a certain level of protection against certain toxic chemicals, particularly hydrophilic antimicrobials (266).

Reduced influx

With respect to reduced influx, *E. coli cmlB* mutants were selected in a single step and confer chloramphenicol resistance (319, 320). Relative to the wild-type parent, strains bearing the *cmlB* mutation are twofold more resistant to both tetracycline (accumulate less drug [108]) and chloramphenicol (320), increase R factor-mediated resistance to both drugs (108, 320), and exhibit a reduced level of OmpF expression (63). In *E. coli*, *cmlB* has been identified as *ompF* (30). In addition, clinical *H. influenzae* isolates bearing a chromosomal chloramphenicol resistance determinant (49, 366) exhibit a decreased expression of a major 40 kDa outer membrane protein (49). Some of these isolates were resistant to multiple antibiotics (51), including fluorinated chloramphenicol analogs (366), and the resistance phenotype was attributed to the loss of a porin (51).

Increased efflux

Decreased influx makes a relatively small contribution to drug resistance. Mutant organisms with decreased porin expression are not much more resistant than their wild-type parental strains, since drugs will eventually enter the cell, only more slowly (189, 268). The principal mechanisms occur at the less selective cytoplasmic membrane. At this point, many specific and nonspecific multidrug efflux pathways (185, 265) function to rid the cell of these antibiotics (see below).

There are four general families of drug efflux systems in bacteria (Table 4): the major facilitator (MF) superfamily (284), composed of proteins with either 12- or 14-TMS regions; the resistance-nodulation-cell division (RND) family (86, 288, 332); the small multidrug resistance (SMR) family (290), which was originally named after the staphylococcal multidrug resistance (Smr) protein (128) and then changed to SMR (290); and the adenosine triphosphate binding cassette (ABC) family (100). Recently the existence of an additional multidrug and toxic compound extrusion (MATE) family has been proposed (45) (Table 4).

The majority of bacterial multidrug efflux proteins are secondary transporters, which catalyze drug/ion antiport exchange reactions, and belong to the MFS, RND, or SMR families. The functioning of these proteins is contingent on proton motive force (PMF), which is composed of chemical/hydrogen ion (ΔpH) and electrical potential/charge ($\Delta\Psi$) components. For proteins like TetA in the MF superfamily, the efflux mechanism does not change membrane

Table 4. Families of multidrug efflux proteins in bacteria[e]

Family[a]	Energy[b]	Organism(s)	Representative member(s)	Substrate profile[c]
Resistance-nodulation-cell division (RND)	PMF	E. coli	AcrB	DET, BD, SDS, ERY, NOV, FUS, TET, MTM, EtBr, ACR, TX100, BS, BLA, CAM, FQ, NAL, OS, TRI, PO, HD
		P. aeruginosa	MexB	PYO, TET, CAM, FQ, BLA, NOV, ERY, FUS, QUO, OS
		N. gonorrhoeae	MtrD	ERY, RIF, TX100, CV, FL, AP, BS, FA
Major facilitator (MF)	PMF	E. coli	EmrB	CCCP, NAL, TCS, THI
		M. smegmatis	LfrA	EtBr, ACR, Qacs, FQ
		S. aureus	NorA	FQ, BD, PUR, CAM, TPP, EtBr, ACR, Qacs, R6G
		B. subtilis	Bmr	CAM, PUR, BD, EtBr, ACR, Qacs, FQ, R6G, TPP, DX
		L. lactis	LmrP	EtBr, TPP, DAU
Staphylococcal multidrug resistance (Smr)/ small multidrug resistance (SMR)[a]	PMF	S. aureus	Smr	Qacs, BD, EtBr, MV, TPP, LC, MTP
		E. coli	EmrE	MTP, BD, EtBr, ERY, MV, TET, SUL
		K. pneumoniae	QacE	Qacs, BD, LC, EtBr, PRO, R6G
		Gram-negative bacteria	QacEΔ1	Similar to QacE
ATP binding cassette (ABC)	ATP	L. lactis	LmrA	EtBr, R6G, R123, DAU, TPP, DX, VB, VC, COL
Multidrug and toxic compound extrusion (MATE)	PMF	V. parahaemolyticus	NorM	NOR, CIP, KAN, EtBr, STM, BER
		E. coli	YdhE	NOR, CIP, ACR, BER, TPP, KAN, STM
		H. influenzae	HI1612[d]	?
		B. subtilis	YojI[d]	?

[a] The staphylococcal multidrug resistance (Smr) protein was first identified in S. aureus (128). Subsequent to this initial report, other homologous proteins have been identified in many different organisms (see references in references 288 and 290), and since most are relatively small, the "S" designation has been changed to "small" (288, 290).

[b] PMF, proton motive force; ATP, adenosine triphosphate hydrolysis.

[c] Abbreviations: ACR, acriflavine; AP, antimicrobial peptides; BD, basic dyes; BER, berberine; BLA, β-lactams; BS, bile salts; CAM, chloramphenicol; CCCP, cyanide m-chlorophenylhydrazone; CIP, ciprofloxacin; COL, colchicine; CV, crystal violet; DET, detergents; DAU, daunomycin; DX, doxorubicin; ERY, erythromycin; EtBr, ethidium bromide; FA, fatty acids; FL, fecal lipids; FQ, fluoroquinolones; HD, household disinfectants; KAN, kanamycin; LC, lipophilic cations; MTM, mitomycin C; MTP, methyltriphenylphosphonium; MV, methyl viologen; NAL, nalidixic acid; NOR, norfloxacin; NOV, novobiocin; OS, organic solvents; PO, pine oil; PRO, proflavine; PUR, puromycin; PYO, pyoverdine; Qacs, quaternary ammonium compounds; QUO, quorums; R123, rhodamine 123; R6G, rhodamine-6G; RIF, rifampin; SDS, sodium dodecyl sulfate; STM, streptomycin; SUL, sulfadiazine; TET, tetracycline; TCS, tetrachlorosalicylanilide; THI, thiolactomycin; TPP, tetraphenylphosphonium; TRI, Triclosan; TX100, Triton X-100; VB, vinblastine; VC, vincristine.

[d] HI1612 and YojI are NorM homologs but are hypothetical proteins (45).

[e] This table was adapted from references 265 and 288.

proton distribution (electroneutral) (see above). However, Smr- (of the SMR family), EmrE- (also an SMR family member), and LmrP (of the MF superfamily)-mediated efflux is electrogenic; two (or more) protons enter for every drug effluxed (38, 129, 409). Nonetheless, efflux in these systems is inhibited by chemicals that destroy PMF, e.g., uncouplers such as 2,4-dinitrophenol (DNP) and carbonyl cyanide m-chlorophenylhydrazone (CCCP). In contrast, efflux by members of the ABC family, primary transporters, is driven by the hydrolysis of ATP (100) and thus is inhibited both by chemicals that interfere either with the functioning of the F_0F_1 ATPase (e.g., N-N'-dichlorohexylcarbodiimide) or PMF (185).

Most multidrug efflux proteins of gram-negative bacteria (see below) are multicomponent systems (Fig. 4) that transverse both the cytoplasmic and outer membranes (194, 264). The protein responsible for substrate recognition is situated within the inner membrane and is linked to an outer membrane component through a member of the membrane fusion protein (MFP) family of peptides (86) (Fig. 4). In this manner, antibiotics present in the cytoplasm (e.g., β-lactams) and drugs that are exported to the periplasm (e.g., tetracycline) are removed from the cell.

The demands of efflux from the cytoplasm are different between gram-negative and gram-positive bacteria. While efflux in gram-negative bacteria can be achieved by single proteins (e.g., the tetracycline efflux proteins that function in the absence of any accessory components [185]) (Fig. 4), these systems will concentrate the drug within the periplasm. From there, the drug will be transported out of the cell. All

Gram-negative bacteria   Gram-positive bacteria

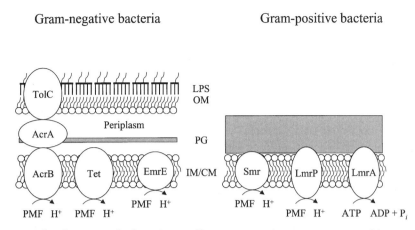

Figure 4. Organization of prokaryotic multiple antibiotic efflux systems. Multicomponent systems like AcrAB/TolC are found in gram-negative bacteria where they connect the cytoplasm directly to the extracellular medium. Other gram-negative efflux proteins include the single membrane-spanning Tet and EmrE systems. In gram-positive bacteria, all of the drug efflux systems (e.g., Smr, LmrP, LmrA) span only a single membrane. Antibiotic efflux in bacteria is powered either by proton motive force or following the hydrolysis of ATP. Abbreviations: PMF, proton motive force; IM, inner membrane; CM, cytoplasmic membrane; OM, outer membrane; PG, peptidoglycan; LPS, lipopolysaccharide.

of the gram-positive MDR proteins need only to span a single membrane, thus their action moves the drug from the cytoplasm directly into the surrounding medium (185, 194, 264). These single-protein systems must function in a manner that will overcome the spontaneous diffusion of the drug(s) back across the cytoplasmic membrane (264, 376) (see below), a much greater possibility with gram-negative bacteria. This feature may explain why multisubunit systems that are linked to the outer membrane are found in gram-negative bacteria.

Multidrug resistance (MDR) systems of gram-negative bacteria

E. coli **and other** *Enterobacteriaceae.* Current estimates suggest the presence of at least 29 known or putative MDR systems in *E. coli* (331), and many appear to have overlapping specificities (119, 194, 214). The same is presumably true of other members of the *Enterobacteriaceae.* Because many of these putative MDR proteins in *E. coli* were identified based solely on their similarity to proteins with known drug efflux activity, it does not necessarily mean that they will function as drug efflux proteins.

Mutations in *acr* increased susceptibility to cationic dyes (e.g., acriflavine, for which the locus was named), detergents, and hydrophobic agents (255, 256) (Table 4). Subsequent sequencing efforts attributed these resistance phenotypes to *acrAB* (formerly *acrAE* [213]) and its homolog *acrEF* (formerly *envCD* [168, 213]). Cells lacking *acrA* are more susceptible to ethidium bromide and novobiocin while those with *acrE* (*envC*) deleted display increased sensitivi-

ties to actinomycin D, vancomycin, and penicillin G (213). AcrB (a member of the RND family) is the inner membrane protein responsible for the recognition of multiple substrates and is linked through AcrA (a MFP family member) to an outer membrane channel, i.e., TolC (see below) (214) (Fig. 4). Since AcrA is a highly asymmetric protein with an elongated shape (415) and promotes membrane fusion (416), it is suggested that this protein is well suited to span the periplasmic space and facilitate the joining of AcrB to TolC (415, 416).

Mutations in the AcrAB system increased the cell's susceptibility to clinically relevant antibiotics (214, 280) and substances normally encountered by *E. coli* (e.g., bile salts and fatty acids) in its natural environment (212) (Table 4). Other conditions of stress, such as entry into stationary phase, growth in 4% ethanol, or osmotic shock, also stimulated expression of *acrAB* (212). Besides many antibiotics that are effluxed by the AcrAB/TolC system in *E. coli*, other chemicals such as organic solvents (400) and household disinfectants, (e.g., pine oil [240] and Triclosan [230]) are also substrates of this pump (10). Since cells lacking *acrA* still exhibit residual drug resistances (212), other homologous loci in addition to *acrEF*, including *yhiVU* (formerly *orfAB* [214]) and *acrD* (214), could have overlapping specificities (212; G. Storz, unpublished data cited in reference 214).

Mutations in the *E. coli tolC* locus confer increased susceptibility to hydrophobic chemicals (253, 401) and have other pleiotropic effects (citations in reference 110). TolC is an outer membrane protein that functions as a putative channel from the periplasm to the extracellular environment for the AcrAB multidrug efflux system (110, 194, 214) and may function with other MDR proteins in *E. coli* (110). The increased antibiotic susceptibility seen in mutants lacking TolC suggests that efflux by the AcrAB complex itself is not efficient enough to keep the drugs in the periplasm.

AcrB has been purified and reconstituted into proteoliposomes where it catalyzed the efflux of fluorescent phospholipids when conditions outside the proteoliposome were acidic (416). This activity was inhibited by known substrates of AcrAB, including bile acids, erythromycin, and cloxacillin but not with chloramphenicol, and was augmented by the addition of AcrA (416). Additionally, when the interior of the proteoliposome was acidic, and in the presence of various drugs, proton efflux was seen (416). These data strongly suggest that AcrB functions via proton antiport (416).

Homologs of AcrAB have also been found in other organisms such as serovar Typhimurium (358) and *H. influenzae* (107). Many of the compounds exported by the *E. coli* AcrAB pump are also substrates of the serovar Typhimurium (176, 269) and *H. influenzae* (337) systems. AcrAB from serovar Typhimurium has a preference for β-lactams with lipophilic side chains (269) and that from *H. influenzae* confers erythromycin resistance (337). Inactivation of the AcrAB multidrug efflux pump in serovar Typhimurium (176) and TolC *Salmonella enteritidis* (356) affects the ability of these organisms to produce disease in experimental animals.

Mutations in the *E. coli emr* (*E. coli* multidrug resistance) locus were identified in cells resistant to CCCP and other hydrophobic chemicals (205) (Table 4). This *E. coli* chromosomal locus comprises three genes, *emrR* (formerly *mprA*, originally identified as a negative regulator of microcin biosynthesis and *proU* expression in *E. coli* [83, 84]), *emrA*, and *emrB* (205). EmrB, a member of the MF superfamily, is composed of 14-TMS regions, functions as an inner membrane multisubstrate binding protein, and is linked to a yet unknown outer membrane component (possibly TolC [unpublished data cited in reference 71]) through EmrA, an MFP family member (288).

Although most MDR systems in *E. coli* are multiprotein complexes, the *emrE* locus, which confers resistance to the oxidative stress agent methyl viologen (*mvrC*) (243, 244) and to monovalent cations such as ethidium bromide (*ebr*) (306, 307), specifies a member of the SMR family of efflux proteins (290) (Table 4). Proteins in this family are characterized by 4-TMS regions and probably function as homomeric complexes (290). With respect to drug efflux, substrates are shuttled from the cytoplasm to the periplasm, and the kinetics of these single membrane-spanning pumps, like that of the Tet protein (376), must function in a manner that will overcome the spontaneous diffusion of lipophilic compounds, such as sulfadiazine, from the periplasm back to the cytoplasm (264). Their exit to the extracellular environment may occur via diffusion, as with the Tet proteins, or by some yet unidentified outer membrane channel.

An unusual characteristic of the EmrE protein is its solubility in organic solvents, which facilitated the study of substrate transport by this protein in functional proteoliposomes (409). The second remarkable feature of this relatively small multidrug transport protein is its ability to confer resistance to many drugs in *Saccharomyces cerevisiae* (408).

The *E. coli* EmrD (257) and Bcr (29, 194) proteins are homologs and members of the MF superfamily with 12-TMS regions. Since EmrD confers resistance only to CCCP and tetrachlorosalicylanilide (TSA, a hydrophobic uncoupler) and Bcr engenders bicyclomycin (29) and sulfathiazole (262) resistance, both proteins are thought to assist cells grown in the presence of uncouplers (194).

K. aerogenes. QacE (ORF1) from *K. aerogenes* is a member of the SMR family of proteins (289). It is specified within an integron on the broad host range plasmid R751 and confers resistance to many structurally dissimilar chemicals (289) (Table 4). Cells expressing QacE are phenotypically similar to strains of both *E. coli* and *S. aureus* that overexpress the *S. aureus* QacC protein (see below). Unlike that of the *S. aureus* QacA system, another protein that confers resistance to Qacs (see below), QacE does not provide resistance to diamidines and chlorhexidine (289). In *E. coli*, QacE mediates resistance in an energy-dependent manner (289).

In addition to *qacE*, the integron on R751 specifies a functional deletion derivative of *qacE*, which is termed *qacE*Δ1 (*orf4*) (289). QacEΔ1 mediates resistance to a narrower spectrum of drugs and at a level lower than that conferred by QacE (289). The QacE/QacEΔ1 (integron-specified on Tn402, Tn5090, Tn5092, and Tn5093 [289 and 311 and references therein]), the *E. coli* EmrE (MvrC/Ebr, specified in the chromosome), *S. aureus* QacC (plasmid-specified), and *P. vulgaris* ORFD (specified in the chromosome) proteins are homologs (289 and references therein).

Vibrio spp. VceAB, a homolog of the *E. coli* EmrAB system, has been identified in *Vibrio cholerae* (71). *vceAB* complements resistance to hydrophobic chemicals in *E. coli* strains lacking either *emrB* or *acrA* but does not restore resistance to hydrophilic antibiotics such as norfloxacin and tetracycline (71). These findings demonstrate that EmrAB, in contrast to AcrAB, has a preference for lipophilic compounds (71, 205).

Active efflux of norfloxacin in *Vibrio parahaemolyticus* has been attributed to the gene product of *norM* (246). In the norfloxacin-hypersusceptible and heterologous host *E. coli*, *norM* conferred resistance to hydrophilic quinolones (e.g., norfloxacin and ciprofloxacin but not to ofloxacin), ethidium, berberine, kanamycin, and streptomycin (246). However, unlike that observed with VceAB, NorM did not confer resistance to hydrophobic quinolone (nalidixic acid) or fluoroquinolone (sparfloxacin) antibiotics (246).

NorM homologs have been identified in *E. coli* (YdhE), *H. influenzae* (also termed YdhE), and *B. subtilis* (YojI) (246). While the *E. coli* YdhE protein complements resistance to some drugs, ethidium resistance remains unaffected in a norfloxacin-hypersusceptible *E. coli* host (246) (Table 4). It was originally suggested that NorM/YdhE were members of the MF family of MDR proteins (246). Further analysis suggested their presence within the multidrug resistance cluster of the new MATE family of proteins (45) (Table 4).

P. aeruginosa. In the clinical setting, infections caused by the opportunistic pathogen *P. aeruginosa* are of great concern because of their resistance to many currently used antibiotics. This property, originally attributed to the low permeability of antibiotics, has more recently been linked to a synergy between decreased permeability and active drug efflux (196, 197).

The link between efflux and resistance in *P. aeruginosa* was first suggested in studies that were initially designed to identify membrane components involved in iron acquisition (301). A segment of chromosomal DNA that restored growth of a *P. aeruginosa* mutant in an iron-deficient medium (301) was later found to specify MexA and MexB, homologs of the *E. coli* AcrEF proteins (301). A third gene within this cistron was later found to encode a 50-kDa outer membrane protein that was initially designated OprK (302) (now termed OprM [127]).

nalB/*cfxB* mutants (325) (the *cfxB* locus is a *nalB* allele [302, 325, 360]), which overexpress OprM, are resistant to many unrelated drugs (126, 197, 302) (Table 5). Insertional inactivation of *oprM*, *mexA*, or *mexB* eliminated the MDR phenotype (126, 127, 301, 410). From these data, it was inferred that these proteins constituted the Mex (multiple efflux) MDR efflux system (302), the organization of which is similar to that of AcrAB/TolC (Fig. 4).

MexB (a member of the RND family) is composed of 12 putative transmembrane regions, is situated within the inner membrane, and is presumably the component that is responsible for substrate recognition (131). It is apparently associated with both MexA (a member of the MFP family of proteins) and OprM (the outer membrane component) through two hydrophilic segments in the periplasm (131).

Three other systems homologous to MexAB-OprM exist in *P. aeruginosa* (Table 5). Unlike the constitutively expressed MexAB-OprM, the other systems are activated only following mutation: MexCD-OprJ in *nfxB* isolates (300), MexEF-OprN in *nfxC* mutants (171), and MexXY-OprM in aminoglycoside-resistant isolates (236; J. R. Aires, C. Vogne, T. Köhler, and P. Plésiat, Abstr. 39th Intersci. Conf. Antimicrob. Agents. Chemother., abstr. 670, 1999).

Mutations in NfxB, the transcriptional repressor of the *mexCD-oprJ* locus, result in OprJ overexpression (278, 279). Relative to wild-type strains, isolates that overexpress OprJ accumulate less chloramphenicol and norfloxacin in an energy dependent manner (300) and are hypersusceptibile to many β-lactams and aminoglycosides (222, 300). This hypersusceptibility is presumably attributed to an alteration in the structure of the bacterial cell membrane (224, 302). While the resistance profile of *nfxB* mutants is similar to that mediated by MexAB-OprM, *nfxB* mutants are resistant to erythromycin (138) (Table 5).

P. aeruginosa nfxC mutants overexpress OprN, a 50-kDa outer membrane protein; exhibit greatly reduced levels of OprD (112, 223); and exhibit an MDR phenotype that includes imipenem (112, 171, 381) (Table 5). The expression of MexEF-OprN and OprD is under the control of the same transcriptional regulator (see below). *nfxC* mutants are also hypersusceptible to many β-lactams and the aminoglycosides. The MexEF-OprN system functions in an energy-dependent manner, but inactivation of *mexE* caused little effect on antibiotic resistance (171).

The chromosomal *P. aeruginosa mexXY* locus was identified in an *E. coli* strain lacking *acrAB* (236). (This same locus has been independently identified as *mexGH* [7] and *amrAB* [399].) The MexXY-mediated MDR phenotype includes aminoglycosides (7, 236, 399) (Table 5). The heterologous host *E. coli* accumulated less ethidium bromide, in an energy-dependent manner, in the presence of MexXY (236). The MDR phenotype of these cells was eliminated when TolC was deleted (236), suggesting the involvement of this outer membrane protein in MexXY-mediated resistance in *E. coli* (236). In a Δ*tolC* background, plasmid-specified *oprM* restored the MDR phenotype, suggesting that MexXY may utilize OprM

Table 5. MDR systems of *Pseudomonas aeruginosa*

Inner membrane component	Membrane fusion protein	Outer membrane component	Substrates[a]	Reference(s)
MexB	MexA	OprM	NAL, FQ, MER, CAM, TET, NOV, MAC, OS, β-lactams[b], PYO, NAL, FUS	302
MexD	MexC	OprJ	4th gen. cephems, NAL, CAM, TET, ERY, OS, FQ	222, 223
MexF	MexE	OprN	TMP, CAM, FQ, OS, TET	223
MexX (MexG, AmrA)	MexY (MexH, AmrB)	OprM[c] (TolC[d])	AM[e], TET, ERY, EtBr, CAM, ACR, FQ	7, 236

[a] Abbreviations according to Table 5 and as follows: 4th gen. cephems, fourth-generation cephems; MAC, macrolides.

[b] In *P. aeruginosa*, MexAB-OprM is the β-lactam efflux system, a property that differentiates it from the other MDR proteins (355).

[c] In one study, OprM was shown to function together with MexXY (236), but in another, OprM was not thought to be part of this efflux system in this host (399).

[d] MexXY can utilize TolC in the heterologous *E. coli* host (236).

[e] MexXY (MexGH, AmrAB) is the aminoglycoside efflux system in *P. aeruginosa*, a property that differentiates it from the other MDR efflux systems in this host (7, 235, 399).

in *P. aeruginosa* and that assembly of the MexXY-OprM system could be achieved in *E. coli* (236). Other studies have suggested that OprM was not associated with MexXY function (399). Like AcrAB in *E. coli* (400), MexAB-OprM, MexCD-OprJ, and MexEF-OprN in *P. aeruginosa* contribute to organic solvent tolerance (198) (Table 5).

Burkholderia **spp.** Multiple antibiotic resistance in *Burkholderia* (*Pseudomonas*) *cepacia*, like *P. aeruginosa*, was also presumed to be attributed to poor antibiotic permeability. A large fragment of chromosomal DNA from a *B. cepacia* clinical isolate engendered resistance to chloramphenicol, trimethoprim, and ciprofloxacin in a susceptible host (53). Subsequent subcloning identified a much smaller piece of DNA, bearing part of a *mexB* homolog and a full-length *oprM* homolog termed *opcM*, that was capable of conferring multidrug resistance (53). This finding suggested that the components of a putative efflux complex in *B. cepacia* resembled the *P. aeruginosa* MexAB-OprM system and that expression of OpcM by itself could contribute to the resistance phenotype (53).

The AmrAB-OprA efflux system of *B. pseudomallei* is unique in that its resistance profile includes only aminoglycosides and macrolides (242). AmrB is similar to AcrD and MexB, AmrA bears homology to MexC, and OprA shows homology to OprJ (242). Deletion of the *amr* locus in a high-level aminoglycoside-resistant clinical isolate abrogated the resistance phenotype (242).

N. gonorrhoeae. That *N. gonorrhoeae* contains a multidrug efflux system was first suggested in studies that characterized single-step multiply resistant mutants (217). Later work suggested a role for the *mtr* (multiple transferable resistance) locus in the resistance phenotype (352). Derepression of the *mtr* locus, specifying MtrR, MtrC, MtrD (133), and MtrE (82), in *N. gonorrhoeae* is caused by mutations in the negative regulator (MtrR) of this locus (345) or in the promoter of *mtrR* (413). Since MtrC is homologous to MexA/AcrA/AcrE and MtrD is an inner membrane protein like MexB/AcrB/AcrF (133), it is surmised that drugs are exported from MtrD to MtrE through MtrC (82, 133). Like the *E. coli* Acr system(s), MtrCDE is able to export environmental toxins (e.g., fecal lipids and bile salts), and it confers macrolide resistance (82, 133, 345). Mutations in *mtrR* and its promoter have been identified in clinical azithromycin-resistant isolates (413). Inactivation of the *mtr* locus increased the susceptibility of *N. gonorrhoeae* to the antimicrobial peptides protegrin-1, found in porcine leukocytes, and LL-37, a product derived from human cells (346 and references therein).

The isolation of *N. gonorrhoeae* clinical isolates that were resistant to fatty acids but were susceptible to erythromycin and Triton X-100 suggested that an *mtr*-independent system was responsible for the resistance phenotype (225). Comparative genome searches identified the *farAB* locus in *N. gonorrhoeae* as a homolog of the *E. coli emrAB* and *V. cholerae vceAB* genes (182). The Far (fatty acid resistance) system mediates resistance to linoleic, palmitic, and oleic acids (182). Since deletion of *mtrE* removed the fatty acid resistance phenotype, it was suggested that both FarAB and MtrCD use the same outer membrane channel (182). However, while MtrR negatively regulates *mtrCDE* expression, it plays a positive role, either directly or indirectly, in the activation of *farAB* (182) (see below).

B. fragilis. *B. fragilis* is intrinsically resistant to many commonly used antibiotics, including fluoroquinolones, aminoglycosides, and β-lactams. Since *B. fragilis* is an obligate anaerobe, resistance to the aminoglycosides can be attributed to decreased uptake of the drug (239) and/or active efflux. Resistance to norfloxacin, EtBr, puromycin, and cetyldimethylethylammonium bromide (CTAB), however, was clearly attributable to active efflux (239). Further experiments demonstrated that reserpine, rescinnamine, and verapamil all increased norfloxacin susceptibility in the *B. fragilis* MDR mutants (239). The substrate specificity of the putative *B. fragilis* MDR protein is similar to that of NorA in *S. aureus*, Bmr in *B. subtilis*, and LfrA in *M. smegmatis* (see below).

Other gram-negative organisms. Less well characterized efflux systems exist in other gram-negative bacteria: *E. aerogenes*, *C. jejuni*, *P. vulgaris*, and *Stenotrophomonas* (*Xanthomonas*) *maltophilia*. Clinical *E. aerogenes* isolates that are resistant to chloramphenicol, tetracycline, and the fluoroquinolones were found to efflux norfloxacin; this mechanism was inhibitable by CCCP (216). Two individual MDR *C. jejuni* mutants, initially selected on cefotaxime or pefloxacin, overexpressed outer membrane proteins not observed in the wild-type parent, and CCCP inhibited antibiotic efflux in these strains (61). An ofloxacin-resistant clinical isolate of *P. vulgaris* accumulated less drug than its wild-type parent, and CCCP restored the bactericidal activity of ofloxacin in this mutant (151). In *S. maltophilia*, an opportunistic nosocomial pathogen, single-step tetracycline-resistant mutants exhibited an MDR phenotype that included resistance to chloramphenicol and quinolones but not to amikacin or β-lactams (13). The resistant mutants exhibited increased expression of a 54-kDa outer membrane protein, and CCCP increased tetracycline accumulation (13). A survey of nine additional unrelated clinical *S. maltophilia* isolates demonstrated that six, which overexpressed a 54-kDa outer membrane protein, exhibited an MDR phenotype (13).

Multidrug resistance (MDR) systems of gram-positive bacteria

In contrast to the dual membranous structure of the gram-negative cell envelope, the gram-positive bacterial membrane is a single lipid bilayer and is surrounded by a much larger and relatively porous layer of peptidoglycan. Efflux in gram-positive organisms is achieved by single membrane-spanning proteins (194).

S. aureus. *The quaternary ammonium compound (Qac) efflux systems.* The first reported MDR efflux protein was the plasmid-specified Qac system from *S. aureus* (372). The *qacA*, *B*, and *C/D* loci confer resistance to cationic lipophilic compounds (210) through an energy-dependent (PMF) mechanism of efflux (159, 201, 372), but each has preferred substrates (201). QacA, composed of 14-TMS segments (287), confers high-level resistances and has the broadest profile (201, 238). It also provides resistance to thrombin-induced platelet microbicidal protein 1, a small cationic peptide, but not to other antimicrobial peptides (175 and references therein). QacB engenders only a low level of resistance to the diamidines and does not afford protection against chlorhexidine (201). Although QacA and QacB bind monovalent cations equally well, the rate of efflux is faster with QacA (238).

The staphylococcal (small) multidrug resistance (Smr) protein. The prototype of the SMR family (290) is the staphylococcal (128) multidrug resistance (Smr) protein (also termed Ebr [ethidium bromide resistance determinant (338)] or QacC/D [128, 159]). (Both QacC and QacD are specified by the same gene, but the expression of each is regulated by different promoters [200].) Smr was first identified in *S. aureus* (128) and presumably functions without any accessory components (128) (Fig. 4). *smr* specifies a 107-amino-acid residue protein consisting of four putative TMS-spanning regions (128, 129). While the resistance profile was initially shown to be quite narrow (200), subsequent experiments in both *S. aureus* and *E. coli* found it to be quite broad (128). Homologs of Smr exist in *E. coli* (EmrE/MvrC [see above]), *P. vulgaris* (an ORF in the *frd* operon), *A. tumefaciens*, and *P. aeruginosa* (citations in references 128 and 129).

Purified Smr has been reconstituted in proteoliposomes, using *E. coli* phospholipids, where it may function as a multimer (129). In this system, transport of methyltriphenylphosphonium and ethidium was demonstrated, and this activity was dependent on both ΔpH and ΔΨ (129).

The NorA efflux protein. The *norA* gene from *S. aureus* was initially identified on the basis of its ability to confer inducible (161) fluoroquinolone resistance in *S. aureus* (385), and it would specify a 388-amino-acid protein composed of 12-TMS regions (411). Its expression in either *S. aureus* or *E. coli* conferred resistance to hydrophilic fluoroquinolones via efflux (385, 411). NorA also mediates ethidium bromide resistance (162) and only a modest increase in hydrophobic quinolone resistance in *S. aureus* (385, 411) but not in *E. coli* (385, 411). Since reserpine, an inhibitor of NorA, does not increase drug sensitivities in a strain lacking *norA*, NorA is the major system responsible for amphipathic cation efflux in *S. aureus* (149) (Table 4).

B. subtilis. Mutants of *B. subtilis* selected for their resistance to rhodamine 6G (a substrate of the mammalian P-glycoprotein efflux system) were later found to exhibit an MDR phenotype (258) (Table 4). The locus that conferred resistance was termed *bmr* (bacterial multidrug resistance) (258) and was later found to engender other resistances (259) (Table 4). Bmr is homologous to the gram-negative TetA-type tetracycline-resistance efflux proteins, and its activity is inhibited by CCCP and pentachlorophenol (another protonophore) (258). Although Bmr resembles P-glycoprotein in its resistance profile (e.g., to ethidium bromide) and in its inhibition by verapamil and reserpine, the two proteins are not homologs (258).

B. subtilis also contains two other well-characterized Bmr-like efflux systems. When overexpressed, Blt (Bmr-like transporter) confers resistance to many of the same drugs as Bmr (5). Bmr, however, is constitutively expressed in wild-type cells, i.e., accounts for basal resistance to ethidium bromide, while Blt is not (5). Moreover, deletion of Blt, unlike that of Bmr, had no effect on drug susceptibility (5), suggesting that it acts as "back-up" system for Bmr.

The third Bmr-like MDR pump in *B. subtilis*, Bmr3, is homologous to the *E. coli* EmrB protein and confers resistance to puromycin, norfloxacin, acriflavine, ethidium bromide, and TPP but not to nalidixic acid or CCCP (277). Bmr3 contributes to basal levels of drug resistance in the wild-type host; however, as the expression of *bmr* increases in late logarithmic phase, that of *bmr3* decreases (277).

S. pneumoniae. The existence of an MDR system(s) in *S. pneumoniae* was first suggested in studies that characterized macrolide-resistant mutants (363). Subsequently, norfloxacin resistance in a laboratory-selected fluoroquinolone-resistant isolate of *S. pneumoniae* and a resistant clinical isolate was reversed by reserpine, an MDR pump inhibitor (41, 42). Increased susceptibility of wild-type *S. pneumoniae* to multiple drugs (norfloxacin, ciprofloxacin, and ethidium bromide) was demonstrated following growth of the host in reserpine (26). A laboratory-selected fluoroquinolone-resistant isolate of *S. pneumoniae* demonstrated an MDR resistance profile, including hydrophilic fluoroquinolones, cetrimide (an antiseptic), acriflavine, tetracycline, and ethidium bromide (414). While these preliminary data suggest that *S. pneumoniae* possesses an endogenous MDR efflux

system(s), no experimental data to show efflux have been reported.

Enterococcus spp. The enterococci are notorious for their broad antibiotic resistance profiles (153) and, like *E. coli* and *N. gonorrhoeae*, are required to survive in conditions of extreme stress, e.g., in the presence of bile salts and fatty acids (209). An initial report demonstrated efflux of ethidium bromide in *Enterococcus hirae* (233). Subsequent studies extended these findings by demonstrating efflux of clinically relevant antibiotics, including norfloxacin and chloramphenicol, in susceptible *E. faecalis* and *E. faecium* (209). The low levels of tetracycline efflux found in some but not other strains of *E. faecalis* suggested that this organism may possess more than one efflux system (209).

Mycobacterium spp. A *M. smegmatis* mutant resistant to ciprofloxacin led to the identification of the *lfrA* gene, which presumably specifies an MDR efflux pump that is homologous to QacA from *S. aureus* (369). Wild-type *M. smegmatis* bearing *lfrA* in *trans* is resistant to multiple drugs (Table 4).

The *M. fortuitum* Tap (tetracycline aminoglycoside resistance) protein confers both low-level aminoglycoside (streptomycin, gentamicin, 2-*N*′-ethylnetilmicin, and 6-*N*′-ethylnetilmicin) and tetracycline resistances in *M. smegmatis* (6). Its homolog from *M. tuberculosis*, however, confers only tetracycline resistance in this heterologous host (6).

L. lactis. Although most bacterial MDR efflux systems are PMF-dependent, initial studies with multidrug-resistant mutants of *L. lactis* identified both PMF- and ATP-dependent systems (36) (Fig. 4). The *L. lactis* LmrP protein confers MDR in *E. coli* (Table 4) and, like that for MDR following expression of NorA and Bmr in *B. subtilis*, this activity was inhibited by reserpine (37). Nigericin, an ionophore that dissipates ΔpH, also inhibited LmrP-mediated EtBr efflux whereas valinomycin, a chemical that destroys ΔΨ, increased this activity (37).

LmrP has been shown to contain multiple substrate-binding sites (308), and in inside-out membrane vesicles prepared from cells overexpressing LmrP, several detergents inhibited LmrP-catalyzed drug efflux (309). LmrP, however, reconstituted in *n*-dodecyl-β-D-maltoside (DDM)-destabilized and preformed *E. coli* liposomes functioned as an PMF-dependent efflux protein (309).

LmrA is homologous to the human MDR1, a protein from *Caenorhabditis elegans*, and the *E. coli*

MsbA protein (388). The *E. coli msbA* gene is essential and was originally identified because of its ability to function as a multicopy suppressor of null mutations in *htrB*, which specifies a lauroyltransferase that participates in lipid A biosynthesis (reference 164 and citations therein). MsbA presumably functions as a transporter of core lipid A molecules across the inner membrane (299) and may also have a role as a glycerophospholipid transporter (see reference 418 and models within for the functions of both HtrB and MsbA). The similarities between LmrA and other known ABC type transporters and its ATP energy requirement place it within the ABC family of MDR efflux systems (100). In transfected human lung fibroblast cells (387), LmrA increased resistance to multiple chemicals, including many that are used in human cancer chemotherapy (387) (Table 4). The LmrA-mediated MDR was inhibited by drugs that block P-glycoprotein function and LmrA was targeted to the plasma membrane of the host (387).

Purified LmrA has been reconstituted in DDM-destabilized liposomes in the presence of *L. lactis* phospholipids (218). With respect to the orientation of LmrA in this system, half of the proteins are situated with their N-termini facing the inside of the liposome, while the others are oriented in the opposite direction (218). Since LmrA in this configuration transports fluorescent phosphatidylethanolamine but not phosphatidylcholine, it is assumed that LmrA transports only specific lipids (218).

There are two possible models by which MDR efflux pumps recognize and efflux multiple drugs: either the substrates are taken up directly from the cytoplasm or they are recognized while partitioned within a membrane bilayer. Since most of the compounds that are effluxed by the MDR proteins have hydrophobic or amphipathic properties, it is presumed that the substrate-binding site would be lipophilic (194, 273). The interaction of a hydrophobic/amphipathic drug with a lipophilic active site might occur more favorably within a hydrophobic environment (i.e., a lipid bilayer/membrane) than within an aqueous environment (i.e., in the cytoplasm) (144, 264, 268, 273).

Studies with P-glycoprotein first demonstrated that this protein was able to efflux substrates that were embedded within the cell membrane (315). Both LmrP and LmrA are also thought to transport drugs from the inner leaflet of the cytoplasmic membrane and not from the cytoplasm directly (38, 39). Experiments have demonstrated a similar mechanism of efflux for the *E. coli* AcrAB (342, 416) and the *P. aeruginosa* MexAB-OprM (273) systems. Since the MexAB-OprM and MexXY (MexGH/AmrAB)-OprM systems also efflux EtBr (273) and aminoglycosides (a purely hydrophilic antibiotic) (7), respectively, these pumps presumably contain recognition components located within both the membrane and the cytoplasm.

C. glutamicum. The cloned *cmr* (corynebacterial multidrug resistance) gene (note that the *C. glutamicum* Cmr protein is not the same as either the plasmid-specified *E. coli* Cmr or *R. fascians* Cmr proteins, which confer chloramphenicol resistance [see above]) engenders resistance to many antibiotics in the *E. coli* heterologous host but not in *C. glutamicum* (156). *cmr* would specify a protein of 459 amino acids in length with 12-TMS (156). These results suggest the presence of a negative regulatory system(s) of *cmr* expression in *C. glutamicum*, which is absent in *E. coli*, or different energetics of drug extrusion in the gram-positive bacterium (156).

The clinical importance of multidrug efflux pumps: decreasing the emergence of resistant strains and reversing resistance mediated by target mutations

Deletion of the *E. coli* AcrAB (276) efflux system and OprM in *P. aeruginosa* (204) reversed clinical high-level fluoroquinolone resistances conferred by mutations in the chromosomal gyrase and topoisomerase genes. The deletion of individual pumps in *P. aeruginosa* did not affect the frequency of mutations that confer levofloxacin resistance (204). Elimination of *mexAB-oprM* and *mexCD-oprJ* (a double mutant), however, did reduce the appearance of levofloxacin-resistant mutants, and removal of *mexAB-oprM*, *mexCD-oprJ*, and *mexEF-oprN* (a triple knockout) abrogated the mutational frequency altogether (204).

Regulation of multidrug resistance in bacteria

Bacterial multidrug resistance can be attributable to the increased synthesis of a single transcriptional activator that acts on a global level (119) (Table 6). The MDR phenotype can also be controlled by repressor/activator proteins that are specified within the same cistron (local regulation) as the resistance genes (Table 6, Fig. 5).

Global regulation of multidrug resistance efflux proteins. *E. coli and the Enterobacteriaceae: the multiple antibiotic resistance (mar) locus.* In *E. coli*, the *mar* regulon represents a salient example of how a single protein (MarA of the *marRAB* operon [66]) can have widespread effects on various other chromosomal genes (reviewed in references 10 and 12). More than 60 different genes appear to be affected by MarA (T. Barbosa and S. B. Levy, submitted for

Table 6. Regulation of bacterial multidrug efflux pumps

System	Activator	Reference(s)
Postively regulated		
E. coli		
AcrAB-TolC	MarA/Rob/SoxS[a]	10
B. subtilis		
Bmr	BmrR, Mta	4, 26
Blt	BltR, Mta	5, 26
N. gonorrhoeae		
MtrCDE	MtrA	328
FarAB	MtrR	182
P. aeruginosa		
MexEF-OprN	MexT[b]	171
P. vulgaris		
?	PqrA	150
K. pneumoniae		
?	RamA	120
E. cloacae		
?	RamA(E), RobA(EC1)	120, 183
Negatively regulated systems		
E. coli		
EmrAB	EmrR	206
AcrAB-TolC	AcrR	211
AcrEF	AcrS	214
P. aeruginosa		
MexAB-OprM	MexR	303
MexCD-OprJ	NfxB	300, 349
N. gonorrhoeae		
MtrCDE	MtrR	134
S. aureus		
QacA	QacR	130

[a] Unless induced, the contribution of SoxS in the activation of MarA, and subsequently AcrAB-TolC, is negligible (citations in reference 10).

[b] MexT also decreases expression of OprD, and this activity is responsible for imipenem resistance (275).

publication), including the synthesis of the AcrAB/TolC efflux system (110, 280) (Table 6), the dominant MDR system in *E. coli* (265, 280). MarA also alters the expression of several membrane proteins (reviewed in references 10 and 12), including OmpF (69), OmpX, and LamB (citations in reference 17). The end result is twofold: the cell's capacity to efflux toxic compounds is increased while at the same time the rate of influx of these chemicals into the cell is decreased.

MarA is the only member of the AraC/XylS family of transcriptional regulators (116) for which a complete crystal structure exists (321), but it is not the sole protein that confers a Mar phenotype in *E. coli*. Overexpression of SoxS, another protein within the AraC/XylS family (117), leads to MDR (reviewed in references 10 and 12).

The *marRAB* operon is negatively regulated by MarR, a member of a newly recognized family of regulatory proteins (12, 359). The repressor activity of MarR can be blocked by many structurally different inducing chemicals including salicylate, plumbagin,

DNP, and menadione (11, 219). The interaction of MarR with an inducer leads to the expression of MarA in whole cells (22, 68, 219, 343).

Another MarR family member is EmrR (MprA [84]), the protein that negatively controls expression of the *emrRA* locus (206) and the *mcb* (microcin producing) operon (84). EmrR also binds several chemicals (44), including CCCP, DNP, and carbonyl cyanide *p*-(trifluoro-methoxy) phenylhydrazone. It is not known whether this interaction alters repressor activity directly (as was found with MarR [11]).

B. subtilis: Mta, a protein that regulates the Bmr and Blt systems. The Mta protein of *B. subtilis*, which is closely related to TipA from *S. lividans* (both members of the MerR family of transcription regulators [361]), is another example of an autogenously regulated transcription activator that controls MDR in a global manner (25) (Table 6). The expression of Mta activates the expression of both the Bmr and Blt (25) systems (Table 6) (see above).

K. pneumoniae and E. cloacae. Susceptible *K. pneumoniae* exhibited an energy-dependent uptake of both chloramphenicol and tetracycline whereas an MDR mutant actively effluxed these drugs (120). The resistance antibiotic multiple (*ram*) locus was identified in single-step chloramphenicol-resistant mutants of *K. pneumoniae*, which also displayed cross-resistance to tetracycline, nalidixic acid, norfloxacin, ampicillin, trimethoprim, and puromycin (120). These mutants were deficient in a large porin, and the resistance phenotype was attributed to a single 113-residue protein (RamA) (120) (Table 6), a MarA homolog.

The *ramA* gene is located proximal to *romA* (reduction of outer membrane protein), a homolog of the *E. coli romA* gene that was initially thought to mediate an MDR phenotype (172, 173). Thus, the organization of these chromosomal regions in *E. cloacae* and *K. pneumoniae* is identical (120) (Fig. 4).

Overexpression of an *E. cloacae* Rob-like protein (a homolog of the *E. coli* Rob protein and termed RobA[EC1]) in both *E. cloacae* and the heterologous host *E. coli* conferred a multidrug resistance phenotype (183). In *E. cloacae* alone, increased levels of resistance to imipenem and moxalactam were observed, attributable to a synergy between high-level cephalosporinase production and reduced porin F expression (183). In an *E. cloacae* strain lacking porins F and D, introduction of RobA(EC1) decreased the host's susceptibility to nalidixic acid, chloramphenicol, and tetracycline (183). Like its counterpart in *E. coli*, overexpression of *E. cloacae* RobA(EC1) mediated MDR via a non-porin-mediated component and through decreased expression of OmpF via *micF* (183).

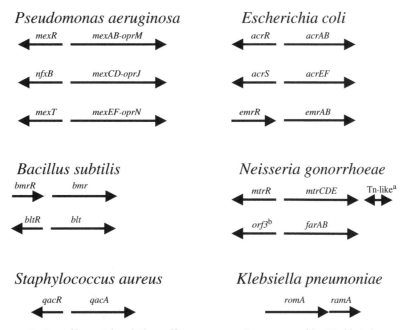

Figure 5. Genetic organization of bacterial multidrug efflux systems. a, A transposon-like (Tn-like) element exists downstream of *mtrE* in *N. gonorrhoeae*, but it is found between *mtrR* and *mtrC* in *N. meningitidis* (2, 82). The arrows represent inverted repeats that are found on both sides of the Tn-like element (82). b, The function of *orf3* in *N. gonorrhoeae* is unknown (182).

P. vulgaris. The *pqrA* was identified in a clinical fluoroquinolone-resistant *P. vulgaris* isolate (150) (Table 6). It would specify an 122-amino-acid protein, with homology to MarA and RamA, that reduced expression of OmpF and conferred resistance to the quinolones, ceftazidime, tetracycline, and chloramphenicol in the heterologous host *E. coli* (150). It is presumed to function in a manner analogous to MarA and RamA (119).

Local regulation of multidrug resistance efflux proteins. *P. aeuruginosa: positively and negatively regulated multidrug efflux systems. mexR* is located distal to the *mexAB-oprM* (303) genes (Fig. 5). It specifies a member of the MarR family of transcriptional regulators (reviewed in references 12, 234, and 359), and it negatively regulates its own expression and that of the *mexAB-oprM* operon (303). While the repressor activity of MarR (11, 219) and other MarR family members (e.g., Ec17kD, a 17-kDa *E. coli* protein of unknown function, and EmrR) is antagonized by salicylate (206, 359), that of MexR is not (303).

Deletion of *mexR* in a wild-type background increased expression of a *mexR∷lacZ* fusion, suggesting that MexR was autoregulated and functioned as a repressor (303; K. Poole, personal communication) (Table 6). Subsequent experiments that characterized 20 independently isolated and laboratory-generated *nalB*-type mutants found that all isolates contained *mexR* mutations (333). Relative to the wild-type par-

ent, these strains overexpressed OprM and a *mexA∷lacZ* fusion (333). Replacement of the wild-type *mexR* with different mutant *mexR* alleles resulted in increased antibiotic resistances and elevated expression of a *mexA∷lacZ* fusion (333). Wild-type *mexR* was also found to repress MexAB-OprM expression in clinical mutants of *P. aeruginosa* (419). These data suggest that a mutated *mexR* was sufficient to confer an MDR phenotype via overexpression of the MexAB-OprM system (333, 419).

The expression of the *P. aeruginosa* MexCD-OprJ MDR efflux system is negatively regulated by NfxB, which is specified by a gene distal to that of *mexCD-oprJ* (300) (Fig. 5 and Table 6). NfxB, a member of the LacR-GalR family of transcriptional repressors (260), reduced expression of a *mexC∷lacZ* fusion in whole cells (300) and negatively regulates its own expression (300, 349).

nfxC-type mutations exhibit increased levels of MexEF-OprN and reduced levels of OprD (see above). Characterization of an Orf upstream of *mexE* resulted in the identification of MexT (Fig. 5), a 304-residue protein with similarity to NahR (170, 275), a member of the LysR family of transcriptional activators (339). The selection of *nfxC*-type mutants is greatly reduced in strains lacking *mexT*, suggesting that this regulator is required for expression of the *nfxC*-type resistance phenotype (170).

Overexpression of plasmid-specified *mexT* increased resistance to imipenem, chloramphenicol, and norfloxacin (275). MexT elevated synthesis of a *mexE::lacZ* fusion (171) and negatively affected expression of *oprD* fusions (171, 275). Thus, MexT apparently acts both as a transcriptional activator and repressor (170, 171, 275) (Table 6).

B. subtilis: control of the Bmr, Blt, and Bmr3 multidrug resistance systems. In *B. subtilis*, genes specifying BmrR or BltR, activate their respective multidrug resistance effector proteins, Bmr and Blt (4, 5) (Fig. 5). BmrR and BltR, like that of Mta (see above), are members of the MerR family of transcription regulators (361).

From early findings, it was realized that the ability of BmrR to bind the *bmr* promoter was increased by chemicals (rhodamine 6G and TPP) that were substrates of the Bmr multidrug transporter (4). The crystal structure of the C-terminal domain of BmrR (BRC, comprising residues 121–279 of the full-length protein) in the absence and presence of TPP has now been determined (417). In the absence of TPP, a single glutamic acid residue is buried within the hydrophobic core of the protein (417). Subsequent to the binding of TPP to BRC, the carboxylic acid portion of the glutamate side chain becomes accessible to the inducer and it participates in hydrogen bonding with TPP (417). This mechanism would explain why BmrR has the ability to interact with many different positively charged substrates (417).

E. coli: negative control of the multidrug AcrAB efflux system. The *E. coli acrR* gene, which negatively controls *acrAB* expression (Fig. 5), is autogenously regulated (211). Like that of AcrAB, its synthesis is induced by general stress conditions such as 4% ethanol, 0.5 M NaCl, and stationary-phase growth, and this property is independent of either MarA or SoxS (211). Since AcrR should repress expression of the AcrAB multidrug efflux system (Table 6), some other yet to be identified factor(s) must be responsible for lack of repression by AcrR of AcrAB under these particular stress conditions (211).

S. aureus: QacR is a negative regulator of multidrug resistance and has multisubstrate binding properties. The organization of the *qacR-qacA* locus (130) is similar to that of *acrR-acrAB* and *bltR-blt* (Fig. 5). QacR does not autogenously regulate its own expression, but structurally diverse compounds reverse the binding of QacR to the *qacA* promoter (130). This interaction relieves QacR-mediated repression of QacA expression (130) (Table 6).

N. gonorrhoeae. Resistance of clinical *N. gonorrhoeae* strains to Triton X-100 (a detergent) and nonoxynol-9 (a spermicide that is structurally similar to Triton X-100) is induced following their exposure to subinhibitory concentrations of these chemicals (328), which stimulate *mtrCDE* expression (328).

That Triton X-100 can further stimulate expression of *mtrCDE* in cells lacking MtrR suggests that a second, MtrR-independent, pathway regulates *mtrCDE* expression (328). Further experiments identified a protein, MtrA (*mtr* Activator), that activated expression of *mtrCDE* (328). Thus, *mtrCDE* is negatively regulated by MtrR and positively regulated by MtrA (328) (Table 6).

OTHER RESISTANCE MECHANISMS

Antibiotic Tolerance

With respect to bactericidal antibiotics (i.e., the β-lactams, glycopeptides, and fluoroquinolones), exposure of a microbe to the drug will result in cell death, and in *S. pneumoniae*, the autolysin LytA is critical to penicillin-induced killing (380). Organisms that are resistant to the lethal effects of these antibiotics will continue to grow even though the antibiotic is present. Microbes that are drug tolerant, however, are neither fully resistant nor completely susceptible (380). These organisms occupy a middle ground; they will not die in the presence of a lethal agent (i.e., retain viability), but they will stop growing (380).

Tolerance to multiple antibiotics in *S. pneumoniae* results from a mutation in the VncS-VncR two-component regulatory system; VncS is presumed to be a sensor histidine kinase/phosphatase and VncR is thought to be the response regulator of VncS (272). These mutants were selected in the presence of penicillin and were later found to be tolerant to vancomycin, aminoglycosides, quinolones, and cephalosporins (272). In the presence of a stimulus (i.e., a bactericidal drug), the phosphatase activity of VncS is presumably activated and VncS acts on VncR to render it inactive and unable to repress the synthesis of a pathway(s) that triggers cell lysis (272). In the absence of VncS, VncR is kept in its active phosphorylated state and functions to repress expression of the lytic pathway(s) (272).

Resistance to Antiseptics and Disinfectants

Efflux-mediated resistance to pine oils and other disinfectants

E. coli chromosomal *mar* mutants are resistant to pine oil or formulations containing this natural product (240). Elimination of AcrAB in these strains reversed the resistance phenotype and rendered cells more susceptible to Qacs and chloroxylenol (240).

Resistance to Qacs has been described above (see sections on multidrug resistance systems of gram-negative [*qacE* in *K. aerogenes*] and gram-positive [*qacA–D* in *S. aureus*] bacteria above). Two new Qac resistance determinants have been identified on integrons (gene cassette-specified) in *E. aerogenes* (*qacF*) (298) and *P. aeruginosa* (*qacG*) (177). QacF conferred resistance to CTAB in *E. coli* (298), and QacG engendered resistance to Qacs and ethidium bromide in the heterologous host (177).

It has recently been suggested that some instances of benzalkonium chloride and oxacillin resistance in *S. aureus* may be linked (8). Relative to the parent strain, MRSA mutants resistant to benzalkonium chloride are also more resistant to many β-lactams (8). While the precise mechanism of coresistance has not been determined, it is thought decreased uptake or increased efflux may play a role in the resistance phenotype (8).

Resistance to Triclosan

Isolation of spontaneous *E. coli* mutants resistant to Triclosan (Ciba-Geigy Corp.), an antibacterial compound found in many consumer products such as toothpaste, hand soap, and deodorant, led to the identification of the chemical's target, enoyl reductase specified by the chromosomal *fabI* gene (231). Confirmatory studies showed that Triclosan inhibited lipid synthesis (231). That enoyl reductase was a target for Triclosan was later supported by the crystal structure of FabI complexed with Triclosan (140, 184) as well as by other biochemical data (141). In *M. smegmatis*, resistance to Triclosan was attributed to mutations in InhA, an enoyl reductase that is 35% identical to the *E. coli* FabI, and an isoniazid-resistant *M. smegmatis* mutant was cross-resistant to Triclosan (229).

Mutants that overexpress MarA, SoxS, or AcrAB in both laboratory and clinical strains are also less susceptible to Triclosan, a property that is linked to its extrusion by the AcrAB efflux pump (230). Similarly, MDR pumps in *P. aeruginosa* also efflux Triclosan (146).

Biofilm Formation

The ineffectiveness of antibiotics is also affected by community structures such as biofilms (72, 73, 263). Antibiotic resistance can be a consequence of poor penetration of the drug into the inner layers of the biofilm (citations in reference 73). Depending on their location within the biofilm, bacteria can exist in different physiologic states (e.g., slow growth) and exhibit reduced susceptibility to different agents (citations in reference 73). Finally, antibiotic resistance in biofilms may be a consequence of an induced phenotype (citations in reference 73) in a manner similar to that seen with the Mar regulon (10, 12). For example, MexAB-OprM is thought to play a role in quorum sensing and cell-to-cell communication (99, 291). Thus, if expression of MexAB-OprM was observed or needed within a biofilm, given that this system is involved in quorum transport (99), then one might expect it to fortuitously contribute to antibiotic resistance.

CONCLUDING REMARKS

Antibiotic-resistant bacteria have historically been problems in urban hospitals. Resistant infectious disease agents, however, now plague community hospitals and communities. This situation is attributed to the extensive use and misuse of antibiotics in people, animals, and agriculture. *S. aureus* isolates with a glycopeptide intermediate-resistant phenotype leading to failed therapy have appeared in Japan (145), Europe, and the United States. Four pediatric deaths, in children lacking established risk factors, have recently been caused by community-acquired MRSA (57), resistant to all β-lactam antibiotics. *S. pneumoniae, M. tuberculosis, N. gonorrhoeae, S. pyogenes,* and *E. coli* are some of the organisms bearing multidrug resistance in the community.

From where have these resistance determinants come? The same resistance genes may occur among gram-negative and gram-positive bacteria, pathogens, and opportunistic pathogens as well as commensal organisms (74). Some resistance genes are probably acquired from antibiotic-producing organisms and function to protect the organisms from their own antibiotic's activity. Through horizontal transfer and evolution (80) these genes have made their way into other bacteria, i.e., those that associate with people.

Clearly, the resistance phenomenon arises from the coming together of two factors, the antibiotic and the resistance gene. Sometimes the use of antibiotics in one area affects another. Unexpectedly, avoparcin, a glycopeptide extensively used as a growth promoter in animal husbandry in Europe, may be the key contributor to the emergence of vancomycin-resistant enterococci (VRE) (188, 386). Virginamycin, a group of streptogramins, has been used for many years as a growth promoter in animal husbandry in the United States and Europe. Synercid, a recently approved drug for clinical use in the treatment of vancomycin-resistant *E. faecium* and MRSA, consists of two similar streptogramins. Its future efficacy is threatened by the existence of *E. faecium* already resistant to the drug (398).

The everninomycins are oligosaccharide antibiotics that were first described over three decades ago

(296, 396). A new derivative (Ziracin [Schering-Plough]) is under clinical trials. Avilamycin, a structural analog of the everninomycins, however, has been used as a growth promoter in animal husbandry (references in reference 1). Although the everninomycins are not yet in clinical practice, there is concern that the previous use of avilamycin will have selected for *E. faecium* and *E. faecalis* with decreased susceptibility to Ziracin (1).

The acquisition of resistance genes from a resistant organism and selective pressure are only a portion of the story. Bacteria also possess intrinsic mechanisms that afford protection. Cytoplasmic regulatory proteins such as MarR, EmrR, and BmrR may sense when the cell is in trouble (195). For MarR, the interaction of the protein with an inducing chemical inactivates its repressor activity, resulting in the increased synthesis of MarA, which functions in a global manner at various places on the chromosome. Alternatively, inducer recognition by BmrR has a more direct (activation of Bmr) downstream effect.

It is clear that resistance traits were not created in response to antibiotics. They have served other purposes over the millennia of bacterial existence. Some may affect microbe-microbe interactions while others may serve very different purposes but are now recruited by bacteria to counteract antibiotics. For instance, the multidrug efflux pumps have been known to have other functions. The MexAB-OprM system in *P. aeruginosa* exports the siderophore pyoverdine (301) and an autoinducer termed *N*-(3-oxododecanoyl)-L-homoserine lactone or PAI-1 (*Pseudomonas* autoinducer 1) (99, 179, 291). The *P. aeruginosa* MexEF-OprN MDR efflux system excretes pyocyanin (171), and in *B. subtilis*, the Blt multidrug efflux pump catalyzes the removal of spermidine from the cell (403). The Acr system of *E. coli*, serovar Typhimurium, and *S. enteritidis* efflux bile acids (176, 356, 375). The *N. gonorrhoeae* MtrCDE system exports toxic environmental (fecal) lipids and may be required for survival on mucosal surfaces (133).

In reviewing the material presented here, it became clear that bacteria have evolved well for survival. The genes and mechanisms of resistance were already there—before people entered their world. All that was needed was the selective advantage offered by our overuse of these natural products. The future for antibiotics can be optimistic if providers and consumers learn from their past mistakes. We should use antibiotics more prudently, in ways to encourage the return and maintenance of the susceptible microbial flora.

Acknowledgments. We are especially grateful to Karen Bush, Ola Sköld, and Göte Swedberg for their valuable comments and suggestions. We also thank Marilyn Roberts, Keith Poole, and Hiroshi Nikaido for sharing unpublished data with us.

Work in our laboratory is supported by National Institute of Health grants GM 51661 and GM 55430 and through funding from the Paralyzed Veterans of America Spinal Cord Research Foundation.

REFERENCES

1. Aarestrup, F. M. 1998. Association between decreased susceptibility to a new antibiotic for treatment of human diseases, everninomicin (SCH 27899), and resistance to an antibiotic used for growth promotion in animals, avilamycin. *Microbial Drug Res.* **4:**137–141.

2. Abadi, F. J. R., P. E. Carter, P. Cash, and T. H. Pennington. 1996. Rifampin resistance in *Neisseria meningitidis* due to alterations in membrane permeability. *Antimicrob. Agents Chemother.* **40:**646–651.

3. Adrian, P. V., M. du Plessis, K. P. Klugman, and S. G. B. Amyes. 1998. New trimethoprim-resistant dihydrofolate reductase cassette, *dfrXV*, inserted in a class 1 integron. *Antimicrob. Agents Chemother.* **42:**2221–2224.

4. Ahmed, M., C. M. Borsch, S. S. Taylor, N. Vazquez-Laslop, and A. A. Neyfakh. 1994. A protein that activates expression of a multidrug efflux transporter upon binding the transporter substrates. *J. Biol. Chem.* **269:**28506–28513.

5. Ahmed, M., L. Lyass, P. N. Markham, S. S. Taylor, N. Vzquez-Laslop, and A. A. Neyfakh. 1995. Two highly similar multidrug transporters of *Bacillus subtilis* whose expression is differentially regulated. *J. Bacteriol.* **177:**3904–3910.

6. Aínsa, J. A., M. C. J. Blokpoel, I. Otal, D. B. Young, K. A. L. De Smet, and C. Martín. 1998. Molecular cloning and characterization of Tap, a putative multidrug efflux pump present in *Mycobacterium fortuitum* and *Mycobacterium tuberculosis*. *J. Bacteriol.* **180:**5836–5843.

7. Aires, J. R., T. Köhler, H. Nikaido, and P. Plésiat. 1999. Involvement of an active efflux system in the natural resistance of *Pseudomonas aeruginosa* to aminoglycosides. *Antimicrob. Agents Chemother.* **43:**2624–2628.

8. Akimitsu, N., H. Hamamoto, R. Inoue, M. Shoji, A. Akamine, K. Takemori, N. Hamasaki, and K. Sekimizu. 1999. Increase in resistance of methicillin-resistant *Staphylococcus aureus* to β-lactams caused by mutations conferring resistance to benzalkonium chloride, a disinfectant widely used in hospitals. *Antimicrob. Agents Chemother.* **43:**3042–3043.

9. Alangaden, G. J., B. N. Kreiswirth, A. Aouad, M. Khetarpal, F. R. Igno, S. L. Moghazeh, E. K. Manavathu, and S. A. Lerner. 1998. Mechanism of resistance to amikacin and kanamycin in *Mycobacterium tuberculosis*. *Antimicrob. Agents Chemother.* **42:**1295–1297.

10. Alekshun, M. N., and M. N. Levy. 1999. The *mar* regulon: multiple resistance to antibiotics and other toxic chemicals. *Trends Microbiol.* **7:**410–413.

11. Alekshun, M. N., and S. B. Levy. 1999. Alteration of the repressor activity of MarR, the negative regulator of the *Escherichia coli marRAB* locus, by multiple chemicals in vitro. *J. Bacteriol.* **181:**4669–4672.

12. Alekshun, M. N., and S. B. Levy. 1997. Regulation of chromosomally mediated multiple antibiotic resistance: the *mar* regulon. *Antimicrob. Agents Chemother.* **10:**2067–2075.

13. Alonso, A., and J. L. Martínez. 1997. Multiple antibiotic resistance in *Stenotrophomonas maltophilia*. *Antimicrob. Agents Chemother.* **41:**1140–1142.

14. Ambler, R. P. 1980. The structure of the β-lactamases. *Philos. Trans. R. Soc. Lond. (Biol.).* **289:**321–331.

15. **Andersson, D. I., and B. R. Levin.** 1999. The biological cost of antibiotic resistance. *Curr. Opin. Microbiol.* **2:**489–493.

16. **Angus, B. L., A. M. Carey, D. A. Caron, A. M. B. Kropinski, and R. E. W. Hancock.** 1982. Outer membrane permeability in *Pseudomonas aeruginosa*: comparison of a wild-type with an antibiotic-supersusceptible mutant. *Antimicrob. Agents Chemother.* **21:**299–309.

17. **Aono, R.** 1998. Improvement of organic solvent tolerance level of *Escherichia coli* by overexpression of stress-responsive genes. *Extremophiles* **2:**239–248.

18. **Arca, P., C. Hardisson, and J. E. Suárez.** 1990. Purification of a glutathione *S*-transferase that mediates fosfomycin resistance in bacteria. *Antimicrob. Agents Chemother.* **34:**844–848.

19. **Arca, P., M. Rico, A. J. Braña, C. J. Villar, C. Hardisson, and J. E. Suárez.** 1988. Formation of an adduct between fosfomycin and glutathione: a new mechanism of antibiotic resistance in bacteria. *Antimicrob. Agents Chemother.* **32:**1552–1556.

20. **Arcangioli, M.-A., S. Leroy-Sétrin, J.-L. Martel, and E. Chaslus-Dancla.** 1999. A new chloramphenicol and florfenicol resistance gene flanked by two integron structures in *Salmonella typhimurium* DT104. *FEMS Microbiol. Lett.* **174:**327–332.

21. **Arias, C. A., M. Martín-Martinez, T. L. Blundell, M. Arthur, P. Courvalin, and P. E. Reynolds.** 1999. Characterization and modelling of VanT: a novel, membrane-bound, serine racemase from vancomycin-resistant *Enterococcus gallinarium* BM4174. *Mol. Microbiol.* **31:**1653–1664.

22. **Ariza, R. R., S. P. Cohen, N. Bachhawat, S. B. Levy, and B. Demple.** 1994. Repressor mutations in the *marRAB* operon that activate oxidative stress genes and multiple antibiotic resistance in *Escherichia coli*. *J. Bacteriol.* **176:**143–148.

23. **Arthur, M., P. E. Reynolds, and P. Courvalin.** 1996. Glycopeptide resistance in the enterococci. *Trends Microbiol.* **4:**401–407.

24. **Asai, T., D. Zaporojets, C. Squires, and C. L. Squires.** 1999. An *Escherichia coli* strain with all chromosomal rRNA operons inactivated: complete exchange of rRNA genes between bacteria. *Proc. Natl. Acad. Sci. USA* **96:**1971–1976.

25. **Baranova, N. N., A. Danchin, and A. A. Neyfakh.** 1999. Mta, a global MerR-type regulator of the *Bacillus subtilis* multidrug-efflux transporters. *Mol. Microbiol.* **31:**1549–1559.

26. **Baranova, N. N., and A. A. Neyfakh.** 1997. Apparent involvement of a multidrug transporter in the fluoroquinolone resistance of *Streptococcus pneumoniae*. *Antimicrob. Agents Chemother.* **41:**1396–1398.

27. **Bellido, F., C. Veuthey, J. Blaser, A. Bauernfeind, and J. C. Pechère.** 1990. Novel resistance to imipenem associated with an altered PBP-4 in a *Pseudomonas aeruginosa* clinical isolate. *J. Antimicrob. Chemother.* **25:**57–68.

28. **Bennett, P. M.** 1991. Transposable elements and transposition in bacteria, p. 323–364, *In* U. N. Streips and R. E. Yasbin (ed.), *Modern Microbial Genetics*. Wiley-Liss, Inc., New York, N.Y.

29. **Bentley, J., L. S. Hyatt, K. Ainley, J. H. Parish, R. B. Herbert, and G. R. White.** 1993. Cloning and sequence analysis of an *Escherichia coli* gene conferring bicyclomycin resistance. *Gene* **127:**117–120.

30. **Berlyn, M. K. B.** 1998. Linkage map of *Escherichia coli* K-12, edition 10: the traditional map. *Microbiol. Mol. Biol. Rev.* **62:**814–984.

31. **Bissonnette, L., S. Champetier, J.-P. Buisson, and P. H. Roy.** 1991. Characterization of the nonenzymatic chloramphenicol resistance (*cmlA*) gene of the In4 integron of Tn*1696*: similarity of the product to transmembrane transport proteins. *J. Bacteriol.* **173:**4493–4502.

32. **Blanc, V., K. Salah-Bey, M. Folcher, and C. J. Thompson.** 1995. Molecular characterization and transcriptional analysis of a multidrug resistance gene cloned from the pristinamycin-producing organism, *Streptomyces pristinaespiralis*. *Mol. Microbiol.* **17:**989–999.

33. **Blattner, F. R., G. Plunkett, C. A. Bloch, N. T. Perna, V. Burland, M. Riley, J. Collado-Vides, J. D. Glasner, C. K. Rode, G. F. Mayhew, J. Gregor, N. W. Davis, H. A. Kirkpatrick, M. A. Goeden, D. J. Rose, B. Mau, and Y. Shao.** 1997. The complete genome sequence of *Escherichia coli* K-12. *Science* **277:**1453–1474.

34. **Blondeau, J. M.** 1999. A review of the comparative *in-vitro* activities of 12 antimicrobial agents, with a focus on five new 'respiratory quinolones'. *J. Antimicrob. Chemother.* **43** (Suppl. B):1–11.

35. **Bohn, C., and P. Bouloc.** 1998. The *Escherichia coli cmlA* gene encodes the multidrug efflux pump Cmr/MdfA and is responsible for isopropyl-β-D-thiogalactopyranoside exclusion and spectinomycin sensitivity. *J. Bacteriol.* **180:**6072–6075.

36. **Bolhuis, H., D. Molenaar, G. Poelarends, H. W. van Veen, B. Poolman, A. J. M. Driessen, and W. N. Konings.** 1994. Proton motive force-driven and ATP-dependent drug extrusion systems in multidrug-resistant *Lactococcus lactis*. *J. Bacteriol.* **176:**6957–6964.

37. **Bolhuis, H., G. Poelarends, H. W. van Veen, B. Poolman, A. J. M. Driessen, and W. N. Konings.** 1995. The lactococcal *lmrP* gene encodes a proton motive force-dependent drug transporter. *J. Biol. Chem.* **270:**26092–26098.

38. **Bolhuis, H., H. W. van Veen, J. R. Brands, M. Putman, B. Poolman, A. J. M. Driessen, and W. N. Konings.** 1996. Energetics and mechanism of drug transport mediated by the lactococcal multidrug transporter LmrP. *J. Biol. Chem.* **271:**24123–24128.

39. **Bolhuis, H., H. W. van Veen, D. Molenaar, B. Poolman, A. J. M. Driessan, and W. N. Konings.** 1996. Multidrug resistance in *Lactococcus lactis*: evidence for ATP-dependent drug extrusion from the inner leaflet of the cytoplasmic membrane. *EMBO J.* **15:**4239–4245.

40. **Bolton, L. F., L. C. Kelley, M. D. Lee, P. J. Fedorka-Cray, and J. J. Maurer.** 1999. Detection of multidrug-resistant *Salmonella enterica* serotype *typhimurium* DT104 based on a gene which confers cross-resistance to florfenicol and chloramphenicol. *J. Clin. Microbiol.* **37:**1348–1351.

41. **Brenwald, N. P., M. J. Gill, and R. Wise.** 1997. The effect of reserpine, an inhibitor of multi-drug efflux pumps, on the in-vitro susceptibilities of fluoroquinolone-resistant strains of *Streptococcus pneumoniae* to norfloxacin. *J. Antimicrob. Chemother.* **40:**458–460.

42. **Brenwald, N. P., M. J. Gill, and R. Wise.** 1998. Prevalence of a putative efflux mechanism among fluoroquinolone-resistant clinical isolates of *Streptococcus pneumoniae*. *Antimicrob. Agents Chemother.* **42:**2032–2035.

43. **Briggs, C. E., and P. M. Fratamico.** 1999. Molecular characterization of an antibiotic resistance gene cluster of *Salmonella typhimurium* DT104. *Antimicrob. Agents Chemother.* **43:**846–849.

44. **Brooun, A., J. J. Tomashek, and K. Lewis.** 1999. Purification and ligand binding of EmrR, a regulator of a multidrug transporter. *J. Bacteriol.* **181:**5131–5133.

45. **Brown, M. H., I. T. Paulsen, and R. A. Skurray.** 1999. The multidrug efflux protein NorM is a prototype of a new family of transporters. *Mol. Microbiol.* **31:**394–395.

46. **Bunny, K. L., R. M. Hall, and H. W. Stokes.** 1995. New mobile gene cassettes containing an aminoglycoside resistance gene, *aacA7*, and a chloramphenicol resistance gene, *catB3*, in

an integron in pBWH301. *Antimicrob. Agents Chemother.* **39:** 686–693.

47. **Burdett, V.** 1996. Tet(M)-promoted release of tetracycline from ribosomes is GTP dependent. *J. Bacteriol.* **178:**3246–3251.

48. **Burns, J. L., D. M. Lien, and L. A. Hedin.** 1989. Isolation and characterization of dihydrofolate reductase from trimethoprim-susceptible and trimethoprim-resistant *Pseudomonas cepacia. Antimicrob. Agents Chemother.* **33:**1247–1251.

49. **Burns, J. L., P. M. Mendelman, J. Levy, T. J. Stull, and A. L. Smith.** 1985. A permeability barrier as a mechanism of chloramphenicol resistance in *Haemophilus influenzae. Antimicrob. Agents Chemother.* **27:**46–54.

50. **Burns, J. L., C. E. Reubens, P. M. Mendelman, and A. L. Smith.** 1986. Cloning and expression in *Escherichia coli* of a gene encoding nonenzymatic chloramphenicol resistance from *Pseudomonas aeruginosa. Antimicrob. Agents Chemother.* **29:**445–450.

51. **Burns, J. L., and A. J. Smith.** 1987. A major outer-membrane protein functions as a porin in *Haemophilus influenzae. J. Gen. Microbiol.* **133:**1273–1277.

52. **Burns, J. L., and A. L. Smith.** 1987. Chloramphenicol accumulation by *Haemophilus influenzae. Antimicrob. Agents Chemother.* **31:**686–690.

53. **Burns, J. L., C. D. Wadsworth, J. J. Barry, and C. P. Goodall.** 1996. Nucleotide sequence analysis of a gene from *Burkholderia* (*Pseudomonas*) *cepacia* encoding an outer membrane lipoprotein involved in multiple antibiotic resistance. *Antimicrob. Agents Chemother.* **40:**307–313.

54. **Bush, K., G. A. Jacoby, and A. A. Medeiros.** 1995. A functional classification scheme for β-lactamases and its correlation with molecular structure. *Antimicrob. Agents Chemother.* **39:** 1211–1233.

55. **Calos, M. P., and J. H. Miller.** 1980. Transposable elements. *Cell* **20:**579–595.

56. **Cameron, F. H., D. J. Groot Obbink, V. P. Ackerman, and R. M. Hall.** 1986. Nucleotide sequence of the AAD(2″) aminoglycoside adenyltransferase determinant *aadB*. Evolutionary relationship of this region with those surrounding *aadA* in R538-1 and *dhfrII* in R388. *Nucleic Acids Res.* **14:**8625–8635.

57. **Centers for Disease Control and Prevention.** 1999. Four pediatric deaths from community-acquired methicillin-resistant *Staphylococcus aureus*—Minnesota and North Dakota, 1997–1999. *Morbid. Mortal. Weekly Rep.* **48:**707–710.

58. **Chaïbi, E. B., D. Sirot, G. Paul, and R. Labia.** 1999. Inhibitor-resistant TEM β-lactamases: phenotypic, genetic, and biochemical characteristics. *J. Antimicrob. Chemother.* **43:**447–458.

59. **Chandrasekaran, S., and D. Lalithakumari.** 1997. Plasmid-mediated rifampicin resistance in *Pseudomonas fluorescens. J. Med. Microbiol.* **47:**197–200.

60. **Charpentier, E., and P. Courvalin.** 1997. Emergence of the trimethoprim resistance gene *dfrD* in *Listeria monocytogenes* BM4293. *Antimicrob. Agents Chemother.* **41:**1124–1136.

61. **Charvalos, E., Y. Tselentis, M. M. Hamzehpour, T. Köhler, and J.-C. Pechere.** 1995. Evidence for an efflux pump in multidrug-resistant *Campylobacter jejuni. Antimicrob. Agents Chemother.* **39:**2019–2022.

62. **Chopra, I.** 1976. Mechanisms of resistance to fusidic acid in *Staphylococcus aureus. J. Gen. Microbiol.* **96:**229–238.

63. **Chopra, I., and S. J. Eccles.** 1978. Diffusion of tetracycline across the outer membrane of *Escherichia coli* K-12: involvement of protein Ia. *Biochem. Biophys. Res. Commun.* **83:**550–557.

64. **Clewell, D. B., S. E. Flannagan, and D. D. Jaworski.** 1995. Unconstrained bacterial promiscuity: the Tn916-Tn1545 family of conjugative transposons. *Trends Microbiol.* **3:**229–236.

65. **Coffey, T. J., C. G. Dowson, M. Daniels, and B. G. Spratt.** 1995. Genetics and molecular biology of β-lactam-resistant pneumococci. *Microb. Drug Resist.* **1:**29–34.

66. **Cohen, S. P., H. Hächler, and S. B. Levy.** 1993. Genetic and functional analysis of the multiple antibiotic resistance (*mar*) locus in *Escherichia coli. J. Bacteriol.* **175:**1484–1492.

67. **Cohen, S. P., D. C. Hooper, J. S. Wolfson, K. S. Souza, L. M. McMurry, and S. B. Levy.** 1988. Endogenous active efflux of norfloxacin in susceptible *Escherichia coli. Antimicrob. Agents Chemother.* **32:**1187–1191.

68. **Cohen, S. P., S. B. Levy, J. Foulds, and J. L. Rosner.** 1993. Salicylate induction of antibiotic resistance in *Escherichia coli*: activation of the *mar* operon and a *mar*-independent pathway. *J. Bacteriol.* **175:**7856–7862.

69. **Cohen, S. P., L. M. McMurry, and S. B. Levy.** 1988. *marA* locus causes decreased expression of OmpF porin in multiple-antibiotic-resistant (Mar) mutants of *Escherichia coli. J. Bacteriol.* **170:**5416–5422.

70. **Collis, C. M., and R. M. Hall.** 1995. Expression of antibiotic resistance genes in the integrated cassettes of integrons. *Antimicrob. Agents Chemother.* **39:**155–162.

71. **Colmer, J. A., J. A. Fralick, and A. N. Hamood.** 1998. Isolation and characterization of a putative multidrug resistance pump from *Vibrio cholerae. Mol. Microbiol.* **27:**63–72.

72. **Costerton, J. W., K. J. Cheng, G. G. Geesey, T. I. Ladd, J. C. Nickel, M. Dasgupta, and T. J. Marrie.** 1987. Bacterial biofilms in nature and disease. *Annu. Rev. Microbiol.* **41:**435–464.

73. **Costerton, J. W., P. S. Stewart, and E. P. Greenberg.** 1999. Bacterial biofilms: a common cause of persistent infections. *Science* **284:**1318–1322.

74. **Courvalin, P.** 1994. Transfer of antibiotic resistance genes between gram-positive and gram-negative bacteria. *Antimicrob. Agents Chemother.* **38:**1447–1451.

75. **Cuchural, G. J., M. H. Malamy, and F. P. Tally.** 1986. β-lactamase-mediated imipenem resistance in *Bacteroides fragilis. Antimicrob. Agents Chemother.* **30:**645–648.

76. **Curtis, N. A. C., R. L. Eisenstadt, S. J. East, R. J. Cornford, L. A. Walker, and A. J. White.** 1988. Iron-regulated outer membrane proteins of *Escherichia coli* K-12 and mechanism of action of catechol-substituted cephalosporins. *Antimicrob. Agents Chemother.* **32:**1879–1886.

77. **Dabbs, E. R.** 1987. Rifampicin inactivation by *Rhodococcus* and *Mycobacterium species. FEMS Microbiol. Lett.* **44:**395–399.

78. **Dabbs, E. R., K. Yazawa, Y. Mikami, M. Miyaji, N. Morisaki, S. Iwasaki, and K. Furihata.** 1995. Ribosylation by mycobacterial strains as a new mechanism of rifampin inactivation. *Antimicrob. Agents Chemother.* **39:**1007–1009.

79. **Dantley, K. A., H. K. Dannelly, and V. Burdett.** 1998. Binding interaction between Tet(M) and the ribosome: requirements for binding. *J. Bacteriol.* **180:**4089–4092.

80. **Davies, J.** 1994. Inactivation of antibiotics and the dissemination of resistance genes. *Science* **264:**375–382.

81. **Davies, J., and G. D. Wright.** 1997. Bacterial resistance to aminoglycoside antibiotics. *Trends Microbiol.* **5:**234–240.

82. **Delahay, R. M., B. D. Robertson, J. T. Balthazar, W. M. Shafer, and C. A. Ison.** 1997. Involvement of the gonococcal MtrE protein in the resistance of *Neisseria gonorrhoeae* to toxic hydrophobic agents. *Microbiology* **143:**2127–2133.

83. **del Castillo, I., J. E. González-Pastor, J. L. San Millán, and F. Moreno.** 1991. Nucleotide sequence of the *Escherichia coli* regulatory gene *mprA* and construction and characterization of *mprA*-deficient mutants. *J. Bacteriol.* **173:**3924–3929.

84. del Castillo, I., J. M. Gómez, and F. Moreno. 1990. *mprA*, an *Escherichia coli* gene that reduces growth-phase-dependent synthesis of microcins B17 and C7 and blocks osmoinduction of *proU* when cloned on a high-copy-number plasmid. *J. Bacteriol.* 172:437–445.

85. Desomer, J., D. Vereecke, M. Crespi, and M. Van Montagu. 1992. The plasmid-encoded chloramphenicol-resistance protein of *Rhodococcus fascians* is homologous to the transmembrane tetracycline efflux proteins. *Mol. Microbiol.* 6:2377–2385.

86. Dinh, T., I. T. Paulsen, and M. H. Saier, Jr. 1994. A family of extracytoplasmic proteins that allow transport of large molecules across the outer membranes of gram-negative bacteria. *J. Bacteriol.* 176:3825–3831.

87. Dittrich, W., M. Betzler, and H. Schrempf. 1991. An amplifiable and deletable chloramphenicol-resistance determinant of *Streptomyces lividans* 1326 encodes a putative transmembrane protein. *Mol. Microbiol.* 5:2789–2797.

88. Dorman, C. J., and T. J. Foster. 1982. Nonenzymatic chloramphenicol resistance determinants specified by plasmids R26 and R55-1 in *Escherichia coli* K-12 do not confer high-level resistance to fluorinated analogs. *Antimicrob. Agents Chemother.* 22:912–914.

89. Dorman, C. J., T. J. Foster, and W. V. Shaw. 1986. Nucleotide sequence of the R26 chloramphenicol resistance determinant and identification of its gene product. *Gene* 41:349–353.

90. Doucet-Populaire, F., P. Trieu-Cuot, I. Dosbaa, A. Andremont, and P. Courvalin. 1991. Inducible transfer of conjugative transposon Tn*1545* from *Enterococcus faecalis* to *Listeria monocytogenes* in the digestive tracts of gnotobiotic mice. *Antimicrob. Agents Chemother.* 35:185–187.

91. Douthwaite, S., J. B. Prince, and H. F. Noller. 1985. Evidence for functional interaction between domain II and V of 23S ribosomal RNA from an erythromycin-resistant mutant. *Proc. Natl. Acad. Sci. USA* 82:8330–8334.

92. Dowson, C. G., A. Hutchison, N. Woodford, A. P. Johnson, R. C. George, and B. G. Spratt. 1990. Penicillin-resistant viridans streptococci have obtained altered penicillin-binding protein genes from penicillin-resistant strains of *Streptococcus pneumoniae*. *Proc. Natl. Acad. Sci. USA* 87:5858–5862.

93. Dyke, K. G. H., S. P. Curnock, M. Golding, and W. C. Noble. 1991. Cloning of the gene conferring resistance to mupirocin in *Staphylococcus aureus*. *FEMS Microbiol. Lett.* 61:195–195.

94. Edgar, R., and E. Bibi. 1997. MdfA, an *Escherichia coli* multidrug resistance protein with an extraordinarily broad spectrum of drug recognition. *J. Bacteriol.* 179:2274–2280.

95. Edgar, R., and E. Bibi. 1999. A single membrane-embedded negative charge is critical for recognizing positively charged drugs by the *Escherichia coli* multidrug resistance protein MdfA. *EMBO J.* 18:822–832.

96. Edwards, D. I. 1993. Nitroimidazole drugs—action and resistance mechanisms. II. Mechanisms of resistance. *J. Antimicrob. Chemother.* 31:201–210.

97. El Solh, N., and J. Allignet. 1998. Staphylococcal resistance to streptogramins and related antibiotics. *Drug Res. Updates* 1:169–175.

98. Etienne, J., G. Gerboud, P. Courvalin, and J. Fleurette. 1989. Plasmid-mediated resistance to fosfomycin in *Staphylococcus epidermidis*. *FEMS Microbiol. Lett.* 52:133–137.

99. Evans, K., L. Passador, R. Srikumar, E. Tsang, J. Nezezon, and K. Poole. 1998. Influence of the MexAB-OprM multidrug efflux system on quorum sensing in *Pseudomonas aeruginosa*. *J. Bacteriol.* 180:5443–5447.

100. Fath, M. J., and R. Kolter. 1993. ABC transporters: bacterial exporters. *Microbiol. Rev.* 57:995–1017.

101. Felsenstein, J. 1989. PHYLIP-Phylogeny Inference Package (Version 3.5). *Cladistics* 5:164–166.

102. Fermér, C., B.-E. Kristiansen, O. Sköld, and G. Swedberg. 1995. Sulfonamide resistance in *Neisseria meningitidis* as defined by site-directed mutagenesis could have its origins in other species. *J. Bacteriol.* 177:4669–4675.

103. Fermér, C., and G. Swedberg. 1997. Adaptation to sulfonamide resistance in *Neisseria meningitidis* may have required compensatory changes to retain enzyme function: kinetic analysis of dihydropteroate synthases from *N. meningitidis* expressed in a knockout mutant of *Escherichia coli*. *J. Bacteriol.* 179:831–837.

104. Ferretti, J. J., K. S. Gilmore, and P. Courvalin. 1986. Nucleotide sequence analysis of the gene specifying the bifunctional 6′-aminoglycoside acetyltransferase 2″-aminoglycoside phosphotranferase enzyme in *Streptococcus faecalis* and identification and cloning of gene regions specifying the two activities. *J. Bacteriol.* 167:631–638.

105. Fines, M., B. Perichon, P. Reynolds, D. F. Sahm, and P. Courvalin. 1999. VanE, a new type of acquired glycopeptide resistance in *Enterococcus faecalis* BM4405. *Antimicrob. Agents Chemother.* 43:2161–2164.

106. Fitzhugh, A. L. 1998. Antibiotic inhibitors of the peptidyl transferase center. 1. Clindamycin as a composite analogue of the transfer RNA fragments L-Pro-Met and the D-ribosyl ring of adenosine. *Bioorg. Med. Chem. Lett.* 8:87–92.

107. Fleischmann, R. D., M. D. Adams, O. White, R. A. Clayton, E. F. Kirkness, A. R. Kerlavage, C. J. Bult, J.-F. Tomb, B. A. Dougherty, J. M. Merrick, K. McKenney, G. Sutton, W. Fitzhugh, C. A. Fields, J. D. Gocayne, J. D. Scott, R. Shirley, L.-I. Liu, A. Glodek, J. M. Kelley, J. F. Weidman, C. A. Phillips, T. Spriggs, E. Hedblom, M. D. Cotton, T. R. Utterback, M. C. Hanna, D. T. Nguyen, D. M. Saudek, R. C. Brandon, L. D. Fine, J. L. Fritchman, J. L. Fuhrmann, N. S. M. Geoghagen, C. L. Gnehm, L. A. McDonald, K. V. Small, C. M. Fraser, H. O. Smith, and J. C. Venter. 1995. Whole-genome random sequencing and assembly of *Haemophilus influenzae* Rd. *Science* 269:469–512.

108. Foster, T. J. 1975. R factor tetracycline and chloramphenicol resistance in *Escherichia coli* K12 *cmlB* mutants. *J. Gen. Microbiol.* 90:303–310.

109. Fraimow, H. S., D. L. Jungknid, D. W. Lander, D. R. Delso, and J. L. Dean. 1994. Urinary tract infection with an *Enterococcus faecalis* isolate that requires vancomycin for growth. *Ann. Int. Med.* 121:22–26.

110. Fralick, J. A. 1996. Evidence that TolC is required for functioning of the Mar/AcrAB efflux pump of *Escherichia coli*. *J. Bacteriol.* 178:5803–5805.

111. Fukuda, H., and K. Hiramatsu. 1999. Primary targets of fluoroquinolones in *Streptococcus pneumoniae*. *Antimicrob. Agents Chemother.* 43:410–412.

112. Fukuda, H., M. Hosaka, K. Hirai, and S. Iyobe. 1990. New norfloxacin resistance gene in *Pseudomonas aeruginosa* PAO. *Antimicrob. Agents Chemother.* 34:1757–1761.

113. Gaffney, D. F., E. Cundliffe, and T. J. Foster. 1981. Chloramphenicol resistance that does not involve chloramphenicol acetyltransferase encoded by plasmids from gram-negative bacteria. *J. Gen. Microbiol.* 125:113–121.

114. Galimand, M., A. Guiyoule, G. Gerbaud, B. Rasoamanana, S. Chanteau, E. Carniel, and P. Courvalin. 1997. Multidrug resistance in *Yersinia pestis* mediated by a transferable plasmid. *N. Engl. J. Med.* 337:677–680.

115. Galimand, M., T. Lambert, G. Gerbaud, and P. Courvalin. 1993. Characterization of the *aac(6′)-Ib* gene encoding an aminoglycoside 6′-N-acetyltransferase in *Pseudomonas aeru-*

ginosa BM2656. *Antimicrob. Agents Chemother.* **37:**1456–1462.

116. Gallegos, M.-T., C. Michan, and J. L. Ramos. 1993. The XylS/AraC family of regulators. *Nucleic Acids Res.* **21:**807–810.

117. Gallegos, M.-T., R. Schleif, A. Bairoch, K. Hofmann, and J. L. Ramos. 1997. AraC/XylS family of transcriptional activators. *Microbiol. Mol. Biol. Rev.* **61:**393–410.

118. García-Lobo, J. M., and J. M. Ortiz. 1982. Tn*2921*, a transposon encoding fosfomycin resistance. *J. Bacteriol.* **151:**477–479.

119. George, A. M. 1996. Multidrug resistance in enteric and other gram-negative bacteria. *FEMS Microbiol. Lett.* **139:**1–10.

120. George, A. M., R. M. Hall, and H. W. Stokes. 1995. Multidrug resistance in *Klebsiella pneumoniae*: a novel gene, *ramA*, confers a multidrug resistance phenotype in *Escherichia coli*. *Microbiol.* **141:**1909–1920.

121. Gibreel, A., and O. Sköld. 1999. Sulfonamide resistance in clinical isolates of *Campylobacter jejuni*: mutational changes in the chromosomal dihydropteroate synthase. *Antimicrob. Agents Chemother.* **43:**2156–2160.

122. Gilbart, J., C. R. Perry, and B. Slocombe. 1993. High-level mupirocin resistance in *Staphylococcus aureus*: evidence for two distinct isoleucyl-tRNA synthetases. *Antimicrob. Agents Chemother.* **37:**32–38.

123. Glynn, M. K., C. Bopp, W. Dewitt, P. Dabney, M. Mokhtar, and F. J. Angulo. 1998. Emergence of multidrug-resistant *Salmonella enterica* serotype typhimurium DT104 infections in the United States. *N. Engl. J. Med.* **338:**1333–1338.

124. Gold, H. S., and R. C. Moellering, Jr. 1996. Antimicrobial-drug resistance. *New Engl. J. Med.* **335:**1445–1453.

125. Goodwin, A., D. Kersulyte, G. Sisson, S. J. Veldhuyzen van Zanten, D. E. Berg, and P. S. Hofman. 1998. Metronidazole resistance in *Helicobacter pylori* is due to null mutations in a gene (*rdxA*) that encodes an oxygen-insensitive NADPH nitroreductase. *Mol. Microbiol.* **28:**383–393.

126. Gotoh, N., N. Itoh, H. Tsujimoto, J. Yamagishi, Y. Oyamada, and T. Nishino. 1994. Isolation of OprM-deficient mutants of *Pseudomonas aeruginosa* by transposon insertion mutagenesis: evidence of involvement in multiple antibiotic resistance. *FEMS Microbiol. Lett.* **122:**267–273.

127. Gotoh, N., H. Tsujimoto, K. Poole, J.-I. Yamagishi, and T. Nishino. 1995. The outer membrane protein OprM of *Pseudomonas aeruginosa* is encoded by *oprK* of the *mexA-mexB-oprK* multidrug resistance operon. *Antimicrob. Agents Chemother.* **39:**2567–2569.

128. Grinius, L., G. Dreguniene, E. B. Goldberg, C. H. Liao, and S. J. Projan. 1992. A staphylococcal multidrug resistance gene product is a member of a new protein family. *Plasmid* **27:**119–129.

129. Grinius, L. L., and E. B. Goldberg. 1994. Bacterial multidrug resistance is due to a single membrane protein which functions as a drug pump. *J. Biol. Chem.* **269:**29998–30004.

130. Grkovic, S., M. H. Brown, N. J. Roberts, I. T. Paulsen, and R. A. Skurray. 1998. QacR is a repressor protein that regulates expression of the *Staphylococcus aureus* multidrug efflux pump QacA. *J. Biol. Chem.* **273:**18665–18673.

131. Guan, L., M. Ehrmann, H. Yoneyama, and T. Nakae. 1999. Membrane topology of the xenobiotic-exporting subunit, MexB, of the MexA, B-OprM extrusion pump in *Pseudomonas aeruginosa*. *J. Biol. Chem.* **274:**10517–10522.

132. Gustafson, J. E., P. V. Candelaria, S. A. Fisher, J. P. Goodridge, T. A. Lichocik, T. M. McWilliams, C. T. D. Price, F. G. O'Brien, and W. B. Grubb. 1999. Growth in the presence of salicylate increases fluoroquinolone resistance in

Staphylococcus aureus. *Antimicrob. Agents Chemother.* **43:**990–992.

133. Hagman, K. E., W. Pan, B. G. Spratt, J. T. Balthazar, R. C. Judd, and W. M. Shafer. 1995. Resistance of *Neisseria gonorrhoeae* to antimicrobial hydrophobic agents is modulated by the *mtrRCDE* efflux system. *Microbiology* **141:**611–622.

134. Hagman, K. E., and W. M. Shafer. 1995. Transcriptional control of the *mtr* efflux system of *Neisseria gonorrhoeae*. *J. Bacteriol.* **177:**4162–4165.

135. Hakenbeck, R., T. Grebe, D. Zähner, and J. B. Stock. 1999. β-lactam reistance in *Streptococcus pneumoniae*: penicillin-binding proteins and non-penicillin-binding proteins. *Mol. Microbiol.* **33:**673–678.

136. Hall, R. M., and C. M. Collis. 1995. Mobile gene cassettes and integrons: capture and spread of genes by site-specific recombination. *Mol. Microbiol.* **15:**593–600.

137. Hampele, I. C., A. D'Arcy, G. E. Dale, D. Kostrewa, J. Nielsen, C. Oefner, M. G. P. Page, H.-J. Schönfeld, D. Strüber, and R. L. Then. 1997. Structure and function of the dihydropteroate synthase from *Staphylococcus aureus*. *J. Mol. Biol.* **268:**21–30.

138. Hamzephour, M. M., J.-C. Pechere, P. Plesiat, and T. Köhler. 1995. OprK and OprM define two genetically distinct multidrug efflux systems in *Pseudomonas aeruginosa*. *Antimicrob. Agents Chemother.* **39:**2392–2396.

139. Hancock, R. E. W. 1997. The bacterial outer membrane as a drug barrier. *Trends Microbiol.* **5:**37–42.

140. Heath, R. J., J. R. Rubin, D. R. Holland, E. Zhang, M. E. Snow, and C. O. Rock. 1999. Mechanism of triclosan inhibition of bacterial fatty acid synthesis. *J. Biol. Chem.* **274:**11110–11114.

141. Heath, R. J., Y.-T. Yu, M. A. Shapiro, E. Olson, and C. O. Rock. 1998. Broad spectrum antimicrobial biocides target the FabI component of fatty acid synthesis. *J. Biol. Chem.* **273:**30316–30320.

142. Henikoff, S., and J. G. Henikoff. 1994. Protein family classification based on searching a database of blocks. *Genomics* **19:**97–107.

143. Hernández-Allés, S., V. J. Benedí, L. Martínez-Martínez, A. Pascual, A. Aguilar, J. M. Tomás, and S. Albertí. 1999. Development of resistance during antimicrobial therapy caused by insertion sequence interruption of porin genes. *Antimicrob. Agents Chemother.* **43:**937–939.

144. Higgins, C. F., and M. M. Gottesman. 1992. Is the multidrug transporter a flippase? *Trends Biochem. Sci.* **17:**18–21.

145. Hiramatsu, K., N. Aritaka, H. Hanaki, S. Kawasaki, Y. Hosoda, S. Hori, Y. Fukuchi, and I. Kobayashi. 1997. Dissemination in Japanese hospitals of strains of *Staphylococcus aureus* heterogeneously resistant to vancomycin. *Lancet* **350:**1670–1673.

146. Hoang, T. T., and H. P. Schweizer. 1999. Characterization of *Pseudomonas aeruginosa* enoyl-acyl carrier protein reductase (FabI): a target for the antimicrobial triclosan and its role in acylated homoserine lactone synthesis. *J. Bacteriol.* **181:**5489–5497.

147. Hooper, D. C. 1999. Mechanisms of fluoroquinolone resistance. *Drug Res. Updates* **2:**38–55.

148. Houvinen, P., L. Sundström, G. Swedberg, and O. Sköld. 1995. Trimethoprim and sulfonamide resistance. *Antimicrob. Agents Chemother.* **39:**279–289.

149. Hsieh, P.-C., S. A. Siegel, B. Rogers, D. Davis, and K. Lewis. 1998. Bacteria lacking a multidrug pump: a sensitive tool for drug discovery. *Proc. Natl. Acad. Sci. USA* **95:**6602–6606.

150. Ishida, H., H. Fuziwara, Y. Kaibori, T. Horiuchi, K. Sato, and Y. Osada. 1995. Cloning of multidrug resistance gene

pqrA from *Proteus vulgaris. Antimicrob. Agents Chemother.* **39**:453–457.

151. **Ito, H., Y. Arakawa, S. Ohsuka, R. Wacharotayankun, N. Kato, and M. Ohta.** 1995. Plasmid-mediated dissemination of the metallo-β-lactamase gene *bla*$_{IMP}$ among clinically isolated strains of *Serratia marcescens. Antimicrob. Agents Chemother.* **39**:824–829.

152. **Jacoby, G., and A. A. Medeiros.** 1991. More extended-spectrum β-lactamases. *Antimicrob. Agents Chemother.* **35**:1697–1704.

153. **Jacoby, G. A.** 1996. Antimicrobial-resistant pathogens in the 1990s. *Annu. Rev. Med.* **47**:169–179.

154. **Jacoby, G. A., and G. L. Archer.** 1991. New mechanisms of bacterial resistance to antimicrobial agents. *N. Engl. J. Med.* **324**:601–612.

155. **Jacoby, G. A., and I. Carreras.** 1990. Activities of β-lactam antibiotics against *Escherichia coli* strains producing extended-spectrum β-lactamases. *Antimicrob. Agents Chemother.* **34**:858–862.

156. **Jäger, W., J. Kalinowski, and A. A. Pühler.** 1997. *Corynebacterium glutamicum* resistance protein conferring multidrug resistance in the heterologous host *Escherichia coli. J. Bacteriol.* **179**:2449–2451.

157. **Jenal, U., T. Rechsteiner, P.-Y. Tan, E. Buhlmann, L. Meile, and T. Leisinger.** 1991. Isoleucyl-tRNA synthetase of *Methanobacterium thermoautotrophicum* Marburg. *J. Biol. Chem.* **266**:10570–10577.

158. **Jenks, P. J., R. L. Ferrero, and A. Labigne.** 1999. The role of the *rdxA* gene in the evolution of metronidazole resistance in *Helicobacter pylori. J. Antimicrob. Chemother.* **43**:753–758.

159. **Jones, I. G., and M. Midgley.** 1985. Expression of a plasmid borne ethidium resistance determinant from *Staphylococcus aureus* in *Escherichia coli*: evidence for an efflux system. *FEMS Microbiol. Lett.* **28**:355–358.

160. **Kaatz, G. W., S. M. Seo, and C. A. Ruble.** 1991. Mechanisms of fluoroquinolone resistance in *Staphylococcus aureus. J. Infect. Dis.* **163**:1080–1086.

161. **Kaatz, G. W., and S. M. Seo.** 1995. Inducible NorA-mediated multidrug resistance in *Staphylococcus aureus. Antimicrob. Agents Chemother.* **39**:2650–2655.

162. **Kaatz, G. W., S. M. Seo, and C. A. Ruble.** 1993. Efflux-mediated fluoroquinolone resistance in *Staphylococcus aureus. Antimicrob. Agents Chemother.* **37**:1086–1094.

163. **Kahan, F. M., J. S. Kahan, P. J. Cassidy, and H. Kroop.** 1974. The mechanism of action of fosfomycin (phosphonomycin). *Ann. N. Y. Acad. Sci.* **235**:364–386.

164. **Karow, M., and C. Georgopoulos.** 1993. The essential *Escherichia coli msbA* gene, a multicopy suppressor of null mutations in the *htrB* gene, is related to the universally conserved family of ATP-dependent translocators. *Mol. Microbiol.* **7**:69–79.

165. **Khodursky, A. B., E. L. Zechiedrich, and N. R. Cozzarelli.** 1995. Topoisomerase IV is a target of quinolones in *Escherichia coli. Proc. Natl. Acad. Sci. USA* **92**:11801–11805.

166. **Kim, E. H., and T. Aoki.** 1996. Sequence analysis of the florfenicol resistance gene encoded on the transferable R-plasmid of a fish pathogen, *Pasteurella piscicida. Microbiol. Immunol.* **40**:665–669.

167. **Kleckner, N.** 1981. Transposable elements in prokaryotes. *Annu. Rev. Genet.* **15**:341–404.

168. **Klein, J. R., B. Henrich, and R. Plapp.** 1991. Molecular analysis and nucleotide sequence of the *envCD* operon of *Escherichia coli. Mol. Gen. Genet.* **230**:230–240.

169. **Knox, J. R.** 1995. Extended-spectrum and inhibitor-resistant TEM-type β-lactamases: mutations, specificity, and the three-

dimensional structure. *Antimicrob. Agents Chemother.* **39**:2593–2601.

170. **Köhler, T., S. F. Epp, L. K. Curty, and J.-C. Pechère.** 1999. Characterization of MexT, the regulator of the MexE-MexF-OprN multidrug efflux system of *Pseudomonas aeruginosa. J. Bacteriol.* **181**:6300–6305.

171. **Köhler, T., M. Michéa-Hamzehpour, U. Henze, N. Gotoh, L. K. Curty, and J.-C. Pechère.** 1997. Characterization of MexE-MexF-OprN, a positively regulated multidrug efflux system of *Pseudomonas aeruginosa. Mol. Microbiol.* **23**:345–354.

172. **Komatsu, T., M. Ohta, N. Kido, Y. Arakawa, H. Ito, and N. Kato.** 1991. Increased resistance of multiple drugs by introduction of the *Enterobacter cloacae romA* gene into OmpF porin-deficient mutants of *Escherichia coli* K-12. *Antimicrob. Agents Chemother.* **35**:2155–2158.

173. **Komatsu, T., M. Ohta, N. Kido, Y. Arakawa, H. Ito, T. Mizuno, and N. Kato.** 1990. Molecular characterization of an *Enterobacter cloacae* gene (*romA*) which pleiotropically inhibits the expression of *Escherichia coli* outer membrane proteins. *J. Bacteriol.* **172**:4082–4089.

174. **Kristiansen, B.-E., P. Rådström, A. Jenkins, E. Ask, B. Facinelli, and O. Sköld.** 1990. Cloning and characterization of a DNA fragment that confers sulfonamide resistance in serogroup B, serotype 15 strain of *Neisseria meningitidis. Antimicrob. Agents Chemother.* **34**:2277–2279.

175. **Kupferwasser, L. I., R. A. Skurray, M. H. Brown, N. Firth, M. R. Yeaman, and A. S. Bayer.** 1999. Plasmid-mediated resistance to thrombin-induced platelet microbicidal protein in staphylococci: role of the *qacA* locus. *Antimicrob. Agents Chemother.* **43**:2395–2399.

176. **Lacroix, F. J. C., A. Cloeckaert, O. Grépinet, C. Pinault, M. Y. Popoff, H. Waxin, and P. Pardon.** 1996. *Salmonella typhimurium acrB*-like gene: identification and role in resistance to biliary salts and detergents and in murine infection. *FEMS Microbiol. Lett.* **135**:161–167.

177. **Laraki, N., M. Galleni, I. Thamm, M. L. Riccio, G. Amicosante, J. M. Frère, and G. M. Rossolini.** 1999. Structure of In*31*, a *bla*$_{IMP}$-containing *Pseudomonas aeruginosa* integron phyletically related to In*5*, which carries an unusual array of gene cassettes. *Antimicrob. Agents Chemother.* **43**:890–907.

178. **Last, J. A., and J. F. Snell.** 1969. Microbiological transformation of oxytetracycline. *Nature* **211**:1002–1003.

179. **Latifi, A., M. K. Winson, M. Foglino, B. W. Bycroft, G. S. A. B. Stewart, A. Lazdunski, and P. Williams.** 1995. Multiple homologues of LuxR and LuxI control expression of virulence determinants and secondary metabolites through quorum sensing in *Pseudomonas aeruginosa* PAO1. *Mol. Microbiol.* **17**:333–343.

180. **Leclercq, R., and P. Courvalin.** 1991. Intrinsic and unusual resistance to macrolide, lincosamide, and streptogramin antibiotics in bacteria. *Antimicrob. Agents Chemother.* **35**:1273–1276.

181. **Leclercq, R., E. Derlot, J. Duval, and P. Courvalin.** 1988. Plasmid-mediated resistance to vancomycin and teicoplanin in *Enterococcus faecium. N. Engl. J. Med.* **319**:157–161.

182. **Lee, E. H., and W. M. Shafer.** 1999. The *farAB*-encoded efflux pump mediates resistance of gonococci to long-chained antibacterial fatty acids. *Mol. Microbiol.* **33**:839–845.

183. **Lee, E.-H., E. Collatz, I. Podglajen, and L. Gutmann.** 1996. A *rob*-like gene of *Enterobacter cloacae* affecting porin synthesis and susceptibility to multiple antibiotics. *Antimicrob. Agents Chemother.* **40**:2029–2033.

184. **Levy, C. W., A. Roujeinikove, S. Sedelnikova, P. J. Baker, A. R. Stuitje, A. R. Slabas, D. W. Rice, and J. B. Rafferty.**

1999. Molecular basis of triclosan activity. *Nature* **398**:383–384.

185. Levy, S. B. 1992. Active efflux mechanisms for antimicrobial resistance. *Antimicrob. Agents Chemother.* **36**:695–703.

186. Levy, S. B. 1992. *The Antibiotic Paradox. How Miracle Drugs Are Destroying the Miracle.* Plenum Publishing, New York, N.Y.

187. Levy, S. B. 1998. The challenge of antibiotic resistance. *Sci. Am.* **278**:46–53.

188. Levy, S. B. 1998. Multidrug resistance—a sign of the times. *N. Engl. J. Med.* **338**:1376–1378.

189. Levy, S. B. 1990. Multiple antibiotic resistance: gene selection, function, and spread, p. 377–386. *In* E. M. Ayoub, G. H. Cassell, W. C. J. Branche, and T. J. Henry (ed.), *Microbial Determinants of Virulence and Host Response.* American Society for Microbiology, Washington, D.C.

190. Levy, S. B., and L. McMurry. 1974. Detection of an inducible membrane protein associated with R-factor-mediated tetracycline resistance. *Biochem. Biophys. Res. Commun.* **27**:1060–1068.

191. Levy, S. B., and L. McMurry. 1978. Plasmid-determined tetracycline resistance involves new transport systems for tetracycline. *Nature* **276**:90–92.

192. Levy, S. B., L. M. McMurry, T. M. Barbosa, V. Burdett, P. Courvalin, W. Hillen, M. C. Roberts, J. I. Rood, and D. E. Taylor. 1999. Nomenclature for new tetracycline resistance determinants. *Antimicrob. Agents Chemother.* **43**:1523–1524.

193. Levy, S. B., L. M. McMurry, V. Burdett, P. Courvalin, W. Hillen, M. C. Roberts, and D. E. Taylor. 1989. Nomenclature for tetracycline resistance determinants. *Antimicrob. Agents Chemother.* **33**:1373–1374.

194. Lewis, K. 1994. Multidrug resistance pumps in bacteria: variations on a theme. *Trends Biochem. Sci.* **19**:119–1123.

195. Lewis, K. 1999. Multidrug resistance: versatile drug sensors of bacterial cells. *Curr. Biol.* **9**:R403–R407.

196. Li, X.-Z., D. M. Livermore, and H. Nikaido. 1994. Role of efflux pump(s) in intrinsic resistance of *Pseudomonas aeruginosa*: resistance to tetracycline, chloramphenicol, and norfloxacin. *Antimicrob. Agents Chemother.* **38**:1732–1741.

197. Li, X.-Z., H. Nikaido, and K. Poole. 1995. Role of MexA-MexB-OprM in antibiotic efflux in *Pseudomonas aeruginosa*. Antimicrob. Agents Chemother. **39**:1948–1953.

198. Li, X.-Z., L. Zhang, and K. Poole. 1998. Role of the multidrug efflux systems of *Pseudomonas aeruginosa* in organic solvent tolerance. *J. Bacteriol.* **180**:2987–2991.

199. Lisitsyn, N. A., E. D. Sverdlov, E. P. Moiseyeva, O. N. Danilevskaya, and V. G. Nikiforov. 1984. Mutation to rifampicin resistance at the beginning of the RNA polymerase β subunit gene in *Escherichia coli*. *Mol. Gen. Genet.* **196**:173–174.

200. Littlejohn, T. G., D. DiBerardino, L. J. Messerotti, S. J. Spiers, and R. A. Skurray. 1990. Structure and evolution of a family of genes encoding antiseptic and disinfectant resistance in *Staphylococcus aureus*. *Gene* **101**:59–66.

201. Littlejohn, T. G., I. T. Paulsen, M. T. Gillespie, J. M. Tennent, M. Midgley, I. G. Jones, A. S. Purewal, and R. T. Skurray. 1992. Substrate specificity and energetics of antiseptic and disinfectant resistance in *Staphylococcus aureus*. *FEMS Microbiol. Lett.* **74**:259–265.

202. Livermore, D. M. 1995. β-lactamases in laboratory and clinical resistance. *Clin. Microbiol. Rev.* **8**:557–584.

203. Livermore, D. M., and J. D. Williams. 1996. β-lactams: mode of action and mechanisms of bacterial resistance, p. 502–578. *In* V. Lorian (ed.), *Antibiotics in Laboratory Medicine*, 4th ed. Williams & Wilkins, Baltimore, Md.

204. Lomovskaya, O., A. Lee, K. Hoshino, H. Ishida, A. Mistry, M. S. Warren, E. Boyer, S. Chamberland, and V. J. Lee. 1999. Use of a genetic approach to evaluate the consequences of inhibition of efflux pumps in *Pseudomonas aeruginosa*. *Antimicrob. Agents Chemother.* **43**:1340–1346.

205. Lomovskaya, O., and K. Lewis. 1992. *emr*, an *Escherichia coli* locus for multidrug resistance. *Proc. Natl. Acad. Sci. USA* **89**:8938–8942.

206. Lomovskaya, O., K. Lewis, and A. Matin. 1995. EmrR is a negative regulator of the *Escherichia coli* multidrug resistance pump EmrAB. *J. Bacteriol.* **177**:2328–2334.

207. Lopez, P., M. Espinosa, B. Greenberg, and S. A. Lacks. 1987. Sulfonamide resistance in *Streptococcus pneumoniae*: DNA sequence of the gene encoding dihydropteroate synthase and characterization of the enzyme. *J. Bacteriol.* **169**:4320–4326.

208. Lovett, P. S. 1996. Translational attenuation regulation of chloramphenicol resistance in bacteria—a review. *Gene* **179**:157–162.

209. Lynch, C., P. Courvalin, and H. Nikaido. 1997. Active efflux of antimicrobial agents in wild-type strains of enterococci. *Antimicrob. Agents Chemother.* **41**:869–871.

210. Lyon, B. R., and R. A. Skurray. 1987. Antimicrobial resistance of *Staphylococcus aureus*: genetic basis. *Microbiol. Rev.* **51**:88–134.

211. Ma, D., M. Alberti, C. Lynch, H. Nikaido, and J. E. Hearst. 1996. The local repressor AcrR plays a modulating role in the regulation of *acrAB* genes of *Escherichia coli* by global stress signals. *Mol. Microbiol.* **19**:101–112.

212. Ma, D., D. N. Cook, M. Alberti, N. G. Pon, H. Nikaido, and J. E. Hearst. 1995. Genes *acrA* and *acrB* encode a stress-induced efflux system of *Escherichia coli*. *Mol. Microbiol.* **16**:45–55.

213. Ma, D., D. N. Cook, M. Alberti, N. G. Pon, H. Nikaido, and J. E. Hearst. 1993. Molecular cloning and characterization of *acrA* and *acrE* genes of *Escherichia coli*. *J. Bacteriol.* **175**:6299–6313.

214. Ma, D., D. N. Cook, J. E. Hearst, and H. Nikaido. 1994. Efflux pumps and drug resistance in gram-negative bacteria. *Trends Microbiol.* **2**:489–493.

215. Malathum, K., and B. E. Murray. 1999. Vancomycin-resistant enterococci: recent advances in genetics, epidemiology and therapeutic options. *Drug Res. Updates* **2**:224–243.

216. Mallea, M., J. Chevalier, C. Bornet, A. Eyraud, A. Davin-Regali, C. Bollet, and J.-M. Pagès. 1998. Porin alteration and active efflux: two *in vivo* drug resistance strategies used by *Enterobacter aerogenes*. *Microbiology* **144**:3003–3009.

217. Maness, M. J., and P. F. Sparling. 1973. Multiple antibiotic resistance due to a single mutation in *Neisseria gonorrhoeae*. *J. Infect. Dis.* **128**:321–330.

218. Margolles, A., M. Putman, H. W. van Veen, and W. N. Konings. 1999. The purified and functionally reconstituted multidrug transporter LmrA of *Lactococcus lactis* mediates the transbilayer movement of specific fluorescent phospholipids. *Biochemistry* **38**:16298–16306.

219. Martin, R. G., and J. L. Rosner. 1995. Binding of purified multiple antibiotic-resistance repressor protein (MarR) to *mar* operator sequences. *Proc. Natl. Acad. Sci. USA* **92**:5456–5460.

220. Martínez-Martínez, L., A. Pascual, S. Hernandez-Alles, D. Alvarez-Diaz, A. I. Suarez, J. Tran, V. J. Benedi, and G. A. Jacoby. 1999. Roles of β-lactamases and porins in activities of carbapenems and cephalosporins against *Klebsiella pneumoniae*. *Antimicrob. Agents Chemother.* **43**:1669–1673.

221. Maskell, J. P., A. M. Sefton, and L. M. C. Hall. 1997. Mechanism of sulfonamide resistance in clinical isolates of *Strep-*

tococcus pneumoniae. Antimicrob. Agents Chemother. **41:** 2121–2126.

222. **Masuda, N., N. Gotoh, S. Ohya, and T. Nishino.** 1996. Quantitative correlation between susceptibility and OprJ production in NfxB mutants of *Pseudomonas aeruginosa. Antimicrob. Agents Chemother.* **40:**909–913.

223. **Masuda, N., E. Sakagawa, and S. Ohya.** 1995. Outer membrane proteins responsible for multiple drug resistance in *Pseudomonas aeruginosa. Antimicrob. Agents Chemother.* **39:**645–649.

224. **McClure, W. R., and C. L. Cech.** 1978. On the mechanism of rifampicin inhibition of RNA synthesis. *J. Biol. Chem.* **253:** 8949–8956.

225. **McFarland, L., T. Mietzner, J. S. Knapp, E. Sandstrom, K. K. Homes, and S. A. Morse.** 1983. Gonococcal susceptibility to fecal lipids can be mediated by an *mtr*-independent mechanism. *J. Clin. Microbiol.* **18:**121–127.

226. **McMurry, L., R. E. Petrucci, Jr., and S. B. Levy.** 1980. Active efflux of tetracycline encoded by four genetically different tetracycline resistance determinants in *Escherichia coli. Proc. Natl. Acad. Sci. USA* **77:**3974–3977.

227. **McMurry, L. M., A. M. George, and S. B. Levy.** 1994. Active efflux of chloramphenicol in susceptible *Escherichia coli* strains and in multiple-antibiotic-resistant (Mar) mutants. *Antimicrob. Agents Chemother.* **38:**542–546.

228. **McMurry, L. M., and S. B. Levy.** 1999. Tetracycline resistance in gram-positive bacteria, p. 660–677. *In* V. A. Fischetti, R. P. Novick, J. J. Ferretti, D. A. Portnoy, and J. J. Rood (ed.), *Gram-Positive Pathogens.* American Society for Microbiology, Washington, D.C.

229. **McMurry, L. M., P. F. McDermott, and S. B. Levy.** 1999. Genetic evidence that InhA of *Mycobacterium smegmatis* is a target for triclosan. *Antimicrob. Agents Chemother.* **43:** 711–713.

230. **McMurry, L. M., M. Oethinger, and S. B. Levy.** 1998. Overexpression of *marA, soxS,* or *acrAB* produces resistance to triclosan in laboratory and clinical strains of *Escherichia coli. FEMS Microbiol. Lett.* **166:**305–309.

231. **McMurry, L. M., M. Oethinger, and S. B. Levy.** 1998. Triclosan targets lipid synthesis. *Nature* **394:**531–532.

232. **Meyers, E., and D. A. Smith.** 1962. Microbial degradation of the tetracyclines. *J. Bacteriol.* **84:**797–802.

233. **Midgley, M.** 1994. Characteristics of an ethidium efflux system in *Enterococcus hirae. FEMS Microbiol. Lett.* **120:**119–124.

234. **Miller, P. F., and M. C. Sulavik.** 1996. Overlaps and parallels in the regulation of intrinsic multiple-antibiotic resistance in *Escherichia coli. Mol. Microbiol.* **21:**441–448.

235. **Mine, T., Y. Morita, A. Kataoka, T. Mizushima, and T. Tsuchiya.** 1998. Evidence for chloramphenicol/H⁺ antiport in Cmr (MdfA) system of *Escherichia coli* and properties of the antiporter. *J. Biochem.* **124:**187–193.

236. **Mine, T., Y. Morita, A. Kataoka, T. Mizushima, and T. Tsuchiya.** 1999. Expression in *Escherichia coli* of a new multidrug efflux pump, MexXY, from *Pseudomonas aeruginosa. Antimicrob. Agents Chemother.* **43:**415–417.

237. **Mingeot-Leclercq, M. P., Y. Glupczynski, and P. M. Tulkens.** 1999. Aminoglycosides: activity and resistance. *Antimicrob. Agents Chemother.* **43:**727–737.

238. **Mitchell, B. A., I. T. Paulsen, M. H. Brown, and R. A. Skurray.** 1999. Bioenergetics of the staphylococcal multidrug export protein QacA. *J. Biol. Chem.* **274:**3541–3548.

239. **Miyamae, S., H. Nikaido, Y. Tanaka, and F. Yoshimura.** 1998. Active efflux of norfloxacin by *Bacteroides fragilis. Antimicrob. Agents Chemother.* **42:**2119–2121.

240. **Moken, M. C., L. M. McMurry, and S. B. Levy.** 1997. Selection of multiple antibiotic resistant (Mar) mutants of *Escherichia coli* using the disinfectant pine oil: roles of the *mar* and *acrAB* loci. *Antimicrob. Agents Chemother.* **41:** 2770–2772.

241. **Mølbak, K., D. L. Baggesen, F. M. Aarestrup, J. M. Ebbesen, J. Engberg, K. Frydendahl, P. Gerner-Smidt, A. M. Petersen, and H. C. Wegener.** 1999. An outbreak of multidrug-resistant, quinolone-resistant *Salmonella enterica* serotype typhimurium DT104. *N. Engl. J. Med.* **341:**1420–1425.

242. **Moore, R. A., D. DeShazer, S. Reckseidler, A. Weissman, and D. E. Woods.** 1999. Efflux-mediated aminoglycoside and macrolide resistance in *Burkholderia pseudomallei. Antimicrob. Agents Chemother.* **43:**465–470.

243. **Morimyo, M.** 1988. Isolation and characterization of methyl viologen-sensitive mutants of *Escherichia coli* K-12. *J. Bacteriol.* **170:**2136–2142.

244. **Morimyo, M., E. Hongo, H. Hama-Inaba, and I. Machida.** 1992. Cloning and characterization of the *mvrC* gene of *Escherichia coli* K-12 which confers resistance against methyl viologen toxicity. *Nucleic Acids Res.* **20:**3159–3165.

245. **Morisaki, N., and S. Iwasaki.** 1993. Inactivated products of rifampicin by pathogenic *Nocardia* spp. : structures of glycosylated and phosphorylated metabolites of rifampicin and 3-formylrifamycin SV. *J. Antibiot.* **46:**1605–1610.

246. **Morita, Y., K. Kodama, S. Shiota, T. Mine, A. Kataoka, T. Mizushima, and T. Tsuchiya.** 1998. NorM, a putative multidrug efflux protein, of *Vibrio parahaemolyticus* and its homolog in *Escherichia coli. Antimicrob. Agents Chemother.* **42:** 1778–1782.

247. **Morosini, M. I., M. C. Negri, B. Shoichet, M. R. Baquero, F. Baquero, and J. Blazquez.** 1998. An extended-spectrum AmpC-type β-lactamase obtained by in vitro antibiotic selection. *FEMS Microbiol. Lett.* **165:**85–90.

248. **Mosher, R. H., D. J. Camp, K. Yang, M. P. Brown, W. V. Shaw, and L. C. Vining.** 1995. Inactivation of chloramphenicol by O-phosphorylation. A novel resistance mechanism in *Streptomyces venezuelae* ISP5230, a chloramphenicol producer. *J. Biol. Chem.* **270:**27000–27006.

249. **Mosher, R. H., N. P. Ranade, H. Schrempf, and L.-C. Vining.** 1990. Chloramphenicol resistance in *Streptomyces*: cloning and characterization of a chloramphenicol hydrolase gene from *Streptomyces venezuelae. J. Gen. Microbiol.* **136:**293–301.

250. **Murray, I. A., and W. V. Shaw.** 1997. O-acetyltransferases for chloramphenicol and other natural products. *Antimicrob. Agents Chemother.* **41:**1–6.

251. **Musser, J. M.** 1995. Antimicrobial agent resistance in mycobacteria: molecular genetic insights. *Clin. Microbiol. Rev.* **8:**496–514.

252. **Nagai, Y., and S. Mitsuhashi.** 1972. New type of R factors incapable of inactivating chloramphenicol. *J. Bacteriol.* **109:** 1–7.

253. **Nagel de Zwaig, R., and S. E. Luria.** 1967. Genetics and physiology of colicin-tolerant mutants of *Escherichia coli. J. Bacteriol.* **94:**1112–1123.

254. **Nagy, E., and J. Földes.** 1991. Inactivation of metronidazole by *Enterococcus faecalis. J. Antimicrob. Chemother.* **27:**63–70.

255. **Nakamura, H.** 1965. Gene-controlled resistance to acriflavine and other basic dyes in *Escherichia coli. J. Bacteriol.* **90:** 8–14.

256. **Nakamura, H.** 1968. Genetic determination of resistance to acriflavin, phenethyl alcohol, and sodium dodecyl sulfate in *Escherichia coli. J. Bacteriol.* **96:**987–996.

257. Naroditskaya, V., M. J. Schlosser, N. Y. Fang, and K. Lewis. 1993. An *E. coli* gene *emrD* is involved in adaptation to low energy shock. *Biochem. Biophys. Res Comm.* **196**:803–809.

258. Neyfakh, A. A., V. Bidnenko, and L. B. Chen. 1991. Efflux-mediated multidrug resistance in *Bacillus subtilis*: similarities and dissimilarities with mammalian systems. *Proc. Natl. Acad. Sci. USA* **88**:4781–4785.

259. Neyfakh, A. A., C. M. Borsch, and G. W. Kaatz. 1993. Fluoroquinolone resistance protein NorA of *Staphylococcus aureus* is a multidrug efflux transporter. *Antimicrob. Agents Chemother.* **37**:128–129.

260. Nguyen, C. C., and M. H. Saier, Jr. 1995. Phylogenetic, structural and functional analyses of the LacI-GalR family of bacterial transcription factors. *FEBS Lett.* **377**:98–102.

261. Nicas, T. I., M. L. Zeckel, and D. K. Braun. 1997. Beyond vancomycin: new therapies to meet the challenge of glycopeptide resistance. *Trends Microbiol.* **5**:240–249.

262. Nichols, B. P., and G. G. Guay. 1989. Gene amplification contributes to sulfonamide resistance in *Escherichia coli*. *Antimicrob. Agents Chemother.* **33**:2042–2048.

263. Nickel, J. C., I. Ruseska, J. B. Wright, and J. W. Costerton. 1985. Tobramycin resistance of *Pseudomonas aeruginosa* cells growing as a biofilm on urinary catheter material. *Antimicrob. Agents Chemother.* **27**:619–624.

264. Nikaido, H. 1996. Multidrug efflux pumps of gram-negative bacteria. *J. Bacteriol.* **178**:5853–5859.

265. Nikaido, H. 1998. Multiple antibiotic resistance and efflux. *Curr. Opin. Microbiol.* **1**:516–523.

266. Nikaido, H. 1989. Outer membrane as a mechanism of antimicrobial resistance. *Antimicrob. Agents Chemother.* **33**:1831–1836.

267. Nikaido, H. 1994. Porins and specific diffusion channels in bacterial outer membranes. *J. Biol. Chem.* **269**:3905–3908.

268. Nikaido, H. 1994. Prevention of drug access to bacterial targets: permeability barriers and active efflux. *Science* **264**:382–387.

269. Nikaido, H., M. Basina, V. Nguyen, and E. Y. Rosenberg. 1998. Multidrug efflux pump AcrAB of *Salmonella typhimurium* excretes only those β-lactam antibiotics containing lipophilic side chains. *J. Bacteriol.* **180**:4686–4692.

270. Nikaido, H., and E. Y. Rosenberg. 1990. Cir and Fiu proteins in the outer membrane of *Escherichia coli* catalyze transport of monomeric catechols: study with β-lactam antibiotics containing catechol and analogous groups. *J. Bacteriol.* **172**:1361–1367.

271. Nilsen, I. W., I. Bakke, A. Vader, Ø. Olsvik, and M. R. El-Gewely. 1996. Isolation of *cmr*, a novel *Escherichia coli* chloramphenicol resistance gene encoding a putative efflux pump. *J. Bacteriol.* **178**:3188–3193.

272. Novak, R., B. Henriques, E. Charpentier, S. Normark, and E. Tuomanen. 1999. Emergence of vancomycin tolerance in *Streptococcus pneumoniae*. *Nature* **399**:590–593.

273. Ocaktan, A., H. Yoneyama, and T. Nakae. 1997. Use of fluorescence probes to monitor function of the subunit proteins of the MexA-MexB-OprM drug extrusion machinery in *Pseudomonas aeruginosa*. *J. Biol. Chem.* **272**:21964–21969.

274. Ochs, M. M., C.-D. Lu, R. E. W. Hancock, and A. T. Abdelal. 1999. Amino acid-mediated induction of the basic amino acid specific outer membrane porin OprD from *Pseudomonas aeruginosa*. *J. Bacteriol.* **181**:5426–5432.

275. Ochs, M. M., M. P. McCusker, M. Bains, and R. E. W. Hancock. 1999. Negative regulation of the *Pseudomonas aeruginosa* outer membrane porin OprD selective for imipenem and basic amino acids. *Antimicrob. Agents Chemother.* **43**:1085–1090.

276. Oethinger, M., W. V. Kern, A. S. Jellen-Ritter, L. M. McMurry, and S. B. Levy. 2000. Ineffectiveness of topoisomerase mutations in mediating clinically significant fluoroquinolone resistance in *Escherichia coli* in the absence of the AcrAB efflux pump. *Antimicrob. Agents Chemother.* **44**:10–13.

277. Ohki, R., and M. Murata. 1997. *bmr3*, a third multidrug transporter gene of *Bacillus subtilis*. *J. Bacteriol.* **179**:1423–1427.

278. Okazaki, T., and K. Hirai. 1992. Cloning and nucleotide sequence of the *Pseudomonas aeruginosa nfxB* gene, conferring resistance to new quinolones. *FEMS Microbiol. Lett.* **76**:197–202.

279. Okazaki, T., S. Iyobe, H. Hashimoto, and K. Hirai. 1991. Cloning and characterization of a DNA fragment that complements the *nfxB* mutation in *Pseudomonas aeruginosa* PAO. *FEMS Microbiol. Lett.* **63**:31–35.

280. Okusu, H., D. Ma, and H. Nikaido. 1996. AcrAB efflux pump plays a major role in the antibiotic resistance phenotype of *Escherichia coli* multiple antibiotic-resistance (Mar) mutants. *J. Bacteriol.* **178**:306–308.

281. Osawa, S., R. Takata, K. Tanaka, and M. Tamaki. 1973. Chloramphenicol resistant mutants of *Bacillus subtilis*. *Mol. Gen. Genet.* **127**:163–173.

282. Padayachee, T., and K. P. Klugman. 1999. Novel expansion of the gene encoding dihydropteroate synthase in trimethoprim-sulfamethoxazole-resistant *Streptococcus pneumoniae*. *Antimicrob. Agents Chemother.* **43**:2225–2230.

283. Palepou, M.-F. I., A. P. Johnson, B. D. Cookson, H. Beattie, A. Charlett, and N. Woodford. 1998. Evaluation of disc diffusion and Etest for determining the susceptibility of *Staphylococcus aureus* to mupirocin. *J. Antimicrob. Chemother.* **42**:577–583.

284. Pao, S. S., I. T. Paulsen, and M. H. Saier, Jr. 1998. Major facilitator superfamily. *Microbiol. Mol. Biol. Rev.* **62**:1–34.

285. Parent, R., and P. H. Roy. 1992. The chloramphenicol acetyltransferase gene of Tn2424: a new breed of *cat*. *J. Bacteriol.* **174**:2891–2897.

286. Patterson, J. E., and M. J. Zervos. 1990. High-level gentamicin resistance in *Enterococcus*: microbiology, genetic basis and epidemiology. *Rev. Infect. Dis.* **12**:644–652.

287. Paulsen, I. T., M. H. Brown, T. G. Littlejohn, B. A. Mitchell, and R. A. Skurray. 1996. Multidrug resistance proteins QacA and QacB from *Staphylococcus aureus*: Membrane topology and identification of residues involved in substrate specificity. *Proc. Natl. Acad. Sci. USA* **93**:3630–3635.

288. Paulsen, I. T., M. H. Brown, and R. A. Skurray. 1996. Proton-dependent multidrug efflux systems. *Microbiol. Rev.* **60**:575–608.

289. Paulsen, I. T., T. G. Littlejohn, P. Rådström, L. Sundström, O. Sköld, G. Swedberg, and R. A. Skurray. 1993. The 3′ conserved segment of integrons contains a gene associated with multidrug resistance to antiseptics. *Antimicrob. Agents Chemother.* **37**:761–768.

290. Paulsen, I. T., R. A. Skurray, R. Tam, M. H. Saier, Jr., R. J. Turner, J. H. Weiner, E. B. Goldberg, and L. L. Grinius. 1996. The SMR family: a novel family of multidrug efflux proteins involved with the efflux of lipophilic drugs. *Mol. Microbiol.* **19**:1167–1175.

291. Pearson, J. P., C. Van Delden, and B. H. Iglewski. 1999. Active efflux and diffusion are involved in transport of *Pseudomonas aeruginosa* cell-to-cell signals. *J. Bacteriol.* **181**:1203–1210.

292. Perichon, B., P. Reynolds, and P. Courvalin. 1997. VanD-type glycopeptide resistant *Enterococcus faecium* BM4339. *Antimicrob. Agents Chemother.* **41**:2016–2018.

293. Perreten, V., F. Schwarz, L. Cresta, M. Boeglin, G. Dasen, and M. Teuber. 1997. Antibiotic resistance spread in food. *Nature* 389:801–802.

294. Philippon, A., R. Labia, and G. Jacoby. 1989. Extended-spectrum β-lactamases. *Antimicrob. Agents Chemother.* 33:1131–1136.

295. Piddock, L. J. V. 1998. Antibacterials—mechanisms of action. *Curr. Opin. Microbiol.* 1:502–508.

296. Piddock, L. J. V. 1998. Fluoroquinolone resistance. *Br. Med. J.* 317:1029–1030.

297. Pierre, J., A. Boisivon, and L. Gutmann. 1990. Alteration of PBP3 entails resistance to imipenem in *Listeria monocytogenes*. *Antimicrob. Agents Chemother.* 34:1695–1698.

298. Ploy, M.-C., P. Courvalin, and T. Lambert. 1998. Characterization of In40 of *Enterobacter aerogenes* BM2688, a class 1 integron with two new gene cassettes, *cmlA2* and *qacF*. *Antimicrob. Agents Chemother.* 42:2557–2563.

299. Polissi, A., and C. Georgopoulos. 1996. Mutational analysis and properties of the *msbA* gene of *Escherichia coli*, coding for an essential ABC family transporter. *Mol. Microbiol.* 20:1221–1233.

300. Poole, K., N. Gotoh, H. Tsujimoto, Q. Zhao, A. Wada, T. Yamasaki, S. Neshat, J.-I. Yamagishi, X.-Z. Li, and T. Nishino. 1996. Overexpression of the *mexC-mexD-oprJ* efflux operon in *nfxB*-type multidrug-resistant strains of *Pseudomonas aeruginosa*. *Mol. Microbiol.* 21:713–724.

301. Poole, K., D. E. Heinrichs, and S. Neshat. 1993. Cloning and sequence analysis of an EnvCD homologue in *Pseudomonas aeruginosa*: regulation by iron and possible involvement in the secretion of the siderophore pyoverdine. *Mol. Microbiol.* 10:529–544.

302. Poole, K., K. Krebes, C. McNally, and S. Neshat. 1993. Multiple antibiotic resistance in *Pseudomonas aeruginosa*: evidence for involvement of an efflux operon. *J. Bacteriol.* 175:7363–7372.

303. Poole, K., K. Tetro, Q. Zhao, S. Neshat, D. E. Heinrichs, and N. Bianco. 1996. Expression of the multidrug resistance operon *mexA-mexB-oprM* in *Pseudomonas aeruginosa*: *mexR* encodes a regulator of operon expression. *Antimicrob. Agents Chemother.* 40:2021–2028.

304. Prammananan, T., P. Sander, B. A. Brown, K. Frischkorn, G. O. Onyi, Y. Zang, E. C. Bottger, and R. J. Wallace, Jr. 1998. A single 16S ribosomal RNA substitution is responsible for resistance to amikacin and other 2-deoxystreptamine aminoglycosides in *Mycobacterium abscessus* and *Mycobacterium chelonae*. *J. Infect. Dis.* 177:1573–1581.

305. Price, C. T. D., F. G. O'Brien, B. P. Shelton, J. R. Warmington, W. B. Grubb, and J. E. Gustafson. 1999. Effects of salicylate and related compounds on fusidic acid MICs in *Staphylococcus aureus*. *J. Antimicrob. Chemother.* 44:57–64.

306. Purewal, A. S. 1991. Nucleotide sequence of the ethidium efflux gene from *Escherichia coli*. *FEMS Microbiol. Lett.* 66:229–232.

307. Purewal, A. S., I. G. Jones, and M. Midgley. 1990. Cloning of the ethidium efflux gene from *Escherichia coli*. *FEMS Microbiol. Lett.* 56:73–76.

308. Putman, M., L. A. Koole, H. W. van Veen, and W. N. Konings. 1999. The secondary multidrug transporter LmrP contains multiple drug interaction sites. *Biochemistry* 38:13900–13905.

309. Putman, M., H. W. van Veen, B. Poolman, and W. N. Konings. 1999. Restrictive use of detergents in the functional reconstitution of the secondary multidrug transporter LmrP. *Biochemistry* 38:1002–1008.

310. Rådström, P., C. Fermér, B.-E. Kristiansen, A. Jenkins, O. Sköld, and G. Swedberg. 1992. Transformational exchanges in the dihydropteroate synthase gene of *Neisseria meningitidis*, a novel mechanism for the acquisition of sulfonamide resistance. *J. Bacteriol.* 174:6386–6393.

311. Rådström, P., O. Sköld, G. Swedberg, J. Flensburg, P. H. Roy, and L. Sundström. 1994. Transposon Tn*5090* of plasmid R751, which carries an integron, is related to Tn7, Mu, and the retroelements. *J. Bacteriol.* 176:3257–3268.

312. Raimondi, A., A. Traverso, and H. Nikaido. 1991. Imipenem- and meropenem-resistant mutants of *Enterobacter cloacae* and *Proteus rettgeri* lack porins. *Antimicrob. Agents Chemother.* 35:1174–1180.

313. Ramsey, M. A., S. F. Bradley, C. A. Kauffman, and T. M. Morton. 1996. Identification of chromosomal location of *mupA* gene encoding low-level mupirocin resistance in staphylococcal isolates. *Antimicrob. Agents Chemother.* 40:2820–2823.

314. Rather, P. N. 1998. Origins of aminoglycoside modifying enzymes. *Drug Res. Updates* 1:285–291.

315. Raviv, Y., H. B. Pollard, E. P. Bruggemann, I. Pastan, and M. M. Gottesman. 1990. Photosensitized labeling of a functional multidrug transporter in living drug-resistant tumor cells. *J. Biol. Chem.* 265:3975–3980.

316. Recchia, G. D., and R. M. Hall. 1995. Gene cassettes: a new class of mobile elements. *Microbiology* 141:3015–3027.

317. Recht, M. I., S. Douthwaite, and J. D. Puglisi. 1999. Basis for prokaryotic specificity of action of aminoglycoside antibiotics. *EMBO J.* 18:3133–3138.

318. Reeve, E. C. 1966. Characteristics of some single-step mutants to chloramphenicol resistance in *Escherichia coli* K12 and their interactions with R-factor genes. *Genet. Res.* 7:281–286.

319. Reeve, E. C., and D. R. Suttlie. 1968. Chromosomal location of a mutation causing chloramphenicol resistance in *Escherichia coli* K12. *Genet. Res.* 11:97–104.

320. Reeve, E. C. R. 1968. Genetic analysis of some mutations causing resistance to tetracycline in *Escherichia coli* K-12. *Genet. Res.* 11:303–309.

321. Rhee, S., R. G. Martin, J. L. Rosner, and D. R. Davies. 1998. A novel DNA-binding motif in MarA: the first structure for an AraC family transcriptional activator. *Proc. Natl. Acad. Sci. USA* 18:10413–10418.

322. Ridley, A., and E. J. Threlfall. 1998. Molecular epidemiology of antibiotic resistance genes in multiresistant epidemic *Salmonella typhimurium* DT 104. *Microb. Drug Resist.* 4:113–118.

323. Roberts, M. C. 1996. Tetracycline resistance determinants: mechanism of action, regulation of expression, genetic mobility, and distribution. *FEMS Microbiol. Rev.* 19:1–24.

324. Roberts, M. C., J. Sutcliffe, P. Courvalin, L. B. Jensen, J. Rood, and H. Seppala. 1999. Nomenclature for macrolide and macrolide-lincosamide-streptogramin B resistance determinants. *Antimicrob. Agents Chemother.* 43:2823–2830.

325. Robillard, N. J., and A. L. Scarpa. 1988. Genetic and physiological characterization of ciprofloxacin resistance in *Pseudomonas aeruginosa* PAO. *Antimicrob. Agents Chemother.* 32:535–539.

326. Rohnson, G. N. 1986. β-lactam antibiotics. *J. Antimicrob. Chemother.* 17:5–36.

327. Ross, J. I., E. A. Eady, J. H. Cove, and W. J. Cunliffe. 1998. 16S rRNA mutation associated with tetracycline resistance is a gram-positive bacterium. *Antimicrob. Agents Chemother.* 52:1702–1705.

328. Rouguette, C., J. B. Harmon, and W. M. Shafer. 1999. Induction of the *mtrCDE*-encoded efflux pump system of *Neisseria gonorrhoeae* requires MtrA, and AraC-like protein. *Mol. Microbiol.* 33:651–658.

329. **Rowe-Magnus, D. A., and D. Mazel.** 1999. Resistance gene capture. *Curr. Opin. Microbiol.* **2:**483–488.

330. **Rubin, R. A., S. B. Levy, R. L. Heinrikson, and F. J. Kézdy.** 1990. Gene duplication in the evolution of the two complementing domains of gram-negative bacterial tetracycline efflux proteins. *Gene* **87:**7–13.

331. **Saier, M. H., Jr., I. T. Paulsen, M. K. Sliwinski, S. S. Pao, R. A. Skurray, and H. Nikaido.** 1998. Evolutionary origins of multidrug and drug-specific efflux pumps in bacteria. *FASEB J.* **12:**265–274.

332. **Saier, M. H., Jr., R. Tam, A. Reizer, and J. Reizer.** 1994. Two novel families of bacterial membrane proteins concerned with nodulation, cell division and transport. *Mol. Microbiol.* **11:**841–847.

333. **Saito, K., H. Yoneyama, and T. Nakae.** 1999. *nalB*-type mutations causing the overexpression of the MexAB-OprM efflux pump are located in the *mexR* gene of the *Pseudomonas aeruginosa* chromosome. *FEMS Microbiol. Lett.* **179:**67–72.

334. **Salyers, A. A., N. B. Shoemaker, A. M. Stevens, and L.-Y. Li.** 1995. Conjugative transposons: a unusual and diverse set of integrated gene transfer elements. *Microbiol. Rev.* **59:**579–590.

335. **Salyers, A. A., B. S. Speer, and N. B. Shoemaker.** 1990. New perspectives in tetracycline resistance. *Mol. Microbiol.* **4:**151–156.

336. **Samuelson, J.** 1999. Why metronidazole is active against both bacteria and parasites. *Antimicrob. Agents Chemother.* **43:**1533–1541.

337. **Sánchez, L., W. Pan, M. Viñas, and H. Nikaido.** 1997. The *acrAB* homolog of *Haemophilus influenzae* codes for a functional multidrug efflux pump. *J. Bacteriol.* **179:**6855–6857.

338. **Sasatsu, M., K. Shima, Y. Shibata, and M. Kono.** 1989. Nucleotide sequence of a gene that encodes resistance to ethidium bromide from a transferable plasmid in *Staphylococcus aureus*. *Nucleic Acids Res.* **17:**101–103.

339. **Schell, M. A.** 1993. Molecular biology of the LysR family of transcriptional regulators. *Annu. Rev. Microbiol.* **47:**597–626.

340. **Schulz, W., and W. Zillig.** 1981. Rifampicin inhibition of RNA synthesis by destabilisation of DNA-RNA polymerase-oligonucleotides-complexes. *Nucleic Acids Res.* **9:**6889–6906.

341. **Scott, J. R.** 1992. Sex and the single circle: conjugative transposition. *J. Bacteriol.* **174:**6005–6010.

342. **Sedgwick, E. G., and P. D. Bragg.** 1996. The role of efflux systems and the cell envelope in fluorescence changes of the lipophilic cation 2-(4-dimethylaminostyryl)-1-ethylpyridinium in *Escherichia coli*. *Biochim. Biophys. Acta* **1278:**205–212.

343. **Seoane, A. S., and S. B. Levy.** 1995. Characterization of MarR, the repressor of the multiple antibiotic resistance (*mar*) operon of *Escherichia coli*. *J. Bacteriol.* **177:**3414–3419.

344. **Severinov, K., M. Soushko, A. Goldfarb, and A. Nikiforov.** 1994. Rif^R mutations in the beginning of the *Escherichia coli rpoB* gene. *Mol. Gen. Genet.* **244:**120–126.

345. **Shafer, W. M., J. T. Balthazar, K. E. Hagman, and S. A. Morse.** 1995. Missense mutations that alter the DNA-binding domain of the MtrR protein occur frequently in rectal isolates of *Neisseria gonorrhoeae* that are resistant to faecal lipids. *Microbiology* **141:**907–911.

346. **Shafer, W. M., X.-D. Qu, A. J. Waring, and R. I. Lehrer.** 1998. Modulation of *Neisseria gonorrhoeae* susceptibility to vertebrate antibacterial peptides due to a member of the resistance/nodulation/division efflux pump family. *Proc. Natl. Acad. Sci. USA* **95:**1829–1833.

347. **Shaw, K. J., P. N. Rather, R. S. Hare, and G. H. Miller.** 1993. Molecular genetics of aminoglycoside resistance genes and familial relationships of the aminoglycoside-modifying enzymes. *Microbiol. Rev.* **57:**138–163.

348. **Shaw, W. V.** 1983. Chloramphenicol acetyltransferase: enzymology and molecular biology. *Crit. Rev. Biochem.* **14:**1–46.

349. **Shiba, T., K. Ishiguro, N. Takemoto, H. Koibuchi, and K. Sugimoto.** 1995. Purification and characterization of the *Pseudomonas aeruginosa* NfxB protein, the negative regulator of the *nfxB* gene. *J. Bacteriol.* **177:**5872–5877.

350. **Sieradzki, K., M. G. Pinho, and A. Tomasz.** 1999. Inactivated *pbp4* in highly glycopeptide-resistant laboratory mutant of *Staphylococcus aureus*. *J. Biol. Chem.* **274:**18942–18946.

351. **Sirot, D., R. Labia, P. Pouedras, C. Chanal-Claris, C. Cerceau, and J. Sirot.** 1998. Inhibitor-resistant OXY-2-derived β-lactamase produced by *Klebsiella oxytoca*. *Antimicrob. Agents Chemother.* **42:**2184-2187.

352. **Sparling, P. F., F. A. Sarubbi, Jr., and E. Blackman.** 1975. Inheritance of low-level resistance to penicillin, tetracycline, and chloramphenicol in *Neisseria gonorrhoeae*. *J. Bacteriol.* **124:**740–749.

353. **Speer, B. S., N. B. Shoemaker, and A. A. Salyers.** 1992. Bacterial resistance to tetracyclines: mechanism, transfer, and clinical significance. *Clin. Microbiol. Rev.* **5:**387–399.

354. **Spratt, B. G.** 1994. Resistance to antibiotics mediated by target alterations. *Science* **264:**388–393.

355. **Srikumar, R., E. Tsang, and K. Poole.** 1999. Contribution of the MexAB-OprM multidrug efflux system to the β-lactam resistance of penicillin-binding protein and β-lactamase-derepressed mutants of *Pseudomonas aeruginosa*. *J. Antimicrob. Chemother.* **44:**537–570.

356. **Stone, B. J., and V. L. Miller.** 1995. *Salmonella enteritidis* has a homologue of *tolC* that is required for virulence in BALB/c mice. *Mol. Microbiol.* **17:**701–712.

357. **Stunt, R. A., A. K. Amyes, C. J. Thomson, D. J. Payne, and S. G. Amyes.** 1998. The production of a novel carbapenem-hydrolysing β-lactamase in *Aeromonas veronii* biovar *sobria*, and its association with imipenem resistance. *J. Antimicrob. Chemother.* **42:**835–836.

358. **Sukupolvi, S., M. Vaara, I. M. Helander, P. Viljanen, and P. H. Mäkelä.** 1984. New *Salmonella typhimurium* mutants with altered outer membrane permeability. *J. Bacteriol.* **159:**704–712.

359. **Sulavik, M. C., L. F. Gambino, and P. F. Miller.** 1995. The MarR repressor of the multiple antibiotic resistance (*mar*) operon in *Escherichia coli*: prototypic member of a family of bacterial regulatory proteins involved in sensing phenolic compounds. *Mol. Med.* **1:**436–446.

360. **Sumita, Y., and M. Fukasawa.** 1996. Meropenem resistance in *Pseudomonas aeruginosa*. *Antimicrob. Agents Chemother.* **42:**47–56.

361. **Summers, A. O.** 1992. Untwist and shout: a heavy metal-responsive transcriptional regulator. *J. Bacteriol.* **174:**3097–3101.

362. **Sutcliffe, J.** 1999. Resistance to macrolides mediated by efflux mechanisms. *Curr. Opin. Anti-infect. Invest. Drugs* **1:**403–412.

363. **Sutcliffe, J., A. Tait-Kamradt, and L. Wondrack.** 1996. *Streptococcus pneumoniae* and *Streptococcus pyogenes* resistant to macrolides but sensitive to clindamycin: a common resistance pattern mediated by an efflux system. *Antimicrob. Agents Chemother.* **40:**1817–1824.

364. **Sutherland, R., R. J. Boon, K. E. Griffin, P. J. Masters, B. Slocombe, and A. R. White.** 1985. Antibacterial activity of

mupirocin (pseudomonic acid), a new antibiotic for topical use. *Antimicrob. Agents Chemother.* **27**:495–498.

365. Swedberg, G., S. Ringertz, and O. Sköld. 1998. Sulfonamide resistance in *Streptococcus pyogenes* is associated with differences in the amino acid sequence of its chromosomal dihydropteroate synthase. *Antimicrob. Agents Chemother.* **42**:1062–1067.

366. Syriopoulou, V. P., A. L. Harding, D. A. Goldmann, and A. L. Smith. 1981. In vitro antibacterial activity of fluorinated analogs of chloramphenicol and thiamphenicol. *Antimicrob. Agents Chemother.* **19**:294–297.

367. Syvanen, M. 1984. The evolutionary implications of mobile genetic elements. *Annu. Rev. Genet.* **18**:271–293.

368. Taber, H. W., J. P. Mueller, P. F. Miller, and A. S. Arrow. 1987. Bacterial uptake of aminoglycoside antibiotics. *Microbiol. Rev.* **51**:439–457.

369. Takiff, H. E., M. Cimino, M. C. Musso, T. Weisbrod, R. Martinez, M. B. Delgado, L. Salazar, B. R. Bloom, and W. R. Jacobs, Jr. 1996. Efflux pump of the proton antiporter family confers low-level fluoroquinolone resistance in *Mycobacterium smegmatis. Proc. Natl. Acad. Sci. USA* **93**:362–366.

370. Tanaka, N., T. Kinoshita, and H. Masukawa. 1968. Mechanism of protein synthesis inhibition by fusidic acid and related antibiotics. *Biochem. Biophys. Res. Commun.* **30**:278–283.

371. Taylor, D. J. 1999. Antimicrobial use in anaimals and its consequences for human health. *Clin. Microbiol. Infect.* **5**:119–124.

372. Tennent, J. M., B. R. Lyon, M. Midgley, I. G. Jones, A. S. Purewal, and R. A. Skurray. 1989. Physical and biochemical characterization of the *qacA* gene encoding antiseptic and disinfectant resistance in *Staphylococcus aureus. J. Gen. Microbiol.* **135**:1–10.

373. Tennigkeit, J., and H. Matzura. 1991. Nucleotide sequence analysis of a chloramphenicol-resistance determinant from *Agrobacterium tumefaciens* and identification of its gene product. *Gene* **98**:113–116.

374. Teuber, M., L. Meile, and F. Schwarz. 1999. Acquired antibiotic resistance in lactic acid bacteria from food. *Antonie Leeuwenhoek* **76**:115–137.

375. Thanassi, D. G., L. W. Cheng, and H. Nikaido. 1997. Active efflux of bile salts by *Escherichia coli. J. Bacteriol.* **179**:2512–2518.

376. Thanassi, D. G., G. S. Suh, and H. Nikaido. 1995. Role of outer membrane barrier in efflux-mediated tetracycline resistance of *Escherichia coli. J. Bacteriol.* **177**:998–1007.

377. Threlfall, E. J., J. A. Frost, L. R. Ward, and B. Rowe. 1994. Epidemic in cattle and humans of *Salmonella typhimurium* DT104 with chromosomally integrated multiple drug resistance. *Vet. Rec.* **134**:577.

378. Threlfall, E. J., J. A. Frost, L. R. Ward, and B. Rowe. 1996. Increasing spectrum of resistance in multiresistant *Salmonella typhimurium. Lancet* **347**:1053–1054.

379. Tomasz, A. 1986. Penicillin-binding proteins and the antibacterial effectiveness of β-lactam antibiotics. *Rev. Infect. Dis.* **8**(Suppl. 3):S260–S278.

380. Tomasz, A., A. Albino, and E. Zanati. 1970. Multiple antibiotic resistance in a bacterium with suppressed autolytic system. *Nature* **227**:138–140.

381. Trias, J., J. Dufresne, R. C. Levesque, and H. Nikaido. 1989. Decreased outer membrane permeability in imipenem-resistant mutants of *Pseudomonas aeruginosa. Antimicrob. Agents Chemother.* **33**:1201–1206.

382. Trias, J., and H. Nikaido. 1990. Protein D2 channel of the *Pseudomonas aeruginosa* outer membrane has a binding site for basic amino acids and peptides. *J. Biol. Chem.* **265**:15680–15684.

383. Tribuddharat, C., and M. Fennewald. 1999. Integron-mediated rifampin resistance in *Pseudomonas aeruginosa. Antimicrob. Agents Chemother.* **43**:960–962.

384. Trinh, S., and G. Reysset. 1996. Detection by PCR of the *nim* genes encoding 5-nitroimidazole resistance in *Bacteroides* spp. *J. Clin. Microbiol.* **34**:2078–2084.

385. Ubukata, K., N. Itoh-Yamashita, and M. Konno. 1989. Cloning and expression of the *norA* gene for fluoroquinolone resistance in *Staphylococcus aureus. Antimicrob. Agents Chemother.* **33**:1535–1539.

386. van den Bogaard, A. E., L. B. Jensen, and E. E. Stobberingh. 1997. Vancomycin-resistant enterococci in turkeys and farmers. *N. Engl. J. Med.* **337**:1558–1559.

387. van Veen, H. W., R. Callaghan, L. Soceneantu, A. Sardini, W. N. Konings, and C. F. Higgins. 1998. A bacterial antibiotic-resistance gene that complements the human multidrug-resistance P-glycoprotein gene. *Nature* **391**:291–295.

388. van Veen, H. W., K. Venema, H. Bolhuis, I. Oussekno, J. Kok, B. Poolman, A. J. M. Driessen, and W. N. Konings. 1996. Multidrug resistance mediated by a bacterial homolog of the human multidrug transporter MDR1. *Proc. Natl. Acad. Sci. USA* **93**:10668–10672.

389. Varon, E., C. Janoir, M.-D. Kitzis, and L. Gutmann. 1999. ParC and GyrA may be interchangeable initial targets of some fluoroquinolones in *Streptococcus pneumoniae. Antimicrob. Agents Chemother.* **43**:302–306.

390. Vedel, G., A. Belaaouaj, R. Labia, A. Philippon, P. Névot, and G. Paul. 1992. Clinical isolates of *Escherichia coli* producing TRI β-lactamases: novel TEM-enzymes conferring resistance to β-lactamase inhibitors. *J. Antimicrob. Chemother.* **30**:449–462.

391. Venkateswaran, P. S., and H. C. Wu. 1972. Isolation and characterization of a phosphonomycin-resistant mutant of *Escherichia coli* K12. *J. Bacteriol.* **110**:935-944.

392. von der Haar, B., S. Walter, S. Schwäpenheuer, and H. Schrempf. 1997. A novel fusidic acid resistance gene from *Streptomyces lividans* 66 encodes a highly specific esterase. *Microbiology* **143**:867–874.

393. Waxman, D. J., and J. L. Strominger. 1983. Penicillin-binding proteins and the mechanism of action of β-lactam antibiotics. *Annu. Rev. Biochem.* **52**:825–869.

394. Wegener, H. C., F. M. Aarestrup, L. B. Jensen, A. M. Hammerum, and F. Bager. 1999. Use of antimicrobial growth promoters in food animals and *Enterococcus faecium* resistance to therapeutic antimicrobial drugs in Europe. *Emerg. Infect. Dis.* **5**:329–335.

395. Wehrli, W. 1983. Rifampin: mechanism of action and resistance. *Rev. Infect. Dis.* **5**(Suppl. 3):S407–S411.

396. Weinstein, M. J., G. H. Wagman, E. M. Oden, G. M. Luedemann, P. Sloane, A. Murawski, and J. Marquez. 1965. Purification and biological studies of everninomicin B. *Antimicrob. Agents Chemother.* **5**:821–827.

397. Weisblum, B. 1998. Macrolide resistance. *Drug Res. Updates* **1**:29–41.

398. Welton, L. A., L. A. Thal, M. B. Perri, S. Donabedian, J. McMahon, J. W. Chow, and M. J. Zervos. 1998. Antimicrobial resistance in enterococci isolated from Turkey flocks fed virginiamycin. *Antimicrob. Agents Chemother.* **42**:705–708.

399. Westbrock-Wadman, S., D. R. Sherman, M. J. Hickey, S. N. Coulter, Y. Q. Zhu, P. Warrener, L. Y. Nguyen, R. M. Shawar, K. R. Folger, and C. K. Stover. 1999. Characterization of a *Pseudomonas aeruginosa* efflux pump contributing to aminoglycoside impermeability. *Antimicrob. Agents Chemother.* **43**:2975–2983.

400. White, D. G., J. D. Goldman, B. Demple, and S. B. Levy. 1997. Role of the *acrAB* locus in organic solvent tolerance mediated by expression of *marA*, *soxS*, or *robA* in *Escherichia coli*. *J. Bacteriol*. **179**:6122–6126.

401. Whitney, E. N. 1971. The *tolC* locus in *Escherichia coli* K12. *Genetics* **67**:39–53.

402. Woodford, N., A. P. Johnson, D. Morrison, and D. C. E. Speller. 1995. Current perspectives on glycopeptide resistance. *Clin. Microbiol. Rev.* **8**:585–615.

403. Woolridge, D. P., N. Vazquez-Laslop, P. N. Markham, M. S. Chevalier, E. W. Gerner, and A. A. Neyfakh. 1997. Efflux of the natural polyamine spermidine facilitated by the *Bacillus subtilis* multidrug transporter Blt. *J. Biol. Chem.* **272**:8864–8866.

404. Wright, G. D. 1999. Aminoglycoside-modifying enzymes. *Curr. Opin. Microbiol.* **2**:499–503.

405. Yamaguchi, A., T. Udagawa, and T. Sawai. 1990. Transport of divalent cations with tetracycline as mediated by the transposon Tn*10*-encoded tetracycline resistance protein. *J. Biol. Chem.* **265**:4809–4813.

406. Yang, Y., P. Wu, and D. M. Livermore. 1990. Biochemical characterization of a β-lactamase that hydrolyzes penems and carbapenems from two *Serratia marcescens* isolates. *Antimicrob. Agents Chemother.* **34**:755–758.

407. Yazawa, K., Y. Mikami, A. Maeda, M. Akao, N. Morisaki, and S. Iwasaki. 1993. Inactivation of rifampin by *Nocardia brasiliensis*. *Antimicrob. Agents Chemother.* **37**:1313–1317.

408. Yelin, R., D. Rotem, and S. Schuldiner. 1999. EmrE, a small *Escherichia coli* multidrug transporter, protects *Saccharomyces cerevisiae* from toxins by sequestration in the vacuole. *J. Bacteriol.* **181**:949–956.

409. Yerushalmi, H., M. Lebendiker, and S. Schyldiner. 1995. EmrE, an *Escherichia coli* 12 kDa multidrug transporter, exchanges toxic cations and H$^+$ and is soluble in organic solvents. *J. Biol. Chem.* **270**:6856–6863.

410. Yoneyama, H., A. Ocaktan, N. Gotoh, T. Nishino, and T. Nakae. 1998. Subunit swapping in the Mex-extrusion pumps in *Pseudomonas aeruginosa*. *Biochem. Biophys. Res. Commun.* **244**:898–902.

411. Yoshida, H., M. Bogaki, S. Nakamura, K. Ubukata, and H. Konno. 1990. Nucleotide sequence and characterization of the *Staphylococcus aureus norA* gene, which confers resistance to quinolones. *J. Bacteriol.* **172**:6942–6949.

412. Yoshimura, F., and H. Nikaido. 1982. Permeability of *Pseudomonas aeruginosa* outer membrane to hydrophilic solutes. *J. Bacteriol.* **153**:636–642.

413. Zarantonelli, L., G. Borthagaray, E. H. Lee, and W. M. Shafer. 1999. Decreased azithromycin susceptibility of *Neisseria gonorrhoeae* due to *mtrR* mutations. *Antimicrob. Agents Chemother.* **43**:2468–2472.

414. Zeller, V., C. Janoir, M.-D. Kitzis, L. Gutmann, and N. J. Moreau. 1997. Active efflux as a mechanism of resistance to ciprofloxacin in *Streptococcus pneumoniae*. *Antimicrob. Agents Chemother.* **41**:1973–1978.

415. Zgurskaya, H. I., and H. Nikaido. 1999. AcrA is a highly asymmetric protein capable of spanning the periplasm. *J. Mol. Biol.* **285**:409–420.

416. Zgurskaya, H. I., and H. Nikaido. 1999. Bypassing the periplasm: reconstitution of the AcrAB multidrug efflux pump of *Escherichia coli*. *Proc. Natl. Acad. Sci. USA* **96**:7190–7195.

417. Zheleznova, E. E., P. N. Markham, A. A. Neyfakh, and R. G. Brennan. 1999. Structural basis of multidrug recognition by BmrR, a transcription activator of a multidrug transporter. *Cell* **96**:353–362.

418. Zhou, Z., K. A. White, A. Polissi, C. Georgopoulos, and C. R. H. Raetz. 1998. Function of *Escherichia coli* MsbA, an essential ABC family transporter, in lipid A and phospholipid biosynthesis. *J. Biol. Chem.* **273**:12466–12475.

419. Ziha-Zarifi, I., C. Llanes, T. Köhler, J.-C. Pechère, and P. Plésiat. 1999. In vivo emergence of multidrug-resistant mutants of *Pseudomonas aeruginosa* overexpressing the active efflux system MexA-MexB-OprM. *Antimicrob. Agents Chemother.* **43**:287–291.

IV. BACTERIA THRIVING IN STRESSFUL ENVIRONMENTS

Bacterial Stress Responses
Edited by G. Storz and R. Hengge-Aronis
©2000 ASM Press, Washington, D.C.

Chapter 23

Thermal Stress in Hyperthermophiles

PURIFICACIÓN LÓPEZ GARCÍA AND PATRICK FORTERRE

Hyperthermophiles, strictly dependent on high temperatures (≥80°C) for optimal growth, must also endure and respond to environmental stress, including heat and cold shock. Except for a few bacteria, most isolated hyperthermophiles belong to the domain Archaea. The study of their response to thermal stress is at its beginnings. Nearly nothing is known about the cold shock response, but the investigation of heat shock has been approached from different perspectives, mostly because of the adaptive and biotechnological interests of hyperthermophiles' molecular components. These are intrinsically adapted to work optimally and stably at their characteristic growth temperatures. Apart from these basic distinctive features, the heat shock response appears to follow general guidelines. Heat shock proteins, which are responsible for thermotolerance, are synthesized. Among them, the archaeal chaperonin (chaperone of the Hsp60 family) is the most abundant and the best studied. Several heat shock proteins have been crystallized in a postgenomics approach that is most useful for the study of a group of organisms for which classic genetic tools are not yet available. Temperature upshifts are also followed by the cytoplasmic accumulation of compatible solutes, and by biofilm formation that act as internal and external protectants, respectively. Transient changes of DNA topology are also observed during heat and cold shock. They attest to the transitory nature of the stress response in these organisms and could be relevant for the transcriptional regulation during the process. Further regulatory aspects during thermal stress in hyperthermophiles, such as posttranscriptional modifications (methylation or phosphorylation), or the role of chaperonin filaments, are among the important directions for future research.

Hyperthermophiles are organisms living optimally at 80°C or above (106). Some can thrive at temperatures around 110°C, the present record being 113°C (11). They have been found almost everywhere on the planet where liquid water is available up to 115°C, from terrestrial solfataras and hot springs to deep marine hydrothermal vents, or geothermally heated environments in the deep subsurface (25, 76, 106). These organisms are specifically adapted to live under such conditions, so that they are rarely able to grow below 60°C. All hyperthermophiles are prokaryotes, and most of the isolated species belong to *Archaea* (106), one of the three phylogenetic domains into which life can be classified (120). In addition to the isolated species, a great diversity of hyperthermophilic organisms has been identified without culturing by direct amplification of their 16S rRNA genes from their natural biotopes (5, 92). Archaea are divided into two major kingdoms, *Crenarchaeota* and *Euryarchaeota*, both of which contain hyperthermophilic species. Recently, two new hyperthermophilic phylotypes were characterized that could constitute a third archaeal kingdom, tentatively named *Korarchaeota* (5, 15) (Fig. 1).

Since their discovery early in the 1970s (12), hyperthermophiles have attracted much attention for several reasons. The most immediate was understanding how their molecular components are adapted not only to endure but to depend on such devilish conditions. From a biotechnological point of view, hyperthermophiles were also a valuable source of stable enzymes and compounds of potential industrial application (20, 71, 117). Additionally, their early branching positions in phylogenetic trees led to the proposal of a hyperthermophilic common ancestor to all extant organisms and thus a hot origin of life (91, 120), opening up a fervent and controversial debate among evolutionary biologists (119). Thanks to these charms, and to the fact that hyperthermophiles possess small genomes (between 1.6 and 3.0 Mb in size),

Purificación López García • División de Microbiología, Facultad de Medicina, Campus de San Juan, Universidad Miguel Hernández, 03550 Alicante, Spain.   **Patrick Forterre** • Institute de Génétique et Microbiologie, Université Paris XI, 91405 Orsay Cedex, France.

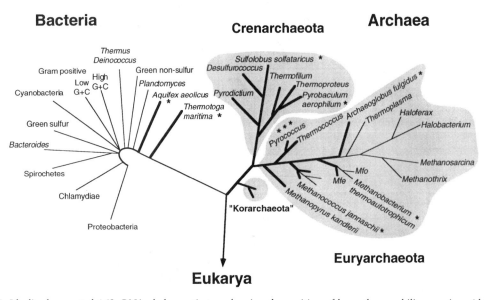

Figure 1. Idealized unrooted 16S rRNA phylogenetic tree showing the position of hyperthermophilic organisms (thick lines). Hyperthermophilic species whose genomes have been completely sequenced, or are being sequenced, are labeled with an asterisk. *Pyrococcus* asterisks correspond to *P. horikoshii*, *P. abyssi*, and *P. furiosus*. Based on references 5, 120.

they found a place very early in genome sequencing projects. Thus, the complete genomic sequence of the archaeon *Methanococcus jannaschii* was among the first to appear (14), and those of *Archaeoglobus fulgidus* (65), *Pyrococcus horikoshii* (60), and the hyperthermophilic bacterium *Aquifex aeolicus* (24) followed (see also Fig. 1). One of the important outcomes of the comparison of the available genome sequences has been the confirmation of *Archaea* as an independent domain, with idiosyncratic genes, eukaryotic-like genes (those involved in replication, transcription, and translation), and bacterial-like genes (metabolism, transport, cell division). At the time of writing, the genome sequences of the archaea *Pyrobaculum aerophilum*, *Aeropyrum pernix*, *Sulfolobus solfataricus*, *Pyrococcus abyssi*, *Pyrococcus furiosus*, and the bacterium *Thermotoga maritima* are either being finished or about to be released. The huge amount of information gathered in this way is expected to be most useful in understanding the biology of these organisms, given that genetic tools to manipulate them are unfortunately not available yet.

HYPERTHERMOPHILES AND ENVIRONMENTAL STRESS

Like other organisms, hyperthermophiles have to adapt to changing environments and respond to stress. The response to environmental stress, especially thermal stress, has been studied better in mesophilic bacteria and eukaryotes (see, for instance, chapters 1 and 3). Underlying these studies was the

idea that stress systems, in which a restricted number of proteins are induced, could serve as simple models to analyze the regulation of gene expression. Heat shock, in particular, has been focus of intense research, although it seems clear that the responses to several different stresses overlap. Thus, heat shock, salt stress, ethanol, starvation, or oxidative stress induce the same set of proteins, called heat shock proteins (21, 49, 75). A general picture of how cells respond to stress is emerging, as shown in Fig. 2, although our knowledge is still fragmentary in many cases.

In general, the response to stress is transitory. A shift in an environmental parameter, such as temperature, is first detected by the cell, which subsequently adjusts gene expression to counteract the physical effects of the external change on its molecular components, finally recovering a steady state (Fig. 2). Sensing mechanisms are not well understood. Obviously, a shift in temperature acts on the physical state of macromolecules. In agreement, alterations in membrane lipids (fluidity) and changes in DNA supercoiling, mRNA as well as protein conformation, and the state of the ribosome seem to be involved in this step (16, 31, 54, 107, 118). A transcriptional response follows in such a way that a set of specific proteins is synthesized. During heat shock, most proteins synthesized are molecular chaperones involved in protein folding (21), as well as proteases involved in degradation of partially denatured proteins (21), proline and disulfide isomerases (21), or specific regulators (44, 49, 87) (see also chapter 1). During cold shock, a transient inhibition of most protein synthesis

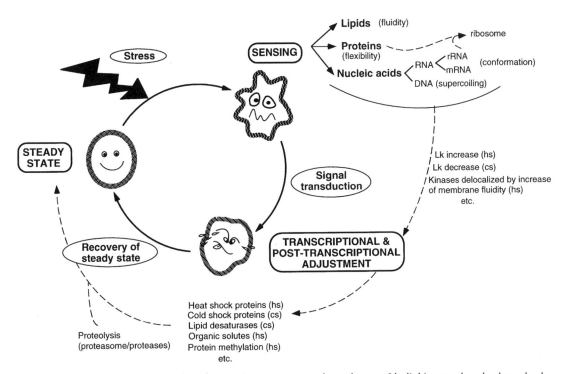

Figure 2. General mechanisms involved in the transient response to thermal stress. Lk, linking number; hs, heat shock; cs, cold shock.

is also observed, and a different set of proteins, called Csps, is mainly induced (40, 41, 107), as are other proteins, such as desaturases (16, 118) (see also chapter 3). In this process (Fig. 2), the transduction of the signal between the sensor and the transcriptional machinery is not fully understood. At any rate, the situation is complex, and possibly, a cross-talk among different signaling mechanisms exists. This would fit with the existence of transcriptional regulatory networks that include positive and negative regulation by specific proteins (44, 49, 87).

In addition, the thermal stress response involves regulation not only at the transcriptional but also at the posttranslational level. Thus, the level and/or pattern of protein methylation changes upon heat shock in eukaryotic histones (29), with a possible role in protection as well as in cellular signaling (2, 29). To further complicate the matter, in addition to lipids, nucleic acids, and proteins, organic solutes, such as the disaccharide trehalose, may prove to be essential players in protein stability and folding during heat shock (103).

Within the domain *Archaea*, the study of heat shock protein induction has been approached in a few mesophiles, including methanogens (18, 19, 48, 73) and halophiles (70). As in mesophilic bacteria and eukaryotes, there are overlapping responses to different stresses (48). In hyperthermophilic archaea, most studies have been aimed at heat shock since, in ad-

dition to the fascination for what is going on at the very border of life at high temperature, the obtainment of clues about protein stabilization has obvious biotechnological potential. Early studies showed that thermotolerance, i.e., the enhanced survival to lethal high temperatures after prolonged heat shock at near-lethal temperatures, could be acquired and that heat shock proteins were essential to this adaptation (50, 109, 110, 114). More recently, heterologous expression of putative heat shock genes found in genome sequences together with crystallographic analysis (hyperthermophilic proteins crystallize relatively well) has become a useful tool to study heat shock proteins from hyperthermophiles (62, 63). Aside from heat shock proteins, we know few things about the thermal stress response in these organisms. Among them, as we will see, and in relation with what is known for other organisms, hyperthermophiles accumulate organic solutes after heat shock, and their DNA supercoiling varies during shocks. In the case of hyperthermophilic bacteria, except for the identification of the heat shock genes in the genome sequences of *A. aeolicus* and *T. maritima*, nearly nothing is known about how these microorganisms behave during stress. Here, we will thus focus mainly on the effect of thermal shocks on hyperthermophilic archaea, to summarize current knowledge and discuss how we can place it in a general framework, as that of Fig. 2. For this, let us first comment on the intrinsic adaptations of hyperthermophiles' macromolecules and how temperature affects them.

INTRINSIC ADAPTATIONS OF HYPERTHERMOPHILES

Lipids

The proper physical state of membranes is essential for all living cells. Under normal growth, membrane lipids are in a liquid crystalline state that ensures an appropriate matrix for proteins and generates a concentration gradient of protons and other solutes across membranes, thus having a profound effect on the bioenergetics of the cell. In bacterial and eukaryotic membranes, glycerol fatty acyl esters, which form a bilayer (the polar headgroups toward the outside, the hydrophobic tails to the interior), are the prevalent lipids. Lipid composition in bacteria is adjusted to maintain a liquid crystalline state regardless of the growth temperature (101). Not surprisingly, the lipid physical state is determinant during heat and cold shock regulation. Heat shock increases membrane fluidity and leads to the expression of heat shock proteins that associate with lipid components increasing rigidification (16, 108, 118), whereas cold shock leads to the synthesis of unsaturated fatty acids, which are more fluid (39, 118).

Archaeal membrane lipids constitute one of the historical distinctive hallmarks of these organisms from bacteria and eukaryotes. They are predominantly ether-linked isoprenoids with a particular stereochemistry (2,3-glycerol ethers instead of bacterial/eukaryotic 1,2-ester links) (59). Isoprenoid ether lipids can be diethers, which are arranged in bilayers, or tetraethers, which span the whole membrane forming monolayers. Archaea thriving in cold environments have more unsaturated lipids, whereas hyperthermophilic archaea tend to possess tetraethers and cycles in isoprenoid chains (115). The increase in tetraether content has indeed been correlated with growth at higher temperatures (105). Furthermore, liposomes composed of tetraethers are more stable and have a much lower proton permeability (34). The dramatic increase of ion permeability with temperature is a major bottleneck for hyperthermophilic life (115, 116). Interestingly, nonthermophilic crenarchaeotes possess cyclic and acyclic tetraether lipids (26). This suggests that lipid composition is the result of the phylogenetic inheritance of the organism plus (current or specific) adaptations to maintain membrane homeostasis. Despite the potential of hyperthermophilic archaea to change their lipid composition, which varies with growth temperature, the study of membrane lipids during heat or cold shock

has not been undertaken yet. The same is true for hyperthermophilic bacteria where lipids present a number of adaptations that somewhat mimic those of archaea, such as the presence of ether links and membrane-spanning lipids (52, 53). The ion permeability of hyperthermophilic bacteria also resembles more closely that of hyperthermophilic archaea than that of mesophilic bacteria (116).

Proteins

If the key word for lipids is fluidity, for proteins it is flexibility, which must be so as to keep an adequate balance between stability and functionality. Because of their thermostability and thermophilicity, proteins from hyperthermophiles have been intensively studied. High-resolution structures, and corresponding primary sequences from genomes, accumulate rapidly, making possible the comparison with their psycro- and mesophilic homologs. This has revealed some general trends for thermophilic adaptation (reviewed recently in references 23, 55, 71, 117). In general, proteins (either meso- or thermophilic) have only marginal stabilities, equivalent to a small number of weak intermolecular interactions. Indeed, at the optimal physiological conditions of each organism, proteins are in corresponding states, i.e., flexibility/rigidity are equivalent (55). In this context, stability should be referred to as the native state although, of course, hyperthermophilic proteins are generally more thermostable and have to cope efficiently with the effect of temperature on folding/unfolding and degradation.

Proteins from hyperthermophiles are not only more stable at high temperature, they are more resistant to a number of extreme conditions such as pH, pressure, salt, organic solvents, and detergents, hence their biotechnological attractiveness. Comparative analyses show that there is no magical recipe for increasing thermostability, but a pool of different possibilities from which each protein chooses an appropriate solution to its needs. Intrinsic and extrinsic forces contribute to the final stability of hyperthermophilic proteins. Among the former, we have sequence and amino acid composition. Phylogenetic and adaptive signals may be difficult to discriminate when comparing homologous protein sequences from evolutionary distant meso- and hyperthermophilic organisms. Thus, the results of this kind of approach have been ambiguous (55, 71, 117). However, a recent and more exhaustive study comparing protein sequences (deduced from complete or large genome sequences) between meso- and hyperthermophilic members of the same genus, *Methanococcus*, points out to some significant changes toward thermophily. These include higher residue volume and hydrophobicity, more charged amino acids (mostly Glu, Arg, Lys), and fewer uncharged polar residues (Ser, Thr, Asn, Gln) (47). Among other intrinsic stabilizing mechanisms are increased packing efficiency (mainly through core hydrophobic interactions), networks of ion pairs and/or hydrogen bonds (including α-helix stabilization), reduction of conformational strain (loop stabilization), and resistance to chemical modification (23, 55, 71, 117). Extrinsic forces comprise the interaction with metabolites and cofactors, protein and salt cytoplasmic concentration, organic solutes, and the interaction with molecular chaperones (71, 117). Therefore, proteins adapt easily to work optimally and stably at high temperatures. In fact, a more important problem for cell biology at high temperature may be the weak stability of many substrates and cofactors. Substrate channeling between enzymes and cytoplasmic protectants is likely a means to avoid thermal inactivation.

The precise effect of heat or cold shock on protein folding has not been studied particularly in hyperthermophiles. Obviously, temperature shifts alter protein flexibility, and cellular mechanisms must exist to avoid irreversible (cold or heat) denaturation. Some have been reported, for instance, the increase in compatible solutes and heat shock proteins (see below). However, other aspects, such as the characterization of additional roles of chaperone availability (after being sequestered by unfolded proteins) in a presumable regulatory network of the stress response, are unknown.

Nucleic Acids

Hyperthermophiles have to face up to two major problems that high temperatures cause in nucleic acids, denaturation and thermodegradation. DNA and RNA in these organisms cope efficiently with them, although RNA is more sensitive to both—denaturation, because of its single-strand structure, and thermodegradation, because of the high reactivity of the 2′ OH of the ribose. As in the case of proteins, extrinsic factors such as cytoplasmic concentrations of salts and organic solutes and binding to proteins act as protectants, but specific factors also exist.

RNA

Among the strategies used by RNA to cope with denaturation are increased GC content, fewer bulged nucleotides in helices, reduction of hairpins, and minimization of alternative folding. Thus, a correlation exists between the growth temperature of an organism and its GC content on rRNA (22, 43). Of course, this correlation does not exist in mRNAs, which may overcome the problem by the coupling between transcription and translation. For tRNAs, however, the case is different, since they have to maintain an appropiate conformation. They achieve this by a number of posttranscriptional modifications. These are already present in moderate thermophiles but are much more frequent and diverse in hyperthermophilic archaea. Normally, they involve extensive methylation (both at base and ribose), thiolation and acetylation, and also the presence of unusual bases containing amino acids, or replacement of N by C atoms in adenine rings (archaeosine) (68). Heat-induced degradation usually follows breaking of the phosphodiester bond after an attack of the 2′ OH of the ribose to the nearby phosphate to form a 2′-3′-cyclic phosphodiester linkage, which could explain why this position is often methylated in these organisms (88).

DNA

Denaturation is not a problem for DNA molecules as long as they are covalently closed. Indeed, no correlation exists between GC content in DNA and thermophily (43). Circular plasmids remain double-stranded at temperatures of at least 107°C (79), and even evidence for dsDNA at 122°C exists (69). Heat-induced depurination and cytosine deamination could have more deleterious effects. However, K^+ and Mg^{2+} salts at physiological concentrations inhibit depurination at high temperature (80). Deamination of cytosine and methyl citosine originates GU and GT mismatches. Powerful DNA repair mechanisms are likely required in these organisms. The study of DNA repair in archaea is just beginning. Interestingly, some hyperthermophilic archaea appear to be highly radioresistant, suggesting an efficient DNA repair system (43). This, and the protection exerted by DNA-binding proteins, could explain the extreme resistance to thermally induced backbone breaks observed in *P. furiosus* (93).

Nevertheless, temperature has an important effect on DNA conformation, because increased temperature tends to unwind the double helix (twist decrease) (28, 32). Consequently, the twist affects the flexibility of the helix axis (supercoiling) in covalently closed molecules, which have a constant linking number (Lk). In this sense, the finding that plasmids isolated from hyperthermophilic archaea range from relaxed to positively supercoiled (17, 77) suggests that these organisms counteract the effect of temperature on the helical path by generating linking excess (37). By contrast, DNA topology in bacteria, eukaryotes, and mesophilic archaea corresponds to negative supercoiling, which is thought to supply the activation energy for opening the two strands in a variety of processes. A linking excess in hyperthermophilic archaea correlates with the presence of reverse gyrase, a unique topoisomerase creating positive supercoils (33, 37, 61). To date, reverse gyrase has been found in all hyperthermophiles, which suggests an essential role in DNA stabilization. However, its function may be restricted to local regions for some organisms. Thus, hyperthermophilic bacteria containing both gyrase (introducing negative supercoils) and reverse gyrase possess negatively supercoiled plasmids (45).

HEAT SHOCK PROTEINS

All organisms studied to date synthesize a number of proteins during heat shock. Most of them facilitate the correct folding of stress-unfolded proteins and, therefore, they are called molecular chaperones. They are also called heat shock proteins (Hsps), although not all proteins induced by heat shock are chaperones, nor are they exclusively induced by heat (21). Molecular chaperones are essential for the acquisition of thermotolerance, but they are also required under normal conditions. According to their approximate size, they are classified in five groups of small size, 12 to 43 kDa (sHsps), 60 kDa (Hsp60), 70 kDa (Hsp70), 90 kDa (Hsp90) and ≥ 100 kDa (Hsp104), which generally correspond to sequence and structural homologs (21).

Hyperthermophilic archaea are not an exception, and they synthesize proteins, as a response to heat shock, that are required for the development of thermotolerance (109, 110, 114). Two major approaches to their study have been followed. Many have been identified in acrylamide protein gels after stress induction, while others have been identified by sequence homology to known Hsps first (Table 1). Eventually, some of them have been purified and analyzed biochemically and structurally. By contrast, cold shock protein induction has not been studied in these organisms yet. Only the crystal structures of the eukaryotic-like translation initiation factor 5A (eIF-5A) from *Pyrobaculum aerophilum* and *M. jannaschii*, which contain one domain homologous to the oligonucleotide-binding domain of the bacterial cold shock protein CspA, have been reported (62, 94). Csps are in general transcriptional activators, in many cases acting as RNA chaperones (40). Whether the archaeal eIF-5A is important to activate transcription during cold shock is unknown.

The Thermosome

The thermosome (also referred to as TF55, rosettasome, archeosome) is the major heat shock pro-

tein found in hyperthermophilic archaea (67, 96, 113). It belongs to the group II of chaperonins (chaperones of the Hsp60 family), being more closely related to the eukaryotic cytosolic TCP-1 (tailless complex polypeptide-1)/CCT (chaperonin-containing TCP-1) than to the bacterial (GroEL), mitochondrial (Hsp60), or chloroplast (Rubisco-subunit-binding protein) homologues (13, 21, 58, 113). The homology extends to the structural level since, while GroEL arranges in two stacked rings of seven-fold symmetry, the archaeal chaperonin, like TCP-1, forms structures of 8 (4, 97) or 9 subunits per ring (67, 113). Its structure was studied by electron microscopy and tridimensional image reconstruction. Recently, the crystal structure of the moderate thermophile *Thermoplasma acidophilum* thermosome has been resolved, revealing a hexadecameric assembly (30). With the exception of *Methanopyrus kandleri*, with only one kind of subunit (4), archaeal thermosomes are heteroligomers composed of two types of subunits, α and β (58, 67, 96). Subunit assembly depends on Mg^{2+} and ATP (67, 98). Interestingly, structural changes are observed after phosphorylation. Phosphorylation is modulated by K^+ and Mg^{2+} and is possibly involved in refolding of denatured proteins (67). The *M. kandleri* thermosome is also exceptional in the sense that its synthesis is not increased upon heat shock (82), which could be related to the very high temperatures at which this organism lives, and because its ATPase activity depends on NH_4^+ (3). Cross-immunological analysis additionally suggests that it could belong to a distinct chaperonin subfamily (3).

The role of the thermosome, and of chaperonins in general, is still being discussed. While it is clear that they are involved in protein folding, additional roles cannot be excluded. In fact, these proteins are extraordinarily abundant in the cell under normal growth (>30 mg/ml in *Sulfolobus shibatae*) and form filaments in vivo. This had led to the proposal of cytoskeletal functions (111, 112). The formation of filaments could serve also to store chaperonin subunits, which would be inactive, the active sites hidden inside filaments. Upon a heat shock, the chaperonins could be released for immediate action (111).

Other Hsps and Heat Shock-Related Proteins

sHsps

Apart from the identification of some induced small proteins after heat shock (Table 1), which could be sHsps, nothing was known about this class of proteins in archaea until very recently. A gene coding for

Table 1. Heat shock genes and proteins detected in hyperthermophiles

Organism	Optimal growth temp (°C)	Size of proteins induced by heat shock (kDa)[a]	Heat shock proteins[b] (reference)		
			Hsp60	Hsp70	Small HSPs
Bacteria					
Thermotoga maritima	80	ND[h]	+ (Gro ES) (13)	+ (46)	+[f]
Aquifex aeolicus	85	ND	+ (GroES) (24)	+ (24)	+ (24)
Archaea					
Crenarchaeota					
Sulfolobus acidocaldarius	75	22, 38, (2X) 64–66, 86 (56)	ND	− (109)	ND
Sulfolobus shibatae B12	83	28, 35, 55[i] (114)	+ (58, 113)	− (109)	ND
Sulfolobus solfataricus	75	27, 32, 49, 55[i], 65, 82 (110)	+ (67)	−[d]	ND
Desulfurococcus sp. strain SY	95	ND	+ (57)	ND[e]	ND
Pyrodictium occultum	105	56[i], 59[i] (96)	+ (96, 97)	ND	ND
Pyrobaculum aerophilum	100	ND	+ (36)	ND	ND
Euryarchaeota					
Methanopyrus kandleri	98	ND	+ (4)	−[d]	ND
Thermococcus sp. strain ES4	76–99	98 (50)	ND	ND	ND
Pyrococcus furiosus	100	ND	+ (13)	ND	ND
Pyrococcus horikoshii	98	ND	+ (60)	− (60)	+[f]
Pyrococcus abyssi	95	ND	+[g]	−[g]	+[g]
Methanococcus jannaschii	85	ND	+ (14)	− (14)	Hsp16.5 (63, 64)
Archaeoglobus fulgidus	83	58.9[i], 59.6[i] (35)	+ (65)	− (65)	+ (65)
Methanobacterium thermoautotrophicum[c]	65	ND	+ (104)	+ (104)	+ (104)
Thermoplasma acidophilum[c]	59	ND	+ (30)	+ (46)	ND

[a] Bands detected in one- or two-dimensional gels of total proteins after heat shock treatment of cells. Proteins corresponding to heat shock genes whose expression is actually induced are included.
[b] Genes identified by sequence homology and/or protein purification.
[c] Moderate thermophiles.
[d] Genes detected by Southern hybridization and/or PCR amplification (42).
[e] Not detected in *D. mobilis* (42).
[f] *T. maritima* and *P. horikoshii* homologs were identified by homology to the *M. jannaschii* sequence at the http://www.tirgr.org and http://www.bio.nite.go.jp/ sites, respectively.
[g] *P. abyssi* genome sequence, http://www.genoscope.cns.fr.
[h] ND, not determined.
[i] Major heat shock polypeptide observed.

a 16.5-kDa sHsp homolog was identified in the complete genome sequence of *M. jannaschii*. After expression in *Escherichia coli*, Mj Hsp16.5 was studied biochemically (64), and its crystal structure resolved (63). It forms a unique spherical oligomer of 24 subunits that protects proteins from thermal aggregation at very high temperatures.

Hsp70

While proteins of the Hsp60 family are ubiquitous, proteins of the Hsp70 family are found among the archaea only in some *Euryarchaeota*, mainly in halophilic methanogens and halophiles, moderate thermophiles, and *A. fulgidus* (Table 1) (19, 109). Efforts to detect the gene in *Crenarchaeota* have been fruitless, and its absence is patent in several genomes

(*Pyrococcus* and *Methanococcus* spp.). This distribution, and its high homology to the gram-positive bacterial *dnaK*, suggests that this gene was acquired by horizontal transfer by some archaea (42, 95). Therefore, most hyperthermophilic archaea are devoid of Hsp70 and, whatever its function, it is properly fulfilled by the other chaperones in the cell.

Hsp90

The highly conserved Hsp90s may play essential roles in evolution because of their dual functions in the response to stress and in the interaction with unstable signal transducers (102). In archaea, the occurrence of this protein has not been reported, except for a region of homology, comprising the ATP-binding domain, to the B subunit of DNA topoisomerase VI (10). Topo VI is characteristic of most archaea and, interestingly, Topo VI-B subunit is slightly induced, or its proteolysis prevented, after prolonged heat shock (77a).

PPI

In addition to the Hsps, peptidyl-prolyl *cis-trans* isomerases (PPI) play an important role in folding and are induced by heat shock. There are two types of distantly related PPIs, cyclophilins (cyclosporine-binding) and FK506-binding proteins (FKBP) (21). As yet another example of a postgenomics application, an FKBP gene homolog was initially detected in *M. jannaschii*, and subsequently, in *Methanococcus thermolithotrophicus*. The purified thermostable protein from the latter was just recently characterized (38).

Proteasome

Proteolytic systems, which degrade unfolded proteins, are as essential as heat shock proteins, which prevent unfolding, during stress. Most bacteria utilize proteases of the families ClpP, HslV, and FtsH, which appear to be absent from hyperthermophilic archaea. The proteasome is the major nonlysosomal proteolytic system in eukaryotes, with a catalytic core of 20S (20S proteasome). It interacts with 19S regulatory ATPase-containing complexes to degrade ubiquitin-bound proteins as a 26S proteasome (21). Genes coding for $\alpha-$ and $\beta-$ 20S proteasome subunits exist in bacteria too, but confined to a very narrow distribution, whereas they are present in all archaeal genomes sequenced so far, including incomplete releases from the crenarchaeon *P. aerophilum* (27, 36). Interestingly, the Lon protease is ubiquitous in the three domains (27). Among hyperthermophiles, the proteasome has been purified from *P. furiosus* (6). The proteasome of the moderate thermophile *T. acidophilum* turned out to be dispensable under normal growth but not under heat shock (100).

DNA SUPERCOILING DURING THERMAL STRESS

Plasmids from hyperthermophilic archaea experience transient changes in supercoiling during heat and cold shocks. This has been studied more in detail using plasmid-bearing strains of *Sulfolobus*. From basal relaxation/slight positive supercoiling, heat shock from 80 to 85 to 87°C induces high positive supercoiling (Lk increase), whereas cold shock from 80 to 65°C generates levels of negative supercoiling (Lk decrease) (77, 77a). Transitory Lk increases and decreases during heat and cold shock, respectively, are general phenomena, since they also occur in mesophilic bacteria (83, 84). Changes in DNA supercoiling are relevant for transcription regulation. Indeed, many promoters of stress-response genes are sensitive to DNA topological variations (31). Thus, the stress-induced DNA supercoiling change triggers the expression of those genes, whereas the transcription of most constitutive genes stops. The control of DNA topology in mesophilic bacteria during stress is partially understood (Fig. 3A). Attempts to unravel the regulation of plasmid topological changes in *Sulfolobus* have been made by analyzing them in parallel to topoisomerase levels and the small DNA-binding protein Sis7. This belongs to the Sso7/Sac7 family, a group of basic, very abundant proteins found in some *Crenarchaeota* members that intercalate in the DNA minor groove, causing unwinding (1, 99), and that may be involved in the generation of negative supercoiling in the absence of a gyrase (78). However, neither Sis7, reverse gyrase, nor the archaeal-specific topoisomerase VI (Topo VI), a powerful decatenase able to relax positive and negative superturns (9, 10), varies importantly during heat and cold shock (77a). Rather, it seems that control of DNA supercoiling during stress in *Sulfolobus* is due to the physical effect of temperature on DNA geometry as well as on topoisomerase activity. Thus, reverse gyrase activity in crude extracts increases at 85°C but is practically abolished at 65°C (77a). A general model on the DNA topology control during thermal stress in these organisms could be hypothesized, as shown in Fig. 3B. In this, a 40-kDa proteolytic product of reverse gyrase, which is homologous to the bacterial ω protein (topoisomerase I), might play an important role. Indeed, this fragment is fairly abundant in vivo (77a), and it may correspond to the 40-kDa fragment that displays a classical ATP-independent relaxation of negative supercoils (86).

Although we have begun to gather some clues about topological changes during stress, much work is needed, not only concerning control of DNA topology, but how it may be connected to the regulation of gene expression during stress. For instance, transcription can proceed independently of template topology at optimal growth temperatures, but negative supercoiling appears essential for transcription at lower temperatures in *Sulfolobus* (8). Transient negative supercoiling during early cold shock steps could thus trigger the expression of specific genes.

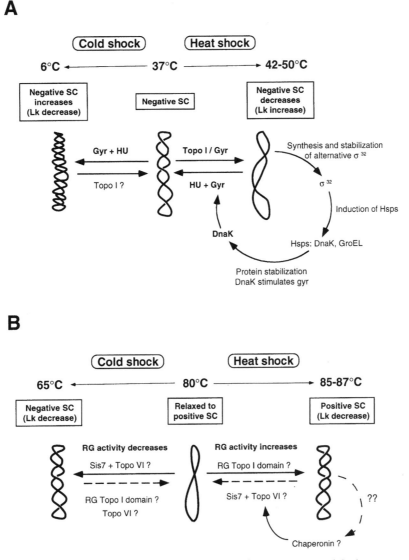

Figure 3. Control of transient DNA topological changes during thermal stress. (A) Mesophilic bacteria (*E. coli*); (B) hyperthermophilic archaea (*Sulfolobus*). Gyr, gyrase; RG, reverse gyrase; Topo I, topoisomerase I (protein ω); Topo VI, topoisomerase VI. Question marks indicate hypothetical candidates for the precise function in each case.

COMPATIBLE SOLUTES AND BIOFILM FORMATION

Hyperthermophilic bacteria and archaea accumulate small organic solutes in response to changes in osmolarity or temperature (20, 72, 81). Nonmethanogenic archaea and bacteria appear to use the same range of organic solutes as thermoprotectants: trehalose, β-mannosylglycerate, glutamate, di-*myo*-inositol-phosphate and related compounds, diglycerol-phosphate (81). Di-*myo*-inositol-phosphate (and its derivatives) is particularly characteristic of hyperthermophilic archaea or bacteria although, in general, the type of solutes accumulated varies de-pending on the organism and the medium composition (81). Moreover, di-*myo*-inositol-phosphate is the solute preferentially accumulated at supraoptimal temperatures, at least in *Archaeoglobus* and *Thermococcus* species (72, 81). However, members of *Crenarchaeota* (*Pyrobaculum*, *Thermoproteus*, *Sulfolobus*) appear to accumulate trehalose (81). In eukaryotes, trehalose plays an important role during heat shock, preventing aggregation of unfolded proteins and helping Hsps to promote correct folding (103). However, persistence of high levels of trehalose may finally interfere with refolding so that not only its synthesis but also its degradation are induced upon heat shock. In hyperthermophilic archaea, it appears that trehalose is imported from the medium (51), but the regulation of this import during the stress response has not been studied yet.

Biofilm formation may be a common stress response mechanism in hyperthermophilic archaea. Different types of stress, including low (55 to 60°C) and high temperatures (≥90°C), induce biofilm formation in *A. fulgidus* growing optimally at 83°C. The biofilm is composed of proteins, polysaccharides, and metals and enhances cell survival presumably by creating a protective environment. This phenomenon has also been observed in *Archaeoglobus profundus*, *M. jannaschii*, and *Methanobacterium thermoautotrophicum* (74).

CONCLUSIONS AND PERSPECTIVES

If the adaptation of life at high temperatures is beginning to be well understood at a molecular level, nearly nothing is known about how hyperthermophiles respond to stress, considering the response to stress as a transient network-regulation-mediated process that leads to maintenance of the homeostasis of the cell (Fig. 2). Only studies about DNA supercoiling have given some insight into the transitory nature of the heat and cold shock response, the latter being practically unexplored in these organisms. Fortunately, despite the inavailability of genetic tools to manipulate these organisms, genome analysis has proven to be an excellent means to overcome, in part, this problem. Thus, many genes potentially involved in the stress response by homology to the better-known bacterial and eukaryal systems have been recently identified. In most cases, their functional analysis is still to be done, but at least we are gathering the pieces that match the general framework depicted in Fig. 2. Their interrelationships remain to be traced in the future.

The first answers to regulatory aspects in the stress response of hyperthermophiles are expected to be those related to heat shock proteins, since they are the best studied. The structure of the archaeal thermosome has been investigated in great detail and its role in vivo critically analyzed. It remains to be tested whether filament formation actually plays a regulatory role in the storage and release of chaperonins after a stress signal. Interestingly, these filaments appear to be linked to membrane fractions (111). It would be, therefore, possible that they act as signal transducers between stress-induced membrane changes (for instance, modifying lipid fluidity, either directly or by delocalizing membrane proteins in interaction with filaments) and the release of operative chaperonin subunits into the cytoplasm. Since phosphorylation induces conformational changes in the protein (67), a possible connection with two-component sensory systems would be worth exploring. Evidence for a possible two-component sensory system in response to phosphate starvation and other stress conditions has been reported for *Sulfolobus acidocaldarius* (90). Finally, the possible cooperation of heat shock proteins with compatible solutes, such as trehalose, which accumulates in *Sulfolobus* and is known to act synergistically with heat shock proteins in other organisms, should be also investigated.

Another aspect to study in the response to stress of hyperthermophilic archaea concerns transcription regulation. Although the transcriptional machinery is relatively well known by now in archaea, due to the interest raised by its similarity to the eukaryal system, transcriptional regulation is much less understood. DNA supercoiling changes during heat and cold shocks, constituting a general event in all organisms, are one of the likely means to adjust transcription. In this sense, changes in supercoiling (sensing by DNA) would attract topoisomerase action, for instance, reverse gyrase during heat shock, and the subsequent Lk change would act at the level of signal transduction (Fig. 2). It would be interesting to see whether the immediate Lk increase following a heat shock induces the synthesis of specific Hsps in an analogous way to the induction experienced by the alternative sigma factor σ^{32} in bacteria (85), or whether cold shock-induced negative supercoiling is actually required to maintain gene expression during adaptation to cold stress. The study of possible interactions of Hsps with topoisomerases could generate clues about how these proteins may, at their turn, contribute to maintain DNA homeostasis. For instance, DnaK stimulates and stabilizes gyrase in bacteria (89). An analogous stabilization by Hsps could be hypothesized for the equivalent step in hyperthermophilic archaea (Fig. 3). Interestingly, induced thermotolerance at the level of reverse gyrase activity linked to heat shock protein synthesis has been recently detected in *Sulfolobus* (77a).

How lipid composition, known to vary as a function of the environmental temperature in all organisms too, and its influence in the permeability to H^+ and Na^+ (thus the bioenergetics of the cell), varies during heat and cold shock is another aspect to study. Many heat shock proteins interact with membranes (16, 118). How may this interaction contribute to sensing the environmental change and transducing the signal to recover homeostasis in hyperthermophiles?

Yet another observation is related to posttranslational modifications. Like eukaryotic histones (29), the archaeal DNA binding protein Sso7d becomes methylated as temperature increases (7). However, methylation alters neither its thermal stability (66) nor the level of DNA unwinding it induces (78). What is the role, if any, of this posttranslational modification?

Finally, even if we do not know a great deal about how hyperthermophiles respond to thermal stress, we are capable now of formulating more precise questions. From what we have already learned, the emerging idea is that the stress response follows similar pathways to that in other organisms, although the precise regulators may be very different. Nevertheless, the discovery of novel mechanisms in hyperthermophilic archaea cannot be excluded. It is hoped that the new approaches to study the molecular biology of these fascinating organisms soon will provide enough data to get better insight on how they keep their homeostasis in the burning environments they have chosen to live in.

REFERENCES

1. Agback, P., H. Baumann, S. Knapp, R. Ladenstein, and T. Hard. 1998. Architecture of nonspecific protein-DNA interactions in the Sso7d-DNA complex. *Nat. Struct. Biol.* **5:**579–584.

2. Aletta, J. M., T. R. Cimato, and M. J. Ettinger. 1998. Protein methylation: a signal event in post-translational modification. *Trends Biochem. Sci.* **23:**89–91.

3. Andrä, S., G. Frey, R. Jaenicke, and K. O. Stetter. 1998. The thermosome from *Methanopyrus kandleri* possesses an NH$_4^+$-dependent ATPase activity. *Eur. J. Biochem.* **255:**93–99.

4. Andrä, S., G. Frey, M. Nistch, W. Baumeister, and K. O. Stetter. 1996. Purification and structural characterization of the thermosome from the hyperthermophilic archaeum *Methanopyrus kandleri*. *FEBS Lett.* **379:**127–131.

5. Barns, S. M., C. F. Delwiche, J. D. Palmer, and N. R. Pace. 1996. Perspectives on archaeal diversity, thermophily and monophyly from environmental rRNA sequences. *Proc. Natl. Acad. Sci. USA* **93:**9188–9193.

6. Bauer, M. W., S. Bauer, and R. M. Kelly. 1997. Purification and characterization of a proteasome from the hyperthermophilic archaeon *Pyrococcus furiosus*. *Appl. Environ. Microbiol.* **63:**1160–1164.

7. Baumann, H., S. Knapp, T. Lundbäck, R. Ladenstein, and T. Härd. 1994. Solution structure and DNA binding properties of a thermostable protein from the archaeon *Sulfolobus solfataricus*. *Struct. Biol.* **1:**808–819.

8. Bell, S. D., C. Jaxel, M. Nadal, P. F. Kosa, and S. P. Jackson. 1998. Temperature, template topology, and factors requirements of archaeal transcription. *Proc. Natl. Acad. Sci. USA* **95:**15218–15222.

9. Bergerat, A. D., D. Gadelle, and P. Forterre. 1994. Purification of a DNA topoisomerase II from the hyperthermophilic archaeon *Sulfolobus shibatae*. *J. Biol. Chem.* **269:**27663–27669.

10. Bergerat, A. D., B. de Massy, D. Gadelle, P. C. Varoutas, A. Nicolas, and P. Forterre. 1997. An atypical topoisomerase II from archaea with implications for meiotic recombination. *Nature* **386:**414–417.

11. Blöchl, E., R. Rachel, S. Burggraf, D. Hafenbradl, H. W. Jannasch, and K. O. Stetter. 1997. *Pyrolobus fumarii*, gen. and sp. nov., represents a novel group of archaea, extending the upper temperature limit for life to 113°C. *Extremophiles* **1:**14–21.

12. Brock, T. D., K. M. Brock, R. T. Belly, and R. L. Weiss. 1972. *Sulfolobus*: a new genus of sulfur-oxidizing bacteria living at low pH and high temperature. *Arch. Microbiol.* **84:**54–68.

13. Brown, J. R., and A. N. Lupas. 1998. What makes a thermophile? *Trends Microbiol.* **6:**349–351.

14. Bult, C. J., et al. 1996. Complete genome sequence of the methanogenic archaeon, *Methanococcus jannaschii*. *Science* **273:**1058–1073.

15. Burggraf, S. P. Heyder, and N. Eis. 1997. A pivotal archaeal group. *Nature* **385:**780.

16. Carratù, L., S. Franceschelli, C. L. Pardini, G. S. Kobayashi, I. Horvath, L. Vigh, and B. Maresca. 1996. Membrane lipid perturbation modifies the set point of the temperature of heat shock response in yeast. *Proc. Natl. Acad. Sci. USA* **93:**3870–3875.

17. Charbonnier, F., and P. Forterre. 1994. Comparison of plasmid DNA topology among mesophilic and thermophilic eubacteria and archaebacteria. *J. Bacteriol.* **176:**1251–1259.

18. Clarens, M., A. J. L. Macario, and E. Conway de Macario. 1995. The archaeal *dnaK-dnaJ* gene cluster: organization and expression in the methanogen *Methanosarcina mazei*. *J. Mol. Biol.* **250:**191–201.

19. Conway de Macario, E., and A. J. L. Macario. 1994. Heat shock response in Archaea. *Trends Biotechnol.* **12:**512–518.

20. da Costa, M. S., H. Santos, and E. A. Galinski. 1998. An overview of the role and diversity of compatible solutes in Bacteria and Archaea. *Adv. Biochem. Eng. Biotechnol.* **61:**117–153.

21. Craig, E. A., B. D. Gambill, and R. J. Nelson. 1993. Heat shock proteins: molecular chaperones of protein biogenesis. *Microbiol. Rev.* **57:**402–414.

22. Daalgard, J. Z., and R. A. Garrett. 1993. Archaeal hyperthermophile genes, p. 535–564. *In* M. Kates, D. J. Kushner, and A. T. Matheson (eds.), *The Biochemistry of Archaea*. Elsevier, Amsterdam, The Netherlands.

23. Danson, M. J., and D. W. Hough. 1998. Structure, function and stability of enzymes from the *Archaea*. *Trends Microbiol.* **6:**307–313.

24. Deckert, G., et al. 1998. The complete genome of the hyperthermophilic bacterium *Aquifex aeolicus*. *Nature* **392:**353–358.

25. DeLong, E. 1998. Archaeal means and extremes. *Science* **280:**542–543.

26. DeLong, E., L. L. King, R. Massana, H. Cittone, A. Murray, C. Schleper, and S. G. Wakeham. 1998. Dibiphytanyl ether lipids in nonthermophilic crenarchaeotes. *Appl. Environ. Microbiol.* **64:**1133–1138.

27. De Mot, R., I. Nagy, J. Walz, and W. Baumeister. 1999. Proteasomes and other self-compartmentalizing proteases in prokaryotes. *Trends Microbiol.* **7:**88–92.

28. Depew, R. E., and J. C. Wang. 1975. Conformational fluctuations of DNA helix. *Proc. Natl. Acad. Sci. USA* **72:**4275–4279.

29. DesRosiers, R., and R. M. Tanguay. 1988. Methylation of *Drosophila* histones at proline, lysine, and arginine residues during heat shock. *J. Biol. Chem.* **263:**4686–4692.

30. Ditzel, L., J. Löwe, D. Stock, K. O. Stetter, H. Huber, R. Huber, and S. Steinbacher. 1998. Crystal structure of the thermosome, the archaeal chaperonin and homolog of CCT. *Cell* **93:**125–138.

31. Dorman, C. J. 1996. Flexible response: DNA supercoiling, transcription and bacterial adaptation to environmental stress. *Trends Microbiol.* **4:**214–216.

32. **Duguet, M.** 1993. The helical repeat of DNA at high temperature. *Nucleic Acids Res.* **21**:463–468.

33. **Duguet, M.** 1995. Reverse gyrase, p. 84–114. *In* F. Eckstein and d. M. J. Lilley (ed.), *Nucleic Acids and Molecular Biology.* Springer-Verlag, Berlin, Germany.

34. **Elferink, M. G., J. G. de Wit, A. J. Driessen, and W. N. Konings.** 1994. Stability and proton-permeability of liposomes composed of archaeal tetraether lipids. *Biochim. Biophys. Acta* **1193**:247–254.

35. **Emmerhoff, O. J., H. P. Klenk, and N. K. Birkeland.** 1998. Characterization and sequence comparison of temperature-regulated chaperonins from the hyperthermophilic archaeon *Archaeoglobus fulgidus. Gene* **215**:431–438.

36. **Fitz-Gibbon, S., A. J. Choi, J. H. Miller, K. O. Stetter, M. I. Simon, R. Swanson, and U.-J. Kim.** 1997. A fosmid-based genomic map and identification of 474 genes of the hyperthermophilic *Pyrobaculum aerophilum. Extremophiles* **1**:36–51.

37. **Forterre, P., A. Bergerat, and P. López-García.** 1996. The unique DNA topology and DNA topoisomerases of hyperthermophilic archaea. *FEMS Microbiol. Rev.* **18**:237–248.

38. **Furutani, M., T. Iida, S. Yamano, K. Kamino, and T. Maruyama.** 1998. Biochemical and genetic characterization of an FK506-sensitive peptidyl prolyl *cis-trans* isomerase from a thermophilic archaeon, *Methanococcus thermolitotrophicus. J. Bacteriol.* **180**:388–394.

39. **Grau, R., and D. de Mendoza.** 1993. Regulation of the synthesis of unsaturated fatty acids by growth temperature in *Bacillus subtilis. Mol. Microbiol.* **8**:535–542.

40. **Graumann, P. L., and M. A. Marahiel.** 1998. A superfamily of proteins that contain the cold-shock domain. *Trends Biochem. Sci.* **23**:286–290.

41. **Graumann, P., T. M. Wendrich, M. H. W. Weber, K. Schöder, and M. A. Marahiel.** 1997. A family of cold shock proteins in *Bacillus subtilis* is essential for cellular growth and for efficient protein synthesis at optimal and low temperature. *Mol. Microbiol.* **25**:741–756.

42. **Gribaldo, S., V. Lumia, R. Creti, E. Conway de Macario, A. Sanangelantoni, and P. Cammarano.** 1999. Discontinuous occurrence of the *hsp70* (*dnaK*) gene among Archaea and sequence features of HSP70 suggest a novel outlook on phylogenies inferred from this protein. *J. Bacteriol.* **181**:434–443.

43. **Grogan, D. W.** 1998. Hyperthermophiles and the problem of DNA instability. *Mol. Microbiol.* **28**:1043–1049.

44. **Gross, C. A.** 1996. Function and regulation of the heat shock proteins, p. 1382–1399. *In* F. C. Neidhart, R. Curtiss III, J. L. Ingraham, E. C. C. Lin, K. B. Low, B. Magasanik, W. S. Reznikoff, M. Riley, M. Schaechter, and H. E. Umbarger (ed.), *Escherichia coli and* Salmonella: *Cellular and Molecular Biology,* 2nd ed. ASM Press, Washington, D.C.

45. **Guipaud, O., E. Marguet, K. M. Noll, C. Bouthier de la Tour, and P. Forterre.** 1997. Both DNA gyrase and reverse gyrase are present in the hyperthermophilic bacterium *Thermotoga maritima. Proc. Natl. Acad. Sci. USA* **94**:10606–10611.

46. **Gupta, R. S.** 1998. What are archaebacteria: life's third domain or monoderm prokaryotes related to gram-positive bacteria? A new proposal for the classification of prokaryotic organisms. *Mol. Microbiol.* **29**:695–707.

47. **Haney, P. J., J. H. Badger, G. L. Buldak, C. I. Reich, C. R. Woese, and G. Olsen.** 1999. Thermal adaptation analyzed by comparison of protein sequences from mesophilic and extremely thermophilic *Methanococcus* species. *Proc. Natl. Acad. Sci. USA* **96**:3578–3583.

48. **Heber, A. M., A. M. Kropinski, and K. F. Jarrell.** 1991. Heat shock response of the archaebacterium *Methanococcus voltae. J. Bacteriol.* **173**:3224–3227.

49. **Hecker, M., W. Schumann, and U. Völker.** 1996. Heat-shock and general stress response in *Bacillus subtilis. Mol. Microbiol.* **19**:417–428.

50. **Holden, J. F., and J. A. Baross.** 1993. Enhanced thermotolerance and temperature—induced changes in protein composition in the hyperthermophilic archaeon ES4. *J. Bacteriol.* **175**:2839–2843.

51. **Horlacher, R., K. B. Xavier, H. Santos, J. DiRuggiero, M. Kossman, and W. Boos.** 1998. Archaeal binding protein-dependent ABC transporter: molecular and biochemical analysis of the trehalose/maltose transport system of the hyperthermophilic archaeon *Thermococcus litoralis. J. Bacteriol.* **180**:680–689.

52. **Huber, R., T. A. Langworthy, H. König, M. Thomm, C. R. Woese, U. B. Sleytr, and K. O. Stetter.** 1986. *Thermotoga maritima* sp. nov. represents a new genus of unique extremely thermophilic eubacteria growing up to 90°C. *Arch. Microbiol.* **144**:324–333.

53. **Huber, R., T. Wilharm, D. Huber, A. Trincone, S. Burggraf, H. Koenig, R. Rachel, I. Rockinger, H. Fricke, and K. O. Stetter.** 1992. *Aquifex pyrophilus,* new genus, new species, represents a novel group of marine hyperthermophilic-oxydizing bacteria. *Syst. Appl. Microbiol.* **15**:340–351.

54. **Hurme, R., and M. Rhen.** 1998. Temperature sensing in bacterial gene regulation—what it all boils down to. *Mol. Microbiol.* **30**:1–6.

55. **Jaenicke, R., and G. Böhm.** 1998. The stability of proteins in extreme environments. *Curr. Opin. Struct. Biol.* **8**:738–748.

56. **Jerez, C.** 1988. The heat shock response in meso- and thermoacidophilic chemolithotrophic bacteria. *FEMS Microbiol. Lett.* **56**:289–294.

57. **Kagawa, H. K., et al.** 1995. Gene of heat shock protein of sulfur-dependent archaeal hyperthermophile *Desulfurococcus. Biochem. Biophys. Res. Commun.* **214**:730–736.

58. **Kagawa, H. K., J. Osipiuk, N. Maltsev, R. Overbeek, E. Quite-Randall, A. Joachimiak, and J. D. Trent.** 1995. The 60 kDa heat shock proteins in the hyperthermophilic archaeon *Sulfolobus shibatae. J. Mol. Biol.* **253**:712–725.

59. **Kates, M.** 1993. Membrane lipids of archaea, p. 261–295. *In* M. Kates, et al. (ed.), *The Biochemistry of Archaea (Archaebacteria).* Elsevier Science Publishers B. V., Amsterdam, The Netherlands.

60. **Kawarabayasi, Y., et al.** 1998. Complete sequence and gene organization of the genome of a hyperthermophilic archaebacterium, *Pyrococcus horikoshii* OT3. *DNA Res.* **5**:55–76.

61. **Kikuchi, A., and K. Asai.** 1984. Reverse gyrase, a topoisomerase which introduces positive superhelical turns into DNA. *Nature* **309**:677–681.

62. **Kim, K. K., L.-W. Hung, H. Yokota, R. Kim, and S.-H. Kim.** 1998. Crystal structures of eukaryotic translation initiation factor 5A from *Methanococcus jannaschii* at 1.8 Å resolution. *Proc. Natl. Acad. Sci. USA* **95**:10419–10424.

63. **Kim, K. K., R. Kim, and S.-H. Kim.** 1998. Crystal structure of a small heat-shock protein. *Nature* **394**:595–599.

64. **Kim, R., K. K. Kim, H. Yokota, and S.-H. Kim.** 1998. Small heat shock protein of *Methanococcus jannaschii,* a hyperthermophile. *Proc. Natl. Acad. Sci. USA* **95**:9129–9133.

65. **Klenk, H.-P., et al.** 1997. The complete genome sequence of the hyperthermophilic, sulphate-reducing archaeon *Archaeoglobus fulgidus. Nature* **390**:364–370.

66. **Knapp, S., A. Karshikoff, K. D. Berndt, P. Christova, B. Atanasov, and R. Ladenstein.** 1996. Thermal unfolding of the DNA-binding protein Sso7d from the hyperthermophile *Sulfolobus solfataricus. J. Mol. Biol.* **264**:1132–1144.

67. **Knapp, S., I. Schmidt-Krey, H. Hebert, T. Bergman, H. Jörnvall, and R. Ladenstein.** 1994. The molecular chaperonin

TF55 from the thermophilic archaeon *Sulfolobus solfataricus*. A biochemical and structural characterization. *J. Mol. Biol.* **242**:397–407.

68. **Kowalak, J. A., J. J. Dalluge, J. A. McCloskey, and K. O. Stetter.** 1994. The role of posttranscriptional modification in stabilization of transfer RNA from hyperthermophiles. *Biochemistry* **33**:7869–7876.

69. **Kozyavkin, S. A., A. V. Pushkin, F. A. Eiserling, K. O. Stetter, J. A. Lake, and A. I. Slesarev.** 1995. DNA enzymology above 100°C. Topoisomerase V unlinks circular DNA at 80–122°C. *J. Biol. Chem.* **23**:13593–13595.

70. **Kuo, Y.-P., D. K. Thompson, A. St. Jean, R. L. Charlebois, and C. J. Daniels.** 1997. Characterization of two heat shock genes from *Haloferax volcanii*: a model system for transcription regulation in the Archaea. *J. Bacteriol.* **179**:6318–6324.

71. **Ladenstein, R., and G. Antranikian.** 1998. Proteins from hyperthermophiles: stability and enzymatic catalysis close to the boiling point of water. *Adv. Biochem. Eng. Biotechnol.* **61**:37–85.

72. **Lamosa, P., L. O. Martins, M. S. da Costa, and H. Santos.** 1998. Effects of temperature, salinity, and medium composition on compatible solute accumulation by *Thermococcus* spp. *Appl. Environ. Microbiol.* **64**:3591–3598.

73. **Lange, M., A. J. L. Macario, B. K. Ahring, and E. Conway de Macario.** 1997. Heat-shock response in *Methanosarcina mazei*. *Curr. Microbiol.* **35**:116–121.

74. **LaPaglia, C., and P. L. Hartzell.** 1997. Stress-induced production of biofilm in the hyperthermophile *Archaeoglobus fulgidus*. *Appl. Environ. Microbiol.* **63**:3158–3163.

75. **Lewis, J. G., R. P. Learmonth, and K. Watson.** 1995. Induction of heat, freezing and salt tolerance by heat and salt shock in *Saccharomyces cerevisiae*. *Microbiology* **141**:687–694.

76. **L'Haridon, A., A.-L. Reysenbach, P. Glénat, D. Prieur, and C. Jeanthon.** 1995. Hot continental biosphere in a continental oil reservoir. *Nature* **377**:223–224.

77. **López-García, P., and P. Forterre.** 1997. DNA topology in hyperthermophilic archaea: reference states and their variation with growth phase, growth temperature, and temperature stresses. *Mol. Microbiol.* **23**:1267–1279.

77a. **López-García, P., and P. Forterre.** 1999. Control of DNA topology during thermal stress in hyperthermophilic Archaea: DNA topoisomerase levels, activities and induced thermotolerance during heat and cold shock in *Sulfolobus*. *Mol. Microbiol.* **33**:766–777.

78. **López-García, P., S. Knapp, R. Ladenstein, and P. Forterre.** 1998. In vitro DNA binding of the archaeal protein Sso7d induces negative supercoiling at temperatures typical for thermophilic growth. *Nucleic Acids Res.* **26**:2322–2328.

79. **Marguet, E., and P. Forterre.** 1994. DNA stability at temperatures typical for hyperthermophiles. *Nucleic Acids Res.* **22**:1681–1686.

80. **Marguet, E., and P. Forterre.** 1998. Protection of DNA by salts against thermodegradation at temperatures typical for hyperthermophiles. *Extremophiles* **2**:115–122.

81. **Martins, L. O., R. Huber, H. Huber, K. O. Stetter, M. da Costa, and H. Santos.** 1997. Organic solutes in hyperthermophilic Archaea. *Appl. Environ. Microbiol.* **63**:896–902.

82. **Minuth, T., M. Henn, K. Rutkat, S. Andra, G. Frey, R. Rachel, K. O. Stetter, and R. Jaenicke.** 1999. The recombinant thermosome from the hyperthermophilic archaeon *Methanopyrus kandleri*: in vitro analysis of its chaperone activity. *Biol. Chem.* **380**:55–62.

83. **Mizushima, T., K. Kataoka, Y. Ogata, R. Inoue, and K. Sekimizu.** 1997. Increase in negative supercoiling of plasmid DNA in *Escherichia coli* exposed to cold shock. *Mol. Microbiol.* **23**:381–386.

84. **Mizushima, T., S. Natori, and K. Sekimizu.** 1993. Relaxation of supercoiled DNA associated with induction of heat shock proteins in *Escherichia coli*. *Mol. Gen. Genet.* **238**:1–5.

85. **Mizushima, T., Y. Ohtsuka, H. Mori, T. Miki, and K. Sekimizu.** 1996. Increase in synthesis and stability of ;$\lambda\gamma\sigma^{32}$ on treatment with inhibitors of DNA gyrase in *Escherichia coli*. *Mol. Gen. Genet.* **253**:297–302.

86. **Nadal, M., E. Couderc, M. Duguet, and C. Jaxel.** 1994. Purification and characterization of reverse gyrase from *Sulfolobus shibatae*. *J. Biol. Chem.* **269**:6255–6263.

87. **Narberhaus, F.** 1999. Negative regulation of bacterial heat shock genes. *Mol. Microbiol.* **31**:1–8.

88. **Noon, K. R., E. Bruenger, and J. A. McCloskey.** 1998. Posttranscriptional modifications in 16S aand 23S of the archaeal hyperthermophile *Sulfolobus solfataricus*. *J. Bacteriol.* **180**:2883–2888.

89. **Ogata, Y., T. Mizushima, K. Kataoka, K. Kita, T. Miki, and K. Sekimizu.** 1996. DnaK heat shock protein of *Escherichia coli* maintains the negative supercoiling of DNA against thermal stress. *J. Biol. Chem.* **271**:29407–29414.

90. **Osorio, G., and C. Jerez.** 1996. Adaptive response of the archaeon *Sulfolobus acidocaldarius* BC65 to phosphate starvation. *Microbiology* **142**:1531–1536.

91. **Pace, N. R.** 1991. Origin of life—facing up to the physical setting. *Cell* **65**:531–533.

92. **Pace, N. R.** 1997. A molecular view of microbial diversity and the biosphere. *Science* **276**:734–740.

93. **Peak, M. J., F. Robb, and J. G. Peak.** 1995. Extreme resistance to thermally induced DNA backbone breaks in the hyperthermophilic archaeon *Pyrococcus furiosus*. *J. Bacteriol.* **177**:6316–6318.

94. **Peat, T. S., J. Newman, G. S. Waldo, J. Berendzen, and T. C. Terwilliger.** 1998. Structure of translation initiation factor 5A from *Pyrobaculum aerophilum* at 1.75 Å resolution. *Structure* **15**:1207–1214.

95. **Philippe, H., K. Budin, and D. Moreira.** 1999. Horizontal transfers confuse the prokaryotic phylogeny based on the HSP70 protein family. *Mol. Microbiol.* **31**:1007–1012.

96. **Phipps, B. M., A. Hoffmann, K. O. Stetter, and W. Baumeister.** 1991. A novel ATPase complex selectively accumulated upon heat shock is a major cellular component of thermophilic archaebacteria. *EMBO J.* **10**:1711–1722.

97. **Phipps, B. M., D. Typske, R. Hegert, S. Volker, A. Hoffman, K. O. Stetter, and W. Baumeister.** 1993. Structure of a molecular chaperone from a thermophilic archaebacterium. *Nature* **361**:475–477.

98. **Quite-Randall, E., J. D. Trent, R. Josephs, and A. Joachimiak.** 1995. Conformational cycle of the archaeosome, aTCP1-like chaperonin from *Sulfolobus shibatae*. *J. Biol. Chem.* **48**:28818–28823.

99. **Robinson, H., Y.-G. Gao, B. S. McCrary, S. P. Edmondson, J. W. Shriver, and A. H.-J. Wang.** 1998. The hyperthermophilic chromosomal protein Sac7d sharply kinks DNA. *Nature* **392**:202–205.

100. **Ruepp, A., C. Eckerskorn, M. Bogyo, and W. Baumeister.** 1998. Proteasome function is dispensable under normal but not under heat shock conditions in *Thermoplasma acidophilum*. *FEBS Lett.* **20**:87–90.

101. **Russell, N. J., and N. Fukunaga.** 1990. A comparison of thermal adaptation of membrane lipids in psycrophilic and thermophilic bacteria. *FEMS Microbiol. Rev.* **75**:171–182.

102. **Rutherford, S. L. and S. Lindquist.** 1998. Hsp90 as a capacitor for morphological evolution. *Nature* **396**:336–342.

103. **Singer, M. A., and S. Lindquist.** 1998. Thermotolerance in *Saccharomyces cerevisiae*: the Yin and Yang of trehalose. *Trends Biotechnol.* **16**:460–468.

104. Smith, D. R., L. A. Doucette-Stamm, C. Deloughery, H. Lee, J. Dubois, T. Aldredge, R. Bashirzadeh, D. Blakely, R. Cook, K. Gilbert, D. Harrison, L. Hoang, P. Keagle, W. Lumm, B. Pothier, D. Qiu, R. Spadafora, R. Vicaire, Y. Wang, J. Wierzbowski, R. Gibson, N. Jiwani, A. Caruso, D. Bush, and J. N. Reeve. 1997. Complete genome sequence of *Methanobacterium thermoautotrophicum* ΔH: functional analysis and comparative genomics. *J. Bacteriol.* 179:7135–7155.

105. Sprott, G. D., M. Meloche, and J. C. Richards. 1991. Proportions of diether, macrocyclic diether, and tetraether lipids in *Methanococcus jannaschii* grown at different temperatures. *J. Bacteriol.* 173:3907–3910.

106. Stetter, K. O. 1990. Hyperthermophilic prokaryotes. *FEMS Microbiol. Rev.* 18:149–158.

107. Thieringer, H. A., P. G. Jones, and M. Inouye. 1998. Cold shock and adaptation. *Bioessays* 20:49–57.

108. Török, Z., I. Horváth, P. Goloubinoff, E. Kovács, A. Glatz, G. Balogh, and L. Vígh. 1997. Evidence for a lipochaperonin: association of active protein-folding GroESL oligomers with lipids can stabilize membranes under heat shock conditions. *Proc. Natl. Acad. Sci. USA* 94:2192–2197.

109. Trent, J. D. 1996. A review of acquired thermotolerance, heat-shock proteins, and molecular chaperones in archaea. *FEMS Microbiol. Rev.* 18:249–258.

110. Trent, J. D., M. Gabrielsen, B. Jensen, J. Neuhard, and J. Olsen. 1994. Acquired thermotolerance and heat shock proteins in thermophiles from the three phylogenetic domains. *J. Bacteriol.* 176:6148–6152.

111. Trent, J. D., H. Kagawa, and T. Yaoi. 1998. The role of chaperonins in vivo: the next frontier. *Ann. N.Y. Acad. Sci.* 851:36–47.

112. Trent, J. D., H. Kagawa, T. Yaoi, E. Olle, and N. J. Zaluzec. 1997. Chaperonin filaments: the archaeal cytoskeleton? *Proc. Natl. Acad. Sci. USA* 94:5383–5388.

113. Trent, J. D., E. Nimmesgern, J. S. Wall, F.-U. Hartl, and A. L. Horwich. 1998. A molecular chaperone from a thermophilic archaebacterium is related to the eukaryotic protein t-complex polypeptide-1. *Nature* 354:490–493.

114. Trent, J. D., J. Osipiuk, and T. Pinkau. 1990. Acquired thermotolerance and heat shock in the extremely thermophilic archaebacterium *Sulfolobus* sp. strain B12. *J. Bacteriol.* 172:1478–1484.

115. van de Vossenberg, J. L. C. M., A. J. M. Driessen, and W. N. Konings. 1991. The essence of being extremophilic: the role of the unique archaeal membrane lipids. *Extremophiles* 2:163–170.

116. van de Vossenberg, J. L. C. M., T. Ubbink-Kok, M. G. L. Elferink, A. J. M. Driessen, and W. N. Konings. 1995. Ion permeability of the cytoplasmic membrane limits the maximum growth temperature of bacteria and archaea. *Mol. Microbiol.* 18:925–932.

117. Vieille, C., D. S. Burdette, and J. G. Zeikus. 1996. Thermozymes. *Biotechnol. Annu. Rev.* 2:1–83.

118. Vigh, L., B. Maresca, and J. L. Harwood. 1998. Does the membrane's physical state control the expression of heat and other genes? *Trends Biochem. Sci.* 23:369–374.

119. Wiegel, J., and M. Adams (eds). 1998. *Thermophiles—The Keys to Molecular Evolution and the Origin of Life?* Taylor & Francis Ltd., London.

120. Woese, C. R., O. Kandler, and M. L. Wheelis. 1990. Towards a natural system of organisms: proposal for the domains *Archaea, Bacteria* and *Eucarya*. *Proc. Natl. Acad. Sci. USA* 87:4576–4579.

Bacterial Stress Responses
Edited by G. Storz and R. Hengge-Aronis
©2000 ASM Press, Washington, D.C.

Chapter 24

The Stress Responses of *Deinococcus radiodurans*

JOHN R. BATTISTA, ASHLEE M. EARL, AND OWEN WHITE

A review of the literature concerning Deinococcus radiodurans *resistance and/or response to ionizing radiation, oxidative stress, UV light, heat, and desiccation was conducted and combined with a search of the* D. radiodurans *genome intended to identify genes encoding homologs of the stress-response proteins of* Escherichia coli. *This analysis revealed three characteristics that distinguish stress responses of* D. radiodurans *from those of* E. coli. *(a) While it appears that* D. radiodurans *expresses a significant response to oxidative stress, it is not possible, at present, to predict how this species regulates this response. The* D. radiodurans *genome does not appear to encode homologs of the regulatory proteins OxyR, SoxR, or SoxS. (b) Even though* D. radiodurans *has a LexA homolog, there is no evidence of an SOS response similar to that seen in* E. coli. *The proteins needed for UV resistance appear to be constitutively expressed. (c) The ability of* D. radiodurans *to survive all forms of stress examined to date requires a functional RecA protein, suggesting that the* D. radiodurans *RecA may have regulatory functions that are sensitive to a wide variety of cellular damage.*

Since its initial description in 1956, *Deinococcus radiodurans* has been considered a poorly understood scientific curiosity. Even though this species is distinguished by an extraordinary capacity to tolerate ionizing radiation, only a limited number of investigators have attempted to study *D. radiodurans* in any detail. As a consequence, our understanding of this species' physiology and biochemistry is rudimentary at best.

Here, we have tried to document what is known of the *D. radiodurans* responses to environmental stress. Although best known for its ionizing radiation resistance, *D. radiodurans* is also resistant to UV light, heat and desiccation. We have also surveyed the *D. radiodurans* genome (73) and identified proteins that, based on sequence similarity, may be involved in various stress responses. The picture that emerges is incomplete. While it is obvious that this organism expresses proteins in response to various stresses, there is no clear indication of how *D. radiodurans* recognizes the stress or regulates the stress response.

SUMMARY DESCRIPTION OF *D. RADIODURANS*

D. radiodurans (9, 58) is a red-pigmented, nonsporulating, nonmotile, spherical bacterium that forms pairs or tetrads in liquid culture. Although this species stains gram positive, it is specifically related to the gram-negative genus *Thermus* (35, 69, 72), both genera forming a phylum within the *Bacteria*. *D. radiodurans* has a complex cell envelope similar to that of gram-negative organisms, with an inner and outer membrane and a relatively thick (~20 nm) peptidoglycan layer. Cells are chemoorganotrophic with respiratory metabolism and grow optimally at 30°C in TGY broth (0.5% tryptone, 0.1% glucose, and 0.3% yeast extract) with aeration. The *D. radiodurans* chromosome is 3.3×10^6 bp long with 67 mol% GC content and has been sequenced in its entirety (73). The genome is segmented, consisting of a 2.64-Mbp chromosome, a 0.41-Mbp megaplasmid, a 0.18-Mbp megaplasmid (designated megaplasmid I and II, respectively), and a 0.045-Mbp plasmid (42). There is believed to be a minimum of four identical copies of the chromosome per stationary-phase cell (29, 34). In exponentially growing cells, the number of chromosome copies increases to as many as 10 per cell (29, 34).

The family *Deinococcaceae* was described, with the genus *Deinococcus* as the type genus (9) and

---

John R. Battista and Ashlee M. Earl • Department of Biological Sciences, Louisiana State University and A & M College, Baton Rouge, LA 70803. **Owen White** • The Institute for Genomic Research, 9712 Medical Center Dr., Rockville, MD 20850.

Deinococcus radiodurans as the type species. Seven species comprise the genus: *D. radiodurans, Deinococcus radiophilus, Deinococcus proteolyticus, Deinococcus radiopugnans, Deinococcus grandis, Deinococcus geothermalis,* and *Deinococcus murrayi* (23, 60). Catalogs of 16S rRNA revealed the deinococci as a distinct phylogenetic lineage in the domain *Bacteria* (74, 75) and established that the genera *Deinococcus* and *Thermus* were related (35). Analyses of a nearly complete 16S rRNA sequence of *D. radiodurans* strain R1 with other complete 16S rRNA sequences established the *Deinococcus-Thermus* phylum (72).

The natural habitat of *Deinococcus* species has not been defined. Only two ecological studies of *Deinococcus* species have been reported (39, 46), and even though these studies were limited to relatively small geographic areas, each concluded that the deinococci were rare but widely distributed soil organisms. Isolates have most frequently been obtained by selecting for ionizing radiation resistance. *Deinococcus* strains have been isolated following γ-irradiation of soil (12, 39, 40), animal feces (59), sewage (37), sawdust (36), processed meats (16, 24), dried foods (41), room dust (12), medical instruments (12), and textiles (40).

It is possible to identify ionizing radiation-resistant bacteria from natural microflora by selecting for desiccation resistance (61). *D. radiodurans* is exceptionally resistant to desiccation (47). Cultures dried onto a glass slide and held for 2 years in a desiccator at 5% humidity exhibit approximately 80% viability (V. Mattimore and J. R. Battista, unpublished observations). The desiccation resistance of *D. radiodurans* and other *Deinococcus* species may provide some insight into the ecology of this genus. Perhaps the dry but viable organism becomes airborne and is distributed randomly within an environment.

RESPONSE TO IONIZING RADIATION AND OXIDATIVE STRESS IN *D. RADIODURANS*

The most startling and distinguishing characteristic of *D. radiodurans* is its ability to tolerate ionizing radiation. The γ-radiation survival curve of log-phase cultures of *D. radiodurans* R1 exhibits a shoulder of resistance to 5,000 grays (Gy) (56), followed by an exponential loss of viability. The ionizing radiation resistance of a culture increases as cells enter stationary phase, the shoulder dose extending to approximately 15,000 Gy (13–15). As cultures enter stationary phase, they become anoxic and cell division stops. Because oxygen potentiates the lethal effects of ionizing radiation, it has been assumed that the lack of oxygen, in part, explains the increased resistance of stationary-phase cultures. The D_{37} dose (i.e., the dose that, on average, is required to inactivate a single colony-forming unit) for *D. radiodurans* R1 is approximately 6,000 Gy, at least 200-fold higher than the D_{37} dose of *Escherichia coli* cultures irradiated under the same conditions. The energy deposited by 6,000 Gy of γ-radiation should introduce approximately 200 DNA double-strand breaks, over 3,000 single-strand breaks, and greater than 1,000 sites of base damage per *D. radiodurans* genome (64). Although there has never been a comprehensive effort to quantify all forms of ionizing radiation-induced DNA damage in *D. radiodurans*, double-strand breaks (47), single-strand breaks (6, 17, 18), and damaged bases (as measured by thymine glycol production) (32, 33) appear at levels predicted by the dose of ionizing radiation administered. In other words, the ionizing radiation resistance of *D. radiodurans* is not the result of a passive process that prevents the introduction of DNA damage. The cell suffers damage and that damage is repaired.

The biochemical details of the *D. radiodurans* response to ionizing radiation are poorly understood. Using two-dimensional gel electrophoresis and silver staining, Tanaka et al. (67) reported changes in the levels of 22 proteins following exposure of *D. radiodurans* to 6,000 Gy of γ-radiation: the intensity of 9 spots increasing and 13 spots decreasing. Pretreating cultures with chloramphenicol before irradiation prevents the observed increase in eight of the nine spots, suggesting that these proteins are generated as part of an overall response to ionizing radiation-induced damage within the cell. It is assumed that the ninth spot appears because the mobility of a constitutively expressed protein is altered by post-translational modification following irradiation. Two of the induced proteins have been tentatively identified as homologs of elongation factor Tu (67) and the *katA* gene product (44) of *E. coli*. The proteins that arise following γ-irradiation are presumably necessary for cell survival, as inhibition of protein synthesis dramatically affects cell survival. Chloramphenicol pretreatment reduces the resistance of *D. radiodurans* to ionizing radiation, preventing the restitution of ionizing radiation-induced single-strand (17, 18) and double-strand (38) DNA breaks and the excision of thymine glycol (32).

The disappearance of 13 proteins from cells exposed to ionizing radiation is an intriguing observation that has not been investigated. There is the possibility that the removal of these proteins may facilitate the cell's recovery postirradiation. It has been suggested that, in response to DNA damage, *D. radiodurans* can inhibit DNA replication until repairs are completed (3, 50). Perhaps one or more of the proteins downregulated following ionizing radiation is involved in that phenomenon.

D. radiodurans is also very resistant to hydrogen peroxide, cells showing only 10% loss of viability when exposed to 40 mM H_2O_2 for 60 min at 30°C (71). Since H_2O_2 induces a spectrum of damage similar to that induced by ionizing radiation, the proteins necessary for ionizing radiation resistance presumably facilitate the repair of H_2O_2-induced damage. Treatment with 10 mM H_2O_2 for 60 min makes that culture 100-fold more resistant to a 20-min exposure to 80 mM H_2O_2 relative to an untreated culture (71). This protective effect is accompanied by a 2.5-fold increase in catalase activity and is eliminated if the culture is incubated in the presence of chloramphenicol. Hydrogen peroxide pretreatment also enhances the resistance of *D. radiodurans* to ionizing radiation, increasing survival of irradiated cultures between 10- and 100-fold (71). This pretreatment is apparently inducing an uncharacterized oxidative stress response that protects against H_2O_2 as well as ionizing radiation, confirming an overlap in the cell's resistance to the lethal effects of both agents.

To our knowledge, there has never been an attempt to investigate the regulation of responses of *D. radiodurans* to oxidative damage, and a search of the *D. radiodurans* genome for regulatory proteins associated with oxidative stress was unsuccessful. There is no evidence that the *D. radiodurans* genome encodes a SoxR or SoxS homolog. There is a 317-amino-acid regulatory protein encoded that is clearly a member of the *lysR* family (62). This putative regulator could be an OxyR homolog, but its homology (27% identity scattered over 310 amino acids) to the *E. coli* OxyR is so limited that it is not possible to assign this protein a function based on sequence similarity alone.

MECHANISMS THAT LIMIT OXYGEN TOXICITY

Catalase and Superoxide Dismutase

The *D. radiodurans* genome encodes two types of catalase and two types of superoxide dismutase (Table 1). The catalases are homologs of the *catX* and *katE* gene products of *Bacillus subtilis*. The superoxide dismutase proteins are homologs of the SodA protein of *E. coli* and the SodC protein of *Brucella abortus*. Strong catalase and superoxide dismutase (SOD) activities are found in *D. radiodurans* cell extracts, and the level of activity can be influenced by exposing cells to a sublethal dose of H_2O_2 (11, 71). The basal level of catalase activity (550 to 600 U/mg of protein)

Table 1. Deinococcal proteins involved in the removal of active oxygen species

Identification	Size[a]	% Similarity[b]
Catalase (*catX*)	536	77 (401/511)
Catalase (*katE*)	686	66 (461/693)
Superoxide dismutase (*sodA*)	201	77 (167/212)
Superoxide dismutase (*sodC*)	182	77 (76/145)

[a] Predicted number of amino acids in the deinococcal protein.
[b] Data in parentheses are ratios of the number of conserved residues to the length of the conserved region.

associated with cultures in exponential growth is quite high (11, 71) relative to *E. coli* K-12 (approximately 5 U/mg of protein), suggesting that *D. radiodurans* is always under some form of intracellular oxidative stress.

Two forms of catalase can be detected in crude extracts of *D. radiodurans* with gels stained for catalase activity. There is an intense, rapidly migrating band, designated KatA, that is the *catX* gene product (44), and a less abundant, slower migrating catalase, designated KatB, that presumably is the *katE* homolog. The KatA protein is induced when *D. radiodurans* cultures are pretreated with H_2O_2, but its activity remains constant as cells enter stationary phase (71). In contrast, the expression of KatB remains unchanged in response to H_2O_2, and increases two- to threefold as cells enter stationary phase, indicating that the expression of these proteins is separately regulated.

Markillie et al. (44) disrupted the *katA* gene and showed that loss of this catalase had little effect on the ionizing radiation resistance of *D. radiodurans* except at doses exceeding 24,000 Gy. *katB* alleles of *D. radiodurans* have not been described.

Cultures of *D. radiodurans* also exhibit approximately sixfold greater levels of SOD activity when compared with *E. coli* (11). Three SOD bands appear on activity stained gels (11). The *sodA* gene product is most abundant (44). Presumably, one of the minor bands is the *sodC* homolog. Since no other SOD-like sequences are found in the *D. radiodurans* genome, it is possible that the third band may be a modified form of one of the other SODs. Genetic inactivation of *sodA* sensitizes *D. radiodurans* to ionizing radiation, but the protective effect of SOD is only seen at doses in excess of 16,000 Gy (44). Since loss of either the *katA* or *sodA* gene product fails to sensitize *D. radiodurans* to doses of γ-radiation that are less than 16,000 Gy, it must be assumed that these proteins contribute little to this species' ionizing radiation resistance at lower doses. As the level of ionizing radiation-induced cellular damage increases, it is obvious that the removal of reactive oxygen species becomes more important to cell survival. It is unlikely that catalase and SOD are protecting DNA from further damage, since extensive DNA damage occurs at lower doses of γ-radiation (3, 49–51). Perhaps catalase and SOD limit the destruction of other cellular molecules, such as lipids or proteins, that are necessary for cell survival but that are relatively unaffected by lower doses of ionizing radiation.

Base Excision Repair

The *D. radiodurans* genome encodes glycosylases necessary for removal of oxidized pyrimidines (three *nth* gene products) and purines (*fpg* and *mutY* gene products), as well as an AP endonuclease (*xthA*) and a deoxyribophophodiesterase (*recJ*) (Table 2). Crude cell extracts of *D. radiodurans* exhibit thymine glycol DNA glycosylase activity (57), formamidopyrimidine (FAPY) DNA glycosylase activity, apurinic/apyrimidinic (AP) endonuclease activity (45), and DNA deoxyribophosphodiesterase (dRPase) activity (57), indicating that these proteins are expressed constitutively. The contribution each protein makes toward alleviating oxidative stress has not yet been assessed.

In addition to the *fpg* and *mutY* gene products, *D. radiodurans* encodes nine possible homologs of the *E. coli* MutT protein (Table 3). MutT dephosphorylates oxidized guanine nucleotides, presumably to prevent their incorporation into the genome during DNA replication (48). Among the collection of proteins listed in Table 3, only three (those encoded by ORF01639, ORF01766, and ORF02527) have similarity over most of the 129 amino acids of the *E. coli* MutT protein. The remaining open reading frames encode proteins that are MutT-like in that

they contain a set of conserved residues characteristic of MutT and other phosphohydrolases (4). Since all of these sequences share less than 45% similarity with *E. coli*'s MutT, it is not possible to assess whether any of these proteins function to dephosphorylate oxidized purine nucleotides without further investigation.

Nucleotide Excision Repair

The *D. radiodurans* R1 genome encodes homologs of all components of the nucleotide excision repair (NER) pathway of *E. coli* (Table 2). Given the degree of similarity, the *D. radiodurans* proteins presumably catalyze the same functions as their *E. coli* counterparts. Consistent with this conclusion, expression of the *E. coli* *uvrA* and *polA* gene products will complement *D. radiodurans* *uvrA* and *pol* muta-

Table 3. Homologs of the *E. coli* MutT protein[a]

Open reading frame[b]	Size[c]	% Similarity[d]
00694	144	55 (32/57)
01193	350	46 (30/70)
01639	192	40 (47/116)
01766	323	45 (47/103)
02527	155	41 (47/113)
02529	194	50 (33/65)
02866	167	52 (31/59)
02919	166	64 (33/51)
03237	194	57 (43/75)

[a] Identified by sequence similarity to their *E. coli* homologs.
[b] Assignments of open reading frames are those of White et al. (73).
[c] Predicted number of amino acids in the deinococcal protein.
[d] Data in parentheses are ratios of the number of conserved residues to the length of the conserved region.

Table 2. Deinococcal proteins potentially involved in repairing oxidative DNA damage[a]

Identification	Size[b]	% Similarity[c]
Base excision repair		
Formamidopyrimidine-DNA glycosylase (*mutM*)	291	51 (148/284)
A/G-specific adenine glycosylase (*mutY*)	363	56 (138/244)
Endonuclease III (*nth*)	259	41 (82/196)
Endonuclease III (*nth*)	225	60 (119/196)
Endonuclease III (*nth*)	338	51 (96/219)
Exodeoxyribonuclease III (*xthA*)	283	51 (115/235)
Single-stranded-DNA-specific exonuclease (*recJ*)	684	49 (253/501)
Nucleotide excision repair		
Excinuclease ABC, subunit A (*uvrA*)	1,016	67 (666/977)
Excinuclease ABC, subunit B (*uvrB*)	730	70 (466/661)
Excinuclease ABC, subunit C (*uvrC*)	617	51 (319/609)
DNA helicase II (*uvrD*)	745	51 (389/739)
DNA polymerase (*polA*)	956	66 (269/401)
DNA ligase (*dnlJ*)	700	51 (380/655)

[a] Identified by sequence similarity to their *E. coli* homologs.
[b] Predicted number of amino acids in the deinococcal protein.
[c] Data in parentheses are ratios of the number of conserved residues to the length of the conserved region.

tions, restoring DNA damage resistance to those strains (1, 26, 27).

uvrA strains of *D. radiodurans* exhibit wild-type ionizing radiation resistance but can be made radiosensitive by genetically inactivating the uncharacterized *irrB* locus (68). Only the *uvrA irrB* double mutant is sensitive to ionizing radiation, suggesting that NER has a redundant role in protecting the cell from γ-radiation-induced DNA damage. Since NER in other species removes modified bases following DNA damage, it is assumed that IrrB is also involved in an excision repair pathway, but this remains to be established.

Recombination Repair

Since multiple copies of the genome provide the cell with a reservoir of genetic information, organisms with higher chromosome multiplicity, like *D. radiodurans* (29, 34), should be more resistant to ionizing radiation-induced DNA damage, provided they have the ability to take advantage of that information. When an exponential phase culture of *D. radiodurans* is exposed to 5,000 Gy of γ-radiation, approximately 150 DNA double-strand breaks (dsbs) are introduced into the chromosome (10). These dsbs are not only repaired without loss of viability, but the chromosome is also reassembled so the linear continuity of the genome is unaffected by either the damage or the repair process. There are no reports that ionizing radiation induces deletions or rearrangements within the *D. radiodurans* genome. *rec* strains of *D. radiodurans* R1 are very sensitive to ionizing radiation and incapable of repairing DNA dsbs (14). The *rec* gene product has been cloned, sequenced, and shown to be a homolog of the *E. coli* RecA protein, sharing 56% identity with the *E. coli* RecA (25). Despite the similarity between the two proteins, the *recA* from *D. radiodurans* will not complement an *E. coli recA* mutant. It has been reported that the expression of the deinococcal RecA protein is toxic to *E. coli* and that even low-level expression is lethal, suggesting a fundamental difference in the activity of the two RecA proteins (25).

Daly and Minton have followed the kinetics of DNA dsb repair in stationary-phase cells exposed to 17.5 kGy of γ-radiation using pulsed-field gel electrophoresis (14). There is a *recA*-independent increase in the molecular weight of chromosomal DNA in the irradiated population during the first 1.5 h postirradiation. A *recA*-dependent increase in molecular weight follows and continues over the next 29 h until the chromosome is restored to its normal size.

RESPONSE TO UV RADIATION IN *D. RADIODURANS*

D. radiodurans exhibits an extraordinary ability to tolerate the lethal effects of a variety of DNA-damaging agents (65, 66), displaying unusually high resistance to the lethal effects of UV light (49, 51). Wild-type *D. radiodurans* cultures in exponential phase growth survive UV doses as high as 1,000 J/m^2 (52). A shoulder of resistance that extends to approximately 500 J/m^2 characterizes the typical UV survival curve of an exponential phase culture. The D$_{37}$ dose is reported to be between 550 and 600 J/m^2. At 500-J/m^2 UV, it is predicted that approximately 1.0% of the irradiated cell's thymine will become part of a pyrimidine dimer (5, 70). Since the *D. radiodurans* genome is 3.3 × 10^6 bp long with a 67% GC content, as many as 5,000 thymine-containing pyrimidine dimers form per genome when the cell is subjected to a nonlethal 500-J/m^2 dose.

Despite the massive levels of UV-induced damage, there is no formal evidence that *D. radiodurans* expresses an SOS-like response (54, 55). Excision of UV-induced pyrimidine dimers from *D. radiodurans* is insensitive to chloramphenicol (5), and the repair proteins needed to survive UV-induced DNA damage appear to be constitutively expressed (2, 21, 22). Hansen (28), however, reported a 30- to 100-fold increase in four proteins, designated α, β, γ, and δ, in UV-irradiated cultures. The identity of these proteins was not determined, and their function, if any, in UV resistance is not known. Curiously, the *D. radiodurans* genome encodes a LexA homolog on megaplasmid I that shares 34% identity with the *E. coli* LexA protein. The residues (Ala-84, Gly-86, Ser-119, and Lys-156) needed for LexA autoproteolysis (63) are conserved in the deinococcal protein, but there is very limited similarity between the deinococcal LexA and the DNA-binding domains found in the amino terminal portion of the *E. coli* LexA.

UV-induced DNA damage is repaired by at least three pathways in *D. radiodurans*: NER, base excision repair (BER), and recombination repair. In contrast to *E. coli uvrA* mutants, *D. radiodurans uvrA* strains exhibit near wild-type resistance to UV light. Functional NER is not necessary for UV resistance in *D. radiodurans* because there is a redundant BER pathway that compensates for the loss. A pyrimidine dimer DNA (PD-DNA) glycosylase, designated endonuclease β, initiates repair of UV-induced damage in NER-defective cells with an efficiency comparable to that of the deinococcal UvrABC exinuclease (22, 53). *D. radiodurans* only becomes sensitive to UV when both repair pathways are inactivated. The *D. radiodurans* genome encodes a putative UV endonuclease that shares 42% similarity with a PD-DNA glycosylase from *Neurospora crassa*. Pending further investigation, it is assumed that this protein is endonuclease β of *D. radiodurans*.

Homologous recombination is critical to cell survival following exposure to UV light. All recombination-defective strains of *D. radiodurans* are very sensitive to UV radiation, failing to exhibit the shoulder of resistance characteristic of the wild-type strain (14, 51, 52).

RESPONSE TO HEAT IN *D. RADIODURANS*

D. radiodurans is resistant to heat (7, 8). Cultures grown at 30°C that have been rapidly shifted to 52°C can be held at this temperature for up to 40 min without loss of viability. By comparison, an *E. coli* culture will exhibit approximately 1% survival after a 40-min incubation at 52°C. The exact reason for this resistance is unknown, but survival requires DNA synthesis and a functional RecA protein (30, 31).

Harada et al. (30, 31) have proposed that the increase in temperature induces single- and double-strand breaks in the *D. radiodurans* genome, and the repair of these lesions is necessary if the cell is to survive. Given the resistance of purified DNA to heat-induced strand breaks (43), this explanation seems unlikely.

If *D. radiodurans* cultures are exposed to 52°C for 30 min and allowed to incubate at 30°C for 4 h, there is a 100-fold increase in the culture's ability to survive a subsequent 100-min exposure to 52°C relative to cultures that did not receive the pretreatment (30). This protective effect is abolished if chloramphenicol is introduced into the incubation, indicating that the 52°C pretreatment induces a protective heat shock response. The nature of this species' heat shock response has not been investigated, but *D. radiodurans* has the typical complement of heat shock proteins (Table 4).

D. RADIODURANS RESPONSE TO DESICCATION

For a vegetative cell, *D. radiodurans* is highly resistant to desiccation. Mattimore and Battista (47) report that the R1 strain survives 6 weeks in a desiccator (relative humidity < 5%) with 85% viability, and there are anecdotal reports of cultures retaining 10% viability after 6 years of desiccation (58). Very little is known concerning the dessiccation resistance of *D. radiodurans* except that this phenomenon re-

Table 4. Deinococcal proteins potentially involved in the heat shock response[a]

Identification	Size[b]	% Similarity[c]
GrpE	197	47 (89/186)
GroEL	580	88 (469/529)
GroES	120	85 (81/95)
DnaK	638	70 (450/636)
RpoD	418	76 (186/241)
ATP-dependent protease (*lon*)	855	67 (526/774)
DnaJ	465	51 (197/379)
ClpA	806	66 (479/720)
ClpB	941	73 (631/851)
ClpC	818	55 (396/709)
ClpP	268	73 (141/190)
ClpX	468	71 (287/399)
HtrA	427	47 (147/304)

[a] Identified by sequence similarity to their *E. coli* homologs.
[b] Predicted number of amino acids in the deinococcal protein.
[c] Data in parentheses are ratios of the number of conserved residues to the length of the conserved region.

quires the ability to repair DNA damage. In fact, when 41 ionizing radiation-sensitive derivatives of *D. radiodurans* were evaluated for the ability to survive 6 weeks of desiccation, all were found to be sensitive to desiccation (47). Like ionizing radiation, dehydration induces a large number of DNA double-strand breaks in bacteria (19, 20, 47). RecA-defective strains are particularly susceptible to the lethal effects of desiccation (47). Presumably the ionizing radiation resistance and desiccation resistance of *D. radiodurans* are functionally interrelated phenomena. When *D. radiodurans* loses the capacity to repair DNA damage, the cell is sensitized to both agents.

CONCLUDING COMMENTS

From the preceding discussion, it is obvious that *D. radiodurans* is highly resistant to the lethal effects of ionizing radiation, H_2O_2, UV radiation, heat, and desiccation. By inference, it must be assumed that this species has evolved mechanisms capable of protecting the cell from each type of stress and of sensing when to employ these mechanisms. Our survey of the literature revealed that *D. radiodurans* expresses at least two stress responses. Proteins are synthesized in response to oxidative stress and heat, and inhibition of protein synthesis sensitizes the cell to these conditions. It is noteworthy that our search of the *D. radiodurans* genome failed to reveal definitive evidence of the presence of regulatory proteins homologous to the OxyR, SoxR, and SoxS proteins of *E. coli*. Presumably, *D. radiodurans* expresses a novel set of regulatory proteins.

Despite the presence of a LexA homolog, there is no evidence of an SOS-like response following the culture's exposure to UV, and we, therefore, must assume that the proteins needed to survive UV-induced intracellular damage are constitutively expressed. The RecA protein, however, appears to play a critical role in cell survival subsequent to each of the stresses discussed here, suggesting two possibilities. (a) Each stress induces a lethal form of DNA damage that can only be repaired by RecA-dependent homologous recombination. The cell succumbs because loss of RecA prevents effective repair of this lesion. Each of the stresses discussed introduces DNA double-strand breaks, either directly as part of the spectrum of damage caused by the agent or indirectly during the DNA repair process. Perhaps double-strand breaks, even at the relatively low levels produced subsequent to the culture's exposure to UV light or heat, are sufficient to kill a *recA*-defective cell. (b) The *D. radiodurans* RecA protein has a global regulatory function analogous to the role of *RecA* in the SOS response of *E. coli*. *D. radiodurans* may need a common subset of gene products to respond to each stress. In this scenario, RecA would either initiate expression of the appropriate genes or activate the necessary gene products. Even though there is no direct evidence that *D. radiodurans* has an SOS response, new protein synthesis has been reported following UV radiation, ionizing radiation, and H_2O_2. It would be informative to determine whether in *D. radiodurans* stress-induced protein synthesis was altered in a *recA* background.

REFERENCES

1. Agostini, H. J., J. D. Carroll, and K. W. Minton. 1997. Identification and characterization of *uvrA*, a DNA repair gene of *Deinococcus radiodurans*. *J. Bacteriol.* **178:**6759–6765.
2. Al-Bakri, G. H., M. W. Mackay, P. A. Whittaker, and B. E. B. Moseley. 1985. Cloning of the DNA repair genes *mtcA, mtcB, uvsC, uvsD, uvsE* and the *leuB* gene from *Deinococcus radiodurans*. *Gene* **33:**305–311.
3. Battista, J. R. 1997. Against all odds: the survival strategies of *Deinococcus radiodurans*. *Annu. Rev. Microbiol.* **51:**203–224.
4. Bessman, M. J., D. N. Frick, and S. F. O'Handley. 1996. The MutT proteins or "Nudix" hydrolases, a family of versatile, widely distributed "housecleaning" enzymes. *J. Biol. Chem.* **271:**25059–25062.
5. Boling, M. E., and J. K. Setlow. 1966. The resistance of *Micrococcus radiodurans* to ultraviolet radiation III. A repair mechanism. *Biochim. Biophys. Acta* **123:**26–33.
6. Bonura, T., C. D. Town, K. C. Smith, and H. S. Kaplan. 1975. The influence of oxygen on the yield of DNA double-strand breaks in X-irradiated *Escherichia coli* K12. *Radiat. Res.* **63:**567–577.
7. Bridges, B. A., M. J. Ashwood-Smith, and R. J. Munson. 1969. Correlation of bacterial sensitivities to ionizing radiation and mild heating. *J. Gen. Microbiol.* **58:**115–124.
8. Bridges, B. A., M. J. Ashwood-Smith, and R. J. Munson. 1969. Susceptibility of mild thermal and of ionizing radiation damage to the same recovery mechanisms in *Escherichia coli*. *Biochem. Biophys. Res. Comm.* **35:**193–196.

9. Brooks, B. W., and R. G. E. Murray. 1981. Nomenclature for "*Micrococcus radiodurans*" and other radiation-resistant cocci: *Deinococcaceae* fam. nov. and *Deinococcus* gen. nov., including five species. *Int. J. Syst. Bacteriol.* **31:**353–360.
10. Burrell, A. D., P. Feldschreiber, and C. J. Dean. 1971. DNA membrane association and the repair of double breaks in X-irradiated *Micrococcus radiodurans*. *Biochim. Biophys. Acta* **247:**38–53.
11. Chou, F. I., and S. T. Tan. 1990. Manganese(II) induces cell division and increases in superoxide dismutase and catalase activities in an aging deinococcal culture. *J. Bacteriol.* **172:**2029–2035.
12. Christensen, E. A., and H. Kristensen. 1981. Radiation-resistance of micro-organisms from air in clean premises. *Acta Path. Microbiol. Scand. Sect. B* **89:**293–301.
13. Daly, M. J., and K. W. Minton. 1995. Interchromosomal recombination in the extremely radioresistant bacterium *Deinococcus radiodurans*. *J. Bacteriol.* **177:**5495–5505.
14. Daly, M. J., L. Ouyang, P. Fuchs, and K. W. Minton. 1994. In vivo damage and *recA*-dependent repair of plasmid and chromosomal DNA in the radiation-resistant bacterium. *J. Bacteriol.* **176:**3608–3517.
15. Daly, M. J., L. Ouyang, and K. W. Minton. 1994. Interplasmidic recombination following irradiation of the radioresistant bacterium *Deinococcus radiodurans*. *J. Bacteriol.* **176:**7506–7515.
16. Davis, N. S., G. J. Silverman, and E. B. Mausurosky. 1963. Radiation resistant, pigmented coccus isolated from Haddock tissue. *J. Bacteriol.* **86:**294–298.
17. Dean, C. J., J. G. Little, and R. W. Serianni. 1970. The control of post irradiation DNA breakdown in *Micrococcus radiodurans*. *Biochem. Biophys. Res. Comm.* **39:**126–134.
18. Dean, C. J., M. G. Ormerod, R. W. Serianni, and P. Alexander. 1969. DNA strand breakage in cells irradiated with X-rays. *Nature* **222:**1042–1044.
19. Dose, K., A. Bieger-Dose, O. Kerz, and M. Gill. 1991. DNA-strand breaks limit survival in extreme dryness. *Orig. Life Evol. Biosphere* **21:**177–187.
20. Dose, K., A. Bieger-Dose, M. Labusch, and M. Gill. 1992. Survival in extreme dryness and DNA-single strand breaks. *Adv. Space Res.* **12:**221–229.
21. Evans, D. M., and B. E. B. Moseley. 1985. Identification and initial characterization of a pyrimidine dimer UV endonuclease (UV endonuclease β) from *Deinococcus radiodurans*; a DNA-repair enzyme that requires manganese ions. *Mutat. Res.* **145:**119–128.
22. Evans, D. M., and B. E. B. Moseley. 1983. Roles of the *uvsC, uvsD, uvsE*, and *mtcA* genes in the two pyrimidine dimer excision repair pathways of *Deinococcus radiodurans*. *J. Bacteriol.* **156:**576–583.
23. Ferreira, A. C., M. F. Nobre, F. A. Rainey, M. T. Silva, R. Wait, J. Burghardt, P. Chung, and M. S. da Costa. 1997. *Deinococcus geothermalis* sp. nov. and *Deinococcus murrayi* sp. nov., two extremely radiation-resistant and slightly thermophilic species from hot springs. *Int. J. Syst. Bacteriol.* **47:**939–947.
24. Grant, I. R., and M. F. Patterson. 1989. A novel radiation resistant *Deinobacter* sp. isolated from irradiated pork. *Lett. Appl. Microbiol.* **8:**21–24.
25. Gutman, P. D., J. D. Carroll, C. I. Masters, and K. W. Minton. 1994. Sequencing, targeted mutagenesis and expression of a *recA* gene required for extreme radioresistance of *Deinococcus radiodurans*. *Gene* **141:**31–37.
26. Gutman, P. D., P. Fuchs, and K. W. Minton. 1994. Restoration of the DNA damage resistance of *Deinococcus radiodur-*

ans DNA polymerase mutants by *Escherichia coli* DNA polymerase I and Klenow fragment. *Mutat. Res.* **314:**87–97.

27. **Gutman, P. D., P. Fuchs, L. Ouyang, and K. W. Minton.** 1993. Identification, sequencing, and targeted mutagenesis of a DNA polymerase gene required for the extreme radioresistance of *Deinococcus radiodurans. J. Bacteriol.* **175:**3581–3590.

28. **Hansen, M. T.** 1980. Four proteins synthesized in response to deoxyribonucleic acid damage in *Micrococcus radiodurans. J. Bacteriol.* **141:**81–86.

29. **Hansen, M. T.** 1978. Multiplicity of genome equivalents in the radiation-resistant bacterium *Micrococcus radiodurans. J. Bacteriol.* **134:**71–75.

30. **Harada, K., and S. Oda.** 1988. Induction of thermotolerance by split dose hyperthermia at 52°C in *Deinococcus radiodurans. Agric. Biol. Chem.* **52:**2391–2396.

31. **Harada, K., A. Uchida, and H. Kadota.** 1984. Post-replication repair of the heat-induced DNA injury in *Deinococcus radiodurans. Agric. Biol. Chem.* **48:**59–65.

32. **Hariharan, P. V., and P. A. Cerutti.** 1972. Formation and repair of γ-ray-induced thymine damage in *Micrococcus radiodurans. J. Mol. Biol.* **66:**65–81.

33. **Hariharan, P. V., and P. A. Cerutti.** 1971. Repair of γ-ray-induced thymine damage in *Micrococcus radiodurans. Nature New Biol.* **229:**247–249.

34. **Harsojo, S., S. Kitayama, and A. Matsuyama.** 1981. Genome multiplicity and radiation resistance in *Micrococcus radiodurans. J. Biochem.* **90:**877–880.

35. **Hensel, R., W. Demharter, O. Kandler, R. M. Kroppenstedt, and E. Stackebrandt.** 1986. Chemotaxonomic and molecular-genetic studies of the genus *Thermus:* evidence for a phylogenetic relationship of *Thermus aquaticus* and *Thermus ruber* to the genus *Deinococcus. Int. J. Syst. Bacteriol.* **36:**444–453.

36. **Ito, H.** 1977. Isolation of *Micrococcus radiodurans* occurring in radurized sawdust culture media of mushroom. *Agric. Biol. Chem.* **41:**35–41.

37. **Ito, H., H. Watanabe, M. Takeshia, and H. Iizuka.** 1983. Isolation and identification of radiation-resistant cocci belonging to the genus *Deinococcus* from sewage sludges and animal feeds. *Agric. Biol. Chem.* **47:**1239–1247.

38. **Kitayama, S., and A. Matsuyama.** 1971. Double-strand scissions in DNA of gamma-irradiated *Micrococcus radiodurans* and their repair during postirradiation incubation. *Agric. Biol. Chem.* **35:**644–652.

39. **Krabbenhoft, K.. L., A. W. Anderson, and P. R. Elliker.** 1965. Ecology of *Micrococcus radiodurans. Appl. Microbiol.* **13:**1030–1037.

40. **Kristensen, H., and E. A. Christensen.** 1981. Radiation-resistant microorganisms isolated from textiles. *Acta Pathol. Microbiol. Scand. Sect. B.* **89:**303–309.

41. **Lewis, N. F.** 1971. Studies on a radio-resistant coccus isolated from Bombay duck (*Harpodon nehereus*). *J. Gen. Microbiol.* **66:**29–35.

42. **Lin, J., R. Qi, C. Aston, J. Jing, T. S. Anantharaman, B. Mishra, O. White, M. J. Daly, K. W. Minton, J. C. Venter, and D. C. Schwartz.** 1999. Whole-genome shotgun optical mapping of *Deinococcus radiodurans. Science* **285:**1558–1562.

43. **Lindahl, T.** 1993. Instability and decay of the primary structure of DNA. *Nature* **362:**709–715.

44. **Markillie, L. M., S. M. Varnum, P. Hradecky, and K.-K. Wong.** 1999. Targeted mutagenesis by duplication insertion in the radioresistant bacterium *Deinococcus radiodurans:* radiation sensitivities of catalase (*katA*) and superoxide dismutase (*sodA*) mutants. *J. Bacteriol.* **181:**666–669.

45. **Masters, C. I., B. E. B. Moseley, and K. W. Minton.** 1991. AP endonuclease and uracil DNA glycosylase activities in *Deinococcus radiodurans. Mutat. Res.* **254:**263–272.

46. **Masters, C. I., R. G. E. Murray, B. E. B. Moseley, and K. W. Minton.** 1991. DNA polymorphisms in new isolates of 'Deinococcus radiopugnans'. *J. Gen. Microbiol.* **137:**1459–1469.

47. **Mattimore, V., and J. R. Battista.** 1996. Radioresistance of *Deinococcus radiodurans:* functions necessary to survive ionizing radiation are also necessary to survive prolonged desiccation. *J. Bacteriol.* **178:**633–637.

48. **Michaels, M. L., and J. H. Miller.** 1992. The GO system protects organisms from the mutagenic effect of the spontaneous lesion 8-hydroxyguanine (7.8-dihydro-8-oxogranine). *J. Bacteriol.* **174:**6321–6325.

49. **Minton, K. W.** 1994. DNA repair in the extremely radioresistant bacterium *Deinococcus radiodurans. Mol. Microbiol.* **13:**9–15.

50. **Minton, K. W.** 1996. Repair of ionizing-radiation damage in the radiation resistant bacterium *Deinococcus radiodurans. Mutat. Res.* **363:**1–7.

51. **Moseley, B. E. B.** 1983. Photobiology and radiobiology of *Micrococcus* (*Deinococcus*) *radiodurans. Photochem. Photobiol. Rev.* **7:**223–275.

52. **Moseley, B. E. B.** 1967. The repair of DNA in *Micrococcus radiodurans* following ultraviolet irradiation. *J. Gen. Microbiol.* **48:**4–24.

53. **Moseley, B. E. B., and D. M. Evans.** 1983. Isolation and properties of strains of *Micrococcus* (*Deinococcus*) *radiodurans* unable to excise ultraviolet light-induced pyrimidine dimers from DNA: evidence of two excision pathways. *J. Gen. Microbiol.* **129:**2437–2445.

54. **Moseley, B. E. B., and H. Laser.** 1965. Repair of x-ray damage in *Micrococcus radiodurans. Proc. R. Soc. Lond.* (*Biol.*). **162:**210–222.

55. **Moseley, B. E. B., and H. Laser.** 1965. Similarity of repair of ionizing and ultra-violet radiation damage in *Micrococcus radiodurans. Nature* **206:**373–375.

56. **Moseley, B. E. B., and A. Mattingly.** 1971. Repair of irradiated transforming deoxyribonucleic acid in wild-type and a radiation-sensitive mutant of *Micrococcus radiodurans. J. Bacteriol.* **105:**976–983.

57. **Mun, C., J. Del Rowe, M. Sandigursky, K. W. Minton, and W. A. Franklin.** 1994. DNA deoxyribophosphodiesterase and an activity that cleaves DNA containing thymine glycol adducts in *Deinococcus radiodurans. Radiat. Res.* **138:**282–285.

58. **Murray, R. G. E.** 1992. The family *Deinococcaceae,* p. 3732–3744. *In* A. Ballows, H. G. Truper, M. Dworkin, W. Harder, and K. H. Scheilefer (ed.), *The Prokaryotes,* 2nd ed, vol. 4. Springer-Verlag, New York, N.Y.

59. **Oyaiza, H., E. Stackebrandt, K. H. Schleifer, W. Ludwig, H. Pohla, H. Ito, A. Hirata, Y. Oyaizu, and K. Komagata.** 1987. A radiation-resistant rod-shaped bacterium, *Deinobacter grandis* gen. nov., sp. nov., with peptidoglycan containing ornithine. *Int. J. Syst. Bacteriol.* **37:**62–67.

60. **Rainey, F. A., M. Nobre, P. Schumann, E. Stackebrandt, and M. da Costa.** 1997. Phylogenetic diversity of the deinococci as determined by 16S ribosomal DNA sequence. *Int. J. Syst. Bacteriol.* **47:**510–514.

61. **Sanders, S. W., and R. B. Maxcy.** 1979. Isolation of radiation-resistant bacteria without exposure to irradiation. *Appl. Env. Microbiol.* **38:**436–439.

62. **Schell, M. A.** 1993. Moleclar biology of the LysR family of transcriptional regulators. *Annu. Rev. Microbiol.* **47:**597–626.

63. **Slilaty, S. N., and J. W. Little.** 1987. Lysine-156 and serine-119 are required for LexA repressor cleavage: a possible mechanism. *Proc. Natl. Acad. Sci. USA* **84:**3987–3991.

64. **Smith, M. D., C. I. Masters, and B. E. B. Moseley.** 1992. Molecular biology of radiation resistant bacteria, p. 258–280. *In* R. A. Herbert and R. J. Sharp (ed.), *Molecular Biology and*

Biotechnology of Extremophiles. Chapman & Hall, New York, N.Y.

65. **Sweet, D. M., and B. E. B. Moseley.** 1974. Accurate repair of ultraviolet-induced damage in *Micrococcus radiodurans. Mutat. Res.* **23:**311–318.

66. **Sweet, D. M., and B. E. B. Moseley.** 1976. The resistance of *Micrococcus radiodurans* to killing and mutation by agents which damage DNA. *Mutat. Res.* **34:**175–186.

67. **Tanaka, A., H. Hirano, M. Kikuchi, S. Kitayama, and H. Watanabe.** 1996. Changes in cellular proteins of *Deinococcus radiodurans* following γ-irradiation. *Radiat. Environ. Biophys.* **35:**95–99.

68. **Udupa, K., P. A. O'Cain, V. Mattimore, and J. R. Battista.** 1994. Novel ionizing radiation-sensitive mutants of *Deinococcus radiodurans. J. Bacteriol.* **176:**7439–7446.

69. **Van den Eynde, H., Y. Van de Peer, H. Vandenabeele, M. Van Bogaert, and R. De Wachter.** 1990. 5S rRNA sequences of myxobacteria and radioresistant bacteria and implications for eubacterial evolution. *Int. J. Syst. Bacteriol.* **40:**399–404.

70. **Varghese, A. J., and R. S. Day.** 1970. Excision of cytosine-thymine adduct from the DNA of ultraviolet-irradiated *Micrococcus radiodurans. Photochem. Photobiol.* **11:**511–517.

71. **Wang, P., and H. E. Schellhorn.** 1995. Induction of resistance to hydrogen peroxide and radiation in *Deinococcus radiodurans. Can. J. Microbiol.* **41:**170–176.

72. **Weisburg, W. G., S. J. Giovannoni, and C. R. Woese.** 1989. The *Deinococcus-Thermus* phylum and the effect of rRNA composition on phylogenetic tree construction. *Syst. Appl. Microbiol.* **11:**128–134.

73. **White, O., J. A. Eisen, J. F. Heidelberg, E. K. Hickey, J. D. Peterson, R. J. Dodson, D. H. Haft, M. L. Gwinn, W. C. Nelson, D. L. Richardson, K. S. Moffet, H. Qin, L. Jiang, W. Pamphile, M. Crosby, M. Shen, J. J. Vamathevan, P. Lam, L. McDonald, T. Utterback, C. Zalewski, K. S. Makarova, L. Aravind, M. J. Daly, K. W. Minton, R. D. Fleishmann, K. A. Ketchum, K. E. Nelson, S. Salzberg, J. C. Venter, and C. M. Fraser.** 1999. Genome sequence of the radioresistant bacterium *Deinococcus radiodurans* R1. *Science* **286:**1571–1577.

74. **Woese, C. R.** 1987. Bacterial evolution. *Microbiol. Rev.* **51:**221–271.

75. **Woese, C. R., E. Stackebrandt, T. J. Macke, and G. E. Fox.** 1985. A phylogenetic definition of the major eubacterial taxa. *Syst. Appl. Microbiol.* **61:**143–151.

Bacterial Stress Responses
Edited by G. Storz and R. Hengge-Aronis
©2000 ASM Press, Washington, D.C.

Chapter 25

Mechanisms of Organic Solvent Tolerance in Bacteria

JASPER KIEBOOM AND JAN A. M. DE BONT

The extreme toxicity of many organic solvents imposes a severe problem in the application of these chemical compounds in biotechnological production processes. The main reason for the toxicity of organic solvents is the accumulation of these molecules in the bacterial cell membrane, causing adverse effects on membrane structure and functioning. However, in the past decade several organic solvent-tolerant bacteria have been isolated from the environment that are able to survive in water-solvent two-phase systems. In these systems the concentrations of organic solvents reach levels that are lethal for normal bacteria. Solvent-tolerant bacteria are able to counterbalance the detrimental effects of organic solvents via several mechanisms that change the structure of the cell envelope. It has now been established that active efflux of solvent from the membrane plays a key role in organic solvent tolerance. We review here the current knowledge about these exceptional bacteria and the mechanisms for their survival in the presence of toxic solvents.

Organic solvents are generally regarded as extremely toxic to living microorganisms and therefore impose an important drawback in the application of these chemical compounds in biotechnology (75). However, in the past decade efforts have been made to isolate microorganisms tolerant to organic solvents. These organisms have potential advantages in either the remediation of highly polluted wastestreams or biocatalytic applications for the production of specialty chemicals. The use of solvent-tolerant bacteria in biocatalysis would allow the introduction of an organic phase to dissolve water-insoluble substrates or to remove toxic products (11).

TOXICITY OF ORGANIC SOLVENTS

Because of their toxicity, organic solvents have been widely used as disinfectants, permeabilization agents, and food preservatives (10, 49). The antimicrobial property of organic solvents correlates with its hydrophobic nature. The hydrophobicity of organic solvents is usually expressed in terms of P_{ow}, which gives the partitioning of a compound over an octanol/water two-phase system. It has been established that the common logarithm of P_{ow} is inversely correlated with the toxicity of the organic solvent (79). This influence of the $\log P_{ow}$ on toxicity toward microorganisms has been described for different classes of organic solvents such as aromatic compounds, alkanes, alcohols, and phenols (38). Organic solvents with a $\log P_{ow}$ value between 1 and 5 are generally regarded as highly toxic for microorganisms (60, 78).

It is generally accepted that the accumulation of organic solvents in the membrane is the main mechanism for toxicity, although other mechanisms cannot be ruled out because of specific chemical properties of each specific compound. The mechanism of membrane toxicity of organic solvents has been reviewed extensively by Sikkema et al. in 1995 (79). To understand the organic solvent tolerance mechanisms of solvent-tolerant bacteria a short overview of organic solvent toxicity is given below.

In most cases, organic solvents, such as aromatic compounds and alcohols, target the bacterial cell membrane in which they preferentially partition (60, 78). De Smet et al. first made this visible in 1978 by means of electron microscopic images of toluene-treated *Escherichia coli* cells (12). While the outer membrane in these cells remained intact, considerable

Jasper Kieboom and Jan A. M. de Bont • Division of Industrial Microbiology, Department of Food Technology and Nutritional Sciences, Wageningen Agricultural University, P.O. Box 8129, 6700 EV Wageningen, The Netherlands.

damage was done to the cytoplasmic membrane, resulting in the permeabilization of the cell membrane. This partitioning of organic solvents prevents the proper functioning of the membrane, which is a selective barrier for solutes between the cell and its external environment. The permeabilization of this barrier impaired the growth of *E. coli* in the presence of toluene and resulted in the leakage of macromolecules such as RNA, phospholipids, and proteins (32, 86). The functioning of the membrane as a selective barrier is especially important for protons and some other ions. In a study by Heipieper et al. in 1991, phenol-treated *E. coli* cells released ATP and potassium ions into medium (19). The loss of ion gradients over the membrane prevents energy transduction in solvent-treated cells. Several other researchers demonstrated that the passive flux of ions destroys the gradients over the membrane, such as the ΔpH and the electrical potential $\Delta\psi$, which dissipates the electron motive force (7, 78).

The accumulation of organic solvents in the membrane of bacteria affects the physicochemical properties of the lipid bilayer by changing lipid ordering and bilayer stability (82). An important aspect of the membrane is its fluidity, which is defined as the reciprocal of the viscosity. Sikkema et al. showed in 1994 that the fluidity increased due to the partitioning of organic solvents in biological membranes. This increased fluidity was the result of hydrophobic interactions of hydrocarbons within the membrane (78, 79). In parallel to this, the accumulation of hydrocarbons resulted in the swelling of the membrane bilayer. As a result of these interactions of solvents with the membrane, its structure and stability were affected (77, 79), resulting in a decreased function of membrane-embedded proteins. This was demonstrated for a variety of organic solvents by measuring cytochrome *c* oxidase activity in artificial membranes (78). In *E. coli* cells it was observed that partitioning of toluene resulted in the total inactivation of the galactose permease system (32).

In conclusion, the accumulation of organic solvents results in the disruption of the membrane structure and bilayer stability by increasing membrane fluidity. These adverse effects on the membrane will cause the loss of membrane functions, leading to cell death.

SOLVENT-TOLERANT BACTERIA

Despite the extreme toxicity of organic solvents, Inoue and Horikoshi in 1989 were able to isolate the first solvent-tolerant *Pseudomonas putida* strain that was able to grow in a two-phase water-toluene system

(26). This discovery was confirmed for several other *P. putida* strains (9, 70, 83) and also for other representatives of the genus *Pseudomonas* (24, 54, 56). Furthermore, solvent tolerance has been found in other gram-negative genera such as *Flavobacterium* (48) and *Alcaligenes* (56). Solvent tolerance was also determined in gram-positive genera such as *Rhodococcus* (61) and *Bacillus* (47; S. Isken and J. A. M. de Bont, poster, International Congress on Extremophiles '98), although this has not been studied in detail yet. In addition to the naturally solvent-tolerant strains, several researchers were able to isolate solvent-tolerant mutants with enhanced resistance properties from nontolerant strains. In this way, mutants of *E. coli* (3), *Pseudomonas aeruginosa* (37), and *P. putida* (76, 80) were obtained. An overview of organic solvent-tolerant bacteria, which are able to withstand organic solvents with a log P_{ow} between 2 and 5, is given in Table 1. It is interesting to note that these naturally solvent-tolerant microorganisms were isolated from various sources such as polluted and nonpolluted soils, domestic wastewater, and the deep sea.

Not surprisingly, many efforts were made to uncover the mechanisms behind solvent tolerance since the discovery of organic solvent-tolerant strains. Because the first mode of action of an organic solvent is to partition in the membrane, researchers initially focused on membrane changes. These membrane changes have also been discussed for nontolerant strains such as *E. coli* (1, 25), but in this review we will focus on solvent tolerance mechanisms of naturally solvent-tolerant gram-negative bacteria. A schematic representation of the solvent tolerance mechanisms in solvent-tolerant bacteria is given in Fig. 1.

CYTOPLASMIC MEMBRANE CHANGES

Several response mechanisms at the level of the cytoplasmic membrane have been observed in reaction to the accumulation of organic solvents. These responses all counteract the change in membrane fluidity and bilayer stability with the purpose of restoring membrane fluidity and membrane functions (82). Mechanisms described include changes in the degree of saturation of membrane lipids (65, 84), changes in the composition of the membrane lipid headgroups (64, 69, 82), changes in the membrane lipid turnover (64), and *cis* into *trans* isomerization of unsaturated membrane lipids (18, 84). Of these mechanisms, the isomerization of the membrane lipids has been studied in detail by several researchers.

The isomerization of the *cis* double bond of an unsaturated fatty acid into the *trans* configuration

Table 1. Overview of organic solvent-tolerant bacteria

Solvent-tolerant strain	Solvent tolerated[a]	Log P_{ow}[b]	Reference
P. putida IH-2000	Toluene	2.5	26
P. putida S12	Toluene	2.5	83
P. putida DOT-T1	Toluene	2.5	70
P. putida Idaho	Toluene	2.5	9
P. putida GM73	Toluene	2.5	36
P. putida KT2442	*p*-Xylene	3.1	15
P. putida F1	*p*-Xylene	3.1	56
P. putida CE2010	Toluene	2.5	56
P. putida PpG1-7T	Toluene	2.5	76
P. putida No.69-3	Heptanol	2.4	80
P. aeruginosa PST-01	Cyclohexane	3.2	55
P. aeruginosa PAO1	Hexane	3.5	42
P. aeruginosa LST-03	Toluene	2.5	54
P. aeruginosa PAK101	Hexane	3.5	37
P. aeruginosa PAK102	*p*-Xylene	3.1	37
P. mendocina K08-1	Dimethylphtalate	2.3	24
P. mendocina LF-1	Dimethylphtalate	2.3	24
Pseudomonas sp. strain LB400	Cyclohexane	3.2	56
A. xylosoxydans A41	*p*-Xylene	3.1	56
A. xylosoxydans subsp. *denitrificans* YO129	Cyclohexane	3.2	56
A. eutrophus H850	Hexane	3.5	56
Flavobacterium sp. strain DS-711	Benzene	2.0	48
Rhodococcus sp. strain 33	Benzene*	2.0	61
Bacillus sp. strain DS-994	Benzene	2.0	47
B. thuringiensis R1	Toluene	2.5	Isken and de Bont[c]
B. mycoides R3	Toluene	2.5	Isken and de Bont[c]
B. cereus R5	Toluene	2.5	Isken and de Bont[c]
E. coli JA300	*p*-Xylene	3.1	3
E. coli OST3121	*p*-Xylene	3.1	3

[a] The tolerance to organic solvents is presented as the ability of the organism to survive in a water-organic solvent two-phase system. However, not in all cases (*) has this criterion been tested rigorously.
[b] Log P_{ow} values as published by Laane et al. in 1986 (38).
[c] Poster, International Congress on Extremophiles '98.

increases the membrane ordering and subsequently decreases membrane fluidity. Isomerization of *cis*-unsaturated fatty acids as a reaction to hydrocarbons has been described by Heipieper et al. in 1992 (20). In *P. putida* P8 cells, grown in the presence of phenols, *cis*-unsaturated fatty acids were isomerized into the *trans* configuration in order to survive high phenol concentrations. *cis* into *trans* isomerization of membrane lipids also takes place in solvent-tolerant strains as a reaction to the addition of an organic solvent phase. In the solvent-tolerant bacterium *P. putida* S12 it was shown that the amount of *trans*-

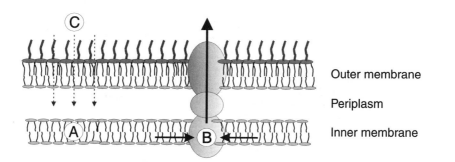

Figure 1. Schematic representation of organic solvent tolerance mechanisms of solvent-tolerant bacteria. To compensate for the accumulation of organic solvents in the membrane, solvent-tolerant bacteria modify the composition of their cytoplasmic membrane (A). This compensation will only be partial, and thus, in a dynamic process, solvents have to be removed continuously from the membrane by the active efflux of the organic solvents (B). In combination with a retarded influx of solvents due to modifications at the outer membrane (C), bacteria are able to withstand the organic solvents.

unsaturated fatty acids corresponds to the survival of the strain in the presence of organic solvents (84). It seems that this energy-independent isomerization (13) of *cis*- into *trans*-unsaturated fatty acids plays an important role in organic solvent tolerance, allowing the cells to survive the initial cell damage by organic solvents.

The *cti* gene responsible for *cis* into *trans* isomerization of membrane lipids in the solvent-sensitive *P. putida* P8 has been isolated by Holtwick et al. by complementing an isomerase-negative mutant. The deduced amino acid sequence of the *cti* gene showed no homology with other proteins (22). In studies by Heipieper et al. it has been shown that *trans*-unsaturated lipids were formed in the presence of chloramphenicol, indicating a constitutive enzyme system for *cis* into *trans* isomerization in *P. putida* P8 (20). This was later confirmed by sequence analysis of the *cti* promoter region, revealing a *Pseudomonas* constitutive *rpoD* promoter sequence in this strain (22).

The *cis-trans* isomerase from *Pseudomonas* sp. strain E-3 seems to be located in the cytosol (58), although the characteristics of the purified isomerase are not typical for cytosolic proteins (57). In a recent study, the *cis-trans* isomerase of *Pseudomonas oleovorans* Gpo12 was purified and determined to be located in the periplasm (63), confirming earlier observations (8). Both isomerases act in vitro only on free unsaturated fatty acids and not on membrane phospholipids (58, 63). However, in vivo *cis*-unsaturated fatty acids are esterified to phospholipids and therefore an additional protein was anticipated to play a role in *cis* into *trans* isomerization (58). In *Pseudomonas* sp. strain E-3 this protein was determined to be located in the cell envelope (58). Moreover, in *P. oleovorans* Gpo12 it was shown that organic solvents induced this protein, and it was speculated that it is a membrane-bound phospholipase (63).

The isomerization of *cis*-unsaturated fatty acids appears to be heme-catalyzed (23). Evidence from isomerase-negative mutants demonstrated the involvement of a protein of the cytochrome *c* type in the isomerization of *cis*-unsaturated fatty acids. It was speculated that this protein was involved in the transport of heme over the cytoplasmic membrane (23). Transposon mutants in this study, which were disrupted in the cytochrome *c* operon, were unable to isomerize *cis*-unsaturated fatty acids. Interestingly, similar observations concerning the involvement of the cytochrome *c* operon in solvent tolerance had been obtained previously with solvent-sensitive transposon mutants of *P. putida* IH-2000. The absence of isomerase activity was not determined in this study (21), but a relation between these two aspects may exist.

Several mutants of *Pseudomonas* became solvent-sensitive because of the lack of isomerase activity (69). However, other strains were shown to be sensitive toward organic solvents and still capable of isomerase activity (65, 69). Therefore, the *cis-trans* isomerization of unsaturated fatty acids is unlikely to be the only mechanism for solvent tolerance. Isomerization of unsaturated fatty acids has also been reported as a reaction to starvation (16), elevated temperatures (69), heavy metals (18), antibiotics (31), and excess salinity (43). Because an increased amount of *trans*-unsaturated fatty acids decreases membrane fluidity, it can be concluded that bacteria use isomerization of the unsaturated lipids to counter the increased fluidity due to general stress conditions, including organic solvents.

OUTER MEMBRANE CHANGES

The majority of the isolated solvent-tolerant bacteria are gram-negative. This observation is in agreement with the fact that gram-negative bacteria are more resistant to deleterious lipophilic compounds (17, 27, 81). No differences between gram-negative and gram-positive bacteria were observed in regard to the critical concentration of organic solvent dissolved in the cytoplasmic membrane (81). Therefore, the existence of an outer membrane in gram-negative bacteria could explain the differences in tolerance toward organic solvents. The outer membrane is considered to be a highly porous shield that allows small hydrophilic molecules to pass via pores, and it has a surprisingly high transfer resistance toward hydrophobic molecules (53). A few studies have described changes at the level of the outer membrane as a result of the presence of organic solvents. Several researchers described the stabilization of the outer membrane by magnesium and calcium ions resulting in an increased solvent tolerance (28, 70, 82). Changes in the outer membrane, as a response to the presence of organic solvents, were described at the level of lipopolysaccharides (LPS) and proteins.

Modification of the outer membrane in gram-negative bacteria has often been related to the LPS composition. Aono and Kobayashi suggested that an increased LPS content in solvent-tolerant mutants of *E. coli* K-12 resulted in a reduction of the cell envelope's hydrophobicity (2). A similar reduction in the hydrophobicity of the cell envelope was observed in the organic solvent-tolerant *P. putida* S12 after adaptation to organic solvents (82). These suggested changes in the LPS composition have indeed been found in the solvent-tolerant *P. putida* Idaho. Exposure of these cells to o-xylene resulted in the elongation of the saccharide chain of the LPS (65).

Other changes at the level of the outer membrane have been related to the membrane-embedded porins. OprL is a peptidoglycan-associated outer membrane lipoprotein (Pal) and was suggested to be involved in the stabilization of the cell envelope and morphology (72, 73). Mutants of *P. putida* DOT-T1 lacking the porin OprL became hypersensitive to organic solvents (72), most likely because their membrane structure was impaired. Similar results were observed in solvent-sensitive mutants of *P. aeruginosa* lacking the porin OprF (39). It was suggested that in *P. aeruginosa* organic solvents normally should enter the cell by passing through the OprF porin. However, OprF and OprL both are bifunctional porins that not only serve as diffusion porin but are also required for maintaining the structural integrity of the bacterial membrane (87). Thus, in analogy with the OprL porin in *P. putida*, solvent sensitivity in *P. aeruginosa* could be due to the decreased stability of the membrane. Similar results concerning the involvement of a porin were obtained by Aono in 1997 (2). A repressed OmpF synthesis resulted in solvent-tolerant mutants of *E. coli* (2). Recently, however, Asako et al. showed that organic solvent tolerance in *E. coli* was independent of OmpF levels and that the higher tolerance toward solvents was due to an elevated level of the *acrAB* efflux system (6). In 1991, Hengge-Aronis and coworkers tested the resistance of *E. coli* to toxic compounds that affect the integrity of the membrane, proteins, and DNA (R. Hengge-Aronis, personal communication). In their experiments stationary-phase cells were shown to exhibit better survival than log-phase cells after treatment with 0.1% toluene. Moreover, *rpoS* mutants were clearly more sensitive to toluene than wild-type cells were. From a present-day perspective, one could speculate that *acrAB* is involved since these genes are stationary phase induced (45) and is regulated by RobA (50), which in turn is under RpoS control (33).

Recently, we have isolated several nonmotile octanol-sensitive transposon mutants of *P. putida* S12. Sequence analysis of these mutants revealed insertion mutations in flagellar genes (J. Kieboom, R. Bruinenberg, and J. A. M. de Bont submitted for publication). Although the exact role of flagellar genes in solvent tolerance has not been determined in detail, it was speculated that a decreased stability of the membrane structure in these mutants resulted in the solvent-sensitive phenotype.

The specific roles of the various porins, the LPS content, the flagellum, and the composition of the outer membrane have not been clarified yet. It has been speculated that alterations at the level of the outer membrane prevent the rapid influx of organic solvent, allowing the cells to adapt to the extreme environment and to induce the necessary resistance mechanisms (11).

ACTIVE EFFLUX OF ORGANIC SOLVENTS

The exceptional resistance of solvent-tolerant bacteria toward a wide variety of deleterious organic solvents cannot be explained solely by the mechanisms as described above. These mechanisms would allow the cells to retard the rapid influx of solvents in the cytoplasmic membrane. Nevertheless, despite the decreased accumulation rate of organic solvents in the membrane, similar equilibrium concentrations will eventually be reached. Therefore, it was anticipated that a longer-lasting active efflux mechanism had to play a critical role in organic solvent tolerance (82). An active efflux system is able to extrude organic solvents from the membrane to decrease the toxin concentration below its critical concentration in the membrane. This allows the survival of the bacterium in the presence of high concentrations of organic solvents.

In 1996, Isken and de Bont were able to determine the active removal of nonmetabolizable ^{14}C-labeled toluene from the solvent-tolerant bacterium *P. putida* S12. It was found that the amount of toluene was 50% lower in toluene-induced cells compared to the amount in noninduced cells. Furthermore, it was determined that the addition of the respiratory chain inhibitor potassium cyanide and the proton conductor carbonyl cyanide *m*-chlorophenylhydrazone increased the accumulation of toluene, suggesting the energy-dependent export of toluene in *P. putida* S12 (29). Similar results were obtained by Ramos et al. in 1997 by measuring the extrusion of ^{14}C-labeled 1,2,4-trichlorobenzene, as a nonmetabolizable analog for toluene, in the solvent-tolerant bacterium *P. putida* DOT-T1 (69). This confirmed the presence of energy-dependent efflux of organic solvents in solvent tolerance.

An attempt to clone the genes for the energy-dependent efflux of toluene in *P. putida* S12 was undertaken (35). Transposon mutants that were sensitive to toluene were constructed, which made it possible to clone and sequence the genes responsible for toluene efflux. The deduced amino acid sequences of the isolated *srp* genes showed strong homology with an efflux mechanism of the resistance-nodulation-cell division (RND) family (52, 62). These results demonstrate that indeed an efflux mechanism in a solvent-tolerant bacterium is involved in the resistance toward uncharged lipophilic compounds such as toluene and xylene.

The involvement of efflux in solvent tolerance was confirmed with the isolation of the *ttgB* gene and the *mepABC* genes in the toluene-tolerant *P. putida* DOT-T1 (71) and in the toluene-tolerant variant of *P. putida* KT2442 (15), respectively. Moreover, Kim et al. reported in 1998 that a transposon insertion in a protein of the RND family resulted in a *P. putida* mutant with a solvent-sensitive phenotype (36).

It is generally accepted that efflux pumps of the RND family are constructed in such a way that they traverse both the cytoplasmic and the outer membrane. Genetic and biochemical evidence supports the notion that the proton motive force is necessary for functioning of RND-type efflux pumps. In the reported efflux systems the inner membrane protein (RND protein) was thought to be a putative 12-transmembrane-segment transporter protein that is attached, via a membrane fusion protein (MFP) belonging to the MFP family, to an outer membrane protein (OMP) (14, 73). The OMP is thought to be an outer membrane channel to circumvent the outer membrane barrier, allowing the pumped molecule to be released into the medium (12). Various toxic substrates are excreted from the cell membrane by RND-type transporters (see Table 2 for an overview of some representatives of the RND family). These efflux mechanisms are involved in the export of antibiotics, metals, organic solvents, and oligosaccharides involved in nodulation signaling (52, 62).

The suggested broad substrate specificity of solvent transporters in solvent-tolerant strains requires that the expression of these pumps be carefully reg-

ulated because their uncontrolled overexpression may lead to the loss of essential metabolic compounds. However, little is known about the physiological regulation of expression. In analogy with multidrug efflux pumps in other organisms, it is to be expected that regulators are located upstream of the efflux operon (62). Indeed, upstream of the *mepABC* operon in *P. putida* KT2442 (15) and the *srpABC* operon in *P. putida* S12 (J. Kieboom, unpublished data), putative regulators were found, but the regulatory system has yet to be uncovered. Induction of RND-type efflux mechanisms has only been studied in some instances. The MexAB-OprM efflux pump in *P. aeruginosa* was not induced by its own substrates (67), whereas the *srpABC* solvent efflux pump in *P. putida* S12 was solely induced by organic solvents (34). With the discovery of several efflux mechanisms involved in organic solvent tolerance, the regulatory mechanisms and induction will be uncovered in the future.

SOLVENT EFFLUX SYSTEMS VERSUS MULTIPLE ANTIBIOTIC RESISTANCE

A close correlation between antibiotic and solvent tolerance has been demonstrated by Aono et al. in 1995. In this study cyclohexane-tolerant mutants of *E. coli* displayed an improved tolerance toward low levels of antibiotics (4). A similar correlation between organic solvent tolerance and multiple drug resistance was observed by Isken et al. in 1997. In the naturally solvent-tolerant strain *P. putida* S12 it was demonstrated that cells, with a solvent-tolerant phe-

Table 2. RND-type efflux systems in gram-negative bacteria involved in efflux of organic solvents and drugs

Efflux protein[a]			Organism	Substrate[b]		Reference(s)
MFP	RND	OMP		Solvent(s)	Antibiotic(s)[c]	
SrpA	SrpB	SrpC	*P. putida*	Toluene, *p*-xylene, octanol, ethylbenzene, propylbenzene, cyclohexane, hexane		34, 35
MepA	MepB	MepC	*P. putida*	Toluene, *p*-xylene, cyclohexane	Tc, PenG, Ery, Nov, Amp	15
TtgA	TtgB	TtgC	*P. putida*	Toluene, *m*-xylene, 1,2,4-trichlorobenzene	Tc, Cm, Amp	75
	Ttg3		*P. putida*	Toluene	ND[d]	36
MexA	MexB	OprM	*P. aeruginosa*	Toluene, *p*-xylene, hexane	Tc, Cm, PenG, Ery, Nov, Carb, Rif	42, 66, 41, 40
MexC	MexD	OprJ	*P. aeruginosa*	*p*-xylene, hexane	Tc, Cm	42, 68
MexE	MexF	OprN	*P. aeruginosa*	*p*-xylene	Cm	42, 46
AcrA	AcrB	TolC	*E. coli*	Hexane, cyclohexane	Tc, Cm, PenG, Ery, Nov, Amp, Rif	85, 44

[a] Abbreviations: MFP, membrane fusion protein; RND, inner membrane (transporter) protein; OMP, outer membrane protein.
[b] Some compounds are expected to be substrates based on solvent-sensitive phenotype or decreased MIC for transposon mutants.
[c] Abbreviations: Tc, tetracycline; Cm, chloramphenicol; PenG, penicillin G; Ery, erythromycin; Nov, novobiocin; Amp, ampicillin; Carb, carbenicillin; Rif, rifampin.
[d] ND, not determined.

notype, were more resistant toward multiple antibiotics. However, cells pregrown in the presence of antibiotic did not show an increased tolerance toward organic solvents (31). These results can be explained in at least three different ways. First, the increased multiple antibiotic resistance can be the result of the active efflux as a result of the induction of the solvent transporter. Second, the solvents can be able to aspecifically induce an antibiotic efflux mechanism. Third, the increased resistance can be the result of other mechanisms induced by organic solvents preventing the influx of the toxic compound.

Indeed, the AcrAB efflux pump of *E. coli* was shown to be involved in both resistance toward antibiotics and solvents (85). It appears that the transcriptional activators *marA* (5), *soxS* (51), and *robA* (50) mediate this phenotype by upregulating the *acrAB* locus. Moreover, a toluene-sensitive transposon mutant of *P. putida* DOT-T1, with an insertional inactivation in the *ttgB* gene, was more susceptible to several antibiotics, including tetracycline, chloramphenicol, and ampicillin (71). Similar results were obtained with an insertion mutant of a toluene-resistant variant of *P. putida* KT2442 (15). Inactivation of the MepABC efflux system resulted in an solvent-sensitive phenotype with a concomitant increase of sensitivity toward antibiotics. It was demonstrated that resistance toward toluene and high concentrations of antibiotics in this strain occurs only in a toluene-tolerant variant and not in wild-type cells (15). Toluene and multiple antibiotic resistance in this strain was suggested to be due to the upregulation of the MepABC efflux system. In *P. aeruginosa* the three efflux mechanisms MexAB-OprM, MexCD-OprJ, and MexEF-OprN are involved in both multiple antibiotic resistance and solvent tolerance (42). Moreover, organic solvent-tolerant mutants of *P. aeruginosa* were multidrug-resistant due to mutations in the regulator MexR (67). This resulted in the overexpression of MexAB-OprM and an increased expression of MexEF-OprN.

Induction mechanisms of efflux mechanisms for antibiotics in gram-negative bacteria have not been studied in detail. Therefore, it is at present difficult to speculate on solvent-induced expression of multiple antibiotic efflux mechanisms.

The solvent transporter SrpABC seems not to be involved in multiple antibiotic resistance. It was observed that a solvent-sensitive mutant of *P. putida* S12, which was inactivated in SrpABC efflux mechanism, was as resistant as the wild-type to antibiotics such as tetracycline, chloramphenicol, and β-lactams (J. Kieboom, unpublished data). These results indicated that other mechanisms have to be involved in multidrug resistance in *P. putida* S12. It seems likely that solvents have induced changes in the outer membrane, resulting in an increased resistance toward antibiotics. This view is supported by our recent observation that a second efflux mechanism is present in this strain. This efflux mechanism is solely involved in the efflux of tetracycline, chloramphenicol, novobiocin, erythromycin, streptomycin, and β-lactams (J. Kieboom, unpublished data).

FUTURE PROSPECTS

Solvent-tolerant bacteria hold considerable promises in the field of applied biotechnology. Two possible applications, namely solvent-tolerant bacteria as cell factories (11, 70) and as the source for solvent-stable enzymes (54, 55), have been demonstrated. With the genetic uncovering of solvent tolerance mechanisms and their regulation, it becomes possible to use solvent-tolerant bacteria as production hosts in biotechnological processes. Furthermore, it has to be seen in the near future what the exact relation is between solvent tolerance and multiple drug resistance.

REFERENCES

1. **Aono, R.** 1998. Improvement of organic solvent tolerance level of *Escherichia coli* by overexpression of stress-responsive genes. *Extremophiles* 2:239–248.
2. **Aono, R., and H. Kobayashi.** 1997. Cell surface properties of organic solvent-tolerant mutants of *Escherichia coli* K-12. *Appl. Environ. Microbiol.* 63:3637–3642.
3. **Aono, R., K. Aibe, A. Inoue, and K. Horikoshi.** 1991. Preparation of organic solvent-tolerant mutants from *Escherichia coli* K12. *Agri. Biol. Chem.* 55:1935–1938.
4. **Aono, R., M. Kobayashi, H. Nakajima, and H. Kobayashi.** 1995. A close correlation between improvement of organic solvent tolerance levels and alteration of resistance towards low levels of multiple antibiotics in *Escherichia coli*. *Biosci. Biotechnol. Biochem.* 59:213–218.
5. **Asako, H., H. Nakajima, K. Kobayashi, M. Kobayashi, and R. Aono.** 1997. Organic solvent tolerance and antibiotic resistance increased by overexpression of *marA* in *Escherichia coli*. *Appl. Environ. Microbiol.* 63:1428–1433.
6. **Asako, H., K. Kobayashi, and R. Aono.** 1999. Organic solvent tolerance of *Escherichia coli* is independent of OmpF levels in the membrane. *Appl. Environ. Microbiol.* 65:294–296.
7. **Cartwright, C. P., J. R. Juroszek, M. J. Beavan, F. M. S. Ruby, S. M. F. de Morais, and A. H. Rose.** 1986. Ethanol dissipates the proton motive force across the plasma membrane of *Saccharomyces cerevisiae*. *J. Gen. Microbiol.* 132:369–377.
8. **Chen, Q.** 1996. Growth of *Pseudomonas oleovorans* in two liquid phase media. Effects of organic solvents and *alk* gene expression on the membrane. Ph.D. thesis, University of Groningen, The Netherlands.
9. **Cruden, D. L., J. H. Wolfram, R. D. Rogers, and D. T. Gibson.** 1992. Physiological properties of a *Pseudomonas* strain which grows with *p*-xylene in a two-phase (organic-aqueous) medium. *Appl. Environ. Microbiol.* 58:2723–2729.

10. Davidson, P. M., and A. L. Branden. 1981. Antimicrobial activity of non-halogenated phenolic compounds. *J. Food Prot.* **44:**623–632.

11. de Bont, J. A. M. 1998. Solvent-tolerant bacteria in biocatalysis. *Trends Biotechnol.* **16:**493–499.

12. de Smet, M. J., J. Kingma, and B. Witholt. 1978. The effect of toluene on the structure and permeability of the outer and cytoplasmic membranes of *Escherichia coli. Biochim. Biophys. Acta* **506:**64–80.

13. Diefenbach, R., and H. Keweloh. 1994. Synthesis of *trans* saturated fatty acids in *Pseudomonas putida* P8 by direct isomerization of the double bond of lipids. *Arch. Microbiol.* **162:**120–125.

14. Dinh, T., I. T. Paulsen, and M. H. Saier, Jr. 1994. A family of extracytoplasmic proteins that allow transport of large molecules across the outer membranes of gram-negative bacteria. *J. Bacteriol.* **176:**3825–3831.

15. Fukumori, F., H. Hirayama, H. Takami, A. Inoue, and K. Horikoshi. 1998. Isolation and transposon mutagenesis of a *Pseudomonas putida* KT2442 toluene-resistant variant: involvement of an efflux system in solvent tolerance. *Extremophiles* **2:**395–400.

16. Guckert, J. B., M. A. Hood, D. C. White. 1986. Phospholipid ester-linked fatty acid profile changes during nutrient deprivation of *Vibrio cholerae*: increase in the *cis/trans* ratio and proportions of cyclopropyl fatty acids. *Appl. Environ. Microbiol.* **52:**794–801.

17. Harrop, A. J., M. D. Hocknull, and M. D. Lilly. 1989. Biotransformation in organic solvents: a difference between gram-positive and gram-negative bacteria. *Biotechnol. Lett.* **11:**807–810.

18. Heipieper, H. J., B. Loffeld, H. Keweloh, and J. A. M. de Bont. 1995. The *cis/trans* isomerisation of unsaturated fatty acids in *Pseudomonas putida* S12: an indicator for environmental stress due to organic compounds. *Chemosphere* **30:**1041–1051.

19. Heipieper, H. J., H. Keweloh, and H. J. Rehm. 1991. Influence of phenols on growth and membrane permeability of free and immobilized *Escherichia coli. Appl. Environ. Microbiol.* **57:**1213–1217.

20. Heipieper, H. J., R. Diefenbach, and H. Keweloh. 1992. Conversion of *cis* unsaturated fatty acids to *trans*, a possible mechanism for the protection of phenol-degrading *Pseudomonas putida* P8 from substrate toxicity. *Appl. Environ. Microbiol.* **58:**1847–1852.

21. Hirayama, H., H. Takami, A. Inoue, and K. Horikoshi. 1998. Isolation and characterization of toluene-sensitive mutants from *Pseudomonas putida* IH-2000. *FEMS Microbiol. Lett.* **169:**219–225.

22. Holtwick, R., F. Meinhart, and H. Keweloh. 1997. Cis-trans isomerization of unsaturated fatty acids: cloning and sequencing of the *cti* gene from *Pseudomonas* P8. *Appl. Environ. Microbiol.* **63:**4292–4297.

23. Holtwick, R., H. Keweloh, and F. Meinhart. 1999. *cis/trans* isomerization of unsaturated fatty acids of *Pseudomonas putida* P8: evidence for a heme protein of the cytochrome *c* type. *Appl. Environ. Microbiol.* **65:**2644–2649.

24. Ikura, Y., Y. Yoshida, and T. Kudo. 1997. Physiological properties of two *Pseudomonas mendocina* strains which assimilate styrene in a two-phase (solvent-aqueous) system under static culture conditions. *J. Ferment. Bioeng.* **83:**604–607.

25. Ingram, L. O. 1977. Changes in the lipid composition of *Escherichia coli* resulting from growth with organic solvents and with food additives. *Appl. Environ. Microbiol.* **33:**1233–1236.

26. Inoue, A., and K. Horikoshi. 1989. A *Pseudomonas* thrives in high concentration of toluene. *Nature* **338:**264–266.

27. Inoue, A., and K. Horikoshi. 1991. Estimation of solvent-tolerance of bacteria by the solvent parameter log *P. J. Ferment. Bioeng.* **71:**194–196.

28. Inoue, A., M. Yahamoto, and K. Horikoshi. 1991. *Pseudomonas putida* which can grow in the presence of toluene. *Appl. Environ. Microbiol.* **57:**1560–1562.

29. Isken, S., and J. A. M. de Bont. 1996. Active efflux of toluene in a solvent-resistant bacterium. *J. Bacteriol.* **178:**6056–6058.

30. Isken, S., and J. A. M. de Bont. 1998. Bacteria tolerant to organic solvents. *Extremophiles* **2:**229–238.

31. Isken, S., P. M. A. C. Santos, and J. A. M. de Bont. 1997. Effect of solvent adaptation on the antibiotic resistance in *Pseudomonas putida* S12. *Appl. Microbiol. Biotechnol.* **48:**642–647.

32. Jackson, R. W., and J. A. de Moss. 1965. Effects of toluene on *Escherichia coli. J. Bacteriol.* **90:**1420–1425.

33. Kakeda, M., C. Ueguchi, H. Yamada, and T. Mizuno. 1995. An *Escherichia coli* curved DNA-binding protein whose expression is affected by the stationary phase-specific sigma factor sigma-s. *Mol. Gen. Genet.* **248:**629–634.

34. Kieboom, J., J. J. Dennis, G. J. Zylstra, and J. A. M. de Bont. 1998. Active efflux of organic solvents in *Pseudomonas putida* S12 is induced by solvents. *J. Bacteriol.* **180:**6769–6772.

35. Kieboom, J., J. J. Dennis, J. A. M. de Bont, and G. J. Zylstra. 1998. Identification and molecular characterization of an efflux pump involved in *Pseudomonas putida* S12 solvent tolerance. *J. Biol. Chem.* **273:**85–91.

36. Kim, K., L. Lee, K. Lee, and D. Lim. 1998. Isolation and characterization of toluene-sensitive mutants from the toluene-resistant bacterium *Pseudomonas putida* GM73. *J. Bacteriol.* **180:**3692–3696.

37. Komatsu, T., K. Moriya, and K. Horikoshi. 1994. Preparation of organic solvent-tolerant mutants from *Pseudomonas aeruginosa* strain PAO1161. *Biosci. Biotech. Biochem.* **58:**1754–1755.

38. Laane, C., S. Boeren, K. Vos, and C. Veeger. 1986. Rules for optimization of biocatalysis in organic solvents. *Biotech. Bioeng.* **30:**81–87.

39. Li, L., T. Komatsu, A. Inoue, and K. Horikoshi. 1995. A toluene-tolerant mutant of *Pseudomonas aeruginosa* lacking the outer membrane protein F. *Biosci. Biotech. Biochem.* **59:**2358–2359.

40. Li, X.-Z., D. Ma, D. M. Livermore, and H. Nikaido. 1994. Role of efflux pump(s) in intrinsic resistance of *Pseudomonas aeruginosa*: active efflux as a contributing factor to beta-lactam resistance. *Antimicrob. Agents Chemother.* **38:**1742–1752.

41. Li, X.-Z., H. Nikaido, and K. Poole. 1995. Role of MexA-MexB-OprM in antibiotic efflux in *Pseudomonas aeruginosa. Antimicrob. Agents Chemother.* **39:**1948–1953.

42. Li, X.-Z., L. Zhang, and K. Poole. 1998. Role of the multidrug efflux systems of *Pseudomonas aeruginosa* in organic solvent tolerance. *J. Bacteriol.* **180:**2987–2991.

43. Loffeld, B., and H. Keweloh. 1996. *cis/trans* isomerization of unsaturated fatty acids as possible control mechanism of membrane fluidity in *Pseudomonas putida* P8. *Lipids* **31:**811–815.

44. Ma, D., D. N. Cook, M. Alberti, N. G. Pon, H. Nikaido, and J. E. Hearst. 1993. Molecular cloning and characterization of *acrA* and *acrE* genes of *Escherichia coli. J. Bacteriol.* **175:**6299–6313.

45. Ma, D., D. N. Cook, M. Alberti, N. G. Pon, H. Nikaido, and J. E. Hearst. 1995. Genes *acrA* and *acrB* encode a stress-induced efflux system of *Escherichia coli. Mol. Microbiol.* **16:**45–55.

46. Masuda, N., E. Sakagawa, and S. Ohya. 1995. Outer membrane proteins responsible for multiple drug resistance in *Pseudomonas aeruginosa*. *Antimicrob. Agents Chemother.* **39**: 645–649.

47. Moriya, K., and K. Horikoshi. 1993. A benzene-tolerant bacterium utilizing sulfur compounds isolated from deep sea. *J. Ferment. Bioeng.* **76**:397–399.

48. Moriya, K., and K. Horikoshi. 1993. Isolation of a benzene-tolerant bacterium and its hydrocarbon degradation. *J. Ferment. Bioeng.* **76**:168–173.

49. Naglak, T. J., D. J. Hettwer, and H. Y. Wang. 1990. Chemical permeabilization of cells for intracellular product release, p. 177–205. *In* J. A. Asenjo (ed.), *Separation Processes in Biotechnology.* Dekker, New York, N.Y.

50. Nakajima, H., K. Kobayashi, M. Kobayashi, H. Asako, and R. Aono. 1995. Overexpression of the *robA* gene increases organic solvent tolerance and multiple antibiotic and heavy metal ion resistance in *Escherichia coli*. *Appl. Environ. Microbiol.* **61**:2302–2307.

51. Nakajima, H., M. Kobayashi, T. Negishi, and R. Aono. 1995. *SoxRS* gene increased the level of organic solvent tolerance in *Escherichia coli*. *Biosci. Biotechnol. Biochem.* **59**:1323–1325.

52. Nikaido, H. 1996. Multidrug efflux pumps of gram-negative bacteria. *J. Bacteriol.* **178**:5853–5859.

53. Nikaido, H., and M. Vaara. 1985. Molecular basis of bacterial outer membrane permeability. *Microbiol. Rev.* **49**:1–32.

54. Ogino, H., K. Miyamoto, and H. Ishikawa. 1994. Organic solvent-tolerant bacterium which secretes an organic solvent-stable lipolytic enzyme. *Appl. Environ. Microbiol.* **60**:3884–3886.

55. Ogino, H., K. Yasui, T. Shiotani, T. Ishihara, and H. Ishikawa. 1995. Organic solvent-tolerant bacterium which secretes an organic solvent-stable proteolytic enzyme. *Appl. Environ. Microbiol.* **61**:4258–4262.

56. Ohta, Y., M. Maeda, T. Kudo, and K. Horikoshi. 1996. Isolation and characterization of solvent-tolerant bacteria which can degrade biphenyl/polychlorinated biphenyls. *J. Gen. Appl. Microbiol.* **42**:349–354.

57. Okuyama, H., A. Ueno, D. Enari, N. Morita, and T. Kusana. 1998. Purification and characterization of 9-hexadecenoic acid *cis-trans* isomerase from *Pseudomonas* sp. strain E-3. *Arch. Microbiol.* **169**:29–35.

58. Okuyama, H., D. Enari, A. Shibahara, K. Yamamoto, and N. Morita. 1996. Identification of activities that catalyze the *cis-trans* isomerization of the double bond of a mono-unsaturated fatty acid in *Pseudomonas* sp. E-3. *Arch. Microbiol.* **165**:415–417.

59. Okuyama, H., S. Sasaki, S. Higashi, and N. Murata. 1990. A *trans*-unsaturated fatty acid in a psychrophilic bacterium, *Vibrio* sp. strain ABE-1. *J. Bacteriol.* **172**:3515–3518.

60. Osborne, S. J., Leaver, J., Turner, M. K., and P. Dunnill. 1990. Correlation of biocatalytic activity in an organic-aqueous two-phase system with solvent concentration in the cell membrane. *Enzyme Microb. Technol.* **12**:281–291.

61. Paje, M. L. F., B. A. Neilan, and I. Couperwhite. 1997. A *Rhodococcus* species that thrives on medium saturated with liquid benzene. *Microbiology* (Reading) **143**:2975–2981.

62. Paulsen, I. T., M. H. Brown, and R. A. Skurray. 1996. Proton-dependent multidrug efflux systems. *Microbiol. Rev.* **60**: 575–608.

63. Pedrotta, V., and B. Witholt. 1999. Isolation and characterization of the *cis-trans*-unsaturated fatty acid isomerase of *Pseudomonas oleovorans* Gpo12. *J. Bacteriol.* **181**:3256–3261.

64. Pinkart, H. C., and D. C. White. 1997. Phospholipid biosynthesis and solvent tolerance in *Pseudomonas putida* strains. *J. Bacteriol.* **179**:4219–4226.

65. Pinkart, H. C., J. W. Wolfram, R. Rogers, and D. C. White. 1996. Cell envelope changes in solvent-tolerant and solvent-sensitive *Pseudomonas putida* strains following exposure to o-xylene. *Appl. Environ. Microbiol.* **62**:1129–1132.

66. Poole, K., K. Krebes, C. McNally, and S. Neshat. 1993. Multiple antibiotic resistance in *Pseudomonas aeruginosa*: evidence for involvement of an efflux operon. *J. Bacteriol.* **175**:7363–7372.

67. Poole, K., K. Tetro, Q. Zhao, S. Neshat, D. E. Heinrichs, and N. Bianco. 1996. Expression of the multidrug resistance operon *mexA-mexB-oprM* in *Pseudomonas aeruginosa*: *mexR* encodes a regulator of operon expression. *Antimicrob. Agents Chemother.* **40**:2021–2028.

68. Poole, K., N. Gotoh, H. Tsujimoto, Q. Zhao, A. Wada, T. Yamasaki, S. Neshat, J.-I. Yamagishi, X.-Z. Li, and T. Nishino. 1996. Overexpression of the *mexC-mexD-OprJ* efflux operon in *nfxB*-type multidrug-resistant strains of *Pseudomonas aeruginosa*. *Mol. Microbiol.* **21**:713–724.

69. Ramos, J. L., E. Duque, J. J. Rodriguez-Herva, P. Godoy, A. Haïdour, F. Reyes, and A. Fernandez-Barrero. 1997. Mechanisms for solvent tolerance in bacteria. *J. Biol. Chem.* **272**: 3887–3890.

70. Ramos, J. L., E. Duque, M. J. Huertas, and A. Haïdour. 1995. Isolation and expansion of the catabolic potential of a *Pseudomonas putida* strain able to grow in the presence of high concentrations of aromatic hydrocarbons. *J. Bacteriol.* **177**: 3911–3916.

71. Ramos, J. L., E. Duque, P. Godoy, and A. Segura. 1998. Efflux pumps involved in toluene tolerance in *Pseudomonas putida* DOT-T1E. *J. Bacteriol.* **180**:3323–3329.

72. Rodriguez-Herva, J. J., and J. L. Ramos. 1996. Characterization of an OprL null mutant of *Pseudomonas putida*. *J. Bacteriol.* **178**:5836–5840.

73. Rodriguez-Herva, J. J., M.-I. Ramos-Gonzales, and J. L. Ramos. 1996. The *Pseudomonas putida* peptidoglycan-associated outer membrane lipoprotein is involved in maintenance of the integrity of the cell envelope. *J. Bacteriol.* **178**:1699–1706.

74. Saier, M. H., Jr., R. Tam, A. Reizer, and J. Reizer. 1994. Two novel families of bacterial membrane proteins concerned with nodulation, cell division and transport. *Mol. Microbiol.* **11**: 841–847.

75. Salter, G. J., and D. B. Kell. Solvent selection for whole cell biotransformation in organic media. *Crit. Rev. Biotechnol.* **15**: 139–177.

76. Shima, H., T. Kudo, and K. Horikoshi. 1991. Isolation of toluene-resistant mutants from *Pseudomonas putida* PpG1 (ATCC 17453). *Agric. Biol. Chem.* **55**:1197–1199.

77. Sikkema, J., B. Poolman, W. N. Konings, and J. A. M. de Bont. 1992. Effects of the membrane action of tetralin on the functional and structural properties of artificial and bacterial membranes. *J. Bacteriol.* **174**:2986–2992.

78. Sikkema, J., J. A. M. de Bont, and B. Poolman. 1994. Interactions of cyclic hydrocarbons with biological membranes. *J. Biol. Chem.* **269**:8022–8028.

79. Sikkema, J., J. A. M. de Bont, and B. Poolman. 1995. Mechanisms of membrane toxicity of cyclic hydrocarbons. *Microbiol. Rev.* **59**:201–222.

80. Tsubata, T., and R. Kurane. 1996. Selection of an *n*-heptanol-resistant bacterium from an organic solvent-sensitive bacterium and characterization of its fatty acids. *Can. J. Microbiol.* **42**:642–646.

81. Vermuë, M., J. Sikkema, A. Verheul, R. Bakker, and J. Tramper. 1993. Toxicity of homologous series of organic solvents for the gram-positive *Arthobacter* and *Norcardia* sp. and the gram-negative *Acinetobacter* and *Pseudomonas* sp. *Biotechnol. Bioeng.* **42**:747–758.

82. **Weber, F. J., and J. A. M. de Bont.** 1996. Adaptation mechanisms of microorganisms to the toxic effects of organic solvents on membranes. *Biochim. Biophys. Acta* **1286:**225–245.

83. **Weber, F. J., L. P. Ooijkaas, R. M. W. Schemen, S. Hartmans, and J. A. M. de Bont.** 1993. Adaptation of *Pseudomonas putida* S12 to high concentrations of styrene and other organic solvents. *Appl. Environ. Microbiol.* **59:**3502–3504.

84. **Weber, F. J., S. Isken, and J. A. M. de Bont.** 1994. *Cis/trans* isomerization of fatty acids as a defense mechanism of *Pseudomonas putida* strains to toxic concentrations of toluene. *Microbiology (NY)* **140:**2013–2017.

85. **White, D. G., J. D. Goldman, B. Demple, and S. B. Levy.** 1997. Role of the *acrAB* locus in organic solvent tolerance mediated by expression of *marA*, *soxS*, or *robA* in *Escherichia coli. J. Bacteriol.* **179:**6122–6126.

86. **Woldringh, C. L.** 1973. Effects of toluene and phenylethyl alcohol on the ultrastructure of *Escherichia coli. J. Bacteriol.* **114:**1359–1361.

87. **Woodruff, W. A., and R. E. W. Hancock.** 1989. *Pseudomonas aeruginosa* outer membrane protein F: structural role and relationship to the *Escherichia coli* OmpA protein. *J. Bacteriol.* **171:**3304–3309.

Bacterial Stress Responses
Edited by G. Storz and R. Hengge-Aronis
©2000 ASM Press, Washington, D.C.

Chapter 26

Bacteria Adapted to Industrial Biotopes:
Metal-Resistant *Ralstonia*

MAX MERGEAY

Metal-resistant bacteria formerly known as Alcalige-nes eutrophus *carry the* czc, ncc, *or* cnr *cation efflux metal-resistance operon and are mostly found in industrial and polluted biotopes, prevailing in metallur-gic wastes. Strain CH34 is the best known among these strains, which have been recently transferred to the* Ralstonia *genus where they form distinct clusters waiting for a definitive assessment at the species level. Here we review the genetic properties (plasmid- and chromosome-borne) of these* R. eutropha-*like bacteria from the perspective of specific adaptation to harsh and poor, nonequilibrated biotopes. Because of their apparent specialization to such biotopes, these strains provide an attractive material to study, both at the genomic and the physiological level, the cellular responses to various types of stress and the regulatory networks that control their expression.*

STRESS AND INDUSTRIAL BIOTOPES

Industrial biotopes refer to ecological niches created by the industrial revolution over the last two centuries, especially to industrial wastes rich in heavy metals, which are often mixed with organic recalci-trant compounds and hydrocarbons. These wastes are deposited as sediments on river banks, fields, or dumping sites (89). Such biotopes may be considered very harsh due to the abundance and regular replen-ishment of toxic compounds, the absence of renew-able vegetation with related strong variations of tem-perature, the limited amounts of organic nutrients, the irregular fluctuations between extreme dryness and moisture, and so forth.

Residents of these biotopes are exposed to a mixture of chronic (heavy metals, food scarcity) and episodic stress (extreme fluctuations in T° [tempera-ture] or water activity, anthropogenic activities) (an-thropogenic specifically refers to man-made situa-tions). These environmental conditions are rarely encountered in "natural" (nonanthropogenic [58, 70]) extremophilic biotopes such as hydrothermal sources or the dry valleys of Antarctica. This combi-nation of stress that is typical of industrial anthro-pogenic activities could likely occur naturally after natural catastrophes such as earthquakes, floods, or volcanic eruptions where specialized (pioneer) mi-crobes and plants are expected to play a crucial role in the recolonization processes.

A STRAIN WITH MULTIPLE RESISTANCE TO HEAVY METALS

Strain CH34 was first found in such an industrial anthropogenic biotope (36, 38), specifically the metal-rich sediments of a zinc factory in Belgium. This bacterium harbors plasmid-borne multiple resis-tance to heavy metals (Cd^{2+}, Co^{2+}, $CrO_4^=$, Cu^{2+}, Hg^{2+}, Ni^{2+}, Pb^{2+}, Tl^+ and Zn^{2+}) and is a facultative hydrogenotroph. The finding that CH34 synthesizes two hydrogenases, one soluble and cytoplasmic and one membrane-bound, was used to classify the strain as *Alcaligenes eutrophus* (38). The taxonomic prox-imity between metal-resistant strains such as CH34 and the *A. eutrophus* hydrogenotrophs was further supported by a variety of genetic and phenotypic tests (24, 38). But, as discussed below, the taxonomic status of *A. eutrophus* and related strains has been

Max Mergeay • Laboratory for Microbiology, Radioactive Waste and Clean-Up Division, Center of Studies for Nuclear Energy, (SCK/CEN), and Environmental Technology, Flemish Institute for Technological Research (VITO), Mol, and Laboratoire de Génétique des Procaryotes, Institut de Biologie Moléculaire et Médicale, Université Libre de Bruxelles, Brussels, Belgium.

recently revisited, and the bacteria have been reclassified in the *Ralstonia* genus (2, 84, 91; J. Goris, P. De Vos, D. Janssens, M. Mergeay, and P. Vandamme, submitted for publication); *Ralstonia metallidurans* is the proposed name for CH34 and related strains (Goris et al., submitted).

Genetic analysis of the CH34 genome has been focused on the heavy-metal-resistance operons, which are located on two megaplasmids, pMOL28 (180 kb) and pMOL30 (240 kb) (38, 45, 47, 62, 74). Conjugation experiments (55), the isolation and characterization of R-prime plasmids (33), the construction of a cosmid library (25), and the development of electroporation assays (76) allowed the circularity of the chromosome to be established and various loci to be mapped. Genetic and physical maps of pMOL28 plasmid and its conjugation-proficient derivative pMOL50 (55) are available (74, 75). Finally, several mobile genetic elements present in the CH34 genome were trapped on appropriate positive selection vectors (mobilizing plasmids [16], resistance to sucrose [21], zinc resistance [9, 30, 77a], and heterologous expression of biphenyl degradation genes [66]) and characterized.

PLASMIDS AND RESISTANCE TO HEAVY METALS

Large conjugative plasmids carrying metal-resistance genes have been identified in various strains. These plasmids were detected via their positive hybridization with specific probes containing the metal-resistance genes *czc* (resistance to Cd^{2+}, Zn^{2+}, and Co^{2+}), *cnr* (resistance to Co^{2+} and Ni^{2+}), and *ncc* (resistance to Ni^{2+}, Cd^{2+}, and Co^{2+}) cloned from strains CH34 and 31A (15, 17, 62, 70, 79). The metal-resistance genes encode tricomponent cation/proton antiporter efflux systems (20, 35, 45–48, 53, 61), which belong to the RND family of proteins involved in resistance, nodulation, and division (32, 34, 56). Table 1 summarizes the metal-resistance genetic determinants carried by large plasmids of strain CH34 and related strains. Further description of the corresponding resistance mechanisms (63) can be found in chapters 10 and 22.

Table 1 also shows that some genes, especially regulatory genes, thus far were found only in *Ralstonia* strains. However, other resistance genes and operons that were first discovered in other bacteria are also present on *Ralstonia* plasmids. The *Ralstonia* plasmids look more specialized in terms of metal-resistance and metal-processing genes. Full sequencing of the pMOL28, pMOL50, and MOL30 plasmids is now in progress and will probably uncover more

genes involved in metal processing (J. Dunn, personal communication).

Heavy-metal-resistance operons such as those residing on the *Ralstonia* plasmids provide good examples of a modular association of genes (40) (Table 1). In the *czc*NICBADRSE (resistance to Cd^{2+}, Zn^{2+}, and Co^{2+}), *ncc*YXHCBAN (resistance to Ni^{2+}, Co^{2+}, and Cd^{2+}), and *cnr*YXHCBA (resistance to Co^{2+} and Ni^{2+}) operons or gene clusters, there is a recognizable module formed by the structural genes, *CBA*, which encode the cation/proton antiporter efflux system (46) (Fig. 1). The same type of module exists in the *Pseudomonas aeruginosa czr*SRCBA operon (resistance to Zn^{2+} and Cd^{2+} [30]). The *YXH* regulatory genes constitute another recognizable module in the *cnr* and *ncc* operons (35, 59, 77a). They form a three-component regulatory unit consisting of a σ factor (CnrH and NccH) (35), a putative membrane-bound anti-σ factor (CnrY and NccY), and a periplasmic metal-sensing protein (CnrX and NccX) (77a). This module is absent from the *czc* operon, which is regulated via a totally different mechanism (15, 27, 87; D. H. Nies, submitted for publication; D. van der Lelie, personal communication) involving the *czc*NI and *czc*DRSE gene clusters. In the latter cluster, *czc*R and *czc*S encode a classical two-component regulatory system (43, 50). This two-component system is also present in the *czr*SRCBA operon (30) and in the various copper-resistance operons known so far; *cop*ABCDRS of *Pseudomonas syringae* (43), *pco*ABCDRSE of *Escherichia coli* (4), and *cop*SRABCD are present on the CH34 pMOL30 plasmid (with divergent transcription of the *RS* and *ABCD* genes) (11, 85). Other examples may be found in Table 1.

Another interesting feature of the resistance operons is the clustering or the strong linkage of genes involved in similar functions. The pMOL30 *cop* cluster contains genes corresponding to different copper-resistant mechanisms: the complex detoxification system encoded by *cop*ABCD genes, which are also represented in *P. syringae* and in *E. coli* (*pco*ABCD) (4, 5), and the ATPase-mediated Cu-efflux system (*cop*F), which has homologs in many other living organisms (49, 52). In pTOM9, the metal-resistance plasmid of *R. metallidurans* 31A, the strong clustering of the *nre* and *ncc* operons, which both encode resistance to nickel, provides another example of this grouping of resistance operons (59, 71).

The precise mechanism of *cop*ABCD-dependent copper resistance in *R. metallidurans* CH34 is still unknown (12, 85; D. van der Lelie, personal communication). *cop*+ *P. syringae* colonies accumulate blue Cu^{2+} ions in the periplasm and outer membrane. At least part of this copper-sequestering activity is due to the copper-binding protein products of the *copABCD* operon that ensure compartmentalization in the periplasm (7). *copC* and *copD* also seem to have a role in copper uptake into the cytoplasm. This function may be explained by the necessary equilibrium between resistance and homoeostasis for an essential trace element (10). In *E. coli*, the *pcoABCD* genes are related to *copABCD* of *P. syringae*, but the resistance mechanism differs from that of the *P. syringae* system and appears to involve efflux rather than sequestration (4, 5).

The sequencing of fragments from large metal-resistance plasmids has also provided the sequence of mobile genetic elements, IS sequences, and transposons (van der Lelie, personal communication). The presence of these elements raises questions about the possible mobility of resistance genes and the possible role played by mobile elements in the modular assembly of metal-resistance operons and operon clusters.

TAXONOMY AND ECOLOGY OF METAL-RESISTANT STRAINS: THE GENUS *RALSTONIA*

It was of interest to know whether metal-resistant strains related to CH34 could be isolated from other industrial metallurgical biotopes. This was tested using *czc* (2, 15, 17) and *ncc* (31, 60, 70) metal-resistance genes as probes. The *czc* probe was mainly used on soils and sediments from metallurgical industries in Belgium and in the Katanga copper belt in the Democratic Republic of Congo (capital, Kinshasa). Table 2 gives a description of one of the Congolese sites (shown in Fig. 2) where a Czc$^+$ strain (i.e., probing positive with *czc*) was found (*R. metallidurans* AB2) (2). The *ncc* probe was used to examine (31, 60, 70) high nickel-resistant strains isolated from natural and anthropogenic nickel-rich soils and sediments. Ncc$^+$ strains were found in both kinds of environments.

Taxonomic studies were then conducted on most of Czc$^+$ and Ncc$^+$ strains. The Czc$^+$ strains were all phenotypically resistant to high concentrations of zinc while the Ncc$^+$ strains displayed resistance to very high concentrations of nickel (up to 40 mM). All of the tested Czc$^+$ and some of the Ncc$^+$ strains clustered in the genus *Ralstonia*, confirming the linkage between *R. eutropha* and CH34 (38). Ncc$^+$ strains coming from natural nickel-rich biotopes as ultramafic New Caledonian soils (58) were preliminarily assigned to the genus *Burkholderia* on the basis of phenotypic tests. With the exception of a few clinical isolates (previously known as CDC IV c-2 strains) that hybridize to the *czc* probe (2) and are now part

of the new genus *Ralstonia paucula* (84), most Czc$^+$ and Ncc$^+$ metal-resistant *Ralstonia* (with zinc or nickel resistance, respectively, as the diagnostic phenotype) were isolated from anthropogenic biotopes (17, 70, 79), mainly metallurgical or metal-processing industries. Yet, these metal-resistant *R. eutropha*-like strains can be assigned to three clusters, which are now under further taxonomic determination at the species level (Goris et al., submitted).

The main cluster includes strain CH34 and about 25 strains isolated from Belgium, the Democratic Republic of Congo, and Germany (2). Many of them contain large plasmids for heavy metal resistance (15), are also resistant to nickel (with the *cnr*-like genes and phenotypes [35]), and are hydrogenotrophs. This cluster, which is closely related to the canonical *R. eutropha* hydrogenotrophs, displays a striking characteristic. Fourteen of its members have a mutator phenotype, which will be detailed below.

A second cluster of Czc$^+$ strains is mainly made of Belgian strains (15, 17, 18), which were found in a zinc-desertified area and in sediments of a galvanization plant. According to an ARDRA taxonomic study (2), this cluster of Czc$^+$ strains is most closely related to the plant pathogenic *Ralstonia solanacearum*.

The genomes of the *Ralstonia* genus seem to shelter interesting groups of specialized functions: (i) hydrogenotrophy, (ii) plant pathogenesis, (iii) multiple resistance to heavy metals, and (iv) not mentioned thus far, the degradation of a variety of aromatic compounds, including chlorinated derivatives (19, 57, 64, 68, 90). At first glance, these functions do not appear compatible with the same niche. However, it may be argued that hydrogen is a product of active vegetation involving legumes and that some specialized rhizospheric bacteria could take advantage of this excess hydrogen produced during nitrogen fixation. *Ralstonia solanacearum* strains are mainly known as phytopathogens. However, many nonvirulent strains that can invade plant tissues, although to a much lower extent than their virulent relatives, have been isolated from a variety of soils (A. Trigalet, personal communication). Nonvirulent phyto- or rhizobacteria able to grow at the expenses of hydrogen and CO_2 could thus exist. If their commensalist plant is confronted with high soil concentrations of heavy metals, it is conceivable that these bacteria capture metal-resistance genes. From the same perspective, it should be mentioned that some well-known *R. eutropha* strains such as JMP134 (19) or A5 (22, 64, 68) can degrade xenobiotics such as chlorinated aromatic compounds. This phenotypic trait may also be seen as an adaptation to inhospitable environments and may have originated as a bacterial

Table 1. Metal-resistance genes in *Ralstonia* plasmids[a]

Plasmid	Operon or cluster	Metal(s)	Mechanism or function	Similar gene(s) (DNA sequence)	Organism or plasmid	Reference(s)
pMOL28	*cnrYXHCBA*	Co²⁺, Ni²⁺				
	cnrYXH		Regulation	*nccYXH*	31A (pTOM9)	35
	cnrH		σ_ECF factor	*nccH*	31A (pTOM9)	35, 77a
	cnrX		Periplasmic sensor	*nccX*		77a
	cnrCBA		Tricomponent efflux of cations	*czcCBA*	CH34 (pMOL30)	
				nccCBA	31A (pTOM9)	30
				czrCBA	*P. aeruginosa*	
				silCBA	*Salmonella*	28
	chrAB	CrO₄=	Sulfate/chromate antiporter efflux	*chr*	*P. aeruginosa*	6, 12; Nies, submitted
	chrC	CrO₄=?		Superoxide dismutase genes	Many genera	Nies, submitted
	merRTPAD (Tn4378)	Hg²⁺	Uptake and reduction	*merRTPAD*	Many genera	16
pMOL30	*czcNICBADRSE*	Cd²⁺, Zn²⁺, Co²⁺				
	czcN		Regulation	*nccN*	pTOM9	87
	czcI		Regulation			15
	czcCBA		Tricomponent efflux of cations	See *cnr*		46, 47
	czcD		Efflux	COT, ZRC	Yeast	1, 47, 87
	czcRS		Two-component regulatory system	*czcRS*	*P. aeruginosa*	30
				pcoRS	*P. syringae*	43, 87
				pcoRS	*E. coli*	4, 87
				copRS	CH34 (pMOL30)	12, 85, 87

Gene/operon	Metal	Function	Similar gene(s)	Organism	Reference(s)
czcE		Unknown	*copH*		van der Lelie, personal communication
pbrRA	Pb²⁺				
pbrR		Regulation	*merR*	Many bacteria	12
pbrA		ATPase-mediated efflux	*cadA* (Cd²⁺)	*Staphylococcus aureus*	12, 85
			pacS (Cu²⁺)	*Synechococcus*	3
			zntA (Zn²⁺)	*E. coli*	12, 85
			copF	CH34 (pMOL30)	15
merRTPAD (Tn4380)	Hg²⁺	Uptake and reduction	*merRTPAD*	Many bacteria	12, 85
copSRABCD	Cu²⁺				
copF copH					
copRS		Two-component regulatory system	see *czcRS*		43, 85
copABCD		Efflux and complexation?	*pcoABCD*	*E. coli*	4, 12, 85
copF		ATPase-mediated efflux?	*copABCD*	*P. syringae*	85
copH		Unknown but needed for full resistance	see *pbrA*	CH34 (pMOL30)	van der Lelie, personal communication
			czcE		
nccYXHCBAN	Ni²⁺, Co²⁺, Cd²⁺			CH34 (pMOL30)	59
nccYXH		Regulation	*cnrYXH*	CH34 (pMOL28)	59
nccCBA		Tricomponent efflux	see *cnrCBA*		59
nccN		Regulation	*czcN*	CH34 (pMOL30)	87
nre	Ni²⁺	Unknown	*nre*	*Klebsiella oxytoca*	71

pTOM9

ᵃ Complete operons or gene clusters are underlined; subdivisions indicate putative modules or functions. Functions deduced from sequences and waiting for physiological confirmation are indicated by question marks.

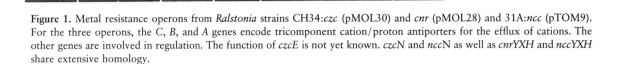

Figure 1. Metal resistance operons from *Ralstonia* strains CH34:*czc* (pMOL30) and *cnr* (pMOL28) and 31A:*ncc* (pTOM9). For the three operons, the *C*, *B*, and *A* genes encode tricomponent cation/proton antiporters for the efflux of cations. The other genes are involved in regulation. The function of *czcE* is not yet known. *czc*N and *ncc*N as well as *cnr*YXH and *ncc*YXH share extensive homology.

response to some phenolic compounds released by plants.

These conjectural proposals could be tested by analyzing the genomes of different *Ralstonia* species. These genomic studies may help to understand how gene functions related to plant colonization or commensalism, chemolithotrophic growth, degradation of xenobiotics, and multiple metal resistance are organized inside the *Ralstonia* genomes and distributed in the species. The studies might also elucidate the reasons why some genes are clustered on plasmids or other mobile genetic elements, as, for instance, appears to be the case for hydrogenotrophy and growth at the expense of nitrate in *R. eutropha* (23, 54, 65) or for metal resistance in CH34 and related strains. The complete sequence of the genome of a typical virulent strain of *R. solanacearum* is now available (C. Boucher, personal communication). Preliminary analysis shows the presence of sequences similar to metal resistance (*czc*) in *R. metallidurans* CH34. This finding should stimulate an extensive analysis of other *Ralstonia* genomes and the use of the *Ralstonia* genus as a model system to study the emergence and genomic dynamics (localization on mobile genetic elements, intra- and intercellular mobility [40], and rearrangements) of specialized functions such as host-plant interactions, pathogenicity, biodegradation, and resistance to heavy metals.

THE MUTATOR PHENOTYPE OF *R. METALLIDURANS* CH34 AND RELATED METAL-RESISTANT STRAINS

When grown at 37°C, cultures of strain CH34 display a peculiar phenotype of strong mortality (reproducible survival of 10^{-4} to 10^{-5}) and a very high percentage (up to 80%) of many types of mutants among the survivors. At 38°C, the growth of the strain is irreversibly impaired while at 36°C there is no detectable mortality or mutagenesis (39, 55). The mutator phenotype is thus still limited to a 1°C window below the lowest lethal T°. Some of the mutants that arise at 37°C carry auxotrophy for one of several amino acids with a special prevalence of lysine-requiring mutants (*lys*) (33). Nevertheless, threonine (*thr*) (39), proline (*pro*), and serine (*ser*) auxotrophs are also found. Other mutants lose the ability to grow under autotrophic conditions (*aut*), or, at the expense of nitrate, under anaerobic conditions, or on one of several carbon sources, such as tyrosine (*tyu*) (13), lactate, or gluconate. Some mutants require high concentrations of ammonium and others lose the resistance to heavy metals (especially those depending on the *czc* operon). Some survivors grow very poorly only on rich medium and form irregular small compact colonies. *lys*, *pro*, *thr*, *tyu*, and *aut* mutations map at loci scattered throughout the chromosome of *R. metallidurans* CH34 (55).

Strains carrying multiple mutations resulting in one or more of these phenotypes are often found. Their occurrence seems to be influenced by the time of exposure at 37°C. Liquid cultures maintained at 37°C accumulate multiple mutations (resulting in two or more of the above-mentioned phenotypes). Mutations can be present in up to 100% of the colonies

Table 2. Description of a site with high content in heavy metals[a]

Metal	Chemical analysis (mg/kg)		
	Total	Extraction EDTA	Extraction H_2O
Cd	1,046	48	30
Cu	49,260	1,350	3
Co	27	9	3
Pb	11,590	1,558	9
Ni	26	<1	<1
Fe	93,690	19	<1
Zn	156,040	4,533	1,027
Tl	<5		

[a] See Fig. 2. The site is a decantation basin (pH 9 to 10) with tailings from a flotation unit at a metallurgical plant (Katanga, Democratic Republic of Congo) located by the underground mine of Kipushi, near Lubumbashi. The bacterial viable counts were as follows: rich medium, 1.4×10^4 CFU/ml; rich medium plus 10 mM Zn^{2+}, 1.8×10^4 CFU/ml; minimal medium (29), 2×10^3 CFU/ml; and minimal medium plus 0.8 mM Cu^{2+}, 2.2×10^3 CFU/ml. The Czc⁺ strain of interest was *R. eutropha*-like strain AB2 (1) with pMOL230 (*czc*; 210 kb).

Figure 2. A decantation basin in the Kipushi metallurgical plant (Democratic Republic of Congo). The sampling place is located on the bank shown on the left. Sample characteristics are reported in Table 2. The picture was taken in December 1990.

that grow when the cultures are plated at 30°C on rich, nonselective media (Y. Markowicz, personal communication). Only some of the colonies that survive at 37°C gain the capacity to better survive or grow at 37°C. This is clearly the case for the *lys* mutants, although their growth remains irreversibly impaired at 38°C. Most of the other mutants (*thr, tyu, aut,* etc.) display no change in temperature sensitivity. When they are recloned at 30°C, a fully permissive temperature, and later grown at 37°C, they behave just like the wild-type parental strain.

Some types of mutants have been studied in more detail. Those that lost the *czc*-encoded metal resistance were cured from pMOL30 and carry a rearranged pMOL28 called pMOL50 (55, 74). Eight changes were identified, mostly deletions, excisions, and insertions, some of which may result from the activity of mobile genetic elements (74). Interestingly, pMOL50 is highly conjugative and efficiently mobilizes the CH34 chromosome. This type of rearranged pMOL28 was regularly observed over a 6-month period but much less frequently thereafter. However, curing of pMOL30 at 37°C was observed regularly.

Lysine auxotrophs are the most frequent class of mutants obtained among the survivors at 37°C, which is likely related to their increased resistance to temperature. Their presence is diagnostic of the mutator phenotype of strain CH34. Up to now, Lys⁻ mutants were never found after transposon mutagenesis or chemical mutagenesis, even when the mutagenesis was followed by extensive ampicillin/carbenicillin enrichment for mutants (55). Genetic analysis suggested that lysine auxotrophy and resistance to temperature could result from a single mutation affecting tightly linked genes (77). The nature of the genetic change leading to these coupled phenotypes is under investigation but may shed new light on previous observations showing a correlation between increased instability of IS elements (20) and a stimulatory effect of sulfur amino acids at 37°C.

A major, yet unresolved question is whether the observed mutations are induced by the exposure at 37°C or preexist that exposure and, for a reason that remains to be elucidated, invade the culture. The two alternatives are reflected in the different acronyms used to designate the phenomenon: "TIMM" for temperature-induced mortality and mutagenesis (86) and "TSP" for "thermospontagenesis" (39, 55). The TIMM induction hypothesis supposes that exposure of the bacteria at 37°C implies a kind of stress-mediated response via the expression of new functions, leading to the generation of specific mutations. Alternatively, one could hypothesize that 37°C is a lethal temperature for wild-type CH34 but not for some families of mutants. Cultures at 37°C would then be enriched in such mutants, similar to the mutant enrichment obtained after ampicillin treatment. This alternative implies that the mutations are present at the permissive temperatures (28 to 30°C) where they may be difficult to screen because of their low frequency of appearance. The wild-type strain might then bear a permanent/constitutive mutator phenotype. In this case the mutations should follow a Luria-Delbruck distribution, which easily could be tested. This hypothesis could be accounted for by more canonical mutators. A diversity of mutants would be generated, among which the mutants observed at 37°C would be the "screenable" fraction.

Remarkably, as mentioned above, this mutator phenotype was observed in 14 different strains originating from different countries and from different anthropogenically polluted biotopes but belonging to the same taxonomic cluster as strain CH34 (2). This suggests that, within the *Ralstonia* genus, the mutator phenotype might be specific to species related to CH34. The 14 strains share the basic features of the mutator phenotype observed in the strain CH34 although there are variations in the spectrum of the mutations (in some strains the dominant Lys⁻ phenotype is, for instance, replaced by Pro⁻), suggesting that the mutagenesis could be related to the survival in harsh environments rather than being purely anecdotal. Indeed, a high rate of mutagenesis (including rearrangements promoted by mobile genetic elements) may help a bacterial population to cope with repeated surges of acute stress by generating variants able to survive in harsh environments (51). In a similar way, the mutator phenotype of these *Ralstonia* strains might be a tool to promote adaptive mutagenesis in the same circumstances (29, 83).

GENE MOBILITY AND FOREIGN GENE EXPRESSION: RELEVANCE FOR ADAPTATION TO STRESSING CONDITIONS

Early on, *R. metallidurans* CH34 was recognized as a good acceptor and expressor of foreign genes. This was especially true for heterologous matings carried out using broad-host-range conjugative plasmids equipped with a proficient transposon such as the prophage mini-Mu (33). Donors for the mating experiments included bacteria from the *Agrobacterium*, *Burkholderia*, *Escherichia*, and *Pseudomonas* genera and thus represented α, β, and γ proteobacteria (33, 37, 55, 69, 88). Most of these matings resulted in heterologous complementations of single mutations. However, the transfer of operational sets of catabolic chromosomal genes was also observed (69), such as the heterologous transfer of the genes for the degradation of meta-hydroxybenzoate or biphenyl. The wild-type CH34 strain is unable to degrade these products. It may be speculated that low restriction proficiency plays a role in the efficiency of heterologous expression of foreign genes in *R. metallidurans* CH34. Low restriction proficiency could be a selective advantage for bacteria living in harsh or polluted environments and helps recruit new catabolic genes or other genes, allowing them to survive in environments where food is scarce or limited to recalcitrant compounds. Gene transfer experiments, in which the recipient CH34 is followed in soils, can be used for microbial ecology studies (14, 80, 82).

R. metallidurans CH34 also has been used extensively as a recipient to illustrate the phenomenon of retrotransfer or plasmid-mediated capture of genes (37), a feature that is characteristic of IncP broad-host-range plasmids (72). The IncP catabolic plasmids pJP4 (19), pSS50 (64), and pBR60 (90) have been found in strains closely related to strain CH34 (2) and can be expressed in metal-resistant strains (23, 66).

In addition to the aforementioned IncP catabolic plasmids, conjugative genetic elements have been detected in the metal-resistant *Ralstonia* and related strains. Large heavy metal plasmids are often conjugative and able to self-transfer at high frequencies although their transfer range is limited (15, 18, 55, 59). A special mention should be made of the large catabolic transposon Tn*4371* (41, 42, 44), which was found in the related strain A5 (2, 64, 68) and has conjugative properties. Tn*4371* hybridized partially with the genome of strain CH34 and could be expressed and integrated in this strain (D. Springael, personal communication).

PERSPECTIVES OF METAL-RESISTANT *RALSTONIA* AS A SYSTEM TO STUDY STRESS RESPONSES IN "CHRONICALLY" STRESSED BIOTOPES

Metal-resistant *Ralstonia* equipped with *czc* or *ncc*-like genes is frequent in industrial environments that could be considered hostile due to the variety and the irregularity of the toxic shocks and the general precariousness of food sources. The biology of these bacteria is clearly still in its infancy, even regarding their main responses to stress such as heat shock, oxidative shock, or iron stress. The heat shock response is detected at temperatures above 33°C. The *dna*K/Hsp70 gene of strain CH34 was recently cloned and sequenced, and an insertional mutant was isolated that displays increased sensitivity to T° as well as other phenotypes normally associated with the inactivation of *dna*K (77). Preliminary studies suggest that it may have retained a mutator phenotype although the temperature window in which mutants can be isolated (34°C) and the spectrum of mutations that accumulate are different (77). Selenium was shown to elicit the synthesis of specific proteins with a striking abundance of a superoxide dismutase (J. Coves, personal communication). Interestingly, one gene of pMOL28 (*chr*C) shows extensive similarity with superoxide dismutase genes from other bacteria (D. H. Nies, submitted for publication), suggesting a role for this family of proteins in resistance mechanisms.

Genes responsible for the biosynthesis and regulation of siderophores have been identified on the CH34 chromosome (25). Experiments to examine a relationship between siderophore overproduction and changes in heavy metal bioavailability are being carried out (26) since heavy metals display a nonnegligible affinity for some siderophores.

Metal-resistant *Ralstonia,* like other soil bacteria adapted to chemical pollutants, has been considered for a variety of environmental applications, including bioremediation (8, 11, 15, 67, 73; Nies, submitted). As recently pointed out (78), the ratio of funding devoted to applied and basic environmental microbiology is currently unbalanced toward applied microbiology. Only integrated studies of the stress-mediated responses, as suggested above for the metal-resistant *Ralstonia*, could help develop controlled bioremediation processes.

Besides the main stress responses (SOS, heat shock, oxidative stress, etc.), a comprehensive study should integrate the key features of the metal-resistant *Ralstonia* and include (i) the biology of their large plasmids, especially for those displaying an efficient conjugative phenotype; (ii) the mutator phenotype and its eventual relationship with SOS repair or other DNA repair pathways; (iii) the diversity of mobile genetic elements (IS, conjugative plasmids, and transposons) (42) and their potential in rearrangements as a consequence of sudden environmental pressures or in relation to the clustering in plasmids of genes involved in the processing of heavy metals; (iv) the emergence and expression of genes involved in the degradation of some families of recalcitrant (chloroaromates) compounds (taking also into account the genetic proximity with catabolic *R. eutropha* strains such as JMP134 [19] and A5 [64, 68]); (v) the regulation of the resistance to heavy metals (homeostasis) and the possible cross-talk between these regulatory mechanisms and central cellular regulatory networks; and (vi) the apparently facilitated acquisition and expression of foreign genes in possible correlation with a very low proficiency in restriction (33, 55, 66, 67, 69, 81, 82). These studies will undoubtedly contribute to our general understanding of the genetic evolution of microbial populations that cope with stringent environmental constraints.

Acknowledgments. Thanks are due to J. M. Collard, P. Corbisier, L. Diels, Q. Dong, M. Faelen, A. Sadouk, D. Springael, S. Taghavi, E. Top, A. Toussaint, D. van der Lelie, and A. Wilmotte for a long-standing collaboration on *Ralstonia/Alcaligenes*; to H.-G. Schlegel for friendly encouragement and 20 years of scientific exchange; to J. Balandreau, G. Maenhaut-Michel, J. Mahillon, J. Shapiro, and F. Taddei for vivid exchanges on stress and mutagenesis; and to B. Borremans, C. Boucher, H. Brim, J. Coves, P. Devos, J. Dunn, M.-E.-T. Hassan, Y. Markowicz, C. Merlin, J.-M. Meyer, D. Nies, S. Silver, S. Taghavi, S. Talbi, C. Tibazarwa, A. Trigalet, P. Vandamme, D. van der Lelie, and S. Wuertz for good discussions and for providing unpublished information or preprints. Special thanks go to A. Toussaint for deeply focused criticisms and comments on this manuscript and for promoting collaboration and exchanges on the genetics of strain CH34 in her laboratories in Rhode-St-Genèse and Grenoble.

This work was supported by the convention between SCK/CEN and ULB, by the bilateral collaboration between France and Belgium (Flemish Community) (TOURNESOL), and by a grant from the Flemish Regional Government (VLAB-ETC-003).

REFERENCES

1. **Anton, A., C. Grosse, J. Reissmann, T. Pribyl, and D. H. Nies.** 1999. CzcD is a heavy metal ion transporter involved in regulation of heavy metal resistance in *Ralstonia* sp. strain CH34. *J. Bacteriol.* 181:6876–6881.
2. **Brim, H., M. Heyndrickx, P. De Vos, A. Wilmotte, D. Springael, H. G. Schlegel, and M. Mergeay.** 1999. Amplified rDNA restriction analysis and further genotypic characterisation of metal-resistant soil bacteria and related facultative hydrogenotrophs. *Syst. Appl. Microbiol.* 22:258–268.
3. **Brocklehurst, K. R., J. L. Hobman, B. Lawley, L. Blank, S. J. Marshall, N. L. Brown, and A. P. Morby.** 1999. ZntR is a Zn(II)-responsive MerR-like transcriptional regulator of *zntA* in *Escherichia coli. Mol. Microbiol.* 31:893–902.
4. **Brown, N. L., D. A. Rouch, and B. T. O. Lee.** 1992. Copper resistance determinants in bacteria. *Plasmid* 27:41–51.
5. **Brown, N. L., S. R. Barrett, J. Camakaris, B. T. O. Lee, and D. A. Rouch.** 1995. Molecular genetics and transport analysis of the copper resistance determinant (*pco*) from *Escherichia coli* plasmid pRJ1004. *Mol. Microbiol.* 17:1153–1166.
6. **Cervantes, C., H. Ohtake, L. Chu, and S. Silver.** 1990. Cloning, nucleotide sequence and expression of the chromate resistance determinant of *Pseudomonas aeruginosa* plasmid pUM505. *J. Bacteriol.* 172:287–291.
7. **Cha J. S., and D. A. Cooksey.** 1991. Copper resistance in *Pseudomonas syringae* mediated by periplasmic and outer membrane proteins. *Proc. Nat. Acad. Sci. USA* 88:8915–8919.
8. **Collard, J. M., P. Corbisier, L. Diels, Q. Dong, C. Jeanthon, M. Mergeay, S. Taghavi, D. van der Lelie, A. Wilmotte, and S. Wuertz.** 1994. Plasmids for heavy metal resistance in *Alcaligenes eutrophus* CH34: mechanisms and applications. *FEMS Microbiol. Rev.* 14:405–414.
9. **Collard, J. M., A. Provoost, S. Taghavi, and M. Mergeay.** 1993. A new type of *Alcaligenes eutrophus* CH34 zinc resistance generated by mutations affecting regulation of the *cnr* cobalt-nickel resistance system. *J. Bacteriol.* 175:779–784.
10. **Cooksey, D. A.** 1994. Molecular mechanisms of copper resistance and accumulation in bacteria. *FEMS Microbiol. Rev.* 14:381–386.
11. **Corbisier, P., E. Thiry, A. Masolijn, and L. Diels.** 1994. Construction and development of metal ion biosensors, p. 151–155. *In* A. K. Campbell, L. J. Kricka, and P. E. Stanley (ed.), *Bioluminescence and Chemiluminescence, Fundamentals and Applied Aspects.* John Wiley & Sons, Chichester, United Kingdom.
12. **Corbisier, P., D. van der Lelie, B. Borremans, A. Provoost, V. de Lorenzo, N. L. Brown, J. R. Lloyd, J. L. Hobman, E. Csöregi, G. Johansson, and B. Mattiasson.** 1999. Whole cell- and protein-based biosensors for the detection of bioavailable heavy metals in environmental samples. *Anal. Chim. Acta* 387:235–244.
13. **David, C., A. Daro, E. Szalai, T. Atarhouch, and M. Mergeay.** 1996. Formation of polymeric pigments in the presence of bacteria and comparison with chemical oxidative coupling II. Catabolism of tyrosine and hydroxyphenylacetic acid by *Alcaligenes eutrophus* CH34 and mutants. *Eur. Polym. J.* 32:669–679.
14. **De Rore, H., E. Top, F. Houwen, M. Mergeay, and W. Verstraete.** 1994. Evolution of heavy metal resistant transconjugants in soil environment with a concomitant selective pressure. *FEMS Microbiol. Ecol.* 14:163–174.
15. **Diels, L., Q. Dong, D. van der Lelie, W. Baeyens, and M. Meageay.** 1995. The *czc* operon of *Alcaligenes eutrophus* CH34: from resistance mechanism to the removal of heavy metals. *J. Ind. Microbiol.* 14:142–153.

16. Diels, L., M. Faelen, M. Mergeay, and D. Nies. 1985. Mercury transposons from plasmids governing multiple resistance to heavy metals in *Alcaligenes eutrophus* CH34. *Arch. Intern. Physiol. Biochem.* **93**:B27–B28.

17. Diels, L., and M. Mergeay. 1990. DNA probe-mediated detection of resistant bacteria from soils highly polluted by heavy metals. *Appl. Environ. Microbiol.* **56**:1485–1491.

18. Diels, L., A. Sadouk, and M. Mergeay. 1989. Large plasmids governing multiple resistances to heavy metals: a genetic approach. *Toxicol. Environ. Chem.* **23**:79–89.

19. Don, R. H., and J. M. Pemberton. 1981. Properties of six pesticide degradation plasmids isolated from *Alcaligenes paradoxus* and *Alcaligenes eutrophus*. *J. Bacteriol.* **145**:681–686.

20. Dong, Q., and M. Mergeay. 1994. Czc/Cnr efflux: a three-component chemiosmotic antiport pathway with a 12-transmembrane-helix protein. *Mol. Microbiol.* **14**:185–186.

21. Dong, Q., A. Sadouk, D. van der Lelie, S. Taghavi, A. Ferhat, J-M. Nuyten, B. Borremans, M. Mergeay, and A. Toussaint. 1992. Cloning and sequencing of IS*1086*, an *Alcaligenes eutrophus* insertion element related to IS*30* and IS*4351*. *J. Bacteriol.* **174**:8133–8138.

22. Faelen, M., C. Merlin, M. Geuskens, M. Mergeay, and A. Toussaint. 1993. Characterization of a temperate phage hosted by *Alcaligenes eutrophus* strain A5. *Res. Microbiol.* **144**:624–631.

23. Friedrich, V., C. Hogrefe, and H. G. Schlegel. 1981. Naturally occurring genetic transfer of hydrogen-oxidizing ability between strains of *Alcaligenes eutrophus*. *J Bacteriol.* **147**:198–205.

24. Gerstenberg, C., B. Friedrich, and H. G. Schlegel. 1982. Physical evidence for plasmids in autotrophic, especially hydrogen-oxidizing bacteria. *Arch. Microbiol.* **133**:90–96.

25. Gilis, A., M. M. Khan, P. Cornelis, J.-M. Meyer, M. Mergeay, and D. van der Lelie. 1996. Siderophore-mediated iron uptake in *Alcaligenes eutrophus* CH34 and identification of *aleB* encoding the ferric-alcaligin E receptor. *J. Bacteriol.* **178**:5499–5507.

26. Gilis, A., P. Corbisier, W. Baeyens, S. Taghavi, M. Mergeay, and D. van der Lelie. 1998. Effect of the siderophore alcaligin E on the bioavailability of Cd to *Alcaligenes eutrophus* CH34. *J. Ind. Microbiol. Biotech.* **20**:61–68.

27. Grosse, C., G. Grass, A. Anton, S. Frank, A. Navarrete Santos, B. Lawley, N. L. Brown, and D. H. Nies. 1999. Transcriptional organisation of the *czc* heavy metal homoeostasis determinant from *Alcaligenes eutrophus*. *J. Bacteriol.* **181**:2385–2393.

28. Gupta, A., K. Matsui, J. F. Lo, and S. Silver. 1999. Molecular basis for resistance to silver cations in *Salmonella*. *Nat. Med.* **5**:183–188.

29. Hall, B. G. 1990. Spontaneous point mutations that occur more often when advantageous than when neutral. *Genetics* **126**:5–16.

30. Hassan, M.-T., D. van der Lelie, D. Springael, U. Romling, N. Ahmed, and M. Mergeay. 1999. Identification of a gene cluster, *czr*, involved in cadmium and zinc resistance in *Pseudomonas aeruginosa*. *Gene* **238**:417–425.

31. Kaur, P., K. Ross, R. A. Siddiqui, and H. G. Schlegel. 1990. Nickel resistance of *Alcaligenes denitrificans* strain 4a-2 is chromosomally coded. *Arch. Microbiol.* **154**:33–138.

32. Köhler, T., M. Michéa-Hamzehpour, U. Henze, G. Naomassa, L. K. Curty, and J. C. Perchère. 1997. Characterization of MexE-MexF-OprN, a positively regulated multidrug efflux system of *Pseudomonas aeruginosa*. *Mol. Microbiol.* **23**:345–354.

33. Lejeune, P., M. Mergeay, F. van Gijsegem, M. Faelen, J. Gerits, and A. Toussaint. 1983. Chromosome transfer and R-prime plasmid formation mediated by plasmid pULB113 (RP4::Mini-Mu) in *Alcaligenes eutrophus* CH34 and *Pseudomonas fluorescens* 6.2. *J. Bacteriol.* **155**:1015–1026.

34. Lo, X.-Z., H. Nikaido, and K. Poole. 1995. Role of MexA-MexB-OprM in antibiotic efflux in *Pseudomonas aeruginosa*. *Antimicrob. Agents Chemother.* **38**:1742–1752.

35. Liesegang, H., K. Lemke, R. A. Siddiqui, and H. G. Schlegel. 1993. Characterization of the inducible nickel and cobalt resistance determinant *cnr* from pMOL28 of *Alcaligenes eutrophus* CH34. *J. Bacteriol.* **175**:767–778.

36. Mergeay, M., C. Houba, and J. Gerits. 1978. Extrachromosomal inheritance controlling resistances to Cd^{++}, Zn^{++} and Co^{++} ions: evidence from curing in a *Pseudomonas*. *Arch. Int. Physiol. Biochim.* **86**:440–441.

37. Mergeay, M., P. Lejeune, A. Sadouk, J. Gerits, and L. Fabry. 1987. Shuttle transfer (or retrotransfer) of chromosomal markers mediated by plasmid pULB113. *Mol. Gen. Genet.* **209**:61–70.

38. Mergeay, M., D. Nies, H. G. Schlegel, J. Gerits, and F. van Gijsegem. 1985. *Alcaligenes eutrophus* CH34, a facultative chemolithotroph displaying plasmid bound resistance to heavy metals. *J. Bacteriol.* **162**:328–334.

39. Mergeay, M., A. Sadouk, L. Diels, M. Faelen, J. Gerits, J. Denecke, and B. Powell. 1987. High level spontaneous mutagenesis revealed by survival at non-optimal temperature in *Alcaligenes eutrophus* CH34. *Arch. Intern. Physiol. Biochem.* **95**:35–36.

40. Merlin, C., J. Mahillon, J. Nesvera, and A. Toussaint. 1999. Gene recruiters and transporters: the modular structure of bacterial mobile elements, p. 363–408. *In* C. M. Thomas, (ed.), *Plasmid Ecology and Biology*. Harwood Academic Publishers, Amsterdam, The Netherlands.

41. Merlin, C., D. Springael, M. Mergeay, and A. Tossaint. 1997. Organisation of the *bph* gene cluster of transposon Tn*4371* encoding enzymes for the degradation of biphenyl and 4-chlorobiphenyl. *Mol. Gen. Genet.* **253**:499–506.

42. Merlin, C., D. Springael, and A. Toussaint. 1999. Tn*4371*: a modular structure encoding a phage-like integrase, a Pseudomonas-like catabolic pathway and RP4/Ti-like transfer. *Plasmid* **41**:40–54.

43. Mills, S. D., C. A. Jasalavich, and D. A. Cooksey. 1993. A two-component regulatory system required for copper inducible expression of the copper resistance operon of *Pseudomonas syringae*. *J. Bacteriol.* **175**:1656–1664.

44. Mouz, S., C. Merlin, D. Springael, and A. Toussaint. 1999. A GntR-like negative regulator of transposon Tn*4371* biphenyl degradation genes. *Mol. Gen. Genet.* **262**:790–799.

45. Nies, D., M. Mergeay, B. Friedrich, and H. G. Schlegel. 1987. Cloning of plasmid genes encoding resistance to cadmium, zinc and cobalt in *Alcaligenes eutrophus* CH34. *J. Bacteriol.* **169**:4865–4868.

46. Nies, D. H. 1995. The cobalt, zinc, and cadmium efflux system CzcABC from *Alcaligenes eutrophus* functions as a cation-proton antiporter in *Escherichia coli*. *J Bacteriol* **177**:2707–2712.

47. Nies, D. H., A. Nies, L. Chu, and S. Silver. 1989. Expression and nucleotide sequences of a plasmid-determined divalent cation efflux system from *Alcaligenes eutrophus*. *Proc. Natl. Acad. Sci. USA* **86**:7351–7355.

48. Nies, D. H., and S. Silver. 1995. Ion efflux systems involved in bacterial metal resistances. *J. Ind. Microbiol.* **14**:186–199.

49. Nucifora, G., L. Chu, T. K. Misra, and S. Silver. 1989. Cadmium resistance from *Staphylococcus aureus* plasmid pI258 *cad*A gene results from a cadmium-efflux ATPase. *Proc. Natl. Acad. Sci. USA* **86**:3544–3548.

50. Parkinson, J. S., and E. C. Kofoid. 1992. Communication modules in bacterial signaling proteins. *Annu. Rev. Genet.* **26:** 71–112.

51. Radman, M., I. Matic, J. Halliday, and F. Taddei. 1994. Editing of homologous recombination by mismatch repair and SOS systems: implications to genome stability. *Philos. Trans. R. Soc.* **347:**97–103.

52. Rensing C., M. Ghosh, and B. P. Rosen. 1999. Families of soft-metal-ion-transporting ATPases. *J. Bacteriol.* **181:**5891–5897.

53. Rensing, C., T. Pribyl, and D. Nies. 1997. New functions for the three subunits of the CzcCBA cation-proton antiporter. *J. Bacteriol.* **179:**6871–6879.

54. Römmermann, D., and B. Friedrich. 1985. Denitrification in *Alcaligenes eutrophus* is plasmid dependent. *J. Bacteriol.* **162:** 852–854.

55. Sadouk A., and M. Mergeay. 1993. Chromosome mapping in *Alcaligenes eutrophus* CH34. *Mol. Gen. Genet.* **240:**181–187.

56. Saier., M. H., R. Tam, A. Reizer, and J. Reizer. 1994. Two novel families of bacterial membrane proteins concerned with nodulation, cell division and transport. *Mol. Microbiol.* **11:** 841–847.

57. Sauret-Ignazi, G., J. Gagnon, C. Beguin, M. Barrelle, Y. Markowicz, J. Pelmont, and A. Toussaint. 1996. Characterisation of a chromosome encoded catechol 1,2-dioxygenase (E.C.1.13.11.1) from *Alcaligenes eutrophus* CH34. *Arch. Microbiol.* **166:**42–50.

58. Schlegel, H. G., J. P. Cosson, and A. J. M. Baker. 1991. Nickel-hyperaccumulating plants provide a niche for nickel-resistant bacteria. *Bot. Acta* **104:**18–25.

59. Schmidt, T., and H. G. Schlegel. 1994. Combined nickel-cobalt-cadmium resistance encoded by the *ncc* locus of *Alcaligenes xylosoxidans* 31A. *J. Bacteriol.* **176:**7045–7054.

60. Schmidt, T., R. D. Stoppel, and H. G. Schlegel. 1991. High-level nickel resistance in *Alcaligenes xylosoxydans* 31A and *Alcaligenes eutrophus* KT02. *Appl. Environ. Microbiol.* **57:**3301–3309.

61. Sensfuss C., and H. G. Schlegel. 1988. Plasmid pMOL28-encoded resistance to nickel is due to specific efflux. *FEMS Microbiol. Lett.* **55:**295–298.

62. Siddiqui, R. A., K. Benthin, and H. G. Schlegel. 1989. Cloning of pMOL28-encoded nickel resistance genes and expression of the genes in *Alcaligenes eutrophus* and *Pseudomonas* spp. *J. Bacteriol.* **171:**5071–5078.

63. Silver S., and L. T. Phung. 1996. Bacterial heavy metal resistance: new surprises. *Annu. Rev. Microbiol.* **50:**753–789.

64. Shields, M. S., S. W. Hooper, and G. S. Sayler. 1985. Plasmid-mediated mineralization of 4-chlorobiphenyl. *J. Bacteriol.* **163:** 882–889.

65. Siedow, A., R. Cramm, R. A. Siddiqui, and B. Friedrich. 1999. A megaplasmid-borne anaerobic ribonucleotide reductase in *Alcaligenes eutrophus* H16. *J. Bacteriol.* **181:**4919–4928.

66. Springael, D., L. Diels, L. Hooyberghs, S. Kreps, and M. Mergeay. 1993. Construction and characterization of heavy metal-resistant haloaromatic-degrading *Alcaligenes eutrophus* strains. *Appl. Environ. Microbiol.* **59:**334–339.

67. Springael, D., L. Diels, and M. Mergeay. 1994. Transfer and expression of PCB-degradative genes into heavy metal resistant *Alcaligenes eutrophus* strains. *Biodegradation* **5:**343–357.

68. Springael, D., S. Kreps, and M. Mergeay. 1993. Identification of a catabolic transposon, Tn*4371*, carrying biphenyl and 4-chlorobiphenyl degradation genes in *Alcaligenes eutrophus* A5. *J. Bacteriol.* **175:**1674–1681.

69. Springael, D., J. van Thor, H. Goorissen, A. Ryngaert, R. de Baere, P. van Hauwe, L. Commandeur, J. Parsons, R. de Wachter, and M. Mergeay. 1996. RP4::Mu3A-mediated in

vivo cloning and transfer of a catabolic pathway. *Microbiology* **142:**3283–3293.

70. Stoppel, R. D., and H. G. Schlegel. 1995. Nickel-resistant bacteria from anthropogenically nickel-polluted and naturally nickel-percolated ecosystems. *Appl. Environ. Microbiol.* **61:** 2276–2285.

71. Stoppel R. D., M. Meyer, and H. G. Schlegel. 1995. The nickel resistance determinant cloned from the enterobacterium *Klebsiella oxytoca*: conjugal transfer, expression, regulation and DNA homologies to various nickel resistant bacteria. *Biometals* **8:**70–79.

72. Szpirer, C., E. Top, M. Couturier, and M. Mergeay. 1999. Retrotransfer or gene capture: a feature of conjugative plasmids, with ecological and evolutionary significance. *Microbiology* **145:**3321–3329.

73. Taghavi, S., M. Mergeay, D. Nies, and D. van der Lelie. 1997. *Alcaligenes eutrophus* as a model system for bacterial interactions with heavy metals in the environment. *Res. Microbiol.* **146:**536–551.

74. Taghavi, S., M. Mergeay, and D. van der Lelie. 1997. Genetic and physical maps of the *Alcaligenes eutrophus* CH34 megaplasmid pMOL28 and its derivative pMOL50 obtained after temperature induced mutagenesis and mortality (TIMM). *Plasmid* **37:**22–34.

75. Taghavi, S., A. Provoost, M. Mergeay, and D. van der Lelie. 1996. Identification of a partition and replication region in the *Alcaligenes eutrophus* megaplasmid pMOL28. *Mol. Gen. Genet.* **250:**169–179.

76. Taghavi, S., D. van der Lelie, and M. Mergeay. 1994. Electroporation of *Alcaligenes eutrophus* with (Mega) plasmids and genomic DNA fragments. *Appl. Environ. Microbiol.* **60:**3585–3591.

77. Talbi, S. 1999. Stress thermique et mutagenèse chez une bactérie du sol résistante aux métaux lourds: *Alcaligenes eutrophus* CH34. Ph.D. thesis. Université Libre de Bruxelles, Brussels, Belgium.

77a.Tibazarwa, C., S. Wuertz, M. Mergeay, L. Wyns, and D. van der Lelie. 2000. Regulation of the *cnr* cobalt and nickel resistance determinant of *Ralstonia eutropha* (*Alcaligenes eutrophus*) CH34. *J. Bacteriol.* **182:**1399–1409.

78. Timmis, K. N. 1999. An urgent need to reassess the balance between fundamental and applied environmental research. *Environ. Microbiol.* **1:**187–188.

79. Timotius, K., and H. G. Schlegel. 1987. Aus Abwässern isolierte nickel-resistente Bakterien. Nachrichten der Akademie der Wissenschaften Göttingen. II. *Math.-Physik. Klasse* **3:**15–23.

80. Top, E., H. De Rore, J. M. Collard, G. Gellens, G. Slobodkina, W. Verstraete, and M. Mergeay. 1995. Retromobilization of heavy metal resistance genes in unpolluted and heavy metal polluted soil. *FEMS Microbiol. Ecol.* **18:**191–203.

81. Top, E., I. De Smet, W. Verstraete, R. Dijkmans, and M. Mergeay. 1994. Exogenous isolation of mobilizing plasmids from polluted soils and sludges. *Appl. Environ. Microbiol.* **60:**831–839.

82. Top, E., M. Mergeay, D. Springael, and W. Verstraete. 1990. Gene escape model: transfer of heavy metal resistance genes from *Escherichia coli* to *Alcaligenes eutrophus* on agar plates and in soil samples. *Appl. Environ. Microbiol.* **56:**2471–2479.

83. Torkelson, J., R. S. Harris, M.-J. Lombardo, J. Nagendran, C. Thulin, and S. M. Rosenberg. 1997. Genome-wide hypermutation in a subpopulation of stationary-phase cells underlies recombination-dependent adaptive mutation. *EMBO J.* **16:** 3303–3311.

84. Vandamme, P., J. Goris, T. Goenye, B. Hoste, D. Janssens, P. De Vos, and E. Falsen. 1999. Assignment of Centers for Dis-

ease Control group IVc-2 to the genus *Ralstonia* as *Ralstonia paucula* sp. nov. *Int. J. Syst. Bacteriol.* **49**:663–669.

85. van der Lelie, D. 1998. Biological interactions: the role of soil bacteria in the bioremediation of heavy metal-polluted soils, p. 31–50. *In* J. Vangronsveld and S. D. Cunningham (ed.), *Metal-Contaminated Soils: In Situ inactivation and Phytorestoration.* Landes-Bioscience, Austin, Texas.

86. van der Lelie, D., A. Sadouk, A. Ferhat, S. Taghavi, A. Toussaint, and M. Mergeay. 1992. Stress and Survival in *Alcaligenes eutrophus* CH34: effects of temperature and genetic rearrangements, p. 27–32. *In* M. J. Gauthier (ed.), *Gene Transfers and Environment.* Springer-Verlag, Berlin, Germany.

87. van der Lelie, D., T. Schwuchow, U. S. Schwidetzky, S. Wuertz, W. Baeyens, M. Mergeay, and D. Nies. 1997. Two-component regulatory system involved in transcriptional control of heavy-metal homoeostasis in *Alcaligenes eutrophus.* *Mol. Microbiol.* **23**:493–503.

88. Waelkens, F., K. Verdickt, L. van Duffel, J. Vanderleyden, A. Vangool, and M. Mergeay. 1987. Complementation by *Agrobacterium tumefaciens* chromosomal genes and its potential use for linkage mapping. *FEMS Microbiol. Lett.* **43**:329–334.

89. Wuertz, S., and M. Mergeay. 1997. The impact of heavy metals on soil microbial communities and their activities, p. 607–642. *In* D. van Elsas, E. Wellington, and J. Trevors (ed.), *Modern Soil Microbiology.* Marcel Dekker, New York, N.Y.

90. Wyndham, R. C., and N. A. Straus. 1988. Chlorobenzoate catabolism and interaction between *Alcaligenes* and *Pseudomonas* species from Bloody Run Creek. *Arch. Microbiol.* **150**:230–236.

91. Yabuuchi, E., Y. Kosako, I. Yano, H. Hotta, and Y. Nishiuchi. 1995. Transfer of two *Burkholderia* and an *Alcaligenes* species to *Ralstonia* gen. nov: proposal of *Ralstonia pickettii* (Ralston, Palleroni & Doudoroff 1973) comb. nov., *Ralstonia solanacearum* (Smith 1896) comb. nov. and *Ralstonia eutropha* (Davis 1969) comb. nov. *Microbiol. Immunol.* **39**:897–904.

V. APPLICATIONS OF STRESS RESPONSE ANALYSIS

Bacterial Stress Responses
Edited by G. Storz and R. Hengge-Aronis
©2000 ASM Press, Washington, D.C.

Chapter 27

A Comparative-Genomic View of the Microbial Stress Response

Eugene V. Koonin, L. Aravind, and Michael Y. Galperin

Comparison of complete bacterial and archaeal genomes provides unprecedented opportunities for the exploration of the evolution of cellular functional systems, including those involved in stress response. Examination of the phylogenetic distribution of these systems shows both striking conservation, as in the case of heat shock genes and, in part, even the respective operons, and notable plasticity as in the case of the anti-sigma factor regulatory systems. Comparative sequence analysis results in the prediction of a number of previously undetected candidate components of stress-response systems, some of which might comprise new pathways. These data are expected to help prioritize targets for future experiments aimed at a comprehensive picture of bacterial and archaeal stress response.

Complete sequences of diverse microbial genomes that have become available in the past several years have largely redefined the task of assessing the evolutionary relationships between different groups of microorganisms. It has become possible not only to compare organisms on the basis of the protein sets encoded in their genomes, but also to identify the proteins that are *not* encoded in each particular genome, and to trace how their absence is compensated for by adaptive changes in the lifestyles of these organisms (40). Perhaps the most important and rather unexpected outcome of the sequencing of the first 20 or so complete bacterial and archaeal genomes is the growing realization that along with the remarkable conservation of a variety of protein domains, horizontal gene transfer and lineage-specific gene loss seem to be the prevalent forces of evolution, at least among prokaryotes.

The systems of stress response, particularly those including heat shock proteins, are generally well conserved throughout the microbial world (see references 19 and 48 for recent reviews). This has been linked to the importance of the functions these systems play in a normal, nonstressed cell (see, e.g., reference 51). On the other hand, many of these genes are nonessential and many are partially redundant, being backed up by alternative homologous or unrelated systems. This provides for significant diversity that seems to be supported by differential gene loss and lateral gene transfer and makes the stress-response systems particularly interesting targets for comparative genomics. The comparative genomics approach supplements the experimental data in two principal ways. First, by comparing the repertoires of conserved proteins involved in stress response in phylogenetically diverse bacteria and archaea, it becomes possible to draw some conclusions regarding possible evolutionary history of these systems. Second, sequence analysis of conserved protein domains in various multidomain proteins allows one to deduce the potential functions of these domains and improves the overall understanding of the roles of the respective proteins in the complex process of cellular response to stress.

At the time of this writing, complete genome sequences of 15 bacteria, 5 archaea, and 1 eukaryotic microorganism were available (3, 15, 21, 23, 26, 32–35, 42, 55, 58, 59, 63, 64, 67, 72, 91, 115, 116, 124). Given the extensive coverage of individual stress response systems in the other chapters of this book, we keep the description of these systems here to a minimum and, instead, concentrate on the analysis of the distribution of these systems in microorganisms that belong to different prokaryotic lineages (98, 100). Genes implicated in stress response in bacteria are numerous, and the levels of characterization of their actual roles in these processes differ widely. Since it would not be practicable to consider them all in one brief chapter, we first discuss the phylogenetic

Eugene V. Koonin, L. Aravind, and Michael Y. Galperin • National Center for Biotechnology Information, National Library of Medicine, National Institutes of Health, Bethesda, MD 20894.

distribution and likely evolutionary histories of some well-characterized components and then consider some more obscure but highly conserved and, by implication, important proteins.

ASSIGNMENT OF ORTHOLOGS

The assessment of the presence or absence of a particular gene (protein) in a given species in fact involves the identification of orthologs, that is, genes directly related by vertical evolutionary descent, as opposed to paralogs, genes within the same genome related by duplication (31). Here, the assignment of likely orthologs in the completely sequenced genomes was based on the clusters of orthologous groups of proteins (COGs, http://www.ncbi.nlm.nih.gov/COG) (40, 70, 121). This approach has been designed to identify orthologous families regardless of the differences in the rates of evolution of different genes, that is, even for the cases when the sequence similarity between likely orthologs is low and might not be statistically significant by itself. Briefly, COGs are identified by detecting and merging cliques of consistent, genome-specific best hits in an all-against-all sequence comparison of proteins from complete microbial genomes. Proteins in each COG are assumed to have evolved from a common ancestor and, generally, may be inferred to possess the same function or very similar functions in different species. However, it should be emphasized that in many cases, the same function in different organisms is carried out by a distantly related or an unrelated protein (27). This phenomenon, referred to as nonorthologous gene displacement, proves to be common in microbial genomes (see references 41 and 69 for further discussion). Therefore, the statement that a particular protein is missing in a given genome means only the absence of a member of the respective COGs, and by no means the absence of the corresponding function.

THE CONSERVED STRESS EFFECTORS—HEAT SHOCK GENES AND OPERONS

Distribution of Heat Shock Genes

The molecular chaperones encoded by heat shock genes, such as the HSP40, HSP60, HSP70, and HSP90 family members, are among the most highly conserved proteins in bacteria and eukaryotes. In particular, the former three are invariably encoded in each sequenced bacterial and eukaryotic genome (Table 1). Therefore, when the first archaeal genome, that of *Methanococcus jannaschii*, was sequenced

(21), it was a major surprise that only HSP60, most closely related to the eukaryotic TCP1 proteins, has been identified among the gene products. Subsequent sequencing of other archaeal genomes showed that this was not an isolated case of gene loss since at least two other euryarchaeal genomes (*Archaeoglobus fulgidus* and *Pyrococcus horikoshii*) and one crenarchaeal genome (*Aeropyrum pernix*) also lack genes for these molecular chaperones (63, 64, 67) (see Table 1). By contrast, these genes are present in *Methanobacterium thermoautotrophicum* and several other *Euryarchaeota*, such as *Methanosarcina barkeri*, *Thermoplasma acidophilum*, and *Halobacterium halobium* (115). These proteins show very high similarity to their bacterial orthologs and appear to be more closely related to them than to eukaryotic orthologs (46). Together with the absence of these chaperones in representatives from both major divisions of the archaea (see also reference 24), this suggests that the heat shock genes have been introduced into some of the archaeal lineages by horizontal transfer from bacteria (see also below). With regard to the chaperone functions in the bulk of the archaea, this seems to be a case of nonorthologous gene displacement whereby the same function is performed by unrelated or distantly related proteins. However, their functional counterparts in the archaea remain to be identified. It is interesting to note in this regard that in some archaeal species, such as *M. jannaschii* and *P. horikoshii*, there is significant expansion of distinct families of predicted ATPases distantly related to the AAA+ class. It cannot be ruled out that these ATPases are involved in specific forms of stress response.

Operon Organization of Heat Shock Genes

Genomes of closely related bacterial species, such as *Mycoplasma genitalium* and *Mycoplasma pneumoniae*, or *Chlamydia trachomatis* and *Chlamydia pneumoniae*, show significant, genome-scale conservation of the gene order (56, 58). At larger phylogenetic distances, however, the conservation of the gene order progressively decreases (90, 122) and involves only a relatively small number of essential operons (25, 40, 68, 99). In this context, it is interesting to note that heat shock genes show considerable conservation of operonic organization in bacteria and those of the archaea that possess them; at the same time, not a single gene pair is conserved universally (Table 1). Thus, the *groESL* operon is seen in all bacterial genomes sequenced to date, with the exception of the two spirochetes, *Borrelia burgdorferi* and *Treponema pallidum* (Table 1). A *Synechocystis* sp. and *Mycobacterium tuberculosis* each possess a second copy of the *groEL* gene, which is unlinked to the *groESL* operon, whereas *C. trachomatis* and *C. pneumoniae* each encode two additional, stand-alone copies of *groEL*.

The genes that encode heat shock proteins GrpE, DnaK, and DnaJ form operons with variable organization in different bacterial and archaeal genomes. The largest operon that has been characterized in detail (128) is seen in *Bacillus subtilis*. In addition to the three genes encoding molecular chaperones, it also includes an upstream gene that encodes the transcriptional regulator HrcA. In other bacterial and archaeal genomes this operon seems to be partially or completely disrupted (Table 1). For example, in *Helicobacter pylori* and in chlamydiae, *hrcA*, *grpE*, and *dnaK* still form an operon, while *dnaJ* is translocated; in *Escherichia coli*, *Haemophilus influenzae*, and *Synechocystis*, the *grpE* gene is translocated, while *dnaK* and *dnaJ* still form an operon. In mycoplasmas and in *Aquifex*, there are only traces of operon arrangement of the heat shock genes (Table 1). Particularly interesting is the arrangement in *Thermotoga maritima* where the main heat shock operon consists of the *hrcA*, *grpE*, and *dnaJ* genes whereas the *dnaK* gene appears to form a separate operon with the gene for the small heat shock protein HSP20 that in other bacteria and archaea is encoded by a stand-alone gene (Table 1). This has been interpreted as an indication of a functional interaction between DnaK and HSP20, although experimental support for this idea is still lacking (84). Another remarkable arrangement of the heat shock genes is seen in *Rickettsia prowazekii* where *grpE* is a part of the *groESL* operon, rather than the *dnaK* operon. Importantly, in archaea, at least in *M. thermoautotrophicum*, the operon organization of the heat shock genes is partially conserved, with the gene order *dnaJ-grpE-dnaK* (115).

Taken together, these observations are most easily interpreted to suggest that the common ancestor of all bacteria possessed both *groESL* and *dnaK* operons, with the latter probably including also the *hcrA* gene and perhaps even the gene for HSP20. During subsequent evolution, these operons apparently have undergone rearrangements that included gene deletions, translocations, and duplications. Furthermore, the *dnaK* operon apparently has been laterally transferred to at least one archaeal lineage, which has been accompanied by its partial disruption and might have been followed by limited dissemination among the archaea. The conclusion that the operon organization of the heat shock genes is an ancestral feature is not, however, logically inevitable. An alternative possibility is that the heat shock operons have evolved in one of the bacterial lineages and, being selected for the advantages of coregulation, have swept the rest of the

prokaryotic world through horizontal transfer. Under this scenario, the mycoplasmas and *Aquifex* have never had those operons although for the other bacteria, operon rearrangement still has to be postulated.

SPECIALIZED SIGMA FACTORS AND THE ANTI-SIGMA-FACTOR SYSTEM

Specialized Sigma Factors

In *E. coli*, certain genes (operons) are transcribed only by the RNA polymerase that contains a particular sigma subunit (47, 52). Thus, RpoH (σ^{32}) is required for the transcription of heat shock genes (47); RpoN (σ^{54}) is responsible, among other things, for nitrogen metabolism (83, 113); and RpoS (σ^{38}, also referred to as KatF) (88) has been shown to be involved in a variety of stress responses in *E. coli* (52, 78). Sigma subunit-dependent regulation also appears to be one of the main mechanisms of transcription regulation in *B. subtilis* and other gram-positive bacteria (49, 50).

With regard to the regulation of the expression of heat shock operons and other stress-response systems, there are at least two unrelated mechanisms. Since the discovery of the controlling inverted repeat of chaperone expression (CIRCE, reference 135) recognized by the HcrA protein, CIRCE-mediated regulation of heat shock genes has been described in a number of phylogenetically diverse bacteria (49). The *E. coli* mechanism, where the expression of heat shock genes is primarily controlled by a single sigma factor (σ^{32}), apparently represents only one, and arguably the simplest, of the many regulatory mechanisms (111, 133). In some bacteria, such as *Bradyrhizobium japonicum*, σ^{32}-mediated and HcrA/CIRCE-mediated regulatory mechanisms appear to function in concert (9), while in other bacteria these mechanisms are responsible for the control of different stress-response systems (see chapter 1 and reference 133 for discussion).

Comparative genome analysis shows that stress-response regulation via specialized sigma factors is prominent in free-living bacteria but has a limited significance in parasitic ones that typically encode only 1 to 4 sigma species, and does not exist in archaea, none of which possess sigma factors (Table 2). Genomes of diverse bacteria encode multiple sigma subunits whose functions have not been experimentally characterized (Table 2). While it is clear that these extra sigma subunits are involved in some forms of transcriptional regulation, their precise functions are difficult to predict, since apparently orthologous sigma factors may be responsible for transmitting different signals in different bacteria. The apparent or-

Table 1. Conservation of heat shock genes and operons in microbial genomes

Organism	HSP10	HSP60	HrcA	GrpE	HSP70	DnaJ	HSP20	HSP90
Bacteria								
Escherichia coli	groES^a*	groEL*	—^b	grpE†	dnaK† hscA ybeW yegD	dnaJ† cbpA	ibpA ibpB	htpG
Haemophilus influenzae	HI0542*	HI0543*	—	HI0071	HI1237† HI0373	HI1238†	—	HI0104
Helicobacter pylori	HP0011*	HP0010*	HP0111†	HP0110†	HP0109†	HP1024 HP1332	—	HP0210
Rickettsia prowazekii	RP627*	RP626*	—	RP629*	RP185† RP200	RP184†	RP273	RP840
Bacillus subtilis	groES*	groEL*	hrcA†	grpE†	dnaK†	dnaJ†	yocM† ydfT ypqA	htpG
Mycoplasma genitalium	MG393*	MG392*	MG205†	MG201†	MG305	MG002 MG019 MG200†	—	—
Mycoplasma pneumoniae	MP268*	MP269*	MP031†	MP035†	MP406	MP152 MP133 MP036†	—	—
Mycobacterium tuberculosis	Rv3418c*	Rv3417c* Rv0440	Rv2374c‡	Rc0351†	Rv0350†	Rv0352† Rv2373c‡	Rv2031c Rv0251c	Rv2299c
Synechocystis sp.	slr2075*	slr2076* sll0416	sll1670	sll0057†	sll0058† slr0086 sll0170† sll1932	sll1011 slr0093 sll0169‡ sll1384 sll1666 sll0897 sll1933	sll1514	sll0430
Borrelia burgdorferi	BB0741	BB0649	—	BB0519†	BB0518† BB0264	BB0517† BB0602 BB0655	—	BB0560
Treponema pallidum	TP1013	TP0030	—	TP0215†	TP0216†	TP0098 TP0843	—	TP0984
Chlamydia trachomatis	CT111*	CT110* CT604 CT755	CT394†	CT395†	CT396†	CT341	—	—
Chlamydia pneumoniae	CPn0135*	CPn0134* CPn0777 CPn0898	CPn0501†	CPn0502†	CPn0503†	CPn0032	—	—
Aquifex aeolicus	aq_2199*	aq_2200*	—	aq_433	aq_996	aq_703 aq_1735	aq_1283	—

	TM0505*	TM0506*	TM0851†	TM0850†	TM0373‡	TM0849†	TM0374‡
Thermotoga maritima							
Archaea							
Methanococcus jannaschii	—	MJ0999	—	—	—	—	MJ0285
Methanobacterium thermoautotrophicum	—	MTH218 MTH794	—	MTH1289†	MTH1290†	MTH1291†	MTH859
Archaeoglobus fulgidus	—	AF1451 AF2238	—	—	—	—	AF1296 AF1971
Pyrococcus horikoshii	—	PH0017	—	—	—	—	PH1842
Aeropyrum pernix	—	APE0907 APE2072	—	—	—	—	APE1950 APE0103
Eukaryota							
Saccharomyces cerevisiae	YOR020c	YBR044c YDL143w YDR188w YDR212w YIL142w YJL008c YJR064w YJL014w YLR259c YJL111w	—	YOR232w	YAL005c YBL075c YBR169c YDL229w YEL030w YER103w YHR064c YJL034w YJR045c YKL073w YLL024c YLR369w YNL209w YPL106c	YER048c YFL016c YFR041c YGL128c YGR285c YIR004w YJL073w YJL162c YJR097w YLR090w YMR214w YNL077w YOR254c YMR161w YNL007c YNL064c YNL227c	YBR072w YDR171w YMR186w YPL240c

[a] The genes that code for the indicated heat shock proteins are listed, based on the comparison of their protein products. The genes are named according to their original authors' designations; these names can be used to retrieve the corresponding protein sequences from the National Center for Biotechnology Information protein database (http://www.ncbi.nlm.nih.gov/Entrez/protein.html). The conserved gene strings (presumed but not always proven to comprise an operon) are indicated by boldface type, and each group is indicated by a specific symbol. ‡, this protein is missing in this particular genome, based on the sequence analysis using the Clusters of Orthologous Groups (http://www.ncbi.nlm.nih.gov/COG) approach. See references 40, 70, and 121 for more details.

[b] —, this protein is missing in this particular genome, which might be carried out by a nonorthologous protein. This does not necessarily mean the absence of the corresponding function.

Table 2. Conservation of sigma subunits in bacterial genomes

Bacteria[a]	Major sigma factors, σ^{70} and σ^{38}	Alternative sigma factors				Total no.	σ^{B} control mechanism		
		RpoH (σ^{32})	RpoE (σ^{24}, σ^{19})	RpoN (σ^{54})	FliA (σ^{28})		Anti-sigma[b]	Anti-anti-sigma	Ser phosphatase
Escherichia coli	RpoD (613)[c] RpoS (330)	RpoH	RpoE FecI	RpoN	FliA	7	—[d]	YrbB	—
Haemophilus influenzae	HI0533 (629)	HI0269	HI0628 HI1459	—	—	4	—	HI1083	—
Helicobacter pylori	HP0088 (671)	—	—	HP0714	HP1032	3	—		—
Rickettsia prowazekii	RP858 (635) RP303 (293)	—	—	—	—	2	—		—
Bacillus subtilis	SigA (371)	SigB SigF SigG	SigH SigV SigW SigX SigY SigZ YhdM YlaC	SigL	SigD SigE SpoIIIC SpoIVCB YkoZ	18	RsbW RsbT SpoIIAB	RsbV RsbS RsbR SpoIIAA YqhA YkoB YojH YrvA	RsbU RsbX SpoIIE YvfP
Mycoplasma genitalium	MG249 (497)	—	—	—	—	1	—	—	—
Mycoplasma pneumoniae	MP485 (499)	—	—	—	—	1	—	—	—
Mycobacterium tuberculosis	Rv2703 (528) Rv2710 (323)	—	Rv0182 Rv0445 Rv0735 Rv1189 Rv1221 Rv2069 Rv3223c Rv3328c Rv3328c Rv3911	—	Rv3286c	13	Rv1364c_2 Rv3287	Rv3687c Rc1365c Rv1904 Rv0516c Rv2638 Rv0941c Rv1364_3	Rv1364c_1

Organism								
Synechocystis sp.	slr0653 (425) slr1689 (369) sll0184 (404) sll0306 (345) sll2012 (318)	slr564	slr545 sll0856 sll0687		slr1861	9	slr1659 slr1856 slr1859 slr1912 ssr1600	slr2031 slr1860 slr1983 sll1365 slr0114
Borrelia burgdorferi	BB0712 (631) BB0771 (266)	—	—	BB0450	—	3	—	—
Treponema pallidum	TP0493 (611) TP1012 (311)	TP0092	TP0111	TP0709	—	5	TP0220 TP0233 TP0540	TP0218 TP0219 TP0854
Chlamydia trachomatis	CT615 (571)	—	CT609	CT061	CT549	3	CT424 CT765	CT588
Chlamydia pneumoniae	CPn0756 (572)	—	CPn0771	CPn0362	CPn0670	3	CPn0511 CPn0909	CPn0793
Aquifex aeolicus	aq_1490 (575) aq_1452 (310)	—	aq_599	aq_1218	—	4	—	—
Thermotoga maritima	TM1451 (399)	TM0534 TM1598	—	TM0902	TM0733	4	TM1081 TM1442	TM0467

[a] No orthologs of bacterial sigma subunits have been identified in archaea; some similarity in DNA-binding motifs of sigma subunits and archaeal transcriptional regulators has been suggested (73).

[b] Anti-sigma factors RsbW and RsbT and their homologs are serine/threonine protein kinases (39, 61, 131).

[c] The lengths of the deduced proteins are given in parentheses.

[d] —, this protein has not been found in this particular genome. See footnotes to Table 1 for more details.

tholog of the nitrogen-regulating sigma subunit RpoN of *E. coli*, for example, is required for the biogenesis of the flagellum and the stalk and for normal cell division in *Caulobacter crescentus* (20). Thus, any conclusions based solely on sequence comparisons of these sigma subunits should be considered tentative. That said, the distribution and the sheer number of different sigma subunits encoded in the genomes of *B. subtilis, M. tuberculosis,* and *Synechocystis* (Table 2) indicate extreme versatility of this regulatory mechanism, much of which remains to be uncovered.

Remarkably, the sigma subunits responsible for the stress response in *E. coli* (RpoS) and *B. subtilis* (SigB) are only distantly related and are regulated by analogous but distinct mechanisms (see chapter 11 and references 39 and 53). Thus, the mechanism for stress-response regulation through proteolytic degradation of the RpoS that has been studied in detail in *E. coli* (see chapter 11 and reference 52) is probably limited to proteobacteria or even gammaproteobacteria only. This mechanism involves the control of the RpoS proteolysis by ClpX/ClpP, mediated by binding to RpoS of the phosphorylated form of the response regulator RssB (also referred to as Hnr and SprE in *E. coli* [106] and MviA in *Salmonella enterica* serovar Typhimurium). RssB is a likely part of a still uncharacterized two-component regulatory system and is phosphorylated on its Asp-58 residue (17). Apparent orthologs of RssB so far have been found only in *Erwinia carotovora* (ExpM) and in unfinished genome sequences of several γ1 and γ2 proteobacteria that belong to the enterobacteria, vibrios, and pseudomonads. Accordingly, the crucial Lys-173 residue of the *E. coli* RpoS that apparently serves as the target for RssB binding (12) is substituted by Glu in *E. coli* RpoD and in all RpoS-like sigma subunits outside of gamma-proteobacteria (M. Y. Galperin, unpublished observations). The second mechanism of regulation of the cellular levels of RpoS that involves stimulation of the *rpoS* gene transcription by the stringent response factor ppGpp (74) appears to be more widespread since orthologs of the *spoT* whose product plays a key role ppGpp turnover (reference 22, see also below) are found in most bacteria, with the exception of several obligate parasites (Table 3).

The Anti-Sigma Regulatory System

The stress response sigma subunit of *B. subtilis*, σ^B, is additionally regulated by the complex anti-sigma-factor system that consists of two paralogous modules, RsbX-RsbS-RsbT-RsbR and RsbU-RsbV-RsbW (see chapter 12). RsbW is a sigma factor antagonist that binds σ^B and prevents σ^B incorporation into the RNA polymerase holoenzyme (60). This binding is regulated through the phosphorylation and dephosphorylation of RsbS and RsbV (39). The level of phosphorylation of each of these proteins is tightly controlled through the interplay between the respective switch kinase RsbT/RsbW and serine phosphatase RsbX/RsbU (1). The RsbX-RsbS-RsbT-RsbR module is primarily responsible for the response to environmental stress signals whereas the RsbU-RsbV-RsbW module responds to intracellular energy stress (1, 39). Interestingly, database searches with the RsbS/RsbV/RsbR sequences reveal in *B. subtilis* at least four additional proteins that, similarly to RsbR, contain the sigma-antagonist domain (which in RsbS /RsbV comprises the entire protein) combined with other domains (Table 2). Of particular interest is the YtvA protein that, in addition to the antagonist domain, contains a PAS domain that is likely to be involved in sensing environmental stimuli, such as light, oxygen, and changes in redox potential (105). The other three uncharacterized proteins share a conserved N-terminal domain with RsbR; this is likely to be a new sensor domain.

Cross-genome comparisons show that orthologs of these anti-sigma system components are encoded in the genomes of such phylogenetically diverse bacteria as *M. tuberculosis, Synechocystis* sp., chlamydiae, *T. pallidum,* and *T. maritima* (Table 2). Furthermore, *M. tuberculosis* and *Synechocystis* possess multiple proteins containing the RsbV domain; direct orthologous relationships between these proteins are difficult to establish. Most of these are small, single-domain proteins, but several contain additional domains likely to be involved in signaling. For example, the serine phosphatase RsbU homologs in *Synechocystis* (slr1860) and both chlamydiae (CT588 and CPn0793 [Table 2]) contain long N-terminal extensions, similar to ones found in N-terminal regions of *B. subtilis* methyl-accepting proteins. These extensions contain two membrane-spanning helices, probable extracellular (periplasmic) regions, and regulatory HAMP (for <u>h</u>istidine kinases, <u>a</u>denylyl cyclases, and <u>m</u>ethyl-accepting <u>p</u>roteins [7]) domains just next to the phosphatase domains. Another RsbU homolog in *Synechocystis*, slr1983, contains in its N-terminal region a CheY-like receiver domain of the two-component system response regulators. Interestingly, the *M. tuberculosis* protein Rv1364c contains all three domains of the anti-sigma system fused within a single large protein. Of further interest is the presence of a single homolog of RsbV in both *E. coli* and *H. influenzae*. This suggests that while the *Proteobacteria* do not have the anti-sigma regulatory circuit as such, they might share some aspects of stress-related signaling with those bacteria that possess this system. Conversely, in *T. pallidum*, there are three *rsbV* homologs and three predicted phosphatases homologous to RsbU but no readily detectable switch kinase. In this case, a non-orthologous displacement with a distantly related or unrelated kinase seems likely.

In *B. subtilis*, the components of the anti-sigma regulatory system form a large operon that also includes the *sigB* gene encoding σ^B itself: *rsbRSTUVW-sigB-rsbX*. In *Synechocystis*, *M. tuberculosis*, and *T. pallidum*, apparent remnants of this operon are seen, whereas in the chlamydiae and in *T. maritima* there is no clustering of the genes encoding this system (Table 2). This situation is reminiscent of the heat shock operon case (see above) and suggests similar evolutionary scenarios that could involve either an ancestral operon organization followed by partial or complete disruption or a relatively late emergence of the operon followed by lateral dissemination and rearrangements.

Taken together, the observations on the organization and phylogenetic distribution of the anti-sigma-factor system indicate that, elaborate as this control mechanism is, it is an ancient and perhaps ancestral bacterial feature. The evolution of this system seems to have involved remarkable diversification, primarily through lineage-specific gene duplication, probably resulting in the corresponding diversification of the response to stress signals.

CONSERVATION OF PROTEOLYTIC SYSTEMS INVOLVED IN STRESS RESPONSE

Proteolysis is the ultimate fate of cellular proteins that have been damaged by stress conditions. This degradation not only clears the potentially deleterious aggregates of unfolded proteins that are formed in a cell but also recycles the amino acids for new polymer chain synthesis. The microbial proteolytic machinery consists of several ATP-dependent proteases that act in the cytoplasm (43) and of ATP-independent forms that are additionally active at the cell surface or in the periplasm.

Recently, a distinct system has been discovered that targets for degradation nascent polypeptides whose synthesis could not be completed due to message damage. This targeting is brought about by an unusual ribonucleoprotein that is composed of 10S RNA *ssrA* and the RNA-binding protein SmpB (62). The RNA simultaneously acts both as an mRNA and a tRNA to add an 11-residue tag to the C terminus of the polypeptide (66). Peptides with this tag are destroyed by C-terminal-specific proteases that include the Clp proteases and FtsH (44, 54). The ubiq-

uity of this targeting system in all bacteria is made clear by the presence of the SmpB orthologs in all the bacteria with completely sequenced genomes (Table 3). Thus, this targeting system appears to be an ancient system that was present even in the common ancestor of all the bacteria. Remarkably, the gene encoding the SmpB ortholog from *T. maritima* (TM0254) contains a frameshift, resulting in the loss of 25 C-terminal amino acid residues. This frameshift has been reported to be authentic (GenBank entry AE001708), and SmpB has been removed from the list of *T. maritima* protein products (91). The *T. maritima ssrA* gene coding for the tmRNA (129) also has not been recognized (91). It seems likely, however, that *T. maritima* SmpB protein is expressed through the programmed reading frameshift mechanism, previously reported in *Thermotoga* (110, 132). In both cases, this slippage apparently occurs at the stretches of A nucleotides, 10 in the case of *tpi-pgk* gene fusion (110) and 6 in the case of *smpB*. The programmed reading frameshift mechanism would allow increased expression of full-length SmpB at elevated temperatures, satisfying the need for increased proteolysis. Existence of this or any other mechanism of SmpB regulation in *Thermotoga* certainly deserves an experimental verification.

Proteolytic systems include some of the most conserved proteins in the bacterial domain. One wing of these systems includes proteins that possess ATPase activity and act as chaperones to the damaged proteins by keeping them from forming aggregates and making them accessible to the proteases. The other wing consists of the various families of proteases that actually degrade the polypeptides (Table 3). Most of these ATPases belong to the AAA+ (ATPases Associated with a variety of cellular Activities) superfamily that contains a core P-loop ATP-binding fold with a characteristic C-terminal α-helical bundle involved in protein-protein interactions (92). Most of these ATPases form a ring-shaped quaternary structure with six or more subunits that acts as a molecular machine with cooperative ATP hydrolysis (30). The proteins like ClpA/B, ClpX, and HslU are standalone ATPases that contain one or two AAA+ modules per polypeptide and associate with separate protease subunits—ClpP—that also form a ring-like structure (126). As shown in Table 3, these are highly conserved in all the bacterial lineages with some cases of gene loss in bacteria with very small genomes, such as the mycoplasmas (loss of ClpP).

There are other members of the AAA+ superfamily that combine the ATPase domain with protease domains in the same polypeptide. One of these protease domains is the Lon protease domain that is present at the C terminus of the AAA+ module in

Table 3. Conservation of proteolytic systems involved in stress response

Organism	tmRNA-binding protein SmpB	ATPase components with or without fused protease domain							Proteases without fused ATPase domains				Other Lon-like proteases
		ClpA	ClpX	HslU (ClpY)	Lon	YifB	FtsH	Sms	ClpP family	HtrA/Do serine protease	Tail-specific protease	S2P protease	
Bacteria													
Escherichia coli	SmpB	ClpA ClpB	ClpX	HslU	Lon	YifB	FtsH	Sms	ClpP SppA SohB	DegQ HtrA DegS	Prc	YaeL	LonB[a]
Haemophilus influenzae	HI0981	HI0859	HI0715	HI0497	HI0462	HI1117	HI1335 HI1465	HI1597	HI0714 HI1541 HI1682	HI1259 HI0945	HI1668	HI0918	HI1324
Helicobacter pylori	HP1444	HP0033 HP0264	HP1374	HP0516	HP1379	HP0792	HP1069 HP0286	HP0223	HP0794 HP1435	HP1019	HP1350	HP0575 HP0980 HP0258	—[c]
Rickettsia prowazekii	RP430	RP036	RP692	RP320	RP450	—	RP043	RP546	RP520 RP525 RP398	RP124 RP186	RP228	RP161	—
Bacillus subtilis	YvaI	ClpC ClpE	ClpX	ClpY	LonA LonB*[a]	—	FtsH YjoB	Sms	ClpP YteI YmfB YqeZ	YkdA YyxA YvtB	YvjB CtpA	SpoIVFB YwhC YydH YluC	YlbL
Mycoplasma genitalium *Mycoplasma pneumoniae*	MG059 MP081	MG355 MP311	— —	— —	MG239 MP505	— —	MG457 MP171	— —	— —	— —	— —	— —	— —
Mycobacterium tuberculosis	Rv3100c	Rv3596c Rv0384c	Rv2457c	—	—	Rv2897c	Rv3610c	Rv3585	Rv2460c Rv2461c Rv0724	Rv1223 Rv0983 Rv0125 Rv3671c Rv1043c	—	Rv2625c Rv0359 Rv2869c	Rv3194c
Synechocystis sp.	slr1639	sll0020 slr1641 slr0156	sll0535	—	—	slr0904	slr1604 slr0228 slr1390 sll1463	slr0448	sll0534 slr0542 slr0165 slr0164 slr0021 sll1703	slr1204 sll1427 sll1679	slr1751 slr0257 slr0008	sll0528 sll0862 slr0643 slr1821	—

Organism													
Borrelia burgdorferi	BB0033	BB0834 BB0369	BB0612	BB0295	BB0613 BB0253	BB0086	BB0789	—	BB0611 BB0757	BB0104	BB0359	BB0118	—
Treponema pallidum	TP0184	TP0508	—	—	TP0524 TP0016*	—	TP0765 TP0330	TP1022	TP0507 TP1041 TP0997	TP0841 TP0773 TP0546	TP0277	TP0600	—
Chlamydia trachomatis	CT076	CT705	—	—	CT344	—	CT841	CT298	CT706 CT431 CT494	CT823	CT441 CT858	CT072	—
Chlamydia pneumoniae	CPn0337	CPn0144 CPn0437	CPn0846	—	CPn0027	—	CPn0998	CPn0053	CPn0847 CPn0520 CPn0613	CPn0979	CPn0555 CPn1016	CPn0344	—
Aquifex aeolicus	aq_287	aq_1296 aq_1672	aq_1337	aq_192	aq_242	aq_291	aq_936	aq_552	aq_1339 aq_1820 aq_814 aq_2080	aq_1450	aq_797	aq_1853 aq_1964	—
Thermotoga maritima	TM0254[b]	TM1391 TM0198	TM0146	TM0522	TM1633 TM1869*	TM0513	TM0580	TM0199	TM0695 TM0916	TM0571	TM0747	TM0890	—
Archaea													
Methanococcus jannaschii	—	—	—	—	MJ1417*	—	—	—	MJ0651 MJ0137 MJ1495	—	—	MJ0392 MJ0611 MJ0971	MJ1318
Methanobacterium thermoautotrophicum	MTH284	—	—	—	MTH785*	—	—	—	MTH806	MTH1813	—	MTH816 MTH986 MTH1368	—
Archaeoglobus fulgidus	—	—	—	—	AF0364*	—	—	—	AF0856 AF1781	—	—	AF0332 AF0053 AF1322	AF0705
Pyrococcus horikoshii	—	—	—	—	PH0452*	—	—	—	PH1510 PH1569 PH0282	—	—	PH0351 PH0256	PH1442
Aeropyrum pernix	—	—	—	—	—	—	—	—	APE2151 APE2212	—	—	APE1967 APE0915 APE0209	APE2279 APE0428

[a] Lon proteases marked with an asterisk comprise the second family of Lon-type proteases having both an ATPase and a protease domain; *E. coli* LonB and its ortholog from *H. influenzae*, while more closely related to the Lon protease of type II (marked with asterisks), show a truncation of the N-terminal ATPase domain.

[b] The *smpB* gene of *T. maritima* (TM0254) contains a frameshift and is probably expressed through the programmed reading frameshift mechanism. See text for more details.

[c] This protein has not been found in this particular genome. See footnotes to Table 1 for more details.

two distinct forms of the Lon-related ATP-dependent protease (Table 3). We discovered a third combination of the Lon protease domain with an AAA+ module (typified by YifB) wherein the protease domain is to the N-terminus of the AAA+ module. The presence of at least one protein with a combination of these two modules in every bacterial lineage studied to date suggests absolute requirement of the Lon-type protease for protein degradation in bacteria. Another such composite protein is FtsH wherein the AAA+ module is combined with a zinc-dependent metalloprotease domain at the C-terminus. FtsH has been shown to complement the Clp system in degradation of C-terminal tagged proteins and is thus far ubiquitous in the bacterial world, with some organisms like *Synechocystis* showing multiple copies of this protein.

The proteases that do not contain an ATPase domain may function either in conjunction with such domains or independently of them. There are several distinct protease families that appear to have analogous roles in different cellular compartments. The ClpP family has representatives in almost all the bacteria and includes ATP-dependent members (ClpP) as well as ATP-independent members such as SohB that possibly functions in the periplasm and SppA (protease VI) (10). While these proteases have an active serine, they are interestingly members of a fold that includes nonprotease members such as enoyl dehydratases (89). The bacterial trypsin-like protease family whose archetype is HtrA is also represented by one or more copies in each of the bacterial genomes with the exception of the mycoplasmas (77, 101, 104). These proteases recognize the C-terminal tail of their substrates via the PDZ (for postsynaptic density protein, disks large, zona occludens) domain (104) and degrade proteins in the periplasm in response to heat shock. Another family of C-terminal proteases—the Prc family—also recognizes substrates using the PDZ domains; however, instead of the serine protease domain, they have the IRBP catalytic domain (65, 101). These proteases are present in all bacteria except for the mycoplasmas and *M. tuberculosis*, which, interestingly, has an expansion of the HtrA-like proteases. We identified two novel families of proteases containing PDZ domains that could act as additional C-terminal proteases on the cell surface or in the periplasm. One of these, the highly conserved S2P family, includes transmembrane metalloproteases such as the *B. subtilis* protease SpoIVFB (Table 3). The other family, represented by YlbL and Rv3194c, combines the PDZ domain with the Lon protease domain, suggesting an analogous mode of action to the HtrA-like proteases. It is worth mentioning that the study of PDZ domains presents not only a number of insights into mechanisms of stress-related protein degradation but also a cautionary tale. "PDZ-like" domains have been described as protein-binding specificity determinants in the Clp/HSP100 family of proteases (75). A more detailed analysis has shown, however, that this portion of the Clp/HSP100 is a conserved part of the ATPase domain itself and, in structural and evolutionary terms, is unrelated to PDZ (92).

Eukaryotes have acquired the orthologs of all these ATPases and proteases from their bacterial endosymbionts, and many of them are known to function in the organelles of bacterial origin (data not shown). The archaea, on the contrary, do not appear to conserve most of these systems, as they have alternative systems, one of which includes the "eukaryote" type proteasomal complex with several distinct AAA+ ATPases. In spite of this, some of the bacterial systems appear to have entered the archaea by way of horizontal transfer, such as HtrA and the ClpA-like ATPase in *M. thermoautotrophicum*. In addition to these cases of horizontal transfer, archaea possess distinct homologs of the conserved bacterial proteins such as ClpP. All the *Euryarchaea* studied possess a class II Lon protease, suggesting that it participates in a conserved activity similar to the bacterial versions. Further, both *Euryarchaea* and *Crenarchaea* possess a large conserved protein with a stand-alone Lon protease domain, e.g., MJ1318, which may represent a hitherto unrecognized proteolytic system of the archaeal domain (Table 3). The *Crenarchaea Sulfolobus* and *Thermoplasma* possess a large multicatalytic protease—the tricorn protease—that forms the basis of a prominent degradative system in these organisms (120). The tricorn protease is an ortholog of the bacterial tail-specific proteases of the Prc family but has a TolB-like β-propeller domain in addition to the PDZ domain at the N terminus. This β-propeller domain may help in either organizing it into the large icosahedral quaternary structure that it assumes in cells or interacting with other proteases (119, 125).

SOS REGULON

The mechanisms of cellular responses to DNA damage are covered in chapter 9 and will not be discussed here in any detail. Besides, a general overview of such systems and the phylogenetic distribution of conserved protein domains in them has been published recently (8). We therefore concentrated here on tracing phylogenetic distribution of the proteins of the SOS response system, the best-studied response mechanism in *E. coli* (38, 107, 114, 118), limiting ourselves to the proteins that are known to be expressed from LexA-dependent promoters (38). The results of such comparison (Table 4) show that while most components of the SOS response system are well conserved in bacteria, they are largely absent in archaea, with the sole exception of *M. thermoautotrophicum*. In *M. thermoautotrophicum*, *uvrB* and *uvrA* genes form an operon, just as they do in *B. subtilis* and *B. burgdorferi*, but not in the other bacteria (Table 4). Moreover, in *M. thermoautotrophicum*, the *uvrBA* operon is directly preceded by the *uvrC* gene, the situation not seen so far in any bacterial genome. This observation strongly suggests the idea that *uvr* genes in *M. thermoautotrophicum* have been acquired relatively recently by horizontal gene transfer.

It should be noted that the conservation in bacteria of such systems as excinuclease UvrABC, Holliday junction DNA helicase RuvAB, cell division protein FtsK, and even the RecA protein itself does not necessarily indicate the presence of the SOS response mechanism in all bacteria. In fact, the LexA protein that regulates transcription of all the genes of the SOS regulon in *E. coli* is present only in bacteria with relatively large genomes, such as *E. coli*, *B. subtilis*, *Synechocystis*, and *M. tuberculosis* (Table 4). The hyperthermophile *T. maritima* and *H. influenzae* are the only two bacteria with genomes smaller than 2 Mb that still code for this protein. To complicate the situation even further, recognition sequence even for orthologous transcription regulators can differ substantially (85). This situation shows the characteristic limitation of the cross-genome comparative analysis: orthologous genes in different genomes are not necessarily regulated by the same signals. The conserved presence of LexA in *B. subtilis*, *Synechocystis*, and *M. tuberculosis* is, however, an indication that some forms of SOS response are likely to be present in these very diverse bacteria.

UNIVERSAL STRESS PROTEIN

The product of the *uspA* gene has been identified as one induced by a variety of stress conditions, such as carbon, nitrogen, and phosphate starvation; oxidative stress; acid shock; and addition of antibiotics (93–95), and designated universal stress protein A. Its overexpression has been shown to alter the cellular protein patterns seen by two-dimensional gel electrophoresis and to increase the negative charge on certain proteins (96), possibly by means of phosphorylation. In addition to *uspA*, the *E. coli* genome contains two more paralogs of this gene, *yiiT* and *yecG*. Close homologs of UspA are easily recognized in the genomes of *H. influenzae*, *B. subtilis*, Synecho-

cystis, and *M. thermoautotrophicum*. Furthermore, a careful analysis of this protein family identifies additional homologs of UspA, encoded in every archaeal genome (81). Since the *E. coli* UspA protein has been reported to have autophosphorylation activity (36), we assume that the proteins of this family are universal nucleotide-binding regulators in prokaryotic cells. Indeed, the structure of one of such protein, solved recently at 1.7-Å resolution, contained bound ATP, indicating that this protein functions as an ATP-regulated molecular switch (134). UspA thus appears to deserve the designation of the "universal stress protein" also in this respect.

Phosphorylation of UspA in vivo has been recently reported to depend on the tyrosine phosphorylation of the product of the *E. coli yihK* gene, also referred to as *o591* (36), *bipA* (29), and *typA* (37). This protein, which is remarkably similar to the elongation factor G, is most likely an autophosphorylating GTPase (37). Close homologs of TypA are present in several bacteria, including *H. influenzae*, *B. subtilis*, *Synechocystis*, *H. pylori*, *R. prowazekii*, and *M. tuberculosis*. The combined presence of *uspA* and *typA* in the first three bacteria suggests that at least these organisms probably possess an *E. coli*-like mechanism of stress response, mediated by UspA.

OTHER STRESS-RESPONSE SYSTEMS

Table 5 shows the phylogenetic distributions of the genes that have been shown to be involved in responses to particular types of stress, such as amino acid starvation, oxidative stress, acid shock, osmotic shock, and addition of antibiotics. While the mechanisms of these responses are discussed in detail in many recent reviews (53, 78, 117) and elsewhere in this book, the phylogenetic distribution has been examined only for a few stress-response systems, such as cold shock response (45, 103, 123), mechanosensitive ion channels (76), and some others. Whenever phylogeny of stress-response systems has been studied, it was able to provide interesting insights into organization and evolution of these systems. Unfortunately, in early studies of the protein patterns induced by a particular kind of stress, identification of stress-induced genes using transcriptional fusions or their products using two-dimensional gel electrophoresis (11, 82) was seldom followed up by DNA or peptide sequencing (4). This makes unequivocal assignment of the stress-responsive genes, identified in those studies, to the complete genome of *E. coli* almost impossible. Here, we have used the lists of better-characterized stress-induced genes, compiled in several comprehensive reviews (13, 14, 79, 117),

Table 4. Conservation of SOS regulon genes in microbial genomes

Organism	Orthologs of *Escherichia coli* SOS-induced genes in other complete genomes											
	lexA	recA	recN	umuC	uvrB	uvrA	uvrD	ruvA	ruvB	polB	dinG	ftsK[a]
Bacteria												
Haemophilus influenzae	HI0749	HI0600	HI0070	—[b]	HI1247	HI0249	HI1188 HI0649	**HI0313**[c]	**HI0312**	—	HI0387	HI1592
Helicobacter pylori	—	HP0153	HP1393	—	HP1114	HP0705	HP1478 HP0911	HP0883	HP1059	—	—	HP1090 HP0066
Rickettsia prowazekii	—	RP761	RP182	—	RP203	RP835	RP447	**RP385**	**RP386**	—	—	RP823
Bacillus subtilis	*lexA*	*recA*	*recN*	*yqjW* *uvrX* *yqjH* *yukA*	*uvrB*	*uvrA*	*yerF* *yjcD*	*ruvA*	*ruvB*	—	*ypvA* *dinG*	SpoIIIE YtpT *yukA*
Mycoplasma genitalium	—	MG339	—	MG360	MG073	MG421	MG244	**MG358**	**MG359**	—	—	—
Mycoplasma pneumoniae	—	MP351	—	MP305	MP621	MP223	MP496 MP497	**MP307**	**MP306**	—	—	—
Mycobacterium tuberculosis	Rv2720	Rv2737c	Rv1696	Rv1537 Rv3056	Rv1633	Rv1638	Rv0949 Rv3201c Rv3198c Rv3202c	**Rv2593c**	**Rv2592c**	—	Rv1329c	Rv2748c Rv3870 Rv3894c Rv1784 Rv3447c Rv0284
Synechocystis sp.	sll1626	sll0569	sll1520	slr0790	sll0459	slr1844	sll1143 sll1121	sll0876	sll0613	—	—	—
Borrelia burgdorferi	—	BB0131	—	—	**BB0836**	**BB0837**	BB0344 BB0607	**BB0023**	**BB0022**	—	—	BB0257
Treponema pallidum	—	TP0692	TP0442	—	TP0116	TP0514	TP0102 TP1028	TP0543	TP0162	—	—	TP0999
Chlamydia trachomatis	—	CT650	—	—	CT586	CT333	CT608	CT501	CT040	—	—	CT739
Chlamydia pneumoniae	—	CPn0762	—	—	CPn0801	CPn0096	CPn0772	CPn0620	CPn0390	—	—	CPn0880
Aquifex aeolicus	—	aq_2150	aq_561	—	aq_1856	aq_686	aq_793	—	—	—	aq_358	—
Thermotoga maritima	TM1082	TM1859	—	—	TM1761	TM0480	TM1238	TM0165	TM1730	—	—	—
Archaea												
Methanococcus jannaschii	—	MJ0254 MJ0869	—	—	—	—	—	—	—	MJ0885	MJ0942	—
Methanobacterium thermoautotrophicum	—	MTH1383 MTH1693	—	—	**MTH442**	**MTH443**	MTH472 MTH511	—	—	MTH1208 +MTH208[d]	MTH1347 MTH1347	—
Archaeoglobus fulgidus	—	AF0351 AF2096	—	—	—	—	—	—	—	AF0497	—	—
Pyrococcus horikoshii	—	PH0119 PH0263	—	—	—	—	—	—	—	PH1947	PH0697	—
Aeropyrum pernix	—	PH0263 APE0119	—	—	—	—	—	—	—	APE2098 APE2098 APE0099 APE2229	APE0322	—

[a] *E. coli* proteins UmuD, DinD, DinI, SulA, and the N-terminal 800 amino acid residues of FtsK (previously referred to as DinH) do not have orthologs in any of the other sequenced genomes; orthologs of the C-terminal part of FtsK are indicated by boldface type.

[b] This protein has not been found in this particular genome. See footnotes to Table 1 for more details.

[c] Adjacent genes, presumed but not always proven to comprise an operon, are indicated by boldface type.

[d] These two distinct proteins of *M. thermoautotrophicum* together correspond to the polymerase of other *Archaea* and probably form a noncovalent complex.

Table 5. Conservation of stress-activated genes in bacteria and archaea

Stress response genes in E. coli or B. subtilis	Presence of homologs of E. coli or B. subtilis stress-activated genes in the complete genomes (no. of paralogs)[a]																
	E. coli	Hinf	Hpyl	Rpro	Bsub	Mgen/Mpne	Mtub	Bbur	Tpal	Ctra/Cpne	Syn	Aaeo	Tmar	Mjan	Mthe	Aful	Phor
Stringent response																	
spoT	2	2	1	—[b]	1	1	1	1	—	—	1	1	1	—	—	—	—
ppGpp-activated																	
himA	4	3	1	3	2	1	1	1	1	1	1	2	1	—	—	—	—
bmp	2*	1*	—	—	1	—	5*	—	—	1*	1*	1	—	—	—	—	—
BS_yvyD	2	1	—	1	1	—	1	1	—	0/1	1	1	1	—	—	—	—
Oxidative stress																	
SoxR/SoxS regulon																	
sodA	2	1	1	1	2	—	1	1	—	1	1	1	—	—	1	—	—
fur	2	1	1	—	3	1	2	1	—	—	3	2	3	—	—	1	—
fpr	2	—	—	—	—	—	2	—	—	—	—	—	—	—	—	—	—
fldA	3	1	1	1	2	—	—	—	1	—	5	—	2	1	2	1	1
acrA	7	1	4	1	2	—	—	1	1	—	5	7	2	2	—	1	—
acrB	7	1	3	1	1	—	—	1	—	1	5	4	1	—	—	—	—
nfo	1	—	—	—	1	1	—	—	—	—	—	1	1	1	1	1	1
acnA	1	—	1	1	1	—	1	—	—	1	1	1	—	—	—	—	—
fumC	1	1	1	1	1	—	1	—	1	1	—	1	1	—	—	—	—
zwf	1	1	1	1	1	—	2	—	1	1	—	1	1	—	—	—	—
ribA	1	1	1	—	1	—	1	—	—	1	1	1	1	—	—	1	—
OxyR regulon																	
katG	1	—	—	—	—	—	1	—	—	—	1	—	—	—	—	—	—
ahpC	1	—	2	1	2	—	2	—	1	1	2	2	1	1	1	1	1
ahpF	2	1	2	2	4	1	1	1	1	1	1	2	2	1	1	1	1
grxA	4	2	1	1	1	—	1	—	1	—	2	—	1	—	1	1	1
dps	1	1	1	1	2	—	—	1	1	—	1	—	—	—	—	—	—
gorA	4	2	—	2	3	1	5	—	—	1	3	1	1	1	1	1	—
Other																	
katE	1	1	2	—	3	—	—	—	—	—	—	—	—	—	—	—	—
sodC	1	—	—	—	1	—	1	—	—	—	2	2	—	—	—	—	—
xthA	1	1	1	2	1	—	1	1	1	—	1	—	—	—	1	1	—

Continued on following page

Table 5. *Continued*

Stress response genes in *E. coli* or *B. subtilis*	Presence of homologs of *E. coli* or *B. subtilis* stress-activated genes in the complete genomes (no. of paralogs)[a]																
	E. coli	*Hinf*	*Hpyl*	*Rpro*	*Bsub*	*Mgen/Mpne*	*Mtub*	*Bbur*	*Tpal*	*Ctra/Cpne*	*Syn*	*Aaeo*	*Tmar*	*Mjan*	*Mthe*	*Aful*	*Phor*
Acid shock																	
gadA	2	—	—	—	—	—	1	—	—	—	1	—	—	1	1	3	1
aceA	1	—	—	—	1	—	2	—	—	—	—	—	—	—	—	—	—
ldcC	5	1	—	—	2	—	1	—	—	—	1	—	—	—	—	—	—
manX	2	—	—	—	1	1	—	1	—	—	—	—	—	—	—	—	—
gatY	4	1	1	—	2	1	1	—	1	—	1	1	1	—	—	—	—
osmY	2	1	—	1	1	—	1	—	—	—	—	—	—	—	—	—	—
yaiQ	1	1	—	—	1	—	1	—	—	—	—	—	—	—	—	—	—
Osmotic Shock																	
osmC	1	—	—	—	3	1	—	—	—	—	—	—	—	—	—	—	—
mscL	1	1	—	—	1	—	1	1	—	—	—	—	—	—	—	—	—
yggB	6	1	3	1	3	—	2	1	1	—	8	2	1	3	1	2	1
kdpD_1	1	—	—	—	—	—	1	—	—	—	1	—	—	—	—	—	1
trkA	1	1	—	—	—	—	2	—	1	—	1	1	1	1	1	1	1
trkH	2	1	—	—	2	1	—	1	1	—	1	1	1	1	1	1	1

[a] The genes for which no orthologs have been found in any other genomes (*tolC*, *inaA*, *yciE*, etc.) were omitted from the table. The asterisks indicates genes whose products are homologous only to the C-terminal flavodoxin reductase domain of the Hmp protein. Abbreviations: *Hinf*, *H. influenzae*; *Hpyl*, *H. pylori*; *Rpro*, *R. prowazekii*; *Bsub*, *B. subtilis*; *Mgen*/*Mpne*, *M. genitalium*/*M. pneumoniae*; *Mtub*, *M. tuberculosis*; *Bbur*, *B. burgdorferi*; *Tpal*, *T. pallidum*; *Ctra*/*Cpne*, *C. trachomatis*/*C. pneumoniae*; *Syn*, *Synechocystis* sp.; *Aaeo*, *A. aeolicus*; *Tmar*, *T. maritima*; *Mjan*, *M. jannaschii*; *Mthe*, *M. thermoautotrophicum*; *Aful*, *A. fulgidus*; *Phor*, *P. horikoshii*.

and compared their phylogenetic distribution using the COG approach. The results of such comparison (Table 5) show several common trends. First, parasitic organisms with smaller genomes usually have fewer stress-response systems than their free-living relatives. Indeed, such human pathogens as *R. prowazekii*, mycoplasmas, or chlamydiae are unlikely to encounter acid or osmotic shock, which may explain virtual absence of these response systems in those parasites. Second, few stress-response systems are present in representatives of both *Bacteria* and *Archaea*, and those that are universal are not necessarily stress-induced in all organisms. It appears that archaea, at least the ones whose genomes are available at this time, are devoid of most stress-response systems that have been found in bacteria. This could be due to the relatively small size of archaeal genomes (assuming the loss of ancestral stress response systems in modern archaeal branches) and/or to the presence of still unidentified archaea-specific systems of shock response. Third, the distribution of the stress-response systems in bacteria shows surprising diversity. The sets of stress proteins in closely related bacteria, such as *M. genitalium* and *M. pneumoniae*, *C. trachomatis* and *C. pneumoniae*, or the two strains of *H. pylori* (not shown), are virtually identical, with the sole exception of the stress-induced *B. subtilis* protein YvyB (28), found in *C. pneumoniae* but not *C. trachomatis* (Table 5). The distribution of stress-related genes in more distantly related bacteria, however, is increasingly patchy and does not seem to correlate with the phylogenetic affinities of the respective organisms. We believe that this patchy distribution of important functions stems from extensive horizontal gene transfer between various branches of the phylogenetic tree. A comparison of the stress response systems found in *A. aeolicus* and *T. maritima*, representatives of ostensibly the two most ancient branches of the bacterial tree, appears to support this idea (Table 5). The difference in abundance of oxidative stress-related proteins between these two organisms can be, of course, attributed to the microaerophily of the former and the obligate anaerobic nature of the latter.

INSIGHTS INTO STRUCTURE AND FUNCTIONS OF STRESS-RESPONSE SYSTEM COMPONENTS THROUGH COMPUTATIONAL ANALYSIS OF PROTEIN DOMAINS

Detailed analysis of protein sequences using recently developed powerful methods, particularly those based on the iterative use of position-specific scoring matrices (2), systematically results in the prediction of the structure and function of previously uncharacterized protein domains (6, 8). A critical aspect of this analysis is the comparison of domain architectures of multidomain proteins that frequently yields completely unexpected insights into their functions and evolution. Specifically, the analysis of the proteins that comprise the stress-response systems in bacteria and archaea has produced a number of functional predictions that await experimental verification and extension. Here we discuss three examples of such predictions.

SpoT/RelA

As indicated above, SpoT/RelA proteins are central components of the bacterial stringent response that are nearly universally present in bacteria and are also found in eukaryotes, probably as the result of horizontal gene transfer. SpoT and RelA are large, highly conserved proteins; until recently, comparative sequence analysis has provided no clues as to their domain organization and possible functional interactions. Our recent detailed examination of these proteins, however, has revealed a fascinatingly complex domain architecture that suggests additional, previously unsuspected functional complexity (Fig. 1). The catalytic domain of SpoT has been shown to belong to the HD superfamily of (predicted) hydrolases (so named after the principal catalytic residues) that also includes bacterial dGTPases, uridylyl transferases (e.g., GlnD), a vast variety of uncharacterized proteins, and as peripheral members, eukaryotic cAMP/cGMP phosphodiesterases (5). The RelA protein that so far has been found only in gammaproteobacteria contains a version of the HD domain in which some of the predicted catalytic residues are replaced, which correlates with the observations that this protein catalyzes only the synthesis but not (at least not at any significant rate) hydrolysis of (p)ppGpp. In addition to the catalytic HD domain, SpoT/RelA proteins contain two domains that are predicted to provide regulatory connections (Fig. 1).

Since all these domains are shared by SpoT and RelA, it appears likely that they primarily regulate the (p)ppGpp synthetase activity of these proteins. The TGS domain (after Threonyl-tRNA synthetase, GTPases, and SpoT) is shared by bacterial threonyl-tRNA synthetases, a distinct family of GTPases (the OBG family), and SpoT/RelA (130). The structure of this domain recently has become available since the structure of the *E. coli* threonyl-tRNA synthetase has been solved (109). The crystal structure suggests that the TGS domain is likely to be involved in RNA binding. Given that (p)ppGpp synthesis activity is stimu-

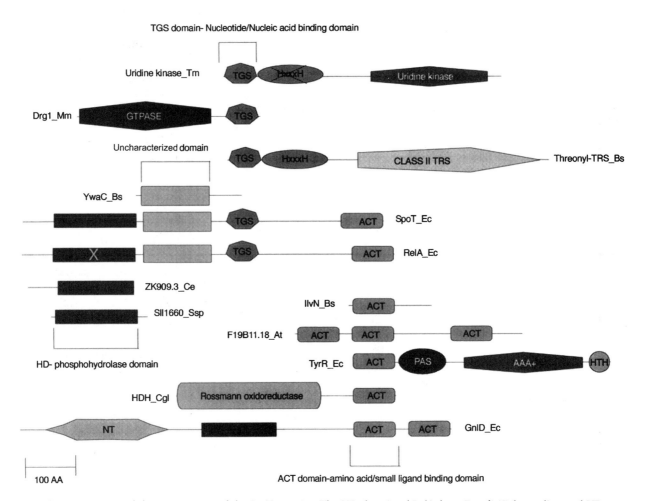

Figure 1. Conserved domain structure of the SpoT proteins. The HD domain of RelA from *E. coli* (5) has a disrupted HD sequence, which is indicated by an 'X.' Other proteins with each of the conserved domains are aligned with respect to SpoT. The following domains are shown: ACT (Aspartokinase, Chorismate mutase, TyrA) domain (6) is predicted to bind amino acids; TGS is the Threonyl-tRNA synthetase (TRS), GTPase (Drg-1), SpoT domain that is also additionally seen in the uridine kinase from certain bacteria (130); the domain denoted by "HxxxH" is a conserved domain with a characteristic histidine dyad that is seen in threonyl-tRNA synthetases, alanyl-tRNA synthetases, and some uridine kinases (130). PAS is ligand-binding and dimerization domain (105), HTH indicates a helix-turn-helix DNA-binding domain. The protein name abbreviations are as follows: Drg-1, a developmentally regulated GTPase; HDH, homoserine dehydrogenase; TyrR, transcriptional regulator of the *aro* operon; GlnD, PII protein uridylyl transferase; IlvN, acetolactate synthase small subunit. The species abbreviations are as follows: At, *Arabidopsis thaliana*; Bs, *Bacillus subtilis*; Ce, *Caenorhabditis elegans*; Cgl, *Corynebacterium glutamicum*; Ec, *Escherichia coli*; Mm, *Mus musculus*; Ssp, *Synechocystis* sp.; Tm, *Thermotoga maritima*.

lated by uncharged tRNAs, it appears likely that the TGS domain of SpoT/RelA mediates this effect through tRNA binding; the role of this domain in GTPases, however, remains unclear.

The C-terminal ACT (for acetolactate synthase, chorismate mutase, and TyrR) domain of SpoT/RelA is a versatile ligand-binding domain that is present primarily in various enzymes and regulators of amino acid metabolism and is involved in allosteric control (6). It appears most likely that in SpoT/RelA, this domain is the primary sensor for stringent control. In the presence of sufficient concentrations of nutrients, the ACT domain probably is bound to an amino acid (its specificity remains unclear; perhaps it could

be broad), which might result in allosteric inhibition of pppGpp synthesis. Under conditions of starvation, the amino acid ligand would be released, resulting in stimulation of the activity. This model suggests a new level of control for (p)ppGpp synthesis, namely direct regulation by amino acids, which to our knowledge has not been discovered so far and is open to experimental testing.

Finally, SpoT/RelA proteins contain a fourth, completely uncharacterized domain that is also represented as stand-alone proteins in *B. subtilis* and *M. tuberculosis* (Fig. 1). Thus, computer analysis of the sequences of the SpoT/RelA proteins sheds completely new light on these key stress-related regulators by identifying distinct catalytic and regulatory domains and suggesting that the regulatory effects exerted by these proteins might be even more complex and multifaceted than previously suspected.

The OsmC Family of Predicted Stress-Related Redox Proteins

The *E. coli* osmC gene and its homologs in *B. subtilis* belong to the RpoS stress regulon and are induced by different forms of stress (particularly high osmotic pressure—hence the gene name) and starvation (18). Iterative database searches indicate that OsmC family proteins are encoded in many bacterial and archaeal genomes, although the phylogenetic distribution is patchy (e.g., the family is not represented in most bacterial parasites and in some archaeal species, such as *M. thermoautotrophicum* and *A. fulgidus*); no eukaryotic homologs of these proteins were detected. Examination of the amino-acid-residue conservation pattern in the multiple alignment of the OsmC family proteins shows that its most notable feature is the presence of two nearly invariant cysteine residues (Fig. 2). This suggests that these two cysteine residues form a labile disulfide bond and, accordingly, the OsmC-like proteins function as antioxidants, which probably is the basis of their role in stress response. This idea is supported by the observation that the Ohr protein from *Xanthomonas campesteris*, an OsmC ortholog (Fig. 2), is necessary for organic hydroperoxide resistance in this organism (86). This protein is specifically induced by organic hydroperoxide and confers resistance to these compounds when expressed in other bacteria, suggesting a similar role of OsmC orthologs in stress response in other bacteria.

The PspE-Like Predicted Rhodanese-Like Enzymes in Stress Response

The *E. coli* PspE protein belongs to the phage-shock operon that is activated, in addition to the filamentous phage infection, by a variety of stress signals. The operon as such is not seen in any other bacterium, and the encoded proteins are either not conserved at all (PspB and PspD) or have detectable orthologs in only a few bacterial and archaeal species (PspA and PspC). By contrast, PSSM-based database searches with the PspE sequence have shown that this protein belongs to a large superfamily of rhodanese (sulfurtransferase)-like proteins (Fig. 3a). The superfamily is represented, frequently by more than one member, in most of the bacterial, archaeal, and eukaryotic genomes. The members of this family are expressed in a variety of stress conditions in both bac-

teria and multicellular eukaryotes. Of interest are the senescence-induced protein Sen1 of *Arabidopsis* (Din1 in *Brassica*) that is related to PspE-like small rhodanese-like proteins from *Synechocystis* (97, 112). This protein appears to be expressed in plants under dark-induced stress and functions in the chloroplasts, suggesting a similar role in the cyanobacteria. Similar small rhodanese-like proteins in eukaryotes (e.g., HSP-67B2 from *Drosophila*) could be a part of a heat stress response (102), while the Acr2 from yeast is involved in resistance to arsenate by way of reduction (87). Rhodanese itself is likely to play a role in cyanide detoxification in addition to its function in the formation of iron-sulfur complexes in metalloproteins (127).

It has been shown, both by a structural comparison and by a PSSM-aided sequence analysis, that rhodanese-like proteins are homologous to the CDC25 family of protein tyrosine phosphatases (57). Most of the proteins in the superfamily contain the (predicted) active cysteine that covalently binds the transferred anion in both rhodanese and tyrosine phosphatases; other members of the family contain an aspartate in place of this catalytic cysteine and the rhodanese-like domain of ubiquitin hydrolases contains threonine (Fig. 3A). Although it can be predicted that the family members containing the catalytic cysteine are active hydrolases or anion transferases, it remains unclear whether the homologous proteins in which this residue is substituted possess enzymatic activity. The studies on a small rhodanese-like protein from *Wolinella succinogenes* show that it acts as a periplasmic sulfide dehydrogenase and uses the same catalytic cysteine involved in anion transferase and hydrolase activity (71). This suggests a possible redox function for rhodanese-like proteins similar to that of the thioredoxin fold proteins. Thus, in stress response, the rhodanese-like proteins could show several alternative catalytic activities that include (a) detoxification of toxic compounds such as arsenate and cyanide by either transferring anions or reducing them; (b) chaperone activity to allow efficient assembly of iron-sulfur complex-containing proteins; and (c) possible dithiol-disulfide redox activity analogous to the one in *W. succinogenes*.

The rhodanese-like domains combine with a variety of other enzymatic and non-enzymatic domains to form multidomain proteins (Fig. 3B). Some of these linked domains include the parvulin-type peptidyl-prolyl isomerase domain (108) and the disulfide isomerase-like thioredoxin domain, suggesting that in these cases the rhodanese-like domain could assist in a chaperone-like function (Fig. 3B). Further, in the actinomycetes, there is a combination of the rhoda-

```
OsmC_Ec_2507088       7 GQAHWEGDIKRGKGTVST-ESGVLNQQ-PYGFNTRFEGEKG--TNPEELIGAHAACFSMALSLMLGEAG-FTPTSID
Ohr_Xc_3098342       11 AHATATGGREGRAVSSDKALDAKLST---PRE-LGGAGGDG--TNPEQLFAAGYAACFIGAMKAVAAQDK-LKLP-GE
orf2_Acni_1657239    10 AHAKATGGRDGRATSSDNILDVQLTV---PKE-MGGMGGG---TNPEQLFAAGYSACFLGAMKFVATRDK-FNIP-KD
YknA_Bs_2632036       6 AKVTARGGRAGHITSDDGVLDFDIVM---PNAKKEGQTG---TNPEQLFAAGYAACFGGALEHVAKEQN-IEID-SE
YklA_Bs_2632034       8 ATVSAVGGREGKVISSDRVLELDVAM---PGTPRAKKLEK--ATNPEQLFAAGYAACFDSALQLVARTER-VKVE-TE
MG454_Mge_1723166    22 TVAQTETGREGSVKTLDG-FQTKLSF---PKPDLSVQT----ENNPEQLFASAYACFSQAVIVVMQQHQ-FSFS-KK
MP454_Mpn_2496453     7 TTAHASAGREGVVQTVDG-FTVSLAF---PKP--GATHQD--KNNPEQLFASAYAGCFSQAVRVVLQQHQ-LQLA-TQ
YhfA_Ec_1176244       3 ARVKWVEGLTFLGESASG-HQILMDGN-----SGDK----APSPMEMVLMAAGGCSAIDVVSILQKGR-QDVVDCE
ORFX_Pa_1888564       3 ARIQWWAGEAMFLGESGSG-HVVVMDGP----PDHGGRNL----GVRPMEMVLIGLGGCTNFDVVSILKKAR-QPVESCE
PH1469_Ph_3257893    11 GKVRWVEGEQFIGRIEGD-KCSVILGE-------GG-------ISPMKLLLLSVAGCTSYDVVMILKKMR-EPIKGLE
PAB1905_Pab_5458120   9 GRVRWIEGEQFIGRIEGD-QCSVILGE-------GG-------ISPMKLLLLSVAGCTAYDVVMILEKMR-EPIKGLE
PH1621_Ph_3258050     2 QFKDMEIRVIGKAVSSTK-TLIKAGNFEIYIDKLGG-----EYPSPLDYTLAALAGCINIVGHMVAKEMG-FEINSLE
PAB2001_Pab_5457976   3 KYKDLEIRVVGKAVSPTK-TLVKTENFEIMVDKLGG-----EHPCPLDYTLAALAGCLNIVGHMVAKDMG-FNIEELE
MG427_Mge_1351579     6 DITAVLNDDSSINAVSDN-FQITLDAR----PKEKSKG-----INPLSAFLAGLAACELATANAMAAAKM-ITLNKAL
MP427_Mpe_2496429     6 DITAVLNEDSSMTAISDQ-FQITLDAR----PKHTAKG-----FGPLAALLSGLAACELATANLMAPAKM-ITINKLL
TM0919_Tma_4981456    1 MQARWIGNMMFHVRTDSN-HDVLMDTK----EEVGGKD----AAPRPLELVLTGLMCCTGMDVVSILRKMKVIDQMKDF
aq_1549_Aae_2983923   2 EVKEVELELSSEATFLSKTSIGEITA--------GEKG-----LNPMELLLVSIGSCSGVDVYHILKKKR-QEVKDIK
PH0818_Ph_3257228    30 AELTWDGNVGSRAKVREFEFSIDTNT------DGFNKG-----PNPTEYLLAALGGCLTVNWGRLIKKMR-LNVESMK
SC5C7.16_Scoe_3560006 7 AHTVWEGNLLEGNGVVTF-DSSGIGE-QPVSWPSRAEQANGK-TSPEELIAAAHSSCFSMALSHGLAGAG-TPPTKLT
YmaD_Bs_2634112       8 LKANWPGNRNDVGTIESGNLITSISI--PKEMDGPGEG-----TNPDEMLLGAAATCYIITLAAMMERSG-LEKEDLQ
MJ0573_Mj_2496050    53 IIVSKDENKEVSAEYLDM-FEALLNVKGLKIHSRGGKGAIKEKISPMDLFLAGLCGCVCIAVGNTLKANN-IDAE-IK
APE2138_Ape_5105837   2 PVIKFFEAKSLVDAEKGK---AVVE--------GGREFELA--TLNPSFMLSMLSACIGKKVAENAGVDK-VSVS-IE
APE2211_Ape_5105911  19 VTAEAVPSGDAVKVTTGGVQIDVYPER----EAGGPERG---LTPLGLLAASLASCEVLMSRLVGRMLG-YNGFDVR
AttW_Atu_5508830     16 AAMGWAGGHTVVIDRPEGKA------------GGLGLG----FNGGQMLALALGGCYCNDLRYVAHERG-VKIEQIA
aq_1515_Aae_2983898   4 KVKQKE-DFHFIGVGPAG-REVPIDAA---DYVGGKGRG----IRPPELLFHSVAGCVGIHLYEALHKEG-KHVEDIE
Rv2923c_Mtu_1723037   1 MTQLWVERTGTRRYIGRSTRGAQVLV-------GSEDVDG--VFTPGELLKIALAACSGMASDQPLARRL-GDDYQAV
consensus/85%           ....h.........s......h...................pP.phhh.uhuuC...s....h...........
```

```
OsmC_Ec_2507088       TTADVSLDKVDAGFAIT------KIALKSEVAVP--GIDASTFDGIIQK--AKAGCPVSQVLKA--EITLDYQLKS- 143
Ohr_Xc_3098342        VSIDSSVGIGQIPGGFG-------IVVELRIAVP--GMDKAELQTLVDK--AHQVCPYSNATRG---NIDVTLTLA 142
orf2_Acni_1657239     AYVEGDVGIGPIPNGFG-------IEVKLHVHLP--GMDTDEAKKLVDA--AHYVCPYSNATRN---NIDVDFEIV 140
YknA_Bs_2632036       IEGQVSLMKDESDGGFK-------IGVTLVVNTK--DLDREKAQELVNA--AHEFCPYSKATRG----NVDVKLELK 136
YklA_Bs_2632034       VTANVSLLKDEADQGYK-------LGVTLQVKGE--GVSASELEALVKK--AHGVCPYSKATSG----NIDVTLEVA 140
MG454_Mge_1723166     PVVSVKVELHQENGLFH-------IKAGVELTTN--SNDQEVGKKLIQK--AHEMCPFSRLIRN----ENFLGLTLN 151
MP454_Mpn_2496453     PIVGVSVELHDQDGLFH-------IKAGVELAIT--GVDQTTAQTVITA--AHAMCPFSRLIKP----ENFLGLTLN 136
YhfA_Ec_1176244       VKLTSERREEAPRLFTH-------INLHFIVTGR--DLKDAAVARAVDLS--AEKYCSVALMLEK--AVNITHSYEVV 132
ORFX_Pa_1888564       AFLEAERADEEPKVFTK-------IHVHFVVKGR--GLKEAQVKRAVELS--AEKYCSASIMLGR-GGVEITHDYEIV 137
PH1469_Ph_3257893     VEIEGVRREEHPRIYKE-------VTIHYKIYG---NVNEKKARRAIELS--QEKYCSASAHLKL-SGTDVKYTLEVI 137
PAB1905_Pab_5458120   VEIEGVRREEHPRIYKE-------VTIHYRIYG---NVDEKKARRAIELS--QEKYCSASAHLKL-SGTNVKYTLEII 135
PH1621_Ph_3258050     IEVNGTFNPAKFKGFDGDRAGFKSIEVTIRVDA---VRVDDETLKEWIKR--VEERCPVSDNLIN--ETPTEITVKKA 143
PAB2001_Pab_5457976   IEVTGIFNPAKFMGLDGERAGFKSVKAIIRVKA---DVDEEKLKEWLKK--VEERCPVSDNLTN--LTPTEVIVEKK 143
MG427_Mge_1351579     ININKGYRLTNPSDGYFG----LRELNIHWEIHS---PNEEEEIKEFIDF--VSKRCPAHNTLHG--TSNFKINISVT 138
MP427_Mpe_2496429     MNVTGSRSTNPTDGYFG----LREINLHWEIHS---PNSETEIKEFIDF--VSKRCPAHNTLQG--VSQLKINVNVT 138
TM0919_Tma_4981456    RIEIEYERTEEHPRIFTK------VHLKYIFKFD-GEPPKDKVEKAVQLS-QEKYCSVSAILKCSSKVTYEIVYEN- 138
aq_1549_Aae_2983923   IFLKGKRREKHPKIYEE-------IEIKYVAVG---KVEEKALEQAVKLS-TEKYCSVLAMVKP--STNLKISWEVK 129
PH0818_Ph_3257228     ITVSGWRSRDEPQLKEI-------RYKVRIVT---NEPEKKILRVKEL--AEKYGTVFNTVGK--EKIKGEVEIV 156
SC5C7.16_Scoe_3560006 TSADVTFQPGEGIKG---------IHLTVEGTVP--GLDNDAFVAAAED--AKKNCPVSQALTG---TTITLSAKLA 141
YmaD_Bs_2634112       MESEGIVNVTKGVFTYK-----KIIHRPSVVLKHDASQDDVALAHKLCKK-AESSCMISRAIQG----NVELQLEAS 144
MJ0573_Mj_2496050     VDGKVEKSFEEGKIKKV-------IINIYVKVDG---DIDKEKLKKLVLE--GSKKCLISNSISC--EIEKNVILE-- 189
APE2138_Ape_5105837   AYVNVDKLLEGKEEIEH-------IVVIRA------PASEEAVLK----GVENCPVFKLIDK--SRVKDVIVEKS 121
APE2211_Ape_5105911   VAVTADVQVAEGLRSLS--------IRYVFKG----VDIDTANLIVSK--VKELCPVYNSLVR-NGVSVEENVEVE 148
AttW_Atu_5508830      VTATLDDLEGVPLVATKA-------VLSVKCET----SDGSDPKALIDH--WASVCTVANSLNR--GIPTIIQVA-- 134
aq_1515_Aae_2983898   IETDAERITDRYPKVFT------KIYLFVKVKG-DVSEEEVKNALDKVIYNPGTCSIAYMVNQ--VAPIEYKVEIL 138
Rv2923c_Mtu_1723037   VKVSGAADRDQERYPL--------IEETMELDLSGLTEDEKERLLVINRAVELACVGRTLKS-GTTVNLEVVDVG 136
consensus/85%         ...ps....................h.h.h.......p......hhp....p..Csh..............hp..
```

Figure 2. Multiple alignment of OsmC homologs. Proteins are designated with (left to right) the protein name, species of origin, and the NCBI protein database gene identification (gi) number. Numbers that precede and follow the aligned regions refer to the numbers of amino acid positions within each protein sequence. The residues are shaded according to the consensus that is shown below the alignment. Uppercase letters on the consensus line indicate conserved amino acid residues in the single-letter code, while lowercase letters indicate conserved classes of amino acids. Inverse shading indicates the predicted active site cysteine residues; light shading includes hydrophobic residues (h) (A, C, F, I, L, M, V, W, Y), aromatic residues (a) (F, H, W, Y), and aliphatic residues (l) (I, L, V); darker shading indicates small residues (s) (A, C, D, N, G, P, S, T, V). Other conserved residues are in bold and are indicated on the consensus line as follows: p, polar residues (C, D, E, H, K, N, Q, R, S, T); e, charged residues (D, E, H, K, R); u, tiny residues (A, G, S); b, bulky residues (E, F, I, K, L, M, Q, R, W, Y). The species abbreviations are as follows: Aae, *Aquifex aeolicus*; Acni, *Acinetobacter* sp.; Ape, *Aeropyrum pernix*; Atu, *Agrobacterium tumefaciens*; Mge, *Mycoplasma genitalium*; Mp, *Mycoplasma pneumoniae*; Mj, *Methanococcus jannaschii*; Mtu, *Mycobacterium tuberculosis*; Pa, *Pseudomonas aeruginosa*; Pab, *Pyrococcus abyssi*; Ph, *Pyrococcus horikoshii*; Xc, *Xanthomonas campesteris*.

```
                              7  LALALVFSLPVFAAEHWIDVRVPE QYQQEHVQ GAINIPLKEVKERIAT 2PDKNDTVKVYC-NAGRQS-GQAKEILSEMGYT HVENA-GGLKDIAMPKVK 103
PspE_Ec_1787567
SEN1_At_2129729              72  SVPVRVARELAQAGYRYLDVRTPD--EFSIGHPT RAINVPYMRVGSGMV12FRKHDEIIIGC-ESQMS-FMASTDLLTAGFT AITDIAGGYVAWTENELP 179
KetR_At_2129628              10  SVSVTVAHDLLLAGHRYLDVRTPE EFSQGHVP1GSINVPYMNRGASGMS12FGQSDNIIVGC-QSGGRS-IKATTDLLHAGFT GVKDIVGGYSAWAKNGFL 118
T13L16.13_At_2708749          8  TIDVNQAQKLLDSGYTFLDVRTVE EFKKGHVD2NVFNVPYFTPGQGE12CNQTDHLILGC-KSIVRS-LHATKFLVSSGFR TVRNMDGGYIAWVNKRFP 117
slr0192_Ssp_1001676          10  VQDLALMLAHPNGDRQLLDVREPH EVEIAALP GFETLSLSEFAQWSGT 4YDGEKETVVLC-HHGIRS-DQMAHWLVDQGFS KVKNVVGGIDAYSRLIDA 109
PARV_At_2246380             117  ELHSKMQQDPVFMDEAQLIDVRNDY EIEIASLP GFKVFPLRQFGTWAPD 4LNPEKDTFVLC-KVGGRS-MQVANWLQSQGFK SVYNITGGIQAYSLKVDP 216
YgaP_Ec_2495651              5  TISPHDAQELIARGAKLIDVRDAD EYLREHIP EADLAPLSVLEQSGLP 1KLRRHEQIIPHC-QAGKRTSNNADKLAAIAAPA EIFLLEDGIDGWKKAGLP 102
slr1261_Ssp_1652937         11  IAPKTLQQLRQQDAVILVDVREPL EFVGEHIT DAYSLPLSRLNPSQLP QAEGKTIVLYC-QSGNRS-GNALQQLRSAGVE GIIHLEGGLLAWKQAGLP 106
HSP-67B2_Dm_123582           2  ATYEQVKDVPNHPDVYLIDVRRKE 1LQQTGFIP ASINIPLDELDKALNL13PEKQSPIIFTC-RSGNRV-LEAEKIAKSGGYS NVVIYKGGWNEWAQKEGL 111
YOR286w_Sc_1420636          35  FDQVRNLVEHPNDKKLLVDVREPK EVKDKVMP TTINIPVNSAPGALGL13PPHDKELIFLC-AKGVRA-KTAEELARSYGYE NTGIYPGGITEWLAKGGA 143
YOR285w_Sc_1279709          27  FEDMKRIVGKHDPNVVLVDVREPS EYSIVHIP ANINVPYRSHPDAFAL13PDSAKELIFLC-ASGKRG-GEAQKVASSHGYS NTSLYPGGMNDWVSHGGD 135
AC4H3.07C_Sp_1723283        62  QVYNLSKRPTGDKSTVLIDVREPD EFKQGAIE TSYNLPVGKIEEAMKL13PVFEDNVVVYC-RSGNRS-TTASDILTKLGYK NIGNYTGWLEWSDKIKS 170
yqhL_Bs_2634888             29  KTLTEEEFRAGYRKAQLIDVREPN EFEGGHIL GARNIPLSQLKQRKNE IRTDKPVYLYC-QNGVRS-GRAAQTLRKNGCT EIYNLKGGFKKWGGKIKA 124
yrkF_Bs_2635098             91  EYEKAHIP GVVHIPLGEVEKRANE LNENDEIYIIC-HSGNRS-EMAARTMKKQGFK NVIYKGGWNEWTGKTE- 185
YbbB_Ec_1786712             4  RHTEQDYRALLIADTPIIDVRAPI EFEHGAMP AAINLPLMNNDERAAV35CLQNPQGILC-ARGQRN-HIVQSWLHAAGID -YPLVEGGYKALRQTAIQ 134
MJ0052_Mj_1498813           94  ITVSELLELIKKEDYIIVDIRSPR EFKEETLP GAILPLFLDDEHALI33LDRDKLIVVFC-ARGMRS-QTMALILQLLGFK -VKRLIGGFKAFKHAVDK 222
GlpE_Ec_41583               7  INVADAHQKLQEKEAVLVDIRDPQ SFAMGHAV QAFHLTNDTLGAFMRD NDFDTPVMVMC-YHGNSS-KGAAQYLLQQGYD VVYSIDGGFEAWQRQFPA 102
YibN_Ec_586732             40  ITRGEATRLINKEDAVVVVDVRQRD DFRKGHIA1SIPSEIKANNVGE 2KHKDKPVIVVGC-GSGMQC-QEPANALTKAGFA QVFVLKEGVAGWAGENLP 138
HP1223_Hp_2314397           3  EDYAISLEEVNFNDFIVVDVNELE EYEELHLP NATLISVNDQEKLADF 2IPDRKVIFVC-RARRA-LDAAKSMHELGYT -PYYLEGNVYDPEKYGFR 99
slr1184_Ssp_1653922        65  RTPAALKKRLLDDAKTTLVDREVK EYQAGHIP GAINIPLRTLSHNLAQ IPPDRKVIFYC-STGYRS-AMAMMALNLLGYE NVLAFSPPTGWQAAGEA 160
ytwF_Bs_2635515             1  STAALKEKIEADEELYLLIDVPEFT EVAEGMIP QAVHIRMGDIPEKMET LPSGDKKVIPIC-RSGNRS-MNVCKYLDEQGFK -TVNVEGGMMAWEGETKP 66
aq_477_Aae_2983145         10  PEEAKKMLEEEKDKVVLLDVTPP EHFQVRIP NSMLIPLDELRYAFQN LPKDKKVIVYC-RIGERS-AFATYFLRQMGYE -AYNLAGGILIWPYEKES 104
aq_1599_Aae_2983956        52  TPKQVVEMIKKGEDVVLLDVRTEA 8YKNSLHIP MDKLFKPENLKKIP-- --LKNDKPVIIYC-RSGARA-IAATFALRSAGFD NVYALKGGIAALADYVTP 151
RP600_Rp_3861144            7  CSTKAYNMLILNNNAFLVDVRTQE EWKQVGIP HLDNKNKVIFLSLQLN12EKIDTAIFFLC-RSGYRS-FIAANFIANIGYK NCYNISDGFEGNNQDKGW 111
Rv0324_Mtu_2193945        119  ITRAELLRRREAGEVTLVDVRPHE EYQAGHIP GAINIPIAELADRLAE LTGDRDIVAYC-RGAYCV-MAPDAVRIARDAG 1EVKRLDDGMLEWRLAGLP 215
Rv1674c_Mtu_2916972       118  ISRDELQARVAAGSVLVDVRPHE EYAAGHLP GAVSIPLDELAERLDE LPSGIDVYVYC-RGAYCV-YAYDALELLRPNG 1SARRLDGGFSEWLAADLP 214
Rv0390_Mtu_1817705          8  TPLQAWEMLSDNPRAVLVDVRCEA EWRFVGVP DLSSLGREVVVEWAT21DQHERPVIFLC-RSGNRS-IGAAEVATEAGIT PAYNVLDGFEGHLDAEGH 124
Rv1066_Mtu_2896703         17  RLAADQVPEAARRGAVLVDVRPQA QRAREGEV PGALVIERNVLEWRCD10VDDDVEWVLLC-GHGYTS-SLAAASLLDLGLH RATDVVGGYRALAAGGVL 122
AF0400_Af_2650233         459  INVFELKEKLEKEDIVILVDVRSEE EFKTRRIE2KVIHIPILELRERLDE IPRDKEIVVVC-AIGLRS-FEASRILKHAGFE KVKILEGGMAFWF---- 551
MTH1096_Mta_2622197       104  RAISVHELREMEGDFFLLDVRKIT DRERFHIE GSEHIWVGDLPDNLDL -IPEKDVVIYC-DSGYKS-TIAASILEGHGFN -VTTVLGGIGAWLRAGYP 197
MTH668_Mta_2621751        162  KHAGHIWRVMEAGDGVFLDVNELG -TISDNPL RALRMLAEHTPE----- 2VTCGRRGSVIC-HGGSRT-RIRAVRAVRELDP -TALGIPTWQPMYMPEEW 253
F42G8.7_Ce_2702453        410  LIAQSEQTSGCPESFCLVDVNDND VYHKSHIA NAYHYDRIMLSRLVYE 7RAEGHLLVIYG-REYER----VAKVLFQRGFK -TVLLKGNTAQFKEQYPC 508
T29A15.190_At_4469021      90  VDVKEAQRLQKENNFVILDVRPEA EYKAGHPP GAINVEMYRLIRETWTA31LDKEAKIIVAC-SSAGTMKPTQNLPEGQQSRS13NVFHLEGGIYTWGKEGLP 230
T24P15.13_At_2673913       52  VNAEEAKQLIAEEGYSVVDVRDKT QFERAHIK SCSHIPLFIYNEDNDI35SSQGVKVIFHC-QEFLRS-AAARSRLEEAGYE 5TSGLQSVKPGTFESVGST 187
ACR2_Sc_2498102             7  SRQLKGLIENQRKDFDQVVVRRE- DFARDHIT NAWHVPVTAQITEKQL12SSQFVKVIFHC-TGAKNR-GPKVAAKFETYLQ 8ESCILVGGFYAWETHCRE 121
YGR203w_Sc_1323363         19  RWMQEGHTTLREPFQVVVRDKT DYMGGHIK DGWHYAYSRLKQDPV14GRGAINVIFHC-MLAQQP-GPSAAMLLLRSLD 7RLWVLRGGFSRWQSVYGD 237
YceA_Ec_1787294           136  LQAAEVNAMLDDPDALFIDVRNHY EYEVGHIK NALEIPADTFREQLPK 6AHKDKKIVMYC-TGGYRC-EKASAWMKHNGFN KVWHIEGGIIEYARKARE 237
ybfQ_Bs_2632519           115  LKPAEFYEKMQDPNTIVIDVRNDY EYDLGHIR -GAVRPDIEAFRELPE 7MLEGKKIILYC-TGGYRC-EKFSGWLVKQGFE DVAQLDGGIVTYGKDPEV 216
YceA-1_Ct_3329074         112  ISPEEWHEKLQENRCLVLDVRNNY EWKIGHFK -NAVLPDIETFREFPD 9DPKVPVMMYC-TGGIRC-ELYSALLLEKGFK EVYHLRGGILQYLEDTHN 213
sll0765_Ssp_1001499       112  VSPQQWNQLLQDPDVVVIDVRNNY EVAIGTFQ -GAVNPCTKKFRQFPD 7QQKNKKVAMFC-TGGIRC-EKASAYLLEEGFA EVYHLRGGILHYLETIAP 215
RP125_Rp_732270           114  IEPKDWDEFITKQNVILLDVRNTY EIDIGTFK -SAINPRTETFKQFPA 7LLQGKKIAMFC-TGGIRC-EKSTSLLKSIGYN EVYHLKGGILKYLEEVPK 215
T7D17.6_At_4895217        197  ETRIGKFK -GAVDPCTTAFRNFPS36PKTLPRIAMYC-TGGIRC-EKASSLLLSQGFE11QVYHLKGGILKYLEEVPK 338
T12M4.1_At_3249095        148  AEHLDGDGKSENKELVLLDVRNLY ETRIGKFE -SENVETLDPEIRQYS11KMKGKNVLMYC-EMASAYIRSKGAG 2NTFQLYGGIQRYLEQFPS 255
ThiI_Ec_1786625           394  EVVVEETVNGFGPNDLIDVRSID EQEDKPLK4DVYFYKLSTKFGD LDQNKTWLLWC-RAGALYLREAGYK NVKVYRP---------- 482
moeB_Ssp_1653525          286  VRELHALMESQEKDFILLDVRNPN EYDIARIP GSVLVPLPDIEEGHGI 7DGHEGILIAHC-KLGG--RS-TKALSLLREAGIE -GINVRAGGTSWSREVDP 387
MoeB2_Mtu_2072679         290  TPRELRDWLDSGRKLALLVDVRPV EWDIVHID GAQLIPKSLINSGEGL 2LPPDRTAVLYC-KTVVRS-AEALAANVKAGSS DAVHLQGGIAEWTRTIDS 387
MoeB1_Mtu_2076691         287  SCDELRTKQQSDQNFLLVDVREPA EFDIAHIP GSILIPKGEIGSAAGL 2LPLDKEIVLYC-KSIIRS-AQALTTLKAAGLH NVKHLDGGIAEWTRTIDS 384
Yhr111w_Sc_529119         329  AFQRIYKDDEFLAKHIFPLDVPSH HYEISHIP EAVNIPIKNLRDMNGD10VEKDSNIVILCRYGNDSQ-LATRLLKDKFGFS NVRDVRGGYFKYIDDIDQ 435
ThtR.1_Ec_401186           7  VGADWLAEHIDDPEIQIIDVRMAS10EYLNGHIP GAVFFDIEALSDHTSP19VNQDKHLIVYD-DGNTFSAPRAWWMLRTFGVE KVSILGGLAGWQRDDLL 132
ThtR.2_Ec_401186          155  VKVTDVLLASHENTAQIIDVRPAA11GLRRGHIP GALNVPWTELVREGEL14VSYDKPIIVSC-GSSVTA-AVVLLALATLDVP NVLKLDGGNSEWGARADL 275
ThtR1.1_Aae_2983547       18  VSPEWLKENLSKKKNLVIEVFGDTQ 1JLVEGHIP GAVLTEKEQWRKMDPE20INNDSEVVIVYD-DGAVYAYWVWFH 5RVGILDGGWENIKKGYP 139
ThtR1.2_Aae_2983547      160  LETNWKYVYENIGKIPIIDVRLPD10AKKYGHIP CSVPFSWEWWVKRDRE20LKPEDEVILFC-FGGTGA-AFLYWVFDVSGHK NMKVYDAEKREWEALNLP 285
ThtR2.1_Aae_2984095       22  SPEELHKLIKEEKNLVIIDVRKSI 1EYWKEHIP GAQWVHFEAFRFPKGG19IDKNTPVVLVD-DGAYAQ-AVVLLALATLLAT 1KVRLYDGGWFNHWKDKDLP 138
ThtR2.2_Aae_2984095      162  ADLDVVKKALNREDVVFLDVRTPE 8WKRLGHIK GAYNPDFDEEHVN23GGGVAAVVRVLT11RVYFLKGGYETFYSEYPE 290
ThtR_Sc_3122964          167  EMFQLVKSGELAKKFNAFDARSLG11DIPSGHIP GTQPLPYGSLLDPET 24LDPSKPTICSC-GTGVSG-VIIKTALELAGVF NVRLYDGGWTEWVLKSGP 296
YnjE.1_Ec_1788054        282  KAEPDFGVKIPAQPQLMLDMEQ-- ARGLLHRQ DASLVSIRSWPEFIGT56GIPDQQVVSFYC-GTGWRA-SETFMYARAMGWK NVSVYDGGWYSSDPKD 427
YnjE.2_Ec_1788054        159  LQQGKEVAETAKPAGDWKVIEAAWGA 2LYLISHIP GADYIDTNEVESEPLW16IRHDTTVVILYG-RDVYAA-ARVAQIMLYAGVK DVRLLDGGWQTWSDAGLP 272
Mih1_Sc_798962            29  NILQNNMCESFYNSCRIIDVRFEY EYTGGHII NSVNIHSRDELEYEF11NTLPTTLLIIHCGHGSSHRG-PSLASHLRNCDRI12DGGYKAVFDNFPE 148
CDC25_Hs_3660166          31  ASVLNGKFANLIKKEDVIDVRYPY EYEGGHIK GAKNLYTTEQILDEFL15GHKRNIIIPHCGEFSGRG-PRMCRYVRERDRL10ELYVLKGGYKEFFMKCQS 144
String_Dm_115912         306  ARLLKGEFSDKVASYRIIDVRYPY EFEGGHIE GAKNLYTTEQILDEFL15GHKRNIIIPHCGEFSGRG-PKMSRFLRNLDRE12EIYLLHNGYKEFFSEHVE 429
DUS5_Hs_2499746            8  RQLRKMLRKEAAARCVVLDVRPYL AFAASNVR QSLNVVNLNVVLVDQGSRHWD-KLREESAARVVLT11RVYFLKGGYETFYSEYPE 139
PTP3_Sc_731478           101  VELGKIIETLPDEKVLLIDVPFEFT EHAKSIIT2IHVCLPSTLLRKNFT20AIDNLRIIIYD-STANQT--ESSVSLPCYGIA12TVSILMCGGFPQFKILFPD 229
UBP4_Sc_577796           195  LISSSANSASSQMEILLIDVRSRL EFNKSHID TKNIICLEPISFKMSY23RNLFKFIILYT-DANEYNVKQQSVLLDILVNH11KIFILESGPGWLKSNYG 325
UBP5_Sc_731042           149  PGKLSSMLHFHGDALLLIDVPRRS EFVRAHIK INECICIDPASFKDSF23RNKFDYIVYYT-DVKTFMTINFDYAFIFFYLM10VPTTLLGGYEKWKKTLHS 280
UBP7_Sc_731044           318  ELFSILSNRVEREKVLLLIDVRIPQ RSAINHIV APNLVNVDPNLLWDKQ23RNKFDYIVYYT-DVKTFMTINFDYAFIFFYLM10VPTTLLGGYEKWKKTLHS 447
consensus/85%                  .............h1DhR..  ......h. .s..hs.........  ......hhh.s..u..s.......h...s.. ....h.suh..a..... 
```

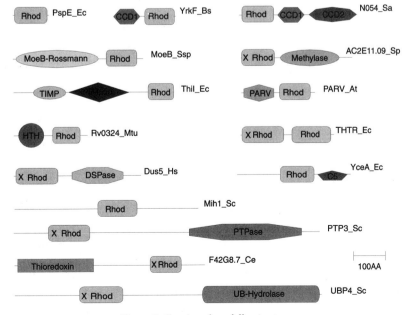

Figure 3. *Continued on following page.*

```
TM0983_Tm_4981522    8 KTLDVREVCPVPDVETKRALQNMKPGEILEVWIDYPMSKERIPETVK-KLGHEVLEIEEVGPSEWKIYIKV  78
AF0554_Af_2650074   17 EVLDIREVCPFTFIETKLKLEEMKSGEILRVIIDHEPAVRDVPRSVE-QEGHEVLSVEKVGEKEWSILIKK  87
YhhP_Ec_586656      10 HTLDALSLRCPEPVMMVRKTVRNMQPGETLLIIADDPATTRDIPGFCT-FMEHELVAKETDGL-PYRYLIRK  79
HI0721_Hi_1176278    8 QTLDTLSLRCPEPVMLVRKNIRHLNDLEILLIIADDPATTRDIPSFCQ-FMDHTLLQCEVEKP-PFKYWVKR  77
YrkI_Bs_1731128      5 KVLDAKSLACPMPIVRTKKAMNELESGQILEVHATDKGAKNDLTAWSK-SGGHDLLEQTDEGD-ILKFWIQK  74
YrkF_Bs_1731125      6 IVLDAKSLACPMPIVKTKKRMKDLKAGEVLEIHATDKGSTADLEAWAK-STGHEYLGTEAEGE-ILRHFLRK  75
H0532_Hsp_2822320    9 ETLDVKSASCPMPVVKTKSAIDDLAEGEILEVLATDSGSMSDIDGWASGTAGVELVDQEEGDD-VYKHYVRK  79
AF0556_Af_2650073    2 VVVDARGSYCPGPLMEMIKTLKQVEVGEVVEVLSSDESSAKDIPEWVK-KAGHELVKVKKEED-YWRIVVRK  71
AF0188_Af_2650456    2 KQVDCIGLYCPEPVFRARKAMEESEVGEIIEILADDPAAESDIPVLVK-KLGQELVEFEKLEDGVLRFVVKI  72
YeeD_Ec_465568       4 KKLDVVTQVCPFFPLIEAKAALAEMVSGDELVIEFDCTQATEAIPQWAA-EEGHAITDYQQIGDAAWSITVQK  74
MJ0990_Mj_2842585    6 KKLDVTGDICPVPVLKTKKALEELNEGEELEVVGDYKPALENIKRFAE-NNGYTVVLAEETES-RFRIVLKK  75
HI0242_Hi_1175198    3 YQLNLTSLRCPIPLLSAKKALKNLDKNDLMLILNISAVENFSIFAE-ENSVALVEQYYASEKEFIVILKK  73
YedF_Ec_401543       8 YRLDMVGEPCPYPAVATLEAMPQLKKKGEILEVVSDCPQSINNIPLDAR-NHGYTVLDIQQDGP-TIRYLIQK  77
aq_1421_Aae_2983825  8 RTLDLSGSLCPLPVVMTSETMRKMEEGQVLKVIICRGEKVPID17LRELEQLVNLGKSFGRMRIIACSGSMELFGLRED  78
N052_Sa_5360841    118 KQFDFRGLQCPGPIVNISKEINNISTGEQIEVTVTDPGFNSDKSWAK-QTGNTLVNLTEEAN-VINAIIQK 187
TM0978_Tm_4981517  112 KFLDMRGQICPVPEITTRKELEKLQPGETLIVMCDYPLSGERITSFSL-REGYEVATEQIGPVTKIYIKKPQ 182
AF0165_Af_2650485   12 HELDCRNMVGPYPVIITKLAMQKVDR---LEVLTNNPPSVRDIEKIAE--KEGWRVEKRRDGE-VWRIKWR  77
MJ0760_Mj_2833569   74 KTINVVGATCPGPIMMVSDMLSKMKNGEILEIICGKNSLTDLTEGLKG--MGNEIIKVEDKGDGTYRILVKK 143
MTH1417_Mta_2622634 73 EEIDVTGETCPGVIIVGDRLSSMEFGMRIKINSESSDVIDDLALSAP-EMKAEVIEKSASHLILEKTDVSR 143
ssl1707_Ssp_1673329  8 ILLDLRGTPCPINFVRTKLKLAQMAPGQCLEVWLDEGEPVEQVPHSLELEGYTEQSLRALEGESAYCLTITV  79
S34615_Sso_479541    5 LRLDLRGKSGDEYILEISKVLVNMKPGDILNIIADKDRIMCTHQLLRN--APRYLFKGDVVDD-HAEISIRR  73
SC6G10.30_Scoe_4539226 389 LVVDALGRRFCPIPVIELAKVIGDVPVGATVRVLSDDEAARLDIPAWCE--MRGQEYVGEEPVEGGTAYLVRR 458
consensus/90%          ..1Dh.u...CP.shh.....h.ph...G..1bl..s.....pph...........h..p.......h...1p.
```

```
aq_401_Aae_2983074    8 DLVIIL ITGPENPKRLPSAFFLASTAAAADMNVV--MYFTGPTLLLKKGVAES  GNTIKDFMELALDNGVQ--FVGGLQSLELNGMTKE   DLAFDVP  LLTPSQALPSLGQAGRVLT 125
Y12804_Ar_2292730     7 RMVFVV --QKRHELFDLMPFSIALTWRQDKNVDVD-LYFMYDGVELLREDFLAG  HQDLLEILNSLLADGAI-VYAGGFCSRACELSAK  DYHPDVQ  VANRQIFHSLMTQRRPVYW 116
HI0576_Hi_1176235     2 RYVIAV KSPIYGKQGAFLAYGFAESLIKKEHEISQ-IFFFQDGVSNGNALVYPA  3VNLQKHWQMFSITYNVP-LHLGVAASQRRGVVDN  9NLAEGFT  IAGLGEFIAASLNADRVIT 125
AF0555_Af_2650077     3 KLAIVL ASGEL--EKVQAASIIASTAATLGEVM---VFATMDGLMAFFKDVVEN16APLFIDVLKQAKEVGNLKVYACGMVMDMLKMSLD  DFVDIFD  1VIGVTKFLGMVEDAKVLFI 127
AF0183_Af_2650453     7 RINGIIIHSGQW--DRIYHAFSIASVLASMGEEVQ--VFLTYWALNICRGEKVF17LRELEQLVNLGKSFGRMRIIACSGSMELFGLRED  ELPEWVD  KVGGLTEV---LGGDNFIF 131
H0526_Hsp_2822319    41 KMSIIS TKGTL--DMAYPPLILASTAAAFGYEVT--VFHTFWGLEILHEERSKN39TATIEELIETSLDMGVE-FQACQMTIDLMDYDED  DFYDGVT  1GVGAAATALQDMAEADIQLL 189
aq_1119_Aae_2983592  73 FKTMLRIITNHL-SVYNFDPFKVHIVVVAHGAGA---KFPLLKLRGKTRWEKEPID  1KAIKAKLEEL-QQYGVE-YVGGITVKFRLLKDK  2DFVKIVP  SGVGAVAH-LQKIGYAYIK 183
MTH1491_Mta_2622607   3 DYRVVFIIDEDDESRVLLLISNVRNLMADLESVRIE-VVAYSMGVNVLRRDSEYS  ----GDVSEL-TGQGVR-FCAGSNTLRASGMDGD  DLLEGVD  2SSGVGHIVRRQTEGWAYIR 112
AF0913_Af_2649702     2 KVVFHL DMD----STSLLELCLANVNNFLNDVPEAEVAVLAN-PKLEDRMRELSERGVE-FFVGNNALTLHGIKIE  DVFDFCE  2PAGVTKLVELQNEGYAYIK 111
aq_389_Aae_2983080    5 RLAIIA TKGTL--DMAYPPLILASVAASLGVETA--VFFTFYGLNIIHKEKVKQ65VASIEELLELCKEADVK-LIPCQMTMDLFGYKRE  DLIDGLE  1PAGATEFFNYVLAADKPMI 179
YrkE_Bs_1731124       7 RTTIVL FSGDY--DKAMAAVIIANGAAAYDHEVT--IFHTFWGFNALRKEELIP44VLTLPQLIEMAQEQGVK-LVACTMTMDLLGLQEK  ELLDDID  YAGVAAYLADAEEGSVNLF 159
N052_Sa_5360841     201 GTTIVL FSGEL--DKAVAAMIIANGAKAAGKDVT--IFHTFWGLNALKKAQSTR44VDSLYSLIDQAIDQDIK-LIACTMSMDVMGISKE  ELRDEVD  YGGVGTYIGHTEQANHNLF 353
AF0187_Af_2650457     3 KLLYVQ TSGIDTPERLYAPFVLAMTAKAMGDDAI--IYFVIKGVTVVKKDNAEK  5FPPLSELMKQARESGVE-MMVGDRSSELLGIDKG  EIVDGVK  IVGAATLNQLVLEADGVLY 118
AF0563_Af_2650056     1 ANSRRE TSGIDTPERLYAPFILAMTAKAMGDDAI--IYFVIKGVTVVKKGNADK  5FPSLKEWMQDAVQSGVE-MMVGDRSSELLGIDKG  EIVEGVK  VVGAATLNQLVLEADGVLY  93
Y10251_Mk_1772578     9 VITVVI SEAPYGQERAYTALRFALTALVEGEEVK--IFLIEDGVFLGKKGQNPD  1VPNYLELLEQCIEQGAE-VKACGPCSKARGLSEE  DFIEGVE  LATMHDLVNWVKESDNVIF 120
MJ0989_Mj_2128041     2 KFTVII TEAPYGKERAYSALRFALTALLEGIEVN--IFLIKDGVFSCKGGNPS  1VPNYLELLKNAIELGAV-VKVCGPCCKARGLKIE  DLIEGAK  LATMHDLIAFVKESDNVVT 113
YheN_Ec_1176234       2 RFAIVV TGPAYGTQQASSAFQAQALIADGHELSS-VFFYREGVYNANQLTSPA  3FDLVRAWQQLNAQHGVA-LNICVAAALRRGVVDE11NLQQGFT  LSGLGALAEASLTCDRVVQ 127
DsrE_Cvi_3228382      2 KFALQI NEGPYQHQASDSAYQFAAKALEKGHEIFR-VFFYHDGVNNSTRLTTPP  3RHIVNRWAELAEQYELD-MVVGVAAAQRRGIVDE11NIHPKFR  ISGLGQLVEAAIQADRLVV 127
MTH1357_Mta_2622464   5 ILTIVV TEGPYRYQYADIAFEMAESAIKNGYDVT--IFLYMDGTHIPKRNQAP  2FPNSAERLRSLVEMCGVK-VTSGIRCSTARGYTCS  2PYIRGVE  IKSVYDLGMVKCWKDSHRVIT 118
MTH1417_Mta_2622530 152 KVLVVQ SNGTGNAERAYATFIFSKAALSMGKDVT--IFLLMDGVSIARKGGAAA  5FPRLDELMAEVIEMGVK-IYVCEMSAQFRGLRED  NMVEGCK  IAGAATF---------- 255
MJ0760_Mj_2833569   158 ELFIIN TTGTGNAEKAYFDFIAKQALEKKGHEVT--IFLLMDGVKFLGDLVKDILSKGVK-IYVCELSAEFRGINEK  NLEEGFE  I---------- 255
YchN_Ec_2506650       3 KIVIVA NGAPYGSESLFNSLRLAIALREQESNLDLRLFLMSDGVTAGLRGQKPG  1GYNIQQMLEILTAQNVP-VKLCKTCTDGRGISTL  PLIDGVE  IGTLVELAQWTLSADKVLT 116
TM0981_Tm_4981520     2 KIGIQV MVPPYTYEDLDTAIKIAEAAMEKGHEVT--LFLFADGVICTNKNIKPI  3NIPQKLVIMEEMLGVK-FEVHICGMDYRGITTD  MIIEGSK  PGEPLANLIATCDRFVIV 116
aq_401_Aae_2983074    7 KIVILM TSGPKTPWRCATPFYIATLMAAQEADVE--IFFNMDGTNLIKKGVAEK16PKTVYDFMKDAKLAGVK-FYSCKQAVDSMGLTEE  DLIPELD  1IVPASEFALRAMMADKLIT 134
aq_390_Aae_2983069   15 TVPFFV4TTGELINPQAGAPFFLATAATTMDYGEV--MVITSEGYLLMDGAK-TYLCTPSDLTDIYKE  2VNKEICD  1ILGGAAFLDKLMSGEYAVI 139
HI0576.1_Hi_2495665   3 KIAFLF RTSPHGTSISREGLDALLAATAFCEPNDIGIFFIDDGVLNLIDNQQPE  3QKDFIRTFKLLDLYDVEQRFICTASLQKFKLDNR  ELILSCE  KIDRSLLLEKLNQAGKLFT 118
YheM_Ec_1176233       3 RIAFVF STAPHGTAAGREGLDALLATSALTDDLA--VFFIADGVFQLLPGQKPD  3ARDYIATFKLLGLYDIECQCWVCAASLRERGLDPQ  1PFVVEAT  PLEADALRRELANYDVILR 118
orf4_Tfe_3790608     14 YLMIVV TTLDMEKSTAIEAVFNRAKAATAEYDVEV-ILTGEGSVIATTHAVKSG  5LRLLYDSMRSAKSAGVK-ISICERSIEWCSLNND  ELIPEID  1VVSSQYIMEERAAPDTYVW 118
MTH1358_Mta_2622465   3 SALIII DRAPYGYENAFSGLYVVIACLNSSLDTD--VLLIEDGVYAAVADQVSE  3YPNVEELTYLIFPEGS--VFVHERSLKERGLEED  DLVEAAE  1IDDTELYEIFRLKRDTAIM 116
TM0980_Tma_4981519    3 KLLFVA YQSPVGSVWVNEAFFTLAMLEGDLEPN--VLLIEEGVSVALSKHYRY  3LLPLSICHVIKIRVGTK-VYAVKQHLEKYRVKL  DENFGAE  VIDEANLPEFLHSFDYVIF 116
DsrF_Cvi_3228383      7 KFMYLN RKAPYGTIYAWEEALEVVLGIAAFDQDVC--VLFLDDGVYQLTRGQDTK  3MKNFSPTYRTLGDYEVRRIYVDRDSLEARGLTQD  DLVEIAF14VIDSARVSELMNESDAVFS 135
DsrH_Cvi_3228384      3 ILHTVN -KSPFERNSLESCLKFATEGAS--------VLLFEDGIYAALAGTRVE  ------SQVTEAL-GRLKLLYVLGPDLKARGFSDE  RVIPGIS  VVDYAGFVDLTTECDTVQA 100
consensus/90%             ..h...h.............sh.h.........h..............h....p........h...s..s..p..h...................h..h...p...................h.
```

Figure 3. Organization and distribution of the rhodanese (Rhod) domain, found in the phage-shock protein PspE. (A) Multiple alignment of the Rhod domain. (B) Domain organization of the Rhod domain-containing proteins. (C) Multiple alignment of the conserved cysteine-containing domain CCD1. (D) Multiple alignment of the conserved cysteine-containing domain CCD2. The protein names and shading patterns in the alignments are as in Fig. 2. Predicted catalytic residues are shown in inverse shading. The Rhod domains with the missing active site cysteine residue are indicated with an "X." Other domain designations are as follows: PARV, parvulin-type peptidyl-prolyl isomerase (108); TIMP, a novel RNA-binding domain (L. Aravind, unpublished); PP-loop, NTPase module with a PP-loop signature (16); HTH, helix-turn-helix DNA-binding domain; DSPase, dual specificity phosphatase; PTPase, phosphotyrosine phosphatase; UB-hydrolase, ubiquitin C-terminal hydrolase domain; C5, a metal-binding domain with 5 conserved cysteine residues (L. Aravind, unpublished); MoeB-Rossmann, a special Rossmann-fold domain of the MoeB class. The protein abbreviations are as follows: THTR, rhodanese (thiosulfate sulfurtransferase); ThiI, thiamin biosynthesis and RNA thiouridine biosynthesis enzyme; MoeB, molybdopterin biosynthesis protein; Dus5, dual specificity phosphatase 5; PspE, phage-shock protein E. The species abbreviations not in Fig. 1 or 2 are as follows: Af, *Archaeoglobus fulgidus*; Ar, *Agrobacterium radiobacter*; Bb, *Borrelia burgdorferi*; Cvi, *Allochromatium vinosum*; Ct, *Chlamydia trachomatis*; Dm, *Drosophila melanogaster*; Hi, *Haemophilus influenzae*; Hp, *Helicobacter pylori*; Hs, *Homo sapiens*; Hsp, *Halobacterium sp.*; Mk, *Methanopyrus kandleri*; Mta, *Methanobacterium thermoautotrophicum*; Rp, *Rickettsia prowazekii*; Sa, *Staphylococcus aureus*; Sc, *Saccharomyces cerevisiae*; Scoe, *Streptomyces coelicolor*; Sp, *Schizosaccharomyces pombe*; Sso, *Sulfolobus solfataricus*; Tfe, *Thiobacillus ferrooxidans*.

nese domain with the DNA-binding helix-turn-helix domain, suggesting that it plays a role in transcriptional regulation, perhaps in response to toxic ligands or specific redox states. Of considerable interest is the combination of the rhodanese domain with two novel domains that are well conserved in the microbial world. Proteins with both these domains sometimes occur in operons that contain a small rhodanese-like protein, further supporting a functional interaction.

Both these domains are characterized by a highly conserved cysteine residue and were accordingly named conserved cysteine-containing domain 1 and 2 (CCD1 and CCD2, Fig. 3C and D). The presence of the conserved cysteine residues and the association with the small rhodanese-like proteins suggest that CCD1 and CCD2 proteins participate in a redox relay system, with the cysteines passing through alternative oxidation states.

The OsmC protein family, the rhodanese-like domain superfamily, and the CCD1/2 families illustrate the general challenge associated with the functional interpretation of genome comparisons. Here we have a highly conserved protein family, with a known three-dimensional structure and the general type of enzymatic activity predicted. It is also known that certain members of each family are stress-induced. However, it remains to be determined experimentally what the specific enzymatic activities are and which are the contexts in which the uncharacterized members of these families function.

CONCLUDING REMARKS

Comparison of the phylogenetic distribution of various stress-related genes (Tables 1 through 5) shows that very few of them, such as *groEL* and *lon*, are universally indispensable for all the living organisms. In several additional cases (*uspA*, *hslV*), two or more families of distantly related proteins appear to perform essentially the same functions. Many of the genes that participate in stress response in *E. coli*, and have been found to be critically important in this organism, are limited in their phylogenetic distribution to proteobacteria, or just enterobacteria. This appears to be the case for the genes encoding such interesting proteins as H-NS, Dps, InaA, and others. There are also clear distinctions in the distribution of stress genes between free-living bacteria such as *E. coli*, *B. subtilis*, and *Synechocystis* sp. with their relatively large genomes and parasitic bacteria such as mycoplasmas, chlamydias, *H. influenzae*, and *H. pylori*, whose smaller genomes encode fewer stress-related proteins. In some cases, there appears to be a division between hyperthermophilic archaea (*M. jannaschii*, *A. fulgidus*, *P. horikoshii*, and *A. pernix*) and all other free-living organisms, including the nonhyperthermophilic archaeon *M. thermoautotrophicum*. The former group, for example, lacks *grpE*, *dnaK*, and *dnaJ* genes, coding for heat shock proteins, *clpA* gene encoding the chaperone subunit of the ATP-dependent protease ClpAP, and the *uvrABC* genes that encode the ATP-dependent excision nuclease involved in DNA repair (Tables 1, 3, and 4). The presence of these systems in *M. thermoautotrophicum* can be explained by the horizontal transfer of the respective genes from bacteria. The conserved operon organization of these genes in *M. thermoautotrophicum* (Tables 1, 3, and 4) appears to support this idea. Other mechanisms of stress response, such as the interplay of multiple sigma subunits (Table 2) or acid shock response (Table 5), appear to be limited to the bacterial domain.

In conclusion, comparative genomics offers us a broad view of the distribution of various stress-response systems in the microbial world. In some cases, *E. coli* turns out to be a representative organism; on other occasions, *E. coli* is an exception whereas *B. subtilis* offers a more universal picture; in yet other cases, neither of them has been studied in sufficient detail and we still have very little understanding of the stress-response mechanisms. Availability of the parasitic genomes should help sort out which stress responses are indeed indispensable for bacteria. Incorporation of archaeal models should shed new light onto the evolution of stress-response systems and make certain conclusions about their universal character (80). Cross-genome sequence analysis of common domains reveals conserved features in their organization and sometimes provides unexpected insights into the functions of stress-response systems, such as RelA/SpoT (Fig. 1), OsmC (Fig. 2), or PspE (Fig. 3), offering new experimentally verifiable hypotheses. Furthermore, a number of functionally uncharacterized paralogs of known stress-response proteins have been identified (the anti-sigma factor system is a good example of this as well as the UspA superfamily), suggesting that unexplored stress-response pathways might exist. In a similar vein, it might be possible to predict additional stress-response genes by analyzing distinct promoters for specialized sigma factors and binding sites for stress-related transcriptional regulators—a subject not covered here but certainly deserving attention. Generally, comparative genomics allows us to transfer experimental information obtained by studying one particular organism into the universal perspective, helping to achieve a better understanding of the functioning and evolution of the microbial world.

REFERENCES

1. Akbar, S., C. M. Kang, T. A. Gaidenko, and C. W. Price. 1997. Modulator protein RsbR regulates environmental signalling in the general stress pathway of *Bacillus subtilis*. *Mol. Microbiol.* **24**:567–578.
2. Altschul, S. F., T. L. Madden, A. A. Schaffer, J. Zhang, Z. Zheng, W. Miller, and D. J. Lipman. 1997. Gapped BLAST and PSI-BLAST—a new generation of protein database search programs. *Nucleic Acids Res.* **25**:3389–3402.
3. Andersson, S. G., A. Zomorodipour, J. O. Andersson, T. Sicheritz-Ponten, U. C. Alsmark, R. M. Podowski, A. K. Naslund, A. S. Eriksson, H. H. Winkler, and C. G. Kurland. 1998. The genome sequence of *Rickettsia prowazekii* and the origin of mitochondria. *Nature* **396**:133–140.
4. Antelmann, H., J. Bernhardt, R. Schmid, H. Mach, U. Volker, and M. Hecker. 1997. First steps from a two-dimensional protein index towards a response-regulation map for *Bacillus subtilis*. *Electrophoresis* **18**:1451–1463.
5. Aravind, L., and E. V. Koonin. 1998. The HD domain defines a new superfamily of metal-dependent phosphohydrolases. *Trends Biochem. Sci.* **23**:469–472.

6. Aravind, L., and E. V. Koonin. 1999. Gleaning non-trivial structural, functional and evolutionary information about proteins by iterative database searches. *J. Mol. Biol.* **287**:1023–1040.

7. Aravind, L., and C. P. Ponting. 1999. The cytoplasmic helical linker domain of receptor histidine kinase and methyl-accepting proteins is common to many prokaryotic signalling proteins. *FEMS Microbiol. Lett.* **176**:111–116.

8. Aravind, L., D. R. Walker, and E. V. Koonin. 1999. Conserved domains in DNA repair proteins and evolution of repair systems. *Nucleic Acids Res.* **27**:1223–1242.

9. Babst, M., H. Hennecke, and H. M. Fischer. 1996. Two different mechanisms are involved in the heat-shock regulation of chaperonin gene expression in *Bradyrhizobium japonicum*. *Mol. Microbiol.* **19**:827–839.

10. Baird, L., B. Lipinska, S. Raina, and C. Georgopoulos. 1991. Identification of the *Escherichia coli sohB* gene, a multicopy suppressor of the HtrA (DegP) null phenotype. *J. Bacteriol.* **173**:5763–5770.

11. Bearson, S., B. Bearson, and J. W. Foster. 1997. Acid stress responses in enterobacteria. *FEMS Microbiol. Lett.* **147**:173–180.

12. Becker, G., E. Klauck, and R. Hengge-Aronis. 1999. Regulation of RpoS proteolysis in *Escherichia coli*: the response regulator RssB is a recognition factor that interacts with the turnover element in RpoS. *Proc. Natl. Acad. Sci. USA* **96**:6439–6444.

13. Bernhardt, J., U. Volker, A. Volker, H. Antelmann, R. Schmid, H. Mach, and M. Hecker. 1997. Specific and general stress proteins in *Bacillus subtilis*—a two-dimensional protein electrophoresis study. *Microbiology* **143**:999–1017.

14. Blankenhorn, D., J. Phillips, and J. L. Slonczewski. 1999. Acid- and base-induced proteins during aerobic and anaerobic growth of *Escherichia coli* revealed by two-dimensional gel electrophoresis. *J. Bacteriol.* **181**:2209–2216.

15. Blattner, F. R., G. Plunkett III, C. A. Bloch, N. T. Perna, V. Burland, M. Riley, J. Collado-Vides, J. D. Glasner, C. K. Rode, G. F. Mayhew, J. Gregor, N. W. Davis, H. A. Kirkpatrick, M. A. Goeden, D. J. Rose, B. Mau, and Y. Shao. 1997. The complete genome sequence of *Escherichia coli* K-12. *Science* **277**:1453–1474.

16. Bork, P., and E. V. Koonin. 1994. A P-loop-like motif in a widespread ATP pyrophosphatase domain: implications for the evolution of sequence motifs and enzyme activity. *Proteins* **20**:347–355.

17. Bouche, S., E. Klauck, D. Fischer, M. Lucassen, K. Jung, and R. Hengge-Aronis. 1998. Regulation of RssB-dependent proteolysis in *Escherichia coli*: a role for acetyl phosphate in a response regulator-controlled process. *Mol. Microbiol.* **27**:787–795.

18. Bouvier, J., S. Gordia, G. Kampmann, R. Lange, R. Hengge-Aronis, and C. Gutierrez. 1998. Interplay between global regulators of *Escherichia coli*: effect of RpoS, Lrp and H-NS on transcription of the gene *osmC*. *Mol. Microbiol.* **28**:971–980.

19. Brown, J. R., and W. F. Doolittle. 1997. Archaea and the prokaryote-to-eukaryote transition. *Microbiol. Mol. Biol. Rev.* **61**:456–502.

20. Brun, Y. V., and L. Shapiro. 1992. A temporally controlled sigma-factor is required for polar morphogenesis and normal cell division in *Caulobacter*. *Genes Dev.* **6**:2395–2408.

21. Bult, C. J., O. White, G. J. Olsen, L. Zhou, R. D. Fleischmann, G. G. Sutton, J. A. Blake, L. M. FitzGerald, R. A. Clayton, J. D. Gocayne, A. R. Kerlavage, B. A. Dougherty, J.-F. Tomb, M. D. Adams, C. I. Reich, R. Overbeek, E. F. Kirkness, K. G. Weinstock, J. M. Merrick, A. Glodek, J. L. Scott, N. S. M. Geoghagen, J. F. Weidman, J. L. Fuhrmann, D. Nguyen, T. R. Utterback, J. M. Kelley, J. D. Peterson, P. W. Sadow, M. C. Hanna, M. D. Cotton, K. M. Roberts, M. A. Hurst, B. P. Kaine, M. Borodovsky, H.-P. Klenk, C. M. Fraser, H. O. Smith, C. R. Woese, and J. C. Venter. 1996. Complete genome sequence of the methanogenic archaeon, *Methanococcus jannaschii*. *Science* **273**:1058–1073.

22. Cashel, M., D. B. Gentry, V. J. Hernandez, and D. Vinella. 1996. The stringent response, p. 1458–1496. *In* F. C. Neidhardt, R. Curtiss III, J. L. Ingraham, E. C. C. Lin, K. B. Low, B. Magasanik, W. S. Reznikoff, M. Riley, M. Schaechter, and H. E. Umbarger (ed.), *Escherichia coli and Salmonella: Cellular and Molecular Biology*, 2nd ed. ASM Press, Washington, D.C.

23. Cole, S. T., R. Brosch, J. Parkhill, T. Garnier, C. Churcher, D. Harris, S. V. Gordon, K. Eiglmeier, S. Gas, C. E. Barry III, F. Tekaia, K. Badcock, D. Basham, D. Brown, T. Chillingworth, R. Connor, R. Davies, K. Devlin, T. Feltwell, S. Gentles, N. Hamlin, S. Holroyd, T. Hornsby, K. Jagels, and B. G. Barrell. 1998. Deciphering the biology of *Mycobacterium tuberculosis* from the complete genome sequence. *Nature* **393**:537–544.

24. Conway de Macario, E., and A. J. Macario. 1994. Heat-shock response in Archaea. *Trends Biotechnol.* **12**:512–518.

25. Dandekar, T., B. Snel, M. Huynen, and P. Bork. 1998. Conservation of gene order: a fingerprint of proteins that physically interact. *Trends Biochem. Sci.* **23**:324–328.

26. Deckert, G., P. V. Warren, T. Gaasterland, W. G. Young, A. L. Lenox, D. E. Graham, R. Overbeek, M. A. Snead, M. Keller, M. Aujay, R. Huber, R. A. Feldman, J. M. Short, G. J. Olsen, and R. V. Swanson. 1998. The complete genome of the hyperthermophilic bacterium *Aquifex aeolicus*. *Nature* **392**:353–358.

27. Doolittle, R. F. 1994. Convergent evolution: the need to be explicit. *Trends Biochem. Sci.* **19**:15–18.

28. Drzewiecki, K., C. Eymann, G. Mittenhuber, and M. Hecker. 1998. The *yvyD* gene of *Bacillus subtilis* is under dual control of sigmaB and sigmaH. *J. Bacteriol.* **180**:6674–6680.

29. Farris, M., A. Grant, T. B. Richardson, and C. D. O'Connor. 1998. BipA: a tyrosine-phosphorylated GTPase that mediates interactions between enteropathogenic *Escherichia coli* (EPEC) and epithelial cells. *Mol. Microbiol.* **28**:265–279.

30. Fasshauer, D., W. K. Eliason, A. T. Brunger, and R. Jahn. 1998. Identification of a minimal core of the synaptic SNARE complex sufficient for reversible assembly and disassembly. *Biochemistry* **37**:10354–10362.

31. Fitch, W. M. 1970. Distinguishing homologous from analogous proteins. *Syst. Zool.* **19**:99–113.

32. Fleischmann, R. D., M. D. Adams, O. White, R. A. Clayton, E. F. Kirkness, A. R. Kerlavage, C. J. Bult, J.-F. Tomb, B. A. Dougherty, J. M. Merrick, K. McKenney, G. G. Sutton, W. FitzHugh, C. Fields, J. D. Gocayne, J. Scott, R. Shirley, L.-I. Liu, A. Glodek, J. M. Kelley, J. F. Weidman, C. A. Phillips, T. Spriggs, E. Hedblom, M. D. Cotton, T. R. Utterback, M. C. Hanna, D. Nguyen, D. M. Saudek, R. C. Brandon, L. D. Fine, J. L. Frichtman, J. L. Fuhrmann, N. S. M. Geoghagen, C. L. Gnehm, L. A. McDonald, K. V. Small, C. M. Fraser, H. O. Smith, and J. C. Venter. 1995. Whole-genome random sequencing and assembly of *Haemophilus influenzae* Rd. *Science* **269**:496–512.

33. Fraser, C. M., S. Casjens, W. M. Huang, G. G. Sutton, R. Clayton, R. Lathigra, O. White, K. A. Ketchum, R. Dodson, E. K. Hickey, M. Gwinn, B. Dougherty, J. F. Tomb, R. D. Fleischmann, D. Richardson, J. Peterson, A. R. Kerlavage, J. Quackenbush, S. Salzberg, M. Hanson, R. van Vugt, N. Palmer, M. D. Adams, J. Gocayne, and J. C. Venter. 1997. Ge-

nomic sequence of a Lyme disease spirochaete, *Borrelia burgdorferi*. *Nature* 390:580–586.

34. Fraser, C. M., J. D. Gocayne, O. White, M. D. Adams, R. A. Clayton, R. D. Fleischmann, C. J. Bult, A. R. Kerlavage, G. Sutton, J. M. Kelley, J. L. Fritchman, J. F. Weidman, K. V. Small, M. Sandusky, J. Fuhrmann, D. Nguyen, T. R. Utterback, D. M. Saudek, C. A. Phillips, J. M. Merrick, J.-F. Tomb, B. A. Dougherty, K. F. Bott, P.-C. Hu, T. S. Lucier, S. N. Peterson, H. O. Smith, C. A. Hutchinson III, and J. C. Venter. 1995. The minimal gene complement of *Mycoplasma genitalium*. *Science* 270:397–403.

35. Fraser, C. M., S. J. Norris, G. M. Weinstock, O. White, G. G. Sutton, R. Dodson, M. Gwinn, E. K. Hickey, R. Clayton, K. A. Ketchum, E. Sodergren, J. M. Hardham, M. P. McLeod, S. Salzberg, J. Peterson, H. Khalak, D. Richardson, J. K. Howell, M. Chidambaram, T. Utterback, L. McDonald, P. Artiach, C. Bowman, M. D. Cotton, C. Fujii, S. Garland, B. Hatch, K. Horst, K. Roberts, M. Sandusky, J. Weidman, H. O. Smith, and J. C. Venter. 1998. Complete genome sequence of *Treponema pallidum*, the syphilis spirochete. *Science* 281:375–388.

36. Freestone, P., T. Nystrom, M. Trinei, and V. Norris. 1997. The universal stress protein, UspA, of *Escherichia coli* is phosphorylated in response to stasis. *J. Mol. Biol.* 274:318–324.

37. Freestone, P., M. Trinei, S. C. Clarke, T. Nystrom, and V. Norris. 1998. Tyrosine phosphorylation in *Escherichia coli*. *J. Mol. Biol.* 279:1045–1051.

38. Friedberg, E. C., G. C. Walker, and W. Siede. 1995. *DNA Repair and Mutagenesis*. ASM Press, Washington, D.C.

39. Gaidenko, T. A., X. Yang, Y. M. Lee, and C. W. Price. 1999. Threonine phosphorylation of modulator protein RsbR governs its ability to regulate a serine kinase in the environmental stress signaling pathway of *Bacillus subtilis*. *J. Mol. Biol.* 288:29–39.

40. Galperin, M. Y., R. L. Tatusov, and E. V. Koonin. 1999. Comparing microbial genomes: how the gene set determines the lifestyle, p. 91–108. *In* R. L. Charlebois (ed.), *Organization of the Prokaryotic Genome*. ASM Press, Washington, D.C.

41. Galperin, M. Y., D. R. Walker, and E. V. Koonin. 1998. Analogous enzymes: independent inventions in enzyme evolution. *Genome Res.* 8:779–790.

42. Goffeau, A., B. G. Barrell, H. Bussey, R. W. Davis, B. Dujon, H. Feldmann, F. Galibert, J. D. Hoheisel, C. Jacq, M. Johnston, E. J. Louis, H. W. Mewes, Y. Murakami, P. Philippsen, H. Tettelin, and S. G. Oliver. 1996. Life with 6000 genes. *Science* 274:546–567.

43. Gottesman, S. 1999. Regulation by proteolysis: developmental switches. *Curr. Opin. Microbiol.* 2:142–147.

44. Gottesman, S., E. Roche, Y. Zhou, and R. T. Sauer. 1998. The ClpXP and ClpAP proteases degrade proteins with carboxy-terminal peptide tails added by the SsrA-tagging system. *Genes Dev.* 12:1338–1347.

45. Graumann, P. L., and M. A. Marahiel. 1998. A superfamily of proteins that contain the cold-shock domain. *Trends Biochem. Sci.* 23:286–290.

46. Gribaldo, S., V. Lumia, R. Creti, E. C. de Macario, A. Sanangelantoni, and P. Cammarano. 1999. Discontinuous occurrence of the *hsp70* (*dnaK*) gene among *Archaea* and sequence features of HSP70 suggest a novel outlook on phylogenies inferred from this protein. *J. Bacteriol.* 181:434–443.

47. Gross, C. A. 1996. Function and regulation of the heat shock proteins, p. 1382–1399. *In* F. C. Neidhardt, R. Curtiss III, J. L. Ingraham, E. C. C. Lin, K. B. Low, B. Magasanik, W. S. Reznikoff, M. Riley, M. Schaechter, and H. E. Umbarger (ed.), *Escherichia coli and Salmonella: Cellular and Molecular Biology*, 2nd ed. ASM Press, Washington, D.C.

48. Gupta, R. S. 1998. What are archaebacteria: life's third domain or monoderm prokaryotes related to gram-positive bacteria? A new proposal for the classification of prokaryotic organisms. *Mol. Microbiol.* 29:695–707.

49. Hecker, M., W. Schumann, and U. Volker. 1996. Heat-shock and general stress response in *Bacillus subtilis*. *Mol. Microbiol.* 19:417–428.

50. Helmann, J. D. 1999. Anti-sigma factors. *Curr. Opin. Microbiol.* 2:135–141.

51. Hengge-Aronis, R. 1996. Back to log phase: sigma S as a global regulator in the osmotic control of gene expression in *Escherichia coli*. *Mol. Microbiol.* 21:887–893.

52. Hengge-Aronis, R. 1996. Regulation of gene expression during entry into stationary phase, p. 1497–1512. *In* F. C. Neidhardt, R. Curtiss III, J. L. Ingraham, E. C. C. Lin, K. B. Low, B. Magasanik, W. S. Reznikoff, M. Riley, M. Schaechter, and H. E. Umbarger (ed.), *Escherichia coli and Salmonella: Cellular and Molecular Biology*, 2nd ed. ASM Press, Washington, D.C.

53. Hengge-Aronis, R. 1999. Interplay of global regulators and cell physiology in the general stress response of *Escherichia coli*. *Curr. Opin. Microbiol.* 2:148–152.

54. Herman, C., D. Thevenet, P. Bouloc, G. C. Walker, and R. D'Ari. 1998. Degradation of carboxy-terminal-tagged cytoplasmic proteins by the *Escherichia coli* protease HflB (FtsH). *Genes Dev.* 12:1348–1355.

55. Himmelreich, R., H. Hilbert, H. Plagens, E. Pirkl, B. C. Li, and R. Herrmann. 1996. Complete sequence analysis of the genome of the bacterium *Mycoplasma pneumoniae*. *Nucleic Acids Res.* 24:4420–4449.

56. Himmelreich, R., H. Plagens, H. Hilbert, B. Reiner, and R. Herrmann. 1997. Comparative analysis of the genomes of the bacteria *Mycoplasma pneumoniae* and *Mycoplasma genitalium*. *Nucleic Acids Res.* 25:701–12.

57. Hofmann, K., P. Bucher, and A. V. Kajava. 1998. A model of Cdc25 phosphatase catalytic domain and Cdk-interaction surface based on the presence of a rhodanese homology domain. *J. Mol. Biol.* 282:195–208.

58. Kalman, S., W. Mitchell, R. Marathe, C. Lammel, J. Fan, R. W. Hyman, L. Olinger, J. Grimwood, R. W. Davis, and R. S. Stephens. 1999. Comparative genomes of *Chlamydia pneumoniae* and *C. trachomatis*. *Nat. Genet.* 21:385–389.

59. Kaneko, T., S. Sato, H. Kotani, A. Tanaka, E. Asamizu, Y. Nakamura, N. Miyajima, M. Hirosawa, M. Sugiura, S. Sasamoto, T. Kimura, T. Hosouchi, A. Matsuno, A. Muraki, N. Nakazaki, K. Naruo, S. Okumura, S. Shimpo, C. Takeuchi, T. Wada, A. Watanabe, M. Yamada, M. Yasuda, and S. Tabata. 1996. Sequence analysis of the genome of the unicellular cyanobacterium *Synechocystis* sp. strain PCC6803. II. Sequence determination of the entire genome and assignment of potential protein-coding regions. *DNA Res.* 3:109–136.

60. Kang, C. M., M. S. Brody, S. Akbar, X. Yang, and C. W. Price. 1996. Homologous pairs of regulatory proteins control activity of *Bacillus subtilis* transcription factor σ^B in response to environmental stress. *J. Bacteriol.* 178:3846–3853.

61. Kang, C. M., K. Vijay, and C. W. Price. 1998. Serine kinase activity of a *Bacillus subtilis* switch protein is required to transduce environmental stress signals but not to activate its target PP2C phosphatase. *Mol. Microbiol.* 30:189–196.

62. Karzai, A. W., M. M. Susskind, and R. T. Sauer. 1999. SmpB, a unique RNA-binding protein essential for the peptide-tagging activity of SsrA (tmRNA). *EMBO J.* 18:3793–3799.

63. Kawarabayasi, Y., Y. Hino, H. Horikawa, S. Yamazaki, Y. Haikawa, K. Jin-no, M. Takahashi, M. Sekine, S. Baba, A. Ankai, H. Kosugi, A. Hosoyama, S. Fukui, Y. Nagai, K. Nishijima, H. Nakazawa, H. Takamiya, S. Masuda, T. Funahashi,

T. Tanaka, Y. Kudoh, J. Yamazaki, N. Kushida, A. Oguchi, K. Aoki, K. Kubota, Y. Nakamura, N. Nomura, Y. Sako, and H. Kikuchi. 1999. Complete genome sequence of an aerobic hyper-thermophilic crenarchaeon, *Aeropyrum pernix* K1. *DNA Res.* 6:83–101, 145–152.

64. Kawarabayasi, Y., M. Sawada, H. Horikawa, Y. Haikawa, Y. Hino, S. Yamamoto, M. Sekine, S. Baba, H. Kosugi, A. Hosoyama, Y. Nagai, M. Sakai, K. Ogura, R. Otsuka, H. Nakazawa, M. Takamiya, Y. Ohfuku, T. Funahashi, T. Tanaka, Y. Kudoh, J. Yamazaki, N. Kushida, A. Oguchi, K. Aoki, and H. Kikuchi. 1998. Complete sequence and gene organization of the genome of a hyper-thermophilic archaebacterium, *Pyrococcus horikoshii* OT3. *DNA Res.* 5:147–155.

65. Keiler, K. C., and R. T. Sauer. 1995. Identification of active site residues of the Tsp protease. *J. Biol. Chem.* 270:28864–28868.

66. Keiler, K. C., P. R. Waller, and R. T. Sauer. 1996. Role of a peptide tagging system in degradation of proteins synthesized from damaged messenger RNA. *Science* 271:990–993.

67. Klenk, H. P., R. A. Clayton, J.-F. Tomb, O. White, K. E. Nelson, K. A. Ketchum, R. J. Dodson, M. Gwinn, E. K. Hickey, J. D. Peterson, D. L. Richardson, A. R. Kerlavage, D. E. Graham, N. C. Kyrpides, R. D. Fleischmann, J. Quackenbush, N. H. Lee, G. G. Sutton, S. Gill, E. F. Kirkness, B. A. Dougherty, K. McKenney, M. D. Adams, B. Loftus, J. D. Peterson, C. I. Reich, L. K. McNeil, J. H. Badger, A. Glodek, L. Zhou, R. Overbeek, J. D. Gocayne, J. F. Weidman, L. McDonald, T. R. Utterback, M. D. Cotton, T. Spriggs, P. Artiach, B. P. Kaine, S. M. Sykes, P. W. Sadow, K. P. D'Andrea, C. Bowman, C. Fujii, S. A. Garland, T. M. Mason, G. J. Olsen, C. M. Fraser, H. O. Smith, C. R. Woese, and J. C. Venter. 1997. The complete genome sequence of the hyperthermophilic, sulphate-reducing archaeon *Archaeoglobus fulgidus*. *Nature* 390:364–370.

68. Koonin, E. V., and M. Y. Galperin. 1997. Prokaryotic genomes: the emerging paradigm of genome-based microbiology. *Curr. Opin. Genet. Dev.* 7:757–763.

69. Koonin, E. V., A. R. Mushegian, and P. Bork. 1996. Nonorthologous gene displacement. *Trends Genet.* 12:334–336.

70. Koonin, E. V., R. L. Tatusov, and M. Y. Galperin. 1998. Beyond the complete genomes: from sequences to structure and function. *Curr. Opin. Struct. Biol.* 8:355–363.

71. Kreis-Kleinschmidt, V., F. Fahrenholz, E. Kojro, and A. Kroger. 1995. Periplasmic sulphide dehydrogenase (Sud) from *Wolinella succinogenes*: isolation, nucleotide sequence of the *sud* gene and its expression in *Escherichia coli*. *Eur. J. Biochem.* 227:137–142.

72. Kunst, F., N. Ogasawara, I. Moszer, A. M. Albertini, G. Alloni, V. Azevedo, M. G. Bertero, P. Bessieres, A. Bolotin, S. Borchert, R. Borriss, L. Boursier, A. Brans, M. Braun, S. C. Brignell, S. Bron, S. Brouillet, C. V. Bruschi, B. Caldwell, V. Capuano, N. M. Carter, S. K. Choi, J. J. Codani, I. F. Connerton, N. J. Cummings, R. A. Daniel, F. Denizot, K. M. Devine, A. Dusterhoft, S. D. Ehrlich, P. T. Emmerson, K. D. Entian, J. Errington, C. Fabret, E. Ferrari, D. Foulger, C. Fritz, M. Fujita, Y. Fujita, Fuma S, A. Galizzi, N. Galleron, S. Y. Ghim, P. Glaser, A. Goffeau, E. J. Golightly, G. Grandi, G. Guiseppi, B. J. Guy, K. Haga, J. Haiech, C. R. Harwood, A. Henaut, H. Hilbert, S. Holsappel, S. Hosono, M. F. Hullo, M. Itaya, L. Jones, B. Joris, D. Karamata, Y. Kasahara, M. Klaerr-Blanchard, C. Klein, Y. Kobayashi, P. Koetter, G. Koningstein, S. Krogh, M. Kumano, K. Kurita, A. Lapidus, S. Lardinois, J. Lauber, V. Lazarevic, S. M. Lee, A. Levine, H. Liu, S. Masuda, C. Mauel, C. Medigue, N. Medina, R. P. Mellado, M. Mizuno, D. Moestl, S. Nakai, M. Noback, D. Noone, M. O'Reilly, K. Ogawa, A. Ogiwara, B. Oudega, S. H. Park, V.

Parro, T. M. Pohl, D. Portetelle, S. Porwollik, A. M. Prescott, E. Presecan, P. Pujic, B. Purnelle, et al. 1997. The complete genome sequence of the gram-positive bacterium *Bacillus subtilis*. *Nature* 390:249–256.

73. Kyrpides, N. C., and C. A. Ouzounis. 1997. Bacterial sigma 70 transcription factor DNA-binding domains in the archaeon *Methanococcus jannaschii*. *J. Mol. Evol.* 45:706–707.

74. Lange, R., D. Fischer, and R. Hengge-Aronis. 1995. Identification of transcriptional start sites and the role of ppGpp in the expression of *rpoS*, the structural gene for the sigma S subunit of RNA polymerase in *Escherichia coli*. *J. Bacteriol.* 177:4676–4680.

75. Levchenko, I., C. K. Smith, N. P. Walsh, R. T. Sauer, and T. A. Baker. 1997. PDZ-like domains mediate binding specificity in the Clp/Hsp100 family of chaperones and protease regulatory subunits. *Cell* 91:939–947.

76. Levina, N., S. Totemeyer, N. R. Stokes, P. Louis, M. A. Jones, and I. R. Booth. 1999. Protection of *Escherichia coli* cells against extreme turgor by activation of MscS and MscL mechanosensitive channels: identification of genes required for MscS activity. *EMBO J.* 18:1730–1737.

77. Lipinska, B., O. Fayet, L. Baird, and C. Georgopoulos. 1989. Identification, characterization, and mapping of the *Escherichia coli htrA* gene, whose product is essential for bacterial growth only at elevated temperatures. *J. Bacteriol.* 171:1574–1584.

78. Loewen, P. C., and R. Hengge-Aronis. 1994. The role of the sigma factor sigma S (KatF) in bacterial global regulation. *Annu. Rev. Microbiol.* 48:53–80.

79. Loewen, P. C., B. Hu, J. Strutinsky, and R. Sparling. 1998. Regulation in the *rpoS* regulon of *Escherichia coli*. *Can. J. Microbiol.* 44:707–717.

80. Macario, A. J., and E. C. de Macario. 1999. The archaeal molecular chaperone machine. Peculiarities and paradoxes. *Genetics* 152:1277–1283.

81. Makarova, K. S., L. Aravind, M. Y. Galperin, N. V. Grishin, R. L. Tatusov, Y. I. Wolf, and E. V. Koonin. 1999. Comparative genomics of the archaea (Euryarchaeota): evolution of conserved protein families, the stable core, and the variable shell. *Genome Res.* 9:608–628.

82. Matin, A. 1992. Genetics of bacterial stress response and its applications. *Ann. N.Y. Acad. Sci.* 665:1–15.

83. Merrick, M. J. 1993. In a class of its own—the RNA polymerase sigma factor sigma 54 (sigma N). *Mol. Microbiol.* 10:903–909.

84. Michelini, E. T., and G. C. Flynn. 1999. The unique chaperone operon of *Thermotoga maritima*: cloning and initial characterization of a functional Hsp70 and small heat shock protein. *J. Bacteriol.* 181:4237–4244.

85. Mironov, A. A., E. V. Koonin, M. A. Roytberg, and M. S. Gelfand. 1999. Computer analysis of transcription regulatory patterns in completely sequenced bacterial genomes. *Nucleic Acids Res.* 27:2981–2989.

86. Mongkolsuk, S., W. Praituan, S. Loprasert, M. Fuangthong, and S. Chamnongpol. 1998. Identification and characterization of a new organic hydroperoxide resistance (*ohr*) gene with a novel pattern of oxidative stress regulation from *Xanthomonas campestris* pv. *phaseoli*. *J. Bacteriol.* 180:2636–2643.

87. Mukhopadhyay, R., and B. P. Rosen. 1998. *Saccharomyces cerevisiae* ACR2 gene encodes an arsenate reductase. *FEMS Microbiol. Lett.* 168:127–136.

88. Mulvey, M. R., and P. C. Loewen. 1989. Nucleotide sequence of *katF* of *Escherichia coli* suggests KatF protein is a novel sigma transcription factor. *Nucleic Acids Res.* 17:9979–9991.

89. Murzin, A. G. 1998. How far divergent evolution goes in proteins. *Curr. Opin. Struct. Biol.* 8:380–387.

90. Mushegian, A. R., and E. V. Koonin. 1996. Gene order is not conserved in bacterial evolution. *Trends Genet.* **12:**289–290.

91. Nelson, K. E., R. A. Clayton, S. R. Gill, M. L. Gwinn, R. J. Dodson, D. H. Haft, E. K. Hickey, J. D. Peterson, W. C. Nelson, K. A. Ketchum, L. McDonald, T. R. Utterback, J. A. Malek, K. D. Linher, M. M. Garrett, A. M. Stewart, M. D. Cotton, M. S. Pratt, C. A. Phillips, D. Richardson, J. Heidelberg, G. G. Sutton, R. D. Fleischmann, J. A. Eisen, O. White, S. L. Salzberg, H. O. Smith, J. C. Venter, and C. M. Fraser. 1999. Evidence for lateral gene transfer between Archaea and bacteria from genome sequence of *Thermotoga maritima*. *Nature* **399:**323–329.

92. Neuwald, A. F., L. Aravind, J. L. Spouge, and E. V. Koonin. 1999. AAA+: a class of chaperone-like ATPases associated with the assembly, operation, and disassembly of protein complexes. *Genome Res.* **9:**27–43.

93. Nystrom, T., and F. C. Neidhardt. 1992. Cloning, mapping and nucleotide sequencing of a gene encoding a universal stress protein in *Escherichia coli*. *Mol. Microbiol.* **6:**3187–3198.

94. Nystrom, T., and F. C. Neidhardt. 1993. Isolation and properties of a mutant of *Escherichia coli* with an insertional inactivation of the *uspA* gene, which encodes a universal stress protein. *J. Bacteriol.* **175:**3949–3956.

95. Nystrom, T., and F. C. Neidhardt. 1994. Expression and role of the universal stress protein, UspA, of *Escherichia coli* during growth arrest. *Mol. Microbiol.* **11:**537–544.

96. Nystrom, T., and F. C. Neidhardt. 1996. Effects of overproducing the universal stress protein, UspA, in *Escherichia coli* K-12. *J. Bacteriol.* **178:**927–930.

97. Oh, S. A., S. Y. Lee, I. K. Chung, C. H. Lee, and H. G. Nam. 1996. A senescence-associated gene of *Arabidopsis thaliana* is distinctively regulated during natural and artificially induced leaf senescence. *Plant Mol. Biol.* **30:**739–754.

98. Olsen, G. J., C. R. Woese, and R. Overbeek. 1994. The winds of (evolutionary) change: breathing new life into microbiology. *J. Bacteriol.* **176:**1–6.

99. Overbeek, R., M. Fonstein, M. D'Souza, G. D. Pusch, and N. Maltsev. 1999. The use of gene clusters to infer functional coupling. *Proc. Natl. Acad. Sci. USA* **96:**2896–2901.

100. Pace, N. R. 1997. A molecular view of microbial diversity and the biosphere. *Science* **276:**734–740.

101. Pallen, M. J., and B. W. Wren. 1997. The HtrA family of serine proteases. *Mol. Microbiol.* **26:**209–221.

102. Pauli, D., C. H. Tonka, and A. Ayme-Southgate. 1988. An unusual split *Drosophila* heat shock gene expressed during embryogenesis, pupation and in testis. *J. Mol. Biol.* **200:**47–53.

103. Phadtare, S., J. Alsina, and M. Inouye. 1999. Cold-shock response and cold-shock proteins. *Curr. Opin. Microbiol.* **2:**175–180.

104. Ponting, C. P. 1997. Evidence for PDZ domains in bacteria, yeast, and plants. *Protein Sci.* **6:**464–468.

105. Ponting, C. P., and L. Aravind. 1997. PAS: a multifunctional domain family comes to light. *Curr. Biol.* **7:**R674–R677.

106. Pratt, L. A., and T. J. Silhavy. 1996. The response regulator SprE controls the stability of RpoS. *Proc. Natl. Acad. Sci. USA* **93:**2488–2492.

107. Radman, M. 1975. SOS repair hypothesis: phenomenology of an inducible DNA repair which is accompanied by mutagenesis, p. 355–367. *In* P. Hanawalt and R. B. Setlow (ed.), *Molecular Mechanisms for Repair of DNA*, vol. A. Plenum Publishing Corp., New York, N.Y.

108. Rudd, K. E., H. J. Sofia, E. V. Koonin, G. Plunkett III, S. Lazar, and P. E. Rouviere. 1995. A new family of peptidylprolyl isomerases. *Trends Biochem. Sci.* **20:**12–14.

109. Sankaranarayanan, R., A. C. Dock-Bregeon, P. Romby, J. Caillet, M. Springer, B. Rees, C. Ehresmann, B. Ehresmann, and D. Moras. 1999. The structure of threonyl-tRNA synthetase-tRNA(Thr) complex enlightens its repressor activity and reveals an essential zinc ion in the active site. *Cell* **97:**371–381.

110. Schurig, H., N. Beaucamp, R. Ostendorp, R. Jaenicke, E. Adler, and J. R. Knowles. 1995. Phosphoglycerate kinase and triosephosphate isomerase from the hyperthermophilic bacterium *Thermotoga maritima* form a covalent bifunctional enzyme complex. *EMBO J.* **14:**442–451.

111. Segal, G., and E. Z. Ron. 1998. Regulation of heat-shock response in bacteria. *Ann. N. Y. Acad. Sci.* **851:**147–151.

112. Shimada, Y., G. J. Wu, and A. Watanabe. 1998. A protein encoded by *din1*, a dark-inducible and senescence-associated gene of radish, can be imported by isolated chloroplasts and has sequence similarity to sulfide dehydrogenase and other small stress proteins. *Plant Cell Physiol.* **39:**139–143.

113. Shingler, V. 1996. Signal sensing by sigma 54-dependent regulators: derepression as a control mechanism. *Mol. Microbiol.* **19:**409–416.

114. Smith, B. T., and G. C. Walker. 1998. Mutagenesis and more: *umuDC* and the *Escherichia coli* SOS response. *Genetics* **148:**1599–1610.

115. Smith, D. R., L. A. Doucette-Stamm, C. Deloughery, H. Lee, J. Dubois, T. Aldredge, R. Bashirzadeh, D. Blakely, R. Cook, K. Gilbert, D. Harrison, L. Hoang, P. Keagle, W. Lumm, B. Pothier, D. Qiu, R. Spadafora, R. Vicaire, Y. Wang, J. Wierzbowski, R. Gibson, N. Jiwani, A. Caruso, D. Bush, H. Safer, D. Patwell, S. Prabhakar, S. McDougall, G. Shimer, A. Goyal, S. Pietrokovsky, G. M. Church, C. J. Daniels, J.-I. Mao, P. Rice, J. Nolling, and J. N. Reeve. 1997. Complete genome sequence of *Methanobacterium thermoautotrophicum* deltaH: functional analysis and comparative genomics. *J. Bacteriol.* **179:**7135–7155.

116. Stephens, R. S., S. Kalman, C. Lammel, J. Fan, R. Marathe, L. Aravind, W. Mitchell, L. Olinger, R. L. Tatusov, Q. Zhao, E. V. Koonin, and R. W. Davis. 1998. Genome sequence of an obligate intracellular pathogen of humans: *Chlamydia trachomatis*. *Science* **282:**754–759.

117. Storz, G., and J. A. Imlay. 1999. Oxidative stress. *Curr. Opin. Microbiol.* **2:**188–194.

118. Taddei, F., M. Vulic, M. Radman, and I. Matic. 1997. Genetic variability and adaptation to stress. *EXS* **83:**271–290.

119. Tamura, N., F. Lottspeich, W. Baumeister, and T. Tamura. 1998. The role of tricorn protease and its aminopeptidase-interacting factors in cellular protein degradation. *Cell* **95:**637–648.

120. Tamura, T., N. Tamura, Z. Cejka, R. Hegerl, F. Lottspeich, and W. Baumeister. 1996. Tricorn protease—the core of a modular proteolytic system. *Science* **274:**1385–1389.

121. Tatusov, R. L., E. V. Koonin, and D. J. Lipman. 1997. A genomic perspective on protein families. *Science* **278:**631–637.

122. Tatusov, R. L., A. R. Mushegian, P. Bork, N. P. Brown, W. S. Hayes, M. Borodovsky, K. E. Rudd, and E. V. Koonin. 1996. Metabolism and evolution of *Haemophilus influenzae* deduced from a whole-genome comparison with *Escherichia coli*. *Curr. Biol.* **6:**279–291.

123. Thieringer, H. A., P. G. Jones, and M. Inouye. 1998. Cold shock and adaptation. *Bioessays* **20:**49–57.

124. Tomb, J.-F., O. White, A. R. Kerlavage, R. A. Clayton, G. G. Sutton, R. F. Fleishmann, K. A. Ketchum, H.-P. Klenk, S. Gill, B. A. Dougherty, K. A. Nelson, J. Quackenbush, L. Zhou, E. F. Kirkness, S. Peterson, B. Loftus, D. Richardson, R. Dodson, H. G. Khalak, A. Glodek, K. McKenney, L. M.

Fitzgerald, N. Lee, M. D. Adams, E. K. Hickey, D. E. Berg, J. D. Gocayne, T. R. Utterback, J. D. Peterson, J. M. Kelley, M. D. Cotton, J. M. Weldman, C. Fujii, C. Bowman, L. Watthey, E. Wallin, W. S. Hayes, M. Borodovsky, P. D. Karp, H. O. Smith, C. M. Fraser, and J. C. Venter. 1997. The complete genome sequence of the gastric pathogen *Helicobacter pylori*. *Nature* 388:539–547.

125. Walz, J., T. Tamura, N. Tamura, R. Grimm, W. Baumeister, and A. J. Koster. 1997. Tricorn protease exists as an icosahedral supermolecule in vivo. *Mol. Cell* 1:59–65.

126. Wang, J., J. A. Hartling, and J. M. Flanagan. 1997. The structure of ClpP at 2.3 Å resolution suggests a model for ATP-dependent proteolysis. *Cell* 91:447–456.

127. Way, J. L., P. Leung, E. Cannon, R. Morgan, C. Tamulinas, J. Leong-Way, L. Baxter, A. Nagi, and C. Chui. 1988. The mechanism of cyanide intoxication and its antagonism. *CIBA Found. Symp.* 140:232–243.

128. Wetzstein, M., U. Volker, J. Dedio, S. Lobau, U. Zuber, M. Schiesswohl, C. Herget, M. Hecker, and W. Schumann. 1992. Cloning, sequencing, and molecular analysis of the *dnaK* locus from *Bacillus subtilis*. *J. Bacteriol.* 174:3300–3310.

129. Williams, K. P. 1999. The tmRNA website. *Nucleic Acids Res.* 27:165–166.

130. Wolf, Y. I., L. Aravind, N. V. Grishin, and E. V. Koonin. 1999. Evolution of aminoacyl-tRNA synthetases—analysis of unique domain architectures and phylogenetic trees reveals a complex history of horizontal gene transfer events. *Genome Res.* 9:689–710.

131. Yang, X., C. M. Kang, M. S. Brody, and C. W. Price. 1996. Opposing pairs of serine protein kinases and phosphatases transmit signals of environmental stress to activate a bacterial transcription factor. *Genes Dev.* 10:2265–2275.

132. Yu, J. S., and K. M. Noll. 1995. The hyperthermophilic bacterium *Thermotoga neapolitana* possesses two isozymes of the 3-phosphoglycerate kinase/triosephosphate isomerase fusion protein. *FEMS Microbiol. Lett.* 131:307–312.

133. Yura, T., and K. Nakahigashi. 1999. Regulation of the heat-shock response. *Curr. Opin. Microbiol.* 2:153–158.

134. Zarembinski, T. I., L. W. Hung, H. J. Mueller-Dieckmann, K. K. Kim, H. Yokota, R. Kim, and S. H. Kim. 1998. Structure-based assignment of the biochemical function of a hypothetical protein: a test case of structural genomics. *Proc. Natl. Acad. Sci. USA* 95:15189–15193.

135. Zuber, U., and W. Schumann. 1994. CIRCE, a novel heat shock element involved in regulation of heat shock operon *dnaK* of *Bacillus subtilis*. *J. Bacteriol.* 176:1359–1363.

Chapter 28

Proteomic Analysis of Bacterial Stress Responses

FREDERICK C. NEIDHARDT AND RUTH A. VANBOGELEN

Bacteria respond to changes in their environment. In the case of beneficial changes—nutritional enrichment, shift to optimal growth temperature, pH, osmotic pressure, or redox potential, for example—growth accelerates as the result of economic adjustments in the cells' proteome. Largely through repression of redundant or unnecessary proteins, resources are diverted to increased synthesis of growth-essential cellular components. In the case of detrimental changes—imposition of any of the nutritional, toxic chemical, or suboptimal physical conditions commonly referred to as stresses—the cells divert resources from growth and preferentially synthesize proteins that counteract or remove the stressful condition and repair any damage. Consequently, growth usually decelerates upon stress but may effect a partial return to the pre-stress growth rate depending on the circumstance. Since the physiological response to stress depends on changes in the protein expression profile, understanding stress responses must involve studies of this profile—a task most directly accomplished through the use of two-dimensional polyacrylamide gel electrophoresis (2-D gels). Work with 2-D gels has in the past taken one or the other of two paths: as an aid in identifying and studying individual proteins involved in stress responses, and as a global monitor of the entire response of the cell to its environment. These two modes of employing the resolving power of 2-D gels currently are coming to fusion, largely as the result of improved, genome-based techniques for identifying protein spots on gels, and as a consequence of improvements in gel matching and scanning, data archiving, and pattern recognition. At the current time, 2-D gels offer the best (only) approach to monitor the protein synthesis during stress responses. Monitoring the responding proteins, identifying them biochemically, and solving the molecular mechanism of their

regulation must in many cases be followed by physiological analysis of the role of the protein in the stress response.

In this chapter we will first emphasize that bacterial reactivity to the environment is more than a triggering of a particular regulon when the cell encounters a specific stress. The cell mounts a complex response involving large portions of its genome to any change in its environment and is constantly adjusting the synthesis rates of individual proteins. We will point out the central role of proteins in stress responses—the fact that it is the adjustment of its complement of active proteins that is the end product of the sensing, signaling, and regulatory processes by which a cell responds to a stress. The investigator interested in bacterial stress responses, therefore, must of necessity be interested in the cell's proteome (the protein complement of the cell's genome) (31).

Two-dimensional polyacrylamide gel electrophoresis (2-D gels) has been the preeminent tool for monitoring changes in the proteome. We will indicate what has been learned from studies using 2-D gels, what factors limit this technique, and what future developments should merge the study of stress responses with general efforts to model cell behavior.

STRESS RESPONSES VERSUS BALANCED GROWTH

When bacterial cells grow with an excess of nutrients at constant temperature, they synthesize all of their protoplasmic constituents at near constant differential rates and divide at a fixed and constant cell mass. Eventually each cellular component increases

Frederick C. Neidhardt • Department of Microbiology and Immunology, University of Michigan Medical School, Ann Arbor, MI 48109-0620. Ruth A. VanBogelen • Molecular Biology Department, Parke-Davis Pharmaceutical Research, Division of Warner-Lambert Company, Ann Arbor, MI 48105.

by the same proportion in each interval of time, achieving the state called balanced growth.

The experience of bacteriologists is that balanced growth is difficult to maintain. It takes only a few hours before stationary phase is entered, and the population becomes one of cells that differ from their parents in average composition, size, and structure. Moreover, long before the onset of stationary phase the cells in a batch liquid culture react to the accumulation of metabolic by-products, the diminution of nutrients, and the increasingly restricted rate of gas exchange. In fact, by rigorous standards, cultures commonly employed in the laboratory may not even achieve balanced growth, or may do so only fleetingly (14). Unless a chemostat is employed or frequent subcultures are made to prevent significant changes in the medium, the *balanced* character of growth is quickly lost, the pattern of gene expression becomes modified, and a changing mix of protein products is synthesized.

The gradual shift in enzyme constitution as a batch culture grows reflects the high reactivity of bacteria; virtually any change in the environment induces changes in the bacteria. In most instances detrimental changes in the growth environment are met by alterations in cellular enzymic makeup that permit continued growth, usually at a reduced rate. Successful bacterial stress responses of this sort involve a shift from one steady state of growth to another. These adaptive patterns of unbalanced growth are responsible for the unusual capacity of bacteria for growth under a variety of ambient conditions.

PROTEOMICS AND STRESS RESPONSES

Adaptive alterations of growth pattern in response to environmental signals take many forms, but at their core is the ability of bacteria to modulate the amount and activity of their protein enzymes. The selective synthesis of proteins is the essence of every stress response. New or enhanced levels of proteins are needed, for example, to scavenge scarce nutrients and utilize new sources of essential elements, to modify or destroy toxic chemical species or to exclude them from the cell (see chapter 10), to degrade denatured proteins (see chapter 1), to recycle nonessential constituents in times of starvation, to expand selected metabolic pathways, to cope with alternative electron acceptors, to protect and repair essential cellular components, to manage aerobic-anaerobic transitions (see chapter 5), and to produce metabolites with specialized protective functions (see chapter 6). The study of bacterial stress responses, therefore, in-

volves a study of their proteome—the subject of proteomics.

Bacteria, as do other cells, modulate physiological functions by adjusting the activity of allosteric proteins through specific ligand binding. Likewise, covalent modification (e.g., phosphorylation, adenylylation, and methylation) of preexisting proteins provides convenient on/off-switching for many enzymes (15). But the dominant feature of bacterial stress responses is a change in protein expression profile (the quantitative array of proteins made under a given circumstance) brought about by induction and repression of individual proteins. Although protein degradation (12) undoubtedly plays a significant role, particularly when growth is severely inhibited, the ability to modulate the synthesis of individual proteins is recognized as the bacterial cell's chief means of adjusting protein levels. The unique metabolism of mRNA in bacteria (reviewed in reference 8), resulting in complete renewal on average every few minutes, establishes a constantly rolling set of instructions for protein synthesis based on up-to-the-minute responses to the environment.

MONITORING PROTEOME EXPRESSION DURING STRESS

The first glimpse into the molecular richness of bacterial stress responses—in fact, the first recognition of the nature of these responses—came with the introduction in 1975 of two-dimensional polyacrylamide gel electrophoresis to resolve complex mixtures of cellular protein (17). This technique introduced a phase shift in bacterial physiology by transforming the basic mode of investigation. Previously the investigator *chose* one or more proteins to study in connection with some phenomenon; henceforth, the investigator could *ask the cell* what proteins should be studied. This approach ushered in the era of global (whole-cell) gene regulation.

For a few investigators the availability of 2-D gel technology provided a kid-in-the-candy-store opportunity. With the use of pulse labeling to monitor rates of synthesis of individual polypeptides on a global scale, the extraordinary modulation of proteome expression was soon discovered. Cultures were grown under many different steady-state conditions, varying the temperature (6) and the chemical nature of the medium (18). Comparisons were made of their protein expression pattern as revealed on 2-D gels. Other cultures were exposed to various toxic agents or were restricted for individual nutrients (27). Pulse labeling at early and advanced times after the stress was applied revealed the changing differential rates of synthesis of proteins resolved on the gels (9, 27).

Without pausing to identify the responding proteins, one after another physical or environmental stress was applied to examine the cells' responses. In this way, the adaptation of *Escherichia coli* to high and low growth temperatures, to different nutrient limitations, to heavy metals, to the transitions between anaerobic and aerobic growth, to redox reagents, to rich nutritional supplementation, etc., was displayed in overwhelming detail (reviewed in reference 24).

Pathway 1: Monitoring Global Changes in Protein Expression Patterns

From the start it was recognized that the information revealed on 2-D gels was valuable even without knowledge of the identity or specific function of the individual polypeptides. An early surprise, for example, was the sheer number of proteins induced (or repressed) during mild stress—the classical studies on bacterial "enzymatic adaptation" had given no hint of this fact.

Some proteins were synthesized uniquely in response to a specific stress condition and could be tagged as good candidates for later physiological and genetic studies to ascertain their function in relation to the stress. Other proteins were found to respond to several stimuli, and some (e.g., UspA) to almost all inhibitory conditions (reviewed in reference 24).

Parallel studies on cultures in balanced growth under a variety of culture conditions suggested tentative interpretations of the complex responses. The synthesis of some proteins—no surprise—depended on the composition of the growth medium. There were clear examples of repression in rich medium—obvious candidates for some role in biosynthesis. Other proteins were induced uniquely by individual carbon and energy sources—obvious candidates for catabolic functions. Still other proteins were synthesized according to simple but unknown rules related in some way to the growth rate of the cells rather than to the chemical nature of the medium (18). Gradually patterns began to emerge. For example, the group of proteins responding to a particular inhibitor could be seen to consist of responders related to slow growth as well as those unique to this inhibitor. Likewise, the protein response that is shared by cells restricted for nitrogen and those restricted for phosphorus could in part be related to early events of stationary phase.

In a similar fashion, proteomic signatures—telltale patterns of protein synthesis that indicate dysfunction of various cellular processes—have been identified and can be used in diagnosing the physiological state of cells subjected to some environmental stress (29). Thus, dysfunction of protein secretion is signaled by the presence of precursors of OmpC and OmpF; chromosome damage leads to, among other responses, increase in RecA; and various inhibited states of ribosomes trigger increases in either heat shock or cold shock proteins (reviewed in reference 29).

Simultaneous induction or repression of a set of proteins cannot be taken as evidence that the proteins are controlled by the same regulatory system. Therefore, the term stimulon was introduced to indicate the group of proteins responding to a given stimulus (16). The distinction was emphasized that membership in a stimulon implied no regulatory mechanism and that the task of the physiologist and molecular geneticist was to ascertain the functional regulons (and operons) constituting a particular stimulon.

By the early 1980s this approach—the first pathway of proteomics, we might call it—had led to the description of many stress-related stimulons and to the discovery of the bacterial heat shock response (13, 33). Through mutant analysis it was established that the heat shock stimulon consisted largely of a set of at least 20 proteins controlled by a single regulatory factor. All this had been accomplished with only hints of the identity of the proteins involved.

A more recently completed study of phosphate limitation and starvation (28) illustrates the usefulness of global analysis of stress responses and indicates the current state of 2-D gel technology. Two cultures were used to examine the response of *E. coli* to suboptimal supply of phosphate by comparison with a third culture growing with ample inorganic phosphate. One was grown with a limiting amount of inorganic phosphate; cellular events were monitored from 10 to 30 and from 30 to 60 min after an inflection in the growth curve that indicated the onset of exhaustion of the phosphate supply. The second was grown with phosphonate as the sole source of phosphorus; this compound can be utilized by an inducible pathway, but only slowly at a growth rate-limiting speed. This culture was monitored during slow, steady-state growth after adaptation to utilize phosphonate. This protocol was designed to distinguish between the proteomic response to phosphate limitation, with attendant entry of the cells into stationary phase, and slow phosphate feeding, which kept the cells growing with a phosphate limitation.

Over 800 protein spots were matched across 20 2-D gels used to resolve proteins pulse-labeled to measure their differential rates of synthesis. Among the cellular responses expected were derepression of the PHO regulon and induction of the phosphonate utilization regulon. The former is a coordinately regulated set of genes (encoding alkaline phosphatase and enzymes of related function) controlled by a DNA-binding protein, PhoB, which is a member of the two-component signaling system, PhoR-PhoB. The latter regulon consists of a set of genes encoding enzymes that convert phosphonate to inorganic phosphate (30).

The results, however, indicated that far more proteins than these were involved. On the basis of experimental (rather than biological) significance, proteins with differential rates differing by a factor of 2 from the control with ample phosphate were scored as part of the PL (phosphate limitation) or PHN (phosphonate) stimulons. By this criterion the PL stimulon included 413 proteins, 208 showing induced synthesis and 205 showing repressed synthesis. The PHN stimulon included 257 proteins, of which 227 were induced and 30 repressed. The aggregate mass of proteins responding to phosphate limitation or to growth on phosphonate accounted for 30 to 40% of the cells' total protein mass. The overlap of the two stimulons included 137 proteins, most (118) of which showed induced synthesis.

What is to be made of the magnitude of this response? It is obvious that one is not dealing here with a simple regulon or two. The 137 proteins responding to both limitation and slow feeding (118 induced and 19 repressed) should presumably be related to the phosphorus restriction and could be considered candidates for members of the PHO regulon, though this would make it one of E. coli's largest regulons. (Some of the identified responders were found to contain a sequence in their promoter region similar to the sequence called the phoB box, suggesting that the regulatory protein PhoB may control more genes than previously discovered.) The 276 proteins responding to phosphate limitation but not to growth on phosphonate should include those preparing the cell to survive stationary phase. Many more were repressed than induced, and of those repressed, many were found in a separate study to be repressed during limitation for nitrogen, strongly supporting the speculation that their response is related to stationary phase.

The large number of proteins responding to phosphate restriction is by now a common observation with stress responses. It might be helpful to consider why this is so. In Table 1 is a list of classes of proteins that might be expected to respond to a particular environmental stress. First, there are the specific proteins directly addressing the main problem at hand. In the case of phosphate nutrition this certainly includes the products of the PHO regulon. Second, there are the more or less general stress response systems that respond to a variety of situations. Good examples are the stringent response network, the heat shock and cold shock stimulons, and catabolite repression. Third, there are proteins with synthesis linked in some way to the overall cellular growth rate, such as proteins of the transcription and translation apparatus, among several others observed but not yet identified. Fourth, partially overlapping with the former category are proteins associated with the transition to stationary phase. Fifth, there are proteins responsible for repair of damage to essential cellular structures independent of the cause of the damage. Finally, there are proteins, perhaps affected only slightly or moderately in synthesis rates, that have no relationship whatsoever to the stress response, but whose synthetic rate is affected by so-called passive control. (The latter in theory include protein products of genes that are not saturated with RNA polymerase.)

Categorizing the responding proteins should not disguise a considerable challenge facing bacteriologists. In many instances the role of a protein in helping the cell cope with a stress turns out to be a puzzle. After biochemical identification of a spot has been made, after the regulatory mechanism has been worked out, there remains the task of assigning and proving the physiological significance of a protein to the cell's stress response.

The phosphate response study reflects the current state of 2-D technology with respect to quantitation of cellular responses. One fact emerges quite clearly from the summary just given of the results obtained in this study. It would be enormously helpful if identifications existed for the 800 proteins monitored, but certainly for the more than 400 responders. With this information tentative assignment of proteins to the six functional categories of responders could be made.

Transcriptional analysis using cDNA hybridization to spot blots on nylon membranes or on glass microarrays can achieve more complete monitoring of gene expression than can currently be achieved by 2-D gels. These high-resolution methods establish a goal for translational monitoring (reviewed in reference 26).

Pathway 2: Identification and Study of Individual Stress Proteins

Not every biologist is thrilled with the approach to stress responses that seems to ignore the identity and function of the responders. Spots on gels are not very gripping to biochemists or, for that matter, to many physiologists and geneticists until enzymatic

Table 1. Nature of proteins belonging to stress stimulons

Class of responders	Nature or function	Example in phosphate limitation stimulon
Specific stress	Specific and direct function to alleviate the particular stress	Alkaline phosphatase
General stress	General stress response	Fur protein
Growth rate	Growth-rate-related proteins, including most proteins involved in transcription and translation	Ribosomal protein L7/L12
Stationary phase	Prepare cells for survival in nongrowth state	H-NS protein
Repair	Damage repair, including damage to proteins, to membrane, and to DNA	GroE protein
Passive	Completely unrelated to stress; induced or repressed, for example, by changed availability of ribosomes or RNA polymerase	Glutamate dehydrogenase (?)

and gene product names can be assigned (see the perspective presented in reference 32).

This skepticism emphasizes one valid point: identification of spots greatly facilitates genetic and physiological interpretations of regulatory observations. For this reason, an intensive effort has been made since the very start of 2-D gel work to identify protein spots on reference gels of *E. coli* (and, for that matter, of many eucaryotic organisms as well). The biochemical, genetic, and physiological means employed to make identifications have been reviewed elsewhere (23). Suffice it to say that hundreds of spot identifications on reference gels of model organisms were made throughout the 1980s. Today, thanks to advances in mass spectrometry and in sequencing very small amounts of proteins, identifying protein spots of interest is relatively simple in any bacterium with a sequenced genome (4, 7, 19, 34).

This second pathway of proteomics—the use of gels for reverse genetics and other reductionist analyses—has renewed the interest in 2-D gels for proteomic studies. Efforts are now under way to identify every spot on reference gels of *E. coli* and to express in vivo and in vitro every open reading frame (ORF) of this cell's genome (20, 23).

PRACTICAL CONSIDERATIONS FOR THE USE OF 2-D GELS IN PROTEOMICS

Currently the two pathways are linked because both use 2-D gels. For global studies of proteins (the first pathway) 2-D gels offer superb capabilities for separation and quantitation of large numbers of proteins. For proteome mapping (the second pathway) 2-D gels offer precise protein purification of large numbers of proteins. However, 2-D gels also constrict the usefulness of each of these pathways:

Current Limitations of 2-D Gels

In the quarter-century since introduction of the 2-D gel technique many improvements have been

made in gel casting, electrophoresis equipment, software for digitalization and storage of images, spot analysis, and gel matching. Nevertheless, challenges to reproducibility and utility remain.

Current formulations for the first dimension are based on isoelectric focusing with ampholines or on immobilized pH gradients. With either approach there are at least two classes of proteins that do not enter the gel system: (i) proteins with basic isoelectric points, and (ii) proteins designed to be maintained in the lipid environment of cellular membranes (e.g., transmembrane proteins). New materials for separation of basic proteins (5) and solubilization of hydrophobic proteins (21) are being investigated but have not advanced to the stage of being used in high-throughput proteome studies.

Second, good separation of proteins on gels requires ample matrix space. Larger gels permit increased protein loading while maintaining good spot separation (35), but because 2-D gels still require much human handling, size becomes a limitation. Loading more protein on the gels is of benefit to both pathways. Completely automated 2-D gel systems are being designed and tested (1). Methods for fractionating complex mixtures of proteins are available, as are zoom gels, i.e., gels with limited pI and/or molecular weight ranges that allow for higher loading of a small portion of a cell's protein complement and thus facilitate identification by mass spectrometry. However, the latter two developments are counterproductive for merging the two pathways because the reassembly of the fractions back into the context of the expressed proteome is a nearly impossible quantitative task.

Partial Inhibition of Growth

There are good reasons for employing partial (e.g., 30, 50, or even 70%) growth inhibition in proteomic studies. For global studies, the continuing slow growth allows one the opportunity to study the response to the agent or condition of interest uncomplicated by changes in the proteome associated with entry into stationary growth. For proteome mapping, partial growth inhibition provides the opportunity for the cells to accumulate more of the newly expressed proteins, thus providing sufficient material for mass spectrometry analysis.

Varying Experimental Conditions

Experimental design is critical in accurately annotating protein phenotypes, stimulons, regulons, and proteomic signatures. Simply said, looking at cells in only a single state reveals very little. True, some information about the phenotype of a protein (expression level, degradation rate, degree of posttranslational modification) can be determined by looking at one condition. Also, stimulons can be defined by comparing only two cellular states (for example, all the proteins induced and repressed by a shift in growth temperature), but little is obtained beyond a list of responding proteins. Questions asking whether the stimulon would be the same if a different pulse-labeling time were chosen, or if a different temperature were employed, reflect the complexity and ambiguity of stimulons. Regulon identification by proteomics relies on comparing mutants (in regulatory genes) with wild-type strains (yes, concern about the pleiotropic effects of a single mutant is warranted) under numerous conditions suspected to activate/deactivate the regulatory protein. Proteomic signatures are even more difficult to establish than phenotypes, stimulons, and regulons. The more experimental conditions that contribute to the database, the more authentic and useful is the set of proteins one establishes as a signature for a particular cellular process.

Image Analysis

Good image analysis is critical for annotation of protein phenotypes, stimulons, regulons, and proteomic signatures. Tracking individual proteins through many 2-D gels representing many conditions can be done with current image analysis systems. These systems are well designed to annotate protein phenotypes and to identify stimulons. They are all designed to convert the image data into numeric data. Finding regulons and proteomic signatures requires additional analysis of images and/or data from images. None of the current commercially available systems are robust enough for high-throughput analysis. This step has been identified by most groups as a major bottleneck in proteomics. Likewise, none provide the means for searching through a database of stored images.

Matching Gel Images

Is a protein spot on this gel really the same protein as a given spot on another gel? Given the number of possible protein components in each proteome, and the density of spots in any given pI and molecular weight region (3), this is an important issue. Although most global studies and proteome mapping studies currently identify each protein only once, most investigators would like verification of each identification for each experiment. If technology for high-throughput analytical determination of proteins were possible—if mass spectrometry (or another approach) could be routinely performed as a third dimension for the gels—the reliability of data from 2-D gels would be greatly improved (19).

Proteome Mapping

Especially when combined with global studies, the identification of all cellular proteins on 2-D gels could provide valuable information about the characterization of proteins. Many posttranslational modifications of proteins occur in bacteria. These modifications are potentially all important for the function/activity of the protein. By identifying the different protein spots on 2-D gels that are encoded by the same gene, the level of such modifications can be monitored in global studies and eventually might be related to a function or activity of that protein.

Organization and Dissemination of Proteomic Information

Large proteome centers are being established throughout the world. Such centers usually include work on both proteomic pathways for each organism chosen for study. However, given the differences in the technical requirements for the two pathways, it is unlikely that a single laboratory will focus on both pathways. So how then could single laboratories contribute to a concerted effort to study the proteome of an organism? The proteome analysis of *E. coli* is an example of how this could be done. Several groups have done global studies of *E. coli*. These studies have been linked by the spot names given to each protein in the Eco2Dbase (25). Two other groups have done proteome mapping of *E. coli* (11, 22). Integration of much of the proteome mapping data with the global data has been done. Cross-referencing can be easily accessed by the scientific community from each protein's SWISS-PROT entry (2). Web maps of 2-D gel images and information databases are available for many organisms (http://www.expasy.ch/ch2d/2d-index.html), and software tools to compare 2-D images over the web exist (10). Databases of mass spectrometry analysis of protein spots from 2-D gels do not currently exist but should be generated and linked to the gel images as the data for the identification. Such databases may in the future play important roles in the characterization of proteins (e.g., posttranslational modification) and in learning how alterations in the proteins relate to functions in the cells.

PERSPECTIVES

There is every reason to bring together the two pathways of proteomics. In the near future cell biology will advance beyond its current information-rich era into an era in which electronic models will be based on this information and will guide experimentation. The information necessary for this new era of synthesis and modeling must include correspondence between the genome and the proteome, and between them and cell behavior. Hence, both the reductionist and the whole-cell aspects of proteomics will be necessary and will be used in concert.

The need for the second pathway was quite evident in the phosphate study described above (28). Because most proteins resolved on the 2-D gels could not be identified, the large amount of response data and information gleaned from that global study could not be linked to the corpus of biochemical, genetic, and physiological data collected over 30 years. On the other hand, the weaknesses of the second proteomic pathway are evidenced in work that does not use the whole-cell approach (22). When the focus is proteome mapping (linking each spot to a gene on the chromosome), the limitation is the fact that cells make very few proteins (fewer than 500) in a quantity that allows identification by current analytical chemistry methods.

The key to merging the two pathways is technology and experimental design, which need to be driven not by each pathway independently, but by the practical considerations needed to merge the two pathways. Foremost among the challenges for technological advances in proteome analysis of stress responses are (i) improvements in the ability to solubilize hydrophobic proteins and to resolve very basic and very small proteins, (ii) new materials and protocols to improve the reproducibility of individual gels, (iii) development of mass spectrophotometric or other means for high-throughput identification of individual protein spots to assist matching gel images, (iv) increased robustness of image analysis to permit high-throughput analysis of very large sets of gel images, (v) identification of posttranslational modifications that produce multiple products from a single gene, and (vi) improved merging of datasets from different laboratories.

Acknowledgments. Work in the laboratory of F.C.N. has been supported by a grant from the National Science Foundation; the most recent members of this laboratory to contribute to these studies are Robert Clark, Mary Hurley, Gretchen McGannon, and Heather Sims.

We thank the genomics, bioinformatics, and analytical chemistry groups at Parke-Davis Pharmaceutical Research for stimulating discussion of proteomics, including James Cavalcoli, Ping Du, Ron Emaus, Ken Greis (currently at Procter and Gamble), Deborah Hogan, Robert Lepley, Joseph Loo, Rachel Loo, Gary McMaster, Brian Moldover, John Rogers, Erin Schiller, Tracy Stevenson, and Jeffrey Thomas.

REFERENCES

1. **Anderson, N. G., and N. L. Anderson.** 1998. Proteome and proteomics: new technologies, new concepts, and new words. *Electrophoresis* **19**:1853–1861.
2. **Bairoch, A., and R. Apweiler.** 1997. The SWISS-PROT protein sequence data bank and its supplement TREMBL. *Nucleic Acids Res.* **24**:31–36.
3. **Cavalcoli, J. D., R. A. VanBogelen, P. C. Andrews, and B. Moldover.** 1997. Unique identification of proteins from small genome organisms: theoretical feasibility of high throughput proteome analysis. *Electrophoresis* **18**:2703–2708.
4. **Courchesne, P. L., and S. D. Patterson.** 1999. Identification of proteins by matrix-assisted laser desorption/ionization mass spectrometry using peptide and fragment ion masses, p. 487–512. *In* A. J. Link (ed.), *2-D Proteome Analysis Protocols.* Humana Press, Totowa, N.J.
5. **Gorg, A., C. Obermaier, G. Boguth, and W. Weiss.** 1999. Recent developments in two-dimensional gel electrophoresis with immobilized pH gradients: wide pH gradients up to pH 12, longer separation distances and simplified procedures. *Electrophoresis* **20**:712–716.
6. **Herendeen, S. L., R. A. VanBogelen, and F. C. Neidhardt.** 1979. Levels of major proteins of *Escherichia coli* during growth at different temperatures. *J. Bacteriol.* **139**:185–194.
7. **Jensen, O. N., M. Wilm, A. Shevchenko, and M. Mann.** 1999. Sample preparation methods for mass spectrometric peptide mapping directly from 2-DE gels, p. 513–530. *In* A. J. Link (ed.), *2-D Proteome Analysis Protocols.* Humana Press, Totowa, N.J.
8. **Kushner, S.** 1996. mRNA decay, p. 849–860. *In* F. C. Neidhardt, R. Curtis III, J. L. Ingraham, E. C. C. Lin, K. B. Low, B. Magasanik, W. S. Reznikoff, M. Riley, M. Schaechter, and H. E. Umbarger (ed.), *Escherichia coli and Salmonella: Cellular and Molecular Biology,* 2nd ed. ASM Press, Washington, D.C.
9. **Lemaux, P. G., S. L. Herendeen, P. L. Bloch, and F. C. Neidhardt.** 1978. Transient rates of synthesis of individual polypeptides in *E. coli* following temperature shifts. *Cell* **13**:427–434.
10. **Lemkin, P. F.** 1999. Comparing 2-D electrophoresis gels across internet databases, p. 393–410. *In* A. J. Link (ed.), *2-D Proteome Analysis Protocols.* Humana Press, Totowa, N.J.
11. **Link, A. J., K. Robinson, and G. M. Church.** 1997. Comparing the predicted and observed properties of proteins encoded in the genome of *Escherichia coli* K-12. *Electrophoresis* **18**:1259–1313.
12. **Miller, C. G.** 1996. Protein degradation, p. 938–954. *In* F. C. Neidhardt, R. Curtis III, J. L. Ingraham, E. C. C. Lin, K. B.

Low, B. Magasanik, W. S. Reznikoff, M. Riley, M. Schaechter, and H. E. Umbarger (ed.), *Escherichia coli and Salmonella: Cellular and Molecular Biology*, 2nd ed. ASM Press, Washington, D.C.

13. **Neidhardt, F. C., and R. A. VanBogelen.** 1981. Positive regulatory gene for temperature-controlled proteins in *Escherichia coli. Biochem. Biophys. Res. Commun.* **100:**894–900.

14. **Neidhardt, F. C., J. L. Ingraham, and M. Schaechter.** 1990. *Physiology of the Bacterial Cell: a Molecular Approach*, p. 1–6 and 215–216. Sinauer Associates, Inc., Sunderland, Mass.

15. **Neidhardt, F. C., J. L. Ingraham, and M. Schaechter.** 1990. *Physiology of the Bacterial Cell: a Molecular Approach*, p. 306–307. Sinauer Associates, Inc., Sunderland, Mass.

16. **Neidhardt, F. C., J. L. Ingraham, and M. Schaechter.** 1990. *Physiology of the Bacterial Cell: a Molecular Approach*, p. 382–383. Sinauer Associates, Inc., Sunderland, Mass.

17. **O'Farrell, P. H.** 1975. High resolution two-dimensional electrophoresis of proteins. *J. Biol. Chem.* **250:**4007–4021.

18. **Pedersen, S., P. L. Bloch, S. Reeh, and F. C. Neidhardt.** 1978. Patterns of protein synthesis in *E. coli*: a catalog of the amount of 140 individual proteins at different growth rates. *Cell* **14:**179–190.

19. **Quadroni, M., and P. James.** 1999. Proteomics and automation. *Electrophoresis* **20:**664–677.

20. **Sankar, P., M. E. Hutton, R. A. VanBogelen, R. L. Clark, and F. C Neidhardt.** 1993. Expression analysis of cloned chromosomal segments of *Escherichia coli. J. Bacteriol.* **175:**5145–5152.

21. **Santoni, V., T. Rabilloud, P. Doumas, D. Rouquie, M. Mansion, S. Kieffer, J. Garin, and M. Rossignol.** 1999. Toward the recovery of hydrophobic proteins on two-dimensional electrophoresis gels. *Electrophoresis* **20:**705–711.

22. **Tonella, L., B. J. Walsh, J. C. Sanchez, K. Ou, M. R. Wilkins, M. Tyler, S. Frutiger, A. A. Gooley, I. Pescaru, R. D. Appel, J. X. Yan, A. Bairoch, C. Hoogland, F. S. Morch, G. J. Hughes, K. L. Williams, and D. F. Hochstrasser.** 1998. '98 *Escherichia coli* SWISS-2DPAGE database update. *Electrophoresis* **19:**1960–1971.

23. **VanBogelen, R. A.** 1999. Generating a bacterial genome inventory: identifying 2-D spots by comigrating products of the genome on 2-D gels, p. 423–430. *In* A. J. Link (ed.), *2-D Proteome Analysis Protocols*. Humana Press, Totowa, N.J.

24. **VanBogelen, R. A., K. Z. Abshire, B. Moldover, E. R. Olson, and F. C. Neidhardt.** 1997. *Escherichia coli* proteome analysis using the gene-protein database. *Electrophoresis* **18:**1243–1251.

25. **VanBogelen, R. A., K. Z. Abshire, A. Pertsemlidis, R. L. Clark and F. C. Neidhardt.** 1996. Gene-protein database of *Escher-*

ichia coli K-12, edition 6, p. 2067–2117. *In* F. C. Neidhardt, R. Curtis III, J. L. Ingraham, E. C. C. Lin, K. B. Low, B. Magasanik, W. S. Reznikoff, M. Riley, M. Schaechter, and H. E. Umbarger (ed.), *Escherichia coli and Salmonella: Cellular and Molecular Biology*, 2nd ed. ASM Press, Washington, D.C.

26. **VanBogelen, R. A., R. M. Blumenthal, K. Greis, T. Tani, and R. Matthews.** 1999. Mapping regulatory networks in microbial cells. *Trends Microbiol.* **7:**320–328.

27. **VanBogelen, R. A., P. M. Kelley, and F. C. Neidhardt.** 1987. Differential induction of heat shock, SOS and oxidation stress regulons and accumulations of nucleotides in *Escherichia coli. J. Bacteriol.* **169:**26–32.

28. **VanBogelen, R. A., E. R. Olson, B. L. Wanner, and F. C. Neidhardt.** 1996. Global analysis of proteins synthesized during phosphorus restriction in *Escherichia coli. J. Bacteriol.* **178:**4344–4366.

29. **VanBogelen, R. A., E. E. Schiller, J. D. Thomas, and F. C. Neidhardt.** 1999. Diagnosis of cellular states of microbial organisms using proteomics. *Electrophoresis* **20:**2149–2159.

30. **Wanner, B. L.** 1996. Phosphorus assimilation and control of the phosphate regulon, p. 1357–1381. *In* F. C. Neidhardt, R. Curtiss III, C. Gross, J. L. Ingraham, E. C. C. Lin, K. B. Low, B. Magasanik, M. Riley, M. Schaechter, and H. E. Umbarger, (ed.), *Escherichia coli and Salmonella: Cellular and Molecular Biology*, ASM Press, Washington, D.C.

31. **Wasinger, V. C., S. J. Cordwell, A. Cerpa-Poljak, J. X. Yan, A. A. Gooley, M. R. Wilkins, M. W. Duncan, R. Harris, K. L. Williams, and I. Humphrey-Smith.** 1995. Progress with geneproduct mapping of the Mollicutes: *Mycoplasma genitalium. Electrophoresis* **16:**1090–1094.

32. **Wilkins, M. R., K. L. Williams, R. D. Appel, and D. F. Hochstrasser.** 1997. *Proteome Research: New Frontiers in Functional Genomics*. Springer-Verlag, Berlin, Germany.

33. **Yamamori, T., and T. Yura.** 1982. Genetic control of heatshock protein synthesis and its bearing on growth and thermal regulation in *Escherichia coli* K12. *Proc. Natl. Acad. Sci. USA* **79:**860–864.

34. **Yates, J. R., III, E. Carmack, L. Hays, A. J. Link, and J. K. Eng.** 1999. Automated protein identification using microcolumn liquid chromatography-tandem mass spectrometry, p. 553–570. *In* A. J. Link (ed.), *2-D Proteome Analysis Protocols*. Humana Press, Totowa, N.J.

35. **Young, D. A., B. P. Voris, E. V. Maytin, and R. A. Colbert.** 1983. Very-high-resolution two-dimensional electrophoretic separation of proteins on giant gels. *Methods Enzymol.* **91:**190–214.

Bacterial Stress Responses
Edited by G. Storz and R. Hengge-Aronis
©2000 ASM Press, Washington, D.C.

Chapter 29

Applications of Stress Responses for Environmental Monitoring and Molecular Toxicology

ROBERT A. LaROSSA AND TINA K. VAN DYK

To maintain a healthy state the bacterial cell seeks to coordinate its activities through a multiplicity of regulatory mechanisms ranging from simple feedback loops and covalent protein modifications to the sometimes baroque, intertwined regulatory cascades that effect patterns of gene expression. The latter responses often occur at the transcriptional level and can be conveniently monitored with gene fusions. Application of gene fusion technology to the monitoring of stressful conditions and the activation of stress responses is the subject of this contribution. The current status of this technology for environmental monitoring and molecular toxicology is explored.

Regulation within a bacterial cell occurs at several levels with a variety of time constants and specificities (75). Among the fastest responses is the almost instantaneous end product inhibition of biosynthetic pathways. In contrast, minutes are needed for transcription and translation, and proteins are usually stable, lasting for hours or days. Certain responses are specific; for example, the histidine operon's attenuation response is a reflection of the intracellular level of histidyl-tRNA (123). Others, such as the stringent response, are more integrative, measuring the deficiency in any aminoacyl-tRNA species relative to the protein biosynthetic demand (21).

In the 1960s and 1970s, many researchers were studying the regulation of a single operon at the transcriptional level or the allosteric interactions of a small molecule with one enzyme or regulatory protein. A few others were taking more holistic approaches by investigating the pleiotropic effects of regulatory mutants and studying the complexities of diauxie. With the isolation of global regulatory mutants altered in genes such as *cya*, *crp*, *hisT*, and *relA*, and the realization that these mutations effected many aspects of cellular metabolism (33, 44, 99, 125), global approaches to regulation were encouraged. Soon after, technology also become available for acquiring the molecular census of the cell; two-dimensional separations of the nucleotide (17) and protein (81) fractions achieved impressive resolution of the constituent molecular species. These technological and conceptual breakthroughs in the study of global regulatory mechanisms set the stage for the applications of bacterial stress responses described in this chapter.

Early applications used bacterial cell viability as tests for toxicity and mutagenicity. For example, inhibition of bacterial bioluminescence following chemical treatment correlates with loss of viability, a classic measure of toxicity; this is the basis of the Microtox assay that was first described in 1977 (19). Another important assay, reversion of *Salmonella enterica* serovar Typhimurium histidine mutants, commonly known as the Ames test, was introduced in 1971 (2). The Ames test depends on induction of a specific response, error-prone repair, to DNA damage to quantitate mutagenicity of chemicals. These early tests incorporated concepts that are reflected in later, more sophisticated measures. The Microtox test relies on bioluminescence, a rapidly generated and easily quantified signal. In the *Salmonella* mutagenicity assay, the cellular envelope was altered by mutation to allow accumulation of toxins in the cytoplasm while interfering cellular activities such as excision repair were eliminated by genetic manipulations. These early tests, both in current use, are not predisposed to identification of the stress that a cell encountered.

Robert A. LaRossa and Tina K. Van Dyk • Biochemical Science and Engineering, Central Research and Development, DuPont Company, Wilmington, DE 19880-0173.

UTILITY OF STRESS RESPONSE MONITORING

The study of stress responses has many practical outlets. One can view activation or repression of gene expression as an indicator of the cell's physiological state (59). Knowledge of the transcriptional status thus reflects the cell's adaptation to its environment. Such readouts could be useful in the control of fermentations (23) or in the safe operation of industrial or domestic wastewater treatment facilities (8) that protect sources of drinking water such as aquifers, rivers, and lakes from contamination. Such monitoring can also be used to qualify materials for entrance into industrial process streams. Similarly, one might be able to exclude harmful materials from entry into the food chain by the application of easy, rapid, reliable, and sensitive testing of animal feed (111a). Furthermore, the presence of toxic materials, such as heavy metals, in the soil is an environmental burden (101); monitoring of remediation is a potential application of stress-responsive gene fusions. Cost and ease are, of course, but two hurdles that must be cleared before tests will be adopted by government and industry.

Stress responses can also be used to characterize bioactive materials. By knowing that a chemical is inhibitory and understanding the biological responses elicited by that chemical, its detailed effects on the cell may be revealed. For example, if a material induces the stringent response (21), it might be an inhibitor of either amino acid biosynthesis or of tRNA aminoacylation. Likewise, if a compound activates the OxyR regulon (36), one would speculate that it may produce reactive oxygen species. In the same manner, induction of the SOS regulon by a chemical suggests that it causes DNA damage (120). Thus, these transcriptional responses can serve as diagnostic tools, allowing modes of inhibitor action to be suggested. These studies, termed molecular toxicology, have accelerated in the past few years.

Such information on the consequences of drug action is of obvious interest for the control of infectious disease through the use of antibiotics. Furthermore, since metabolic pathways and cellular structures are often common to bacteria, fungi, plants, and animals, studies of the interactions between bacteria and other inhibitory chemicals are also worthwhile. Indeed, several "blockbuster" herbicides have been extensively studied in photosynthetic and enteric bacteria; these bacteriological studies identified biochemical sites of inhibition (4, 29, 64), physiological cascades (65) and stress responses (108) resulting from herbicide application.

POTENTIAL SIGNATURE MOLECULES

The adaptation of bacteria such as *Escherichia coli* to their environment is most remarkable; both cellular barriers and cytoplasmic content change markedly as the surrounding milieu is altered. For example, the lipid component of membranes is tailored to ensure membrane fluidity. Similarly, protein and small molecule content varies with culture conditions. Hence, one can suggest the environment that the cell has encountered from the analysis of cellular pools of small and/or macromolecules. Thus, the potential for deducing the cellular effects of chemical action by analyzing those induced cellular stress responses is clear. The keys are (a) to discover "signature molecules" indicative of a particular stress, (b) to devise facile assays that reflect the intracellular concentrations of those molecules, and (c) to exploit those assays for environmental or other monitoring purposes.

Means for analyzing cellular pools have been acquired steadily. Early work focused on pool measurements of amino acids, glycolytic intermediates, and enzymes, the latter using specific activities in crude extracts as the means of obtaining a titer. Immunoassays provided other means for measuring the content of either specific small molecules or individual proteins. As noted previously, such work was expanded by the successful resolution of the cellular nucleotide (17) and protein (81) pools into their individual components by two-dimensional separation methodologies followed by quantitation of each species. Although much of this work originated with analysis of pools derived from exponentially growing wild-type cultures, the technologies were soon applied to the study of mutant and inhibitor effects. These analyses led to romantic appellations such as "magic spots" for the highly phosphorylated guanylates that accumulate after amino acid starvation (21). Several such small-molecule alarmones (99) have been identified, providing one class of signature molecules. However, the need to develop specific methods to quantitate each of these small molecules has limited practical applications.

The mRNA species transcribed from stress-activated promoters constitute another set of signature molecules, while the corresponding translation products comprise a further series of markers. The levels of these species are often measured directly. Any specific RNA species can be quantitated by Northern analysis or primer extension; both techniques rely on hybridization intensity to measure RNA content. Polypeptide content can be measured by antibody-dependent methods such as immunoprecipitation and Western blotting. Alternatively, the entire protein content can be separated by two-dimensional gel electrophoresis before quantitation of individual species. Presently, however, only about 350 polypeptides, or less than 10 %, have been identified with this method (113). Thus, application of protein-based technologies lags behind nucleic-acid-based detection systems.

REPORTER GENE CASSETTES

Further technological advances involved the construction of gene fusions (1, 20); such fusions allow convenient, though indirect, measurements of specific mRNA species. This approach was attractive due to the short mRNA half-life or mRNA lability found in *E. coli* and other bacteria. The realization that several such fusions could be grown, challenged, and assayed in parallel set the stage for the first comprehensive gene expression profiling experiment that identified mitomycin C-activated genes of *E. coli* (58). Such fusion strains, expected to mirror the responsiveness of wild-type cells to environmental challenges, provide a genetically engineered solution for environmental monitoring. The reporter gene technology originally developed for fundamental studies of gene expression thus has provided a convenient assay to monitor stress responses at the transcriptional level (84, 109). Indeed, most practical applications of stress response monitoring have employed gene fusions, although other approaches such as reverse transcription-PCR are beginning to be utilized in fundamental studies (53, 72).

Several genes and operons have been fused to promoters by in vitro recombinant DNA techniques or in vivo transposon-based strategies. Among them are *E. coli lacZ*, *Pseudomonas putida xylE*, *E. coli phoA*, *E. coli galK*, *E. coli uidA*, *cat* derived from a drug-resistance plasmid, *nptII* from the transposon Tn5, insect *luc*, coelenterate *gfp*, and *luxAB* and *luxCDABE* from a variety of luminescent bacterial species (73). Each reporter gene or operon has attendant advantages, caveats, and limitations when used to define patterns of gene expression (Table 1). *lacZ*, encoding β-galactosidase, is probably the most commonly used reporter due to the early and continued development of the *lac* genetic system (6). *lac*-based selections and screens using histochemical and other indicators are routine, fueling a continuing stream of innovations. A major drawback is that quantitative measurement uses an enzymatic assay requiring cell growth, cell disruption, substrate addition, and product measurement. This suite of operations requires considerable skill and cost in terms of personnel and reagents. Nonetheless, the *lacZ* reporter has been in-

corporated in many cellular biosensors. In contrast, several other reporters have not been widely applied.

Practical use of gene fusion-based biosensors benefits from facile assays of the reporter gene product. Thus, reporters yielding fluorescent or luminescent readouts continue to be most attractive. One such reporter gene, *gfp* (24), encodes a fluorescent protein, thus obviating the need for cell disruption and enzyme assay. However, questions surrounding heterologous expression, such as the rate at which the native protein conformation is achieved after the completion of translation, are legitimate concerns for the utilization of any cell-based *gfp* assay. The *gfp* fusions require expression of the nonfluorescent translation product, its proper attainment of tertiary structure, and an autocatalytic protein modification to achieve a fluorescent state (50). That this last step is particularly sluggish in bacteria (34) limits the utility of *gfp* fusions for studies of bacterial gene expression. Mutagenesis of *gfp* to yield faster-folding variants of green fluorescent protein has resulted in improvements (32); however, the rate is still too slow to allow real-time monitoring of gene expression changes. Few applications of bacterial stress monitoring use *gfp* reporters; however, its use is more common for gene expression monitoring in yeast.

The reporter genes whose products produce visible light are frequently used in developing microbial biosensors. Like *gfp*, the *luxCDABE* reporter produces a cellular signal, thereby eliminating need for cell disruption and enzymatic assay. Expression of the five *lux* genes in the presence of oxygen, ATP, and reducing power results in continuous bioluminescence (71) (Fig. 1). Several luminescent bacteria are sources of *luxCDABE* operons for gene expression monitoring. The gene products of some of these operons, including those derived from *Vibrio fischeri*, *Vibrio harveyi*, and *Photobacterium phosphoreum*, display thermal instabilities at temperatures in the growth range of the typical host bacterial species (25). Marked improvements in reporter performance are achieved by the use of the *Photorhabdus luminescens luxCDABE* genes encoding significantly more robust gene products that are active when expressed in *E. coli* at temperatures up to 42°C (103). As an alternative to the five-gene *luxCDABE* operon, two genes, *luxAB* that encode the heterodimeric luciferase enzyme, are also used as a reporter of gene expression. In this case, light production occurs when a long straight-chain aldehyde substrate, such as *n*-nonanal or *n*-decanal, is added to the cells (25). These lipophilic aldehydes readily diffuse across bacterial membranes, thus making cell lysis unnecessary. Other light-producing reporters are the *luc*-encoded insect luciferases (18). Despite their shared name, the bac-

Table 1. Some reporter gene systems[a]

Gene	Reporter(s)	Substrate	Product(s)	Comments
cat	Chloramphenicol acetyltransferase	Chloramphenicol and acetyl-CoA	CoA and acetyl-chloramphenicol	Sensitive radioactive enzyme assay, chromogenic enzyme assay
galK	Galactokinase	Galactose	Galactose-1-P	Sensitive radioactive enzyme assay
gfp	Green fluorescent protein	Primary translation product	Fluorescent protein	Slow autocatalytic reaction, cellular assay
inaZ	Ice nucleation protein	Water	Ice	Raises freezing points of solutions
lacZ (bgaB)	β-Galactosidase	Lactose	Glucose and galactose	Wide range of selections and screens
		ONPG	Galactose and ONP	ONP is yellow while ONPG is colorless; sensitive enzyme assay
luc	Insect luciferase	X-Gal	Blue precipitate	Colony stain
		D-Luciferin	Light (hν)	Highly sensitive enzymatic assay; cell lysis not required at low pH
luxAB	Bacterial luciferase	Long-chain aldehyde	Long-chain acid and light (hν)	Cellular or highly sensitive enzymatic assay; cell lysis not required
luxCDABE	Bacterial bioluminescence	Central metabolism	Light (hν)	Cellular assay; see Fig. 1 for details
nptII	Neomycin phosphotransferase	Kanamycin and ATP	Kanamycin-P and ADP	
phoA	Alkaline phosphatase	R-O-P	R-OH and HO-P	Active, dimeric enzyme forms only in periplasm
		XP	Blue precipitate	Colony stain
uidA	β-Glucuronidase	β-Glucuronide	D-Glucuronate	
		X-Gluc	Blue precipitate	Colony stain
xylE	Catechol-2,3-dioxygenase	Catechol	2-hydroxy semialdehyde	Product is yellow; suitable for colony staining and enzyme assay

[a] Abbreviations: CoA, coenzyme A; ONPG, o-nitrophenyl-β-D-galactopyranoside; ONP, o-nitrophenol; X-Gal, 5-bromo-4-chloro-3-indolyl-β-D-galactopyranoside; R-O-P, orthophosphoric monoester; R-OH, alcohol; HO-P, phosphoric acid; XP, 5-bromo-4-chloro-3-indolyl phosphate; X-Gluc, 5-bromo-4-chloro-3-indoyl-β-D-glucuronic acid.

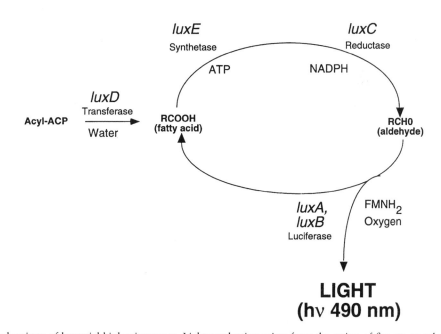

Figure 1. Biochemistry of bacterial bioluminescence. Light production arises from the action of five structural gene products, those encoded by *luxA*, *luxB*, *luxC*, *luxD*, and *luxE* (25). The *luxD* product generates a long-chain fatty acid by hydrolysis of acylated acyl-carrier protein. The long-chain fatty acid is activated in an ATP-dependent reaction to an acyl-enzyme conjugate by the action of the *luxE* product. The acyl-enzyme (*luxE* product) serves as a transfer agent, donating the acyl group to the *luxC* product. The resultant acyl-LuxC binary complex is then reduced to the long-chain aldehyde utilizing NADPH as an electron pair and proton donor. Light production is catalyzed by the heterodimeric luciferase, the product of *luxA* and *luxB*. The energy for light emission is provided by the aldehyde to fatty acid conversion and reduced flavin mononucleotide oxidation.

terial and insect luciferases are unrelated enzymes. The luciferin substrate of the insect luciferase, which is much more costly than the substrate for bacterial luciferase, passes through bacterial membranes only at low pH (22). Thus, for developing inexpensive and facile tests, the *luxCDABE* or *luxAB* reporters are often chosen.

EXPLOITED GENE FUSIONS

A variety of bacterial strains containing many kinds of gene fusions have been developed as cellular biosensors. Table 2 lists such biosensors that are based on stress responses and have been applied to environmental monitoring or molecular toxicology. It is evident that both specific and global regulatory mechanisms have been exploited. Details on some of these applications are given below.

DNA Damage Assessment with Tests Based on the SOS Response

Development of cellular biosensors for detection of agents causing DNA damage is a very active area in the application of stress responses. Fusions to genes of the SOS regulon and to genes of the adaptive

response regulon have been used (Table 2). The latter fusions have not been extensively characterized, while several of the former are quite advanced. The SOS-Chromotest (88) uses a *lacZ* fusion to *sfiA* (*sulA*) in an *E. coli* host strain that is deleted for the *lac* operon so that elevated β-galactosidase activity is a measure of increased induction of the SOS response to DNA damage. The host strain also contains mutations in *uvrA*, which destroys the excision repair pathway, slowing the correct repair of DNA lesions, thus prolonging the SOS response, and *rfa*, which disrupts lipopolysaccharide biosynthesis, thus allowing improved permeability of many molecules (88). This constellation of "accessory" mutations mimics those used to increase the utility of the *Salmonella* histidine reversion assays (2, 3). The SOS-Chromotest test has been extensively used to characterize the ability of chemicals to induce the SOS response, and the results have a good correlation (87) with results from the *Salmonella* histidine reversion assays (3, 69). This correlation, coupled with the hours needed to complete the SOS-Chromotest rather than days required for the histidine reversion assay, makes the gene fusion test most attractive.

More recently, several groups have taken advantage of the simplicity of measuring gene expression by following light emission; they have developed

Table 2. Stress-responsive gene fusions used for environmental monitoring or molecular toxicology[a]

Stress response	Bacterium	*lacZ* fusion(s) (reference)	*lux* fusion(s) (reference)	Other fusions (reference)
SOS (DNA damage)	E. coli	dinD (84) recA (78) sfi (88)	cda (86, 91) λ lysogen (68) recA (33a, 111a, 118) recA mutant (74) umuDC (57; T. Imaeda, O. Asami, and M. Hirai, abstract, 216th ACS National Meeting) uvrA (118)	λ lysogen-*luc* (66)
	Pseudomonas aeruginosa		recA (43)	
	S. enterica serovar Typhimurium	umuDC (79)	recN mutant of E. coli (106, 116)	umuDC-*luc* (92)
Adaptive response (DNA alkylation)	E. coli	ada (84)	alkA (118)	
Heat shock	Bacillus subtilis	clpC-bgaB (93)		clpB-*gfp* (23)
	E. coli	clpB (84) groE (84)	dnaK (110, 111a) grpE (110, 111a) lon (111)	dnaK-*gfp* (23)
OxyR (oxidative stress)	E. coli	katG (84)	katG (11, 111a)	
SoxS/MarA/Rob (superoxide or antibiotic stress)	E. coli	micF (84) nfo (84) soi28 (84) zwf (84)	inaA (83) micF (82a)	
σ^S-dependent stress response	E. coli		osmY (83) xthA (109) yciG (111a)	
Acid pH	S. enterica serovar Typhimurium	aniG (84)		
FadR (fatty acid metabolism)	E. coli	uspA (84)	fabA (109) uspA (112)	
Stringent response (amino acid starvation)	E. coli		his (109)	
cAMP-CRP (carbon limitation)	E. coli	rpoS (84)	lac (82)	
Carbon limitation	P. fluorescens	? (115)		
Phosphate limitation	E. coli		phoA (109)	
	P. fluorescens		pho (61)	
	P. putida	? (40)		
Nitrogen limitation	E. coli		glnA (109)	
	P. fluorescens		? (56)	
Iron availability	P. fluorescens and Pseudomonas syringae			pvd-inaZ (67)
Sulfometuron methyl stress	E. coli		o513 (111a)	
Quorum sensing	E. coli		luxR (5, 102)	
Aluminum	E. coli		fliC (48)	
Arsenite and antimonite	E. coli	arsB (41) arsR (90)	arsD (89)	ars-*luc* (105)
	S. aureus		arsB (30)	ars-*luc* (105)
Cadmium	B. subtilis			cadA-*luc* (104)
	P. fluorescens		? (95)	
	S. aureus		cadA (30)	cadA-*luc* (104)
Copper	A. eutrophus		copA (35)	
Chromate	A. eutrophus		chr (35)	
Lead	A. eutrophus		pbrRA (35)	
Mercury	A. eutrophus		mer (35)	
	E. coli	merR (84)	mer (52, 94)	mer-*luc* (117)
Nickel	E. coli		celF (49)	

[a] ? indicates that the gene to which the reporter is fused has not been identified.

biosensor strains containing *lux* or *luc* fusions to SOS-responsive genes that respond with increased bioluminescence to chemical agents that damage DNA (Table 2). Van Dyk *et al.* first described *luxCDABE* fusions to the *E. coli recA* and *uvrA* promoters in 1994 (109). These biosensors have been extensively characterized (7, 9, 10, 12, 13, 118) and several applications, such as testing contaminated soil (33a) and monitoring the efficacy of wastewater treatment plant action (118), have been demonstrated. The Vitotox test also makes use of a *luxCDABE* fusion to an SOS-regulated gene for detection of mutagens (106, 116). Here, a fusion to a mutant *E. coli recN* promoter is placed in the *Salmonella* histidine reversion strains that, like the host strain for the SOS-Chromotest, contain mutations in excision repair (*uvrB*) and lipopolysaccharide synthesis (*rfa*). Also like the SOS-Chromotest, the Vitotox test shows good correlation with the *his* reversion measures (106), yet has the advantages of *luxCDABE* fusion tests versus assays relying on β-galactosidase quantitation.

Because bacterial cells often lack enzymes that transform chemicals from promutagens to mutagens, it is useful to test compounds that have undergone in vitro activation reactions. Such reactions usually use rat liver extracts from animals that have been previously treated with an inducer of metabolic enzymes. Methods to transform promutagens to mutagens, such as that described by Maron and Ames (69), can be readily adapted to liquid assays for use with strains containing gene fusions. This additional step has been successfully added to many bacterial tests detecting the SOS response (10, 88, 96, 106, 111a). Typical results showing the effect of metabolic activation are depicted in Fig. 2. The *E. coli* strain used in this example contains a mutation in *tolC*, which encodes an outer membrane channel for efflux pumps; detection of the SOS response induced by the metabolically activated 2-aminoanthracene is not possible in an otherwise isogenic *tolC*⁺ strain (111a). Although both the *tolC* mutation and the *rfa* mutation increase host cell sensitivity to many chemicals, the *tolC* mutation works by preventing efflux of chemicals from the cytoplasm while the *rfa* mutation allows more facile access into the cell. Future applications may benefit by using a *rfa tolC* background because such double mutants are even more sensitive to chemicals than either single mutant (45).

Figure 2 further illustrates the kinetics of the bioluminescent response reported by the *recA*:: *luxCDABE* fusion following treatment with a mutagen. Light production from the untreated culture increases slightly during the course of the experiment due to growth of the biosensor strain. In the culture treated with activated 2-aminoanthracene, a lag time

during which no difference in bioluminescence between it and the untreated culture is initially seen. This lag, which is typical of other stress-responsive *luxCDABE* fusion strains (107), is attributed to the time required for generating the stress signal and for subsequent transcription and translation of the *luxCDABE* reporter. Following the lag, a large increase in the bioluminescent signal from the biosensor treated with the activated chemical is observed, indicating increased gene expression. Such kinetic data are readily obtained when the *lux* reporter is used because light output can be monitored in real time without cell disruption.

Detection of the Heat Shock Response

The induction of the heat shock response is a useful indicator of biological stress in a wide range of organisms, both prokaryotic and eukaryotic (37, 46). Thus, bacterial biosensors reporting induction of this universal response are effective indicators of environmental stresses. Biosensor strains containing genetic fusions of *E. coli* heat shock responsive promoters to the *luxCDABE* reporter respond with increased light production to many stresses (110). Metals, organic solvents, herbicides, fungicides, weak acids, detergents, UV irradiation, as well as other classes of toxicants, induce this stress response to various degrees (111, 112), thus suggesting that induction of the heat shock response is a general indication of adverse environmental conditions. The bioluminescent response from these fusion strains is induced at sublethal concentrations of the environmental insults, thereby providing an advantage in its low end toxicant detection limit over systems that rely on cell death (10); a similar advantage in sensitivity is evident in work on chemical induction of *E. coli* heat shock proteins analyzed by sodium dodecyl sulfate-polyacrylamide gel electrophoresis (SDS-PAGE) (80) or by two-dimensional PAGE (16). The sensitivity of *E. coli* strains carrying heat shock promoter-*luxCDABE* fusions to hydrophobic chemicals is further enhanced by introduction of a *tolC* mutation that enhances toxicant accumulation (110). This class of biosensors provides an alternative to general toxicity tests with the advantages of improved sensitivity and response time.

Starvation Responsive Sensors

Development of nutrient limitation-responsive fusions is an area of recent activity. In *E. coli*, fusions of the *V. fischeri luxCDABE* operon have been made to the *lac* promoter to indicate carbon starvation, the *phoA* promoter to signal phosphate starvation, the

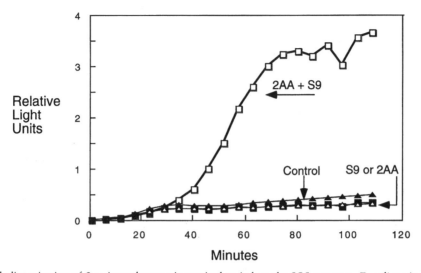

Figure 2. Metabolic activation of 2-aminoanthracene is required to induce the SOS response. *E. coli* strain DPD2222 (Van Dyk et al., submitted) containing a chromosomal *recA-lux* fusion in a *tolC* host was treated with 50 μg/ml 2-aminoanthracene with and without pretreatment with a rat liver postmitochondrial S9 fraction. Kinetics of light production (relative light units) after addition of compound to the cell culture are plotted. The solid triangles represent the control with 0.25% dimethyl sulfoxide added, which was also present in all other samples. The open triangles represent the culture that contains the S9 extract at 1.5%. The solid squares represent the cell culture to which unactivated 2-aminoanthracene was added. The open squares represent the cell culture amended with 2-aminoanthracene that had been incubated with a mixture of 30% S9 extract and cofactors (Van Dyk et al., submitted) for 10 min before addition to the biosensor cells.

glnA2 promoter to reflect nitrogen starvation, and the *his* promoter separated from the *his* attenuator to detect starvation for any amino acid through triggering of the stringent response (109). These *E. coli* fusions have not been extensively tested in environmental conditions; however, similar fusions in the soil bacteria *Pseudomonas fluorescens* and *P. putida* that respond to nitrogen, phosphate, or carbon limitation have been demonstrated to report on nutrient availability in the environment. A transposon containing a promoterless *V. fischeri luxAB* cassette has been used to identify nitrogen- and phosphate-regulated fusions of *P. fluorescens* (60). The bioluminescence from one such nitrogen limitation-responsive strain, in which *luxAB* is fused to a gene without homology to known genes, is expressed at ammonia concentrations below 10 to 90 μM in cell culture; however, when this strain is introduced into soil, nitrogen starvation is signaled only if the soil is amended with carbon and phosphorous sources (56). This underscores that cells must be metabolically competent, having sufficient ATP and reducing power, if bioluminescence is to be observed. One of the phosphate starvation-inducible fusions of *P. fluorescens* contains *luxAB* driven by a promoter with sequences typical of Pho regulon members (61). This biosensor is useful to distinguish environmental conditions where phosphate limitation occurs in the vicinity of barley roots (61). Likewise, *lacZ* fusions to phosphate-regulated genes in *P. putida* are useful to characterize growth

conditions in the rhizosphere and in bulk soil or sand (40). Similarly, a *P. fluorescens* strain with a *lacZ* fusion, which had been found from a screen for response to carbon limitation, is useful to characterize the availability of carbon in soil (115). When test conditions have been optimized, this class of starvation-responsive biosensors will be useful to characterize the nutritional stresses microorganisms encounter in aqueous as well as soil environments.

Biosensors for Detection of Toxic Metals

Monitoring specific stress responses to toxic metals was an early, and continues to be an important, application of gene fusion technology. In contrast to the examples given thus far, where regulatory circuits controlling expression of multiple genes have been utilized, metal-responsive sensors typically exploit more specific regulatory circuits. In the case of a recently described lead-responsive *pbrA-lux* gene fusion in *Alcaligenes eutrophus*, the product of a regulatory gene, *pbrR,* is responsible for lead-specific induction (35). Metal-responsive sensors are often sensitive to very low concentrations of metals; for example, antimonite and arsenite biosensor strains utilizing *lux* or *lacZ* reporters can detect concentrations of these metals as low as 10^{-15} M (89, 90). Some metal-responsive sensors are specifically induced by one metal; for example, an *E. coli* strain carrying a *mer-luc* fusion, which can detect 10^{-16} M mercury, is not induced by zinc, cobalt, copper, or manganese ions (117). In contrast, others metal-responsive sensors are induced by a limited group of toxic metals, such as *cadA-luc* fusion-containing strains that respond to cadmium, lead, and antimony (104). The utility of some of these metal-responsive biosensors has been demonstrated in field conditions. For example, mercury-contaminated pond water induces a *mer-lux* fusion; the levels of mercury determined by the bioluminescent strain are in reasonable agreement with those determined by analytical chemistry, suggesting that such fusions are useful sensors of bioavailable mercury in aqueous environments (94). Bioavailable toxic metal contaminants in solids such as incinerator fly ash and soil samples are similarly detected with metal-responsive biosensors (31). This class of biosensors will continue to find utility in assessing the risk of contaminated materials or environments, as well as in monitoring the efficacy of clean-up efforts.

STRESS FINGERPRINTS FROM PANELS OF GENE FUSIONS

Thousands of transcriptional units exist in a bacterium, each presumably responsive to one or more control circuits. Bacteria thus integrate a wealth of information as they tune themselves to perform optimally in their environment. How much of that information can be accessed? How much is needed for developing useful applications? While answers to these questions may eventually emerge, current practice utilizes two basic discovery platforms (75). In the first, global regulatory mechanisms are uncovered by mutant selection and analysis; this avenue leads to the definition of regulons, sets of genes each controlled by a master regulatory gene. The second approach defines an alternative gene grouping, termed a stimulon. This identifies all genes induced by a single environmental challenge. Knowledge of stimulons and regulons (76), coupled with the complete sequencing of experimentally tractable bacterial genomes (15), has allowed an engineering approach to be applied. Hence, panels of gene fusions thought to cover all important regulatory circuits have been assembled, allowing generation of molecular toxicological fingerprints of chemical action.

The well-studied bacterium *E. coli* is the host strain for gene fusions forming two panels of stress-responsive biosensors. In one, a group of 15 *lacZ* fusion strains make up the "ProTox" test (84); the induction activity of numerous chemicals has been tested with this panel and characteristic patterns of stress induction are found for various classes of chemicals. Similarly, panels of *lux* genetic fusions, each strain of which monitors a different stress response, have been described (10, 109; Van Dyk et al., submitted). In a like manner to the "Pro-Tox" *lacZ* fusion panel, the pattern of gene expression, or "stress fingerprint," induced by a chemical or environmental sample yields information on the physiological effects of that insult. Furthermore, these stress fingerprints allow grouping of chemicals into classes that elicit comparable transcriptional responses. Four examples of stress fingerprints induced by chemicals with distinct mechanisms of toxicity are shown in Fig. 3. Ethanol treatment of a panel of six strains (Fig. 3, panel A) resulted in increased expression from only the *grpE:: lux* fusion, a member of the heat shock regulon. This result is consistent with the known potent induction of the heat shock response by ethanol (110, 114, 122). A DNA gyrase inhibitor, nalidixic acid, also induced a somewhat specific pattern of increased expression (Fig. 3, panel B); the fusion to *recA*, a member of the SOS regulon, was very highly induced, while there was a low level of induction of the *grpE-lux* fusion. In contrast to these chemicals with fairly specific stress fingerprints, hydrogen peroxide induced elevated expression of three of the six panel members (Fig. 3, panel C). A reasonable interpretation of this characteristic fingerprint is that hydrogen peroxide damages DNA, as indicated by induction of the SOS response, and protein, as reported by induction of the heat shock response, through a peroxide-mediated mechanism as sensed by the elevated expression of a member of the OxyR regulon. A fungicide, captan, which is a thiol-reactive molecule, also induced multiple stress responses (Fig. 3, panel D); five of the six stress responses monitored by the panel were elevated, consistent with its nonspecific toxic effect on the cell. Nevertheless, the very potent induction of the heat shock response suggests that captan severely damages protein. Induction of the SOS response was also observed, consistent with the mutagenic scoring of captan with both the SOS-Chromotest and the Ames test (87). It is also worth noting that while hydrogen peroxide and captan both appear to have nonspecific toxic effects, the modes of their biological effects were, nevertheless, distinguished by the induction patterns obtained with this small set of six stress-responsive *lux* fusion strains.

Other such panels containing bioluminescent biosensors responsive to protein denaturation, DNA damage, oxidative damage caused by hydrogen peroxide or reactive oxygen species, membrane synthesis limitations, various starvation conditions, and other stresses have been shown to be useful in evaluating influent and effluent from wastewater treatment plants (7, 10, 12) and in analyzing the modes of ac-

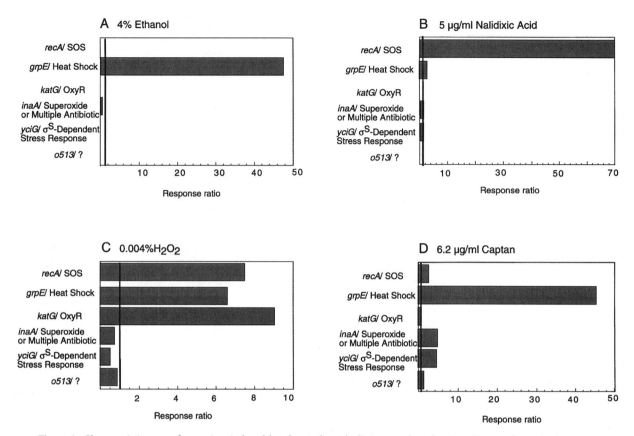

Figure 3. Characteristic stress fingerprints induced by chemicals with distinct modes of action. Six members of a biosensor panel, each containing a *lux* fusion to a gene from a distinct regulon (Van Dyk et al., submitted), were individually grown in Luria-Bertani medium to mid-exponential phase and treated with 4% ethanol (panel A), 5 μg/ml nalidixic acid (panel B), 0.004% hydrogen peroxide (panel C), or 6.2 μg/ml of captan (panel D). Following quantitation of bioluminescence, the response ratios, which are the ratio of the relative light units of the chemically treated cell culture divided by the relative light units of the untreated culture (107), were calculated at the point of maximal response. This was at 40 to 45 min for the strains containing *lux* fusions to *grpE*, *katG*, *inaA*, and *yciG* and at 110 min for the *recA-lux* and the *o513-lux* fusion strains. Response ratios greater than 1.0, which is indicated by the thick line, represent induction of gene expression.

tion of antibacterial treatments and antibiotics (42, 82, 82a, 83, 85). The wealth of information on the physiological basis of *E. coli* stress responses often allows a rational deduction of the biological effects of chemicals from observed induction patterns. However, caution and further experimentation may be necessary because many genes in *E. coli* are subject to control by multiple regulatory circuits and all regulatory circuits controlling expression of particular genes may not be known. Dissection of multiple regulation is possible by formation of gene fusions using only parts of the upstream regulatory regions, such as separation of the *his* promoter (63) controlled by the stringent response (124) from the *his* operon attenuator responsible for histidine starvation-specific control (62, 123). Use of bacterial strains containing chromosomal mutations in regulatory genes is also a useful method for discovering the regulatory circuits responsible for stress-induced gene expression. For example, a membrane-active antibiotic, polymyxin,

induces elevated expression of a *micF-lux* fusion; interpretation of this response was not initially possible because of the multiple regulatory proteins effecting *micF* expression. However, use of strains mutant in *soxRS*, *marA*, or *robA* clearly demonstrated the unexpected role of *robA* in the polymyxin-mediated *micF* response (82a). An alternative analysis of induction patterns is provided by empirical approaches, such as use of clustering algorithms; such methods allow chemicals to be grouped without requiring knowledge of the regulatory significance of the individual panel members (14).

Despite the success of these panels in yielding molecular toxicological information, a challenge to the assumption that all important stress responses have already been characterized has recently arisen. The identification of many unexpected gene fusions (108) induced by the well-studied, and supposedly well-understood, chemical challenge imposed by the herbicide sulfometuron methyl (64) suggests that there are stress responses still to be discovered, even in *E. coli*. Such work resonates with earlier studies of Christman et al. (27) where it was shown that hydrogen peroxide induced both members of the OxyR regulon and many other proteins controlled by unknown regulatory circuits. Likewise, the work of Matin (16), who demonstrated induction of many previously uncharacterized proteins after exposure to various pollutant molecules, sounds a similar chord. Together, these studies suggest that current panel approaches are incomplete, perhaps revealing only a portion of the story. Recent developments suggest comprehensive gene profiling will vastly improve selection of fusions for inclusion in panels of biosensors.

SUPPLYING CELLULAR BIOSENSORS

The simplest way to use cellular biosensors containing gene fusions is by contacting a chemical or environmental sample with actively growing cells containing the fusion of interest. However, growing fresh bacterial cultures for each experiment is not optimal for high-throughput or repetitive applications, may result in reproducibility issues, and is not practical for field applications. Thus, several alternatives have been explored; these include freezing, immobilization, lyophilization, continuous culture, and construction of integrated circuits incorporating microbial cells.

A relatively simple upgrade to daily cultivation is the use of frozen culture aliquots. This approach has been successfully demonstrated with an *E. coli* bioluminescent sensor that is competent for gene expression assays immediately after thawing, having been previously stored at −80°C using glycerol as a cryoprotectant (100). Although such a method makes laboratory experiments more convenient by separating tester strain preparation from exposure, it probably will not be useful for field applications.

Two other common methods of handling bacteria, lyophilization and continuous culture, have also proven to be useful sources of *lux* fusion strains for testing purposes. Lyophilization has been successfully applied to several metal responsive cellular biosensor strains (31, 104, 105). Surprisingly, lyophilization of an *E. coli* strain containing a heat shock promoter-*lux* fusion did not overwhelmingly stress the cells because they could be rehydrated and used immediately to detect a chemically induced heat shock response (119). Continuous cultivation of this same *E. coli* biosensor in minibioreactors has also allowed reproducible detection of the stress response (47). Continuous culture makes it possible to consider on-line ap-

plications, such as monitoring wastewater treatment facilities, while lyophilization can extend testing from the laboratory out to the field.

Immobilization by entrapment in a carrier material, an approach that has often been applied to isolated enzymes, has also succeeded with cellular biosensors. Strontium alginate immobilization of a *lux* fusion-containing bacterial strain enables its use as a probe for wastestreams (51). This immobilized biosensor has been tested under conditions that simulate subsurface environments (121) and groundwater with various pH regimens (70). In another example of immobilization, calcium alginate beads, harboring an SOS responsive *lux* fusion strain, stored in a $CaCl_2$ solution at 4°C were found to give useful DNA damage induction responses for up to 1 month after formation (33a). Likewise, calcium alginate immobilization of a copper-responsive biosensor results in superior stability of the biosensor relative to the immobilization of the same biosensor in agarose (35). Combination of immobilized cells and light detection equipment is possible for sensors that produce visible light as a signal; in one such case, an *E. coli luc* fusion strain is immobilized on the end of a fiber optic monitoring device (55). Such immobilized biosensor strains will allow more facile field applications.

A particularly interesting recent development is the bioluminescent bioreporter integrated circuit, or BBIC (97). In this approach, which takes advantage of the cellular signal generated by the five-gene *luxCDABE* reporter, a bioluminescent biosensor strain is deposited onto a microluminometer fabricated within an integrated circuit. The light produced by the biosensor is detected by the integrated circuit, which then processes and communicates the results. The BBIC, thus, has the capability of functioning as a complete stand-alone sensor (98).

CONCLUSIONS

Exploitation of stress-responsive regulatory circuits is in its infancy. Providing impetus for advances are several recent milestones, including the complete sequencing of several microbial genomes. Such sequencing capacity can be used to precisely map transposition events and gene fusions, allowing the rapid correlation of phenotype with precise genotype. Such sequence-based mapping of interesting gene fusions and insertions is already common (77, 108). Furthermore, complete genomic sequence data make possible formation of ordered arrays of DNA containing each open reading frame of a species (38) or complementary oligonucleotides (26) that serve as capture reagents. These reagents allow comprehensive analysis of the individual mRNA species present in any RNA sample. Such analyses have been reported using yeast as an experimental system (54); the application of such technology to the study of *E. coli* and other bacteria has been both foreshadowed (28, 39) and is being realized by several groups. It is envisioned that such comprehensive, array-based hybridization tests will become an important discovery tool to identify genes whose expression is severely altered by adverse environmental conditions. The resulting hypotheses will need to be tested; to this end, gene fusions will allow both independent verification of hypotheses and convenient cellular biosensors for facile, routine, and continuous monitoring purposes. It is likely that these tools will continue to stimulate the integrative approach to biology that is a hallmark of the stress response field.

REFERENCES

1. **Ames, B. N., P. E. Hartman, and F. Jacob.** 1963. Chromosomal alterations affecting the regulation of histidine enzymes in *Salmonella. Mol. Biol.* 7:23–42.

2. **Ames, B. N., F. D. Lee, and W. E. Durston.** 1973. An improved bacterial test system for the detection and clasification of mutagens and carcinogens. *Proc. Natl. Acad. Sci. USA* 70:782–786.

3. **Ames, B. N., J. McCann, and E. Yamasaki.** 1975. Methods for detecting carcinogens and mutagens with the *Salmonella/mammalian*-microsome mutagenicity test. *Mutat. Res.* 31:347–364.

4. **Arntzen, C. J., and J. H. Duesing.** 1984. Chloroplast-encoded herbicide resistance, p. 273–294. *In* F. Ahmand, K. Downey, J. Schultz, and R. W. Voellmy (ed.), *Advances in Gene Technology: Molecular Genetics of Plants and Animals.* Academic Press, New York, N.Y.

5. **Batchelor, S. E., M. Cooper, S. R. Chhabra, L. A. Glover, G. S. A. B. Stewart, P. Williams, and J. I. Prosser.** 1997. Cell density-regulated recovery of starved biofilm populations of ammonia-oxidizing bacteria. *Appl. Environ. Microbiol.* 63:2281–2286.

6. **Beckwith, J.** 1996. The operon: an historical account, p. 1227–1231. *In* F. C. Neidhardt, R. Curtiss III, J. L. Ingraham, E. C. C. Lin, K. B. Low, B. Magasanik, W. S. Reznikoff, M. Riley, M. Schaechter, and H. E. Umbarger (ed.), *Escherichia coli and Salmonella: Cellular and Molecular Biology,* 2nd ed. ASM Press, Washington, D.C.

7. **Belkin, S.** 1998. A panel of stress-responsive luminous bacteria for monitoring wastewater toxicity, p. 247–258. *In* R. A. LaRossa (ed.), *Methods in Molecular Biology: Bioluminescence Methods and Protocols,* vol. 102. Humana Press Inc, Totowa, N.H.

8. **Belkin, S.** 1998. Stress-responsive luminous bacteria for toxicity and genotoxicity monitoring, p. 171–183. *In* P. G. Wells, K. Lee, and C. Blaise (ed.), *Microscale Testing in Aquatic Toxicology.* CRC, Boca Raton, Fla.

9. **Belkin, S., and R. A. LaRossa.** 1997. Biotechnological applications of microbial stress responses: new trends in environmental monitoring, p. 565–570. *In* M. T. Martins, M. I. Z. Sato, J. M. Tiedje, L. C. N. Hagler, J. Döbereiner, and P. S. Sanchez (ed.), *Progress in Microbial Ecology.* Brazilian Society for Microbiology, São Paulo, Brazil.

10. **Belkin, S., D. R. Smulski, S. Dadon, A. C. Vollmer, T. K. Van Dyk, and R. A. LaRossa.** 1997. A panel of stress-responsive luminous bacteria for the detection of selected classes of toxicants. *Water Res.* 31:3009–3016.

11. **Belkin, S., D. R. Smulski, A. C. Vollmer, T. K. Van Dyk, and R. A. LaRossa.** 1996. Oxidative stress detection with *Escherichia coli* harboring a *katG′::lux* fusion. *Appl. Environ. Microbiol.* 62:2252–2256.

12. **Belkin, S., T. K. Van Dyk, A. C. Vollmer, D. R. Smulski, and R. A. LaRossa.** 1996. Monitoring subtoxic environmental hazards by stress-responsive luminous bacteria. *Environ. Toxicol. Water Qual.* 11:179–185.

13. **Belkin, S., A. C. Vollmer, T. K. Van Dyk, D. R. Smulski, T. R. Reed, and R. A. LaRossa.** 1994. Oxidative and DNA damaging agents induce luminescence in *E. coli* harboring *lux* fusions to stress promoters, p. 509–512. *In* A. K. Cambell, L. J. Kricka, and P. E. Stanley (ed.), *Bioluminescence and Chemiluminescence: Fundamentals and Applied Aspects.* John Wiley & Sons, Chichester, England.

14. **Ben-Israel, O., H. Ben-Israel, and S. Ulitzur.** 1998. Identification of quantification of toxic chemicals by use of *Escherichia coli* carrying *lux* genes fused to stress promoters. *Appl. Environ. Microbiol.* 64:4346–4352.

15. **Blattner, F. R., G. Plunkett III, C. A. Bloch, N. T. Perna, V. Burland, M. Riley, J. Collado-Vides, J. D. Glasner, C. K. Rode, G. F. Mayhew, J. Gregor, N. W. Davis, H. A. Kirkpatrick, M. A. Goeden, D. J. Rose, B. Mau, and Y. Shao.** 1997. The complete genome sequence of *Escherichia coli* K-12. *Science* 277:1453–1462.

16. **Blom, A., W. Harder, and A. Matin.** 1992. Unique and overlapping pollutant stress proteins of *Escherichia coli. Appl. Environ. Microbiol.* 58:331–334.

17. **Bochner, B. R., and B. N. Ames.** 1982. Complete analysis of cellular nucleotides by two-dimensional thin layer chromatography. *J. Biol. Chem.* 257:9759–9769.

18. **Bronstein, I., J. Fortin, P. E. Stanley, G. S. A. B. Stewart, and L. J. Kricka.** 1994. Chemiluminescent and bioluminescent reporter gene assays. *Anal. Biochem.* 219:169–181.

19. **Bulich, A. A.** 1982. A practical and reliable method for monitoring the toxicity of aquatic samples. *Process Biochem.* 17:45–47.

20. **Casadaban, M. J., and S. N. Cohen.** 1980. Lactose genes fused to exogenous promoters in one step using a new *lac* bacteriophage: in vivo probe for transcriptional control sequences. *Proc. Natl. Acad. Sci USA* 76:4530–4533.

21. **Cashel, M., D. R. Gentry, V. J. Hernandez, and D. Vinella.** 1996. The stringent response, p. 1458–1496. *In* F. C. Neidhardt, R. Curtiss III, J. L. Ingraham, E. C. C. Lin, K. B. Low, B. Magasanik, W. S. Reznikoff, M. Riley, M. Schaechter, and H. F. Umbarger (ed.), *Escherichia coli and Salmonella: Cellular and Molecular Biology,* 2nd ed. ASM Press, Washington, D.C.

22. **Cebolla, A., M. E. Vázquez, and A. J. Palomares.** 1995. Expression vectors for the use of eukaryotic luciferases as bacterial markers with different colors of luminescence. *Appl. Environ. Microbiol.* 61:660–668.

23. **Cha, H. J., R. Srivastava, V. N. Vakharia, G. Rao, and W. E. Bentley.** 1999. Green fluorescent protein as a noninvasive stress probe in resting *Escherichia coli* cells. *Appl. Environ. Microbiol.* 65:409–414.

24. **Chalfie, M., Y. Tu, G. Euskircheng, W. W. Ward, and D. C. Prasher.** 1994. Green fluorescent protein as marker for gene expression. *Science* 263:802–805.

25. **Chatterjee, J., and E. A. Meighen.** 1995. Biotechnological applications of bacterial bioluminescence (*lux*) genes. *Photochem. Photobiol.* 62:641–650.

26. Chee, M., R. Yang, E. Hubbell, A. Berno, X. C. Huang, D. Stern, J. Winkler, D. J. Lockhart, M. S. Morris, and S. P. Fodor. 1996. Accessing genetic information with high-density DNA arrays. *Science* **274**:610–614.

27. Christman, M. F., R. W. Morgan, F. S. Jacobson, and B. N. Ames. 1985. Positive control of a regulon for defenses against oxidative stress and some heat-shock proteins in *Salmonella typhimurium*. *Cell* **41**:735–762.

28. Chuang, S.-E., D. Daniels, and F. R. Blattner. 1993. Global regulation of gene expression in *Escherichia coli*. *J. Bacteriol.* **175**:2026–2036.

29. Comai, L., L. C. Sen, and D. M. Stalker. 1983. An altered *aroA* gene product confers resistance to the herbicide glyphosate. *Science* **221**:370–371.

30. Corbisier, P., G. Ji, G. Nuyts, M. Mergeay, and S. Silver. 1993. *luxAB* gene fusions with the arsenic and cadmium resistance operons of *Staphylococcus aureus* plasmid pI258. *FEMS Microbiol. Lett.* **110**:231–238.

31. Corbisier, P., E. Thiry, and L. Diels. 1996. Bacterial biosensors for the toxicity assessment of solid wastes. *Environ. Toxicol. Water Qual.* **11**:171–177.

32. Cormack, B. P., R. H. Valdivia, and S. Falkow. 1996. FACS-optimized mutants of the green fluorescent protein (GFP). *Gene* **173**:33–38.

33. Cortese, R., R. A. Landsberg, R. A. von der Haar, H. E. Umbarger, and B. N. Ames. 1974. Pleiotropy of *hisT* mutants blocked in pseudouridine synthesis in tRNA: leucine and isoleucine-valine operons. *Proc. Natl. Acad. Sci. USA* **71**:1857–1861.

33a. Davidov, Y., D. R. Smulski, T. K. Van Dyk, A. C. Vollmer, D. A. Elsemore, R. A. LaRossa, and S. Belkin. Assessment of SOS promoter::*lux* fusions for genotoxicity detection. *Mutat. Res.*, in press.

34. Davis, D. F., W. W. Ward, and M. W. Cutler. 1994. Posttranslational chromophore formation in recombinant GFP from *E. coli* requires oxygen, p. 596–599. *In* A. K. Cambell, L. J. Kricka, and P. E. Stanley (ed.), *Bioluminescence and Chemiluminescence: Fundamentals and Applied Aspects*. John Wiley & Sons, Chichester, England.

35. de Lorenzo, V., N. L. Brown, J. R. Lloyd, J. L. Hobman, E. Csöregi, G. Johansson, B. Mattiasson, D. van der Lelie, L. Diels, and P. Corbisier. 1999. Whole cell- and protein-based biosensors for the detection of bioavailable heavy metals in environmental samples. *Anal. Chim. Acta* **387**:235–244.

36. Demple, B. 1991. Regulation of bacterial oxidative stress genes. *Annu. Rev. Genet.* **25**:315–337.

37. de Pomerai, D. I. 1996. Heat-shock proteins as biomarkers of pollution. *Hum. Exp. Toxicol.* **15**:279–285.

38. DeRisi, J. L., V. R. Iyer, and P. O. Brown. 1997. Exploring the metabolic and genetic control of gene expression on a genomic scale. *Science* **278**:680–686.

39. de Saizieu, A., U. Certa, J. Warrington, C. Gray, W. Keck, and J. Mous. 1998. Bacterial transcript imaging by hybridization of total RNA to oligonucleotide arrays. *Nat. Biotechnol.* **16**:45–48.

40. de Weger, L. A., L. C. Dekkers, A. J. van der Bij, and B. J. J. Lugtenberg. 1994. Use of phosphate-reporter bacteria to study phosphate limitation in the rhizosphere and in bulk soil. *Mol. Plant-Microbe Interact.* **7**:32–38.

41. Diorio, C., J. Cai, J. Marmor, R. Shinder, and M. S. DuBow. 1995. An *Escherichia coli* chromosomal *ars* operon homolog is functional in arsenic detoxification and is conserved in gram-negative bacteria. *J. Bacteriol.* **177**:2050–2056.

42. Dukan, S., S. Dadon, D. R. Smulski, and S. Belkin. 1996. Hypochlorous acid activates the heat shock and *soxRS* systems of *Escherichia coli*. *Appl. Environ. Microbiol.* **62**:4003–4008.

43. Elasri, M. O., and R. V. Miller. 1998. A *Pseudomonas aeruginosa* biosensor responds to exposure to ultraviolet radiation. *Appl. Microbiol. Biotechnol.* **50**:455–458.

44. Emmer, M., B. de Crombrugghe, I. Pastan, and R. Perlman. 1970. Cyclic AMP receptor protein of *E. coli*: its role in the synthesis of inducible enzymes. *Proc. Natl. Acad. Sci. USA* **66**:480–487.

45. Fralick, J. E., and L. L. Burns-Keliher. 1994. Additive effect of *tolC* and *rfa* mutations on the hydrophobic barrier of the outer membrane of *Escherichia coli* K-12. *J. Bacteriol.* **176**:6404–6406.

46. Goering, P. L. 1995. Stress proteins. Molecular biomarkers of chemical exposure and toxicity. *Environ. Sci. Res.* **50**:217–227.

47. Gu, M. B., P. S. Dhurjati, T. K. Van Dyk, and R. A. LaRossa. 1996. A miniature bioreactor for sensing toxicity using recombinant bioluminescent *Escherichia coli* cells. *Biotechnol. Prog.* **12**:393–397.

48. Guzzo, A., C. Diorio, and M. S. DuBow. 1991. Transcription of the *Escherichia coli fliC* gene is regulated by metal ions. *Appl. Environ. Microbiol.* **57**:2255–2259.

49. Guzzo, A., and M. S. DuBow. 1994. A *luxAB* transcriptional fusion to the cryptic *celF* gene of *Escherichia coli* displays increased luminescence in the presence of nickel. *Mol. Gen. Genet.* **242**:455–460.

50. Heim, R., D. C. Prasher, and R. Y. Tsien. 1994. Wavelength mutations and posttranslational autooxidation of green fluorescent protein. *Proc. Natl. Acad. Sci. USA* **91**:12501–12504.

51. Heitzer, A., K. Malachowsky, J. E. Thonnard, P. R. Beinkowski, D. White, and G. S. Sayler. 1994. Optical biosensor for environmental on-line monitoring of naphthalene and salicylate bioavailability with an immobilized bioluminescent catabolic reporter bacterium. *Appl. Environ. Microbiol.* **60**:1487–1494.

52. Holmes, D. S., S. K. Dubey, and S. Gangolli. 1994. Development of biosensors for the detection of mercury and copper ions. *Environ. Geochem. Health* **16**:229–233.

53. Holmstrøm, K., T. Tolker-Nielsen, and S. Molin. 1999. Physiological states of individual *Salmonella typhimurium* cells monitored by in situ reverse transcription-PCR. *J. Bacteriol.* **181**:1733–1738.

54. Holstege, F. C. P., E. G. Jennings, J. J. Wyrick, T. I. Lee, C. J. Hengartner, M. R. Green, T. R. Golub, E. S. Lander, and R. A. Young. 1998. Dissecting the regulatory circuitry of a eukaryotic genome. *Cell* **95**:717–728.

55. Ikariyama, Y., S. Nishiguchi, T. Koyama, E. Kobatake, M. Aizawa, M. Tsuda, and T. Nakazawa. 1997. Fiber-optic-based biomonitoring of benzene derivatives by recombinant *E. coli* bearing luciferase gene-fused TOL-plasmid immobilized on the fiber-optic end. *Anal. Chem.* **69**:2600–2605.

56. Jensen, L. E., L. Kragelund, and O. Nybroe. 1998. Expression of a nitrogen regulated *lux* gene fusion in *Pseudomonas fluorescens* DF57 studied in pure culture and in soil. *FEMS Microbiol. Ecol.* **25**:23–32.

57. Justus, T., and S. M. Thomas. 1998. Construction of a *umuC'-luxAB* plasmid for the detection of mutagenic DNA repair via luminescence. *Mutat. Res.* **398**:131–142.

58. Kenyon, C. J., and G. C. Walker. 1980. DNA damaging agents stimulate gene expression at specific loci in *Escherichia coli*. *Proc. Natl. Acad. Sci. USA* **77**:2819–2823.

59. Konstantinov, K. B. 1996. Monitoring and control of the physiological state of cell cultures. *Biotechnol. Bioeng.* **52**:271–289.

60. Kragelund, L., B. Christoffersen, O. Nybroe, and F. J. de Bruijn. 1995. Isolation of *lux* reporter gene fusions in *Pseu-*

domonas fluorescens DF57 inducible by nitrogen or phosphorus starvation. *FEMS Microbiol. Ecol.* **17**:95–106.

61. Kragelund, L., C. Hosbond, and O. Nybroe. 1997. Distribution of metabolic activity and phosphate starvation response of *lux*-tagged *Pseudomonas fluorescens* reporter bacteria in the barley rhizosphere. *Appl. Environ. Microbiol.* **63**:4920–4928.

62. Landick, R., C. L. Turnbough Jr., and C. Yanofsky. 1996. Transcription attenuation, p. 1263–1286. *In* F. C. Neidhardt, R. Curtiss III, J. L. Ingraham, E. C. C. Lin, K. B. Low, B. Magasanik, W. S. Reznikoff, M. Riley, M. Schaechter, and H. E. Umbarger (ed.), *Escherichia coli and Salmonella: Cellular and Molecular Biology*, 2nd ed. ASM Press, Washington, D.C.

63. LaRossa, R. A., W. R. Majarian, and T. K. Van Dyk. 1997. U.S. patent US 5,683,868.

64. LaRossa, R. A., and J. V. Schloss. 1984. The sulfonylurea herbicide sulfometuron methyl is an extremely potent and selective inhibitor of acetolactate synthase in *Salmomella typhimurium. J. Biol. Chem.* **259**:8753–8757.

65. LaRossa, R. A., and T. K. Van Dyk. 1987. Metabolic mayhem caused by 2-ketoacid imbalances. *Bioessays* **7**:125–130.

66. Lee, S., M. Suzuki, M. Kumagai, H. Ikeda, E. Tamiya, and I. Karube. 1992. Bioluminescence detection system of mutagen using firefly luciferase genes introduced in *Escherichia coli* lysogenic strain. *Anal. Chem.* **64**:1755–1759.

67. Loper, J. E., and S. E. Lindow. 1994. A biological sensor for iron available to bacteria in their habitats on plant surfaces. *Appl. Environ. Microbiol.* **60**:1934–1941.

68. Maillard, K. I., M. J. Benekik, and R. C. Willson. 1996. Rapid detection of mutagens by induction of luciferase-bearing prophage in *Escherichia coli. Environ. Sci. Technol.* **30**:2478–2483.

69. Maron, D. M., and B. N. Ames. 1983. Revised methods for the *Salmonella* mutagenicity test. *Mutat. Res.* **148**:25–34.

70. Matrubutham, U., J. E. Thonnard, and G. S. Sayler. 1997. Bioluminescence induction response and survival of the bioreprorter bacterium *Pseudomonas fluorescens* HK44 in nutrient-deprived conditions. *Appl. Microbiol. Biotechnol.* **47**:604–609.

71. Meighen, E. A. 1991. Molecular biology of bacterial bioluminescence. *Microbiol. Rev.* **55**:123–142.

72. Michán, C., M. Manchado, G. Dorado, and C. Pueyo. 1999. In vivo transcription of the *Escherichia coli oxyR* regulon as a function of growth phase and in reponse to oxidative stress. *J. Bacteriol.* **181**:2759–2764.

73. Miller, J. H. 1992. *A Short Course in Bacterial Genetics.* Cold Spring Harbor Press, Plainview, N.Y.

74. Mizumoto, M., T. Izumi, and K. Nakamura. 1998. Detection of DNA damaging agents using altered DNA sequence of the *Escherichia coli recA* gene. *Mizu Kankyo Gakkaishi* **21**:347–352.

75. Neidhardt, F. C., J. L. Ingraham, and M. Schaechter. 1990. *Physiology of the Bacterial Cell: A Molecular Approach.* Sinauer Associates, Inc., Sunderland, Mass.

76. Neidhardt, F. C., and M. A. Savageau. 1996. Regulation beyond the operon, p. 1310–1324. *In* F. C. Neidhardt, R. Curtiss III, J. L. Ingraham, E. C. C. Lin, K. B. Low, B. Magasanik, W. S. Reznikoff, M. Riley, M. Schaechter, and H. E. Umbarger (ed.), *Escherichia coli and Salmonella: Cellular and Molecular Biology*, 2nd ed. ASM Press, Washington, D.C.

77. Nichols, B. P., O. Shafiq, and V. Meiners. 1998. Sequence analysis of Tn*10* insertion sites in a collection of *Escherichia coli* strains used for genetic mapping and strain construction. *J. Bacteriol.* **180**:6408–6411.

78. Nunoshiba, T., and H. Nishioka. 1991. 'Rec-lac test' for detecting SOS-inducing activity of environmental genotoxic substances. *Mutat. Res.* **254**:71–77.

79. Oda, Y., S. Nakamura, I. Oki, T. Kato, and H. Shinagawa. 1985. Evaluation of the new system (umu-test) for the detection of environmental mutagens and carcinogens. *Mutat. Res.* **147**:219–229.

80. Ödberg-Ferragut, C., M. Espigares, and D. Dive. 1991. Stress protein synthesis, a potential toxicity marker in *Escherichia coli. Ecotoxicol. Environ. Saf.* **21**:275–282.

81. O'Farrell, P. H. 1975. High-resolution two-dimensional electrophoresis of proteins. *J. Biol. Chem.* **250**:4007–4021.

82. Oh, J.-T., Y. Cajal, P. S. Dhurjati, T. K. Van Dyk, and M. K. Jain. 1998. Cecropins induce the hyperosmotic stress response in *Escherichia coli. Biochim. Biophys. Acta* **1415**:235–245.

82a. Oh, J.-T., Y. Cajal, E. M. Skowronska, S. Belkin, J. Chen, T. K. Van Dyk, M. Sasser, and M. K. Jain. 2000. Cationic peptide antimicrobials induce selective *micF* and *osmY* transcriptional response in *Escherichia coli. Biochim. Biophys. Acta* **1463**:43–54.

83. Oh, J.-T., T. K. Van Dyk, Y. Cajal, P. S. Dhurjati, M. Sasser, and M. K. Jain. 1998. Osmotic stress in viable *Escherichia coli* as the basis for the antibiotic response by polymyxin B. *Biochem. Biophys. Res. Commun.* **246**:619–623.

84. Orser, C. S., F. C. F. Foong, S. R. Capaldi, J. Nalezny, W. MacKay, M. Benjamin, and S. B. Farr. 1995. Use of prokaryotic stress promoters as indicators of the mechanisms of chemical toxicity. *In Vitro Toxicol.* **8**:71–85.

85. Pedahzur, R., H. I. Shuval, and S. Ulitzur. 1997. Silver and hydrogen peroxide as potential drinking water disinfectants: their bactericidal effects and possible modes of action. *Water Sci. Tech.* **35**:87–93.

86. Ptitsyn, L. R., G. Horneck, O. Komova, S. Kozubek, E. A. Krasavin, M. Bonev, and R. Rettberg. 1997. A biosensor for environmental genotoxin screening based on an SOS *lux* assay in recombinant *Escherichia coli* cells. *Appl. Environ. Microbiol.* **63**:4377–4384.

87. Quillardet, P., and M. Hofnung. 1993. The SOS chromotest: a review. *Mutatat. Res.* **297**:235–279.

88. Quillardet, P., O. Huisman, R. D'Ari, and M. Hofnung. 1982. SOS chromotest, a direct assay of induction of an SOS function in *Escherichia coli* K-12 to measure genotoxicity. *Proc. Natl. Acad. Sci. USA* **79**:5971–5975.

89. Ramanathan, S., W. Shi, B. Rosen, and S. Daunert. 1997. Sensing antimonite and arsenite at the subattomole level with genetically engineeried bioluminescent bacteria. *Anal. Chem.* **69**:3380–3384.

90. Ramanathan, S., W. Shi, B. P. Rosen, and S. Daunert. 1998. Bacteria-based chemiluminescence sensing system using β-galactosidase under the control of the ArsR regulatory protein of the *ars* operon. *Anal. Chim. Acta* **369**:189–195.

91. Rettberg, P., C. Baumstark-Khan, K. Bandel, L. R. Ptitsyn, and G. Horneck. 1999. Microscale application of the SOS-LUX-TEST as biosensor for genotoxic agents. *Anal. Chim. Acta* **387**:289–296.

92. Schmid, C., G. Reifferscheid, R. K. Zahn, and M. Bachmann. 1997. Increase of sensitivity and validity of the SOS/umu-test after replacement of the β-galactosidase reporter gene with luciferase. *Mutat. Res.* **394**:9–16.

93. Schrogel, O., and R. Allmansberger. 1997. Optimization of the BgaB reporter system: determination of transcriptional regulation of stress responsive genes in *Bacillus subtilis. FEMS Microbiol. Lett.* **153**:237–243.

94. Selifonova, O., R. Burlage, and T. Barkay. 1993. Bioluminescent sensors for detection of bioavailable Hg(II) in the environment. *Appl. Environ. Microbiol.* **59**:3086–3090.

95. Sendrowski, H., and C. C. Tebbe. 1998. Use of novel genetically modified reporter bacteria for the detection of soil-bound contaminants. *Med. Fac. Landbouww. Univ. Gent.* **63/4:**1893–1896.

96. Shimada, T., Y. Oda, H. Yamazaki, M. Mimura, and F. P. Guengerich. 1994. SOS function tests for studies of chemical carcinogenesis using *Salmonella typhimurium* TA1535/pSK1002, NM2009, and NM3009. *Methods Mol. Genet.* **5:**342–355.

97. Simpson, M., G. Sayler, D. Nivens, S. Ripp, M. Paulus, and G. Jellison. 1998. Bioluminescent bioreporter integrated circuits (BBICs). *Proc. SPIE-Int. Soc. Opt. Eng.* **3328:**202–212.

98. Simpson, M. L., G. S. Sayler, B. M. Applegate, S. Ripp, D. E. Nivens, M. J. Paulus, and G. E. Jellison, Jr. 1998. Bioluminescent-bioreporter integrated circuits form novel whole-cell biosensors. *Trends Biotechnol.* **16:**332–338.

99. Stephens, J. C., S. W. Artz, and B. N. Ames. 1975. Guanosine 5′-diphosphate 3′-diphosphate (ppGpp): postive effector for histidine operon transcription and general signal for amino acid deficiency. *Proc. Natl. Acad. Sci. USA* **72:**4389–4393.

100. Sticher, P., M. C. M. Jaspers, K. Stemmler, H. Harms, A. J. B. Zehnder, and J. R. van der Meer. 1997. Development and characterization of a whole-cell bioluminescent sensor for bioavailable middle-chain alkanes in contaminated groundwater samples. *Appl. Environ. Microbiol.* **63:**4053–4060.

101. Summers, A. O. 1992. The hard stuff: metal in bioremediation. *Curr. Opin. Biotechnol.* **3:**271–276.

102. Swift, S., M. K. Winson, P. F. Chan, N. J. Bainton, M. Birdsall, P. J. Reeves, C. E. Rees, S. R. Chhabra, P. J. Hill, J. P. Throup, B. W. Bycroft, G. P. C. Salmond, P. Williams, and G. S. A. B. Stewart. 1993. A novel strategy for the isolation of *luxI* homologues: evidence for the widespread distribution of a LuxR:LuxI superfamily in enteric bacteria. *Mol. Microbiol.* **10:**511–520.

103. Szittner, R., and E. Meighen. 1990. Nucleotide sequence, expression, and properties of luciferase coded by *lux* genes from a terrestrial bacterium. *J. Biol. Chem.* **265:**16581–16587.

104. Tauriainen, S., M. Karp, W. Chang, and M. Virta. 1998. Luminescent bacterial sensor for cadmium and lead. *Biosensors Bioelectronics* **13:**931–938.

105. Tauriainen, S., M. Karp, W. Chang, and M. Virta. 1997. Recombinant luminescent bacteria for measuring bioavailable arsenite and antimonite. *Appl. Environ. Microbiol.* **63:**4456–4461.

106. van der Lelie, D., L. Regniers, B. Borremans, A. Provoost, and L. Verschaeve. 1997. The VITOTOX test, an SOS bioluminescence *Salmonella typhimurium* test to measure genotoxicity kinetics. *Mutat. Res.* **389:**279–290.

107. Van Dyk, T. K. 1998. Stress detection using bioluminescent reporters of the heat shock response, p. 153–160. *In* R. A. LaRossa (ed.), *Methods in Molecular Biology: Bioluminescence Methods and Protocols*, vol. 102. Humana Press Inc., Totowa, N.J.

108. Van Dyk, T. K., B. L. Ayers, R. W. Morgan, and R. A. LaRossa. 1998. Constricted flux through the branched-chain amino acid biosynthetic enzyme acetolacatate synthase triggers elevated expression of genes regulated by *rpoS* and internal acidification. *J. Bacteriol.* **180:**785–792.

109. Van Dyk, T. K., S. Belkin, A. C. Vollmer, D. R. Smulski, T. R. Reed, and R. A. LaRossa. 1994. Fusions of *Vibrio fischeri lux* genes to *Escherichia coli* stress promoters: detection of environmental stress, p. 147–150. *In* A. K. Cambell, L. J. Kricka, and P. E. Stanley (ed.), *Bioluminescence and Chemi-*

luminescence: Fundamentals and Applied Aspects. John Wiley & Sons, Chichester, England.

110. Van Dyk, T. K., W. R. Majarian, K. B. Konstantinov, R. M. Young, P. S. Dhurjati, and R. A. LaRossa. 1994. Rapid and sensitive pollutant detection by induction of heat shock gene-bioluminescence gene fusions. *Appl. Environ. Microbiol.* **60:**1414–1420.

111. Van Dyk, T. K., T. R. Reed, A. C. Vollmer, and R. A. LaRossa. 1995. Synergistic induction of the heat shock response in *Escherichia coli* by simultaneous treatment with chemical inducers. *J. Bacteriol.* **177:**6001–6004.

111a. Van Dyk, T. K., D. R. Smulski, D. A. Elsemore, R. A. LaRossa, and R. W. Morgan. A panel of bioluminescent biosensors for characterization of chemically induced bacterial stress responses. *ACS Symp. Ser.*, in press.

112. Van Dyk, T. K., D. R. Smulski, T. R. Reed, S. Belkin, A. C. Vollmer, and R. A. LaRossa. 1995. Responses to toxicants of an *Escherichia coli* strain carrying a *uspA′*::*lux* genetic fusion and an *E. coli* strain carrying a *grpE′*::*lux* genetic fusion are similar. *Appl. Environ. Microbiol.* **61:**4124–4127.

113. VanBogelen, R., K. Z. Abshire, A. Pertsemlidis, R. L. Clark, and F. C. Neidhardt. 1996. Gene-protein database of *Escherichia coli* K-12, edition 6, p. 2067–2117. *In* F. C. Neidhardt, R. Curtiss III, J. L. Ingraham, E. C. C. Lin, K. B. Low, B. Magasanik, W. S. Reznikoff, M. Riley, M. Schaechter, and H. E. Umbarger (ed.), *Escherichia coli and Salmonella: Cellular and Molecular Biology*, 2nd ed. ASM Press, Washington, D.C.

114. VanBogelen, R. A., P. M. Kelley, and F. C. Neidhardt. 1987. Differential induction of heat shock, SOS, and oxidation stress regulons and accumulation of nucleotides in *Escherichia coli. J. Bacteriol.* **169:**26–32.

115. van Overbeek, L. S., J. D. van Elsas, and J. A. van Veen. 1997. *Pseudomonas fluorescens* Tn5-B20 mutant RA92 responds to carbon limitation in soil. *FEMS Microbiol. Ecol.* **24:**57–71.

116. Verschaeve, L., J. Van Gompel, L. Thilemans, L. Regniers, P. Vanparys, and D. van der Lelie. 1999. VITOTOX® bacterial genotoxicity and toxicity test for the rapid screening of chemicals. *Environ. Mol. Mutagen.* **33:**240–248.

117. Virta, M., J. Lampinen, and M. Karp. 1995. A luminescence-based mercury biosensor. *Anal. Chem.* **67:**667–669.

118. Vollmer, A. C., S. Belkin, D. R. Smulski, T. K. Van Dyk, and R. A. LaRossa. 1997. Detection of DNA damage by use of *Escherichia coli* carrying *recA′*::*lux*, *uvrA′*::*lux* or *alkA′*::*lux* reporter plasmids. *Appl. Environ. Microbiol.* **63:**2566–2571.

119. Wagner, L. W., and T. K. Van Dyk. 1998. Cryopreservation and reawakening, p. 123–127. *In* R. A. LaRossa (ed.), *Methods in Molecular Biology: Bioluminescence Methods and Protocols*, vol. 102. Humana Press Inc., Totowa, N.J.

120. Walker, G. C. 1996. The SOS response of *Escherichia coli*, p. 1400–1416. *In* F. C. Neidhardt, R. Curtiss III, J. L. Ingraham, E. C. C. Lin, K. B. Low, B. Magasanik, W. S. Reznikoff, M. Riley, M. Schaechter, and H. E. Umbarger (ed.), *Escherichia coli and Salmonella: Cellular and Molecular Biology*, 2nd ed. ASM Press, Washington, D.C.

121. Webb, O. F., P. R. Bienkowski, U. Matrubutham, F. A. Evans, A. Heitzer, and G. S. Sayler. 1997. Kinetics and response of a *Pseudomonas fluorescens* HK44 biosensor. *Biotechnol. Bioeng.* **54:**491–502.

122. Welch, W. 1990. The mammalian stress response: cell physiology and biochemistry of stress proteins, p. 223–278. *In* R. I. Morimoto, A. Tissières, and C. Georgopoulos (ed.), *Stress Proteins in Biology and Medicine.* Cold Spring Harbor Laboratory Press, Cold Spring Harbor, N.Y.

123. **Winkler, M. E.** 1996. Biosynthesis of histidine, p. 485–505. *In* F. C. Neidhardt, R. Curtiss III, J. L. Ingraham, E. C. C. Lin, K. B. Low, B. Magasanik, W. S. Reznikoff, M. Riley, M. Schaechter, and H. E. Umbarger (ed.), *Escherichia coli and Salmonella: Cellular and Molecular Biology*, 2nd ed. ASM Press, Washington, D.C.

124. **Winkler, M. E., J. Roth, and P. E. Hartman.** 1978. Promoter- and attenuator-related metabolic regulation of the *Salmonella typhimurium* histidine operon. *J. Bacteriol.* **133:**830–843.

125. **Zubay, G., D. O. Schwartz, and J. Beckwith.** 1970. Mechanism of activation of catabolite-sensitive genes: a positive control system. *Proc. Natl. Acad. Sci. USA* **66:**104–110.

INDEX